Postharvest Pathology of Fresh Horticultural Produce

Postharvest Pathology of Fresh Horticultural Produce

Edited by

Lluís Palou
and
Joseph L. Smilanick

CRC Press
Taylor & Francis Group
Boca Raton London New York

CRC Press is an imprint of the
Taylor & Francis Group, an **informa** business

CRC Press
Taylor & Francis Group
6000 Broken Sound Parkway NW, Suite 300
Boca Raton, FL 33487-2742

CRC Press is an imprint of Taylor & Francis Group, an Informa business

Printed on acid-free paper

International Standard Book Number-13: 978-1-1386-3083-3 (Hardback)

Library of Congress Cataloging-in-Publication Data

Names: Palou, Lluís, editor. | Smilanick, Joseph L., editor.
Title: Postharvest pathology of fresh horticultural produce / edited by Lluís Palou and Joseph L. Smilanick.
Description: Boca Raton : CRC Press, [2020] | Includes bibliographical references and index. |
Summary: "Describes the most important postharvest diseases of fruits and vegetables, symptoms, key aspects related to infection and epidemiology, and both conventional and innovative management strategies. Fruit diseases, molecular insights in pathogenicity, quiescent infections, mycotoxins detection and control, safety and security issues are discussed"– Provided by publisher.
Identifiers: LCCN 2019023295 (print) | LCCN 2019023296 (ebook) | ISBN 9781138630833 (hardback) | ISBN 9781315209180 (ebook)
Subjects: LCSH: Horticultural crops–Postharvest diseases and injuries.
Classification: LCC SB319.5 .P68 2020 (print) | LCC SB319.5 (ebook) | DDC 635–dc23
LC record available at https://lccn.loc.gov/2019023295
LC ebook record available at https://lccn.loc.gov/2019023296

Visit the Taylor & Francis Web site at
www.taylorandfrancis.com

and the CRC Press Web site at
www.crcpress.com

Contents

Preface

Our intention with this book is to provide a comprehensive resource of information about the biology and control of postharvest diseases of many fresh horticultural products. For this reason, sources cited include appropriate literature of any age, rather than only the most recent. The information in this book covers a wide range of topics related to postharvest pathology of fresh fruits and vegetables. The book is structured in three different sections. In Section I—Postharvest Diseases of Fresh Horticultural Produce—the most important microbial pathogens causing postharvest decay, the symptoms they cause (including color plates), their economic importance, the epidemiology and significant preharvest and postharvest aspects affecting disease incidence, and conventional methods for commercial control are described for citrus fruits, pome fruits, stone fruits, table grapes, kiwifruit, strawberries, pomegranates, persimmons, loquats, avocados, papayas, bananas, Solanaceae and Cucurbitaceae crops, and leafy vegetables. In Section II—General Aspects of Infection Causing Postharvest Disease—chapters are devoted to the molecular insights in the pathogenicity of necrotrophic fungi, mechanisms of fungal quiescence, and detection and control of mycotoxins. In Section III—Novel Technologies to Control Postharvest Decay of Fruits and Vegetables—the most relevant methods for postharvest disease control alternative to conventional chemical fungicides are discussed. These include biological control with antagonistic microorganisms, heat treatments, UV-C light, low-toxicity acids and salts, plant-derived compounds including extracts and essential oils, antifungal edible coatings with especial reference to chitosan and *Aloe* spp. gels, antimicrobial peptides and proteins, and elicitors for induced host resistance.

We hope to appeal to both research and commercial audiences. Graduate and undergraduate students with interest in postharvest sciences should also find this book of value. The best managed postharvest programs address research into the biology and control of the significant diseases, but also include other aspects, such as employing major cultivars in their work, an understanding of many aspects of the commercial horticultural practices and marketing, and they serve as an important resource when regulatory issues are raised concerning postharvest diseases. They understand there are some producers who avoid the use of conventional fungicides, others that employ them consistently, and still others that produce for both conventional and organic markets, and strive to conduct their research to understand and serve the needs of both. Another very important role for a postharvest researcher is to generate unbiased information regarding the effectiveness of the many new postharvest products and processes introduced commercially. As public employees, they are uniquely qualified for this valuable task.

We are truly grateful to the authors that have contributed chapters to the book. All the contributors are renowned scientists who have had outstanding careers and have attained a high level of expertise on specialized topics within postharvest pathology. We also sincerely acknowledge the CRC Press editors Steven Zollo and Laura Piedrahita and the CRC Press production team for their support and guidance with the publication of the book.

The editors: Lluís Palou and Joe Smilanick

Editor Biography

Prof. Lluís Palou completed a PhD in Agricultural Engineering with major emphasis on Postharvest Plant Pathology in 2002 (University of Lleida, Catalonia, Spain). In 2003 he created the Pathology Laboratory at the Postharvest Technology Center (CTP), Valencian Institute of Agrarian Research (IVIA), Montcada, Valencia, Spain. Since 2017 he has been a research professor and the head of the CTP. His main research interests are applied research and extension on general postharvest pathology of horticultural produce, mainly Mediterranean fruits such as citrus, stone fruits, pomegranate, or persimmon; integrated disease management; and alternatives to conventional fungicides. He has directed and evaluated Spanish and international research projects, supervised four doctoral theses, trained national and international students, and collaborated with research groups in the USA, Brazil, Mexico, Turkey, Tunisia, and Australia. To date he has published over 75 peer-reviewed research articles (WOS *h*-index of 25), 16 book chapters or invited reviews, and 58 extension or technical articles.

Dr. Joseph L. Smilanick completed a PhD in Plant Pathology in 1984 under the direction of Joseph W. Eckert in the Fawcett Postharvest Laboratory of the University of California, Riverside. He was a Research Plant Pathologist with the USDA-ARS from 1983 to 2014. Located in Fresno and later in Parlier, California, his primary specialty has been the biology and control of postharvest plant pathogens of citrus fruit and table grapes, with a thorough understanding of the production and handling of these and other tree and vine crops. He authored or co-authored approximately 250 scientific and technical publications, many produced in collaboration with other scientists located in Israel, Spain, Italy, Turkey, Uruguay, Argentina, New Zealand, Australia, and

Mexico. Now a consultant, he currently reviews journal and book chapter manuscripts, manages commercial research projects, conducts crop loss investigations, presents educational seminars in shelf life and food safety, and consults on technical and regulatory issues.

Contributors

Noam Alkan
Department of Postharvest Science of Fresh Produce
Agricultural Research Organization Volcani Center
Rishon LeZion, Israel

Dionisio G. Alvindia
Philippine Center for Postharvest Development and Mechanization (PHilMech)
Science City of Muñoz, Nueva Ecija, Philippines

Joseph Arul
Department of Food Science and Horticultural Research Centre
Laval University
Quebec, Canada

Ana-Rosa Ballester
Institute of Agrochemistry and Food Technology (IATA)
Spanish National Research Council (CSIC)
Paterna, Valencia, Spain

Silvia Bautista-Baþos
National Polythecnical Institute (IPN) Biotic Products Development Center (CEPROBI)
San Isidro, Morelos, Mexico

Lúcia Raquel Ramos Berger
Laboratory of Food Microbiology Department of Nutrition
Federal University of Paraíba
João Pessoa, Brazil

Yang Bi
College of Food Science and Engineering
Gansu Agricultural University
Lanzhou, China

Shifeng Cao
College of Food Science and Technology
Nanjing Agricultural University
Nanjing, PR China

Carla Casals
Institute of Agrifood Research and Technology (IRTA)

XaRTA-Postharvest
Edifici Fruitcentre, Parc Cientìfic
i Tecnològic Agroalimentari
de Lleida, Parc de
Gardeny Lleida, Catalonia

Salvador Castillo

Department of Food Technology
University Miguel Hernández
Orihuela, Alicante, Spain

Marie Thérèse Charlesèse

Saint-Jean-sur-Richelieu Research and
Development Centre
Agriculture and Agri-Food Canada
Saint-Jean-sur-Richelieu
Quebec, Canada

María Coca

Centre for Research in Agricultural
Genomics (CRAG)
CSIC-IRTA-UAB-UB, Campus UAB
Cerdanyola del Vallès, Barcelona, Spain

Marina Collina

Department of Agricultural Science
University of Bologna
Viale Fanin 46, Bologna, Italy

Rosalie Daniel

New South Wales Department
of Primary Industries
Gosford, Australia

Salvatore D'Aquino

Institute of Sciences of Food
Production
National Research Council
Traversa La Crucca 3, Sassari, Italy

Antonieta De Cal

National Institute for Agricultural
and Food Research and
Technology (INIA)
Madrid, Spain

Samir Droby

Department of Postharvest Science
Institute of Postharvest and Food
Sciences
Agricultural Research Organization (ARO)
the Volcani Center, Rishon
LeZihon, Israel

Arturo Duarte-Sierra

Department of Biotechnology and
Food Sciences, Sonora Technological
Institute
Ciudad Obregón, Sonora, Mexico

Philip A. G. Elmer

The New Zealand Institute for Plant &
Food Research Ltd
Waikato Mail Centre Hamilton,
New Zealand

Arno Erasmus

Wonderful Citrus Delano,
CA, USA

Deena Errampalli

Agriculture and Agri-Food Canada
Canada

Kerry Everett

The New Zealand Institute for Plant &
Food Research Ltd.,
Auckland, New Zealand

Elazar Fallik

Department of Postharvest Science of
Fresh Produce
Agricultural Research Organization (ARO)
the Volcani Center, Rishon
LeZiyyon, Israel

Erica Feliziani

Department of Agricultural
Food and Environmental
Sciences
Marche Polytechnic University
Via Brecce Bianche
Ancona, Italy

Mónica Gandía

Institute of Agrochemistry and Food
Technology (IATA),
Spanish National Research
Council (CSIC)
Paterna, Valencia, Spain

Sandra Garrigues

Institute of Agrochemistry and Food
Technology (IATA)
Spanish National Research
Council (CSIC)
Paterna, Valencia, Spain

John B. Golding

New South Wales Department of
Primary Industries
Gosford, Australia

Luis González-Candelas

Institute of Agrochemistry and Food
Technology (IATA),
Spanish National Research
Council (CSIC)
Paterna, Valencia, Spain

Alberto González-León

Food and Development Research
Center (CIAD, A.C.)
Hermosillo, Sonora, Mexico

Tania Elisa González-Soto

Food and Development Research
Center (CIAD, A.C.)
Hermosillo, Sonora, Mexico

Fabián Guillén

Department of Food Technology
University Miguel Hernández
Orihuela, Alicante, Spain

Zoran Ilic'

Faculty of Agriculture in Lešak
University of Pristina in Kosovska
Mitrovica Lešak, Serbia

Antonio Ippolito

Department of Soil, Plant and Food
Sciences
University of Bari Aldo Moro
Bari, Italy

M. Haïssam Jijakli

Integrated and Urban Plant Pathology
Laboratory
Gembloux Agro-Bio Tech, ULg
Gembloux, Belgium

George Karaoglanidis

Laboratory of Plant Pathology, Faculty
of Agriculture,
Aristotle University of Thessaloniki
Thessaloniki, Greece

Anjani M. Karunaratne

Department of Botany
Faculty of Science
University of Peradeniya
Peradeniya, Sri Lanka

Pervin Kinay-Teksür

Department of Plant Protection
Faculty of Agriculture
Ege University Bornova, Izmir, Turkey

Lise Korsten

Department of Microbiology and Plant
Pathology, Faculty of Natural and
Agricultural Sciences

University of Pretoria, Pretoria,
South Africa

Paloma Manzanares

Institute of Agrochemistry and
Food Technology (IATA)
Spanish National Research
Council (CSIC)
Paterna, Valencia, Spain

Jose F. Marcos

Institute of Agrochemistry and
Food Technology (IATA),
Spanish National Research
Council (CSIC)
Paterna, Valencia, Spain

Marta Mari

CRIOF, Department of Agricultural
Science
University of Bologna
Cadriano, Bologna, Italy

Anna Marín

Research Institute of Food
Engineering for Development
(IIAD),
Department of Food Technology
Polytechnical University of
Valencia (UPV)
Valencia, Spain

Domingo Martínez-Romero

Department of Food Technology
University Miguel Hernández
Orihuela, Alicante, Spain

Buddhie S. Nanayakkara

Department of Botany, Faculty of
Science
University of Peradeniya
Peradeniya, Sri Lanka

Amedeo Palma

Institute of Sciences of Food
Production
National Research Council
Sassari, Italy

Lluís Palou

Laboratory of Pathology
Postharvest Technology Center (CTP)
Valencian Institute of Agrarian
Research (IVIA)
Montcada, Valencia, Spain

María B. Pérez-Gago

Postharvest Technology Center (CTP)
Valencian Institute of Agrarian
Research (IVIA)
Montcada, Valencia, Spain

Dov Prusky

Department of Postharvest Science of
Fresh Produce
Agricultural Research Organization
(ARO)
the Volcani Center
Rishon LeZion, Israel

Lucia Ramos

The New Zealand Institute for Plant &
Food Research Ltd.,
Auckland, New Zealand

Margarita de Lorena Ramos-García

Faculty of Nutrition
Morelos State Authonomical
University (UAEM)
Cuernavaca, Morelos, Mexico

Gianfranco Romanazzi

Department of Agricultural
Food and Environmental Sciences

Marche Polytechnic University
Via Brecce Bianche, Ancona, Italy

Simona Marianna Sanzani

Department of Soil Plant and Food Sciences
University of Bari Aldo Moro
Bari, Italy

María Serrano

Department of Food Biology
University Miguel Hernández
Orihuela, Alicante, Spain

Dharini Sivakumar

Phytochemical Food Network Group
Department of Crop Sciences
Tshwane University of Technology
Pretoria, South Africa

Joseph L. Smilanick

Private consultant, formerly USDA-ARS
Kingsburg, CA, USA

Evandro Leite de Souza

Laboratory of Food Microbiology
Department of Nutrition
Federal University of Paraíba
João Pessoa, Brazil

Davide Spadaro

Department of Agricultural
Forestry and Food Sciences (DISAFA)
University of Torino
Grugliasco, Torino, Italy

Subbaraman Sriram

Plant Pathology
ICAR-Indian Institute of Horticultural Research
Hessaraghatta, Bengaluru, India

Darisi Venkata Sudhakar Rao

Postharvest Technology & Agricultural Engineering
ICAR-Indian Institute of Horticultural Research
Hessaraghatta, Bengaluru, India

Neus Teixidó

Institute of Agrifood Research and Technology (IRTA)
XaRTA-Postharvest, Edifici Fruitcentre
Parc CientUÇ fic i TecnolOEgic Agroalimentari de Lleida, Parc de Gardeny
Lleida, Catalonia

Len Tesoriero

New South Wales Department of Primary Industries
Gosford, Australia

Martín Ernesto Tiznado-Hernández

Food and Development Research Center (CIAD, A.C.)
Hermosillo, Sonora, Mexico

Rosario Torres

Institute of Agrifood Research and Technology (IRTA)
XaRTA-Postharvest, Edifici Fruitcentre
Parc Cientìfic i Tecnològic Agroalimentari de Lleida, Parc de Gardeny
Lleida, Catalonia

Rosalba Troncoso-Rojas

Food and Development Research Center (CIAD, A.C.)
Hermosillo, Sonora, Mexico

Nikos Tzortzakis

Department of Agricultural Sciences
Biotechnology and Food Science

Cyprus University of Technology
Limassol, Cyprus

Josep Usall

Institute of Agrifood Research and Technology (IRTA), XaRTA-Postharvest
Edifici Fruitcentre, Parc Cientìfic i Tecnològic Agroalimentari de Lleida, Parc de Gardeny
Lleida, Catalonia

Daniel Valero

Department of Food Technology
University Miguel Hernández
Orihuela, Alicante, Spain

Juan Miguel Valverde

Department of Food Technology
University Miguel Hernández
Orihuela, Alicante, Spain

Rosa Isela Ventura-Aguilar

National Council of Science and Technology (CONACYT)
National Polythecnical Institute (IPN)
Biotic Products Development Center (CEPROBI)
San Isidro, Morelos, Mexico

Antonio Vicent

Mycology Unit
Plant Protection and Biotechnology Center
Valencian Institute of Agrarian Research (IVIA)
Montcada, Valencia, Spain

Junjie Wang

College of Biological Science and Engineering
North Minzu University
Yinchuan, China

Michael Wisniewski

USDA-ARS
Appalachian Fruit Research Station
Kearneysville, USA

Huali Xue

College of Science
Gansu Agricultural University
Lanzhou, China

Pedro J. Zapata

Department of Food Technology
University Miguel HernA˜ndez
Orihuela, Alicante, Spain

Carmit Ziv

Department of Postharvest Science of Fresh Produce
Agricultural Research Organization
Volcani Center
Rishon LeZion, Israel

Section I

POSTHARVEST DISEASES OF FRESH HORTICULTURAL PRODUCE

Chapter 1

Citrus Fruits

Joseph L. Smilanick
Private consultant, formerly USDA-ARS, Kingsburg, CA, USA

Arno Erasmus
Wonderful Citrus, Delano, CA, USA

Lluís Palou
Laboratory of Pathology, Postharvest Technology Center (CTP), Valencian Institute of Agrarian Research (IVIA), Montcada, Valencia, Spain

Abbreviations

2,4-D	2,4-dichlorophenoxyacetic acid
AZX	Azoxystrobin
CFU	Colony forming units
EC	Emulsifiable concentrate
EU	European Union
FAO	Food and Agriculture Organization of the United Nations
FLU	Fludioxonil
FRAC	Fungicide Resistance Action Committee
GRAS	Generally recognized as safe
IMZ	Imazalil
MRL	Maximum residue limit
OPP	Ortho-phenylphenol
PAL	Phenylalanine ammonia-lyase
PAA	Peroxyacetic acid
PCZ	Propiconazole
PYR	Pyrimethanil
ROS	Reactive oxygen species
SOPP	Sodium ortho-phenylphenate
SS	Sulfate salt
TBZ	Thiabendazole
USA	United States of America

1 Introduction

Citrus fruits (*Citrus* spp., Rutaceae) are among the most important fruits produced for human consumption in the world and rank first in terms of value of international trade. Total worldwide production of fresh citrus exceeded 124 million tons in 2016, according to the following distribution: 67 million tons of oranges (*Citrus sinensis* L.); 33 million tons of mandarins or tangerines (*Citrus reticulata* Blanco), including clementines (*Citrus clementina* hort. ex Tanaka), Satsumas (*Citrus unshiu* Marcow.), and a variety of hybrid mandarins; 16 million tons of lemons (*Citrus limon* [L.] Burm. f.) and limes (*Citrus aurantiifolia* [Christm.]); and 8 million tons of grapefruits (*Citrus paradisi* Macfad.) (FAO, 2017). In 2016, the most important citrus-producing countries were China, Brazil, India, the United States of America (USA), Spain, Mexico, Egypt, Turkey, Iran, Italy, Argentina, South Africa, and Morocco, among others. In terms of international trade, Spain is the leading country, with 4.1 million tons of exports of fresh produce in 2016 (FAO, 2017).

We interpret postharvest diseases of citrus to be those caused by microbial pathogens that colonize or otherwise harm fruit, and do not include those disorders of physiological or abiotic origin. Most postharvest diseases of citrus fruits are caused by filamentous fungi, and the common names of the diseases they cause are based on the symptoms they produce. They have been classically divided into two groups according to the time when infection predominantly occurs (Eckert and Eaks, 1989). The first group of postharvest diseases is those that initiated when the pathogen infects the fruit before harvest, termed *preharvest infections*. These infections are called "latent" (when not visible) or "quiescent" (when the inactive infection is visible), and they do not cause significant disease until after harvest. The second group of postharvest diseases is those that initiated when infection occurs just before, during, or after harvest. Most of these pathogens infect through rind wounds and disease progresses immediately after infection. Wound pathogens infect mature fruit through rind injuries or bruises inflicted in the field near harvest or during harvest, postharvest handling in the packinghouse, transportation, or when marketed. In contrast, latent pathogens infect the fruit in the field during the growing season and remain inactive until they resume growth after harvest because of significant changes in the fruit properties and environmental conditions. The relative importance of each type of these diseases varies, but the most important factor is the climate of the citrus-producing area. In typical summer-rainfall production areas, such as Florida, Brazil, or Southeast Asia, diseases from preharvest infections are usually high. In contrast, in areas of sparse summer rainfall or Mediterranean-type climate areas, such as Spain and other Mediterranean countries, California, Australia, and most citrus areas in South Africa, the incidence of postharvest diseases is typically lower and diseases from harvest or postharvest infections, especially green and blue molds caused by *Penicillium* spp., cause most losses.

Regardless of the climatic area, losses of citrus fruits due to postharvest diseases are quite variable and dependent on the variety, tree age and condition, weather conditions during the growing and harvest season, the extent of physical injury to the fruit during harvest and subsequent operations, pathogen inoculum density, the effectiveness of antifungal treatments, sorting/grading operations after harvest, and the postharvest environment. In a summary of 12 yr of inspections of commercial shipments of California and Florida citrus fruits in the New York produce market, Penicillium decays, sour rot, and stem-end rot were present in 30, 9, and 5%, respectively, of the inspected shipments (Ceponis et al., 1986). In another study in California, the total percentage of fruit lost to decay during both storage and subsequent marketing was about 8% in three lemon packinghouses (Bancroft et al., 1984). Green mold losses in California are typically 2–4% during ethylene degreening, but can exceed 30% in disease-conducive years when heavy rains occur before harvest, or when numerous split fruit or other rind injuries occur and become infected on trees before harvest (Smilanick et al., 2006a). In Spain, Tuset (1987) estimated that Penicillium molds, black rot, gray mold, anthracnose, and sour rot accounted for 55–80, 8–16, 8–15, 2.5–6, and 2–3%, respectively, of total postharvest decay observed in oranges and mandarins during the marketing season. Pelser (1977a) reported that Penicillium molds accounted for about 75% of the decay observed in South African 'Valencia' oranges shipped to London. In Florida, among samples of several cultivars of untreated oranges collected over a 5-yr period in this high-rainfall area, stem-end rot was the most common disease, and it infected 13–42% of the fruit after storage at 21°C for 3 wk (Smoot, 1977). The fruit examined in all of these reports were from commercial packinghouses that employed sanitation measures, fungicides, and temperature management regimes to minimize decay; losses among fruit handled without these measures can be much higher. Among smaller growers, especially organic growers and some of those in underdeveloped countries, where efficient transportation, refrigeration, and chemical treatments may be less available, losses are typically much higher and can make export sales infeasible (Kader, 2005).

The percentage of fruit lost to postharvest diseases is only one measure of their economic impact. Citrus postharvest diseases can also cause significant losses for growers, packers, shippers, and consumers by harming future sales. Inspection standard thresholds for decay are low; typically, if the number of decayed fruits exceeds 0.5%, the grade and price of the shipment will be reduced. The unsightly appearance of rotten fruit, although the incidence may be comparatively low, repels wholesale buyers who may as a consequence abandon the affected producer and seek other sources. In addition to the losses of individual fruit, propagules produced from lesions on decaying fruit contaminate the surrounding environment, initiating new cycles of decay, and requiring the remaining healthy fruit to be cleaned and repackaged. If the losses occur after shipment, the producer is usually billed for these added costs, which can exceed their returns (Smilanick et al., 2006a). Furthermore, the loss of one piece of fruit after harvest can be more than twice as costly as the loss of one before harvest since expenses that begin at harvest typically surpass those to produce the fruit. For example, between 51 and 67% of the total cost of growing, harvesting, packing, marketing, and shipping California oranges and lemons occurred after the fruit were harvested (Eckert and Eaks, 1989).

The ubiquitous occurrence of virulent and aggressive pathogens, the employment of relatively warm storage conditions to avoid chilling injury (but favor rapid pathogen growth), and minimal resistance of citrus fruits to infection make management of postharvest diseases of paramount importance for producers and marketers of citrus fruits, particularly if export sales or long-term storage are employed. In this chapter, the major diseases and actions to manage them will be described.

2 Diseases from Postharvest Fruit Infection

2.1 Green and Blue Molds

Green mold, caused by *Penicillium digitatum* (Pers.: Fr.) Sacc., is the most important postharvest disease of citrus fruits produced in areas with a Mediterranean-type climate, characterized by scant summer rainfall. Blue mold, caused by *Penicillium italicum* Wehmer, is typically of lesser overall importance since it grows slower than *P. digitatum* at ambient temperatures, but may become the major problem under certain conditions. Penicillium molds are also important in production areas with abundant summer rainfall, where the total incidence of postharvest decay is higher and diseases caused by preharvest fruit infections are predominant (Eckert and Eaks, 1989; Palou, 2014). A third *Penicillium* sp., *Penicillium ulaiense* H.M. Hsieh, H.J. Su & Tzean, the cause of whisker mold, was first found in California in mixed infections with *P. digitatum* in stored citrus fruits (Holmes et al., 1994). It resembles *P. italicum*, especially on obverse colony color, but grows much slower and shows paler reverse colors on all media (Frisvad and Samson, 2004). The economic importance of whisker mold is very modest because *P. ulaiense* is considerably less virulent than *P. digitatum* and *P. italicum*, which decay citrus fruits about three to five times faster (Holmes et al., 1994). The disease has been recently reported in countries such as Japan (Tashiro et al., 2012), Tunisia (Rouissi et al., 2015), Pakistan (Khan et al., 2017), Korea (Park et al., 2018), and Spain (Palou and Taberner, 2019).

Infections of *P. digitatum* and *P. italicum* start in rind wounds and produce a water-soaked, soft, circular area seen by the naked eye after about 3 d of incubation at temperatures around 20°C. Both are necrotrophic pathogens and produce hydrolytic enzymes that macerate the peel tissue during disease development (Barkai-Golan and Karadavid, 1991). As the fungus grows, aerial white mycelium develops in the center of the lesion and expands radially. Depending on the inoculum load, sporulation begins after 3–5 d at room temperatures (15–28°C) and also expands radially, forming a colored layer of conidia with a velutinous texture. In the case of green mold, after 7–8 d, the central area of the lesion is olive green surrounded by a broad band of dense, non-sporulating white mycelium limited by moderately firm decaying peel. With blue mold, the central sporulating area is blue or bluish green surrounded by a very narrow band of non-sporulating white mycelium limited by a broad band of soft, water-soaked peel (Figure 1.1a; Palou, 2014). Eventually, the entire surface of the fruit is completely covered with conidia and the fruit begins to shrink. Conidiophores of *P. digitatum* are terverticillate, irregularly branched, with whorls of three to six solitary, cylindrical phialides. Conidia are typically ellipsoidal, of 3.5–8.0 × 3.0–4.0 μm in size. The conidial apparatus of *P. italicum* consists of asymmetric penicilli bearing tangled chains of conidia. Conidiophores are terverticillate, hyaline, with stipes bearing three to six cylindrical phialides. Conidia are elliptical or subglobose, smooth, of 4.0–5.0 × 2.5–3.5 μm in size. *Penicillium digitatum* is the first phytopathogenic *Penicillium* species whose complete and mitochondrial genomes have been entirely sequenced (Sun et al., 2011; Marcet-Houben et al., 2012), and the phylome is available at the public database PhylomeDB (www.phylomedb.org). Recently, specific primers for rapid detection of *P. digitatum* have been developed (Chen et al., 2017).

Conidia of *P. digitatum* and *P. italicum* that are formed on fruit rotting on the ground in the grove or in packinghouse facilities are transported by air currents to sound fruit. The conidia do not germinate on the surface of fruit until the peel is wounded. These wounds are mainly inflicted at harvest and after harvest, but some infections can occur in the field in injuries produced by wind or insects. Usually, fruit infected less than 3 d before harvest cannot be detected and are harvested as sound (Eckert and Eaks, 1989). Free water and nutrients are required for conidia to germinate. Conidia situated in injuries that penetrate into the albedo of the peel, a depth of 2 mm or more, or those in ruptured oil glands, usually cause irreversible infection within 48 hr at 20–25°C (Smilanick et al., 2006a). Germination and subsequent hyphal growth is stimulated by citrus volatiles present in the rind essential oils and is favored by specific temperature and moisture conditions. Work by Droby et al. (2008) showed that germination and germ tube elongation in *P. digitatum* and *P. italicum* responded primarily to limonene and myrcene, respectively, which are the first and second most abundant compounds in the peel oil. Furthermore, when transgenic oranges with reduced limonene content in the peel were challenged with *P. digitatum*, the incidence and severity of green mold were markedly reduced (Rodríguez et al., 2015). The optimum temperature for germination and growth of both species is 25°C, but while green mold is more frequent on fruit at room temperatures, blue mold can be more prevalent among fruit stored below 10°C, and particularly at 3–5°C, the commercial cold storage temperatures for oranges or mandarins. *Penicillium italicum* is active at lower temperatures and water activities than *P. digitatum*. For instance, at 4°C, lesions caused by

P. italicum were visible 16 d after the inoculation of 'Valencia' oranges, while those of *P. digitatum* did not appear until 23 d had elapsed (Plaza et al., 2004). Conversely, green mold can overgrow blue mold in a mixed infection on fruit at room temperatures and is typically more common. Blue mold, in contrast to green mold, can spread by contact from fruit to fruit, resulting in pockets of decay involving several diseased fruits (Barmore and Brown, 1982). An important problem associated with the occurrence of green and blue molds in loose-fill cartons or boxes is a condition termed "soilage", caused by conidia from a diseased fruit dispersing onto adjacent fruit in the same container (Smilanick et al., 2006a).

Development of green and blue molds in infected citrus fruits is mediated by complex interactions between pathogen virulence mechanisms and host defense responses. While the responses triggered by the fruit host lead to natural disease resistance in immature fruit, all citrus fruits become progressively susceptible to decay as they age. All commercial citrus species and cultivars are susceptible to green and blue molds, although they can show different degrees of susceptibility. In general, clementines and other mandarins are more susceptible than other citrus species (Palou, 2014). The presence of constitutive or preformed antifungal compounds in the rind is one of the biochemical mechanisms of the fruit host to resist infection by *Penicillium* spp. Among them, citral, *p*-coumaric acid, and several polymethoxyflavones and flavanones have shown significant antifungal activity (Ortuño et al., 2011). The synthesis of these substances is primarily regulated by the activity of the enzyme phenylalanine ammonia-lyase (PAL), and their concentration declines as the fruit ages (González-Candelas et al., 2010). Other resistance mechanisms induced in the infection site by rind wounding and/or fungal infection include lignin synthesis, which creates a physical barrier, phytoalexins or other secondary metabolites, and pathogenesis-related proteins (PRP) (Ballester et al., 2006, 2010; Zhu et al., 2017). Phytoalexins, such as coumarins scoparone and scopoletin, have significant antifungal activity against *P. digitatum* and *P. italicum* and can also be induced by some physical or chemical postharvest treatments via the enhancement of PAL activity (Venditti et al., 2005; Rojas-Argudo et al., 2012; Ballester and Lafuente, 2017). Chitinases and β-1,3-glucanases are common PRP produced by citrus fruits that inhibit mycelial growth by damaging fungal cell walls (Pavoncello et al., 2001). During infection and fruit colonization, *P. digitatum* and *P. italicum* are able to suppress the burst of reactive oxygen species (ROS) with accumulation of hydrogen peroxide (H_2O_2) that occurs in citrus fruits tissue as the precursor step of most of these disease resistance mechanisms. ROS metabolism and H_2O_2 production are regulated by enzymes, such as PAL, superoxide dismutase, catalase, ascorbate peroxidase, or glutathione reductase, whose activity in the fruit peel can be modulated by the application of postharvest treatments of different nature (Macarisin et al., 2010; Lu et al., 2013; Fallanaj et al., 2016). The effect of such treatments on induction of disease resistance is clearly regulated by the fruit host maturity and the environmental conditions (Vilanova et al., 2012).

One of the most important factors determining the final incidence of green and blue molds is the amount of rind wounds, injuries, and mechanical damage inflicted to the fruit during harvest, transportation, and postharvest handling. The fruits should be harvested by well-trained teams of workers. In arid production areas, they should clip the stems as short as possible and never pull the fruit from the trees because it creates a wound that could later become infected. In summer-rainfall production areas, the fruits are pulled or snapped from the trees to reduce the stem-end rot incidence that can often exceed that of Penicillium molds (Smilanick et al., 2006a). Harvest in the early morning, when fruit turgor is high, is generally avoided because excessive rind turgidity in the oil glands favors their rupture during handling that can cause oleocellosis and opportunities for infection (Eckert and Eaks, 1989). Early-season citrus fruits degreened with ethylene after harvest, particularly if done at 20–22°C (common in arid-climate production areas) and not protected by a prior fungicide application, may have a high incidence of green mold because these temperatures are conducive to decay development. Furthermore, degreening can significantly increase the severity of Penicillium molds because ethylene accelerates senescence of the fruit peel. This effect, however, depends on initial rind color and fruit maturity and it is more pronounced when less green (more mature) fruit are degreened (Moscoso-Ramírez and Palou, 2014). In contrast, in Florida and other humid citrus production areas, where degreening is performed at temperatures surrounding 30°C, the process exerts a curing effect that reduces green and blue molds by wound lignification (Brown, 1973; Plaza et al., 2003b; Nunes et al., 2007). In any case, unlike other postharvest pathogens, ethylene alone has little direct impact on the growth of *P. digitatum* and *P. italicum*, and its effects on decay are only indirect to the fruit host (Porat et al., 1999b; Moscoso-Ramírez and Palou, 2014). Early studies also report benefits from removal of ethylene from the room atmosphere during citrus long-term cold storage (McGlasson and Eaks, 1972; Wills et al., 1999). Besides fruit and packinghouse sanitation and specific postharvest fungicide treatments, cold storage is an important tool to control green and blue molds. Low temperatures directly delay fungal growth and indirectly contribute to the maintenance of fruit disease resistance since it reduces the fruit metabolic activity and delay its senescence. As previously mentioned, blue mold incidence can exceed that of green mold on cold-stored fruit because *P. italicum* grows better than *P. digitatum* at low temperatures.

2.2 Sour Rot

Second in importance to Penicillium decay among the wound-initiated diseases of citrus fruits, sour rot is caused by *Geotrichum citri-aurantii* (Ferraris) E.E. Butler (Butler et al., 1988), formerly identified as *G. candidum.* Although some authors prefer this term be retained and sour rot isolates from citrus termed "citrus type" (Nakamura et al., 2001), the name *G. citri-aurantii* is currently preferred. In this book, the convention "one fungus = one name" has been adopted (Taylor, 2011). Before the adoption of this more rational nomenclature, the teleomorph state of the species was known as *Galactomyces citri-aurantii* E.E. Butler. McKay et al. (2012a) stated that sexual reproduction is not rare and seems to have a major role in the life cycle of this pathogen. All *Geotrichum* spp. can produce a pseudomycelium and grow rapidly by fission in liquid culture. They have septate mycelia that readily fragment into asexual arthroconidia, which are the primary means of reproduction. *Geotrichum citri-aurantii* can be differentiated from *G. candidum*, which is a pathogen of many fresh fruit and processed products but not of fresh citrus fruits, by their morphology when cultured in lemon juice and polymerase chain reaction amplification using species-specific primers from endo-polygalacturonase and β-tubulin genes (Nakamura et al., 2001, 2008; McKay et al., 2012a). *Geotrichum citri-aurantii* grows in lemon juice as elongated cells that fragment into numerous individual conidia, while *G. candidum* grows as small filamentous colonies with irregular branching and few individual conidia (McKay et al., 2012a).

Geotrichum citri-aurantii is widely distributed in citrus grove soils (Eckert, 1959; Butler et al., 1965, 1988; Suprapta et al., 1995). Arthroconidia are disseminated by water splash from rain or irrigation, by airborne soil particles, or by insects to the surface of the fruit, where they cause infections at sites of injury (Roth, 1967; Baudoin and Eckert, 1982; Huang, 1991). It does not readily infect through shallow injuries, especially on immature fruit. Fruits located lower on trees are most likely to be contaminated with this fungus. Brown (1979) observed that more inoculum of *G. citri-aurantii* was present on scarred than smooth fruit, reflecting the accumulation of soil particles on the former. Postharvest gibberellic acid applications, which delay peel aging, reduced sour rot incidence on lemons (Coggins et al., 1992) but failed to do so on Navel oranges (Cunningham and Taverner, 2007). Arthroconidia of *G. citri-aurantii* will only germinate when water activity is high, unlike the more tolerant conidia of *P. digitatum* and *P. italicum* (Plaza et al., 2003a). Even with ripe fruit, the fungus may not develop an active decay lesion unless the peel has a relatively high water content and the inoculated fruit is held in a water-saturated atmosphere (Baudoin and Eckert, 1982, 1985a, 1985b; Suprapta et al., 1996), or if the fruit is submerged in water before inoculation (Cohen et al., 1991). Inoculation with a mixture of *G. citri-aurantii* arthroconidia and *P. digitatum* conidia increases the development of sour rot (Morris, 1982). Because of this synergy, treatments that control green mold can affect marked reductions in sour rot, even if the treatment has no activity alone on *G. citri-aurantii*. Cunningham and Taverner (2007) simultaneously inoculated Navel oranges with both pathogens to evaluate treatment effectiveness to control both diseases since mixed infections occur quite frequently within packinghouses.

Sour rot is exacerbated by storage at 10°C or higher of fruits that are very mature, harvested after prolonged wet periods, or treated with ethylene gas (Savastano and Fawcett, 1929). Temperatures below 10°C greatly suppress its development, and its growth stops at 6°C (Eckert and Eaks, 1989; Plaza et al., 2004). Sour rot is a major problem on lemons and mandarins, which are particularly susceptible to infection (Nazerian and Alian, 2013). Sour rot lesions are soft, gelatinous, penetrate deeply into the fruit, and do not alter the color of colonized tissue. Its incipient infections and rind-colored lesions are difficult to eliminate, so the infected fruit may be inadvertently marketed, where it then develops and spreads rapidly among healthy fruit (Smith, 1917). The pathogen produces a full complement of enzymes that digest the host tissue into a liquid that can drip on to underlying fruit or be dispersed by insects, resulting in rapid spread of the disease. Under high relative humidity (RH), symptoms on advanced lesions include the formation of a yeasty, wrinkled layer of whitish mycelium (Figure 1.1b) (Timmer et al., 2000). On packinglines, the disintegrating lesions cause extensive contamination, so sanitation is critical to its management. In some facilities, high-pressure washing with chlorinated water is done just after the fruits are placed on packinglines to disintegrate lesion tissue so they can be more easily observed and eliminated during subsequent grading. Thermal treatments, sodium hypochlorite, chlorine dioxide, ozone, peracetic acid, and quaternary ammonium compounds can be used to decontaminate equipment and facilities (Smilanick et al., 1999, 2002; Smilanick and Mansour, 2007; Diaz et al., 2015).

2.3 Miscellaneous Diseases

Other citrus wound pathogens that are infrequent or cause minor economic losses are *Mucor piriformis* A. Fisch., the most common cause of Mucor rot; *Rhizopus stolonifer* (Ehrenb.) Vuill. (syn.: *R. nigricans* Ehrenb.), cause of Rhizopus rot; *Trichoderma viride* Pers. (syn.: *T. lignorum* (Tode) Harz), cause of Trichoderma rot; *Aspergillus niger* Tiegh, the most common cause of Aspergillus rot; and *Cladosporium* spp., cause of Cladosporium rot.

Recent reports indicate that Mucor rot is an important emerging postharvest disease of mandarin fruit in California. It was found in 11 of 15 mandarin lots with an average of almost 50% of the total decay observed. About 92% of the *Mucor* spp. isolated were molecularly identified as *M. piriformis*. Other pathogenic *Mucor* spp., although generally less virulent, were *M. circinelloides* Tiegh., *M. racemosus f. racemosus* Fresen., *M. hiemalis* Wehmer, and *M. mucedo* L. (Saito et al., 2016). In another study in California, Mucor rot caused 27% of decay observed on mandarins stored at 4–5°C for several weeks after harvest (Saito and Xiao, 2017c). Earlier reports in Spain found *R. stolonifer* as the cause of 1–3% of total postharvest decay on non-refrigerated oranges and mandarins (Tuset, 1987). Infection by these pathogens occurs mainly in the field from highly decomposed organic matter in the soil. Inoculum is also present in the air, especially under high humidity conditions. Soil dirt or residual rotten fruit in field packages or boxes can be the inoculum source to infect healthy fruit through rind microwounds or injuries. Symptoms of decay caused by both species are similar. They produce a very soft, watery rot that can encompass the entire whole fruit. Early symptoms can be confused with those of sour rot. Under high temperature and moisture, dark brown soft lesions expand rapidly and a characteristic coarse, long, white to gray mycelium bearing large, black, spherical sporangia covers the fruit. Although these diseases are infrequent if citrus postharvest handling and sanitation are adequate, they can be especially devastating due to the formation of nests of decay in piles of bins or boxes in the packinghouse (Figure 1.1c). *Mucor*- or *Rhizopus*-infected citrus fruits collapse and exudates that leak from decayed tissues carry inoculum that may easily infect adjacent healthy fruit by the action of pectolytic enzymes (Tuset, 1987). The most important difference between the two species is that while *R. stolonifer* does not grow at temperatures lower than 5°C, *M. piriformis* is able to grow, although slowly, at temperatures in the range of 0–5°C, thus causing decay on commercially cold-stored oranges or mandarins. Rhizopus rot can affect cold-stored lemons or grapefruits, which are typically stored at temperatures beyond 8–10°C.

Trichoderma rot is typically more important in lemons than other citrus species. The reason is that the growth of *T. viride* below 5°C is minimal, and lemons are typically stored at 10°C to avoid chilling injury, whereas oranges and mandarins are usually stored at 3–5°C. The pathogen is ubiquitous in the soil and contaminates the fruit through rain splash or soiled field containers. Since it is able to grow on wood products, its incidence was higher when wood boxes (instead of current plastic materials) were commonly used for citrus handling and storage. Deep rind wounds and release of oil gland contents facilitate infection of contaminated fruit, which is more frequent in the stem or stylar ends of the fruit. Effective infection occurs preferentially in mature fruit near harvest or those harvested exceptionally late in the season. Rotten fruit turns cocoa brown and has a firm texture and a characteristic coconut-like odor. Decay lesions enlarge slowly and, under humid conditions, white mycelium and yellow green to dark green masses of conidia are abundantly produced within 10 d at 25°C, especially if the fruit is exposed to light. If juice from decayed fruit reaches surrounding healthy fruit, nests of decay can occur. Fruit cold-stored for long periods (>30 d) are more prompt to disease, especially when the ventilation of storage rooms is not optimal. Imazalil and other benzimidazole fungicides are not effective against Trichoderma rot (Tuset, 1987; Timmer et al., 2000).

Aspergillus rot causes significant losses only on citrus fruits kept at very high temperatures (27–32°C) for relatively long periods. Since this is a very rare postharvest practice in modern citrus packinghouses, the disease is of minor importance in developed countries. Different *Aspergillus* spp. can be the causal agent, but the most frequent is *A. niger,* which causes a characteristic soft black rot. Conidia of the pathogen are abundantly produced in many different plant substrates in the grove and easily transported by air currents to the fruit. Conidia only infect fruit through rind injuries in a similar way to *Penicillium* spp. Lesions are circular, water-soaked, discolored areas that resemble those of *P. italicum* or *G. citri-aurantii* until they become sunken and darker and covered by dense masses of black conidia. It can spread to adjacent healthy fruit causing nests of very soft and watery decayed tissues. Minimum and optimal temperatures for pathogen growth are 15 and 32°C, respectively (Eckert and Eaks, 1989; Timmer et al., 2000).

Airborne spores of *Cladosporium* spp. are present at extremely high levels in the environment and on fruit surfaces in citrus groves and packinghouses worldwide (Palou et al., 2001b, 2001c; Fischer et al., 2009; Moubasher et al., 2016). However, these species are typically saprophytic fungi developing in any type of plant substrate, mainly dry twigs and branches. Moreover, they are commonly among the fungi, together with *Capnodium* spp. and *Alternaria* spp., that comprise sooty mold on areas of citrus branches, leaves, and fruits, where honeydew has been largely excreted by different types of insects (Tuset, 1984; Snowdon, 1990). Citrus fruits are readily contaminated with conidia in both the grove and the packinghouse, but postharvest infection and subsequent development of Cladosporium rot only occur occasionally on overmature or senescent fruits or on fruits that are intensively affected by rind damage of a different etiology. Two *Cladosporium* spp., viz. *C. herbarum* (Pers.) Link and *C. cladosporioides* (Fresen.) G.A. de Vries, have been sporadically reported as weak wound pathogens causing postharvest decay of citrus fruits. In Spain, *C. herbarum* caused superficial, firm, dark brown lesions in peel areas (mainly in the stem and stylar ends) of oranges and lemons affected by physiological disorders, such as rind breakdown or water spots. Under humid conditions, mycelia and grayish to dark green masses of conidia almost completely covered the lesions (Tuset, 1987). In Japan, *C. cladosporioides* caused a large number of superficial, tiny, black spots on Satsuma mandarins grown in heated greenhouses and shipped to local markets. The mandarins did not rot, but could not be sold because of their unpleasant appearance. This cosmetic disease was named as sooty spot (Tashiro et al., 2013). In our laboratory at the IVIA (Valencia, Spain), we isolated *C. cladosporioides* growing abundantly on the surface of senescent long-term cold-stored mandarins (Figure 1.1d), but when pathogenicity tests were conducted by artificial wound inoculation of drops of a high-density conidia suspension on recently harvested mature healthy fruit, no decay developed and Koch's postulates were not fulfilled.

Figure 1.1 a. Green and blue molds of orange caused by *Penicillium digitatum* and *Penicillium italicum*, respectively. b. Sour rot of mandarin caused by *Geotrichum citri-aurantii*. c. Nest of Mucor rot caused by *Mucor piriformis* on stored mandarins. d. Cladosporium rot of mandarin caused by *Cladosporium cladosporioides*.

3 Diseases from Preharvest Fruit Infection

3.1 Stem-End Rots

Stem-end rots cause dark brown rind lesions that initiate at the stem end of the fruit; these progress toward the calyx end as the causal pathogens grow. The diseases are Diplodia stem-end rot, caused by *Lasiodiplodia theobromae* (Pat.) Griffon & Maubl. (syn.: *Diplodia natalensis* Pole-Evans), and Phomopsis stem-end rot, caused by *Phomopsis citri* H.S. Fawc. Previously to the convention "one fungus = one name", the teleomorph states of these fungi were known as *Botryosphaeria rhodina* (Berk. & M.A. Curtis) Arx and *Diaporthe citri* F.A. Wolf, respectively. A third pathogen, *Alternaria alternata* (Fr.) Keissl., can also cause stem-end rot symptoms, in addition to causing internal rot, calyx infections, and lesions on the rind.

Phomopsis spp. also cause melanose, a widely distributed disease causing small spots or scab-like lesions on citrus fruits and leaves (Gopal et al., 2014). Species of *Diaporthe*, the perfect stage of *Phomopsis*, causing melanose and stem-end rot diseases of *Citrus* spp. were revised to three species of *Diaporthe* that infect citrus: *D. citri, D. cytosporella*, and *D. foeniculina* (Udayanga et al., 2014). *Diaporthe citri* occurs on citrus throughout the citrus-growing regions of the world. *Diaporthe cytosporella* is found on citrus in Europe and California. *Diaporthe foeniculina*, including the synonym *D. neotheicola*, is recognized as a species with an extensive host range, including citrus. The taxonomy of *L. theobromae* isolates causing

Diplodia stem-end rot has not been thoroughly elucidated. Alves et al. (2008) reported that *L. theobromae* is an unspecialized plant pathogen that causes numerous diseases on hundreds of hosts that is composed of a diverse complex of different cryptic species. Phillips et al. (2013) developed a key based on morphological features to identify *L. theobromae*, although they stated morphological features alone were inadequate to confidently define genera or identify species in Botryosphaeriaceae.

Diplodia and Phomopsis stem-end rots are the principal postharvest diseases of citrus in areas with rainfall during fruit development. Incidence can exceed 50% (Smoot, 1977; Dantas et al., 2003). These diseases are rare or absent in production areas with dry growing seasons. In summer-rainfall production areas, such as the Gulf states of the USA, the West Indies, southeastern Asia, and Brazil, they are often the most important postharvest diseases. The conidia of *L. theobromae* and *P. citri* are produced in pycnidia that develop on dead wood in trees. They move by splashing water onto developing fruits and initiate incipient infections in the "button" (calyx + disc) of the fruit (Brown and Wilson, 1968). *Lasiodiplodia theobromae* and *P. citri* do not aggressively attack the button during fruit growth, but remain quiescent or grow saprophytically on necrotic tissue on the disc and the inner surface of the calyx. After harvest, the button senescences and begins to separate from the fruit. Fungi quiescent in the button during the growing season resume active growth and penetrate through the abscission zone into internal tissues of the fruit (Brown and Wilson, 1968). They colonize and darken the tissues of the peel and central axis (pith) of the fruit, causing the typical stem-end rot symptoms. Diplodia and Phomopsis stem-end rots can be distinguished by their symptoms. In Diplodia stem-end rot, the advancing front of the lesion appears irregular, with finger-like projections into the healthy tissue (Figure 1.2a), while lesions of Phomopsis stem-end rot grow as a uniform, dark brown front that advances down the rind (Figure 1.2b).

Although *L. theobromae* and *P. citri* are often present together in the calyx of the fruit at harvest, the incidence of each disease is more or less seasonal. Diplodia stem-end rot is most common early in the fall harvest season, whereas Phomopsis stem-end rot appears later in the fall or winter and spring months. Diplodia stem-end rot is prominent early because it develops faster than Phomopsis stem-end rot at the higher ambient temperatures that prevail at that time of the year. Furthermore, the incidence of Diplodia stem-end rot is exacerbated by ethylene degreening used to remove green color from the peel of early season fruit (Brooks, 1944; McCornack, 1972a, 1972b; Porat et al., 1999a). The recommended temperature for this operation is 28–29°C in Florida and 21–22°C in California, reflecting physiological differences in fruit grown under different climatic conditions. Ethylene stimulates the growth of *L. theobromae* (Brown and Lee, 1993), and increases polygalacturonase and cellulase activity in the fruit that hastens abscission of the stem-end button, a condition that increases stem-end rot (Brown and Burns, 1998). As the season progresses, ambient temperatures decline and ethylene degreening is discontinued because the peel develops sufficient color naturally. The lower temperature and absence of ethylene favor the development of *P. citri*, so Phomopsis stem-end rot becomes the principal stem-end rot rather than Diplodia stem-end rot (Smoot et al., 1983).

3.2 Anthracnose

Until recently, postharvest anthracnose of citrus fruits was thought to be caused only by the fungus *Colletotrichum gloeosporioides* (Penz.) Penz. & Sacc. (*C. gloeosporioides sensu stricto*). Nevertheless, other *Colletotrichum* spp., such as *C. constrictum* Damm, P.F. Cannon, Crous, P.R. Johnst. & B. Weir, *C. truncatum* (Schwein.) Andrus & W.D. Moore, *C. fructicola* Prihastuti, L. Cai & K.D. Hyde, or *C. novae-zelandiae* Damm, P.F. Cannon, Crous, P.R. Johnst. & B. Weir, were also associated with citrus fruits decay worldwide (Huang et al., 2013; Guarnaccia et al., 2017). On the other hand, other *Colletotrichum* spp., such as *C. acutatum* J.H. Simmonds or *C. karstii* Y.L. Yang, Zuo Y. Liu, K.D. Hyde & L. Cai, are also present in citrus groves and can cause disease in leaves and twigs, but have not been reported as causal agents of fruit postharvest disease. Moreover, *C. acutatum* can also infect flower petals and cause postbloom fruit drop in the orchard (Timmer et al., 1998; Ramos et al., 2016; Silva et al., 2017). Species of *Colletotrichum* affecting cultivated citrus plants belong to four different species complexes (*C. gloeosporioides, C. acutatum, C. boninense*, and *C. truncatum*) and can currently be properly identified using molecular techniques, such as the analysis of the DNA regions ITS, GAPDH, calmodulin, β-tubulin, chitin synthase 1, or histone 3 (Aiello et al., 2014; Guarnaccia et al., 2017).

Symptoms of postharvest anthracnose in weakened fruit are firm and dry brown to black spots as small as 1.5 mm in diameter that can slowly enlarge at ambient temperatures. Under humid conditions, pink to salmon conidial masses appear on the lesion surface. Distinctive lesions on ethylene-degreened fruit are larger, firm, flat, and silver gray, with a leathery texture (Timmer at al., 2000). As the decay advances, lesions become darker and can affect much of the rind, leading eventually to a brown to grey black soft rot (Figure 1.2c).

Postharvest anthracnose has been reported in citrus production areas worldwide (Smoot et al., 1983; Tuset, 1984; Huang et al., 2013; Honger et al., 2016), but it is more important in summer-rainfall areas, where prolonged wet periods favor the production and dispersal of inoculum and the incidence of fruit latent infections in the field, particularly if significant rains occur later in the season than normal. Conidia of *C. gloeosporioides* are produced abundantly in acervuli on dead plant parts and are spread over short distances by rain splash, heavy dew, and overhead irrigation to the developing fruits. Ascospores, although less numerous, are airborne and consequently are significant in long distance dispersal. The spores germinate on the fruit surface, giving rise to appressoria. As the season progresses and the fruit begins to mature, some of the appressoria germinate and send out infection hyphae that penetrate a short distance into the peel. However, most of the appressoria remain in an ungerminated state on the surface of the fruit (Timmer et al., 2000; Smilanick et al., 2006a). Although considerable variation can be observed for isolates of *C. gloeosporioides* from different locations, optimal environmental conditions that favor spore germination and appressorium formation are temperatures surrounding 25°C and free water or RH higher than 95%. Spores can be readily inactivated under sunlight, low temperatures, and dry weather (Siddiqui and Ali, 2014). *Colletotrichum gloeosporioides* is a weak pathogen on citrus fruits and, in general, anthracnose is a minor problem because these latent or quiescent infections only develop after harvest, mainly on fruit injured by other agents or with a naturally weakened rind (harvested late in the season, overripe, senescent, or cold-stored for too long periods). The exceptions, however, are early-season mandarins, tangerines, or oranges that are subjected to degreening treatments with exogenous ethylene to give the fruit an attractive orange-colored appearance (Timmer et al., 2000). Ethylene treatment causes a loss of chlorophyll and an increase in carotenoids in the fruit, and accelerates senescence of the peel, making it more susceptible to invasion by the infection hyphae of the pathogen (Brown, 1975). Likewise, ethylene stimulates germination of conidia and the formation of appressoria, and can also stimulate germination of appressoria (Brown, 1992). Therefore, anthracnose is clearly exacerbated in degreened fruit and can be conducive to significant economic losses in early citrus cultivars. Procedures advisable to reduce this risk include to pick the fruit with high initial rind color index (not excessively green-colored), sort the fruit by color in the packingline to maximize the uniformity of initial rind color, maintain ethylene concentration and exposure time to the minimum levels required for effective degreening, and treat with a fungicide and dry the fruit before degreening.

3.3 Brown Rot

Caused by several *Phytophthora* spp., brown rot of citrus fruits usually develops from infections that take place in the grove prior to harvest. Its control is important because it often causes economically significant losses and because of quarantines to limit its introduction on imported fruit (Adaskaveg and Förster, 2014). Many species of *Phytophthora* have been associated with brown rot, and although the symptoms caused by all are indistinguishable, their etiology varies (Graham et al., 1998). In California, *P. citrophthora* and *P. parasitica* probably account for 90% of the brown rot infections (Feld et al., 1979), although recent surveys reported that *P. citrophthora, P. parasitica* (syn. *P. nicotianae*), *P. syringae*, and *P. hibernalis* were isolated from infected fruit (Adaskaveg et al., 2015). *Phytophthora syringae*, the cause of quarantines in 2013 to stop the import of infected fruit into China from California, is considered of limited distribution and minor importance as compared to *P. citrophthora* or *P. parasitica*. In Florida, the restricted occurrence of severe brown rot on fruit was attributed to the limited distribution of *P. citrophthora* in these areas (Whiteside, 1970). In both areas, *P. parasitica* is the major cause of Phytophthora disease of the roots and trunks of the tree, but this species is less pathogenic to fruit than *P. citrophthora*. However, epidemics in 1994–1997 in Florida, which were associated with repeated rainfall events, were caused by *P. palmivora* and *P. nicotianae* (Graham et al., 1998). *Phytophthora nicotianae* was common in soil and confined to the lowest 1 m of the canopy. In contrast, *P. palmivora* typically infected fruit above 1 m in the canopy. It was primarily found on fallen immature fruit and its populations in soil were smaller than those of *P. nicotianae*.

Motile zoospores of *Phytophthora* spp. that develop on the surface of the moist soil or fallen infected fruit are splashed by rain onto fruit hanging on the lower skirt of the tree. Epidemics are associated with repeated rain events over a period of days, or wet foggy weather with temperatures from 18 to 25°C, although conducive temperatures for some Florida isolates were 23–30°C (Zitko et al., 1991). Most of the infected fruit develop on the tree within 1 m of the soil surface, although fruit higher in the tree may be infected as a result of wind-driven rains or in groves with a heavy cover crop (Graham et al., 1998). The zoospores of

Phytophthora may infect a fruit on its surface at any point; hence, this decay is not necessarily associated with the stem or stylar end of the fruit. The susceptibility of the fruit increases with maturity, so the greatest incidence of infection occurs just before or during the harvest period. Infected fruits often fall from trees earlier than healthy ones, and some managers halt harvest until the infected fruits have fallen to minimize their inclusion with healthy fruit.

Fruit infected with Phytophthora brown rot has a characteristic pungent rancid odor, which immediately distinguishes this disease from fruit afflicted with the stem-end rots. The most serious aspect of brown rot is that the fruit infected before harvest may be inspected and graded before symptoms of the disease are visible. Infected fruits therefore become mixed with sound fruits in storage or packages shipped to market (Klotz and DeWolfe, 1961). Tan spots become visible on the fruit after 3 d incubation at 25°C or after 10 d at 10°C. Under optimum conditions for decay development, the entire fruit becomes tan to brown in about 7 d, but the hyphae of the fungus may not be visible on the surface of the fruit. The texture of the lesions is leathery and the fruit remains firm (Figure 1.2d). A delicate white growth of mycelium does form on the surface of fruit stored under very high humidity conditions. Furthermore, the disease may spread from fruit to fruit by contact under the normal, relatively warm conditions of lemon storage (Klotz and DeWolfe, 1961). Fruits infected with *Phytophthora* are readily colonized by the wound pathogens *Penicillium* spp. and *G. citri-aurantii*, which may transform the firm brown rot into a soft watery rot.

3.4 Black Rot (Alternaria Rot)

Citrus trees are affected by three diseases caused by *Alternaria alternata* (Fr.) Keissl.: brown spot of tangerines, leaf spot of rough lemon, and black rot (Timmer et al., 2003). In contrast to brown spot and leaf spot, black rot is primarily a postharvest problem. It can occur in the rind, blossom end, stem end, or internal rot and can be found in most citrus production areas. The taxonomy of *Alternaria* pathogens affecting citrus is very diverse and incompletely known (Simmons, 1999; Peever et al., 2002; Troncoso-Rojas and Tiznado-Hernández, 2014), and it is likely that the literature citing *A. citri* as the cause of black rot would identify the pathogen as *A. alternata* today (Timmer et al., 2003). Peever et al. (2004) advocated collapsing all small-conidia, citrus-associated isolates of *Alternaria* into a single phylogenetic species, *A. alternata*. Isolates causing black rot do not produce the distinct host-specific toxins characteristic of virulent *A. alternata* isolates that cause serious defoliation, widespread necrosis, and halos around brown spots on mandarin fruit and foliage (Timmer et al., 2003; Garganese et al., 2016). Numerous *A. alternata* isolates from citrus fruits or leaves and even other hosts caused black rot and rind lesions on citrus fruits, indicating those causing black rot show little or no host specificity (Timmer et al., 2003; Sauer et al., 2015). Isshiki et al. (2001) reported isolates that cause black rot, all producing macerating enzymes since mutants incapable of endopolygalacturonase production were incapable of causing black rot.

Postharvest black rot was considered of minor importance in California, until recent surveys of mandarin fruit decay by Saito and Xiao (2017a). It comprised more than 53–83% of the decayed fruit in large-scale surveys of packinghouses conducted in 2015 and 2016, and was the most prevalent disease (Saito and Xiao,

2017c). This prevalence was among fruit entering the packinghouse before sorting; after sorting and cold storage, it comprised 12–15% of the decayed fruit. Postharvest black rot infects the fruit before harvest by various routes. Although colonization of single infected fruit continues after harvest, it does not contaminate facilities or spread significantly from fruit to fruit after harvest. On mandarin fruit, it most often causes large, dark brown lesions with a leathery texture on the rind (Figure 1.2e, left). It can also cause a stem-end infection when the pathogen invades through the stem end of the fruit before harvest and grows into the peel and juice sacs in a manner similar to that observed with Diplodia and Phomopsis stem-end rots (Bartholomew, 1926; Joly, 1967; Brown and McCornack, 1972; Schiffmann-Nadel et al., 1981; Isshiki et al., 2003). Infections of this type occur primarily on lemons. It can also enter at the blossom end of the fruit, commonly through cracks near the opening of the Navel oranges, where it invades internally into the juice sacks and pith of the central columella of the fruit (Figure 1.2e, right).

In contrast to Diplodia and Phomopsis stem-end rots, infection by *A. alternata* conidia is not dependent upon rainfall for dispersal during the growing season, so it can occur in all areas of citrus production. *Alternaria alternata* grows saprophytically on dead plant materials in the citrus grove and the conidia are transported by air currents, rather than by rain, to the flower or the developing fruit. However, moisture is needed for the abundant production and germination of conidia. The calyx cup ("button") is an effective receptacle for collecting airborne spores, which then become trapped under the sepals as the fruit enlarges. Removal of the calyx at harvest by snap picking, which reduces Diplodia and Phomopsis stem-end rots, does not reduce black rot incidence because *A. alternata* invades more deeply into both the calyx and the underlying tissue (Bartholomew, 1926; Pelser, 1977b; Isshiki et al., 2003). *Alternaria alternata*, like the stem-end rot pathogens, can remain quiescent and resume growth after harvest. On Navel oranges, it can also develop as a saprophyte on the necrotic style of the flower and, thereby, gain entrance to the stylar end of the fruit before harvest and colonize the interior of the fruit. It can also colonize cracks in the rind at the stylar end of the Navel oranges. Infected fruits develop mature color earlier than those not infected, and are occasionally removed manually at this time to prevent their inclusion with the healthy fruits at harvest.

Postharvest black rot can be a serious market problem because, unlike the other stem-end rots, the fungus may grow abundantly in the central axis of the fruit without any external symptoms that would be obvious to the buyer. *Alternaria alternata* is a slower-growing pathogen than either *L. theobromae* or *P. citri* and therefore, internal infections of black rot primarily become a problem among overmature oranges or mandarins and on fruits after long storage (Smoot et al., 1983). Black rot is a significant disease in California particularly among lemons, where stem-end infections develop, occasionally among Navel oranges, when cracks on the blossom end of mature fruit become infected, and commonly on the rind of mandarin oranges. Diplodia and Phomopsis stem-end rots are rare in California because of the scant precipitation during the growing season (Bartholomew, 1926; Harvey, 1946). Frost injury in the grove predisposed grapefruit to Alternaria stem-end rot during storage (Schiffmann-Nadel et al., 1975). The disease also is a major problem in the long-term storage (10–12 wk) of 'Valencia' oranges at 1°C in Florida because the low temperature and the fungicide treatments suppress *L. theobromae* and *P. citri* to a much greater extent than *A. alternata* (Smoot, 1969;

Brown and McCornack, 1972). Black rot may be exacerbated by low-temperature storage of cultivars that are sensitive to chilling. Israeli workers reported that stem-end rots in grapefruit incited by *Alternaria, Phomopsis*, and *Fusarium* were more severe at 6–8°C than at 10–12°C (Schiffmann-Nadel, 1969; Schiffmann-Nadel et al., 1981). Greater growth of the pathogen at the higher temperature was a less important factor in disease development than the loss of host resistance (incipient chilling injury) at the lower temperature. Alternaria black rot can be managed by fungicide applications before harvest. Other than fungicides, 2,4-dichlorophenoxy acetic acid (2,4-D) both before and after harvest has been used to increase resistance to infection. Applications of this growth regulator to flowers or developing fruit reduce splitting and subsequent infections through the stylar end (Klotz, 1973; Stander et al., 2017). In California, it is also routinely mixed with storage waxes applied to lemons to retard senescence of the calyx and entry of the pathogen.

3.5 Miscellaneous Diseases

Other citrus postharvest diseases originating from preharvest infections that are less important in terms of causing general economic losses are gray mold, cottony rot, and Fusarium rot. Postharvest diseases, such as Dothiorella rot, caused by *Botryosphaeria dothidea* (Moug. ex Fr.) Ces. & de Not.; Pleospora rot, caused by *Pleospora herbarum* (Pers.) Rabenh. ex Ces. & De Not.; pink rot, caused by *Trichothecium roseum* (Pers.) Link; and Septoria spot, caused by *Septoria citri* Pass., have been occasionally cited affecting *Citrus* spp. (Smoot et al., 1983; Tuset, 1987; Timmer et al., 2000). Fungi such as *Aureobasidium* spp., *Epicoccum* spp., *Phoma* spp., *Pyronema* spp., *Sordaria* spp., and *Ulocladium* spp. have also been found in infected mandarins, but at very low frequencies and often as secondary or tertiary infections without distinctive symptomatology. When tested, these fungi failed pathogenicity tests (Saito and Xiao, 2017c).

Postharvest gray mold of citrus fruits is caused by the polyphagous fungus *Botrytis cinerea* Pers. It was found in early work in Spain that this pathogen caused 2–15% of total postharvest rots on oranges and mandarins after 2 mon of storage at 2–4°C (Tuset, 1984). In California, it recently caused up to 30% of total decay on mandarins stored at 4–5°C for several weeks, and is considered as an emerging postharvest pathogen of mandarin fruit in that producing area (Saito and Xiao, 2017c). Traditionally, however, frequency of gray mold has been higher on lemons than on oranges or mandarins (Smoot et al., 1983). Typically, inoculum of *B. cinerea* in citrus orchards is produced at large quantities both saprophytically and parasitically on citrus trees, fallen fruit, and many other hosts, especially during foggy, drizzly weather, or humid conditions, which favor sporulation. Conidia are easily spread by rain, wind, and insects and may infect flowers causing quiescent infections in the stem end of the fruit that develop after harvest. Under especially favorable conditions for fungal development, infected blossoms or young fruits may drop in the field. Citrus fruit infections can also occur at any place of the surface through rind wounds of various origins or by contact with infected flower tissues that adhere to adjacent fruit (Eckert and Eaks, 1989; Timmer et al., 2000). After harvest, disease symptoms on lemons are first firm, drab brown lesions that later become leathery, pliable, and dark brown. On oranges, affected surface is medium to dark yellowish brown. Under conditions of high temperature and humidity, lesions considerably enlarge and characteristic white mycelia later covered with tufts of gray, granular-appearing conidia develop on the lesion surface (Figure 1.2f). Its incidence may be especially high on long-term cold-stored citrus fruits since *B. cinerea* is able to grow at temperatures as low as 0°C. When fruits are bulk-stored in cartons, boxes, or bins, the pathogen can spread by contact from infected to adjacent healthy fruit causing "nests" of decay (Smoot et al., 1983; Eckert and Eaks, 1989). Recent research in California showed that many phenotypes of *B. cinerea* isolated from mandarins were highly resistant to common citrus postharvest fungicides, such as fludioxonil, pyrimethanil, and thiabendazole (Saito and Xiao, 2017b).

Cottony or Sclerotinia rot of citrus is caused by *Sclerotinia sclerotiorum* (Lib.) de Bary, a necrotrophic and non-host-specific fungus that, besides fruit, can also infect twigs, bark, and roots of citrus trees (Polizzi et al., 2011). Cottony rot is widely distributed in many of the citrus-growing areas of the world, including Spain (Tuset and Martí, 1988), California (Eckert and Eaks, 1989), and Pakistan (Hanif et al., 2016), and although it affects all citrus species and cultivars, decay losses have been typically more important on stored lemons. Nonetheless, the widespread practice of clean cultivation and the abandonment of cover cropping and green manuring in lemon orchards have substantially reduced the incidence of the disease because they are the main source of sclerotia production (Timmer et al., 2000). After a long dormant period, sclerotia germinate on the soil under humid conditions to form apothecia, from which ascospores are released and dispersed by air currents to fruit on the tree. Rain splash of soil onto fruit or field harvest bags, buckets, or boxes can also contribute to inoculum dissemination. Fruit infection occurs mainly in the stem and stylar ends or through deep rind wounds. Similarly to gray mold, infected flower remnants in contact with fruit surfaces can also be a route of infection (Tuset, 1987; Eckert and Eaks, 1989). Field infections commonly remain latent and after harvest, during the early stages of fungal development, *S. sclerotiorum* secretes cell wall-degrading enzymes such as polygalacturonases, exo-β-1,3-glucanases, xylanases, and cellulases that cause discoloration and slow softening and leathering of the rind (Oliveira et al., 2013). Symptoms on lemons in a dry atmosphere are yellowish brown soft lesions. In a moist atmosphere, the lesions are rapidly covered with a characteristic white, fluffy, cottony mycelium and, after several weeks, large, irregularly shaped, black sclerotia are formed. Major losses due to this disease arise from decay "nests" produced by fruit contact during prolonged cold storage of lemons, which, due to chilling injury risks, is conducted at temperatures of 10–15°C (Smoot et al., 1983; Eckert and Eaks, 1989).

Species of the genus *Fusarium* that were described in early reports as causal agents of Fusarium rot of citrus fruits include *F. oxysporum* Schltdl., *F. moniliforme* J. Sheld., *F. lateritium* Nees, *F. fructigenum* Fr., *F. culmorum* (W.G. Sm.) Sacc., and *F. solani* (Mart.) Sacc. (Tuset, 1984; Schiffmann-Nadel et al., 1987; Eckert and Eaks, 1989). Among them, the latter species seems to be currently preponderant or more active, at least in some producing areas, as recently reported by authors from Egypt (Abd-Elsalam et al., 2015; Youssef et al., 2017) or China (Fu et al., 2017). Fusarium rot can affect all citrus species and cultivars, but its incidence is usually higher on grapefruits and oranges, especially on those of the Navel group, because disease at the fruit stylar end can initiate from incomplete closure of the navel. *Fusarium* spp. are common saprophytic soil inhabitants developing on plant debris and producing abundant macroconidia and chlamydospores that disperse by water

Figure 1.2 a. Stem-end rot of orange caused by *Lasiodiplodia theobromae*. b. Stem-end rot of orange caused by *Phomopsis citri*. c. Anthracnose of orange caused by *Colletotrichum gloeosporioides*. d. Brown rot of lemon caused by *Phythophthora* sp. e. External (mandarin, left) and internal (orange, right) black rot caused by *Alternaria alternata*. f. Nest of gray mold caused by *Botrytis cinerea* on stored mandarins.

or wind to reach the fruits on the tree. Infections may occur in the stem and stylar ends, but also in any part of the fruit surface through rind wounds or injuries. These infections typically remain quiescent and only develop after harvest during long-term storage. In general, *Fusarium* spp. are weak pathogens of citrus tissues and latent infections only develop on senescent or weakened fruit. External lesions first resemble those of anthracnose or black rot, starting as a soft brown spot that becomes leathery, sunken, and dark brown in color. Under humid conditions, they slowly become covered by white mycelium that may turn to beige or pink depending on the causal *Fusarium* sp. An internal decay originated from infections in the stem or stylar ends can also occur in the central axis of the fruit, which becomes decomposed with a reddish-brown discoloration. In general, if citrus fruits are sound, with the button in good condition, and stored at the usual temperatures of 3–5°C for oranges and mandarins, the incidence of Fusarium rot is low and no specific postharvest control measures are needed (Smoot et al., 1983; Tuset, 1987; Timmer et al., 2000).

4 Commercial Disease Management

Postharvest fungicides form part of the integrated packinghouse strategy to minimize diseases such as green and blue molds as well as sour rot that require fruit rind wounds to initiate infections. Other elements of a decay management strategy are to minimize wounds, apply treatments promptly after harvest, and store and ship at appropriate temperatures. This strategy includes actions to control pathogens, such as disinfecting the fruit and its environment, inhibition of conidia in wounds, and to protect the fruit from future infections by means such as deposition of a fungicide residue (Eckert, 1990). The efficacy of fungicide application is determined by the potency of the active ingredient and its formulation, timing (infection age), and application method, as well as fruit vitality. All currently implemented postharvest fungicides are registered for the control of green and blue molds, while a few have action against sour rot and diseases caused by latent pathogens, such as Diplodia and Phomopsis stem-end rots and anthracnose. A ranking of fungicide active ingredients for the control of green mold is presented in Table 1.1.

Table 1.1 Ranking of fungicide effectiveness for the control of green mold caused by *Penicillium digitatum*

Ranking	*Fungicide*	*Curative control*	*Protective control*	*Sporulation control*	*Other target diseases*
1	Imazalil (IMZ)	Excellent	Good	Excellent	Alternaria rot Diplodia rot
2	Thiabendazole (TBZ)	Excellent	Fair	Good	Diplodia rot Anthracnose
3	Sodium ortho-phenylphenate (SOPP)	Excellent	Poor	Fair	...
4	Pyrimethanil (PYR)	Excellent	Poor	Poor	...
5	Fludioxonil (FLU)	Fair to good	Fair to good	Fair to poor	Diplodia rot
6	Propiconazole (PCZ)	Fair	Poor	Poor	Sour rot
7	Azoxystrobin (AZX)	Poor	Fair	Fair to poor	...

4.1 Fungicides

4.1.1 Imazalil

1-[2-(2,4-dichlorophenyl)-2-(2-propenyloxy-ethyl)]-1*H*-imidazole

Mode of action: Demethylation inhibitor of ergosterol biosynthesis (Siegel and Ragsdale, 1978; Siegel, 1981).

Since its introduction in the late 1970s (Laville et al., 1977), the usage of imazalil (IMZ) grew to it being the first choice in most packinghouses. This is due to having three distinct attributes; both curative and protective action, and inhibition of the sporulation of pathogenic *Penicillium* spp. on infected fruit. Sporulation produces abundant conidia and their release increases airborne inoculum and conidia that deposit on adjacent fruit, termed "soilage" that requires costly and time-consuming cleaning to remove. Although not well-documented, control of Alternaria rot, stem-end rots, and sour rot was also claimed (Laville et al., 1977; McCornack and Brown, 1977). The introduction of IMZ to the South African citrus industry was associated with a 50% reduction in decay losses (Pelser and La Grange, 1981). In the European Union (EU), this active ingredient is currently under revision due to uncertainties about the toxicity of some IMZ metabolites and a decision on its legal status is expected by 2020.

IMZ is available in two distinct formulations. The emulsifiable concentrate (EC) formulation with limited water solubility was used in the majority of research projects employing IMZ (Eckert, 1977; Schirra et al., 1996, 1997; Smilanick et al., 1997b, 2005; Cabras et al., 1999; D'Aquino et al., 2006; Dore et al., 2009, 2010). The IMZ sulfate salt (SS) formulation is better suited for aqueous applications due to its water solubility and superior effectiveness in aqueous treatments (Sepulveda et al., 2015). In contrast to the SS formulation, the EC formulation is better suited for wax applications. IMZ SS applied in an aqueous dip solution has better curative than protective activity, although sporulation inhibition is less efficient than the IMZ EC formulation (Erasmus et al., 2011). In contrast, IMZ EC applied in wax has better protective than curative activity, with excellent sporulation inhibition (Njombolwana et al., 2013). The double application of both IMZ formulations, with the first application of an aqueous solution of SS followed by the EC in wax, was shown to give the best results in terms of curative and protective control of green mold as well as sporulation inhibition (Njombolwana et al., 2013; Sepulveda et al., 2015), and has become a common commercial practice worldwide.

4.1.2 Thiabendazole

4-(1*H*-1,3-benzimidazol-2-yl)-1,3-thiazole

Mode of action: Thiabendazole (TBZ), like other benzimidazole fungicides, inhibits the polymerization of microtubules, which causes fungal cell division to cease (Clemons and Sisler, 1971; Hammerschlag and Sisler, 1973).

TBZ has been in use since the early 1960s (Harding, 1962). It has curative (Schirra et al., 2008) and protective (Brown, 1977) action as well as modest inhibition of *P. digitatum* sporulation (Ladaniya, 2008). Efficacy to control Diplodia stem-end rot and anthracnose on citrus has also been shown (Wardowski and Brown, 1993). Aqueous application of TBZ provides better curative control compared to protective control, while, in contrast, TBZ applied in wax provided superior inhibition of sporulation (Kellerman et al., 2014). An important complementary effect of TBZ is its ability to reduce chilling injury (Hordijk et al., 2013). The limited solubility of postharvest formulations of TBZ cause it to be very difficult to keep in suspension as it precipitates rapidly in both aqueous and wax suspensions.

4.1.3 Pyrimethanil

4,6-dimethyl-*N*-phenylpyrimidin-2-amine

Mode of action: Inhibits methionine biosynthesis (Fritz et al., 1997) and the secretion of cell wall-degrading enzymes (Daniels and Lucas, 1995).

Pyrimethanil (PYR) was introduced to the citrus postharvest industry in the first decade of the 2000s (Smilanick et al., 2006b). Good curative control of green mold can be expected from PYR applied in aqueous dip treatments, while protective activity and sporulation inhibition are minimal (Kellerman et al., 2018). When inoculated after PYR treatment of as high as 1000 mg/L, green mold was not controlled. In contrast, when inoculated many hours before PYR treatment, a solution of 500 mg/L effectively controlled green mold. Modest sporulation inhibition was observed with residue levels higher than 4 mg/kg. Increasing PYR solution temperature increases PYR residue content (Smilanick et al., 2006b). The formulation of PYR is a suspension, and therefore will cause challenges in aqueous and wax application due to precipitation.

4.1.4 Fludioxonil

4-(2,2-difluoro-1,3-benzodioxol-4-yl)-1*H*-pyrrole-3-carbonitrile

Mode of action: Inhibits osmoregulatory signal function, causing inhibition of spore germination, germ tube elongation, and mycelial growth (Rosslenbroich and Stuebler, 2000).

Aqueous dip treatments of fludioxonil (FLU) resulted in acceptable curative control of green mold 24 hr after inoculation (D'Aquino et al., 2013). Work done by Kanetis et al. (2007) indicated that FLU when applied in aqueous dip treatment had some activity against younger infections (<12 hr), but not as good as IMZ and PYR. However, both IMZ and PYR could control infections from inoculations done of 21 hr before treatment, while FLU could not. When FLU residues were increased to >2 mg/kg, by means of its application in heated solutions (50°C), nearly 100% of 24 hr-old green mold infections were controlled (Schirra et al., 2005). Kanetis et al. (2007) showed that FLU was able to inhibit sporulation of *P. digitatum*, although the specific residue level was not indicated. This fungicide is also formulated in a suspension and this can cause solubility problems in aqueous and wax treatments.

4.1.5 Azoxystrobin

Methyl (2*E*)-2-{2-[6-(2-cyanophenoxy) pyrimidin-4-yloxy]phenyl}-3-methoxyacrylate

Mode of action: Inhibits mitochondrial respiration (Mansfield and Wiggins, 1990).

This fungicide does not very effectively control green mold but has modest antisporulant action that can be beneficial in a postharvest management program (Bushong and Timmer, 2000). Although registered separately in some countries, azoxystrobin (AZX) is commonly used in conjunction with FLU, especially in the USA. The combination of FLU and AZX had a synergistic effect in controlling green mold when applied in an aqueous spray (Kanetis et al., 2007). This combination had protective action (about 60% control) when applied in a dip treatment, but its performance to cure green mold infections when applied 24 hr after inoculation was modest (<50% control) (Erasmus et al., 2015a). Formulations of these two active ingredients are also suspensions and therefore are difficult to maintain in suspension in high-volume aqueous treatments.

4.1.6 Sodium Ortho-Phenylphenate and Ortho-Phenylphenol

Sodium [1,1'-biphenyl]-2-olate and [1,1'-Biphenyl]-2-ol

Mode of action: Interfere with the cell wall and enzymes essential for cell metabolism; also thickening of constituents in the cytosol (Koleci et al., 2007).

These fungicides have been in use since 1965 (McCornack, 1970). Although still registered in most citrus-producing countries, sodium ortho-phenylphenate (SOPP) or ortho-phenylphenol (OPP) is rarely used. Both these formulations leave the residue OPP. SOPP is soluble in water when the pH is maintained between 11.5 and 12.5. In solutions of lower pH than this, the product is less soluble and it becomes phototoxic, while in solutions of higher pH than recommended, decay control effectiveness decreases (McCornack, 1970). The risk of phytotoxicity and vigilant solution management renders this product a less desirable choice than other fungicides (Dezman et al., 1986). Nevertheless, it was shown to be an excellent alternative in an IMZ resistance management program (Erasmus et al., 2015b).

4.1.7 Propiconazole

1-((2-(2,4-dichlorophenyl)-4-propyl-1,3-dioxolan-2-yl) methyl)-1*H*-1,2,4-triazole

Mode of action: Demethylation inhibitor of ergosterol biosynthesis (Leroux, 2003; Calonne et al., 2012).

Propiconazole (PCZ) is currently used to control sour rot and green mold in many countries. However, its non-renewal as an active ingredient for postharvest applications in the EU countries was voted in October 2018, and it will be definitively banned in that area by 2020. When applied in a high-volume aqueous treatment, it controls young infections (12–16 hr) of green mold and sour rot (McKay et al., 2012b). Cross-resistance in pathogen populations is possible between PCZ and IMZ since both inhibit demethylation in the ergosterol biosynthesis pathway (McKay et al., 2012c). It is currently the only fungicide with a nearly worldwide maximum residue limit (MRL) available to control sour rot, so it is a compelling option to consider when sour rot is prevalent. More research is necessary to determine the most effective use of this fungicide. Sour rot control effectiveness, expressed as percent reduction in its incidence, on clementine, lemon, and orange fruit when the PCZ drench treatment was applied to 6 hr-old infections was >90% on all. On the older infections of 14, 18, and 24 hr-old, it was 76, 87, and 96%, respectively (Mamba et al., 2018). The same group found that control was already <50% on an 18 hr-old infection on clementine and <40 and <70% on 24 hr-old infections on lemon and orange, respectively. This fungicide is also formulated in a suspension, which causes challenges to keep it in aqueous or wax suspension.

4.1.8 Potassium Phosphite (K_3O_3P)

Mode of action: Enhances the plant defense system (Smillie et al., 1989) and/or interferes with the pathogen phosphate-starvation response (Carswell et al., 1996; Förster et al., 1998).

Potassium phosphite is exempt from tolerance in the USA (Adaskaveg et al., 2015), but MRLs are still lacking in many prominent export countries. It is commonly applied before harvest in many countries, primarily for plant nutrition purposes. The most important attribute of this compound is its ability to control brown rot caused by *Phytophthora* spp. at relatively low rates (Adaskaveg et al., 2015). For this application, it could replace the organophosphorous aluminum salt Fosetyl-Al, which is still in use as a postharvest fungicide. Heated phosphite solutions were more effective than ambient solutions, in most cases. At higher rates, especially when heated, it has good to excellent curative control action against green and blue molds, sour rot, and Phomopsis stem-end rot, and improves the performance of fungicides when they are used with it (Cerioni et al., 2013a, 2013b, 2017).

4.1.9 Other Active Ingredients

Other fungicides evaluated to control citrus postharvest diseases and registered for this use in some producing countries include myclobutanil, thiophanate methyl, trifloxystrobin, cyprodinil, and Fosetyl-Al. The use of prochloraz and guazatine is rapidly declining due to their elimination on MRL lists of many citrus-importing countries.

4.2 Application Methods

4.2.1 High-Volume Aqueous Applications

These types of treatments are mostly overhead drench- or waterfall-type treatments or the fruit pass through aqueous solutions in tanks. Infection age, temperature, exposure time, fungicide concentration, and in some cases, solution pH play pivotal roles that influence residue content and decay control efficacy.

4.2.1.1 Bin Drench **Definition**: Harvest bins or crates containing fruit are transferred on a conveyer underneath a waterfall-type system that dumps circulating fungicide solution in high volumes onto the fruit, which then drains through the binned fruit into a circulation tank from where the solution is pumped back to the delivery system above the fruit bins. An alternative bin drench method is truck drenching, where trailers of fruit in harvest bins behind trucks are treated within a drive-through drenching system before delivering the fruit to the packinghouse.

Janisiewicz et al. (2005) described a type of bin drench system, but they used it to apply biological control agents on apples. Brown et al. (1988) described a conveyer bin drench system while investigating a truck bin drench system. In many cases, stacks of two or three bins are drenched. Residue levels can fluctuate from the top to the bottom bin, where the bottom usually reflects lower levels than the top bin. Drenching fruit before degreening can significantly reduce Diplodia stem-end rot compared to untreated fruit or those treated after degreening (Brown et al., 1988). This author stated that concentration management, exposure time, and continuous solution sanitation are required for effective residue loading and to limit the buildup of pathogen contaminants such as fungal propagules causing green mold and sour rot. Smilanick et al. (2006c) reported that decay could be controlled up to 60% in an ambient (13°C) bin drench solution of TBZ and by about 80% in a heated solution (41°C). In practice, bin drench solutions are typically not heated. Although bin drenching is widely used in the citrus industries of Spain, South Africa, Australia, Florida, or Turkey, to name just a few, very little research has been published on this topic. An experimental small-scale drench of IMZ EC applied to fruit crates reduced green mold incidence to <10% compared to >90% among untreated fruit (Erasmus et al., 2011). Increasing the flow volume increased residue content and improved green mold control. By means of an improved experimental drench unit employed in South Africa, Kellerman et al. (2014) investigated the effect of exposure time and TBZ concentration. Exposure time did not have a significant influence on efficacy and residue content, whereas concentration did. Increasing TBZ concentration by 1000 mg/L resulted in a >1 mg/kg increase in residue content. Longer exposure times (28–56 s) controlled infections and sporulation better than shorter time (14 s) with fruit inoculated 24 hr or longer before drench treatment with a combination of PYR and TBZ (Christie, 2016). In relation to decay control, factors of fruit orientation (infection site facing upwards or downwards), action (curative or protective), and exposure time were most important. Infection sites that faced upward toward the pouring solution were about 20% better controlled compared to infections facing downward away from the pouring solution (Christie, 2016). Drench treatment with TBZ was more curative (33.9–92.1% control) than protective (6.0–39.8% control) (Kellerman et al., 2018). Prolonging exposure of the fruit to the drench solution from 30 to 60–90 s improved both curative and protective control (Kellerman et al., 2018). These workers focused on PYR and found that drenching younger infections (inoculated 6 hr before treatment) were more effectively controlled than older ones (inoculated 24 hr before treatment); drenching with fungicide mixtures (various combinations of IMZ, TBZ, PYR, and guazatine) had better curative and protective activity compared to using single active ingredients, and that IMZ had exceptional curative and protective activity even when used alone.

4.2.1.2 Dip **Definition**: Fruits are transferred on the packingline by means of conveyers into a tank containing an aqueous fungicide solution where they are submerged for a certain amount of time before they are conveyed out, where excess solution is removed by water elimination rollers consisting of combinations of cylindrical sponges and/or brushes and air applied at high velocity.

Dip tanks are extensively used in South African packinghouses to apply IMZ SS (78%; Erasmus et al., 2011), although this application is less popular in other major citrus-producing countries such as Spain. Fungicides such as TBZ, FLU, and PYR are typically not applied in a dip tank due to the difficulty of keeping these products in suspension. The majority of research done on IMZ and other fungicides is mostly with dip application and specifically with the IMZ EC formulation (Eckert, 1977; Schirra et al., 1996, 1997, 2005; Smilanick et al., 1997b, 2005, 2006b; Cabras et al., 1999; D'Aquino et al., 2006; Dore et al., 2009, 2010).

When studying dip application from a commercial perspective the factor of post-dip brushing needs to be considered as part of the treatment. Brushing after dip treatment can reduce a potential IMZ residue by up to 90% at a solution pH of 3; if the pH was increased to 6, where the fungicide is less water soluble, the loss was reduced to 60% (Erasmus et al., 2015a). This did not affect curative control as much as sporulation inhibition, which was lost at lower residue levels (<0.5 mg/kg at pH 3 and 1.3–2.3 mg/kg at pH 6 depending on exposure time). Dip application resulted in better curative control at similar residue levels compared with wax treatment with the IMZ EC formulation (Smilanick et al., 1997b). Residue content on fruit dipped for 60 s in a 20°C IMZ EC solution was more than double than that in an IMZ SS solution regardless of concentration (Erasmus et al., 2011). Curative efficacy of green mold infections made by fungicide-sensitive *P. digitatum* isolates by these treatments was similar or slightly better in the IMZ EC treatment and more so on IMZ-resistant isolates. Dip application gave excellent curative control of green mold when using most of the available fungicides such as IMZ, PYR, and TBZ (Smilanick et al., 2006b; Erasmus et al., 2011; Kellerman et al., 2014, 2018). Sporulation control was not satisfactory by dip treatments of the fungicides FLU, IMZ, PYR, and TBZ (Smilanick et al., 2006b; Kanetis et al., 2007; Kellerman et al., 2014, 2018). This may be due to dip application loading inadequate residue levels; natural wax thickness on the rind is irregular and does not facilitate deposition of a uniform residue needed to get good sporulation control (Brown and Dezman, 1990).

4.2.1.3 Flooder **Definition**: A flooder is a recirculating high-volume low-pressure laminar waterfall application on fruit over rotating brushes, although some variations employ roller conveyers instead of brushes. The number of waterfalls or weirs may vary from one to as many as nine. The flooder may also be referred to as an in-line drench or heated IMZ flooder.

Very little research has been done on this application. In the late 1960s, McCornack and Brown (1967) use a flood treatment, but it is unclear whether it was a bin drench or flooder on a packingline. Wardowski et al. (1974) described a flooder application that deposited adequate residues (>0.30 mg/kg) of TBZ and controlled natural infections of green mold on orange fruit. PYR applied in a flooder effectively controlled green mold (Smilanick et al., 2006b). Kanetis et al. (2007) showed that PYR, FLU, and AZX had effective curative control when applied with a flooder. Sour rot and green mold were adequately controlled with a flooder treatment of PCZ (McKay et al., 2012b). Flooder treatments of IMZ SS at either 250 or 500 mg/L resulted in excellent curative control (85% or higher) and even better protective control (>90%) (Erasmus, unpublished data). In the same study, it was shown that one weir (waterfall) provided >75% curative control of green mold, although three to five weirs were more consistently effective and control levels approached 90%. Flooder treatments of 8 s with 250 mg/L IMZ SS at 45°C and pH 3 gave curative control higher than 89% in most cases (Savage, 2017). Protective control is slightly weaker or similar under these treatment conditions. This was achieved with IMZ residue levels of 0.31–0.78 mg/kg, increasing the pH to 4 resulted in residue levels of about 1.00 mg/kg with higher and more consistent levels of control. This research group showed that IMZ residue levels of 1.29–2.47 mg/kg loaded by means of flooder treatment inhibited sporulation to levels lower than 10%. They also showed that, under these treatment conditions, solution temperatures ≥55°C and pH ≥5 can result in exceeding the MRL (5.00 mg/kg) of most countries.

4.2.2 Low-Volume Non-Recirculating Spray Applications

4.2.2.1 Aqueous Sprays **Definition**: Aqueous solutions or suspensions of fungicide applied through nozzles at low volume under high-to-medium pressure onto fruit rolling over rotating polyethylene brushes.

Non-recirculating sprays are often used to apply a water-soluble storage wax (0.5–5% wax in water) in combination with fungicides for fruit destined for long-term storage in California. Although not as effective as a dip treatment, a low-volume spray application of PCZ can give close to 70% control of sour rot (McKay et al., 2012b). This method also gave some level of control when the combinations FLU/AZX, TBZ/FLU, or IMZ/PYR were applied to control green mold, but not as effective as a flooder treatment (Kanetis et al., 2007). Although a higher IMZ residue level (6.03 mg/kg) was loaded when sprayed with 1000 mg/L, compared to 2.39 mg/kg loaded after a dip in 500 mg/L, green mold control was similar or better with the dip treatment (Erasmus et al., 2011).

4.2.2.2 Pack Wax Sprays **Definition**: Fungicides are mixed with a viscous emulsion of waxy solids (single or combinations of polyethylene, shellac, wood resin, and/or carnauba) applied via a non-recirculating system through nozzles, drip, or controlled disc atomizers onto fruit rolling over rotating brushes of polyethylene or horse hair or a combination of these.

Wax is applied to citrus fruits to improve their shine and reduce weight loss from the loss of water. In so doing, waxes preserve quality and act as a carrier for fungicides (Hall and Sorenson, 2006; Palou et al., 2015). Hall (1981) and Palou et al. (2015) published thorough descriptions of the composition and components of waxes and other citrus fruit coatings. Fungicide (AZX, FLU, IMZ, PYR, and TBZ) application in wax generally gives better protective action and mediocre to poor curative action (Kanetis et al., 2007; Njombolwana et al., 2013; Kellerman et al., 2014, 2018). IMZ applied in wax gave excellent sporulation control (Njombolwana et al., 2013).

4.3 Sanitizing Recirculating Solutions

The requirement to sanitize recirculating fungicide solutions is becoming more and more prominent due to the potential of unwanted microbial populations developing in packinghouse solutions and increased focus on food safety. Countries such as the USA have implemented laws to ensure safety throughout the food chain that affect the postharvest management of horticultural produce including citrus fruits (US FDA, 2011). Currently, very few chemical sanitation options are compatible with postharvest fungicides. The popular sanitizer chlorine is compatible only with FLU and TBZ of the postharvest fungicides in use today. However, the combination of H_2O_2/peroxyacetic acid (PAA) is compatible with AZX, FLU, IMZ, PCZ, PYR, and TBZ (Kanetis et al., 2008; McKay et al., 2012b). Arthroconidia of *G. citri-aurantii* were reduced from >5000 to <156 CFU/mL when chlorine and PAA were used to sanitize a drench solution of PYR, TBZ, and 2,4-D contaminated with such arthroconidia (Christie, 2016).

Table 1.2 Suggested options for combinations of fungicide active ingredients and application methods for specific outcomes regarding the control of postharvest diseases indicated

Approach		*Drench*	*Flooder*	*Pack wax*	*Storage wax*
Conservative	Fungicides	TBZ + Chlorine	IMZ + TBZ[1] + PAA	IMZ	IMZ
	Target	• Green and blue molds • Diplodia rot	• Green and blue molds • Diplodia rot • Alternaria rot	• Green and blue molds	• Green and blue molds
	Benefit	• Curative control • Rind fortification	• Curative control • Protective control • Rind fortification	• Protective control • Sporulation control	• Protective control • Sporulation control
Resistance management	Fungicides	TBZ + PYR + PAA	IMZ + FLU[2] + PAA	IMZ	IMZ
	Target	• Green and blue molds • Diplodia rot	• Green and blue molds • Diplodia rot • Alternaria rot	• Green and blue molds	• Green and blue molds
	Benefit	• Curative control • Rind fortification	• Curative control • Protective control	• Protective control • Sporulation control	• Protective control • Sporulation control
High performance	Fungicides	TBZ + IMZ[3] + PCZ[4] + PAA	IMZ + PYR + PAA	IMZ	IMZ
	Target	• Green and blue molds • Sour rot • Diplodia rot	• Green and blue molds • Alternaria rot	• Green and blue molds	• Green and blue molds
	Benefit	• Curative control • Sporulation control • Rind fortification	• Curative control • Protective control	• Protective control • Sporulation control	• Protective control • Sporulation control

(*Continued*)

Table 1.2 (Cont.)

Approach		*Drench*	*Flooder*	*Pack wax*	*Storage wax*
Sour rot management	Fungicides	TBZ + PCZ[3] + PAA	IMZ + PCZ[5] + PAA	IMZ	IMZ
	Target	• Green and blue molds • Sour rot • Diplodia rot	• Green and blue molds • Sour rot • Alternaria rot	• Green and blue molds	• Green and blue molds
	Benefit	• Curative control • Sporulation control • Rind fortification	• Curative control • Protective control	• Protective control • Sporulation control	• Protective control • Sporulation control

[1]TBZ can be applied in the flooder or storage wax, flooder performance will be better.

[2]FLU can be applied in the flooder or storage wax, flooder performance will be better.

[3]By applying IMZ and/or PCZ before and after a delay or degreening, the risk for resistance development is increased.

[4]If no bin drench available, PCZ can be applied in the flooder or storage wax, flooder will give the best performance.

[5]PCZ can be applied in the flooder or storage wax, flooder will give the best performance.

4.4 Sequence and Combinations of Application Methods and Fungicides

When IMZ is used as a case study to investigate the concept of precision application, meaning aiming for a combination of targeted effects, the following has been established by recent studies. IMZ applied in high-volume aqueous treatments followed by brushing will result in excellent curative action despite loading very low residues (<0.50 mg/kg) (Erasmus et al., 2015a). These low residue levels will not be adequate for protective or sporulation control where >2.00 mg/kg is needed. Adequate residues for sporulation control can be obtained by adding IMZ to the wax treatment, but IMZ in a pack wax treatment only will not give adequate curative control (Smilanick et al., 1997b; Njombolwana et al., 2013). The double application (high-volume aqueous plus pack wax) of IMZ will hit the targets of curative activity, protective activity, and sporulation control of green mold. Fungicide resistance has been widely documented, especially on the older fungicides, such as OPP, TBZ, and IMZ. Although IMZ can achieve all targeted effects described here, it is not sustainable to use a fungicide with a single active ingredient season after season. Currently and especially for green mold control, the citrus industry has a variety of fungicides to its disposal with various modes of action that can be combined to manage resistance development or give more potent control. Taking all above-mentioned research into account and adding some practical experience, ideal combination of fungicides and application methods are suggested for specific outcomes (Table 1.2).

4.5 Factors Further Affecting Fungicide Efficacy

4.5.1 Effective Residue

Until 1995, the state of Florida enforced minimum residue levels for fungicides on fresh citrus (FDOC, 2018). The Florida minimum requirement was 1/20 of the MRL of at least one of the fungicides used on citrus fruits destined for the USA domestic market (McCornack, 1970). This relates to 0.5 mg/kg for the USA MRL of IMZ (10 mg/kg). This would ensure that a consistent level of decay control was achieved. In recent years, much attention has been paid to residue levels and retailers, especially those in Europe, are demanding residues of fewer pesticide active ingredients be present on produce, as well as lower levels of those with no scientific basis to support these requirements (Melo et al., 2013; Sonntag et al., 2016). Several researchers have specifically or indirectly investigated what an effective residue level would be for a specific fungicide to control, mostly green mold. For example, some information is available for IMZ (Smilanick et al., 1997b; Njombolwana et al., 2013; Erasmus et al., 2015a, b), FLU (Schirra et al., 2005), PYR (Smilanick et al., 2006b; Kellerman et al., 2018), and TBZ (Smilanick et al., 2008; Kellerman et al., 2014), and it is summarized in Table 1.3. Residue levels of 1.00–2.00 mg/kg of FLU, IMZ, PYR, or TBZ will give good to excellent curative control applied by means of high-volume aqueous treatments. Higher levels (2.00–4.00 mg/kg) are necessary for protective or sporulation control in the case of green mold.

Table 1.3 Effective residue levels in fruit (mg/kg fresh weight) of common fungicides used for the control of citrus green mold caused by *Penicillium digitatum*

	For 50% control		*For 75% control*		*For 100% control*		*For sporulation control*
Fungicide	*Curative*	*Protective*	*Curative*	*Protective*	*Curative*	*Protective*	
Imazalil (IMZ)	0.15–0.37[a]	0.11–0.35[a]	–	–	≥1.00[e]	–	≥2.00[f,i]
Thiabendazole (TBZ)	0.03–0.11[b]	–	0.06–0.22[b]	–	>0.97[g]	–	≈3.00[b]
Pyrimethanil (PYR)	0.14–0.53[c]	0.97–1.43[c]	0.47–1.16[c]	2.30–4.75[c]	>0.83[g] 1.00–2.00[h]	–	>4.00[h]
Fludioxonil (FLU)	–	–	≈0.50[d]	–	≥0.80[d]	–	2.00–4.00[i]

a Erasmus et al. (2015b)
b Kellerman et al. (2014)
c Kellerman et al. (2018)
d Schirra et al. (2005)
e Erasmus et al. (2015a)
f Smilanick et al. (1997b)
g Smilanick et al. (2008)
h Smilanick et al. (2006b)
i Ladaniya (2008) and
j Njombolwana et al. (2013)

4.5.2 Infection Age and Fruit Susceptibility

Mostly during the harvest process, fresh wounds are inflicted on citrus rinds that considerably increase the risk of infection by decay pathogens (Rose et al., 1951). It takes *P. digitatum* as short a period as 4 hr from germination to infection (Plaza et al., 2003a). A few recent studies investigated the effect of infection age on green mold control on various citrus types and harvest batches. Drenching 6 hr-old infections compared to 24 hr-old infections of green mold with PYR on 'Valencia' orange fruit resulted in 5–12 times better curative control depending on exposure time (57.3–69.8% control for 6 hr and 4.9–14.6% control for 24 hr) (Kellerman et al., 2018). In the same study, much lower control levels (≤7.0%) were achieved on Navel orange fruit compared to those on 'Valencia' orange fruit. Curative control of green mold by IMZ dip treatments on clementine mandarin fruit was only effective on 6–12 hr-old infections depending on wound size, where 18–36 hr-old infections were controlled on Navel orange fruit with the same treatment (Erasmus et al., 2015a). This was confirmed in another study by Kellerman et al. (2016), where effective control levels were achieved on 24 hr-old infections on Navel orange and lemon fruit, but not on clementine fruit regardless of increased IMZ pH level, solution temperature, or exposure time. Kanetis et al. (2007) tested AZX, FLU, IMZ, and PYR as single fungicide aqueous spray treatments on a green mold infection age range from 9 to 21 hr on lemon fruit. All four fungicides were able to control the 9 hr-old infection; IMZ and PYR could control all infections up to 21 hr, while AZX and FLU showed a steady decline as age increase where control with AZX and FLU was reduced by around 40 and 20%, respectively, at 21 hr. When TBZ was combined with FLU, control levels were similar to those of IMZ and PYR. When AZX and FLU were combined control levels, especially of the older infections, improved but not as well as when IMZ and PYR were combined. Interestingly, in this study, IMZ was able to control 9- and 12-hr infections of *P. digitatum* isolates resistant to IMZ, as well as AZX and PYR, and better than FLU. Christie (2016) inoculated two different batches each of Satsuma mandarin, lemon, and Navel orange fruit with *P. digitatum* conidia at a time range of 0–54 hr prior to treatment with a combination of PYR and TBZ. A threshold of <90% green mold will be used for loss of control in the discussion that follows below. According to their prediction, control was lost at about 25 and 33 hr-old infections on the two respective Satsuma mandarin batches, with the first sign of control loss already observed at 12 hr. On lemon, control was predicted to be lost after about 30 and 45 hr for the two respective batches, with the first sign of control loss at 18 hr. Lastly, on Navel orange, they predicted to lose control at about 30 hr for both batches with no sign of control loss before 30 hr. Curability of infection before the predicted loss of control varied the most on Satsuma and this variability was very small on Navel orange fruit.

Brown (1979) mentioned that fruit becomes more susceptible to sour rot as fruit maturity increases, while differences in green mold control on different fruit harvest batches of the same citrus type were also observed in other studies (Njombolwana et al., 2013; Kellerman et al., 2014; Erasmus et al., 2015a). The industry rule to treat fruit within 24 hr after harvest is clearly validated, but to maintain consistent and improved control, a more sophisticated approach may be required. Factors to consider are fruit maturity at the time of harvest, time from harvest to treatment, fungicide active ingredients used in treatment, and application type. When treating overmature fruit or soft citrus species, such as Satsuma or clementine mandarins, or using fungicides such as AZX, FLU, or PCZ, the time between harvest and the first treatment should be as brief as possible, preferably less than 12 hr to be consistently effective. Prompt treatment would be more feasible with a bin drench rather than a packinghouse treatment. Packinghouses tend to accumulate fruit for a day or longer before processing them due to logistical and practical reasons, which may push the fruit out of the 24-hr window for optimal fungicide effectiveness. In contrast, fruit can be treated in harvest bins by drenching trailers of them on arrival at the packinghouse, or when they are unloaded for storage and the fruit removed sometime later for packingline processing. The two older fungicides, IMZ and TBZ, and the newer PYR have "kick back" action in that they cure 24 hr-old green mold infections, so they are good choices for bin drenching. Most producers would harvest and transport fruit to the packinghouse on day 1 and then stage fruit to be processed on day 2. This means that if there is no drenching, the fruit would be treated within 18–24 hr or even longer after harvest depending on the dump rate (speed of the packingline). If a bin drencher is utilized, this time could be shorted to 6–18 hr, depending on the diligence of the postharvest team.

The inoculation method could explain why D'Aquino et al. (2013) achieved much better curative control of a 24 hr-old green mold infection with FLU compared to the poor control seen in the Erasmus et al. (2015b) study. D'Aquino and coworkers wounded fruit first and then dipped it in a 10^5 conidia/mL suspension versus dipping the wound-inducer into a 10^6 conidia/mL suspension and deposing the spores directly into the fresh wound as used by Erasmus and coworkers. Kanetis et al. (2007) showed very good curative green mold control with FLU using a small inoculation wound, but the infection age only ranged from 13 to 15 hr.

4.5.3 Wound Size and Inoculum Load

The largest threat to the fruit in terms of postharvest decay losses in Mediterranean-type climate citrus-growing areas are infections caused during harvest mostly by the green and blue molds and sour rot pathogens. To test the efficacy of fungicides properly, these natural infections need to be simulated in a laboratory.

First, the wound size should be considered. Large wounds could be caused by clippers or crushing the fruit in orchard bins. Smaller wounds are caused by long stems, this happens when the harvester cuts the stem too far from the fruit. These stumps can be as sharp as a nail and as the fruit gets transferred and handled, these will induce multiple wounds on the rind of neighboring fruit. Small wounds are also caused by insects, such as fruit flies and moths. Brown et al. (1983) showed that higher levels of IMZ were loaded in wounded tissue compared to the intact rind. Mean control levels of various fungicides (including TBZ and IMZ) were more than double on larger wounds (87%; 1 mm diameter) compared to smaller wounds (41%; 0.3 mm diameter). Segments with a single large wound of 2 mm wide × 2 mm deep loaded higher IMZ residues (10.19 mg/kg) compared to segments with three smaller wounds of 0.5 mm wide × 2 mm deep (9.06 mg/kg) and both these were higher than intact rind segments (7.91 mg/kg) (Erasmus et al., 2015a). Loss of control was observed on older infections of these larger wounds compared to the smaller wounds (18 vs. 12 hr on clementine mandarins and 42 vs. 30 hr on Navel oranges, for large and small wounds, respectively). It is therefore the small wounds that should be studied since infections originating from larger wounds should be more easily controlled, and because they are large, they have a better chance of being eliminated by manual or machine grading during packinghouse processing.

The second factor to consider is inoculum load. Among other factors determining the interaction between citrus fruits and the pathogens *P. digitatum* and *P. italicum*, inoculum load in rind wounds was discussed by Palou (2014). Early studies showed that the amount of conidia in the rind wound is linearly related with the incidence of infection, provided that the fruit is mature (susceptible to disease) and the temperature is appropriate for infection (20–25°C). In these conditions, inoculation of 50 and 500 spores of *P. digitatum* into rind wounds of oranges produced effective infection on 10 and 65% of the fruit, respectively (Eckert and Eaks, 1989). Similarly, the number of arthroconidia of *G. citri-aurantii* necessary for 50% active lesions in artificially inoculated rind wounds of lemons was determined by Baudoin and Eckert (1982). They observed that this value was greatly dependent on the fruit physiological age (harvest time) and water status.

For evaluation of the efficacy of postharvest antifungal treatments to control green and blue molds, inoculum densities of 10^6 conidia/mL of *P. digitatum* or *P. italicum* were recommended by Eckert and Brown (1986). These concentrations can be obtained with a conidial suspension with an optical density of 0.1 at 425 nm and ensure close to 100% infection on untreated fruit. Kellerman et al. (2018) explored inoculum load (10^3, 10^4, 10^5, and 10^6 conidia/mL) for curative green mold control with PYR applied by means of a drench treatment on lemon, Satsuma mandarin, or Navel orange fruit 24 hr prior to treatment. Infection levels on untreated fruit of just over 50% were observed on lemon and Satsuma and <10% on Navel orange for the 10^3 inoculum load, which resulted to >90% control on all three citrus types on treated fruit. Infection levels on untreated fruit of about 90% were achieved with the 10^4 inoculum load for lemon and Satsuma mandarin and just under 50% for Navel orange with control levels between 80 and 90% on treated fruit for all three citrus types. The 10^5 inoculum load caused >95% infection on untreated fruit of lemon and Satsuma and just over 80% on Navel oranges with control levels of >75% for untreated lemon and Navel and 50% for Satsuma. The highest load of 10^6 conidia/mL resulted in close to or 100% infection for all three untreated citrus types and control levels varied from <53, <33, and <63% for treated lemon, Satsuma mandarins, and Navel oranges, respectively. These results indicate that lower inoculum loads (10^4–10^5 conidia/mL) should be considered when using soft citrus and higher levels (10^5–10^6 conidia/mL) for oranges and lemons.

4.6 Fungicide Resistance

Fungicide resistance has been reported to most of the current registered fungicides active against green and blue molds (Holmes and Eckert, 1999; Kinay et al., 2007): SOPP (Harding, 1962), TBZ (Harding, 1972), guazatine (Wild, 1983), IMZ (Eckert, 1987), and FLU (Kim et al., 2015). Resistance to FLU and PYR was reported to be naturally occurring in packinghouse populations (Kanetis et al., 2008), and currently, there are commercial reports of FLU and PYR resistance occurring in packinghouse environments (Erasmus, unpublished data). With the exception of IMZ and PCZ, all other registered fungicides for green and blue molds control fall in different Fungicide Resistance Action Committee (FRAC) groups, indicating different modes of action. *Penicillium digitatum* fungicide-resistant populations can be managed with a strategy that either alternate or combine fungicides from different FRAC groups. Resistant levels are usually high when fruits are stored for extended periods, either as a marketing strategy or because of poor sales, and reworked again where sporulating green mold is dumped in the packinghouse environment. A country such as South Africa, which mainly does export sales, has fewer problems with fungicide resistance because fresh fruits are exported promptly after harvest and fruit storage is not a common practice. Another sure way to enhance the development of fungicide resistance in the packinghouse environment is by the application of a postharvest fungicide in the orchard. This was the case with TBZ, where benomyl (which falls in the same FRAC group) was applied for citrus black spot control and this resulted in TBZ resistance in green mold populations in South African citrus orchards. The implementation of IMZ had a significant practical impact on decay management when TBZ was rendered ineffective against green mold (Pelser and La Grange, 1981).

4.7 Alternatives to Conventional Fungicides

Palou et al. (2008) wrote a thorough review on the subject of the alternative control of citrus green and blue molds. Many alternatives using physical, chemical, and biological control approaches were evaluated to control postharvest decay of horticultural produce, and these are thoroughly covered in Section III of this book. Heat treatments, irradiation, and complimentary practices such as cold storage, ozonation of storage atmosphere, as well as controlled and modified atmosphere storage or packaging are all physical ways to control green and blue molds, but these vary in efficacy and present risk of physiological injury to the fruit. Food additives, generally recognized as safe (GRAS) compounds, essential oils, peptides, proteins, and chitosan are all alternative or supplemental chemical compounds with variation in efficacy and consistency in terms of green and blue molds control (Palou et al., 2008; Palou, 2014). Since the recent cancellation of the fungicides guazatine and PCZ in important citrus-producing countries, increasing interest has arisen on the evaluation of many of these alternatives for sour rot control (Karim et al., 2017; Wu et al., 2017; Dou et al., 2018). Despite the considerable research for the last 30 yr, the industrial usage of microbial antagonists as postharvest biological control agents has been sporadic, mainly due to the reduced and inconsistent performance of biocontrol products when used under commercial conditions (Droby et al., 2016).

GRAS salts such as sodium bicarbonate, sodium carbonate, and potassium sorbate stand out as alternatives that are more frequently used in the citrus industry worldwide. Heated (around 40°C) treatments of 4–6% sodium carbonate for 1–2 min in a dip tank can significantly control green mold on oranges (Smilanick et al., 1997a). Ambient treatments of 2–4% sodium bicarbonate for 150 s can also significantly control green and blue molds (Palou et al., 2001a). Brief potassium sorbate treatments at 0.5–3% in ambient or heated solutions can significantly reduce sour rot or green mold incidence (Smilanick et al., 2008; Montesinos-Herrero et al., 2009). These three GRAS salts can complement the efficacy of conventional fungicide treatments, particularly when fungicide resistance develops, and more research in this area is needed. In this sense, these and other GRAS and natural compounds are also evaluated as antifungal ingredients of edible coatings for postharvest treatment of citrus fruits (Palou et al., 2015). Some commercial coatings, especially containing potassium sorbate, have been developed by specialized postharvest companies and are currently available for the citrus industry.

Acknowledgments

The authors would like to thank all research funding agencies and private companies in California, South Africa, and Spain that funded and supported research on this topic.

References

Abd-Elsalam, K.A., Youssef, K. and Almoammar, H. 2015. First morphogenetic identification of *Fusarium solani* isolated from orange fruit in Egypt. *Phyton-International Journal of Experimental Botany* 84, 128–131.

Adaskaveg, J.E. and Förster, H. 2014. Integrated strategies for management of Phytopthora brown rot of citrus in the United States. pp. 123–131. *In*: D. Prusky and M. Gullino (eds.). *Post-Harvest Pathology. Plant Pathology in the 21st Century (Contributions to the 9th International Congress)*, vol. 7. Springer, New York, NY.

Adaskaveg, J.E., Hao, W. and Förster, H. 2015. Postharvest strategies for managing Phytophthora brown rot of citrus using potassium phosphite in combination with heat treatments. *Plant Disease* 99, 1477–1482.

Aiello, D., Carrieri, R., Guarnaccia, V., Vitale, A., Lahoz, E. and Polizzi, G. 2014. Characterization and pathogenicity of *Colletotrichum gloeosporioides* and *C. karstii* causing preharvest disease on *Citrus sinensis* in Italy. *Journal of Phytopathology* 163, 168–177.

Alves, A., Crous, P.W., Correia, A. and Phillips, A.J.L. 2008. Morphological and molecular data reveal cryptic speciation in *Lasiodiplodia theobromae*. *Fungal Diversity* 28, 1–13.

Ballester, A.R., Izquierdo, A., Lafuente, M.T. and González-Candelas, L. 2010. Biochemical and molecular characterization of induced resistance against *Penicillium digitatum* in citrus fruit. *Postharvest Biology and Technology* 56, 31–38.

Ballester, A.R. and Lafuente, M.T. 2017. LED blue light-induced changes in phenolics and ethylene in citrus fruit: Implication in elicited resistance against *Penicillium digitatum* infection. *Food Chemistry* 218, 575–583.

Ballester, A.R., Lafuente, M.T. and González-Candelas, L. 2006. Spatial study of antioxidant enzymes, peroxidase and phenylalanine ammonia-lyase in the citrus fruit-*Penicillium digitatum* interaction. *Postharvest Biology and Technology* 39, 115–124.

Bancroft, M.N., Gardner, P.D., Eckert, J.W. and Baritelle, J.L. 1984. Comparison of decay control strategies in California lemon packinghouses. *Plant Disease* 68, 24–28.

Barkai-Golan, R. and Karadavid, R. 1991. Cellulolytic activity of *Penicillium digitatum* and *P. italicum* related to fungal growth and to pathogenesis in citrus fruits. *Journal of Phytopathology* 131, 65–72.

Barmore, C.R. and Brown, G.E. 1982. Spread of *Penicillium digitatum* and *Penicillium italicum* during contact between citrus fruits. *Phytopathology* 72, 116–120.

Bartholomew, E.T. 1926. Alternaria rot of lemons. California Agricultural Experiment Station Bulletin 408, University of California, Berkely, CA.

Baudoin, A.B.A.M. and Eckert, J.W. 1982. Factors influencing the susceptibility of lemons to infection by *Geotrichum candidum*. *Phytopathology* 72, 1592–1597.

Baudoin, A.B.A.M. and Eckert, J.W. 1985a. Development of resistance against *Geotrichum candidum* in lemon peel injuries. *Phytopathology* 74, 174–179.

Baudoin, A.B.A.M. and Eckert, J.W. 1985b. Influence of preformed characteristics of lemon peel on susceptibility to *Geotrichum candidum*. *Physiological Plant Pathology* 26, 1–163.

Brooks, C. 1944. Stem-end rot of oranges and factors affecting its control. *Journal of Agricultural Research* 68, 363–381.

Brown, G.E. 1973. Development of green mold in degreened oranges. *Phytopathology* 63, 1104–1107.

Brown, G.E. 1975. Factors affecting postharvest development of *Colletotrichum gloeosporioides* in citrus fruits. *Phytopathology* 65, 404–409.

Brown, G.E. 1977. Application of benzimidazole fungicides for citrus decay control. *Proceedings of the International Society of Citriculture* 1, 273–277.

Brown, G.E. 1979. Biology and control of *Geotrichum candidum* the cause of citrus sour rot. *Proceedings of the Florida State Horticultural Society* 92, 186–189.

Brown, G.E. 1992. Factors affecting the occurrence of anthracnose on Florida citrus fruit. *Proceedings of the International Society of Citriculture* 7, 1044–1048.

Brown, G.E. and Burns, J.K. 1998. Enhanced activity of abscission enzymes predisposes oranges to invasion by *Diplodia natalensis* during ethylene degreening. *Postharvest Biology and Technology* 14, 217–227.

Brown, G.E. and Dezman, D.J. 1990. Uptake of imazalil by citrus fruit after postharvest application and the effect of residue distribution on sporulation of *Penicillium digitatum*. *Plant Disease* 74, 927–930.

Brown, G.E. and Lee, H.S. 1993. Interactions of ethylene with citrus stem-end rot caused by *Diplodia natalensis*. *Phytopathology* 83, 909–912.

Brown, G.E., Mawk, P. and Craig, J.O. 1988. Pallet treatment with benomyl of citrus fruit on trucks for control of diplodia stem-end rot. *Proceedings of the Florida State Horticultural Society* 101, 187–190.

Brown, G.E. and McCornack, A.A. 1972. Decay caused by *Alternaria citri* in Florida citrus fruit. *Plant Disease Reporter* 56, 909–912.

Brown, G.E., Nagy, S. and Maraulja, M. 1983. Residues from postharvest nonrecovery spray applications of imazalil to oranges and effects on green mold caused by *Penicillium digitatum*. *Plant Disease* 67, 954–957.

Brown, G.E. and Wilson, W.C. 1968. Mode of entry of *Diplodia natalensis* and *Phomopsis citri* into Florida oranges. *Phytopathology* 58, 736–739.

Bushong, P.M. and Timmer, L.W. 2000. Evaluation of postinfection control of citrus scab and melanose with benomyl, fenbuconazole, and azoxystrobin. *Plant Disease* 84, 1246–1249.

Butler, E.E., Fogle, D. and Miranda, M. 1988. *Galactomyces citri-aurantii*, a newly found teleomorphe of *Geotrichum citri-aurantii*, the cause of sour rot of citrus fruit. *Mycotaxon* 33, 197–212.

Butler, E.E., Webster, R.K. and Eckert, J.W. 1965. Taxonomy, pathogenicity, and physiological properties of the fungus causing sour rot of citrus. *Phytopathology* 55, 1262–1268.

Cabras, P., Schirra, M., Pirisi, F.M., Garau, V.L. and Angioni, A. 1999. Factors affecting imazalil and thiabendazole uptake and persistence in citrus fruits following dip treatments. *Journal of Agricultural and Food Chemistry* 47, 3352–3354.

Calonne, M., Sahraoui, A.L.-H., Campagnac, E., Debiane, D., Laruelle, F., Grandmougin-Ferjani, A. and Fontaine, J. 2012. Propiconazole inhibits the sterol 14a-demethylase in *Glomus irregulare* like in phytopathogenic fungi. *Chemosphere* 87, 376–383.

Carswell, C., Grant, B.R., Theodorou, M.E., Harris, J., Niere, J.O. and Plaxton, W.C. 1996. The fungicide phosphonate disrupts the phosphate-starvation response in *Brassica nigra* seedlings. *Plant Physiology* 110, 105–110.

Ceponis, M.J., Cappellini, R.A. and Lightner, G.W. 1986. Disorders in citrus shipments to the New York market, 1972–1984. *Plant Disease* 70, 1162–1165.

Cerioni, L., Bennasar, P.B., Lazarte, D., Sepulveda, M., Smilanick, J.L., Ramallo, J. and Rapisarda, V. 2017. Conventional and reduced-risk fungicides to control postharvest Diplodia and Phomopsis stem-end rot on lemons. *Scientia Horticulturae* 225, 783–787.

Cerioni, L., Rapisarda, V.A., Doctor, J., Fikkert, S., Ruiz, T., Fassel, R. and Smilanick, J.L. 2013a. Use of phosphite salts in laboratory and semicommercial tests to control citrus postharvest decay. *Plant Disease* 97, 201–212.

Cerioni, L., Sepulveda, M., Rubio Ames, Z., Volentini, S.I., Rodríguez-Montelongo, L., Smilanick, J.L., Ramallo, J. and Rapisarda, V.A. 2013b. Control of lemon postharvest diseases by low-toxicity salts combined with hydrogen peroxide and heat. *Postharvest Biology and Technology* 83, 17–21.

Chen, K., Tian, Z., Wang, L. and Long, C.A. 2017. Development of specific primers based on the genomes of *Penicillium* spp. for rapid detection of *Penicillium digitatum* among fungal isolates in citrus. *European Journal of Plant Pathology* 149, 201–209.

Christie, C. 2016. Optimisation of postharvest drench application of fungicides on citrus fruit. M.Sc. Thesis, Stellenbosch University, Stellenbosch, South Africa. Available at: http://hdl.handle.net/10019.1/98775

Clemons, G.P. and Sisler, H.D. 1971. Localization of the site of action of a fungitoxic benomyl derivative. *Pesticide Biochemistry and Physiology* 1, 32–43.

Coggins, C.W., Anthony, M.F. and Fritts, R. 1992. Postharvest use of gibb on lemons. *California Citrograph* 78, 11–14.

Cohen, E., Coggins, C.W. and Eckert, J.W. 1991. Predisposition of citrus fruits to sour rot when submerged in water. *Plant Disease* 75, 166–168.

Cunningham, N.M. and Taverner, P.D. 2007. Efficacy of integrated postharvest treatments against mixed inoculations of *Penicillium digitatum* and *Geotrichum citri-aurantii* in 'Leng' navel oranges (*Citrus sinensis*). *New Zealand Journal of Crop and Horticultural Science* 35, 187–192.

D'Aquino, S., Palma, A., Angioni, A. and Schirra, M. 2013. Residue levels and efficacy of fludioxonil and thiabendazole in controlling postharvest green mold decay in citrus fruit when applied in combination with sodium bicarbonate. *Journal of Agricultural and Food Chemistry* 61, 296–306.

D'Aquino, S., Schirra, M., Palma, A., Angioni, A., Cabras, P. and Migheli, Q. 2006. Residue levels and effectiveness of pyrimethanil *vs.* imazalil when using heated postharvest dip treatments for control of *Penicillium* decay on citrus fruit. *Journal of Agricultural and Food Chemistry* 54, 4721–4726.

Daniels, A. and Lucas, J.A. 1995. Mode of action of the anilinopyrimidine fungicide pyrimethanil. 1. *In-vivo* activity against *Botrytis fabae* on broad bean (*Vicia faba*) leaves. *Pesticide Science* 45, 33–41.

Dantas, S.A.F., Oliveira, S.M.A., Michereff, S.J., Nascimento, L.C., Gurgel, L.M.S. and Pessoa, W.R.L.S. 2003. Doenças fúngicas pós-colheita em mamões e laranjas comercializados na Central de Abastecimento do Recife. *Fitopatologia Brasileira* 28, 528–533 (In Portuguese with English abstract).

Dezman, D.J., Nagy, S. and Brown, G.E. 1986. Postharvest fungal decay control chemicals: Treatments and residues in citrus fruits. *Residue Reviews* 97, 37–92.

Diaz, L.E., del Campo, R.M. and Mara, H. 2015. In vitro effectiveness of disinfectants on *Geotrichum citri-aurantii* spore viability. *Acta Horticulturae* 1065, 1571–1575.

Dore, A., Molinu, M.G., Venditti, T. and D'hallewin, G. 2009. Immersion of lemons into imazalil mixtures heated at 50°C alters the cuticle and promotes permeation of imazalil into rind wounds. *Journal of Agricultural and Food Chemistry* 57, 623–631.

Dore, A., Molinu, M.G., Venditti, T. and D'hallewin, G. 2010. Sodium bicarbonate induces crystalline wax generation, activates host-resistance, and increases imazalil level in rind wounds of oranges, improving the control of green mold during storage. *Journal of Agricultural and Food Chemistry* 58, 7297–7304.

Dou, S., Ouyang, Q., You, K., Qian, J. and Tao, N. 2018. An inclusion complex of thymol into b-cyclodextrin and its antifungal activity against *Geotrichum citri-aurantii. Postharvest Biology and Technology* 138, 31–36.

Droby, S., Eick, A., Macarisin, D., Cohen, E., Rafael, G., Stange, R., McColum, G., Dudai, N., Nasser, A., Wisniewski, M. and Shapira, R. 2008. Role of citrus volatiles in host recognition, germination and growth of *Penicillium digitatum* and *Penicillium italicum. Postharvest Biology and Technology* 49, 386–396.

Droby, S., Wisniewski, M., Teixidó, N., Spadaro, D. and Jijakli, M.H. 2016. The science, development, and commercialization of postharvest biocontrol products. *Postharvest Biology and Technology* 122, 22–29.

Eckert, J.W. 1959. Lemon sour rot. *California Citrograph* 45, 30–31, 35, 36.

Eckert, J.W. 1977. Control of postharvest diseases. pp. 269–352. *In*: M.R. Siegel and H.D. Sisler (eds.). *Antifungal Compounds*, 1st. Marcel Dekker Incorporated, New York, NY.

Eckert, J.W. 1987. *Penicillium digitatum* biotypes with reduced sensitivity to imazalil. *Phytopathology* 77, 1728.

Eckert, J.W. 1990. Recent developments in the chemical control of postharvest diseases. *Acta Horticulturae* 269, 477–494.

Eckert, J.W. and Brown, G.E. 1986. Evaluation of postharvest fungicide treatments for citrus fruits. pp. 92–97. *In*: K.D. Hickey (ed.). *Methods for Evaluating Pesticides for Control of Plant Pathogens.* American Phytopathological Society Press, St. Paul, MN.

Eckert, J.W. and Eaks, I.L. 1989. Postharvest disorders and diseases of citrus fruits. pp. 179–260. *In*: W. Reuter, E.C. Calavan and G.E. Carman (eds.). *The Citrus Industry.* Division of Agriculture and Natural Resources, University of California Press, Berkeley, CA.

Erasmus, A., Lennox, C.L., Jordaan, H., Smilanick, J.L., Lesar, K. and Fourie, P.H. 2011. Imazalil residue loading and green mould control in citrus packhouses. *Postharvest Biology and Technology* 62, 193–203.

Erasmus, A., Lennox, C.L., Korsten, L., Lesar, K. and Fourie, P.H. 2015a. Imazalil resistance in *Penicillium digitatum* and *P. italicum* causing citrus postharvest green and blue mould: Impact and options. *Postharvest Biology and Technology* 107, 66–76.
Erasmus, A., Lennox, C.L., Njombolwana, N.S., Lesar, K. and Fourie, P.H. 2015b. Curative control of citrus green mould by imazalil as influenced by infection age, wound size, fruit exposure time, solution pH and fruit brushing after treatment. *Postharvest Biology and Technology* 101, 26–36.
Fallanaj, F., Ippolito, A., Ligorio, A., Garganese, F., Zavanella, C. and Sanzani, S.M. 2016. Electrolyzed sodium bicarbonate inhibits *Penicillium digitatum* and induces defence responses against green mould in citrus fruit. *Postharvest Biology and Technology* 115, 18–29.
FAO (Food and Agriculture Organization of the United Nations). 2017. Citrus Fruit, Fresh and Processed, Statistical Bulletin 2016. Available at: www.fao.org/3/a-i8092e.pdf
FDOC (State of Florida Department of Citrus). 2018. Official rules affecting the Florida citrus industry. FDOC, Bartow, FL. Available at: https://fdocgrower.app.box.com/s/xn65jdjfsf93p1952no8a2sknpiv17k5/file/187106877628
Feld, S.J., Menge. J.A. and Pehrson, J.E. 1979. Brown rot of citrus: A review of the disease. *Citrograph* 64, 101–106.
Fischer, I.H., Lourenço, S.A., Spósito, M.B. and Amorim, L. 2009. Characterisation of the fungal population in citrus packing houses. *European Journal of Plant Pathology* 123, 449–460.
Förster, H., Adaskaveg, J.E., Kim, D.H. and Stanghellini, M.E. 1998. Effect of phosphite on tomato and pepper plants and on susceptibility of peppers to Phytophthora root and crown rot in hydroponic culture. *Plant Disease* 82, 1165–1170.
Frisvad, J.C. and Samson, R.A. 2004. Polyphasic taxonomy of *Penicillium* subgenus *Penicillium*. A guide to identification of food and air-borne terverticillate Penicillia and their mycotoxins. *Studies in Mycology* 49, 1–174.
Fritz, R., Lanen, C., Colas, V. and Leroux, P. 1997. Inhibition of methionine biosynthesis in *Botrytis cinerea* by the anilinopyrimidine fungicide pyrimethanil. *Pesticide Science* 49, 40–46.
Fu, Y.P., Liu, Z.H. and Li, Y. 2017. *Fusarium solani*: A new pathogen causing postharvest lemon rot in Changchun, China. *Plant Disease* 101, 1548.
Garganese, F., Schena, L., Siciliano, I., Prigigallo, M.I., Spadaro, D., De Grassi, A., Ippolito, A. and Sanzani, S.M. 2016. Characterization of citrus associated *Alternaria* species in Mediterranean areas. *PLoS ONE* 11, e0163255.
González-Candelas, L., Alamar, S., Sánchez-Torres, P., Zacarías, L. and Marcos, J.F. 2010. A transcriptomic approach highlights induction of secondary metabolism in citrus fruit in response to *Penicillium digitatum* infection. *BMC Plant Biology* 10, 194.
Gopal, K., Mukunda Lakshmi, L., Sarada, G., Nagalakshmi, T., Gouri Sankar, T., Gopi, V. and Ramana, K.T.V. 2014. Citrus melanose (*Diaporthe citri* Wolf): A Review. *International Journal of Current Microbiology and Applied Sciences* 4, 113–124.
Graham, J.H., Timmer, L.W., Drouillard, D.L. and Peever, T.L. 1998. Characterization of *Phytophthora* spp. causing outbreaks of citrus brown rot in Florida. *Phytopathology* 88, 724–729.

Guarnaccia, V., Groenewald, J.Z., Polizzi, G. and Crous, P.W. 2017. High species diversity in *Colletotrichum* associated with citrus diseases in Europe. *Persoonia* 39, 32–50.

Hall, D.J. 1981. Innovations in citrus waxing - An overview. *Proceedings of the Florida State Horticultural Society* 94, 258–263.

Hall, D.J. and Sorenson, D. 2006. Washing, waxing and color-adding. pp. 421–450. *In*: W.F. Wardowski, W.M. Miller, D.J. Hall and W. Grierson (eds.). *Fresh Citrus Fruits*. Florida Science Source, Florida, FL.

Hammerschlag, R.D. and Sisler, H.D. 1973. Benomyl and methyl-2-benzimidazolecarbamate (MBC): Biochemical, cytological and chemical aspects of toxicity to *Ustilago maydis* and *Saccharomyces cerevesiae*. *Pesticide Biochemistry and Physiology* 3, 42–54.

Hanif, S., Hafeez, R., Akram, W., Ashfaq, M. and Ali, A. 2016. First report of Sclerotinia fruit rot of *Citrus paradisi* caused by *Sclerotinia sclerotiorum* in Pakistan. *Plant Disease* 100, 863.

Harding, P.R. 1962. Differential sensitivity to sodium orthophenylphenate by biphenyl-sensitive and biphenyl-resistant strains of *Penicillium digitatum*. *Plant Disease Reporter* 46, 100–104.

Harding, P.R. 1972. Differential sensitivity to thiabendazole by strains of *Penicillium italicum* and *P. digitatum*. *Plant Disease Reporter* 56, 256–260.

Harvey, E.M. 1946. Changes in Lemons during Storage as Affected by Air Circulation and Ventilation. USDA Technical Bulletin No. 908. Available at: http://agecon search.umn.edu/record/169915/files/tb908.pdf

Holmes, G.J. and Eckert, J.W. 1999. Sensitivity of *Penicillium digitatum* and *P. italicum* to postharvest citrus fungicides in California. *Phytopathology* 89, 716–721.

Holmes, G.J., Eckert, J.W. and Pitt, J.I. 1994. Revised description of *Penicillium ulaiense* and its role as a pathogen of citrus fruits. *Phytopathology* 84, 719–727.

Honger, J.O., Offei, S.K., Oduro, K.A., Odamtten, G.T. and Nyaku, S.T. 2016. Identification and molecular characterisation of *Colletotrichum* species from avocado, citrus and pawpaw in Ghana. *South African Journal of Plant and Soil* 33, 177–185.

Hordijk, J., Cronjé, P.J.R. and Opara, L. 2013. Postharvest application of thiabendazole reduces chilling injury of citrus fruits. *Acta Horticulturae* 1007, 119–125.

Huang, F., Chen, G.Q., Hou, X., Fu, Y.S., Cai, L., Hyde, K.D. and Li, H.Y. 2013. *Colletotrichum* species associated with cultivated citrus in China. *Fungal Diversity* 61, 61–74.

Huang, J.K. 1991. A study on the original inoculum of citrus sour rot (*Geotrichum candidum*). *Proceedings of the International Citrus Symposium Guangzhou China*, 623.

Isshiki, A., Akimitsu, K., Yamamoto, M. and Yamamoto, H. 2001. Endopolygalacturonase is essential for citrus black rot caused by *Alternaria citri* but not brown spot caused by *Alternaria alternata*. *Molecular Plant-Microbe Interactions* 14, 749–757.

Isshiki, A., Ohtani, K., Kyo, M., Yamamoto, H. and Akimitsu, K. 2003. Green fluorescent detection of fungal colonization and endopolygalaturonase gene expression in the interaction of *Alternaria citri* with citrus. *Phytopathology* 93, 768–773.

Janisiewicz, W.J., Peterson, D.L., Yoder, K.S. and Miller, S.S. 2005. Experimental bin drenching system for testing biocontrol agents for control of postharvest decay of apples. *Plant Disease* 89, 487–490.

Joly, P. 1967. The black decays of citrus fruits caused by *Alternaria*. *Fruits* 22, 89–95.

Kader, A.A. 2005. Increasing food availability by reducing postharvest losses of fresh produce. *Acta Horticulturae* 682, 2169–2175.

Kanetis, L., Förster, H. and Adaskaveg, J.E. 2007. Comparative efficacy of the new postharvest fungicides azoxystrobin, fludioxonil, and pyrimethanil for managing citrus green mold. *Plant Disease* 91, 1502–1511.

Kanetis, L., Förster, H., Jones, C.A., Borkovich, K.A. and Adaskaveg, J.E. 2008. Characterization of genetic and biochemical mechanisms of fludioxonil and pyrimethanil resistance in field isolates of *Penicillium digitatum*. *Phytopathology* 98, 205–214.

Karim, H., Boubaker, H., Askarne, L., Cherifi, K., Lakhtar, H., Msanda, F., Boudyach, E. H. and Ait Ben, A.A. 2017. Use of Cistus aqueous extracts as botanical fungicides in the control of Citrus sour rot. *Microbial Pathogenesis* 104, 263–267.

Kellerman, M., Erasmus, A., Cronjé, P.J. and Fourie, P.H. 2014. Thiabendazole residue loading in dip, drench and wax coating applications to control green mould and chilling injury on citrus fruit. *Postharvest Biology and Technology* 96, 78–87.

Kellerman, M., Joubert, J., Erasmus, A. and Fourie, P.H. 2016. The effect of temperature, exposure time and pH on imazalil residue loading and green mould control on citrus through dip application. *Postharvest Biology and Technology* 121, 159–164.

Kellerman, M., Liebenberg, E., Njombolwana, N., Erasmus, A. and Fourie, P.H. 2018. Postharvest dip, drench and wax coating application of pyrimethanil on citrus fruit: Residue loading and green mould control. *Postharvest Biology and Technology* 103, 115–129.

Khan, A., Subhani, M.N., Chattha, M.B., Anwar, W. and Nawaz, K. 2017. First report of whisker mold of citrus (*Citrus sinensis*) caused by *Penicillium ulaiense* in Pakistan. *Plant Disease* 101, 1042.

Kim, Y.K., Saito, S. and Xiao, C.L. 2015. Occurrence of fludioxonil resistance in *Penicillium digitatum* from citrus in California. *Plant Disease* 99, 1447.

Kinay, P., Mansour, M.F., Mlikota Gabler, F., Margosan, D.A. and Smilanick, J.L. 2007. Characterization of fungicide-resistant isolates of *Penicillium digitatum* collected in California. *Crop Protection* 26, 647–656.

Klotz, L.J. 1973. *Color Handbook of Citrus Diseases*, 4th ed. University of California, Division of Agricultural Sciences, Berkeley, CA.

Klotz, L.J. and DeWolfe, T.A. 1961. Brown rot contact infection of citrus fruits prior to hot water treatment. *Plant Disease Reporter* 45, 264–267.

Koleci, X., Quinn, P.J., Çela, M. and Malaj, Z. 2007. The place of disinfection in the control of infectious diseases. *Albanian Journal of Natural Technology and Science* 12, 139–156.

Ladaniya, M.S. 2008. Postharvest diseases and their management. pp. 417–450. *In*: *Citrus Fruit - Biology, Technology and Evaluation*. Academic Press, San Diego, CA.

Laville, E.Y., Harding, P.R., Dagan, Y., Rafael, G., Kraght, A.J. and Rippon, L.E. 1977. Studies on imazalil as potential treatment for control of citrus fruit decay. *Proceedings of the International Society of Citriculture* 1, 269–273.

Leroux, P. 2003. Modes d'action des produits phytosanitaires sur les organisms pathogènes des plantes. *Comptes Rendus Biologies* 326, 9–21 (In French with English abstract).

Lu, L., Lu, H., Wu, C., Fang, W., Yu, C., Ye, C., Shi, Y., Yu, T. and Zheng, X. 2013. *Rhodosporidium paludigenum* induces resistance and defense-related responses against *Penicillium digitatum* in citrus fruit. *Postharvest Biology and Technology* 85, 196–202.

Macarisin, D., Droby, S., Bauchan, G. and Wisniewski, M. 2010. Superoxide anion and hydrogen peroxide in the yeast antagonist-fruit interaction: A new role for reactive oxygen species in postharvest biocontrol? *Postharvest Biology and Technology* 58, 194–202.

Mamba, L.C., Meitz-Hopkins, J.C., Fourie, P.H., du Plooy, W. and Lennox, C.L. 2018. Citrus sour rot management by propiconazole drench application in South Africa. p. 105. *In*: *Abstracts Book of the 10th Citrus Research Symposium*. Drakensberg, South Africa.

Mansfield, R.W. and Wiggins, T.E. 1990. Photoaffinity labelling of the b-methoxyacrylate binding site in bovine heart mitochondrial cytochrome bc1 complex. *Biochimica et Biophysica Acta (BBA) - Bioenergetics* 1015, 109–115.

Marcet-Houben, M., Ballester, A.R., de la Fuente, B., Harries, E., Marcos, J.F., Gonzalez-Candelas, L. and Gabaldon, T. 2012. Genome sequence of the necrotrophic fungus *Penicillium digitatum*, the main postharvest pathogen of citrus. *BMC Genomics* 13, 646.

McCornack, A.A. 1970. Status of postharvest fungicides for citrus fruit. *Proceedings of the Florida State Horticultural Society* 83, 229–232.

McCornack, A.A. 1972a. Factors affecting decay and peel injury in temples. *Proceedings of the Florida State Horticultural Society* 85, 232–235.

McCornack, A.A. 1972b. Effect of ethylene degreening on decay of Florida citrus fruit. *Proceedings of the Florida State Horticultural Society* 84, 270–272.

McCornack, A.A. and Brown, G.E. 1967. Thiabendazole. An experimental fungicide for fresh citrus fruit. *Proceedings of the Florida State Horticultural Society* 83, 225–228.

McCornack, A.A. and Brown, G.E. 1977. Decay control of Florida citrus fruits with imazalil. *Proceedings of the Florida State Horticultural Society* 90, 141–144.

McGlasson, W.B. and Eaks, I.L. 1972. A role for ethylene in the development of wastage and off-flavors in stored "Valencia" oranges. *HortScience* 7, 80–81.

McKay, A.H., Förster, H. and Adaskaveg, J.E. 2012a. Distinguishing *Galactomyces citri-aurantii* from *G. geotrichum* and characterizing population structure of the two postharvest sour rot pathogens of fruit crops in California. *Phytopathology* 102, 528–538.

McKay, A.H., Förster, H. and Adaskaveg, J.E. 2012b. Efficacy and application strategies for propiconazole as a new postharvest fungicide for managing sour rot and green mold of citrus fruit. *Plant Disease* 96, 235–242.

McKay, A.H., Förster, H. and Adaskaveg, J.E. 2012c. Toxicity and resistance potential of selected fungicides to *Galactomyces* and *Penicillium* spp. causing postharvest fruit decays of citrus and other crops. *Plant Disease* 96, 87–96.

Melo, O., Engler, A., Nauelhual, L., Cofre, G. and Barrena, J. 2013. Do sanitary, phytosanitary, and quality-related standards affect international trade? Evidence from Chilean fruit exports. *World Development* 54, 350–359.

Montesinos-Herrero, C., del Río, M.A., Pastor, C., Brunetti, O. and Palou, L. 2009. Evaluation of brief potassium sorbate dips to control postharvest penicillium decay on major citrus species and cultivars. *Postharvest Biology and Technology* 52, 117–125.

Morris, S.C. 1982. Synergism of *Geotrichum candidum* and *Penicillium digitatum* in infected citrus fruit. *Phytopathology* 72, 1336–1339.

Moscoso-Ramírez, P.A. and Palou, L. 2014. Effect of ethylene degreening on the development of postharvest penicillium molds and fruit quality of early season citrus fruit. *Postharvest Biology and Technology* 91, 1–8.

Moubasher, A.H., Abdel-Sater, M.A. and Soliman, Z. 2016. Biodiversity and molecular characterization of yeast and filamentous fungi in the air of citrus and grapevine plantations in Assiut area, Egypt. *Micosphere* 7, 236–261.

Nakamura, M., Suprapta, D.N. and Iwai, H. 2008. Differentiation of pathogenic and nonpathogenic isolates of *Geotrichum candidum* sensu Suprapta et al. (1995) on citrus fruit based on PCR-RFLP analysis of rDNA ITS and PCR using specific primers designed in polygalacturonase genes. *Mycoscience* 49, 155–158.

Nakamura, M., Suprapta, D.N., Iwai, H. and Arai, K. 2001. Comparison of endo-polygalacturonase activities of citrus and non-citrus races of *Geotrichum candidum*, and cloning and expression of the corresponding genes. *Molecular Plant Pathology* 2, 265–274.

Nazerian, E. and Alian, Y.M. 2013. Association of *Geotrichum citri-aurantii* with citrus fruits decay in Iran. *International Journal of Agronomy and Plant Production* 4, 1839–1843.

Njombolwana, N.S., Erasmus, A. and Fourie, P.H. 2013. Evaluation of curative and protective control of *Penicillium digitatum* following imazalil application in wax coating. *Postharvest Biology and Technology* 77, 102–110.

Nunes, C., Usall, J., Manso, T., Torres, R., Olmo, M. and García, J.M. 2007. Effect of high temperature treatments on growth of *Penicillium* spp. and their development on 'Valencia' oranges. *Food Science and Technology International* 13, 63–68.

Oliveira, M.B., Barbosa, S.C. and Petrofeza, S. 2013. Comparative *in vitro* and *in planta* analyses of extracellular enzymes secreted by the pathogenic fungus *Sclerotinia sclerotiorum*. *Genetics and Molecular Research* 12, 1796–1807.

Ortuño, A., Díaz, L., Álvarez, N., Porras, I., García-Lidón, A. and Del Río, J.A. 2011. Comparative study of flavonoid and scoparone accumulation in different *Citrus* species and their susceptibility to *Penicillium digitatum*. *Food Chemistry* 125, 232–239.

Palou, L. 2014. *Penicillium digitatum, Penicillium italicum* (Green mold, Blue mold). pp. 45–102. *In*: S. Bautista-Baños (ed.). *Postharvest Decay. Control Strategies.* Academic Press, Elsevier Inc., London, UK.

Palou, L., Smilanick, J.L. and Droby, S. 2008. Alternatives to conventional fungicides for the control of citrus postharvest green and blue moulds. *Stewart Postharvest Review* 2:2, 1–16.

Palou, L., Smilanick, J.L., Usall, J. and Viñas, I. 2001a. Control of postharvest blue and green molds of oranges by hot water, sodium carbonate, and sodium bicarbonate. *Plant Disease* 85, 371–376.

Palou, L. and Taberner, V. 2019. First report of *Penicillium ulaiense* causing postharvest whisker mold of oranges (*Citrus sinensis*) in Spain. *Plant Disease* 103, 153.

Palou, L., Usall, J., Pons, J., Cerdá, M.C. and Viñas, I. 2001b. Fungal populations in Tarragona citrus packinghouses. *Investigación Agraria: Producción y Protección Vegetal* 16, 447–462 (in Spanish with English summary).

Palou, L., Usall, J., Pons, J. and Viñas, I. 2001c. Fruit epiphyte and environmental fungal populations in 'Clemenules' mandarin orchards in Tarragona (Spain). *Investigación Agraria: Producción y Protección Vegetal* 16, 257–272 (in Spanish with English summary).

Palou, L., Valencia-Chamorro, S.A. and Pérez-Gago, M.B. 2015. Antifungal edible coatings for fresh citrus fruit: A review. *Coatings* 5, 962–986.

Park, J.H., Hyun, J.W., Park, M.J. and Choi, Y.J. 2018. First report of whisker mold as a postharvest disease caused by *Penicillium ulaiense* on citrus (*Citrus unshiu*) in Korea. *Plant Disease* 102, 2643.

Pavoncello, D., Lurie, S., Droby, S. and Porat, R. 2001. A hot water treatment induces resistance to *Penicillium digitatum* and promotes the accumulation of heat shock and pathogenesis-related proteins in grapefruit flavedo. *Physiologia Plantarum* 111, 17–22.

Peever, T.L., Ibáñez, A., Akimitsu, K. and Timmer, L.W. 2002. Worldwide phylogeography of the citrus brown spot pathogen, *Alternaria alternata*. *Phytopathology* 92, 794–802.

Peever, T.L., Su, G., Carpenter-Boggs, L. and Timmer, L.W. 2004. Molecular systematics of citrus associated *Alternaria* species. *Mycologia* 96, 119–134.

Pelser, P.DuT. 1977a. Postharvest handling of South African citrus fruit. *Proceeding of the International Society of Citriculture* 1, 244–249.

Pelser, P.DuT. 1977b. Development of Alternaria and Diplodia rots in stored grapefruit and lemons as influenced by snap-picking. *Citrus and Subtropical Fruit Journal* 519, 5, 7, 9.

Pelser, P.DuT. and La Grange, J.M. 1981. Latest developments in the control of postharvest decay of citrus fruit in South Africa. *Proceedings of the International Society of Citriculture* 2, 812–814.

Phillips, A.J.L., Alves, A., Abdollahzadeh, J., Slippers, B., Wingfield, M.J., Groenewald, Z. and Crous, P.W. 2013. The Botryosphaeriaceae: Genera and species known from culture. *Studies in Mycology* 76, 51–167.

Plaza, P., Usall, J., Teixidó, N. and Viñas, I. 2003a. Effect of water activity and temperature on germination and growth of *Penicillium digitatum, P. italicum* and *Geotrichum candidum*. *Journal of Applied Microbiology* 94, 549–554.

Plaza, P., Usall, J., Teixidó, N. and Viñas, I. 2004. Effect of water activity and temperature on competing abilities of common postharvest citrus fungi. *International Journal of Food Microbiology* 90, 75–82.

Plaza, P., Usall, J., Torres, R., Lamarca, N., Asensio, A. and Viñas, I. 2003b. Control of green and blue molds by curing on oranges during ambient and cold storage. *Postharvest Biology and Technology* 28, 195–198.

Polizzi, G., Aiello, D., Scuderi, G. and Cirvilleri, G. 2011. First report of Sclerotinia stem and twig blight caused by *Sclerotinia sclerotiorum* on *Citrus volkameriana* rootstock in Italy. *Plant Disease* 95, 1030–1031.

Porat, R., Lers, A., Dori, S., Cohen, L., Weiss, B., Daus, A., Wilson, C.L. and Droby, S. 1999a. Induction of chitinase and β-1,3-endoglucanase proteins by

UV irradiation and wounding in grapefruit peel tissue. *Phytoparasitica* 27, 233–238.

Porat, R., Weiss, B., Cohen, L., Daus, A., Goren, R. and Droby, S. 1999b. Effects of ethylene and 1-methylcyclopropene on the postharvest qualities of 'Shamouti' oranges. *Postharvest Biology and Technology* 15, 155–163.

Ramos, A.P., Talhinhas, P., Sreenivasaprasad, S. and Oliveira, H. 2016. Characterization of *Colletotrichum gloeosporioides*, as the main causal agent of citrus anthracnose, and *C. karstii* as species preferentially associated with lemon twig dieback in Portugal. *Phytoparasitica* 44, 549–561.

Rodríguez, A., Shimada, T., Cervera, M., Redondo, A., Alquézar, B., Rodrigo, M. J., Zacarías, L., Palou, L., López, M.M. and Peña, L. 2015. Resistance to pathogens in terpene down-regulated orange fruits inversely correlates with the accumulation of D-limonene in peel oil glands. *Plant Signaling Behaviour* 10, e1028704.

Rojas-Argudo, C., Palou, L., Bermejo, A., Cano, A., del Río, M.A. and González-Mas, M.C. 2012. Effect of X-ray irradiation on nutritional and antifungal bioactive compounds of 'Clemenules' clementine mandarins. *Postharvest Biology and Technology* 68, 47–53.

Rose, D.H., Cook, H.T. and Redit, W.H. 1951. Harvesting, Handling, and Transportation of Citrus Fruits, 1st ed. United States Department of Agriculture, Washington, DC.

Rosslenbroich, H.J. and Stuebler, D. 2000. *Botrytis cinerea* - History of chemical control and novel fungicides for its management. *Crop Protection* 19, 557–561.

Roth, G. 1967. Citrus fruit decay in South Africa caused by *Penicillium digitatum* Sacc. *Phytopathology Zeitschrift* 58, 383–396.

Rouissi, W., Cherif, M., Ligorio, A., Ippolito, A. and Sanzani, S.M. 2015. First report of *Penicillium ulaiense* causing whisker mould on stored citrus fruit in Tunisia. *Journal of Plant Pathology* 97, 402.

Saito, S., Michailides, T.J. and Xiao, C.L. 2016. Mucor rot - An emerging postharvest disease of mandarin fruit caused by *Mucor piriformis* and other *Mucor* spp. in California. *Plant Disease* 100, 1054–1063.

Saito, S. and Xiao, C.L. 2017a. Emerging post-harvest fruit rot diseases of mandarins. *Citrograph* 8, 44–47.

Saito, S. and Xiao, C.L. 2017b. Fungicide resistant phenotypes in *Botrytis cinerea* from mandarin and their impact on control of gray mold on stored mandarin fruit in California. *Phytopathology* 107, S5.20.

Saito, S. and Xiao, C.L. 2017c. Prevalence of postharvest diseases of mandarin fruit in California. *Plant Health Progress* 18, 204–210.

Sauer, A.V., Hugo, J., Tozze, H.J. Jr., Sposito, M.B., Feichtenberger, E., Colauto, N. B., Paccola-Meirelles, L.D., Nelson, S. and Massola, N.S. Jr. 2015. Characterization of *Alternaria alternata* causal agent of brown spot in *Citrus* spp. *Agronomy Science and Biotechnology* 1, 45–51.

Savage, S. 2017. Further optimisation of in-line aqueous application of imazalil to control citrus green mould caused by *Penicillium digitatum*. M.Sc. Thesis, Stellenbosch University, Stellenbosch, South Africa. Available at: http://hdl.handle.net/10019.1/100931

Savastano, G. and Fawcett, H.S. 1929. A study of decay in citrus fruit produced by inoculations with known mixtures of fungi at different temperatures. *Journal of Agricultural Research* 39, 163–198.

Schiffmann-Nadel, M. 1969. Stem-end rot diseases in refrigerated and non-refrigerated storage of citrus fruit. *Proceeding of the International Society of Citriculture* 3, 1295–1299.

Schiffmann-Nadel, M., Chalutz, E., Waks, J., Lomaniec, E. and Yoffe, A.Z. 1987. Increase of Fusarium rot in stored citrus fruit. *Journal of Phytopathology* 120, 154–157.

Schiffmann-Nadel, M., Waks, J. and Chalutz, E. 1975. Frost injury predisposes grapefruit to storage rots. *Phytopathology* 65, 630.

Schiffmann-Nadel, M., Waks, J., Gutter, Y. and Chalutz, E. 1981. Alternaria rot of citrus fruit. *Proceeding of the International Society of Citriculture* 2, 791–793.

Schirra, M., Cabras, P., Angioni, A., D'hallewin, G., Ruggiu, R. and Minelli, E.V. 1997. Effect of heated solutions on decay control and residues of imazalil in lemons. *Journal of Agricultural and Food Chemistry* 45, 4127–4130.

Schirra, M., Cabras, P., Angioni, A. and Melis, M. 1996. Residue level of imazalil fungicide in lemons following prestorage dip treatment at 20 and 50°C. *Journal of Agricultural and Food Chemistry* 44, 2865–2869.

Schirra, M., D'Aquino, S., Palma, A., Angioni, A. and Cabras, P. 2008. Factors affecting the synergy of thiabendazole, sodium bicarbonate, and heat to control postharvest green mold of citrus fruit. *Journal of Agricultural and Food Chemistry* 56, 10793–10798.

Schirra, M., D'Aquino, S., Palma, A., Marceddu, S., Angioni, A., Cabras, P., Scherm, B. and Migheli, Q. 2005. Residue level, persistence, and storage performance of citrus fruit treated with fludioxonil. *Journal of Agricultural and Food Chemistry* 53, 6718–6724.

Sepulveda, M., Cuevas, I.I., Smilanick, J.L., Cerioni, L., Rapisarda, V.A. and Ramallo, J. 2015. Improvement in imazalil treatments in commercial packinglines to control green mold on lemon fruit. *Scientia Horticulturae* 192, 387–390.

Siddiqui, Y. and Ali, A. 2014. *Colletotrichum gloeosporioides* (Anthracnose). pp. 337–371. *In*: S. Bautista-Baños (ed.). *Postharvest Decay. Control Strategies*. Academic Press, Elsevier Inc., London, UK.

Siegel, M.R. 1981. Sterol-inhibiting fungicides: Effects on sterol biosynthesis and sites of action. *Plant Disease* 65, 986–989.

Siegel, M.R. and Ragsdale, N.N. 1978. Antifungal mode of action of imazalil. *Pesticide Biochemistry and Physiology* 9, 48–56.

Silva, A.O., Savi, D.C., Gomes, F.B., Gos, F.M.W.R., Silva Jr, G.J. and Glienke, C. 2017. Identification of *Colletotrichum* species associated with postbloom fruit drop in Brazil through GAPDH sequencing analysis and multiplex PCR. *European Journal of Plant Pathology* 147, 731–748.

Simmons, E.G. 1999. *Alternaria* themes and variations. Classification of citrus pathogens. *Mycotaxon* 48, 263–323.

Smilanick, J.L., Aiyabei, J., Mlikota Gabler, F., Doctor, J. and Sorenson, D. 2002. Quantification of the toxicity of aqueous chlorine to spores of *Penicillium digitatum* and *Geotrichum citri-aurantii*. *Plant Disease* 86, 509–514.

Smilanick, J.L., Brown, G.E. and Eckert, J.W. 2006a. The biology and control of postharvest diseases. pp. 339–396. *In*: W.F. Wardowski, W.M. Miller, D. J. Hall and W. Grierson (eds.). *Fresh Citrus Fruits*. Florida Science Source, Florida, FL.

Smilanick, J.L., Crisosto, C.H. and Mlikota, F. 1999. Postharvest use of ozone on fresh fruit. *Perishables Handling Quarterly* 99, 10–14.
Smilanick, J.L., Mackey, B.E., Reese, R., Usall, J. and Margosan, D.A. 1997a. Influence of concentration of soda ash, temperature, and immersion period on the control of postharvest green mold of oranges. *Plant Disease* 81, 379–382.
Smilanick, J.L., Mansour, M., Gabler, F.M. and Goodwine, W. 2006b. The effectiveness of pyrimethanil to inhibit germination of *Penicillium digitatum* and to control citrus green mold after harvest. *Postharvest Biology and Technology* 42, 75–85.
Smilanick, J.L. and Mansour, M.F. 2007. Influence of temperature and humidity on survival of *Penicillium digitatum* and *Geotrichum citri-aurantii*. *Plant Disease* 91, 990–996.
Smilanick, J.L., Mansour, M.F., Gabler, F.M. and Sorenson, D. 2008. Control of citrus postharvest green mold and sour rot by potassium sorbate combined with heat and fungicides. *Postharvest Biology and Technology* 47, 226–238.
Smilanick, J.L., Mansour, M.F., Margosan, D.A., Gabler, F.M. and Goodwin, W.R. 2005. Influence of pH and $NaHCO_3$ on effectiveness of imazalil to inhibit germination of *Penicillium digitatum* and to control postharvest green mold on citrus fruit. *Plant Disease* 89, 640–648.
Smilanick, J.L., Mansour, M.F. and Sorenson, D. 2006c. Pre- and postharvest treatments to control green mold of citrus fruit during ethylene degreening. *Plant Disease* 90, 89–96.
Smilanick, J.L., Michael, I.F., Mansour, M.F., Mackey, B.E., Margosan, D.A., Flores, D. and Weist, C.F. 1997b. Improved control of green mold of citrus with imazalil in warm water compared with its use in wax. *Plant Disease* 81, 1299–1304.
Smillie, R., Grant, B.R. and Guest, D. 1989. The mode of action of phosphite: Evidence for both direct and indirect modes of action on three *Phytophthora* spp. in plants. *Phytopathology* 79, 921–926.
Smith, C.O. 1917. Sour rot of lemon in California. *Phytopathology* 7, 37–41.
Smoot, J.J. 1969. Decay of Florida citrus fruits stored in controlled atmospheres and in air. *Proceeding of the International Society of Citriculture* 3, 1285–1289.
Smoot, J.J. 1977. Factors affecting market diseases of Florida citrus fruits. *Proceedings of the International Society of Citriculture* 2, 250–254.
Smoot, J.J., Houck, L.G. and Johnson, H.B. 1983. Market Diseases of Citrus and other Subtropical Fruits. 2nd ed. USDA-ARS Agriculture Handbook No. 398.
Snowdon, A.L. 1990. Sooty mould. p. 86. *In: A Color Atlas of Post-Harvest Diseases and Disorders of Fruit and Vegetables, vol. 1: General Introduction and Fruits.* Wolfe Scientific, London, UK.
Sonntag, W., Theuvsen, L., Kersting, V. and Otter, V. 2016. Have industrialized countries shut the door and left the key inside? Rethinking the role of private standards in the international fruit trade. *International Food and Agribusiness Management Review* 19, 151–170.
Stander, O.P.J., Gilbert, M.J., Moore, S.D., Kirkman, W. and Schutte, G.C. 2017. Benefits of reducing the size of the navel-end opening in 'Navel' sweet oranges (*Citrus sinensis*). *Crop Protection* 96, 123–129.
Sun, X., Li, H. and Yu, D. 2011. Complete mitochondrial genome sequence of the phytopathogenic fungus *Penicillium digitatum* and comparative analysis of closely related species. *FEMS Microbiology Letters* 323, 29–34.

Suprapta, D.N., Arai, K. and Iwai, H. 1995. Distribution of *Geotrichum candidum* citrus race in citrus groves and non-citrus fields in Japan. *Mycoscience* 36, 277–282.

Suprapta, D.N., Arai, K., Iwai, H. and Matsuo, T. 1996. Change in susceptibility of satsuma mandarin fruit to sour rot pathogen (*Geotrichum candidum* citrus race) with relation to biochemical changes during maturation and storage. *Mycoscience* 37, 209–216.

Tashiro, N., Manabe, K. and Ide, Y. 2012. First report of whisker mold, a postharvest disease on citrus caused by *Penicillium ulaiense* (in Japan). *Journal of General Plant Pathology* 78, 140–144.

Tashiro, N., Noguchi, A., Ide, Y. and Kuchiki, F. 2013. Sooty spot caused by *Cladosporium cladosporioides* in postharvest Satsuma mandarin grown in heated greenhouses. *Journal of General Plant Pathology* 79, 158–161.

Taylor, J.W. 2011. One Fungus = One Name: DNA and fungal nomenclature twenty years after PCR. *IMA Fungus* 2, 113–120.

Timmer, L.W., Brown, G.E. and Zitko, S.E. 1998. The role of *Colletotrichum* spp. in postharvest anthracnose of citrus and survival of *C. acutatum* on fruit. *Plant Disease* 82, 415–418.

Timmer, L.W., Garnsey, S.M. and Graham, J.H. (eds.). 2000. *Compendium of Citrus Diseases*, 2nd ed. APS Press, St. Paul, MN.

Timmer, L.W., Peever, P.L., Solel, Z. and Kazuya Akimitsu, K. 2003. Alternaria diseases of citrus – Novel pathosystems. *Phytopathologia Mediterranea* 42, 99–112.

Troncoso-Rojas, R. and Tiznado-Hernández, M.E. 2014. *Alternaria alternata* (Black rot, Black spot). pp. 147–187. *In*: S. Bautista-Baños (ed.). *Postharvest Decay. Control Strategies*. Academic Press, Elsevier Inc., London, UK.

Tuset, J.J. 1984. Post-harvest citrus diseases in Spain: Problems and solutions. *Proceedings of the International Society of Citriculture* 5, 508–511.

Tuset, J.J. 1987. *Podredumbres de los Frutos Cítricos*. Conselleria d'Agricultura i Pesca, Generalitat Valenciana, Valencia, Spain (in Spanish).

Tuset, J.J. and Martí, M.C. 1988. *Sclerotinia sclerotiorum* - A new decay of citrus fruit in Spain. *Proceedings of the International Society of Citriculture* 6, 1451–1459.

Udayanga, D., Castlebury, L.A., Rossman, A.Y. and Hyde, K.D. 2014. Species limits in *Diaporthe*: Molecular re-assessment of *D. citri, D. cytosporella, D. foeniculina* and *D. rudis*. *Persoonia* 32, 83–101.

US FDA (United States Food and Drug Administration). 2011. FDA Food Safety Modernization Act. Available at: www.gpo.gov/fdsys/pkg/PLAW-111publ353/pdf/PLAW-111publ353.pdf

Venditti, T., Molinu, M.G., Dore, A., Agabbio, M. and D'hallewin, G. 2005. Sodium carbonate treatment induces scoparone accumulation, structural changes, and alkalinization in the albedo of wounded *Citrus* fruits. *Journal of Agricultural and Food Chemistry* 53, 3510–3518.

Vilanova, L., Viñas, I., Torres, R., Usall, J., Jauset, A.M. and Teixidó, N. 2012. Infection capacities in the orange-pathogen relationship: Compatible (*Penicillium digitatum*) and incompatible (*Penicillium expansum*) interactions. *Food Microbiology* 29, 56–66.

Wardowski, W.F. and Brown, G.E. 1993. *Postharvest Decay Control Recommendations for Florida Citrus Fruit*. Florida Cooperative Extension Service, Circular 359-A. Institute of Food and Agricultural Sciences, University of Florida, Gainesville, FL.

Wardowski, W.F., Hayward, F.W. and Dennis, J.D. 1974. A flood-recovery TBZ fungicide treatment system for citrus fruits. *Proceedings of the Florida State Horticultural Society* 87, 241–243.

Whiteside, J.O. 1970. Factors contributing to the restricted occurrence of citrus brown rot in Florida. *Plant Disease Reporter* 54, 608–612.

Wild, B.L. 1983. Double resistance by citrus green mould *Penicillium digitatum* to the fungicides guazatine and benomyl. *Annals of Applied Biology* 103, 237–241.

Wills, R.B.H., Ku, V.V.V., Shohet, D. and Kim, G.H. 1999. Importance of low ethylene levels to delay senescence of non-climacteric fruit and vegetables. *Australian Journal of Experimental Agriculture* 39, 221–224.

Wu, Y., Duan, X., Jing, G., OuYang, Q. and Tao, N. 2017. Cinnamaldehyde inhibits the mycelial growth of *Geotrichum citri-aurantii* and induces defense responses against sour rot in citrus fruit. *Postharvest Biology and Technology* 129, 23–28.

Youssef, K., Hashim, A.F., Rubina, M., Alghuthaymi, M.A. and Abd-Elsalam, K.A. 2017. Fungicidal efficacy of chemically-produced copper nanoparticles against *Penicillium digitatum* and *Fusarium solani* on citrus fruit. *Philippine Agricultural Scientist* 100, 69–78.

Zhu, C., Sheng, D., Wu, X., Wang, M., Hu, X., Li, H. and Yu, D. 2017. Identification of secondary metabolite biosynthetic gene clusters associated with the infection of citrus fruit by *Penicillium digitatum*. *Postharvest Biology and Technology* 134, 17–21.

Zitko, S.E., Timmer, L.W. and Sandler, H.A. 1991. Isolation of *Phytophthora palmivora* pathogenic to citrus in Florida. *Plant Disease* 75, 532–535.

Chapter 2

Pome Fruits

Davide Spadaro

Department of Agricultural, Forestry and Food Sciences (DISAFA), University of Torino, Largo Braccini, Grugliasco, Torino, Italy

Centre of Competence for the Innovation in the Agro-Environmental Sector (AGROINNOVA), University of Torino, Largo Braccini, Grugliasco, Torino, Italy

Rosario Torres

Institute of Agrifood Research and Technology (IRTA), XaRTA-Postharvest, Edifici Fruitcentre, Parc Científic i Tecnològic Agroalimentari de Lleida, Parc de Gardeny, Lleida, Catalonia

Deena Errampalli

Agriculture and Agri-Food Canada, Vineland Station, Canada

Kerry Everett and Lucia Ramos

The New Zealand Institute for Plant & Food Research Ltd., Private Bag, Auckland, New Zealand

Marta Mari

CRIOF, Department of Agricultural Science, University of Bologna, Cadriano, Bologna, Italy

POSTHARVEST PATHOLOGY

Abbreviations

1-MCP	1-Methylcyclopropene
a_w	Water activity
BLAST	Basic local alignment search tool
CA	Controlled atmosphere
CYA	Czapek-dox yeast autolysate agar
DPA	Diphenylamine
EC_{50}	Effective concentration for 50% inhibition
EC_{95}	Effective concentration for 95% inhibition
EF1-alpha	Elongation factor 1-alpha
EU	European Union
IPM	Integrated pest management
ITS	Internal transcribed spacer
MEA	Malt extract agar
PCR	Polymerase chain reaction
PDA	Potato dextrose agar
RAPD	Random amplification of polymorphic DNA
rDNA	ribosomal DNA
RH	Relative humidity
UK	United Kingdom
USA	United States of America

1 General Introduction

Around 89 million tons of apples were produced worldwide across over 5 million hectares in 2016 (FAOSTAT, 2018). China produces more than 44 million tons of apples, followed by the United States of America (USA), Poland, Turkey, India, Iran, and Italy. Apple cultivars in commercial production change slowly, and the most popular apple cultivars are 'Golden Delicious,' 'Red Delicious,' 'Gala,' 'Fuji,' 'Granny Smith,' 'Jonagold,' and 'Idared.' Over 27 million tons of pears are produced annually worldwide on 1.5 million hectares. Also, for pears, China is the largest producer, with around 11 million tons in 2016, followed by Argentina, the USA, Italy, and Turkey. The principal pear cultivars are 'Abate Fétel,' 'Beurré Bosc,' 'Conference,' 'Coscia,' 'd'Anjou,' 'Doyenneé du Comice,' 'Packham's Triumph,' and 'Williams Bon Crétien.' Besides apples and European pears, other pome fruit include Asian pear, cotoneaster, hawthorn, quince, loquat, hawthorn, and firethorn.

2 Main Diseases

2.1 *Blue Mold (*Penicillium *spp.)*

Blue mold or soft rot is a common storage disease in apples (Figure 2.1a) and pears (Figure 2.1b) for the fresh market. It is the most important postharvest disease in all apple-producing countries worldwide. Blue mold is one of the most

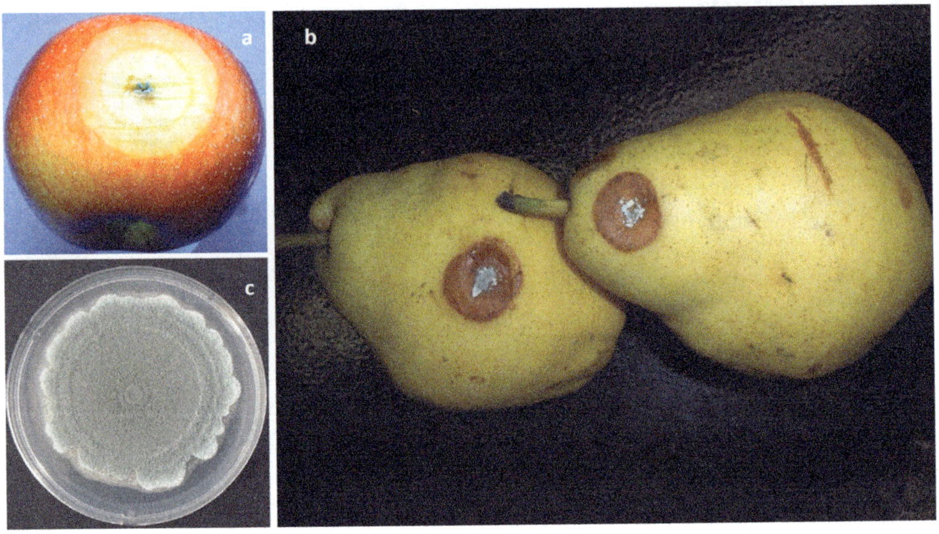

Figure 2.1 Blue mold on: a. Apple and b. Pear. c. Colony of *Penicillium expansum.*

destructive rots of stored apples and annual economic losses are estimated at 4.5–5 million USD. Postharvest losses up to 50% of total production were reported in developing countries, where sanitation and refrigeration are lacking. The blue mold disease is caused by *Penicillium* spp., including *Penicillium expansum* Link. Some isolates of *P. expansum* produce the mycotoxin patulin, with a potential damage to public health (Li et al., 2015). Patulin is heat resistant, particularly in an acidic environment and, as a result, the toxin persists in shelf-stable apple products, such as juices, concentrates, jellies, and baby foods. Since the juices and baby foods are consumed by infants and children, Canada, the USA, and the European Union (EU) have set regulatory limits of patulin as 50 μg/L or less in apple juices, including the apple juice portion of any juice blends or drinks, and unfermented apple cider (US FDA, 2001; HealthCanada, 2013). On a world scale, *P. expansum* is one of the most important pathogens of pome fruit and its infection may occur during the growing season, at harvest, in storage, or transit (Dennis, 1983).

The first symptoms appear as soft, watery lesions that are light brown in color and, as the lesion ages, the conidia of *Penicillium* spp. turn blue–green on the surface. Sporulation does not usually develop while fruits are kept in controlled atmosphere (CA) storage, but can develop after CA rooms are opened. The lesions expand rapidly between 20 and 25°C, they have a very sharp margin between diseased and healthy tissues, and the diseased tissue can be readily separated from healthy tissue, leaving it like a "bowl." The surface of older lesions may be covered with blue–green spores and conidial tufts of *P. expansum*. Because of the soft, watery consistency of the rotted tissue, this disease is often referred to as soft rot. Decayed fruit has an earthy and musty odor. The presence of blue–green spore masses at the decayed area and associated musty odor are the positive diagnostic indication of blue mold.

Since its first description by Link in 1809, historically, the identification of species in the genus *Penicillium* relied on their morphological characteristics. Morphological characterization of *P. expansum* was based on the macroscopic study, growth rate, color texture, topography of the colony using two standard media, namely potato dextrose agar (PDA) and Czapek Dox, and microscopic study, observation of hyphae and conidia under a light microscope (Figure 1c). The conidia are small, ranging in size between 4.5 and 5.0 μm in diameter, spherical to subspherical or elliptical, and borne in chains on conidiophores. The branching of the conidiophores is an important part of species identification. Conidiophores are dense brush-like conidia-bearing structures. Conidia are produced in dry chains from the tips of flask-shaped phialides, with the youngest spore at the base of the chain, and are nearly always green. On the basis of the type of spore-bearing systems (penicilli), *P. expansum* was classified as terverticillate, branches that bear a second order of branches, bearing, in turn, a cluster of phialides. However, distinguishing different species of *Penicillium* based on morphological characteristics alone is difficult. With the recent advances in nucleic acid-based technologies, molecular characterization was carried out with molecular markers (RAPD) and DNA sequences of internal transcribed spacer (ITS) or β-tubulin gene (Peterson, 2000; Czerwinski et al., 2009). A polyphasic approach to the taxonomy based on morphology, growth pattern, ecology, extrolites, and partial β-tubulin sequences placed *P. expansum, P. marinum* Frisvad & Samson, and *P. sclerotigenum* W. Yamam. in the section *penicillium* and series *expansa* (Frisvad and Samson, 2004). Blue mold is reported to be caused by over 17 *Penicillium* spp., those have been isolated from naturally infected pome fruit showing blue mold symptoms, but *P. expansum* was the most common species (Sanderson and Spotts, 1995; Sutton et al., 2014). The species *P. expansum* could be differentiated from other *Penicillium* spp. that attack pome fruits by its rate of growth in apple fruit. When apples are wounded, inoculated with conidia, and incubated at 20–22°C, *P. expansum* usually produced a decayed area 30–40 mm in diameter after 8–10 d, whereas decayed areas caused by other species are often less than half as large (Sutton et al., 2014). Some strains of *P. expansum* produce patulin, a heat-resistant mycotoxin, in apple fruit. When apples carrying some strains of *P. expansum* are used for juice or fresh cider, patulin concentrations can exceed limits considered safe (50 μg/L) for human consumption.

Penicillium is considered a wound pathogen, as wounds on the apple fruit are required to infect the host. Unlike other fungi, *Penicillium* does not produce appressoria and therefore is unable to penetrate the intact epidermis of fruits and vegetables. Poor-quality fruits with surface defects, such as wounds caused by insects or storm damage, are the most common entry points for *P. expansum* in the orchard during the preharvest period. Infection may occur in the damaged areas near the lenticels or wounds caused during apple harvest. In some cases, the pathogen may infect unwounded fruits via stem-end or calyx-end entry. *Penicillium* spp. may gain entry through the lesions formed by other pathogens, such as *Gloeosporium, Mucor,* or *Phytophthora*. Temperature and relative humidity (RH) are the two factors that influence the progress of blue mold on the fruit. The pathogen grows even at low temperatures of 0–4°C, but rapid growth occurs when the fruit is transferred to warmer temperatures.

2.2 *Gray Mold* (Botrytis cinerea)

Gray mold is one of the most important postharvest diseases of pome fruits, especially in cold-stored pears and, less commonly, apples. The disease is found worldwide, wherever these fruits are grown and stored, and develops more rapidly at cold storage temperatures than other postharvest decay. Losses as high as 20–60% of total production due to gray mold are reported after an extended period of storage (Sutton et al., 2014), especially on fruits that were not treated with fungicides prior to storage. In addition, gray mold can cause large losses because of the ability to spread from decayed to surrounding healthy fruits through fruit-to-fruit contact during storage. These losses are higher in developing countries due to the lack of adequate handling and refrigeration facilities. Because of the lack of precise statistics, however, it is hard to determinate the extent of real losses, which may vary depending on the commodity and the producing country. Even so, gray mold is considered an important postharvest pathogen because the environmental conditions during storage and/or shipment are favorable to pathogen development and to infect healthy fruits through fruit-to-fruit contact (Droby and Lichter, 2007).

Gray mold causes a firm and pale tan rot without sharp margins that becomes covered by ash-gray spore masses (Figure 2.2a). Depending on cultivar

Figure 2.2 Gray mold on: a. 'Conference' pear, b. Through lenticels of pear, and c. Red apple. d. Nesting of gray mold on 'Blanquilla' pears. (Courtesy of XaRTA-Postharvest, IRTA).

and storage conditions, the decayed area appears light brown to dark brown. The mold originates primarily from infection through wounds, such as punctures, cracks, and bruises. These could be produced at harvest and during postharvest handling, but also from stem punctures, especially in 'd'Anjou' pear, or less often from the calyx end (Sutton et al., 2014). This mold can also penetrate fruit through their lenticels (Figure 2b). On red apples, dark areas may persist around lenticels even after the fruits are completely decayed, and these give the fruit a speckled appearance (Figure 2c). A fine white or gray-white mycelium may appear over the rotted area and, when the fruits are held at high humidity, gray spore masses usually appear; less sporulation occurs at cold storage temperatures (Jones and Aldwinckle, 1990). As the decay enlarges, the rot becomes darker as it progresses, but the edges often remain with a paler color. The rotted area may become wrinkled and often appear baked after removal from storage. The decayed area is spongy and diseased tissue is not readily separable from the healthy tissue, but after an extended storage, advanced decayed fruits may become soft (Xiao and Kim, 2008). Spread from fruit to fruit, as secondary infection, is rapid and causes significant losses after a long period of storage, producing nests of infection (Figure 2d). Generally, gray mold does not have a distinct odor but, in advanced stages, the rotted fruit may have a sweet and fermented cedar-like smell (Xiao and Kim, 2008).

The causal organism of gray mold is the fungus *Botrytis cinerea* Pers. that also has a teleomorph form (Coley-Smith et al., 1980), which is placed in Sclerotiniaceae of Helotiales. The teleomorph state is seldom found in nature, but apothecia can be readily obtained in laboratory (Faretra et al., 1988). The species name, *B. cinerea*, is derived from the Latin for "grapes like ashes": the "grapes" refers to the bunching of the fungal spores on their conidiophores, and "ashes" just refers to the grayish color of the spore masses. *Botrytis cinerea* is characterized by abundant hyaline conidia (asexual spores) borne on gray, branching tree-like conidiophores. On PDA, czapek-dox yeast autolysate agar (CYA) or malt extract agar (MEA), colonies covering the whole Petri dish are floccose, with growth sometimes patchy or irregular, mycelium white becoming gray to dark gray as conidiogenesis proceeds; the reverse color is pale to gray. Conidiophores borne from aerial hyphae, stipes of indeterminate length each bearing terminally an irregular cluster of short branches, 10–30 μm long, with swollen spherical apices, 8–10 μm in diameter. Conidia borne singly from these apices on denticles (small pegs), round or ellipsoidal, 8–12 μm long, smooth walled, not released at maturity. Short branches and the production of conidia on pegs from spherical swellings are characteristic of *Botrytis* spp. (Pitt and Hocking, 1999).

Botrytis cinerea is a necrotrophic fungus that affects many plant species, including cultivated apple (*Malus* × *domestica* Borkh), and is worldwide present (Coley-Smith et al., 1980). Growth of *B. cinerea* depends on the temperature and is rather variable, with minima from −2 to 5°C, maxima 28–35°C, and optima 22–25°C (Domsch et al., 1980). Temperature determines how fast infection occurs and, together with RH, are the key environmental parameters that promote *Botrytis* development. Mycelial growth and secondary infection through fruit-to-fruit spread is enhanced by high RH. However, little sporulation occurs at cold storage temperatures (Jones and Aldwinckle, 1990). This fungus can grow from pH 2 to 8 (Jarvis, 1977), at O_2 concentration lower than 1%. It produces black, irregular-shaped sclerotia when grows in culture. Sclerotia are not found on decayed fruit in storage but may form on windfall fruit on the orchard floor (Jones and Aldwinckle, 1990).

Isolates from pome fruit are highly variable: some isolates sporulate heavily in culture and others produce only sclerotia or sclerotia with a few conidia. In general, sporulation is enhanced by exposure to light that includes the near-ultraviolet spectrum. The fungus also produces highly resistant sclerotia as survival structures in older cultures. It overwinters as sclerotia or intact mycelia, both germinating in spring to produce conidiophores. The conidia are dispersed by wind and rainwater and cause new infections.

Botrytis cinerea is a broad-spectrum pathogen (Sommer, 1985) that infects numerous host plants causing gray mold on various plant parts. After infection and death of host tissues, the pathogen can survive and sporulate as saprophyte on the necrotic tissue or produce long-term survival structures, such as sclerotia (Holz et al., 2007). Gray mold is rarely seen in the field, although the fungus is a common saprophyte on decaying organic matter on the orchard floor. The largest damage inflicted by *Botrytis* spp. is recorded for senescing fruit tissue after harvest, since it is the most vulnerable to infection, and the pathogen effectively develops and grows at temperatures just before freezing. Most infections result from the contamination of wounds with spores that have spread from decaying plant material in the orchard by air dispersion, but are more commonly water-dispersed in flumes and drenches and from contaminated bins or containers. The disease can spread quickly through infected fruits, which is an additional inoculum to infect adjacent healthy fruits during storage, producing nests or pockets of decayed fruits, and can result in important losses from a single original infection. *Botrytis cinerea* can also infect stems, especially on fruits with thick stems, and cause stem-end gray mold. At harvest, these fruits are detached from the tree and fresh wounds at the stem are exposed to potential contamination by this pathogen.

2.3 Bitter Rot and Other Diseases Caused by Colletotrichum *spp.*

Recently, there have been several reports of bitter rot of apple caused by *Colletotrichum acutatum* J.H. Simmonds in Brazil (Braganca et al., 2016), Uruguay (Alaniz et al., 2012), Croatia (Ivic et al., 2013b), Italy (Mari et al., 2012), Iran (Arzanlou et al., 2015), Indonesia (Dharmaputra et al., 2013), Norway (Sundelin et al., 2015; Børve and Stensvand, 2017), Serbia (Grahovac et al., 2012), the Czech Republic (Vichova et al., 2012), Kentucky, USA (Munir et al., 2016), or Latvia (Volkova and Juhnevica-Radenkova, 2015). Older records are from Canada (Biggs and Miller, 2001), China (Zhang et al., 2008, 2009; Fu et al., 2013), New Zealand (Johnston et al., 2005; Boyd-Wilson et al., 2014), Japan (Sakai et al., 2002; Tsushima and Yukita, 2002), southeastern USA (Shi et al., 1996), and South Carolina, USA (Gonzalez and Sutton, 2004). The proliferation of new records for *C. acutatum* in the past decade could be because the species *Colletotrichum gloeosporioides* (Penz.) Penz. & Sacc. is less tolerant than *C. acutatum* to benomyl (Gonzalez and Sutton, 2005). Fungicides in the group to which benomyl belongs, the benzimidazoles, have been commonly used in apple orchards (Barkai-Golan, 2001). It is possible that the widespread use of these fungicides, and other changes in fungicide use, has altered the composition of *Colletotrichum* spp. present on apple orchards. For instance, in New Zealand, the older records are exclusively of *C. gloeosporioides* (Kirk, 1906; Dingley, 1969; Brook, 1972, 1977, 1981). *Colletotrichum acutatum* was not reported as a disease of apple until 1997 (Johnston and Jones, 1997), even though the first deposition of a culture from a domestic property into the national collection was in 1964 (Anonymous, 2017). By 2005, it was reported as the major cause of bitter rot of apple in New Zealand (Johnston et al., 2005). Likewise, in Norway, where it has recently been reported as a serious disease (Stensvand et al., 2012), examination of herbarium records showed that it was isolated from apples in 1957 (Sundelin et al., 2015). However, it has been suggested in Italy that *C. acutatum* is a new disease that was first observed on strawberries (Bosshard, 1997). The pathogen was removed from the quarantine list in 2008 because it had become established in the EU (Mari et al., 2012). *Colletotrichum acutatum* has a wide host range (Peres et al., 2005; Freeman, 2008) and a host shift from strawberries could be the reason for the recent proliferation of new reports in some parts of Europe (Mari et al., 2012).

The economic cost of bitter rot disease is largely borne by the orchard manager. Fungicides are recommended to be applied at 10- to 14-d intervals throughout the season for good control (Sugar, 2014). The cost of postharvest losses is less well documented, but occasional losses from postharvest diseases of 15–25% have been reported (Rosenberger, 2014). There have been calculations of the economic cost of postharvest apple diseases to the USA at 4.4 million USD/yr (Rosenberger, 1997). In spray trials, conducted on 'Royal Gala' in the Waikato growing region of New Zealand, incidence of *C. acutatum* in unsprayed controls at harvest ranged from 7.9 to 26% (Boyd-Wilson et al., 2014; Everett et al., 2015). Incidence in unsprayed controls after 15 wk cool storage plus 7 d at 20°C for Waikato apple fruit harvested during the 2008/2009 season was 69%. In the southern states and mid-Atlantic USA there have been reports of 50–80% losses in unsprayed orchards (Camilo et al., 1988; Jurick et al., 2011), as well as significant losses during postharvest storage (Struble and Keitt, 1950; Barkai-Golan, 2001). Although fungicide application results in a significant cost of bitter rot disease, the exclusion of this measure would be commercially non-viable.

Apple fruit are infected by *C. acutatum* in the field, where they first express as small black spots (1–2 mm diameter), which can be surrounded by reddish tissue (Figures 2.3a,c), usually on the sun-exposed side. These spots first become depressed and then expand to form tan to dark brown to black, usually circular lesions (Figures 2.3b,f). Finally, lesions develop acervuli that produce distinctive orange spore masses in concentric rings (Figure 2.3d). *Colletotrichum acutatum* can also infect the distal end to cause an eye rot (Figure 2.3e).

Severely infected fruit can drop off or remain as mummified fruit in the canopy. Apple fruit can be latently infected in the field without symptoms, and the disease then expresses in the postharvest period. Symptoms caused by *C. gloeosporioides* are almost identical. The only observable differences are the color of the conidia produced in the acervuli; for this species, the color is cream to salmon to pale pink (Sutton, 1990). Black perithecia can also form in lesions on apple fruit infected by *C. gloeosporioides*, but these are not observed on apples infected by *C. acutatum*. Conidia on lesions associated with perithecia can be orange to salmon, but when orange conidia are present, these lesions can be distinguished from those of *C. acutatum* by the presence of perithecia. Postharvest symptoms differ slightly compared with those expressed in the field. Lesions are usually circular, can converge and overlap, and are tan in color (Figures 2.4a–f). More advanced symptoms result in the production of masses of orange-colored conidia (Figure 2.5, top) in a concentric ring pattern.

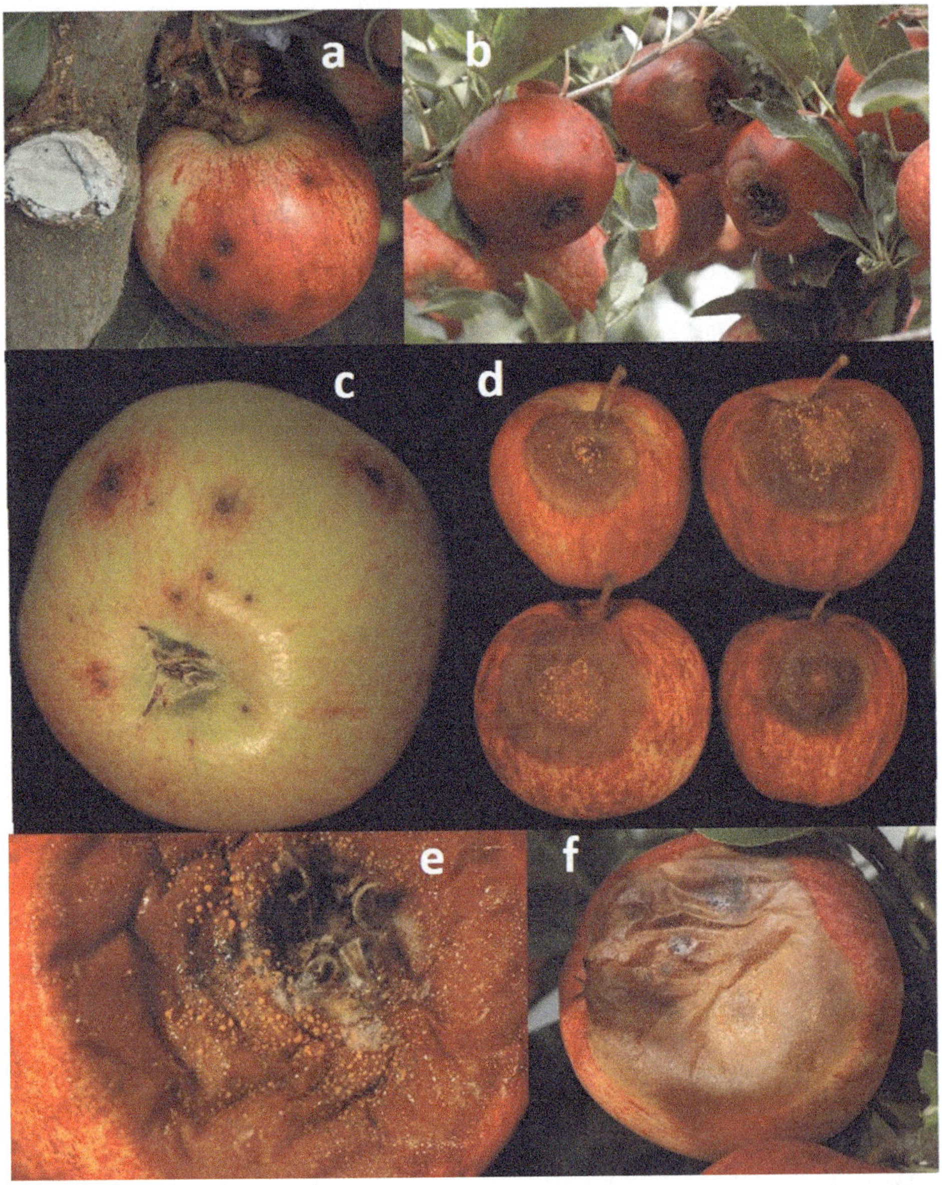

Figure 2.3 Field symptoms of bitter rot caused by *Colletotrichum acutatum*. a. Small (2–4 mm in diameter) dark sunken spots, b. Large tan to dark brown lesions, c. Spots surrounded by reddish tissue, d. Lesions producing acervuli and concentric rings of orange-colored conidia, e. Eye rot, and f. A large lesion without conidia.

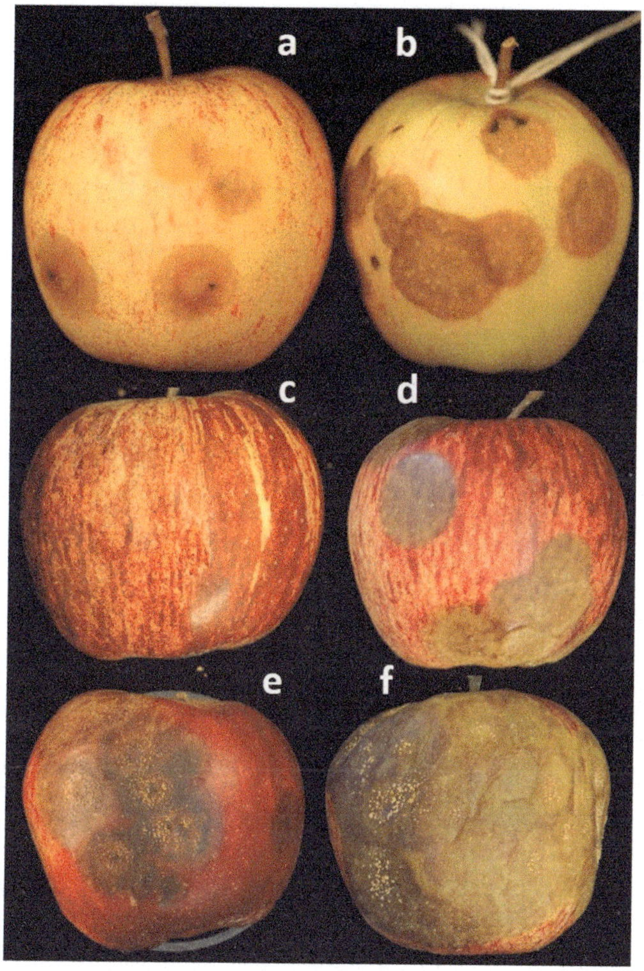

Figure 2.4 Typical postharvest bitter rot symptoms on 'Royal Gala' apples caused by *Colletotrichum acutatum* from the Waikato apple-growing region of New Zealand after: a,b. 12 wk and d. 15 wk of cool storage followed by e,f. 7 d at room temperature. c. A symptomless fruit is shown for comparison.

Recently, the *Colletotrichum* group of fungi has undergone extensive taxonomic revision (Cannon et al., 2012). Using multi-locus sequence analysis of randomly selected gene regions, the fungus *C. acutatum* has been split into 29 new phylospecies (Damm et al., 2012), and *C. gloeosporioides* into 22 (Weir et al., 2012). Currently, there is no convincing evidence that there are important biological differences between phylospecies (Everett, 2014). The new species names are possibly describing population variation within a species, and thus are no longer useful species concepts for plant pathologists. For these reasons, the species' complex names are used here. In the orchard and during the postharvest phase, bitter rot

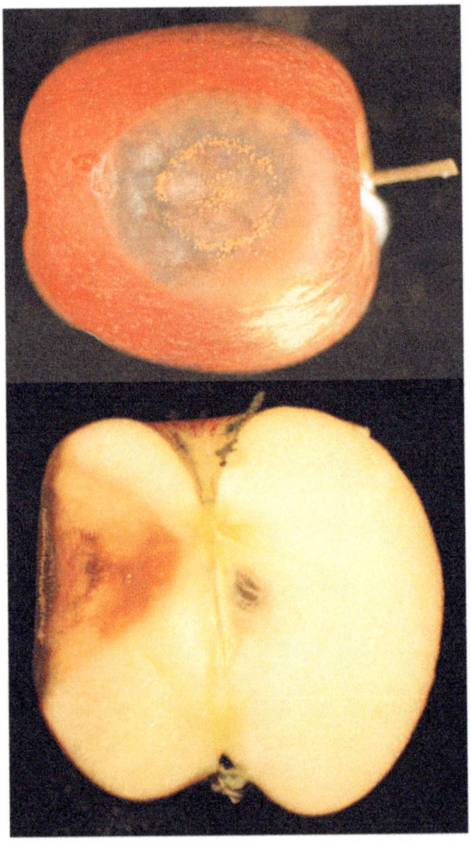

Figure 2.5 External (top) and internal (V-shaped in section) (bottom) lesion following wound-inoculation of 'Royal Gala' apple fruit with *Colletotrichum acutatum* in the orchard.

caused by *C. acutatum* and *C. gloeosporioides* can be distinguished from other apple rots by the presence of orange- or salmon-colored conidia (Figure 2.5, top). Both species of *Colletotrichum* penetrate the apple fruit to form a distinctive V-shape rot through the fruit flesh (Figure 2.5, bottom), which can be used as a valuable diagnostic (Sugar, 2014). Following isolation, *C. acutatum* can be distinguished from *C. gloeosporioides* on the basis of a slower rate of colony growth on PDA, fusiform conidia compared with the cylindrical conidia with obtuse ends of *C. gloeosporioides* (Dyko and Mordue, 1979), and an orange to red color on the underside of colonies grown on PDA (Shi et al., 1996; Mari et al., 2012). Sequencing the ITS region of ribosomal DNA (rDNA) can also be used to identify these two species (Phoulivong et al., 2010). Although there is no evidence that there are biological differences between phylospecies, there are differences between *C. acutatum* and *C. gloeosporioides*. For instance, *C. gloeosporioides* is less tolerant to low temperatures, whereas *C. acutatum* is less tolerant to high temperatures (Everett, 2003; Gonzalez and Sutton, 2005). *Colletotrichum acutatum* is also more tolerant to benomyl (Gonzalez and Sutton, 2005), grows more slowly on PDA, and has differently shaped conidia (Sugar, 2014). The sexual stage of *C. gloeosporioides* has also been shown to be important in the disease cycle. However, the sexual stage of *C. acutatum* has rarely been observed; it was first reported from the laboratory in 2001 (Guerber and Correll, 2001). Since then, there have been two reports of the sexual stage from the field, on blueberry fruits in Norway (Talgo et al., 2007) and on apple leaves in Brazil (Berton, 2007).

Initial infections of *C. gloeosporioides* on apples have been shown to start in spring from cankers on twigs and mummified fruits in the canopy. Either ascospores or conidia are released during rain to symptomlessly infect all stages of fruits, leaves, and twigs. Ascospores are found throughout the season (Struble and Keitt, 1950; Brook, 1977; Sutton and Shane, 1983) and have been shown to traverse distances over 60 m in apple orchards (Sutton and Shane, 1983). Infections of twigs result in the formation of cankers and, in the later part of the season, infections of fruits can lead to disease that expresses in the orchard or remain symptomless until expression during the postharvest storage phase. The fungus overwinters in mummified fruits in the canopy, rotten fruits that fall to the ground, and in cankers on twigs in the tree (Struble and Keitt, 1950; Atkinson, 1971; Brook, 1977; Sutton and Shane, 1983; Snowdon, 1990). Although symptoms on fruits in the orchard usually appear when fruits are close to harvest, fruits are susceptible to infection by *C. gloeosporioides* at all stages of maturity (Taylor, 1971; Noe and Starkey, 1982). Apple tissue is infected by conidia that deposit as a mass in mucilage during rain. Germination of conidia autoinhibits until rain disperses the mass, whereupon spores germinate to form dark appressoria that are cemented to the surface by mucilage and a short infection peg (Brook, 1981; Skipp et al., 1995). Appressoria remain latent until conditions are favorable for infection, whereupon the infection peg extends into the plant tissue and proliferating hyphae cause a lesion or a canker (Atkinson, 1971; Brook, 1981; Skipp et al., 1995). *Colletotrichum acutatum* has been shown to overwinter in apple buds (Børve and Stensvand, 2007; Everett et al., 2018). Although inoculum is also present on twigs, in New Zealand, twigs were less extensively colonized than buds (Everett et al., 2010). In Norway, buds were less frequently infected than twigs, but were considered an important source of inoculum for infections in spring. *Colletotrichum acutatum* was isolated from symptomless leaves in New Zealand throughout the season, but leaves were more frequently infected during summer, while in Norway, isolations were only possible from leaves collected in summer (Børve and Stensvand, 2017; Everett et al., 2018). A temperature threshold of 15°C and a concurrent 72-hr wetness period were required for infection of 'Royal Gala' apple fruits (Everett et al., 2018). It was also demonstrated that inoculum was splashed up from the ground during summer and from the apple canopy during spring. On the basis of this evidence, a putative disease cycle was suggested for New Zealand (Everett et al., 2018). In summary, inoculum released from opening buds during spring infects leaves that remain in the canopy, and contaminates and infects bud scales and flower petals that fall to the ground. Fallen twigs, decomposing leaves, and rotten fruits from the previous season add to this inoculum in and on the ground. This inoculum is rain-splash dispersed to infect fruits, buds, and leaves during summer when the mean daily temperature exceeds 15°C and is coincident with a 72-hr wetness period. There was some evidence that inoculum was also released from twig cankers. Mummified fruits remaining in the canopy have been reported as an inoculum source for other crops infected by *C. acutatum* (Wilson et al., 1992; Moral and Trapero, 2012; Samuelian et al., 2012) and also for *C. gloeosporioides* on apple (Sugar, 2014).

2.4 Bull's Eye Rot (*Gloeosporium spp.*)

Bull's eye rot, previously known as Gloeosporium rot (Snowdon, 1990), is one of the most frequent and damaging disease occurring in stored pome fruits produced in areas such as Northern regions of Europe (Maxin et al., 2014; Michalecka et al., 2016), North-West USA (Henriquez et al., 2004), and several regions of Canada (de Jong et al., 2001), Brazil (Spolti et al., 2010), Chile (Soto-Alvear et al., 2013), and Australia (Cunnington, 2004). The incidence of disease varies considerably; in humid areas, with abundant rain and persistent fog close to harvest, the economic losses can be as high as 40% for 'Golden Delicious' apple (Sholberg and Haag, 1996) and over 50% for 'Bosc' pear fruit (Lennox et al., 2004). Among apple, 'Cripps Pink', a highly regarded apple cultivar, is extremely susceptible to bull's eye rot and the incidence of disease can exceed 70% after extended storage (Figure 2.6; Cameldi et al., 2016b).

In the field, *Neofabraea* spp., especially *N. perennans* Kienholz and *N. malicorticis* (Cordley) H.S. Jacks., can produce anthracnose and perennial cankers on the branches and twigs of apple and pear trees. *Neofabraea vagabunda* (Desm.) P.R. Johnst. is considered a weak pathogen for wood tree, living as saprophyte on the dead bark, pruning snags, and fruiting spurs (Henriquez et al., 2006), while *N. kienholzii* (Seifert, Spotts & Lévesque) Spotts has been mainly isolated from apple and pear fruits (Spotts et al., 2009). These pathogens infect the fruit in the field, which remain symptomless until harvest, developing lesions only after several months of storage. Initially, the symptoms are restricted to lenticels; however, they can also occur on wounds and around the stem or calyx. Bull's eye rot appears as singular spherical lesions, with well-defined margins. The diseased

Figure 2.6 Bull's eye rot on 'Cripps Pink' apple.

areas are usually flat or slightly concave, dark and firm with a light brown center; the rotten tissue is easily separated from the healthy part and conidiomata can be present in old lesions (Soto-Alvear et al., 2013).

According to the recently updated nomenclature (Seifert, 2013), the disease is caused by different fungal species belonging to genus *Neofabraea*, of which four species have been described as pathogenic for pome fruit worldwide. *Neofabraea vagabunda* (syn.: *N. alba* [E.J. Guthrie] Verkley) is the main species known in continental Europe and reported as minor pathogen of apple in North America. *Neofabraea perennans* was associated most frequently with bull's eye rot occurring in humid areas of the West coast of the USA; however, in Europe, it has been detected in the Netherlands (Verkley, 1999), Germany, and the United Kingdom (UK) (de Jong et al., 2001). The occurrence of *N. malicorticis* was connected with the disease symptoms on the West coast of the USA (Kienholz, 1951) and Canada (de Jong et al., 2001), as well as in Europe (Cameldi et al., 2016a). The fourth species, formerly described by de Jong et al. (2001) as *Neofabraea* sp. *nov.*, was detected in samples from Canada and Portugal, and apples and pears from Oregon and Australia. In 2009, Spotts et al. (2009) provided the complete description of this species and proposed the name *Cryptosporiopsis kienholzii* Seifert reporting its presence on apples and pears in Washington State. Finally, after Seifert (2013), the name *N. kienholzii* Seifert, Spotts & Levesque was established for this fungus. Morphological differences among bull's eye pathogens are subtle and often overlap, making specific identification difficult and undependable (Henriquez et al., 2004). The variation among members of the genus *Neofabraea* is slight and the identification using conventional criteria, such as the shape and dimensions of conidia, the features of conidiomata, and the colony morphology, is challenging. Moreover, in the past, there was a lack of agreement regarding the genus to which these fungi belong, partially solved by a revision of members *Pezicula, Dermea*, and *Neofabraea* (Abeln et al., 2000), and the phylogenetic analysis of member *Neofabraea* (de Jong et al., 2001). Currently, the exact identification of *Neofabraea* spp. requires molecular techniques, such as polymerase chain reaction (PCR) fingerprinting and DNA sequencing of ITS rDNA region, mitochondrial small subunit rDNA, beta-tubulin, and elongation factor 1-alpha (EF1-alpha) genes. Among them, the sequence of ITS rDNA region may not be a sufficient approach; in fact, *N. malicorticis* and *N. perennans* have an identical sequence (de Jong et al., 2001), while EF1-alpha seems to be a suitable marker for the identification of *Neofabraea* spp. (Pesicova et al., 2017).

The studies about *Neofabraea* spp. biology are still poor and limited; some of the reasons for this are the slow growth of the fungus and the difficulty of obtaining a sufficient number of conidia in artificial media for in vitro experiments and artificial inoculations. Recently, a tomato puree-based medium was found suitable for the in vitro production of *N. vagabunda* conidia, after 14 d of incubation at 15°C (Cameldi et al., 2017), also revealing that the relatively low temperatures (lower than 20°C) are fundamental for conidia production. Previously, a discrepancy in growth optimum temperature was found between *N. vagabunda* and *N. malicorticis*, ranging between 18–20°C and 18–22°C, respectively (Senula, 1985); however, all four *Neofabraea* spp. are capable of growing at 0°C, supporting the statement that bull's eye rot cannot be prevented by storage at low temperatures (Hortova et al., 2014). Micro- and macroconidia production was observed in fungal pathogens causing bull's eye rot (Spotts et al., 2009), in particular, the variation in *N. vagabunda* conidia size could be a result of incubation temperature. Cameldi et al. (2017) found that conidia of larger dimensions were formed from colonies grown at optimal temperatures for sporulation than at less favorable temperatures. The hypothesis, also reported in other pathosystems (Tian and Bertolini, 1999), is that conidia of smaller size produced by *N. vagabunda*, in a warm environment, are less aggressive to fruit than large conidia produced at low temperature. Many pathogens have developed mechanisms to modify ambient pH and ensure a good activity of their pathogenesis enzymes (Bi et al., 2016). The secretion of alkalizing compounds caused an increase of pH (+1.2 units) at the margin of *N. vagabunda* colonies after 14 d incubation. These results were also partially confirmed on artificially inoculated 'Cripps Pink' apple, where infected tissues showed higher pH values than tissues of non-inoculated fruit (Cameldi et al., 2017).

In the orchard, conidia of *Neofabraea* spp. are produced on cankers or dead bark tissue; the highest sporulation occurs at the end of summer and during fall (Henriquez et al., 2006). In general, conidia are splash-dispersed to fruit surface and remain in a quiescent stage for some months after harvest. The knowledge is limited about the state of conidia in this time: they can be germinated showing a germ tube or remain ungerminated, into lenticels or under wax layers (Neri, personal communication). The timing of fruit harvest affected the incidence of bull's eye rot of 'Bosc' (Henriquez et al., 2008), 'Anjou,' and 'Barlett' pear (Spotts, 1985b), revealing the highest levels of disease in late-harvested fruits. Likewise, in 'Cripps Pink' apple, a late maturing variety, the percentage of disease increased delaying the harvest from 2 to 4 wk after commercial harvest (Cameldi et al., 2016a). Henriquez et al. (2008) studied the influence of wetting on bull's eye rot development showing that the disease incidence was independent of wetness duration longer than 0.5 hr, the minimum considered. Because the duration of wetting necessary for infection was very short, the authors concluded that water might have a greater impact on conidial dispersal than on infections itself. Few investigations refer to the mechanism regulating the end of pathogen quiescent stage in stored apple. Preliminary data allow hypothesizing the involvement of organic acids, mainly malic acid, in the appearance of bull's eye rot symptoms. In in vitro trials, *N. vagabunda* was strongly inhibited by malic acid with an EC_{95} value (28 mg/g dry matter) very close to the concentration detected in unripe apple (34.8 mg/g). While its EC_{50} value (13.9 mg/g) was comparable to the malic acid quantified in fruit 8 wk after harvest (14.9 mg/g) when the first symptoms of bull's eye rot appeared (Cameldi, 2015).

2.5 Brown Rot (*Monilinia* spp.)

Brown rot caused by *Monilinia fructigena* (Aderh. & Ruhland) Honey represents one of the limiting factors for apple production, leading to economic losses both in the orchard and storage (Figure 7; Byrde and Willetts, 1977). Other two species, *M. laxa* (Aderh. & Ruhland) Honey and *M. fructicola* (G. Winter) Honey, have been occasionally detected on pome fruits (Batra, 1991). Moreover, a new species, *M. polystroma* (G.C.M. Leeuwen) Kohn, was reported on European apples (Petroczy and Palkovics, 2009). The entity of losses varies from year to year, largely influenced by weather conditions; however, in well-managed orchards under integrated production, the incidence of disease usually remains below 10% (Xu

Figure 2.7 Symptoms of brown rot caused by *Monilinia polystroma* on 'Cripps Pink' apple.

et al., 2001). In contrast, considerable yield losses, up to 46%, have been reported from organic apple production in Hungary (Holb, 2004), suggesting different dynamics of brown rot development in orchards with different management systems. In storage, brown rot incidence was up to 5% in the UK (Berrie, 1989) and 4.1% in Denmark on 'Ingrid Marie' apples after 100 d storage at 2°C (Maxin et al., 2012). *Monilinia fructigena* was also confirmed as the dominant pathogen in stored 'Idared' apples in Croatia, occurring with an incidence of 3.1% after 90 d at 0°C (Ivic et al., 2013a).

The primary and most frequent symptoms of brown rot occur on fruit, although *M. fructigena* also causes symptoms as blossom blight and cankers on stems of apple and pear trees (Jones and Aldwinckle, 1990). Fruits affected by brown rot show similar symptoms in the field and storage; initially, the lesions, starting from a wound, are brown, circular, and firm, subsequently covered by large cream or gray–brown tufts of mycelium and conidia not always arranged in concentric rings. When RH is low or the fruits are not ripe, no or rare mycelium and conidial tufts develop (CABI, 2017). Initially, symptoms caused by *M. fructicola* and *M. laxa* cannot be distinguished from brown rot caused by *M. fructigena*. Only the appearance of gray-colored conidia pustules on the fruit lesions allows a partial differentiation of *M. fructicola* and *M. laxa* from *M. fructigena*, which produces distinctly buff-colored tufts. Likewise, symptoms caused on apple by *M. polystroma* and *M. fructigena* were indistinguishable 5 d after inoculation, but the differences were noticeable after 1 mon incubation; *M. polystroma* formed an extensive and discontinuous black stromatal layer over the host surface, while fruit tissue stromatization caused by *M. fructigena* was sporadic (Vasic et al., 2016).

The identification of the causal agent causing brown rot in the field is not possible using symptomology. Laboratory analyses based on culture characteristics and morphology, including color of the colony, growth rate, sporulation, conidia size, and germ tube features, are required, although they are not always sufficient to distinguish between the four *Monilinia* spp. (Poniatowska et al., 2013), because these features are frequently related to the single isolate. Recently, during a survey of Italian *Monilinia* populations, at least three morphotypes were found for each of the four pathogenic species (Beda et al., 2016), rendering a molecular test necessary for the correct identification of the isolate (Hughes et al., 2000). The analysis of the ribosomal ITS region is mainly used for the molecular detection of *Monilinia* spp. isolated from stone and pome fruits (Fulton et al., 1999). Côté et al. (2004) developed an identification method using multiplex PCR that does not required isolation of the fungal agent and it is therefore significantly faster than identification procedures based on quantitative characteristics. Afterwards, an automated DNA extraction method combined with a multiplex PCR based on TaqMan chemistry has been developed for fast, convenient, and reliable detection of the four *Monilinia* spp. limiting false negatives (Van Brouwershaven et al., 2010). The size of *M. fructigena* conidia is bigger (24 × 12 μm) than that of the other three *Monilinia* spp.; however, their production was sparse on artificial medium as PDA or cherry decoction agar (Poniatowska et al., 2013). A better sporification occurred in apple fruit and the conidia size was greater than that obtained on V8 medium (Vasic et al., 2016). The germination of conidia is markedly influenced by the interaction of temperature and water activity (a_W). For *M. fructigena*, the optimum range for conidia germination is within 15–30°C; moreover, depending on temperature, the pathogen can germinate at a_w from 0.90 to 0.99 (Casals et al., 2010).

Monilinia fructigena is considered a wound parasite infecting apple fruit only via wounds, although infection by conidia through lenticels has been observed in the laboratory (Sharma and Kaul, 1990). The pathogen overwinters mainly in/on diseased mummified fruits attached to the tree or located on the ground. The conidia of *M. fructigena* are not actively discharged, but only by air currents and wind. When temperatures, day length, and RH are suitable, tufts of conidiophores form sporodochia in infected tissues on the surface of the mummified fruit. The conidia are transported by wind, water, or insects to young fruits. Occasionally, scab lesions or sites of insect damage can be the initial court of infection, such as growth cracks (Holb and Scherm, 2008). Apple may become infected at harvest time and then fruit rots can develop during the postharvest period. Apparently, healthy fruits can be contaminated with spores and develop brown rot symptoms during storage and marketing. The incidence of disease is greatly affected by fruit maturity and wound age (Xu and Robinson, 2000), suggesting the importance to protect fruits, especially near-mature fruits, from being wounded.

2.6 Storage Apple Scab (Venturia inaequalis)

Storage scab of cultivated apple is caused by *Venturia inaequalis* (Cooke) G. Winter (Bratley, 1940; Hockey, 1953; Schwabe, 1980). The appearance of scab lesions on apple fruit in storage reduces the quality of the fruit and economic losses from storage scab results directly from fruit infection. Scabby apples may be infected by secondary pathogens in storage, thus causing further losses.

Scab infections that occur very late in the season often do not show symptoms at harvest or packing (Figure 2.8). In the orchard, the entire fruit surface of a young apple is susceptible to infection. An apple becomes more resistant as the

Figure 2.8 Storage scab on 'Golden Delicious' apple.

fruit matures. Lesions that appear late in the season are small, sunken, and dark, and because of their small size, they are sometimes referred as "pin head" or "pin point" scab (Bratley, 1940; Hockey, 1953). The pin point scab appears shortly before harvest. Pin point scab is frequently followed by "storage scab." The symptoms of storage scab on apples may become visible sometimes after a period of 10–12 wk in storage. The brownish-black lesions ranging from 0.1 to 4 mm in diameter are shiny due to intact cuticle, circular and sunken and sometimes are referred to as "ink spot" because of the color. The storage scab lesions are very dark as compared to velvety appearance of lesions on young apple fruit in the orchard.

The storage scab lesions do not produce conidia in storage and therefore the identification of the pathogen presented here is based on apple scab disease in the orchard. The fungal pathogen that causes apple scab, *V. inaequalis*, has two states: the perfect (sexual) or saprophytic state (previously named as *V. inaequalis*) and the conidial (imperfect) or parasitic state (previously named as *Spilocaea pomi* Fr.; syn.: *Fusicladium dendriticum* [Wallr.] Fuckel). The pathogen, *V. inaequalis*, infects only *Malus* species, including apples and crab apples. The fungus belongs to the subclass Pleosporomycetidae, order Pleosporales, and family Venturiaceae. The fungus is bipolar heterothallic and produces pseudothecia (ascocarps) in a stroma in overwintering leaves of fruit on orchard floor. Pseudothecia are solitary, dark brown to black, spherical with a short beak. The asci (6–12 × 55–75 μm) within the pseudothecium are cylindrical and eight spored. Ascospores (5–8 × 11–15 μm) are yellowish brown to tan, slightly unequally once septate with the upper cell, shorter and wider than the lower cell. The species name derives from the unequal size of the two cells in the ascospore. The parasitic state, previously known as *S. pomi* or *F. dendriticum*, was placed in the subdivision Deuteromycota, class Hyphomycetes, and order Moniliales. The brown mycelium produces short erect, closely septate, conidiophores. Conidiophores produce conidia successively giving characteristics abscission ridges or annelations (*Spilocaea*) on the conidiophores. Conidia (6–9 × 12–22 μm) are single, terminal, obpyriform-obclavate, and pale to olive brown, smooth, one or two celled. The scab spots on apples that develop in storage have no conidial production.

The fungus *V. inaequalis* overwinters in fallen infected leaves on the orchard floor. In early spring, when the overwintered leaves become wet by rain or heavy dew, ascospores are released from pseudothecia and disseminated by wind. The ascospores germinate in the presence of free moisture and form primary infections on freshly growing leaf surfaces. The spores germinate if the surface remains wet with over 95% RH for a period of time (18–23 hr at 6°C to 6–9 hr at 16–25°C). Ascospores continue to mature and discharge over a period of 5–9 wk. Following spore germination, the hyphae forms an appressorium and a penetration peg and penetrate the leaf surface and become stablished between the epidermis and the cuticle. The fungus forms subcuticular stroma that will continue to produce conidia on conidiophores in a lesion. The cuticle ruptures and exposes mycelium and release conidia. Conidia are the principal inoculum that will cause subsequent infections and build up apple scab disease in the orchard. Conidia have the same germination requirement as ascospores, but require longer periods of wetness. In case of storage scab, the infection occurs on apple late in the season and the storage scab symptoms become visible only after 10–12 wk in storage.

3 Minor Diseases

*3.1 Black Rot (*Alternaria *spp.)*

Black rot, caused by *Alternaria* spp., may mainly occur on apples, but also on pears, from all producing areas. Fungal spores, produced on dead or weakened plant tissues, contaminate fruits in the orchard. The amount of decay developing from this wound parasite on fruits in storage depends on the number of wounds and the strength of the fruits (Pierson et al., 1971). Besides being deleterious on causing food spoilage and being a potent airborne allergen, *Alternaria* spp. are of particular interest to mycotoxicologists because they can produce approximately 30 metabolites with possible toxicity (Robiglio and Lopez, 1995). The first *Alternaria* toxins were isolated and characterized in the years 1953–1986 (EFSA, 2011). Because of their possible harmful effects, *Alternaria* toxins are of concern for public health. However, currently, there are no regulations on *Alternaria* toxins in food and feed in Europe or in other regions of the world. *Alternaria alternata* (Fr.) Keissl. is the most common *Alternaria* sp. in harvested fruits and vegetables, and the most important mycotoxin-producing species.

Black rot is characterized by brown–black, dry and firm, round and shallow lesions on both pear (Figure 2.9a) and apple (Figure 2.9b). Lesions can appear around wounds or injuries or at the calyx or stem insertion; in general, rot shows a slow advancement, especially in cold storage. Lesion starts on the skin of the fruit, but it advances in depth toward the center of the fruit and become spongy. Sometimes, infections occur at the calyx and in the depression around the stem or in the wound where the stem is torn out at harvest. In conditions of high RH, the lesion area may appear dark-colored mycelium, which allow to distinguish this rot from those caused by *Cladosporium herbarum* (Pers.) Link or *Cadophora* spp. (Sutton et al., 2014).

Figure 2.9 Black mold on: a. 'Flor d'hivern' pear and b. 'Golden Delicious' apple. c. Morphology of conidia of *Alternaria* sp. (Courtesy of XaRTA-Postharvest, IRTA).

Alternaria alternata is probably the species more reported from a very wide range of food. Fungi in the genus *Alternaria* are easy to recognize by the morphology of their large conidia (Figure 9c). More problems are related to conduct an accurate classification at the species level, due to the variability of forms present in *Alternaria* spp. In general, conidia show a pale to light brown color, several vertical, transverse, and even oblique septa; habitually, long and branching chains contained many conidia, obclavate, obpyriform ovoid or ellipsoidal, often rostrate with a short conical or cylindrical beak. The phenotype of colonies on synthetic media depends on the media, but in general colonies are plane, dark (olive brown or dark gray) or gray colored, and with the reverse from gray, brown to black. Optimum growth temperature of *Alternaria* spp. is near 25°C, with maximum 31–32°C and minimum from −5 to 6.5°C. The minimum a_w for growth at 25°C is 0.88 and optimal growth occurs at pH 4–5.4, although the pH range of growth is 2.7–8.0. *Alternaria alternata* is able to grow in oxygen concentration as low as 0.25% (v/v) in N_2 and its growth rate is proportional to oxygen concentration (Pitt and Hocking, 1999).

In contrast to other *Alternaria* spp., *A. alternata* penetrates predominantly through wounds. Conidia are easily disseminated by wind and water and are usually present in the microbiota on apple and pear orchards because they can persist as mycelium or conidia that survive in waste of plant material and on soil throughout the growing season (Sutton et al., 2014). Under disease-favorable conditions, conidia from few initial infections can possibly produce a great number of new lesions and important damage in a relatively short period of time, because *Alternaria* spp. have a life system quite flexibly in response to the environmental conditions.

3.2 Soft Rots (Rhizopus spp. and Mucor spp.)

Soft rot is a fungal disease caused by both mucoral fungi: *Rhizopus* spp., which causes Rhizopus rot, and *Mucor piriformis* A. Fisch., which causes Mucor rot. Symptoms, life cycle of the pathogens, and control strategies are quite similar for both diseases. However, the characteristics related with their identification and physiology are slightly different. The rot appearance in pome fruits is less constant and regular than other main diseases as blue or gray mold. Soft rot is not one of the most important causes of postharvest disease on pome fruits; however, losses caused by this disease can be serious and achieve up to 50% of stored fruits affected (Usall et al., 2013). The disease is commonly observed in packinghouses after cold storage and occasionally in the field.

Soft rot is primarily originated through wounds or microwounds on the skin of the fruit. The main symptoms of the rot are a light brown color, soft, and watery tissue that usually has a sour smell (Figure 2.10a,b). The decay rapidly extends to the entire fruit producing internal decomposition and juice containing sporangiospores, which can cause additional infections. In the initials stages of infection development, it is already possible to observe a white filament mass crowned by globular sporangia that quickly becomes of black color and is completely visible by human eye. Sporangiophores may protrude through cracks and fruit lenticels in the skin. However, a microscopic examination is needed to determinate if the disease is caused by *Rhizopus* spp. or *Mucor* spp., since only *Rhizopus* spp. produce rhizoids.

In Petri dish, the *Rhizopus* spp. sporangium shows a black color, usually at the margins of the plate. Sporangiophores born in groups of three to five from clusters of rhizoids and stipe unbranched, robust and up to 3 mm long with brown walls. Columellae are roughly spherical and up to 200 μm in diameter. The size of sporangiospores is 8–20 μm in long axis, color is pale brownish and walls are striated (Pitt and Hocking, 1999). Colonies of *M. piriformis* grown in Petri dish are low and sparse (Figure 10c). The mycelium is colorless, overall color buff from sporangia. On MEA, colonies with 35–60 mm diameter are discrete with gray color or brownish. Sporangiosphores on CYA born from the hyphae surface, stipes short, broad, and often are sympodially branched and with encrusted walls (Pitt and Hocking, 1999). The distinctive feature between the genera *Rhizopus* and *Mucor* is the presence or absence of root-like rhizoids. *Mucor piriformis* is classified as a psychrophile, being the minimum and maximum growth temperatures close to 0 and 26°C, respectively. The optimum temperature for mycelial growth and sporangiospores germination is 20°C (Sutton et al., 2014). For *Rhizopus* spp., the optimum temperature is around 25°C and the pathogen can grow from 4.5–5 to 37°C. Both fungi reproduce by forming sexual and asexual spores (sporangiospores). Sporangiospores are produced inside the sporangium, supported by a large apophysate columella atop a long stalk, the sporangiophore. The vegetative mycelium is white. In sexual reproduction, a dark zygospore is produced at the point where two compatible mycelia fuse. Upon germination, a zygospore produces colonies

Figure 2.10 a. Mucor rot on 'Conference' pear. b. Rhizopus rot on 'Blanquilla' pear. c. Detail of rhizoids of *Rhizopus* sp. on apple skin. d. Nest of soft rot on 'Golden Delicious' apples. (Courtesy of XaRTA-Postharvest, IRTA).

that are genetically different from either parent. However, zygospores are not common in nature (Sutton et al., 2014).

Both pathogens are soilborne and the sporangiospores can survive in the orchard soil, although the presence of conidia in the air is also common. The incidence of pome fruits affected by mucoral fungi in the field is quite low. Pathogens enter packinghouses through infected soil or organic material adhering to field boxes and bins. Secondary infections spread in cold storage mainly in pears and less in apples (Sutton et al., 2014). Disease occurs mainly on ripe fruits; even more on overripe fruits or fruits affected by freezing injury. If fruits have been handled carefully along the processes of harvest, packing, and transport to market, the risk of mechanical damage decreases and the disease is unlike to occur. Drench solution and dump-tank water, together with cold storage chambers, are the main source of inoculum for fruit infection in the postharvest period, so they must be exhaustively disinfected to avoid infections by secondary inoculum. The disease is spread by fruit–fruit contact or by means of the sporangiospores. From rotted fruits, healthy fruits are easily infected by juices produced during the decomposition of the tissues, which contributes to increase the incidence of infected fruits causing nests of rotten fruit (Figure 2.10d).

3.3 Side Rots (Cadophora spp.)

Side rot caused by *Cadophora malorum* (Kidd & Beaumont) W. Gams (syns.: *Phialophora malorum* [Kidd & Beaumont] McColloch, *Sporotrichum malorum* Kidd & Beaumont, and *Sporotrichum carpogenum* Rühle) is considered to be a minor disease of apples. However, in the Pacific Northwest USA and Europe, it can cause occasional economic losses on pears (Sugar, 2014). *Cadophora* sp. was first described as an apple pathogen in the UK (Kidd and Beaumont, 1924) and a pathogen of pears in Chicago, USA (Ramsey, 1937). A new species, *Cadophora luteo-olivacea* (J.F.H. Beyma) T.C. Harr. & McNew, was recently isolated from 'Conference' pears in the Netherlands (Wenneker et al., 2016c).

On pears, *C. malorum* causes a dark brown circular to oval lesion usually less than 2 cm in diameter after more than 12 wk of cold storage (Sugar, 2014). Lesions are slightly depressed and diseased material is easily removed from healthy tissue to leave a round crater. The affected tissue under the skin is dark brown to black and can be spongy, especially if the skin is broken. The symptoms are indistinguishable from lesions caused by *Cladosporium herbarum* and *Alternaria* spp., and only isolations and identification of the cultures can separate these diseases (Snowdon, 1990; Sugar, 2014). Lesions caused by *C. luteo-olivacea* were similar in color and shape, but with slightly smaller spots ranging from 0.5 to 1.0 cm in diameter (Figure 11a). *Cadophora malorum* produces cultures that are "woolly textured and gray, with a white margin,

Figure 2.11 a. 'Conference' pears showing typical dark olive circular spots caused by *Cadophora luteo-olivacea*. b. Culture morphology of *C. luteo-olivacea* growing on potato dextrose agar. (Courtesy of Marcel Wenneker).

appearing black from below" (Sugar, 2014). Conidia are produced in phialides that are produced singly along hyphae and conidia can accumulate in mucilaginous masses. Conidia are hyaline, unicellular, and 1.5–4 × 4–8 μm (Sugar, 2014). Cultures of *C. luteo-olivacea*, in contrast, vary in color from white to gray and olive black (Figure 2.11b) (Wenneker et al., 2016c). Conidia are ellipsoid or elongate and slightly thinner and shorter, measuring 2.2–2.3 × 4.9–6.5 μm. Sequencing the ITS region of the rDNA and comparing with known sequences using basic local alignment search tool (BLAST) analysis is also a useful method for identification of this fungus. There are two reports of PCR primers designed to detect *C. malorum* and *C. luteo-olivacea* (Spadaro et al., 2011; Wenneker et al., 2016c).

Examination of colonization of wounds confirmed that *C. malorum* was a weak pathogen, and was only able to invade when its competitor, *C. herbarum*, was controlled in the orchard by benomyl sprays (Sugar and Powers, 1986), as *C. malorum* was shown to be resistant to this fungicide (Bertrand et al., 1977). Infection occurred in pear immersion tanks during packing (Sugar and Spotts, 1993a), and was also observed in the orchard (Bertrand et al., 1977), possibly entering the host through lenticels and microcracks in the cuticle. Spores were found in the air and on fruit surfaces at harvest, but not during the growing season (Sugar and Spotts, 1993a), suggesting infection at harvest. Examination of lenticels after harvest failed to demonstrate entry by this means. Instead, increased hydrostatic pressure was shown to facilitate infections of pears at the bottom of submersion tanks, possibly through microwounds induced by pear-to-pear contact during picking and movement through the packingline (Sugar and Spotts, 1993b). Late picked fruits were also more susceptible to rots caused by *C. malorum* (Bertrand et al., 1977; Spotts, 1985a).

Cadophora malorum was readily isolated in semi-selective media from soil from affected pear orchards. Inoculum apparently increased in soil to reach relatively high levels close to the time of pear harvest (Sugar and Spotts, 1992). The fungus was found to increase in soil for 6 mon after addition of fruits to the soil (Sugar and Spotts, 1992). This suggests that *C. malorum* was a weak saprophyte, and was only able to colonize the fruit after breakdown by more aggressive pathogens. Short-lived cankers were induced following wound inoculations in the tree (Sugar and Spotts, 1992), and isolations from inoculated bark were possible for up to a year. Cankers producing *C. malorum* spores were not found in commercial pear orchards (Sugar, 2014), and were probably not important in the disease cycle of this fungus. Bark was probably not a primary inoculum source. Later, preliminary results using quantitative PCR and specific primers showed that necrotic pear leaves fallen to the ground were a potential source of inoculum for *C. luteo-olivacea* (Kohl et al., 2015; Wenneker et al., 2016a), as were mummies and cankers, and other plant species present on the orchard floor (e.g., weeds and grass). It was also possible that these tissues were contaminated with spores from another inoculum source, such as soil.

3.4 Core Rots (*Fusarium spp.*)

Apple core rot infections can be caused by several fungal pathogens in the orchard or during postharvest. Due to the infection being internal, diseases remain undetectable until the product is processed or consumed. Among core rots, dry and wet core rots are generally more important because they affect the fruit flesh, whereas moldy core remains limited to the seed cavity and does not affect the flavor (Carpenter, 1942; Sutton et al., 2014). Up to 8% of the postharvest losses were attributed to dry and wet core rots in South Africa (Serdani et al., 1998). In Australia, Wong (1996) reported up to 20% losses in the 1980s, while more recently Brown (2010) mentioned new cases after a long period of low incidence of the disease.

Depending on the symptoms and development of the disease in the core and flesh of the fruit, core diseases can be named as moldy, wet, and dry core rots (Jijakli and Lepoivre, 2004; McLeod et al., 2008; Maxin, 2012), and more recently as browning rot (Gao et al., 2013). Different symptoms occur depending on the causal organism, as well as on spore concentration (Gao et al., 2013). For example, at lower concentrations, *Alternaria* spp. and *Cladosporium* spp. can cause core browning (when the carpel becomes brown but without mycelial growth in the core), while at higher concentrations, the lesions could develop into moldy core (Gao et al., 2013). The disease is called moldy core when the growth of the fungus is restricted to the locules or endocarp. Moldy core can develop into dry core rot at a later stage when the disease invades the flesh (mesocarp) and degenerates the colonized tissue. Apple tissue affected by dry core rot remains firm. Wet core rots are characterized by soft, watery lesions and subsequent decay of the fruit (McLeod et al., 2008; Gao et al., 2013). Alternaria core rots are characterized by a dark mold in the center of the seed cavity (Snowdon, 1990). The fungus can grow into the mesocarp of the fruit, and causes a dry core rot. Because Alternaria core rots are already described in more detail in the Section 3.1, this section focuses on core rots produced by *Fusarium* spp. The first symptoms of Fusarium core rot occur in the seed cavity, where a white to reddish mycelial layer develops with a brown soft wet rot surrounding the pericarp (Figure 2.12). The fungus then infects the apple cortex and watery lesions spread into the fruit flesh, finally colonizing the fruit, covering it with mycelia, and causing the flesh to decay (Sorensen et al., 2009; Maxin, 2012; Sanzani et al., 2013).

Figure 2.12 Postharvest symptoms of wet core rot caused by *Fusarium* sp. after natural infection of 'Jazz' apple. (Courtesy of Kerry Everett).

Apple core infections have been frequently associated with species from the genera *Alternaria, Fusarium, Penicillium, Aspergillus, Pezicula, Colletotrichum, Pleospora, Coniothyrium, Phoma, Trichothecium, Cladosporium, Epicoccum, Trichoderma*, and *Stemphylium* (Manning and Burmeister, 2000; Archer, 2002; Niem et al., 2007; Gao et al., 2013; Riad and Ziedan, 2015; Soliman et al., 2015; Abdelfattah et al., 2016; Wenneker et al., 2016b). Among those, *A. alternata* was one of the most frequently found causal agents of core rots (Ellis and Barrat, 1983; Robiglio and Lopez, 1995; Archer, 2002; Niem et al., 2007); however, other *Alternaria* spp. have also been attributed as causal agents (Serdani et al., 1998, 2002; Kang et al., 2002; Gao et al., 2013; Abdelfattah et al., 2016). Predominating species causing dry core rot belong to the genera *Alternaria, Epicoccum*, and *Phoma* (Serdani et al., 1998; Archer, 2002; Jijakli and Lepoivre, 2004; Brown, 2010; Maxin, 2012; Gao et al., 2013), whereas for wet core rot *Penicillium, Pezicula*, and *Fusarium* species were most commonly reported (Combrink et al., 1985; Archer, 2002; Van der Walt et al., 2010; Gao et al., 2013; Sanzani et al., 2013; Wenneker et al., 2016b). Focusing on *Fusarium* spp., although several species were isolated from decayed tissue with core wet rot symptoms (Gao et al., 2013; Kou et al., 2014; Almeida and Lopes, 2015; Gualandri et al., 2015), *Fusarium avenaceum* (Fr.) Sacc. has been frequently reported in recent years as the causal agent of core rot diseases in several countries, including Slovenia (Sorensen et al., 2009), Italy (Sanzani et al., 2013), and the Netherlands (Wenneker et al., 2016b), as well as of other related fruit rot diseases in Greece (Konstantinou et al., 2011), Serbia (Grahovac et al., 2014), and the USA (Kou et al., 2014). *Fusarium* spp. can be grown in vitro on PDA to describe colony morphology and pigmentation, which are useful criteria for identification (Samson et al., 1995; Leslie and Summerell, 2006). Colonies of *F. avenaceum* on PDA are initially pigmented by a light yellow or white color floccose aerial mycelium, which later turns peach-colored or reddish-brown and felty. On the reverse side, the colony is characterized by a yellow to reddish-brown color (Samson et al., 1995). For further morphological identification, the fungus can be cultured on synthetic nutrient-poor agar, carnation leaf-piece agar, or carrot agar, and incubated at 22–25°C (Samson et al., 1995; Leslie and Summerell, 2006; Sever et al., 2012). Slimy masses of spores are presented in distinctive, orange-colored sporodochia. Macroconidia (35–89 × 3.5–4 μm) are narrowly fusiform, sickle-shaped to straight, with narrow ends, 4–7 septa, and an elongated apical cell and footcell, and are produced in mono- and polyphialides. Microconidia (6–30 × 2.5–4.5 μm) can be observed and, if present, these are fusiform and 0–3 septate. Formation of chlamydospores can sometimes occur on conidia but not in the mycelium (Samson et al., 1995). Other *Fusarium* spp. described in the literature associated with core rots or fruit rot of apple are: *F. proliferatum* (Matsush.) Nirenberg, *F. equiseti* (Corda) Sacc., *F. solani* (Mart.) Sacc., *F. tricinctum* (Corda) Sacc., *F. lateritium* Nees, (Snowdon, 1990; Konstantinou et al., 2011; Gao et al., 2013), and, more recently, *F. decemcellulare* Brick (Lee et al., 2017). *Fusarium* spp. on apple can also be identified by sequencing the ITS regions of rDNA and analyzed using BLAST (Sever et al., 2012; Sanzani et al., 2013).

Fusarium infections in the orchard are generally enhanced by wet conditions, especially during the blossom time (Maxin, 2012). The spectrum of different *Fusarium* spp. from season to season is dependent on weather conditions as described in cereals and maize (Doohan et al., 2003; Logrieco et al., 2003; Leslie and Summerell, 2006) where, for example, *F. avenaceum* occurs more frequently in temperate regions or cold years (Uhlig et al., 2007; Goertz, 2010), whereas *F. proliferatum* is favored by lower precipitation (i.e., dry conditions) and high temperatures (Vigier et al., 1997; Logrieco et al., 2002; Goertz, 2010).

Fusarium spp. are ubiquitous fungi that survive mostly in mycelial form, in perithecia (the sexual stage, which only plays a role for some species), as chlamydospores in the soil or in organic debris. Propagules can be dispersed by wind or rain (Agrios, 2005). On apple, core rot infections occur in the field at blossom time when fungal spores invade flowers, for example, in water drops, and through the calyx to finally penetrate the core cavity (Combrink et al., 1985; Manning and Burmeister, 2000; Maxin, 2012). *Fusarium* spp. are considered to enter the core through this mechanism in the majority of cases (Jones and Aldwinckle, 1990; Maxin, 2012). A temperature drop after blossom could also delay the closing of the fructification channel, resulting in an extended time period for infection (Gualandri et al., 2015). Core rot infections could also occur during different periods of the season (Brown, 2010) or fruit processing, as it has been reported for *A. alternata* (Niem et al., 2007). Core rot diseases can also enhance premature fruit fall (Brown, 2010). *Tarsonemus* mites are considered to be involved in the epidemiology of apple core rots, performing as vectors carrying fungal spores into the core (Michailides et al., 1994; Van der Walt, 2009; Van der Walt et al., 2010). They have been positively correlated with dry core rot in 'Red Delicious' fruits (Van der Walt et al., 2010). Core rot infections remain in the core cavity until, once established, the fungus colonizes the developing fruit flesh (Ellis and Barrat, 1983; Wong, 1996). However, this further colonization of the fruit will depend on the resistance or susceptibility of the cultivar (Niem et al., 2007; Peter, 2018). During processing, storage atmosphere conditions and inoculum-contaminated dipping solutions could enhance the development of core rot diseases. Treatment with diphenylamine (DPA) during immersion of the fruit after harvest could enhance wet core rots due to the lower surface tension supporting the penetration of the solution and thus spores into the core (Combrink and Ginsburg, 1973; Spotts et al., 1988).

*3.5 White Haze (*Tilletiopsis *spp.)*

White haze has been reported on apples cultivated in Europe, particularly in the Netherlands (Boekhout et al., 2006), Italy (Lindner and Baric, 2006), Germany (Weber and Zabel, 2011), and Croatia (Prencipe et al., 2016). Since 1999, symptoms of white haze have been reported on apples prior to harvest or during storage (Lindner and Baric, 2006), particularly during ultra-low oxygen storage (Baric et al., 2010). The marketability of fruits with white haze is decreased, as the fungus can cause considerable economic losses (Lindner, 2009).

White haze is considered an apple cosmetic disorder, characterized by imperfection of the apple skin. White haze is caused by extensive mycelial growth of anamorphic smut fungi over the surface of apple fruits. It is evident as a whitish to grayish waxy and opaque superficial thin layer of filamentous fungal growth, adhering to the cuticle, sometimes associated with russeting (Figure 2.13a). It is randomly distributed over the fruit skin as spots, particularly around the peduncle and the stem end. Fruits with more symptoms are in the shadow side and in the lower branches of the tree.

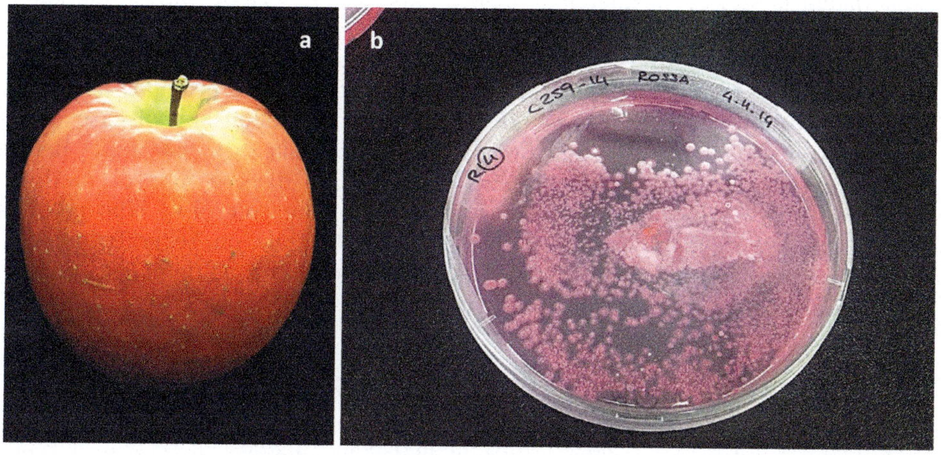

Figure 2.13 a. White haze on apple. b. Spore fall of *Tilletiopsis pallescens* on rose bengal chloramphenicol agar. (Courtesy of AGROINNOVA and DISAFA, University of Torino).

White haze is caused by species of *Tilletiopsis*, ballistosporous fungi of the class Basidiomycetes and order Tilletiales (Figure 13b; Boekhout et al., 2006). Morphologically, the fungi showed thin-walled hyaline hyphae without clamp connections, but had sterigmata bearing the ballistospores. Analysis of DNA sequences (D1/D2 and ITS) showed that nine distinct *Tilletiopsis* taxa are associated with white haze in the Netherlands (Boekhout et al., 2006). Baric et al. (2010) showed that six haplotypes were identified, which could be assigned to the species *Tilletiopsis pallescens* Gokhale and *Tilletiopsis washingtonensis* Nyland, together with a lineage in the Entylomataceae. Members of the genus *Tilletiopsis* are yeast-like fungi, which are saprophytes in nature and colonize the leaf and fruit surface as epiphytes. These fungi naturally occur in the apple phylloplane microflora as yeast-like forms. The ballistospores can germinate into the filamentous molds in response to particular environmental conditions with the ability to overgrow the apple carposphere.

White haze has an erratic incidence over the years and it has been reported on different apple cultivars: 'Elstar,' 'Topaz,' 'Braeburn,' 'Golden Delicious,' 'Gala,' 'Red Delicious,' 'Fuji,' and 'Cripps Pink,' among others. Fungal growth is promoted by high RH during the cropping season and high amount of rainfall during harvest. Hail protection nets could favor longer periods of leaf wetness and higher incidence of white haze. The disorder seems to be promoted at low temperatures. Fertile soils, high density of plants in the orchard, and east–west orientation of the rows can favor white haze.

4 Disease Management

Postharvest diseases of pome fruits can be caused by wound opportunistic pathogens, such as *P. expansum* or *B. cinerea*, or by latent pathogens, able to cause latent or quiescent infections before harvest. The ecology and the etiology of the target disease must be understood in order to develop an effective disease management strategy. Postharvest rots resulting from preharvest latent infections can be managed with preharvest practices that promote a healthy crop, reduce conducive environments for pathogen infection and disease development, and minimize the amount of pathogenic inocula that may infect before harvest. There is a need for sustainable and safe disease control methods against postharvest decays of apples. For example, the control of blue mold in apples that are destined for fresh fruit market depends on several practices: good farm management, appropriate long-term storage conditions, postharvest treatments, and packinghouse sanitation.

4.1 Preharvest Management of Postharvest Diseases

Good cultural practices should be implemented before harvest to reduce blue and gray molds. These include irrigation practices that minimize wetness duration (i.e., drip irrigation), balanced nitrogen fertilization, and canopy management (pruning). Foliar fertilizers should be avoided to reduce white haze. White haze is also reduced by equilibrated winter pruning, with low number of branches left, and green pruning. Anthracnose and bitter rots can be greatly reduced by removing perennial cankers in the orchard, as tree cankers are the main source of conidial infection. Monitoring premature fruit drop in the orchard (Brown, 2010) could provide some indication of moldy core rot infections, because this is a symptom.

In the orchard, an extensive nutritional study conducted in New Zealand (Everett et al., 2016) showed a strong correlation between boron content in the fruit and bitter rot caused by *C. acutatum*; fruit flesh containing 0.25 mg boron per 100 g of apple tissue had the least bitter rot (calculated as the area under a disease curve using lesion diameter), while samples with 0.14 mg boron per 100 g tissue had the most bitter rot. If boron levels in fruit are not at least 0.24 mg per 100 g, then boron added to the soil or applied to the foliage may reduce the incidence of bitter rot disease. Although nitrogen was related to disease, it was also significantly related to fruit maturity, and all maturity indicators (°Brix, starch pattern index, and mean weight) were also related to disease (Everett et al., 2016).

The use of fungicides may reduce the amount of fruit decay before and after harvest and reduce inoculum levels of the target pathogens. Preharvest fungicide applications could reduce postharvest blue mold in storage. Partial control of bull's eye rot can be achieved by repeated applications of fungicides, such as captan, phenyl-pirroles, and strobilurines, during the 2 mon preceding harvest (Giraud and Bompeix, 2012). Besides fungicides, some biocontrol agents can also be applied in preharvest to control postharvest diseases. The biofungicide Boni Protect (Manica, Italy), based on two antagonistic strains of *Aureobasidium pullulans*, is used as preharvest application to control wound pathogens developing on apples during storage.

Control of primary as well as secondary scab inoculum is necessary to reduce storage scab incidence. Chemical fungicides play an important role in controlling primary scab inoculum during the growing season. Since storage scab infections occur late in the growing season, late season chemical sprays are required to control secondary scab infection when longer wetting periods persist. In South Africa, curative fungicides controlled storage scab under conditions of low inoculum potential, but protective fungicides were required every 3–4 wk to control apples inoculated with moderate and high inoculum (Schwabe, 1980). In Canada, some control of storage scab was achieved by regular summer fungicide applications and postharvest dip treatments, with several fungicides were found to be ineffective for storage scab control (Ross and Lokhart, 1960; Northover, 1975).

To control bitter rot, fungicide applications every 10–14 d throughout the season, from when leaves first develop until harvest, are recommended. In addition, removal of inoculum sources, such as mummified fruits, cankers, or dead wood and twigs (Sutton, 1990), reduces disease. Timing of fungicide application can be aided by a model developed by Crusius et al. (2002) for management of Glomerella leaf spot in Brazil on apple (Sugar, 2014). There have been reports that this type of preventative fungicide program is ineffective against *C. acutatum* (Everett et al., 2015; Munir et al., 2016). A series of spray timing trials were conducted in New Zealand against natural infections of *C. acutatum* for three seasons (2007–2010) (Boyd-Wilson et al., 2014; Everett et al., 2015). Spraying eight times during the season with a low concentration of calcium chloride (0.33 kg per 100 L) provided control as effective as fungicide applications in the field, and some control of disease that developed after harvest (Boyd-Wilson et al., 2014). Biggs (1999) also showed good control of *C. acutatum* by application of calcium chloride to inoculated fruits. Fungicides applied in early summer in two seasons (2007–2008 and 2008–2009) effectively controlled field expression of the disease (Everett et al., 2015). Fungicides applied in late summer in both seasons did not control field expression, presumably because they were applied after infection had occurred, but did control postharvest expression when it was examined in 2008–2009 (Everett et al., 2015).

Fungicides have been reported to control infection of core rots when applied over the flowering period (Wong, 1996; Reuveni and Prusky, 2007; Brown, 2010). However, after flowering fungicides were no longer effective because they could not penetrate into the core of the newly formed fruits (Ellis and Barrat, 1983; Manning and Burmeister, 2000).

Tarsonemus mite control programs could also be a helpful measure to control the entry of spores into the locule cavity (Serdani, 2006). The use of varieties with small or closed calyx openings is an effective practice to avoid fungal infections (Miller, 1959; Spotts, 1990). For example, 'Red Delicious' apples are characterized by a large open sinus and are more susceptible to core rot diseases, whereas the calyx in 'Granny Smith' fruits is closed and resistant to core rot (Spotts et al., 1988; Serdani et al., 1998; Van der Walt, 2009). The size of the calyx opening could also reduce the presence of *Tarsonemus* mites in the core and thus the arrival of spores. Van der Walt et al. (2010) found a high occurrence of mites in 'Red Delicious' cores compared with 'Granny Smith' apples, where mites were only observed in the calyx tube.

At harvest, workers should be very careful with fruit handling, to avoid skin injuries and discard wounded fruits. All picking bins should be cleaned before the beginning of the season, and soil build up on the bins during picking should be avoided. Fruits should not get in contact with soil, which can bear several postharvest pathogens.

4.2 Postharvest Management

Postharvest decay control practices should also be considered as part of an integrated pest management (IPM) strategy to control pathogens (Errampalli, 2014; Sutton et al., 2014; Janisiewicz and Jurick, 2017).

Fruit handling methods that minimize wounds and bruising after harvest are essential for controlling blue mold in apples. Rough areas in the grading and packinglines, including dust and debris, which could abrade the fruit, must be eliminated. Rotting apples stuck to the storage bins and spores adhering to packing store walls serve as sources of inoculum. Sanitation of bins and storage room walls prior to harvest can significantly reduce inoculum levels for *P. expansum* (Sanderson and Spotts, 1995). Plastic bins should be preferred to wooden bins, as they carry less pathogen inoculum and reduce apple bruising and abrasion produced by contact with the sides of the bins. The contamination of prestorage drenches, dips, or sprays should be minimized by washing fruit and bins in fresh or sanitized water before further treatment. Following storage and prior to packing, the treatment of water flumes and dump tanks on packinglines with a biocide (e.g., hypochlorite or ozone) or steam, or by appropriate filtration, will kill conidia that could inoculate wounded fruits during the packaging process. Rollers, brushes, and belts must not become contaminated with spores, and should be thoroughly cleaned as necessary. Rinsing fruits in fresh or sanitized water, after fruits with rots have been removed on the sorting table, will help to minimize further contamination and infection of fruits.

Postharvest handling practices should focus on maintaining a healthy fruit physiology, while minimizing decay losses. Pome fruits with an active metabolism show considerable resistance to microbial infection and decay, whereas stressed or senescent pome fruits are prone to disease. Any environment that slows microbial activity and maintains fruit quality will help in reducing the amount of decay. Physical methods that maintain the vitality of the crop include temperature management and modification of the atmospheres using reduced oxygen and elevated carbon dioxide. As pome fruits can withstand extreme cold storage temperature, it is a high value for prevention of fungal decay, but it is not enough to slow down development of some pathogens such as *B. cinerea*.

Synthetic fungicides, when permitted, are the primary means to control postharvest diseases. Properly applied treatments prevent or impede the development of pathogens and are generally economical. Postharvest application of chemical fungicides, such as thiophanate methyl, fludioxonil, or pyrimethanil, provides excellent control of blue mold and other postharvest diseases, as well. Isolates of *Penicillium* and *Botrytis* that are resistant to some of these fungicides are present in some packing houses, restricting their effectiveness (Rosenberger and Meyer, 1985). Several reasons, such as the growing public concern over risks for the human health and the environmental pollution associated with pesticide usage in orchards, the development of fungicide resistant strains of postharvest pathogens, and the lack of re-registration of some of the most effective fungicides, have encouraged the search of alternative approaches. Late-season and postharvest use of fungicides on apples is under scrutiny by the European Commission, because of the increasing demand for fruit without pesticide residues (Poulsen et al., 2009); therefore, the tendency toward alternative strategies is encouraged.

For a decade, 1-methylcyclopropene (1-MCP), a synthetic cyclic olefin that blocks an ethylene-binding receptor (Sisler and Serek, 1997), has been used on stored apples to extend fruit firmness during storage and marketing (Saftner et al., 2003). A reduction of bull's eye rot by 1-MCP was observed on 'd'Anjou' pears (Spotts et al., 2007), 'Elstar' apple stored for 6 mon in CA (Maxin and Weber, 2001), and 'Cripps Pink' apple after 5 mon of storage at 2°C. An integrated strategy providing a combination of two fungicide preharvest treatments and one postharvest 1-MCP treatment achieved a better control (>87.5%) than the single treatments tested alone (Cameldi et al., 2016a).

Alternative approaches to fungicide sprays include biological control with bacteria and yeast; natural compounds, such as plant extracts, essential oils, volatiles, and chitosan; heat treatments; and induced resistance with compounds, such as methyl jasmonate or acibenzolar-S-methyl.

Much effort has been devoted to the development of biocontrol for postharvest diseases of apples. Several biocontrol agents have been developed, proving to be effective in controlled and semi-commercial conditions against postharvest pathogens. The main mechanisms of action explored include antibiosis, mycoparasitism, biofilm formation, production of lytic enzymes, induced resistance, competition for limiting nutrients and space, and the role of oxidative stress (Spadaro and Droby, 2016). A biofungicide based on *Pseudomonas syringae* was registered in the USA for postharvest application on pome fruit (Janisiewicz and Peterson, 2004). Among the yeast-based products, one based on *Candida oleophila* (Liu et al., 2013) and another based on *Cryptococcus albidus* (Janisiewicz and Korsten, 2002) were commercialized for some years. A product based on another strain of *C. oleophila* (Nexy, Agrauxine, Lesaffre Plant Care, France) has recently been registered throughout the EU (Lahlali et al., 2011). Several other biocontrol agents have been isolated, characterized, and developed to control postharvest diseases of pome fruit (El-Ghaouth et al., 1998; Viñas et al., 1998; Janisiewicz et al., 2001; Vero et al., 2002; Qin et al., 2004; Chan and Tian, 2005; Spadaro et al., 2008, 2010). Anyway, commercialization of biofungicides has been slow (Droby et al., 2016). Reasons for this could be related to low profitability due to limited market for postharvest biofungicides, difficulties in market penetration and perception of industry, high profile over food safety issues for apples in public debates, small enterprises with low economic resources, and lack of eradicant activity toward latent infections.

Janisiewicz et al. (2003) used a yeast biocontrol agent in combination with 4 d hot air (38°C) and storage in 1-MCP under CA conditions against bitter rot. While 1-MCP alone did not control *C. acutatum* bitter rot on wound-inoculated apples, in combination with the heat and the biocontrol agent, rots were reduced by 50–70% compared with the untreated control. When 2% sodium bicarbonate was combined with another yeast-based biocontrol agent, bitter rot control was greatly improved and the disease was almost completely eliminated (Conway et al., 2004, 2005).

Several natural products have been evaluated for their efficacy against postharvest diseases of pome fruit. Plants, particularly, provide a wide range of compounds with antimicrobial activity, including aromatics, essential oils, glucosinolates, or extracts. Hexanal, an ester that provides the characteristic apple aroma, was very effective in reducing blue mold incidence on apples (Lua et al., 2006). Similarly, trans-2-hexanal also showed a similar activity (Neri et al., 2006). Allyl isothiocyanates are volatile products of the hydrolysis of glucosinolates, typically present in brassicas, able to reduce the incidence of blue mold of pears (Mari et al., 2002). Essential oils are obtained by extraction from aromatic plant material. The biological activity of essential oils and their constituents can be fungistatic or fungicidal, depending on the concentration used (Spadaro and Gullino, 2014). Some essential oils extracted from Mediterranean plants demonstrated their antifungal activity against *B. cinerea* and *P. expansum* on different cultivars of apple (Lopez-Reyes et al., 2010). Garlic extract applied to apples reduced blue and gray molds, and in combination with clove essential oil, further increased its efficacy (Daniel et al., 2015). Among the generally recognized as safe substances, sodium bicarbonate, ethanol, calcium chloride, or chitosan have been used to control gray or blue mold on apples, alone (Conway et al., 2004; Bautista-Baños et al., 2006) or in combination with biocontrol agents and heat treatments (Spadaro et al., 2004; Janisiewicz and Conway, 2010).

Heat treatments may be applied as hot dry air (Conway et al., 1999), hot water dip (Maxin, 2012), or short hot water rinse and brushing (Fallik, 2004) to control postharvest decays of apples. Heat treatments have a direct biocidal effect on the pathogen, but can also induce resistance in the host, for example, by up-regulating genes involved in heat shock (Spadoni et al., 2015). Fruits exposed to 40°C for 24 hr in combination with CA storage reduced bitter rot on fruits wound inoculated with *C. gloeosporioides* by 73 and 65% on 'Aroma' and 'Ingrid Marie' apples, respectively, stored for 4 mon (Tahir et al., 2009). An in-line hot water treatment was also tested, and dipping in water at 50–54°C for 3 min or rinsing in water at 55°C for 20–25 s effectively controlled rots caused by *C. acutatum* following 14 wk cold storage and 14 d at room temperature (Maxin, 2012). Although control by hot water treatment is a feasible option against fungal storage-rot organisms, the control of some fungal pathogens in the core, such as *Fusarium* spp., is often difficult (Maxin, 2012; Maxin et al., 2014). In addition, treatments of pome fruits with chemical inducers, such as salycilic acid, jasmonates, β-amino butyric acid, or acybenzolar-S-methyl, are able to activate plant defense responses, including accumulation of reactive oxygen species and pathogenesis-related proteins (Quaglia et al., 2011; Yin et al., 2013).

Much of the research focused on the development of a single treatment to control postharvest diseases of apples. The hurdle concept (Janisiewicz and Jurick, 2017) states that each additional treatment is able to reduce the postharvest losses by a certain amount, but an additive, or even a synergistic effect, can be obtained by combination of different treatments. Integrated management of postharvest diseases of pome fruits with one or more of the above-mentioned control methods will help in achieving a sustainable control, which is safe to humans and the environment.

5 Physiological Disorders

5.1 Scald

Scald is a term loosely applied to a group of skin disorders of apple and pear. Usually, it develops after the fruits have been removed from storage to warmer temperatures, although extended months of refrigeration can also render visible alteration on stored fruits (Lurie and Watkins, 2012). Superficial scald represents a limiting factor for prolonged storage of several apple cultivars ('Granny Smith,' 'Fuji,' and 'Golden Delicious') and pear ('Abate Fetel'). Generally, after 2–3 mon at 0°C and 1 wk of shelf life, 'Granny Smith' apples show superficial scald symptoms (Figure 2.14a). Moreover, 'Abate Fetel' pears cannot be stored at −0.5°C, in normal atmosphere, longer than 4 mon, because the fruits become sensitive to the disorder (Figure 2.14b) (Rizzolo et al., 2010). Potentially scald can destroy the market value of millions tons of apple and pear annually. Powell and Fulton (1903) described the scald for the first time, identified as a problem associated with the advent of refrigerated storage, while the term "superficial scald" was coined by Tiller (1929). Afterwards, the study of scald has received much attention by researchers and various aspects of etiology and control have been reviewed (Smock, 1961; Wang and Dilley, 2000; Lurie and Watkins, 2012).

The symptoms ascribed to scald can take several related forms, including brown or black patches on the fruit skin but also rugose and lenticel scald (Wilkinson and Fidler, 1973), especially on the shaded side. The margins of lesion are diffused and the flesh remaining firm and of eating quality. In lenticel scald, the damage is predominantly around the lenticels, so that it seems a spotting rather than a blotchy disorder. Other variants of scald are sunscald, skin darkening during storage (Lurie et al., 1991), stem-end browning, and senescent scald (Wilkinson and Fidler, 1973), the latter not correlated to chilling injury.

In general, scald is positively correlated with the length of refrigerated storage, it is more prevalent in fruits harvested early, exhibiting a partial chlorophyll degradation and, within the single fruit, the shaded part is more susceptible than that exposed to sun. Most, but not all, apple and pear cultivars are affected by scald, suggesting a genetic basis for the disorder. The environmental conditions during the growing season influences scald susceptibility; warm, lightness, and dry before harvest can favor the disorder, while the air temperature below 10°C for at

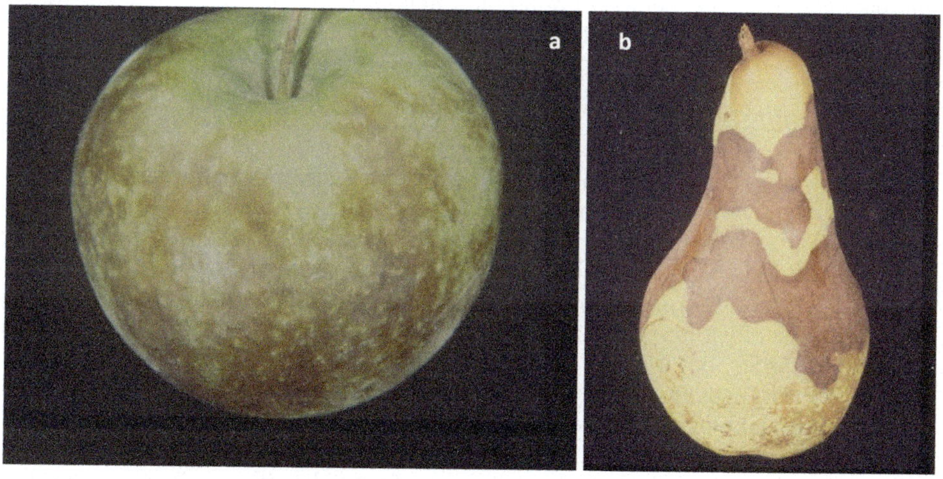

Figure 2.14 a. Superficial scald on 'Granny Smith' apple. b. Soft scald on 'Abate Fétel' pear.

least 150 hr before harvest reduces the incidence of scald. This may be related to the nature and amount of wax components developed on the fruit during storage (Emongor et al., 1994). Moreover, the mineral nutrition can affect the appearance of scald; apples with low calcium levels develop the disorder more often than those with high levels. The scald is the result of biochemical reactions that happen after harvest in the fruit skin, starting from the farnesyl diphosphate, a precursor of the acylic sesquiterpene α-farnesene. Concentrations of α-farnesene are negligible at the preclimacteric stage, increasing as ethylene concentrations grow during ripening (Barden and Bramlage, 1994). The control of α-farnesene synthesis is not fully elucidated, but it seems associated with ethylene production. Huelin and Murray (1966) provided the evidence that the development of the scald is associated to α-farnesene oxidation products, such as conjugated trienols, which are more stable compounds than conjugated trienes. The concentration of these oxidation products in the fruit skin could determine the cell damage and the appearance of browning; however, the critical threshold is different, depending the on cultivar. The oxidative process is slowed down by natural antioxidants (ascorbic acid, flavonoids, tocopherol, etc.) present in the fruit skin, although the relation between natural antioxidants and scald still requires more investigations. In addition to this hypothesis, called α-farnesene hypothesis, Knee and Hatfield (1981) suggested that α-farnesene only has a scald promoting role, as participant in a chain reaction rather than as an initiator. Similarly, Rupasinghe et al. (2000) revealed that the conversion of α-farnesene to conjugated trienols could be a secondary event in scald development.

In the past, ventilation and oil wraps were the only methods available for successful scald prevention (Brooks et al., 1919). Later, the discover of the antioxidants, such as DPA and ethoxyquin, was an important contribution to the disorder control (Smock, 1955). However, concerns about postharvest treatments have contributed to ban the use of DPA in the EU countries and phase out in others. Currently, DPA has been replaced by 1-MCP, a widely registered inhibitor of ethylene receptor, which exerts its effect indirectly by inhibition of phytormone synthesis. 1-MCP is frequently associated with newer CA strategies, such as ultra-low oxygen, approaching anaerobic compensation point, hypobaric, or initial low oxygen storage (0.04 kPa O_2 for 10 d prior to storage in 0.7 kPa O_2) (Lurie and Watkins, 2012).

5.2 Bitter Pit

Bitter pit is one of the most important physiological disorders of apples and pears. Most apple cultivars are susceptible, including 'Golden Delicious,' 'Granny Smith,' 'Cox's Orange Pippin,' 'Honeycrisp,' 'Jonathan,' and 'Fuji' (Ferree and Warrington, 2003). Within a cultivar, bitter pit is related to orchard management, harvest date, and climate (Ferguson and Watkins, 1989). Influence of climate is partially related to its effect on calcium concentration in the fruit. It is predominantly a storage disorder, but it could be sometimes found before harvest.

Bitter pit usually develops only after harvest (Perring, 1986), with small brown depressions on the apple skin, associated with the collapse of flesh, due to accumulation of dead cells. Necrotic lesions are round (1–4 mm), spongy, and dry, with a bitter tasting, and mainly located in the fruit calyx (Figure 2.15). In severely affected fruit, the lesions may coalesce into blotches.

Figure 2.15 Bitter pit on apple. (Courtesy of AGROINNOVA and DISAFA, University of Torino).

Bitter pit has a complex etiology, but it is considered a nutritional disorder, associated with imbalance in minerals, and particularly to calcium deficiency. Calcium has the role to stabilize cell membrane and reduce its permeability, preventing water stress. Insoluble calcium fraction is inactive and bound to the cell wall, while water-soluble calcium is physiologically active (Jemric et al., 2016). In bitter-pitted fruits, calcium is mainly water insoluble and it accumulates around the pits on the fruit skin (Val et al., 2008). Calcium deficiency leads to cell collapse and tissue browning, due to lower activity of superoxide dismutase, peroxidase, catalase, and polyphenol oxidase. Bitter pit is also associated to high levels of nitrogen in the fruit, and with an imbalance with magnesium levels. Bitter pit is favored by excess nitrogen supply (Terblanche et al., 1980) and water stress (Lotter et al., 1985). Risk of bitter pit is higher in hot dry climates. Furthermore, high concentration of gibberellins late in the season can increase the permeability of cell membrane, thus causing water stress and favoring bitter pit (Saure, 1996; Silveira et al., 2012). Incidence of bitter pit is higher on vigorous or moderate rootstocks, while it is lower on less vigorous rootstocks (Ben, 1995). Bitter pit is a common problem of acidic soils. Excessive pruning and high temperatures during the growing season can increase bitter pit incidence. In susceptible cultivars, fruits harvested before they are fully mature are more susceptible to bitter pit. Larger fruits are more susceptible to bitter pit than are small fruits, as calcium is diluted in bigger fruits.

Preharvest management practices to increase calcium levels include lime applications to the soil, after a soil analysis of the calcium content. Foliar applications of calcium (calcium chloride, calcium nitrate, calcium carbonate, and calcium-amino acids) after full bloom increase the firmness and reduce the bitter pit susceptibility. Nitrogen fertilization and pruning should be equilibrated in order to avoid excessive tree vigor. Fruits from severely thinned trees contain less calcium and more potassium, being more susceptible to bitter pit (Wojcik et al., 2001). Excessive leaf growth could increase the incidence of bitter pit (Sutton et al., 2014). Irregular irrigation can favor bitter pit, particularly during fruit formation (Ferguson et al., 1999). Some schemes based on fruit or leaf sampling for calcium content have been developed to predict the risk of bitter pit development. Fruits should be harvested just before the strong increase in the synthesis of ethylene (Prange et al., 2011). Postharvest treatment of fruits with calcium chloride can reduce bitter pit development (Sharples and Johnson, 1976). Moreover, rapid cooling, low temperature and high humidity storage, and CA can delay the occurrence of bitter pit, and/or reduce its incidence (Watkins et al., 2004). Treatment with 1-MCP can contribute to reduce the incidence of bitter pit.

Acknowledgments

Work carried out with the contribution of the LIFE financial instrument of the EU for the project "Low pesticide IPM in sustainable and safe fruit production" (Contract No. LIFE13 ENV/HR/000580) and with the contribution of Fondazione CRC for the project "SMART APPLE - Innovative and SMART technologies for sustainable APPLE production".

References

Abdelfattah, A., Wisniewski, M., Droby, S. and Schena, L. 2016. Spatial and compositional variation in the fungal communities of organic and conventionally grown apple fruit at the consumer point-of-purchase. *Horticulture Research* 3, 16047.

Abeln, E.C.A., de Pagter, M.A. and Verkley, G.J.M. 2000. Phylogeny of *Pezicula, Dermea* and *Neofabraea* inferred from partial SSU and the Internal Transcribed Spacer Region of the Nuclear Ribosomal Gene Cluster. *Mycologia* 92, 685–693.

Agrios, G.N. 2005. *Plant Pathology*, 5th ed. Academic Press, San Diego, CA.

Alaniz, S., Hernandez, L., Damasco, D. and Mondino, P. 2012. First report of *Colletotrichum acutatum* and *C. fragariae* causing bitter rot of apple in Uruguay. *Plant Disease* 96, 458.

Almeida, C.M.M. and Lopes, M.M. 2015. Fungi isolated from traditional and exotic apple varieties from Portugal and patulin production. *Journal of Pharmacy and Nutrition Sciences* 5, 30–37.

Anonymous. 2017. Systematics Collections Data. Available at: https://scd.landcareresearch.co.nz/specimen/ICMP_1765

Archer, C. 2002. *The Use of Honeybees as a Transfer Vector for Control of Core Rot in Apples.* Rural Industries Research and Development Corporation, Publication No. 02/046, Barton, Australia.

Arzanlou, M., Bakhshi, M., Karimi, K. and Torbati, M. 2015. Multigene phylogeny reveals three new records of *Colletotrichum* spp. and several new host records for the mycobiota of Iran. *Journal of Plant Protection Research* 55, 198–211.

Atkinson, J.D. 1971. *Diseases of Fruit Trees in New Zealand.* New Zealand Department of Scientific and Industrial Research, Information Series No. 81. Auckland, New Zealand.

Barden, C.L. and Bramlage, W.J. 1994. Relationships of antioxidants in apple peel to changes in α-farnesene and conjugated trienes during storage, and to superficial scald development after storage. *Postharvest Biology and Technology* 4, 23–33.

Baric, S., Lindner, L., Marschall, K. and Dalla Via, J. 2010. Haplotye diversity of *Tilletiopsis* spp. causing white haze in apple orchards in Northern Italy. *Plant Pathology* 59, 535–541.

Barkai-Golan, R. 2001. Postharvest disease summary. pp. 291–310. *In*: R. Barkai-Golan (ed.). *Postharvest Diseases of Fruits and Vegetables, Development and Control.* Elsevier Science, Amsterdam, The Netherlands.

Batra, L.R. 1991. *World Species of Monilinia (Fungi): Their Ecology, Biosystematics and Control.* Mycologia Memoir, Vol. 16. Lubrecht & Cramer Ltd., Berlin, Germany.

Bautista-Baños, S., Hernandez-Lauzardo, A.N., Velazquez-del Valle, M.G., Hernandez-Lopez, M., Barka, E.A., Bosquez-Molina, E. and Wilson, C.L. 2006. Chitosan as a potential natural compound to control pre and postharvest diseases of horticultural commodities. *Crop Protection* 25, 108–118.

Beda, E., Martini, C., Di Francesco, A. and Mari, M. 2016. *Monilinia polystroma*: Nuovo patogeno delle produzioni di drupacee e pomacee. *Frutticoltura* 5, 32–36. (In Italian with English abstract).

Ben, J. 1995. Influence of rootstock on mineral content and storage of apple fruits. *Acta Horticulturae* 383, 353–357.

Berrie, A.M. 1989.Storage rots of apple and pear in South East England 1980–88: Incidence and fungicide resistance. pp. 229–239. *In*: C. Gessler (ed.). *Integrated Control of Pome Fruit Diseases.* IOBC, Locarno, Switzerland.

Berton, O. 2007. First report of the teleomorph of *Colletotrichum acutatum* on apple leaves in Brazil. *Agropecuaria Catarinense* 20, 82–84.

Bertrand, P.F., MacSwan, I.C., Rackham, R.L. and Moore, B.J. 1977. An outbreak of side rot in Bosc pears in Oregon. *Plant Disease Reporter* 61, 890–893.

Bi, F., Barad, S., Ment, D., Luria, N., Dubey, A., Casado, V., Glam, N., Diaz Minguez, J., Espeso, E.A., Fluhr, R. and Prusky, D. 2016. Carbon regulation of environmental pH by secreted small molecules that modulate pathogenicity in phytopathogenic fungi. *Molecular Plant Pathology* 17, 1178–1195.

Biggs, A.R. 1999. Effects of calcium salts on apple bitter rot caused by two *Colletotrichum* spp. *Plant Disease* 83, 1001–1005.

Biggs, A.R. and Miller, S.S. 2001. Relative susceptibility of selected apple cultivars to *Colletotrichum acutatum*. *Plant Disease* 85, 657–660.

Boekhout, T., Gildemacher, P., Theelen, B., Muller, W.H., Heijne, B. and Lutz, M. 2006. Extensive colonization of apples by smut anamorphs causes a new postharvest disorder. *FEMS Yeast Research* 6, 63–76.

Børve, J. and Stensvand, A. 2007. *Colletotrichum acutatum* found on apple buds in Norway. *Plant Health Progress*. Available at: www.plantmanagementnetwork.org/pub/php/brief/2007/apple/Cacutatum.pdf

Børve, J. and Stensvand, A. 2017. *Colletotrichum acutatum* occurs asymptomatically on apple leaves. *European Journal of Plant Pathology* 147, 943–948.

Bosshard, E. 1997. Why is *Colletotrichum acutatum* a quarantine organism, and *C. gloeosporioides* and *C. fragariae* are not? *Acta Horticulturae* 439, 799–802.

Boyd-Wilson, K.S.H., Butler, R.C., Alspach, P.A., Everett, K.R., Pushparajah, I.P.S. and Walter, M. 2014. Compounds alone and in combination with yeasts to control *Colletotrichum acutatum* in apples. *Australasian Plant Pathology* 43, 703–714.

Braganca, C.A.D., Damm, U., Baroncelli, R., Massola, N.S. and Crous, P.W. 2016. Species of the *Colletotrichum acutatum* complex associated with anthracnose diseases of fruit in Brazil. *Fungal Biology* 120, 547–561.

Bratley, C.O. 1940. Development of scab on stored apple. *Phytopathology* 30, 174–178.

Brook, P.J. 1972. The apple bitter rot epidemic in Auckland. *Orchardist of New Zealand* 45, 238, 241.

Brook, P.J. 1977. *Glomerella cingulata* and bitter rot of apple. *New Zealand Journal of Agricultural Research* 20, 547–555.

Brook, P.J. 1981. Protective function of an ultraviolet-absorbing compound associated with conidia of *Glomerella cingulata*. *New Zealand Journal of Botany* 19, 299–304.

Brooks, C., Cooley, J.S. and Fisher, D.F. 1919. Apple scald. *Journal of Agricultural Research* 16, 195–217.

Brown, G. 2010. Fuji and Red Delicious growers beware: Core rot chemicals no longer available. *Australian Fruitgrower* 4, 16–19.

Byrde, R.J.W. and Willetts, H.J. 1977. *The Brown Rot Fungi of Fruit*. Pergamon Press, New York.

CABI (Centre for Agriculture and Biosciences International). 2017. *Monilinia fructigena* (brown rot). Invasive Species Compendium. Available at: www.cabi.org/isc/datasheet/34747

Cameldi, I. 2015. Apple latent infection caused by *Neofabraea alba*: Host-pathogen interaction and disease management. PhD. Thesis, University of Bologna, Bologna, Italy.

Cameldi, I., Neri, F., Menghini, M., Pirondi, A., Nanni, I., Collina, M. and Mari, M. 2017. Characterization of *Neofabraea vagabunda* isolates causing apple bull's eye rot in Italy (Emilia-Romagna Region). *Plant Pathology* 66, 1432–1444.

Cameldi, I., Neri, F., Ventrucci, D., Ceredi, G., Muzzi, E. and Mari, M. 2016a. Influence of harvest date on bull's eye rot of 'Cripps Pink' apple and control chemical strategies. *Plant Disease* 100, 2287–2293.

Cameldi, I., Pirondi, A., Neri, F., Collina, M. and Mari, M. 2016b. First report of apple bull's eye rot caused by *Neofabraea malicorticis* in Italy. *Plant Disease* 100, 2532–2533.

Camilo, A.P., Lamb, R.C. and Aldwinckle, H.S. 1988. Genetic resistance to bitter rot incited by *Glomerella cingulata* (Stoneman.) Spaulding and Schrenk in apple (*Malus domestica* (Borkh.)). *Acta Horticulturae* 232, 37–45.

Cannon, P.F., Damm, U., Johnston, P.R. and Weir, B.S. 2012. *Colletotrichum* – Current status and future directions. *Studies in Mycology* 73, 181–213.

Carpenter, J.B. 1942. Moldy core of apples in Wisconsin. *Phytopathology* 32, 896–900.

Casals, C., Viñas, I., Torres, R., Griera, C. and Usall, J. 2010. Effect of temperature and water activity on in vitro germination of *Monilinia* spp. *Journal of Applied Microbiology* 108, 47–54.

Chan, Z.L. and Tian, S.P. 2005. Interaction of antagonistic yeasts against postharvest pathogens of apple fruit and possible mode of action. *Postharvest Biology and Technology* 36, 215–223.

Coley-Smith, J.R., Verhoeff, K. and Jarvis, W.R. 1980. *The Biology of Botrytis*. Academic Press, San Diego, CA.
Combrink, J.C. and Ginsburg, L. 1973. Core rot in starking apples – A preliminary investigation into the origin and control. *Deciduous Fruit Grower* 23, 16–19.
Combrink, J.C., Kotze, J.M., Wehner, F.C. and Grobbelaar, C.J. 1985. Fungi associated with core rot of starking apples in South Africa. *Phytophylactica* 17, 81–83.
Conway, W.S., Janisiewicz, W.J., Klein, J.D. and Sams, C.E. 1999. Strategy for combining heat treatment, calcium infiltration, and biological control to reduce postharvest decay of 'Gala' apples. *HortScience* 34, 700–704.
Conway, W.S., Leverentz, B., Janisiewicz, W.J., Blodgett, A.B., Saftner, R.A. and Camp, M.J. 2004. Integrating heat treatment, biocontrol and sodium bicarbonate to reduce postharvest decay of apple caused by *Colletotrichum acutatum* and *Penicillium expansum*. *Postharvest Biology and Technology* 34, 11–20.
Conway, W.S., Leverentz, B., Janisiewicz, W.J., Saftner, R.A. and Camp, M.J. 2005. Improving biocontrol using antagonist mixtures with heat and/or sodium bicarbonate to control postharvest decay of apple fruit. *Postharvest Biology and Technology* 36, 235–244.
Côté, M.J., Tardif, M.C. and Meldrum, A.J. 2004. Identification of *Monilinia fructigena, M.fructicola, M. laxa,* and *M. polystroma* on inoculated and naturally infected fruit using multiplex PCR. *Plant Disease* 88, 1219–1225.
Crusius, L.U., Forcelini, C.A., Sanhueza, R.M.V. and Fernandes, J.M.C. 2002. Epidemiology of apple leaf spot. *Fitopatologia Brasileira* 27, 65–70.
Cunnington, J.H. 2004. Three *Neofabraea* species on pome fruit in Australia. *Australasian Plant Pathology* 33, 453–454.
Czerwinski, M., Popovic, V. and Errampalli, D. 2009. Multi-site mutations in Beta-tubulin gene sequences are involved in thiabendazole resistance in *Penicillium* species collected from Southern Ontario. *Canadian Journal of Plant Pathology* 31, 52.
Damm, U., Cannon, P.F., Woudenberg, J.H.C. and Crous, P.W. 2012. The *Colletotrichum acutatum* species complex. *Studies in Mycology* 73, 37–113.
Daniel, C.K., Lennox, C.L. and Vries, F.A. 2015. In vivo application of garlic extracts in combination with clove oil to prevent postharvest decay caused by *Botrytis cinerea, Penicillium expansum* and *Neofabraea alba* on apples. *Postharvest Biology and Technology* 99, 88–92.
de Jong, S.N., Levesque, C.A., Verkley, G.J.M., Abeln, E.C.A., Rahe, J.E. and Braun, P.G. 2001. Phylogenetic relationships among *Neofabraea* species causing tree cankers and bull's-eye rot of apple based on DNA sequencing of ITS nuclear rDNA, mitochondrial rDNA, and the beta-tubulin gene. *Mycological Research* 105, 658–669.
Dennis, C. 1983. *Postharvest Pathology of Fruits and Vegetables*. Academic Press., London, UK.
Dharmaputra, O.S., Putri, A.S.R. and Dewi, A.U. 2013. Mycobiota of apple fruit: Effects on bitter rot caused by *Colletotrichum acutatum*. *Acta Horticulturae* 975, 223–229.
Dingley, J.M. 1969. *Records of Plant Diseases in New Zealand*. Bulletin of the New Zealand Department of Scientific and Industrial Research No. 192. Wellington, New Zealand.

Domsch, K.H., Gams, W. and Anderson, T.H. 1980. *Compendium of Soil Fungi*. Vol. 2. Academic Press, London, UK.

Doohan, F.M., Brennan, J. and Cooke, B.M. 2003. Influence of climatic factors on *Fusarium* species pathogenic to cereals. *European Journal of Plant Pathology* 109, 755–768.

Droby, S. and Lichter, A. 2007. Post-harvest *Botrytis* infection: Etiology, development and management. pp. 349–367. *In*: Y. Elad, B. Williamson, P. Tudzynski and N. Delen (eds.). *Botrytis: Biology, Pathology and Control*. Springer, Dordrecht, The Netherlands.

Droby, S., Wisniewski, M., Teixidó, N., Spadaro, D. and Jijakli, M.H. 2016. The science, development, and commercialization of postharvest biocontrol products. *Postharvest Biology and Technology* 122, 22–29.

Dyko, B.J. and Mordue, J.E.M. 1979. *Colletotrichum acutatum*. CMI Descriptions of Pathogenic Fungi and Bacteria No. 630. Commonwealth Mycological Institute, Kew, UK.

EFSA (European Food Safety Authority). 2011. Scientific opinion on the risks for animal and public health related to the presence of *Alternaria* toxins in feed and food. *EFSA Journal* 9, 2407–2504.

El-Ghaouth, A., Wilson, C.L. and Wisniewski, M. 1998. Ultrastructural and cytochemical aspects of the biological control of *Botrytis cinerea* by *Candida saitoana* in apple fruit. *Phytopathology* 88, 282–291.

Ellis, M.A. and Barrat, J.G. 1983. Colonization of Delicious apple fruits by *Alternaria* spp. and effect of fungicide sprays on moldy-core. *Plant Disease* 67, 150–152.

Emongor, V.E., Murr, D.P. and Lougheed, E.C. 1994. Preharvest factors that predispose apples to superficial scald. *Postharvest Biology and Technology* 4, 289–300.

Errampalli, D. 2014. *Penicillium expansum* (blue mold). pp. 189–231. *In*: S. Bautista-Baños (ed.). *Postharvest Decay: Control Strategies*. Academic Press, Elsevier, London, UK.

Everett, K.R. 2003. The effect of low temperatures on *Colletotrichum acutatum* and *Colletotrichum gloeosporioides*, fungi causing body rots of avocado in New Zealand. *Australasian Plant Pathology* 32, 441–448.

Everett, K.R. 2014. Anthracnose and stem-end rots of tropical and subtropical fruit-new names for old foes. pp. 55–70. *In*: D. Prusky and M.L. Gullino (eds.). *Post-Harvest Pathology. Plant Pathology in the 21st Century, Contributions to the 10th International Congress, ICPP 2013*. Springer International Publishing, Cham, Switzerland.

Everett, K.R., Pushparajah, I.P.S., Taylor, J.T., Timudo-Torrevilla, O.E., Spiers, T.M., Ah Chee, A., Shaw, P.W. and Wallis, D.R. 2015. Evaluation of fungicides for control of bitter and sprinkler rots on apple fruit. *New Zealand Plant Protection* 68, 264–274.

Everett, K.R., Pushparajah, I.P.S., Timudo-Torrevilla, O.E., Ah Chee, A., Scheper, R. W.A., Shaw, P.W., Spiers, T.M., Taylor, J.T., Wallis, D.R. and Wood, P. 2018. Infection criteria, inoculum sources and splash dispersal pattern of *Colletotrichum acutatum* causing bitter rot of apples in New Zealand. *European Journal of Plant Protection* 152, 367–383.

Everett, K.R., Timudo-Torrevilla, O.E., Pushparajah, I.P.S., Scheper, R.W.A., Shaw, P.W., Spiers, T.M., Ah Chee, A., Taylor, J.T., Wood, P., Wallis, D.R. and Manning, M.A.

2010. Inoculum sources for *Colletotrichum acutatum* causing summer rot of apple. *The Orchardist* 83, 18–22.
Everett, K.R., Timudo-Torrevilla, O.E., Scheper, R.W.A., Shaw, P.W., Wood, P.N., Mundy, D.C. and Wallis, D.R. 2016. The influence of nutrition, maturity and canopy density on the incidence of apple bitter rot. *New Zealand Plant Protection* 69, 99–110.
Fallik, E. 2004. Prestorage hot water treatments (immersion, rinsing and brushing). *Postharvest Biology and Technology* 32, 125–134.
FAOSTAT (Food and Agriculture Organization of the United Nations, Statistics Division). 2018. Crop data. Available at: www.fao.org/faostat/en/#data/QC
Faretra, F., Antonacci, E. and Pollastro, S. 1988. Sexual-behavior and mating system of *Botryotinia fuckeliana*, teleomorph of *Botrytis cinerea*. *Journal of General Microbiology* 134, 2543–2550.
Ferguson, I.B., Volz, R. and Woolf, A. 1999. Preharvest factors affecting physiological disorders of fruit. *Postharvest Biology and Technology* 15, 255–262.
Ferguson, I.B. and Watkins, C.B. 1989. Bitter pit in apple fruit. *Horticultural Reviews* 11, 289–355.
Ferree, D.C. and Warrington, I.J. 2003. *Apples. Botany, Production and Uses.* CABI Publishing, Wallingford, UK.
Freeman, S. 2008. Management, survival strategies, and host range of *Colletotrichum acutatum* on strawberry. *HortScience* 43, 66–68.
Frisvad, J.C. and Samson, R.A. 2004. Polyphasic taxonomy of *Penicillium* subgenus *Penicillium* – A guide to identification of food and air-borne terverticillate Penicillia and their mycotoxins. *Studies in Mycology* 49, 1–174.
Fu, D., Zhuang, J., Zhang, R. and Sun, G. 2013. Optimization of ISSR-PCR reaction system and genetic diversity analysis of *Colletotrichum* species causing bitter rot of apple. *Acta Phytophylacica Sinica* 40, 231–236.
Fulton, C.E., van Leeuwen, G.C.M. and Brown, A.E. 1999. Genetic variation among and within *Monilinia* species causing brown rot of stone and pome fruits. *European Journal of Plant Pathology* 105, 495–500.
Gao, L.L., Zhang, Q., Sun, X.Y., Jiang, L., Zhang, R., Sun, G.Y., Zha, Y.L. and Biggs, A.R. 2013. Etiology of moldy core, core browning, and core rot of Fuji apple in China. *Plant Disease* 97, 510–516.
Giraud, M. and Bompeix, G. 2012. Postharvest diseases of pome fruits in Europe: Perspective for integrated control. *IOBC-WPRS Bulletin* 84, 257–263.
Goertz, A. 2010. Auftreten der Fusarium-Kolbenfäule im Maisanbau in Deutschland und Maßnahmen zur Vermeidung der Mykotoxinbelastung in Maiskörnern Ph. D. Thesis. Rheinischen Friedrich-Wilhelms-Universität Bonn, Bonn, Germany (In German with English abstract).
Gonzalez, E. and Sutton, T.B. 2004. Population diversity within isolates of *Colletotrichum* spp. causing Glomerella leaf spot and bitter rot of apples in three orchards in North Carolina. *Plant Disease* 88, 1335–1340.
Gonzalez, E. and Sutton, T.B. 2005. Differentiation of isolates of *Glomerella cingulata* and *Colletotrichum* spp. associated with Glomerella leaf spot and bitter rot of apples using growth rate, response to temperature, and benomyl sensitivity. *Plant Health Progress* Available at: www.plantmanagementnetwork.org/pub/php/research/2005/apple/differentiate.pdf

Grahovac, M.S., Balaz, J.S., Grahovac, J.A., Dodic, J.M., Tanovic, B.B., Hrustic, J.G. and Tadijan, I.Z. 2014. Screening of antagonistic activity of selected microorganisms against apple rot pathogens. *Romanian Biotechnological Letters* 19, 8959–8965.

Grahovac, M.S., Indic, D., Balaz, J., Vukovic, S., Tanovic, B., Hrustic, J. and Tanaskovic, S. 2012. Phytopathogenic fungi from *Colletotrichum* genus on fruit species. *Biljni Lekar (Plant Doctor)* 40, 28–38.

Gualandri, V., Concini, M.D., Branz, A. and Angeli, G. 2015. Apple, for the rot of the heart and vital prevention. *Informatore Agrario* 71, 44–46.

Guerber, J.C. and Correll, J.C. 2001. Characterization of *Glomerella acutata*, the teleomorph of *Colletotrichum acutatum*. *Mycologia* 93, 216–229.

HealthCanada. 2013. Consultation on Health Canada's Proposed Standard (Maximum Level) for the Presence of the Mycotoxin Patulin in Apple Juice and Unfermented Apple Cider. Available at www.hc-sc.gc.ca/fn-an/consult/max-mycotox-pat/index-eng.php.

Henriquez, J.L., Sugar, D. and Spotts, R.A. 2004. Etiology of bull's eye rot of pear caused by *Neofabraea* spp. in Oregon, Washington, and California. *Plant Disease* 88, 1134–1138.

Henriquez, J.L., Sugar, D. and Spotts, R.A. 2006. Induction of cankers on pear tree branches by *Neofabraea alba* and *N. perennans*, and fungicide effects on conidial production on cankers. *Plant Disease* 90, 481–486.

Henriquez, J.L., Sugar, D. and Spotts, R.A. 2008. Effects of environmental factors and cultural practices on bull's eye rot of pear. *Plant Disease* 92, 421–424.

Hockey, J.F. 1953. *Studies in Fruit Diseases, IX. Apple Scab*. Vol. 519. Canadian Department of Agricultural Publications, Otawa, Canada.

Holb, I.J. 2004. Yield loss and disease development of *Monilinia fructigena* (Aderh. & Ruhl.) Honey in organic apple orchard. *Journal of Agricultural Science* 15, 6–8.

Holb, I.J. and Scherm, H. 2008. Quantitative relationships between different injury factors and development of brown rot caused by *Monilinia fructigena* in integrated and organic apple orchards. *Phytopathology* 98, 79–86.

Holz, G., Cortze, S. and Williamson, B. 2007. The ecology of *Botrytis* on plant surfaces. pp. 9–27. *In*: Y. Elad, B. Williamson, P. Tudzynski and N. Delen (eds.). *Botrytis: Biology, Pathology and Control*. Springer, Dordrecht, The Netherlands.

Hortova, B., Novotny, D. and Erban, T. 2014. Physiological characteristics and pathogenicity of eight *Neofabraea* isolates from apples in Czechia. *European Journal of Horticultural Science* 79, 327–334.

Huelin, F.E. and Murray, K.E. 1966. α-farnesene in the natural coating of apples. *Nature* 210, 1260–1261.

Hughes, K.J.D., Fulton, C.E., McReynolds, D. and Lane, C.R. 2000. Development of new PCR primers for identification of *Monilinia* species. *OEPP/EPPO Bulletin* 30, 507–511.

Ivic, D., Sever, Z. and Milicevic, T. 2013a. Estimation of economic loss due to postharvest diseases of apple (cv. Idared) during four seasons. *Pomologia Croatica* 19, 51–61.

Ivic, D., Voncina, D., Sever, Z., Simon, S. and Pejic, I. 2013b. Identification of *Colletotrichum* species causing bitter rot of apple and pear in Croatia. *Journal of Phytopathology* 161, 284–286.

Janisiewicz, W.J. and Conway, W.S. 2010. Combining biological control with physical and chemical treatments to control fruit decay after harvest. *Stewart Postharvest Review* 6:3, 1–16.

Janisiewicz, W.J. and Jurick, W.M. II. 2017. Sustainable approaches to control postharvest diseases of apples. pp. 307–336. *In*: K. Evans (ed.). *Achieving Sustainable Cultivation of Apples*. Burleigh Dodds Science Publishing, Cambridge, UK.

Janisiewicz, W.J. and Korsten, L. 2002. Biological control of postharvest diseases of fruits. *Annual Reviews of Phytopathology* 40, 411–441.

Janisiewicz, W.J., Leverentz, B., Conway, W.S., Saftner, R.A., Reed, A.N. and Camp, M.J. 2003. Control of bitter rot and blue mold of apples by integrating heat and antagonist treatments on 1-MCP-treated fruit stored under controlled atmosphere conditions. *Postharvest Biology and Technology* 29, 129–143.

Janisiewicz, W.J. and Peterson, D.L. 2004. Susceptibility of the stem pull area of mechanically harvested apples to blue mold decay and its control with a biocontrol agent. *Plant Disease* 88, 662–664.

Janisiewicz, W.J., Tworkoski, T.J. and Kurtzman, C.P. 2001. Biocontrol potential of *Metchnikowia pulcherrima* strains against blue mold of apple. *Phytopathology* 91, 1098–1108.

Jarvis, W.R. 1977. Botryotinia *and* Botrytis *Species: Taxonomy, Physiology and Pathogenicity: A Guide to the Literature*. Canada Department of Agriculture, Research Brand, Otawa, Canada.

Jemric, T., Fruk, I., Fruk, M., Radman, S., Sinkovic, L. and Fruk, G. 2016. Bitter pit in apples: Pre- and postharvest factors: A review. *Spanish Journal of Agricultural Research* 14, e08R01.

Jijakli, M.H. and Lepoivre, P. 2004. State of the art and challenges of post-harvest disease management in apples. *Fruit and Vegetable Diseases* 1, 59–94.

Johnston, P.R. and Jones, D. 1997. Relationships among *Colletotrichum* isolates from fruit-rots assessed using rDNA sequences. *Mycologia* 89, 420–430.

Johnston, P.R., Pennycook, S.R. and Manning, M.A. 2005. Taxonomy of fruit-rotting fungal pathogens: What's really out there? *New Zealand Plant Protection* 58, 42–46.

Jones, A.L. and Aldwinckle, H.S. 1990. *Compendium of Apple and Pear Diseases*. The American Phytopathological Society Press, St. Paul, MN.

Jurick, W.M. II, Janisiewicz, W.J., Saftner, R.A., Vico, I., Gaskins, V.L., Park, E., Forsline, P.L., Fazio, G. and Conway, W.S. 2011. Identification of wild apple germplasm (*Malus* spp.) accessions with resistance to the postharvest decay pathogens *Penicillium expansum* and *Colletotrichum acutatum*. *Plant Breeding* 130, 481–486.

Kang, J.C., Crous, P.W., McHau, G.R.A., Serdani, M. and Song, S.M. 2002. Phylogenetic analysis of *Alternaria* spp. associated with apple core rot and citrus black rot in South Africa. *Mycological Research* 106, 1151–1162.

Kidd, M.N. and Beaumont, A. 1924. Apple rot fungi in storage. *Transactions of the British Mycological Society* 10, 98–118.

Kienholz, J.R. 1951. The bull's eye rot problem of apples and pears. *Proceedings of Oregon Horticultural Society* 43, 75–77.

Kirk, T.W. 1906. Principal fungus diseases of the year. *Report of the Department of Agriculture, New Zealand* 14, 341–365.

Knee, M. and Hatfield, S.G.S. 1981. Benefits of ethylene removal during apple storage. *Annals of Applied Biology* 98, 157–165.

Kohl, J., Wenneker, M., Haas, L.D., Anbergen, R., Lombaers, C.H. and Kastelein, P. 2015. Sources of *Neofabraea* spp. and *Cadophora* spp. in Dutch apple and pear orchards. *IOBC/WPRS Bulletin* 110, 143–145.

Konstantinou, S., Karaoglanidis, G.S., Bardas, G.A., Minas, I.S., Doukas, E. and Markoglou, A.N. 2011. Postharvest fruit rots of apple in Greece: Pathogen incidence and relationships between fruit quality parameters, cultivar susceptibility, and patulin production. *Plant Disease* 95, 666–672.

Kou, L.P., Gaskins, V.L., Luo, Y.G. and Jurick, W.M. II. 2014. First report of *Fusarium avenaceum* causing postharvest decay of 'Gala' apple fruit in the United States. *Plant Disease* 98, 690.

Lahlali, R., Raffaele, B. and Jijakli, M.H. 2011. UV protectants for *Candida oleophila* (strain O), a biocontrol agent of postharvest fruit diseases. *Plant Pathology* 60, 288–295.

Lee, S.Y., Park, S.J., Lee, J.J., Back, C.H., Ten, L.N., Kang, I.K. and Jung, H.Y. 2017. First report of fruit rot caused by *Fusarium decemcellulare* in apples in Korea. *The Korean Journal of Mycology* 45, 54–62.

Lennox, C.L., Spotts, R.A. and Booyse, M. 2004. Postharvest decay of d'Anjou pears from the Pacific Northwest and control with a thiabendazole drench. *Plant Disease* 88, 474–478.

Leslie, J.F. and Summerell, B.A. 2006. *The Fusarium Laboratory Manual*, 1st ed. Blackwell Publishing Ltd., Oxford, UK.

Li, B., Zong, Y., Du, Z., Chen, Y., Zhang, Z., Qin, G., Zhao, W. and Tian, S. 2015. Genomic characterization reveals insights into patulin biosynthesis and pathogenicity in *Penicillium* species. *Molecular Plant-Microbe Interactions* 28, 635–647.

Lindner, L. 2009. 'Patina bianca', rugginosita', fumaggine: l'Alto Adige si interroga sui sistemi per prevenire i danni alle mele. *Frutticoltura* 71, 58–62. (In Italian).

Lindner, L. and Baric, S. 2006. White haze: A new disorder of apple caused by fungi colonizing the fruit surface. *Informatore Fitopatologico* 56, 60–63. (In Italian with English abstract).

Liu, J., Sui, Y., Wisniewski, M., Droby, S. and Liu, Y. 2013. Review: Utilization of antagonistic yeasts to manage postharvest fungal diseases of fruit. *International Journal of Food Microbiology* 167, 153–160.

Logrieco, A., Bottalico, A., Mule, G., Moretti, A. and Perrone, G. 2003. Epidemiology of toxigenic fungi and their associated mycotoxins for some Mediterranean crops. *European Journal of Plant Pathology* 109, 645–667.

Logrieco, A., Mule, G., Moretti, A. and Bottalico, A. 2002. Toxigenic *Fusarium* species and mycotoxins associated with maize ear rot in Europe. *European Journal of Plant Pathology* 108, 597–609.

Lopez-Reyes, J.G., Spadaro, D., Gullino, M.L. and Garibaldi, A. 2010. Efficacy of plant essential oils on postharvest control of rot caused by fungi on four cultivars of apples in vivo. *Flavour and Fragrance Journal* 25, 171–177.

Lotter, J.D.V., Beukes, D.J. and Weber, H.W. 1985. Growth and quality of apples as affected by different irrigation treatments. *Journal of Horticultural Science & Biotechnology* 60, 181–192.

Lua, F., Jun, S., Beaudry, R.M. and Hildebrand, P.D. 2006. Effect of hexanal vapor on spore viability of *Penicillium expansum*, lesion development on whole apples and fruit volatile biosynthesis. *Journal of Food Science* 71, M105–M109.

Lurie, S., Pesis, E. and Ben-Arie, R. 1991. Darkening of sunscald on apples in storage is a non-enzymatic and non-oxidative process. *Postharvest Biology and Technology* 1, 119–125.

Lurie, S. and Watkins, C.B. 2012. Superficial scald, its etiology and control. *Postharvest Biology and Technology* 65, 44–60.

Manning, M.A. and Burmeister, D. 2000. *Strategies to Reduce Postharvest Rots in Pacific Rose Apples. Internal Report to ENZA*. The Horticulture & Food Research Institute of New Zealand Ltd., Auckland, New Zealand.

Mari, M., Guidarelli, M., Martini, C. and Spadoni, A. 2012. First report of *Colletotrichum acutatum* causing bitter rot on apple in Italy. *Plant Disease* 96, 144.

Mari, M., Leoni, O., Iori, R. and Cembali, T. 2002. Antifungal vapour-phase activity of allyl-isothiocyanate against *Penicillium expansum* on pears. *Plant Pathology* 51, 231–236.

Maxin, P. 2012. Improving apple quality by hot water treatment. M.Sc. Thesis, Aarhus University, Aarhus, Denmark.

Maxin, P. and Weber, R.W. 2001. Control of *Phacidiopycnis washingtonensis* storage rot of apples by hot-water treatments without the ethylene inhibitor 1-MCP. *Journal of Plant Diseases and Protection* 118, 222–224.

Maxin, P., Weber, R.W.S., Pedersen, H.L. and Williams, M. 2012. Control of a wide range of storage rots in naturally infected apples by hot-water dipping and rinsing. *Postharvest Biology and Technology* 70, 25–31.

Maxin, P., Williams, M. and Weber, R.W.S. 2014. Control of fungal storage rots of apples by hot-water treatments: A northern European perspective. *Erwerbs-Obstbau* 56, 25–34.

McLeod, A., Smit, F.J., Walt, L.V.D. and Mostert, L. 2008. Apple core rot diseases past and present. *SA Fruit Journal* 7, 21–24.

Michailides, T.J., Morgan, D.P., Mitcham, E. and Crisosto, C.H. 1994. Occurrence of moldy core and core rot of Fuji apple in California. *Central Valley Postharvest Newsletter* 3, 6–9.

Michalecka, M., Bryk, H., Poniatowska, A. and Pulawska, J. 2016. Identification of *Neofabraea* species causing bull's eye rot of apple in Poland and their direct detection in apple fruit using multiplex PCR. *Plant Pathology* 65, 643–654.

Miller, P.M. 1959. Open calyx tubes as a factor contributing to carpel discoloration and decay of apples. *Phytopathology* 49, 520–523.

Moral, J. and Trapero, A. 2012. Mummified fruit as a source of inoculum and disease dynamics of olive anthracnose caused by *Colletotrichum* spp. *Phytopathology* 102, 982–989.

Munir, M., Amsden, B., Dixon, E., Vaillancourt, L. and Gauthier, N.A.W. 2016. Characterization of *Colletotrichum* species causing bitter rot of apple in Kentucky orchards. *Plant Disease* 100, 2194–2203.

Neri, F., Mari, M., Menniti, A.M., Brigati, S. and Bertolini, P. 2006. Control of *Penicillium expansum* in pears and apples by trans-2-hexenal vapours. *Postharvest Biology and Technology* 41, 101–108.

Niem, J., Miyara, I., Ettedgui, Y., Reuveni, M., Flaishman, M. and Prusky, D. 2007. Core rot development in Red Delicious apples is affected by susceptibility of the seed locule to *Alternaria alternata* colonization. *Phytopathology* 97, 1415–1421.

Noe, J.P. and Starkey, T.E. 1982. Relationship of apple fruit maturity and inoculum concentration to infection by *Glomerella cingulata*. *Plant Disease* 66, 379–381.

Northover, J. 1975. Preharvest NAA and postharvest fungicide treatments of apples against natural storage scab and blue mold infections. *Plant Disease Reporter* 59, 969–971.

Peres, N.A., Timmer, L.W., Adaskaveg, J.E. and Correll, J.C. 2005. Lifestyles of *Colletotrichum acutatum*. *Plant Disease* 89, 784–796.

Perring, M.A. 1986. Incidence of bitter pit in relation to the calcium content of apples – Problems and paradoxes, a review. *Journal of the Science of Food and Agriculture* 37, 591–606.

Pesicova, K., Kolarik, M., Hortova, B. and Novotny, D. 2017. Diversity and identification of *Neofabraea* species causing bull's eye rot in the Czech Republic. *European Journal of Plant Pathology* 147, 683–693.

Peter, K.A. 2018. Moldy Core Of Apple. PennState Extension, College of Agricultural Sciences. The Pennsylvania State University. Available at: http://extension.psu.edu/plants/tree-fruit/news/2017/moldy-core-of-apple

Peterson, S.W. 2000. Phylogenetic analysis of *Penicillium* species based on ITS and LSU-rDNA nucleotide sequences. pp. 163–178. *In*: R.A. Samson and J.I. Pitt (eds.). *Integration of Modern Taxonomic Methods for Penicillium and Aspergillus Classification*. Harwood Academic Publishers, Amsterdam, The Netherlands.

Petroczy, M. and Palkovics, L. 2009. First report of *Monilia polystroma* on apple in Hungary. *European Journal of Plant Pathology* 125, 343–347.

Phoulivong, S., Cai, L., Chen, H., McKenzie, E.H.C., Abdelsalam, K., Chukeatirote, E. and Hyde, K.D. 2010. *Colletotrichum gloeosporioides* is not a common pathogen on tropical fruits. *Fungal Diversity* 44, 33–43.

Pierson, C.F., Ceponis, M.J. and McColloch, L.P. 1971. Market Diseases of Apples, Pears and Quinces. United States Department of Agriculture, Agricultural Research Service, Agricultural Handbook No. 376.

Pitt, J.I. and Hocking, A.D. 1999. *Fungi and Food Spoilage*, 2nd ed. Aspen Publishers Inc., New York.

Poniatowska, A., Michalecka, M. and Bielenin, A. 2013. Characteristic of *Monilinia* spp. fungi causing brown rot of pome and stone fruits in Poland. *European Journal of Plant Pathology* 135, 855–865.

Poulsen, M.E., Naef, A., Gasser, S., Christen, D. and Rasmussen, P.H. 2009. Influence of different disease control pesticide strategies on multiple pesticide residue levels in apple. *Journal of Horticultural Science and Biotechnology* 84, 58–61.

Powell, H.G. and Fulton, S.H. 1903. *The Apple in Cold Storage*. United States Department of Agriculture, Bureau Plant Industry, Bulletin No. 48. Washington. Available at: www.biodiversitylibrary.org/item/27541#page/519/mode/1up

Prange, R., Delong, J., Nichols, D. and Harrison, P. 2011. Effect of fruit maturity on the incidence of bitter pit, senescent breakdown, and other post-harvest disorders in 'Honeycrisp'^{TM} apple. *Journal of Horticultural Science & Biotechnology* 86, 245–248.

Prencipe, S., Spadaro, D., Fruk, G. and Jemric, T. 2016. First report of *Tilletiopsis pallescens* causing white haze on apple in Croatia. *Plant Disease* 100, 225.
Qin, G.Z., Tian, S.P. and Xu, Y. 2004. Biocontrol of postharvest diseases on sweet cherries by four antagonistic yeasts in different storage conditions. *Postharvest Biology and Technology* 31, 51–58.
Quaglia, M., Ederli, L., Pasqualini, S. and Zazzerini, A. 2011. Biological control agents and chemical inducers of resistance for postharvest control of *Penicillium expansum* Link. on apple fruit. *Postharvest Biology and Technology* 59, 307–315.
Ramsey, G.B. 1937. Fruit and vegetable diseases on the Chicago market in 1936. *Plant Disease Reporter Supplement* 101, 81–96.
Reuveni, M. and Prusky, D. 2007. Improved control of moldy-core decay (*Alternaria alternata*) in Red Delicious apple fruit by mixtures of DMI fungicides and captan. *European Journal of Plant Pathology* 118, 349–357.
Riad, E.M.R.S. and Ziedan, E.S.H. 2015. First record of core rot disease on apple fruit cv. Anna 106 local cultivar in Egypt. *International Journal of Agricultural Technology* 11, 1371–1380.
Rizzolo, A., Grassi, M. and Vanoli, M. 2010. La conservazione delle pere in AC dinamica: Scambi gassosi e incidenza delle fisiopatie. *Frutticoltura* 72, 38–44. (In Italian with English abstract).
Robiglio, A.L. and Lopez, S.E. 1995. Mycotoxin production by *Alternaria alternata* strains isolated from Red Delicious apples in Argentina. *International Journal of Food Microbiology* 24, 413–417.
Rosenberger, D.A. 1997. Recent research and changing options for controlling postharvest decays of apples. pp. 42–53. *In*: Proceedings of the Harvesting, Handling, and Storage Workshop 1997. Cornell University, Ithaca, NY.
Rosenberger, D.A. 2014. Postharvest diseases. p. 75. *In*: T.B. Sutton, H.S. Aldwinckle, A.M. Agnello and J.F. Walgenbach (eds.). *Compendium of Apple and Pear Diseases and Pests*, 2nd ed. The American Phytopathological Society Press, St. Paul, MN.
Rosenberger, D.A. and Meyer, F.W. 1985. Negatively correlated cross-resistance to diphenylamine in benomyl-resistant *Penicillium expansum*. pp. 74–79. *In*: Proceedings of the Harvesting, Handling, and Storage Workshop 1985, Cornell University, Ithaca, NY.
Ross, R.G. and Lokhart, C.L. 1960. Influence of orchard fungicides and storage scab of apple. *Canadian Plant Disease Survey* 40, 10–19.
Rupasinghe, H.P.V., Murr, D.P., Paliyath, G. and Skog, L. 2000. Inhibitory effect of 1-MCP on ripening and superficial scald development in 'McIntosh' and 'Delicious' apples. *Journal of Horticultural Science & Biotechnology* 75, 271–276.
Saftner, R.A., Bai, J.H., Abbott, J.A. and Lee, Y.S. 2003. Sanitary dips with calcium propionate, calcium chloride, or a calcium amino acid chelate maintain quality and shelf stability of fresh-cut honeydew chunks. *Postharvest Biology and Technology* 29, 257–269.
Sakai, H., Yoshioka, M., Horigome, M., Urushibara, T. and Shiraishi, T. 2002. Distribution of *Colletotrichum acutatum* causing bitter rot of apple in Gunma Prefecture. *Annual Report of the Kanto-Tosan Plant Protection Society*, 55–57.
Samson, R.A., Hoekstra, E.S. and Frisvad, J.C. 1995. *Introduction to Food Borne Fungi*. Centraalbureau voor Schimmelcultures, Baarm, The Netherlands.

Samuelian, S.K., Greer, L.A., Savocchia, S. and Steel, C.C. 2012. Overwintering and presence of *Colletotrichum acutatum* (ripe rot) on mummified bunches, dormant wood, developing tissues and mature berries of *Vitis vinifera*. *Vitis* 51, 33–37.

Sanderson, P.G. and Spotts, R.A. 1995. Postharvest decay of winter pear and apple fruit caused by species of *Penicillium*. *Phytopathology* 85, 103–110.

Sanzani, S.M., Cariddi, C., Roccotelli, A., Garganese, F., Fallanaj, F. and Ippolito, A. 2013. First report of *Gibberella avenacea* causing wet apple core rot in Italy. *Journal of Plant Pathology* 95, 217.

Saure, M.C. 1996. Reassessment of the role of calcium in development of bitter pit in apple. *Australian Journal of Plant Physiology* 23, 237–243.

Schwabe, W.F.S. 1980. Prevention of storage scab of apples. *Phytophylactica* 12, 209–211.

Seifert, K.A. 2013. *Neofabraea kienholzii* (Seifert, Spotts & Levesque). Nomenclatural novelties. *Index Fungorum* 28, 1.

Senula, A. 1985. Untersuchungen zur morphologie und physiologie von *Cryptosporiopsis malicorticis* (Cordl) Nannf. und *Phlyctena vagabunda* Desm. *Archivies of Phytopathology and Plant Protection* 21, 273–286. (In German with English abstract).

Serdani, M. 2006. Shedding light on core rots. Dry core rot develops in the orchard; wet core rot is a postharvest disease. Good Fruit Grower. Available at: www.goodfruit.com/shedding-light-on-core-rots/.

Serdani, M., Crous, P.W., Holz, G. and Petrini, O. 1998. Endophytic fungi associated with core rot of apples in South Africa, with specific reference to *Alternaria* species. *Sydowia* 50, 257–271.

Serdani, M., Kang, J.C., Andersen, B. and Crous, P.W. 2002. Characterisation of *Alternaria* species-groups associated with core rot of apples in South Africa. *Mycological Research* 106, 561–569.

Sever, Z., Ivic, D., Kos, T. and Milicevic, T. 2012. Identification of *Fusarium* species isolated from stored apple fruit in Croatia. *Archives of Industrial Hygiene and Toxicology* 63, 463–470.

Sharma, R.L. and Kaul, J.L. 1990. Mode of entry and histophatological changes induced by *Monilinia* species in apple fruit. *Indian Phytopathology* 43, 113–115.

Sharples, R.O. and Johnson, D.S. 1976. Post-harvest treatments for the control of storage disorders of apples. *Annals of Applied Biology* 83, 157–167.

Shi, Y., Correll, J.C., Guerber, J.C. and Rom, C.R. 1996. Frequency of *Colletotrichum* species causing bitter rot of apple in the southeastern United States. *Plant Disease* 80, 692–696.

Sholberg, P.L. and Haag, P.D. 1996. Incidence of postharvest pathogens of stored apples in British Columbia. *Canadian Journal of Plant Pathology* 18, 81–85.

Silveira, J.P.G., Talamini do Amarante, C.V., Steffens, C.A., Miqueloto, A. and Katsurayama, J.M. 2012. Inhibition of gibberellin synthesis reduces vegetative growth of apple trees and provides control of bitter pit in the fruit. *Revista Brasileira de Fruticultura* 34, 328–335. (In Portuguese with English abstract).

Sisler, E.C. and Serek, M. 1997. Inhibitors of ethylene responses in plants at the receptor level: Recent developments. *Physiologia Plantarum* 100, 577–582.

Skipp, R.A., Beever, R.E., Sharrock, K.R., Rikkerink, E.H.A. and Templeton, M.D. 1995. Colletotrichum. pp. 119–144. *In*: U.S. Singh, R.P. Singh and K. Kohmoto (eds.). *Pathogenesis and Host Specificity in Plant Diseases. Histopathological, Biochemical, Genetic and Molecular Bases*. Elsevier Science Ltd., Oxford, UK.

Smock, R.M. 1955. A new method of scald control. *American Fruit Grower* 75, 20.

Smock, R.M. 1961. Methods of scald control on the apple. *Bulletin of Cornell Agricultural Experimental Station* No. 970.

Snowdon, A.L. 1990. *A Colour Atlas of Post-Harvest Diseases & Disorders of Fruits & Vegetables*. Vol. 1: General Introduction & Fruits. Wolfe Scientific Ltd., London, UK.

Soliman, S., Li, X.Z., Shao, S., Behar, M., Svircev, A.M., Tsao, R. and Zhou, T. 2015. Potential mycotoxin contamination risks of apple products associated with fungal flora of apple core. *Food Control* 47, 585–591.

Sommer, N.F. 1985. Role of controlled environments in suppression of postharvest diseases. *Canadian Journal of Plant Pathology* 7, 331–339.

Sorensen, J.L., Phipps, R.K., Nielsen, K.F., Schroers, H.J., Frank, J. and Thrane, U. 2009. Analysis of *Fusarium avenaceum* metabolites produced during wet apple core rot. *Journal of Agricultural and Food Chemistry* 57, 1632–1639.

Soto-Alvear, S., Lolas, M., Rosales, I.M., Chavez, E.R. and Latorre, B.A. 2013. Characterization of the bull's eye rot of apple in Chile. *Plant Disease* 97, 485–490.

Spadaro, D., Ciavorella, A., Dianpeng, Z., Garibaldi, A. and Gullino, M.L. 2010. Effect of culture media and pH on the biomass production and biocontrol efficacy of a *Metschnikowia pulcherrima* strain to be used as a biofungicide for postharvest disease control. *Canadian Journal of Microbiology* 56, 128–137.

Spadaro, D. and Droby, S. 2016. Development of biocontrol products for postharvest diseases of fruit: The importance of elucidating the mechanisms of action of yeast antagonists. *Trends in Food Science & Technology* 47, 39–49.

Spadaro, D., Garibaldi., A. and Gullino, M.L. 2004. Control of *Botrytis cinerea* and *Penicillium expansum* on apple combining a biocontrol agent with hot water dipping and acibenzolar-S-methyl, baking soda, or ethanol application. *Postharvest Biology and Technology* 33, 141–151.

Spadaro, D. and Gullino, M.L. 2014. Use of essential oils to control postharvest rots on pome and stone fruit. pp. 101–110. *In*: D. Prusky and M.L. Gullino (eds.). *Post-harvest Pathology. Plant Pathology in the 21st Century, Contributions to the 10th International Congress, ICPP 2013*. Springer International Publishing, Cham, Switzerland.

Spadaro, D., Pellegrino, C., Garibaldi, A. and Gullino, M.L. 2011. Development of SCAR primers for the detection of *Cadophora luteo-olivacea* on kiwifruit and pome fruit and of *Cadophora malorum* on pome fruit. *Phytopathologia Mediterranea* 50, 430–441.

Spadaro, D., Sabetta, W., Acquadro, A., Portis, E., Garibaldi, A. and Gullino, M.L. 2008. Use of AFLP for differentiation of *Metschnikowia putcherrima* strains for postharvest disease biological control. *Microbiological Research* 163, 523–530.

Spadoni, A., Guidarelli, M., Phillips, J., Mari, M. and Wisniewski, M. 2015. Transcriptional profiling of apple fruit in response to heat treatment: Involvement of a defense response during *Penicillium expansum* infection. *Postharvest Biology and Technology* 101, 37–48.

Spolti, P., Valdebenito-Sanhueza, R.M. and Del Ponte, E.M. 2010. Semi-selective medium for recovery and quantification of *Cryptosporiopsis perennans* from apple fruits. *Ciencia Rural* 40, 661–665. (In Portuguese with English abstract).

Spotts, R.A. 1985a. Effect of preharvest pear fruit maturity on decay resistance. *Plant Disease* 69, 388–390.

Spotts, R.A. 1985b. Environmental-factors affecting conidial survival of five pear decay fungi. *Plant Disease* 69, 391–392.

Spotts, R.A. 1990. Moldy core and core rot. pp. 29–30. *In*: A.L. Jones and H.S. Aldwinckle (eds.). *Compendium of Apple and Pear Diseases.* The American Phytopathological Society Press, St. Paul, MN.

Spotts, R.A., Holmes, R.J. and Washington, W.S. 1988. Factors affecting wet core rot of apples. *Australasian Plant Pathology* 17, 53–57.

Spotts, R.A., Seifert, K.A., Wallis, K.M., Sugar, D., Xiao, C.L., Serdani, M. and Henriquez, J.L. 2009. Description of *Cryptosporiopsis kienholzii* and species profiles of *Neofabraea* in major pome fruit growing districts in the Pacific Northwest USA. *Mycological Research* 113, 1301–1311.

Spotts, R.A., Sholberg, P.L., Randall, P., Serdani, M. and Chen, P.M. 2007. Effects of 1-MCP and hexanal on decay of d'Anjou pear fruit in long-term cold storage. *Postharvest Biology and Technology* 44, 101–106.

Stensvand, A., Talgo, V., Blystad, D.R. and Sletten, A. 2012. Problematic root, stem and shoot diseases on apple in Norway. *IOBC/WPRS Bulletin* 84, 109–111.

Struble, F.B. and Keitt, G. 1950. Variability in inheritance in *Glomerella cingulata* (Stonem.) S. and V.S. from apple. *American Journal of Botany* 37, 563–576.

Sugar, D. 2014. Side rot. pp. 81–82. *In*: T.B. Sutton, H.S. Aldwinckle, A.M. Agnello and J.F. Walgenbach (eds.). *Compendium of Apple and Pear Diseases and Pests,* 2nd ed. The American Phytopathological Society, St. Paul, MN.

Sugar, D. and Powers, K. 1986. Interactions among fungi causing postharvest decay of pear. *Plant Disease* 70, 1132–1134.

Sugar, D. and Spotts, R.A. 1992. Sources of inoculum of *Phialophora malorum*, causal agent of side rot of pear. *Phytopathology* 82, 735–738.

Sugar, D. and Spotts, R.A. 1993a. Dispersal of inoculum of *Phialophora malorum* in pear orchards and inoculum redistribution in pear immersion tanks. *Plant Disease* 77, 47–49.

Sugar, D. and Spotts, R.A. 1993b. The importance of wounds in infection of pear fruit by *Phialophora malorum* and the role of hydrostatic pressure in spore penetration of wounds. *Phytopathology* 83, 1083–1086.

Sundelin, T., Stromeng, G.M., Gjaerum, H.B., Amby, D.B., Orstad, K., Jensen, B., Lund, O.S. and Stensvand, A. 2015. A revision of the history of the *Colletotrichum acutatum* species complex in the Nordic countries based on herbarium specimens. *FEMS Microbiology Letters* 362, fnv130.

Sutton, T.B. 1990. Bitter rot. p. 20. *In*: A.L. Jones and H.S. Aldwinckle (eds.). *Compendium of Apple and Pear Diseases.* The American Phytopathological Society, St. Paul, MN.

Sutton, T.B., Aldwinckel, H.S., Agnello, A.M. and Walgenbach, J.F. 2014. *Compendium of Apple and Pear Diseases and Pests,* 2nd ed. The American Phytopathological Society, St. Paul, MN.

Sutton, T.B. and Shane, W.W. 1983. Epidemiology of the perfect stage of *Glomerella cingulata* on apples. *Phytopathology* 73, 1179–1183.

Tahir, I.I., Johansson, E. and Olsson, M.E. 2009. Improvement of apple quality and storability by a combination of heat treatment and controlled atmosphere storage. *HortScience* 44, 1648–1654.

Talgo, V., Aamot, H.U., Stromeng, G.M., Klemsdal, S.S. and Stensvand, A. 2007. *Glomerella acutata* on highbush blueberry (*Vaccinium corymbosum* L.) in Norway. *Plant Health Progress*. Available at: www.plantmanagementnetwork.org/pub/php/research/2007/blueberry/Gacutata.pdf

Taylor, J. 1971. Epidemiology and symptomology of apple bitter rot. *Phytopathology* 61, 1028–1029.

Terblanche, J.H., Gurgen, K.H. and Hesebeck, J. 1980. An integrated approach to orchard nutrition and bitter pit control. pp. 71–82. *In*: D. Atkinson, J. E. Jackson, R.O. Sharples and W.M. Waller (eds.). *Mineral Nutrition of Fruit Trees: Studies in the Agricultural and Food Sciences*. Butterworths, Sevenoaks, UK.

Tian, S.P. and Bertolini, P. 1999. Effect of temperature during conidial formation of *Monilinia laxa* on conidial size, germination and infection of stored nectarines. *Journal of Phytopathology* 147, 635–641.

Tiller, L.W. 1929. Cold storage of fruit. *Bulletin of the New Zealand Department of Industrial Research* No. 16. Wellington, New Zealand.

Tsushima, Y. and Yukita, K. 2002. Isolates of apple bitter rot fungi collected from apple orchards in Aomori. *Annual Report of the Society of Plant Protection of North Japan* 53, 118–121.

Uhlig, S., Jestoi, M. and Parikka, P. 2007. *Fusarium avenaceum* – The North European situation. *International Journal of Food Microbiology* 119, 17–24.

Usall, J., Torres, R., Viñas, I., Abadias, M. and Teixidó, N. 2013. Principales enfermedades de poscosecha y su control. pp. 253–255. *In*: I. Viñas, J. Usall and J. Graell (eds.). *Poscosecha de Pera, Manzana y Melocotón*. Mundi-Prensa, Madrid, Spain. (In Spanish).

US FDA (United States Food and Drug Administration). 2001. Guidance for Industry: Juice HACCP; Small Entity Compliance Guide. U. S. Regulatory Requirements for Irradiating Foods. Available at: www.fda.gov/Food/GuidanceRegulation/GuidanceDocumentsRegulatoryInformation/Juice/ucm072637.htm

Val, J., Gracia, M.A., Monge, E. and Blanco, A. 2008. Visual detection of calcium by GBHA staining in bitter pit-affected apples. *Food Science and Technology International* 14, 315–319.

Van Brouwershaven, I.R., Bruil, M.L., van Leeuwen, G.C.M. and Kox, L.F.F. 2010. A real-time (TaqMan) PCR assay to differentiate *Monilinia fructicola* from other brown rot fungi of fruit crops. *Plant Pathology* 59, 548–555.

Van der Walt, L. 2009. Characterisation of mites and *Penicillium* species associated with apple core rot diseases. M.Sc. Thesis, Stellenbosch University, Stellenbosch, South Africa.

Van der Walt, L., Spotts, R.A., Visagie, C.M., Jacobs, K., Smit, F.J. and McLeod, A. 2010. *Penicillium* species associated with preharvest wet core rot in South Africa and their pathogenicity on apple. *Plant Disease* 94, 666–675.

Vasic, M., Duduk, N., Vico, I., Rancic, D., Pajic, V. and Backhouse, D. 2016. Comparative study of *Monilinia fructigena* and *Monilia polystroma* on morphological features, RFLP analysis, pathogenicity and histopathology. *European Journal of Plant Pathology* 144, 15–30.

Verkley, G.J.M. 1999. A monograph of the genus *Pezicula* and its anamorphs. *Studies in Mycology* 44, 1–176.

Vero, S., Mondino, P., Burgueno, J., Soubes, M. and Wisniewski, M. 2002. Characterization of biocontrol activity of two yeast strains from Uruguay against blue mold of apple. *Postharvest Biology and Technology* 26, 91–98.

Vichova, J., Stankova, B. and Pokorny, R. 2012. First report of *Colletotrichum acutatum* on tomato and apple fruits in the Czech Republic. *Plant Disease* 96, 769–770.

Vigier, B., Reid, L.M., Seifert, K.A., Stewart, D.W. and Hamilton, R.I. 1997. Distribution and prediction of *Fusarium* species associated with maize ear rot in Ontario. *Canadian Journal of Plant Pathology* 19, 60–65.

Viñas, I., Usall, J., Teixidó, N. and Sanchis, V. 1998. Biological control of major postharvest pathogens on apple with. *Candida Sake. International Journal of Food Microbiology* 40, 9–16.

Volkova, J. and Juhnevica-Radenkova, K. 2015. Bitter rot of apple: Different causal agents, two diseases. pp. 149–152. *In*: Proceedings of the Zinatniski Praktiska Konference Lidzsvarota Lauksaimnieciba, Jelgava, Latvia.

Wang, Z.Y. and Dilley, D.R. 2000. Initial low oxygen stress controls superficial scald of apples. *Postharvest Biology and Technology* 18, 201–213.

Watkins, C.B., Nock, J.F., Weis, S.A., Jayanty, S. and Beaudry, R.M. 2004. Storage temperature, diphenylamine, and pre-storage delay effects on soft scald, soggy breakdown and bitter pit of 'Honeycrisp' apples. *Postharvest Biology and Technology* 32, 213–221.

Weber, R.W.S. and Zabel, D. 2011. White haze and scarf skin, two little-known cosmetic defects of apples in Northern Germany. *European Journal of Horticultural Science* 76, 45–50.

Weir, B.S., Johnston, P.R. and Damm, U. 2012. The *Colletotrichum gloeosporioides* species complex. *Studies in Mycology* 73, 115–180.

Wenneker, M., Kohl, J., van Leeuwen, P., Pham, K. and van Schaik, A. 2016a. Control of postharvest storage rots of apples and pears in The Netherlands. *Acta Horticulturae* 1144, 189–194.

Wenneker, M., Pham, K.T.K., Lemmers, M.E.C., Boer, F.A.D., Lans, A.M.V.D., Leeuwen, P.J.V., Hollinger, T.C. and Thomma, B.P.H.J. 2016b. First report of *Fusarium avenaceum* causing wet core rot of 'Elstar' apples in The Netherlands. *Plant Disease* 100, 1501–1502.

Wenneker, M., Pham, K.T.K., Lemmers, M.E.C., Boer, F.A.D., Leeuwen, P.J.V., Hollinger, T.C., Groenenboom-de Haas, B.H. and Kohl, J. 2016c. First report of *Cadophora luteo-olivacea* causing side rot on 'Conference' pears in The Netherlands. *Plant Disease* 100, 2162.

Wilkinson, B.G. and Fidler, J.C. 1973. Injuries to the skin of the fruit. pp. 67–80. *In*: J. C. Fidler, B.G. Wilkinson, K.L. Edney and R.O. Sharples (eds.). *The Biology of Apple and Pear Storage*. Commonwealth Agricultural Bureaux, London, UK.

Wilson, L.L., Madden, L.V. and Ellis, M.A. 1992. Overwinter survival of *Colletotrichum acutatum* in infected strawberry fruit in Ohio. *Plant Disease* 76, 948–950.

Wojcik, P., Rutkowski, K. and Treder, W. 2001. Quality and storability of 'Gala' apples as affected by crop load. *Folia Horticulturae* 13, 89–96.

Wong, J. 1996. *Control of Core Rots in Red Delicious Apples*. Horticulture Research and Development Corporation, Report AP011. Gordon, NSW, Australia.

Xiao, C.L. and Kim, Y.K. 2008. Postharvest fruit rots in apples caused by *Botrytis cinerea, Phacidiopycnis washingtonensis*, and *Sphaeropsis pyriputrescens*. *Plant Health Progress*. Available at: www.plantmanagementnetwork.org/pub/php/diagnosticguide/2008/apple/apple.pdf

Xu, X.M. and Robinson, J.D. 2000. Epidemiology of brown rot (*Monilinia fructigena*) on apple: Infection of fruits by conidia. *Plant Pathology* 49, 201–206.

Xu, X.M., Robinson, J.D., Berrie, A.M. and Harris, D.C. 2001. Spatio-temporal dynamics of brown rot (*Monilinia fructigena*) on apple and pear. *Plant Pathology* 50, 569–578.

Yin, L., Zou, Y., Li, M., Ke, X., Li, C., Liang, D. and Ma, F. 2013. Resistance of *Malus* plants to Diplocarpon mali infection is associated with the antioxidant system and defense signaling pathways. *Physiological and Molecular Plant Pathology* 84, 146–152.

Zhang, R., Wang, S., Cui, J. and Sun, G. 2009. Identification of pathogens causing apple fruit bitter rot in Shaanxi and He'nan provinces. *Scientia Agricultura Sinica* 42, 3224–3229. (In Chinese with English abstract).

Zhang, R., Wang, S.F., Cui, J.Q., Sun, G.Y. and Gleason, M.L. 2008. First report of bitter rot caused by *Colletotrichum acutatum* on apple in China. *Plant Disease* 92, 1474.

Chapter 3

Stone Fruits

Marta Mari

CRIOF, Department of Agricultural Science, University of Bologna, Cadriano, Bologna, Italy

Davide Spadaro

Department of Agricultural, Forestry and Food Sciences (DISAFA) and Agroinnova, Centre of Competence for the Innovation in the Agroenvironmental Sector, University of Torino, Largo Braccini, Grugliasco, Torino, Italy

Carla Casals

Institute of Agrifood Research and Technology (IRTA), XaRTA-Postharvest, Edifici Fruitcentre, Parc Científic i Tecnològic Agroalimentari de Lleida, Parc de Gardeny, Lleida, Catalonia

Marina Collina

Department of Agricultural Science, University of Bologna, Viale Fanin, Bologna, Italy

Antonieta De Cal

National Institute for Agricultural and Food Research and Technology (INIA), Madrid, Spain

Josep Usall

Institute of Agrifood Research and Technology (IRTA), XaRTA-Postharvest, Edifici Fruitcentre, Parc Científic i Tecnològic Agroalimentari de Lleida, Parc de Gardeny, Lleida, Catalonia

POSTHARVEST PATHOLOGY

Abbreviations

ACC	1-aminocyclopropane-1-carboxylic acid
ACO	ACC-oxidase
CA	Controlled atmosphere
CI	Chilling injury
CYA	Czapek yeast extract agar
EPPO	European and Mediterranean Plant Protection Organization
ITS	Internal transcribed spacer
MEA	Malt extract agar
PDA	Potato dextrose agar
1-MCP	1-methylcyclopropene
PG	Polygalacturonase
USA	United States of America

1 General Introduction

About 40 million tons of stone fruits are produced annually worldwide, from an area of 5 million hectares. The main economically important cultivated members of the genus *Prunus* are peach, nectarine, plum, cherry, and apricot. In 2016, China produced over 50% of the world's peaches and plums, while the United States of America (USA), Italy, Spain, and Greece produced over 17 million tons of stone fruits (FAOSTAT, 2018). Stone fruits are revered as delicious and healthy summer fruits in most temperate regions of the world; however, in many countries, their consumption has remained relatively static over the last 10–20 yr (USDA FAS, 2014). Stone fruits are characterized by a relatively short postharvest life, passing quickly from ideal ripeness to overmature fruit, depending in part on temperature and handling practices (Kader and Mitchell, 1989). Moreover, fruit maturity at harvest will determine their postharvest life: advanced maturity at harvest increases fruit susceptibility to decay and reduces the shelf life, while the fruit of minimal maturity at harvest can have less flavor and excessive firmness, harming their consumer acceptance.

The fruit losses after harvest are mainly due to biotic diseases caused by fungal pathogens such as *Monilinia* spp., *Rhizopus* spp., *Mucor* spp., and *Botrytis cinerea* Pers.; however, other minor pathogens can cause disease including *Geotrichum candidum* Link, *Penicillium* spp., *Alternaria* spp., and *Cladosporium* spp. Furthermore, the storage of fruit at low temperature (2–7°C) can favor the appearance of physiological disorders collectively termed chilling injuries (CIs) (Crisosto et al., 1999).

2 Main Diseases

2.1 Brown Rot (*Monilinia spp.*)

Brown rot caused by *Monilinia* spp. is the most economically important fungal disease of stone fruits, causing substantial preharvest and postharvest losses (Byrde and Willetts, 1977). Brown rot is present in all stone fruit areas of the world (De Cal and Melgarejo, 2000). Direct yield losses result from infection of flowers (termed flower and twig blight) and fruits. Fruits can rot in the field, at harvest, and after harvest (Villarino et al., 2012). Fruits can be infected at any time but their susceptibility to infection increases with maturation, partly due to decreased content of chlorogenic acid and its isomer, neochlorogenic acid, in ripe fruit near harvest (Villarino et al., 2011). However, postharvest losses, which routinely occur during storage and transport, are typically more severe than preharvest losses (Hong et al., 1997) and have a greater economic impact on producers, since fruit losses at the consumer level can negatively affect consumer satisfaction and their propensity to consume the fruit in the future. Brown rot incidence may be as high as 80% when the climatic conditions are favorable for disease development, especially in the case of late varieties (Larena et al., 2005) and in commercial organic orchards, despite the use of copper-based fungicides.

Brown rot occurs in peach (*Prunus persica* [L.] Batsch), nectarine (*Prunus persica* var. *nectarina* Maxim), cherry (*Prunus avium* L.), plum (*Prunus domestica* L.), and apricot (*Prunus armeniaca* L.). It can be caused at least by four *Monilinia* species: *M. laxa* (Aderh. & Ruhland) Honey, *M. fructicola* (G. Winter) Honey, *M. fructigena* (Aderh. & Ruhland) Honey (Byrde and Willetts, 1977), and *M. polystroma* (G.C.M. Leeuwen) Kohn (Martini et al., 2014). *Monilinia fructicola* is classified by the European and Mediterranean Plant Protection Organization (EPPO) as an A2 quarantine pest (EPPO, 2009), because it has only been identified in several EPPO member countries. Furthermore, the relative frequency of *M. fructicola* has increased over the years to coexist at the same level with *M. laxa*, and it has partially displaced *M. fructigena* as a pathogen on stone fruits (Villarino et al., 2013; Papavasileiou et al., 2015). *Monilinia fructicola* is the most virulent of these species, causing larger fruit lesions and shorter periods of both incubation and latency (Villarino et al., 2016).

Brown rot symptoms are blossom and twig blight and fruit rot (Figure 3.1). Flower blight begins in spring from infection of the anthers of the flower, which continue to the pistil and the ovary. Branches and twigs can also be infected. From the outbreaks on infected branches, elliptical or fusiform cankers occur, characterized by massive production of gums and abundant sporulation. Fruit can be infected at any time during their development, but normally large lesions do not develop until the fruit begins to mature.

Figure 3.1 Brown rot caused by *Monilinia fructicola* on peach.

Traditionally, *Monilinia* spp. were differentiated based on morphological and cultural traits, which require at least 9–10 d (De Cal and Melgarejo, 1999). Molecular techniques could overcome many of the shortcomings of the conventional assays, especially polymerase chain reaction (PCR) (Côté et al., 2004; Gell et al., 2007). The analytical sensitivity (detection limit) of these conventional PCR assays is sufficiently low when using pure cultures of *Monilinia* spp., but is too high for routine detection on fruit samples. Recently, some quantitative PCR(qPCR)-based methods were reported that detected visible and latent brown rot infections in the blossoms and fruit of nectarine trees and could distinguish the causal *Monilinia* spp. (García-Benítez et al., 2017).

Monilinia mycelium in culture is hyaline initially, later developing dark and irregular stromata, which contains abundant melanin (De Cal and Melgarejo, 1994). Macroconidia are produced in moniliform chains that are single or dichotomously branched (Byrde and Willetts, 1977; Mordue, 1979a, 1979b, 1979c) (Figure 3.2). These macroconidia are unicellular, hyaline, and limonate, and their dimensions (11.5–21 × 8–13 mm) depend on the temperature and culture medium (Biggs and Northover, 1985). Microconidia, 2 mm in diameter, are produced on phialids, which are found in many cases in fruiting bodies similar to pycnidia (pseudopycnidia). Microconidia are produced in culture media and on mummies. The microconidia do not germinate; it seems that they have a spermatid function for the formation of apothecia. The apothecia are orange, with the shape of a cup, and 5–20 mm in diameter. The asci are 102–215 × 3–13 mm and the unicellular and ovoid ascospores are 6–15 × 4–8 mm. However, sexual reproduction and vegetative compatibility groups have only been reported for *M. fructicola* (De Cal et al., 2014), which suggests that the genetic variability of *M. fructicola* could also be maintained by sexual and/or parasexual recombination.

The infection process begins when *Monilinia* conidia germinate on flower or fruit surfaces to produce germ tubes and/or appressoria, which then penetrate the surface depending on the prevailing environmental conditions (Lee and Bostock,

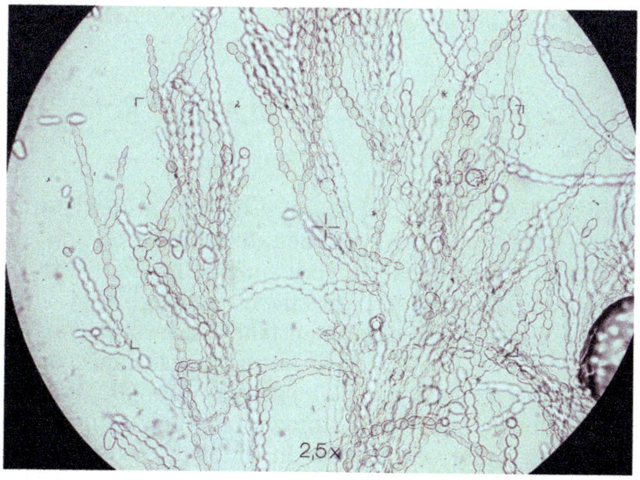

Figure 3.2 Macroconidia of *Monilinia fructicola* on single or dichotomically branched moniliform chains under a microscope (×25).

2006; García-Benítez et al., 2016). *Monilinia* colonizes the epidermis and mesocarp of fruit with thin and thick hyphae, causing the collapse and disruption of epidermal and mesocarpic cells, lysogenic cavities in the subepidermis and mesocarp, and total degradation of the cuticle and epidermis (García-Benítez et al., 2016). When climatic conditions are unfavorable, fruit infections can remain dormant until the season and fruit maturity advance, keeping the pathogen inactive from early spring until a later optimal time for disease development (Byrde and Willetts, 1977; García-Benítez et al., 2016). It has been shown that there is a positive correlation between the number of latent infections of *Monilinia* spp. and the incidence of postharvest brown rot (Emery et al., 2000; Luo and Michailides, 2001; Gell et al., 2008). The incidence of postharvest rot can also be predicted from time of the first pathogenic conidia observed on the fruit surface or in the air and the occurrence of the first latent infection (Villarino et al., 2012). Two fundamental phases were distinguished in the development of peach brown rot caused by any of *Monilinia* species: flower blight and fruit brown rot. Primary inoculum was the mycelium on mummies found in the tree or soil, or on infected shoots and branches (Gell et al., 2009; Villarino et al., 2010; Casals et al., 2015), or ascospores from apothecia in the case of *M. fructicola* (Biggs and Northover, 1985). A positive correlation was reported between the number of mummies found in trees and postharvest brown rot (Villarino et al., 2010). The secondary dispersion is due to a new production of conidia on infected flowers, twig blights, cankers, thinning fruits, or rotten fruits. Conidia are dispersed by air, water, or insects. A correlation was observed between the number of *Monilinia* conidia per m^3 of air and the number of mummies and branches infected with the pathogen present in the orchard soil (Villarino et al., 2010).

The climatic conditions are critical to disease development, with the temperature and period of wetness being the factors that most influence penetration and infection by *Monilinia* spp. (Gell et al., 2008; Villarino et al., 2012). In the case of infection of immature fruit, which generally leads to infections that remain dormant, these two climatic factors explain more than 90% of their incidence (Gell et al., 2008), with the wetness period more influential than temperature. Temperatures of 15–25°C favor the development of the fungus, although the infection can take place between 5 and 30°C (Tamm and Flückiger, 1993; Hong et al., 1997).

2.2 *Rhizopus Rot (*Rhizopus *spp.) and Mucor Rot (Mucor piriformis)*

Rhizopus stolonifer (Ehrenb.) Vuill. and *Mucor piriformis* A. Fisch. are both members of the Mucoraceae family, and the symptoms they cause on stone fruits, their life cycles, and their control strategies are very similar. These diseases are commonly called Rhizopus rot and Mucor rot, respectively, although they are also known as soft rots. Both pathogens cause very important diseases on stone fruits in many regions, and they could be the second in terms of causing economic losses, just after brown rot. Losses caused by *Rhizopus* spp. and *M. piriformis* are mostly observed in packinghouses after cold storage, although sometimes they occur in the field, especially after long wet periods. In addition to stone fruits, they can also infect many fruits and vegetables, including potato, eggplant, watermelon, melon cucumber, sweet potato, grape, strawberry, raspberry, blackberry, apple, and pear (Pitt and Hocking, 2009).

When *Rhizopus* spp. or *M. piriformis* grow on stone fruit, they cause circular, brown, soft lesions that rapidly extend to the entire fruit, causing internal decomposition and a liquid exudation. After the initial stages of infection, a white filament mass crowned by globular sporangia soon appears; the sporangia then become black in color and readily visible to the naked eye. However, a microscopic examination is needed to determine if the disease is caused by *Rhizopus* spp. or *Mucor* spp., since only *Rhizopus* spp. produce rhizoids.

Conidia typically survive on plant material that resides on the soil surface in stone fruit orchards, although the presence of conidia in the air is common. These conidia can be easily deposited on healthy fruits present both in the tree and the ground. It was thought that *Rhizopus* spp. were strictly wound parasites, since they typically penetrate the fruit tissue through fresh wounds or microwounds before successful infection. However, recent studies indicated that *R. stolonifer* can penetrate directly into unwounded stone fruits, probably due to the production of esterase enzymes (Baggio et al., 2016). On the intact fruit surface, conidia require external nutrient sources to germinate, and when external nutrients are not available, the pathogen is unable to produce the specific infection enzymes. The incidence of Rhizopus or Mucor rot in the field is generally low, and they primarily infect fruits on the ground and rarely those in trees (Figure 3.3). Disease occurs mainly after harvest on ripe fruits with high sugar contents. Moreover, fruit susceptibility depends on mechanical injuries that occur during harvest and marketing. During the postharvest period, conidia of these fungi can also survive when adhered on wood packing boxes used at harvest or even in the atmosphere of cold storage rooms. This implies that boxes and postharvest facilities, especially cold storage rooms, must be rigorously cleaned and disinfected before the storage of new fruits, especially if Rhizopus or Mucor rots were detected after storage. Water used in dump tanks must be disinfected to avoid new infections during fruit cleaning processes. If harvested fruits are stored under favorable conditions for disease development, the disease will spread rapidly from infected fruits to healthy adjacent fruits causing nests of rotten fruits, since fruit flesh disintegrates and

Figure 3.3 Fruit located on the tree affected by *Mucor* sp.

produces juices with degrading enzymes. Although it has been reported that soft rots spread rapidly at temperatures higher than 5°C, Mucor rot could also develop at 0°C, the standard temperature used for stone fruit cold storage.

On Czapek yeast extract agar (CYA), colonies of *R. stolonifer* are low and sparse and rapidly cover the whole Petri dish at temperatures of 20–25°C. Sporangia are up to 3 mm long with brown walls and are borne in groups of three to five from clusters of rhizoids and unbranched, robust stipes. The columella is roughly spherical and up to 200 µm in diameter. The size of the sporangiospore is 8–20 µm in long axis, color is pale brownish, and walls are striated (Pitt and Hocking, 2009). Most studies indicate that *R. stolonifer* is able to grow from 4.5–5 to 37°C and that the optimum temperature is around 25°C (Pitt and Hocking, 2009).

The colonies of *M. piriformis* grown on CYA are low and sparse, spreading across the Petri dish. The mycelium is colorless, but the overall color of the colony with developing sporangia is buff or pale yellow brown. On malt extract agar (MEA), colonies of 35–60 mm in diameter are discrete and gray or brownish in color. The fungus does not grow at 37°C or higher. Sporangiospores on CYA borne from the hyphae surface. They have short stipes, are broad, often are sympodially branched, and have encrusted walls. The sporangium is up to 150 µm in diameter, with spinulose walls. The columella is spheroidal to short cylindrical in shape, with diameter or length of 25–80 µm. Chlamydoconidia are uncommon and zygospores are not formed in pure culture (Pitt and Hocking, 2009). *Mucor piriformis* is classified as psychrophilic, with minimum and maximum growth temperatures of about 0 and 26°C, respectively. The optimum temperature is 20–21°C (Michailides and Spotts, 1990).

*2.3 Gray Mold (*Botrytis cinerea*)*

Gray mold, caused by *B. cinerea,* is an important postharvest disease of stone fruits in all producing areas. It usually affects only 1–2% of fruits, but under conducive conditions and especially on particularly susceptible cultivars, the magnitude of losses may be much higher (Fourie and Holz, 1994). Epidemics are sporadic, but have been reported to reduce yields by 70% or more in California (Ogawa et al., 1995), Greece (Thomidis et al., 2009), China (Wang and Wang, 2001), and South Africa (Fourie and Holz, 1995). In South Africa, it is considered to be the most important postharvest decay of stone fruits (Fourie et al., 2002). Moreover, green fruit rot, caused by *B. cinerea,* has been found in Japan, Italy, France, South Africa, Australia, New Zealand, and the USA (Fourie and Holz, 1995).

Initially, pale brown circular spots develop on the surface of the infected fruit. Later, the spots become darker and larger and, under conditions of high relative humidity (RH), become covered with a white mold. At low RH, gray–brown conidiophores and conidia are produced. Infected flesh is soft. At early infection stages, symptoms of gray mold are similar to those of brown rot, and inexperienced observers can confuse the two.

The pathogen can be easily identified based on morphological and cultural characteristics (Hennebert, 1973). Sclerotia size and form and conidium size are useful in delimiting some species of *Botrytis,* but many species are morphologically similar and growing conditions significantly influence these variations (Beever and Weeds, 2004). *Botrytis cinerea* forms grayish brown, branched conidiophores with a swollen basal cell. Oval or elliptical conidia are attached to the ampullae by a sterigma. Flat sclerotia adhering to the substrate are produced. The apothecial disk of the teleomorph is 1–5 mm in diameter.

DNA-based molecular identification has been recently used to define *Botrytis* species. Phylogenetic analyses using the sequences of the nuclear genes glyceraldehyde-3-phosphate dehydrogenase, heat-shock protein 60, and DNA-dependent RNA polymerase subunit II have been used to corroborate the identification of *Botrytis* species (Lorenzini and Zapparoli, 2014). The necrosis and ethylene-inducing protein (NEP) genes 1 and 2 could be used to design specific primers to identify some species of *Botrytis* (Staats et al., 2005).

Botrytis cinerea overwinters as sclerotia in the soil or as mycelium in plant debris. Conidia are dispersed by wind and rain and can infect senescent blossoms, later resulting in gray mold when the fruit ripens. The fungus can also cause green fruit rot on blossoms and immature fruit (Ogawa et al., 1995). Conidia can also infect through wounds during harvest, handling, and storage. Gray mold can become prevalent on fruits after prolonged cold storage. The appearance of infections assumes different morphologies depending on the degree of tissue hydration and RH (Fourie and Holz, 1995). The prevalence and severity of gray mold increase on stressed fruits, such as on hot water-treated peaches and nectarines (Smith and Anderson, 1975). Decay can also diffuse from a diseased fruit to an adjacent healthy fruit, resulting in nesting.

High humidity and rain during blossom time can favor the development of *B. cinerea*. Sclerotia germinate over a wide range of temperatures to form conidia and, in particular conditions, apothecia. Under optimal conditions (15–20°C, high humidity, and light), conidia can form in less than 8 hr, and are mainly released during the day. Several weeds can become infected during the early season and may contribute inoculum of *B. cinerea* in stone fruit orchards, increasing the prevalence of gray mold on early maturing cultivars (Fourie et al., 2002). Gray mold can occur during storage if the fruit has been contaminated through harvest and handling wounds. Fruit susceptibility increases if storage exceeds 3–4 wk, particularly on late season cultivars. The processes of germination and infection are highly dependent on humidity and wetness. In apricot, during wet spring weather, the fruit may suffer blossom blight and rotting of young fruit. The pathogen develops on immature fruits and at low temperature, as low as −4°C, and may also cause large economic losses.

3 Minor Diseases

3.1 Sour Rot (*Geotrichum candidum*)

Sour rot occurs in many fresh market horticultural products, such as cantaloupe (Wade and Morris, 1982), tomato (Eckert and Ogawa, 1988), carrot (Horita and Hatta, 2016), loquat (Hafeez et al., 2015), and other fruits (Pitt and Hocking, 2009). On stone fruits, the disease is a problem primarily in humid production regions (Eckert and Ogawa, 1988). Detected for the first time in California in peaches and nectarines (Burton and Wright, 1969; Wells, 1977), the incidence of sour rot has increased in recent years (Michailides et al., 2004). Although California region is characterized by an atypical and unfavorable climate for sour rot, the irrigation practices can lead to high humidity promoting the pathogen development (Yaghmour et al., 2012). Fruit losses are generally low when the disease develops prior to harvest, but they can be very high during storage, transport, and market display (Adaskaveg et al., 2007). Amorim et al. (2008) investigated the incidence of postharvest diseases in the wholesale market of Sao Paulo (Brazil) and found that 0.5% of assessed stone fruits was infected by *G. candidum*. Diseases similar to sour rot may be caused by other yeasts and fungi that have not been characterized (Michailides et al., 2004). No data are available on the incidence of sour rot in European and Asian stone fruit production areas.

Initially, a dark brown, wet discoloration of skin appears on the fruit surface that quickly develops into a wet, mushy decay of the entire fruit. A white mycelium covers the lesions, frequently accompanied by a creamy exudate (Wells, 1977). Decay can penetrate to the pit and completely degrade the fruit (Yaghmour et al., 2012). Fruits infected by *G. candidum* are easily recognized since they have a characteristic odor, ranging from yeasty to vinegary, which gives the common name to the disease. When the decay lesions are originated close to the stylar end, the juice stream from them dissolves the cuticle, epidermis, and some of the flesh, creating distinct furrows in the tissue (Michailides et al., 2004).

Geotrichum candidum is a dimorphic fungus that grows both by the production of hyphae that reproduce by arthrospores, and from single cells that reproduce by budding. It is currently considered to be a true yeast (Wouters et al., 2002). On potato dextrose agar (PDA), at 20°C, colonies of *G. candidum* grow up to 30 mm in diameter within 72 hr; they are white, smooth with a fruit odor, and the reverse is pale yellow. On Leonian's agar (Booth, 1971), the growth is more rapid than on PDA, the colonies appear very thin, white, and without reverse coloration. Hyphae rapidly disarticulate arthrospores (2.5–11 × 3–3.5 μm) and the large aseptate or sparingly septate hyphae abundant among disarticulating hyphae can be observed. The pathogen is a yeast that is widely present in numerous environments and hosts, including dairy products and humans. Molecular techniques are routinely used in clinical laboratories for its identification: sequence analysis of universal barcoding markers, such as internal transcribed spacer regions (ITS 1 and ITS 2), large subunit (LSU) D1 and D2 domains of the ribosomal DNA (rDNA), and more recently matrix-assisted laser desorption/ionization – time of flight mass spectrophotometry (MALDI-TOF MS) (Bader et al., 2011).

The soil is the main source of *G. candidum* inoculum in stone fruit orchards. The pathogen living in the soil showed a high variation in the propagule density, often positively correlated with the number of fallen fruits that remain on the ground until harvest (Yaghmour et al., 2012). Moreover, leaves and fruit surfaces can be contaminated in the orchard by the propagules that can easily reach the lowest parts of the trees from the soil. Important vectors of sour rot are nitidulid beetles (*Carpophilus* spp.) that transmit *G. candidum* from decayed fruits to wounded and unwounded fruits. The visitation of diseased fruits on the trees or orchard floor by insects provides a pathway for the spread of the arthrospores, which are mainly carried on the surface of the body of the insect to healthy fruits on trees. When fruits are transported to the packinghouse, pathogen propagules present in soil debris and on the fruit surface can contaminate the packingline and infect healthy fruits. The pathogen can infect firm, immature fruits, especially those with wounds, although it grows slowly until the fruit ripens. Infections in mature, soft fruits develop very rapidly.

3.2 *Blue Mold (*Penicillium *spp.)*

Blue mold of stone fruits is a minor disease, occurring in different countries (Smith and Anderson, 1975; Morales, 1984). Only in senescent fruits, particularly plums and cherries, stored for long periods, the incidence exceeds 10%. Blue mold is less common on apricots, peaches, and nectarines. Although in general, the incidence of blue mold is low, this disease comprises a health risk to consumers because *Penicillium expansum* Link. can produce the mycotoxin patulin that can be found in juices made from infected apricot and peach fruits (Spadaro et al., 2009).

Blue mold lesions are circular, soft, and watery. Mycelium growing from lesions is white initially, but becomes blue when the pathogen sporulates. On cherries and plums, infection causes small cracks in the skin to appear, while on peaches and nectarines, pale brown spots occur that are accompanied by pulp softening. Later, a dense mass of conidia forms on the surface of lesions on peaches, nectarines, and apricots. On plums and cherries, sporulation on lesions occurs in concentric circles around the point of infection (Figure 3.4).

The *Penicillium* genus contains 354 accepted species. The polyphasic species concept adopted for *Penicillium* spp. includes morphology, multigene phylogenetics, and extrolite profiling. The sequence of β-tubulin (*BenA*) is one of the best options as an identification marker for *Penicillium* spp. Other sequences that could help in the identification are ITS, calmodulin, and RNA polymerase II (RPB2) (Visagie et al., 2014). Together with the presence of blue mold, a musty odor on the fruit can help identify *Penicillium* spp.

Blue mold is mainly caused by *P. expansum* (Pitt and Hocking, 2009), but other *Penicillium* spp. can also cause it. Some studies reported of *Penicillium crustosum* Thom infected peach (Restuccia et al., 2006), *Penicillium chrysogenum* Thom infected black plum (Eseigbe and Bankole, 1996), and *Penicillium digitatum* (Pers.) Sacc. infected nectarine (Navarro et al., 2011). The pathogenicity of these four *Penicillium* spp. on plum was also reported (Louw and Korsten, 2016).

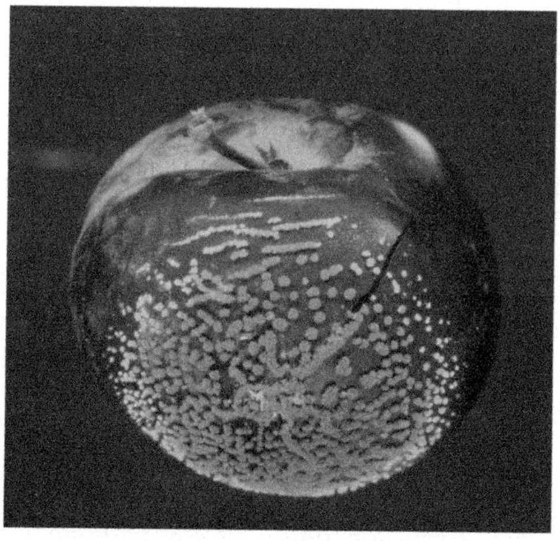

Figure 3.4 Blue mold caused by *Penicillium* spp. on plum.

Abundant conidia of *P. expansum* can be produced on various fruits, other organic substrates, or in the soil. The pathogen overwinters on dead plant material. Airborne conidia can contaminate surfaces of fruit-handling equipment. Conidia must be deposited within wounds to establish infections. *Penicillium expansum* can spread by contact infections from infected to healthy fruits to form nests of rotten fruits within containers, but the short storage period of stone fruits usually limits the development of nesting.

Blue mold occurs in peach, nectarine, and plum after one or more weeks of storage at 0°C. The minimum temperature for the growth of *P. expansum* is −3 to −2°C. Even after several weeks of storage, its incidence is usually lower than that of brown rot or gray mold. In sweet cherries, blue mold is common during storage if fruits are harvested after cold and moist weather. Cherries are particularly vulnerable since they can suffer cracking after rain, which facilitates pathogen penetration. Blue mold incidence is higher on fruits with physiological disorders and on overmature, senescent fruits. In plums, the severity of bruising has been related to the incidence of blue mold (Ceponis and Friedman, 1957).

In commercial practice, blue mold is not controlled with chemicals because its incidence is typically very low, particularly on fruits marketed just after harvest or stored for brief periods.

3.3 *Alternaria Rot* (Alternaria *spp.)*

Alternaria rot caused by *Alternaria alternata* (Fr.) Keissler occurs with some frequency on plums and cherries, while it is occasionally found on peaches, nectarines, and apricots. Alternaria rot symptoms are usually visible only after long storage. The disease has been reported in the USA (McPhee, 1980), Canada, Mexico, Italy (Tonini and Capriotti, 1996), India, China (Zhang et al., 1995), Japan (Inoue and Nasu, 2000), Argentina, Libya, and Australia. Outbreaks of Alternaria rot are increasing in the USA due to the emergence and selection of *A. alternata* strains resistant to common fungicides used in the orchard (Yang et al., 2015).

Lesions, first brown and later black, are circular, dry, firm, and slightly sunken, with a well-defined margin. Under high humidity, a mat of olive green conidia covers the lesions, which are firm and lightly sunken in most stone fruits. In some apricot cultivars, red spots have been reported on the upper layers of the fruit skin because the pathogen had died back after penetration (Larsen et al., 1980). In sweet cherries, the rot is cone-shaped extending towards the stone, while the rot is superficial in sour cherries (Snowdon, 1990). In humid years, the symptoms may occur in the orchard, but they are more frequent during storage.

Microscopic observations may be necessary to distinguish Alternaria rot caused by *A. alternata* from Cladosporium rot. *Alternaria alternata* can be easily identified by its fusiform septate conidia (20–63 × 9–17 μm) that are produced in chains. The conidial beak is short, up to one-third as long as the conidium. Conidia are formed in long, often branched chains, arise singly or in groups, and are straight or flexuous. Conidiophores are simple or branched and straight or flexuous (Simmons, 2007). Molecular identification of isolates employed sequencing ITS and a small fragment of the elongation factor 1 alpha gene (EF1a) (White et al., 1990).

Alternaria alternata is saprophytic or weakly pathogenic on many plants. The fungus overwinters on plant residues in the orchard. Wind disseminates its airborne conidia. New infections occur through microwounds, caused by surface cracks occurring during fruit expansion or by mechanical injuries at harvest and/or processing. Skin cracking or splitting of cherries in the orchard facilitate pathogen penetration. On sweet cherries, the pathogen attacks the dying tissue of aborted fruits and proceeds into healthy fruits as they mature. During storage, the pathogen infects mature fruits from infected spurs. On apricots, the pathogen can penetrate sound fruits through the stigma after bloom, later progressing into the ripening fruit or through the stomata causing small spots (Larsen et al., 1980). On apricot, the disease is common if unseasonable rain causes fruit splitting. On dark plums, *A. alternata* can colonize sunburn-injured tissues and develop before harvest, and these lesions then enlarge after harvest. *Alternaria* spp. can also produce lesions that cause a rot or moldiness of the fruit heart called "moldy heart" or "core rot" of peaches (Thomidis et al., 2007; Pose et al., 2010). Infected peaches do not show any external symptoms, but develop an internal black or gray mycelium in the pulp and stone, respectively. Infected fruits do not show apparent external symptoms in the orchard and are thus harvested. Alternaria rot may also occur on dried fruits, if they were infected before the drying process.

In spore trap studies, the concentration of *A. alternata* conidia in the air increases during and after rain or foggy conditions. Alternaria rot commonly develops close to insect feeding wounds. Since *A. alternata* grows at low temperatures, it can cause damage during refrigerated storage (Tournas and Stack, 2001). The decay is common even under storage temperatures close to 0°C, but the slow rate of growth generally limits apparent damage unless fruits are held over 2–3 wk at low temperatures.

3.4 Cladosporium Rot (Cladosporium spp.)

Cladosporium rot caused by *Cladosporium* spp. is considered a secondary disease of stone fruits. These fungi are weak pathogens infecting fruit damaged by rain and rough handling. Plums that have been shaken from the trees for harvest (Michailides et al., 1987) or sweet cherries damaged by weather conditions causing cracking on the skin are highly susceptible to this disease. The incidence of the disease is never high; however, in a survey at the wholesale market of Sao Paulo (Brazil), Cladosporium rot was found in 4.5% (peach) and 7.5% (plum) of fruit sampled (Amorim et al., 2008). After harvest, the disease generally appears only after prolonged storage or transit. In addition, Cladosporium rots can be present in fruit treated with fungicides or biological control agents effective against the main fruit pathogens of peaches and nectarines, *Monilinia* spp. and *B. cinerea*, after prolonged storage (Mari, unpublished data).

Decayed areas are small and associated with injuries such as cracks, cuts, bruises, or others. In the early stages, the lesions are shallow and overlaid with a white mycelium, which assumes a velvety aspect due to the appearance of dark green spores (Figure 3.5). In the worst cases, the decay can extend from the surface to the pit, forming a dark cone-shape mass of rotted tissues, easily separable from healthy tissues (Barkai-Golan, 2001). On cherries, the decayed area often affects only the lining of the crack or wound around which it centers. On peaches, the spots usually are less than 2–3 cm in diameter (Harvey et al., 1972).

Figure 3.5 Cladosporium rot on apricot.

Cladosporium herbarum (Pers.) Link. is the species that causes Cladosporium rot on stone fruits in storage, while *Cladosporium carpophilum* Thüm. causes scab in twigs, leaves, and fruits of peach and nectarine trees. In culture, *Cladosporium* spp. are readily recognized. In particular, colonies of *C. herbarum* on MEA have a diameter of 18–32 mm after 7 d at 25°C, with a velutinous to lightly floccose aspect, plane or slightly wrinkled, olive colored; on the reverse side, the colonies are olive gray to dark greenish gray. Conidiophores are sparsely and irregularly branched. Conidia are ellipsoidal to cylindrical, usually not septate, 4–8 mm wide, with distinctly roughened walls (Pitt and Hocking, 2009). Because *C. herbarum* has long been the best known of the *Cladosporium* spp., it is likely that some reports on this species really refer to others, especially *C. cladosporioides*. However, molecular identification of *C. herbarum* is possible by the application of csM13 fingerprinting or ITS-amplified rDNA restriction analysis (ARDRA) (Guimaraes et al., 2011).

Cladosporium herbarum survives on dead plant material in the soil and produces abundant conidia that are readily disseminated by air; in fact, it generally constitutes the most common component of the air spores (Gregory, 1973). The pathogen was also recovered in the inner tissues of stone fruit mummies, more frequently isolated from tree mummies than from ground mummies, where it could represent an antagonist of *M. fructicola* (Hong et al., 2000). Conidia of *Cladosporium* were also found in the environment and on different surfaces within two Spanish packinghouses. It was, with *Penicillium* spp., the most prevalent genus recovered in all of the packinghouse areas tested (Bernat et al., 2017). The pathogen enters the fruit through wounds; however, its growth in the host is limited to a small area around the penetration point. It develops rapidly when fruits are stored at room temperature, but slowly when fruits are kept at temperatures as low as 0°C.

4 Disease Management

Postharvest disease management in stone fruits is mainly focused on the control of brown rot caused by *Monilinia* spp. For this reason, most of the strategies described below are related to the control of this disease. However, some specific means for controlling other postharvest diseases will be also described. Commonly, postharvest disease management in stone fruits is based on a program of fungicide applications in the field. However, in recent years, workers have focused on developing complementary preharvest and postharvest strategies, with goals to improve their efficacy, reduce the number of chemical applications, and minimize the presence of residues on fruit.

4.1 Control of Postharvest Diseases in the Field

No cultivars immune to brown rot are available, only some partially tolerant cultivars exist, and most commercial cultivars are susceptible or very susceptible to infection by *Monilinia* spp., in particular, those fruits with soft skin. Cultivars harvested early in the season can escape infection, whereas those harvested later are exposed to greater inoculum pressure and to climatic conditions more favorable to disease. Brown rot tolerance in some cultivars seems to be associated with a higher content of phenolic compounds, a greater thickness of the epidermis, and more waxes in the cuticle (Feliciano et al., 1987; Gradziel and Wang, 1993; Walter et al., 2004; Villarino et al., 2011). Consequently, all control strategies used in the field to reduce fruit losses are a combination of cultural practices and fungicide applications.

Among cultural practices, the elimination of all diseased parts of the host (mummies, shoots, branches, and fruits) that are sources of primary and secondary inoculum is essential to reduce potential inoculum, especially mummies and aborted fruits found in trees (Villarino et al., 2010). Mummies located on the tree can be a source of viable conidia into the next growing season, while conidia from mummies located on the ground are viable for a shorter period of time (Casals et al., 2015). However, in areas where the sexual form of *M. fructicola* has been found, mummies located on the ground can produce apothecia. Therefore, it is a good practice in winter during pruning or at the beginning of spring to remove mummified fruits located on trees and the ground, as well any wood colonized by this pathogen. This practice will reduce the presence of inoculum and consequently will contribute to controlling decay caused by *Monilinia* spp. Moreover, removing infected fruits should continue throughout the fruit-growing season, especially 10–15 d before harvest when fruit are more susceptible to infection and rapid disease development, to minimize the production and dissemination of conidia to healthy fruits. In the case of contamination with *G. candidum*, the fallen fruits should be buried to depths between 10 and 20 cm to reduce arthrospore vitality and sour rot incidence (Yaghmour et al., 2012).

Another tool available in the field to minimize fungal infections is to maintain an optimum microclimate in the tree canopy by pruning to improve ventilation and achieve higher sunlight and solar radiation to avoid humid conditions that favor the development of *Monilinia* spp. and other fungi (Michailides and Morgan, 1997). Irrigation to reduce tree water stress and increase tolerance to infection is also a recommended practice (Ogawa et al., 1995). However, excessive irrigation should be avoided to lessen fruit cracking in the orchard and reduce the appearance of Alternaria rot. Disease management practices in the field should also include attention to tree fertilization and other nutritional elements needed by the crop. Applications of chemical elements such as silica or calcium have significantly reduced brown rot. Elmer et al. (2006) found that the incidence of fruits infected by *Monilinia* spp. was significantly reduced after the application of six preharvest foliar treatments of calcium, and this practice is currently employed by stone fruit growers in New Zealand. Calcium sprays could also be useful to prevent cracking or splitting in cherries, particularly after rain, limiting indirectly Monilinia, Alternaria, and Penicillium rots. After harvest, the use of clean boxes, avoidance of exposing fruit to the sun, and careful handling of fruit to minimize injuries inflicted during harvest and transportation are without doubt sound practices that will reduce the incidence of postharvest diseases.

The main strategy used to control brown rot and other diseases is based on the applications of fungicides in the field using an established program. Treatments will be initiated earlier or later depending on the climatic conditions. In areas with a wet spring, fungicide applications begin at flowering and do not stop until harvest, with an average of 14 treatments per season. In contrast, in producing areas that are drier during flowering, fruit set and pit hardening, an environment that does not favor pathogen infection and development, treatments are mostly initiated 30 or 45 d before harvest, with an average of 3–4 treatments per season. Typically, synthetic chemical fungicides are used for these treatments, although sometimes biocontrol products are also used.

Among the fungicides, a wide range of active ingredients are currently available worldwide to be applied in the field to control brown rot: boscalid, cyproconazole, cyprodinil, difenoconazole, fenbuconazole, fenparazamine, fenhexamid, fludioxonil, fluopyram, iprodione, pyraclostrobin, tebuconazole, among others. For each active ingredient, the authorization for application in the field, doses, the re-entry or security period, maximum residue limit, and the maximum number of treatments that can be applied depends on the applicable legislation in each country. Most of these products are used to control brown rot, but some of them have proven effective for controlling Rhizopus rot and other important diseases that affect stone fruits. For example, fungicides can prevent blossom infections caused by *B. cinerea* in the orchard, and the "reduced-risk" fungicides fenhexamid and fludioxonil are highly effective for gray mold management. However, fungicides, such as tebuconazole, efficiently target *Monilinia* spp., but were less effective to control *B. cinerea* (Förster et al., 2007). In addition, fungicides routinely applied to control brown rot have unintentionally selected for resistance in both *Monilinia* spp. and *A. alternata* populations; indeed, strains resistant to succinate dehydrogenase inhibitor (SDHI), methyl benzimidazole carbamate (MBC), and quinone outside inhibitors (QoI) fungicides are currently reported in the USA and other countries (Fan et al., 2015; Martini et al., 2016). For brown rot control, fungicides could be applied at specific calendar dates or by using a prediction model of infection risk, which are currently already available in some producing areas, such as California, New Zealand, and Catalonia. To minimize the presence of residues on fruit, it is possible to select the active ingredients according to their degradation rates. Therefore, active ingredients that need more time to degrade on fruit will be applied before others that degrade faster. In addition, use of a good rotation program is recommended to reduce the risk of the development and proliferation of resistant strains (Brent and Hollomon, 2007).

Consumers demand fruits that are produced under environmentally friendly practices and with minimal exposure to synthetic chemicals that may be a risk to consumer health. Consequently, legislation has become progressively more restrictive in relation to the number of active ingredients available and their presence as chemical residues on fruits. Furthermore, commercial food distributors also limit the number and concentration of residues on fruits that exceed the legislated limits. This situation presents a challenge for fruit management and marketing, and extra efforts are necessary to make products more competitive. For example, to address this problem, the Catalan government created in 2011 the Fruit.Net Program that promotes cooperative work among all personnel involved in fruit production, providing valuable technical assistance to implement alternatives to chemical control and to introduce scientific advances from the wide research network.

Biocontrol products are increasingly included or integrated into conventional strategies designed to control pests and diseases, especially to control brown rot caused by *Monilinia* spp. Under conventional agriculture, the tendency is to include biocontrol products as complementary treatments to chemical fungicides, so one important requirement is that they have to be compatible with chemical products. Currently, only Serenade Max, containing *Bacillus subtilis,* is approved in several countries for field application to control brown rot.

4.2 Control of Postharvest Diseases after Harvest

Seasons characterized by prolonged wet periods with moderate temperatures favor the infection and development of *Monilinia* spp. Under these conditions and for mid or late season varieties, the chemical treatments applied in the field can be insufficient to control brown rot effectively and losses after harvest can be high. The adoption of good agricultural practices during the postharvest phase is a prerequisite to produce a high-quality fruit product for marketing. Hence, cooling harvested fruits to 0–0.5°C as soon as possible after harvest is an important step to reduce the incidence of Monilinia rot and other diseases, such as those caused by *Cladosporium* spp., *Alternaria* spp., or *Penicillium* spp. Rapid cooling is also important for fruits already infected in the field, although brown rot development will only be slowed and not stopped. New infections by *Monilinia* spp. rarely occur on stone fruits during the cold storage (Bernat et al., 2017). However, the adoption of effective sanitization practices in the packinghouse could be a useful tool to reduce inoculum of pathogens present in the atmosphere and on surfaces within the packinghouse environment, such as spores of *Penicillium* spp. Wells (1977) showed that stone fruit packinghouses using poor sanitization practices had more sour rot caused by *G. candidum* than packinghouses that used good sanitization practices. Prompt cooling and storage of fruit at 0°C are good disease management practices since the minimum temperature for spore germination, growth, and infection of *G. candidum* is about 2°C (Adaskaveg et al., 2007). In addition, equipment should be routinely disinfected using sanitizing treatments.

Postharvest chemical treatments are recommended when climatic conditions are favorable for disease and control practices in the field are insufficient. In this case, the current strategy is the use of synthetic fungicides applied in the packinghouse by the use of a water dump tank. These postharvest treatments are relatively inexpensive, easy to apply, and show curative and preventive activity against established and new infections, respectively. The active substances authorized depend on the producing country, but the most used are fludioxonil, iprodione, and fenhexamid. Although fludioxonil is very effective to control brown rot, Rhizopus rot, and gray mold of stone fruits, in semi-commercial trials, this fungicide failed to control sour rot caused by *G. candidum* on 'Summer Fire' nectarines. Some food additives (potassium sorbate, sodium benzoate, and 2-deoxy-D-glucose) also failed to satisfactorily control sour rot (Palou et al., 2009). In contrast, propiconazole, specifically formulated for postharvest use as Mentor 45WP and registered for the USA market, controlled brown rot effectively (Adaskaveg et al., 2007).

Alternative treatments to chemical products have been developed recently, mainly to control brown rot. The strategies studied employed physical, chemical, and biological treatments, alone or in combinations of these treatments (Usall et al., 2015). Among physical treatments, heat treatments are effective to control postharvest diseases by direct pathogen inhibition and stimulation of host defense mechanisms (Usall et al., 2016). To heat fruit, one approach is a hot water treatment. These treatments can be used now because they do not require regulatory approval and are easily employed to treat freshly harvested fruits. Currently, they are commercially applied to stone fruit, in particular to organic fruit. The incidence of fruits infected by *Monilinia* spp. was reduced by 50%, without affecting the quality of peaches or nectarines dipped in hot water at 60°C for 40 s (Casals et al., 2010a). Previous studies showed that the combination of hot water treatment with other strategies, such as the use of antifungals based on natural compounds, was more effective than these treatments applied separately and was promising (Margosan et al., 1997). Curing is another method to apply heat based on fruit exposure under high temperature and RH for short periods. Heat transfer is done by the water condensation on fruit surfaces. Exposure to 50°C and RH higher than 95% for 2 hr completely controlled the incidence of brown rot on a wide range of peach and nectarine cultivars. Moreover, it was observed that curing fruit under these conditions improved fruit firmness, since it was higher on treated than on non-treated fruits (Casals et al., 2012). Other possibilities for heating fruit are technologies based on radiofrequency or microwaves. The dielectric properties of fruits allow rapid heating of them in a continuous process. These technologies have also been thoroughly investigated, with promising results (Casals et al., 2010b; Sisquella et al., 2013a, 2014). Controlled atmosphere (CA), based on high concentrations of carbon dioxide and low concentrations of oxygen, is another physical treatment that reduces both CIs and diseases such as Alternaria rot and gray mold of cherries (Snowdon, 1990). Currently, CA is applied commercially in containers for long-distance transportation of stone fruits such as plum (Singh and Singh, 2013).

Chemical treatments of low toxicity have also been thoroughly studied in recent years. A wide range of substances has been evaluated for the control of *Monilinia* spp. and other pathogens on stone fruits, such as *Rhizopus* spp. The list of substances evaluated is large and, among others, includes chitosan, sodium bicarbonate, peracetic acid, and others (Casals et al., 2010a; Sisquella et al., 2013b; Karaca et al., 2014). In general, these substances are evaluated as curative treatments; however, chitosan was also evaluated as a preventive treatment. Chitosan applied at 1% by dipping fruit for 1 min reduced the incidence of new infections by 50% (Casals et al., 2012). Currently, few of these substances are registered for postharvest use.

Treatments with biological control agents have been investigated as alternatives to chemical products applied after the harvest of stone fruits for more than 30 yr and numerous studies reported they were effective (Karabulut and Baykal, 2003; Mercier and Jimenez, 2004; Larena et al., 2005; Mari et al., 2007; Zhou et al., 2008; Yánez-Mendizábal et al., 2011). Despite the huge effort to develop biocontrol products, currently in the USA only one, Bio-Save® (Jet Harvest Solutions, Apopka, FL) containing *Pseudomonas syringae*, is approved for postharvest application to cherry fruit to control diseases caused by *Penicillium* spp. and *Botrytis* spp.

5 Chilling Injury

After harvest, stone fruits deteriorate rapidly at ambient temperature; therefore, cold storage is necessary to slow this process and minimize decay development. The optimal storage temperature is about 0°C, but often these fruits are stored between 2.2 and 7.6°C. Termed the "killing temperature zone," these higher than optimal temperatures negatively affect fruit quality and favor the appearance of CI. The symptoms mainly develop during the shelf-life period, after 20–25 d of cold storage. The problem is not apparent until these fruits reach customers. It causes significant economic losses of fruits both shipped to distant markets and stored before sale (Crisosto et al., 1999). CI is one of the major reasons for low consumption of stone fruits, in particular peaches and nectarines. The susceptibility of stone fruits to CI is greatly influenced by the genetic background of the cultivar (Peace et al., 2006) and triggered by a combination of low temperatures and storage periods (Crisosto et al., 1999). Although other factors can affect the appearance of CI symptoms, fruit maturity and some orchard practices, such as fertilization and irrigation regimes that directly influence fruit size, are especially important and were thoroughly reviewed by Lurie and Crisosto (2005).

Chilling-injured fruits can show a variety of symptoms: mealiness or woolliness, leatheriness, internal browning, and internal reddening. In more advanced maturity stages, particularly in white flesh peach cultivars, flesh tissue separation and cavity formation can be frequently observed. Nevertheless, not all authors agree concerning the influence of fruit ripening on CI appearance. Some researchers found a greater incidence of mealiness in fruits harvested at an immature stage (Kailasapathy and Melton, 1992; Fernández-Trujillo et al., 1998), while others consider ripe fruits more susceptible to mealiness (Salunkhe and Desai, 1984). This is probably related to genetic characteristics of peach and nectarine cultivars; however, in all cases, flavor is compromised approximately 5 d prior to visual CI symptoms (Crisosto and Labavitch, 2002). Mealiness or woolliness is characterized by a lack of juiciness and a mealy texture (Lill et al., 1989). Both peaches and nectarines exhibit the disorder, although some variations in susceptibility between different cultivars were reported (Crisosto et al., 1999). In plums and apricots, the symptoms are slightly different, with a gel-like region forming near the stone (Taylor et al., 1993) (Figure 3.6). In all cases, the exterior of the fruit is apparently sound and fruit weight and firmness are not affected; only tasting the fruit reveals its lack of juice, absence of flavor, and dryness, which make it inedible and lead to consumer dissatisfaction. Leatheriness symptoms are similar to mealiness symptoms, but the dry texture of leatheriness is associated with firm fruit. In leathery peaches, the mesocarp parenchyma cells have collapsed, while the intercellular space has continued to increase. In addition, at the ultrastructural level, observations include dissolution of the middle lamella, cell separation, irregular thickening of the primary wall, and plasmolysis of the mesocarp parenchyma cells (Luza et al., 1992). Internal browning, frequently associated with the pit cavity, is characterized by an internal discoloration of peach tissue (flesh browning) that may extend to the rest of the fruit, excluding a zone of about 1 cm

Figure 3.6 Symptoms of chilling injury in plum.

below the skin that remains unaffected (Manganaris et al., 2006). Even though this symptom is often seen in mealy and leathery fruits, it may also occur in the absence of mealiness; however, only a few cultivars develop flesh browning without mealiness. Internal reddening, with flesh bleeding, is an important cause of rejection of canning peaches for esthetic reasons. It appears as an intense red color that develops in the flesh radiating from the pit. Flesh bleeding, however, does not affect fruit taste. In cultivars with abundant red pigments, flesh bleeding is associated with fruit maturity and not to injuries caused by storage at low temperatures (Lurie and Crisosto, 2005).

Although the physiological basis of CI has been studied in detail in peach (Lurie and Crisosto, 2005), the exact mechanism by which CI affects stone fruits has not been fully elucidated. In general, it is known that the disorder involves loss of membrane integrity and ion leakage from cells with changes in enzyme activities (Brummel et al., 2004). Moreover, the study of genetic control of these traits reported that mealiness, browning, and bleeding is probably by major genes (Ogundiwin et al., 2007). Clingstone non-melting flesh cultivars do not develop mealiness, but some of these may have a short market life because they are very susceptible to leatheriness or browning. In addition, genetic traits that were associated with internal browning were highly heritable in the population, with relatively small seasonal or rootstock effects (Peace et al., 2006).

Mealiness develops not during cold storage but during subsequent shelf life (Buescher and Furmanski, 1978). The disorder is the result of gel formation due to high-molecular weight-pectin binding and a low degree of esterification in combination with calcium, extracellular water, and solutes that pass through the cell wall (Sonego et al., 1995). The activity of enzymes such as pectin methylesterase, exo-polygalacturonase (exo-PG), and endo-PG changes during cold storage in different ways, depending on the enzyme, although a reduction in endo-PG activity during cold storage has been commonly observed (Zhou et al., 2000b). It is assumed that endo-PG plays a qualitative role in the development of mealiness, as a functional endo-PG gene must be present for the expression of other genes that influence mealiness susceptibility (Peace et al., 2006). Furthermore, other enzymes have been studied in relation to mealiness in peach. Endo-1,4-glucanase activity increased after fruit storage at chilling temperatures (Zhou et al., 2000a), while endo-1,4-β-mannase, β-galactosidase, and α-arabinosidase had lower activity in mealy fruits than in ripe juicy fruits (Brummel et al., 2004).

Leatheriness, another CI symptom, can develop in early harvested fruits after a few days at 20°C following cold storage. Ju et al. (2000) reported that 'Huangjin' peaches failed to soften after storage at low temperatures. In these fruits, failure to ripen might have been due to the failure of ethylene synthesis from very low levels of 1-aminocyclopropane-1-carboxylic acid (ACC) content and ACC-oxidase (ACO) activity. However, both ACC content and ACO activity were similar in both juicy and mealy fruits, indicating that ethylene synthesis was not impaired in the mealy fruit. In addition, reduced activity of PG and β-galactosidase was observed in fruits affected by leatheriness. Leatheriness and mealiness are both symptoms of CI, but they must be considered different disorders where early harvested fruits are more susceptible to the former and late harvested fruits are more susceptible to the latter.

Internal browning, associated with senescence and tissue deterioration, develops in a range of temperatures between 2 and 5°C, while its development is reduced at 0°C (Ceretta et al., 2000). Generally, phenols and polyphenol oxidase are present in separate compartments in fruit cells, but when they can be brought into contact, the enzymes act on phenolic substrates and the disorder occurs. The browning potential of peach is related to the amount of total phenolic compounds present in the fruit and associated with polyphenol oxidase activity (Kader and Chordas, 1984).

The causes of flesh bleeding are still not fully elucidated; the disorder is visible in halved peaches and nectarines stored for long periods. The entire area around the pit is reddish in appearance, and this can extend throughout the flesh when the disorder is severe. The red pigments are presumably anthocyanins spreading through fruit flesh. Abidi et al. (2015) found a negative correlation between high concentrations of phenolics, antioxidants, or sugars and relatively low or intermediate susceptibility to flesh bleeding.

The development of techniques to induce chilling tolerance in stone fruits is an urgent need during their storage and transportation. Several physical and chemical treatments have been found to be effective in delaying and limiting fruit CI and they have been widely investigated (Lurie and Crisosto, 2005). Among physical treatments, cold storage of peaches and nectarines in CA with elevated CO_2 and low O_2 concentrations was found to delay or prevent woolliness. However, after the use of 15% CO_2, a slight off-flavor that persisted through additional shelf life of peaches was detected (Truque et al., 2012). The use of intermittent thermal conditioning has been proposed as a technique to prevent and alleviate CI; however, it can encourage undesirable alterations in fruit sensory characteristics and favor the appearance of browning without enabling longer storage life in non-melting peach cultivars (Ferrer-Mairal et al., 2012). Recently, hot air treatment, an environmentally friendly technology, reduced CI in peach fruit by enhancing the antioxidant system activity, stimulating accumulation of heat shock proteins, and maintaining membrane integrity (Wang et al., 2014). Among chemical treatments, some substances have been proposed as protectants against CI. Cao et al. (2016) investigated the effect of an exogenous melatonin treatment on CI production in harvested peach fruit during cold storage. While peach fruits without melatonin treatment developed CI symptoms after 7 d of cold storage, the CI symptoms were significantly reduced in peaches treated with 100 or 200 μM of melatonin. The chilling tolerance of harvested peaches induced by melatonin could be associated with higher levels of polyamine, γ-aminobutyric acid, and proline, confirming the relationship between the increase of the polyamine content and enhancement of chilling tolerance in harvested horticultural crops. In recent years, oxalic acid, a natural organic acid present in plants, has received much attention for postharvest fruit management since it could increase resistance to CI, in particular, in 'Baifeng' peaches. A treatment with oxalic acid (5 mM) for 10 min significantly reduced the subsequent development of internal browning of cold-stored peaches (Jin et al., 2014). Results from different studies are not in agreement regarding the effects of 1-methylcyclopropene (1-MCP), an ethylene perception inhibitor widely used to reduce ethylene production in horticultural crops, on the appearance of CI symptoms. 1-MCP treatment (0.5 mL/L) retarded internal browning and flesh mealiness in 'Baifeng' peaches (Jin et al., 2011) and 'Maria Aurelia' nectarines (Özkaya et al., 2016). Conversely, other researchers found that treatments with 0.5 or 0.9 mL/L of 1-MCP increased CI symptoms in 'Elberta' or 'Chiripá' peaches (Fan et al., 2002; Girardi et al., 2005). These contradictory results may be related to the different fruit cultivars and 1-MCP concentrations used in these studies.

References

Abidi, W., Cantín, C.M., Jiménez, S., Giménez, R., Moreno, M.A. and Gogorcena, Y. 2015. Influence of antioxidant compounds, total sugars and genetic background on the chilling injury susceptibility of a non-melting peach (*Prunus persica* (L.) Batsch) progeny. *Journal of the Science of Food and Agriculture* 95, 51–358.

Adaskaveg, J.E., Förster, H., Driever, G. and Crisosto, C.H. 2007. Update on Mentor 45WP - a 'new tool' for postharvest management of sour rot of stone fruit in the 2007 season. *Central Valley Postharvest Newsletter* 16, 1–8.

Amorim, L., Martins, M.C., Lourenco, S.A., Gutierrez, A.S.D., Abreu, F.M. and Goncalves, F.P. 2008. Stone fruit injuries and damage at the wholesale market of Sao Paulo, Brazil. *Postharvest Biology and Technology* 47, 353–357.

Bader, O., Weig, M., Taverne-Ghadwal, L., Lugert, R., Gross, U. and Kuhns, M. 2011. Improved clinical laboratory identification of human pathogenic yeasts by matrix-assisted laser desorption ionization time-of-flight mass spectrometry. *Clinical Microbiology Infection* 17, 1359–1365.

Baggio, J.S., Gonçalves, F.P., Lourenço, S.A., Tanaka, F.A.O., Pascholati, S.F. and Amorim, L. 2016. Direct penetration of *Rhizopus stolonifer* into stone fruits causing Rhizopus rot. *Plant Pathology* 65, 633–642.

Barkai-Golan, R. 2001. *Postharvest Diseases of Fruits and Vegetables. Development and Control.* Elsevier, Amsterdam, The Netherlands.

Beever, R.E. and Weeds, P.L. 2004. Taxonomy and genetic variation of *Botrytis* and *Botryotinia*. pp. 29–52. *In*: Y. Elad, B. Williamson, P. Tudzynski and N. Delen (eds.). *Botrytis: Biology, Pathology and Control.* Kluwer Academic Publisher, Dordrecht, The Netherlands.

Bernat, M., Segarra, J., Casals, C., Teixidó, N. and Usall, J. 2017. Identification of fungal population in the environment and on surface of stone fruit packinghouses. *European Journal of Plant Pathology* 148, 723–731.

Biggs, A.R. and Northover, J. 1985. Inoculum sources for *Monilinia fructicola* in Ontario peach orchards. *Canadian Journal of Plant Pathology* 7, 302–307.

Booth, C. 1971. Fungal culture media. pp. 44–94. *In*: C. Booth (ed.). *Methods in Microbiology*, vol. 4. Academic Press, London, UK.

Brent, K.J. and Hollomon, D.W. 2007. *Fungicide Resistance in Crop Pathogens: How Can It Be Managed?* 2nd ed. Fungicide Resistance Action Committee (FRAC), Monograph No. 1, Brussels, Belgium.

Brummel, D.A., Dal Cin, V., Lurie, S., Crisosto, C.H. and Labavitch, J.M. 2004. Cell wall metabolism during the development of chilling injury in cold-stored peach fruit: Association of mealiness with arrested disassembly of cell wall pectins. *Journal of Experiment in Botany* 55, 2041–2052.

Buescher, R. and Furmanski, R.J. 1978. Role the pectinesterase and polygalacturonase in the formation of woolliness in peaches. *Journal of Food Science* 43, 264–266.

Burton, C.L. and Wright, W.R. 1969. Sour rot disease of peaches on the market. *Plant Disease Reporter* 53, 580–582.

Byrde, R.J. and Willetts, H.J. 1977. *The Brown Rot Fungi of Fruit - Their Biology and Control.* Pergamon Press, Oxford, UK.

Cao, S., Song, C., Shao, J., Bian, K., Chen, W. and Yang, Z. 2016. Exogenous melatonin treatment increases chilling tolerance and induces defense response in harvested peach fruit during cold storage. *Journal of Agricultural and Food Chemistry* 64, 5215–5222.

Casals, C., Elmer, P.A.G., Viñas, I., Teixidó, N., Sisquella, M. and Usall, J. 2012. The combination of curing with either chitosan or *Bacillus subtilis* CPA-8 to control brown rot infections caused by *Monilinia fructicola. Postharvest Biology and Technology* 64, 126–132.

Casals, C., Segarra, J., De Cal, A., Lamarca, N. and Usall, J. 2015. Overwintering of *Monilinia* spp. on mummified stone fruit. *Journal of Phytopathology* 163, 160–167.

Casals, C., Teixidó, N., Viñas, I., Silvera, E., Lamarca, N. and Usall, J. 2010a. Combination of hot water, *Bacillus subtilis* CPA-8 and sodium bicarbonate treatments to control postharvest brown rot on peaches and nectarines. *European Journal of Plant Pathology* 128, 51–63.

Casals, C., Viñas, I., Landl, A., Picouet, P., Torres, R. and Usall, J. 2010b. Application of radio frequency heating to control brown rot on peaches and nectarines. *Postharvest Biology and Technology* 58, 218–224.

Ceponis, M.J. and Friedman, B.A. 1957. Effect of bruising injury and storage temperature upon decay and discoloration of fresh, Idaho-grown Italian prunes on the New York City market. *Plant Disease Reporter* 41, 491–492.

Ceretta, M., Antunes, P.L., Brackmann, A. and Nakasu, B.H. 2000. Controlled atmosphere storage of the peach cultivar Eldorado. *Ciência Rural* 30, 73–79 (In Portuguese with English abstract).

Côté, M.-J., Tardif, M.-C. and Meldrum, A.J. 2004. Identification of *Monilinia fructigena, M. fructicola, M. laxa,* and *Monilia polystroma* on inoculated and naturally infected fruit using multiplex PCR. *Plant Disease* 88, 1219–1225.

Crisosto, C.H. and Labavitch, J.M. 2002. Developing a quantitative method to evaluate peach (*Prunus persica*) flesh mealiness. *Postharvest Biology and Technology* 25, 151–158.

Crisosto, C.H., Mitchell, F.G. and Ju, Z. 1999. Susceptibility to chilling injury of peach, nectarine, and plum cultivars grown in California. *HortScience* 34, 1116–1118.

De Cal, A., Egüen, B. and Melgarejo, P. 2014. Vegetative compatibility groups and sexual reproduction among Spanish *Monilinia fructicola* isolates obtained from peach and nectarine orchards, but not *Monilinia laxa. Fungal Biology* 118, 484–494.

De Cal, A. and Melgarejo, P. 1994. Effects of *Penicillium frequentans* and its antibiotics on unmelanizated hyphae of *Monilinia laxa*. *Phytopathology* 84, 1010–1014.

De Cal, A. and Melgarejo, P. 1999. Effects of long-wave UV light on *Monilinia* growth and identification of species. *Plant Disease* 83, 62–65.

De Cal, A. and Melgarejo, P. 2000. Momificado de los frutales de hueso (*Monilinia* spp.). pp. 66–67. *In*: E. Montesinos, P. Melgarejo, M.A. Cambra and J. Pinochet (eds.). *Enfermedades de los Frutales de Pepita y de Hueso*. Mundi-Prensa, Madrid, Spain (In Spanish).

Eckert, J.W. and Ogawa, J.M. 1988. The chemical control of postharvest diseases: Deciduous fruits, berries, vegetables and root tuber crops. *Annual Review of Phytopathology* 26, 433–469.

Elmer, P.A.G., Spiers, T.M. and Wood, P.N. 2006. Effects of pre-harvest foliar calcium sprays on fruit calcium levels and brown rot of peaches. *Crop Protection* 1, 11–18.

Emery, K.M., Michailides, T.J. and Scherm, H. 2000. Incidence of latent infection of immature peach fruit by *Monilinia fructicola* and relationship to brown rot in Georgia. *Plant Disease* 84, 853–857.

EPPO (European and Mediterranean Plant Protection Organization). 2009. List of A2 pests regulated as quarantine pests in the EPPO region. Available at: www.eppo.org/QUARANTINE/listA2.htm.

Eseigbe, D.A. and Bankole, S.A. 1996. Fungi associated with post-harvest rot of black plum (*Vitex doniana*) in Nigeria. *Mycopathologia* 136, 109–114.

Fan, X., Argenta, L. and Mattheis, J.P. 2002. Interactive effects of 1-MCP and temperature on 'Elberta' peach quality. *HortScience* 37, 134–138.

Fan, Z., Yang, J.H., Fan, F., Luo, C.-X. and Schnabel, G. 2015. Fitness and competitive ability of *Alternaria alternata* field isolates with resistance to SDHI, QoI, and MBC fungicides. *Plant Disease* 99, 1744–1750.

FAOSTAT (Food and Agriculture Organization of the United Nations, Statistics Division). 2018. Available at: http://faostat3.fao.org/browse/Q/QC/E.

Feliciano, A., Feliciano, A.J. and Ogawa, J.M. 1987. *Monilinia fructicola* resistance in the peach cultivar Bolinha. *Phytopathology* 77, 776–780.

Fernández-Trujillo, J.P., Cano, A. and Artés, F. 1998. Physiological changes in peaches related to chilling injury and ripening. *Postharvest Biology and Technology* 13, 109–119.

Ferrer-Mairal, A., Remón, S., Peiró, J.M. and Oria, R. 2012. Effects of intermittent conditioning on the color and enzymatic activity of peaches during controlled atmosphere storage. *Journal of Food Biochemistry* 36, 129–138.

Förster, H., Driever, G.F., Thompson, D.C. and Adaskaveg, J.E. 2007. Postharvest decay management for stone fruit crops in California using the "reduced-risk" fungicides fludioxonil and fenhexamid. *Plant Disease* 91, 209–215.

Fourie, J.F. and Holz, G. 1994. Infection of plum and nectarine flowers by *Botrytis cinerea*. *Plant Pathology* 43, 309–315.

Fourie, J.F. and Holz, G. 1995. Initial infection processes by *Botrytis cinerea* on nectarine and plum fruit and the development of decay. *Phytopathology* 85, 82–87.

Fourie, P.H., Holz, G. and Calitz, F.J. 2002. Occurrence of *Botrytis cinerea* and *Monilinia laxa* on nectarine and plum in Western Cape orchards, South Africa. *Australasian Plant Pathology* 31, 197–204.

García-Benítez, C., Melgarejo, P. and De Cal, A. 2017. Detection of latent *Monilinia* infections in nectarine flowers and fruit by qPCR. *Plant Disease* 101, 1002–1008.

García-Benítez, C., Melgarejo, P., De Cal, A. and Fontaniella, B. 2016. Microscopic analyses of latent and visible *Monilinia fructicola* infections in nectarines. *PloS ONE* 11:8, e0160675.

Gell, I., Cubero, J. and Melgarejo, P. 2007. Two different PCR approaches for universal diagnosis of brown rot and identification of *Monilinia* spp. in stone fruit trees. *Journal of Applied Microbiology* 103, 2629–2637.

Gell, I., De Cal, A., Torres, R., Usall, J. and Melgarejo, P. 2008. Relationship between the incidence of latent infections caused by *Monilinia* spp. and the incidence of brown rot of peach fruit: Factors affecting latent infection. *European Journal of Plant Pathology* 121, 487–498.

Gell, I., De Cal, A., Torres, R., Usall, J. and Melgarejo, P. 2009. Conidial density of *Monilinia* spp. on peach fruit surfaces in relation to the incidences of latent infections and brown rot. *European Journal of Plant Pathology* 123, 415–424.

Girardi, C.L., Adriana, R., Corrent, A.R., Lucchetta, L., Zanuzo, M.R., Da Costa, T.S., Brackmann, A., Twyman, R.M., Nora, F.R., Nora, L., Silva, J.A. and Rombaldi, C.V. 2005. Effect of ethylene, intermittent warming and controlled atmosphere on postharvest quality and the occurrence of woolliness in peach (*Prunus persica* cv. Chiripà) during cold storage. *Postharvest Biology and Technology* 38, 25–33.

Gradziel, T.M. and Wang, D. 1993. Evaluation of brown rot resistance and its relation to enzymatic browning in clingstone peach germplasm. *Journal of the American Society for Horticultural Science* 118, 675–679.

Gregory, P.H. 1973. *The Microbiology of the Atmosphere* 2nd ed. Leonard Hill, London, UK.

Guimaraes, J.B., Chambel, L., Melzoch, K., Pereira, P. and Tenreiro, R. 2011. *Cladosporium* sp. from phylloplane: A diversity evaluation on a continental ecosystem. *Mycosphere* 2, 191–201.

Hafeez, R., Akhtar, N., Shoaib, A., Bashir, U., Haider, M.S. and Awan, Z.A. 2015. First report of *Geotrichum candidum* from Pakistan causing postharvest sour rot in loquat (*Eriobotrya japonica*). *Journal of Animal and Plant Sciences* 25, 1737–1740.

Harvey, J.M., Smith, W.L. and Kaufman, J. 1972. *Market Diseases of Stone Fruits: Cherries, Peaches and Plums*. Agricultural Research Service, United States Department of Agriculture, Washington, DC.

Hennebert, G.L. 1973. *Botrytis* and *Botrytis*-like genera. *Persoonia* 7, 183–204.

Hong, C.X., Holtz, B.A., Morgan, D.P. and Michailides, T.J. 1997. Significance of thinned fruit as a source of secondary inoculum of *Monilinia fructicola* in California nectarine orchards. *Plant Disease* 81, 519–524.

Hong, C.X., Michailides, T.J. and Holtz, B.A. 2000. Mycoflora of stone fruit mummies in California orchards. *Plant Disease* 84, 417–422.

Horita, H. and Hatta, Y. 2016. Sour rot of carrot caused by *Geotrichum candidum* in Japan. *Journal of General Plant Pathology* 82, 65–68.

Inoue, K. and Nasu, H. 2000. Black spot of peach caused by *Alternaria alternata* (Fr.) Keissler. *Journal of General Plant Pathology* 66, 18–22.

Jin, P., Shang, H., Chen, J., Zhu, H., Zhao, Y. and Zheng, Y. 2011. Effect of 1-methylcyclopropene on chilling injury and quality of peach fruit during cold storage. *Journal of Food Science* 76, S485–S491.

Jin, P., Zhu, H., Wang, L., Shan, T. and Zheng, Y. 2014. Oxalic acid alleviates chilling injury in peach fruit by regulating energy metabolism and fatty acid contents. *Food Chemistry* 161, 87–93.

Ju, Z., Duan, Y. and Ju, Z. 2000. Leatheriness and mealiness of peaches in relation to fruit maturity and storage temperature. *The Journal of Horticultural Science and Biotechnology* 75, 86–91.

Kader, A. and Mitchell, F.G. 1989. Maturity and quality. pp. 191–196. *In*: J.H. LaRue and R.S. Johnson (eds.). *Plum and Nectarine: Growing and Handling for Fresh Market.* University of California, Division of Agriculture and Natural Resources. Publication No. 331, Oakland, CA.

Kader, A.A. and Chordas, A. 1984. Evaluating the browning potential of peaches. *California Agriculture* 38, 14–15.

Kailasapathy, K. and Melton, L.D. 1992. Woolliness in stone fruit. *Asian Food Journal* 7, 13–16.

Karabulut, O.A. and Baykal, N. 2003. Integrated control of postharvest disease of peaches with a yeast antagonist, hot water and modified atmosphere packaging. *Crop Protection* 23, 431–435.

Karaca, H., Pérez-Gago, M.B., Taberner, V. and Palou, L. 2014. Evaluating food additives as antifungal agents against *Monilinia fructicola* in vitro and in hydroxypropyl methylcellulose-lipid composite edible coatings for plums. *International Journal of Food Microbiology* 179, 72–79.

Larena, I., Torres, R., De Cal, A., Liñan, M., Melgarejo, P., Domenichini, P., Bellini, A., Mandrin, J.F., Lichou, J., de Eribe, X.O. and Usall, J. 2005. Biological control of postharvest brown rot (*Monilinia* spp.) of peaches by field applications of *Epicoccum nigrum*. *Biological Control* 32, 305–310.

Larsen, H.J., Covey, R.P. and Fischer, W.R. 1980. A red spot fruit blemish in apricots. *Phytopathology* 70, 139–142.

Lee, M.-H. and Bostock, R.M. 2006. Induction, regulation, and role in pathogenesis of appressoria in *Monilinia fructicola*. *Phytopathology* 96, 1072–1080.

Lill, R.E., O'Donoghue, E.M. and King, G.A. 1989. Postharvest physiology of peaches and nectarines. *Horticultural Reviews* 11, 413–452.

Lorenzini, M. and Zapparoli, G. 2014. An isolate morphologically and phylogenetically distinct from *Botrytis cinerea* obtained from withered grapes possibly represents a new species of *Botrytis*. *Plant Pathology* 63, 1326–1335.

Louw, J.P. and Korsten, L. 2016. Postharvest decay of nectarine and plum caused by *Penicillium* spp. *European Journal of Plant Pathology* 146, 779–791.

Luo, Y. and Michailides, T.J. 2001. Factors affecting latent infection of prune fruit by *Monilinia fructicola*. *Phytopathology* 91, 864–872.

Lurie, S. and Crisosto, C.H. 2005. Chilling injury in peach and nectarine. *Postharvest Biology and Technology* 37, 195–208.

Luza, J.G., Van Gorsel, R., Polito, V.S. and Kader, A.A. 1992. Chilling injury in peaches: A cytochemical and ultrastructural cell wall study. *Journal of the American Society for Horticultural Science* 117, 114–118.

Manganaris, G.A., Vailakakis, M., Diamantidis, G. and Mignani, I. 2006. Cell wall physicochemical aspects of peach fruit. *Postharvest Biology and Technology* 39, 69–74.

Margosan, D.A., Smilanick, J.L., Simmons, G.F. and Henson, D.J. 1997. Combination of hot water and ethanol to control postharvest decay of peaches and nectarines. *Plant Disease* 81, 1405–1409.

Mari, M., Torres, R., Casalini, L., Lamarca, N., Mandrin, J.F., Lichou, J., Larena, I., De Cal, M.A., Melgarejo, P. and Usall, J. 2007. Control of post-harvest brown rot on nectarine by *Epicoccum nigrum* and physico-chemical treatments. *Journal of the Science of Food and Agriculture* 87, 1271–1277.

Martini, C., Guidarelli, M., Di Francesco, A., Ceredi, G. and Mari, M. 2016. Characterization of thiophanate methyl resistance in Italian *Monilinia fructicola* isolates. *Journal of Plant Pathology* 98, 453–462.

Martini, C., Lantos, A., Di Francesco, A., Guidarelli, M., D'Aquino, S. and Baraldi, E. 2014. First report of Asiatic brown rot caused by *Monilinia polystroma* on peach in Italy. *Plant Disease* 98, 1595.

McPhee, W.J. 1980. Some characteristics of *Alternaria alternata* strains resistant to iprodione. *Plant Disease* 64, 847–849.

Mercier, J. and Jimenez, J.I. 2004. Control of fungal decay of apples and peaches by the biofumigant fungus *Muscodor albus*. *Postharvest Biology and Technology* 31, 1–8.

Michailides, T.J. and Morgan, D.P. 1997. Influence of fruit-to-fruit contact on the susceptibility of French prune to infection by *Monilinia fructicola*. *Plant Disease* 81, 1416–1424.

Michailides, T.J., Morgan, D.P. and Day, K.R. 2004. First report of sour rot of California peaches and nectarines caused by yeasts. *Plant Disease* 88, 222.

Michailides, T.J., Ogawa, J.M. and Sholberg, P.L. 1987. Chemical control of fungi causing decay of fresh prunes during storage. *Plant Disease* 71, 14–17.

Michailides, T.J. and Spotts, R.A. 1990. Postharvest diseases of pome and stone fruits caused by *Mucor piriformis* in the Pacific Northwest and California. *Plant Disease* 74, 537–543.

Morales, M.A.R. 1984. Manejo de post-cosecha de nectarinos y duraznos. *Revista Fruticola* 5, 56–59 (In Spanish).

Mordue, J.E.M. 1979a. Sclerotinia fructicola. *CMI Descriptions of Pathogenic Fungi and Bacteria* 616, 1–2.

Mordue, J.E.M. 1979b. Sclerotinia laxa. *CMI Descriptions of Pathogenic Fungi and Bacteria* 619, 1–2.

Mordue, J.E.M. 1979c. Sclerotinia fructigena. *CMI Descriptions of Pathogenic Fungi and Bacteria* 617, 1–2.

Navarro, D., Díaz-Mula, H.M., Guillén, F., Zapata, P.J., Castillo, S., Serrano, M., Valero, D. and Martínez-Romero, D. 2011. Reduction of nectarine decay caused by *Rhizopus stolonifer, Botrytis cinerea* and *Penicillium digitatum* with *Aloe vera* gel alone or with the addition of thymol. *International Journal of Food Microbiology* 151, 241–246.

Ogawa, J.M., Zehr, E.I., Bird, G.W., Ritchie, D.F., Uriu, K. and Uyemoto, J.K. 1995. *Compendium of Stone Fruit Diseases*. APS Press, St. Paul, MN.

Ogundiwin, E.A., Peace, C.P., Gradziel, T.M., Dandekar, A.M., Bliss, F.A. and Crisosto, C.H. 2007. Molecular genetic dissection of chilling injury in peach fruit. *Acta Horticulturae* 738, 633–638.

Özkaya, O., Yildirim, D., Dundar, O. and Tukel, S.S. 2016. Effects of 1-methylcyclopropene (1-MCP) and modified atmosphere packaging on postharvest storage quality of nectarine fruit. *Scientia Horticulturae* 198, 454–461.

Palou, L., Smilanick, J.L. and Crisosto, C.H. 2009. Evaluation of food additives as alternative or complementary chemicals to conventional fungicides for the control of major postharvest diseases of stone fruit. *Journal of Food Protection* 72, 1037–1046.

Papavasileiou, A., Testempasis, S., Michailides, T.J. and Karaoglanidis, G.S. 2015. Frequency of brown rot fungi on blossoms and fruit in stone fruit orchards in Greece. *Plant Pathology* 64, 416–424.

Peace, C.P., Crisosto, C.H., Garner, D., Dandekar, A.M., Gradziel, T. and Bliss, F.A. 2006. Genetic control of internal breakdown in peach. *Acta Horticulturae* 713, 489–496.

Pitt, J.I. and Hocking, A.D. 2009. *Fungi and Food Spoilage* 3rd ed. Springer, Dordrecht, The Netherlands.

Pose, G.N., Ludemann, V., Fernandez, D., Segura, J.A. and Pinto, V.F. 2010. *Alternaria* species associated with 'moldy heart' on peaches in Argentina. *Tropical Plant Pathology* 35, 174–177.

Restuccia, C., Giusino, F., Licciardello, F., Randazzo, C., Caggia, C. and Muratore, G. 2006. Biological control of peach fungal pathogens by commercial products and indigenous yeasts. *Journal of Food Protection* 69, 2465–2470.

Salunkhe, D.K. and Desai, B.B. 1984. *Postharvest Biotechnology of Fruits*. CRC Press, Boca Raton, FL.

Simmons, E.G. 2007. *Alternaria: An Identification Manual*. CBS Fungal Biodiversity Centre, Utrecht, The Netherlands.

Singh, S.P. and Singh, Z. 2013. Controlled and modified atmospheres influence chilling injury, fruit quality and antioxidative system of Japanese plums (*Prunus salicina* Lindell). *International Journal of Food Science and Technology* 48, 363–374.

Sisquella, M., Casals, C., Picouet, P., Viñas, V., Torres, R. and Usall, J. 2013a. Immersion of fruit in water to improve radio frequency treatment to control brown rot in stone fruit. *Postharvest Biology and Technology* 80, 31–36.

Sisquella, M., Casals, C., Viñas, I., Teixidó, N. and Usall, J. 2013b. Combination of peracetic acid and hot water treatment to control postharvest brown rot on peaches and nectarines. *Postharvest Biology and Technology* 83, 1–8.

Sisquella, M., Picouet, P., Viñas, I., Teixidó, N., Segarra, J. and Usall, J. 2014. Improvement of microwave treatment with immersion of fruit in water to control brown rot in stone fruit. *Innovative Food Science and Emerging Technologies* 26, 168–175.

Smith, W.L. and Anderson, R.E. 1975. Decay control of peaches and nectarines during and after controlled atmosphere and air storage. *Journal of the American Society for Horticultural Science* 100, 84–86.

Snowdon, A.L. 1990. *A Color Atlas of Post-Harvest Diseases and Disorders of Fruits and Vegetables. Vol. 1. General Introduction and Fruits*. CRC Press, Boca Raton, FL.

Sonego, L., Ben-Arie, R., Raynal, J. and Pech, J.C. 1995. Biochemical and physical evaluation of textural characteristics of nectarines exhibiting woolly breakdown – NMR imaging, X-ray computed tomography and pectin composition. *Postharvest Biology and Technology* 5, 187–198.

Spadaro, D., Garibaldi, A. and Gullino, M.L. 2009. Occurrence of patulin and its dietary intake through pear, peach and apricot juices in Italy. *Food Additives and Contaminants Part B* 1, 134–139.

Staats, M., van Baarlen, P. and van Kan, J.A. 2005. Molecular phylogeny of the plant pathogenic genus *Botrytis* and the evolution of host specificity. *Molecular Biology Evolution* 22, 333–346.

Tamm, L. and Flückiger, W. 1993. Influence of temperature and moisture on growth, spore production, and conidial germination of *Monilinia laxa*. *Phytopathology* 83, 1321–1326.

Taylor, M.A., Rabe, E., Jacobs, G. and Dodd, M.C. 1993. Physiological and anatomical changes associated with ripening in the inner and outer mesocarp of cold stored 'Songold' plums and concomitant development of internal disorders. *Journal of Horticultural Science* 68, 911–918.

Thomidis, T., Michailides, T.J. and Exadaktylou, E. 2009. Contribution of pathogens to peach fruit rot in Northern Greece and their sensitivity to iprodione, carbendazim, thiophanate-methyl and tebuconazole fungicides. *Journal of Phytopathology* 157, 194–200.

Thomidis, T., Michailides, T.J. and Karayiannis, I. 2007. Fruit core rot of the peach cultivar 'Fayette' caused by *Alternaria alternata* in Greece. *Australian Journal of Experimental Agriculture* 47, 1476–1479.

Tonini, G. and Capriotti, M. 1996. Postharvest damage to stone fruits and prevention. *Informatore Agrario* 52, 32–45 (In Italian with English abstract).

Tournas, V.H. and Stack, M.E. 2001. Production of alternariol and alternariol methyl ether by *Alternaria alternata* grown on fruits at various temperatures. *Journal of Food Protection* 64, 528–532.

Truque, E., Aguayo, E., Artés-Hernández, F., Gómez, P. and Artés, F. 2012. Controlled atmosphere for the export of 'Miraflores' peaches. *Acta Horticulturae* 962, 585–590.

Usall, J., Casals, C., Sisquella, M., Palou, L. and De Cal, A. 2015. Alternative technologies to control postharvest diseases of stone fruits. *Stewart Postharvest Review* 4, 2–6.

Usall, J., Ippolito, A., Sisquella, M. and Neri, F. 2016. Physical treatments to control postharvest diseases of fresh fruits and vegetables. *Postharvest Biology and Technology* 122, 30–40.

USDA FAS (United States Department of Agriculture, Foreign Agricultural Services, Europe Offices). 2014. EU-28 Stone Fruit Annual 2014, Global Agricultural Information Network (GAIN) Reports. Available at: https://gain.fas.usda.gov/Recent%20GAIN%20Publications/Stone%20Fruit%20Annual_Madrid_EU-28_8-21-2015.pdf

Villarino, M., Egüen, B., Lamarca, N., Usall, J., Segarra, J., Melgarejo, P. and De Cal, A. 2013. Occurrence of *Monilinia laxa* and *M. fructigena* after introduction of *M. fructicola* in peach orchards in Spain. *European Journal of Plant Pathology* 137, 835–845.

Villarino, M., Melgarejo, P. and De Cal, A. 2016. Growth and aggressiveness factors affecting *Monilinia* spp. survival peaches. *International Journal of Food Microbiology* 224, 22–27.

Villarino, M., Melgarejo, P., Usall, J., Segarra, J. and De Cal, A. 2010. Primary inoculum sources of *Monilinia* spp. in Spanish peach orchards and their relative importance in brown rot. *Plant Disease* 94, 1048–1054.

Villarino, M., Melgarejo, P., Usall, J., Segarra, J., Lamarca, N. and De Cal, A. 2012. Secondary inoculum dynamics of *Monilinia* spp. and relationship to the incidence of postharvest brown rot in peaches and the weather conditions during the growing season. *European Journal of Plant Pathology* 133, 585–598.

Villarino, M., Sandin-España, P., Melgarejo, P. and De Cal, A. 2011. High chlorogenic and neochlorogenic acid levels in immature peach reduce *Monilinia laxa* infection by interfering with fungal melanin biosynthesis. *Journal of Agricultural and Food Chemistry* 59, 3205–3213.

Visagie, C.M., Houbraken, J., Frisvad, J.C., Hong, S.-B., Klaassen, C.H.W., Perrone, G., Seifert, K.A., Varga, J., Yaguchi, T. and Samson, R.A. 2014. Identification and nomenclature of the genus *Penicillium*. *Studies in Mycology* 78, 343–371.

Wade, N.L. and Morris, S.C. 1982. Causes and control of cantaloupe postharvest wastage in Australia. *Plant Disease* 66, 549–552.

Walter, M., Gillian, F.M., Fraser, J.A., Frampton, C.M., Boyd-Wilson, K.S.H. and Perry, J.H. 2004. Methods of screening apricot fruit for resistance to brown rot caused by *Monilinia* spp. *Australasian Plant Pathology* 33, 541–547.

Wang, K., Shao, X., Gong, Y., Xu, F. and Wang, H. 2014. Effects of postharvest hot air treatment on gene expression associated with ascorbic acid metabolism in peach fruit. *Plant Molecular Biology Report* 32, 881–887.

Wang, O. and Wang, J. 2001. The occurrence of peach grey mold disease and its control. *South China Fruits* 30, 38–39 (In Chinese with English abstract).

Wells, J.M. 1977. Sour rot of peaches caused by *Monilinia implicata* and *Geotrichum candidum*. *Phytopathology* 67, 404–408.

White, T.J., Bruns, T., Lee, S. and Taylor, J. 1990. Amplification and direct sequencing of fungi ribosomal RNA genes for phylogenetics. pp. 315–322. *In*: M.A. Innis, D.H. Gelfand, J.J. Sninsky and T.J. White (eds.). *PCR Protocols. A Guide to Methods and Applications.* Academic Press, San Diego, CA.

Wouters, J.T.M., Ayad, E.H.E., Hugenholtz, J. and Smit, G. 2002. Microbes from raw milk for fermented dairy products. *International Dairy Journal* 12, 91–109.

Yaghmour, M.A., Bostock, R.M., Morgan, D.P. and Michailides, T.J. 2012. Biology and source of inoculum of *Geotrichum candidum* causing sour rot of peach and nectarine fruit in California. *Plant Disease* 96, 204–210.

Yánez-Mendizábal, V., Usall, J., Viñas, I., Casals, C., Marín, S., Solsona, C. and Teixidó, N. 2011. Potential of a new strain of *Bacillus subtilis* CPA-8 to control the major postharvest diseases of fruit. *Biocontrol Science & Technology* 9, 409–426.

Yang, J.H., Brannen, P.M. and Schnabel, G. 2015. Resistance in *Alternaria alternata* to SDHI fungicides causes rare disease outbreak in peach orchards. *Plant Disease* 99, 65–70.

Zhang, Z., Wang, J. and Li, Y. 1995. Study on the occurrence of peach black spot (*Alternaria alternata*) and disease dissemination. *Journal of Agricultural University of Hebei* 18, 53–58 (In Chinese with English abstract).

Zhou, H.W., Lurie, S., Lers, A., Khalchitski, A., Sonego, L. and Ben Arie, R. 2000a. Delayed storage and controlled atmosphere storage of nectarines: Two strategies to prevent woolliness. *Postharvest Biology and Technology* 18, 133–141.

Zhou, H.W., Sonego, L., Khalchitski, A., Ben Arie, R., Lers, A. and Lurie, S. 2000b. Cell wall enzymes and cell wall changes in 'Flavortop' nectarine: MRNA abundance, enzyme activity, and changes in pectic and neutral polymers during ripening and in woolly fruit. *Journal of the American Society for Horticultural Science* 125, 603–637.

Zhou, T., Schneider, K.E. and Li, X.Z. 2008. Development of biocontrol agents from food microbial isolates for controlling post-harvest peach brown rot caused by *Monilinia fructicola*. *International Journal of Food Microbiology* 126, 180–185.

Chapter 4

Table Grape, Kiwifruit, and Strawberry

Gianfranco Romanazzi

Department of Agricultural, Food and Environmental Sciences, Marche Polytechnic University, Ancona, Italy

Philip A.G. Elmer

The New Zealand Institute for Plant & Food Research Limited, Waikato Mail Centre, Hamilton, New Zealand

Erica Feliziani

Department of Agricultural, Food and Environmental Sciences, Marche Polytechnic University, Ancona, Italy

Abbreviations

BCAs	Biological control agents
CA	Controlled atmosphere
EU	European Union
GRAS	Generally recognized as safe
MA	Modified atmosphere
PAL	Phenylalanine ammonia-lyase
Psa	*Pseudomonas syringae pv. actinidiae*
RH	Relative humidity
SO_2	Sulfur dioxide
TA	Titratable acidity
TSS	Total soluble solids
TSS	Total soluble solids
USA	United States of America

1 Introduction

This chapter describes the cause of postharvest rots of table grapes, kiwifruits, and strawberries. Details on the current control strategies and alternative management practices will only be touched upon briefly, and readers are encouraged to refer to Section III of the book for more information on them.

2 Table Grape

2.1 Botanical Classification

Grapevines belong to the genus *Vitis* of the family Vitaceae, which includes climbing plants with very small flowers arranged in clusters. Species of *Vitis* are found throughout Asia, Europe, and America. Eurasia has only one endemic species of *Vitis, Vitis vinifera,* while North America and China have many endemic species. Grapevines are believed to have been domesticated in the Caucasus during the Neolithic period and their cultivation was associated with human settlements in Syria, Palestine, northern Egypt, and Mesopotamia. The North American species were critical to save the European cultivars from the *Phylloxera* epidemics in the mid-19th century, for the breeding of resistant rootstocks. Of the North American species, *V. labrusca,* which is represented by the cultivar 'Isabella', is one of the major genetic resources for hybrid table grapes. For example, 'Concord' (*V. labrusca* × *V. vinifera*) is resistant to the major diseases including downy mildew, which is caused by *Plasmopara viticola* (Berk. & M.A. Curtis) Berl. & De Toni. The goals of table grape breeding include selection for seedless cultivars, specific harvest maturity dates, high yields, reduced need for viticultural practices, good storage potential, and favorable berry characteristics such as large size, elongated shape, uniform color, thin skin, and firm texture. Seedless berries are anatomically stenospermocarpic and have seed remnants. There are very few true parthenocarpic cultivars and they have limited commercial significance (Lichter and Romanazzi, 2019).

The grapevine inflorescence has been described as a complex, highly modified branched system with reduced shoots and flowers (Jackson, 2014). The rachis is composed of a basal stem (the peduncle), two main branches, and numerous sub-branches that terminate in pedicels. Flowers develop individually on the pedicels. The grapevine fruit is a berry. The fruit wall (pericarp and skin) surrounds the fleshy portion of the berry (pulp). The skin of table grapes is typically thinner than that of wine grapes. This results in better texture, but can result in cracking and increased disease susceptibility in the vineyard and during storage. The epidermal layers of the skin are coated with epicuticular waxes that have complex structures and give the skin its natural, non-shiny appearance, termed "bloom". Handling before and after harvest can leave shiny marks on the bloom, which reduce the fresh appearance of the berries and thus their market value.

The maturity of any particular cultivar spans a period of a few weeks, which can be extended by geographic diversity and viticultural practices. In many "seedless" grapes, the plant growth regulators gibberellin and cytokinin are used to obtain reasonable or enhanced berry size. However, overuse of plant growth regulators can have negative effects on berry properties, such as shatter, reduced flavor (Zoffoli et al., 2009a), and reduced berry set and yield in subsequent seasons. Ethylene, which is generated by application of ethephon (a common commercial formulation is Ethrel®, Bayer Crop Science), and more recently, s-abscisic acid (a common commercial formulation is ProTone®, Valent BioSciences) are used to enhance the red color of some cultivars, such as 'Flame Seedless' and 'Crimson Seedless' (Cantín et al., 2007; Peppi and Fidelibus, 2008). The weight of a single table grape berry is typically 6–10 g, with clusters of table grapes typically weighing 400–700 g, although much larger sizes occur (e.g., cvs. 'Italia' and 'Redglobe'). Controlling cluster size and compactness are two critical labor-intensive viticultural tasks that can affect diseases before and after harvest, as a tighter and more compacted cluster can favor microlesions that facilitate infections by *Botrytis cinerea* Pers., and/or spread of the pathogen from an infected berry to its close neighbors. After decay caused by *B. cinerea*, the second most important postharvest problem of table grapes is the physiological disorder referred to as rachis browning (Lichter, 2016). The primary factor that influences rachis browning during storage is desiccation and this is the reason why grapes have to be stored at very high humidity (Lichter and Romanazzi, 2019).

2.2 Postharvest Diseases of Table Grapes

Gray mold is caused by *B. cinerea*. Within the project "One Fungus: One Name" at the XVI *Botrytis* Symposium held in Locorotondo (Bari, Italy) in 2013, there was unanimous support for retaining the older name of the imperfect stage (*B. cinerea*) and discard the name of the teleomorph stage (*Botryotinia fuckeliana*) (Johnston et al., 2014). *Botrytis cinerea* has a worldwide distribution and it is considered the second most important fungal plant pathogen (Dean et al., 2012). Gray mold is by far the most important postharvest disease of table grapes (Pearson and Goheen, 1988). However, in subtropical Mediterranean climates with dry summers, *Aspergillus niger* Tiegh., *Rhizopus* spp., *Alternaria* spp., *Penicillium* spp., and the "sour rot complex" of pathogens can also cause major bunch rot problems (Luvisi et al., 1992).

2.2.1 Gray Mold

Gray mold is the most important postharvest disease of numerous fruit and vegetable crops, including table grapes (Pearson and Goheen, 1988; Romanazzi et al., 2016b). The market tolerance to gray mold incidence is very low (typically 0.5% of symptomatic berries). The pathogen can spread among berries. If no control measures are applied, it can reach incidence of 50% or more within 1 mon of cold storage, or lead to the loss of entire packages in a few days of storage under ambient conditions (shelf life). Therefore, disease control strategies must be implemented throughout the season and after harvest to prevent gray mold outbreaks.

Infections mainly start from wounds to the berries that occur in the field or during and after harvest. Infections are first visible as small lesions that are a few millimeters in diameter (Figure 4.1a). As the internal tissues of the berry are colonized, brown lesions develop on the berry skin. The lesions enlarge and are subsequently covered by white to gray mycelia and later by gray conidia, which are the typical signs of gray mold (Figure 4.1b). The infection can spread from diseased to healthy berries, which leads to a mass of mycelia and decayed berries, a process that is termed "nesting". A characteristic symptom of postharvest gray mold of grape berries is the separation of the skin from the inner pulp. This is due to production of skin degrading enzymes by *B. cinerea* under the skin but not within the pulp. The skin slips away easily when the berry is touched, which is a condition termed "slip skin".

Botrytis cinerea grows best at high humidity and moderate temperatures, although it can grow across a wide temperature range, from –0.5 to 32°C. Thus, *B. cinerea* can develop during cold storage of table grapes when the host is less metabolically active, although its growth is much slower than at higher temperatures (Romanazzi et al., 2016b).

The mycelia of *B. cinerea* are branched, septate, and hyaline to brown in color. Conidiophores arise directly from the mycelium or the sclerotia and are tall, slender, and irregularly branched in the terminal portion, with apical cells that are enlarged or rounded and bear clusters of conidia simultaneously on short denticles. The conidia are smooth, hyaline or gray in color, oval-shaped, with a mean length of 10 μm and a mean width of 5 μm. The sclerotia, which are survival structures, are often present. They are formed from mycelial branches that fuse together to form a globular mass, which is at first hyaline but later turns brown or black due to deposition of melanin pigments in the outer rind. The melanized layer and β-glucans encasing the internal mycelia can protect the sclerotia from desiccation, ultraviolet (UV) radiation, and microbial attack over long periods (Williamson et al., 2007). When environmental conditions are favorable (e.g., temperature and moisture), the sclerotia can germinate. They produce conidiophores, or less commonly after sexual recombination, an elongated stipe with a terminal apothecium. Chlamydospores are short-term survival structures that can help the fungus to overcome unfavorable periods that can be encountered on plant surfaces. They are formed from hyphal cells and are liberated when the mycelia disintegrate (Holz et al., 2007).

During the sexual reproductive cycle of *B. cinerea*, numerous microconidia are observed in all crosses that form apothecia. Microconidia provide an alternative propagule for the survival of this fungus during adverse conditions. In general, they are found in aging cultures or in those contaminated by other organisms and

Figure 4.1 a. Initial gray mold infections (top left). b. Gray mold infection with sporification of *Botrytis cinerea* (top right). c. Leak disease caused by *Rhizopus stolonifer* (bottom left). d. Black mold with distinctive black *Aspergillus niger* spore masses (bottom center). e. Infections on the same berry, starting from a wound, of black mold (top of the berry), gray mold (middle of the berry), and blue mold caused by *Penicillium* sp. (bottom of the berry) (bottom right).

in association with sclerotia. Microconidia develop from germ tubes that are produced by the macroconidia, endogenously inside empty cells of mature hyphae, and from appressoria and sclerotia (Fukumori et al., 2004; Holz et al., 2007). During sexual reproduction of *B. cinerea*, the microconidia function as spermatia (Fukumori et al., 2004; Williamson et al., 2007). In laboratory studies, sexual bodies can be formed during mating of a sclerotium of one isolate with microconidia from another isolate. The apothecia that form on sclerotia are at first deeply concave or funnel-shaped; as the heads expand, they became flat and ultimately convex. They are from 10 to 15 mm in height and from 5 to 7 mm in cap diameter. Asci arise in a fascicle from the base of the caps of apothecia. These are cylindrical, hyaline, and evanescent at maturity and contained eight ascospores. The ascospores are hyaline, one-celled, and ovoid or ellipsoid in shape (Fukumori et al., 2004).

Botrytis cinerea is present in many habitats as mycelia, microconidia and macroconidia, chlamydospores, sclerotia, or apothecia with ascospores. These are dispersed by diverse means, but the most important propagule, the macroconidia, are mainly disseminated by dry air currents (Holz et al., 2007). In temperate climates, the sclerotia germinate in spring and are the primary source of inoculum (Williamson et al., 2007). *Botrytis cinerea* can also survive unfavorable periods as mycelia in infected crop debris on the ground, and in dead leaves and flowers, and mummified inside the canopy, and sometimes inside seeds. With its extremely wide host range, *B. cinerea* has a worldwide distribution. The most serious losses occur in temperate areas with cooler temperatures and greater wetness. The presence of water or high relative humidity (RH>95%) is required for conidia to germinate. The minimum temperature for growth is 0°C, the optimum temperature is around 20°C, and above 32°C growth can be very restricted. Therefore, gray mold develops well during cold storage of fruits and vegetables (Romanazzi and Feliziani, 2014).

In the vineyard, *B. cinerea* infects dead flower tissues and remains latent until environmental conditions become favorable for infection of the developing bunch. The inoculum load is higher in vineyards with poor air circulation and in bunches where the berries are tightly clustered. The pathogen starts actively growing at veraison, when the berry cuticle becomes thinner, and the sugar content increases. Infections progress rapidly inside tight berry clusters. When berries enlarge, microcracks can develop around the pedicel, which provide excellent entry points for penetration of *B. cinerea*. Other factors that contribute to disease development are humid microclimate, dense canopy, and thin berry cuticle. After veraison, disease development progresses rapidly when the environmental conditions of "two 15s" occur (i.e., a minimum temperature of 15°C and at least 15 hr of continuous moisture on the berries). The disease is more severe when the berries are damaged in the field by hail, during bunch cleaning, or at harvest (Figure 4.1e). It is also promoted by other pathogens, such as powdery mildew, caused by *Erysiphe necator* Schwein., or by pests, such as the worm *Lobesia botrana* or *Frankliniella occidentalis* and other thrips. Severe infections can also occur after heavy rain, which in some cultivars (e.g., 'Victoria') can lead to a circular crack around the pedicel. To minimize damage during postharvest handling, clusters are often packed directly in the vineyard (Romanazzi et al., 2016b).

Because gray mold is such an important problem during table grape storage, management practices have to be applied to every fruit that is stored for more than 1 or 2 wk. Although in many production areas a large fraction of the crop is used for local consumption, table grapes are also marketed over long distances. For example, in the United States (USA), California accounts for approximately 90% of the national table grape production and large quantities are shipped to the East Coast or are exported to Canada, China, and many other countries. Even if the crop is locally consumed, it is often kept in cold storage for up to 2 or 3 mon to extend the marketing period. As an alternative to cold storage, grapes of some cultivars are kept on the vines for longer periods. For example, in southern Italy and California, mature table grapes, which are protected by plastic tunnels or canopy covers (to avoid rain damage), can be left on the vines for 2–3 mon and sometimes even up to Christmas (Romanazzi et al., 2016b).

Some cultivars, such as 'Redglobe', can be stored for 3 mon or longer (Lichter et al., 2008), while the quality of others, such as 'Flame Seedless', declines quickly over time, as the fruit is more prone to decay development. Grape berries with injuries are highly susceptible to gray mold, but those with few natural pores exhibit mechanical resistance to infection by conidia of *B. cinerea*. In addition, the thickness of the skin, including the epidermis and the epicuticular wax layer, has been correlated with resistance to gray mold (Mlikota Gabler et al., 2003). In a comparison of *V. vinifera* cultivars, only very few (e.g., 'Emperor' and 'Autumn Black') were moderately resistant to *B. cinerea*, whereas other *Vitis* spp. (e.g., *V. rotundifolia* and *V. labrusca*) were highly resistant. This comparison included 42 cultivars and lines. Cluster compactness or chemical components, which are commonly associated with resistance, were not found to be important factors in susceptibility to *B. cinerea*. Because the cuticle is considered to be of major importance in resistance, the recent development of a sensor that measures cuticle properties may enhance efforts to develop resistant cultivars (Herzog et al., 2015; Lichter and Romanazzi, 2019).

A review of the effects of viticultural practices on gray mold that focused on wine grapes (Elmer and Michailides, 2007) indicated that *B. cinerea* infections are facilitated by a dense canopy that creates a more favorable microclimate for disease development. In addition, compact clusters are more likely to be more conducive to gray mold and removing excess grapes within clusters can significantly reduce decay. Leaf removal around grape clusters provides better air circulation and light exposure and ensures more effective coverage of the fruit during application of fungicides and biofungicides. Moreover, ultraviolet C (UV-C) irradiation increases the content of the phytoalexin *trans*-resveratrol (Romanazzi et al., 2006), which might contribute to reduced sensitivity to fungal pathogens of those parts of the fruit and of the canopy that are exposed to the sunlight.

In southern Italy, almost all table grape vineyards are sprayed two to four times a season with calcium chloride, from bunch closure to the harvest, over the last couple of decades. Calcium chloride is now also used in several other table-grape-producing areas of the world, because improved control of gray mold decay has been observed. Calcium strengthens the plant tissues by ligating to pectin, which makes the berries less susceptible to degradation by the tissue-macerating enzymes of *B. cinerea*. However, while calcium applications can be very good practice for late-harvested cultivars, it can delay ripening in early cultivars (Romanazzi et al., 2012). Regular applications of calcium (e.g., four to five times from veraison) or reduced nitrogen fertilization in wine grape vineyards can contribute to gray mold control (Elmer and Michailides, 2007), although these effects can be cultivar-specific and it is not clear if these treatments will also work with table grapes.

Fungicides are commonly applied for preharvest and postharvest gray mold control in temperate climates with rainfall during the season (Leroux, 2007). In areas with dry climates, such as in some parts of the Mediterranean basin and California, they are mainly applied before bunch closure and after veraison to prevent development of postharvest decay (Smilanick et al., 2010; Romanazzi et al., 2016b). The active ingredients of the main fungicides are boscalid, cyprodinil, fenhexamid, fenpyrazamine, fluxapyroxad, fludioxonil, fluopyram, mepanipyrim, pyraclostrobin, and pyrimethanil, or combinations of these (Fillinger and Walker, 2016), and these can be highly effective in reducing the incidence of gray mold. However, in several areas, rapid development of resistance in *B. cinerea* to almost all of these active ingredients has been observed (De Miccolis Angelini et al., 2014; Li et al., 2014; Amiri et al., 2018). Thus, there is an urgent need for alternatives to synthetic fungicides that can complement, or even replace, their use.

Biocontrol agents, such as *Bacillus subtilis, Bacillus amyloliquefaciens* subsp. *plantarum, Aureobasidium pullulans, Metschnikowia fructicola,* and *Ulocladium oudemansii,* have been formulated for the control of gray mold in several countries, and when used in integrated programs, they can provide effective gray mold management (Calvo-Garrido et al., 2013; Liu et al., 2013; Rotolo et al., 2016). It is likely that microbial-based biocontrol agents and antimicrobial natural products (collectively referred to as biofungicides) integrated with viticultural control practices will be more widely used in the future close to harvest, the most effective timing for gray mold control. This is because residues of conventional fungicide would not meet the requirements by retailers in many countries to reduce or eliminate detectable fungicide residues on the fruit (Wurms et al., 2011a).

Control of insect pests is a continuous effort in table grape vineyards, including such as the moths *L. botrana* and *F. occidentalis,* thrips, mealybugs, *Drosophila suzukii* and other fruit flies, and additional local pests. Their control is important in a decay control program because insects can be effective carriers of the conidia of *B. cinerea* and other pathogens, including species of *Penicillium, Rhizopus,* and *Aspergillus,* and especially sour rot agents.

Many of the postharvest decay control practices for table grapes have been discussed in recent publications (Zoffoli and Latorre, 2011; Chervin et al., 2012; Romanazzi et al., 2016b; Lichter and Romanazzi, 2019). Some practices are primarily aimed at reducing physiological disorders, but these in turn can have direct or indirect effects on the incidence of postharvest decay. Table grapes are harvested when the sugar content exceeds 15°Brix, depending on the cultivar. Two basic technologies that are commonly used commercially are refrigeration and sulfur dioxide (SO_2) (Nelson, 1985). Fumigation with SO_2 during initial forced-air precooling followed by weekly applications during cold storage, primarily to kill conidia and aerial mycelium of *B. cinerea* on table grapes, is a standard practice in California. In Chile, in the vineyard just after harvest, a pulse of SO_2 is injected into the cartons. The final packaging of grapes is often done in the vineyard or packinghouse. Most packaging is done in vented cluster bags, polycarbonate clamshell containers, or punnets. The fruit is stored in plastic totes or returnable plastic containers if the packaging is carried out after storage. For extended storage, SO_2 sheets, which release the gas during storage, are placed on top of the clusters that are then surrounded by a plastic liner to contain the gas. SO_2 sheets are also often placed inside a paper pad, which allows SO_2 to diffuse through, and to control the humidity in the carton. These also have a liner to contain the SO_2 released from the sheets upon absorption of water. One variation is the use of a laminated plastic liner that releases SO_2 to the interior of the package to the grapes. Within 6 hr of harvest, forced-air cooling is used to cool the grapes to 0–4°C. This requires the pallets to be stacked in two rows in front of a powerful fan that forces cold air into the vented cartons. Forced-air cooling takes 4–15 hr, depending on the packaging material used. Instead of using liners, the whole pallet can be wrapped in "stretch" polyethylene after forced-air cooling (Lichter et al., 2016), a practice that shortens the cooling time significantly. A different approach is practiced in California, where grapes are fumigated during forced-air cooling, typically with 400 μL/Lxhr. This CxT value is the multiplication product of the SO_2 concentration (μL/L) and the fumigation time (hr). For example, application of 100 μL/L SO_2 for 4 hr provides a CxT value of 400 μL/Lxhr. This is followed by weekly fumigation with a CxT of 200 μL/Lxhr. Damage due to excess SO_2 is primarily evident as bleaching around the pedicel or in mechanical injuries and natural openings among red cultivars. Hairline cracking is another symptom, which can occur as longitudinal fine cracks that appear after storage (Zoffoli et al., 2009b). White grape cultivars can also show browning when exposed to excess SO_2 (Nelson, 1985). SO_2 damage is more common when a break in the cold chain occurs (Romanazzi et al., 2016b). Fumigated grapes must not exceed sulfite residues of 10 mg/kg, the maximum residue limit.

Grapes under cold storage are maintained between 0 and −1°C at an RH of 90–95%. Lower temperatures can freeze the rachis and subsequently the berries. Cold transport is mainly done in marine containers. The container should be pre-cooled to the dew point when loading the pallets to prevent condensation, which can have detrimental consequences for decay development. Ideally, grapes should be loaded with the container connected to the packinghouse through a sleeve. The length of storage varies significantly, which is mainly based on the grape cultivar, but also on the incidence of latent infections. For example, 'Redglobe' grapes can be stored for 100–120 d, while 'Thompson Seedless' and 'Crimson Seedless' grapes can be stored for about 60 d, and 'Flame Seedless' grapes for about 40 d (Zoffoli and Latorre, 2011; Lichter et al., 2016; Romanazzi et al., 2016b).

In California, more than half of the organic grape crop is stored in the presence of ozone (Feliziani et al., 2014). Ozone is an oxidizing agent that has fungistatic activity and can, thus, inhibit the growth of *B. cinerea*. Ozone gas is also employed for the storage of some conventionally grown table grapes. It also oxidizes residues of several fungicides that are commonly applied before harvest (Mlikota Gabler et al., 2010). However, low fungicide residues might instead allow the pathogen to develop, which would result in significant losses throughout the handling chain (Romanazzi et al., 2016b). Furthermore, ozone can oxidize and turn to yellow the white liners that are placed into the boxes to avoid contact of bunches with the wood and this is objected to by retailers (Lichter and Romanazzi, 2019).

A long list of treatments have been evaluated in recent years in the search for effective potential alternatives to synthetic fungicides and SO_2 for the control of postharvest gray mold of table grapes (Lichter et al., 2006; Romanazzi et al., 2012, 2016b) and some of these are now used commercially. These include preharvest treatments that can be used in organic farming, such as biofungicides and calcium chloride, and several postharvest treatments. Among the postharvest treatments, ozone is the most promising. Many of these alternatives were studied only at the laboratory scale and decay control after storage did not meet current commercial standards. Other studies have used combinations of treatments that have been based on changing current practices (e.g., postharvest wet treatments) or that required costly equipment (e.g., ozone generators). There is interest by growers for natural compounds from plants (e.g., plant essential oils) or of animal origin (e.g., chitosan). Some of these are being marketed as plant defense boosters or biostimulants and are often not registered for use as plant protection products (Romanazzi et al., 2016a; Gutiérrez-Gamboa et al., 2019). In New Zealand, two products are registered for use against Botrytis rots (not just in grapes): MidiZen®, which is based on soybean oil, and ARMOUR-Zen®, which is based on chitosan (Wurms et al., 2011a). Recently, in Italy, the application of microwave technology in the packinghouses applied for few minutes to bunches in commercial plastic boxes that were sealed with semi-permeable film provided good decay control after storage (Lichter and Romanazzi, 2019). However, further research is required to demonstrate the efficacy under commercial conditions and packinghouses have to commit to investment in this new technology. Other promising alternatives are certain essential oils (e.g., thyme and fennel oils) (Abdollahi et al., 2012) and their synthetic derivatives (e.g., carvacrol) (Martínez-Romero et al., 2007), which have reduced postharvest gray mold of table grapes in laboratory studies. Recently, a product based on Australian tea-tree oil, known as Timorex Gold®, was registered to control Botrytis rot in several countries. However, the effects of volatile essential oils on gray mold can be short-lived if they are not applied in a slow-release form. A formulation of the synthetic terpenes eugenol, geraniol, and thymol (3Logy®; Sipcam Italia, Italy) that are essential oil components was recently registered in several countries for preharvest application to control gray mold of table grapes.

2.2.2 Other Diseases

Relative to the great importance of gray mold, other decays of table grapes are usually of minor importance. Additionally, in the presence of gray mold, other decays are often not observed or are masked. However, under certain conditions or when *B. cinerea* is well controlled, other pathogens can be present at commercially significant levels.

2.2.2.1 Blue Mold

Penicillium expansum Link, the causal agent of blue mold, is a common pathogen of numerous fruits, especially apples and pears. It has also been reported as a commercial problem for table grapes (Franck et al., 2005). On white grapes, a soft, brown decay develops that is first covered by white hyphae and then by bluish spores (Figure 4.1e). In addition to *P. expansum*, other species can be involved, including *Penicillium chrysogenum* Thom (Donoso and Latorre, 2006). The application of SO_2 very effectively controls *P. expansum*, although after storage of more than 60 d, infections can appear on senescent tissues or at very high RH (Franck et al., 2005). Fungicides that control *B. cinerea* are usually also effective against *P. expansum*, with some exceptions (e.g., fenhexamid). Fruit injuries that arise mainly from the pedicels where the berries were detached can be a major site of infection. Decay severity was reported to be low in immature berries and overmature berries were resistant to infection, possibly because a water activity below 0.96 restricts the growth of *P. expansum* (Donoso and Latorre, 2006). A temporary rise in temperature during cold storage can result in development of blue mold. *Penicillium expansum* can grow at low temperatures and become a significant commercial problem after extended storage of grapes (Zoffoli and Latorre, 2011).

2.2.2.2 Alternaria and Cladosporium Rots

Alternaria rot is caused by *Alternaria alternata* (Fr.) Keissl. and can occur on table grapes after storage, although at low incidence. Generally, the pathogen is outcompeted by *B. cinerea*. However, after storage of 'Redglobe' grapes for 90–120 d with continuous SO_2 release, black lesions of different sizes caused by *A. alternata* can still develop on the skin and can become the dominant cause of spoilage (Zutahy et al., 2008). These black lesions are dry and firm and the fungus can sporulate on the surface. Occasional consignments in South

Africa were reported to have a high incidence of such infection (Swart and Holz, 1994). All of the bunch tissues contained latent *A. alternata*, with the incidence of infection peaking at bunch closure. Mature bunches were asymptomatic despite being colonized by the fungus and there was no apparent relation between inoculum level and postharvest rot incidence.

Long storage periods might also be conducive to the development of Cladosporium rot, which is caused by *Cladosporium herbarum* (Pers.) Link or *Cladosporium cladosporioides* (Fresen.) G.A. de Vries, and is characterized by superficial velvet-like, dark green lesions at the pedicel end of the berries (Zoffoli and Latorre, 2011). Cladosporium rot caused large losses in the 2016–2017 table grape season in California (Joseph L. Smilanick, personal communication).

2.2.2.3 Rhizopus Rot Infections by *Rhizopus stolonifer* (Ehrenb.) Vuill. are often observed in vineyards as brown berries with black stripes that protrude from the berry skin along the axis of the berry, giving the disease the common name of "black mold" (Lisker et al., 1996; Latorre et al., 2002). In packaged table grapes, the berries can be covered by abundant white to gray hyphae that bear black sporangia (Figure 4.1c) and leakage of juice (termed "leak") frequently occurs from the bottom of the packages. Rhizopus rot occurs mainly on table grapes for local distribution that are not subjected to cold storage with SO_2. Air-borne sporangiospores disseminated from nearby peach orchards have been reported to provide an inoculum source for table grapes, although this fungus is also abundantly present in most soils (Lisker et al., 1996). In contrast to *B. cinerea* and *A. alternata, R. stolonifer* is more resistant to disinfection with ethanol (Mlikota Gabler et al., 2004). Storage temperatures below 5°C will stop the growth of *R. stolonifer*, but not of *Mucor* spp., which can cause similar symptoms.

2.2.2.4 Black Rot and Black Mold Several closely related species in *Aspergillus* sec. *Nigri* are abundant in table grape vineyards in semi-arid warm climates, such as the eastern part of the Mediterranean basin, including *A. niger* and *Aspergillus carbonarius* (Bainier) Thom (Guzev et al., 2006). *Aspergillus carbonarius* is recognized by its large conidia and its production of the mycotoxin ochratoxin A. In vineyards, infected table grapes develop brown water-soaked lesions that are eventually covered by an abundance of black conidia that can be spread by air movement to neighboring berries (Figure 4.1d,e). The incidence of *A. niger* infections can be high when grapes are maintained after harvest for 7 d at 20°C. *Aspergillus* spp. infections do not occur at low temperatures and the incidence of infected fruits is reduced during cold storage (Guzev et al., 2008). *Aspergillus carbonarius* cultures exposed to SO_2 fail to sporulate even after transfer to fresh medium (Guzev et al., 2008). In contrast to *B. cinerea* and *A. alternata, A. niger* is more resistant to disinfection with ethanol (Mlikota Gabler et al., 2004).

2.2.2.5 Melting Decay and Breakdown Disorder "Melting decay" and "breakdown disorder" are terms that are used for a phenomenon that occurs mainly in California on 'Redglobe' table grapes (Morgan and Michailides, 2004; Hashim-Buckey et al., 2008). This decay is similar to the "soft tissue breakdown" that was described in South Africa (Witbooi et al., 2010). Symptoms include discolored berries that split and lose their integrity. According to Morgan and Michailides (2004), the symptoms appear only after 2 wk of cold storage, while according to Hashim-Buckey et al. (2008), the decay can also be observed after veraison. The main organisms isolated from this melting decay have been *B. subtilis* and *Cryptoccocus laurentii*, as well as *A. pullulans* and other yeasts. These microorganisms are probably not sensitive to SO_2, but melting decay was apparently partly solved when the infected vineyards were replanted (Joseph L. Smilanick, personal communication). New molecular tools for in-depth analysis of microbiological communities associated with melting decay and breakdown disorder should result in a better understanding of these disorders and their causal agents.

3 Kiwifruit

3.1 History, Species, and Cultivars

Cultivated kiwifruits of the family Actinidiaceae are valued worldwide for their sensory characteristics, nutritional value including high vitamin C content (Nishiyama et al., 2004), dietary fiber, and other compounds beneficial to human health (Stonehouse et al., 2012). The fruit has also been used in traditional medicine to improve health and well-being (Ferguson and Ferguson, 2003; Lee et al., 2015). In addition, long storage life (from 6 to 9 mon) means the fruit can be consumed throughout the year in many parts of the world (Minas et al., 2010; Ferguson, 2015; Cotrut et al., 2016).

Historically, much of the kiwifruits grown outside China originated from seed that had been brought into New Zealand in the early 20th century (Costa and Ferguson, 2015). The green-fleshed *A. chinensis* var. *deliciosa* 'Hayward' (A. Chev.) A. Chev. (formerly *Actinidia deliciosa* [A. Chev.] C.F. Liang and A.R. Ferguson) originated from this single batch of seed and formed the basis of kiwifruit production in California, China, Chile, Greece, France, Italy, and New Zealand. Globally, kiwifruit production is currently estimated to be at least 2.4 million tons per year with China alone producing half the total production. Most of China's production is consumed domestically with less than 1% exported fresh (Ferguson, 2015). Overall, global kiwifruit production only represents 0.2–0.3% of fresh fruit production worldwide, but its contribution to the economy of New Zealand is very significant, earning over NZ$ 1 billion per year (Costa and Ferguson, 2015).

Fruits from the genus *Actinidia* are incredibly diverse in size, appearance, flavor, and nutrient composition and new varieties with green, yellow, and red flesh being developed and commercialized. Examples of hairless, yellow-fleshed kiwifruit include 'Hort16A' (*A. chinensis* var. *chinensis* Planch.), marketed as Zespri® Gold Kiwifruit, *A. chinensis* var. *chinensis* 'Zesy002', informally known as Gold3 (marketed under the name Zespri® SunGold Kiwifruit), 'Jintao', 'Soreli', and 'Dori'. The most widely planted variety of *A. chinensis* Planch. var. *rufopulpa* (C.F. Liang et R.H Huang) (C.F. Liang et A.R. Ferguson) in China is 'Hongyang' (Wang, 2011; Li et al., 2015), which has distinctive red flesh around the core and comprises nearly two-thirds of the total commercial kiwifruit production in China (Ferguson, 2015). New red-fleshed varieties are also in the process of commercialization in China (Wang et al., 2015c) and New Zealand (e.g., 'Hort22D') (Seal et al., 2016).

Kiwifruits belonging to *A. arguta* differ markedly from those of *A. chinensis* var. *deliciosa* or *A. chinensis* var. *chinensis* and are often referred to as kiwiberry or hardy kiwi. Although the fruits are small (grape sized), they are prized for their sweetness (Jung et al., 2016), diversity of flesh and skin color (Cossio et al., 2015), smooth, often edible thin skins, high vitamin C content, intense flavor, and greater cold tolerance; attributes that have increased commercial interest in this species worldwide (Jung et al., 2016).

3.2 Postharvest Diseases of Actinidia spp.

3.2.1 *A. chinensis* var. *deliciosa*

'Hayward' is the most widely planted green-fleshed kiwifruit in the world but there are three other green-fleshed cultivars that have been commercialized in China: 'Qinmei', 'Xuxiang', and 'Miliang No. 1', and further information on these varieties can be found in Ferguson (2015). A new green-fleshed cultivar, 'Zesh004' (marketed as Zespri® Sweet Green Kiwifruit), is under development in New Zealand, but total production remains small, while its commercial viability is being evaluated. 'Hayward' are harvested when they are unripe (minimum total soluble solids [TSS] content of 6.2%) and firm (6–8 kg force) and this has been established as the optimum for subsequent storage life (Burdon et al., 2013). Ripening then occurs slowly during cold storage in normal atmospheres at 0±1°C (Feng et al., 2006). Controlled atmosphere (CA) cold storage in 2% oxygen and 5% carbon dioxide has been shown to extend the storage life of kiwifruit for up to 8 mon. While extended storage periods help coordinate marketing of the fruit, it has been associated with fruit losses resulting from physiological disorders (e.g., chilling injury and physiological pitting) and fruit rots caused by several fungal species. The most important postharvest rot affecting 'Hayward' is Botrytis stem-end rot or gray mold caused by the fungus *B. cinerea* with the first symptoms appearing as early as 1 mon after cold storage at 0±1°C (Michailides and Elmer, 2000; Spadaro et al., 2010).

3.2.1.1 Botrytis Stem-End Rot

Botrytis stem-end rot symptoms do not appear in the field but are usually first visible after 3–4 wk of cold storage at 0°C (Figure 4.2a, b). In New Zealand, the long delay between initial fruit infection at the picking wound at harvest and first symptom expression has been shown to be a consequence of slow *B. cinerea* development (0.2 mm/d) through the fruit tissues in cold storage (Sharrock and Hallett, 1992). Fruit surface wounds caused by poor handling practices can also be entry points for *B. cinerea*, which may cause "side rots", but these are usually infrequent because of improved postharvest and packinghouse handling practices. Externally, the skin of *Botrytis*-infected fruit, extending down from the stem end usually appears darker than the healthy part of the fruit (creating the so-called "tide-mark" appearance). In addition, white to gray mycelium, sclerotia, and conidiophores bearing conidia can also develop on infected fruits (Figure 4.2a). When cut open, infected tissue appears dark green and water-soaked (Figure 4.2b). Crop losses caused by *B. cinerea* can be divided into direct and indirect losses. Direct losses are those with the typical, clearly visible rot that extends down from the pedicel or stem end of the fruit. Indirect losses also occur when *Botrytis*-infected fruit produce ethylene, thereby causing premature fruit softening in apparently healthy fruit (Brook,

1992; Niklis et al., 1997; Manning et al., 2010). The presence of even one infected fruit in a tray can accelerate the softening of healthy fruits and increase the incidence of ripe rots caused by other fungi when fruits are removed from cold storage (Brook, 1992). Percentage crop loss varies considerably and is influenced by differences in cold storage systems (Brigati et al., 2003), from year to year, from orchard to orchard (Pennycook and Manning, 1992; Manning et al., 2010), from region to region, and from country to country (Michailides and Elmer, 2000). Reported losses were 2–32% in New Zealand (Michailides and Elmer, 2000), 2–45% in Italy (Prencipe et al., 2016), 3–20% in Chile (Díaz et al., 2014a; Elfar et al., 2017), and 1–1.5% in Greece (Kulakiotu et al., 2004). Crop losses reported in Korea and China were less precise and it was described as a "major pathogen" in Korea (Kim et al., 2015b) and a "significant postharvest disease" in China (Chen et al., 2015). Additional costs may also be incurred as a result of labor-intensive re-sorting and re-packing at onshore and offshore facilities (Elmer and Michailides, 2007).

Many pre- and postharvest factors influence the incidence of Botrytis stem-end rot and have been described in detailed reviews (Michailides and Elmer, 2000; Elmer and Michailides, 2007; Manning et al., 2010). In summary, *B. cinerea* infection of petals and anthers during flowering provides inoculum for sepal and receptacle infections that remain latent in the stem end until later in cold storage in California (Michailides and Morgan, 1996; Michailides and Elmer, 2000; Michailides et al., 2010). The primary infection pathway in New Zealand begins with infection of picking wounds at harvest from *B. cinerea* conidia contaminating the fruit surface (Elmer et al., 1995, 1997; Manning et al., 2010). The third type of infection occurs when fruit-to-fruit contact ("nesting") enables the spread of *B. cinerea* from one fruit to another in the same tray or container during cold storage (Minas et al., 2010). In contrast to many other exporting countries, Botrytis stem-end rot in New Zealand has been successfully managed using pre- and postharvest non-fungicide based cultural control strategies, including summer and winter pruning to create open canopies and postharvest curing of fruit. Their effectiveness is such that this pathogen is no longer regarded as economically important to the industry (Manning et al., 2010). The postharvest curing process in New Zealand involves holding harvested 'Hayward' fruits in bins in covered areas at ambient temperatures for at least 2 d before packing and cold storage (Pennycook and Manning, 1992; Bautista-Baños et al., 1997; Manning et al., 2010; Mari et al., 2015). In contrast, in Italy, the use of curing at ambient temperatures for 48–72 hr represented "considerable logistical" challenges and packinghouse-related problems mostly due to the need for the 'Hayward' crop to be harvested within a few days to avoid autumn frosts (Tonini et al., 1999). Furthermore, the practice of holding fruits at ambient temperature prior to cold storage was often found not to be effective in several kiwifruit regions. The researchers then proposed a modified curing process where the fruit was actually cured at normal atmospheres in the cold storage by gradually reducing temperature from 10 to 0°C over a period of 12–14 d. The results of this modification were impressive. Botrytis rot incidence among the non-cured fruits was 25%, and reduced to 15 and 2%, respectively, among those cured for 2 d (at ambient, 8–14°C) or cured for 14 d in the cold storage.

3.2.1.2 Non-Botrytis Rots In Italy, the skin pitting disease of 'Hayward' caused by *Cadophora luteo-olivacea* (J.F.H. Beyma) T.C. Harr. & McNew (syn. *Phialophora* sp.) was initially thought of as a sporadic problem affecting fruits in the south of the

Figure 4.2 a. Gray mold infections on kiwifruit starting from the pedicel picking wound (top left). b. Progression of *Botrytis cinerea* infection in the fruit flesh (top right) with the distinctive boundary visible between infected and healthy flesh ("tide mark"). c. Blue mold infection of kiwifruit (bottom left). d. Gray mold infection on 'Hongyang' kiwifruit (bottom right).

country, but in early 2000, up to 30% of fruits were reported to be infected after 100–120 d of cold storage (Spadaro et al., 2010). A 5-yr project to gain a greater understanding of factors affecting *C. luteo-olivacea* concluded that before cold storage, the dry matter was inversely correlated while nitrogen content was positively correlated with *C. luteo-olivacea* skin pitting outbreaks. Furthermore, when CA conditions were delayed from 8–12 to 25 d or more, *C. luteo-olivacea* skin pitting symptoms were absent (Spadaro et al., 2010).

The majority of non-Botrytis rots in 'Hayward' appear when fruits have become overripe (Brook, 1992; Michailides and Elmer, 2000) and the fungi reported to be associated with fruit postharvest rots worldwide include species of *Botryosphaeria, Cryptosporiopsis, Penicillium* (Figure 4.2c), *Phomopsis, Colletotrichum, Alternaria, Sclerotinia, Fusarium,* and *Phoma*. However, *Botryosphaeria* sp. and *Phomopsis* sp. stem-end rots are generally regarded as the two most important postharvest fungal pathogens of kiwifruit in China (Li et al., 2015; Zhou et al., 2015) and Korea (Kim et al., 2015b). The other fungal rots occur sporadically and are considered minor pathogens on 'Hayward' and will not be discussed further in this chapter.

3.2.1.2.1 Botryosphaeria *spp.* Three species of *Botryosphaeria*, namely, *B. dothidea* (Moug. ex Fr.) Ces. & De Not., *B. parva* Pennycook and Samuels (syn.: *Neofusicoccum parvum* [Pennycook & Samuels] Crous, Slippers & A.J.L. Phillips), and *B. stevensii* Shoemaker, were identified as causing postharvest rots in 'Hayward' in New Zealand and were associated with fruits that were physiologically ripe and ready to consume (Hawthorne et al., 1982; Brook, 1992), and also in fruits that were "disagreeably overripe" (Pennycook, 1985). Interestingly, of the these three *Botryosphaeria* spp. studied, none was able to grow in vitro at typical cold storage temperatures (0.3°C), and hence, these belong to the group often referred to as "ripe rots" (Brook, 1992). In other important kiwifruit-growing regions, such as Korea and China, *B. dothidea* was regarded as one of the most important pathogens causing postharvest rots (Koh et al., 2003; Li et al., 2016a). Taxonomically, the genus *Botryosphaeria* is complex and molecular studies have identified that the taxa previously grouped under *B. dothidea* in New Zealand includes *B. dothidea, B. lutea, B. parva,* and *B. ribis* (Johnston et al., 2005), but not all of these species have been isolated from postharvest kiwifruit rots.

Typical symptoms caused by *Botryosphaeria* spp. appear as a distinctive, single expanding lesion on the side of the fruit immediately after removal from extended periods of cold storage, or following a period of incubation at ambient temperatures after cold storage. Infected fruits have a much reduced shelf life, and exhibit deterioration of fruit flesh quality and the development of unpleasant odors and flavor (Pennycook, 1985; Pennycook and Samuels, 1985). Although currently in New Zealand, *Botryosphaeria* is not generally regarded as an economically important postharvest pathogen, up to 15% of fruits were infected in fruit samples in the Te Puke region over three decades ago (Pennycook, 1985). In contrast, *B. dothidea*, the predominant postharvest pathogen in Korea (Koh et al., 2003), was usually associated with overripe fruits and substandard cold storage conditions (Koh et al., 2005).

3.2.1.2.2 Phomopsis Stem-End Rot Several species of *Phomopsis* (Sacc.) Bubak have been described as causing significant postharvest rots at the stem end of kiwifruit and are considered to be among the major fruit rot pathogens of kiwifruit in Korea (Lee et al., 2001; Koh et al., 2005; Kim et al., 2015a) and China (Li et al., 2016a). Symptoms may vary slightly from country to country, but typically the brown pubescent skin at the stem end becomes soft and lighter in color than the adjacent firm healthy tissue, and a watery exudate may also be produced. When the skin was peeled back, the flesh appears water-soaked, soft, disorganized, lighter green in color than the healthy tissue (Lee et al., 2001; Koh et al., 2005), and a fermented sour odor may be produced on severely affected fruits (Auger et al., 2013). The teleomorph state of *Phomopsis* (*Diaporthe actinidiae* N.F. Sommer & Beraha) was reported by Lee et al. (2001) from up to 40% of isolations from rotted fruits after cold storage (Koh et al., 2003). To avoid possible confusion when naming this pathogen, we will use the term *Phomopsis* sp. (anamorph state of *Diaporthe*) where the asexual state has been identified and *Diaporthe* sp. (telomorph state of *Phomopsis*) will be used where the sexual state was positively identified from samples of rotted kiwifruit. Multiple species of *Diaporthe* were isolated from postharvest rots and identified from long-term stored 'Hayward' kiwifruits in Chile, namely, *D. novem* J.M. Santos, Vrandecic & A.J.L. Phillips, *D. ambigua* Nitschke, *D. australafricana* Crous & Van Niekerk, and *D. viticola* Nitschke (Díaz et al., 2014a, 2014b). Surprisingly, *D. actinidiae*, commonly described in the literature as a postharvest pathogen of kiwifruit, was not found in this study. In a separate study, *D. ambigua* was isolated and identified from cold-stored 'Hayward' kiwifruits in Chile (Auger et al., 2013). In China, samples of *A. chinensis* var. *deliciosa* 'Miliang No. 1' and *A. chinensis* var. *chinensis* 'Jinyan' were cold stored in CA for 4 mon at 1°C then given a period of ambient incubation for 14 d. At the end of this period of postharvest treatment, 27% ('Miliang No. 1') and 33% ('Jinyan') developed typical Diaporthe stem-end rot symptoms. Morphological and molecular identification by DNA sequencing confirmed that the cause of these stem-end rots was *D. actinidiae* (Li et al., 2017). Interestingly, *D. actinidiae* was excluded as the cause of kiwifruit stem-end rot in Italy and morphological, DNA sequencing, and pathogenicity tests could not identify the species responsible (Luongo et al., 2011). A new species of *Diaporthe*, which was associated with rots of yellow-fleshed 'Jinyan' kiwifruit, was identified as *D. lithocarpus* Speg. (Li et al., 2016a). In New Zealand, postharvest storage conditions may affect the incidence of Phomopsis stem-end rot in 'Hayward'. When harvested fruits were cold stored in CA conditions (2% O_2 and 5% CO_2) for up to 24 wk, then packed in export trays and cold stored for a further 7 wk, *Phomopsis* spp. stem-end rot incidence was 13 and 78% for 1992 and 1991, respectively, compared with 44 and 7% Botrytis stem-end rot in the same years (Manning and Lallu, 1997). On the basis of these findings, recommendations were made to kiwifruit producers in New Zealand to avoid water condensation on fruit surfaces in CA stores, and to conduct effective prestorage curing of the fruit.

3.2.1.2.3 Blue Mold Blue mold, caused by *P. expansum* (Figure 4.2c), causes postharvest losses in a very wide range of fruits, including kiwifruit, although it is not regarded as economically important as gray mold caused by *B. cinerea* (Chen et al., 2015). Factors known to contribute to postharvest rots of green-fleshed kiwifruit caused by *P. expansum* include poor postharvest handling practices that provide wound sites for infection, overmature fruit, and long-term storage conditions (Zhu et al., 2016).

3.2.2 *A. chinensis* var. *chinensis*

'Hort16A' (marketed as Zespri® Gold Kiwifruit), 'Zesy002' (marketed as Zespri® SunGold Kiwifruit), 'Jintao', 'Soreli', and 'Dori' have been gaining in popularity worldwide because of their stunning visual appearance when cut open, flavor profiles, and improved sweetness, especially when compared to 'Hayward'. A similar range of postharvest pathogens that cause postharvest rots in 'Hayward' described above, also affect varieties that belong to *A. chinensis* var. *chinensis*, although some preharvest symptoms occur in some of the yellow-fleshed varieties that do not occur in 'Hayward' (Manning et al., 2003). In New Zealand, one or more species of *Phomopsis* spp. and *C. actinidiae* (Johnston et al., 2004) were described as the primary fungal pathogens of 'Hort16A' when assessed after 12 wk of cold storage (Manning et al., 2003; Fullerton et al., 2007). These fungi are present on the fruit surface before harvest, but disease lesions do not develop until after prolonged periods of cold storage (Wurms et al., 2007). Fungal pitting caused by *Cryptosporiopsis actinidiae* P.R. Johnst., M.A. Manning & X. Meier was reported to be the most prevalent postharvest rot of 'Hort16A', with fruit losses of up to 10% after 18 wk of cold storage in air (Wurms et al., 2011b). However, losses as high as 30% were reported in 2001 and 2002 by some exporters of 'Hort16A', with an estimated cost of at least NZ$ 80 million (Beresford, 2005). Clearly, there was an urgent need to review existing knowledge of the two primary pathogens responsible at that time and a 2-yr study (2002–2004) was undertaken to better understand the *Phomopsis* and *Cryptosporiopsis* disease cycles in an unsprayed, organically managed block of mature vines of 'Hort16A' in the Te Puke region of New Zealand. Fruit losses, measured after 18 wk of cold storage in normal atmosphere and CA were divided into four categories, namely, blossom-end rot (a rot that originates and spreads from the stigma end of the fruit), stem-end rot (originating from the pedicel end of the fruit), and side rots and fungal pits (discrete, depressed pits usually on the side of the fruit). The relative percentage of fruits with blossom-end rot or side rot, where *Phomopsis* spp. was isolated, was 40 and 10%, respectively. The relative percentage of fruits with fungal pits or side rots, where *C. actinidiae* was isolated, was 11 and 20%, respectively (Fullerton et al., 2007). Incidence of typical stem-end rot, caused by *Phomopsis* spp., was low at about 5%. The authors concluded that *C. actinidiae* infected 'Hort16A' mostly during the flowering period, but could also infect the fruit at any stage during fruit development and remain symptomless. No specific growth stage of susceptibility was identified for *Phomopsis* spp., indicating that fruits were susceptible to infection throughout fruit development. Furthermore, both of these fungi were able to remain in a symptomless (latent) state, up to and after harvest.

Storage conditions were also reported to have a major effect on rot expression in 'Hort16A' (Manning and Lallu, 1997; Patterson et al., 2003), especially on fruits cold stored in CA conditions. In contrast to 'Hayward', chilling injury increased in 'Hort16A' fruits when stored for long durations at 1.5°C and especially if the fruits were immature at the time of picking and cold storage (Maguire et al., 2005). Chilling injury was described as "a major precursor" for the expression of fungal rots in 'Hort16A' (Fullerton et al., 2007), representing a significant stress factor that altered the physiological state of the fruit, allowing latent infections to activate and become symptomatic. Other less frequently isolated fungal pathogens of 'Hort16A' in New Zealand were described; examples include *B. cinerea, Cylindrocarpon* cf. *candidum* (Link) Wollenw., *Diaporthe* spp., *Phoma exigua* Desm., *Colletotrichum acutatum* J.H. Simmonds, and *Fusarium* spp. (Manning et al., 2003). Pre- and postharvest research on the ecology, epidemiology, and control of fruit rots of 'Hort16A' came to an abrupt halt in New Zealand following the devastating outbreak of bacterial canker caused by *Pseudomonas syringae* pv. *actinidiae* (Psa), biovar 3, in November 2010 (Everett et al., 2011; Vanneste, 2012). The impact of this new bacterial pathogen on 'Hort16A' in New Zealand was described as "brutal" (Vanneste, 2017), and as a direct consequence of this incursion, more than 80% of the 'Hort16A' plantings in New Zealand have now been grafted over to a new yellow-fleshed cultivar 'Zesy002' (marketed under the name Zespri® SunGold Kiwifruit), informally known as Gold3 (Costa and Ferguson, 2015). Importantly, this new cultivar has a greater degree of tolerance to Psa biovar 3 than 'Hort16A' and is leading the economic recovery of kiwifruit production in New Zealand.

Some researchers have suggested that yellow-fleshed cultivars may be more susceptible to postharvest fruit rots than green-fleshed cultivars (Kwon et al., 2011), but very few comparative studies have been reported. In one study, a new yellow-fleshed cultivar, 'Haegeum' (originally introduced in Korea in 2007), was compared with 'Hayward' (Kim et al., 2015b). Not only was 'Haegeum' significantly less susceptible to postharvest rots than 'Hayward', with fruit losses of 3 and 19%, respectively, but this cultivar was also less susceptible to bacterial canker caused by Psa. In China, the most widely planted of a number of yellow-fleshed cultivars is 'Huayou' (a hybrid between *A. chinensis* var. *chinensis* and *A. chinensis* var. *deliciosa* (Costa and Ferguson, 2015), but literature searches found no information related to percentage crop loss caused by specific postharvest rot pathogens. In one study, *D. lithocarpus* was isolated from postharvest rotted fruit of 'Jinyan', another yellow-fleshed kiwifruit from China (Li et al., 2016a). The same researchers also reported isolating, for the first time, *Pestalotiopsis microspora* (Speg.) Bat. & Peres from fruits rotted after harvest from the Hubei Province (Li et al., 2016b).

3.2.3 *A. chinensis* var. *rufopulpa*

'Hongyang', the first red-fleshed cultivar developed from *A. chinensis* var. *rufopulpa* (Wang, 2011; Li et al., 2015), is very popular with Chinese consumers (Zhou et al., 2015), with 17,833 ha under production representing nearly 16% of the total Chinese area in kiwifruit (Costa and Ferguson, 2015). A 3-yr survey (2010–2013) identified three fungal species belonging to the *Botryosphaeriaceae* family (*B. dothidea, Lasiodiplodia theobromae* [Pat.] Griffon & Maubl., and *N. parvum*) to be the primary causal organisms of postharvest rots of 'Hongyang' in Sichuan Province (Zhou et al., 2015). Other significant postharvest rots of 'Hongyang' included *B. cinerea* (Figure 4.2d) and *P. expansum* (Figure 4.3a), with up to 37% fruit infection reported, primarily caused by *B. cinerea* (Tang et al., 2015). Artificial inoculation studies in New Zealand have also demonstrated that 'Hongyang' is also susceptible to *Phomopsis* sp. (Figure 4.3b), *Cryptosporiopsis* sp. (Figure 4.3c), and *C. acutatum* (Figure 4.3d).

The novelty and visual appeal of red-fleshed cultivars to consumers worldwide have prompted many plant breeders to begin screening their *Actinidia* collections for this genetic trait. However, to be commercially successful, a red-fleshed cultivar must meet certain criteria, including significant and consistent red pigmentation, acceptable fruit size, high yield, tolerance of long storage, superior eating quality, and disease resistance (Seal et al., 2016). In New Zealand, a selection of *A. chinensis* 'Hort22D' was identified with an intense red center, but subsequent field evaluation has shown that this cultivar was highly susceptible to Psa biovar 3 and its future is now uncertain (Stephen Hoyte, personal communication). The focus of research, especially in China, has been on identifying factors that affect the intensity of red pigmentation and to improve storage life in both air and CA storage conditions. Little has been published about postharvest rots of red-fleshed cultivars. However, if Botrytis rot was to become established as a persistent and perennial problem for producers in China, the wealth of information describing Botrytis rot in other kiwifruit will be invaluable to the development of new strategies to control Botrytis rot in 'Hongyang' in China.

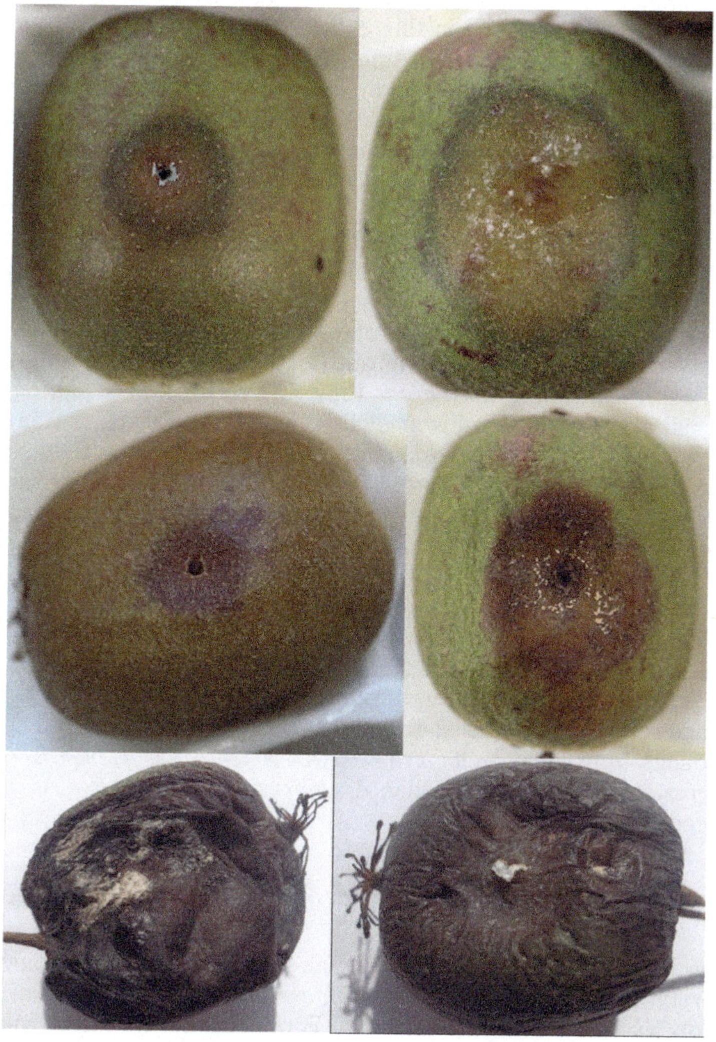

Figure 4.3 a. Blue mold infection by *Penicillium expansum* on 'Hongyang' kiwifruit (top left). Artificial infection on 'Hongyang' kiwifruit of: b. *Phomopsis* sp. (top right), c. *Cryptosporiopsis* sp. (center left), and d. *Colletotrichum acutatum* (center right). Symptoms of natural infection on *Actinidia arguta* of: e. Gray mold (bottom left) and f. Blue mold (bottom right).

3.2.4 *Actinidia arguta*

Kiwifruit belonging to *A. arguta* differs markedly compared to that belonging to *A. chinensis* var. *deliciosa* and *A. chinensis* var. *chinensis* and is referred to as kiwiberry or hardy kiwi. Successful commercialization of kiwiberry has been hampered by short storage life (about 1 mon), compared with those of *A. chinensis* var. *deliciosa* (6–8 mon) and *A. chinensis* var. *chinensis* (4–6 mon). There have been few reports of postharvest rots associated with *A. arguta* fruit during storage and only three publications are described in this chapter. In 2008, in Italy, up to 5% of fruit samples of *A. arguta* Ananasnaya ('Anna red') developed browning and depressions on the fruit surface and decay in the flesh, which was caused by *B. cinerea* (Figure 4.3e) (Romanazzi, 2009). A survey of postharvest losses in cold-stored *A. arguta* described lesions as soft, watery, pale yellow to light brown in color with masses of blue–green conidiophores on 10–15% of fruits sampled from Jilin Province, China, in 2014. Subsequent tests confirmed that the cause was *P. expansum* (Wang et al., 2015b), which was found but not yet reported in Italy (Figure 4.3f). In a related study, 5–10% of "ripe rots" were reported after cold storage at 4°C, with lesions described as soft, brown, slightly water-soaked, and with abundant white to pink mycelium. Morphological and molecular identification and pathogenicity tests confirmed that the cause was *Fusarium acuminatum* Ellis & Everh., a new pathogen responsible for postharvest decay of kiwifruit in China (Wang et al., 2015a). With the exception of *B. cinerea, A. arguta* does not appear to suffer from the usual range of postharvest rots more commonly associated with *A. chinensis* var. *deliciosa, A. chinensis* var. *chinensis*, or *A. chinensis* var. *rufopulpa* cultivars, and this absence of postharvest diseases was described as a real advantage for the future commercialization of this kiwifruit (Hassall et al., 1998). With new selections of *A. arguta*, especially those with significantly longer storage life (e.g., 'Mansu', from Korea (Jung et al., 2016), more postharvest diseases may be reported. In what could be described as an ironic twist, it was quickly discovered that *A. arguta* in New Zealand had a high degree of resistance to Psa biovar 3, compared with the commercial green- and yellow-fleshed kiwifruits, and this highly valuable attribute has been used in new breeding programs, where seedlings from interspecific crosses (*A. chinensis* var. *chinensis* × *A. arguta*) exhibited greater resistance to Psa than *A. chinensis* var. *chinensis* × *A. chinensis* var. *chinensis* seedlings (Hoyte et al., 2015).

3.3 Psa and Postharvest Rot Relationships

It has been noted that the pandemic outbreak of the highly virulent biovar 3 of Psa changed everything for kiwifruit production, even threatening the future of kiwifruit industries in many countries (Costa and Ferguson, 2015). Furthermore, recent evidence suggests that fruits harvested from Psa-infected vineyards have more Botrytis stem-end rot than fruits from healthy vineyards (Prencipe et al., 2016). They reported, after a 2-yr study (2013–2014), located in the Piedmont region of Italy, that fruits from Psa-infected orchards had significantly higher dry matter content, lower calcium content, and lower percentage of titratable acidity (TA) at harvest and during storage, that firmness was significantly lower, and that TSS of the fruit was significantly higher than fruits from apparently healthy vines that were free of Psa. The authors concluded that biophysical and biochemical changes had occurred in the fruit, as a consequence of preharvest Psa infection in the canopy, significantly increasing the susceptibility of 'Hayward' fruits to postharvest Botrytis rot development. This has potential postharvest implications for other green-, yellow-, and red-fleshed cultivars of *Actinidia* and further research is required to better understand the complex relationships between preharvest canopy infection and fruit susceptibility to postharvest pathogens.

3.4 Conclusions

A wide range of fungal pathogens often infects commercialized kiwifruit before harvest, remain symptomless and cause significant postharvest losses after a period of time in normal and CA cold storage. The most important postharvest pathogen we described was *B. cinerea*, the cause of Botrytis stem-end rot, especially in 'Hayward' in Italy, the USA, Chile, and Greece. In China, statistics relating to commercial cultivars and postharvest crop losses were difficult to find, but *B. cinerea* was reported to cause significant postharvest losses of the red-fleshed variety 'Hongyang'. Overall, it is not surprising that *B. cinerea* has been responsible for economically significant losses in commercialized *Actinidia* cultivars given that conducive conditions exist for Botrytis rot in many kiwifruit vineyards before harvest and throughout the postharvest chain, including inadequate cold storage, poor handling practices, high RH, a slowly senescing fruit, and increasing TSS (Romanazzi and Droby, 2016). Many pre- and postharvest factors affecting the occurrence of *B. cinerea* infections in kiwifruit have been extensively reviewed (Michailides and Elmer, 2000; Elmer and Michailides, 2007; Manning et al., 2010; Romanazzi and Droby, 2016). Successful management of *B. cinerea* has been accomplished in New Zealand through a range of pre- and postharvest non-chemical and cultural control practices in 'Hayward'. In contrast, fungicides are still used as a preharvest gray mold management tool in several other countries, at times combined with modified curing practices (e.g., Italy), but problems in 'Hayward' persist, which seems oddly out of step with the New Zealand experience. Climatic differences most likely play a role, as highlighted in a comparative study by Michailides and Elmer (2000) of *B. cinerea* epidemiology in a cool temperate (New Zealand) versus hot, dry climate (California). Pathogen ecology and epidemiology in these two climatic regions differed. The two cornerstones of New Zealand's successful gray mold management program are preharvest vine management, to increase canopy aeration, and postharvest curing by holding fruits in bins in covered areas at ambient temperatures for at least 2 d before packing and cold storage (Pennycook and Manning, 1992; Bautista-Baños et al., 1997; Manning et al., 2010; Mari et al., 2015). However, curing at ambient temperatures represented considerable logistical problems in some kiwifruit regions in Italy and was often not very effective (Tonini et al., 1999). Modifications were made to the New Zealand curing practice, in Italian cold stores, to adapt to local needs and this change was reported to be successful. However, preharvest fungicides may still be used in Italy to achieve commercially acceptable postharvest Botrytis stem-end rot control.

The two so-called ripe rot pathogens responsible for globally significant postharvest losses in 'Hayward' are *Phomopsis* spp. stem-end rot and *Botryosphaeria* spp. body rots. If postharvest conditions are optimized and storage is not extended beyond the physiological life of the fruit, these rots are generally regarded as not being problematic in New Zealand, Italy, California, or Greece. In contrast, *Phomopsis/Diaporthe* spp. stem-end rot has been reported to be an issue in 'Hayward' in Chile, which argues for more fundamental research to identify the host, pathogen, and environmental conditions that allow these latent infections to occur in 'Hayward' in Chile.

Generally, the yellow-fleshed cultivars of *Actinidia* are more prone to *Phomopsis* spp. rots within their shorter (4–6 mon) "optimum" storage time compared with that of the longer stored 'Hayward', where exporters may extend the storage time beyond 6 mon. The type of cold storage system was shown to affect the incidence of *Phomopsis* spp. rots. Fruits emerging from CA cold storage had significantly more *Phomopsis* spp. rots compared with fruits that had been stored in normal atmosphere cold stores. CA cold stores are often preferred because fruit firmness is retained for longer periods compared with those in normal atmosphere cold storage, but further studies are needed to identify the biological and physiological basis for greater *Phomopsis* spp. stem-end rot incidence in CA cold stores.

Botryosphaeria spp. rots were reported to occur sporadically in the yellow-fleshed 'Hort16A' in New Zealand and also in 'Hayward' in California, Chile, Greece, and Italy. In contrast, in China, this pathogen has caused significant postharvest rots, mostly affecting the red-fleshed cultivar 'Hongyang'. This may be a function of 'Hongyang' being highly susceptible to this pathogen and/or being stored beyond its physiological life. In addition, factors such as suboptimal fruit ripening conditions after cold storage and during fruit marketing, including excessive heat and high RH, may play a role in ripe rot occurrence.

3.5 Looking Forward

'Hayward' will probably remain the most important cultivar of *A. chinensis* var. *deliciosa,* which will be traded globally in the near future, because the fruit is large, has good flavor, and stores remarkably well (Ferguson, 2015). New, green-fleshed cultivars are being planted, such as 'Summer 3373', Green Light® (Italy and Chile), and 'Zesh004' (marketed as Zespri® Sweet Green Kiwifruit) in New Zealand, but production and commercialization of these are in the early stages.

The global outbreak of Psa biovar 3 had a devastating economic and social impact (Vanneste, 2017) and essentially destroyed 'Hort16A' production in several kiwifruit-growing regions (e.g., New Zealand and Italy). Ferguson (2015) suggested that even the viability of 'Jintao' and 'Soreli' in Italy (Scortichini et al., 2012) and 'Hongyang' in China may be threatened by Psa. With this new challenge, global breeding programs will need to shift their emphasis from long storage, flavor, sweetness, fruit size, and yields to incorporating a degree of resistance to Psa. As new Psa-resistant cultivars are introduced, new postharvest pathogen complexes may emerge.

4 Strawberry

4.1 Botanical Classification

Cultivated strawberries, *Fragaria* × *ananassa*, belonging to the Rosaceae family, are the product of an interspecific cross between two wild species, *Fragaria chiloensis*, originated from Chile, and *Fragaria virginiana*, from North America. The edible part of strawberry, even though it is commonly called a fruit, is not technically a real fruit, because it is derived from the enlargement of the receptacle of the flower. The real fruits are the achenes, disseminated into the enlarged receptacle; however, to avoid any misunderstanding, we will refer to the enlarged receptacle as strawberry fruit. Thanks to breeding activities and advanced techniques of cultivation, actual strawberry fruits are much larger compared with the wild species and their appearance, flavor, texture, and color are appreciated all over the world, as indicated by global strawberry production, which continues to increase. Global strawberry production has increased by 59% from 5.8 million tons in 2007 to 9.2 million tons in 2017 (FAOSTAT, 2019), and this growth is predicted to increase over the next decade. Despite this growth, there are many challenges ahead, particularly because strawberries are very delicate and highly susceptible to fungal pathogens responsible for postharvest rots. Strawberries are also highly susceptible to mechanical injuries due to poor handling practices, damage during transportation after harvest, and rapid water loss after harvest.

4.2 Postharvest Diseases of Strawberry

4.2.1 Gray Mold

The general properties of *B. cinerea* were described earlier in the table grape section of this review. In strawberries, the disease cycle starts with conidia or resistance structures (e.g., sclerotia and overwintering mycelium) germinating in response to conducive environmental conditions with temperatures between 15 and 25°C and periods of rainfall (Fillinger and Walker, 2016), both events typical of spring in warm temperate regions, when strawberries are flowering. The infection starts from the flowers; the growth of mycelium into various flower parts gives rise to later fruit rots (Bristow et al., 1986; Romanazzi et al., 2016b) (Figure 4.4a,b). Lesions become visible on ripening fruit 2–4 wk after infection or they can remain latent until cold storage, where the higher RH and lower temperatures are conducive for disease development. In cold storage, enzymatic processes are slowed, including those involved in host defenses. After the primary infection before harvest, subsequent disease spread occurs by conidia dispersed in the crop by wind, splashing rain, or irrigation water (Holz et al., 2007). During the season, epidemics of gray mold are characterized by continuous infection cycles. The main symptoms of gray mold are the formation of brown-colored, soft lesions on the fruit surface, which slowly enlarge and develop a fluffy mycelium. These are initially white but become gray once conidia are produced; hence, the common name of the disease is gray mold. During cold storage, the disease spreads quickly from one infected berry to the next adjacent healthy one, a phenomenon known as "nesting", which can result in the destruction of entire batches of fruits. Although disease spread is slower during cold storage compared to room temperature, losses during cold storage can be significant.

4.2.1.1 Cultivation Practices Affecting Strawberry Postharvest Gray Mold

Preharvest practices can influence the incidence and causal pathogens responsible for strawberry postharvest decay. Poor agronomic choices can also facilitate the spread of pathogen inoculum in the field. For example, during strawberry planting, the choice of cultivar and planting distance and site, such as within greenhouses or cultivation in open fields, can influence other management factors, such as irrigation and harvest date. Among cultivars susceptible to postharvest decay, disease incidence can be facilitated by the locale, such as cultivation near rivers or other water sources that are moist and stagnant, especially during spring. Drip irrigation is preferred over using overhead sprinklers or lateral move irrigation. Drip irrigation minimizes close proximity of the fruits to saturated, stagnant water, which can occur when strawberries are cultivated in raised beds and plastic mulch. In some cases, fertilization is carried out with irrigation using a system termed fertigation, where not exceeding

Figure 4.4 a. Gray mold caused by *Botrytis cinerea* infecting strawberry fruit during storage, with necrotized infected flower residues (top left). b. Gray mold on strawberry fruit under field conditions (top right). c. Soft rot caused by *Rhizopus stolonifer* affecting strawberry fruit during storage (bottom left). d. Blue mold on strawberry fruit during storage (bottom right).

the recommended quantity of nitrogen is recommended because excessive nitrogen favors the disease. Spacing between rows is usually managed without the need for grass, thereby reducing competition for nutrients and water, and improved aeration is also another benefit. Aeration is especially important within greenhouses regardless of their models and fans are recommended to avoid excess of humidity. It is very common to use straw mulching between crop rows to control weeds and to reduce soil being splashed onto fruit surfaces due to rainfall. At the end of every season, crop residues must be removed and destroyed by deep burial or burning since they usually harbor inoculum structures, such as conidia, mycelia, and sclerotia, which enable the pathogen to successfully overwinter. Once spring and mild temperatures occur, sclerotia of *B. cinerea* germinate, giving start to a new wave of gray mold infections (Figure 4.4b). The time of harvest is an important factor that affects subsequent storage fruit decay and overripe fruits are more susceptible to infection. Berries are very delicate fruits and mechanical injuries at harvest are potential entry points for several postharvest decay pathogens. To reduce the risk of mechanical injury at harvest, fruit picking and packaging is often carried out in the field. After harvest, rapid cooling and cold storage with no breaks in the cold chain can extend the shelf life of the strawberries (Romanazzi et al., 2016b).

4.2.1.2 Fungicides Used to Control Strawberry Postharvest Gray Mold Strawberries are considered a valuable crop, and growers can receive higher returns per hectare compared with other crops, such as cereals. However, the fruit must be visually perfect and the inputs required from growers are significantly greater. In conventional production, repeated applications of fungicides are applied up to harvest. For example, in strawberries the productive cycle is typically 25 d from flowering to harvest, and approximately five fungicides are applied, depending on climatic conditions. After harvest, no fungicides are applied in both "organic" and "Integrated Pest Management" systems. Therefore, the effectiveness of fungicides applied in the field before harvest should persist into cold storage, when enzymatic processes, including for plant defense mechanisms, are slowed down by low temperatures. Gray mold is the main target of fungicides applied in strawberry fields. However, *B. cinerea* is considered a high-risk pathogen for the development of fungicide resistance (Myresiotis et al., 2007). Careful use of a limited number of active ingredients and care with respect to rates and timing of their application are important to reduce the development of pathogen strains resistant to fungicides (Amiri et al., 2018). There are many fungicides registered for strawberry plants and these are classified by the Fungicide Resistance Action Committee (FRAC) according to their mechanisms of action. For example, the fungicide fenhexamid can be applied as late as 3 d before harvest and its effectiveness persists during the subsequent fruit storage, as it has been observed on treated table grapes (Smilanick et al., 2010). However, the concerns of consumers and government agencies about detectable fungicide residues on/in food has encouraged growers to use alternatives to conventional fungicides, which are more environmentally sustainable, and to seek other cultural control methods or at least to combine the application of multiple control strategies for effective control of gray mold. In some cases, the effectiveness of individual cultural control methods and/or biofungicides may not equal that of conventional fungicides, but postharvest decay control can be achieved if these practices are integrated with conventional fungicides. Moreover, alternating pest management strategies between conventional and alternative means can have additional important benefits since it can decrease the risk of the development of pathogen fungicide resistance.

4.2.1.3 Alternatives to Conventional Fungicides Alternatives to conventional synthetic fungicides are generally grouped into three categories: physical methods, biological control agents (BCAs), and substances generally recognized as safe (GRAS) (Palou et al., 2016). GRAS substances are allowed under "organic" agriculture rules since their effects on the environment and on human health are lower than those of synthetic fungicides and this status is certified by the United States Food and Drug Administration in the USA or by the European Food Safety Authority (EFSA) in the European Union (EU). They also include disinfectants and sanitizing agents, resistance inducers, and natural products. Disinfecting agents minimize both populations of postharvest pathogens and the risks of contaminating bacterial human pathogens on the commodities and within packing/processing facilities, both improving storage life of products and improving food safety (Feliziani et al., 2016). Resistance inducers stimulate the plants own defense systems and constrain attacks by pathogens (Latif et al., 2013).

4.2.1.3.1 Host Resistance Inducers Resistance inducers are compounds that protect plant tissues from pathogen infection or development since they induce a defense response in the plant (Amil-Ruiz et al., 2011). Chitosan acts as a resistance inducer, stimulating fruit defense pathways (Landi et al., 2014; Romanazzi et al., 2018), and in addition, it has antimicrobial properties (Feliziani et al., 2013). Chitosan, which is a natural biopolymer, was successfully used to control postharvest decay of strawberries, either when applied onto fruits (Romanazzi et al., 2013) or when sprayed during the season in the field (Feliziani et al., 2015). In addition, chitosan has film-forming properties that create a coating on fruit surface acting as a mechanical barrier to pathogens (Romanazzi et al., 2017). These properties make chitosan an attractive alternative to conventional fungicides and two different sources of chitosan have been formulated into a commercial product (ARMOUR-Zen®) and registered for use to control gray mold in grapes (Reglinski et al., 2010; Calvo-Garrido et al., 2013). Benzothiadiazole, which is a functional analogue of salicylic acid (Lawton et al., 1996), can either be applied to fruit after harvest (Romanazzi et al., 2013) or to the canopy during the growing season to prolong the storage of strawberries (Feliziani et al., 2015). Induced resistance in fruit tissues occurs by modulating the expressions of several genes, including those associated with the plant immune system, hormone metabolism, systemic acquired resistance, and biotic and abiotic stress (Landi et al., 2017).

4.2.1.3.2 Coatings The use of edible coatings has gained more interest among researchers and food industries (Chiabrando and Giacalone, 2015; Pérez-Gallardo et al., 2015). These film-forming coatings are reported to reinforce the mechanical resistance of fruits and vegetables to pathogen infection and decrease water loss via transpiration (Falguera et al., 2011; Calvo-Garrido et al., 2014; Flores-López et al., 2016). In addition, these coatings may incorporate antimicrobial compounds that further protect the fresh produce from microbial infection (Peretto et al., 2014). On strawberry plants and fruits, edible coatings were experimentally applied during the growing season as a liquid solution, which then solidified on the fruit surfaces and reduced fruit damage from mechanical injuries (Feliziani et al., 2015).

4.2.1.3.3 Biocontrol Agents BCAs developed for use in fruit crops are microorganisms applied to plant and fruit surfaces to act as antagonists of pathogens that cause pre- and postharvest decay. BCAs have a range of modes of action, including competition for nutrients and space, mycoparasitism, production of volatile compounds, induced resistance, or through secretion of extracellular enzymes or compounds with antimicrobial activity (Spadaro and Droby, 2016). Several good reviews on the use of BCAs to control plant pathogens, especially *B. cinerea*, have been published and readers are encouraged to investigate them for further details (Sharma et al., 2009; Spadaro and Droby, 2016). Research on the effectiveness of BCAs to control postharvest decay of strawberry fruit has a long tradition, starting more than 40 yr ago with studies on the inhibitory activity of *Trichoderma* spp. against *B. cinerea* causing gray mold (Tronsmo and Dennis, 1977). Twenty years later, a Canadian research group studied the potential for controlling fruit rots in commercial raspberry plantings (Yu and Sutton, 1997). In initial experiments, they sprayed suspensions of the BCA, the fungus *Gliocladium roseum* Bainier, to flowers and to the above ground portions of raspberry plants to control *B. cinerea*. Further research optimized the delivery of *G. roseum* to raspberry flowers through honeybees and bumblebees (Yu and Sutton, 1997). Similar research was conducted to disseminate *Trichoderma harzianum* Rifai to strawberry flowers to control gray mold (Kovach et al., 2000). On strawberry, the use of BCAs for disease management has been reported for a wide range of microorganisms, including *Trichoderma* spp. (Tronsmo and Dennis, 1977; Kovach et al., 2000; Robinson-Boyer et al., 2009), *Candida* spp. (Lima et al., 1997; Huang et al., 2011), *A. pullulans* (Lima et al., 1997; Adikaram et al., 2002), *Kloeckera apiculata* (Long and Yuan, 2009), *Lactobacillus plantarum* (Zamani-Zadeha et al., 2014), *Pichia carribbica* (Zhao et al., 2012), *Hanseniaspora uvarum* (Cai et al., 2015; Qin et al., 2017), *M. fructicola* (Karabulut et al., 2004), *C. laurentii* (Wei et al., 2014), *B. subtilis* (Robinson-Boyer et al., 2009), and *U. oudemansii*. The latter, formulated and registered into BOTRY-Zen® (Thomidis et al., 2015), and *Ulocladium atrum* Preuss (Boff et al., 2002a, 2002b) have been successfully tested on strawberry and shown to reduce postharvest decay. The application of BCAs in the field can be made with the BCA alone or in combination with other means of control. For example, after Zhang and co-workers confirmed the effectiveness of BCAs within the genus *Rhodotorula* to reduce postharvest gray mold of strawberry (Zhang et al., 2007), they combined their application with phytic acid (Zhang et al., 2013), salicylic acid (Zhang et al., 2010), and edible coatings (Zhang et al., 2014).

4.2.1.3.4 Disinfecting Agents Ozone at constant low levels prolonged the storage life of blackberry (Barth et al., 1995), raspberry (Giuggioli et al., 2015), blueberry (Crowe et al., 2012), and cranberry (Norton et al., 1968) without injury, but not that of strawberry (Pérez et al., 1999; Nadas et al., 2003) where it reduced fruit quality and did not reduce gray mold incidence. This is likely because *B. cinerea* infections in strawberry are often latent, initiated before harvest, and within the strawberry flesh, where ozone could not reach them. Fumigation of berry fruits has been carried out with compounds of natural origin. A triple fumigation of 30 min each with 2 mg/L acetic acid vapors reduced decay of strawberry by *B. cinerea* by 56% (Hassenberg et al., 2010). Essential oil fumigation and essential oils in solution or formulated in coatings were used to manage postharvest decay of strawberries (Feliziani and Romanazzi, 2016). Indeed, interest in research into the use of essential oils for disease management has increased since they are natural compounds with good antimicrobial properties. It is important that the safety and influence of essential oil treatments on organoleptic properties of fruit should be evaluated (Servili et al., 2017).

CA is obtained during the storage within cold rooms through the control of the atmosphere gaseous composition. Alternatively, modified atmosphere (MA) is obtained when the atmosphere composition is changed by the respiration of the fruit and the permeability of the packaging material. Typically, in both CA and MA, the oxygen content is reduced and carbon dioxide increased. Raspberries cold stored in an MA had better quality compared to the control fruits, with higher TSS and TA and better retention of their original color (Moor et al., 2014). A storage atmosphere with an elevated oxygen content of 40 kPa (approximately double that of air) or higher, increased the antioxidant capacity and postharvest life of strawberry fruit (Ayala-Zavala et al., 2007).

Strawberries treated after harvest or during the season from flowering to maturity with a mixture of organic acids and calcium had less postharvest decay (gray mold and Rhizopus rot) compared to control fruits treated with water (Feliziani et al., 2015). An investigation at the molecular level revealed that this commercial formulation induced resistance to pathogens in the strawberry tissues, stimulating the expression of key phenylpropanoid pathway genes (Landi et al., 2014).

4.2.1.3.5 Physical Treatments Antifungal physical treatments are those that employ temperature or pressure to control postharvest fruit decay (Usall et al., 2016). Rapid cooling and prompt cold storage after harvest can reduce postharvest decay and prolong berry shelf life. Storage at hypobaric pressure (less than atmospheric pressure) reduced postharvest gray mold and Rhizopus rot and induced resistance in strawberry tissues that were treated for 4 hr at 50 kPa, approximately one-half normal atmospheric pressure at sea level (Romanazzi et al., 2001; Hashmi et al., 2013). The activity of the defense-related enzymes, such as phenylalanine ammonia-lyase (PAL), chitinase, and peroxidase increased in strawberry tissues exposed to the hypobaric environment compared to untreated samples. Other physical methods include sterilizing the external berry surface with UV light. Strawberry fruit treated with UV-C doses of 0.50 and 1.00 kJ/m^2 had significantly less postharvest gray mold after either artificial or natural *B. cinerea* infection, in comparison with the non-irradiated controls (Nigro et al., 2000). In addition, UV-C doses increased the activity of PAL in the strawberry tissues 12 hr after irradiation, which indicated activation of metabolic pathways related to the biosynthesis of phenolic compounds had occurred.

4.2.2 Other Diseases

Rhizopus stolonifer and *Mucor piriformis* A. Fisch. are the two pathogens responsible for soft rots, sometimes referred to as "leak diseases". This name derives from the disease-specific symptoms. Enzymes of these pathogens disrupt the middle lamellae of the fruit cells and, as a consequence, the plant tissues become soft and the cell contents leak out. This leak is extremely infectious during fruit storage and facilitates widespread dispersal of the pathogens through entire fruit lots. In addition, within fruit clamshells, *R. stolonifer* produces stolons, which are filamentous, rapidly growing mycelial extensions that easily infect nearby healthy fruits (Figure 4.4c). In the case of *M. piriformis*, cold storage does not completely inhibit fungal growth, even at temperatures of 0–5°C. Conversely, the growth of *R. stolonifer* usually stops at storage temperatures below 5°C (Sommer, 1982). In the field, both diseases can spread quickly, especially during rainy, warm weather. Fungicide applications are usually recommended in the field as soon as the first leak symptoms occur, but very few fungicides that are registered effectively control these pathogens. Distinguishing between the two soft rot pathogens is difficult both in the lab and in the field; however, some morphological and physiological differences exist to aid with identification. *Mucor piriformis* sporangia have a visible film of viscous liquid on the surface that makes them sticky to aid dispersal by invertebrates, whereas *R. stolonifer* sporangia have a dry appearance. After infection with *R. stolonifer*, the fruit surface can be covered with thin, fluffy mycelium. Eventually, sporangiophores develop with dark masses of black sporangia at their tips, which can cover the entire surface of the fruit (Bautista-Baños et al., 2014).

Other postharvest pathogens of strawberry and other berry fruits are *Colletotrichum* spp., *Penicillium* spp., *Aspergillus* spp., *Alternaria* spp., and *Cladosporium* spp. These are considered minor fruit rots of limited or only occasional economic importance. *Colletotrichum acutatum* is responsible for strawberry anthracnose disease. The typical symptoms on the berry surface appear as brown sunken spots that in the middle of the lesion produce an orange lactate of conidia, the ideal means for the fungus widespread.

The fungus *P. expansum*, causal agent of blue mold, produces blue-grayish masses of dust-like conidia (Figure 4.4d). *Aspergillus* spp., *Alternaria* spp., and *Cladosporium* spp. are not considered economically important in strawberries, but the incidence of these pathogens has increased recently among those cultivars with resistance to gray mold (Romanazzi et al., 2019). Losses vary according to environmental conditions during harvest, timing of fungicide applications, cultivar, and storage and transport conditions. Generally, few or no control strategies have been developed for these secondary diseases and most are suppressed by refrigeration and good crop hygiene.

5 Concluding Remarks

Storage technology and fruit quality and shelf life have improved greatly in recent decades, but this progress has not included substantial improvements in the management of postharvest decay, in general, and gray mold, in particular. Preharvest fungicides are still used in many kiwifruit-growing countries, and the sustainability of this practice must be questioned with the introduction of strict, new EU regulations on pesticide use in food crops (Lamichhane et al., 2016, 2017). New Zealand's Zespri Group Limited (the world's largest marketer of kiwifruit) does not allow any fungicide to be applied after petal fall in green- and yellow-fleshed cultivars to ensure fruit remain pesticide residue-free (Elaine Gould, personal communication). Therefore, the combination of stricter regulations on fungicide use, loss of fungicide chemistry due to voluntary withdrawal by chemical companies, pathogen resistance development to several key fungicide groups, and an increasing demand for certified "organic" fruits indicate significantly reduced fungicide use in food crops in the future, including kiwifruit. To compensate for reduced fungicide inputs, there will be a need for greater employment of biofungicides and non-chemical management tools to effectively manage postharvest rots in kiwifruit. In kiwifruit systems, where synthetic fungicide inputs are reduced, there may be emergence of kiwifruit pathogens that were previously regarded as minor; therefore, kiwifruit researchers need to remain vigilant for new and unusual postharvest diseases. Early screening in specific kiwifruit-growing regions in breeding programs to determine the susceptibility of selections to significant postharvest pathogens is of great value in identifying those that would be promising cultivars. Cultivars with highly desirable consumer and marketing traits, but an unacceptably high number of rots, in the absence of fungicide, should be abandoned early in the commercialization process. This is because it is unlikely that conventional fungicides will be freely available and may disappear from spray recommendations in many countries in the near future. In the short-to-medium term, it is expected that there will be a continual withdrawal of key pre- and postharvest fungicides from the market place, especially in Europe, as a result of their prohibition by regulatory agencies and tougher EU regulations. This in turn will lead to the need for effective alternatives to synthetic fungicides to be identified and implemented into fully integrated control strategies, which combine cultural, biological, physical, chemical, and nutrient management practices with partial host resistance into a truly holistic postharvest disease control program for kiwifruit. Finally, the fundamental and applied research effort directed at all aspects of Botrytis stem-end rot in kiwifruit over the last two decades in several countries serves as an excellent model for developing effective management practices of non-Botrytis rots in *Actinidia* spp. cultivars.

Botrytis cinerea remains the most important postharvest pathogen of table grape and strawberry. The fungus produces abundant inoculum in the field that can both infect before and after harvest, when fruit defenses become weaker and conditions are optimal for the pathogen. Regular preharvest treatment with fungicides can be effective at reducing pre- and postharvest gray mold, but resistance by this pathogen to commonly used active ingredients is common. Combined with concerns by consumers and retailers about the human and environmental risks of fungicide residues on fruit, have led to a reduction in the number of active ingredients being applied. This may favor an increase in the amount of pathogenic inoculum on fruit surface and/or in the frequency of latent infections that can later develop during storage, transportation, and shelf life in the market and at the consumers' homes. Therefore, the need for effective alternatives to synthetic fungicides has become more urgent. BCAs have effectively controlled gray mold of strawberries in many studies, but reliance on their use alone is unwise because their effectiveness can vary. Therefore, their use must be integrated with other disease management practices, such as preharvest practices, including strategic early use of fungicides to reduce inoculum pressure, cultural control methods, plant nutrition, and activators of host defenses during the growing season. These actions, when complemented with postharvest treatments, may approach similar levels of fruit quality and disease control obtained using fungicide treatments alone. Shtienberg (2007) cited good examples where these fully integrated approaches were successful, which may serve as a model for their increased development and use to ensure that high quality and safe fruits arrive at the consumers' tables.

Acknowledgments

We are grateful to Drs. Tony Reglinski, Stephen Hoyte, Pia Rheinlander (The New Zealand Institute for Plant & Food Research Limited), and Elaine Gould (Zespri International Limited) for their valuable comments on early drafts of the kiwifruit section of this manuscript, and to Amnon Lichter (Volcani Center, Israel) for critical revision of the table grape section. We also thank the Science Publications Office of Plant & Food Research Ltd. for their valuable editorial comments on the kiwifruit section of this chapter.

References

Abdollahi, A., Hassani, A., Ghosta, Y., Bernousi, I., Meshkatalsadat, M.H., Shabani, R. and Ziaee, S.M. 2012. Evaluation of essential oils for maintaining postharvest quality of Thompson seedless table grape. *Natural Product Research* 26, 77–83.

Adikaram, N.K.B., Joyce, D.C. and Terry, L.A. 2002. Biocontrol activity and induced resistance as a possible mode of action for *Aureobasidium pullulans* against grey mould of strawberry fruit. *Australasian Plant Pathology* 31, 223–229.

Amil-Ruiz, F., Blanco-Portales, R., Muñoz-Blanco, J. and Caballero, J.L. 2011. The strawberry plant defense mechanism: A molecular review. *Plant and Cell Physiology* 52, 1873–1903.

Amiri, A., Zuniga, A.I. and Peres, N.A. 2018. Potential impact of populations drift on *Botrytis* occurrence and resistance to multi- and single-site fungicides in Florida southern highbush blueberry fields. *Plant Disease* 102, 2142–2148.

Auger, J., Pérez, I. and Esterio, M. 2013. *Diaporthe ambigua* associated with post-harvest fruit rot of kiwifruit in Chile. *Plant Disease* 97, 843.

Ayala-Zavala, J.F., Shiow, Y.W., Chien, Y.W. and González-Aguilar, G.A. 2007. High oxygen treatment increases antioxidant capacity and postharvest life of strawberry fruit. *Food Technology and Biotechnology* 45, 166.

Barth, M.M., Zhou, C., Mercier, J. and Payne, F.A. 1995. Ozone storage effects on anthocyanin content and fungal growth in blackberries. *Journal of Food Science* 60, 1286–1288.

Bautista-Baños, S., Bosquez-Molina, E. and Barrera-Necha, L.L. 2014. *Rhizopus stolonifer* (Soft rot). pp. 1–44. *In*: S. Bautista-Baños (ed.). *Postharvest Decay. Control Strategies*. Academic Press. Elsevier Inc., London, UK.

Bautista-Baños, S., Long, P.G. and Ganesh, S. 1997. Curing of kiwifruit for control of postharvest infection by *Botrytis cinerea*. *Postharvest Biology and Technology* 12, 137–145.

Beresford, R.M. 2005. Reducing the impact of fruit spoilage diseases on New Zealand's fruit export industries. *Proceedings of the Fifty Eighth New Zealand Plant Protection Conference* 58, 37–41.

Boff, P., Köhl, J., Gerlagh, M. and de Kraker, J. 2002a. Biocontrol of grey mould by *Ulocladium atrum* applied at different flower and fruit stages of strawberry. *BioControl* 47, 193–206.

Boff, P., Köhl, J., Jansen, M., Horsten, P.J.F.M., Lombaers-van der Plas, C. and Gerlagh, M. 2002b. Biological control of gray mold with *Ulocladium atrum* in annual strawberry crops. *Plant Disease* 86, 3, 220–224.

Brigati, S., Gregori, R., Neri, F. and Pratella, G.C. 2003. New procedures for 'curing' and controlled atmosphere (CA) storage to control *Botrytis cinerea* in kiwifruit. *Acta Horticulturae* 610, 283–288.

Bristow, P.R., McNicol, R.J. and Williamson, B. 1986. Infection of strawberry flowers by *Botrytis cinerea* and its relevance to grey mould development. *Annals of Applied Biology* 109, 545–554.

Brook, P.J. 1992. Botrytis stem-end rot and other storage diseases of kiwifruit – A review. *Acta Horticulturae* 297, 545–550.

Burdon, J., Lallu, N., Pidakala, P. and Barnett, A. 2013. Soluble solids accumulation and postharvest performance of 'Hayward' kiwifruit. *Postharvest Biology and Technology* 80, 1–8.

Cai, Z., Yang, R., Xiao, H., Qin, X. and Si, L. 2015. Effect of preharvest application of *Hanseniaspora uvarum* on postharvest diseases in strawberries. *Postharvest Biology and Technology* 100, 52–58.

Calvo-Garrido, C., Elmer, P.A.G., Parry, F.J., Viñas, I., Usall, J., Torres, R., Agnew, R.H. and Teixidó, N. 2014. Mode of action of a fatty acid-based natural product to control *Botrytis cinerea* in grapes. *Journal of Applied Microbiology* 116, 967–979.

Calvo-Garrido, C., Elmer, P.A.G., Viñas, I., Usall, J., Bartra, E. and Teixidó, N. 2013. Biological control of botrytis bunch rot in organic wine grapes with the yeast antagonist *Candida sake* CPA-1. *Plant Pathology* 62, 510–519.

Cantín, C.M., Fidelibus, M.W. and Crisosto, C.H. 2007. Application of abscisic acid (ABA) at veraison advanced red color development and maintained postharvest quality of 'Crimson Seedless' grapes. *Postharvest Biology and Technology* 46, 237–241.

Chen, H.Z., Cheng, Z., Wisniewski, M., Liu, Y.S. and Liu, J. 2015. Ecofriendly hot water treatment reduces postharvest decay and elicits defense response in kiwifruit. *Environmental Science and Pollution Research* 22, 15037–15045.

Chervin, C., Aked, J. and Crisosto, C.H. 2012. Grapes. pp. 187–211. *In*: D. Rees, G. Farrell and J. Orchard (eds.). *Crop Post-Harvest: Science and Technology*. Wiley-Blackwell, Oxford, UK.

Chiabrando, V. and Giacalone, G. 2015. Anthocyanins, phenolics and antioxidant capacity after fresh storage of blueberry treated with edible coatings. *International Journal of Food Sciences and Nutrition* 66, 248–253.

Cossio, F., Debersaques, F. and Latocha, P. 2015. Kiwiberry (*Actinidia arguta*): New perspectives for a great future. *Acta Horticulturae* 1095, 423–434.

Costa, G. and Ferguson, A.R. 2015. Bacterial canker of kiwifruit: Response to a threat. *Acta Horticulturae* 1095, 27–40.

Cotrut, R.C., Stanica, F. and Cimpeanu, S.M. 2016. Influence of cold storage on fruit quality of some kiwifruit genotypes organically produced. *Romanian Biotechnological Letters* 21, 12110–12115.

Crowe, K.M., Bushway, A. and Davis-Dentici, K. 2012. Impact of postharvest treatments, chlorine and ozone, coupled with low-temperature frozen storage on the antimicrobial quality of lowbush blueberries (*Vaccinium angustifolium*). *LWT – Food Science and Technology* 47, 213–215.

De Miccolis Angelini, R.M., Rotolo, C., Masiello, M., Gerin, D., Pollastro, S. and Faretra, F. 2014. Occurrence of fungicide resistance in populations of *Botryotinia fuckeliana* (*Botrytis cinerea*) on table grape and strawberry in southern Italy. *Pest Management Science* 70, 1785–1796.

Dean, R., van Kan, J.A.L., Pretorius, Z.A., Hammond-Kosack, K.E., Di Pietro, A., Spanu, P.D., Rudd, J.J., Dickman, M., Kahmann, R., Ellis, J. and Foster, G.D. 2012. The top 10 fungal pathogens in molecular plant pathology. *Molecular Plant Pathology* 13, 414–430.

Díaz, G.A., Latorre, B.A., Jara, S., Ferrada, E., Naranjo, P., Rodríguez, J. and Zoffoli, J. P. 2014a. First report of *Diaporthe novem* causing postharvest rot of kiwifruit during controlled atmosphere storage in Chile. *Plant Disease* 98, 1274–1275.

Díaz, G.A., Latorre, B.A., Jara, S., Ferrada, E., Naranjo, P. and Zoffoli, J.P. 2014b. *Diaporthe* species identified from postharvest rot on kiwifruits during long term storage. *Phytopathology* 104, 33.

Donoso, A. and Latorre, B.A. 2006. Characterization of blue mold caused by *Penicillium* spp. in cold stored table grapes. *Ciencia e Investigación Agraria* 33, 119–130 (In Spanish with English abstract).

Elfar, K., Riquelme, D., Zoffoli, J.P. and Latorre, B.A. 2017. First report of *Botrytis prunorum* causing fruit rot on kiwifruit in Chile. *Plant Disease* 101, 388–389.

Elmer, P.A.G., Gaunt, R.E., Boyd-Wilson, K.S.H., Pyke, N.B. and Fermaud, M. 1995. External contamination of kiwifruit by *Botrytis cinerea*: An important source of inoculum for fruit infection. *Proceedings of the Forty Eighth New Zealand Plant Protection Conference* 48, 95–100.

Elmer, P.A.G. and Michailides, J.T. 2007. Epidemiology of *Botrytis cinerea* in orchard and vine crops. pp. 243–272. *In*: Y. Elad, B. Williamson, P. Tudzynski and N. Delen (eds.). Botrytis: *Biology, Pathology and Control*. Springer, Dordrecht, The Netherlands.

Elmer, P.A.G., Whelan, H.G. and Pyke, N.B. 1997. Relationship between *Botrytis cinerea* inoculum in kiwi fruit vines, contamination of the fruit surface at harvest and stem end rot in cool storage. *Acta Horticulturae* 444, 713–717.

Everett, K.R., Taylor, R.K., Romberg, M.K., Rees-George, J., Fullerton, R.A., Vanneste, J.L. and Manning, M.A. 2011. First report of *Pseudomonas syringae* pv. *actinidiae* causing kiwifruit bacterial canker in New Zealand. *Australasian Plant Disease Notes* 6, 67–71.

Falguera, V., Quintero, J.P., Jiménez, A., Muñoz, J.A. and Ibarz, A. 2011. Edible films and coatings: Structures, active functions and trends in their use. *Trends in Food Science and Technology* 22, 292–303.

FAOSTAT (Food and Agriculture Organization of the United Nations, Statistics Division). 2019. Food and agricultural commodities production. Available at: www.fao.org/faostat/en/#data/QC

Feliziani, E., Landi, L. and Romanazzi, G. 2015. Preharvest treatments with chitosan and other alternatives to conventional fungicides to control postharvest decay of strawberry. *Carbohydrate Polymers* 132, 111–117.

Feliziani, E., Lichter, A., Smilanick, J.L. and Ippolito, A. 2016. Disinfecting agents for controlling fruit and vegetable diseases after harvest. *Postharvest Biology and Technology* 122, 53–69.

Feliziani, E. and Romanazzi, G. 2016. Postharvest decay of strawberry fruit: Etiology, epidemiology, and disease management. *Journal of Berry Research* 6, 47–63.

Feliziani, E., Romanazzi, G. and Smilanick, J.L. 2014. Application of low concentration of ozone during cold storage of table grapes. *Postharvest Biology and Technology* 93, 38–48.

Feliziani, E., Santini, M., Landi, L. and Romanazzi, G. 2013. Pre and postharvest treatment with alternatives to synthetic fungicides to control postharvest decay of sweet cherry. *Postharvest Biology and Technology* 78, 133–138.

Feng, J.Q., Maguire, K.M. and MacKay, B.R. 2006. Discriminating batches of 'Hayward' kiwifruit for storage potential. *Postharvest Biology and Technology* 41, 128–134.

Ferguson, A.R. 2015. Kiwifruit in the world – 2014. *Acta Horticulturae* 1096, 33–46.

Ferguson, A.R. and Ferguson, L.R. 2003. Are kiwifruit really good for you? *Acta Horticulturae* 610, 131–138.

Fillinger, S. and Walker, A.-S. 2016. Chemical control and resistance management of *Botrytis* diseases. pp. 188–214. *In*: S. Fillinger and Y. Elad (eds.). Botrytis – *The Fungus, the Pathogen and Its Management in Agricultural Systems*. Springer International Publishing, Cham, Switzerland.

Flores-López, M.L., Cerqueira, M.A., Jasso de Rodríguez, D. and Vicente, A.A. 2016. Perspectives on utilization of edible coatings and nano-laminate coatings for extension of postharvest storage of fruits and vegetables. *Food Engineering Review* 8, 292–305.

Franck, J., Latorre, B.A., Torres, R. and Zoffoli, J.P. 2005. The effect of preharvest fungicide and postharvest sulfur dioxide use on postharvest decay of table grapes caused by *Penicillium expansum. Postharvest Biology and Technology* 37, 20–30.

Fukumori, Y., Nakajima, M. and Akutsu, K. 2004. Microconidia act the role as spermatia in the sexual reproduction of *Botrytis cinerea. Journal of General Plant Pathology* 70, 256–260.

Fullerton, R.A., Rheinlander, P.A., Manning, M.A. and Casonato, S.G. 2007. Infection and rot expression by *Cryptosporiopsis actinidiae* and *Phomopsis* sp. in kiwifruit, *Actinidia chinensis* 'Hort16A'. *Acta Horticulturae* 753, 653–660.

Giuggioli, N.R., Briano, R., Girgenti, V. and Peano, C. 2015. Quality effect of ozone treatment for the red raspberry storage. *Chemical Engineering Transactions* 44, 25–30.

Gutiérrez-Gamboa, G., Romanazzi, G., Garde-Cerdán, T. and Pérez-Álvarez, E.P. 2019. A review on the use of biostimulants in the vineyard for improved grape and wine quality: Effects on prevention of grapevine diseases. *Journal of the Science of Food and Agriculture* 99, 1001–1009.

Guzev, L., Danshin, A., Zahavi, T., Ovadia, A. and Lichter, A. 2008. The effects of cold storage of table grapes, sulphur dioxide and ethanol on species of black *Aspergillus* producing ochratoxin A. *International Journal of Food Science and Technology* 43, 1187–1194.

Guzev, L., Danshin, A., Ziv, S. and Lichter, A. 2006. Occurrence of ochratoxin A producing fungi in wine and table grapes in Israel. *International Journal of Food Microbiology* 111, S67–S71.

Hashim-Buckey, J., Mlikota Gabler, F., Mansour, M.F., Schrader, P., Pryor, M., Margosan, D.A. and Smilanick, J.L. 2008. Effectiveness of preharvest applications of fungicides on preharvest bunch rot and postharvest sour rot of 'Redglobe' grapes. *Proceedings of the 2nd Annual National Viticulture Research Conference*, 35–36.

Hashmi, M.S., East, A.R., Palmer, J.S. and Heyes, J.A. 2013. Hypobaric treatment stimulates defence-related enzymes in strawberry. *Postharvest Biology and Technology* 85, 77–82.

Hassall, A.K., Pringle, G.J. and MacRae, E.A. 1998. Development, maturation, and postharvest responses of *Actinidia arguta* (Sieb. et Zucc.) Planch. ex Miq. fruit. *New Zealand Journal of Crop and Horticultural Science* 26, 95–108.

Hassenberg, K., Geyer, M. and Herppich, W.B. 2010. Effect of acetic acid vapour on the natural microflora and *Botrytis cinerea* of strawberries. *European Journal of Horticultural Science* 75, 141–146.

Hawthorne, B.T., Rees-George, J. and Samuels, G.J. 1982. Fungi associated with leaf spots and post-harvest fruit rots of kiwifruit (*Actinidia chinensis*) in New-Zealand. *New Zealand Journal of Botany* 20, 143–150.

Herzog, K., Wind, R. and Töpfer, R. 2015. Impedance of the grape berry cuticle as a novel phenotypic trait to estimate resistance to *Botrytis cinerea. Sensors* 15, 12498–12512.

Holz, G., Coertze, S. and Williamson, B. 2007. The ecology of *Botrytis* on plant surfaces. pp. 9–27. *In*: Y. Elad, B. Williamson, P. Tudzynski and N. Delen (eds.). Botrytis: *Biology, Pathology and Control*. Springer, Dordrecht, The Netherlands.

Hoyte, S., Reglinski, T., Elmer, P., Mauchline, N., Stannard, K., Casonato, S., Chee, A.A., Parry, F., Taylor, J., Wurms, K., Yu, J., Cornish, D. and Parry, J. 2015. Developing and using bioassays to screen for Psa resistance in New Zealand kiwifruit. *Acta Horticulturae* 1095, 171–180.

Huang, R., Li, G.Q., Zhang, J., Yang, L., Che, H.J., Jiang, D.H. and Huang, H.C. 2011. Control of postharvest Botrytis fruit rot of strawberry by volatile organic compounds of *Candida intermedia*. *Phytopathology* 101, 859–869.

Jackson, S.R. 2014. *Wine Science: Principles and Application*. Academic Press, Tokyo, Japan.

Johnston, P.R., Manning, M.A., Meier, X., Park, D. and Fullerton, R.A. 2004. *Cryptosporiopsis actinidiae* sp nov. *Mycotaxon* 89, 131–136.

Johnston, P.R., Pennycook, S.R. and Manning, M.A. 2005. Taxonomy of fruit-rotting fungal pathogens: What's really out there? *Proceedings of the Fifty Eighth New Zealand Plant Protection Conference* 58, 42–46.

Johnston, P.R., Seifert, K.A., Stone, J.K., Rossman, A.Y. and Marvanova, L. 2014. Recommendations on generic names competing for use in Leotiomycetes (Ascomycota). *IMA Fungus* 5, 91–120.

Jung, B.J., Cho, H.S., Park, M. and Cho, Y.S. 2016. 'Masu', a hardy kiwifruit (*Actinidia arguta* Planch. et Miq.) cultivar with improved storage life. *Korean Journal of Horticultural Science and Technology* 34, 755–760.

Karabulut, O.A., Tezcan, H., Daus, A., Cohen, L., Wiess, B. and Droby, S. 2004. Control of preharvest and postharvest fruit rot in strawberry by *Metschnikowia fructicola*. *Biocontrol Science and Technology* 14, 513–521.

Kim, G.H., Koh, Y.J., Jung, J.S. and Hur, J.S. 2015a. Control of postharvest fruit rot diseases of kiwifruit by antagonistic bacterium *Bacillus subtilis*. *Acta Horticulturae* 1096, 377–382.

Kim, G.H., Song, K.I., Koh, Y.J., Jung, J.S. and Cho, Y.S. 2015b. Reactions of Korean new cultivar 'Haegeum' against major diseases of kiwifruit. *Acta Horticulturae* 1096, 403–408.

Koh, Y.J., Hur, J.S. and Jung, J.S. 2005. Postharvest fruit rots of kiwifruit (*Actinidia deliciosa*) in Korea. *New Zealand Journal of Crop and Horticultural Science* 33, 303–310.

Koh, Y.J., Jung, J.S. and Hur, J.S. 2003. Current status of occurrence of major diseases on kiwifruits and their control in Korea. *Acta Horticulturae* 610, 437–443.

Kovach, J., Petzoldt, R. and Harman, G.E. 2000. Use of honey bees and bumble bees to disseminate *Trichoderma harzianum* to strawberries for *Botrytis* control. *Biological Control* 18, 235–242.

Kulakiotu, E.K., Thanassoulopoulos, C.C. and Sfakiotakis, E.M. 2004. Postharvest biological control of *Botrytis cinerea* on kiwifruit by volatiles of 'Isabella' grapes. *Phytopathology* 94, 1280–1285.

Kwon, J.H., Cheon, M.G., Kim, J. and Kwack, Y.B. 2011. Black rot of kiwifruit caused by *Alternaria alternata* in Korea. *Plant Pathology Journal* 27, 298.

Lamichhane, J.R., Bischoff-Schaefer, M., Bluemel, S., Dachbrodt-Saaydeh, S. and Dreux, L. 2017. Identifying obstacles and ranking common biological control research priorities for Europe to manage most economically important pests in arable, vegetable and perennial crops. *Pest Management Science* 73, 14–21.

Lamichhane, J.R., Dachbrodt-Saaydeh, S., Kudsk, P. and Messean, A. 2016. Toward a reduced reliance on conventional pesticides in European agriculture. *Plant Disease* 100, 10–24.

Landi, L., De Miccolis Angelini, R.M., Pollastro, S., Feliziani, E., Faretra, F. and Romanazzi, G. 2017. Global transcriptome analysis and identification of differentially expressed genes in strawberry after pre-harvest application of BTH and chitosan. *Frontiers in Plant Sciences* 8, 235.

Landi, L., Feliziani, E. and Romanazzi, G. 2014. Expression of defense genes in strawberry fruit treated with different resistance inducers. *Journal of Agricultural and Food Chemistry* 62, 3047–3056.

Latif, N.S.A., Wake, G.C., Reglinski, T., Elmer, P.A.G. and Taylor, J.T. 2013. Modeling induced resistance to plant disease using a dynamical systems approach. *Frontiers in Plant Sciences* 4, 19.

Latorre, B.A., Viertel, S.C. and Spadaro, I. 2002. Severe outbreaks of bunch rots caused by *Rhizopus stolonifer* and *Aspergillus niger* on table grapes in Chile. *Plant Disease* 86, 815.

Lawton, K., Friedrich, L., Hunt, M., Weymann, K., Delaney, T., Kessmann, H., Staub, T. and Ryals, J. 1996. Benzothiadiazole induces disease resistance in *Arabidopsis* by activation of the systemic acquired resistance signal transduction pathway. *Plant Journal* 10, 71–82.

Lee, I., Im, S., Jin, C.R., Heo, H.J., Cho, Y.S., Baik, M.Y. and Kim, D.O. 2015. Effect of maturity stage at harvest on antioxidant capacity and total phenolics in kiwifruits (*Actinidia* spp.) grown in Korea. *Horticulture Environment and Biotechnology* 56, 841–848.

Lee, J.G., Lee, D.H., Park, S.Y., Hur, J.S. and Koh, Y.J. 2001. First report of *Diaporthe actinidiae*, the causal organisms of stem-end rot of kiwifruit in Korea. *Plant Pathology Journal* 17, 110–113.

Leroux, P. 2007. Chemical control of *Botrytis* and its resistance to chemical fungicides. pp. 195–222. *In*: Y. Elad, B. Williamson, P. Tudzynski and N. Delen (eds.). Botrytis: *Biology, Pathology and Control*. Springer, Dordrecht, The Netherlands.

Li, H.J., Zhu, Y.Q., Luo, F.Y., He, H.Y., Yuan, H.Y., Gao, J., Zeng, X.D. and Huang, C. 2015. Use of controlled atmospheres to maintain postharvest quality and improve storage stability of a novel red-fleshed kiwifruit (*Actinidia chinensis* Planch. var. *rufopulpa* [C.F. Liang et R.H. Huang] C.F. Liang et A.R. Ferguson). *Journal of Food Processing and Preservation* 39, 907–914.

Li, L., Pan, H., Chen, M.Y. and Zhong, C.H. 2016a. First report of *Diaporthe lithocarpus* causing postharvest rot of kiwifruit in Sichuan province, China. *Plant Disease* 100, 2327.

Li, L., Pan, H., Chen, M.Y. and Zhong, C.H. 2016b. First report of *Pestalotiopsis microspora* causing postharvest rot of kiwifruit in Hubei province, China. *Plant Disease* 100, 2161.

Li, L., Pan, H., Liu, M.Y. and Zhong, C.H. 2017. First report of *Diaporthe actinidiae* causing stem end rot of kiwifruit during post-harvest in China. *Plant Disease* 101, 1054.

Li, X., Fernández-Ortuño, D., Chen, S., Grabke, A., Luo, C.-X., Bridges, W.C. and Schnabel, G. 2014. Location-specific fungicide resistance profiles and evidence for stepwise accumulation of resistance in *Botrytis cinerea*. *Plant Disease* 98, 1066–1074.

Lichter, A. 2016. Rachis browning in table grapes. *Australian Journal of Grape and Wine Research* 22, 161–168.
Lichter, A., Kaplunov, T., Zutahy, Y. and Lurie, S. 2016. Unique techniques developed in Israel for short-and long-term storage of table grapes. *Israel Journal of Plant Science* 63, 2–6.
Lichter, A., Mlikota Gabler, F. and Smilanick, J.L. 2006. Control of spoilage in table grapes. *Stewart Postharvest Review* 6, 1.
Lichter, A. and Romanazzi, G. 2019. Postharvest diseases of table grapes. *In*: D. Prusky and J.E. Adaskaveg (eds.). *Postharvest Pathology of Fruits and Vegetables*. APS Press, MN. In press.
Lichter, A., Zutahy, Y., Kaplunov, T. and Lurie, S. 2008. Comparison of storage of grapes with SO_2 releasing pads in boxes with internal liners or external wrapping. *HortTechnology* 18, 206–214.
Lima, G., Ippolito, A., Nigro, F. and Salerno, M. 1997. Effectiveness of *Aureobasidium pullulans* and *Candida oleophila* against postharvest strawberry rots. *Postharvest Biology and Technology* 10, 169–178.
Lisker, N., Keren-Shacham, Z., Sarig, P., Zutkhi, Y. and Ben-Arie, R. 1996. The biology and pathology of the fungus *Rhizopus stolonifer*, cause of black mould disease of table grapes in Israel. *Plant Pathology* 45, 1099–1109.
Liu, J., Sui, Y., Wisniewski, M., Droby, S. and Liu, Y. 2013. Review: Utilization of antagonistic yeasts to manage postharvest fungal diseases of fruit. *International Journal of Food Microbiology* 167, 153–160.
Long, C.A. and Yuan, G. 2009. *Kloeckera apiculata* strain (34-9) to control *Botrytis cinerea* during the pre- and postharvest handling of strawberries. *Annals of Microbiology* 59, 77–81.
Luongo, L., Santori, A., Riccioni, L. and Belisario, A. 2011. *Phomopsis* sp. associated with post-harvest fruit rot of kiwifruit in Italy. *Journal of Plant Pathology* 93, 205–209.
Luvisi, D., Shorey, H., Smilanick, J.L., Thompson, J., Gump, B.H. and Knutson, J. 1992. *Sulfur Dioxide Fumigation of Table Grapes*. Division of Agriculture and Natural Resources, University of California, Bulletin 1932, Oakland, CA.
Manning, M.A. and Lallu, N. 1997. Fungal diseases of kiwifruit stored in controlled atmosphere conditions in New Zealand. *Acta Horticulturae* 444, 725–731.
Manning, M.A., Meier, X. and Olsen, T.L. 2003. Fungi associated with fruit rots of *Actinidia chinensis* 'Hort16A' in New Zealand. *New Zealand Journal of Crop and Horticultural Science* 31, 315–324.
Manning, M.A., Pak, H.A. and Beresford, R.M. 2010. Non-fungicidal control of Botrytis storage rot in New Zealand kiwifruit through pre- and postharvest crop management. pp. 183–196. *In*: D. Prusky and M.L. Gullino (eds.). *Post-Harvest Pathology, Plant Pathology in the 21st Century*, Vol. 2. Springer Science+Business Media B.V., Dordrecht, The Netherlands.
Mari, M., Spadoni, A. and Ceredi, G. 2015. Alternative technologies to control postharvest diseases of kiwifruit. *Stewart Postharvest Review* 11, 1.
Martínez-Romero, D., Guillén, F., Valverde, J.M., Bailén, G., Zapata, P.J., Serrano, M., Castillo, S. and Valero, D. 2007. Influence of carvacrol on survival of *Botrytis cinerea* inoculated in table grapes. *International Journal of Food Microbiology* 115, 144–148.

Michailides, T.J. and Elmer, P.A.G. 2000. Botrytis gray mold of kiwifruit caused by *Botrytis cinerea* in the United States and New Zealand. *Plant Disease* 84, 208–223.

Michailides, T.J. and Morgan, D.P. 1996. Using incidence of *Botrytis cinerea* in kiwifruit sepals and receptacles to predict gray mold decay in storage. *Plant Disease* 80, 248–254.

Michailides, T.J., Morgan, D.P. and Luo, Y. 2010. Epidemiological assessments and postharvest disease incidence. pp. 69–88. *In*: D. Prusky and M.L. Gullino (eds.). *Post-Harvest Pathology, Plant Pathology in the 21st Century*, Vol. 2. Springer Science+Business Media B.V., Dordrecht, The Netherlands.

Minas, I.S., Karaoglanidis, G.S., Manganaris, G.A. and Vasilakakis, M. 2010. Effect of ozone application during cold storage of kiwifruit on the development of stem-end rot caused by *Botrytis cinerea*. *Postharvest Biology and Technology* 58, 203–210.

Mlikota Gabler, F., Mansour, M.F., Smilanick, J.L. and Mackey, B.E. 2004. Survival of spores of *Rhizopus stolonifer, Aspergillus niger, Botrytis cinerea* and *Alternaria alternata* after exposure to ethanol solutions at various temperatures. *Journal of Applied Microbiology* 96, 1354–1360.

Mlikota Gabler, F., Smilanick, J.L., Mansour, M., Ramming, D.W. and Mackey, B.E. 2003. Correlations of morphological, anatomical, and chemical features of grape berries with resistance to *Botrytis cinerea*. *Phytopathology* 93, 1263–1273.

Mlikota Gabler, F., Smilanick, J.L., Mansour, M.F. and Karaca, H. 2010. Influence of fumigation with high concentrations of ozone gas on postharvest gray mold and fungicide residues on table grapes. *Postharvest Biology and Technology* 55, 85–90.

Moor, U., Poldma, P., Tonutare, T., Moor, A. and Starast, M. 2014. The effect of modified atmosphere storage on the postharvest quality of the raspberry 'Polka'. *Agronomy Research* 12, 745–752.

Morgan, D.P. and Michailides, T.J. 2004. First report of melting decay of 'Red Globe' grapes in California. *Plant Disease* 88, 1047.

Myresiotis, C.K., Karaoglanidis, G.S. and Tzavella-Klonari, K. 2007. Resistance of *Botrytis cinerea* isolates from vegetable crops to anilinopyrimidine, phenylpyrrole, hydroxyanilide, benzimidazole, and dicarboximide fungicides. *Plant Disease* 91, 407–413.

Nadas, A., Olmo, M. and García, J.M. 2003. Growth of *Botrytis cinerea* and strawberry quality in ozone-enriched atmospheres. *Journal of Food Science* 68, 1798–1802.

Nelson, K.E. 1985. *Harvesting and Handling California Table Grapes for Market*. Division of Agriculture and Natural Resources, University of California, Bulletin 1913, Oakland, CA.

Nigro, F., Ippolito, A., Lattanzio, V., Di Venere, D. and Salerno, M. 2000. Effect of Ultraviolet-C light on postharvest decay of strawberry. *Journal of Plant Pathology* 82, 29–37.

Niklis, N., Sfakiotakis, E. and Thanassoulopoulos, C.C. 1997. Ethylene production of *Botrytis cinerea*, kiwi fruit and Botrytis rotted kiwi fruit under several storage temperatures. *Acta Horticulturae* 444, 733–737.

Nishiyama, I., Yamashita, Y., Yamanaka, M., Shimohashi, A., Fukuda, T. and Oota, T. 2004. Varietal difference in vitamin C content in the fruit of kiwifruit and other *Actinidia* species. *Journal of Agricultural and Food Chemistry* 52, 5472–5475.

Norton, J.S., Charig, A.J. and Demoranville, I.R. 1968. The effect of ozone on storage of cranberries. *Journal of American Society Horticultural Science* 93, 792–796.

Palou, L., Ali, A., Fallik, E. and Romanazzi, G. 2016. GRAS, plant- and animal-derived compounds as alternatives to conventional fungicides for the control of postharvest diseases of fresh horticultural produce. *Postharvest Biology and Technology* 122, 41–52.

Patterson, K., Burdon, J. and Lallu, N. 2003. 'Hort16A' kiwifruit: Progress and issues with commercialisation. *Acta Horticulturae* 610, 267–273.

Pearson, R.C. and Goheen, A.C. 1988. *Compendium of Grape Diseases.* APS Press, MN.

Pennycook, S.R. 1985. Fungal fruit rots of *Actinidia deliciosa* (kiwifruit). *New Zealand Journal of Experimental Agriculture* 13, 289–299.

Pennycook, S.R. and Manning, M.A. 1992. Picking wound curing to reduce Botrytis storage rot of kiwifruit. *New Zealand Journal of Crop and Horticultural Science* 20, 357–360.

Pennycook, S.R. and Samuels, G.J. 1985. *Botryosphaeria* and *Fusicoccum* species associated with ripe fruit rot of *Actinidia deliciosa* (kiwifruit) in New Zealand. *Mycotaxon* 24, 445–458.

Peppi, M.C. and Fidelibus, M.W. 2008. Effects of forchlorfenuron and abscisic acid on the quality of 'Flame Seedless' grapes. *HortScience* 43, 173–176.

Peretto, G., Du, W.X., Avena-Bustillos, R.J., Berrios, J.D.J., Sambo, P. and McHugh, T.H. 2014. Optimization of antimicrobial and physical properties of alginate coatings containing carvacrol and methyl cinnamate for strawberry application. *Journal of Agricultural and Food Chemistry* 62, 984–990.

Pérez, A.G., Sanz, C., Ríos, J.J., Olías, R. and Olías, J.M. 1999. Effects of ozone treatment on postharvest strawberry quality. *Journal of Agricultural and Food Chemistry* 47, 1652–1656.

Pérez-Gallardo, A., García-Almendárez, B., Barbosa-Cánovas, G., Pimentel-González, D., Reyes-González, L.R. and Regalado, C. 2015. Effect of starch-beeswax coatings on quality parameters of blackberries (*Rubus* spp.). *Journal of Food Science and Technology* 52, 5601–5610.

Prencipe, S., Nari, L., Vittone, G., Gullino, M.L. and Spadaro, D. 2016. Effect of bacterial canker caused by *Pseudomonas syringae* pv. *actinidiae* on postharvest quality and rots of kiwifruit 'Hayward'. *Postharvest Biology and Technology* 113, 119–124.

Qin, X., Xiao, H., Cheng, X., Zhou, H. and Si, L. 2017. *Hanseniaspora uvarum* prolongs shelf life of strawberry via volatile production. *Food Microbiology* 63, 205–212.

Reglinski, T., Elmer, P.A.G., Taylor, J.T., Wood, P.N. and Hoyte, S.M. 2010. Inhibition of *Botrytis cinerea* growth and suppression of botrytis bunch rot in grapes using chitosan. *Plant Pathology* 59, 882–890.

Robinson-Boyer, L., Jeger, M.J., Xu, X. and Jeffries, P. 2009. Management of strawberry grey mould using mixtures of biocontrol agents with different mechanisms of action. *Biocontrol Science and Technology* 19, 1051–1065.

Romanazzi, G. 2009. Gray mold infection of *Actinidia arguta* in Italy. *Plant Disease* 93, 1221.

Romanazzi, G. and Droby, S. 2016. Control strategies for postharvest grey mould on fruit crops. pp. 217–228. In: S. Fillinger and Y. Elad (eds.). Botrytis – *The Fungus, the Pathogen and Its Management in Agricultural Systems.* Springer International Publishing, Cham, Switzerland.

Romanazzi, G. and Feliziani, E. 2014. *Botrytis cinerea.* pp. 131–146. *In*: S. Bautista-Baños (ed.). *Postharvest Decay. Control Strategies.* Academic Press. Elsevier Inc., London, UK.

Romanazzi, G., Feliziani, E., Bautista-Baños, S. and Sivakumar, D. 2017. Shelf life extension of fresh fruit and vegetables by chitosan treatment. *Critical Reviews in Food Science and Nutrition* 57, 579–601.

Romanazzi, G., Feliziani, E. and Peres, N.A. 2019. Postharvest diseases of strawberries. *In*: D. Prusky and J.E. Adaskaveg (eds.). *Postharvest Pathology of Fruits and Vegetables*. APS Press, MN. In press.

Romanazzi, G., Feliziani, E., Santini, M. and Landi, L. 2013. Effectiveness of postharvest treatment with chitosan and other resistance inducers in the control of storage decay of strawberry. *Postharvest Biology and Technology* 75, 24–27.

Romanazzi, G., Feliziani, E. and Sivakumar, D. 2018. Chitosan, a biopolymer with triple action on postharvest decay of fruit and vegetables: Eliciting, antimicrobial and film-forming properties. *Frontiers in Microbiology* 9, 2745.

Romanazzi, G., Lichter, A., Mlikota Gabler, F. and Smilanick, J.L. 2012. Recent advances on the use of natural and safe alternatives to conventional methods to control postharvest gray mold of table grapes. *Postharvest Biology and Technology* 63, 141–147.

Romanazzi, G., Mlikota Gabler, F. and Smilanick, J.L. 2006. Preharvest chitosan and postharvest UV-C irradiation treatments suppress gray mold of table grapes. *Plant Disease* 90, 445–450.

Romanazzi, G., Nigro, F., Ippolito, A. and Salerno, M. 2001. Effect of short hypobaric treatments on postharvest rots of sweet cherries, strawberries and table grapes. *Postharvest Biology and Technology* 22, 1–6.

Romanazzi, G., Sanzani, S.M., Bi, Y., Tian, S., Gutiérrez-Martínez, P. and Alkan, N. 2016a. Induced resistance to control postharvest decay of fruit and vegetables. *Postharvest Biology and Technology* 122, 82–94.

Romanazzi, G., Smilanick, J.L., Feliziani, E. and Droby, S. 2016b. Integrated management of postharvest gray mold on fruit crops. *Postharvest Biology and Technology* 113, 69–76.

Rotolo, C., De Miccolis Angelini, R.M., Pollastro, S. and Faretra, F. 2016. A TaqMan-based qPCR assay for quantitative detection of the biocontrol agents *Bacillus subtilis* strain QST713 and *Bacillus amyloliquefaciens* subsp. *plantarum* strain D747. *BioControl* 61, 91–101.

Scortichini, M., Marcelletti, S., Ferrante, P., Petriccione, M. and Firrao, G. 2012. *Pseudomonas syringae* pv. *actinidiae*: A re-emerging, multi-faceted, pandemic pathogen. *Molecular Plant Pathology* 13, 631–640.

Seal, A., McGhie, T., Boldingh, H., Rees, J. and Blackmore, A. 2016. The effect of pollen donor on fruit weight, seed weight and red colour development in 'Hort22D'. *New Zealand Journal of Crop and Horticultural Science* 44, 1–12.

Servili, A., Feliziani, E. and Romanazzi, G. 2017. Exposure to volatiles of essential oils alone or under hypobaric treatment to control postharvest gray mold of table grapes. *Postharvest Biology and Technology* 133, 36–40.

Sharma, R.R., Singh, D. and Singh, R. 2009. Biological control of postharvest diseases of fruits and vegetables by microbial antagonists: A review. *Biological Control* 50, 205–221.

Sharrock, K.R. and Hallett, I.C. 1992. Physiology of *Botrytis* infection in kiwifruit. *Acta Horticulturae* 2, 551–558.

Shtienberg, D. 2007. Rational management of *Botrytis*-incited diseases: Integration of control measures and use of warning systems. pp. 335–347. *In*: Y. Elad, B. Williamson, P. Tudzynski and N. Delen (eds.). Botrytis: *Biology, Pathology and Control*. Springer, Dordrecht, The Netherlands.

Smilanick, J.L., Mansour, M.F., Mlikota Gabler, F., Margosan, D.A. and Hashim-Buckey, J. 2010. Control of postharvest gray mold of table grapes in the San Joaquin Valley of California by fungicides applied during the growing season. *Plant Disease* 94, 250–257.

Sommer, N.F. 1982. Postharvest handling practices and postharvest diseases of fruit. *Plant Disease* 66, 357–364.

Spadaro, D. and Droby, S. 2016. Development of biocontrol products for postharvest diseases of fruit: The importance of elucidating the mechanisms of action of yeast antagonists. *Trends in Food Science and Technology* 47, 39–49.

Spadaro, D., Galliano, A., Pellegrino, C., Gilardi, G., Garibaldi, A. and Gullino, M.L. 2010. Dry matter, mineral composition, and commercial storage practices influence the development of skin pitting caused by *Cadophora luteo-olivacea* on kiwifruit 'Hayward'. *Journal of Plant Pathology* 92, 349–356.

Stonehouse, W., Gammon, C.S., Beck, K.L., Conlon, C.A., von Hurst, P.R. and Kruger, R. 2012. Kiwifruit: Our daily prescription for health. *Canadian Journal of Physiology and Pharmacology* 91, 442–447.

Swart, A.E. and Holz, G. 1994. Colonization of table grape bunches by *Alternaria alternata* and rot of cold-stored grapes. *South African Journal of Enology and Viticulture* 15, 19–25.

Tang, J.M., Liu, Y.Q., Li, H.H., Wang, L.M., Huang, K. and Chen, Z.X. 2015. Combining an antagonistic yeast with harpin treatment to control postharvest decay of kiwifruit. *Biological Control* 89, 61–67.

Thomidis, T., Pantazis, S., Navrozidis, E. and Karagiannidis, N. 2015. Biological control of fruit rots on strawberry and grape by BOTRY-Zen. *New Zealand Journal of Crop and Horticultural Science* 43, 68–72.

Tonini, G., Barberini, K., Bassi, F. and Proni, R. 1999. Effects of new curing and controlled atmosphere storage technology on *Botrytis* rots and flesh firmness in kiwifruit. *Acta Horticulturae* 498, 285–291.

Tronsmo, A. and Dennis, C. 1977. The use of *Trichoderma* species to control strawberry fruit rots. *Netherlands Journal of Plant Pathology* 83, 449–455.

Usall, J., Ippolito, A., Sisquella, M. and Neri, F. 2016. Physical treatments to control postharvest pathogens. *Postharvest Biology and Technology* 122, 30–40.

Vanneste, J.L. 2012. *Pseudomonas syringae* pv. *actinidiae* (Psa): A threat to the New Zealand and global kiwifruit industry. *New Zealand Journal of Crop and Horticultural Science* 40, 265–267.

Vanneste, J.L. 2017. The scientific, economic and social impacts of the New Zealand outbreak of bacterial canker of kiwifruit (*Pseudomonas syringae* pv. *actinidiae*). *Annual Review of Phytopathogy* 55, 377–399.

Wang, C.W., Ai, J., Fan, S.T., Lv, H.Y., Qin, H.Y., Yang, Y.M. and Liu, Y.X. 2015a. *Fusarium acuminatum*: A new pathogen causing postharvest rot on stored kiwifruit in China. *Plant Disease* 99, 1644.

Wang, C.W., Ai, J., Lv, H.Y., Qin, H.Y., Yang, Y.M., Liu, Y.X. and Fan, S.T. 2015b. First report of *Penicillium expansum* causing postharvest decay on stored kiwifruit (*Actinidia arguta*) in China. *Plant Disease* 99, 1037.

Wang, L.H., Zheng, X.Q., Zhuang, Q.G., Dong, G.Y., Xie, Y. and Li, M.Z. 2015c. Collection of wild germplasm resources and breeding of red-fleshed kiwifruit. *Acta Horticulturae* 1096, 155–160.

Wang, Z. 2011. Stability of red flesh color in kiwifruit: From breeding to postharvest. *Acta Horticulturae* 913, 117–122.

Wei, Y., Mao, S. and Tu, K. 2014. Effect of preharvest spraying *Cryptococcus laurentii* on postharvest decay and quality of strawberry. *Biological Control* 73, 68–74.

Williamson, B., Tudzynski, B., Tudzynski, P. and van Kan, J.A.L. 2007. *Botrytis cinerea*: The cause of grey mould disease. *Molecular Plant Pathology* 8, 561–580.

Witbooi, W.R., Fourie, J.F. and Taylor, M.A. 2010. Information on the little understood problem of soft tissue breakdown in cold stored table grapes. *South African Fruit Journal* 9, 46–49.

Wurms, K., Chee, A., Elmer, P., Agnew, R., Wood, P. and Chee, A.A. 2011a. Developing new biologically-based products for control of botrytis bunch rot. Part 1: Developing a new natural product for mid-season botrytis control – NP2 moves closer to the market. *Wine and Viticulture Journal* 26, 64–72.

Wurms, K., Wei, C., Ah-Chee, A., Rees-George, J. and Merima, B. 2011b. Down regulation of putative defence-associated transcripts correlates with ripe rot symptoms on kiwifruit (*Actinidia chinensis*) kiwifruit defence responses and ripe rots. *Journal of Phytopathology* 159, 435–442.

Wurms, K.V., Reglinski, T., Friel, E.N., Wang, M. and Chynoweth, R. 2007. Postharvest biochemical changes in 'Hort16A' kiwifruit: Effects of fungal inoculation and storage environment. *Acta Horticulturae* 753, 677–684.

Yu, H. and Sutton, J.C. 1997. Effectiveness of bumblebees and honeybees for delivering inoculum of *Gliocladium roseum* to raspberry flowers to control *Botrytis cinerea. Biological Control* 10, 113–122.

Zamani-Zadeha, M., Soleimanian-Zadb, S., Sheikh-Zeinoddina, M. and Goli, S.A.H. 2014. Integration of *Lactobacillus plantarum* A7 with thyme and cumin essential oils as a potential biocontrol tool for gray mold rot on strawberry fruit. *Postharvest Biology and Technology* 92, 149–156.

Zhang, H., Ge, L., Chen, K., Zhao, L. and Zhang, X. 2014. Enhanced biocontrol activity of *Rhodotorula mucilaginosa* cultured in media containing chitosan against postharvest diseases in strawberries: Possible mechanisms underlying the effect. *Journal of Agricultural and Food Chemistry* 62, 4214–4224.

Zhang, H., Ma, L., Jiang, S., Lin, H., Zhang, X., Ge, L. and Xu, Z. 2010. Enhancement of biocontrol efficacy of *Rhodotorula glutinis* by salicyclic acid against gray mold spoilage of strawberries. *International Journal of Food Microbiology* 141, 122–125.

Zhang, H., Wang, L., Dong, Y., Jiang, S., Cao, J. and Meng, R. 2007. Postharvest biological control of gray mold decay of strawberry with *Rhodotorula glutinis. Biological Control* 40, 287–292.

Zhang, H., Yang, Q., Lin, H., Ren, X., Zhao, L. and Hou, J. 2013. Phytic acid enhances biocontrol efficacy of *Rhodotorula mucilaginosa* against postharvest gray mold spoilage and natural spoilage of strawberries. *LWT - Food Science and Technology* 52, 110–115.

Zhao, L., Hongyin, Z., Li, J., Cui, J., Zhang, X. and Ren, X. 2012. Enhancement of biocontrol efficacy of *Pichia carribbica* to postharvest diseases of strawberries by addition of trehalose to the growth medium. *International Journal of Molecular Sciences* 13, 3916–3932.

Zhou, Y., Gong, G.S., Cui, Y.L., Zhang, D.X., Chang, X.L., Hu, R.P., Liu, N. and Sun, X.F. 2015. Identification of *Botryosphaeriaceae* species causing kiwifruit rot in Sichuan province, China. *Plant Disease* 99, 699–708.

Zhu, Y., Yu, J., Brecht, J.K., Jiang, T. and Zheng, X. 2016. Pre-harvest application of oxalic acid increases quality and resistance to *Penicillium expansum* in kiwifruit during postharvest storage. *Food Chemistry* 190, 537–543.

Zoffoli, J.P. and Latorre, B. 2011. Table grape (*Vitis vinifera* L.). pp. 179–212. *In*: E.M. Yahia (ed.). *Postharvest Biology and Technology of Tropical and Subtropical Fruits*, Vol. 3. Woodhead Publishing, Cambridge, UK.

Zoffoli, J.P., Latorre, B.A. and Naranjo, P. 2009a. Preharvest applications of growth regulators and their effect on postharvest quality of table grapes during cold storage. *Postharvest Biology and Technology* 51, 183–192.

Zoffoli, J.P., Latorre, B.A., Rodríguez, J. and Aguilera, J.M. 2009b. Biological indicators to estimate the prevalence of gray mold and hairline cracks on table grapes cv. Thompson Seedless after cold storage. *Postharvest Biology and Technology* 52, 126–133.

Zutahy, Y., Lichter, A., Kaplunov, T. and Lurie, S. 2008. Extended storage of 'Red Globe' grapes in modified SO_2 generating pads. *Postharvest Biology and Technology* 50, 12–17.

Chapter 5

Pomegranate, Persimmon, and Loquat

Lluís Palou

Laboratory of Pathology, Postharvest Technology Center (CTP), Valencian Institute of Agrarian Research (IVIA), Montcada, Valencia, Spain

Pervin Kinay-Teksür

Department of Plant Protection, Faculty of Agriculture, Ege University, Bornova, Izmir, Turkey

Shifeng Cao

College of Food Science and Technology, Nanjing Agricultural University, Nanjing, PR China

George Karaoglanidis

Laboratory of Plant Pathology, Faculty of Agriculture, Aristotle University of Thessaloniki, Thessaloniki, Greece

Antonio Vicent

Mycology Unit, Plant Protection and Biotechnology Center, Valencian Institute of Agrarian Research (IVIA), Montcada, Valencia, Spain

Abbreviations

1-MCP	1-methylcyclopropene
ABS	Alternaria black spot
CA	Controlled atmosphere
CPPU	Forchlorfenuron
EU	European Union
FLU	Fludioxonil
GA	Gibberellic acid
GRAS	Generally recognized as safe
ITS	Internal transcribed spacer
MAP	Modified atmosphere packaging
MeJA	Methyl jasmonate
PS	Potassium sorbate
QoIs	Quinone outer inhibitor fungicides
RH	Relative humidity
SAR	Systemic acquired resistance
SBC	Sodium bicarbonate
SBIs	Sterol biosynthesis inhibitors fungicides
SC	Sodium carbonate
USA	United States of America

1 Introduction

Pomegranate, persimmon, and loquat are traditional minor fruit crops of increasing economic importance in subtropical production areas. Pomegranate (*Punica granatum* L., Myrtales, Lythraceae, Punicoideae) originated and currently is mainly cultivated in the Middle East (India [first world producer with about 1.8 million tons in 2015], Iran, Iraq, Afghanistan, Turkey, Egypt, Israel, etc.), but also in the Mediterranean Basin (Spain, Morocco, Tunisia, Greece, Cyprus, etc.) and the United States of America (USA). It is also of increasing importance in the southern hemisphere (Peru, Chile, South Africa, Australia, etc.), as a consequence of increasing consumer demand worldwide (Erkan and Kader, 2011; POMASA, 2015). Pomegranate fruit is a fleshy berry with yellowish to dark red leathery rind that contains the edible juicy arils, which themselves contain the woody seeds. Arils are highly nutritious and a source of phenolic antioxidants and health-promoting bioactive compounds (anthocyanins, punicalagin, ellagic acid derivatives, etc.) (Gil et al., 2000). Consumption is mainly as whole fresh fruit, extracted juice, or minimally processed products (whole arils in packages). Marketing is increasing of pomegranate-derived nutraceuticals, dietary or health supplements, and processed products, such as wine or candy (Erkan and Kader, 2011; Faria and Calhau, 2011).

Persimmon (*Diospyros kaki* Thunb., Ericales, Ebenaceae; sweet, Asian, or Japanese persimmon) originated in China, where there is by far the largest production worldwide of almost 4 million tons in 2014 (more than 70% of total world production), which is followed by South Korea and Japan. Non-Asian producing countries include Brazil, Azerbaijan, Italy, Israel, New Zealand, and Spain, among others (FAOSTAT,

2014). Global consumption is increasing, especially outside of Asia where the potential health benefits associated with persimmon phytochemicals are being increasingly recognized (George and Redpath, 2008). The persimmon fruit is round, of different sizes, depending on the cultivar, with thin yellow to orange peel that becomes intense red as the fruit ripens, and it retains a persistent green leafy calyx at the stem end. While some persimmon cultivars are non-astringent, others are astringent and inedible at harvest. Fresh fruit in the latter condition can only be consumed soft when naturally over ripened or after the application of postharvest deastringency treatments, such as exposure to CO_2 or ethanol. Consumption can also be as dried fruit and processed products such as jams, pastries, or liquors are also manufactured, especially in Asian countries (Woolf and Ben-Arie, 2011).

Loquat (*Eriobotrya japonica* Lindl., Rosales, Rosaceae, Amygdaloideae; Chinese plum) originated and is widely cultivated in subtropical regions of southern China, which is the largest producer worldwide with about 1 million tons grown annually on about 170,000 ha. Other important producing areas are Japan, northern India, Pakistan, Brazil, and Mediterranean countries, including Spain, Italy, Turkey, and Israel (Liu et al., 2016). The fruit is round or oval in shape, with a thin tough skin. Ripe loquat fruit flesh is soft and juicy, varying from white to deep orange. The fresh fruit has nutritional and commercial importance in terms of its high content of bioactive compounds. After harvest, the loquat fruit is perishable and susceptible to mechanical injury and microbial decay. Other uses include processed products such as jams or liquors and prescription in traditional Asian medicine, due to benefits to human health associated with loquat consumption (Cao and Zheng, 2008; Pareek et al., 2014).

2 Postharvest Diseases of Pomegranate

2.1 Gray Mold

Postharvest gray mold of pomegranate fruit is caused by *Botrytis cinerea* Pers. This fungus has a wide host range and causes pre- and postharvest crop losses worldwide. *Botrytis cinerea* has been classified as the second most damaging phytopathogen worldwide (Dean et al., 2012), being pathogenic on more than 580 plant genera (Elad et al., 2016). Gray mold caused by *B. cinerea* is considered as the most economically important postharvest disease of pomegranate fruit in many producing areas worldwide (Tedford et al., 2005; Bardas et al., 2009; Pala et al., 2009; Palou et al., 2013d).

Botrytis cinerea invades plant tissues directly or through wounds (Elad et al., 2005). In the case of pomegranate, *B. cinerea* may infect flower parts during the blooming period, which leads to the occurrence of fruit latent infections. Airborne spores produced after winter from infected plant tissues or other inoculum sources present in the orchard (weeds, sclerotia in the soil, etc.) spread freely through dry air currents to deposit on open flowers, where in the presence of a film of water-containing nutrients, they readily infect the stamens or other senescent flower tissues. It has been observed in many different crops that the pathogen preferentially uses dehiscent flower parts to invade the host (Viret et al., 2004). In addition, the pathogen can also infect the crown (calyx containing stamens and pistil) of recently formed pomegranate fruit and remain latent until after harvest (Palou et al., 2007a).

Infection may also occur through fruit surface wounds created by mechanical or insect injuries when fruits are immature. On the other hand, penetration through natural openings and even direct penetration of the fruit peel by conidia have also been observed for many *Botrytis*-host combinations (Holz et al., 2007). Once attached to the host surface, the conidium germinates under moist conditions and produces a germ tube that develops into an appressorium that penetrates the host surface (van Kan, 2005). Conidia, ascospores, sclerotia, chlamydospores, and mycelia can be inoculum sources of the pathogen. *Botrytis cinerea* typically produces many sclerotia on the organic matter of plant debris and the soil, which are the most important structures for pathogen survival during winter (Holz et al., 2007).

Pomegranate gray mold symptoms from direct, wound, or latent infections typically start as small, superficial, rounded, soft, and decolorized spots that, if relative humidity (RH) is high, readily enlarge into extended dark brown lesions (Figure 5.1a). The pathogen sporulates on the lesion surface to form a typical gray coating of spores that can cover the whole fruit surface depending on the storage temperature (Figure 5.1b). Since symptoms from latent infections usually start in the fruit crown (calyx), the disease is often termed Botrytis crown rot. *Botrytis cinerea* is able to grow and sporulate even at 0°C if the humidity is high, although cold storage considerably delays its growth. Gray mold develops on pomegranate fruit stored at the usual commercial storage conditions of 5–10°C and RH>90% (Palou et al., 2007b). Furthermore, the pathogen is able to infect stored healthy fruits through mycelia and conidia that spread from infected fruits present in the same box or package to adjacent healthy fruits, causing "nests" of decay (Kinay-Teksür, 2015). The extent of gray mold disease symptoms is strongly influenced by fruit maturity and senescence and favored by prolonged storage, during which fruit natural antifungal compounds and defense mechanisms decline or become ineffective (Prusky et al., 2013). Many factors, such as sugars, pH, organic acids, and amino acids, play an important role on decay development and are influenced by fruit ripening during storage (Elad and Evensen, 1995; Blanco-Ulate et al., 2016).

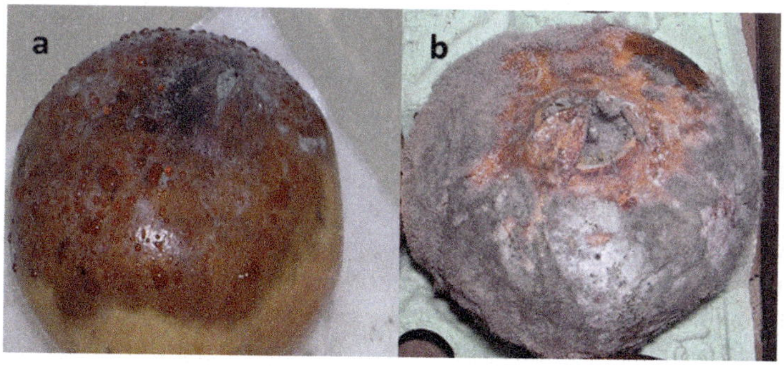

Figure 5.1 a. Brownish discoloration of the crown area of a pomegranate fruit infected by *Botrytis cinerea*. b. Sporulation of *B. cinerea* and symptoms of gray mold on pomegranate fruit.

2.2 *Alternaria* Black Heart and Black Spot

2.2.1 Black Heart

Black heart, also known as heart rot, is one of the most common diseases of pomegranate fruit with a worldwide distribution. The disease occurs in semi-arid and Mediterranean production areas, in countries such as Spain (Vicent et al., 2016), Greece (Tziros et al., 2008), Italy (Faedda et al., 2015), Israel (Ezra et al., 2015b), Turkey (Pala et al., 2009), California (Michailides et al., 2008), India (Jamadar et al., 2011), and South Africa (Munhuweyi et al., 2016). In a recent study conducted in Greece and Cyprus, it was found that black heart was the second and third most important preharvest and postharvest rot, respectively, of pomegranate fruit (Kanetis et al., 2015). Similarly, infection rates in 'Wonderful' pomegranate orchards have been reported to reach 10% or even higher in California and to exceed 20% in Cyprus (Michailides et al., 2008; Kahramanoglu et al., 2014).

Until recently, black heart of pomegranate fruit was referred to in the literature as either undefined at the species level (Zhang and McCarthy, 2012) or attributed to *Alternaria alternata* (Fr.:Fr.) Keissl. and other *Alternaria* spp. (Michailides et al., 2008; Tziros et al., 2008). Kanetis et al. (2015) recently identified the causal agents of black heart using the sequence of the *endoPG* gene as a molecular marker and suggested that the disease is caused in Greece and Cyprus by a complex of small-spored *Alternaria* that includes *A. alternata, A. tenuissima*, and *A. arborescens*. Among these species, *A. alternata* was isolated at a frequency of around 50%, while *A. tenuissima* and *A. arborescens* occurred at frequencies of around 30 and 20%, respectively. In California, *A. alternata* and *A. arborescens* were identified as the causal species through recent molecular work with internal transcribed spacer (ITS), glyceraldehyde-3-phosphate dehydrogenase (*gapdh*), and OPA 10-2 gene region sequences (Luo et al., 2017).

The absence of obvious external symptoms leads to a reduction of consumers' confidence since very often the infected fruits escape detection during sorting at harvest and packing, so they can be purchased to end-consumers or enter the juice production process (Zhang and McCarthy, 2012). The use of infected fruits for juice production may lead to contamination with multiple toxins. In recent reports, most of the *Alternaria* spp. isolates collected in Greece and Cyprus from infected pomegranates produced a range of *Alternaria* mycotoxins such as alternariol, alternariol monomethyl ether, and tentoxin, both in vitro and in vivo, in artificially inoculated or naturally infected fruits (Kanetis et al., 2015; Myresiotis et al., 2015). However, no *Alternaria* mycotoxins were detected in any commercial pomegranate juice tested in a survey conducted in Greece (Myresiotis et al., 2015).

Symptoms of the disease consist of a partial or complete internal rot of the fruit arils that usually starts from the calyx end of the fruit. Initially, infected arils are brown and soft but progressively become covered by a dark gray to black mass of mold (Michailides et al., 2008) (Figure 5.2a). Severely affected fruits weigh less and show rough exocarp, a pale red color, and an asymmetrical shape. Fruits with light-to-moderate lesions, an indication of more recent infection development, appear externally healthy and the leathery rind remains firm. However, an experienced eye is able to recognize the abnormally darker skin color on infected fruit (Figure 5.2b). In addition, the infected fruit emits a hollow sound when struck, while the healthy fruit emits a dull sound (Ezra et al., 2015b).

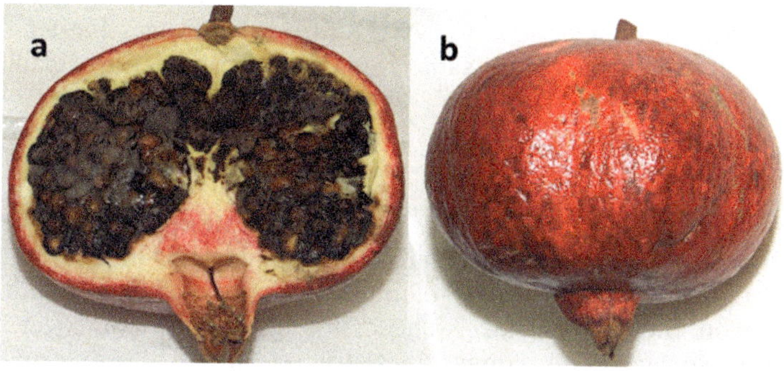

Figure 5.2 a. Black heart of pomegranate fruit infected by *Alternaria* sp. b. Pomegranate fruit with darker color skin and absence of other external symptoms, despite infection by *Alternaria* sp. causing black heart.

Most of the infections causing black heart occur during the bloom period (Michailides et al., 2008; Ezra et al., 2015a). Ezra et al. (2015a) showed that over 90% of pomegranate blossoms are inhabited by *Alternaria* spp. They found that the fungus penetrates the host through open flowers and gains entrance from the stigmata. It continues its growth through the pistil and the tunnel and colonizes the lower loculus, where it remains latent until the onset of fruit ripening. Then, mycelial growth resumes in the lower loculus and the rot of arils becomes evident. The extent of tissue colonized by the fungus expands to the upper part of the loculus and the entire fruit may decay (Ezra et al., 2015b). Similarly, a detailed epidemiological study conducted in California showed that infection is mostly likely to occur at the onset of the bloom (Puckett et al., 2014). According to this study, the infection pathway of *Alternaria* spp. is through the style and pollen tube to the fruit locules. The time of maximum susceptibility to fruit infection was determined by a series of artificial inoculations conducted at three different periods from the onset of bloom until 3 wk after fruit set. The maximum incidence of black heart was observed when artificial inoculation was conducted at bloom. Both Puckett et al. (2014) and Ezra et al. (2015b) found that the frequency of pistil colonization by *Alternaria* spp. was significantly higher than the frequency of fruit that showed heart rot symptoms, suggesting that infection of the pistil does not necessarily lead to disease development. This was recently confirmed in California (Luo et al., 2017) and Spain (Vicent and Palou, unpublished data), where heart rot did not consistently develop after spray inoculations of open pomegranate flowers with aqueous suspensions of high concentrations of conidia of *Alternaria* spp. Factors that inhibit disease development after pistil colonization by the fungus may include the environmental conditions, the position of the fruit in the canopy, and tree physiology, which is also linked with cultivar susceptibility (Ezra et al., 2015b). Pomegranate cultivars differ in their susceptibility to the disease, although the basis of these differences is currently unknown. Palou et al. (2013d) showed that the local Spanish cultivar 'Mollar de Elche' did not develop the disease, while the cultivar 'Wonderful' and other red varieties grown in the same area were susceptible. In Cyprus, Kahramanoglu et al. (2014) showed that among the cultivars 'Acco,' 'Herskovitz,' and 'Wonderful,' the former was the most susceptible and 'Wonderful' the most resistant. Similarly, the incidence of disease in 'Acco' fruit was higher than that in 'Wonderful' fruit cultivated in the same regions of Greece (Karaoglanidis, unpublished data).

2.2.2 Black Spot

In addition to black heart, *Alternaria* spp., and particularly *A. alternata*, can cause other symptoms on pomegranate fruits and leaves. Infections of the fruit are associated with the appearance of numerous, small, black round spots and corky lesions on the rind surface. These necrotic spots can cover more than 50% of the fruit surface. The damage is restricted only to the external surface of the fruit and no internal symptoms occur. Similar black spots can appear on the leaves of the tree, but in this case, the spots are irregular in shape. The infected leaves turn to chlorotic and abscission of some of them may occur. The disease has been reported in several Mediterranean countries such as Israel, Turkey, Egypt, and Spain (Ezra et al., 2010; Gat et al., 2012; Ammar and El-Naggar, 2014; Berbegal et al., 2014), and India (Archana and Jamadar, 2014).

An etiological study of the disease in Israel revealed that it was caused by some specific strains of *A. alternata* that differed from those causing black heart (Gat et al., 2012). Pathogenicity tests on pomegranate blossoms, leaves, and fruits showed that these strains were highly virulent on leaves and the rind surface. A molecular marker was developed enabling the discrimination of these specific black spot strains from other *A. alternata* strains that occur saprophytically on pomegranate leaves or fruits or from strains causing black heart (Gat et al., 2012).

2.3 Aspergillus Black Rot

Black rot caused by *Aspergillus* spp. is a common disease of pomegranate fruit that occurs in most pomegranate cultivation areas around the Mediterranean Basin, California, India, South Africa, and Saudi Arabia (D'Aquino et al., 2009; Jamadar et al., 2011; Yehia, 2013; Palou et al., 2013d; Kanetis et al., 2015; Munhuweyi et al., 2016). The disease is often confused with the decay caused by *Alternaria* spp. due to the black discoloration of the infected arils. Recently, in Greece and Cyprus, Kanetis et al. (2015) reported that black rot was by far the most common preharvest rot of pomegranate fruit, although it could be found at low frequencies as a postharvest disease as well. However, black rot is considered to be the most important postharvest disease of pomegranate in India (Kumar et al., 2016). Identification of *Aspergillus* spp. using a polymerase chain reaction–restriction fragment length polymorphism method and partial sequencing of the calmodulin gene revealed that more than 95% of the isolates from both countries were *A. niger* van Tiegh, with the remainder identified as *A. tubingensis* R. Mosseray (Kanetis et al., 2015). Besides *A. niger* and other *Aspergillus* spp. in the section *Nigri*, other species identified as causal agents of Aspergillus rot include *A. flavus* L., *A. niveus* Blochwitz, *A. versicolor* (Vuill.) Tirab, *A. nidulans* (Eidam) G. Winter, *A. clavatus* Desm., and *A. parasiticus* Speare (Kanwar and Thakur, 1972; Nallathambi and Umamaheswari, 2009; Pala et al., 2009; Jamadar et al., 2011; Palou et al., 2013d). Mycotoxin contamination from *Aspergillus* spp. in fruit with black rot may occur in pomegranate fruits or their processed products. Recently, in Greece and Cyprus, more than 20% of the *A. niger* and *A. tubingensis* isolates from infected pomegranate fruits produced either ochratoxin A or fumonisin B_2 (Kanetis et al., 2015).

Fruit rot caused by *Aspergillus* spp. is associated with decay of the arils, where they become covered by the black spores of the pathogen. The internal appearance of infected fruit is similar to that with *Alternaria* spp. infections, and this has led to confusion about the etiology and naming of black heart disease of pomegranate (Puckett et al., 2014). However, there are some marked differences between the two diseases. Although external symptoms of black rot may also initially be absent, the rot is much softer compared to that caused by *Alternaria* spp. and in advanced stages of disease development, symptoms extend to the outer surface of the fruit causing pale red or most often brown discoloration of the rind and juice exudates (Ammar and El-Naggar, 2014) (Figure 5.3a). Under conditions of ambient temperatures and high RH, the infected fruit surface is covered by black masses of fungal conidiophores and conidia (Yehia, 2013). The area where sporulation occurs on the rind is usually surrounded by an outer zone of soft and discolored rind (Figure 5.3b).

Aspergillus spp. causing black rot of pomegranate are considered to be mainly wound pathogens. Dry conidia produced saprophytically on many different plant substrates are readily disseminated by air currents in large amounts and gain entrance into the fruit through wounds caused on the rind by several biotic or abiotic agents, such as insects, bird pecking, physiological fruit cracking, russeting, or mechanical injuries (Adaskaveg and Michailides, 2013; Yehia, 2013). However, there is some evidence in the literature suggesting that infections by *Aspergillus* spp. can also occur through flowers or young fruit during the bloom or the first stages of fruit development. Rainy conditions are considered to be a basic prerequisite for these infections (Ammar and El-Naggar, 2014). Decay develops rapidly on fruits stored at ambient temperatures and new infections in rind wounds inflicted at harvest and during fruit handling can occur from airborne spores produced from infected moldy stamens (Labuda et al., 2004; Palou et al., 2013d). Rapid

Figure 5.3 a. Soft rot, brownish discoloration, and juicy exudates on pomegranate fruit infected by *Aspergillus* sp. section *Nigri*. b. Sporulation of *Aspergillus* sp. section *Nigri* causing black rot on the surface of an infected pomegranate fruit.

and appropriate cold storage is a key factor to prevent black rot since the growth of many *Aspergillus* spp. is greatly inhibited below 15°C.

2.4 Pilidiella Fruit Rot

The ascomycete *Pilidiella granati* Sacc. (syn.: *Coniella granati* [Sacc.] Petr. & Syd.) causes an important fruit rot of pomegranate worldwide (Hebert and Clayton, 1963; Tziros and Tzavella-Klonari, 2008; Michailides et al., 2010; Palou et al., 2010; Levy et al., 2011; Sharma and Tegta, 2011; Mirabolfathy et al., 2012; Thomidis, 2015; KC and Vallad, 2016; Mincuzzi et al., 2016; Munhuweyi et al., 2016; Cintora-Martínez et al., 2017; Yang et al., 2017; Lennox et al., 2018). According to these reports, depending on the area and the season, Pilidiella fruit rot, also sometimes termed dry rot, can appear before or after harvest and may cause yield losses of up to 50%. The pathogen can also cause blight and cankers on the crown and shoots of pomegranate trees (Çeliker et al., 2012; Chen et al., 2014; Pollastro et al., 2016; Lennox et al., 2018).

Typically, the pathogen initiates the disease as latent infections on the fruit surface in the orchard and decay occurs after harvest or later in cold storage. Disease symptoms occur as small, circular, soft lesions that rapidly expand at ambient temperatures to encompass the entire fruit, including the internal rind, membranes, and arils, which turn brown and leak juice. The pathogen produces globose yellowish pycnidia on the surface of the lesions that progressively become dark brown and black (Figure 5.4a). The conidia of the fungus are one-celled, hyaline to olivaceous brown, ellipsoid, apex obtuse, base truncate, with a mucoid appendage along the side of each conidium, which average 9–16 × 3–4.5 μm in size (Alvarez et al., 2016).

Pilidiella granati can cause crown, trunk, and stem cankers in pomegranate trees that often become inoculum sources for fruit infections. Other sources are infected shoots, leaves, and fruit that abscise and fall to the soil surface. Pycnidia,

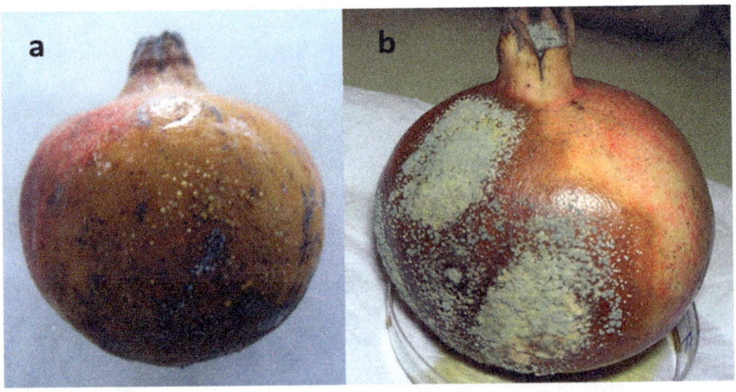

Figure 5.4 a. dry rot and yellowish young pycnidia produced by *Pilidiella granati* on the surface of an infected pomegranate fruit. b. Sporulating lesions of blue mold caused by *Penicillium* sp. on a pomegranate fruit. The edge of the lesions is a well-limited non-sporulated brown and soft discolored area.

located in fruit mummies, left in the orchard, appear to be the main overwintering propagules of the pathogen (Thomidis, 2015). Warm temperatures and high humidity favor conidial production, their dispersal by water splash, and the occurrence of fruit infections. Infections on the surface of young fruits usually remain latent, but wound infections on more mature fruits, particularly if followed by warm and rainy periods, can decay fruits in the orchard and produce new fruit mummies (Sharma and Tegta, 2011). After harvest, latent infections can decay pomegranates even when they are stored at the usual commercial temperature of 5°C, especially if the fruit rind condition is poor or storage has been excessively prolonged (Palou et al., 2013d). Optimum temperatures for mycelial growth and conidial germination of *P. granati* in vitro were 25–30°C, with complete inhibition that occurred only at 2–4°C and 35°C (Thomidis, 2015). At optimal growth temperatures, affected fruit decay completely and tend to leak during storage, causing additional infections by contact and nests of decay (Adaskaveg and Michailides, 2013). Therefore, bulk storage of pomegranates within bins in packinghouses should be minimized.

2.5 Blue/Green Molds

Several species of *Penicillium* can cause blue or green mold on pomegranate fruit. *Penicillium expansum* L., *P. sclerotiorum* J.F.H. Beyma, *P. glabrum* (Wehmer) Westling, *P. implicatum* Biourge, and *P. chrysogenum* Thom have been identified as pathogens causing blue or green mold (Labuda et al., 2004; Pala et al., 2009; Palou et al., 2010; Jamadar et al., 2011; Hammerschmidt et al., 2012; Özgüven et al., 2012; Khokhar et al., 2013; Thomidis, 2014). Typically, decay caused by *Penicillium* spp. appears in rind wounds among fruits stored for long periods. Its incidence can be high. For example, blue mold symptoms were observed on up to 10–20% of pomegranates in a survey of commercial markets in Lahore, Pakistan (Khokhar et al., 2013), up to 20% of cold-stored 'Mollar de Elche' pomegranates in Spanish packinghouses (Palou et al., 2010), or up to 25% of cold-stored 'Primosole' pomegranates in a study conducted in Italy (D'Aquino et al., 2009).

Initial symptoms of the disease appear as small, round, collapsed, and clearly delimited water-soaked areas on the fruit surface. Depending on the causal species, masses of blue or green conidia develop from the center to the edge of the lesions (Figure 5.4b). Radial growth and sporulation occur rapidly at ambient temperatures of 20–25°C and high RH (Palou et al., 2010). In some cases, infections can start from conidia produced from moldy stamens in the crown that subsequently deposit in nearby peel injuries (Labuda et al., 2004). At advanced disease stages, the mycelium of the pathogen can grow inside the fruit through the connective tissue, which becomes dark brown and water-soaked. Decay can then extend to the edible arils that disintegrate into a watery rot.

Penicillium spp. are general saprophytes on plant debris in the soil and senescent plant tissues where they produce large powdery masses of conidia that are readily dispersed by air currents. Airborne conidia infect pomegranate fruits in the orchard during the growing period and at harvest through peel wounds and bruises of varied etiology (mechanical injuries, sunburn, physiological cracking, insects, birds, harvesting punctures, etc.). Infected rind wounds can develop to decay fruit in the field (Thomidis, 2014), but many near-harvest infections only develop after harvest. Postharvest decay also often occurs when conidia present on the fruit surface or in packinghouse facilities infect rind wounds inflicted during fruit handling and/or storage. In this case, the physical and physiological condition of the fruit peel is a key factor influencing the susceptibility of pomegranate to Penicillium molds. Palou et al. (2013d) found in Spain that the incidence of blue mold was only very high after cold storage periods at 5°C exceeding 5 mon and this was associated with infection by *Penicillium* spp. of rind microcracks and injuries caused by rind senescence and/or chilling injury (peel surface scald, pitting, dehydration, and/or shriveling). It was also observed that among different pathogenic *Penicillium* spp. infecting 'Mollar de Elche' pomegranates, *P. expansum* was the most adapted to grow at the usual commercial temperature of 5°C (Palou et al., 2013d). As in other studies (D'Aquino et al., 2010), the incidence of blue mold on cold-stored fruit considerably increased after shelf-life periods of several days at 20°C.

2.6 Other Diseases

Anthracnose, caused by *Colletotrichum gloeosporioides* (Penz.) Penz. & Sacc., has been occasionally reported as an economically important postharvest disease of pomegranate, especially in India. The frequency of anthracnose was found to reach nearly 21% in Maharashtra, the leading pomegranate producer state in India (Joshi et al., 2014). The pathogen can infect the fruit in the orchard at any stage of development during the growing season and infections may decay fruit in the field or, more often, remain latent until the symptoms appear after harvest. Symptoms are dark brown depressed spots often located in the calyx area. Severe early infections can cause blotchy colored, corky fruit that may crack while still on the tree (Jamadar et al., 2011). In Turkey, *C. gloeosporioides* was found infecting the leaves of pomegranate trees, which can be an important inoculum source for fruit infection (Uysal and Kurt, 2018). In general, cultural practices and fungicide programs properly designed for field control of gray mold and Alternaria rots during the fruit-growing season will also effectively control pomegranate anthracnose.

Rhizopus stolonifer (Ehrenb.) Vuill. is a virulent pathogen that can cause severe soft rot on pomegranate fruit. The occurrence of this disease under commercial conditions is not high, but it can cause serious losses when it occurs. The fungus invades pomegranate fruits through peel wounds and the infecting hyphae secrete pectinolytic enzymes that break down the middle lamellae of the infected tissue, causing cell disintegration and a very soft, watery rot. Later, a white or gray fluffy fungal mycelium and black fruiting bodies appear on decayed fruit. Decay develops extremely rapidly at ambient temperatures causing extensive nests of decay among bulk-stored fruits when exudates and juices quickly spread from disintegrating infected fruit to healthy fruit (Munhuweyi et al., 2016). Therefore, fruits from the field should be processed, or at least cold-stored, as soon as possible after their arrival at the packinghouse.

Other fungal pathogens occasionally causing postharvest decay of pomegranate include *Cytospora punicae* Sacc. (Mincuzzi et al., 2017; Venter et al., 2017), *Cladosporium* spp., *Trichoderma* spp., *Phomopsis* spp., *Pestalotiopsis versicolor* (Speg.) Stey, *Aureobasidium pullulans* (de Bary) G. Arnaud, and *Elsinoë punicae* Bitanc. & Jenkins (Snowdon, 1990; Jamadar et al., 2011; Thomidis, 2014).

2.7 Chemical and Integrated Control

Since postharvest diseases of pomegranate fruit are commonly initiated by fungal infections in the orchard (latent infections of flowers and/or young fruit or fruit wound infections through rind injuries), disease management should focus on integrated approaches that encompass preharvest, harvest, and postharvest factors affecting infection and disease development to schedule appropriate control practices.

2.7.1 Preharvest Management

In general, the most important factor for an effective reduction of field latent infections by major pomegranate pathogens such as *B. cinerea, A. alternata*, or *P. granati* is a reduction in the production and dispersal of fungal inoculum. This can be partially accomplished by chemical fungicide applications and adoption of orchard sanitation practices such as removal of decayed fruit, mummies, infected stems, twigs and leaves, as well as dead or senescent prunings and plant materials (Adaskaveg and Michailides, 2013). However, these measures are not always economically feasible and may be of little value to control pathogens that can grow and reproduce on a wide range of hosts and substrates and, therefore, can be ubiquitous in orchards. In the case of wound pathogens such as *Aspergillus* spp. or *Penicillium* spp., the most important preharvest factor is the preservation of the integrity of the fruit rind. Protection from wind and bird and insect control may reduce the number of rind punctures and injuries that enable infections by wound pathogens. Furthermore, appropriate irrigation and fertilization regimes are key factors to minimize cracking or microcracking on mature fruit near harvest. The benefits of avoiding excessive ambient moisture from irrigation and the need for balanced nitrogen fertilization to reduce cracking are well-known phenomena (Adaskaveg and Michailides, 2013). Calcium nutrition is of great importance to preserve peel quality during the storage of pomegranate. Spray application of calcium to pomegranate trees, particularly calcium chloride, during the growing season effectively increased fruit calcium content and retarded decay of stored fruit (El-Kassas et al., 1995).

Results about the effectiveness of preharvest fungicide applications to control pomegranate postharvest diseases have been contradictory. Work in Turkey showed that fungicide sprays during blooming time and 1–2 wk before harvest with active ingredients such as a mixture of pyraclostrobin + boscalid (Bellis®, BASF Crop Protection), registered on pomegranate to control *Alternaria* spp., reduced the number of latent infections by *B. cinerea* and *Alternaria* spp., and significantly inhibited decay on fruits sprayed before harvest and packed in modified atmosphere packaging (MAP) for 2 mon. These preharvest fungicide programs also effectively reduced pomegranate fruit microbial load and disease pressure (Kinay-Teksür et al., 2012, 2014; Şen et al., 2013; Kinay-Teksür, 2015). In Israel, a mixture of cyprodinil + fludioxonil (FLU) (Switch®, Syngenta), applied at a concentration of 0.1% four times in 1-wk intervals from the beginning of flowering until the end of fruit set, significantly reduced the incidence of postharvest crown rots caused by *B. cinerea* and *Penicillium* spp. (Porat et al., 2015). In India, dry rot caused by *P. granati* was effectively controlled by periodic field applications of carbendazim or mancozeb (Sharma and Tegta, 2011). In other studies, in contrast, preharvest fungicide sprays did not reduce postharvest disease caused by *B. cinerea* or *P. granati*. In Greece, applications of azoxystrobin, thiophanate methyl, tebuconazole, cyproconazole, or a mixture of pyraclostrobin + boscalid did not reduce postharvest decay (Thomidis, 2014).

In the case of black heart, caused by *Alternaria* spp., preharvest fungicide applications done before the opening of the calyx (BBCH 59) or after petal fall (BBCH 67) are not effective because the flower is still closed or the pathogen had already colonized the stigmata and the style, which were then protected by the stamens. Fungicide sprays cannot reach potential infection sites at these stages, even when systemic fungicides are used. The phenological stage of open flowers (BBCH 61) appears to be the only time window for the fungicides to reach the flowers and protect them from infections by *Alternaria* spp. Many commercial fungicides can protect pomegranate flowers from *Alternaria* infections. Some fungicide groups, such as quinone outer inhibitor fungicides (QoIs), sterol biosynthesis inhibitors fungicides (SBIs) or dicarboximides, also have translaminar activity and might exhibit curative activity (Oliver and Hewitt, 2014). Nevertheless, field trials in Israel and Spain with contact and systemic fungicides sprayed in spring did not effectively control black heart (Ezra et al., 2015b; Vicent et al., 2017). Flowers are irregularly distributed in the canopy of pomegranate trees and are inadequately treated by current equipment used for the application of fungicides. Consequently, spray coverage on the stigmata and the style of the flower, where *Alternaria* infections are initiated, is much less than optimal in most cases. In addition, pomegranate flowers emerge sequentially in spring during the blooming period, which, depending on the cultivar and other agronomical factors, can extend for more than 1 mon. Considering that the average duration of the open flower stage (BBCH61) is of only about 6 d (Melgarejo et al., 1997), to be effective, excessively frequent fungicide sprays would be needed, which would not be cost-effective. The use of plant growth regulators and specific cultural practices, such as pruning or irrigation, might be explored to synchronize flowering and facilitate the concentration of fungicide applications in a shorter and more feasible period of time. Likewise, further improvements in coverage of flowers by spraying systems could improve fungicide effectiveness. In general, treatments after fruit set are ineffective because blossom tissues are protected by the fruit crown and the fungicide cannot reach the infection sites (Adaskaveg and Michailides, 2013). Approaches that could be evaluated as alternatives to chemical fungicides for pomegranate disease control include the field application of biological control agents or systemic acquired resistance (SAR) compounds, although little has been done in these areas.

2.7.2 Postharvest Management

Despite its apparent toughness, pomegranate rind is very sensitive to physical damage, so it is important to conduct harvest and postharvest operations as gently as possible to avoid peel injuries and bruises where fungal infections can be initiated. Harvest teams should be especially aware of this fact.

Postharvest handling practices to control postharvest diseases of pomegranate in different production areas vary according to national legislations regarding the use of fungicides. While they are used in some countries such as the USA, Israel, and India, no registered active ingredients are currently available in the European Union (EU) or other countries such as Turkey. The fungicide FLU (Scholar®, Syngenta), classified as a reduced risk pesticide, was approved in 2005 by the United States Environmental Protection Agency for use on pomegranates with a maximum residue limit of 5 ppm (Tedford et al., 2005). Since then, it has been registered in many other countries. In Spain, it received an emergency exemption registration for use during the 2019 season. FLU is a contact fungicide without systemic activity mainly recommended to control gray mold caused by *B. cinerea*, but it also controls *Penicillium* ssp., *Alternaria* spp. and *Aspergillus* spp. (D'Aquino et al., 2009; Porat et al., 2015). Postharvest fungicide application is most effective by dipping the fruit into the fungicide solution in order to reach latent fungal infections present in the blossom tissues protected by the fruit crown (Adaskaveg and Förster, 2002). On the other hand, FLU is also effective against relatively recent infections initiated during harvesting or postharvest handling of pomegranates, although its efficacy decreases when the infections occur earlier than 24 h before treatment (Tedford et al., 2005; D'Aquino et al., 2009). In California, FLU is commonly used in combination with chlorine to kill suspended conidia present in the fungicide recycling solutions. These sanitation treatments are especially useful to reduce decay caused by wound pathogens such as *Penicillium* spp., *Aspergillus* spp., and *Rhizopus* spp. (Adaskaveg and Förster, 2000). Other researchers showed that FLU solutions heated to temperatures of about 50°C were more effective than non-heated solutions to control pomegranate decay caused by *B. cinerea* or *Penicillium* spp. (Palou et al., 2007a; D'Aquino et al., 2009). Other active ingredients registered for postharvest application to pomegranates, but infrequently used compared to FLU, are fenhexamid, registered in the USA, and prochloraz, registered in Israel. The latter seems to be more effective than FLU to control *P. granati*, but it is not labeled for use on fruit destined for the EU markets (Nerya and Levin, 2015).

Research efforts to develop alternative decay control methods and integrated disease management strategies for pomegranate have increased in recent years because of the unavailability of fungicides for postharvest use in some countries and the problematic issues of fungicide resistance and consumer concerns related to the use of synthetic fungicides. Currently, in some countries such as Turkey or the EU, the lack of postharvest fungicides compels the growers to handle the fruit in the packingline without applying any kind of washing or sanitation treatments. This is because wetting of the tissues where latent infections are located can stimulate disease development and even cause the fermentation of stored fruits (Kinay-Teksür, 2015). In these cases, appropriate handling employs gentle and effective manual sorting to discard all fruits with sites for fungal infection, such as those with peel injuries or disorders: sun burn, splits and cracks, or husk scald. Sorted pomegranates should be properly packaged according to destination markets and promptly stored at low temperature and high RH to preserve rind integrity. Optimal storage conditions for many commercial pomegranate cultivars to reduce fruit chilling injury, dehydration, and weight loss are 5–8°C and 90–95% RH (Erkan and Kader, 2011). Acceptable alternatives to conventional fungicides for postharvest disease control are physical, biological, or low-toxicity chemical treatments. These are usually applied in combinations to exploit synergistic activities to achieve efficacy similar to that of conventional fungicides. In an evaluation of food additives or generally recognized as safe (GRAS) compounds as alternatives for gray mold control applied after harvest to 'Wonderful' pomegranates, potassium sorbate (PS) was superior in effectiveness to sodium bicarbonate (SBC) and sodium carbonate (SC), and the combination of PS treatment (3 min dip in 3% solution at 21°C) and controlled atmosphere (CA) storage was as effective as the application of heated FLU

(30 s dip in 0.6 g/L at 49°C) (Palou et al., 2007a). Immersion of 'Sheshi-kab' pomegranates in hot water at 50°C for 3 min followed by a 3 min dip in 1–2 mM salicylic acid or 1–2% calcium chloride at 20°C significantly reduced postharvest decay compared to control fruit (Moradinezhad and Khayyat, 2014). In recent work, the application of edible coatings or casted films composed of chitosan and oregano essential oils to artificially inoculated pomegranates effectively controlled gray mold (Munhuweyi et al., 2017). CA and MAP are physical treatments that effectively reduced postharvest decay of cold-stored pomegranates by pathogens such as *B. cinerea, Penicillium* spp., or *P. granati* (Hess-Pierce and Kader, 2003; Kinay-Teksür, 2015). CA and MAP prevent decay by both direct effects on the pathogen and indirect effects on the fruit host, providing that water does not condense on the surface of the stored fruit. Palou et al. (2016a) recently reported that exposure to very high CO_2 concentrations (50 or 95%) at 20°C and 90% RH for brief periods (48 h) effectively prevented gray mold development on pomegranates artificially inoculated with *B. cinerea.*

3 Postharvest Diseases of Persimmon

3.1 Alternaria Black Spot

Alternaria black spot (ABS), caused by *A. alternata*, is one of the most important postharvest diseases of persimmon worldwide. It was isolated and identified in Israel in the early 1980s, where it caused considerable economic losses of persimmons during long-term commercial cold storage (up to 3 mon at 0°C) (Prusky et al., 1981). In 1990, the disease was reported in the comprehensive atlas of postharvest diseases and disorders by Snowdon (1990). Later, first reports of the occurrence of the disease and proper pathogen identification showed that it was in Brazil (Cia et al., 2003), South Korea (Kwon et al., 2004), Turkey (Kurt et al., 2010), and Spain (Palou et al., 2012). A characteristic symptom is the appearance of numerous, superficial, small, firm, round black spots. These are located mainly under or near the calyx, or irregularly distributed throughout the fruit surface. At ambient temperatures, these small black spots can enlarge rapidly and lesions of various sizes may coexist on the same fruit (Figure 5.5). During storage at high temperature and humidity conditions (optimum of 25–28°C and RH>90%), short dark mycelia can grow and become visible in the central part of the lesion, eventually producing abundant olive green to black conidia. The ABS causal organism, *A. alternata*, is a common saprophytic and pathogenic ascomycete that can colonize a wide variety of substrates and plant hosts, including many fruit species. It is also one of the most important postharvest fungal pathogens of horticultural produce worldwide, and its production of mycotoxins such as alternariol, tentoxin, or alternariol monomethyl ether has been repeatedly reported (Rotem, 1998; Kanetis et al., 2015). However, to the best of our knowledge, the capacity of strains causing disease on persimmon fruit to produce mycotoxins has not been studied.

Early epidemiological studies in Israel with persimmon cv. 'Triumph' (Prusky et al., 1981) showed that conidia of *A. alternata* are produced profusely during the entire season in the orchard from infected leaves and twigs, but also from saprophytic colonies on dead organic matter in trees and on the soil. Conidia are easily

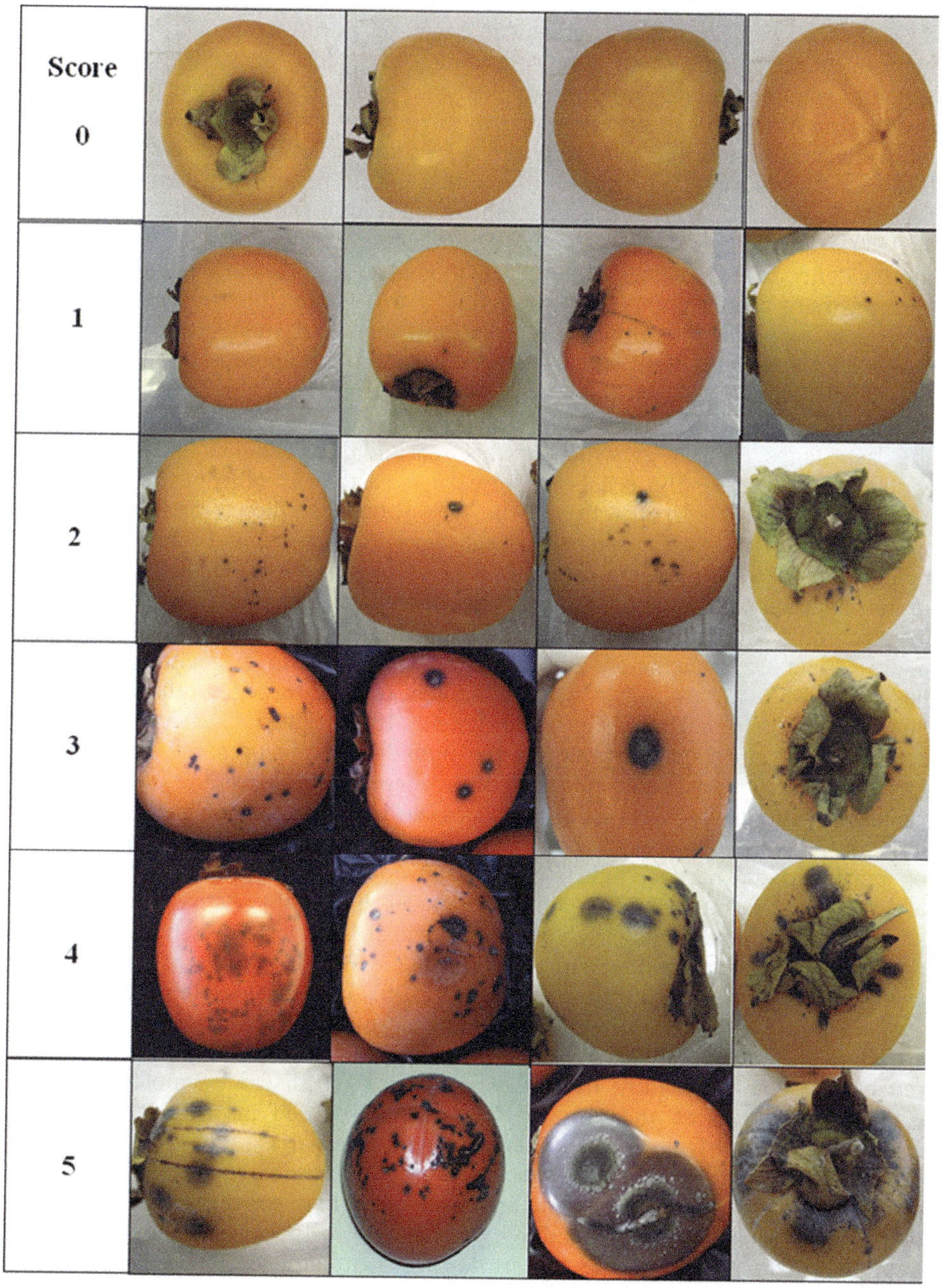

Figure 5.5 Photographs of the quantitative severity index (score 0–5) of Alternaria black spot developed for 'Rojo Brillante' persimmons (Palou et al., 2015). The index illustrates the different types, locations, and intensities of symptoms caused by *Alternaria alternata* on the surface of infected persimmon fruit, from very small rounded black spots to large sporulating lesions.

dispersed by wind or splashing water. Depending on the time of the year, conidia deposit on flowers, developing immature fruits, or mature fruits near harvest. Primary infections in fruit occur after germination of conidia followed by either direct penetration of the cuticle or through peel microwounds or injuries. Formation of appressoria has not been observed and the hyphae of the fungus colonize only the intercellular spaces of the most superficial layers of cells (first three or four layers of the skin), causing them to collapse. Since the germination of conidia requires free water on the fruit surface, wetness and high humidity in the orchard (RH>80%) is required for infection. Warm weather (temperature>20°C) increases the incidence of infections. Although infections can occur at any point of the fruit surface, there are differences in susceptibility to ABS on the fruit surface. These differences are due to physiological dissimilarities that influence the stem end (calyx end) and the bottom end (stylar end) of the fruit. There is a greater production of ethylene and CO_2 at the stem end during early stages of fruit development, which leads to reduced chlorophyll levels and a propensity for peel cracking during the last stages of fruit growth. For this reason, the incidence of ABS is often higher under or near the sepals in the calyx area than in other areas of the fruit (Biton et al., 2014b). Irrespective of the type (wound or direct) or the location of the infection in the fruit surface, incipient superficial infections in immature developing fruits remain quiescent and only develop after harvest. This is probably due to nutrient deficiency or the high levels of constitutive fungistatic compounds naturally present in immature fruit. In contrast, infections that take place near harvest in mature fruits are immediately active, although *A. alternata* in healthy, living plant tissues grows relatively slowly and the ambient temperatures in autumn (far below 20°C) are not optimal for its growth (Rotem, 1998). The incidence of these late active infections depends largely on the humidity in the orchard. In addition to increasing the abundance of inoculum of *A. alternata* and providing conditions that facilitate infection, intense rainfall or excessive irrigation before harvest increases fruit turgidity and the occurrence of microwounds or small cracks in the stem end of mature fruit. These significantly enhance opportunities for fungal penetration and increase the number of infections (Biton et al., 2014b). Active infections initiated in microwounds in the calyx area, which are not visible at harvest, are the most important cause of postharvest losses due to ABS in Israel and Spain (Prusky et al., 1981; Biton et al., 2014b; Palou et al., 2015). In contrast, latent infections are the main cause of postharvest losses in more humid persimmon production areas or in particularly, warm and very rainy seasons. In these conditions, recent infections in mature fruits develop rapidly so disease symptoms become visible before harvest and affected fruits are discarded in the orchard (Perez et al., 1995; Lee et al., 2013). Besides latent and calyx microcrack infections, other potential infections causing postharvest ABS can originate in peel wounds or bruises inflicted during harvest and even after harvest, during fruit transportation and/or packinghouse handling. This is due to the often numerous conidia of *A. alternata* contaminating the fruit surface at harvest. Palou et al. (2016c) recently reported that, although in general, the results were highly variable depending on the season, similar numbers of airborne conidia of *Alternaria* spp. were trapped in an orchard of ‘Rojo Brillante’ persimmons in Spain during the entire season. The incidence of latent infections was constantly low during the whole season or reached a peak at the beginning of October. This cultivar is typically harvested in mid November. In contrast, Prusky et al. (1981) reported a continuous increase in the number of latent infections in 'Triumph' persimmons from fruit set until harvest.

Postharvest factors affecting the final fruit market losses due to ABS include delays to place harvested persimmons in cold storage, availability and efficacy of postharvest antifungal treatments, proper sanitation of packinghouse facilities, duration and conditions of commercial fruit storage in the packinghouse, application of postharvest technologies, such as deastringency treatments with CO_2 or 1-methylcyclopropene (1-MCP) treatments, and duration and conditions of both transportation to the market place and market shelf life. Typically, persimmon fruits are stored in ambient atmospheres at 0–1°C and RH>90% for various periods depending on the tolerance of the cultivar to low temperatures. Although the pathogen can grow under these environmental conditions, they slow the development of ABS symptoms effectively as long as the good physical and physiological condition of the fruit is maintained. However, visible black spots may develop on senescent fruit or fruit softened by chilling injury induced by excessively prolonged cold storage periods. For this reason, ABS incidence usually limits commercial storage of 'Triumph' persimmons to about 3 mon in Israel (Prusky et al., 1981, 1997). In Spain, storage is typically about 40 d for 'Rojo Brillante' persimmons previously treated with the ethylene inhibitor 1-MCP, which significantly prevents chilling injury (Salvador et al., 2004; Palou et al., 2015). Black spots enlarge rapidly when cold-stored persimmons are transferred to ambient temperatures for marketing. The best method to assess the severity of persimmon ABS is to quantify the percentage of the fruit surface area covered with black spots. Since the affected area and the fruit surface area are determined by the shape, volume, and dimensions of the fruit, specific disease severity indexes are needed for each persimmon cultivar. Indexes can be based on diagrammatic scales, such as those developed in Israel for 'Triumph' persimmon (Perez et al., 1995), or objective categorical scales, such as that developed in Spain for 'Rojo Brillante' fruit (Figure 5.5; Palou et al., 2015).

3.2 Anthracnose

Anthracnose, caused by *Colletotrichum* spp., is one of the most important fungal diseases of persimmon in Asia, where the crop originated. In China, Japan, and South Korea, the leading persimmon-producing countries in the world, but also in other producing areas, such as Brazil, the disease has been reported to cause significant reductions in yield due to infections of leaves, shoots, and branches, and also losses in the field due to fallen fruit or symptomatic fruit that must be discarded at harvest (Zhang, 2008; Kwon et al., 2013; Blood et al., 2015). However, latent anthracnose infections in the field that remain asymptomatic at harvest may also cause significant losses after harvest.

The species most frequently identified as the causal agent of persimmon anthracnose is *C. horii* B. Weir & P.R. Johnst., a member of the *C. gloeosporioides* sensu lato species complex that has been reported to cause both pre- and postharvest anthracnose (Weir and Johnston, 2010; Yu et al., 2013; Hassan et al., 2018). This fungus, however, had been previously named *C. gloeosporioides* (*C. gloeosporioides* sensu stricto) and also with the synonymic name *Gloeosporium kaki* Hori. Other *Colletotrichum* species that have been reported to cause postharvest persimmon anthracnose in different growing areas are *C. acutatum* J.H. Simmonds (Williamson and Sutton, 2010; Kwon and Kim, 2011), *C. gloeosporioides* sensu stricto (Palou et al., 2013c), and, very recently, *C. siamense* (Chang et al., 2018; Hassan et al., 2018). Moreover, the species *C. karstii* was recently found to cause anthracnose in persimmon leaves (Wang et al., 2016). Although *C. horii* is morphologically very close to *C. gloeosporioides* sensu stricto, identification by newer molecular techniques supported a separate taxonomical classification. Currently, *Colletotrichum* spp. are molecularly identified by analyzing sequences of the DNA regions ITS, actin, glutamine synthetase, calmodulin, β-tubulin, or *gapdh* (Weir et al., 2012; Silva et al., 2017). Cultured on potato dextrose agar, colonies of *C. horii* are grayish, velvety, with concentric zonation and abundant yellowish-orange central conidial masses. Maximum mycelial growth and sporulation occur at around 25°C with inhibition above 30°C. Conidiophores are short-cylindrical, congregated, produced in acervuli, fasciculate, and straight or occasionally geniculate. Conidia (4.5–6.5 × 16.5–22.5 μm) are holoblastic, cylindrical, straight or slightly curved, and non-septate. Appressoria are dark, brownish, smooth, and globose. The optimal pH for conidia germination and appressoria formation is 5–6, although both occur over a wide pH range of 2–9 (Xie et al., 2010; Kwon et al., 2013). A genomic library of *C. horii* has been constructed and characterized (Sun et al., 2012).

Typical symptoms of persimmon anthracnose slightly vary depending on the *Colletotrichum* sp. causing disease. Symptoms of anthracnose caused by *C. horii* include one or many small, round or oval, slightly depressed, superficial black spots that can be present anywhere on fruit surface (Figure 5.6a). At ambient temperatures and under humidity, the lesions enlarge and grayish and pale orange conidial masses are produced. Conversely, under dry conditions, lesions can be sunken and central longitudinal cracks may appear (Xie et al., 2010). Symptoms of anthracnose caused by *C. gloeosporioides* sensu stricto or *C. acutatum* are also superficial firm, round, brownish or blackish spots with a well-defined perimeter line, but they are usually more numerous, smaller, and not as depressed as those caused by *C. horii* (Figure 5.6b). Under favorable temperature and moisture conditions, adjacent spots coalesce and numerous black acervuli with salmon-colored conidial masses are produced on the lesion surface (Kwon and Kim, 2011; Figure 5.6c).

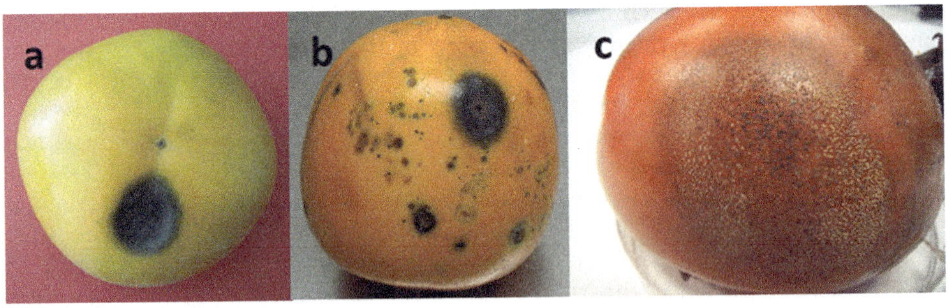

Figure 5.6 Symptoms of anthracnose of persimmon caused by: a. *Colletotrichum horii* and b. *Colletotrichum gloeosporioides* sensu stricto. c. Black acervuli and salmon-colored conidial masses of *C. gloeosporioides* sensu stricto growing in a mature persimmon fruit.

Epidemiology of *C. horii* in persimmon orchards has been studied in China and South Korea (Zhang, 2008; Xie et al., 2010; Kwon et al., 2013). The pathogen overwinters mainly as mycelia in infected living twigs and buds. Typical rainy weather from April to July favors the abundant production of conidia (primary inoculum) and their dissemination by rain splash and wind to fruits or new shoots. Symptoms of preharvest anthracnose comprise blight of young twigs and shoots and leaf defoliation. Although possible, conidia production is considerably less on dead plant tissues than on living tissues. Conidia contaminating the fruit surface can germinate and form dark appressoria that produce infection pegs that penetrate the fruit cuticle. Depending on persimmon cultivar, fruit condition, and environmental conditions, these infections may cause visible disease symptoms in the field, even leading to fruit drop from trees, or they remain in a latent state and cause visible disease only after harvest. New infected shoots and fallen fruits are a source of conidia (secondary inoculum) that may produce new latent infections on young fruits or wound infections in injuries or microcracks on fruits approaching maturity. The incidence of postharvest anthracnose will depend on the proportion of these infections that are asymptomatic at the time of harvest.

3.3 Other Diseases

Although ABS and anthracnose are the most economically important postharvest diseases of persimmon worldwide, other minor postharvest fruit rots have been reported in different persimmon-producing areas. To date, significant economic losses due to these diseases are only occasional, since their incidence is limited to fruit subjected to sporadic and particularly adverse growing and/or handling conditions. Among them, the most frequent include stem-end rots, usually caused by field latent infections of different fungal pathogens, gray mold, caused by latent or wound infections of *B. cinerea*, and several diseases, caused by common postharvest wound pathogens. Common wound pathogens reported were *P. expansum*, causing blue mold of persimmon in South Korea (Kwon et al., 2006) or Spain (Palou et al., 2015), *Cladosporium cladosporiodes* (Fresen.) G.A. de Vries, causing surface black spots in South Korea (Kwon and Park, 2003) and Spain (Palou et al., 2015), and *R. stolonifer*, causing soft and watery rot in Brazil (Cia et al., 2003).

3.3.1 Stem-End Rots

Besides *A. alternata* and *Colletotrichum* spp., which can both also affect this part of the fruit, the most frequent pathogens associated with postharvest stem-end rots of persimmon are *Pestalotiopsis* spp., some species in the Botryosphaeriaceae family, and *Phacidiopycnis washingtonensis* Xiao & J.D. Rogers, which was described in Italy as the causal agent of a disease observed in persimmons in cold storage (Garibaldi et al., 2010). The most common *Pestalotiopsis* sp. causing persimmon stem-end rot is *P. diospyri* (Syd. & P. Syd.) Rib. Souza. It was reported in New Zealand (Goh et al., 1991), South Korea (Kwon et al., 2004), and Spain (Blanco et al., 2008). Other species reported to occur are *P. longiseta* (Speg.) K. Dai & Ts. Kobay. and *P. foedans* (Sacc. & Ellis) Steyaert in Japan (Taguchi et al., 2001), and *P. clavispora* in Spain (Palou et al., 2015). In a report from

Brazil, *Pestalotiopsis* sp. was observed, but not identified at the species level (Cia et al., 2003). Postharvest persimmon stem-end disease, caused by *Pestalotiopsis* spp., typically begins as dark brown to black circular dry spots located in the calyx under the sepals. At ambient temperatures, the spots enlarge and coalesce to form a continuous lesion that covers the entire stem-end area. Then the pathogen produces very compact cottony white mycelia with numerous visible globose black acervuli (Figure 5.7a).

Species in the Botryosphaeriaceae family that have been identified as the cause of persimmon stem-end rot include *Lasiodiplodia theobromae* (Pat.) Griffon & Maubl., reported in Spain, Israel, the USA, Japan, and Australia (DAFF, 2004; Palou et al., 2013a), and *Neofusicoccum mediterraneum* Crous, M.J. Wingf. & A.J.L. Phillips and *Neofusicoccum luteum* (Pennycook & Samuels) Crous, Slippers & A.J.L. Phillips, both reported in Spain (Palou et al., 2013a). Besides fruit stem-end rots, these pathogens are known for causing cankers, blight, or dieback in a large variety of plant hosts. Usual symptoms in persimmon fruit are irregular brownish, soft lesions surrounding the fruit calyx that, at ambient conditions, enlarge and evolve to be dark gray to black. These later become covered with abundant cottony mycelium that is whitish to gray, and produce numerous globose, black pycnidia (Figures 5.7b,c,d; Palou et al., 2013a).

The incidence of postharvest stem-end rot of persimmon caused by latent infections of *Pestalotiopsis* spp., *L. theobromae, N. mediterraneum*, or *N. luteum* is

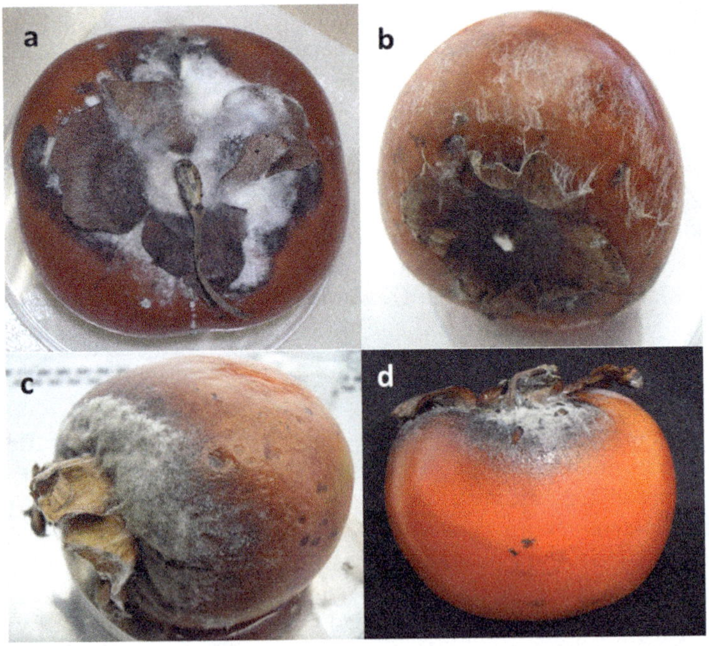

Figure 5.7 Symptoms of stem-end rot of persimmon fruit caused by: a. *Pestalotiopsis clavispora*, b. *Lasiodiplodia theobromae*, c. *Neofusicoccum mediterraneum*, and d. *Neofusicoccum luteum*.

influenced by the environmental conditions in the field during fruit growth. In general, these pathogens overwinter in woody parts of the tree or in pruning debris in the soil, and the production and dispersal of conidia occur primarily during rainy weather and at high temperatures. Rain and its consequent water-splash over short distances are the principle means of conidial dispersal and fruit contamination, although some inoculum is disseminated by the wind. The role of near harvest infections of stem-end microcracks on the fruit has not been documented for these pathogens but, presumably, it could be important (Palou et al., 2013a). Likewise, limited information is available regarding postharvest factors affecting the development of these pathogens. Commercial storage at 0–1°C should greatly inhibit their development, although they were able to grow very slowly in persimmons inoculated artificially and stored at 5°C (Palou et al., 2015).

3.3.2 Gray Mold

Postharvest gray mold of persimmon has been reported in South Korea (Kwon et al., 1999) and Spain (Palou et al., 2015). Its symptoms are large soft lesions where abundant gray mycelium and conidia of *B. cinerea* develop under ambient temperatures and high RH (Figure 5.8). The pathogen can cause nests of decay by the contact of infected with healthy fruits. In New Zealand, field infections of *B. cinerea* during fruit set were associated with a cosmetic disorder known as scarring, that is, dark corky superficial spots caused by the hypersensitive reaction of the fruit (Rheinländer et al., 2013).

3.4 Chemical and Integrated Control

3.4.1 Preharvest Management

The causal agent of ABS disease of persimmon, *A. alternata*, is a polyphagous necrotrophic fungus that can grow and reproduce on a wide range of hosts and substrates. Therefore, inoculum reduction measures appear to be of little value for

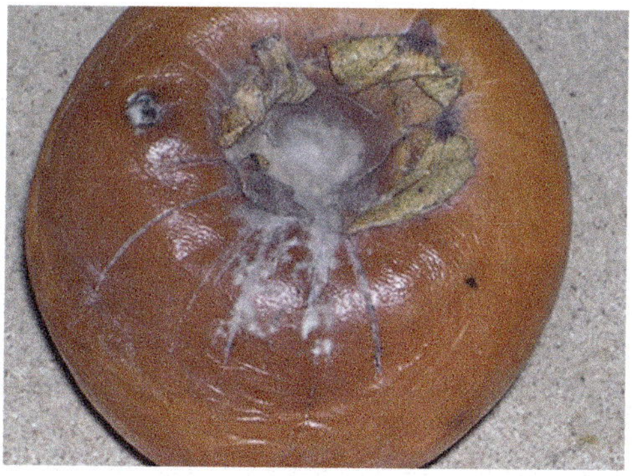

Figure 5.8 Gray mold lesion on persimmon fruit originated from a stem-end infection of *Botrytis cinerea*.

the management of ABS in persimmon orchards. Cultural measures that increase ventilation and reduce orchard humidity may help to create a microclimate that is less favorable for *A. alternata*, although this species is well adapted to develop epidemics even in arid conditions. When designing spray programs to control ABS on persimmon, it is important to consider that most infections by *A. alternata* are initiated in the area beneath the fruit calyx, where rains cause conditions in this micro-environment to be more favorable for conidia to germinate. Moreover, beneath the calyx peel, cracks frequently occur due to more advanced maturity of the stem end that facilitates the penetration of *A. alternata* into the fruit (Biton et al., 2014b). This particular area of the fruit is poorly contacted by fungicides applied by any of the spray systems currently used in the orchard. This poor coverage harms the efficacy and efficiency of solutions applied before harvest for disease control purposes.

The activation of latent infections of *A. alternata* and the enhancement of fruit susceptibility to infection are strongly affected by the physiological processes associated with ripening, which can be modulated by the application of exogenous hormones (Biton et al., 2014a). Plant growth regulators were extensively evaluated in Israel for the control of ABS on 'Triumph' persimmon. The application of gibberellic acid (GA) increased the cell wall cellulose content in persimmon fruit and inhibited maceration enzymes produced by *A. alternata* during the infection process (Eshel et al., 2000). Perez et al. (1995) evaluated GA treatments to delay fruit senescence, enhance calyx erectness, and subsequently, reduce disease severity. One spray application of GA 30 d before harvest was sufficient to increase calyx erectness and reduce ABS severity. Two additional GA sprays, applied 20 and 30 d prior to harvest, improved disease control and inhibited fruit softening. In a subsequent study, Kobiler et al. (2011) evaluated the combination of field treatments with a plant growth regulator and postharvest dips of sodium troclosene in 'Triumph' persimmon. The cytokinin-like CPPU (forchlorfenuron) at 0.5–1.0 mg/L applied 30 d after fruit set reduced ABS incidence and delayed fruit softening after cold storage without influencing fruit maturity at harvest. However, increasing the CPPU concentration to 2.5 mg/L inhibited fruit maturity. Biton et al. (2014a) evaluated different spray programs with a mixture of gibberellin and benzyl adenine at monthly intervals. The application of both plant growth regulators delayed fruit maturation, decreased the area on the fruit surface with cracks, and reduced ABS incidence after 3 mon of storage. Earlier experiments in Israel indicated that field applications of contact fungicides, such as dithiocarbamates, did not control ABS effectively. Wash off due to frequent and somewhat intense rains, typical of the Mediterranean area, dramatically affected the persistence of contact fungicides. The calyx area of the fruit is seldom reached by contact fungicides, which can neither penetrate nor redistribute into plant tissues. Fungicides that are systemic or have translaminar movement are more appropriate choices to control ABS. Infections by *A. alternata* in persimmon fruit are mainly found under the cuticle and in wound cracks in the epidermis. Systemic fungicides can reach these infection sites and exhibit curative activity. Perez et al. (1995) reported that the percentage of unmarketable fruits due to ABS was reduced when a GA spray program was supplemented with applications of the dicarboximide fungicide iprodione applied between 40 and 15 d before harvest. Likewise, Kobiler et al. (2011) significantly increased the percentage of marketable fruits by a single spray of polyoxin B applied 14 d before harvest. Many curative fungicides are available for the control of Alternaria rots in fruit tree crops (Oliver and Hewitt, 2014). However, regulations on fungicide registration and maximum residue levels do not allow their use in persimmon in many countries. Field treatments with biological control agents or SAR activators for ABS control should be examined.

Appropriate preharvest disease management can reduce the incidence of persimmon anthracnose in the field, mainly caused by *C. horii*, but can also contribute to control postharvest anthracnose, also caused by *C. horii*, but also by other *Colletotrichum* species, such as *C. gloeosporioides* and *C. acutatum*. Inoculum reduction by removing affected twigs, fruit, and leaf litter is recommended for the management of anthracnose. Improving orchard ventilation by using wide tree spacing, setting rows oriented to dominant winds, and appropriate pruning and training may also assist in reducing disease severity in persimmon orchards.

In addition to cultural practices, the primary means to control *Colletotrichum* spp. is the application of fungicides in the field. Spray programs are basically preventive because fungicides are less effective once anthracnose symptoms develop. Spray timing is critical for the control of anthracnose because the occurrence of this disease is highly dependent on weather conditions, especially rainfall. In experiments conducted in South Korea, two to eight sprays with difenoconazole and QoIs effectively controlled persimmon anthracnose. These were applied from late May or early June and concluded in July or August (Kwon et al., 2013). Likewise, Jeon et al. (2015) evaluated fungicide programs to control persimmon anthracnose including two to eight applications of dithianon, metconazole, or propineb. The disease was effectively and equally controlled with four or eight fungicide sprays, so the four-spray program from early June to late August was recommended.

3.4.2 Postharvest Management

Currently, the use of conventional chemical fungicides for the postharvest treatment of persimmon is not permitted in many producing areas such as Turkey, Israel, or most EU countries. In Spain, emergency registrations of gaseous pyrimethanil (Deccopyr Pot, 30% a.i., Decco Iberica Post-Cosecha SAU, Paterna, Valencia, Spain) have been released since the 2017 season. Therefore, postharvest disease control management in these areas relies on the evaluation of alternative, environmentally friendly chemical control methods or sanitation measures that complement treatments applied before harvest and cold storage after harvest in integrated control strategies. Cold storage under commercial conditions of 0–1°C and 90–95% RH reduce fungal growth and preserve fruit firmness and quality, which positively contributes to the maintenance of fruit resistance to ABS and other postharvest diseases. This effect can be enhanced by the application of other postharvest technologies such as 1-MCP treatments (Salvador et al., 2004), MAP, and CA storage (Prusky et al., 1997; Neuwald et al., 2009).

Kobiler et al. (2011) reported that postharvest applications (dip, spray, or fogging) of the sanitizer sodium troclosene, which contains about 60% chlorine, significantly reduced ABS on 'Triumph' persimmons and were synergistic with fungicide applications before harvest. In work conducted in Spain, hot water dips at temperatures from 55 to 75°C and dips in 3% aqueous solutions of GRAS salts such as PS, sodium benzoate, SC, and SBC had no curative activity to control ABS on artificially inoculated 'Rojo Brillante' persimmons (Palou, unpublished data). Likewise, further recent research showed that the disease was not satisfactorily reduced on fruit exposed for 48 hr at 20°C to atmospheres containing 15, 50, or 95 kPa CO_2, 35 or 45 kPa O_2, or a mixture of 30 kPa O_2 + 70 kPa CO_2 (Palou and Taberner, 2018).

4 Postharvest Diseases of Loquat

4.1 Anthracnose

Anthracnose caused by *Colletotrichum* ssp. has been described as one of the most serious postharvest diseases affecting loquat fruit worldwide. Reports from East Asian countries, the most important loquat producers in the world, identified *C. acutatum* as the causal agent of the disease (Gu et al., 2007; Liu et al., 2007), but *C. gloeosporioides* sensu stricto has also been reported in other loquat-producing areas such as Spain (Palou et al., 2016b; Figure 5.9a), the USA (Horst, 2013), and Pakistan (Naz et al., 2017). Epidemiological factors influencing the incidence of latent infections in the field and disease development after harvest are similar to those described for anthracnose of persimmon fruit in this chapter. Abundant rains, high humidity, and mild temperatures are key factors for inoculum production and dispersal. Optimum values for in vitro sporulation of *C. acutatum* were 18–26°C and pH 9–10. For conidial germination, these values were 20–26°C, pH 6, and 100% RH (Liu et al., 2007). Infected fruits initially have small black spots in the area of infection, then these enlarge gradually and the entire fruit becomes brown (Cao and Zheng, 2008). If the environmental humidity is high enough, the whole fruit may rot and the fungal fruiting bodies form on the rotten surface.

4.2 Gray Mold

Botrytis cinerea can infect loquat fruit during its entire growth, storage, and marketing periods, since it is an aggressive airborne pathogen with a necrotrophic lifestyle. However, postharvest gray mold develops more often from field latent infections and can be economically important on cold-stored fruits (Palou et al., 2016b). Symptoms start as soft, circular, discolored spots that rapidly enlarge at ambient temperatures covering the entire fruit of typical gray sporulating mycelia (Figure 5.9b). Besides postharvest losses, *B. cinerea* can also cause losses in the field prior to harvest mainly due to blossom blight, in which the flower spike wilts and the fruit dies (Sun et al., 2009).

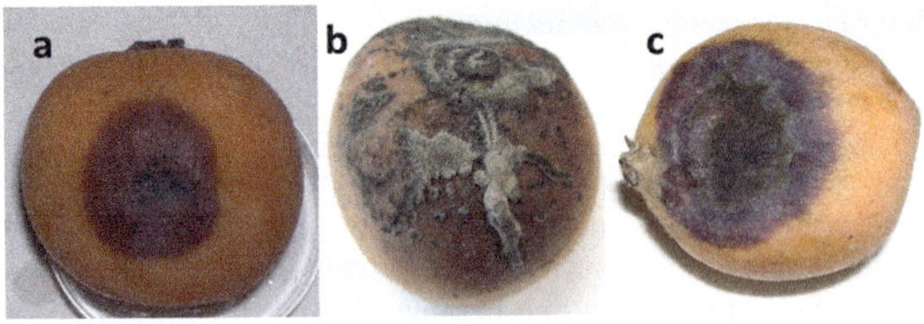

Figure 5.9 a. Anthracnose lesion caused by *Colletotrichum gloeosporioides* on a Spanish-grown loquat fruit. b.Symptoms of advanced gray mold caused by *Botrytis cinerea* on loquat fruit. c.Sporulated lesion of black spot caused by *Alternaria alternata* on the surface of a loquat fruit.

4.3 Black Spot

Black spot caused by *A. alternata* was identified as the most frequent postharvest disease that developed from latent infections on 'Algerie' loquats grown in Spain. Although with lower frequency, the pathogen also caused disease on artificially wounded fruit (Palou et al., 2016b). Presence of *Alternaria* spp. causing leaf spots or fruit rot on loquat was also reported in Greece (Tziros, 2013), Iran (Mirhosseini et al., 2015), Palestine (Batta, 2005), China (Gu et al., 2007), and Taiwan (Ko et al., 2010). These workers reported that the spots were mainly caused by field infections of *A. alternata*, which remained quiescent until harvest and then visible symptoms appeared after harvest. Typically, these infections developed marginally during cold storage, and expanded further during fruit shelf life or marketing periods (Palou et al., 2016b). Symptoms on loquat fruit are black, circular, dry spots of variable size located on any part of the fruit surface that, after some time at ambient temperatures and high RH, develop an external dark mycelium and olive green to black conidia (Figure 5.9c).

4.4 Other Diseases

Species of the Botryosphaeriaceae such as *L. theobromae* (syn.: *Diplodia natalensis* Pole-Evans) (Tian et al., 2011) and *Diplodia seriata* De Not. (Palou et al., 2013e, 2016b) can also cause postharvest disease of loquat fruit. Their origin is typically from latent infections in the field. Although the symptoms occur most frequently at the stem end of the fruit; they can also occur occasionally on the stylar end or fruit sides. These fungi rapidly grow unevenly through the rind and produce finger-like projections of brown tissue in the infected fruit.

Other pathogens causing loquat postharvest disease from latent or wound infections that were isolated, identified, and characterized in terms of pathogenicity and disease development in Spain include *P. expansum, P. clavispora*, and *Diaporthe phaseolorum* (Cooke & Ellis) Sacc. (Palou et al., 2013b, 2016b). *Penicillium expansum* caused the most frequent disease on 'Algerie' loquats, artificially wounded and incubated at 20°C, and it was also capable of growing on fruit stored at 5°C (Palou et al., 2016b). Another species of *Pestalotiopsis, P. eriobotryfolia*, was described as a pathogen of loquat fruit in China (Gu et al., 2007).

Pseudomonas syringae pv. *eriobotryae* is the causal agent of stem canker of loquat, an economically important tree disease distributed widely throughout the loquat cultivation areas. It can also infect loquat fruit after harvest through wounds made during fruit handling (Lin et al., 1999).

4.5 Chemical and Integrated Control

4.5.1 Preharvest Management

Effective orchard sanitation practices such as the removal of dead wood and rotten fruit are recommended to reduce fungal inoculum levels and the incidence of postharvest disease caused by latent infections, especially those of *Colletotrichum* spp. and *Alternaria* spp. Nevertheless, since these cultural practices may be excessively costly, the application of fungicides in the field is the most common decay control practice. In the Mediterranean Basin, most fungicide treatments in the orchard are primarily applied to control scab, caused by *Fusicladium eriobotryae* (Cavara) Sacc, and they could be useful against pathogens causing postharvest decay. This pathogen causes necrotic spots on young twigs, leaves, and fruit. Scab lesions on loquat fruit may coalesce to cover most of the rind and affected fruits are not acceptable for the fresh market or industrial processing. The pathogen has a relatively long incubation period and symptoms do not appear until several weeks after infection (González-Domínguez et al., 2014a). Copper compounds are usually applied during flowering, followed by SBIs combined with dithiocarbamates or captan. Up to five fungicide sprays per year are applied for loquat scab control in some areas of Spain (González-Domínguez et al., 2014b).

4.5.2 Postharvest Management

Refrigeration can reduce the severity of decay and extend the postharvest life of loquat. However, low temperature regimes, such as 2–3 wk of storage at 1–5°C, increase the severity of chilling injury symptoms in red-fleshed fruit, with symptoms of flesh leatheriness, stuck peel, and firm and juiceless flesh (Zheng et al., 2000). Therefore, other postharvest technologies to complement cold storage are employed, such as CA (10–12 kPa O_2 + 1–2 kPa CO_2) and MAP based on microperforated plastic films, to preserve loquat fruit quality after harvest (Pareek et al., 2014). In addition, depending on the regulations of each loquat-producing country, synthetic chemical fungicides may or may not be applied after harvest to control postharvest decay. Their use is banned in Turkey, Spain and other EU countries (Palou et al., 2016b), while postharvest fungicides are approved and used in China and other countries. However, growing concerns about public health and the environment, and increasing resistance of many fungi to commonly used fungicides have stimulated the search for alternative methods (Wisniewski et al., 2016).

Low-toxicity chemicals and biological control are two alternative control methods that have been evaluated to control loquat postharvest diseases, especially anthracnose. Cao et al. (2008c, 2008d, 2014) reported that treatments with the plant regulator methyl jasmonate (MeJA) effectively inhibit loquat postharvest anthracnose caused by *C. acutatum*. Treatment of loquat fruits with MeJA significantly reduced disease incidence and reduced lesion sizes compared to control fruit. MeJA treatment significantly increased the activity of several enzymes and the biosynthesis of polyamines, which was related to induction of disease resistance. Furthermore, in in vitro tests, MeJA inhibited spore germination, germ tube elongation, and mycelial growth of *C. acutatum*. Similarly, the ethylene inhibitor 1-MCP was effective to indirectly reduce decay by delaying loquat fruit senescence and consequently maintaining natural resistance of the fruit (Cao and Zheng, 2010). In other work, postharvest applications of the GRAS compound ethanol at 300 μL/L significantly inhibited anthracnose and maintained the overall quality of loquat fruits. Ethanol treatment enhanced the activity of defense-related enzymes in *C. acutatum*-inoculated fruits and effectively reduced spore germination and mycelial growth of *C. acutatum* in vitro (Wang et al., 2015). Yan et al. (2008) reported that the application of the pathogenesis-related protein chitinase expressed in the yeast *Komagataella pastoris* (syn. *Pichia pastoris*) to loquats wound inoculated with *B. cinerea* significantly reduced the severity of gray mold, although the extent of the reduction was dependent on the timing of pathogen inoculation and chitinase treatment.

Several microorganisms were tested as biological control agents for the control of loquat postharvest diseases. Research with the yeast *Pichia membranifaciens* indicated that application of washed cell suspensions controlled loquat anthracnose better than this yeast applied in culture broth at the same concentration, although the efficacy depended on the concentration. Treatment with autoclaved cell cultures or culture filtrates did not control the disease (Cao et al., 2008a). Treatment of loquats with *Bacillus cereus* AR156 resulted in lower incidence and severity of anthracnose than on untreated fruits through the induction and priming of defense responses in the fruit (Wang et al., 2014).

Integrated disease control strategies utilizing biological control and physical or chemical alternative control methods have also been investigated on loquat to find synergistic activities that improve control of the disease. A beneficial effect of combining MeJA applications with the antagonistic yeast *P. membranifaciens* to control loquat postharvest anthracnose has been demonstrated (Cao et al., 2009). The combination induced higher activities of defense-related enzymes and greater inhibition of spore germination and germ tube elongation of *C. acutatum* than either treatment alone. Similar synergistic activity for direct (an adverse effect on the pathogen) and indirect (induction of host resistance) control of loquat anthracnose caused by *C. acutatum* was observed with a combination of *P. membranifaciens* with $CaCl_2$ applications (Cao et al., 2008b). Other work indicated that heat treatment (hot air at 38°C for 36 hr) and the yeast *Pichia guilliermondii*, either alone or in combination, significantly reduced both natural decay of loquat fruits and the incidence and severity of anthracnose on artificially inoculated fruits (Liu et al., 2010).

Acknowledgments

The authors thank all the public agencies and private enterprises from Spain, Turkey, China, and Greece that have funded research in this topic.

References

Adaskaveg, J.E. and Förster, H. 2000. Postharvest decays of pomegranates and their management in California. Available at: http://ir4.rutgers.edu/FoodUse/PerfData/229.pdf

Adaskaveg, J.E. and Förster, H. 2002. Management of gray mold of pomegranates caused by *Botrytis cinerea* using two reduced-risk fungicides, fludioxonil and fenhexamid. *Phytopathology* 93, S127.

Adaskaveg, J.E. and Michailides, T.J. 2013. Diseases. pp. 33–37. *In*: University of California Statewide Integrated Pest Management Program (ed.). *UC IPM Pest Management Guidelines for Agriculture - Pomegranate*. University of California Statewide Integrated Pest Management Program, Agriculture and Natural Resources No. 3474. ANR Communication Services, Richmond, CA.

Alvarez, L.V., Groenewald, J.Z. and Crous, P.W. 2016. Revising the Schizoparmaceae: *Coniella* and its synonyms *Pilidiella* and *Schizoparme*. *Studies in Mycology* 85, 1–34.

Ammar, M.I. and El-Naggar, M.A. 2014. Screening and characterization of fungi and their associated mycotoxins in some fruit crops. *International Journal of Advanced Research* 2, 1216–1227.

Archana, B.C. and Jamadar, M.M. 2014. Management of leaf spot and fruit spot/rot of pomegranate (*Punica granatum* L.) caused by *Alternaria alternata* (Fr.) Keissler. *Karnataka Journal of Agricultural Science* 27, 247–249.

Bardas, G.A., Tzelepis, G.D., Lotos, L. and Karaoglanidis, G.S. 2009. First report of *Botrytis cinerea* causing gray mold of pomegranate (*Punica granatum*) in Greece. *Plant Disease* 93, 1346.

Batta, Y. 2005. Control of *Alternaria* spot disease on loquat (*Eriobotrya japonica* Lindl.) using detached fruits and leaf disk assay. *An-Najah University Journal for Research - Natural Sciences* 19, 69–82.

Berbegal, M., López-Cortés, I., Salazar, D., Gramaje, D., Pérez-Sierra, A., García-Jiménez, J. and Armengol, J. 2014. First report of Alternaria black spot of pomegranate caused by *Alternaria alternata* in Spain. *Plant Disease* 98, 689.

Biton, E., Kobiler, I., Feygenberg, O., Yaari, M., Friedman, H. and Prusky, D. 2014a. Control of Alternaria black spot in persimmon fruit by a mixture of gibberellin and benzyl adenine, and its mode of action. *Postharvest Biology and Technology* 94, 82–88.

Biton, E., Kobiler, I., Feygenberg, O., Yaari, M., Kaplunov, T., Ackerman, M., Friedman, H. and Prusky, D. 2014b. The mechanism of differential susceptibility to Alternaria black spot, caused by *Alternaria alternata*, of stem-and bottom-end tissues of persimmon fruit. *Postharvest Biology and Technology* 94, 74–81.

Blanco, C., Romero, F. and De los Santos, B. 2008. First report of *Pestalotiopsis dyospiri* causing calyx blight of sweet persimmon in Huelva province (Southwestern Spain). *Plant Disease* 92, 176.

Blanco-Ulate, B., Vincenti, E., Cantu, D. and Powell, A.L.T. 2016. Ripening of tomato fruit and susceptibility to *Botrytis cinerea*. pp. 387–412. *In*: S. Fillenger and Y. Elad (eds.). *Botrytis – The Fungus, the Pathogen and Its Management in Agricultural Systems*. Springer, New York, NY.

Blood, R.R.Y., Rozwalka, L.C. and De Mio, L.L.M. 2015. Persimmon anthracnose caused by *Colletotrichum horii*: Incidence in branches, leaves, flowers and fruits in the field. *Revista Brasileira de Fruticultura* 37, 335–345.

Cao, S.F., Cai, Y.T., Yang, Z.F., Joyce, D.C. and Zheng, Y.H. 2014. Effect of MeJA treatment on polyamine, energy status and anthracnose rot of loquat fruit. *Food Chemistry* 145, 86–89.

Cao, S.F. and Zheng, Y.H. 2008. Postharvest biology and handling of loquat fruit. *Stewart Postharvest Review* 4, 1–5.

Cao, S.F. and Zheng, Y.H. 2010. Effect of 1-methylcyclopropene on anthracnose rot caused by *Colletotrichum acutatum* and disease resistance in loquat fruit. *Journal of the Science of Food and Agriculture* 90, 2289–2294.

Cao, S.F., Zheng, Y.H., Tang, S.S., Jin, P. and Wang, K.T. 2008a. Biological control of postharvest anthracnose rot of loquat by *Pichia membranifaciens*. *The Journal of Horticultural Science and Biotechnology* 83, 816–820.

Cao, S.F., Zheng, Y.H., Tang, S.S. and Wang, K.T. 2008b. Improved control of anthracnose rot in loquat fruit by a combination treatment of *Pichia membranifaciens* with $CaCl_2$. *International Journal of Food Microbiology* 126, 216–220.

Cao, S.F., Zheng, Y.H., Tang, S.S., Wang, K.T. and Rui, H.J. 2009. Effect of yeast antagonist in combination with MeJA treatment on postharvest anthracnose rot of loquat fruit. *Biological Control* 50, 73–77.

Cao, S.F., Zheng, Y.H., Yang, Z.F., Tang, S.S. and Jin, P. 2008c. Control of anthracnose rot and quality deterioration in loquat fruit with methyl jasmonate. *Journal of the Science of Food and Agriculture* 88, 1598–1602.

Cao, S.F., Zheng, Y.H., Yang, Z.F., Tang, S.S., Jin, P., Wang, K.T. and Wang, X.M. 2008d. Effect of methyl jasmonate on inhibition of *Colletotrichum acutatum* infection in loquat fruit and the possible mechanism. *Postharvest Biology and Technology* 49, 301–307.

Çeliker, N.M., Uysal, A., Çetinel, B. and Poyraz, D. 2012. Crown rot on pomegranate caused by *Coniella granati* in Turkey. *Australasian Plant Disease Notes* 7, 161–162.

Chang, T., Hassan, O., Jeon, J.Y., Shin, J.S., Oh, N.K. and Lim, T.H. 2018. First report of anthracnose of persimmon (*Diospyros kaki* L. f.) caused by *Colletotrichum siamense* in Korea. *Plant Disease* 102, 443.

Chen, Y., Shao, D.D., Zhang, A.F., Yang, X., Zhou, M. and Xu, Y.L. 2014. First report of a fruit rot and twig blight on pomegranate (*Punica granatum*) caused by *Pilidiella granati* in Anhui Province of China. *Plant Disease* 98, 695.

Cia, P., Benato, E.A., Sigrist, J.M., Sarantopoulos, C., Oliveira, L.M. and Padula, M. 2003. In vitro effect of modified atmosphere on mycelial growth of persimmon pathogens and on the control of Rhizopus rot in "Fuyu'. *Summa Phytopathologica* 29, 266–274.

Cintora-Martínez, E.A., Leyva-Mir, S.G., Ayala-Escobar, V., Ávila-Quezada, G.D., Camacho-Tapia, M. and Tovar-Pedraza, J.M. 2017. Pomegranate fruit rot caused by *Pilidiella granati* in Mexico. *Australasian Plant Disease Notes* 12, 1–4.

D'Aquino, S., Palma, A., Schirra, M., Continella, A., Tribulato, E. and La Malfa, S. 2010. Influence of film wrapping and fludioxonil application on quality of pomegranate fruit. *Postharvest Biology and Technology* 55, 121–128.

D'Aquino, S., Schirra, M., Palma, A., Angioni, A., Cabras, P., Gentile, A. and Tribulato, E. 2009. Effectiveness of fludioxonil in control storage decay on pomegranate fruit. *Acta Horticulturae* 818, 313–318.

DAFF (Department of Agriculture, Fisheries and Forestry, Australian Government). 2004. Final import policy: Persimmon fruit (*Diospyros kaki* L.) from Japan, Korea and Israel. Available at: www.agriculture.gov.au/SiteCollectionDocuments/ba/plant/ungroupeddocs/persimmon_final.pdf

Dean, R., van Kan, J.A.L., Pretorius, Z.A., Hammond-Kosack, K.E., Di Pietro, A., Spanu, P.D., Rudd, J.J., Dickman, M., Kahmann, R., Ellis, J. and Foster, G.D. 2012. The top 10 fungal pathogens in molecular plant pathology. *Molecular Plant Pathology* 13, 414–430.

Elad, Y. and Evensen, K. 1995. Physiological aspects of resistance to *Botrytis cinerea*. *Phytopathology* 85, 637–643.

Elad, Y., Pertot, I., Cotes Prado, A.M. and Stewart, A. 2016. Plant hosts of *Botrytis* spp. pp. 413–486. *In*: S. Fillenger and Y. Elad (eds.). *Botrytis – The Fungus, the Pathogen and Its Management in Agricultural Systems*. Springer, New York, NY.

Elad, Y., Williamson, B., Tudzynski, P. and Delen, N. 2005. *Botrytis* spp. and diseases they cause in agricultural systems – An introduction. pp. 1–8. *In*: Y. Elad,

B. Williamson, P. Tudzynski and N. Delen (eds.). *Botrytis: Biology, Pathology and Control*. Springer, New York, NY.

El-Kassas, S.E., Mahmoud, H.M., Salhy, A.M. and Mohamed, S.A. 1995. Effect of GA3 and calcium treatments on the keeping quality and storability of Manfalouty pomegranate fruits. *Assuit Journal of Agricultural Sciences* 26, 217–230.

Erkan, M. and Kader, A.A. 2011. Pomegranate (*Punica granatum* L.). pp. 287–311. *In*: E.M. Yahia (ed.). *Postharvest Biology and Technology of Tropical and Subtropical Fruits, vol. 4. Mangosteen to White Sapote*. Woodhead Publishing Limited, Cambridge, UK.

Eshel, D., Ben Arie, R., Dinoor, A. and Prusky, D. 2000. Resistance of gibberellin-treated persimmon fruit to *Alternaria alternata* arises from the reduced ability of the fungus to produce endo-1,4-β-glucanase. *Phytopathology* 90, 1256–1262.

Ezra, D., Gat, T., Skovorodnikova, Y., Vardi, Y. and Kosto, I. 2010. First report of Alternaria black spot of pomegranate caused by *Alternaria alternata* in Israel. *Australasian Plant Disease Notes* 5, 1–2.

Ezra, D., Kirshner, B., Gat, T., Shteinberg, D. and Kosto, I. 2015a. Heart rot of pomegranate, when and how does the pathogen cause the disease?. *Acta Horticulturae* 1089, 167–171.

Ezra, D., Kirshner, B., Hershcovich, M., Shteinberg, D. and Kosto, I. 2015b. Heart rot of pomegranate: Disease etiology and the events leading to development of symptoms. *Plant Disease* 99, 496–501.

Faedda, R., Granata, G., MassiminoCocuzza, G.E., Lo Giudice, V., Audoly, G., Pane, A. and Cacciola, S.O. 2015. First report of heart rot of pomegranate (*Punica granatum*) caused by *Alternaria alternata* in Italy. *Plant Disease* 99, 1446.

FAOSTAT (Food and Agriculture Organization of the United Nations, Statistics Division). 2014. Food and agricultural commodities production. Available at: www.fao.org/faostat/en/#data/QC

Faria, A. and Calhau, C. 2011. The bioactivity of pomegranate: Impact on health and disease. *Critical Reviews in Food Science and Nutrition* 51, 626–634.

Garibaldi, A., Bertetti, D., Amatulli, M.T. and Gullino, M.L. 2010. First report of postharvest fruit rot in persimmon caused by *Phacidiopycnis washingtonensis* in Italy. *Plant Disease* 94, 788.

Gat, T., Liarzi, O., Skovorodnikova, Y. and Ezra, D. 2012. Characterization of *Alternaria alternata* causing black spot disease of pomegranate in Israel using a molecular marker. *Plant Disease* 96, 1513–1518.

George, A.P. and Redpath, S. 2008. Health and medical benefits of persimmon fruit: A review. *Advances in Horticultural Science* 22, 244–249.

Gil, M.I., Tomás-Barberán, F.A., Hess-Pierce, B., Holcroft, D.M. and Kader, A. 2000. Antioxidant activity of pomegranate juice and its relationship with phenolic composition and processing. *Journal of Agricultural and Food Chemistry* 48, 4581–4589.

Goh, H.H., Lyons, S.N. and Mowat, A.W. 1991. Preliminary observations of diseases on persimmon (*Diospyros kaki*) and their potential effect on fruit quality. *Proceedings of the New Zealand Weed and Pest Control Conference* 44, 265–267.

González-Domínguez, E., Armengol, J. and Rossi, V. 2014a. Development and validation of a weather-based model for predicting infection of loquat fruit by *Fusicladium eriobotryae*. *PLoS ONE* 9, e107547.

González-Domínguez, E., Rodríguez-Reina, J.M., García-Jiménez, J. and Armengol, J. 2014b. Evaluation of fungicides to control loquat scab caused by *Fusicladium eriobotryae*. *Plant Health Progress* 15, 15–18.

Gu, H., Liu, A.Y., Chen, W.X., Feng, S.J. and Shi, J.Y. 2007. Development and control of postharvest diseases of loquat fruit. *Acta Horticulturae* 750, 437–443.

Hammerschmidt, L., Wray, V., Lin, W., Kamilova, E., Proksch, P. and Aly, A.H. 2012. New styrylpyrones from the fungal endophyte *Penicillium glabrum* isolated from *Punica granatum*. *Phytochemistry Letters* 5, 600–603.

Hassan, O., Jeon, J.Y., Chang, T., Shin, J.S., Oh, N.K. and Lee, Y.S. 2018. Molecular and morphological characterization of *Colletotrichum* species in the *Colletotrichum gloeosporioides* complex associated with persimmon anthracnose in South Korea. *Plant Disease* 102, 1015–1024.

Hebert, T.T. and Clayton, C.N. 1963. Pomegranate fruit rot caused by *Coniella granati*. *Plant Disease Reporter* 47, 222–223.

Hess-Pierce, B. and Kader, A.A. 2003. Responses of 'Wonderful' pomegranates to controlled atmospheres. *Acta Horticulturae* 600, 751–757.

Holz, G., Coertze, S. and Williamson, B. 2007. The ecology of *Botrytis* on plant surface. pp. 9–27. *In*: Y. Elad, B. Williamson, P. Tudzynski and N. Delen (eds.) *Botrytis: Biology, Pathology and Control*. Kluwer Academic Publishers, Dordrecht, The Netherlands.

Horst, R.K. 2013. *Field Manual of Diseases on Trees and Shrubs*. Springer Science & Business Media, New York, NY.

Jamadar, M.M., Sataraddi, A.R., Patil, P.V., Jawadagi, R.S. and Patil, D.R. 2011. Status of pomegranate diseases of Northern Karnataka in India. *Acta Horticulturae* 890, 501–507.

Jeon, C.W., Kwon, Y., Lee, J.H. and Kwak, Y.S. 2015. Improvement and effectiveness for chemical control protocol of sweet persimmon anthracnose disease. *The Korean Journal of Pesticide Science* 19, 312–316.

Joshi, S., Sawant, D.M. and Gaikwad, A.P. 2014. Isolate variations in *Colletotrichum gloeosporioides* infecting pomegranate. *The Journal of Plant Protection Sciences* 6, 21–26.

Kahramanoglu, I., Usanmaz, S. and Nizam, I. 2014. Incidence of heart rot in pomegranate fruits caused by *Alternaria* spp. in Cyprus. *African Journal of Agricultural Research* 9, 905–907.

Kanetis, L., Testempasis, S., Goulas, V., Samuel, S., Myresiotis, C. and Karaoglanidis, G.S. 2015. Identification and mycotoxigenic capacity of fungi associated with pre- and postharvest fruit rots of pomegranates in Greece and Cyprus. *International Journal of Food Microbiology* 208, 84–92.

Kanwar, Z.S. and Thakur, D.P. 1972. Post-harvest fungal diseases of pomegranate in Haryana. *Indian Society of Mycology and Plant Pathology* 2, 189–191.

KC, A.N. and Vallad, G.E. 2016. First report of *Pilidiella granati* causing fruit rot and leaf spots on pomegranate in Florida. *Plant Disease* 100, 1238.

Khokhar, I., Bajwa, R. and Nasim, G. 2013. New report of *Penicillium implicatum* causing a postharvest rot of pomegranate fruit in Pakistan. *Australasian Plant Disease Notes* 8, 39–41.

Kinay-Teksür, P. 2015. Alternative technologies to control postharvest diseases of pomegranate. *Stewart Postharvest Review* 4, 3.

Kinay-Teksür, P., Şen, F. and Yildiz, F. 2014. The Prevention of fungal decays and improving fruit quality with some pre and postharvest treatments on pomegranate. *Fruit Report of Project Tubitak, Tovag* 110 O 624, 158.

Kinay-Teksür, P., Şen, F., Yildiz, F., Hizaler, K. and Celik, Y. 2012. The effect of pre and postharvest treatments on decay development of 'Hicaz' pomegranates (*Punica granatum* L. var. Hicaz). Abstracts Book of the 7th International Postharvest Symposium. Kuala Lumpur, Malaysia. Abstract No. 350.

Ko, Y., Liu, C.W., Chen, C.C., Yao, K.S., Maruthasalam, S. and Lin, C.H. 2010. First report of fruit rot of loquat caused by an *Alternaria* sp. in Taiwan. *Plant Disease* 94, 481.

Kobiler, I., Akerman, M., Huberman, L. and Prusky, D. 2011. Integration of pre- and postharvest treatments for the control of black spot caused by *Alternaria alternata* in stored persimmon fruit. *Postharvest Biology and Technology* 59, 166–171.

Kumar, V., Soni, R. and Nathawat, B.D.S. 2016. Eco-friendly management of Aspergillus rot of pomegranate (*Punica granatum*). *Advances in Life Sciences* 5, 2859–2865.

Kurt, S., Soylu, E.M. and Soylu, S. 2010. First report of black spot disease caused by *Alternaria alternata* in persimmon fruits in Turkey. *Plant Disease* 94, 1069.

Kwon, J.H., Ahn, G.H. and Park, C.S. 2004. Leaf blight of sweet persimmon tree in the field and fruit rot in the storage caused by *Pestalotia diospyri. The Korean Journal of Mycology* 32, 130–133.

Kwon, J.H., Jeong, S.G., Hong, S.B., Chae, Y.S. and Park, C.S. 2006. Occurrence of blue mold on sweet persimmon (*Diospyros kaki*) caused by *Penicillium expansum. Research in Plant Disease* 12, 290–293.

Kwon, J.H., Kang, S.W. and Park, C.S. 1999. Persimmon gray mold caused by *Botrytis cinerea. Plant Diseases in Agriculture* 5, 55–57.

Kwon, J.H. and Kim, J. 2011. First report of fruit black spot of *Diospyros kaki* caused by *Colletotrichum acutatum* in Korea. *Journal of Plant Pathology* 27, 100.

Kwon, J.H., Kim, J., Choi, O., Gang, G.H., Han, S. and Kwak, Y.S. 2013. Anthracnose caused by *Colletotrichum horii* on sweet persimmon in Korea: Dissemination of conidia and disease development. *Journal of Phytopathology* 161, 497–502.

Kwon, J.H. and Park, C.S. 2003. Sooty mold of persimmon (*Diospyros kaki*) caused by *Cladosporium cladosporioides* in Korea. *The Plant Pathology Journal* 19, 266–268.

Labuda, R., Hudec, K., Piecková, E., Mezey, J., Bohovic, R., Mátéová, S. and Lukác, S.S. 2004. *Penicillium implicatum* causes a destructive rot of pomegranate fruits. *Mycopathologia* 157, 217–223.

Lee, J.H., Kim, J. and Kwak, Y.-S. 2013. First report of black spot disease caused by *Alternaria alternata* on sweet persimmon fruits. *Mycobiology* 41, 67–169.

Lennox, C.L., Mostert, L., Venter, E., Laubscher, W. and Meitz-Hopkins, C. 2018. First report of *Coniella granati* (Sacc.) fruit rot and dieback on pomegranate in the Western Cape of South Africa. *Plant Disease* 102, 821.

Levy, E., Elkind, G., Ben-Arie, R. and Ben-Ze'ev, I.S. 2011. First report of *Coniella granati* causing pomegranate fruit rot in Israel. *Phytoparasitica* 39, 403–405.

Lin, S., Sharpe, R.H. and Janick, J. 1999. Loquat: Botany and horticulture. *Horticultural Reviews* 23, 233–276.

Liu, A.Y., Chen, W.X., Gu, H., Shi, J.Y. and Li, J. 2007. Biological characteristic of pathogenic fungus causing anthracnose of loquat fruit. *Acta Horticulturae* 750, 465–470.

Liu, F.J., Tu, K., Shao, X.F., Zhao, Y., Tu, S.C., Su, J., Hou, Y.P. and Zou, X.R. 2010. Effect of hot air treatment in combination with *Pichia guilliermondii* on postharvest anthracnose rot of loquat fruit. *Postharvest Biology and Technology* 58, 65–71.

Liu, Y., Zhang, W., Xu, C. and Li, X. 2016. Biological activities of extracts from loquat (*Eriobotrya japonica* Lindl.): A review. *International Journal of Molecular Sciences* 17, 1983.

Luo, Y., Hou, L., Förster, H., Pryor, B. and Adaskaveg, J.E. 2017. Identification of *Alternaria* species causing fruit rot of pomegranates in California. *Plant Disease* 101, 421–427.

Melgarejo, P., Martínez-Valero, R., Guillamón, J.M. and Amorós, M.M. 1997. Phenological stages of the pomegranate tree (*Punica granatum* L.). *Annals of Applied Biology* 130, 135–140.

Michailides, T.J., Morgan, D., Quist, M. and Reyes, H. 2008. Infection of pomegranate by *Alternaria* spp. causing black heart. *Phytopathology* 98, 105.

Michailides, T.J., Puckett, R. and Morgan, D.P. 2010. Pomegranate decay caused by *Pilidiella granati* in California. *Phytopathology* 100, 83.

Mincuzzi, A., Garganese, F., Ippolito, A. and Sanzani, S.M. 2016. First report of *Pilidiella granati* causing postharvest fruit rot on pomegranate in Southern Italy. *Journal of Plant Pathology* 98, 377.

Mincuzzi, A., Sanzani, S.M., Garganese, F., Ligorio, A. and Ippolito, A. 2017. First report of *Cytospora punicae* causing fruit rot on pomegranate in Italy. *Journal of Plant Pathology* 99, 302.

Mirabolfathy, M., Groenewald, J.Z. and Crous, P.W. 2012. First report of *Pilidiella granati* causing dieback and fruit rot of pomegranate (*Punica granatum*) in Iran. *Plant Disease* 96, 461.

Mirhosseini, H.A., Babaeizad, V. and Basavand, E. 2015. Identification and detection of agent of loquat leaf spot and fruit rot in north of Iran. *Journal on New Biological Reports* 4, 135–138.

Moradinezhad, F. and Khayyat, M. 2014. Effects of intermittent warming and prestorage treatments (hot water, salicylic acid, calcium chloride) on postharvest life of pomegranate fruit cv. 'Shishe-Kab' during long-term cold storage. *International Journal of Horticultural Science and Technology* 1, 43–51.

Munhuweyi, K., Caleb, O.J., Lennox, C.L., van Reenen, A.J. and Opara, U.L. 2017. In vitro and in vivo antifungal activity of chitosan-essential oils against pomegranate fruit pathogens. *Postharvest Biology and Technology* 129, 9–22.

Munhuweyi, K., Lennox, C.L., Meitz-Hopkins, J.C., Caleb, O.J. and Opara, U.L. 2016. Major diseases of pomegranate (*Punica granatum* L.), their causes and management - A review. *Scientia Horticulturae* 211, 126–139.

Myresiotis, C., Testempasis, S., Vryzas, Z., Karaoglanidis, G.S. and Papadopoulou-Mourkidou, E. 2015. Determination of mycotoxins in pomegranate fruits and juices using a QuEChERS-based method. *Food Chemistry* 182, 81–88.

Nallathambi, P. and Umamaheswari, C. 2009. Detection of aflatoxins in pomegranate arils infected by *Aspergillus* species. *Indian Phytopathology* 62, 178–182.

Naz, F., Abbas, M.F., Rauf, C.A., Tariq, A., Mumtaz, A., Irshad, G., Shaheen, F.A. and Hassan, I. 2017. First report of *Colletotrichum gloeosporioides* causing anthracnose on loquat in Pakistan. *Plant Disease* 101, 1550.

Nerya, O. and Levin, A. 2015. Innovative treatment of pomegranates from harvest to market. *Acta Horticulturae* 1089, 489–493.

Neuwald, D.A., Streif, J., Sestari, I., Giehl, R.F.H., Weber, A. and Brackmann, A. 2009. Quality of 'Fuyu' persimmon during modified atmosphere storage. *Acta Horticulturae* 833, 227–232.

Oliver, R.P. and Hewitt, H.G. 2014. *Fungicides in Crop Protection*. 2nd ed. CABI, Wallingford, UK.

Özgüven, A.I., Yilmaz, C. and Keles, D. 2012. Pomegranate biodiversity and horticultural management. *Acta Horticulturae* 940, 21–28.

Pala, H., Tatli, A., Yilmaz, C. and Ozguven, A.I. 2009. Important diseases of pomegranate fruit and control possibilities in Turkey. *Acta Horticulturae* 818, 285–289.

Palou, L., Crisosto, C.H. and Garner, D. 2007a. Combination of postharvest antifungal chemical treatments and controlled atmosphere storage to control gray mold and improve storability of 'Wonderful' pomegranates. *Postharvest Biology and Technology* 43, 133–142.

Palou, L., Guardado, A. and Montesinos-Herrero, C. 2010. First report of *Penicillium* spp. and *Pilidiella granati* causing postharvest fruit rot of pomegranate in Spain. *New Disease Reports* 22, 21.

Palou, L., Montesinos-Herrero, C., Besada, C. and Taberner, V. 2013a. Postharvest fruit rot of persimmon (*Diospyros kaki*) in Spain caused by *Lasiodiplodia theobromae* and *Neofusicoccum* spp. *Journal of Phytopathology* 161, 625–631.

Palou, L., Montesinos-Herrero, C. and del Río, M.A. 2007b. Incidence of postharvest gray mold on 'Mollar de Elche' pomegranates. *Phytopathology* 97, 89.

Palou, L., Montesinos-Herrero, C., Tarazona, I., Besada, C. and Taberner, V. 2015. Incidence and etiology of postharvest fungal diseases of persimmon (*Diospyros kaki* Thunb. cv. Rojo Brillante) in Spain. *Plant Disease* 99, 1416–1425.

Palou, L., Montesinos-Herrero, C., Tarazona, I. and Taberner, V. 2013c. Postharvest anthracnose of persimmon fruit caused by *Colletotrichum gloeosporioides* first reported in Spain. *Plant Disease* 97, 691.

Palou, L., Montesinos-Herreros, C., Guardado, A. and Taberner, V. 2013b. First report of *Pestalotiopsis clavispora* causing postharvest fruit rot of loquat in Spain. *Journal of Plant Pathology* 95, 69.

Palou, L., Rosales, R., Montesinos-Herrero, C. and Taberner, V. 2016a. Short-term exposure to high CO_2 and O_2 atmospheres to inhibit postharvest gray mold of pomegranate fruit. *Plant Disease* 100, 424–430.

Palou, L., Sánchez-Torres, P., Montesinos-Herrero, C. and Taberner, V. 2016b. Incidence and etiology of postharvest fungal diseases of loquat fruit (*Eriobotrya japonica* (Thunb.) Lindl. cv. 'Algerie') in Alacant province (Spain). *European Journal of Plant Pathology* 146, 847–860.

Palou, L. and Taberner, V. 2018. Evaluation of short postharvest gaseous treatments for inhibition of Alternaria black spot of persimmon 'Rojo Brillante'. *In*: Proceedings of the VI International Symposium on Persimmon. Valencia, Spain. *Acta Horticulturae* 1195, 243–249

Palou, L., Taberner, V., Guardado, A., del Río, M.A. and Montesinos-Herrero, C. 2013d. Incidence and etiology of postharvest fungal diseases of pomegranate (*Punica granatum* cv. Mollar de Elche) in Spain. *Phytopathologia Mediterranea* 52, 478–489.

Palou, L., Taberner, V., Guardado, A. and Montesinos-Herrero, C. 2012. First report of *Alternaria alternata* causing postharvest black spot of persimmon in Spain. *Australasian Plant Disease Notes* 7, 41–42.

Palou, L., Taberner, V., Mira, J.L. and Vicent, A. 2016c. Assessment of latent infections of *Alternaria alternata* in fruit of persimmon cv. Rojo Brillante cultivated in the Valencia area (Spain). Abstracts Book of the VI International Symposium on Persimmon. Valencia, Spain. p. 55.

Palou, L., Taberner, V. and Montesinos-Herreros, C. 2013e. First report of *Diplodia seriata* causing loquat fruit rot in Spain. *Plant Disease* 97, 421.

Pareek, S., Benkeblia, N., Janick, J., Cao, S. and Yahia, E.M. 2014. Postharvest physiology and technology of loquat (*Eriobotrya japonica* Lindl.) fruit. *Journal of the Science of Food and Agriculture* 94, 1495–1504.

Perez, A., Ben-Arie, R., Dinoor, A., Genizi, A. and Prusky, D. 1995. Prevention of black spot disease in persimmon fruit by gibberellic acid and iprodione treatments. *Phytopathology* 85, 221–225.

Pollastro, S., Dongiovanni, C., Gerin, D., Pollastro, P., Fumarola, G., De Miccolis, A. and Faretra, F. 2016. First report of *Coniella granati* (Sacc.) Petr. & Syd. as a causal agent of pomegranate crown rot in Southern Italy. *Plant Disease* 100, 1498.

POMASA (Pomegranate Producers Association of South Africa). 2015. Pomegranate market analysis. Available at: www.sapomegranate.co.za/wp-content/uploads/2013/02/Pomegrante-Market-Analysis_Nina-Goosen.pdf

Porat, R., Mayuoni-Kirshenbaum, L., Lichter, A., Ezra, D. and Kosto, I. 2015. Reduction of crown rots development during storage of pomegranate fruit by preharvest and postharvest fungicidal treatments. *Acta Horticulturae* 1089, 173–177.

Prusky, D., Alkan, N. and Mengiste, T. 2013. Quiescent and necrotrophic lifestyle choice during postharvest disease development. *Annual Review of Phytopathology* 51, 155–176.

Prusky, D., Ben-Arie, R. and Guelfat-Reich, S. 1981. Etiology and histology of Alternaria rot of persimmon fruits. *Phytopathology* 71, 1124–1128.

Prusky, D., Perez, Y., Zutkhi, Y. and Ben-Arie, R. 1997. Effect of modified atmosphere for control of black spot, caused by *Alternaria alternata*, on stored persimmon fruits. *Phytopathology* 87, 203–208.

Puckett, R., Morgan, D., Luna, M. and Michailides, T.J. 2014. Understanding black heart and Botrytis gray mold of pomegranate and management of these diseases. Available at: http://ir4.rutgers.edu/Fooduse/PerfData/3797.pdf

Rheinländer, P.A., Sutherland, P.W. and Fullerton, R.A. 2013. Fruit infection and disease cycle of *Botrytis cinerea* causing cosmetic scarring in persimmon fruit (*Diospyros kaki* Linn.). *Australasian Plant Pathology* 42, 551–560.

Rotem, J. 1998. *The Genus Alternaria. Biology, Epidemiology and Pathogenicity.* 2nd ed. American Phytopathological Society Press, St. Paul, MN.

Salvador, A., Arnal, L., Monterde, A. and Cuquerella, J. 2004. Reduction of chilling-injury symptoms in persimmon cv. "Rojo Brillante" by 1-MCP. *Postharvest Biology and Technology* 33, 285–291.

Şen, F., Altun, A. and Kinay-Teksür, P. 2013. Effects of different modified atmosphere packing on storage quality and decay development of 'Hicaznar' pomegranates (*Punica granatum* L.). *Acta Horticulturae* 1012, 971–978.

Sharma, R.L. and Tegta, R.K. 2011. Incidence of dry rot of pomegranate in Himachal Pradesh and its management. *Acta Horticulturae* 890, 491–499.

Silva, A.O., Savi, D.C., Gomes, F.B., Gos, F.M.W.R., Silva Jr, G.J. and Glienke, C. 2017. Identification of *Colletotrichum* species associated with postbloom fruit drop in Brazil through GAPDH sequencing analysis and multiplex PCR. *European Journal of Plant Pathology* 147, 731–748.

Snowdon, A.L. 1990. *A Colour Atlas of Post-Harvest Diseases and Disorders of Fruits and Vegetables. Vol. 1: General Introduction and Fruits.* CRC Press, Boca Raton, FL.

Sun, H., Zhang, J.Z., Zhang, L.L. and Hyde, K.D. 2012. Construction and characterization of a genomic library of *Colletotrichum horii* from persimmon. *Plant Pathology & Quarantine* 2, 1–8.

Sun, H., Zhang, L.L., Zhang, J.Z. and Hu, D.W. 2009. Identification of causal agent of blossom blight of loquat in Zhejiang province, China. *Journal of Zhejiang University - Agricultural and Life Sciences* 35, 237–242.

Taguchi, Y., Watanabe, H., Akita, S. and Hyakumachi, M. 2001. Occurrence and control of ripe rot symptoms on persimmon fruit. *Japanese Journal of Phytopathology* 67, 33–41.

Tedford, E.C., Adaskaveg, J.E. and Ott, A.J. 2005. Impact of Scholar (a new post-harvest fungicide) on the California pomegranate industry. *Plant Health Progress.* Available at: www.plantmanagementnetwork.org.

Thomidis, T. 2014. Fruit rots of pomegranate (cv. Wonderful) in Greece. *Australasian Plant Pathology* 43, 583–588.

Thomidis, T. 2015. Pathogenicity and characterization of *Pilidiella granati* causing pomegranate diseases in Greece. *European Journal of Plant Pathology* 141, 45–50.

Tian, S., Quin, G. and Li, B. 2011. Loquat (*Eriobotrya japonica* L). pp. 424–442. *In*: E. M. Yahia (ed.) *Postharvest Biology and Technology of Tropical and Subtropical Fruits, vol. 3. Cocona to Mango.* Woodhead Publishing Limited, Cambridge, UK.

Tziros, G.T. 2013. *Alternaria alternata* causes leaf spot and fruit rot on loquat (*Eriobotrya japonica*) in Greece. *Australasian Plant Disease Notes* 8, 123–124.

Tziros, G.T., Lagopodi, A.L. and Tzavella-Klonari, K. 2008. *Alternaria alternata* fruit rot of pomegranate (*Punica granatum*) in Greece. *Plant Pathology* 57, 379.

Tziros, G.T. and Tzavella-Klonari, K. 2008. Pomegranate fruit rot caused by *Coniella granati* confirmed in Greece. *Plant Pathology* 57, 783.

Uysal, A. and Kurt, S. 2018. *Colletotrichum gloeosporioides* causing anthracnose on pomegranate in Turkey. *Australasian Plant Disease Notes* 13, 19.

van Kan, J.A.L. 2005. Infection strategies of *Botrytis cinerea*. *Acta Horticulturae* 669, 77–90.

Venter, E., Lennox, C.L. and Meitz-Hopkins, J.C. 2017. First report of *Cytospora punicae* causing post-harvest fruit rot on pomegranate in South Africa. *Plant Disease* 101, 631.

Vicent, A., Bartual, J., Mira, J.L., Beltrán, V. and Palou, L. 2017. Control of black heart of cv. 'Wonderful' pomegranate caused by *Alternaria alternata* in Spain. p. 89. *In*: Abstracts Book of IV International Symposium on Pomegranate and Minor Mediterranean Fruits. September 18-22, 2017, Elche, Spain.

Vicent, A., Mira, J.L., Bartual, J., Beltrán, V., Taberner, V. and Palou, L. 2016. First report of black heart of pomegranate caused by *Alternaria alternata* in Spain. *Plant Disease* 100, 1952.

Viret, O., Keller, M., Jaudzems, V.G. and Cole, F.M. 2004. *Botrytis cinerea* infection of grape flowers: Light and electron microscopical studies of infection sites. *Phytopathology* 94, 850–857.

Wang, J., Ai, C.X., Yu, X.M., An, M., Sun, S. and Gao, R. 2016. First report of *Colletotrichum karstii* causing anthracnose on persimmon leaves in China. *Plant Disease* 100, 532.

Wang, K.T., Cao, S.F., Di, Y.Q., Liao, Y.X. and Zheng, Y.H. 2015. Effect of ethanol treatment on disease resistance against anthracnose rot in postharvest loquat fruit. *Scientia Horticulturae* 188, 115–121.

Wang, X., Wang, L., Wang, J., Jin, P., Liu, H. and Zheng, Y. 2014. *Bacillus cereus* AR156-induced resistance to *Colletotrichum acutatum* is associated with priming of defense responses in loquat fruit. *PLoS ONE* 9, e112494.

Weir, B.S. and Johnston, P.R. 2010. Characterisation and neotypification of *Gloeosporium kaki* Hori as *Colletotrichum horii* nom. nov. *Mycotaxon* 111, 209–219.

Weir, B.S., Johnston, P.R. and Damm, U. 2012. The *Colletotrichum gloeosporioides* species complex. *Studies in Mycology* 73, 115–180.

Williamson, S.M. and Sutton, T.B. 2010. First report of anthracnose caused by *Colletotrichum acutatum* on persimmon fruit in the United States. *Plant Disease* 94, 634.

Wisniewski, M., Droby, S., Norelli, J., Liu, J. and Schena, L. 2016. Alternative management technologies for postharvest disease control: The journey from simplicity to complexity. *Postharvest Biology and Technology* 122, 3–10.

Woolf, A.B. and Ben-Arie, R. 2011. Persimmon (*Diospyros kaki* L.). pp. 166–193. *In*: E. M. Yahia (ed.). *Postharvest Biology and Technology of Tropical and Subtropical Fruits, Vol. 4. Mangosteen to White Sapote*. Woodhead Publishing Limited, Cambridge, UK.

Xie, L., Zhang, J.Z., Cai, L. and Hyde, K.D. 2010. Biology of *Colletotrichum horii*, the causal agent of persimmon anthracnose. *Mycology* 1, 242–253.

Yan, R.X., Ding, D.F., Guan, W.Q., Hou, J.H. and Li, M.G. 2008. Control of grey mould rot of loquat with chitinase expressed in *Pichia pastoris*. *Crop Protection* 27, 1312–1317.

Yang, X., Hameed, U., Zhang, A.-F., Zang, H.-Y., Gu, C.-Y., Chen, Y. and Xu, Y.-L. 2017. Development of a nested-PCR assay for the rapid detection of *Pilidiella granati* in pomegranate fruit. *Scientific Reports* 7, 1–8.

Yehia, H.M. 2013. Heart rot caused by *Aspergillus niger* through splitting in leathery skin of pomegranate fruit. *African Journal of Microbiological Research* 7, 834–837.

Yu, X., Qin, Z., Ai, C., Wang, H. and An, M. 2013. Isolation and identification of pathogen of 'Jirou' persimmon anthracnose disease from Yishui, Shandong, China. *Acta Horticulturae* 996, 271–276.

Zhang, J.Z. 2008. Anthracnose of persimmon caused by *Colletotrichum gloeosporioides* in China. *The Asian and Australasian Journal of Plant Science and Biotechnology* 2, 50–54.

Zhang, L. and McCarthy, M.J. 2012. Black heart characterization and detection in pomegranate using NMR relaxometry and MR imaging. *Postharvest Biology and Technology* 67, 96–101.

Zheng, Y.H., Li, S.Y. and Xi, Y.F. 2000. Changes of cell wall substances in relation to flesh woodiness in cold-stored loquat fruits. *Acta Phytophysiologica Sinica* 26, 306–310.

Chapter 6

Avocado

Silvia Bautista-Baños

Instituto Politécnico Nacional. Centro de Desarrollo de Productos Bióticos (CEPROBI), San Isidro, CEPROBI 8, Yautepec, Morelos, México

Rosa Isela Ventura-Aguilar

CONACYT, Instituto Politécnico Nacional-CEPROBI, San Isidro, CEPROBI 8, Yautepec, Morelos, México

Margarita de Lorena Ramos-García

Facultad de Nutrición, Universidad Autónoma del Estado de Morelos, Col. Los Volcanes, Cuernavaca, Morelos, México

Abbreviations

DNA	Deoxyribonucleic acid
DPPH	2, 2-diphenyl-1-pycrylhydrazyl
ABTS	2, 2-azinobis-(3-ethylbenzothiazoline- 6-sulfonic acid)
FAO	Food and Agriculture Organization of the United Nations
FRAP	Ferric reducing antioxidant power
HDL cholesterol	High-density lipoprotein cholesterol
LDL cholesterol	Low-density lipoprotein cholesterol
GAE	Gallic acid equivalents
ppm	Parts per million
RH	Relative humidity

1 Introduction

The avocado (*Persea americana* Mill.) is a species native to Mexico and Central America. Its cultivation has increased in recent years, currently occurring in 69 countries. It is listed as the fourth most important fruit in the world because of its good organoleptic characteristics and high nutritional value. It belongs to the Lauraceae family, which includes around 2200 species. The avocado is considered the most important species of this family due to its high global production (Anaya and Burgos, 2015; Urquiza et al., 2015).

1.1 Production Worldwide

In recent years, the worldwide production area of avocado has increased considerably. In 2006, around 394,048 ha were cultivated, with a production of 3,709,147 tons per year. In 2016, the production of avocado fruit was greater than 5.6 million tons, with a cultivated area of 583,978 ha, a production increase of more than 50%. More than 90% of the production is of the 'Hass' variety. Mexico ranks first in avocado production (1,889,354 tons), followed by the Dominican Republic (601,349 tons), Peru (455,394 tons), Colombia (309,431 tons), and Indonesia (304,938 tons) (Rubí-Arriaga et al., 2013; FAO, 2016) (Figure 6.1).

1.2 Production in Mexico

Avocado production in Mexico has increased in the last 7 yr (Figure 6.2), due to consumer demand and the profitability of the crop. About 70% of national production is destined for fresh consumption, 19% for the domestic processing industry, and 12% for export. The state of Michoacán ranks first in the production and cultivated area of avocado worldwide (153,000 ha). Approximately 1,000,000 tons are produced there, which is equivalent to 90% of the national production. Nayarit ranks second with 26,000 tons (2.5%), followed by Morelos with 25,000 tons (2.2%) and the State of Mexico with 21,000 tons (2%) (Rubí-Arriaga et al., 2013; FAO, 2016).

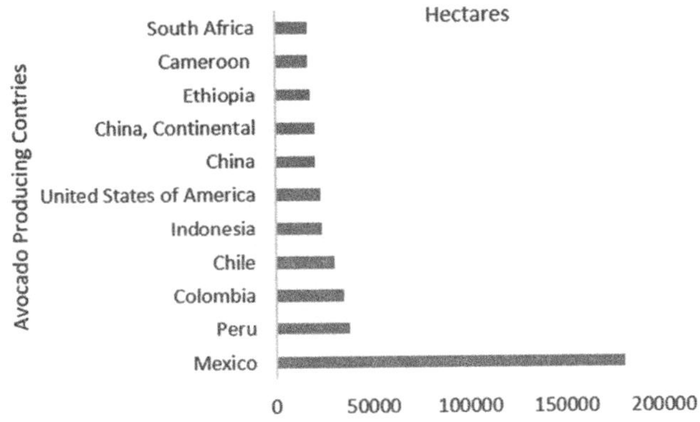

Figure 6.1 Main producing countries of avocado fruit.

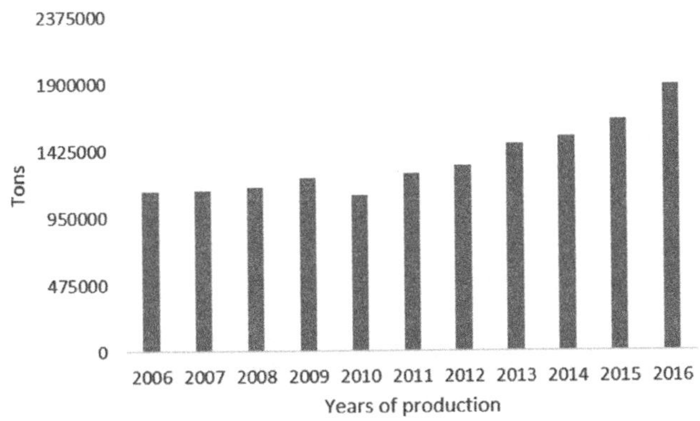

Figure 6.2 Annual avocado production in Mexico in the 2006–2016 period.

2 Botanical Characteristics

The avocado is a species that is cultivated throughout the year. It belongs to the order Laurales and to the family Lauraceae. It is a perennial tree, which can reach a height of up to 20 m. However, for commercial plantations, it is limited to 5 m in height to facilitate the cultural work of the crop. The diameter of the tree canopy is about 25 m, with branches that are glabrous or with trichomes and short apical buds. The leaves are distributed along the branches and rarely grouped at the end of the branches. The leaf blades are elliptical, 7.5–23.7 × 3.9–10.0 cm in size, and can be glabrous or pubescent. The stem is thick, cylindrical, erect, and woody; it has branches with dense foliage, its bark is rough and sometimes furrowed longitudinally. The taproot is branched and can exceed 1-m deep in the soil, while the secondary and tertiary roots are distributed superficially in the first 60 cm. The flowers are grouped in bunches of long stemmed inflorescence, which can mature over the course of 6 months. In this species, female and male flowers are developed, which are functional at different times. Each tree can produce up to 1 million flowers of which between 0.01 and 1.0% become fruit. The fruit is a berry of 4–12 cm in length and varies in shape according to the cultivar. The skin can be smooth or rough and when the fruit is ripe it is usually dark green to black, the color of the pulp can be ivory, yellow, light yellow, bright yellow, light green, or green. It has a central seed generally with a smooth surface and large in size that can weigh between 150 and 350 g (Ferrer-Pereira, 2012; Pérez et al., 2015; Guzmán et al., 2017).

3 Nutritional Content and Functional Characteristics

Avocado fruit, which is also referred to as 'butter fruit', is eaten throughout the world and has gained recognition for its high nutritional value (Takenaga et al., 2008). The edible portion consists of a thick pale-yellow pulp that is rich in fats. The seed comprises 13–18% of the fruit weight and is a by-product that is generally not utilized (Ejiofor et al., 2018). The fruit has an olive-green peel that contains significant amounts of minerals and compounds that prevent lipid oxidation, such as phenols (Rotta et al., 2016). Numerous reports describe the presence of significant amounts of carbohydrates, protein, total lipid, and ash in avocado (Table 6.1). These compounds are assimilated during the digestion process and relate to different functions in the human body. For example, carbohydrates supply most of the energy and carbon needed for the biosynthesis of proteins, nucleic acids, and lipids, and they are the second most abundant group of compounds in avocados after lipids. Avocado carbohydrates are contained mainly in the seed, with the pulp and peel of the avocado each having similar and lesser contents. Proteins have a variety of different biochemical functions. Specifically, they have a catalytic function (enzymes), are structural, they protect the body against foreign substances, cells, and viruses, and serve as nutritional and amino acid sources, among other functions (Bohinski, 1998). The avocado fruit is used as a high-energy food source due to the lipid content of the pulp, of which about 72% are monounsaturated fatty acids such as oleic, eicosenoic, and palmitoleic acids, 15% are saturated fatty acids, i.e., palmitic, myristic, and stearic acids, and last 13% are polyunsaturated fatty acids (linoleic acid). The composition of these fatty acids varies depending on the cultivars, the maturity stage, the anatomical region of the fruit, and the geographic location of plant growth (Araújo et al., 2018). Additionally, clinical research on the effects of avocado consumption such as on cardiovascular health, weight management, blood glucose control, serum lipids, and healthy living, has been reported by Pieterse et al. (2003) and Dreher and Davenport (2013), among others. In this respect, Carranza et al. (1995) undertook an assessment of humans partaking of a diet rich in monounsaturated fatty acids using the avocado as their major source of calories over a 4-wk period. The results indicated that dyslipidemia patients significantly reduced their total cholesterol and LDL-cholesterol levels.

Table 6.1 Proximate composition of pulp, seed, and peel of avocado per 100 g of weight fresh

Nutrient	*Pulp*		*Seed*		*Peel*
Energy	130.21–167 kcal		-		62.18 kcal
Water	72.33–79.37 g		15.1 g		6.505 g
Protein	1.11–1.96 g		15.55 g		2.162 g
Lipid	10.15–15.41 g		17.9 g		1.221 g
Carbohydrate	7.3–8.9 g		49.03 g		11.345 g
Ash	0.6–2.31 g		2.26 g		0.543 g
Fiber, total dietary	6.8 g		5.9–6.2 g		19.124 g
Sugars, total	0.3 g				
Minerals		**Anti-nutritional Constituents**			
Calcium, Ca	13 mg	Tannin	6.98 mg	Phenols	62.136 mg GAE
Iron, Fe	0.61 mg	Total oxalate	14.98 mg		
Magnesium, Mg	29 mg	Phytic acid	3.18 mg		
Phosphorus, P	54 mg				
Potassium, K	507 mg				
Sodium, Na	8 mg				
Zinc, Zn	0.68 mg				
Vitamins					
Vitamin C (total ascorbic acid)	8.8 mg				
Thiamin	0.075 mg				
Riboflavin	0.143 mg				
Niacin	1.912 mg				
Vitamin B-6	0.287 mg				
Folate	89 µg				
Vitamin A	7 µg				
Vitamin E (alpha-tocopherol)	1.97 mg				
Vitamin K (phylloquinone)	21 µg				
Lipids					
Fatty acids, total saturated	2.126 g				

(Continued)

Table 6.1 (Cont.)

Nutrient	*Pulp*	*Seed*	*Peel*
Fatty acids, total monounsaturated	9.799 g		
Fatty acids, total polyunsaturated	1.816 g		

GAE = Gallic acid equivalents.
Sources: Mooz et al. (2012); USDA-ARS (2014); Rotta et al. (2016); Ejiofor et al. (2018); Krumreich et al. (2018).

3.1 Functional Compounds

3.1.1 Phytosterols, Terpenoids, and Unsaturated Fatty Acids

Phytosterols are substances of plant origin with a structure very similar to that of cholesterol. They act in the body to inhibit intestinal cholesterol absorption and decrease hepatic cholesterol synthesis. Avocados contain substantial amounts of phytosterols, especially in the lipid fraction. The main representative is β-sitosterol, but stigmasterol is also present (0.02–6%) (Duarte et al., 2016). The phytosterol content depends on the cultivar, the maturity stage, and the geographic location of the avocado plant, as mentioned by Mooz et al. (2012). Calderón-Vázquez et al. (2013) determined that the β-sitosterol content is affected by the year of harvest and ranges from 146 to 672 μg/g. On the other hand, terpenes in the form of antioxidant molecules are important because they react with free radicals by partitioning themselves into fatty membranes by virtue of their long carbon side chain. Perhaps the most commonly studied of the terpene antioxidants are the tocotrienols and tocopherols. Nevertheless, there are also other compounds such as lycopersene (Dillard and German, 2000; Abaide et al., 2017). Particularly in avocado, compounds such as vitamin E or α-tocopherol (6.98 to a high of 44.94 μg/g) are influenced by both location and year of harvest as reported by Calderón-Vázquez et al. (2013).

Finally, the presence of fatty acids in diet is important since they are associated with major diseases such as cancers, diabetes, and cardiovascular disease. However, not all of them have negative effects on health. For instance, monounsaturated fatty acids or n-6 polyunsaturated fatty acids reduce LDL (the 'bad') cholesterol and thus reduce the risk of developing cardiovascular diseases. Furthermore, unsaturated fatty acids such as linoleic acid, oleic acid (73.88 g per 100 g), and palmitoleic acid (11.35 g per 100 g) that are characteristic fatty acids found in avocado, also slightly raise HDL (the 'good') cholesterol.

3.1.2 Phenols and Antioxidant Capacity

The 'Hass' and 'Fuerte' avocado peel and seed extracts have an abundance of phenolic compounds with high antioxidant capacity (Table 6.2) making them interesting to the food, cosmetics, and pharmaceutical industries; paradoxically these are by-products and they are not totally exploited (Tremocoldi et al., 2018). A total

Table 6.2 Phenols and antioxidant capacity of peel and seed of 'Hass' and 'Fuerte' avocados

		Antioxidant capacity		
	Total phenols[a] *(mg GAE/g)*	*DPPH*[b] *(µmol/g)*	*ABTS*[b] *(µmol/g)*	*FRAP*[c] *(µmol* Fe^{2+}*/g)*
'Hass' peel	63.5±7.2	310±36.9	791.5±35.9	1175.1±102.9
'Hass' seed	57.3±2.7	410.7±35.8	645.8±17.9	656.9±26
'Fuerte' peel	120.3±7.8	420.5±23.2	1004.5±52	1881.4±75.3
'Fuerte' seed	59.2±6.9	464.9±32.7	580.8±31	931.7±65.6

[a] Expressed as mg gallic acid equivalents (GAE) per g of avocado peel and seed (lyophilized).
[b] Expressed as µmol Trolox equivalents per g of avocado peel and seed (lyophilized).
[c] Expressed as µmol Fe^{2+} per g of avocado peel and seed (lyophilized).
DPPH: 2,2-diphenyl-1-pycrylhydrazyl; ABTS: 2,2-azinobis-(3-ethylbenzothiazoline-6-sulfonic acid); FRAP: ferric reducing antioxidant power.
Values are the means of three replicates ± standard deviation.
Source: Tremocoldi et al. (2018).

of 61 phenolic compounds were identified in avocado peel. These were classified into procyanidins (epicatechin, quinone methide), flavonoids (quercetin glucuronide, quercetin 3-glucoside, and kaempferol O-glucosyl-rhamnoside), hydroxybenzoic (gentisic acid and benzoic acid) and hydroxycinnamic acids [caffeic acid, p-coumaric acid, and 4-O-caffeoylquinic acid (cryptochlorogenic acid)], among others. These make up the most representative groups (Figueroa et al., 2018). In addition, procyanidin B_1, catechin, trans-5-O-caffeoyl-D-quinic acid, perseitol, quinic acid, citric acid, penstemide, tyrosol hexoside, and chlorogenic acid were present in the seed. Finally, seven hydroxycinnamic acid derivatives (sinapic acid-C-hexoside, p-coumaric acid glucoside and one of its isomers, ferulic acid glucoside and its isomer, p-coumaric acid rutinoside and coumaric acid) and one hydroxybenzoic acid derivative (octyl gallate) were identified in the avocado pulp. Based on this result, it is possible to conclude that avocado peel and seed are the parts of the fruit with the highest phenol content (López-Cobo et al., 2016). With regard to the antioxidant capacity as quantified by ferric reducing antioxidant power (FRAP) and 2-azinobis (3-ethylbenzothiazoline-6-sulfonic acid) radical (ABTS), Tremocoldi et al. (2018) reported that it was considerably higher in peel than in seed, while a converse behavior was observed when using the 2,2-diphenyl-1-picrylhydrazyl radical (DPPH). This might be because this radical is not specific for phenols, and consequently may underestimate the antioxidant capacity. Antioxidants are highly valued in the human diet because of their capability to reduce the action of free radicals during the inflammatory process.

3.1.3 Fiber

Barbosa-Martín et al. (2016) found that the total fiber content in avocado seed and pulp was similar at 5–6%, with 76% of this being insoluble dietary fiber. This fiber type in the diet has several beneficial properties, including an increase of fecal mass, an improvement of intestinal transit, the retention of bile acids, and decreased serum cholesterol levels, among others (Calderón-Vázquez et al., 2013).

4 Postharvest Diseases

Market surveys worldwide identified the magnitude of losses caused by postharvest rots of avocado during storage, marketing, and retail sales, and how these diseases can limit the commercialization of avocado fruit. Overall, avocado can be infected in the field by fungi, bacteria, and viruses, although fungi are the principal cause of postharvest diseases (Marais, 2004).

In general, the literature indicates that worldwide the most important fungal pathogens are those from the genus *Colletotrichum*, including *Colletotrichum gloeosporioides* (Penz.) Penz. & Sacc. (*C. gloeosporioides* sensu estricto); *C acutatum* J.H. Simmonds; *C. boninense* Moriwaki, Toy. Sato & Tsukib.; *C. fructicola* Prihastuti, L. Cai & K.D. Hyde; *C. hymenocallidis* Yan L. Yang, Zuo Y. Liu, K.D. Hyde & L. Cai; *C. siamense* Prihast., L. Cai & K.D. Hyde; *C. tropicale* Rojas, Rehner & Samuels; *C. karstii* Y.L. Yang, Zuo Y. Liu, K.D. Hyde & L. Cai; *C. perseae* sp. nov.; *C. aenigma* B.S. Weir & P.R. Johnst.; *C. alienum* B.S. Weir & P.R. Johnst.; *C. theobromicola* Delacr.; and *C. nupharicola* D.A. Johnson, Carris & J.D. Rogers (Table 6.3).

Disease symptoms occurring on avocado fruit caused by *Colletotrichum* spp. are commonly called anthracnose (Figure 6.3a,b). Anthracnose has been regarded as the most important postharvest avocado disease in terms of the economic losses it causes. A market survey conducted in South Africa reported an anthracnose incidence of 80% caused by *C. gloeosporioides* in overripe fruit (Sanders et al., 2000). In the same study, 300 isolates of this fungus were isolated from 'Fuerte' avocados harvested throughout different production areas of the country. In Hawaii, *C. gloeosporioides* is considered the most problematic pathogen in the production areas of avocado (Nelson, 2008). In a survey conducted in three avocado producing locations in Kenia, Mutembei (2009) isolated fungi from four different avocado cultivars and *C. gloeosporioides* was the most frequently isolated (53.3%), followed by *Fusarium* spp. (16.4%). Other minor pathogens were isolated, such as species of *Cladosporium* (9.0%), *Penicillium* (8.6%), *Botryosphaeria* (8.5%), *Pestalotiopsis* (2.2%), and *Rhizopus* (2.2%). In another study conducted in 'Hall' avocado orchards in Taiwan during a 2-yr period, the incidence of stem-end rot, body rot or fruit rot spot, and anthracnose in the first and second years were about 33, 32, and 47%, and 20, 17, and 30%, respectively (Ni et al., 2011). In a field survey done in Southern Ethiopia, anthracnose was the main postharvest disease, with an incidence of about 45, 32, and 24% in three different production areas, respectively (Mekonnen et al., 2015).

Another important disease is stem-end rot (Figure 6.3c,d), caused by a number of fungal pathogens from different genera and species, i.e., *Dothiorella* spp., *Phomopsis perseae* Zerova, *Lasiodiplodia theobromae* (Pat.) Griffon & Maubl., *Botryosphaeria* sp., and *Fusarium* sp., among others (Stovold and Dirou, 2004; Pérez-Jiménez, 2008; Ni et al., 2009; Valencia et al., 2011; Garibaldi et al., 2012; Twizeyimana et al., 2013; Akgül et al., 2016; Kimaru et al., 2018).

Avocado blight or avocado scab is a disease caused by the fungus *Sphaceloma perseae* Jenkins. It is considered an important disease found in various countries of

Table 6.3 Postharvest diseases, causal agents, and overall symptoms on avocado fruit

Disease common name	*Causal microorganism*	*Symptoms*	*Reference*
Anthracnose, pepper spot	*Colletotrichum gloeosporioides, C acutatum, C. boninense, C. fructicola C. hymenocallidis, C. siamense, C. tropicale, C. karstii, C. perseae* sp. nov., *C. aenigma, C. alienum, C. theobromicola, C. nupharicola*	Brown and black rounded spots and sunken lesions on the fruit surface, enlarged rapidly in the skin and fruit flesh. Salmon-colored spore masses developed at advanced stages of this disease	Stovold and Dirou (2004); Nelson (2008); Pérez-Jiménez (2008); Giblin et al. (2010); Ni et al. (2011); Silva-Rojas and Avila Quezada (2011); Freeman et al. (2013); Akgül et al. (2016); Trinidad-Ángel et al. (2017); Sharma et al. (2017); Fuentes-Aragón et al. (2018)
Cercospora/Pseudocercospora spot	*Pseudocercospora purpurea*	Fruits show small scattered sunken spots, initially brown, becoming purple with time. The fungus may invade the flesh	Pérez-Jiménez (2008)
Phytophthora fruit rot	*Phytophthora cinnamomi*	Black soft decay, in a roundish blotch	UC (2014)
Pink rot	*Trichothecium roseum*	Profuse salmon-colored sporulation, soft lesions	Sharma et al. (2016)
Soft rot fruit	*Erwinia* sp.	Typical rancid smell. External metallic color. Brown flesh, softened and liquid	Stovold and Dirou (2004); DAF (2014)
Avocado blight/ avocado scab	*Sphaceloma perseae*	Corky fruit lesions, brownish with a scabby and cracking appearance	Marroquín-Pimentel (1999); Pegg et al. (2009)
Stem-end rot	*Botryosphaeria/ Fusicoccum, Dothiorella* spp., *Phomopsis perseae, Lasiodiplodia theobromae, Botryosphaeria dothidea, B. rhodina,*	Disease begins in the abscission scar of the fruit, advancing externally and internally with a well-defined margin, showing brown and black	Hartill (1991); Stovold and Dirou (2004); Pérez-Jiménez (2008); Ni et al. (2009); Valencia et al. (2011); Garibaldi

(*Continued*)

Table 6.3 (Cont.)

Disease common name	*Causal microorganism*	*Symptoms*	*Reference*
	B. parva, *B. ribis*, *Fusarium* sp., *Nectria pseudotrichia*, *Alternaria* sp., *Rhizopus stolonifer*, *Pestalotiopsis versicolor*, *P. clavispora*, *P. icrospora*, *Neufosicoccum luteum*, *N. austral*, *N. mangiferae*	color. No superficial development of mycelia. Internally, mycelia are shown in the avocado flesh	et al. (2012); Twizeyimana et al. (2013); Akgül et al. (2016); Montealegre et al. (2016); Kimaru et al. (2018)

Figure 6.3 The two most important postharvest diseases of avocado. a. External and b. Internal symptoms of anthracnose. c. External and d. Internal symptoms of stem-end rot.

the Americas, including Mexico and the United States, where it is mainly in the state of Florida, and in some areas of Africa (Marroquín-Pimentel, 1999; Pegg et al., 2009). The initial symptoms are mostly superficial and cosmetic, affecting the appearance of the peel, but later it affects the avocado pulp quality during storage.

The genera of pathogenic bacteria that affect avocado fruit during storage are few. Pegg et al. (2009) reported that the genus *Erwinia*, specifically *E. carotovora*, caused disease. Overall, symptoms include a gray-to-black tender lesion, with darkened metallic bright color on the infected area of the fruit and an intense rot smelling.

5 Pre- and Postharvest Factors Influencing the Incidence of Postharvest Rots

A broad range of factors can contribute to the incidence of storage diseases caused by fungi in avocado fruit. The environmental conditions in the field, the agronomic practices employed during fruit production, and the handling practices from harvest further influence the incidence and severity of anthracnose, which, as already mentioned, is the main postharvest disease of this fruit (Table 6.4).

Rainfall incidence and prolonged wet periods on the avocado surface were associated with a high incidence of anthracnose caused by *C. gloeosporioides* in 'Fuerte' fruit (Peterson, 1978). Other studies indicated that the selection of rootstock also influenced the appearance of anthracnose during storage of 'Hass' avocados (Willingham et al., 2001b, 2006; Marques et al., 2003). These workers reported that the rootstocks 'Duke 6' and 'Duke 7' were associated with a higher incidence of this disease in the harvested fruit compared to the rootstock 'Velvick'. In addition, other rootstocks including 'Anderson 8', 'Anderson 10', 'Nabal', and 'Parida' were also tested, although no influence on anthracnose was detected. In other studies (Penter and Stassen, 2000; Willingham, 2006), nitrogen and calcium fertilization was evaluated with the conclusion that

Table 6.4 Environmental conditions and agronomic practices influencing disease incidence during storage of avocado fruit

Cultivar	*Preharvest factor*	*Postharvest disease effect*	*Reference*
'Fuerte'	Rainfall duration during a 3-yr period season	Higher anthracnose incidence during the maximum rainfall period and harvest month	Peterson (1978)
'Hass'	Rootstocks: 'Duke 6' and 'Velvick'	Higher infection of anthracnose (ca. 41%) in fruit grafted on 'Duke 6' than on 'Velvick' rootstock (ca. 12%)	Willingham et al. (2001b)
'Hass'	Rootstocks: 'Duke 7', clonal 'Velvick' and 'Velvick'	Significantly more body rots in 'Duke 7' avocados and in those stored at 5°C for 5 wk	Marques et al. (2003)
'Hass'	Rootstocks: 'Duke 6' and 'Velvick' and nitrogen fertilization: ammonium and nitrate, both at 13.3 and 26.6%	Overall, the rootstock had more influence on disease incidence (37%) and severity (64%) than mineral nutrition	Willingham et al. (2006)
'Hass'	Tree vigor based on previous disease incidence by *Phythopthora cinnamomi*: Non-vigorous vs. vigorous trees	Overall, more anthracnose infections in fruit from vigorous trees in two seasons (95.2 vs. 41.7%)	Willingham et al. (2004)
'Pinkerton'	Calcium application: 1. Calcium Dextrolac (2 L/ha) 2. Calcimax (0.5%) 3. Basfoliar Calcium (4.5 L/ha) 4. Caltrac (4 L/ha)	Calcimax and Caltrac significantly reduced anthracnose by approximately 90%	Penter and Stassen (2000)

preharvest application of calcium was associated with greater reduction of anthracnose incidence compared to nitrogen in 'Pinkerton' and 'Hass' avocados.

Regarding postharvest management of anthracnose, storage temperature is considered the main factor influencing the growth of the causal pathogen at this stage. Temperature regimes can be used to reduce postharvest decay. According to Hopkirk et al. (1994), fruit ripened at 20°C had fewer postharvest rots if they had been previously stored at 4 or 6°C than if they had been stored at either lower or higher temperatures. In addition, both body rots and stem-end rots caused by *Botryosphaeria* spp. and *Colletotrichum* spp. were reduced. In another study, anthracnose incidence was lower during fruit ripening at 20°C when avocados were stored before ripening at 5.5°C for 30 d than if ripened immediately (Everett, 2003).

The source of inoculum was also investigated by Hartill and Everett (2002). In this study, the harvesting method influenced storage disease incidence. Plucking the fruit rather that clipping them resulted in notably higher presence of *C. acutatum*. Likewise, sterilized clippers reduced the incidence of disease caused by *Botryosphaeria parva* Pennycook & Samuels.

As described above, avocados are very susceptible to infection by numerous pathogens. Symptoms from some of these infections are more evident during fruit development and symptoms from others during storage of the fruit. However, according to Londsale (1992), in all cases infection takes place in the orchard, thus most control measures should be applied during this stage.

6 Postharvest Handling and Technologies for Preserving Avocados

Unlike many other fruits, the ripening or softening of avocados does not occur during maturation on the tree, but it occurs from 3 to 5 d after the fruit has been harvested (Ozdemir and Topuz, 2004). Ripening occurs optimally between 21 and 27°C and temperatures in excess of 30°C cause irregular ripening and darkening of the flesh (Arpaia et al., 2018). In Mexico, avocados are harvested manually, leaving its peduncle flush with the skin, at a physiological maturity indicated by at least 21.0% of dry matter in the flesh. Subsequently, they are placed in a clean plastic box, selected, and transported in chilled vehicles to a storage house or packinghouse where the fruits are precooled at 12–16°C for up to 24 hr (NMX-FF-016-1995, 1995). Commercially, forced-air cooling is the most common method used in packinghouses to cool the fruit, while new alternatives have been proposed (Figure 6.4). Among these, Chen et al. (2017) reported the use of a cold shock treatment involving cold air or ice water as a modified precooling approach. This rapidly lowers the internal temperature of agricultural products. This treatment inhibits peel discoloration and firmness loss, suppresses polygalacturonase enzymes and endo-β-1,4-glucanase activities, and reduces the ethylene production rate (0–5 μL/kg/h for 6 d) and the respiration rate (40–60 mg/kg/h for 6 d) in avocados.

Following precooling, avocados are graded for color and size, packed in new cardboard cartons or plastic boxes (10–13 kg), and cold stored, typically at 5°C. Alternatively, they are sent to wholesale and retail national markets in chilled vehicles, depending on the final market (Figure 6.4). Because avocados are stored at low temperatures, ripening can be delayed for up to 4–6 wk. Therefore, when avocado demand is high, ripening can be induced by the use of exogenous ethylene and storage

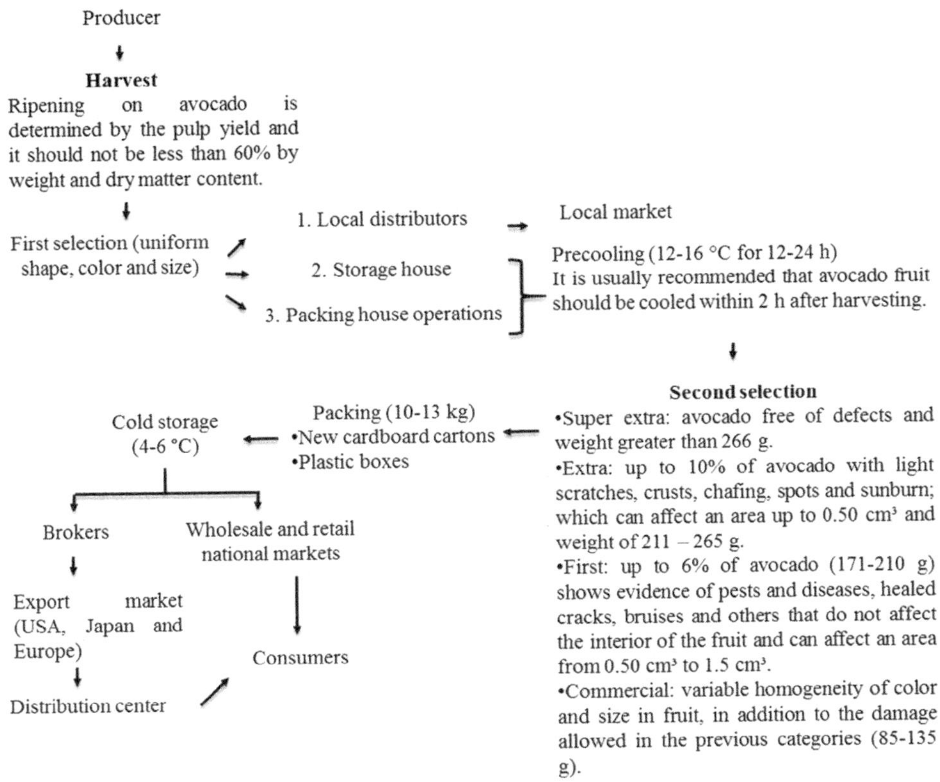

Figure 6.4 Whole supply chain during postharvest handling of avocado (NMX-FF-016-SCFI-2016 (Norma Mexicana), 2016).

at higher temperatures of 10 to 21°C. Mazhar et al. (2018) reported that avocados can be dipped in 1000 μL/L of the ethylene-releasing compound ethephon, using a commercial product such as Ethrel® (480 g/L), followed by air drying. Following this, the fruit can be kept in a shelf-life room at 20°C and 85% relative humidity (RH) until the required stage of firmness is achieved. Ethylene plays a key role in the ripening of avocados, with softening being far more influenced by ethylene than skin color changes (Gwanpua et al., 2018). Additionally, postharvest handling and ripening also influence the fatty acid content and dry matter of avocado (Ozdemir and Topuz, 2004). These authors observed that the dry matter content decreased in 'Fuerte' avocados (28.2–25.0%) but increased slightly in 'Hass' avocados (25.4–26.5%) after harvest. In contrast, the proportion of other fatty acids, such as oleic acid (47.2–59.3%), and the nutritional value did not change significantly.

Despite the avocado being highly valued for its nutritional value and sensory quality, research to develop technologies to prolong its quality after harvest has been insufficient. This is due to the fact, among other things, that the main consuming countries forbid the use of chemical compounds such as fungicides, waxes, coatings, or other technologies that could represent a health risk or affect the process of ripening of the fruit. Consequently, cold storage represents one of the most widely used technologies to extend the postharvest life of avocados by the delay of ripening-associated softening. However, poor handling accelerates the avocado decomposition process. Glowacz et al. (2017) investigated the incidence of chilling injury and found that its incidence was substantially reduced, by approximately 20%, when avocados of commercial maturity were treated with methyl jasmonate and methyl salicylate vapors for 24 hr at 20°C, then kept at 2°C for 21 d, followed by 6–7 d of shelf life at 20°C. These conditions simulated commercially applied quarantine treatments for fruit flies and the supply chain. According to these authors, the mode of action of methyl jasmonate and methyl salicylate is based on the ability of these treatments to alter the membrane integrity and the fatty acid content/composition by downregulating the activity of the lipoxygenase enzyme.

Another product that contributes to retain firmness in cold-stored avocados is 1-methylcyclopropene (1-MCP), which maintains the integrity of the avocado cell wall. Defilippi et al. (2018) reported that treatment with 1-MCP (300 nL/L, SmartFresh®, Agrofresh) for 24 hr at 20°C and 45% RH followed by storage at 5°C and 92% RH for 25 d, then subsequent ripening at 20°C and 45% RH until the ready-to-eat stage, delayed softening and increased the ethylene production rate during ripening. This regimen caused 'Hass' avocados (32% average dry matter) to ripen 4 d later than the control fruit (stored at 5°C and 92% RH for 25 d with subsequent ripening at 20°C and 45% RH until reaching the ready-to-eat stage). However, although the fruit did not soften during cold storage for 25 d, once at room temperature (20°C), the softening and ethylene production occurred faster than in the control group. Ortiz-Viedma et al. (2018) showed through a microstructure analysis that avocado softening during ripening was the result of the release of endogenous lipids and components of the cell wall, and that 1-MCP retarded this process. They reported that an application of 1-MCP extended avocado shelf life by 6 d compared with the control group.

Postharvest technologies such as active packaging and edible coatings, among others, have been used to preserve avocados. Gaona-Forero et al. (2018) used a package made of rigid polypropylene terephthalate (PET) that contained sodium polyacrylate-cotton as a moisture absorber with 160–200 g of whole avocado fruit at 12°C. It was not successful, however, because fruits packaged with moisture absorber lost 2.1% more weight than those of the control group. In addition, Handayani et al. (2018) applied edible coatings from cassava peel starch with the addition of bay leaf extract on green avocados stored at room temperature. They observed a 4% lower weight loss in coated avocados than in non-treated ones and retained a stable peel color for up to 8 d. Although there is interest in the development of postharvest technologies to preserve avocado quality, currently they are not used for marketing this product around the world.

7 Control of Postharvest Rots

7.1 Synthetic Fungicides

The use of fungicides is the primary commercial method used to control the main postharvest fungal diseases of avocado (anthracnose and stem-end rots). Currently, the preference is to apply all synthetic chemicals before harvest, and principally to fruits aimed for export markets (Table 6.5). Currently, copper (mainly copper

Table 6.5 Preharvest application of fungicides worldwide for controlling postharvest diseases during storage of avocado fruit

Country	*Disease/Causal agent*	*Chemical*	*Level of control*	*Reference*
Australia	*Colletotrichum gloeosporioides*, *C. acutatum*	1. Amistar®, 2. Flint®, Stroby®	Reduction of anthracnose disease by 66 and 74% with the first two fungicides.	Willingham et al. (2001a)
		1. NaturalGreen, 2. Ecocarb, 3. Aminogro, 4. Product A, 5. Kasil, 6. Serenade, 7. Rainshield, 8. Copper and azoxystrobin, 9. Control	Overall, greater percentage of marketable fruit with the combination of copper and azoxystrobin.	Smith et al. (2011)
Brazil	*Colletotrichum* sp., *Lasiodiplodia* sp. and *Fusicoccum* sp.	1. Azoxystrobin, 2. Thiabendazole, 3. Difenoconazole+ azoxystrobin, 4. Acibenzolar-S-methyl+ azoxystrobin, 5. Control, 6. Acibenzolar-S-methyl, 7. Calcium chloride	Average fruit rots from 54 to 75%.	Fischer et al. (2018)
Colombia	*Glomerella* sp. and *Cercospora* sp.	1. Net (tulle), 2. Pesticide (cypermethrine), 3. Fungicide (mancozeb), 4. Pesticide (abamectine), 5. Control, 6. Net (muslin)	Control depending on the production area. In trial 1, all treatments reduced disease severity by 29%. In trial 2, severity was reduced by 35% with mancozeb and abamectine	Reina-Noreña et al. (2015)
Mexico	*Sphaceloma persea*, *Cercospora purpurea*, *Colletotrichum gloeosporioides*	Bordeaux mixture (1:1:100)	–	Fucikovsky and Luna (1987)
		Copper hydroxide (300 g), folpet (200 g), benomyl (60 g) and copper oxychloride (300 g)	National and export production increased.	Vidales et al. (2005)

(*Continued*)

Table 6.5 (Cont.)

Country	*Disease/Causal agent*	*Chemical*	*Level of control*	*Reference*
		1. Azoxystrobin-fludioxonil, 2. Azoxystrobin, 3. Pyraclostrobin, 4. Copper sulphate	Infection less than 13% compared to 60% in untreated avocados	Herrera-González et al. (2017)
New Zealand	Stem-end rots, body rots	Benomyl and copper	Higher number of fungicide applications = lesser number of stem-end rots	Everett et al. (2007a)
		Pristine® (pyraclostrobin + boscalid), Shirlane®, Tilt™ (propiconazole) EC, Nufilm™-17, Kocide® Opti., Champ™DP1	Effectiveness depended on the application time, season, and growing location	Everett and Pushparajah (2008), Everett et al. (2011)
		Copper	Effectiveness depended on season and number of applications throughout the year	Sorensen (2017)
South Africa	Black spot, sooty blotch	Copper oxychloride, captafol	Significant reduction of sooty blotch	Smith et al. (1987)
		Copper oxychloride, copper ammonium carbonate, cyproconazole, flusilazol, triadimenol, benomyl	Fruit free of infection with the fungicide copper oxychloride	Londsale (1992)
		Copper oxychloride Benomyl + copper oxychloride Benomyl + *Bacillus subtilis* Copper oxyclhoride + *B. Subtilis*	Effectiveness depended on the application time, season, growing location, and treatment	Korsten et al. (1997)
		14 fungicides, applied alone or combined	Disease index: Copper + Sporekill® (didecyldimethyl ammonium chloride) = 26.4%	Manicom and Schoeman (2010)

(*Continued*)

Table 6.5 (Cont.)

Country	*Disease/Causal agent*	*Chemical*	*Level of control*	*Reference*
			Copper + Breaktrhu® (polyether trisiloxane) = 40.1% Copper + oil = 40.5% Copper + Ortiva® (azoxystrobin) = 42% Copper = 66%	
USA	Cercospora rot and anthracnose	Abound®, Heritage®, Kocide®, Copper hydroxide	–	Palmateer et al. (2006)
		Abound® (azoxystrobin), Tilt™ (propiconazole), Pristine® (pyraclostrobin + boscalid), Cuprofix® Ultra 40 (basic copper sulphate)	93 and 92% control of Cercospora rot and anthracnose, respectively, with the fungicide Pristine®, 76 and 87% with Abound®, 69 and 73% with Tilt®, and 44 and 71% with Cuprofix®	Pérez-Martínez and Monterroso (2015)

– Not indicated

oxychloride) is considered worldwide the best approved fungicide to control these postharvest rots (Manicom and Schoeman, 2010; Everett et al., 2011). Other fungicides that also significantly control these diseases include pyraclostrobin/boscalid (Pristine®), fluazinam (Shirlan®), and azoxystrobin (Abound®), among others (Everett et al., 2011; Pérez-Martínez and Monterroso, 2015), but their effectiveness to control postharvest rots is also influenced by the avocado production area, frequency of application, season, and cultivar (Everett and Pushparajah, 2008; Pérez-Martínez and Monterroso, 2015; Daneel et al., 2017).

Combinations of the fungicides (copper oxychloride or benomyl copper-oxychloride) with a biological control agent, the antagonist *Bacillus subtilis*, were evaluated in different avocado groves during a 3-yr period in South Africa (Korsten et al., 1997). Control of black spot, caused by *Pseudocercospora purpurea* (Cooke) Deighton, was variable, according to the producing region, year of treatment application, and type of treatment applied.

Regarding fungicides applied after harvest, prochloraz (Sportak®) has been evaluated repeatedly, in part due to its efficacy, and also because some countries of Europe accept its use with a maximum residue tolerance of 2 ppm (Le Roux et al., 1985; Fischer et al., 2011; Daneel et al., 2017). This fungicide combined with a commercial wax effectively controlled avocado rots in South Africa (Everett and Korsten, 1996); however, its application did not proceed beyond the experimental stage. Similarly, dipping 'Hass' and 'Fuerte' avocados in prochloraz combined with citric acid nearly completely controlled anthracnose and stem-end rots

Table 6.6 Application of synthetic fungicides during the postharvest handling of avocado fruit

Postharvest disease	*Chemical*	*Level of control*	*Reference*
Anthracnose	1. Benomyl (0.2%) 2. Prochloraz (0.2%) 3. Benomyl (0.05%) + prochloraz (0.05%) 4. Control	Decay index on 'Pinkerton' 2.6 1.8 1.9 9.0	McMillan and Narayanan (1990)
Anthracnose	1. Omega® (prochloraz) in wax 2. Omega® dip 3. Omega® dip + wax	External and internal severity on 'Fuerte': 2.1 and 2.0 2.3 and 3.0 3.0 and 3.8	Everett and Korsten (1996)
Anthracnose	1. Azoxistrobin, 2. Benzalkonium chloride, 3. Chlorine dioxide, 4. Ecolife, 5. Sodium hypochlorite, 6. Imazalil, 7. Prochloraz, 8. Thiabendazole	More effective when prochloraz followed by imazalil on 'Hass'. After 14 d storage, disease incidence was 10.8 and 38.1%, respectively	Fischer et al. (2011)
Anthracnose	1. Prochloraz EC and SC (both at 200 and 810 ppm) 2. Prochloraz + hydrochloric acid (HCl) (50 nM) 3. HCl alone (50 nM)	No significant disease control on 'Hass', more effective on 'Fuerte'	Mavuso and van Nierkek (2011)
Anthracnose	1. Prochloraz (800 ppm) 2. Phrocholraz + citric acid (50 nM)	Higher percentage of fruit of 'Fuerte' and 'Hass' free of anthracnose and stem-end rots, compared with the control	Mavuso and van Nierkek (2013)
Anthracnose, stem-end rots	1. Imazalil sulphate 2. Propiconazole 3. Pyrimethanil 4. Fludioxonil 5. Prochloraz	Prochloraz (180 mL) was the most effective followed by fludioxonil (170 g) and propiconazole (240 mL)	Daneel et al. (2017)
Stem-end rots	1. Omega® in wax 2. Omega® dip 3. Omega® dip + wax 4. Control	External and internal severity on 'Fuerte': 1.8 and 1.8 3.2 and 5.9 3.6 and 4.5 2.6 and 2.2	Everett and Korsten (1996)

(*Continued*)

Table 6.6 (Cont.)

Postharvest disease	*Chemical*	*Level of control*	*Reference*
Numerous fruit rots	1.Pristine® (boscalid + pyraclostrobin) 2. Sportak®	Control depended on the production area, but overall the fungicide Pristine® was more effective. No effect during long-term storage	Everett et al. (2007b), Everett and Pushparajah (2008)

(Mavuso and van Nierkek, 2013). In addition, these authors found a marked reduction in fungicide residues and costs when prochloraz was combined with citric acid. Another potential postharvest fungicide is pyraclostrobin/boscalid (Pristine®), but it cannot be used at present since it is not registered for 'Hass' avocados (Everett et al., 2007b). In the later work, results were variable with this fungicide. Its efficacy was influenced by the harvest areas within New Zealand avocado orchards, and the length of storage of the fruit (28, 46, and 56 d) (Everett and Pushparajah, 2008). Table 6.6 shows a summary of the fungicides applied during the handling of avocado fruit, and their level of control of the main postharvest diseases.

7.2 Other Alternatives under Experimentation

7.2.1 Antagonists

Most biocontrol strategies in avocado are directed to the use of bacterial strains and antagonistic yeasts that produce antibiotics or toxins, compete for nutrients, or produce lytic enzymes that affect germinating fungal hyphae or enzymes that degrade the fungal cell wall. It is important to highlight that the search of biocontrol agents to control postharvest pathogens has been difficult because the pathogen penetrates the fruit cuticle and the infective hyphae are protected by the tissue of the host fruit itself (Yakoby et al., 2001). In spite of this, however, some antagonists have controlled postharvest diseases of avocado when applied both before and after harvest (Korsten and De Jager, 1995).

The antagonists most often used to control anthracnose of avocado are of the genus *Bacillus* (Table 6.7), due to their ease of formulation and storage. They have also the ability to form spores, which enables their survival in harsh environments, and they do not deposit visible residues on the surface of the fruit (Korsten and Kotzé, 1992; Van Dyk et al., 1997; Cazorla et al., 2007; Guardado-Valdivia et al., 2018).

7.2.2 Plant Derivatives (Plant Extracts and Essential Oils)

Extracts and essential oils derived from plants are considered to be promising natural alternatives to conventional fungicides, because they inhibit the growth of

Table 6.7 Microbial antagonists used to control postharvest diseases of avocado fruit

Cultivar	*Antagonist*	*Concentration (cell/mL)*	*Postharvest pathogen*	*Reference*
'Fuerte'	*Bacillus subtilis*	1×10^4	*Dothiorella* sp.	Korsten and Kotzé (1992)
	B. licheniformis	1×10^2	*Colletotrichum* sp.	
	B. cereus	1×10^4		
	B. subtilis	1×10^7	*Colletotrichum* sp. *Dothiorella* sp.	Van Dyk et al. (1997)
'Hass'	*B. amyloliquefaciens*	1×10^9	*Colletotrichum gloeosporioides*	Canizal (2017)
	Wickerhamomyces anomalus	1×10^8		
	B. amyloliquefaciens	1×10^4		Martínez (2015)
	B. cereus	3×10^6	*C. gloeosporioides*	Lemus-Soriano and Pérez-Aguilar (2017)
	B. subtilis	3×10^6		
	Burkholderia sp.	3×10^6		
	Picha anomala	4×10^7	*C. gloeosporioides* *C. acutatum*	Campos (2014)

several phytopathogenic fungi. These compounds are gaining interest because they are relatively safe and have been well accepted by consumers (Bill et al., 2016). The preventive application of natural compounds to avocado fruit reduces both postharvest disease incidence and severity. In the case of 'Hass' and 'Gem' avocado fruits, extracts *of Moringa oleifera, Larrea tridentate*, and *Equisetum arvense* have been evaluated, alone or in combination with essential oils, mainly to control the development of *C. gloeosporioides* (Table 6.8) and promising results have been reported, with reductions in the incidence of anthracnose to 95% (Lemus-Soriano and Pérez-Aguilar, 2017; Tesfay et al., 2017).

The mechanisms of action of these compounds can vary depending on the type of plant species and the components extracted. For example, in *C. gloeosporioides*, according to Pérez et al. (2011), the secondary metabolites of the plants can be toxic, which is attributed to the oxidation of fungal compounds since they can penetrate the plasma membrane of the pathogen, then combine with DNA and denature plasma proteins. In addition, they break the membrane through the lipophilic compounds or they can form ion channels in the microbial membrane. Bill et al. (2016) stated that the plant essential oils have bioactivity during the vapor phase, a feature that makes them attractive as potential fumigants for the protection of stored agricultural products.

Table 6.8 Derivatives of plants used to control postharvest diseases of avocado fruit

Cultivar	*Treatment*	*Concentration*	*Postharvest pathogen*	*Reduction of incidence (%)*	*Reference*
'Gem'	*Moringa oleifera* extract	10%	*Colletotrichum gloeosporioides*	80.79	Tesfay et al. (2017)
			Alternaria sp.	89.8	
'Hass'	*Thymus vulgaris* essential oil	66.7 mL	*C. gloeosporioides*	88	Bill et al. (2015)
		10%		90	Bill et al. (2016)
	Larrea tridentata extract + *Ricinus communis* essential oil	90% 10%		26.25	Lemus-Soriano and Pérez-Aguilar (2017)
	Larrea tridentata + *Citrus* sp. extracts	52.5% 7.5%		95	
	Equisetum arvense extract + *Ricinus communis* essential oil	90% 10%		36.25	
	Equisetum arvense + *Citrus* sp. extracts	52.5% 7.5%		45	
	Thymus vulgaris essential oil	1%		60	Bill et al. (2014)
	Aloe vera	2%		40	

8 Conclusions

The incidence of diseases caused by microorganisms during the storage of avocado fruit can become a serious issue for those professionals involved in the production, sale, and consumption of this crop. Overall, the same causal fungi have been reported throughout the harvesting seasons, years, and production areas, including several species of *Colletotrichum*, and those of the fungal complex causing the disease termed stem-end rot. Management strategies have been focused in the application of fungicides in the field, because the infection process is mainly initiated at this stage, although the symptoms can be visible after harvest and storage of the fruit. Postharvest use of chemical fungicides has been also common, but worldwide regulations are increasingly prohibiting the use of these compounds once the avocados are harvested. The perception that chemical fungicides may cause environmental harm or pose a risk to human health has made the development of alternatives important. The application of other means for controlling these pathogens has been tested during both fruit growth and after harvest. Because infections occur during flowering and fruit set periods, the application of control alternatives should be focused at these times. The combination of low doses of synthetic fungicides with other alternatives such as those mentioned in this review could be tested. Currently, the environmental issue should always be considered.

References

Abaide, E.R., Zabot, G.L., Tres, M.V., Martins, R.F., Fagundez, J.L., Nunes, L.F., Druzian, S., Soares, J.F., Dal Prá, V., Silva, J.R.F., Kuhn, R.C. and Mazutti, M.A. 2017. Yield, composition, and antioxidant activity of avocado pulp oil extracted by pressurized fluids. *Food and Bioproducts Processing* 102, 289–298.

Akgül, D.S., Awan, Q.N., Güler, P.G. and Önelge, N. 2016. First report of anthracnose and stem end rot disease caused by *Colletotrichum gloeosporioides* and *Neofusicoccum australe* on avocado fruits in Turkey. *Plant Disease* 100, 1792.

Anaya, C. and Burgos, A.L. 2015. Energy consumption in the management of avocado orchards in Michoacan, Mexico. *Revista Chapingo Serie Horticultura* 21, 5–20.

Araújo, R.G., Rodríguez-Jasso, R.M., Ruiz, H.A., Pintado, M.M.E. and Aguilar, C.N. 2018. Avocado by-products: Nutritional and functional properties. *Trends in Food Science and Technology* 80, 51–60.

Arpaia, M.L., Collin, S., Sievert, J. and Obenland, D. 2018. 'Hass' avocado quality as influenced by temperature and ethylene prior to and during final ripening. *Postharvest Biology and Technology* 140, 76–84.

Barbosa-Martín, E., Chel-Guerrero, L., González-Mondragón, E. and Betancur-Ancona, D. 2016. Chemical and technological properties of avocado (*Persea americana* Mill.) seed fibrous residues. *Food and Bioproducts Processing* 100, 457–463.

Bill, M., Obianom, P.C., Korsten, L., Mavuso, Z., van Rooyen, Z. and Sivakumar, D. 2016. Using plant volatiles in biodegradable sachets and exposure of volatiles under pallet covers to control postharvest decay in avocados. *South African Avocado Growers' Association Yearbook* 39, 44–49.

Bill, M., Sivakumar, D., Korsten, L. and Thompson, K. 2014. The efficacy of combined application of edible coatings and thyme oil in inducing resistance components in avocado (*Persea americana* Mill.) against anthracnose during post-harvest storage. *Crop Protection* 64, 159–167.

Bill, M., Sivakumar, D., van Rooyan, Z. and Mavuso, Z.S. 2015. New methods of postharvest disease control: Using thyme oil fumigation. *South African Avocado Growers' Association Yearbook* 38, 98–104.

Bohinski, R.C. 1998. *Bioquímica*, 5th ed. E. Elizondo and R. Bautista (eds.). Pearson Editorial, Mexico. (In Spanish).

Calderón-Vázquez, C., Durbin, M.L., Ashworth, V., Tommasini, L., Meyer, K. and Clegg, M. 2013. Quantitative genetic analysis of three important nutritive traits in the fruit of avocado. *Journal of the American Society for Horticultural Science* 138, 283–289.

Campos, A. 2014. Identificación de levaduras y evaluación de su capacidad para inhibir a *Colletotrichum* spp. causante de la antracnosis en aguacate (*Persea americana* Mill). MSc. Thesis. Instituto Politécnico Nacional (IPN), Centro de Desarrollo de

Productos Bióticos (Ceprobi). Yautepec, Morelos, Mexico. Available at: https://tesis.ipn.mx/handle/123456789/14687 (In Spanish with English abstract).

Canizal, A. 2017. Aplicación de microorganismos antagonistas nativos contra la antracnosis del aguacate en Morelos. MSc. Thesis Instituto Politécnico Nacional (IPN), Centro de Desarrollo de Productos Bióticos (Ceprobi). Yautepec, Morelos, Mexico. Available at: https://tesis.ipn.mx/handle/123456789/24311 (In Spanish with English abstract).

Carranza, J., Alvizouri, M., Alvarado, M.R., Chávez, F., Gómez, M. and Herrera, J.E. 1995. Effects of avocado on the level of blood lipids in patients with phenotype II and IV dyslipidemias. *Archivos del Instituto de Cardiologia de México* 65, 342–348.

Cazorla, F.M., Romero, D., Pérez-García, A., Lugtenberg, B.J., de Vicente, A. and Bloemberg, G. 2007. Isolation and characterization of antagonistic *Bacillus subtilis* strains from the avocado rhizoplane displaying biocontrol activity. *Journal of Applied Microbiology* 103, 1950–1959.

Chen, J., Liu, X., Li, F., Li, Y. and Yuan, D. 2017. Cold shock treatment extends shelf life of naturally ripened or ethylene-ripened avocado fruits. *PLoS ONE* 12, e0189991.

DAF (Department of Agriculture and Fisheries). 2014. Queensland Government. Queensland, Australia. Available at: www.daf.qld.gov.au/business-priorities/plants/fruit-and-vegetables/a-z-list-of-horticultural-diseases-and-disorders/bacterial-soft-rots

Daneel, M.S., Sippel, A., de Beer, M.S. and Mlimi, J. 2017. Phrocloraz – A look into factors affecting the variability of residue levels on avocado fruit. *South African Avocado Growers' Association Yearbook* 40, 72–77.

Defilippi, B.G., Ejsmentewicz, T., Covarrubias, M.P., Gudenschwager, O. and Campos-Vargas, R. 2018. Changes in cell wall pectins and their relation to postharvest mesocarp softening of 'Hass' avocados (*Persea americana* Mill.). *Plant Physiology and Biochemistry* 128, 142–151.

Dillard, C.J. and German, J.B. 2000. Phytochemicals: Nutraceuticals and human health. *Journal of the Science of Food and Agriculture* 80, 1744–1756.

Dreher, M.L. and Davenport, A.J. 2013. Hass avocado composition and potential health effects. *Critical Reviews in Food Science and Nutrition* 53, 738–750.

Duarte, P.F., Chaves, M.A., Borges, C.D. and Mendonca, C.R.B. 2016. Avocado: Characteristics, health benefits and uses. *Ciência Rural* 46, 747–754.

Ejiofor, N.C., Ezeagu, I.E., Ayoola, M. and Umera, E.A. 2018. Determination of the chemical composition of avocado (*Persea Americana*) seed. *Advances in Food Technology and Nutrional Sciences Open Journal Special Edition* 2, S51-S55.

Everett, K.R. 2003. The effect of low temperatures on *Colletotrichum acutatum* and *Colletotrichum gloeosporioides* causing body rots of avocados in New Zealand. *Australasian Plant Pathology* 32, 441–448.

Everett, K.R., Boyd, L.M., Pak, H.A. and Cutting, J.G.M. 2007a. Calcium, fungicide sprays and canopy density influence postharvest rots of avocado. *Australasian Plant Pathology* 36, 22–31.

Everett, K.R. and Korsten, L. 1996. Postharvest rots of avocados: Improved chemical control by using different application methods. Proceedings of the 49th New Zealand Plant Protection Conference 37–46.

Everett, K.R., Pak, H.A., Pushparajah, I.P.S., Taylor, J.T., Astill, M.S., King, D. B. and Chee, A.A. 2011. Field evaluation of fungicides to control postharvest rots of avocados in New Zealand. *New Zealand Plant Protection* 64, 112–118.

Everett, K.R. and Pushparajah, I.P.S. 2008. Postharvest fungicide treatments and their effect on long-term storage of avocados from three growing regions in New Zealand. *New Zealand Avocado Growers' Association Annual Research Report* 8, 85–92.

Everett, K.R., Timudo-Torrevilla, O.E. and Pushparajah, I.H.S. 2007b. Evaluation of a new chemical and two biological control agents for postharvest rot control in avocados. *New Zealand Avocado Growers' Association Annual Research Report* 7, 73–78.

FAO (Food and Agriculture Organization of the United Nations). 2016. Crops and livestock products. Available at: www.fao.org/faostat/es/#data/TP

Ferrer-Pereira, H. 2012. Aportes al conocimiento taxonómico del género *Persea* (Lauraceae) en Venezuela. *Hoehnea* 39, 435–478.

Figueroa, J.G., Borrás-Linares, I., Lozano-Sánchez, J. and Segura-Carretero, A. 2018. Comprehensive identification of bioactive compounds of avocado peel by liquid chromatography coupled to ultra-high-definition accurate-mass Q-TOF. *Food Chemistry* 245, 707–716.

Fischer, H.I., de Noraes, M.F., de Arruda, P.M.C., de Souza, F.M., Sodário, C.J.C. and Firmino, A.C. 2018. Effect of conventional alternative products on postharvest disease control in avocados. *Revista Brasileira de Fruticultura* 40, 1–10.

Fischer, I.H., Tosse, J.H.J., de Arruda, M.C. and Massola, J.N.S. 2011. Postharvest of 'Fuerte' and 'Hass' avocados: Physical and chemical characteristics, damages and control of diseases. *Semina: Ciências Agrárias* 32, 209–220. (In Portuguese with English abstract).

Freeman, S., Horowits-Brown, S., Afabador-Kafuri, L. and Maymon, D.M. 2013. *Colletotrichum*: Host specificity and pathogenicity on selected tropical and subtropical crops. *Acta Horticulturae* 975, 209–216.

Fucikovsky, L. and Luna, I. 1987. Avocado fruit diseases and their control in Mexico. *South African Avocado Growers' Association Yearbook* 10, 119–121.

Fuentes-Aragón, D., Juárez-Vázquez, S.B., Vargas-Hernández, M. and Silva-Rojas, H. V. 2018. *Colletotrichum fructicola*, a member of *Colletotrichum gloeosporioides* sensu lato, is the causal agent of anthracnose and soft rot in avocado fruits cv. 'Hass'. *Mycobiology* 46, 92–100.

Gaona-Forero, A., Agudelo-Rodríguez, G., Herrera, A.O. and Castellanos, D.A. 2018. Modeling and simulation of an active packaging system with moisture adsorption for fresh produce. Application in 'Hass' avocado. *Food Packaging and Shelf Life* 17, 187–195.

Garibaldi, A., Bertetti, D., Amatulli, M.T., Cardinale, J. and Gullino, M.L. 2012. First report of postharvest fruit rot in avocado (*Persea americana*) caused by *Lasiodiplodia thebromae* in Italy. *Plant Disease* 96, 460.

Giblin, F.R., Coates, L.M. and Irwin, J.A.G. 2010. Pathogenic diversity of avocado and mango isolates of *Colletotrichum gloeosporioides* causing anthracnose and pepper spot in Australia. *Australasian Plant Pathology* 30, 50–62.

Glowacz, M., Bill, M., Tinyane, P.P. and Sivakumar, D. 2017. Maintaining postharvest quality of cold stored 'Hass' avocados by altering the fatty acids content and composition with the use of natural volatile compounds–methyl jasmonate and methyl salicylate. *Journal of the Science of Food and Agriculture* 97, 5186–5193.

Guardado-Valdivia, L., Tovar-Pérez, E., Chacón-López, A., López-García, U., Gutiérrez-Martínez, P., Stoll, A. and Aguilera, S. 2018. Identification and characterization of a new *Bacillus atrophaeus* strain B5 as biocontrol agent of postharvest anthracnose disease in soursop (*Annona muricata*) and avocado (*Persea americana*). *Microbiological Research* 210, 26–32.

Guzmán, L.F., Machida-Hirano, R., Borrayo, E., Cortés-Cruz, M., Espíndola-Barquera, M. and Heredia-García, E. 2017. Genetic structure and selection of a core collection for long term conservation of avocado in Mexico. *Frontiers in Plant Science* 8, 1–10.

Gwanpua, S.G., Qian, Z. and East, A.R. 2018. Modelling ethylene regulated changes in 'Hass' avocado quality. *Postharvest Biology and Technology* 136, 12–22.

Handayani, M.N., Karlina, S., Sugiarti, Y. and Cakrawati, D. 2018. Application of edible coating from cassava peel - bay leaf on avocado. *Journal of Physics: Conference Series* 1013, 012168.

Hartill, W.F.T. and Everett, K.R. 2002. Inoculum sources and infection pathways of pathogens causing stem-end rots of 'Hass' avocado (*Persea americana*). *New Zealand Journal of Crop and Horticultural Science* 30, 249–260.

Herrera-González, J.A., Zapién-Ramos, J.E., Vite-Hernández, M.Y. and Mercado-Silva, E. 2017. Aplicación de fungicidas en precosecha que controlan enfermedades postcosecha en aguacate 'Hass' en Michoacán. Memorias del V Congreso Latinoamericano del Aguacate. Ciudad Guzmán, Jalisco, México. pp. 126–134. (In Spanish with English abstract).

Hopkirk, G., White, A., Beever, D.J. and Forbes, S.K. 1994. Influence of postharvest temperatures and the rate of fruit ripening on internal postharvest rots and disorders of New Zealand 'Hass' avocado fruit. *New Zealand Journal of Crop and Horticultural Science* 22, 305–311.

Kimaru, S.K., Monda, E., Cheruiyot, R.C., Mbaka, J. and Alakonya, A. 2018. Morphological and molecular identification of the causal agent of anthracnose disease of avocado in Kenya. *International Journal of Microbiology* 4568520.

Korsten, L. and De Jager, E.E. 1995. Mode of action of *Bacillus subtilis* for control of avocado postharvest pathogens. *South African Avocado Growers' Association Yearbook* 18, 124–130.

Korsten, L., De Villiers, E.E., Wehner, F.C. and Kotzé, J.M. 1997. Field sprays of *Bacillus subtilis* and fungicides control of preharvest fruit diseases of avocado in South Africa. *Plant Disease* 81, 455–459.

Korsten, L. and Kotzé, J. 1992. Postharvest biological control of avocado postharvest diseases. Procedings of the Second World Avocado Congress. Orange, CA. pp. 473–478.

Krumreich, F.D., Borges, C.D., Mendonça, C.R., Jansen-Alves, C. and Zambiazi, R.C. 2018. Bioactive compounds and quality parameters of avocado oil obtained by different processes. *Food Chemistry* 257, 376–381.

Le Roux, A.W.G., Wentzel, R.C. and Roose, C. 1985. Efficacy of prochloraz treatments for post harvest disease control in avocados. *South African Avocado Growers' Association Yearbook* 8, 44–45.

Lemus-Soriano, B.A. and Pérez-Aguilar, D.A. 2017. Manejo de la antracnosis del aguacate con biofungicidas. Memorias del V Congreso Latinoamericano del Aguacate. Ciudad Guzmán, Jalisco, México. pp. 135–139. (In Spanish with English abstract).

Londsale, J.H. 1992. Evaluation of systemic fungicides as pre-harvest treatments of avocados. *South African Avocado Growers' Association Yearbook* 15, 35–38.

López-Cobo, A., Gómez-Caravaca, A.M., Pasini, F., Caboni, M.F., Segura-Carretero, A. and Fernández-Gutiérrez, A. 2016. HPLC-DAD-ESI-QTOF-MS and HPLC-FLD-MS as valuable tools for the determination of phenolic and other polar compounds in the edible part and by-products of avocado. *LWT - Food Science and Technology* 73, 505–513.

Manicom, B.Q. and Schoeman, M.H. 2010. Evaluation of fungicides for Pseudocercospora control - final report. *South African Avocado Growers' Association Yearbook* 33, 70–73.

Marais, L.J. 2004. Avocado diseases of major importance worldwide and their management. pp. 1–36. *In*: S.A.M.H. Naqvi (ed.). *Diseases of Fruits and Vegetables. Diagnosis and Management*, Vol. II. Kluwer Academic Publisher, New York, NY.

Marques, J.R., Hofman, P.J. and Wearing, A.H. 2003. Rootstocks influence 'Hass' avocado fruit quality and fruit minerals. *The Journal of Horticultural Science and Biotechnology* 78, 673–679.

Marroquín-Pimentel, F.J. 1999. Factores que favorecen la incidencia de roña (*Sphaceloma perseae* Jenk.) en el cultivo del aguacate (*Persea americana* Mill.) 'Hass', en tres regiones agroclimáticas de Michoacán, México. *Revista Chapingo Serie Horticultura* 5, 309–312. (In Spanish).

Martínez, Y. 2015. Identificación y caracterización de bacterias antagonistas del patógeno de aguacate: *Colletotrichum gloeosporioides*. MSc. Thesis. Instituto Politécnico Nacional (IPN), Centro de Desarrollo de Productos Bióticos (Ceprobi). Yautepec, Morelos, Mexico. Available at: https://tesis.ipn.mx/handle/123456789/16193

Mavuso, Z.S. and van Nierkek, J.M. 2011. Development of a more effective post-harvest treatment for the control of post-harvest diseases of avocado fruit - Results from 2010. *South African Avocado Growers' Association Yearbook* 34, 37–41.

Mavuso, Z.S. and van Nierkek, J.M. 2013. Development of a more effective postharvest treatment for the control of postharvest diseases of avocado fruit. *South African Avocado Growers' Association Yearbook* 36, 23–26.

Mazhar, M., Joyce, D., Hofman, P. and Vu, N. 2018. Factors contributing to increased bruise expression in avocado (*Persea americana* M.) cv. 'Hass' fruit. *Postharvest Biology and Technology* 143, 58–67.

Mekonnen, Y., Alemayehu Chala, A. and Alemayehu, S. 2015. Prevalence of *Colletotrichum* spp. infecting fruits in Southern Ethiopia. *World Journal of Agricultural Sciences* 11, 34–41.

Montealegre, J.R., Ramírez, M., Riquelme, D., Armengol, J., León, M. and Pérez, L. M. 2016. First report of *Neofusicoccum australe* in Chile causing avocado stem-end rot. *Plant Disease* 100, 2532.

Mooz, E.D., Gaino, N.M., Shimano, M.Y.H., Amancio, R.D. and Spoto, M.H.F. 2012. Physical and chemical characterization of the pulp of different varieties of avocado targeting oil extraction potential. *Ciência e Tecnologia de Alimentos* 32, 274–280.

Mutembei, M.M. 2009. Agronomic practices and postharvest managements of anthracnose in avocado. MSc. Thesis. Faculty of Agriculture, University of Nairobi, Nairobi, Kenya.

Nelson, S. 2008. Anthracnose of avocado. PD-58. Cooperative Extension Service, College of Tropical Agriculture and Human Resources (CTAHR), University of Hawaii at Manoa, Honolulu, Hawaii, USA. Available at: www.ctahr.hawaii.edu/oc/freepubs/pdf/PD-58.pdf

Ni, H.F., Chuang, M.F., Hsu, S.L., Lai, S.Y. and Yang, H.R. 2011. Survey of *Botryosphaeria* spp., casual agents of postharvest disease of avocado, in Taiwan. *Journal of Taiwan Agricultural Research* 60, 157–166.

Ni, H.F., Liou, R.F., Hung, T.H., Chen, R.S. and Yang, H.R. 2009. First report of a fruit rot disease of avocado caused by *Neofusicoccum mangiferae. Plant Disease* 93, 760.

NMX-FF-016-1995 (Norma Mexicana). 1995. Aguacate *(Persea Americana* Mill*)*. México. Available at: www.colpos.mx/bancodenormas/nmexicanas/NMX-FF-016-1995.PDF

NMX-FF-016-SCFI-2016 (Norma Mexicana). 2016. Non industrialized food products for human consumption - fresh fruit Hass Avocado (*Persea americana* Mill.) - Specifications. Mexico. Available at: www.ciamac.org/assets/files/Articulos/nmx-ff-016-scfi-2016.pdf

Ortiz-Viedma, J., Rodriguez, A.L., Vega, C., Osorio, F., Defillipi, B., Ferreira, R. and Saavedra, J. 2018. Textural, flow and viscoelastic properties of Hass avocado (*Persea americana* Mill.) during ripening under refrigeration conditions. *Journal of Food Engineering* 219, 62–70.

Ozdemir, F. and Topuz, A. 2004. Changes in dry matter, oil content and fatty acids composition of avocado during harvesting time and post-harvesting ripening period. *Food Chemistry* 86, 79–83.

Palmateer, A.J., Ploetz, R.C. and Harmon, P.F. 2006. Florida Plant Disease Management Guide: Avocado (*Persea americana*). PP-233. Institute of Food and Agricultural Sciences (IFAS) Extension, University of Florida, Gainesville, FL, USA. Available at: http://ufdcimages.uflib.ufl.edu/UF/00/05/38/71/00029/PG13400.pdf

Pegg, K., Coates, L. and Dann, E. 2009. Avocado. pp. 49–64. *In*: T. Cook, D. Persley and S. House (eds.). *Diseases in Fruit Crops.* CSIRO Publishing, Clayton, Victoria, Australia.

Penter, M.G. and Stassen, P.J.C. 2000. The effect of pre- and postharvest calcium applications on the postharvest quality of Pinkerton avocados. *South African Avocado Growers' Association Yearbook* 23, 1–7.

Pérez, A., Rojas, S.J., Chamorro, L. and Pérez, P. 2011. Evaluation of the antifungi activity of *Melia azedarach* on isolate of *Colletotrichum* spp. *Revista Colombiana de Ciencia Animal* 3, 309–320. (In Spanish with English abstract).

Pérez, S., Avila, G. and Coto, O. 2015. El aguacatero (*Persea Americana* Mill). *Cultivos Tropicales* 36, 111–123. (In Spanish with English abstract).

Pérez-Jiménez, R.M. 2008. Significant avocado diseases caused by fungi and oomycetes. *The European Journal of Plant Science and Biotechnology* 2, 1– 24.

Pérez-Martínez, J. and Monterroso, A. 2015. Semi-commercial evaluation of fungicides on 'Hall' avocados to control cercospora and anthracnose. Final Report. R & D Department at Brooks Tropicals, LLC., Homestead, FL, USA. Available at: http://ir4.rutgers.edu/Fooduse/PerfData/4158.pdf

Peterson, R.A. 1978. Susceptibility of Fuerte avocado fruit at various stages of growth, to infection by anthracnose and stem-end rot fungi. *Australian Journal of Experimental Agriculture and Animal Husbandry* 18, 158–160.

Pieterse, Z., Jerling, J. and Oosthuizen, W. 2003. Avocados (monounsaturated fatty acids), weight loss and serum lipids. *South African Avocado Growers' Association Yearbook* 26, 65–71.

Reina-Noreña, J., Mayorag-Cobos, M.J., Caldas-Herrera, S.J., Rodríguez-Valenzuela, J. and Varón-Devia, E. 2015. The spot problem in avocado (*Persea Americana* Mill.) plantations of northern Tolima, Colombia. *Corpoica Ciencia y Tecnología Agropecuaria* 16, 265–278. (In Spanish with English abstract).

Rotta, E.M., De Morais, D.R., Biondo, P.B.F., Dos Santos, V.J., Matsushita, M. and Visentainer, J.V. 2016. Use of avocado peel (*Persea americana*) in tea formulation: A functional product containing phenolic compounds with antioxidant activity. *Acta Scientiarum-Technology* 38, 23–29.

Rubí-Arriaga, M., Franco-Malvaíz, M., Rebollar-Rebollar, S., Bobadilla-Soto, E., Martínez-De La Cruz, I. and Siles-Hernández, Y. 2013. Situación actual del cultivo del aguacate (*Persea Americana* Mill.) en el estado de México, México. *Tropical and Subtropical Agroecosystems* 16, 93–101. (In Spanish with English abstract).

Sanders, G.M., Korsten, L. and Wehner, F.C. 2000. Market survey of post-harvest diseases and incidence of *Colletotrichum gloeosporioides* on avocado and mango fruit in South Africa. *Tropical Science* 40, 192–198.

Sharma, G., Maymon, M. and Freeman, S. 2016. First detailed report of *Trichothecium roseum* causing post-harvest pink rot of avocado in Israel. *Plant Disease* 100, 856.

Sharma, G., Maymon, M. and Freeman, S. 2017. Epidemiology, pathology and identification of *Colletotrichum* including a novel species associated with avocado (*Persea americana*) anthracnose in Israel. *Scientific Reports* 7, 15839.

Smith, E.M., Kotze, J.M. and Wehner, F.C. 1987. Occurrence and control of avocado sooty blotch. *South Africa Avocado Growers' Association Yearbook* 10, 11–113.

Smith, L.A., Dann, E.K., Leonardi, J., Dean, J.R. and Cooke, W. 2011. Exploring non-traditional products for management of postharvest anthracnose and stem end rot in avocado. Proceedings of the VII World Avocado Congress. Cairns, Queensland, Australia. Available at: www.avocadosource.com/WAC7/Section_02/SmithLuke2011b.pdf

Sorensen, S. 2017. Review of fruit rots in the New Zealand avocado industry. New Zealand Avocado, Avocado Industry Council Ltd., Tauranga, New Zealand. Available at: https://industry.nzavocado.co.nz/resources/4564836/August_2017_Review_of_Fruit_Rots_in_the_New_Zealand_Avocado_Industry_On_Orchard.pdf

Stovold, G.E. and Dirou, J.F. 2004. Avocado diseases. Agfact H6.AB. 5. NSW Agriculture, NSW Department of Primary Industries, New South Wales, Australia. Available at: www.dpi.nsw.gov.au/_data/assets/pdf_file/0006/120012/avocado-diseases.pdf

Takenaga, F., Matsuyama, K., Abe, S., Torii, Y. and Itoh, S. 2008. Lipid and fatty acid composition of mesocarp and seed of avocado fruits harvested at northern range in Japan. *Journal of Oleo Science* 57, 591–597.

Tesfay, S.Z., Magwaza, L.S. and Mditshwa, A. 2017. Carboxyl methylcellulose (CMC) containing moringa plant extracts as new postharvest organic edible coating for avocado (*Persea americana* Mill.) fruit. *South African Avocado Growers' Association Yearbook* 40, 108–116.

Tremocoldi, M.A., Rosalen, P.L., Franchin, M., Massarioli, A.P., Denny, C., Daiuto, É. R., Paschoal, J.A.R., Melo, P.S. and De Alencar, S.M. 2018. Exploration of avocado by-products as natural sources of bioactive compounds. *PLoS ONE* 13, e0192577.

Trinidad-Ángel, E., Ascencio-Valle, F.D.J., Ulloa, J.A., Ramírez-Ramírez, J.C., Ragazzo-Sánchez, Calderón-Santoyo, M. and Bautista Rosales, P.U. 2017. Identification and characterization of *Colletotrichum* spp. causing anthracnose in avocado Nayarit, Mexico. *Revista Mexicana de Ciencias Agrícolas* 19, 3953–3964.

Twizeyimana, M., Förster, H., McDonald, V., Wang, D.H., Adaskaveg, J.E. and Eskalen, A. 2013. Identification and pathogenicity of fungal pathogens associated with stem-end rot of avocado in California. *Plant Disease* 97, 1580–1584.

UC (University of California). 2014. Avocado. Causes of fruit damage. Statewide Integrated Management Program, Agriculture and Natural Resources, University of California, Davis, CA, USA. Available at: http://ipm.ucanr.edu/PMG/C008/m008fpfrtdmg.html

Urquiza, L.S., Rebollar, S., Callejas, N., Hernández, J. and Gómez, G. 2015. Análisis de viabilidad económica para la producción comercial de aguacate Hass. *Revista Mexicana de Agronegocios* 36, 1325–1338. (In Spanish with English abstract).

USDA-ARS (United States Department of Agriculture – Agricultural Research Service). 2014. Food Composition Databases, Show Foods List. Available at: https://ndb.nal.usda.gov/ndb/foods/show/09038

Valencia, A.L., Torres, R. and Latorre, B.A. 2011. First report of *Pestalotiopsis clavispora* and *Pestalotiopsis* spp. causing postharvest stem end rot of avocado in Chile. *Plant Disease* 95, 492.

Van Dyk, K., de Villiers, E.E. and Korsten, L. 1997. Alternative control of avocado post-harvest diseases. *South African Avocado Growers' Association Yearbook* 20, 109–112.

Vidales, J.A., Coria, V.M., Anguiano, J. and Alcántar, J.J. 2005. Control integrado de la antracnosis del aguacate. Instituto Nacional de Investigaciones Forestales, Agrícolas y Pecuarias. Centro de Investigación Regional del Pacífico Centro, Campo Experimental Uruapan. Uruapan, Michoacán, México.

Willingham, S.L., Coates, L.M., Cooke, A.W. and Dean, J.R. 2004. Tree vigour influences disease susceptibility of 'Hass' avocado fruits. *Australasian Plant Pathology* 33, 17–21.

Willingham, S.L., Pegg, K.G., Anderson, J.M., Cooke, A.W., Dean, J.R., Giblin, F.R. and Coates, L.M. 2006. Effects of rootstock and nitrogen fertiliser on postharvest anthracnose development in Hass avocado. *Australasian Plant Pathology* 35, 619–629.

Willingham, S.L., Pegg, K.G., Coates, L.M., Cooke, A.W., Dean, J.R., Langdon, P.W. B. and Beasley, D.R. 2001a. Field management of avocado postharvest diseases. *Acta Horticulturae* 553, 435–438.

Willingham, S.L., Pegg, K.G., Cooke, A.W., Coates, L.M., Langdon, P.W.B. and Dean, J.R. 2001b. Rootstock influences postharvest anthracnose development in 'Hass' avocado. *Australian Journal of Agricultural Research* 52, 1017–1022.

Yakoby, N., Zhou, R., Kobiler, A., Dinoor, A. and Prusky, D. 2001. Development of *Colletotrichum gloeosporioides* restriction enzyme-mediated integration mutants as biocontrol agents against anthracnose disease in avocado fruits. *Phytopathology* 91, 143–148.

Chapter 7

Papaya

Subbaraman Sriram
Plant Pathology, ICAR-Indian Institute of Horticultural Research, Hessaraghatta, Bengaluru, India

Darisi Venkata Sudhakar Rao
Postharvest Technology & Agricultural Engineering, ICAR-Indian Institute of Horticultural Research, Hessaraghatta, Bengaluru, India

Abbreviations

CA	Controlled atmosphere
EBDC	Ethylene bis dithiocarbamate
EFF	Enhanced freshness formulation
FADH	Forced air dry heat
HWT	Hot water treatment
ITCs	Isothiocyanates
LDPE	Low-density polyethylene
MA	Modified atmosphere
MT	Metric tons
USA	United States of America
USDA-APHIS	United States Department of Agriculture – Animal and Plant Health Inspection Service
US EPA	United States Environmental Protection Agency
VH	Vapor heat

1 Introduction

Papaya (*Carica papaya* L.), a member of Caricaceae, is an important fruit crop. It is native to tropical America, mainly Guatemala, and its culture is now popular worldwide in the tropics and sub-tropics (Singh, 1990). Next to banana, it helps farmers in sub-tropics to get a good revenue. Besides providing calcium, it is a good source of vitamins A and C. Papain, an alkaloid present in raw fruits of papaya, is an important ingredient in medicine and food preparations in many countries. It is consumed not only as a fresh fruit when ripe, but also in the form of processed postharvest products such as juices, jam, candies, and dried fruit (Villegas, 1997; González-Aguilar et al., 2008). Raw unripe fruit is also consumed (typically cooked like a vegetable).

As per the Department of Agriculture and Cooperation, Government of India, India produced 6,108,000 metric tons (MT) of papaya on 136,000 ha during the season 2016–2017, with Andhra Pradesh and Gujarat leading in the production (1.28 and 1.24 million MT, respectively), while productivity in Tamil Nadu was high (229.74 MT/ha). The export of papaya (as fresh or dried fruits) from India to other countries was 12,773 MT, valuing of 7.23 million US$ during 2016–2017 with most exported to the United Arab Emirates and Saudi Arabia (Anonymous, 2017). Countries in Asia, South America, North Central America, and Africa are the leading producers of papaya. The world production of papaya was estimated to be 13.02 million MT per year (FAOSTAT, 2017).

Fungi, bacteria, nematodes, and viruses can infect papaya and incur yield loss. Anthracnose caused by *Colletotrichum gloeosporioides* (Penz.) Penz. & Sacc. causes serious fruit losses during transit and storage and is considered the most serious yield constraint worldwide. As observed by Eckert and Ogawa (1985), mostly synthetic chemicals have been used to manage the postharvest diseases of fruits. These are perceived by some consumers as a risk to human and environmental health, and preferably their use should be avoided or minimized. Furthermore, the development of resistance to fungicides among postharvest pathogens also poses a serious threat to their continued use (Spalding, 1982; Spotts and Cervantes, 1986). Hence, the development of user-friendly and safer management practices is essential for postharvest disease management. Biological control agents (Wilson and Wisniewski, 1989), especially yeasts of fructo-plane origin, were found to be effective tools in the management of postharvest diseases of fruits (Sharma et al., 2009).

2 Postharvest Diseases of Papaya

Postharvest diseases reduce the market quality of papaya and cause losses during shipment. Losses up to 10–40% during road shipment and 5–30% during air transport are common (Fatima et al., 2006). Earlier inspection reports of papaya shipments arrived in New York terminal markets revealed that anthracnose rot caused by *C. gloeosporioides* affected the 62% of shipments, while other diseases such as Rhizopus rot, stem-end rot, and gray mold affected about 35% of the shipments (Cappellini et al., 1988). Californian inspections also showed that 73% of the inspected cartons had decay and mold growth and 52% had sunken defects (Paull et al., 1997). The statistical figures on postharvest losses caused by decay in papaya clearly demonstrate the scale and importance of this problem. Fruit rots include surface rots, stem-end rots, and internal fruit infections. Papaya fruits can be affected by various postharvest diseases, *viz.* anthracnose, stem-end rot, chocolate rot, Fusarium rot, Aspergillus rot, Rhizopus rot, and others.

2.1 Fruit Surface Rots

Surface rots in fruits are caused by fungal pathogens (which infect the fruits while they are green and yet to be plucked from tree) or by pathogens (which enter through wounds either before harvest or after storage). Anthracnose, chocolate spot, Cercospora black spot, and Phytophthora fruit rot occur while the fruits are borne on the tree, whereas weak pathogens (such as species of *Mycosphaerella, Phomopsis, Alternaria, Stemphylium, Fusarium,* and *Guignardia*) enter through wounds (Hunter and Buddenhagen, 1972).

2.1.1 Anthracnose

It is caused by the hemibiotrophic fungus *C. gloeosporioides*. There are around 600 synonyms of *C. gloeosporioides*, among them *Gloeosporium papayae* Henn. and *C. papayae* Henn., reported to cause anthracnose of papaya (Arx, 1957). *Colletotrichum gloeosporioides* has been reported to be a pathogen of 470 genera of plants, including papaya (Alahakoon et al., 1994; Simmonds, 1965; Sutton, 1980; Hartill, 1992). It is common in all papaya-growing areas. Ratnam and Neema (1967) made the first report describing its occurrence on papaya from Bihar (India). Srivastava et al. (1964) reported that *G. papayae* and *C. papayae* caused fruit rot of papaya. Arx (1957) made a detailed study of the species of the genus *Colletotrichum* and assigned the ascogenous state of *C. gloeosporioides* as *Glomerella cingulata* (Stoneman) Spauld. & H. Schrenk. Saccardo (1884) and Potebnia (1910) placed *Colletotrichum* in Melanconiales and Acervulales, respectively. *Colletotrichum gloeosporioides* produces aggregated perithecia, which are globose, dark brown to black, with periphysate ostioles along with paraphyses. The asci, which bear eight ascospores, are clavate or cylindrical. Ascospores are hyaline, unicellular, oval, cylindrical, or fusiform.

The asexual stage is represented by acervuli that are round or oblong, colored black or pink, and with or without setae. The conidiophores measure about 20–30 μm in length. Conidia are hyaline or pink when in mass, straight rounded at both ends, measure 11–32 x 4–55 μm in size, and usually with two, or rarely one, oil globules. Acervuli develop profusely on diseased parts of the plant, such as petioles, leaves, and fruits (Sattar and Malik, 1939). They are irregular in size, appear as brown to black dots on the leaves, and occur on both leaf surfaces. Setae are common on twigs but not on fruits. The acervuli, when mature, exude pink masses of conidia under moist conditions.

Infections start in the field when conidia germinate and form infection pegs and appressoria, followed by a quiescent stage until fruit reaches the climacteric phase of ripening, then the pathogen shifts from biotrophic to necrotrophic phase. The necrotrophic phase starts with a water-soaked lesion that expands and later becomes a light-brown lesion (Figure 7.1a). Conidia are formed on asexual-spore-bearing structures called acervuli. Internal tissues in the infected area appear grayish white and later on, turn to brown. Late-stage infections may affect the entire fruit and develop pink masses of conidia on the lesion surface if humidity is high enough (Figure 7.1b). A callose layer forms in the infected parenchyma tissues, which enables lifting the infected tissue free from the fruit as a plug (Stanghellini and Aragaki, 1966). Fungicide applications after the shift to the necrotic phase have limited efficacy to control the disease because by that time the pathogen has become established within the tissue and has damaged the fruit. Although fungicide applications can reduce the number of new infections, it is difficult to stop the damage caused by the pathogens that have infected and penetrated the host before ripening. *Colletotrichum* occasionally causes chocolate spot as well, where it forms minute, superficial, reddish-brown lesions (Figure 7.1).

2.1.2 Dry Rot

Dry rot of papaya is caused by *Mycosphaerella* sp. Like anthracnose pathogen, this fungus cannot penetrate the fruit cuticle and needs mechanical injuries to cause infection. The symptoms include initial wrinkles on the outer peel followed by brown lesions with translucent margins. A hard layer of parenchyma tissue develops, which can be easily separated from the epidermal portion of the papaya fruit. Asexual stage of this fungus was earlier described as *Ascochyta caricae* Pat. (Chowdhury, 1950; Hunter and Buddenhagen, 1972) and *A. caricae-papayae* Tarr (Chau and Alvarez, 1979), but later Punithalingam (1980) described it as *Phoma caricae-papayae* (Tarr) Punith. This pathogen is also known to cause stem-end rot.

2.1.3 Phomopsis Rot

Fruit rot due to *Phomopsis* spp. is also termed wet rot. After infection, the lesions first become soft and translucent and later on, pycnidia will be produced at the center of the lesion. Because the affected area is always wet, the lesion spreads quickly and forms a cavity. The infected tissue can be removed easily from the rest of the fruit (Hunter and Buddenhagen, 1972).

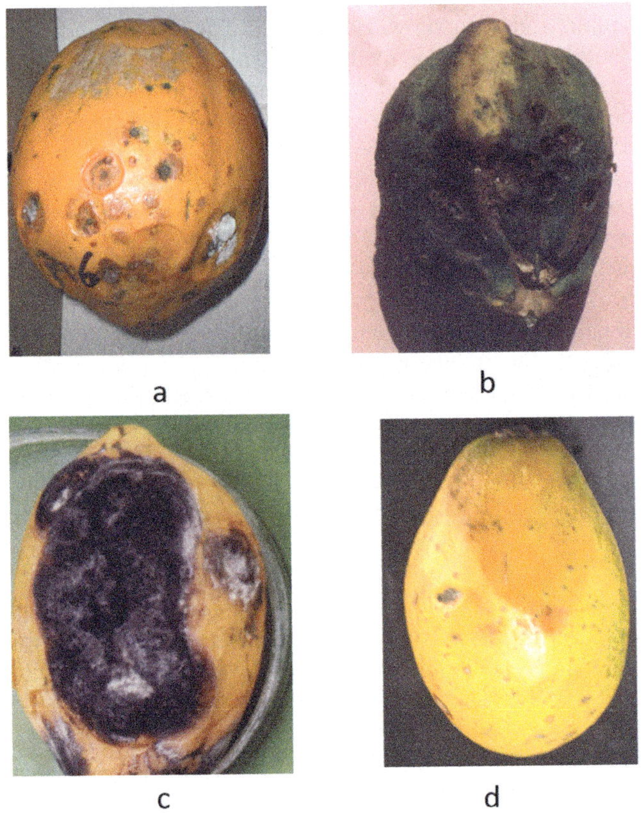

Figure 7.1 Symptoms of postharvest diseases of papaya. a. Early stage of anthracnose. b. Late stage of anthracnose. c. Alternaria rot. d. Stem-end rot.

2.1.4 Charcoal Rot

Charcoal rot is caused by *Macrophomina phaseolina* (Tassi) Goid. Fully mature, ripe fruits are more susceptible to this rot than the raw green fruits. The pathogen enters through wounds and infects the fruit. The pycnidia appear brown or black. This pathogen also produces sclerotia, which are colored black and has a smooth surface. Because of its characteristic black sclerotial and pycnidial bodies, the disease has been named charcoal rot.

2.1.5 Alternaria Rot

Fruit rot and brown spots by *Alternaria alternata* (Fr.) Keissl. is distinct from others by the formation of black spore masses in ring-link patterns within the infected area of the lesions (Figure 7.1c). The lesions are limited to the surface of the fruit and not much damage is seen on the pulp. Infection by *Alternaria* on senescing petioles at the time of harvest provides the major inoculum source. Routine plant protection measures and removal of old leaves help in managing this disease.

2.2 Stem-End Rots

After the harvest, postharvest pathogens enter through the cut wounds at the peduncle. They can also enter through the crevices between peduncle and flesh. Stem-end rot can be caused by species of *Ascochyta, Botryodiplodia, Phomopsis, Fusarium, Mycosphaerella*, and *Guignardia*, and by *A. alternata, C. gloeosporioides, Stemphylium lycopersici* (Enjoji) W. Yamam., *Rhizopus stolonifer* (Ehrenb.) Vuill. (syn.: *Rhizopus nigricans* Ehrenb.), and others (Hunter and Buddenhagen, 1972; Chau and Alvarez, 1983; Alvarez and Nishijima, 1987). Sometimes mixed infections also occur. Infection by *Mycosphaerella* results in a translucent zone followed by slight browning at peduncle. Later, the affected tissue becomes dark, wrinkled, and dry. *Botryodiplodia* infection results in a wide margin of water-soaked lesion with a rough surface (Figure 7.1d). Later, affected portions become blackish with an irregular pattern of pycnidia production. Infection by *S. lycopersici* results in reddish-brown discoloration of tissues with bright-purple margin (Tandon, 1967; Prasad and Verma, 1970). *Phomopsis* infection causes wrinkles in the fruit skin with green to yellow lesions. The infected portion can be easily lifted from the remaining part of the fruit. *Rhizopus* causes the most serious infection because the pathogen invades through wounds, its growth progresses rapidly, and it spreads to other fruits in the storage container.

2.3 Other Minor Fruit Rot Pathogens

Some pathogens causing stem-end rots can also infect through other parts of the fruit, causing minor postharvest diseases of papaya. Infections by *S. lycopersici* cause lesions with reddish brown to purple margins with dense dark-green spore masses with a gray mycelium. Association of *Fusarium* spp. with dry fruit rot has been observed (Hunter and Buddenhagen, 1972), and involvement of *Fusarium* spp. such as *F. solani* (Mart.) Sacc., *F. roseum* Link, *F. moniliforme* J. Sheld., and *F. oxysporum* Schltdl. has been reported (Tandon, 1967; Shukla et al., 1978; Arya et al., 1986). Other minor pathogens include *Guignardia* spp., which enter through cut wounds. *Guignardia* infections increased when hot water treatment was prolonged beyond the recommended 30 min. *Rhizopus stolonifer* entering through peel wounds causes a watery and soft rot that results in collapse of the fruit and exudate of water and emission of foul smell (Srivastava et al., 1964; Tandon and Mishra, 1969; Sarwar and Kamal, 1971; Pathak et al., 1976). *Pythium aphanidermatum* (Edson) Fitzp. also causes soft fruit rot (Trujillo and Hine, 1965). Fatima et al. (2006) reported 37 fungal species that caused postharvest rots in papaya. Tandon (1967) reported the involvement of many fungal species such as *Alternaria tenuis* Nees, *Trichothecium roseum* (Pers.) Link, *Phytophthora parasitica* Dastur, *Aspergillus* spp., and *Cladosporium cucumerinum* Ellis & Arthur in addition to the above-mentioned causal agents of papaya fruit rot. Tewari et al. (1988) reported a soft rot caused by *Ceratocystis paradoxa* (Dade) C. Moreau [syn.: *Thielaviopsis paradoxa* (De Seynes) Höhn.] in papaya.

2.4 Pulp Rot or Internal Fruit Infections

Sometimes a 'smut'-like growth due to fungal infection is seen when blossom-end rot affected fruits are not properly covered and *Cladosporium, Penicillium*, and *Fusarium* species were found in the cavities. Two bacterial species, *Erwinia herbicola* and *Enterobacter cloacae*, were also reported to colonize the internal tissues. The infection by *E. herbicola* causes violet or purple streaks, whereas infection by *E. cloacae* results in yellowing of the internal tissues (Nelson and Alvarez, 1980; Alvarez and Nishijima, 1987).

3 Control of Postharvest Diseases

3.1 Preharvest Measures

Because the origin of most of the postharvest pathogens is the field, control measures also should start in the field. To reduce the inoculum level, contact fungicides as preventive measure can be used (Eckert and Ogawa, 1985). Contact fungicides such as mancozeb, zineb, and chlorothalonil can be used at 2 g/L along with stickers before rainy periods. The removal of infected and senescent leaves and infected fruits is important in stopping the spread of the inoculum. The control of anthracnose on papaya was achieved by preharvest sprays with the dithiocarbamates metiram 80% wettable powder at 200 g/100 L of water or propineb 70% at 200 g/100 L of water, or copper oxychloride 50% wettable powder at 400 g/100 L of water applied at 7 to 10 d intervals (Kamal and Agbari, 1985).

Debysingh et al. (2018) demonstrated that preharvest treatments with 2% of an enhanced freshness formulation (EFF), with hexanal as the main active ingredient, delayed the onset of fruit ripening. The treatment contributed to maintain postharvest fruit quality because it delayed color development and maturity, which resulted in a reduction of the incidence of postharvest diseases.

3.2 Measures After Harvest

Although preharvest sprays can reduce field infections, entry of postharvest pathogens through the peduncle after fruit picking cannot be ruled out and hence, for disease control it is necessary to apply postharvest treatments such as chemical fungicides or other alternative methods, as described below.

Orchard sanitation is the most important consideration to reduce the field pathogen inoculum. The discarded and infected fruits should be removed immediately to control spread and survival of the pathogen (Alvarez and Nishijima, 1987). Postharvest temperature management and packinghouse hygiene and decontamination of packingline equipment are essential to prevent the build-up of inoculum and re-infection of the fruits subjected to disease control measures. There should be utmost care during harvest and postharvest handling to minimize the physical damage to fruit. The mechanical injuries caused during different packinghouse operations provide excellent opportunities to the entry of wound pathogens that causes serious problems of fruit rots. The use of plastic liners for field bins, proper fruit arrangement and cushioning material in the box, and avoiding impact and compression damages during handling could be effective ways to reduce fruit decay problems caused by secondary pathogens (Paull et al., 1997).

3.2.1 Temperature Management

Postharvest temperature management is a key factor to prevent the development of latent infections after harvest of papaya and also to reduce the growth and development of wound pathogens. The conditions favorable for slow ripening of fruit (15–18°C) allow the latent fungi to grow appreciably. Therefore, the rapid postharvest cooling of fruit to about 13°C before cold storage or transport and then the faster ripening of fruit at 22.5 to 27.5°C can be helpful to reduce the losses caused by fungal diseases (Singh and Sudhakar Rao, 2011).

Temperature fluctuations during ground- or air-shipment of papayas can increase marketing losses owing to more decay incidence and poor fruit quality. A study on the effects of simulated commercial temperature regimes during air-transport on the quality of papaya fruit throughout the handling chain showed that fruit handled under a fluctuating cold or warm temperature regime lost more weight; developed objectionable color; were softer and more shriveled; had more decay; and had lower soluble solids, acidity, and ascorbic acid contents than papayas handled in a semi-constant temperature regime (12°C for 52 hr, 8°C for 24 hr, and 7 d at 20°C) (Nunes et al., 2006). 'Redy Lady' papayas handled in a fluctuating cold temperature regime (at 8–15°C for 76 hr) that were briefly exposed to 1°C for 2 hr developed chilling injury symptoms during a subsequent ripening period of 7 d at 20°C. Therefore, proper temperature management without any significant deviations from the recommended conditions is crucial to provide consumers with a high-quality fruit and limit the possibility of rejection of papaya consignments.

3.2.2 Fungicides

Synthetic fungicides belonging to different groups have been reported for the management of postharvest diseases of papaya. They include benzimidazole (thiabendazole and benomyl), imidazole (prochloraz), ethylene bis-diothiocarbamate (EBDC), strobilurin (azoxystrobin), and benzonitrile (chlorothalonil) groups.

A wide range of fungicides (including carbendazim (0.1%), thiophanate methyl (0.1%), chlorothalonil (0.2%), benomyl (0.2%), zineb (0.2%), thiabendazole (0.1%), copper oxychloride (0.2%), prochloraz (0.1%), hexaconazole, propiconazole, and tricyclazole) have been reported to inhibit postharvest pathogens, especially *Colletotrichum* spp. (Washathi and Bhargava, 2000; Patel and Joshi, 2002; Suseela Bhai et al., 2003; Ashoka, 2005). Many of them have been tested in vivo as well. Registration by the United States Environmental Protection Agency (US EPA) permits the use of the fungicides azoxystrobin, chlorothalonil, mancozeb, maneb, prochloraz, and thiabendazole and the efficacy of some of these fungicides to control postharvest diseases of papaya has been summarized (Bautista-Baños et al., 2013). In other studies, postharvest treatments with the active ingredients prochloraz and propiconazole were effective to control anthracnose in papaya (Ong et al., 2013). Postharvest application of thiabendazole at 2 g/L was also found effective to reduce papaya decay (Couey and Farias, 1979). Henriod et al. (2016) found that prochloraz can be replaced with fludioxonil as a postharvest fungicide for the management of anthracnose and stem-end rot in papaya. In recent years, the use of combination products has become a practice instead of using single fungicide molecules to avoid the development of fungicide resistance (Ramallo et al., 2019).

However, the use of chemical fungicides for the management of postharvest diseases, especially as postharvest treatments, led to important issues related to chemical residues and their potential adverse effects on human health and the environment. Many of the chemicals that had been used in postharvest treatment earlier are either not recommended or banned now. Therefore, alternative treatments have also been evaluated for papaya disease control.

3.2.3 Use of Plant Extracts and Essential Oils

As Papavizas (1973) suggested, management of plant diseases should be with minimum interference to biological equilibrium. To achieve this, the integration of many approaches is needed. Plant extracts have been reported to have potential to inhibit many plant pathogens (Ark and Thompson, 1959). For example, the antifungal activity of tulsi leaf extract (Godara and Pathak, 1995), garlic oil and its synthetic form diallyl disulphide (Shirshikar, 2002), mint (Shekhawat and Prasad, 1971), polyalthia (Jetti et al., 1987), eucalyptus oil, castor oil, ginger extract, turmeric, and lantana leaves (Chauhan and Joshi, 1990) to inhibit postharvest pathogens affecting papaya such as *C. gloeosporioides, Gignardia, Alternaria*, and other species has been reported. Similarly, inhibition of *C. gloeosporioides* by extracts of *Allium sativum, Azadirachta indica, Ocimum sanctum, Pongamia pinnata*, and *Vitex negundo* (Chavan, 1996), extracts of medicinal plants (Escopalao and Silvestre, 1996; Raheja and Thakore, 2002), extracts of *Pongamia pinnata* and *Cathranthus roseus* (Shirshikar, 2002), and leaf extracts of gando baval, Bhoy ringni, and ginger (Patel and Joshi, 2001) has also been reported.

A wide range of essential oils, such as those from thyme, Mexican lime, lemongrass, cinnamon, mint, lavender, and castor oil, have proved to be effective against anthracnose caused by *C. gloeosporioides* on papaya fruit (Maqbool et al., 2011; Ali et al., 2015; Sarkhosh et al., 2017). The incorporation of essential oils of thyme (0.1%) and Mexican lime (0.05%) to fruit-coating materials reduced the incidence of anthracnose and Rhizopus rot in 'Maradol' papayas (Bosquez-Molina et al., 2010). Papaya fruits treated with both essential oils resulted in reduced decay caused by *C. gloeosporioides* and *R. stolonifer* by up to 50 and 40%, respectively, compared with the 100% infection observed in non-treated papayas. Dip treatment of fruits with clove essential oil (50 μg/L) showed a higher efficacy than cinnamon oil in reducing natural fungal infections in papaya (Barrera-Necha et al., 2008). Antifungal compounds identified in ginger oil were superior in yield and composition to those from ginger extract, which translated into higher antifungal activity of the ginger oil. Further, a composite coating formulated with ginger oil and gum arabic was effective in maintaining the quality of papaya fruits in terms of firmness, peel color, soluble solids content, and titratable acidity. Ginger oil (2.0%) combined with 10% gum arabic was an effective postharvest treatment to control anthracnose and maintain quality of 'Eksotika II' papaya fruits (Ali et al., 2016).

3.2.4 Irradiation

Few studies on evaluating the effect of irradiation on the postharvest quality of papaya fruits concluded that ripening could be delayed with higher fruit firmness during storage. The effect of irradiation depended on the dose, ripening stage, storage temperature, and duration of exposure (Meirelles and Melges, 2004). These authors observed that an irradiation dose lower than 0.75 kGy could not reduce anthracnose incidence in 'Solo' papaya. Cia et al. (2007) found that gamma irradiation above 1.0 kGy was required for anthracnose management. Too high irradiation doses, however, can induce phytotoxicities and negatively affect the quality of the fruits.

3.2.5 Biocontrol Agents

Efficacy of the filamentous fungus *Trichoderma* spp. to inhibit many postharvest pathogens, especially *Colletotrichum* spp., has been documented widely (Deshmukh and Raut, 1992; Mederios and Menezes, 1994; Majumdar and Pathak, 1995; Rocha and de Oliveira, 1998; Gud, 2001; Patel and Joshi, 2001; Santha Kumari, 2002; Shirshikar, 2002; Ashoka, 2005). de Capdevile et al. (2007) observed that one isolate of the fungus *Cryptococcus magnus* could delay the postharvest rot in papaya.

Baños-Guevara et al. (2004) reported the in vitro efficacy of two isolates of the bacterium *Bacillus subtilis* against papaya anthracnose. Osman et al. (2011) reported that *Bacillus amyloliquefaciens* effectively controlled anthracnose and Phomopsis rots on papaya fruits that were pre-treated with 1-methylcyclopropene. The treatment delayed the ripening and maintained the firmness of fruits. Shi et al. (2009) observed that *Pseudomonas putida* (MPG1) provided significant control of *Phytophthora nicotianae* Breda de Haan infection in papaya.

Gamagae et al. (2003) evaluated the efficacy of *Candida oleophila* to control anthracnose of papaya. Furthermore, the combined application of this yeast with a sodium bicarbonate (2%) incorporated wax coating resulted in a significant reduction of anthracnose incidence and severity on naturally infected fruits stored at 13.5°C and 95% RH for 14 d and for additional 2 d under simulated marketing conditions (Gamagae et al., 2004). In these conditions, the survival of *C. oleophila* in 2% sodium bicarbonate-incorporated wax coating was 100 and 90% for 60 min and 14 d, respectively. Sriram and Poornachandra (2013) identified *Candida tropicalis* and *Alcaligenes feacalis* as bioagents as pre-harvest and postharvest sprays for the management of mango anthracnose.

3.2.6 Salts

Sivakumar et al. (2002) examined the effect of ammonium carbonate (3%) or sodium bicarbonate (2%) in aqueous solution or incorporated the same into a wax formulation on anthracnose severity in inoculated or naturally infected papaya fruits. Both salts had significant effects, but that of ammonium carbonate was greater than that of sodium bicarbonate in controlling anthracnose. Ammonium carbonate (3%) incorporated into the wax formulation effectively reduced anthracnose incidence by 70% in naturally infected papaya and extended the storage life by maintaining the firmness, color, and overall quality of fruits stored at a low temperature (13.5°C and 95% RH for 21 d followed by 2 d under marketing conditions).

The effect of pre- and postharvest application of calcium was studied by Madani et al. (2014), who reported that 1.5 and 2% calcium chloride significantly reduced anthracnose incidence and severity on papaya during 5 wk of storage. Mahmud et al. (2008) found that 'Eksotika II' papaya fruits infiltrated (vacuum) with 2.5% calcium chloride showed lowest incidence of anthracnose.

3.2.7 Hot Water Treatment

Hot water treatment (HWT) delays papaya fruit ripening, but has also positive effects on reducing postharvest decay. HWT, combined with water vapor treatment, reduced *Botryodiplodia* infection in papaya (Yaguchi and Nakamura, 1993). Maharaj and Sankat (1990) used HWT at 48°C for 20 min or hot water plus benomyl (1.5 g/L) at 52°C for 2 min to control anthracnose caused by *C. gloeosporioides.* HWT at 48–50°C for 20 min has also been reported as part of integrated management practices to control postharvest anthracnose of papaya (Allong et al., 2000). Aragaki et al. (1981) observed that HWT of papaya fruit reduced *Phytophthora* infections. HWT of papaya fruit at 54°C for 4 min had a pronounced effect on reducing inoculum of *C. gloeosporioides* in the fruit peel and significantly reduced the incidence of anthracnose and stem-end rot (Li et al., 2013). In this work, HWT also reduced fruit ripeness to a certain extent and induced changes in the wax arrangement on the surface of treated fruits, causing the wax to melt. The cracks and most stomata appeared to be partially or completely plugged by the melted wax, thereby providing a mechanical barrier against wound pathogens. Together, these results confirmed that HWT could reduce disease incidence and induce disease resistance and thus maintain postharvest quality during storage and prolong the shelf life of papaya fruits.

The duration and temperature of HWT depends on import or export regulations. A single HWT dip (49°C for 20 min) has been very promising to control postharvest diseases in papaya fruits without detrimental effect on fruit quality. Postharvest quarantine treatments aimed to control fruit flies and other pests, such as forced air dry heat (FADH) and vapor heat (VH), also provide some control of postharvest diseases. Moreover, the single HWT dip before or after FADH and VH treatments can provide additional disease control to the same level than that provided by the combination of the fungicide thiabendazole with FADH and VH treatments (Nishijima et al., 1992). This single HWT dip was modified into a double hot water immersion (Couey et al., 1984; Couey and Hayes, 1986) and was also accepted by the USDA-APHIS as a quarantine treatment in 1990. Export to the United States (USA) mainland requires papaya fruit to be one fourth ripe, disinfested, subjected to a double HWT immersion (42°C for 30 min followed by 49° C for 20 min), and then immediately cooled to less than 30°C with ambient water dips or sprays. Care should be taken to avoid excessive heating and delayed post-treatment cooling. Otherwise, the ripening process will be inhibited and result in scald.

On the other hand, the combination of forced hot air treatment at 47.5 to 48.5°C and prochloraz was found effective in the control of stem-end rot and other fruit rots (Lay-Yee et al., 1998).

3.2.8 Chitosan and Isothiocyanates

Chitosan is a polymer derived from chitin present in crustaceans. It is de-acetylated form of N-acetyl glucosamine. Chitosan anti-microbial properties are well documented. Combination of chitosan with a papaya seed extract at 1.5% reduced anthracnose of papaya by 50% (Bautista-Baños et al., 2003). Lakshmi Marpudi et al. (2011) reported a reduction in postharvest diseases of papaya with a combination of chitosan, papaya leaf extract, and *Aloe* gel, while Sivakumar et al. (2005) reported that the combination of chitosan with ammonium carbonate and sodium carbonate reduced infections by *C. gloeosporioides.* In another research, anthracnose disease was effectively controlled in vitro using chitosan treatment, whereas calcium at 2.5% combined with chitosan demonstrated the best effect in maintaining fruit firmness and controlling anthracnose incidence on papaya fruits (Al Eryani-Raquup et al., 2009). Ali et al. (2010) also studied both the in vitro and in vivo antifungal activity of chitosan against *C. gloeosporioides*. They found that chitosan at concentrations of 1.5 and 2.0% inhibited fungal mycelial growth in culture medium by 90–100%. In vivo studies showed that the application of 1.5 and 2.0% chitosan coatings to papaya fruits not only reduced fruit decay but also delayed the onset of disease symptoms by 3–4 wk during 5 wk of storage at 12°C and slowed down the subsequent disease development.

Ramos-García et al. (2010) observed that isothiocyanates (ITCs), which are present in plants belonging to families including Capparaceae, Brassicaceae, Koeberliniaceae, Moringaceae, Resedaceae, and Tovariaceae, were effective in inhibiting papaya anthracnose.

3.2.9 Modified and Controlled Atmospheres

Modified (MA) and controlled (CA) atmospheres used during storage help in minimizing the postharvest diseases in many crops. MA packaging (individual fruits) of 'Solo' papayas with low-density polyethylene (LDPE) or Pebax-C® film minimized fruit spoilage by alleviating chilling injury in fruits stored at 13° C for 30 d (Singh and Sudhakar Rao, 2005a). Similarly, individual shrink wrapping of 'Solo' papayas (pre-treated with 250 ppm of the fungicide prochloraz) with Cryovac® D-955 film alleviated chilling injury and reduced anthracnose (10%) in fruits stored at 13°C for 30 d followed by un-wrapping and ripening at 20°C (Singh and Sudhakar Rao, 2005b). Minimum decay in 'Solo' papaya fruits due to storage in a mixture of gases (5% CO_2 and 1% O_2, at 15.4°C) was reported in early work by Spalding and Reeder (1974). CA storage at 16°C for 35 d protected 'Tainung 1' papaya fruits (Maharaj and Sankat, 1990). Rohani and Zaipun (2007) also reported benefits from CA storage of 'Eksotika' papayas. However, de Oliveira et al. (2004) reported the adverse effect, an increased incidence of anthracnose and other fungal spots, of controlled atmosphere (6% CO_2 and 3% O_2 at 10°C) during storage of 'Sunrise Solo' and 'Golden' papaya fruits. González-Aguilar et al. (2003) observed an enhancement in the efficacy of methyl jasmonate (10^{-5} M at 20°C for 16 hr) when combined with a low-density polyethylene film to protect 'Sunrise' papayas from anthracnose and rot caused by *Alternaria* during storage at 10°C for 32 d.

Hypobaric storage has also been suggested as a CA system to extend storage life and reduce decay incidence in papaya. The exposure of papaya fruits to subatmospheric pressure (20 mm Hg at 10°C and 90–98% RH) for 18–21 d during shipment in hypobaric containers from Hawaii to Los Angeles and New York inhibited both fruit ripening and disease development (Alvarez, 1980). The post-hypobaric storage did not affect the ripening process after removal from the containers. Disease incidence was significantly reduced in the hypobaric-stored fruits; these fruits had 63% less peduncle infection, 55% less stem-end rot, and 45% fewer fruit surface lesions than those stored in a refrigerated container at normal atmospheric pressure (Alvarez, 1980). The inhibitory effect of hypobaric storage on disease development in papaya was further confirmed by Chau and Alvarez (1983). Fruits artificially inoculated with *C. gloeosporioides*, held at 15 mm Hg at 10°C for 3 wk, and then ripened at ambient conditions for 5 d showed less anthracnose than the control fruits. A significant delay in the infection progress on fruits under hypobaric conditions has been reported (Chau and Alvarez, 1983).

4 Conclusion

Most of the postharvest pathogens causing papaya decay infect the fruit in the field and develop disease during storage. Hence, more precautions need to be taken in the preharvest stage. The most important papaya postharvest pathogen *C. gloeosporioides* is a hemi-biotroph. It continues as biotroph after its infection of green fruit and goes into quiescent stage or latent phase until the fruit enters into the ripening phase. Once the fruit ripens, the pathogen shifts to the necrotrophic phase, where necrosis and rotting are initiated, causing maximum damage. Other weak pathogens infect the ripening fruit owing to ready availability of sugars. Though preharvest measures are important, substantial progress is yet to be made in this regard. However, considerable efforts have been devoted to the development of CA or MA use during storage and transit so as to minimize postharvest quality loss. Many chemical fungicides have been identified that are effective to control postharvest pathogens. Combination of preharvest measures with ecofriendly post-harvest approaches constitutes the best elements for the integrated management of postharvest diseases of papaya.

References

Al Eryani-Raquup, A., Mahmud, T.M.M., Syed-Omer, S.R., Mohamed-Zaki, A.R. and Al Eryani, A.R. 2009. Effect of calcium and chitosan treatments on controlling anthracnose and post harvest quality of papaya. *International Journal of Agricultural Research* 4, 53–68.

Alahakoon, P.W., Brown, A.E. and Sreenivasaprasad, S. 1994. Cross-infection potential of genetic groups of *Colletotrichum gloeosporioides* on tropical fruits. *Physiological and Molecular Plant Pathology* 44, 93–103.

Ali, A., Hei, G.H. and Keat, Y.W. 2016. Efficacy of ginger oil and extract combined with gum arabic on anthracnose and quality of papaya fruit during cold storage. *Journal of Food Science and Technology* 53, 1435–1444.

Ali, A., Muhammad, M.T.M., Sijam, K. and Siddiqui, Y. 2010. Potential of chitosan coating in delaying the postharvest anthracnose (*Colletotrichum gloeosporioides* Penz.) of Eksotika II papaya. *International Journal of Food Science and Technology* 45, 2134–2140.

Ali, A., Tan, W.P. and Mustafa, M.A. 2015. Application of lemongrass oil in vapour phase for the effective control of anthracnose of 'Sekaki' papaya. *Journal of Applied Microbiology* 118, 1456–1464.

Allong, R., Wickham, L. and Majeed Mohammed, D. 2000. Effect of cultivar, hot water treatment and storage conditions on quality of fresh-cut papaya (*Carica papaya* L.). *Journal of Applied Horticulture* 2, 15–18.

Alvarez, A.M. 1980. Improved marketability of fresh papaya by shipment in hypobaric containers. *HortScience* 15, 517–518.

Alvarez, A.M. and Nishijima, W.T. 1987. Postharvest diseases of papaya. *Plant Disease* 71, 681–686.

Anonymous. 2017. *Horticultural Statistics at a Glance 2017.* Horticulture Statistics Division, Department of Agriculture, Cooperation & Farmers Welfare, Ministry of Agriculture & Farmers Welfare, Government of India. New Delhi, India. p. 514.

Aragaki, M., Kimoto, W.S. and Uchida, J.Y. 1981. Limitations of hot water treatment in the control of Phytophthora fruit rot of papaya. *Plant Disease* 65, 744–745.

Ark, P.A. and Thompson, J.P. 1959. Control of certain diseases of plant with antibiotics and garlic (*Allium sativum* L.). *Plant Disease Reporter* 43, 276.

Arx, J.A. 1957. Die arten der gattung Colletotrichum Cda. *Phytopathologische Zeitschrift* 29, 413–468.

Arya, A., Lal, B., Agarwal, R. and Srivastava, R.C. 1986. Some new fruit rot diseases II: Symptomatology and host range. *Journal of the Indian Botanical Society* 69, 269.

Ashoka, S. 2005. Studies on fungal pathogens of vanilla with special reference to *Colletotrichum gloeosporioides* (Penz.) Penz. and Sacc. M.Sc. (Agri.) Thesis, University of Agricultural Sciences, Dharwad, Karnataka, India.

Baños-Guevara, P.E., Zavaleta-Mejía, E., Colinas-León, M.T., Luna-Romero, I. and Gutiérrez-Alonso, J.G. 2004. Control biológico de *Colletotrichum gloeosporioides* [(Penz.) Penz. y Sacc.] en papaya Maradol roja (*Carica papaya* L.) y fisiología postcosecha de frutos infectados. *Mexican Journal of Phytopathology* 22, 198–205. (In Spanish with English abstract).

Barrera-Necha, L.L., Bautista-Baños, S., Flores-Moctezuma, H.E. and Estudillo, A.R. 2008. Efficacy of essential oils on the conidial germination, growth of *Colletotrichum gloeosporioides* (Penz.) Penz. and Sacc. and control of postharvest diseases in papaya (*Carica papaya* L.). *Plant Pathology Journal* 7, 174–178.

Bautista-Baños, S., Hernández-López, M., Bosquez-Molina, E. and Wilson, C.L. 2003. Effect of chitosan and plant extracts on growth of *Colletotrichum gloeosporioides*, anthracnose levels and quality of papaya fruit. *Crop Protection* 22, 1087–1092.

Bautista-Baños, S., Sivakumar, D., Bello Pérez, A., Villanueva Arce, R. and Hernández-López, M. 2013. A review of the management alternatives for controlling fungi on papaya fruit during the postharvest supply chain. *Crop Protection* 49, 8–20.

Bosquez-Molina, E., Ronquillo-de Jesús, E., Bautista-Baños, S., Verde-Calvo, J.R. and Morales-López, J. 2010. Inhibitory effect of essential oils against *Colletotrichum gloeosporioides* and *Rhizopus stolonifer* in stored papaya fruit and their possible application in coatings. *Postharvest Biology and Technology* 57, 132–137.

Cappellini, R.A., Ceponis, M.J. and Lightner, G.W. 1988. Disorders in apricot and papaya shipments to the New York market, 1972–1985. *Plant Disease* 72, 366–368.

Chau, K.F. and Alvarez, A.M. 1979. Role of *Mycosphaerella ascospores* in stem-end rot of papaya fruit. *Phytopathology* 69, 500–503.

Chau, K.F. and Alvarez, A.M. 1983. Postharvest fruit rot of papaya caused by *Stemphylium lycopersici. Plant Disease* 67, 1279–1281.

Chauhan, H.L. and Joshi, H.U. 1990. Evaluation of phyto-extracts for control of mango fruit anthracnose. Proceedings of National Symposium: Botanical Pesticides in Integrated Pest Management. Rajahmundry, India. pp. 455–459.

Chavan, M. 1996. Effect of plant extracts on some important pathogenic fungi of Konkan region. M.Sc. (Agri.) Thesis, KKV, Dapoli, Maharashtra, India.

Chowdhury, S. 1950. A fruit rot of papaya caused by *Ascochyta caricae* Pat. *Transactions of British Mycological Society* 33, 317–322.

Cia, P., Pascholati, S.F., Benato, E.A., Camili, E.C. and Santos, C.A. 2007. Effects of gamma and UV-C irradiation on the postharvest control of papaya anthracnose. *Postharvest Biology and Technology* 43, 366–373.

Couey, H.M., Alvarez, A.M. and Nelson, M.G. 1984. Comparison of hot water spray and immersion treatments for control of postharvest decay of papaya. *Plant Disease* 68, 436–437.

Couey, H.M. and Farias, G. 1979. Control of postharvest decay of papaya. *Horticultural Science* 14, 719–721.

Couey, H.M. and Hayes, C.F. 1986. Quarantine procedures for Hawaiian papaya using fruit selection and a two–stage hot–water immersion. *Journal of Economic Entomology* 79, 1307–1314.

de Capdevile, G., Texeira Jr., S.M., Pereira, S.J.R., de Paula, M.S., Rodrigues, C.A. and Goncalves, T.F.A. 2007. Selection and testing of epiphytic yeasts to control anthracnose in post-harvest of papaya fruit. *Scientia Horticulturae* 111, 179–185.

de Oliveira, M.J., Rocha, L.N. and Agosthinho, C.S. 2004. Padrao de ocorrencia de doenças em mamao armazenado sob atmosfera controlada. *Revista Brasileria de Fruticultura* 26, 547–549. (In Portuguese with English abstract).

Debysingh, N., Wickham, L.D., Mohammed, M., Legall, G., Paliyath, G. and Subramanian, J. 2018. Effects of pre- and post-harvest treatments with hexanal formulations on time to ripening and shelf life of papaya (*Carica papaya* L.) fruits. *Tropical Agriculture* 95, 36–41.

Deshmukh, P.P. and Raut, J.G. 1992. Antagonism by *Trichoderma* spp. on five plant pathogenic fungi. *New Agriculturist* 3, 127–130.

Eckert, J.W. and Ogawa, J.M. 1985. The chemical control of postharvest diseases: Subtropical and tropical fruits. *Annual Review of Phytopathology* 23, 421–454.

Escopalao, W.M. and Silvestre, J.C. 1996. Evaluation of fifteen medicinal plants extracts for fungicidal property against *Colletotrichum gloeosporioides* Penz. causing anthracnose on mango fruits. *Philippine Phytopathology* 32, 130.

FAOSTAT (Food and Agriculture Organization of the United Nations, Statistics Division). 2017. Food and agricultural commodities production. Available at: www.fao.org/faostat/en/#data/QC

Fatima, S., Javeed, Z. and Ade, A. 2006. *Post-Harvest Rots of Fruits*. Discovery Publishing House, New Delhi, India.

Gamagae, S.U., Sivakumar, D. and Wijesundera, R.L.C. 2004. Evaluation of post-harvest application of sodium bicarbonate–incorporated wax formulation and *Candida oleophila* for the control of anthracnose of papaya. *Crop Protection* 23, 575–579.

Gamagae, S.U., Sivakumar, D., Wilson-Wijeratnam, R.S. and Wijesundra, R.L.C. 2003. Use of sodium bicarbonate and *Candida oleophila* to control anthracnose in papaya during storage. *Crop Protection* 22, 775–779.

Godara, S.L. and Pathak, V.N. 1995. Effect of plant extracts on postharvest rotting of sweet orange fruit. Proceedings of the Global Conference on Advances in Research in Plant Diseases and Their Management. RCA, Udaipur, Rajasthan, India. p. 172.

González-Aguilar, G.A., Buta, J.G. and Wang, C.Y. 2003. Methyl jasmonate and modified atmosphere packaging (MAP) reduce decay and maintain postharvest quality of papaya 'Sunrise'. *Postharvest Biology and Technology* 28, 361–370.

González-Aguilar, G.A., Valenzuela-Soto, E., Lizardi-Mendoza, J., Goycoolea, F., Martínez-Téllez, M.A., Villegas-Ochoa, M.A., Monroy-García, I.N. and Ayala-

Zavala, J.F. 2008. Effect of chitosan coating in preventing deterioration and preserving the quality of fresh-cut papaya 'Maradol'. *Journal of the Science of Food and Agriculture* 89, 15–23.

Gud, M.B. 2001. Studies on fungal diseases of mango (*Mangifera indica* L.) fruits. M. Sc. (Agri.) Thesis, KKV, Dapoli, Maharashtra, India.

Hartill, W.F.T. 1992. Post-harvest rots of avocados in New Zealand and their control. Proceedings of the Brighton Crop Protection Conference, Brighton, UK. pp. 1157–1162.

Henriod, R., Diczbalis, Y., Sole, D., Stice, K.N. and Tora, L. 2016. Investigation into various fungicides and alternative solutions for controlling postharvest diseases in papaya fruit. *Acta Horticulturae* 1111, 113–119.

Hunter, J.E. and Buddenhagen, I.W. 1972. Incidence, epidemiology and control of fruit diseases of papaya in Hawaii. *Journal of Tropical Agriculture (Trinidad)* 49, 61–72.

Jetti, A., Iyengar, D.S., Rao, S.N. and Bhalerao, U.T. 1987. Antifungal spectrum of leaf extracts of *Polyalthia longifolia*. *Pesticides* 20, 29.

Kamal, M. and Agbari, A.A.A. 1985. *Manual of Plant Diseases in the Yemen Arab Republic*. Precision Press, London, UK.

Lakshmi Marpudi, S., Abirami, L.S.S., Pushkala, R. and Srividya, N. 2011. Enhancement of storage life and quality maintenance of papaya fruits using *Aloe vera* based antimicrobial coating. *Indian Journal of Biotechnology* 10, 83–89.

Lay-Yee, M., Clare, G.K., Petry, R.J., Fullerton, R.A. and Gunson, A. 1998. Quality and disease incidence of 'Waimanalo Solo' papaya following forced-air heat treatments. *HortScience* 33, 878–880.

Li, X., Zhu, X., Zhao, N. and Fu, D. 2013. Effect of hot water treatment on anthracnose disease in papaya fruit and its possible mechanism. *Postharvest Biology and Technology* 86, 437–446.

Madani, B., Mohamed, M.T.M., Biggs, A.R., Kadir, J., Awang, Y., Tayebimeigooni, A. and Roodbar-Shojaei, T. 2014. Effect of pre-harvest calcium chloride applications on fruit calcium level and post-harvest anthracnose disease of papaya. *Crop Protection* 55, 55–60.

Maharaj, R. and Sankat, C.K. 1990. Storability of papayas under refrigerated and controlled atmosphere. *Acta Horticulturae* 269, 375–385.

Mahmud, T.M.M., Eryani-Raqeeb, A.A., Sayed-Omar, S.R., Zaki, A.R.M. and Al Eryani, A.R. 2008. Effects of different concentrations and applications of calcium on storage life and physiochemical characteristics of papaya (*Carica papaya* L.). *American Journal of Agricultural and Biological Sciences* 3, 526–533.

Majumdar, V.L. and Pathak, V.N. 1995. Biocontrol of postharvest diseases of guava fruits by *Trichoderma* spp. *Acta Botanica Indica* 23, 263–267.

Maqbool, M., Ali, A., Alderson, P.G., Mohamed, M.T.M., Siddiqui, Y. and Zahid, N. 2011. Postharvest application of gum arabic and essential oils for controlling anthracnose and quality of banana and papaya during cold storage. *Postharvest Biology and Technology* 62, 71–76.

Mederios, S.A.F. and Menezes, M. 1994. Antagonistic potential of some fungi to *Colletotrichum gloeosporioides* causal agent of cashew tree (*Anacardium occidentale*) anthracnose. *Fitopatologia Brasileira* 19, 84–89.

Meirelles, R.D.A.P. and Melges, W.J.M. 2004. Gamma radiation in papaya harvested at three stages of maturation. *Scientia Agricola* 61, 146–150.

Nelson, M.N. and Alvarez, A.M. 1980. Purple stain of *Carica papaya*. *Plant Disease* 64, 93–95.

Nishijima, K.A., Miura, K., Armstrong, J.W., Brown, S.A. and Hu, B.K.S. 1992. Effect of forced, hot-air treatment of papaya fruit on fruit quality and incidence of postharvest diseases. *Plant Disease* 76, 723–727.

Nunes, M.C.N., Emond, J.P. and Brecht, J.K. 2006. Brief deviations from set point temperatures during normal airport handling operations negatively affect the quality of papaya (*Carica papaya*) fruit. *Postharvest Biology and Technology* 41, 328–340.

Ong, M.K., Kazi, F.K., Forney, C.F. and Ali, A. 2013. Effect of gaseous ozone on papaya anthracnose. *Food Bioprocess Technology* 6, 2996–3005.

Osman, S.M., Sivakumar, D. and Korsten, L. 2011. Effect of biocontrol agent *Bacillus amyloliquefaciens* and 1-methyl cyclopropene on the control of postharvest diseases and maintenance of fruit quality. *Crop Protection* 30, 173–178.

Papavizas, G.C. 1973. Status of biological control of soil borne pathogen. *Soil Biology and Biochemistry* 5, 709.

Patel, K.D. and Joshi, K.R. 2001. Antagonistic effect of some bioagents in vitro against *Colletotrichum gloeosporioides* Penz. and Sacc. the causal agent of leaf spot of turmeric. *Journal of Mycology and Plant Pathology* 31, 126.

Patel, K.D. and Joshi, K.R. 2002. Efficacy of different fungicides against *Colletotrichum gloeosporioides* Penz. and Sacc. the causing leaf spot of turmeric. *Journal of Mycology and Plant Pathology* 32, 413–414.

Pathak, V.N., Goyal, J.P. and Bhatanagar, L.G. 1976. Effect of chemical and hot water treatment on *Fusarium* and *Rhizopus* rots of papaya. *Indian Phytopathology* 29, 210–211.

Paull, R.E., Nishijima, W., Reyes, M. and Cavaletto, C. 1997. Postharvest handling and losses during marketing of papaya (*Carica papaya* L.). *Postharvest Biology and Technology* 11, 165–179.

Potebnia, A. 1910. Beitage Dar. Micromycetam flora miltel Russlands. *Annals of Mycology* 8, 42–98.

Prasad, S.S. and Verma, R.A.B. 1970. Fungal diseases of papaya fruit in Bihar. *Indian Phytopathology* 23, 722–724.

Punithalingam, E. 1980. A combination in *Phoma* for *Ascochyta caricae-papayae*. *Transactions of British Mycological Society* 75, 340.

Raheja, S. and Thakore, B.B.L. 2002. Effect of physical factor, plant extracts and bio-agent on *Colletotrichum gloeosporioides* Penz., the causal organism of anthracnose of Yam. *Journal of Mycology and Plant Pathology* 32, 293–294.

Ramallo, A.C., Olmedo, G.M., Ramallo, J., Cerioni, L. and Rapisarda, V.A. 2019. Effectiveness of an ametoctradin-dimethomorph formulation to control brown rot on postharvest lemons. *Scientia Horticulturae* 246, 574–577.

Ramos-García, M., Bautista-Baños, S., Troncoso-Rojas, R., Bosquez-Molina, E., Alia-Tejacal, I., Guillén-Sánchez, D. and Gutiérrez-Martínez, P. 2010. Papaya postharvest handling in México: Use of chitosan and isothiocynanates to control postharvest diseases. *Fresh Produce* 4 (Special Issue 1), 21–28.

Ratnam, C.V. and Neema, K.G. 1967. Studies on market diseases of fruits and vegetables. *The Andhra Agricultural Journal* 14, 60–65.

Rocha, J.R.S. and de Oliveira, N.T. 1998. Biocontrol of *Colletotrichum gloeosporioides* anthracnose agent on passion fruits (*Passiflora duulis*) with *Trichoderma koningi*. *Phytopathologica* 24, 272–275.

Rohani, M.Y. and Zaipun, M.Z. 2007. MA storage and transportation of 'Eksotika' papaya. *Acta Horticulturae* 740, 303–311.

Saccardo, P.A. 1884. Sylloge Fungorum: Sylloge Sphaeropsidearum et Melanconiearum. *Sylloge Fungorum* 3, 1–840.

Santha Kumari, P. 2002. Biocontrol of anthracnose of black pepper. *Journal of Mycology and Plant Pathology* 32, 358.

Sarkhosh, A., Schaffer, B., Vargas, A.I., Palmateer, A.J., Lopez, P., Soleymani, A. and Farzaneh, M. 2017. Antifungal activity of five plant-extracted essential oils against anthracnose in papaya fruit. *Biological Agriculture & Horticulture* 34, 18–26.

Sarwar, M. and Kamal, M. 1971. Studies on fruit rot of papaya caused by *Rhizopus oryzae*. *Pakistan Journal of Scientific and Industrial Research* 14, 234–236.

Sattar, A. and Malik, S.A. 1939. Some studies on anthracnose of mango caused by *Glomerella cingulate* (Stonem). Spauld. Sch. *Colletotrichum gloeosporioides* Penz. *Indian Journal of Agricultural Sciences* 1, 511–521.

Sharma, R.R., Singh, D. and Singh, R. 2009. Biological control of postharvest diseases of fruits and vegetables by microbial antagonists: A review. *Biological Control* 50, 205–221.

Shekhawat, P.S. and Prasad, R. 1971. Antifungal properties of some plant extracts in inhibition of spore germination. *Indian Phytopathology* 24, 800–802.

Shi, J., Liu, A., Li, X., Feng, S. and Chen, W. 2009. Identification of endophytic bacterial strain MGP1 selected from papaya and its biocontrol effects on pathogens infecting harvested papaya fruit. *Journal of Food Science and Agriculture* 90, 227–232.

Shirshikar, G.S. 2002. Studies on fruit rots of mango (*Mangifera indica* L.) caused by *Botryodiplodia theobromae* Pat. and *Colletotrichum gloeosporioides* Penz. and their management. M.Sc. (Agri) Thesis, KKV, Dapoli, Maharashtra, India.

Shukla, D.N., Bhargava, S.N. and Singh, N.K. 1978. A note on stalk rots of papaya on fruits. *Indian Journal of Horticulture* 5, 282–283.

Simmonds, J.H. 1965. A study of the species of *Colletotrichum* causing ripe fruit rots in Queensland. *Queensland Journal of Agricultural and Animal Sciences* 22, 437–459.

Singh, I.D. 1990. *Papaya*. Oxford and IBH Publishing Company Private Limited, New Delhi, India.

Singh, S.P. and Sudhakar Rao, D.V. 2005a. Effect of modified atmosphere packaging on the alleviation of chilling injury and dietary antioxidants levels in 'Solo' papaya during cold storage. *European Journal of Horticultural Science* 70, 246–252.

Singh, S.P. and Sudhakar Rao, D.V. 2005b. Quality assurance of papaya by shrink film wrapping during storage and ripening. *Journal of Food Science and Technology* 42, 523–525.

Singh, S.P. and Sudhakar Rao, D.V. 2011. Papaya (*Carica papaya* L.). pp. 86–124. *In*: E.M. Yahia (ed.). *Postharvest Biology and Technology of Tropical and Subtropical Fruits, Volume 4: Mangosteen to White Sapote*. Woodhead Publishing Limited, Cambridge, UK.

Sivakumar, D., Hewarathgamagae, N.K., Wijeratnam, R.S.W. and Wijesundera, R.L.C. 2002. Effect of ammonium carbonate and sodiumbicarbonate on anthracnose of papaya. *Phytoparasitica* 30, 1–7.

Sivakumar, D., Sultanbawa, Y., Ranasingh, N., Kumara, P. and Wijesundera, R.L.C. 2005. Effect of the combined application of chitosan and carbonate salts on the incidence of anthracnose and on the quality of papaya during storage. *The Journal of Horticultural Science and Biotechnology* 80, 447–452.

Spalding, D.H. 1982. Resistance to mango pathogens to fungicides used to control postharvest diseases. *Plant Disease* 66, 1185–1186.

Spalding, D.H. and Reeder, W.P. 1974. Current status of controlled atmosphere storage of four tropical fruits. *Proceedings of the Florida State Horticultural Society* 87, 334–337.

Spotts, R.A. and Cervantes, L.A. 1986. Populations, pathogenicity and benomyl resistance of *Botrytis* spp., *Penicillium* spp., *Mucor piriformis* in packing houses. *Plant Disease* 70, 106.

Sriram, S. and Poornachandra, S.R. 2013. Biological control of postharvest mango fruit rot caused by *Colletotrichum gloeosporioides* and *Diplodia natalensis* with *Candida tropicalis* and *Alcaligenes feacalis*. *Indian Phytopathology* 66, 375–380.

Srivastava, M.P., Tandon, R.N., Bilgrami, K.S. and Ghosh, A.K. 1964. Studies on fungal diseases of some tropical fruits. I. A list of fungi isolated from fruits and fruit trees. *Phytopathology* 50, 250–261.

Stanghellini, M.E. and Aragaki, M. 1966. Relation of periderm formation and callose deposition to anthracnose resistance in papaya fruit. *Phytopathology* 56, 444–450.

Suseela Bhai, R., Ishwara Bhat, A. and Anandaraj, M. 2003. Premature yellowing and bean shedding in vanilla (*Vanilla planifolia* Andrews). Proceedings of the Symposium on Resent Development in the Diagnosis and Management of Plant Diseases for Meeting Global Challenges. Dharwad, Karnataka, India. pp. 88–89.

Sutton, B.C. 1980. *The Coelomycetes. Fungi Imperfecti with Pycnidia, Acervuli and Stromata.* Commonwealth Mycological Institute, Kew, Surrey, UK.

Tandon, R.N. 1967. Final technical report PL 480 scheme (FG-IN-133). University of Allahabad, Uttar Pradesh, India.

Tandon, R.N. and Mishra, A.N. 1969. Fruit rot diseases of *Carica papaya* and *Musa parasitica* caused by *Rhizopus stolonifer*. *Indian Phytopathology* 22, 334–341.

Tewari, D.K., Srivastava, R.C., Katiyar, N. and Lal, B. 1988. Chemical control of *Thielaviopsis* rots of papaya. *Indian Phytopathology* 41, 491–492.

Trujillo, E.E. and Hine, R.B. 1965. The role of papaya residues in papaya root rot caused by *Pythium aphanidermatum* and *Phytopthora parasitica*. *Phytopathology* 55, 1293–1298.

Villegas, V.N. 1997. *Carica papaya* L. p.446. *In*: E.W.M. Verheij and R.E. Coronel (eds.). *Plant Resources of South-East Asia, No. 2: Edible Fruits and Nuts.* Prosea Foundation, Bogor, Indonesia..

Washathi, A.D. and Bhargava, P.K. 2000. Effect of blight (*Colletotrichum dematium*) on seed yield of chickpea (*Cicer arietum*). *Indian Journal of Agricultural Sciences* 20, 45–47.

Wilson, C.L. and Wisniewski, M.S. 1989. Biological control of postharvest diseases of fruits and vegetables: An emerging technology. *Annual Review of Phytopathology* 27, 425–441.

Yaguchi, Y. and Nakamura, S. 1993. Effect of hot-water and vapor-heat treatments on the control of stem-end rot of papaya. *Japanese Journal of Tropical Agriculture* 37, 167–170.

Chapter 8

Banana and Plantain

Dionisio G. Alvindia

Philippine Center for Postharvest Development and Mechanization (PHilMech), Science City of Muñoz, Nueva Ecija, Philippines

College of Science Center for Natural Sciences and Environmental Research (CENSER), De La Salle University, Malate Manila, Philippines

Abbreviations

LDPE	Low-density polyethylene
RH	Relative humidity
NCA	Nonchemical control approach
MA	Modified atmosphere
CA	Controlled atmosphere
HWT	Hot water treatment
UV-C	Ultraviolet-C
MCA	Microbial control agent
CE	Cinnamon extracts

1 Introduction

The climacteric nature of banana and plantain (*Musa* x *paradisiaca*) (Musaceae) makes them fragile and highly perishable. Owing to this inherent property, the primary goals of banana research scientists are to reduce losses in quantity and quality and to maintain safety between harvest and consumption. Determining and maintaining the best fruit maturity at harvest, fruit care, postharvest treatment and extra care in handling during transport and packing are some of the significant requirements to obtain high quality fruits. Preharvest factor likewise plays a major role in the performance of banana fruit after harvest, for producing quality fruits, i.e. large size, blemish-free, long green life, absence of postharvest diseases, etc. Some of the preharvest factors that significantly affect the postharvest performance of banana fruit are cultivar, climatic and soil conditions, cultural management, the number of functional leaves on the plant at fruit development, and postharvest practices.

'Cavendish' is the popular export banana cultivar coming mainly from developing countries in Latin America, the West Indies, Southeast Asia, and Africa. Significant losses, however, occur mainly during handling and transport to final market destinations and also because of fruit decay due to postharvest diseases. For instance, 86 and 83% of pesticide-free banana from the Philippines are infected with crown rot and anthracnose, respectively (Alvindia et al., 2000a). Such high incidence of postharvest diseases negatively impacts the market value of bananas because the consumers refuse to buy such fruit. In most cases, the country of destination refuses to accept fruit with high incidence of decay/contamination and immediately sends back the cargo to the country of origin. This chapter emphasizes the interrelated relationships of cultural, production, and postharvest practices in preserving the overall quality of banana fruit.

2 Factors of Consideration at Harvest

2.1 Fruit Maturity and Harvest Index

Fruit maturity is an important postharvest criterion for banana because the stage of maturation greatly influences the green life, shelf life, and eating quality. Every fruit attains its fully positive characteristics(e.g. flavor, taste, and color) if harvested at optimum time. Fruits harvested at an advanced stage of maturity are not suitable for long-distance transportation owing to their shorter storage life (Kader, 1994).

Standards of fruit maturity depend on the market destinations wherein a period of green life is required by the fruit before ripening takes place. Basically, fruit intended for local markets are harvested at full maturity for immediate ripening and marketing, while those for export markets, requiring long distance transport mainly by ship, are harvested at 75% maturity (Morton, 1987). Another maturity index is the general appearance of the fruit. For local markets, banana fruits are harvested when the ridges on the surface of fingers change from angular to round. For export destinations, fruits are harvested with the three-quarters round index, referring to fruit with pronounced ridges but with convex planes between them. The accuracy and consistency of the three-quarter round index is difficult to maintain in subtropical countries. Several different external and internal fruit characteristics can be used to determine plantain maturity. These include fruit diameter, age of the bunch, angularity of the fruit, length of the fruit, and peel color (Johnson et al., 1998).

Another method for estimating banana and plantain maturity is to record the age of the bunch after flowering. The time the fruit bunch first becomes visible (shooting) is recorded. Bunches can be tagged with different-colored ribbons during their shooting period, and subsequently harvested after the appropriate time for the particular cultivar, based on the season of the year and prior experience. The color of the ribbons is changed weekly to coincide with the time of shooting and subsequently with the age of the bunch (Johnson et al., 1998).

One very useful fruit-harvesting criterion used commercially is the age of the bunch after emergence from the pseudo stem, which refers to the day in which the first complete hand of fruit is visible. In some plantations, banana bunches are marked with color straps according to weeks of age, cushioning each hand from the other by using plastic film, paper, or other soft, non-abrasive materials such as leaves. The maturity of hands in a bunch varies at a given age, with those hands at the proximal end of the stem being more mature than those at the distal end. Generally, maturity of the entire stem is estimated using the second hand from the proximal end measuring the length and caliper grade. Many plantations in the Philippines use a combination of phenology and caliper measurement of finger diameter for detection of harvestable fruits.

2.2 Harvest Care

The whole fruit bunch must be protected with a cushioning material, such as a plastic bag or soft fabric, to avoid the latex dropping before cutting the stem. The bunch must be secured and pushed up by another worker while cutting is done to avoid bunch dropping or touching the ground. If possible, it is recommended that banana bunches must be individually covered with plastic until dehanding begins in the packinghouse to minimize mechanical damage and to protect the bunches from latex. If this is not possible, the whole bunch should be carried with care using soft fabric material, as padding, thus avoiding the contact of the latex with the peel. Soft fabric tissue stained with latex, however, becomes rough and abrasive to banana peel. Each bunch of banana is individually placed upright onto a trailer that is padded. Padding is also placed on each side to prevent any rubbing between bunches.

Some measures are done to reduce desiccation and the evapotranspiration rate of the fruit to prolong freshness and shelf life for a couple of days. These measures include keeping bunches away from direct sunlight, stacking bunches under shades that shield them from the sun, and protecting piles of fruit by covering them with leafs of banana or any fabric material regularly moistened with water (Alvindia et al., 2000b).

3 Factors Affecting Development of Postharvest Diseases

3.1 Inspection

It is foremost that the fruits are inspected for finger fullness and length, blemishes, scars, insect attacks, and decay upon arrival to the packinghouse. To assure compliance with maximum quality requirements, only fruits that meet these requirements are packed and those with lower quality can be sent for local market or processing (Ahmad and Siddiqui, 2016).

3.2 Washing

Fruits in bunches arriving the packing stations are usually hung on a metallic hook, if present, or carefully moved in a flat table for dehanding. Subsequently, the dehanded fruits are unloaded in a tank filled with water. During unloading, water is useful to protect each hand from rubbing together and for washing of the entire hands. A latex coagulator is usually added in the first water tank. A fungicide and a healing-bleach substance are added to the second tank. Likewise, sodium hypochlorite is commonly added to water to reduce inoculum of decay organisms. Also, it is important not to fill the tanks to full capacity to avoid spilling water (Lassois et al., 2010).

If water is not available, bunches are gently unloaded over a flat surface covered with spongy, soft materials (e.g., cardboard, fabric, leaves). Foreign bodies such as leaves, dirt, impurities, or flowers are carefully removed. It is very important that the materials covering the table must be changed very often.

3.3 Dehanding/Clustering

Dehanding should be carried out with a small, clean, sharp, stainless steel, straight-bladed knife, making a smooth cut as close as possible to the stem. Once dehanded, any undersized, diseased, and damaged fingers are removed. After dehanding, the fruits are placed with the crown facing downwards onto a layer of leaves to allow latex to drain. Then the hand can be placed on a packing wheel or into a water through/conveyor system where it is sorted and graded for size and quality. When the hands are further parted or clustered into a group of fingers, extra care should be taken not to cause scars on the fingers and crown.

3.4 Packing

Bananas are packed as whole hands, part hands, or clusters in cardboard cartons with plastic liners. Plastic slip-sheets are used between full hands and absorbent paper is placed in the bottom of the carton. Cartons are staked onto pallets for ease of pickup and delivery for transport. To ensure a net weight of 13 kg each when cartons of bananas reach the markets, they are usually packed to a weight of 13.5 kg to 13.7 kg in order to allow for any weight loss.

3.5 Packaging

The arrangement of fruits in a box has to be horizontal in two rows, keeping the crown-end towards the box side and the fruit tip towards the center of the box. While packing in single layer, the hands should be placed in the vertical position by keeping their tips up and crown downside. Cushioning pads or paper should be placed at the bottom of the box and fruits may be covered in low-density polyethylene (LDPE) liners of 0.025 mm inside the box to create a modified atmosphere (MA). The benefits of film packaging include ease of handling (consumer package); protection from injuries; reduction in water loss, shrinkage, and wilting; reduction in decay by MA; reduction in physiological disorders (chilling injury); and retardation of ripening and senescence processes. Harmful effects of film packaging include enhancement of decay due to excess humidity and generation of off-flavors and/or off-odors (El-Ramady et al., 2015).

Packaging can influence temperature management and water loss of the produce. To reduce water loss, plastic wraps and liners may be used, but restricted ventilation can result in problems of low oxygen, high carbon dioxide, or an accumulation of water. This can be improved by perforation of the films. Ethylene absorbers could be used to extend green life and shelf life of fruit at ambient temperature. Their use has considerable economic importance in countries where cold storage is not readily available or expensive.

3.6 Transport and Ripening

A delay in ripening of green bananas is accomplished by a controlled temperature of 13 to 14°C and a relative humidity (RH) of 85 to 95% during shipping and storage. Lower temperatures cause chilling injury that may result in dull, gray skin color, poor ripening and poor conversion of starch to sugars, poor flavor development, and an increased susceptibility to decay. Generally, fruits are kept at 13.3 to 15.6°C and 80 to 85% RH after removal from storage and during delivery to markets to avoid rapid spoilage. Increased fruit weight loss (or loss of water), ethylene production, and cellular respiration are triggered by lower RH. Temperatures above 25°C shorten the duration of the pre-climacteric stage and fruit quality is altered because of modification of metabolism during ripening.

4 Postharvest Diseases and Their Symptoms

Banana is affected by several fungal diseases at all stages of its life. Being highly perishable, losses are attributed mainly to postharvest diseases, which cause serious qualitative and quantitative losses of fruits. Fungal-decayed fruits have no market value. There are many postharvest diseases of banana and plantain, and the most economically important include crown rot, finger rot, anthracnose, and cigar-end rot.

4.1 Crown Rot

It is the most important postharvest disease of banana. It is a disease complex and can be caused by several fungi, including *Lasiodiplodia theobromae* (Pat.) Griffon & Maubl. (Ogawa, 1970; Johanson and Blazquez, 1992; Alvindia et al., 2000b), *Colletotrichum musae* (Berk. & M.A. Curtis) Arx (Finlay and Brown, 1993), *Thielaviopsis paradoxa* (De Seynes) Höhn (Alvindia et al., 2002), and a complex of *Fusarium* spp. (Knight et al., 1977; Jimenez et al., 1993; Alvindia et al., 2000b, 2002; Hirata et al., 2001). Recent studies showed that the fungus *Fusarium musae* Van Hove, Waalwijk, Munaut, Logrieco & Moretti is frequently found associated with banana crown rot (Molnár et al., 2015; Kamel et al., 2016). The species was a separate, sister species of *Fusarium verticillioides* (Sacc.) Nirenberg sensu stricto, which is also frequently found associated with banana crown rot (Van Hove et al., 2011; Molnár et al., 2015; Kamel et al., 2016). To date, *F. musae* has been isolated from banana produced by several Latin-American countries (Mexico, Panama, Ecuador, etc.), the Canary Islands (Spain), and the Philippines, but not from banana produced by African countries (Van Hove et al., 2011; Kamel et al., 2016; Triest et al., 2016). *Fusarium musae* was a human pathogen with keratitis cases (i.e., eye infections) from the multistate, contact-lens-associated outbreak in the United States as well as superficial infections such as sinusitis (Triest and Hendrickx, 2016).

The fungi infect the crown through fresh wounds created after trimming the crown of the banana hand into a crescent shape. When the hands are cut from the stems, the massive open wound is an ideal weak spot for crown rot fungi to enter and grow. Additionally, the wounds resulting from the clustering of banana hands are the main portal of infection by crown rot pathogens because more surface area of the crown has open wounds. Fungal spores present on the fruit in the field are carried (after bunch harvesting) into the packinghouse. Spores follow the fruit into de-latexing baths, where they are drawn deeply into the vulnerable wound in the crown tissue exposed by dehanding. Symptoms of crown rot include softening and blackening of tissues at the cut crown surface; white, gray, or pink mold may form on the surface of the cut crown, infected tissue within the crown and fruit that turns black, and the rot may advance into the finger stalk and fingers (Figure 8.1a). When the hands are further parted or clustered into a group of fingers, the crown area would have more fresh cut sections, exposing more vulnerable potential infection sites for crown-rot-causing pathogens (Figure 8.1b).

Figure 8.1 a. Blackening of tissues at the crescent-shaped crown surface with mycelial growth produced on the surface of the cut crown. The infected tissue has turned black and decay has advanced into the finger stalk (finger stalk rot) and fingers (finger rot). The disease-causing agent is a complex of pathogens, including *Colletotrichum musae, Fusarium verticillioides, Lasiodiplodia theobromae,* and *Thielaviopsis paradoxa.* b. Banana clusters have a large fresh cut surface area that provides infection opportunities for crown-rot-causing pathogens. c. Details of necrotic finger stalk rot infections occurred through injuries caused by flexing or force exerted upwards or downwards. These injuries may occur during harvest, packing, or transport.

4.2 Finger Stalk Rot

A rot of the fruit pedicel is caused by fungal invasion through damaged tissue of the stalk. The disease can be caused by fungi that may or may not be the same ones as the crown-rot pathogens. The pedicels become necrotic, usually starting in green life when the pedicels or stalks are injured through flexing or force exerted upwards or downwards, and these injuries may occur in harvest, packing, or transport (Figure 8.1c). In the early days of banana trade, finger stalk rot was an important problem because the fruits were transported on whole bunches and the pedicels were susceptible to rot as a result of mechanical injuries (Meredith, 1971). The disease has been given many names such as black end, finger drop, neck rot, and finger stalk rot. If the disease originates from the surface of the cut crown, formation of white, gray, dark, or pink mycelia can occur. Soft, infected tissue turns black and can extend into the finger stalks, where an external fungal mycelium may form. The whole pedicel is attacked and so weakened that the fingers drop when handled. As is the case with crown rot, finger stalk rot is a disease complex caused largely by *T. paradoxa* and *C. musae* (Alvindia et al., 2000b, 2002).

4.3 Finger Rot

Finger rot normally begins at the tip of one of the fingers, peel wound sites, or blemishes. Within few days, the entire finger may be soft, dark brown, and pimples with minute black pycnidia become apparent (Williamson and Tadon, 1966). Rotting speed depends on the causal pathogen. *Lasiodiplodia theobromae* and *T. paradoxa* are the most active causal pathogens of finger rot. Decay symptoms of infection in wounded fruits are dark brown to black spots visible within two days; few days later, the entire finger becomes soft with maceration of fruit pulp (Alvindia et al., 2002b) (Figure 8.1a). Furthermore, *Phomopsis* spp., *Fusarium oxysporum* Schltdl., and *F. verticillioides* can also be considered active pathogens on wounded banana. On the other hand, *L. theobromae* can infect and cause decay on unwounded fruits and it is capable of macerating the flesh of the fruit (Alvindia et al., 2002b).

4.4 Anthracnose

Anthracnose is a widespread disease of banana and plantain, especially in export markets with prolonged period of transport before the fruits are sold. Anthracnose is caused by the fungus *C. musae* and it harms fruit quality and marketability (Su et al., 2011). *Colletotrichum musae* is also one of the causal agents of crown rot and finger stalk rot of bananas (Alvindia et al., 2000b; Sangeetha et al., 2010) (Figure 8.2a). Symptoms of anthracnose appear on green banana fruits as brown to black diamond-shaped lesions. Orange or salmon-colored rings of spore masses may be observed on severely infected bananas. On yellowing fruits, brown spots initially appear, which later become sunken and covered with orange masses of spores that develop on the lesions (Figure 8.2b). At the onset, small, circular, black spots develop on the affected fruit. Then, these spots enlarge in size and turn brown in color. The infected skin or crown of the fruit turns black, shrivels, and becomes covered with pink spore masses. Finally, the disease affects the whole crown or finger and sometimes spreads to the whole bunch. Anthracnose infection can cause premature ripening and shriveling of the fruit, blackening/withering of the pedicel, and dropping of the fingers from the hands. On green fruit, the peel has dark-brown to black diamond-shaped lesions. The lesions are sunken with pale margins.

Usually, the inoculum of the causal pathogen adheres onto the intact fruit surface in the field, then produces appressoria and these may remain dormant during fruit development. When fruit approach maturity, the inoculum resumes growth and causes typical lesions on ripe fruit. High RH and temperature favor disease development. The conidia of the fungus survive in dead and decaying leaves and fruit, and are dispersed by water, wind, and insects. The conidia adhere to the fructoplane and start germinating when the fruit starts ripening. After harvest, new infection sites can occur through fruit blemishes and/or wounds in the peel

Figure 8.2 a. Symptoms of anthracnose caused by *Colletotrichum musae* on the crown and finger stalk of clustered bananas. b. Characteristic symptoms of anthracnose on yellowing fruit with formation of brown to black diamond-shaped lesions that become sunken and covered with orange spore masses.

incurred during handling and transport, or by fresh cuts in the crown during dehanding/clustering. Thereafter, conidia of the fungus germinate and initiate colonization and expression of symptoms.

4.5 Freckle

Freckle disease occurs on several species and varieties in Musaceae (Jones and Alcorn, 1982). The causal agent induces freckling on the fruit, causing a series of black, raised spots with a sand-paper-like texture; this is due to the protruding pycnidia and/or ascomata (Figure 8.3a,b). The causal agent of banana freckle is reported as *Guignardia musae* Racib. (Aa, 1973) and its anamorph as *Phyllosticta musarum* (Cooke) Aa (Aa, 1973; Aa and Vanev, 2002). A few widely scattered spots may be seen 2–4 wk after bunch emergence, or in some cases, dense aggregates appear in the form of streaks or circles. The severity of the disease increases as the fruit matures. Individual spots first appear as minute reddish-brown flecks surrounded by a halo of dark-green, water-soaked tissue. Dense aggregations of spots

Figure 8.3 a. Banana freckle symptoms caused by *Guignardia musae* include minute black and raised spots on the fruit surface. b. Close-up of the symptoms and signs of freckle. Note the minute black flecks surrounded by a halo of brownish, water-soaked tissue with a sand paper-like texture due to the protruding pycnidia and/or ascomata of the pathogen.

follow and large areas of the fruit surface may become black with a rough texture similar to sandpaper (Meredith, 1968). The eating qualities of the fruit are not affected since the disease is confined to only the fructoplane and/or crown area. The disease affects the aesthetic value of the fruit, especially in export market. It is very important that 'Cavendish' fruits are marketed as blemish-free as possible to satisfy fastidious consumers (Meredith, 1968).

Removing infected leaves and bagging bunches immediately after the male bud removal is a common practice in the Philippines to help prevent fruit infection. During disease outbreaks, plants are usually pruned weekly or biweekly. However, owing to the need of maintaining the functional leaves during fruit development, deleafing causes problems. The growth and yield of banana depends on the maintenance of leaves until the complete development of the fruit. Heavy infections of freckle can also lead to the premature death of older leaves of some cultivars and they are reported to significantly reduce yields (Tsai et al., 1993). Freckle is controlled in 'Cavendish' plantations in the Philippines with the same fungicides that control black leaf streak/black Sigatoka.

5 Management of Postharvest Diseases

The commercial control of postharvest diseases of banana, particularly crown rot, is generally performed with the application of synthetic fungicides. Routine postharvest fungicide treatment is still the most used crown rot control method (Lassois, 2014). Ideally, the control of crown rot starts in the field with the regular removal of leaf trash. Proper field sanitation can greatly reduce the number of crown rot fungi spores present, as well as not keeping rotting fruits or plant waste materials near the packing station. Equally important is a preventive approach to inhibit fruit decay. Complete removal of floral parts harboring numerous disease-causing inocula, in the field or in the packinghouse, before the dehanding operation reduce the risk of contamination of the washing baths (Finlay et al., 1992; Marin et al., 1996; Krauss and Johanson, 2000). Regular changing of water can reduce spore accumulation in the water baths serving as source for contamination of newly trimmed crowns. In most possible ways, the contamination of water baths with plant debris such as dried pistils, leaves, trimming waste, etc. have to be avoided. As a common practice, plant debris is scooped out of the water baths. Moreover, bath water is treated with chlorine to hamper contamination (Shillingford, 1977), but the efficacy of this practice is controversial for fruit decay control (Slabaugh and Grove, 1982). The chlorine concentration is regularly adjusted to offset large losses in it that occur by volatilization or through redox reactions with latex or other organic matter in the washing baths (Shillingford, 1977). Furthermore, hygiene and sanitation of the packinghouse and the adjoining facilities is a foremost protective measure to keep the crowns of freshly trimmed bananas away from all inoculum sources. Trimmings from the crown/stalks and the rejected fruits must be kept away from the packinghouse. It has been shown that trimming clusters in a clean environment rather than in the field can reduce crown rot incidence by 50% (Finlay and Brown, 1993). A clean and sharp stainless steel blade is highly recommended for trimming/clustering banana. A contaminated trimming knife could spread inocula from the peel into the crown tissues (Lukezic et al., 1967). A dull knife causing serrated and rough edges on trimmed crowns, or ripping them off the hands, significantly increases the level of fruit contamination because the tissue fragments on the surface of the crowns dry out and quickly become senescent, thus providing an ideal site for infection (Finlay and Brown, 1993). The tip of the trimming knife must be rounded to avoid causing banana fruit wounds (Krauss and Johanson, 2000). A wider area retained after crown trimming containing as much tissue as possible appears to increase crown resistance to decay and seldom leads to the spread of decay into the fruit pedicels (Muirhead and Jones, 2000).

Synthetic fungicides such as azoxystrobin, carbendazim, prochloraz, propiconazole, and thiabendazole are commonly used in many banana-producing countries for effective management of banana postharvest diseases, but their repeated application is perceived by many as harmful to human health and the environment. Furthermore, they become ineffective after prolonged use owing to the buildup of fungicide resistance in pathogen populations. For instance, to ensure low or zero incidence of crown rot and/or anthracnose diseases, some packinghouses in the Philippines double the dose of fungicides, thus leaving high traces of residues on the fruit and affecting the overall image of fresh banana from the Philippines.

The increasing restrictions on the use of synthetic chemicals for postharvest application may ultimately put this practice to an end, since many active substances registered for postharvest treatment have been banned by national regulations (Chillet and de Lapeyre de Bellaire, 1996). Consumers are now becoming highly selective with respect to food safety concerns, and demand is increasing for banana without postharvest synthetic chemical treatments. For instance, Japan and South Korea are studying the imposition of stricter pesticide residue limit for fresh bananas. Hence, alternatives to synthetic chemical fungicides are therefore being considered. The required interventions should be nonchemical or of very low toxicity, as demanded by the consumers.

5.1 Environmental Factors

Extending green life of harvested bananas is one of the primary concerns, particularly for export markets where transport of the fruit for long periods is required. Green life is significant in the decay process because the relationship previously described between the degree of banana ripeness and susceptibility to disease (Muirhead and Jones, 2000). Environmental factors such as temperature, RH, and atmospheric gaseous composition affect green life and disease development. These factors may directly affect the biology of the pathogens and favor green life of the fruit by slowing down fruit respiration. These environmental parameters can thus be modified to carefully manage banana postharvest diseases (Krauss and Johanson, 2000). The main strategy to extend fruit green life is to provide low temperatures during storage and transport. This minimizes fruit respiration and retards the development of disease, particularly crown rot. Right after packing, banana in boxes are placed in a cold room or shipping container awaiting transit, which is a common practice for small and corporate banana growers in the Philippines and other countries. Refrigerating the fruits in boxes to the lowest possible temperature without inducing physiological disorders or chilling injuries is a critical action. Commonly, 13°C is the lowest temperature that should be used because at temperatures below 13°C chilling injury may occur. Temperatures below 12°C may cause peel browning owing to oxidization (Muirhead and Jones, 2000). Temperature influences the fungal growth; hence, containers designed for maritime shipping of bananas are adjusted to 13–14°C. Generally, the growth of fungi is slowed at low temperatures. However, temperatures of 13–14°C are not low enough to deter fungal infection and colonization because many common phytopathogenic fungi can grow at temperatures within the range of 10–35°C. For example, crown rot incidence and severity are higher in bananas exposed to temperatures over 16°C (Slabaugh and Grove, 1982).

RH influences the quality of banana fruit because green life is markedly reduced at low RH (30 to 40%) as a result of ethylene production in the fruit peel tissue (Peacock, 1973). High RH, which is essential to ensure a long green life, hinders transpiratory water loss from the fruit. In fact, there is an inverse relationship between fruit water loss and RH of the storage environment, which can affect the occurrence of some physiological disorders and alter the uniformity in ripening (Kader, 1985).

The composition of the atmosphere around bananas during shipping can be manipulated to slow metabolic activity. MA and controlled atmosphere (CA) storage systems involve altering and maintaining an atmospheric gas composition in the banana environment different from the ambient air atmosphere. The underlying principle of MA and CA lies in reducing the fruit respiration rate and ethylene synthesis during transport and storage. MA or hermetic storage is generally achieved by keeping the fruit inside a polyethylene bag. Since there is a sealed condition, the fruit surrounding environment is modified over time. The modification of the atmosphere depends on the fruit respiration, the volume of the gas around the fruit, and the permeability of the enclosing plastic bag. Under MA conditions, the O_2 concentration falls over time and the concentration of CO_2 increases owing to banana respiration. Other gases, such as ethylene, may accumulate. It was shown that crown rot was partially controlled by packing bananas in MA (Bastiaanse et al., 2010). High CO_2 (>15%) and low O_2 (<1%) levels inhibit the growth of many fungi (Al Zaemey et al., 1994). Unfortunately, bananas cannot be stored under these conditions because major negative alterations generally occur during ripening when the atmospheric CO_2 level is above 7 to 12% or when the O_2 level is below 1 to 2% (Win et al., 2007). Another way to modify gas exchange is to coat the fruit peel with a wax, which blocks the stomatal pores, thus reducing gas exchange (Banks, 1984). Meanwhile, CA is obtained artificially by injecting nitrogen into the storage rooms and does not depend on fruit respiration.

A recent approach involves a combined MA and CA storage that modified both the micro- and macro-environments. Because there is an inverse relationship between respiration rates and postharvest life of fresh banana, the effects of combining MA and CA were beneficial (Goswami and Mangaraj, 2011). In other work, a synergistic effect between MA and a potassium-permanganate-based ethylene scrubber extended green life and restricted softening during ripening of the banana (Chauhan et al., 2006).

5.2 Hot Water Treatment (HWT)

Dipping 'Cavendish' bananas in water at 55°C for 2 min partially controlled crown rot and anthracnose latent infections caused by *C. musae* (Burden, 1968). Dipping bananas at 45–50°C for between 15 and 30 min controlled crown rot caused by *Chalara paradoxa* (De Seynes) Sacc., *Fusarium proliferatum* (Matsush.) Nirenberg, and *C. musae* without affecting the quality of 'Santa Catarina Prata' and 'Williams Cavendish' fruits (Reyes et al., 1998). Although HWT provides some degree of control of crown rot, its efficacy is not comparable with that of fungicides. Variability in disease severity among fruit after HWT treatments and within a treatment was significant and the reasons these occurred were not apparent (López-Cabrera and Marrero-Domínguez, 1998; Reyes et al., 1998). This could be explained by the ineffectiveness of HWT in killing dormant spores, although the treatment delayed mycelial growth and conidial germination. HWT is fungistatic and has no residual effect to prevent development of fungi after treatment. Mycelium and spore suspensions of pathogens dipped in hot water at 50°C for 20 min were later able to develop on banana and eventually caused crown rot (Alvindia, 2012a). Nevertheless, the effectiveness of HWT to control crown rot disease and maintain postharvest quality of banana cvs. 'Buñgulan' (Alvindia, 2012a) and 'Berangan' (Amin et al., 2012; Mirshekari et al., 2015) was demonstrated. The optimal temperature and exposure time for this application were 50°C and 20 min. However, this treatment may not be applicable for postharvest handling of other banana varieties. Beyond exposures of 20 min, HWT caused fruit injury such as scalding, shriveling, and failure to soften. Hence, the temperature and exposure time determined for cv. 'Buñgulan' may cause mild skin scald to cvs. 'Santa Catarina Prata' and 'Williams Cavendish' (Reyes et al., 1998), suggesting that the application of HWT in banana is highly dependent on crop variety. Universally effective temperature and exposure time to control fruit postharvest pathogens with HWT are not available (Barkai-Golan and Phillips, 1991).

5.3 Irradiation with Ultraviolet-C (UV-C) or Gamma Rays

UV-C irradiation, defined as light of 100 to 280 nm, is one of the presently available nonchemical strategies to control postharvest diseases (Canale et al., 2011; Darras et al., 2012). The control of storage rots by UV-C irradiation could be credited to its direct germicidal effect as well as its ability to stimulate defense responses in tissues of the host (Darras et al., 2012). UV-C irradiation is highly effective in controlling the anthracnose disease of banana (Bokhari et al., 2013). However, the peel of banana seems to be too sensitive to UV-C for the use of this technique for postharvest control of pathogens of the crown rot complex such as *C. musae* (Joas, 1997 as cited in Lassois et al., 2010). Further reports noted that UV-C irradiation could alter the banana peel (Marriott and Palmer, 1980).

Irradiation with gamma rays shows similar limitations. The maximum dose tolerated by the fruit is likely around 0.5 kGy (Wills et al., 1998). Moreover, the high cost of the treatments and low consumer acceptance are hampering the further development of postharvest fruit irradiation (Wills et al., 1998). On the other hand, Kanapathipillai et al. (1987) reported that gamma ray treatment (38 min at 4 kGy) inhibited spore germination, the formation of *C. musae* appressoria, and all fungal development on the surface of banana pieces. However, the effects of gamma rays on fruit quality were not elucidated.

5.4 Microbial Control Agents (MCAs)

Several reports indicated that MCAs are suitable alternatives to fungicides for postharvest decay control on banana. Harvested bananas are easy to treat with MCAs since the postharvest environment (e.g., storage container, temperature, and RH) is easily controlled (Spotts and Sanderson, 1994). Bacteria, yeast, and fungi have been extensively investigated as MCAs to replace conventional agrochemical fungicides. MCAs can be 'native antagonists' that dwell on the fruit or 'alien antagonists' that are not native to banana fruit. The advantages of using 'native' MCAs are their pre-existing adaption to the fruit as a natural substrate (Alvindia and Natsuaki, 2008, 2009). *Bacillus amyloliquefaciens* (Alvindia and Natsuaki, 2009), *Bacillus subtilis* (Fu et al., 2010), *Burkholderia cepacia* (De Costa and Erabadupitiya, 2005), *Burkholderia spinose* (De Costa et al., 2008), *Pantoea agglomerans* and *Flavobacterium* sp. (Gunasinghe et al., 2004), *Pichia anomala* and *Candida oleophila* (Lassois et al., 2008), and *Pseudomonas syringae* (Williamson et al., 2008) are some of the bacteria and yeast reported to control postharvest banana diseases. Antagonistic *Clonostachys byssicola, Curvularia pallescens* and *Penicillium oxalicum* (Alvindia and Natsuaki, 2008), *Trichoderma asperellum, T. harzianum, T. koningii* and *T. viride* (Mortuza and Ilag, 1999; Alvindia and Natsuaki, 2008; Adebesin et al., 2009; Sangeetha et al., 2009), *Gliocladium* spp., *Gliomatrix murorum*, and *Paecilomyces* spp. (Krauss et al., 1998) are some of the filamentous fungi reported to control postharvest banana diseases. Moreover, strains of the yeast *Rhodotorula glutinis, Cryptococcus albidus*, and *Cryptococcus laurentii* (Postmaster et al., 1997) were reported to be effective MCAs to control postharvest diseases of banana. A commercial biocontrol product based on *Bacillus amyloliquefaciens* is currently available in the Philippines for postharvest treatment of banana destined for local and export markets (Yoeki AS®, Eco-Anzen Corp., Davao City, Philippines).

5.5 Salts and Organic Acids

Postharvest dip treatments with the sanitizer sodium hypochlorite (NaOCl) (at 5 g/L), or the salts sodium bicarbonate ($NaHCO_3$), calcium chloride ($CaCl_2$), and sodium chloride (NaCl), all at a concentration of 6 g/L, reduced the incidence of crown rot by 67, 62, 33, and 38%, respectively (Alvindia et al., 2004). Ameliorating the salts with a surfactant increased their fungistatic activity against pathogens and reduced banana decay by 7–75%. The surfactant strengthened the adhesiveness of the salts, provided uniform distribution on the treated surface, and prevented their crystallization on the surface of the fruit (Alvindia and Natsuaki, 2007). The practical use and application of salts with a surfactant to control postharvest disease has restrictions because banana fruits are very sensitive to their combined effect. Salts combined with a nonionic surfactant may cause red and brown 'burn spots' on the fruit peels, which later become dark brown to black as the storage period progresses. It was recommended that salts used with a surfactant must not exceed 5 g/L to avert toxicity to the fruit (Alvindia and Natsuaki, 2007). $CaCl_2$ treatment has been shown to increase the shelf life of banana, mainly through making cell walls less susceptible to pathogens and softening enzymes; however, pressure infiltration of $CaCl_2$ accelerated ripening, disease development, and reduced firmness (Perera and Karunaratne, 2002). Treatment with $NaHCO_3$ at 300 mM reduced anthracnose development on banana fruits by 55% (De Costa and Gunawardhana, 2012). Treatments with 0.2% acetic acid (Ethugala and Karunaratne, 2002), 0.05% citric acid, or 0.1% ascorbic acid (Perera and Karunaratne, 2001), each tested separately, significantly reduced disease incidence on bananas. Malic, citric, oxalic, and maleic acids at concentrations of 1–3% significantly reduced the growth of *C. musae* (Al Zaemey et al., 1994).

5.6 Natural Substances

Presently, natural substances are probably the most researched alternatives to synthetic fungicides to control postharvest diseases of banana. Extracts of *Cinnamomum zeylanicum* and *Syzygium aromaticum* (Ranasinghe et al., 2002; Win et al., 2007; Maqbool et al., 2010), *Calotropis procera, Vitex negundo, Lantana camara, Azadirachta indica, Ficus religiosa, Ocimum sanctum, Thuja orientalis, Argemone mexicana, Achyranthes aspera, Datura fastuosa*, and *Ricinus communis* (Singh et al., 1993), *Cymbopogon nardus, Cymbopogon flexuosus*, and *Ocimum basilicum* (Anthony et al., 2003), *Solanum torvum, Jatropha curcas*, and *Emblica officinalis* (Thangavelu et al., 2004) highly inhibited the in vitro mycelial growth of fungal pathogens causing postharvest crown rot of banana. Their inhibitory effect was directly related to the quantity of extract added to the culture medium.

Cinnamon extracts (CE) and clove oil (*S. aromaticum*) were effective alternative treatments to control anthracnose when used at concentrations of 0.03–0.11% (v/v) (Ranasinghe et al., 2002). Win et al. (2007) also showed that CE reduced crown rot, increased green life, and had no negative effects on banana postharvest quality. The mechanisms of control of the plant extracts are reported to be fungistatic and/or fungitoxic. In general, they inhibit conidial germination and mycelial growth of a wide range of pathogens causing postharvest disease. In some cases, the level of control provided by these natural substances is not sufficient to meet market requirements and, in other cases, phytotoxic effects on fruit have been described. Extracts with fungitoxic properties may include volatile constituents (Kumar and Tripathi, 1991; Doubrava et al., 1998).

Another natural substance with antifungal properties is chitosan. Used as coating of banana fruits at 1%, it prolonged storage life for up to 27 d at 13°C and 90–95% RH (Wattanakorn, 2003). The inhibitory effect of oligochitosan (20 g/L), prepared by enzymatic hydrolysis of chitosan polymer, on *C. musae* is due to the combination of a direct antifungal effect on the pathogen and an indirect effect, whereby the activities of defense-related enzymes in the banana fruit are enhanced (Meng et al., 2012).

5.7 Integration of Nonchemical Control Approaches (NCAs)

It is often true that no single method is effective indefinitely to control a disease. The previously described stand-alone nonchemical treatments provide some degree of control of banana postharvest diseases, but its efficacy is not comparable with that of fungicides, especially when used under commercial conditions (Alvindia, 2013a). One of the significant problems noted when implementing some of the aforementioned control methods is inconsistent performance and variability in disease severity among and within treatments. Therefore, it is essential to implement sound integrated control strategies throughout the commodity channel considering the complexity of factors affecting each postharvest disease and the difficulty of its control. A possible option to upgrade the level of protection is the integration of different methods of control. Reports indicate that a reliable and efficient level of disease control can be reached by the use of two or more alternative means in an integrated approach (Janisiewicz and Korsten, 2002).

The combined effect of a chemical dip and/or an edible coating and/or CA exposure extended the shelf-life period and quality of fresh-cut bananas (Bico et al., 2010). Combining *Cymbopogon citratus* extract (lemongrass) and $NaHCO_3$ significantly improved control of crown decay of 'Embul' bananas compared to lemongrass or $NaHCO_3$ alone (Abeywickrama et al., 2004). Gum arabic (GA) alone had no antifungal activity, but 10% GA supplemented with 0.4% CE was effective to control anthracnose (Maqbool et al., 2011). Citrus seed extract in combination with a wax-based adjuvant provided an organic alternative with a level of crown rot control equal to that of the current standard commercial fungicide treatment (Demerutis et al., 2008). Conversely, amalgamation of some nonchemical alternatives can produce negative effects on decay control efficacy. For example, CE alone did not harm banana fruit quality, but CE in combination with HWT caused unacceptable peel browning (Win et al., 2007).

Postharvest dip treatments of banana fruits with MCAs after HWT significantly improved the efficacy of decay control. HWT in combination with *Burkholderia cepacia* improved control of anthracnose, crown rot, and blossom-end rot than using either of the two treatments alone (De Costa and Erabadupitiya, 2005). Similarly, on bananas treated with HWT and dipped into a *B. amyloliquefaciens* cell suspension, the mycelial growth of crown rot decay fungi was inhibited by 82.61% when compared with reductions of 66.68% by *B. amyloliquefaciens* and 37.90% by HWT as single treatments. Postharvest application of HWT combined with *B. amyloliquefaciens* reduced fruit decay by 90% control, which was comparable to a 93% reduction provided by a synthetic fungicide (Alvindia, 2013c). Additionally, *Trichoderma harzianum* combined with HWT reduced fruit decay by 93%, which was significantly similar to a 95% reduction provided by a fungicide treatment (Alvindia and Acda, 2012).

The combination of MCAs with salts as a safe and effective means to improve the biocontrol activity to control postharvest diseases of fruits has been previously reported (Gamagae et al., 2003; Bastiaanse et al., 2010; Geng et al., 2011; Liu et al., 2011). The combined treatment *B. amyloliquefaciens* + Na_2CO_3 caused the in vitro mycelial growth of the banana pathogens *C. musae, F. verticillioides, L. theobromae*, and *T. paradoxa* to be reduced by 88%; *B. amyloliquefaciens* + $NaHCO_3$ reduced it by 83%; and *B. amyloliquefaciens* + NaOCl reduced it by 52% (Alvindia, 2013b). Crown rot was reduced after fruits were dipped for 30 min in *B. amyloliquefaciens* + $NaHCO_3$ by 93%, in *B. amyloliquefaciens* + Na_2CO_3 by 70%, or in *B. amyloliquefaciens* + NaOCl by 9% (Alvindia, 2013b). Moreover, the compatibility of *T. harzianum* with inorganic salts at a concentration of 1% (w/v) was established. The efficacy of *T. harzianum* to inhibit the mycelial growth of the crown rot pathogens was enhanced by 10–13% with the addition of inorganic salts. As postharvest dip treatment, *T. harzianum* + $NaHCO_3$ reduced the incidence of crown rot by 92%, an efficacy similar to that of synthetic fungicides. On the other hand, *T. harzianum* combined with Na_2CO_3 or NaOCl had no additive effect in the control of crown rot (Alvindia, 2012b).

Dipping banana fruits into salt solutions following HWT likewise yielded promising results. Banana fruits dipped for 30 min in 1% (w/v) Na_2CO_3 following HWT (50°C, 20 min) maintained the overall quality of fruit and reduced the incidence of crown rot by 88%, which was comparable with that of treatment with a fungicide (Alvindia, 2013a). The efficacy of integrating three control measures (*Candida oleophila*, calcium chloride, and MA) yielded a 53% increase in efficacy against crown rot when compared with their efficacy as individual treatments (Bastiaanse et al., 2010).

6 Conclusion

Worldwide concerns about the environment, consumers' health, and the development of pathogenic fungal strains resistant to fungicides are becoming compelling reasons to continuously seek alternatives to the use of synthetic fungicides to control postharvest diseases. This has led to a considerable amount of research on the use of NCAs, such as MCAs, natural substances, salts, UV-C, and HWT, used either singly or in combinations, in an effort to replace synthetic fungicides for decay control while maintaining the overall quality of fruit. To date, the success of NCAs, however, remains limited owing to several reasons. Among these are the variable efficacy of these products, an incomplete understanding of the mechanisms of control, and a lack of general strategies of how to adapt NCAs to a commercial setting with the same convenience of the synthetic fungicides that are applied in the packinghouses today. The diversity, variability in the composition, and pathogenicity of the fungal complexes involved in the development of banana postharvest diseases, particularly crown rot, also complicate the commercial adoption of NCAs. Moreover, the limited spectrum of many NCAs against the fungal complexes causing disease and the risks of phytotoxicity are issues that must be addressed. It is also very important to enhance the overall understanding of the physiological mechanisms of banana involved in the induction of fruit resistance to postharvest disease pathogens with the aim of improving control.

NCAs in commercial settings currently do not match the efficacy of fungicide treatments. However, the results of some semi-commercial trials have shown that postharvest diseases can be effectively managed by combining different NCAs that would eventually lead to the development of an effective integrated NCA in the future. Examples of these approaches are combining MCAs with inorganic salts and/or HWT.

The association of *F. musae* with crown rot disease is another significant concern. The occurrence of *F. musae* as a human pathogen suggests that banana crown rot may be a potential public health concern of importance (Triest and Hendrickx, 2016). Continuous improvement in cultural and production practices adjoined to postharvest research are recommended to manage postharvest diseases of banana. Specifically, crown rot research should enhance the overall understanding of this disease and thus lead to the development of an effective integrated control strategy (Lassois et al., 2010).

References

Aa, H.V.D. 1973. Studies in *Phyllosticta* 1. *Studies in Mycology* 5, 1–110.

Aa, H. A. van der and Vanev, S. 2002. *A Revision of the Species Described in* Phyllosticta. Centraalbureau voor Schimmelcultures (CBS), Utrecht, The Netherlands.

Abeywickrama, K., Kularathna, L., Sarananda, K. and Abeygunawardena, D. 2004. *Cymbopogon citratus* (lemongrass) and citral a+b spray treatments alone or in combination with sodium bicarbonate in controlling crown rot in Embul banana (*Musa acuminata* AAB). *Tropical Agricultural Research and Extension* 7, 104–111.

Adebesin, A., Odebode, C.A. and Ayodele, A.M. 2009. Control of post-harvest rots of banana fruits by conidia and culture filtrates of *Trichoderma asperellum*. *Journal of Plant Protection Research* 49, 302–308.

Ahmad, M.S. and Siddiqui, M.W. 2015. Factors affecting postharvest quality of fresh fruits. pp. 7–32. *In*: *Postharvest Quality Assurance of Fruits: Practical Approaches for Developing Countries*. Springer, New York, NY.

Al Zaemey, A.B., Magan, N. and Thompson, A.K. 1994. *In vitro* studies of the effect of environmental conditions on the anthracnose pathogen of bananas. *Colletotrichum Musae. International Biodeterioration & Biodegradation* 33, 369–381.

Alvindia, D.G. 2012a. Revisiting hot water treatments in controlling crown rot of banana cv. Buñgulan. *Crop Protection* 33, 59–64.

Alvindia, D.G. 2012b. Sodium bicarbonate enhances efficacy of *Trichoderma harzianum* DGA01 in controlling crown rot of banana. *Journal of General Plant Pathology* 79, 136–144.

Alvindia, D.G. 2013a. An integrated approach with hot water treatment and salt in the control of crown rot disease and preservation of quality in banana. *International Journal of Pest Management* 59, 271–278.

Alvindia, D.G. 2013b. Enhancing the bioefficacy of *Bacillus amyloliquefaciens* DGA14 with inorganic salts for the control of banana crown rot. *Crop Protection* 51, 1–6.

Alvindia, D.G. 2013c. Improving control of crown rot disease and quality of pesticide-free banana fruit by combining *Bacillus amyloliquefaciens* DGA14 and hot water treatment. *European Journal of Plant Pathology* 136, 183–191.

Alvindia, D.G. and Acda, M.A. 2012. An integrated approach with *Trichoderma harzianum* DGA01 and hot water treatment on control of crown rot disease and retention of overall quality in banana. *Biocontrol Science and Technology* 22, 1021–1033.

Alvindia, D.G., Kobayashi, T., Natsuaki, K.T. and Tanda, S. 2004. Inhibitory influence of inorganic salts on banana pathogens and preliminary application to control crown rot. *Journal of General Plant Pathology* 70, 61–65.

Alvindia, D.G., Kobayashi, T., Yaguchi, Y. and Natsuaki, K.T. 2000a. Evaluation of cultural and postharvest practices in relation to fruit quality problems in Philippine non-chemical bananas. *Japanese Journal of Tropical Agriculture* 44, 178–185.

Alvindia, D.G., Kobayashi, T., Yaguchi, Y. and Natsuaki, K.T. 2000b. Symptoms and the associated fungi of postharvest diseases on non-chemical bananas imported from the Philippines. *Japanese Journal of Tropical Agriculture* 44, 87–93.

Alvindia, D.G., Kobayashi, T., Yaguchi, Y. and Natsuaki, K.T. 2002. Pathogenicity of fungi isolated from non-chemical bananas. *Japanese Journal of Tropical Agriculture* 46, 215–223.

Alvindia, D.G. and Natsuaki, K.T. 2007. Control of crown rot-causing fungal pathogens of banana by inorganic salts and a surfactant. *Crop Protection* 26, 667–1673.

Alvindia, D.G. and Natsuaki, K.T. 2008. Evaluation of fungal epiphytes isolated from banana fruit surfaces for biocontrol of banana crown rot disease. *Crop Protection* 27, 1200–1207.

Alvindia, D.G. and Natsuaki, K.T. 2009. Biocontrol activities of *Bacillus amyloliquefaciens* DGA14 isolated from banana fruit surface against banana crown rot-causing pathogens. *Crop Protection* 28, 236–242.

Amin, M., Ding, P. and Ghazali, H.M. 2012. Effect of hot water dip treatment on postharvest anthracnose of banana var. Berangan. *African Journal of Agricultural Research* 7, 6–10.

Anthony, S., Abeywickrama, K., Wilson, C. and Wijeratnam, S. 2003. The effect of spraying essential oils of *Cymbopogon nardus, Cymbopogon flexuosus* and *Ocimum basilicum* on postharvest diseases and storage life of Embul banana. *Journal of Horticultural Science and Biotechnology* 78, 780–785.

Banks, N.H. 1984. Studies of the banana fruit surface in relation to the effects of Tal Pro-long coating on gaseous exchange. *Scientia Horticulturae* 24, 279–286.

Barkai-Golan, R. and Phillips, D.J. 1991. Postharvest heat treatments of fresh fruits and vegetables for decay control. *Plant Disease* 75, 1085–1089.

Bastiaanse, H., de Lapeyre de Bellaire, L., Lassois, L., Misson, C. and Jijakli, H. 2010. Integrated control of crown rot of banana with *Candida oleophila* strain O, calcium chloride and modified atmosphere packaging. *Biological Control* 53, 100–107.

Bico, S.L.S., Raposo, M.F., de Morais, R.M.S.C. and de Morais, A.M.M.B. 2010. Chemical dips and edible coatings to retard softening and browning of fresh-cut banana. *International Journal of Postharvest Technology and Innovation* 2, 13–24.

Bokhari, N.A., Siddiqui, I., Parveen, K., Siddique, I., Rizwana, H. and Soliman, D.A.W. 2013. Management of anthracnose of banana by UV irradiation. *The Journal of Animal & Plant Sciences* 23, 1211–1214.

Burden, O.J. 1968. Reduction of banana anthracnose following hot water treatment of green fruit. *Queensland Journal of Agricultural and Animal Science* 25, 135–143.

Canale, M.C., Benato, E.A. and Cia, P. 2011. In vitro effect of UV-C irradiation on *Guignardia citricarpa* and on postharvest control of citrus black spot. *Tropical Plant Pathology* 36, 356–361.

Chauhan, O.P., Raju, P.S., Dasgupta, D.K. and Bawa, A.S. 2006. Instrumental textural changes in banana (var. Pachbale) during ripening under active and passive modified atmosphere. *International Journal of Food Properties* 9, 237–253.

Chillet, M. and de Lapeyre de Bellaire, L. 1996. Conditionnement en polybag pour le contrôle de l'anthracnose de blessures des bananes. *Fruits* 51, 163–172. (in French with English abstract).

Darras, A.I., Demopoulos, V. and Tiniakou, C. 2012. UV-C irradiation induces defence responses and improves vase-life of cut gerbera flowers. *Postharvest Biology and Technology* 64, 168–174.

De Costa, D.M. and Erabadupitiya, H.R.U.T. 2005. An integrated method to control postharvest diseases of banana using a member of the *Burkholderia cepacia* complex. *Postharvest Biology and Technology* 36, 31–39.

De Costa, D.M. and Gunawardhana, H.M.D.M. 2012. Effects of sodium bicarbonate on pathogenicity of *Colletotrichum musae* and potential for controlling postharvest diseases of banana. *Postharvest Biology and Technology* 68, 54–63.

De Costa, D.M., Zahra, A.R.F., Kalpage, M.D. and Rajapakshe, E.M.G. 2008. Effectiveness and molecular characterization of *Burkholderia spinosa*, a prospective biocontrol agent for controlling postharvest diseases of banana. *Biological Control* 47, 257–267.

Demerutis, C., Quiros, L., Martinuz, A., Alvarado, E., Williams, R.N. and Ellis, M.A. 2008. Evaluation of an organic treatment for post-harvest control of crown rot of banana. *Ecological Engineering* 34, 324–327.

Doubrava, N.S., Dean, R.A. and Kuc, J. 1998. Induction of systemic resistance to anthracnose caused by *Coletotrichum legenarium* in cucumber by oxalate and extract from spinach and rhubarb leaves. *Physiological and Molecular Plant Pathology* 33, 69–79.

El-Ramady, H.R., Domokos-Szabolcsy, E., Abdalla, N.A., Taha, H.S. and Fári, M. 2015. Postharvest management of fruits and vegetables storage. pp. 66–152. *In*: E. Lichtfouse (ed.). *Sustainable Agriculture Reviews*, vol. 15. Springer International Publishing, Cham, Switzerland.

Ethugala, C.V. and Karunaratne, A.M. 2002. Disease incidence, physicochemical changes and taste of bananas treated with acetic acid or vinegar. *Fruits* 57, 11–18.

Finlay, A.R. and Brown, A.E. 1993. The relative importance of *Colletotrichum musae* as a crown-rot pathogen on Windward Island bananas. *Plant Pathology* 42, 67–74.

Finlay, A.R., Lubin, C. and Brown, A.E. 1992. The banana stalk as a source of inoculum of fungal pathogens which cause crown rot. *Tropical Science* 32, 343–352.

Fu, G., Huang, S.L., Ye, Y.F., Wu, Y.G., Cen, Z.L. and Lin, S.H. 2010. Characterization of a bacterial biocontrol strain B106 and its efficacy in controlling banana leaf spot and post-harvest anthracnose diseases. *Biological Control* 55, 1–10.

Gamagae, S.U., Sivakumar, D., Wilson, R.S.W. and WIjesundera, R.L.C. 2003. Use of sodium bicarbonate and *Candida oleophila* to control anthracnose in papaya during storage. *Crop Protection* 22, 775–779.

Geng, P., Chen, S., Hu, M., Rizwan-ul-Haq, M., Lai, K., Qu, F. and Zhang, Y. 2011. Combination of *Kluyveromyces marxianus* and sodium bicarbonate for controlling green mold of citrus fruit. *International Journal of Food Microbiology* 151, 190–194.

Goswami, T.K. and Mangaraj, S. 2011. Advances in polymeric materials for modified atmosphere packaging (MAP). pp. 163–242. *In*: M. Lagaron (ed.). *Multifunctional and Nanoreinforced Polymers for Food Packaging*. Woodhead Publishing Limited, Cambridge, UK.

Gunasinghe, R.N., Ikiriwatte, C.J. and Karunaratne, A.M. 2004. The use of *Pantoea agglomerans* and *Flavobacterium* sp. to control banana pathogens. *Journal of Horticultural Science and Biotechnology* 79, 1002–1006.

Hirata, T., Kimishima, E., Aoki, T., Nirenberg, H.I. and O'Donnell, K. 2001. Morphological and molecular characterization of *Fusarium verticillioides* from rotten banana imported into Japan. *Mycoscience* 42, 155–166.

Janisiewicz, W.J. and Korsten, L. 2002. Biological control of postharvest diseases of fruits. *Annual Review of Phytopathology* 40, 411–441.

Jimenez, M., Logrieco, A. and Bottalico, A. 1993. Occurrence and pathogenicity of *Fusarium* species in banana fruits. *Journal of Phytopathology* 137, 214–220.

Joas, J. 1997. Programme qualité de la banane d'exportation. p. 2. *In*: *Rapport d'Essai Sur l'Emploi Des UV En Post Récolte.* Rapport interne. Cirad-Flhor, Fort de France, Martinique, France (in French).

Johanson, A. and Blazquez, B. 1992. Fungi associated with banana crown rot on field packed fruit from the Windward Islands and assessment of their sensitivity to fungicides thiabendazole, prochloraz and imazalil. *Crop Protection* 11, 79–83.

Johnson, P.N.T., Brennan, J.G. and Addo-Yobo, F.Y. 1998. Air-drying characteristics of plantain (*Musa* AAB). *Journal of Food Engineering* 31, 233–242.

Jones, D.H. and Alcorn, J.L. 1982. Freckle and black Sigatoka diseases of banana in far North Queensland. *Australasian Plant Pathology* 11, 7–9.

Kader, A.A. 1985. Quality factors. Definition and evaluation for fresh horticultural crops. pp. 118–121. *In*: A. Kader (ed.). *Postharvest Technology of Horticultural Crops.* CRC Press, Boca Raton, FL.

Kader, A.A. 1994. Fruit maturity, ripening and quality relationships. *Perishables Handling Newsletter* 80, 2.

Kamel, M.A.M., Cortesi, P. and Saracchi, M. 2016. Etiological agents of crown rot of organic bananas in Dominican Republic. *Postharvest Biology and Technology* 120, 112–120.

Kanapathipillai, V.S., Ahmad, R. and Mahamad, M.I. 1987. The effect of sterile filtrates of *Trichoderna* spp. and *Penicillium* spp. and 4 KGy irradiation on the spore germination of *Colletotrichum musae*. pp. 283–292. *In*: K.G. Singh, P. L. Manalo, S.S. Sastrontomo, K.C. Chan, L.G. Lim, A.N. Ganapathi, M.A. A. Rahim, P.S.S. Durai and M.C. Doss (eds.). *Movements of Pests and Control Strategies.* ASEAN Plant and Quarantine Centre and Training Institute, Kuala Lumpur, Malaysia.

Knight, C., Cutts, D.F. and Colhoun, J. 1977. The role of *Fusarium semitectum* in causing crown rot of bananas. *Phytopathologische Zeitschrift* 89, 170–176.

Krauss, U., Bidwell, R. and Ince, J. 1998. Isolation and preliminary evaluation of mycoparasites as biocontrol agents of crown rot of banana. *Biological Control* 13, 111–119.

Krauss, U. and Johanson, A. 2000. Recent advances in the control of crown rot of banana in the Windward Islands. *Crop Protection* 19, 151–160.

Kumar, A. and Tripathi, S.C. 1991. Evaluation of the leaf juice of some higher plants for their toxicity against soil borne pathogens. *Plant Soil* 132, 297–301.

Lassois, L. 2014. Crown rot disease of bananas. Available at: http://hdl.handle.net/2268/169795

Lassois, L., de Lapeyre de Bellaire, L. and Jijakli, M.H. 2008. Biological control of crown rot of bananas with *Pichia anomala* strain K and *Candida oleophila* strain O. *Biological Control* 45, 410–418.

Lassois, L., Jijakli, M.H., Chillet, M. and de Lapeyre de Bellaire, L. 2010. Crown rot of bananas: Preharvest factors involved in postharvest disease development and integrated control methods. *Plant Disease* 94, 648–658.

Liu, Y., Chen, Z., Liu, Y., Wang, X., Luo, C., Nie, Y. and Wang, K. 2011. Enhancing bio-efficacy of *Bacillus subtilis* with sodium bicarbonate for the control of ring rot in pear during storage. *Biological Control* 57, 110–117.

López-Cabrera, J. and Marrero-Domínguez, A. 1998. Use of hot water dips to control the incidence of banana crown rot. *Acta Horticulturae* 490, 563–570.

Lukezic, F.L., Kaiser, W.J. and Martinez, M.M. 1967. The incidence of crown rot of boxed bananas in relation to microbial populations of the crown tissues. *Canadian Journal of Botany* 45, 413–421.

Maqbool, M., Ali, A. and Alderson, P.G. 2010. Effect of cinnamon oil on incidence of anthracnose disease and postharvest quality of bananas during storage. *International Journal of Agriculture & Biology* 12, 516–520.

Maqbool, M., Ali, A., Alderson, P.G., Mohamed, M.T.M., Siddiqui, Y. and Zahid, N. 2011. Postharvest application of gum arabic and essential oils for controlling anthracnose and quality of banana and papaya during cold storage. *Postharvest Biology and Technology* 62, 71–76.

Marin, D.H., Sutton, T.B., Blankenship, S.M. and Swallow, W.H. 1996. Pathogenicity of fungi associated with crown rot of bananas in Latin America on Grande Naine and disease-resistant hybrid bananas. *Plant Disease* 80, 525–528.

Marriott, J. and Palmer, J.K. 1980. Bananas - Physiology and biochemistry of storage and ripening for optimum quality. *Critical Reviews in Food Science and Nutrition* 13, 41–88.

Meng, X.C., Tang, Y.X., Zhang, A.Y., Huang, X.M. and Zhang, Z.Q. 2012. Effect of oligochitosan on development of *Colletotrichum musae* in vitro and in situ and its role in protection of banana fruits. *Fruits* 67, 147–155.

Meredith, D.S. 1968. Freckle diseases of banana in Hawaii caused by *Phyllosticta musarum* (Cke.) Petr. *Annals of Applied Biology* 62, 329–340.

Meredith, D.S. 1971. Transport and storage diseases of bananas: Biology and control. *Tropical Agriculture* 48, 35–50.

Mirshekari, A., Ding, P. and Ghazali, H.M. 2015. Enzymatic activity and microstructural changes of hot water treated banana during ripening. *Journal of Agricultural Science and Technology* 17, 949–962.

Molnár, O., Bartók, T. and Szési, A. 2015. Occurrence of *Fusarium verticillioides* and *Fusarium musae* on banana fruits marketed in Hungary. *Acta Microbiologica Et Immunologica Hungarica* 62, 109–119.

Morton, J. F. 1987. Banana. pp. 29–46. *In*: J.F. Morton (ed.). *Fruits of Warm Climates.* Florida Flair Books, Miami, FL.

Mortuza, M.G. and Ilag, L.L. 1999. Potential for biocontrol of *Lasiodiplodia theobromae* (Pat.) Griff. & Maulb. in banana fruits by *Trichoderma* species. *Biological Control* 15, 235–240.

Muirhead, I.F. and Jones, D.R. 2000. Fungal diseases of banana fruit. Postharvest diseases. pp. 190–206. *In*: D.R. Jones (ed.). *Diseases of Banana, Abaca and Ensete.* CABI Publishing, Wallingford, UK.

Ogawa, J.M. 1970. Postharvest diseases of bananas in China (Taiwan). *FAO Plant Protection Bulletin* 18, 31–42.

Peacock, B.C. 1973. Effect of *Colletotrichum musae* infections on the preclimacteric life of bananas. *Queensland Journal of Agricultural and Animal Science* 30, 239–246.

Perera, O.D.A.N. and Karunaratne, A.M. 2001. Response of bananas to postharvest acid treatments. *The Journal of Horticultural Science and Biotechnology* 76, 70–76.

Perera, O.D.A.N. and Karunaratne, A.M. 2002. Postharvest calcium chloride treatments do not help to increase shelf-life of bananas. *Fruits* 57, 87–94.

Postmaster, A., Kuo, J., Sivasithamparam, K. and Turner, D.W. 1997. Interaction between *Colletotrichum musae* and antagonistic microorganisms on the surface of banana leaf discs. *Scientia Horticulturae* 71, 113–125.

Ranasinghe, L., Jayawardena, B. and Abeywickrama, K. 2002. Fungicidal activity of essential oils of *Cinnamomum zeylanicum* (L.) and *Syzygium aromaticum* (L.) Merr et L.M. Perry against crown rot and anthracnose pathogens isolated from banana. *Letters in Applied Microbiology* 35, 208–211.

Reyes, M.E.Q., Nishijima, W. and Paull, R.E. 1998. Control of crown rot in 'Santa Catarina Prata' and 'Williams' banana with hot water treatments. *Postharvest Biology Technology* 14, 71–75.

Sangeetha, G., Usharani, S. and Muthukumar, A. 2009. Biocontrol with *Trichoderma* species for the management of postharvest crown rot of banana. *Phytopathologia Mediterranea* 48, 214–225.

Sangeetha, G., Usharani, S. and Muthukumar, A. 2010. Significance of *Lasiodiplodia theobromae* and *Colletotrichum musae* in causing crown rot in banana and their reaction on some commercial banana cultivars. *Indian Journal of Horticulture* 67, 21–25.

Shillingford, C.A. 1977. Control of banana fruit rots and of fungi that contaminate washing water. *Tropical Science* 19, 197–203.

Singh, H.N.P., Prasad, M.M. and Sinha, K.K. 1993. Efficacy of leaf extracts of some medicinal plants against disease development in banana. *Letters in Applied Microbiology* 17, 269–271.

Slabaugh, W.R. and Grove, M.D. 1982. Postharvest diseases of bananas and their control. *Plant Disease* 66, 746–750.

Spotts, R.A. and Sanderson, P.G. 1994. The postharvest environment and biological control. pp. 43–56. *In*: C.L. Wilson and M.E. Wisneiwski (eds.). *Biological Control of Postharvest Diseases - Theory and Practice.* CRC Press, Boca Raton, FL.

Su, Y.Y., Noireung, P., Liu, F., Hyde, K.D., Moslem, M.A., Bahkali, A.H., Abd-Elsalam, K.A. and Cai, L. 2011. Epitypification of *Colletotrichum musae*, the causative agent of banana anthracnose. *Mycoscience* 52, 376–382.

Thangavelu, R., Sundararaju, P. and Sathiamoorthy, S. 2004. Management of anthracnose disease of banana caused by *Colletotrichum musae* using plant extracts. *The Journal of Horticultural Science and Biotechnology* 79, 664–668.

Triest, D. and Hendrickx, M. 2016. Postharvest disease of banana caused by *Fusarium musae*: A public health concern? *PLoS Pathogens* 12, e1005940.

Triest, D., Piérard, D., De Cremer, K. and Hendrickx, M. 2016. *Fusarium musae* infected banana fruits as potential source of human fusariosis: May occur more frequently than we might think and hypotheses about infection. *Communicative & Integrative Biology* 9, e1162934.

Tsai, Y.P., Chen, H.P. and Liu, S.H. 1993. Freckle disease of banana in Taiwan. pp. 298–307. *In*: R.V Valmayor, S.C. Hwang, R. Ploetz, S.C. Lee and N.V. Roa (eds.). *Proceedings of the International Symposium on Recent Developments in Banana Cultivation Technology,* Chiuju, Pingtung, Taiwan. INIBAP- ASPNET, Los Baños, Laguna, Philippines.

Van Hove, F., Waalwijk, C., Logrieco, A., Munaut, F. and Moretti, A. 2011. *Gibberella musae* (*Fusarium musae*) sp. nov., a recently discovered species from banana is sister to *F. verticillioides*. *Mycologia* 103, 570–585.

Wattanakorn, C. 2003. Effect of Konjac Gum Coatings on Storage Time of Banana cv. Kluai Hom Thong. M.Sc Thesis. King Mongkut's University of Technology Thonburi, Bangkok, Thailand.

Williamson, D. and Tadon, R.N. 1966. Some pathological studies on *Botryodiplodia theobromae* Pat. causing banana rot. *Mycopathologia et Mycologia Applicata* 29, 245–263.

Williamson, S.M., Guzmán, M., Marin, D.H., Anas, O., Jin, X. and Sutton, T.B. 2008. Evaluation of *Pseudomonas syringae* strain ESC-11 for biocontrol of crown rot and anthracnose of banana. *Biological Control* 46, 279–286.

Wills, R., McGlasson, W.B., Graham, D. and Joyce, D.C. 1998. *Postharvest. An Introduction to the Physiology and Handling of Fruit, Vegetables and Ornamentals.* CABI Publishing, Wallingford, UK.

Win, N., Jitareerat, P., Kanlayanarat, S. and Sangchote, S. 2007. Effects of cinnamon extract, chitosan coating, hot water treatment and their combinations on crown rot disease and quality of banana fruit. *Postharvest Biology and Technology* 45, 333–340.

Chapter 9

Solanaceae and Cucurbitaceae Crops

Nikos Tzortzakis

Department of Agricultural Sciences, Biotechnology and Food Science, Cyprus University of Technology, Limassol, Cyprus

Noam Alkan

Dept. of Postharvest Science of Fresh Produce, Agricultural Research Organization, Volcani Center, Rishon LeZion, Israel

Carmit Ziv

Dept. of Postharvest Science of Fresh Produce, Agricultural Research Organization, Volcani Center, Rishon LeZion, Israel

Lise Korsten

Department of Microbiology and Plant Pathology, Faculty of Natural and Agricultural Sciences, University of Pretoria, Pretoria, South Africa

Abbreviations

AA	Acetic acid
ABA	Abscisic acid
CA	Controlled atmosphere
$CaCl_2$	Calcium chloride
CHEM	Chemical/fungicide
CHI	Chitosan
ClO_2	Chlorine dioxide
CO_2	Carbon dioxide
DCA	Dynamic control atmospheres
DNA	Deoxyribonucleic acid
Ecc	*Erwinia carotovora* subsp. *carotovora*
EO	Essential oil
ETH	Ethanol
EU	European Union
GI	Gamma irradiation
GMO	Genetically modified organism
GRAS	Generally recognized as safe
HIN	Hinokitiol
HT	Heat treatment
HWT	Hot water treatment
LP	Low-pressure
LT	Low temperatures
MA	Modified atmosphere
MJ	Methyl jasmonate
NAA	h-naphthalene acetic acid
PIR	Plant-induced resistance
RH	Relative humidity
ROS	Reactive oxygen species
SOPP	Sodium *ortho*-phenylphenate
TBZ	Thiabendazole
TLV-LTEL	Lower threshold limit value-long-term exposure limit
US FDA	United States Food and Drug Administration
USA	United States of America
UV	Ultraviolet
VIN	Vinegar

1 Introduction

Postharvest losses of fresh produce are a critical contributor to global food losses and waste (FAO, 2011), and these losses have been of growing concern in both the food industry as well as the Food and Agricultural Organization (FAO) of the United Nations in recent years. According to FAO, food production should increase by 70% in the next 50 yr to support the growing world population. In order to meet this demand in the future and to ensure global food security without increasing the burden on the natural environment, postharvest losses must be significantly reduced (Kader, 2005; Hodges et al., 2011).

Perishable fruit vegetables are characterized by high water and nutritional contents and a nearly neutral pH, which make them an ideal habitat for microorganisms. Furthermore, since perishable fresh produce are stored at a high relative humidity (>85% RH) to reduce water loss and conserve quality, during short- to medium-term storage they are highly susceptible to attack by pathogenic microorganisms, especially where damage to the skin has occurred (Agrios, 2005; Bouraz and Özcan, 2006; Pitt and Hocking, 2009). The degree of postharvest losses of fruits and vegetables due to attack by microorganisms worldwide is estimated to be between 10 and 30% of the total crop yield, with significantly higher losses in developing countries (Agrios, 2005).

Mainly fungi, with some bacteria, are responsible for postharvest decay of fresh produce. Those pathogens contaminate produce in the field during transport and/or during handling and packaging causing significant economic losses. When estimating postharvest disease losses, it is important to consider both losses of whole pieces of produce as well as reductions in produce quality, which occurs due to diseases rendering a reduction of the fruit value. For example, the presence of minor or major defects may exclude sale as a whole fresh product but the produce may still be suitable at a lower price for processing purposes, such as tomato paste and other tomato products. Another issue to consider is the fact that diseased produce poses a potential health risk since several fungi infecting the fruits such as *Alternaria, Penicillium, Aspergillus*, and *Fusarium* are known to produce mycotoxins under certain conditions (Tournas, 2005). This is mainly when the produce is used in the production of processed food or animal feed and very rarely in fresh whole produce.

Fruit vegetables production is often challenged by an array of plant diseases exacerbated by a warm and humid climate. These conditions are particularly favorable for the development of fruit rots, both in the field and during handling and shipping. Infections after harvest are an important factor to consider when planning, for example, the handling strategy of cucumber fruits. Since there is a limited number and quantity of fungicides that can be applied after harvest, it is essential to minimize infections during storage and shipping by proper handling, sanitation, and temperature management.

The use of synthetic fungicides for the control of the postharvest pathogens of fruit vegetables is significantly limited for several reasons: (i) the occurrence of fungicide-resistant strains (Leroux, 2004); (ii) a great increase in consumer concerns about pesticide contamination of food and environment, particularly the hazard to human health they may pose (Janisiewicz and Korsten, 2002; Vitale et al., 2012); and (iii) the high costs of discovery, synthesis, and registration of new chemicals. Therefore, there is a need for alternatives to synthetic fungicides.

The families Solanaceae and Cucurbitaceae are the widest cultivated fruit vegetable crops worldwide and they are well adapted to cultivation in intensive production systems, such as greenhouses and hydroponic production. Eggplants (*Solanum melongena* L.), tomatoes (*Solanum lycopersicum* L.), and peppers (*Capsicum annuum* L.) belong to the Solanaceae family and their fruits are widely consumed due to their nutritional value and culinary quality. Tomato is considered a key horticultural crop providing from one medium tomato about 25 mg of vitamin C per 100 g of fresh weight and 47% of vitamin B, 33% of vitamin A, and 0.3–0.6% minerals (calcium, potassium, magnesium, iron) of the daily dietary requirements of an average person. Consumption of tomato on a regular basis can prevent deficiencies in vitamins and minerals (Rama and Narasimham, 2003). Eggplant, besides its vitamins and mineral content, has been associated with cell membrane protection because it contains nasunin, a potent antioxidant and free radical scavenger (Matsubara et al., 2005). Sweet pepper, also termed bell pepper, is a popular vegetable crop. The FAO reported that the production of green chilies and bell peepers in 2013 was slightly more than 31 million tons worldwide (FAO, 2015). Pepper is considered a great source of various phytochemicals that are beneficial to human health, including vitamin C, phenolic compounds, flavonoids, and carotenoids (Zhuang et al., 2012). Nutritional quality of all of these products is influenced by cultivar, maturity stage, agronomic practices, and postharvest treatment of the fruit (Rama and Narasimham, 2003; Zhuang et al., 2012).

The gourd family, Cucurbitaceae, encompasses over 900 species of plants known collectively as gourds or cucurbits, with five vegetable crops of worldwide importance belonging to three genera (Paris et al., 2017). These are: (i) *Cucumis* (cucumbers, muskmelons); (ii) *Cucurbita* (pumpkins, squash); and (iii) *Citrullus* (watermelons). They rank among the top 10 in economic importance among the vegetable crops of the world, while several others in this family have regional importance. Cucurbits are also grown for use as ornaments or containers, and some are used for medicinal applications and other purposes. Some wild cucurbits have potential economic value. Cucumber (*Cucumis sativus* L.) originated from India and it is the most widely cultivated fruit vegetable worldwide, after tomato and watermelon. The production of cucumber is the second largest of all cucurbits, second to watermelon. In 2005, 42 million tons were produced (Staub et al., 2008). Cucumber fruit freshness and high water content are crucial aspects to be maintained during their storage; therefore, high RH is necessary to maintain the quality of the fruits. These storage conditions favorable to product quality can unfortunately favor infections by opportunistic pathogens. Muskmelon (*Cucumis melo* L.) is an important horticultural crop with a worldwide production of 27.3 million metric tons, with China, Iran, Turkey, Egypt, and the USA accounting for 68% of the world production (FAO, 2013). Due to its high nutrient value, that includes β-carotene, muskmelons are among the most commonly consumed fresh fruits in the USA (Lester and Eischen, 1996). Postharvest physiological changes occur in melons because they are a climacteric fruit that ripen after harvest in response to ethylene (Zhou et al., 2012). At the later stage of ripening, tissue softening in melon occurs (Li et al., 2013) causing it to have a short shelf life, reduced fruit quality, and reduced resistance to pathogens. Tissue softening in fruits is commonly associated with changes in the structure, composition, and linkages between cell wall polysaccharides (Ortiz et al., 2011). Watermelon (*Citrullus lanatus* L.) is the third most popular fruit vegetable in the world (Guner and Wehner, 2004). Its postharvest quality is retained by careful handling and transport, proper grading, and cool temperature storage. Nondestructive quality determination of watermelon has been a challenge for its customers since it has a different structure from other fruits. Red-fleshed watermelons have a high lycopene content (Tadmor et al., 2005). Small amounts of phytoene, phytofluene, ζ-carotene, α-carotene, lutein, zeaxanthin, and violaxanthin were also reported in red-fleshed cultivars (Liu et al., 2012a). The major pigments in orange-fleshed cultivar were prolycopene, phytoene, ζ-carotene, β-carotene, and traces of lycopene (Tadmor et al., 2005). Additionally, the consumption of fresh-cut watermelon has increased at a rate of 20–30% annually for several years (Fonseca et al., 2004). Pumpkins and squashes [*Cucurbita pepo* L., *Cucurbita moschata* (Duchesne ex Lam., Duchesne ex Poir.), and *Cucurbita maxima* (Duchesne in Lam.)] vary considerably in nutrient content depending on their cultivation, environment, and species. For example, *C. maxima* has higher carbohydrate, protein, fat, fiber, amino acid, and arginine contents than *C. pepo* or *C. moschata* (Kim et al., 2012). On the other hand, pumpkin fruits contain over 60% of starch in dry matter (Stevenson et al., 2005) and are a food crop targeted for enrichment with pro-vitamin A carotenoids (Ribeiro et al., 2015).

2 Important Postharvest Diseases of Solanaceae and Cucurbitaceae

The following are selected pathogens of great importance causing postharvest disease on both Solanaceae and Cucurbitaceae species. Significant postharvest pathogens for the most important commodities in each family are listed in Subsections 2.9.1 and 2.9.2.

2.1 Alternaria alternata *(Fr.) Keissl.*

Alternaria alternata (Fr.) Keissl. (originally *A. tenuis* Nees) is present worldwide causing Alternaria rot in all solanaceous fruits as well as in cucurbits (Pitt and Hocking, 1997a; Tournas, 2005). *Alternaria alternata* is classified as fungi imperfecti since its sexual stage has not been identified. Its characteristic conidia are ovate, divided by transverse and vertical walls, with minimal development of apical extensions. The hyphae and conidiophores are light brown and septate (Thomma, 2003; Troncoso-Rojas and Tiznado-Hernández, 2014). *Alternaria alternata* causes black spot in tomatoes, peppers, and eggplants (Snowdon, 1988).

Symptoms

In Solanaceae, the disease begins as small, multiple circular, slightly sunken dark brown to black spots. The spots enlarge into sharply sunken lesions that can reach several centimeters in diameter (Pitt and Hocking, 1997a). When the disease progresses, gray to olive-colored green mold is apparent within the colonized tissue. In tomatoes *A. alternata* causes a water-soaked decay, while in pepper the lesions are firm (Figure 9.1). In eggplant, the diseased tissue within lesions becomes tan or grayish-tan in color and spongy (Tournas, 2005). In cucurbits, *A. alternata* is the most important disease of rock melons (cantaloupes) producing dark brown to black lesions that eventually invade the flesh, forming firm, adherent areas (Pitt and Hocking, 1997a). *Alternaria alternata* also infects pumpkin and winter squash (Figure 9.2), causing cream-colored spots that are covered with gray or olive-colored mycelium when the decay is advanced (Tournas, 2005).

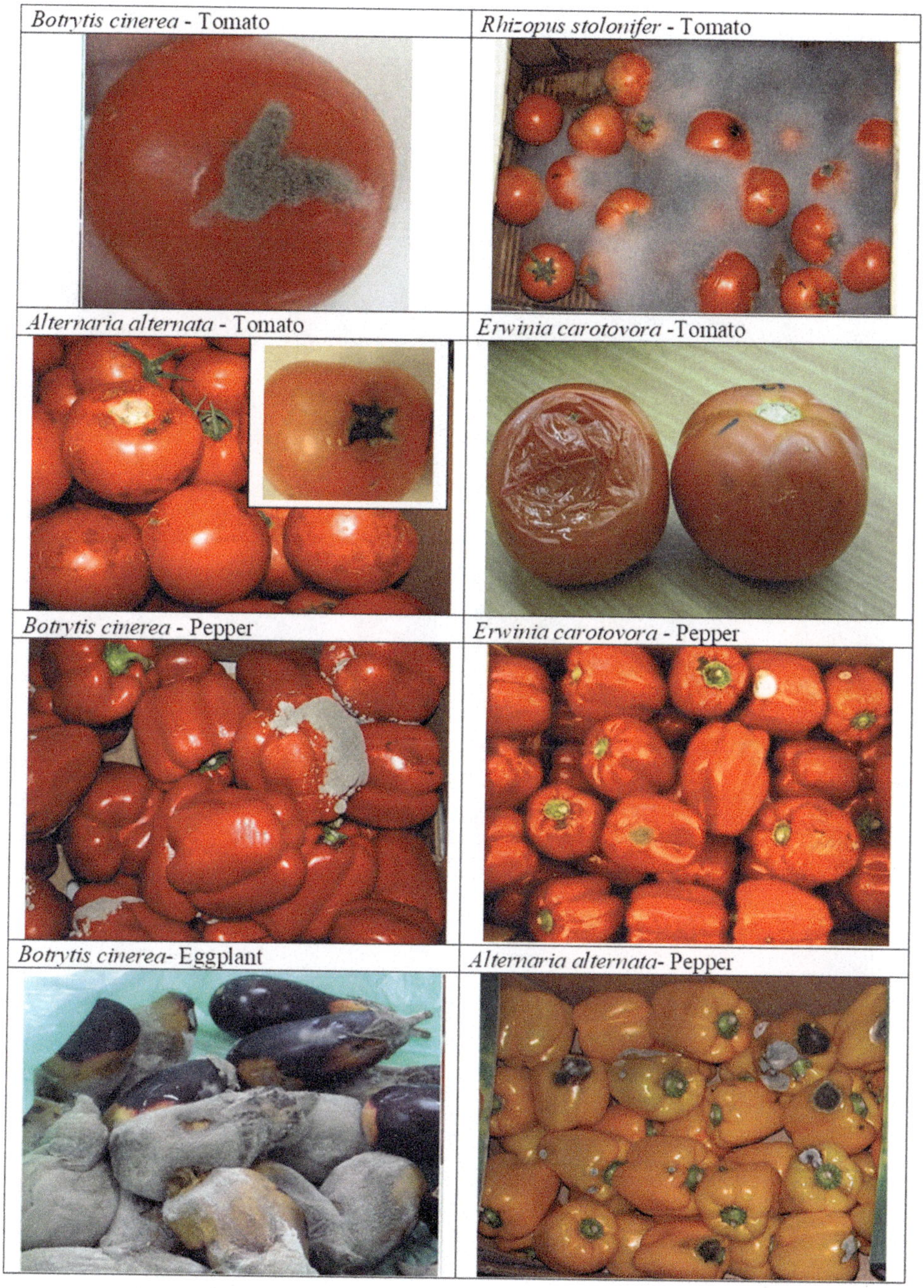

Figure 9.1 The main postharvest diseases of tomato, pepper, and eggplant (Photos by Dr. E. Fallik and Dr. N. Tzortzakis).

Figure 9.2 The main postharvest diseases of cucumber, melon, and squash (Photos by Dr. E. Fallik, Dr. N. Tzortzakis, and Dr. C. Ziv).

Epidemiology

Alternaria alternata is a necrotrophic, mostly opportunistic pathogen. Alternaria rot primarily affects mature fruit by entering the tissue through wounds, cracks, or other surface injuries. The fungus rapidly colonizes tissue that has been weakened by unfavorable growing conditions or by chilling injury that occur during storage (Snowdon, 1988; Pitt and Hocking, 1997a; Thomma, 2003). Infection occurs in the field and is quiescent until conditions favor fungal growth, especially when the fruit ripen (Thomma, 2003). Fruit spots are not always visible during harvest and develop 3–5 d after harvest and during ripening (Snowdon, 1988). The pathogen can survive in the field as mycelium or spores on decaying plant debris for a considerable time and it is spread by wind and splashing rain (Thomma, 2003). *Alternaria alternata* largely depends on toxin production for the colonization of its host (Chung, 2012; Tsuge et al., 2013). These host-specific phytotoxins are chemically diverse but generally target basic cellular processes, and they can pose a health risk to humans and livestock (Thomma, 2003; Chung, 2012).

Management

Alternaria rots develop at all standard handling temperatures and can be avoided only by rapid marketing, combined with proper handling (Pitt and Hocking, 1997a; McGrath, 2004). While there are no other effective control means with cucurbits (McGrath, 2004), with Solanaceae the use of synthetic fungicides before and after harvest is quite effective (Troncoso-Rojas and Tiznado-Hernández, 2014), in addition to the use of resistant tomato cultivars (Cota et al., 2007).

2.2 Botrytis cinerea *Pers.*

Botrytis cinerea Pers. causes gray mold in many cucurbit fruits such as melons, cucumbers, squash, and pumpkin, and in many Solanaceae fruits such as tomatoes, peppers, and eggplants (Snowdon, 1988; Pitt and Hocking, 1997a; Tournas, 2005). It is a generalist necrotrophic pathogen that causes major postharvest losses and is considered the second most important fungal pathogen of plants (Dean et al., 2012; Fillinger and Elad, 2016). The asexual stage of *B. cinerea* produces gray mycelium and branched conidiophores that produce colorless or gray, one-celled, ovoid conidia (Agrios, 2005). In response to nutrient limitation and extreme low temperature, the fungus produces black sclerotia, which are hard and very resistant structures (Markellou, 1999). The sexual (teleomorph) stage of the fungus [previously known as *Botryotinia fuckeliana* (de Bary) Whetzel] produce ascospores in an apothecium (Agrios, 2005; Romanazzi and Feliziani, 2014).

Symptoms

At the beginning of the infection, a darker circular area is visible where the fruit tissues are softer than the other fruit parts. The decay appears as a well-defined and initially water-soaked spot, then the lesion becomes brown in color and penetrates deeply and rapidly into the tissue. Subsequently, under humid conditions, a grayish or brownish-gray, granular, furry mold layer develops on the surface of decaying colonized tissue as the fungus sporulates producing abundant white to gray conidia (Figures 9.1 and 9.2) (Agrios, 2005; Tournas, 2005; Romanazzi and Feliziani, 2014).

Epidemiology

Infections by *B. cinerea* begin with conidia that are released in damp weather and carried by moving air (Snowdon, 1988). The conidia can remain as latent inoculum on plant surface until a thin film of water is formed that facilitates their germination followed by penetration into the fruit tissue (Elad et al., 2007). In addition, the sclerotia often germinate to produce aerial mycelia which can directly infect plant tissues (Agrios, 2005). Fungal infection occurs in the field, exploiting natural openings, and through blossom end, or during harvest time through mechanical wounds such as the cut stem. During storage, infection may occur through mechanical or chilling injuries or by contact with aerial mycelia. After ripening the quiescent fungi activate the rapid colonization of the tissue and cause extensive breakdown of the commodity (Fillinger and Elad, 2016). Optimal temperatures for fungal growth are 18–23°C. However, some growth will occur even at cold storage temperatures (usually 0–5°C), when fruit resistance is decreased and the pathogen is still capable of growing (Romanazzi and Feliziani, 2014). During active host tissue colonization, *B. cinerea* secrets cell wall degrading enzymes and produces toxins (Elad et al., 2007).

Management

Proper field sanitation including soil sterilization before planting is required since the pathogen persists in the soil and in crop debris. The use of clean or treated seed, in addition to well-controlled irrigation and ventilation to prevent the buildup of excessive moisture, and timed fungicide sprays should help reducing the inoculum. Moreover, careful handling during harvest to prevent fruit wounding reduces the incidence of gray mold (Snowdon, 1988; Fillinger and Elad, 2016).

2.3 Colletotrichum *spp.*

Colletotrichum spp. are among the most prevalent fungal pathogens of plants and are important pathogens of fruits and vegetables (Tournas, 2005; Cannon et al., 2012). The anamorph stage of these fungi was known as *Glomerella. Colletotrichum coccodes* (Wallr.) S. Hughes and *C. gloeosporioides* (Penz.) Penz. & Sacc. cause anthracnose in cucumbers, squash, pumpkin, peppers, and tomatoes (Tournas, 2005). *Colletotrichum capsici* (Syd. & P. Syd.) E.J. Butler & Bisby causes anthracnose in Solanaceae (Manandhar et al., 1995; Shenoy et al., 2007) while *C. lagenarium* (Pass.) Ellis & Halst. causes anthracnose in cucurbits (Pitt and Hocking, 1997a; Tournas, 2005).

Symptoms

Anthracnose in cucumber, pumpkin, and squash is manifested by water-soaked lesions with pinkish *Colletotrichum* conidia that are produced in acervuli under high humidity.

In watermelons, this disease forms circular or elongate welts which are initially dark green and later become brown, disfiguring the melon surface (Pitt and Hocking, 1997a; Tournas, 2005). In peppers, anthracnose is an important postharvest disease (Kim et al., 1986; Manandhar et al., 1995; Oh et al., 1998) and causes delimited, often sunken necrotic tan to black lesions on the fruit. As the infection progresses, pink-colored conidia appear either scattered or in concentric rings within the lesions (Agrios, 2005; Cannon et al., 2012). In tomatoes, the symptoms are circular with slightly sunken lesions with small black specks in the center.

Epidemiology

Colletotrichum spp. are seed-borne pathogens and can survive in the soil and on infected plants debris. Both the asexual conidia and sexual ascospores are spread via air transmission, water-splash, insects, and agricultural handling. The conidia germinate and produce appressoria that facilitate the penetration of the fungus into the host without a need for wounding. The infective hyphae penetrate the cuticle and in some cases epidermal cells. Initial infection usually occurs prior to harvest, when the fruit is less mature and its tissues are more resistant to rapid colonization by the pathogen. The fungus enters a biotrophic or quiescent phase on the young fruit. During ripening, the fungus enters a necrotrophic phase that results in significant postharvest losses (Prusky and Plumbley, 1992; Alkan et al., 2015). Symptoms become visible later only in ripened fruits when tissue colonization proceeds. Temperatures of 20 to 30°C and high RH (>85%) promote disease development (Cannon et al., 2012).

Management

Anthracnose can cause serious fruit losses during storage if not controlled. Field sanitation combined with seed treatment, crop rotation, improved air circulation, and fungicide applications from bloom to harvest are necessary to control this disease. In addition, postharvest treatments by fumigation or by dipping or spraying of the fruit are sometimes required in combination with fast fruit cooling and storage at low temperature.

2.4 Erwinia carotovora *subsp.* carotovora (Ecc)

Erwinia carotovora subsp. *carotovora* (*Ecc*) is one of the most important and widespread bacterial pathogens. It causes bacterial soft rot diseases in a large variety of vegetables such as green peppers, paprika, cucumber, and tomato (Tournas, 2005; Bhat et al., 2010). *Ecc* are soil-born, gram negative, facultative anaerobic bacteria. They are motile straight rods, 0.5–1 × 1–3 microns in size with peritrichous flagella (Bhat et al., 2010).

Symptoms

Bacterial spoilage causes rapid fruit rot that is characterized by a soft, watery, and slimy appearance of the fruit tissue, without much discoloration (Figure 9.1), but often accompanied by an offensive odor (Tournas, 2005; Bhat et al., 2010).

Epidemiology

Preharvest infection by the bacteria occurs especially if free water is present, during rainy weather, or when soil containing the bacteria is splashed onto susceptible fruit. The bacteria can enter the fruit through any wound, crack, or opening, such as through the cut stem at harvest time. Tissue damaged by insect feeding is particularly susceptible. Contaminated wash water in postharvest cleaning operations may also rapidly spread this disease. Warm and moist weather increases bacterial soft rot incidence. The optimal temperature for bacterial soft rot development is 27 to 30°C, but most *Erwinia* spp. also grow well during storage at cold temperatures (Snowdon, 1988; Tournas, 2005; Bhat et al., 2010). *Ecc* can metabolize many vegetable sugars and alcohols that are not utilized by other bacterial species. Once within the plant tissue, the bacterium produces copious amounts of plant cell wall degrading enzymes such as pectinase, which disrupt host cell integrity and promote decay (Toth and Birch, 2005).

Management

Currently, the primary methods for controlling postharvest bacterial soft rot are biocides, such as hypochlorite and formaldehyde solutions (Bhat et al., 2010). However, the disease can also be managed by agronomic practices that include controlling insects, picking fruits when conditions are dry, and avoiding injuries during handling. If fruits are washed after harvest, the water should be properly chlorinated and the fruits should be dried as quickly as possible after washing (Snowdon, 1988; Tournas, 2005). Holding the fruits at 0–10°C with low air humidity also prevents the bacterial soft rot (Bhat et al., 2010).

2.5 Penicillium *spp.*

Penicillium spp. are ascomycete fungi that can cause soft rot in fruits and vegetables when sanitation and refrigeration are lacking. *Penicillium expansum* Link causes blue mold in cucurbits and tomato while *P. digitatum* (Pers.) Sacc. causes green mold in melon and tomato (Pitt and Hocking, 1997b; Andersen and Frisvad, 2004; Bankole et al., 2004; Errampalli, 2014; Palou, 2014).

Symptoms

The disease symptoms appear as soft, light brown, watery lesions on the fruit vegetable. As the lesions age, they become covered with blue-green spores. Decayed fruits have an earthy, musty odor (Errampalli, 2014). Some *Penicillium* spp. can produce mycotoxins while growing on stored fruit vegetables like patulin and citrinin that are produced by *P. expansum* infecting tomatoes (Harwig et al., 1979; Andersen and Frisvad, 2004; Tournas, 2005; Errampalli, 2014).

Epidemiology

Penicillium spp. are wound pathogens that infect the host through skin damaged by birds, insects, chilling injury, or prolonged storage. Contamination with *Penicillium* spores usually occurs before harvest or from contaminated storage facilities. The asexual conidia are distributed through the air, germinate in wounds, and then colonize the fruit. Rot mostly develops during storage where the fungus secretes cell wall degrading enzymes that decompose the fruit tissue. The fungus can produce abundant spores that easily survive from season to season (Pitt and Hocking, 1997b; Errampalli, 2014; Palou, 2014).

Management

Sanitation with chlorine compounds or other sanitizers such as peracetic acid, careful handling to minimize injuries, and cooling of the fruits immediately after harvest to a low storage temperature, sometimes in combination with a controlled atmosphere (CA), will significantly control the disease. The application of fungicides is common but hampered by the development of resistance in *P. expansum* (McGrath, 2004; Errampalli, 2014).

2.6 Rhizoctonia solani *J.G. Kühn*

Rhizoctonia solani J.G. Kühn is a basidiomycete soil-borne pathogen that causes belly rot disease in various fruit vegetables (Snowdon, 1988; Tournas, 2005). It is one of the most common fruit rots of cucumbers (Snowdon, 1988; Pitt and Hocking, 1997a; McGrath, 2004) and tomatoes (Fajola, 1979; Coates and Johnson, 1997).

Symptoms

In cucumbers, belly rot appears as a dark brown, water-soaked decay on the side of the fruit that contacts the soil, followed by a yellowish-brown discoloration of the fruit surface. In tomatoes, soil rot develops in green fruits and appears as small circular brown spots. In ripe tomato fruit, the decayed areas enlarge and the reddish brown spots are moderately firm, with a water-soaked border (Snowdon, 1988).

Epidemiology

Infection occurs in the field when there is a direct contact between the fruit and the soil. Wounds or fruit injuries are not required for penetration of the pathogen. High humidity and rainy weather promote development of the disease (Snowdon, 1988; Pitt and Hocking, 1997a). Belly rot develops rapidly at ambient temperature and the entire fruit can rot within several days.

Management

Creating a barrier between the fruit and the soil by stakes or trellises in addition to seed treatment and preharvest sprays of fungicides will reduce infections. Postharvest treatment with fungicides by fumigation in addition to holding the fruit at 10°C will reduce disease development during transportation and storage (Snowdon, 1988; McGrath, 2004).

2.7 *Rhizopus stolonifer* (Ehrenb.) Vuill. and Other *Rhizopus* spp.

Rhizopus stolonifer (Ehrenb.) Vuill. and other *Rhizopus* spp. are zygomycete fungi that can attack almost any fruit or vegetable causing soft rot (also called black mold or Rhizopus rot). In particular, *R. stolonifer* is an important and common postharvest pathogen of tomatoes and cucumbers (Pitt and Hocking, 1997a; Bautista-Baños et al., 2014).

Symptoms

Rhizopus soft rot is manifested by a watery decay of the infected tissue. Grayish-white masses of mycelium develop over the wounded area, which eventually turns black due to the production of the characteristic spores forming on sporangia (Figure 9.1). In cucurbits (cucumber, melon, squash, zucchini, pumpkin, etc.), the initial infection appears as yellowish-brown water-soaked irregular spots with a fairly distinct boundary. The spots develop into sunken lesions (Figure 9.2). The diseased tissue has water-soaked appearance that softens and collapses (Tournas, 2005; Bautista-Baños et al., 2014). In tomatoes, Rhizopus soft rot is manifested by a watery decay (Figure 9.1) and a fermented odor (Tournas, 2005). In pepper, *Rhizopus* spp. initially causes small water-soaked spots which quickly enlarge, but they do not become discolored, and a clear liquid is released from diseased tissues. In eggplant, the fruits show water-soaked areas, brownish liquid, and characteristic odor (Bautista-Baños et al., 2014).

Epidemiology

Rhizopus stolonifer spores are naturally found in soil, plant debris, and air. They are dispersed by wind and by mites and insects. Infection of *R. stolonifer* starts at points of injury in wounded areas of the fruit tissue or through the broken stem and spreads rapidly from infected to healthy fruits (Tournas, 2005). High temperature and RH and fresh wounds promote disease development. For most fruits and vegetables, extensive symptoms of infection caused by this fungus may be seen 3–6 d after infection. Infection by *R. stolonifer* involves enzymatic activities of polygalacturonase and other macerating enzymes such as xylanase, cellulose, and amylase, which are important for the colonization of fruit and result in the rapid softening of the tissue (Bautista-Baños et al., 2014).

Management

The primary approaches to control Rhizopus rot are by sanitation both in the field and after harvest, as well as careful handling to prevent wounding of the fruit tissue, drying wet surfaces of the fruit, and fast cooling the fruit (McGrath, 2004; Bautista-Baños et al., 2014). The application of synthetic fungicides to control Rhizopus rot after harvest is problematic due to concerns about residue hazards to human health and the environment. In addition, their action do not control infection of new injures that may be present after their application. Disinfection of packingline surfaces and wash water with chlorine (HOCl) or other sanitizers is an important means to prevent Rhizopus rot. Ozone (O_3) is another sanitation tool for controlling decay caused by *R. stolonifer* (Bautista-Baños et al., 2014).

2.8 Sclerotinia sclerotiorum *(Lib.) de Bary*

Sclerotinia sclerotiorum (Lib.) de Bary is a white mold ascomycete fungus. This broad host-range, soil-borne plant pathogen (Bolton et al., 2006) causes Sclerotinia rot, also called watery soft rot or cottony rot (Snowdon, 1988; Kora et al., 2003; Tournas, 2005; Saharan and Mehta, 2008). While *S. sclerotiorum* can cause postharvest disease in tomatoes (Pitt and Hocking, 1997a), peppers (Jeon et al., 2006), eggplants (Coates and Johnson, 1997), and cucumbers (McGrath, 2004), it is not considered to be a major pathogen of these fruits during storage (Snowdon, 1988; Tournas, 2005).

Symptoms

Watery soft rot starts as lesions on the fruit surface, which subsequently develop patches of fluffy white mycelium, often with sclerotia, the most obvious sign of this pathogen. Infected fruit tissue becomes soft and watery. In cucumber, water-like droplets, similar to water dew formations, can be observed on the fruits, and after about 3–5 d the fruits are completely rotted (Figure 9.2) (Agrios, 2005; Bolton et al., 2006).

Epidemiology

The sclerotium of *S. sclerotiorum* is a melanized hyphal aggregate that can remain viable over long periods of time in the soil or plant debris and facilitate its long-term survival. The sclerotia can germinate myceliogenically to produce hyphae that directly infect plant tissues, or carpogenically to form apothecia that produce ascospores, which serve as the primary means of infection. The ascospores are ejected into the air, a process that is facilitated by water splashes, then germinate to produce mycelium that can penetrate the cuticle of the host plant using enzymes or mechanical force via appressoria-infection cushions, unless penetration occurs through natural openings such as lenticels or stomata. High temperature and humidity and fresh wounds facilitate infection. Once the disease has been initiated in the host, infection can spread to adjacent fruits through direct contact (Bolton et al., 2006).

Management

Control of this disease is obtained by good preharvest sanitation practices, removal of diseased plants as a source of inoculum, and holding the fruit as low as at 7°C to slow down the growth of the pathogen (McGrath, 2004). Application of fungicides is still the major control method for Sclerotinia diseases (Bolton et al., 2006) since preharvest infections of *S. sclerotiorum* are difficult to control due to the survival of sclerotia between growing seasons and the lack of resistance in the major crops (McGrath, 2004),

2.9 Postharvest Disease Development

Many fruit infections by fungal pathogens occur before harvest. Those fruits will develop disease symptoms after harvest and during storage or shelf life. Fungal pathogens penetrate into the fruit through wounds or natural openings (lenticels, stem end, flowers) (Buzby et al., 2014), by living endophytically in the stem end

(Johnson et al., 1992), or by direct penetration through the fruit cuticle (Deising et al., 2000; Prusky et al., 2013). This direct penetration could be achieved by enzymatically degrading host cuticle and cell wall (Lipinski et al., 2013) or by producing an infection peg from an appressorium that applies enormous mechanical pressure to affect penetration of the cuticle (Deising et al., 2000).

After penetration, the pathogenic fungus may remain quiescent for months awaiting the right conditions to cause disease. Those conditions occur when the fruit ripens and senesces. The period from penetration to the initiation of rotting symptoms is termed quiescence (Prusky et al., 2013). Most of the postharvest fungal pathogens have been reported to live quiescently in their hosts until the fruits ripen; they include: *Botrytis, Alternaria, Sclerotinia, Colletotrichum, Monilinia, Lasiodiplodia, Phomopsis*, and *Botryospheria* (Prusky et al., 1981; Adaskaveg et al., 2000; Prins et al., 2000). During penetration and quiescence, the fungus cannot be detected by simple external examination.

During this quiescent stage, postharvest fungi adopt different lifestyles. The hemibiotrophic fungi such as *Colletotrichum* and *Phytophthora* live quiescently as biotrophs in specialized structures in the unripe fruit cells without killing the host's cells (O'Connell et al., 2012; Alkan et al., 2015). Conversely, fully necrotrophic fungi such as *Botrytis, Alternaria* and others remain restricted 1 to 3 cells within tissues of the unripe fruit in a quiescent stage without damaging the surrounding tissue (Cantu et al., 2008).

During ripening several physiological processes occur, such as activation of ethylene biosynthesis (Seymour et al., 2013), cell-wall loosening (Brummell et al., 1999; Huckelhoven, 2007), soluble sugar accumulation, pH changes (Prusky et al., 2013), cuticular changes (Bargel and Neinhuis, 2005), decline of preformed and inducible antifungal compounds and secondary metabolites (Prusky et al., 2013), reduced reactive oxygen species (ROS) (Alkan and Fortes, 2015), and a decline in inducible host defense responses (Beno-Moualem and Prusky, 2000). Most of those changes are governed by hormonal signals including ethylene, abscisic acid (ABA), jasmonic acid, and salicylic acid, which increase during fruit ripening (Giovannoni, 2001; Seymour et al., 2013). Interestingly, similar phytohormones are regulated in the host defense response to pathogens (Blanco-Ulate et al., 2013; Alkan et al., 2015).

The pathogen could sense the changes that occur during ripening, and switch to the necrotrophic lifestyle and cause decay. During this aggressive stage, the postharvest fungi live as necrotrophs. At this stage, the fungus kills fruit cells by secreting pathogenicity factors, cell wall degrading enzymes, and toxins, then metabolizes the released cell nutrients, followed by mycelial growth and the initiation of fruit decay lesions (Prusky et al., 2013).

2.9.1 Solanaceae (Tomato, Pepper, Eggplant)

Fruit vegetables of the family Solanaceae include tomato, pepper, and eggplant. Of these, tomato is a climacteric fruit with high respiration rates and ethylene production during ripening stages. Several pathogens can be found attacking Solanaceae fruits. *Botrytis cinerea, A. alternata, E. carotovora, R. stolonifer, Mucor* spp., *Geotrichum candidum* Link, *Fusarium* spp., *Phytopthora* spp., and *R. solani* are among the most important (Figure 9.1). The main postharvest pathogens of tomatoes, peppers, and eggplants cold-stored at 10–12°C are the fungi *B. cinerea* and *A. alternata* causing gray mold and black spot, respectively. In addition, bacterial soft rot is caused by the species *Pseudomonas* and *Erwinia*, such as *E. carotovora*. Infection typically occurs via wounds or through the cut stem at harvest time. During postharvest handling, washing can spread infections if microbial contamination is not controlled. Other diseases of solanaceous fruits include anthracnose (*Colletotrichum* spp.), Cladosporium rot (*Cladosporium* spp.), cottony leak (*Pythium* spp.), Fusarium rot, Phoma rot, Phomopsis rot, Phytophthora rot, Pleospora rot, Rhizopus rot, sour rot (*G. candidum*), and watery soft rot (*Sclerotinia* spp.) (Snowdon, 1991). The spectrum of diseases that occur is related to the climate in different countries. *Botrytis cinerea* generally predominates as the major cause of loss, with a lower prevalence of *R. stolonifer, G. candidum*, and *E. carotovora* (McColloch et al., 1982). The main diseases, the pathogens that cause them, and sanitation means employed for their control on Solanaceae fruits are presented in Table 9.1.

2.9.2 Cucurbitaceae (Cucumber, Melon, Water Melon, Squash, Pumpkin)

Fruit vegetables of the Cucurbitaceae family include cucumber, melon, watermelon, and squash-pumpkins. Of these, melon and watermelon are climacteric fruits. In this family, several pathogens that cause postharvest losses include *B. cinerea, A. alternata, E. carotovora, R. stolonifer, Didymella bryoniae* (Fuckel) Rehm, *Penicillium* spp., *Colletotrichum orbiculare* (Berk.) Arx, *Pythium* spp., *Lasiodiplodia theobromae* (Pat.) Griffon & Maubl. (syn.: *Botryodiplodia theobromae* Pat.), *Pseudomonas syringae, S. sclerotiorum, Cladosporium cucumerinum* Ellis & Arthur, *R. solani*, and *Fusarium* spp. (Figure 9.2). The main decay pathogens of Curcubitaceae fruits cold-stored at 10–12°C are *B. cinerea, S. sclerotiorum, A. alternata*, and *E. carotovora.*

One of the most important diseases of cucurbits is black rot/gummy stem blight caused by *D. bryoniae*, although gray mold rot caused by *B. cinerea* is also important. Both pathogens infect fruits via flowers, through the cut stem at harvest time or through wounds. Therefore, sanitation in the field and during and after harvest is essential to minimize the number of infections. To reduce the source of inoculum in the field, plant debris should be removed because fungi can proliferate on them. Additionally, since infection by soft rot bacteria depends on wet conditions, the fruit should be picked while dry. Temperature limitations for disease control in cucurbits during refrigeration are restrictive due to the produce susceptibility to chilling injury.

Other diseases leading to postharvest losses of cucurbits are anthracnose (*Colletotrichum* ssp.), blue mold (*Penicillium* ssp.), Choanephora rot (*Choanephora cucurbitarum* (Berk. & Ravenel) Thaxt.), Cladosporium rot, Myrothecium rot, pink mold rot [*Trichothecium roseum* (Pers.) Link], Rhizopus rot, Sclerotium rot (*Sclerotium rolfsii* Sacc.), soil rot (*R. solani*), sour rot (*G. candidum*), and watery soft rot (*Sclerotinia* spp.). Bacterial soft rot, caused by *Pseudomonas* spp., is particularly difficult to control (Snowdon, 1991).

On watermelon, common fungal diseases that cause peel decay after harvest include black rot (*D. bryoniae*), anthracnose (*C. orbiculare*), Phytophthora fruit rot (*Phytophthora capsici* Leonian), Fusarium rot (*Fusarium* ssp.), and stem-end rot (*L. theobromae*). The most common postharvest bacterial disease is soft rot (*E. carotovora*) (Anonymous, 2003).

Table 9.1 Postharvest diseases of Solanaceae

	Disease	*Pathogen*	*Control*	*Reference*
Tomato	Gray mold	*B. cinerea*	HT	Fallik et al. (2002)
		B. cinerea	HT	Zong et al. (2010)
		B. cinerea	O_3	Tzortzakis et al. (2007b)
		B. cinerea	O_3	Tuffi et al. (2012)
		B. cinerea	EO	Plotto et al. (2003)
		B. cinerea	EO	Tzortzakis (2009a)
		B. cinerea	AA	Sholberg and Gaunce (1995)
		B. cinerea	CHI+EO	Guerra et al. (2015)
	Black mold	*A. alternata*	O_3	Tzortzakis et al. (2008)
		A. alternata	EO	Feng and Zheng (2007)
		A. alternata	HT	Barkai-Golan (1973)
		A. alternata	$CaCl_2$	Wang et al. (2010a)
		A. arborescens	EO	Plotto et al. (2003)
	Anthracnose	*C. coccodes*	O_3	Tzortzakis et al. (2008)
		C. coccodes	EO	Tzortzakis (2009a, 2010)
		C. coccodes	ETH, VIN	Tzortzakis (2010)
	Sour rot	*G. candidum*	EO	Plotto et al. (2003)
		G. candidum	CHEM	Snowdon (1991)
	Rhizopus rot	*R. stolonifer*	EO	Plotto et al. (2003)
		R. stolonifer	UV	Wilson et al. (1997)
		R. stolonifer	HT+*P. guilliermondii*	Zhao et al. (2010)
		R. stolonifer	HT, LT	Snowdon (1991)
		R. stolonifer	CHI+EO	Guerra et al. (2015)
	Bacterial soft rot	*Erwinia* spp.	ClO_2	Mahovic et al. (2007)
	Penicillium rot	*P. expansum*	O_3	Tuffi et al. (2012)
		P. expansum	CHI+EO	Guerra et al. (2015)
Pepper	Gray mold, ghost spots	*B. cinerea*	HT	Fallik et al. (1996)
		B. cinerea	EO	Tzortzakis (2009a)
		B. cinerea	MJ, EO, VIN	Tzortzakis et al. (2016)
		B. cinerea	H_2O_2	Fallik et al. (1994)
		B. cinerea	$KHCO_3$	Fallik et al. (1997)
		B. cinerea	CHI	El Ghaouth et al. (1997)
	Black mold	*A. alternata*	H_2O_2	Fallik et al. (1994)
		A. alternata	$KHCO_3$	Fallik et al. (1997)
		A. alternata	HT	Fallik et al. (1996)
	Bacterial soft rot	*E. carotovora*	*B. amyloliquefaciens*	Zhao et al. (2013)
		E. carotovora	GI	Jeong et al. (2016)
	Rhizopus rot	*R. stolonifer*	HT, LT	Snowdon (1991)
	Watery soft rot	*S. sclerotiorum*	–	Na

(*Continued*)

Table 9.1 (Cont.)

	Disease	*Pathogen*	*Control*	*Reference*
	Sour rot	*G. candidum*	CHEM	Snowdon (1991)
	Anthracnose	*C. capsici*	CHI	Edirisinghe et al. (2014)
		C. coccodes	EO	Tzortzakis (2009a)
	Fusarium rot	*F. lactis;*	EO	Adisa and Lekunze (1986)
		F. equiseti	LT, CHEM	Snowdon (1991)
		F. oxysporum		
	Penicillium rot	*Penicillium* spp.	O_3	Zhao and Cranston (1995)
	Aspergillus rot	*Aspergillus* spp.	O_3	Zhao and Cranston (1995)
Egg-plant	Gray mold	*B. cinerea*	EO	Stavropoulou et al. (2014)
		B. cinerea	H_2O_2	Fallik et al. (1994)
		B. cinerea	HIN	Fallik and Grinberg (1992)
	Black mold	*A. alternata*	H_2O_2	Fallik et al. (1994)
		A. alternata	HIN	Fallik and Grinberg (1992)
		A. alternata	NAA	Temkin-Gorodeiski et al. (1993)

EO: essential oil; HT: heat treatment; ETH: ethanol; O_3: ozone; CHI: chitosan; VIN: vinegar; H_2O_2: hydrogen peroxide; NAA: h-naphthalene acetic acid; AA: acetic acid; $CaCl_2$: calcium chloride; UV: ultraviolet; MJ: methyl jasmonate; ClO_2: chlorine dioxide; $KHCO_3$: potassium bicarbonate; HIN: hinokitiol; GI: gamma irradiation; LT: low temperatures; CHEM: chemical/fungicide; Na: no available studies.

The main diseases, the pathogens that cause them, and means for their control on Cucurbitaceae fruit are presented in Table 9.2.

2.10 Comparative Means for Sanitation and Control Applied to Solanaceae and Cucurbitaceae

Poor postharvest handling, such as a broken cold chain, mechanical injuries, and unsuitable packaging materials, increases postharvest losses due to accelerated senescence, faster pathogen growth, increased water loss, and other processes that are associated with the deterioration of fresh produce. Thus, to meet the market demand for quality and prolonged shelf life of fresh produce, postharvest diseases should be controlled by various means (Choi et al., 2015). To extend the duration of storage, postharvest treatment protocols are necessary, either applied individually or in combinations in order to maintain the produce quality after harvest (Aghdam et al., 2013). Several means are employed to prolonged fruit resistance to pathogens, including cold storage, CA and modified atmosphere (MA) storage,

Table 9.2 Postharvest diseases of Cucurbitaceae

	Disease	*Pathogen*	*Control*	*Reference*
Cucumber	Black mold	*A. alternata*	–	Na
	Belly rot	*R. solani*	Preharvest sprays	Barkai-Golan (2001)
	Cottony leak	*Pythium* spp.	*Bacillus cereus*	Smith et al. (1993)
	Rhizopus soft rot	*R. stolonifer*	–	Na
	Anthracnose	*C. orbiculare*	–	Na
	Gray mold	*B. cinerea*	EO	Tzortzakis et al. (unpublished data)
	Blue mold	*Penicillium* spp.	–	Na
	Stem-end rot	*B. theobromae*	–	Na
	Bacterial soft rot	*E. carotovora*	–	Na
	Bacterial spot	*P. syringae*	*Trichoderma* -PIR	Gal-Hemed et al. (2011)
	Scab	*C. cucumerinum*	–	Na
Melon	Fusarium rot	*F. solani*	HT	Fallik et al. (2000)
		F. solani	HT	Yuan et al. (2013)
		F. solani	Sanosil-25	Aharoni et al. (1994)
		F. semitectum	*Bacillus subtilis*	Yang et al. (2006)
	Gray mold	*B. cinerea*	Sanosil-25	Aharoni et al. (1994)
	Sclerotinia rot	*Sclerotinia* sp.	–	Na
	Botryodiplodia rot	*Botryodiplodia* sp.	–	Na
	Black mold	*A. alternata*	HT	Fallik et al. (2000)
		A. alternata	HT	Yuan et al. (2013)
		A. alternata	O_3	Spalding (1968)
		A. alternata	Sanosil-25	Aharoni et al. (1994)
		A. alternata	*B. subtilis*	Wang et al. (2010b)
	Penicillium rot	*Penicillium* spp.	O_3	Spalding (1968)
	Cladosporium rot	*Cladosporium* spp.	–	Na

(*Continued*)

Table 9.2 (Cont.)

	Disease	Pathogen	Control	Reference
	Rhizopus soft rot	*R. stolonifer*	HT	Yuan et al. (2013)
	Pink rot	*T. roseum*	HT	Yuan et al. (2013)
		T. roseum	*B. subtilis*	Yang et al. (2006)
Water-melon	Black rot	*D. bryoniae*	LT	Zitter (1992)
	Anthracnose	*C. orbiculare*	–	Na
	Phytophthora rot	*P. capsici*	–	Na
	Fusarium rot	*Fusarium* spp.	Resistant cultivars	Zitter (1998)
	Stem-end rot	*L. theobromae*	–	Na
	Bacterial soft rot	*E. carotovora*	–	Na
Squash	Black rot	*D. bryoniae*	LT	Zitter (1992)
	Anthracnose	*C. orbiculare*	–	Na

Sanosil-25 (48% hydrogen peroxide + silver salts); HT: heat treatment; O_3: ozone; EO: essential oil; PIR: plant-induced resistance; LT: low temperatures; Na: no available studies.

hypobaric pressure, growth regulators, and calcium application (Tables 9.1 and 9.2). Chemical control with fungicides directly inhibits pathogens and can be applied before or after harvest. Other control means that could be of use as alternatives to fungicides include generally recognized as safe (GRAS) compounds (hydrogen peroxide, ozone, acetic acid, bicarbonate and carbonate salts, etc.), natural chemical compounds (acetaldehyde, ethanol, jasmonates, essential oils and plant extracts, glucosinolates, chitosan, *Aloe vera*, etc.), and several physical means including heat treatment, ionizing radiation, and ultraviolet illumination (Barkai-Golan, 2001; Aghdam et al., 2013).

Chemical treatments, cold storage, and modified atmosphere storage techniques are the primary means for controlling postharvest decay of fruit vegetables (Pramila and Dubey, 2004; Agrios, 2005). Several chemical fungicides, such as imazalil, dichloran, sodium ortho-phenylphenate (SOPP), thiabendazole (TBZ), benomyl, etc., have been used with variable effectiveness for postharvest fresh produce sanitation and disease control (Barkai-Golan, 2001). Obstacles and constraints arising over the use of chemicals generated the need for alternatives to fungicides for the control of postharvest diseases. The use of chemicals to control postharvest decay is limited in most countries, with increasing reductions in maximum residue levels allowed and the withdrawal of a number of fungicides for postharvest application in production areas such as the European Union (EU). This has become a major driver for the development of alternative disease control methods (Wilson et al., 1991; Tzortzakis, 2009b).

Alternative strategies to chemical treatments to control postharvest diseases have been investigated within the last 20 yr, and include: (1) induction of fruit resistance (Romanazzi et al., 2016); (2) use of plant or animal products with fungicidal activity; (3) physical treatments (thermal, pressure, CA, MA); and (4) application of antagonistic microorganisms (Spadaro and Gullino, 2004). Although there has been significant progress in this research, most of the alternative methods of control have not been effective or consistent enough to be commercialized (Droby et al., 2009).

Chlorine and chlorine-based products are the main preservative means for postharvest sanitation for fresh commodities. Numerous studies have confirmed a limited effectiveness of chlorination in controlling pathogenic bacteria (Beuchat et al., 1998; Pirovani et al., 2000), viruses (Tyrrell et al., 1995), and protozoan cysts (Korich et al., 1990), particularly at high pH. Therefore, microbial contamination remains high despite the adoption of the current 'best practices' for commodity sanitation (fruit dipping, controlled temperature, precooling, etc.). Moreover, the use of chlorinated compounds at effective concentrations [20 to 200 ppm free chlorine (Beuchat et al., 1998)] can generate off-flavors and off-odors in treated fresh produce (Hassenberg et al., 2008). The generated trihalomethanes and haloacetic acids can be harmful to both humans and the environment (Xu, 1999; Han et al., 2002). Consequently, for example, the EU legislation (e.g., EEC 2092/91 and Biocides Directive 98/8/EC) imposes strict guidelines governing the use of chlorine- and bromine-based sanitizers with the goal of moving toward a complete ban of their use in the future.

Calcium and growth regulator applications suppress decay development indirectly, by retarding ripening and senescence processes in fruit vegetables to maintain the natural resistance of young tissues (Barkai-Golan, 2001). Delay of deterioration by the synthetic auxin, h-naphthalene acetic acid (NAA) was studied in eggplants (Temkin-Gorodeiski et al., 1993). Wang et al. (2010a) reported that calcium chloride ($CaCl_2$) effectiveness to control *A. alternata* on tomato fruits improved when $CaCl_2$ was combined with microbial antagonists (*Rhodosporidium paludigenum*). However, both calcium and growth regulators received less research attention compared to alternatives based on sanitation technologies.

UV-C irradiation has been tested in recent years as a postharvest treatment to delay fungal growth and/or product senescence in a variety of fresh produce, including tomato, pepper, and watermelon (Mercier et al., 2001; Charles et al., 2008a, 2009; Artés-Hernández et al., 2010). In stored tomatoes, induced resistance by UV-C was shown to be related to phytoalexins accumulation (Charles et al., 2008b, 2009), including phenolic compounds and the formation of a biochemical barrier containing lignin and suberine, which were associated with resistance to *B. cinerea* (Charles et al., 2008a). Moreover, UV-C caused an up-regulation of genes in tomato fruit involved in signal transduction, defense response, and metabolism (Liu et al., 2011). Mercier et al. (2001) reported that bell pepper decay was reduced by inactivating *B. cinerea* conidia in fruit wounds by UV-C doses of 2.2–2.4 kJ/m. Watermelon irradiated with low doses of UV-C (1.6 and 2.8 kJ/m) were stored successfully up to 11 d at 5°C, whereas moderate-to-high doses of UV-C shortened the storage time up to 27% (Artés-Hernández et al., 2010).

Among various nonchemical approaches, thermal treatments of hot water treatment (HWT) or heat treatment (HT) are among the most effective and promising methods, especially for organically grown fruit vegetable crops, to control postharvest decay in an environmentally friendly way (Zong et al., 2010; Jemric et al., 2011; Liu et al., 2012b). Postharvest decay control by HWT involves effects on both plant pathogen and plant host (Liu et al., 2012b). HWT can both decrease fruit decay and maintain fruit quality (Fallik et al., 1996, 1999). With regard to the mechanisms of control of postharvest decay by HWT, several studies indicated that HWT induces a defense response within fruits and vegetables, subsequently inhibiting the growth of the pathogen throughout host tissues (Fallik et al., 1996). Heat treatment affects cell wall structure/components and enzymatic activity, including hydrolase activity (Vicente et al., 2005). Postharvest losses of Galia melon (Fallik et al., 2000; Zhou et al., 2015) and muskmelon (Yuan et al., 2013) were reduced by HWT applications. Moreover, HWT effectively controlled gray mold and black rot in sweet red pepper (Fallik et al., 1996) and gray mold in tomato (Fallik et al., 2002; Zong et al., 2010).

Hypobaric or low-pressure (LP) storage was applied to delay ripening and senescence in fruits and vegetables in order to extent their postharvest life. This is accomplished with subatmospheric pressure that reduces oxygen levels and removes by-products of metabolism, such as ethylene and carbon dioxide.

MA packaging or CA technologies decrease oxygen (O_2) and elevate carbon dioxide (CO_2). They reduce fruit respiration which delays postharvest ripening, fruit senescence, and prolongs the retention of fruit quality during storage (Wills et al., 1998; Fagundes et al., 2015). An exception to the positive relationship between ripening and disease susceptibility is bacterial soft rot of tomatoes caused by *E. carotovora*, which develops even more rapidly in green fruits than in mature ones (Parsons and Spalding, 1972). The effect of MA/CA is primarily to slow fruit respiration rates to retard fruit ripening and senescence, which prolongs the pathogen resistance characteristic of less mature fruit. In addition to the indirect effect of maintaining host resistance and keeping fruit in a superior physiological condition, low oxygen levels or high carbon dioxide also can directly inhibit disease by suppressing various stages of the pathogen growth or its enzymatic activities (Barkai-Golan, 2001).

Ozone (O_3) is a well-known strong oxidizing agent that has been used by the food industry as an antimicrobial agent for many years. In 2001, it received GRAS status (US FDA, 2001) (Tzortzakis et al., 2007b). Ozone is generally safer to use (lower threshold limit value–long-term exposure limit; TLV-LTEL as 0.06–0.1 ppm over during an 8 hr working day) than many other alternatives (Pryor, 2001; Tzortzakis et al., 2007a) and can be cost-effectively generated and controlled onsite. Additionally, ozone can be applied in gaseous or aqueous phases. Ozone could serve as a fungicide on both vegetative (mycelium growth) and reproductive stages (spores germination/production) of several pathogens including *B. cinerea, C. coccodes*, and *A. alternata* (Tzortzakis et al., 2007b, 2008, 2011). Atmospheric ozone-enrichment induced gene expression and profiling shifts in protein complement in tomato fruit (Tzortzakis et al., 2011; 2013). Additionally, several studies confirmed the ozone effectiveness for fresh produce preservation and maintenance of fruit quality in tomato (Tzortzakis et al., 2007a, 2007b, 2008; Tuffi et al., 2012), pepper (Zhao and Cranston, 1995; Han et al., 2002; Glowacz et al., 2015), cucumber (Li et al., 2014; Glowacz et al., 2015), and zucchini (Glowacz et al., 2015).

3 Future Direction and Challenges in Fruit Vegetable Storage

Control of opportunistic pathogens during fresh produce storage is a challenge. Despite the use of fungicides by the producers during crop growth as well as the developing of new chemicals with apparently fewer potentially negative characteristics, producers are under heavy pressure to reduce or eliminate their application.

New and expanding trends in food and agriculture for chemical-free techniques have prompted the research of alternative means or possible future options that work either alone or in conjunction with synthetics fungicides. These approaches are numerous and include physical means such as HT, HWT, irradiation, low temperatures, hypobaric pressure, MA and CA, and dynamic control atmospheres (DCA), biological means such as biological control by antagonistic microorganisms, and chemical means such as applications of edible coatings or GRAS compounds such as sanitizers, salts, volatiles, and others (Feng et al., 2008; Droby et al., 2009; Tzortzakis, 2009b; Wang et al., 2010a). Recently, postharvest pathology research is focusing on inducing resistance of individual hosts to respond to the pathogen attack. Products such as jasmonates, chitosan, salicylic acid, and harpin among others are under study (Romanazzi et al., 2016).

Depending on the particular pathogen and the affected fruit vegetable, control may vary; however, there are no silver bullets. The integration of one or more different strategies to achieve better disease control is efficient and should be pursued. Additionally, the specific strategies for protection of fruit from pathogens but also for fruit quality maintenance (or even quality improvement) should be optimized for specific species and even for different varieties and types of produce (i.e., beefsteak tomato and cherry tomato; bell pepper and long pepper; green and colored peppers, etc.). For instance, application of sage essential oil in different concentrations to immature and mature tomatoes affects their carotenoids levels and fruit softening differently (Tzortzakis, unpublished data). Today, more than ever, due to the public demand to reduce the use of pesticides, there is an urgent need for efforts to improve cultivars resistance to postharvest diseases and to enhance retention of their quality during prolonged storage. For this to happen, classical breeders should be guided and assisted to screen germplasm for specific fruit traits of value during storage. However, classical breeding alone is time consuming and other techniques can be employed. The emerging fast and efficient approaches of introducing new traits to crops like TALEN and crispr/cas as genome editing systems (Zhang et al., 2013; Nemudryi et al., 2014; Clasen et al., 2016), without introducing foreign DNA and thus not considered as genetically modified organisms (GMO), are probably more promising with regard to the development of improved/resistant cultivars than other approaches. Nevertheless, for this approach to be feasible, a significant basic research of fruit-pathogen interactions and the basic understanding of the physiological changes that fruits undergo in storage are still required to provide the targets for such a mutagenesis approach. In addition, screening for desired traits (like cold/pathogen resistance or new biocontrol agents) from the wild germplasm is still not exhausted, especially if searching for new traits is conducted in unexplored ecosystems such as the marine environment (Gal-Hemed et al., 2011; Fei et al., 2015). These efforts should be combined with the development of improved diagnostic tools to accurately detect and identify plant pathogens that will facilitate disease management (Ahmad et al., 2012). More specifically, to identify biomarkers of very early stages of pathogen inoculation or quiescent microbial infection of the fruit, in order to implement the specific control means as early as possible, preferably in the field before harvest. While these new approaches are still at early stages of development, the rapid technological advances most probably will be employed in the field of postharvest disease management.

References

Adaskaveg, J.E., Förster, H. and Thompson, D.F. 2000. Identification and etiology of visible quiescent infections of *Monilinia fructicola* and *Botrytis cinerea* in sweet cherry fruit. *Plant Disease* 84, 328–333.

Adisa, V.A. and Lekunze, J.K. 1986. Fruit rots of *Capsicum annuum* and *C. frutescens* in Nigeria. Their control by oils. *Fitopatologia Brasileira* 11, 817–822.

Aghdam, M.S., Sevillano, L., Flores, F.B. and Bodbodak, S. 2013. Heat shock proteins as biochemical markers for postharvest chilling stress in fruits and vegetables. *Scientia Horticulturae* 160, 54–64.

Agrios, G.N. 2005. Plant Pathology, 5th ed. Elsevier Academic Press, Burlington, MA.

Aharoni, Y., Copel, A. and Fallik, E. 1994. The use of hydrogen peroxide to control postharvest decay on 'Galia' melons. *Annals of Applied Biology* 125, 189–193.

Ahmad, F., Babalola, O.O. and Tak, H.I. 2012. Potential of maldi-tof mass spectrometry as a rapid detection technique in plant pathology: Identification of plant-associated microorganisms. *Analytical and Bioanalytical Chemistry* 404, 1247–1255.

Alkan, N. and Fortes, A.M. 2015. Insights into molecular and metabolic events associated with fruit response to post-harvest fungal pathogens. *Frontiers in Plant Science* 6, article 889.

Alkan, N., Friedlander, G., Ment, D., Prusky, D. and Fluhr, R. 2015. Simultaneous transcriptome analysis of *Colletotrichum gloeosporioides* and tomato fruit pathosystem reveals novel fungal pathogenicity and fruit defense strategies. *The New Phytologist* 205, 801–815.

Andersen, B. and Frisvad, J.C. 2004. Natural occurrence of fungi and fungal metabolites in moldy tomatoes. *Journal of Agricultural and Food Chemistry* 52, 7507–7513.

Anonymous, 2003. Watermelon, Postharvest Care and Market Preparation. New Guyana National Agricultural Research Institute, Technical Bulletin No. 6. Available at: http://pdf.usaid.gov/pdf_docs/Pnacy822.pdf

Artés-Hernández, F., Robles, P.A., Gómez, P.A., Tomós-Callejas, A.T. and Artés, F. 2010. Low UV-C illumination for keeping overall quality of fresh-cut watermelon. *Postharvest Biology and Technology* 55, 114–120.

Bankole, S.A., Ogunsanwo, B.M. and Mabekoje, O.O. 2004. Natural occurrence of molds and aflatoxin B1 in melon seeds from markets in Nigeria. *Food and Chemical Toxicology* 42, 1309–1314.

Bargel, H. and Neinhuis, C. 2005. Tomato (*Lycopersicon esculentum* Mill.) fruit growth and ripening as related to the biomechanical properties of fruit skin and isolated cuticle. *Journal of Experimental Botany* 56, 1049–1060.

Barkai-Golan, R. 1973. Postharvest heat treatment to control *Alternaria tenuis* Auct. rot in tomato. *Phytopathologia Mediterranea* 12, 108–111.

Barkai-Golan, R. 2001. *Postharvest Diseases of Fruits and Vegetables. Development and Control*. Elsevier B.V., Amsterdam, The Netherlands.

Bautista-Baños, S., Bosquez-Molina, E. and Barrera-Necha, L.L. 2014. *Rhizopus stolonifer* (soft rot). pp. 1–44. *In*: S. Bautista-Baños (ed.). *Postharvest Decay. Control Strategies*. Academic Press, Elsevier, London, UK.

Beno-Moualem, D. and Prusky, D. 2000. Early events during quiescent infection development by *Colletotrichum gloeosporioides* in unripe avocado fruits. *Phytopathology* 90, 553–559.

Beuchat, L.R., Nail, B.V., Adler, B.B. and Clavero, M.R.S. 1998. Efficacy of spray application of chlorinated water in killing pathogenic bacteria on raw apples, tomatoes, and lettuce. *Journal of Food Protection* 61, 1305–1311.

Bhat, K.A., Masood, S.D., Bhat, N.A., Bhat, M.A., Razvi, S.M., Mir, M.R., Akhtar, S., Wani, N. and Habib, M. 2010. Current status of post harvest soft rot in vegetables: A review. *Asian Journal of Plant Sciences* 9, 200–208.

Blanco-Ulate, B., Vincenti, E., Powell, A.L.T. and Cantu, D. 2013. Tomato transcriptome and mutant analyses suggest a role for plant stress hormones in the interaction between fruit and *Botrytis cinerea*. *Frontiers in Plant Science* 4, article 142.

Bolton, M.D., Thomma, B.P. and Nelson, B.D. 2006. *Sclerotinia sclerotiorum* (Lib.) de Bary: Biology and molecular traits of a cosmopolitan pathogen. *Molecular Plant Pathology* 7, 1–16.

Bouraz, N. and Özcan, M. 2006. Inhibition of phytopathogenic fungi by essential oil, hydrosol, ground material and extract of summer savory (*Satureja hortensis* L.) growing wild in Turkey. *International Journal of Food Microbiology* 107, 238–242.

Brummell, D.A., Harpster, M.H., Civello, P.M., Palys, J.M., Bennett, A.B. and Dunsmuir, P. 1999. Modification of expansin protein abundance in tomato fruit alters softening and cell wall polymer metabolism during ripening. *Plant Cell* 11, 2203–2216.

Buzby, J.C., Wells, H.F. and Hyman, J. 2014. The estimated amount, value, and calories of postharvest food losses at the retail and consumer levels in the United States. USDA-ARS Economic Information Bulletin No. 121. Available at: www.ers.usda.gov/webdocs/publications/43833/43680_eib121.pdf

Cannon, P.F., Damm, U., Johnston, P.R. and Weir, B.S. 2012. *Colletotrichum* - current status and future directions. *Studies in Mycology* 73, 181–213.

Cantu, D., Vicente, A.R., Greve, L.C., Dewey, F.M., Bennett, A.B., Labavitch, J.M. and Powell, A.L.T. 2008. The intersection between cell wall disassembly, ripening, and fruit susceptibility to *Botrytis cinerea*. *Proceedings of the National Academy of Sciences of the United States* 105, 859–864.

Charles, M.T., Goulet, A. and Arul, J. 2008a. Physiological basis of UV-C induced resistance to *Botrytis cinerea* in tomato fruit: IV. Biochemical modification of structural barriers. *Postharvest Biology and Technology* 47, 41–53.

Charles, M.T., Mercier, J., Makhlouf, J. and Arul, J. 2008b. Physiological basis of UV-C-induced resistance to *Botrytis cinerea* in tomato fruit: I. Role of pre- and post-challenge accumulation of the phytoalexin-rishitin. *Postharvest Biology and Technology* 47, 10–20.

Charles, M.T., Tano, K., Asselin, A. and Aruln, J. 2009. Physiological basis of UV-C induced resistance to *Botrytis cinerea* in tomato fruit. V. Constitutive defense enzymes and inducible pathogenesis-related proteins. *Postharvest Biology and Technology* 51, 414–424.

Choi, D.S., Park, S.H., Choi, S.R., Kim, J.S. and Chun, H.H. 2015. The combined effects of ultraviolet-C irradiation and modified atmosphere packaging for inactivating *Salmonella enterica* serovar *Typhimurium* and extending the shelf life of cherry tomatoes during cold storage. *Food Packaging and Shelf Life* 3, 19–30.

Chung, K.R. 2012. Stress response and pathogenicity of the necrotrophic fungal pathogen. *Alternaria alternata. Scientifica* 2012, article 635431.

Clasen, B.M., Stoddard, T.J., Luo, S., Demorest, Z.L., Li, J., Cedrone, F., Tibebu, R., Davison, S., Ray, E.E., Daulhac, A., Coffman, A., Yabandith, A., Retterath, A., Haun, W., Baltes, N.J., Mathis, L., Voytas, D.F. and Zhang, F. 2016. Improving cold storage and processing traits in potato through targeted gene knockout. *Plant Biotechnology Journal* 14, 169–176.

Coates, L. and Johnson, G. 1997. Post harvest diseases of fruit and vegetables. pp. 533–547. *In*: J.F. Brown and H.J. Ogle (eds.). *Plant Pathogens and Plant Diseases*. Rockvale Publications, Armidale, Australia.

Cota, I.E., Troncoso-Rojas, R., Sotelo-Mundo, R., Sánchez-Estrada, A. and Tiznado-Hernández, M.E. 2007. Chitinase and β-1,3-glucanase enzymatic activities in response to infection by *Alternaria alternata* evaluated in two stages of development in different tomato fruit varieties. *Scientia Horticulturae* 112, 42–50.

Dean, R., Van Kan, J.A., Pretorius, Z.A., Hammond-Kosack, K.E., Di Pietro, A., Spanu, P.D., Rudd, J.J., Dickman, M., Kahmann, R., Ellis, J. and Foster, G.D. 2012. The top 10 fungal pathogens in molecular plant pathology. *Molecular Plant Pathology* 13, 414–430.

Deising, H.B., Werner, S. and Wernitz, M. 2000. The role of fungal appressoria in plant infection. *Microbes and Infection* 2, 1631–1641.

Droby, S., Wisniewski, M., Macarisin, D. and Wilson, C. 2009. Twenty years of postharvest biocontrol research: Is it time for a new paradigm? *Postharvest Biology and Technology* 52, 137–145.

Edirisinghe, M., Ali, A., Maqbool, M. and Alderson, P.G. 2014. Chitosan controls postharvest anthracnose in bell pepper by activating defense-related enzymes. *Journal of Food Science and Technology* 51(12), 4078-4083.

El Ghaouth, A., Arul, J., Wilson, C. and Benhamou, N. 1997. Biochemical and cytochemical aspects of the interactions of chitosan and *Botrytis cinerea* in bell pepper fruit. *Postharvest Biology and Technology* 12, 183–194.

Elad, Y., Williamson, B., Tudzynski, P. and Delen, N. 2007. Botrytis: *Biology, Pathology and Control*. Springer, Dordrecht, The Netherlands.

Errampalli, D. 2014. *Penicillium expansum* (blue mold). pp. 189–231. *In*: S. Bautista-Baños (ed.). *Postharvest Decay. Control Strategies*. Academic Press, Elsevier, London, UK.

Fagundes, C., Moraes, K., Pérez-Gago, M.B., Palou, L., Maraschin, M. and Monteiro, A.R. 2015. Effect of active modified atmosphere and cold storage on the postharvest quality of cherry tomatoes. *Postharvest Biology and Technology* 109, 73–81.

Fajola, A.O. 1979. The post-harvest fruit rots of tomato (*Lycopersicum esculentum*) in Nigeria. *The Journal Nahrung* 23, 105–109.

Fallik, E., Aharoni, Y., Copel, A., Rodov, V., Tuvia-Alkalai, S., Horev, B., Yekutieli, O., Wiseblum, A. and Regev, R. 2000. Reduction of postharvest losses of Galia melon by a short hot-water rinse. *Plant Pathology* 49, 333–338.

Fallik, E., Aharoni, Y., Grinberg, S., Copel, A. and Klein, J.D. 1994. Postharvest hydrogen peroxide treatment inhibits decay in eggplant and sweet red pepper. *Crop Protection* 13, 451–454.

Fallik, E. and Grinberg, S. 1992. Hinokitiol: A natural substance that controls postharvest diseases in eggplant and pepper fruits. *Postharvest Biology and Technology* 2, 137–144.

Fallik, E., Grinberg, S., Alkalai, S. and Lurie, S. 1996. The effectiveness of postharvest hot water dipping on the control of gray and black molds in sweet red pepper (*Capsicum annuum*). *Plant Pathology* 45, 644–649.

Fallik, E., Grinberg, S., Alkalai, S., Yekutieli, O., Wiseblum, A., Regev, R., Beres, H. and BarLev, E. 1999. A unique rapid hot water treatment to improve storage quality of sweet pepper. *Postharvest Biology and Technology* 15, 25–32.

Fallik, E., Grinberg, S. and Ziv, O. 1997. Potassium bicarbonate reduces postharvest decay development on bell pepper fruits. *Journal of Horticultural Science* 72, 35–41.

Fallik, E., Ilic, Z., Tuvia-Alkalai, S., Copel, A. and Polevaya, Y. 2002. A short hot water rinsing and brushing reduces chilling injury and enhance resistance against *Botrytis cinerea* in fresh harvested tomato. *Advances in Horticultural Science* 16, 3–6.

FAO (Food and Agriculture Organization of the United Nations). 2011. Global food losses and food waste – Extent, causes and prevention. Available at: www.fao.org/docrep/014/mb060e/mb060e00.pdf

FAOSTAT (Food and Agriculture Organization of the United Nations, Statistics Division). 2013. Food and Agricultural Commodities Production. Available at: www.fao.org/faostat/en/#data/QC

FAOSTAT (Food and Agriculture Organization of the United Nations, Statistics Division). 2015. Available at: http://faostat3.fao.org/compare/E.

Fei, J., Wang, Y.S., Jiang, Z.Y., Cheng, H. and Zhang, J.D. 2015. Identification of cold tolerance genes from leaves of mangrove plant *Kandelia obovata* by suppression subtractive hybridization. *Ecotoxicology* 24, 1686–1696.

Feng, W. and Zheng, X. 2007. Essential oils to control *Alternaria alternata* in vitro and in vivo. *Food Control* 18, 1126–1130.

Feng, W., Zheng, X.D., Chen, J.P. and Yang, Y. 2008. Combination of cassia oil with magnesium sulphate for control of postharvest storage rots of cherry tomatoes. *Crop Protection* 27, 112–117.

Fillinger, S. and Elad, Y. 2016. Botrytis – *The Fungus, the Pathogen and Its Management in Agricultural Systems*. Springer International Publishing, Switzerland.

Fonseca, J.M., Rushing, J.W. and Testin, R.F. 2004. The anaerobic compensation point for fresh-cut watermelon and implications for postprocess handling. *Postharvest Biology and Technology* 39, 562–566.

Gal-Hemed, I., Atanasova, L., Komon-Zelazowska, M., Druzhinina, I.S., Viterbo, A. and Yarden, O. 2011. Marine isolates of *Trichoderma* spp. as potential halotolerant agents of biological control for arid-zone agriculture. *Applied and Environmental Microbiology* 77, 5100–5109.

Giovannoni, J. 2001. Molecular biology of fruit maturation and ripening. *Annual Review of Plant Physiology* 52, 725–749.

Glowacz, M., Colgan, R. and Rees, D. 2015. Influence of continuous exposure to gaseous ozone on the quality of red bell peppers, cucumbers and zucchini. *Postharvest Biology and Technology* 99, 1–8.

Guerra, I.C., De Oliveira, P.D., Pontes, A.L., Lucio, A.S., Tavares, J.F., Barbosa-Filho, J.M., Madruga, M.S. and De Souza, E.L. 2015. Coatings comprising chitosan and *Mentha piperita* or *Mentha* x *villosa* Huds essential oils to prevent common postharvest mold infections and maintain the quality of cherry tomato fruit. *International Journal of Food Microbiology* 214, 168–178.

Guner, N. and Wehner, T.C. 2004. The genes of watermelon. *HortScience* 39, 1175–1182.

Han, Y., Floros, J.D., Linton, R.H., Nielsen, S.S. and Nelson, P.E. 2002. Response surface modeling for the inactivation of *E.coli* O157: H7on green peppers (*Capsicum annuum*) by ozone gas treatment. *Journal of Food Science* 67, 1188–1193.

Harwig, J., Scott, P.M., Stoltz, D.R. and Blanchfield, B.J. 1979. Toxins of molds from decaying tomato fruit. *Applied and Environmental Microbiology* 38, 267–274.

Hassenberg, K., Frohling, A., Geyer, M., Schluter, O. and Herppich, W.B. 2008. Ozonated wash water for inhibition of *Pectobacterium carotovorum* on carrots and the effect on the physiological behaviour of produce. *European Journal of Horticultural Science* 73, S37-S42.

Hodges, R.J., Buzby, J.C. and Bennett, B. 2011. Postharvest losses and waste in developed and less developed countries: Opportunities to improve resource use. *The Journal of Agricultural Science* 149, 37–45.

Huckelhoven, R. 2007. Cell wall - associated mechanisms of disease resistance and susceptibility. *Annual Review of Phytopathology* 45, 101–127.

Janisiewicz, W.J. and Korsten, L. 2002. Biological control of postharvest diseases of fruits. *Annual Review of Phytopathology* 40, 411–441.

Jemric, T., Ivic, D., Fruk, G., Matijas, H.S., Cvjetkovic, B., Bupic, M. and Pavkovic, B. 2011. Reduction of postharvest decay of peach and nectarine caused by *Monilinia laxa* using hot water dipping. *Food and Bioprocess Technology* 4, 149–154.

Jeon, Y.J., Kwon, H.W., Nam, J.S. and Kim, S.H. 2006. Characterization of *Sclerotinia sclerotiorum* isolated from paprika. *Mycobiology* 34, 154–157.

Jeong, R.D., Chu, E.H., Park, D.H. and Park, H.J. 2016. Control of postharvest bacterial soft rot by gamma irradiation and its potential modes of action. *The Plant Pathology Journal* 32, 157–161.

Johnson, G.I., Mead, A.J., Cooke, A.W. and Dean, J.R. 1992. Mango stem end rot pathogens - fruit infection by endophytic colonization of the inflorescence and pedicel. *Annals of Applied Biology* 120, 225–234.

Kader, A.A. 2005. Increasing food availability by reducing postharvest losses of fresh produce. *Acta Horticulturae* 682, 2169–2176.

Kim, M.Y., Kim, E.J., Kim, Y.-N., Choi, C. and Lee, B.-H. 2012. Comparison of the chemical compositions and nutritive values of various pumpkin (Cucurbitaceae) species and parts. *Nutrition Research and Practice* 6, 21–27.

Kim, W.G., Cho, E.K. and Lee, E.J. 1986. Two strains of *Colletotrichum gloeosporioides* Penz. causing anthracnose on pepper fruits. *Korean Journal of Plant Pathology* 2, 107–113.

Kora, C., Mcdonald, M.R. and Boland, G.J. 2003. Sclerotinia rot of carrot: An example of phenological adaptation and bicyclic development by *Sclerotinia sclerotiorum*. *Plant Disease* 87, 456–470.

Korich, D.G., Mead, J.R., Madore, M.S., Sinclair, N.A. and Sterling, C.R. 1990. Effects of ozone, chlorine dioxide, chlorine, and monochloramine on *Cryptosporidium parvum* oocyst viability. *Applied and Environmental Microbiology* 56, 1423–1428.

Leroux, P. 2004. Chemical control of *Botrytis* and its resistance to chemicals fungicides. pp. 195–222. *In*: Y. Elad, B. Williamson, P. Tudzynski and N. Delen (eds.). Botrytis: *Biology, Pathology and Control.* Kluwer Academic Publishers, Dordrecht, The Netherlands.

Lester, G.E. and Eischen, F. 1996. Beta-carotene content of postharvest orange-fleshed muskmelon fruit Effect of cultivar, growing location and fruit size. *Plant Foods for Human Nutrition* 49, 191–197.

Li, F., Ma, B., He, Q., Lv, C., Wang, B. and Tian, H. 2013. Non-destructive detection of firmness of Hami melon by hyperspectral imaging technique. *Acta Photonica Sinica* 42, 592–595.

Li, J., Yan, S., Wang, Q., Li, Y. and Wang, Q. 2014. Effects of ionized air treatments on postharvest physiology and quality of fresh cucumber. *Journal of Food Processing and Preservation* 38, 271–277.

Lipinski, B., Hanson, C., Lomax, J., Kitinoja, L., Raite, W. and Searchinger, T. 2013. Reducing Food Loss and Waste. World Resources Institute, UNEP, Washington, DC. Available at: www.wri.org/sites/default/files/reducing_food_loss_and_waste.pdf.

Liu, C., Cai, L., Han, X. and Ying, T. 2011. Temporary effect of postharvest UV-C irradiation on gene expression profile in tomato fruit. *Gene* 486, 56–64.

Liu, C.H., Zhang, H.Y., Dai, Z.Y., Liu, X., Liu, Y., Deng, X., Chen, F. and Xu, J. 2012a. Volatile chemical and carotenoid profiles in watermelons [*Citrullus vulgaris* (Thunb.) schrad (Cucurbitaceae)] with different flesh colors. *Food Science and Biotechnology* 21, 531–541.

Liu, J., Sui, Y., Wisniewski, M., Droby, S., Tian, S., Norelli, J. and Hershkovitz, V. 2012b. Effect of heat treatment on inhibition of *Monilinia fructicola* and induction of disease resistance in peach fruit. *Postharvest Biology and Technology* 65, 61–68.

Mahovic, M.J., Tenney, J.D. and Bartz, J.A. 2007. Applications of chlorine dioxide gas for control of bacterial soft rot in tomatoes. *Plant Disease* 91, 1316–1320.

Manandhar, J.B., Hartman, G.L. and Wang, T.C. 1995. Conidial germination and appressorial formation of *Colletotrichum capsici* and *C. gloeosporioides* isolates from pepper. *Plant Disease* 79, 361–366.

Markellou, E. 1999. Biological control of gray mold (*Botrytis cinerea* Pers: Fr.) of greenhouse-grown tomatoes in Crete using *Bacillus brevis*. Ph.D. Thesis, University of Aberdeen, Aberdeen, UK.

Matsubara, K., Kaneyuki, T., Miyake, T. and Mori, M. 2005. Antiangiogenic activity of nasunin, an antioxidant anthocyanin, in eggplant peels. *Journal of Agricultural and Food Chemistry* 53, 6272–6275.

McColloch, L.P., Cook, H.T. and Wright, W.R. 1982. *Market Diseases of Tomatoes, Peppers and Eggplants.* USDA-ARS Agriculture Handbook No. 28, Washington, DC.

McGrath, M.T. 2004. Diseases of cucurbits and their management. pp. 455–510. *In*: S.A.M.H. Naqvi (ed.). *Diseases of Fruits and Vegetables. Diagnosis and Management*, vol. 1. Kluwer Academic Publishers, Dordrecht, The Netherlands.

Mercier, J., Baka, M., Reddy, B., Corcuff, R. and Arul, J. 2001. Shortwave ultraviolet irradiation for control of decay caused by *Botrytis cinerea* in bell pepper: Induced resistance and germicidal effects. *Journal of the American Society for Horticultural Science* 126, 128–133.

Nemudryi, A.A., Valetdinova, K.R., Medvedev, S.P. and Zakian, S.M. 2014. Talen and crispr/cas genome editing systems: Tools of discovery. *Acta Naturae* 6, 19–40.

O'Connell, R.J., Thon, M.R., Hacquard, S., Amyotte, S.G., Kleemann, J., Torres, M.F., Damm, U., Buiate, E.A., Epstein, L., Alkan, N., et al. 2012. Lifestyle transitions in plant pathogenic *Colletotrichum* fungi deciphered by genome and transcriptome analyses. *Nature Genetics* 44, 1060–1065.

Oh, B.-J.U.N., Kim, K.I.D. and Kim, Y.S. 1998. A microscopic characterization of the infection of green and red pepper fruits by an isolate of *Colletotrichum gloeosporioides*. *Journal of Phytopathology* 146, 301–303.

Ortiz, A., Graell, J. and Lara, I. 2011. Preharvest calcium applications inhibit some cell wall-modifying enzyme activities and delay cell wall disassembly at commercial harvest of 'Fuji Kiku-8' apples. *Postharvest Biology and Technology* 62, 161–167.

Palou, L. 2014. *Penicillium digitatum, Penicillium italicum* (green mold, blue mold). pp. 45–102. *In*: S. Bautista-Baños (ed.). *Postharvest Decay. Control Strategies.* Academic Press, Elsevier, London, UK.

Paris, H.S., Tadmor, Y. and Schaffer, A.A. 2017. Curcubitaceae: Melons, squash, cucumber. *Encyclopedia of Applied Plant Sciences*, 3, 209-217.

Parsons, C.S. and Spalding, D.H. 1972. Influence of a controlled atmosphere, temperature, and ripeness on bacterial soft rot of tomatoes. *Journal of the American Society for Horticultural Science* 97, 297–299.

Pirovani, M., Güemes, D., Di Pentima, J. and Tessi, M. 2000. Survival of *Salmonella hadar* after washing disinfection of minimally processed spinach. *Letters in Applied Microbiology* 31, 143–148.

Pitt, J.I. and Hocking, A.D. 1997a. Spoilage of fresh and perishable foods. pp. 469–487. *In*: J.I. Pitt and A.D. Hocking (eds.). *Fungi and Food Spoilage.* Springer, Boston, MA.

Pitt, J.I. and Hocking, A.D. 1997b. *Penicillium* and related genera. pp. 203–238. *In*: J. I. Pitt and A.D. Hocking (eds.). *Fungi and Food Spoilage.* Springer, Boston, MA.

Pitt, J.I. and Hocking, A.D. 2009. *Fungi and Food Spoilage.* Springer, Boston, MA.

Plotto, A., Roberts, D.D. and Roberts, R.G. 2003. Evaluation of plant essential oils as natural postharvest disease control of tomato (*Lycopersicon esculentum*). *Acta Horticulturae* 628, 737–745.

Pramila, T. and Dubey, N.K. 2004. Exploitation of natural products as an alternative strategy to control postharvest fungal rotting of fruit and vegetables. *Postharvest Biology and Technology* 32, 235–245.

Prins, T.W., Tudzynski, P., von Tiedemann, A., Tudzynski, B., ten Have, A., Hansen, M.E., Tenberge, K. and van Kan, J.A.L. 2000. Infections strategies of *Botrytis cinerea* and related necrotrophic pathogens. pp. 33–65. *In*: J. W. Kronstad (ed.). *Fungal Pathology.* Kluwer Academic Publishers, Dordrecht, The Netherlands.

Prusky, D., Alkan, N., Mengiste, T. and Fluhr, R. 2013. Quiescent and necrotrophic lifestyle choice during postharvest disease development. *Annual Review of Phytopathology* 51, 155–176.

Prusky, D., Benarie, R. and Guelfatreich, S. 1981. Etiology and histology of Alternaria rot of persimmon fruits. *Phytopathology* 71, 1124–1128.

Prusky, D. and Plumbley, R.A. 1992. Quiescent infections of *Colletotrichum* in tropical and subtropical fruits. pp. 289–307. *In*: J.A. Bailey and M.J. Jeger (eds.). Colletotrichum: *Biology, Pathology and Control.* CAB International, Wallingford, UK.

Pryor, A. 2001. Ozone as a gaseous soil treatment agent as an alternative to methyl bromide for pre-plant fumigation. Proceedings of the International Ozone Association PAG Conference, Newport Beach, CA.

Rama, M.V. and Narasimham, P. 2003. Potatoes and related crops. Fruits of the Solanaceae. pp. 4666–4674. *In*: B. Caballero (ed.). *Encyclopedia of Food Sciences and Nutrition*, 2nd ed. Academic Press, Oxford, UK.

Ribeiro, E.M.G., Chitchumroonchokchai, C., de Carvalho, L.M.J., de Moura, F.F., de Carvalho, J.L.V. and Failla, M.L. 2015. Effect of style of home cooking on retention and bioaccessibility of pro-vitamin A carotenoids in biofortified pumpkin (*Cucurbita moschata* Duch.). *Food Research International* 77, 620–626.

Romanazzi, G. and Feliziani, E. 2014. *Botrytis cinerea* (gray mold). pp. 131–146. *In*: S. Bautista-Baños (ed.). *Postharvest Decay. Control Strategies*. Academic Press, Elsevier, London, UK.

Romanazzi, G., Sanzani, S.M., Bi, Y., Tian, S., Martínez, P.G. and Alkan, N. 2016. Induced resistance to control postharvest decay of fruit and vegetables. *Postharvest Biology and Technology* 122, 82–94.

Saharan, G.S. and Mehta, N. 2008. *Sclerotinia Diseases of Crop Plants: Biology, Ecology and Disease Management*. Springer, Dordrecht, The Netherlands.

Seymour, G.B., Østergaard, L., Chapman, N.H., Knapp, S. and Martin, C. 2013. Fruit development and ripening. *Annual Review of Plant Biology* 64, 219–241.

Shenoy, B.D., Jeewon, R., Lam, W.H., Bhat, D.J., Than, P.P., Taylor, P.W.J. and Hyde, K.D. 2007. Morpho-molecular characterisation and epitypification of *Colletotrichum capsici* (glomerellaceae, sordariomycetes), the causative agent of anthracnose in chilli. *Fungal Diversity* 27, 197–211.

Sholberg, P.L. and Gaunce, A.P. 1995. Fumigation of fruit with acetic acid to prevent postharvest decay. *Plant Disease* 82, 689–693.

Smith, K.P., Havey, M.J. and Handelsman, J. 1993. Suppression of cottony leak of cucumber with *Bacillus cereus* strain UW85. *Plant Disease* 77, 139–142.

Snowdon, A.L. 1988. Post-harvest decay: Review of the nature and causes of postharvest deterioration in fruits and vegetables, with special reference to those in international trade. pp. 585–602. *In*: D.R. Houghton, R.N. Smith and H.O. W. Eggins (eds.). *Biodeterioration* 7. Elsevier Science Publishers Ltd., New York, NY.

Snowdon, A.L. 1991. *A Color Atlas of Post-Harvest Diseases and Disorders of Fruits and Vegetables*, vol. 2 Vegetables. Wolfe Scientific, London, UK.

Spadaro, D. and Gullino, M.L. 2004. State of the art and future prospects of the biological control of postharvest fruit diseases. *International Journal of Food Microbiology* 91, 185–194.

Spalding, D.H., 1968. Effects of ozone atmospheres on spoilage of fruits and vegetables after harvest. Marketing Research Report 756. Agricultural Research Service-United States Department of Agriculture, p. 11.

Staub, J.E., Robbins, M.D. and Wehner, T.C. 2008. Cucumber. pp. 241–282. *In*: J. Prohens-Tomás and F. Nuez (eds.). *Vegetables I: Asteraceae, Brassicaceae, Chenopodicaceae, and Cucurbitaceae*. Springer-Verlag, New York.

Stavropoulou, A., Loulakakis, K., Magan, N. and Tzortzakis, N. 2014. *Origanum dictamnus* oil vapour suppresses the development of gray mold in eggplant fruit and in vitro. *BIOMed Research International* 2014, article 562679.

Stevenson, D.G., Yoo, S.H., Hurst, P. and Jane, J. 2005. Structural and physicochemical characteristics of winter squash (*Cucurbita maxima* D.) fruit starches at harvest. *Carbohydrate Polymers* 59, 153–163.

Tadmor, Y., King, S., Levi, A., Davis, A., Meir, A., Wasserman, B., Hirschberg, J. and Lewinsohn, E. 2005. Comparative fruit colouration in watermelon and tomato. *Food Research International* 38, 837–841.

Temkin-Gorodeiski, N., Shapiro, B., Grinberg, S., Rosenberger, I. and Fallik, E. 1993. Postharvest treatments to control eggplant deterioration during storage. *Journal of Horticultural Science* 68, 689–693.

Thomma, B.P.H.J. 2003. *Alternaria* spp.: From general saprophyte to specific parasite. *Molecular Plant Pathology* 4, 225–236.

Toth, I.K. and Birch, P.R. 2005. Rotting softly and stealthily. *Current Opinion in Plant Biology* 8, 424–429.

Tournas, V.H. 2005. Spoilage of vegetable crops by bacteria and fungi and related health hazards. *Critical Reviews in Microbiology* 31, 33–44.

Troncoso-Rojas, R. and Tiznado-Hernández, M.E. 2014. *Alternaria alternata* (black rot, black spot). pp. 147–187. *In*: S. Bautista-Baños (ed.). *Postharvest Decay. Control Strategies.* Academic Press, Elsevier, London, UK.

Tsuge, T., Harimoto, Y., Akimitsu, K., Ohtani, K., Kodama, M., Akagi, Y., Egusa, M., Yamamoto, M. and Otani, H. 2013. Host-selective toxins produced by the plant pathogenic fungus *Alternaria alternata*. *FEMS Microbiology Reviews* 37, 44–66.

Tuffi, R., Lovino, R., Canese, S., Cafiero, L.M. and Vitali, F. 2012. Effects of exposure to gaseous ozone and negative air ions on control of epiphytic flora and the development of *Botrytis cinerea* and *Penicillium expansum* during cold storage of strawberries and tomatoes. *Italian Journal of Food Science* 24, 102–114.

Tyrrell, S., Rippey, S. and Watkins, W. 1995. Inactivation of bacterial and viral indicators in secondary sewage effluents, using chlorine and ozone. *Water Research* 29, 2483–2490.

Tzortzakis, N., Chrysargyris, A., Sivakumar, D. and Loulakakis, K. 2016. Vapour or dipping applications of methyl jasmonate, vinegar and sage oil for pepper fruit sanitation towards gray mold. *Postharvest Biology and Technology* 18, 120–127.

Tzortzakis, N.G. 2009a. Impact of cinnamon oil-enrichment on microbial spoilage of fresh produce. *Innovative Food Science and Emerging Technologies* 10, 97–102.

Tzortzakis, N.G. 2009b. Essential oil: Innovative tool to improve the preservation of fresh produce-A review. *Fresh Produce* 3, 87–97.

Tzortzakis, N.G., Borland, A., Singleton, I. and Barnes, J. 2007a. Impact of atmospheric ozone-enrichment on quality-related attributes of tomato fruit. *Postharvest Biology and Technology* 45, 317–326.

Tzortzakis, N.G., Singleton, I. and Barnes, J. 2007b. Deployment of low-level ozone-enrichment for the preservation of chilled fresh produce. *Postharvest Biology and Technology* 43, 261–270.

Tzortzakis, N.G., Singleton, I. and Barnes, J. 2008. Impacts of low-level atmospheric ozone-enrichment on black spot and anthracnose rot of tomato fruit. *Postharvest Biology and Technology* 47, 1–9.

Tzortzakis, N.G. 2010. Ethanol, vinegar and Origanum vulgare oil vapour suppress the development of anthracnose rot in tomato fruit. *International Journal of Food Microbiology*, 142, 14–18.

Tzortzakis, N.G., Taybi, T., Roberts, R., Singleton, I., Borland, A. and Barnes, J. 2011. Low-level atmospheric ozone exposure induces protection against *Botrytis cinerea* with down regulation of ethylene-, jasmonate- and pathogenesis- related genes in tomato fruit. *Postharvest Biology and Technology*, 61, 152–159.

Tzortzakis, N.G., Taybi, T., Antony, E., Singleton, I., Borland, A. and Barnes, J. 2013. Profiling shifts in protein complement in tomato fruit induced by atmospheric ozone-enrichment and/or wound-inoculation with *Botrytis cinerea. Postharvest Biology and Technology*, 78, 67–75.

US FDA (United States Food and Drug Administration) 2001. Secondary direct food additives permitted in food for human consumption, final rule. *Federal Register* 66(123), 33829–33830.

Vicente, A.R., Costa, M.L., Martínez, G.A., Chaves, A.R. and Civello, P.M. 2005. Effect of heat treatments on cell wall degradation and softening in strawberry fruit. *Postharvest Biology and Technology* 38, 213–222.

Vitale, A., Cirvilleri, G., Panebiance, A., Epifani, F., Perrone, G. and Polizzi, G. 2012. Molecular characterization and pathogenicity of *Aspergillus* Sect. *Nigri* causing Aspergillus vine cracking of table grapes in Italy. *European Journal of Plant Pathology* 132, 483–487.

Wang, Y., Ren, X., Song, X., Yu, T., Lu, H., Wang, P., Wang, J. and Zheng, X.D. 2010a. Control of postharvest decay on cherry tomatoes by marine yeast *Rhodosporidium paludigenum* and calcium chloride. *Journal of Applied Microbiology* 109, 651–656.

Wang, Y., Xu, Z., Zhu, P., Liu, Y., Zhang, Z., Mastuda, Y., Toyoda, H. and Xu, L. 2010b. Postharvest biological control of melon pathogens using *Bacillus subtilis* EXWB1. *Journal of Plant Pathology* 92, 645–652.

Wills, R.B.H., McGlasson, B., Graham, D. and Joyce, D. 1998. *Postharvest: An Introduction to the Physiology and Handling of Fruits, Vegetables and Ornamentals.* University of New South Wales Press, Sydney, Australia.

Wilson, C.L., El-Ghaouth, A., Upchurch, B., Stevens, C., Khan, V., Droby, S. and Chalutz, E. 1997. Using an on-line UV-C apparatus to treat harvested fruit for controlling postharvest decay. *HortTechnology* 7, 278–282.

Wilson, C.L., Wisniewski, M.E., Biles, C.L., McLaughlin, R., Chalutz, E. and Droby, S. 1991. Biological control of post-harvest diseases of fruits and vegetables: Alternatives to synthetic fungicides. *Crop Protection* 10, 172–177.

Xu, L. 1999. Use of ozone to improve the safety of fresh fruits and vegetables. *Food Technology* 53, 58–63.

Yang, D.M., Bi, Y., Chen, X.R., Ge, Y.H. and Zhao, J. 2006. Biological control of postharvest diseases with *Bacillus subtilis* (B1 strain) on muskmelons (*Cucumis melo* L. cv. *Yindi*). *Acta Horticulturae* 712, 735–739.

Yuan, L., Bi, Y., Ge, Y., Wang, Y., Liu, Y. and Li, G. 2013. Postharvest hot water dipping reduces decay by inducing disease resistance and maintaining firmness in muskmelon (*Cucumis melo* L.) fruit. *Scientia Horticulturae* 161, 101–110.

Zhang, Y., Zhang, F., Li, X., Baller, J.A., Qi, Y., Starker, C.G., Bogdanove, A.J. and Voytas, D.F. 2013. Transcription activator-like effector nucleases enable efficient plant genome engineering. *Plant Physiology* 161, 20–27.

Zhao, J. and Cranston, P. 1995. Microbial decontamination of black pepper by ozone and the effect of the treatment on volatile oil constituents of the spice. *Journal of the Science of Food and Agriculture* 68, 11–18.

Zhao, Y., Li, P., Huang, K., Wang, Y., Hu, H. and Sun, Y. 2013. Control of postharvest soft rot caused by *Erwinia carotovora* of vegetables by a strain of *Bacillus amyloliquefaciens* and its potential modes of action. *World Journal of Microbiology and Biotechnology* 29, 411–420.
Zhao, Y., Tu, K., Tu, S., Liu, M., Su, J. and Hou, Y.P. 2010. A combination of heat treatment and *Pichia guilliermondii* prevents cherry tomato spoilage by fungi. *International Journal of Food Microbiology* 137, 106–110.
Zhou, R., Wang, X., Hu, Y., Zhang, G., Yang, P. and Huang, B. 2015. Reduction in Hami melon (*Cucumis melo* var. *saccharinus*) softening caused by transport vibration by using hot water and shellac coating. *Postharvest Biology and Technology* 110, 214–223.
Zhou, R.J., Qiao, Y.J., Wang, H.H. and Chen, S.L. 2012. Effect of high-energy electron beam irradiation on physiological quality of fresh-cut Hami melon. *Journal of Nuclear Agricultural Sciences* 26, 300–305.
Zhuang, Y., Chen, L., Sun, L. and Cao, J. 2012. Bioactive characteristics and antioxidant activities of nine peppers. *Journal of Functional Foods* 4, 331–338.
Zitter, T.A. 1992. Fruit Rots of Squash and Pumpkins. Cornell University, Fact Sheet p. 732.10. Available at: http://vegetablemdonline.ppath.cornell.edu/factsheets/Cucurbit_FrtRots.htm
Zitter, T.A. 1998. Fusarium Diseases of Curcubits. Cornell University, Fact Sheet p. 733.00. Available at: http://vegetablemdonline.ppath.cornell.edu/factsheets/Cucurbits_Fusarium.htm
Zong, Y., Liu, J., Li, B., Qin, G. and Tian, S. 2010. Effects of yeast antagonists in combination with hot water treatment on postharvest diseases of tomato fruit. *Biological Control* 54, 316–321.

Chapter 10

Leafy Vegetables

John Golding, Len Tesoriero and Rosalie Daniel
New South Wales Department of Primary Industries, Gosford, NSW, Australia

Abbreviations

CO_2	Carbon dioxide
FRAC	Fungicide Resistance Activity Committee
Gp	Group
O_2	Oxygen
RH	Relative humidity

1 Introduction

Leafy vegetables, which include lettuce, spinach, and Brassicaceae crops, are important vegetables around the world. Leafy vegetables, also called leafy greens, salad greens, pot herbs, vegetable greens, or simply greens, are plant leaves eaten as a vegetable which are consumed for convenience and high nutrient value.

Postharvest diseases are responsible for the wastage of up to 50% of fresh leafy vegetables, particularly where preharvest and postharvest management is inadequate (Ambuko et al., 2017). Although leafy vegetables can be afflicted with a wide range of diseases caused by fungi, viruses, and bacteria, postharvest diseases are more likely to be caused by fungal and bacterial pathogens. Infection can occur prior to harvest or during harvest, grading, packaging, distribution, and storage.

The development and expression of postharvest diseases in leafy vegetables is a complex interaction of the pathogen, the host, and preharvest and postharvest conditions. Growing conditions in the field can significantly contribute to the level of postharvest disease (Nunes, 2008). For example, leafy vegetables grown under hydroponics with protective structures often have fewer leaf diseases than those grown under field conditions (Gruda, 2005). Reducing disease load and risk during all growing stages of the crop is essential to minimize postharvest disease losses (Coates et al., 1997).

The level of natural resistance of leafy vegetables to infection by microbial pathogens declines progressively after harvest. This is due to the removal of the produce from the main plant and roots, breakdown of chlorophyll, and the naturally high rate of general senescence of leafy vegetables after harvest (Pogson and Morris, 2004; Koukounaras, 2009). Postharvest handling and storage conditions can be managed to reduce these risks. For example, a move toward field packing of leafy vegetables can minimize handling and mechanical damage, thereby reducing the potential entry points for pathogens. In addition, improvements in storage technology (such as vacuum cooling) have also improved storage life and quality of leafy vegetables.

Understanding the pathogens involved in postharvest diseases and the conditions favoring disease development is essential to develop methods for effective disease management. There are ample, thorough, and informative publications on many postharvest pathogens and disorders and potential options for their management (Snowdon, 1992; Narayanasamy, 2005; Koike et al., 2007). In this chapter, we will not reiterate what has been published, but provide new information on selected postharvest pathogens of leafy vegetables including lettuce, crucifers (leafy brassicas

Brassica chinensis and *B. oleraceae*), Chinese cabbage (*B. pekinensis*), leafy mustards (various species: *B. juncea, B. campestris, B. alba, B. napus, B. nigra, B. rapa*), arugula (*Eruca vesicaria* subsp. *sativa*), wild rocket (*Diplotaxis tenuifolia*), silverbeet (or Swiss chard, *Beta vulgaris* subsp. *cicla*), and chard (baby beetroot leaves, *Beta vulgaris*). This chapter describes the cause, symptoms, disease cycle, and management of a range of important leafy vegetable diseases that are observed or expressed after harvest. These diseases include: bacterial leaf spots and rots; lettuce varnish spot and marginal leaf blight; black rot of brassicas; bacterial soft rots; major fungal leaf spots and rots (such as Alternaria black spot, gray leaf spot, anthracnose, Stemphylium leaf spot, Cladosporium leaf spot, and Cercospora leaf spot); white rust (white blister); gray mold (Botrytis rot); downy mildews; powdery mildew; and Sclerotinia rot.

In addition to postharvest diseases, leafy vegetables can also develop storage disorders that are not caused by pathogens. Several key storage disorders will be briefly described in this chapter because the symptoms of these physiological disorders can sometimes be confused with pathological infections. The prevention of these storage disorders is critical not only to reduce waste but also they are often sites for secondary microbial infection.

2 Postharvest Management

While preharvest infection is a common attribute of the development of postharvest leafy vegetable diseases, disease progression can often be minimized with good postharvest practices. Reduced mechanical damage during harvest, grading, and packaging reduces injury and sites for postharvest infection. While different leafy vegetables have different postharvest handling and storage requirements, there are some critical postharvest practices to minimize the development of diseases in leafy vegetables. The most important factor to maintain quality and storage life is postharvest temperature management. It is important to retard the respiration and senescence of harvested produce and reduce pathogen growth by rapidly lowering temperatures after harvest. In general, brassicas and leafy vegetables should be stored at 0°C. The optimal storage conditions of some selected leafy vegetables are presented in Table 10.1.

In addition, sanitation of the washing water is essential to avoid the spread of postharvest pathogens during the washing process. There is a range of specific recommendations for the postharvest handling, packaging, and storage of different types of leafy vegetables. These are widely published and should be followed to minimize the risk of developing postharvest diseases (USDA-ARS, 2016).

3 Bacterial Leaf Spots and Rots

3.1 Bacterial Spot of Lettuce

Causes

Bacterial leaf spot of lettuce is caused by *Xanthomonas campestris* pv. *vitians* (Brown) Dye.

Table 10.1 Optimal storage conditions of selected leafy vegetables (from USDA-ARS, 2016).

	Temperature (°C)	*Relative humidity (%)*	*Sensitivity to ethylene*	*Controlled atmosphere considerations*
Swiss chard (*Beta vulgaris* var. *cycla*)	0	95 to 98	Yes	2 to 3% CO_2 and 10% O_2
Lettuce (*Lactuca sativa*)	0	95 to 98	Yes	1 to 3% O_2
Bok choy (*Brassica campestris* ssp. *chinensis*)	0 to 5	>95	–	0.5 to 1.5% O_2
Chinese cabbage *Brassica rapa* (*B. campestris*) ssp. *pekinensis*	0	98	Yes	1 to 2% O_2 and 2 to 6% CO_2
Rocket salad, roquette, arugula, rucola, or rugula (*Eruca vesicaria* ssp. *sativa*)	0 to 2	95 to 100	Yes	–

Symptoms

Field infections of bacterial spot appear as black greasy lesions that are often delimited by veins and mostly develop on older leaves. As the infection progresses, these spots coalesce leading to leaf collapse. Secondary decay organisms such as soft rot bacteria also colonize damaged leaf tissue in transit and cause postharvest losses.

Disease Cycle

In the field, bacteria can be seed-borne and survive on crop residues or as epiphytic populations on certain weeds such as wild lettuce (Barak et al., 2001; Toussaint et al., 2012). Bacterial spread and infection are favored by wet conditions.

Management

The primary management tool against bacterial leaf spot of lettuce is to use pathogen-free seed. Field infections can also be managed by irrigation practices that minimize periods of leaf wetness and by ensuring crop residues and weeds are ploughed-in or removed (Subbarao et al., 1997; Fayette et al., 2017). Maintain the postharvest cool chain to slow the development of the disease in transit and storage, but even at the recommended storage temperatures, infections can spread.

3.2 Lettuce Varnish Spot and Marginal Leaf Blight

Causes

Pseudomonas viridiflava, P. cichorii, and *P. marginalis.*

Symptoms

These diseases are more common in head lettuce but can also affect leafy lettuce varieties under wet conditions. Varnish spot appears as brown shiny spots or streaks along the leaf veins (Figure 10.1). Marginal leaf blight is self-descriptive but can lead to symptoms similar to varnish spot where all three pathogen species are present. Veins can become darkened, and a soft watery rot occurs that can extend down to the butt of the lettuce during transit and storage (Figure 10.1).

Disease Cycle

Often all three bacteria can be associated with the infected tissue with either of the two disease symptoms (Davis et al., 1997). These bacteria are all common leaf surface inhabitants on a wide range of plants and in soils. They infect lettuce leaves through wounds or natural openings such as hydathodes. In postharvest storage, ethylene damage can weaken the tissue and predispose the leaves to infection (Grogan et al., 1977).

Management

Prevent physical damage to leaves in the field and at harvest. Schedule irrigation to minimize leaf wetness (Pauwelyn et al., 2010). Good postharvest practices, such as good handling and storage conditions (e.g., low ethylene levels in storage), also reduce the severity of these diseases.

3.3 Bacterial Leaf Blight and Leaf Spot of Swiss Chard and Chard

Causes

Pseudomonas syringae pv. *aptata.* This bacterium is mostly significant as a pathogen of sugar beet, but it can also cause postharvest losses when baby-leaf chard is infected.

Figure 10.1 a. Lettuce varnish spot (left). b. Marginal leaf blight (right) symptoms in lettuce.

Symptoms

Seed infections cause seedling blight while secondary spread causes dark water-soaked circular or ellipsoid (3–8 mm diameter) leaf spots with a dark border where their centers turn tan-colored. These spots eventually coalesce, causing large areas of the leaf surface to turn necrotic.

Disease Cycle

This pathogen can be seed-borne and spreads with irrigation water and water splash. Disease symptoms are mostly observed during wet cool or mild conditions. Differentiation of this bacterial pathovar from similar strains has been difficult, and it is not clear if those records from a range of other hosts, such as cereal grain crops, legumes, solanaceous vegetables, lettuce, and melons, can also cause bacterial blight of sugar beet and chard. If they do cross-infect beets and chards, then it has a very wide host range. For instance, Morris et al. (2000) showed that cantaloupe isolates from France also affect sugar beet.

Management

The primary mode of management is to decontaminate seed. In addition, traditional copper sprays have been used to limit bacterial spread.

3.4 Black Rot of Brassicas; Bacterial Leaf Spot of Wild Rocket, Broccolini; and Bacterial Blight of Arugula and Broccoli Raab (Rappini)

Causes

Black rot is caused by *Xanthomonas campestris* pv. *campestris* and affects a wide range of brassica crops. Bacterial leaf spot of wild rocket is caused by *X. campestris* pv. *raphani* (Pernezny et al., 2007). Bacterial leaf spot of broccolini and other brassicas is caused by *Pseudomonas syringae* pv. *maculicola* and bacterial blight of arugula, broccoli, and broccoli raab is caused by *Pseudomonas syringae* pv. *alisalensis* (Bull et al., 2004) (Figure 10.2).

Symptoms

The symptoms of black rot are yellow or brown V-shaped lesions when infection has entered from the leaf margin (Figure 10.2). Symptoms of bacterial blight of broccoli raab and arugula initially are small water-soaked flecks on the lower foliage, which expand and develop a yellow halo. Lesions coalesce forming large necrotic patches. Wild rocket affected by bacterial leaf spot develops similar fleck symptoms that rarely exceed 1 mm in diameter.

Disease Cycle

Xanthomonas campestris is seed-borne and is spread in water, mechanically, and by insects. *Pseudomonas syringae* pv. *alisalensis* has a wide host range in the Brassicaceae, *Solanum* spp., and some grass species (Cintas et al., 2002). They found some strains

Figure 10.2 a. Black rot of cabbage (top left). b. Bacterial leaf spot of wild rocket (top right). c,d. Black rot of leafy Brassicas (middle and lower left). e. Bacterial leaf spot of arugula (lower right).

were pathogenic when artificially inoculated on a range of brassicas, tomato, oats, and brome grass.

Management

As the bacteria are spread by water, avoiding overhead irrigation is a simple management tool. Noting the wider host range of *P. syringae* pv. *alisalensis*, it is essential to consider the appropriate choice of rotation crops such as grass cover crops to minimize its survival in the field.

3.5 Bacterial Soft Rots

Causes

Soft rot enterobacteria are the most important causative agents of soft rot diseases of leafy vegetables during crop production and postharvest storage. They are comprised of a number of bacteria that were formerly known as pectinolytic species and subspecies of the genus *Erwinia*. This group has undergone several revisions in their nomenclature as well as the descriptions of new taxa since the advent of multilocus phylogenetic analyses of gene sequences as well as analyses of electrophoretic patterns of repetitive DNA fragments (Adeolu et al., 2016). The current and respective former names are *Pectobacterium* spp. (formally *E. carotovora*) and *Dickeya* spp. (formally *E. chrysanthemi*). Most of these species have broad host ranges. One particular example is a soft rot disease of brassicas that is caused by *P. carotovorum* subsp. *odoriferum* (Oskiera et al., 2017). This bacterium also occurs in a range of stored vegetables (Waleron et al., 2013). Another pathogen, *P. aroidearum*, was more recently described as causing soft rot of Chinese cabbage in China (Xie et al., 2018). This pathogen was previously described from strains occurring on certain monocots and potatoes (Nabhan et al., 2013). Certain *Pseudomonas* spp. (such as *P. marginalis* and *P. viridiflava*) can also cause or be associated with bacterial soft rots on various leafy host plants (also see lettuce varnish spot).

Symptoms

Characteristic symptoms are a slimy wet rot with associated leaf yellowing and brown to black soft mushy lesions (Figure 10.3) which is accompanied by an unpleasant odor.

Disease Cycle

Pectobacteria are aggressive necrotrophs that secrete cell-wall-degrading enzymes as their primary virulence determinants. They macerate host tissue through the coordinated production of pectinases, cellulases, hemicellulases, and proteinases through a type II secretion system (Cianciotto and White, 2017). In addition, they secrete proteins that promote plant cell death (Davidsson et al., 2013). These bacteria can colonize latently and use quorum-sensing mechanisms to activate their arsenal of degrading enzymes (Pérombelon, 2002). The disease thrives under warm and humid conditions. It is not completely clear how these soft rot bacteria survive between growing seasons, although they can survive on decomposing plant material. They are not thought to overwinter for long periods and only survive in soil for

Figure 10.3 Bacterial soft rot symptoms. a. Pak choi with macerated petiole tissue (top left). b. Head lettuce (top right). c. Individual lettuce leaf showing vein browning and areas of tissue maceration (center left). d. Bok choi with leaf yellowing and petiole maceration (center right). e. Baby spinach with dark areas of leaf maceration (bottom).

weeks or months. They can be spread with irrigation water via aerosols and rain splash, mechanically on equipment and footwear, and by insects such as flies. After harvest, soft rot bacteria are often secondary invaders of wounds or lesions caused by pests, pathogens, and certain physiological disorders such as tip-burn or damaged tissue caused by other environmental or nutritional factors.

Management

Strict crop sanitation and hygiene is important during cropping to avoid plant injury and bacterial spreading. Other management options include scheduling crops to allow complete disintegration of crop residues before planting, reducing plant density, and improving soil drainage or rotation with nonhosts such as cereals. Postharvest losses can be reduced by not harvesting crops when they are wet, frequent disinfection of cutting knives, using disinfected wash water where appropriate, rapid cooling, and maintenance of a cool chain during transit and storage.

4 Fungal Leaf Spots and Rots

Many fungi can cause leaf spots on leafy vegetables that can then affect their quality and shelf life following harvest. The following are descriptions of the most important fungal leaf spots and rots in leafy vegetables.

4.1 Alternaria Black Spot, Gray Leaf Spot

Causes

A number of species of the fungal genus *Alternaria* can affect vegetable brassicas and rocket. Two species are reported as the main causes: *Alternaria brassicicola* (Schwein.) Wiltshire and *A. brassicae* (Berk.) Sacc. Any of a complex of strains of *A. alternata* (Fr.) Keissl., *A. arborescens* E.G. Simmons, and *A. tenuissima* (Nees) Wiltshire have also been shown to cause dark leaf lesions on these vegetables. Another species, *A. japonica* Yoshii, occurs on rocket and radish.

Symptoms

Typical symptoms are black or gray necrotic lesions often with yellow margins on seedlings, leaves, stems, flower heads, and seed pods (Figure 10.4). Circular leaf spot symptoms on brassica vegetables caused by *A. brassicicola* and *A. brassicae* are often indistinguishable, while *A. japonica* causes dark lesions on wild and cultivated rocket spreading to veins and petioles. Disease symptoms are often expressed on the older leaves first.

Disease Cycle

Alternaria spp. can be seed-borne and grow as saprophytes on organic matter. When growing on plants, they produce chains of conidia that become airborne or are dispersed in water. Airborne dispersal of conidia generally occurs with rising temperatures and decreasing relative humidity (RH). Infection occurs through

Figure 10.4 a. Alternaria spots on leafy brassica with concentric ring patterns (top left). b. Dark leaf spots on Bok choi with "shot-hole" symptoms (top right). c. Yellowing of kale leaves associated with necrotic spots and patches (bottom).

wounded tissue, through stomata, or directly through the epidermis. The release of a range of phytotoxic metabolites from the fungus damages cells and allows the pathogen to invade. Some of these toxins have been shown to be important for virulence and/or pathogenesis while some have been reported as carcinogens of mammalian tissue (Tsuge et al., 2013; Wu et al., 2014; Meena et al., 2017). Brassica weeds and crop residues can also harbor these fungal infections and spores. While *Alternaria* species have adapted to different climatic conditions, they generally do not tolerate low temperatures. Secondary fungi and bacteria can infect black spot damaged tissue causing further breakdown during storage and transport to market.

Management

Some brassicas have moderate resistance to *Alternaria* pathogens which are thought to be associated with the thickness of the waxy layer on leaf surfaces, stomatal density, and other factors (Nowicki et al., 2012). Incorporation of infected brassica crop residues, a three-year rotation with non-brassica crops, and control of brassica weeds are thought to reduce primary infections.

Scheduling crops to avoid periods of warm and humid periods can be useful in some growing regions. Seed disinfection or chemical treatments of seed or in

Figure 10.5 Anthracnose symptoms. a. Lettuce anthracnose (left) showing typical elongated spots on the underside of a leaf with brown margins. b. Spinach anthracnose (right) with black setose acervuli sporulating in the center of leaf spots.

the field can also reduce the risk of disease. The maintenance of the postharvest cool-chain is important to reduce the risks of disease expression after harvest.

4.2 Anthracnose

Causes

Colletotrichum higginsianum Sacc. affects Chinese cabbage and turnip; *Colletotrichum dematium* f. *spinaciae* (Ellis & Halst.) Arx affects spinach, and *Microdochium panattonianum* (Berl.) B. Sutton, Galea & T.V. Price occurs on lettuce.

Symptoms

On brassicas, leaf spots are small and dry, with a pale center which may become perforated (Figure 10.5). During transit and storage spots may coalesce causing large necrotic areas and leaves to yellow. The symptoms of spinach anthracnose are water-soaked leaf lesions that turn brown, papery, and covered with black fungal acervuli with spiky setae. As the disease progresses, whole leaves turn yellow. In lettuce, anthracnose produces small (2–5 mm) water-soaked circular or elliptical spots with pale to brown (to black) and papery centers that can drop out giving a shot-hole appearance. Spots usually occur along the midrib on lower leaf surfaces fanning out onto the leaf blade. Secondary bacterial rots and soft rots invade infected tissue during transit and storage.

Disease Cycle

Colletotrichum higginsianum overwinters on brassica crop residues and volunteers as well as related weeds. It may also be seed-borne. Infection occurs under moist conditions, generally in mild or warm conditions prior to harvest, but storage rots can occur at lower temperatures. It has been shown that *C. dematium* f. *spinaciae* has a similar disease cycle on spinach and may also be seed-borne (Correll et al., 1993). While *M. panattonianum* infects plants under cool wet conditions and produces microsclerotia that survive in the soil (Patterson and Grogan, 1991). It also infects common weeds such as prickly lettuce (*Lactuca serriola*) and related crops such as endive, but it is not known to be seed-borne.

Management

Infected brassica or spinach seed can be hot-water treated. Cultural management such as good sanitation through crop rotation, weed control, careful irrigation scheduling to minimize the periods of leaf wetness, and good soil drainage all contribute to reducing disease risks from these pathogens. Fungicides such as the demethylation inhibitor (DMI) chemical group member, prochloraz, have been used in management programs (Wicks et al., 1994).

4.3 Stemphylium Leaf Spot

Causes

Stemphylium leaf spot occurs in the USA, Europe, and Australia. The disease is caused by the fungus *Stemphylium botryosum* Wallr. Isolates from spinach are host-specific to spinach (Koike et al., 2001).

Symptoms

Following infection, small, circular to oval gray-green-colored spots, 2–6 mm in diameter, develop on leaves. These spots become larger and darker brown as the disease progresses. Old lesions appear dry and papery in texture (Figure 10.6). This disease can be differentiated from other foliar diseases of spinach by the absence of signs of fungal growth. However, symptoms can be confused with damage due to chemical burn.

Disease Cycle

Little is known about the disease cycle of Stemphylium leaf spot. The pathogen is seed-borne (Hernandez-Perez and du Toit, 2006). In the field the disease spreads slowly. Lesions on leaves make affected spinach undesirable for marketing.

Management

Several studies have shown that hot water and chlorine seed treatments are effective in reducing seed-borne inoculum (du Toit and Hernandez-Perez, 2005). Overhead irrigation can exacerbate the disease.

Figure 10.6 Stemphylium leaf spots with brown papery centers.

4.4 Cladosporium Leaf Spot

Causes

Cladosporium leaf spot of spinach is caused by the fungus *Cladosporium variabile* (Cooke) G.A. de Vries.

Symptoms

Leaf spot due to *Cladosporium* infection initially develops as round, brown lesions, generally less than 10 mm in diameter. Over time, dark green mycelium and spores develop in the center of the lesions (Figure 10.7). The occurrence of fungal structures on the lesion enables the disease to be distinguished from Stemphylium leaf spot.

Disease Cycle

Details of the disease cycle of Cladosporium leaf spot are lacking. The fungus is seed-borne (Hernandez-Perez and du Toit, 2006) and disease is favored by cool, wet weather. The conidia on leaves are dispersed by wind and by water splash.

Management

Management of Cladosporium leaf spot is similar to that of Stemphylium leaf spot.

Figure 10.7 a. Pale leaf spots with darker concentric rings caused by *Cladosporium variable* on spinach on lower (left). b. Upper leaf surfaces (right).

4.5 Cercospora Leaf Spots

Causes

Cercospora leaf spot is caused by the fungus *Cercospora beticola* Sacc. and is a major disease affecting silverbeet, chard, and to a lesser extent spinach. *Cercospora lactucae-sativae* Sawada (syn.: *C. longissima* Cooke & Ellis) affects lettuce, endive, and related plants. Cercospora leaf spot of brassicas is caused by *C. brassicicola* Henn., while *C. acetosella* Ellis affects sorrel.

Symptoms

The first symptoms on silverbeet and chard appear as small, white spots on the surface of the leaf (Figure 10.8). The white spots typically have a white halo around them and become gray as the disease progresses. Typical symptoms on lettuce are brown circular spots up to 15 mm in diameter with a white center and sometimes with a chlorotic halo (Figures 10.8c, d). Spots can coalesce causing large necrotic areas on leaves. Older leaves generally display more pronounced symptoms (Figure 10.8e).

Disease Cycle

Cercospora beticola can be seed-borne. Cercospora leaf spot of beet is most severe under conditions of intermittent leaf wetness and high RH. It is dispersed with irrigation water and it overwinters as sclerotia in infected leaves which survive in soil for up to 2 yr. *Cercospora lactucae-sativae* also spreads with water splash and is

Figure 10.8 Cercospora leaf spots. a. *Cercospora lactucae-sativae on lettuce* (top left). b. *Cercospora acetosella* on sorrel (top right). c. *Cercospora beticola* on chard (center left). d. *Cercospora brassicicola* on leafy brassica (center right). e. Cercospora leaf spot on swiss chard (bottom).

favored by higher temperatures (25–35°C) and high humidity or leaf wetness during rainy periods.

Management

Resistant chard cultivars are available, but seed treatments reduce the risk of seed-borne infections. Seed treatment, sanitation practices, crop rotation, weed management, and fungicides are used for disease management in the field. There is a wide range of fungicides with efficacy against *Cercospora* spp. Availability differs with various country registrations. The following fungicides are listed under their Fungicide Resistance Activity Committee (FRAC) chemical group and code numbers (in parentheses): dicarboximides (Gp2); demethylation inhibitors (Gp3); succinate dehydrogenase inhibitors (Gp7); strobilurins (Gp11); pyrimidines (Gp9); phenylpyrrols (Gp12); benzimidazoles (Gp1); and dithiocarbamates (Gp M3).

4.6 White Rust (White Blister)

Causes

White rust is a disease caused by the oomycetes *Albugo* spp. They primarily affect spinach (*A. occidentalis* G.W. Wilson and *A. ipomoeae-aquaticae* Sawada) and crucifer vegetables including rocket, radish, and broccoli [*A. candida* (Pers. ex J.F. Gmel.) Kuntze].

Symptoms

Symptoms first appear as small white blisters on the underside of leaves and then on the upper side (Figure 10.9). The disease can also affect broccoli florets.

Disease Cycle

Infection typically occurs in the field prior to harvest but can significantly affect postharvest quality and marketability.

Management

Broccoli should not be stored wet, as free moisture can increase stem rots and splitting as well as increase diseases in florets. Chemical control means are similar to those listed next for downy mildews (Section 4.8).

4.7 Gray Mold (Botrytis Rot)

Causes

Gray mold is caused by the fungus *Botrytis cinerea* Pers. and it is a common disease of lettuce, but also affects brassicas. It can also be a significant problem in stored vegetables.

Symptoms

In brassica vegetables, brown, water-soaked lesions from 10–20 mm to much larger legions develop on stems and leaves. The fungus may be observed as pale gray

Figure 10.9 White rust (white blister) symptoms. a. *Albugo ipomoeae-aquaticae* (top left) on water spinach. b. *Albugo candida* on leafy brassica on (top right). c. Postharvest development of white blister on leafy brassica (lower left). d. *A. candida* on rocket (lower right).

fluffy mycelium in the center of lesions, or more generally over the vegetable. Symptoms are typically associated with senescence or physical damage.

Water-soaked lesions develop on affected lettuce. As the disease progresses, the fungus produces profuse pale gray, fluffy mycelium, becoming darker as it sporulates. The lettuce becomes soft and mushy. Black sclerotia may form on diseased plant parts.

Disease Cycle

Botrytis infection of vegetables occurs in the field. The fungus survives between crops as a saprophyte on crop debris and weeds and as sclerotia in soil. Conidia are windborne. Conidia germinate in the presence of free water and infected damaged or senescing tissues. The fungus then colonizes the healthy stems and leaves. Once established, the fungus may cause disease symptoms in the field or in storage. Disease development is favored by cool temperatures, free moisture, and high RH. When environmental conditions are favorable, gray mold can also occur in greenhouse and field-grown lettuces.

Management

Damage caused by chemical burns, pests and other pathogens, environmental stresses, nutritional imbalances, and poor handling can predispose vegetables to infection. Preventing injury and providing sufficient nutrition can reduce the chance of infection occurring. Reducing leaf wetness by avoiding or reducing sprinkler irrigation can make conditions less favorable for infection to occur in the field. Packing lettuces when the leaves and heads are wet can also increase the risk of postharvest disease development. Therefore, pack the lettuces with no excess free water. In addition, removing the outer, older, infected leaves can reduce incidence of gray mold in brassica vegetables. Fungicide programs are also effective management tools, but care must be taken with fungicide groups and rotations because of the high genetic variability of *B. cinerea* and potential for development of fungicide resistance (Kim et al., 2016). Keep the product at the recommended storage temperature to reduce the spread of the disease.

4.8 Downy Mildews

Causes

Downy mildew is the devastating disease threatening sustainable spinach production worldwide (Choudhury et al., 2017). Downy mildew of lettuce is considered the most significant foliar disease of lettuce globally (Koike et al., 2007). Downy mildew is caused by obligate oomycetes. Downy mildew diseases are common worldwide, affecting brassica, spinach, beet, and lettuce crops. While they are often significant at the establishment stage, these diseases can affect marketability if they develop after harvest. In particular, downy mildew is a preharvest disease of rocket, but symptoms can be more obvious following harvest and washing.

On crucifer vegetables it is caused by *Hyaloperonospora parasitica* (Pers.) Constant. [syn.: *Peronospora parasitica* (Pers.) Tul.], while on lettuce it is caused by *Bremia lactucae* Regel. The fungus affects all lettuce and other species (*Lactuca* spp.), as well as artichoke and other plants in the Asteraceae family. *Bremia lactucae* consists of multiple races or pathotypes that are characterized by inoculating a set of lettuce varieties carrying different resistance genes.

On spinach, downy mildew is caused by *Peronospora effusa* (Grev.) Rabenh. (syn.: *P. farinosa* f.sp. *spinaciae*). This pathogen occurs in all spinach growing regions, and co-occurs with other closely related downy mildew species. There are several races for which specific resistance genes have been found. A related oomycete, *P. farinosa* f.sp. *betae* Byford, affects silverbeet and chard.

Symptoms

Early symptoms are typically observed as irregular yellow lesions on the upper surface of leaves (Figure 10.10a). The lesions may appear angular between the major leaf veins. On the underside of the leaf, white "downy" fungal growth can be seen (Figure 10.10c). If growing conditions are cool and humid, sporulation can occur on the upper leaf surface. The center of older lesions becomes papery, and dark specks may be observed (Figure 10.10b). Older leaves are typically affected first. Infection may also be systemic, resulting in gray or black flecking and streaking of internal tissues.

Disease Cycle

Humid, cool (10–15°C) conditions favor sporulation, germination, and infection of downy mildew pathogens. Lesion expansion is favored by milder temperatures (around 20°C). Sporangia are dispersed by wind and by water splash. In the field, inoculum typically comes from surrounding infected plants, and in some areas, oospores. The pathogen may survive as oospores in crop residue and in soil.

Figure 10.10 Downy mildew symptoms. a. White tufts of *Peronospora effusa* on undersurface of baby spinach leaf (top left). b. Downy mildew of Swiss Chard (top right). c. *Bremia lactucae* on lettuce (lower).

Management

Infection in the field can be managed by improving ventilation around the plants, irrigating early in the day so that foliage can be readily dried, and by fungicide application. Resistant lettuce and spinach cultivars are available. But the development of new races creates an ongoing challenge for breeders and growers. Fungicides are available to manage these diseases in the field, and little can be done after harvest to control the spread of downy mildew except storing at the recommended temperature. A range of different chemical groups are available subject to use regulations in different countries. Common FRAC activity groups are: phenylamides (Gp4); strobilurins (Gp11); carbamates (Gp28); carboxylic acid amides (Gp40); benzamides (Gp43); oxathiopiprolin (Gp49); as well as copper (Gp M1) and dithiocarbamates (Gp M3).

4.9 Powdery Mildew

Causes

Powdery mildew of lettuce is caused by the obligate biotrophic ascomycete *Golovinomyces cichoracearum* (DC.) V.P. Heluta (syn.: *Erysiphe cichoracearum* DC.). The fungus has a broad host range including *Lactuca* spp., wild Asteraceae and cultivated lettuce species, artichoke, chicory, and endive (Lebeda and Mieslerova, 2011). Powdery mildew of lettuce in Korea is caused by *Podosphaera fusca* (Fr.) U. Braun & Shishkoff. Powdery mildews are cosmopolitan fungi affecting almost 10,000 plant species. Lettuce powdery mildew is considered a minor disease but can significantly impact quality and marketability. Powdery mildew of brassicas is caused by *Erysiphe cruciferarum* Opiz ex L. Junell, while powdery mildew of beets and chard is caused by *E. polygoni* DC. or *E. heraclei* DC. An excellent review of lettuce powdery mildew is given by Lebeda and Mieslerova (2011).

Symptoms

Powdery mildew appears as patches of white powdery growth on the upper and lower leaf surface and can cover the entire leaf (Figure 10.11). Older leaves and mature plants are often affected first, becoming chlorotic then necrotic. Symptoms can develop rapidly.

Disease Cycle

Powdery mildew more commonly affects mature plants rendering lettuce unmarketable. White mycelium and powdery spores can develop on both sides of the leaf.

Management

Crop sanitation including the proper disposal of infected plant material can lower disease incidence in production areas. Crop rotation is also an important cultural control of powdery mildew as the pathogen can persist in the soil. Continually growing the same crop without rotation will encourage the formation of sclerotia in soil that can persist for long periods. Nutrient management has been shown to affect disease development, where high nitrogen fertilizer rates can favor infection. There are a number of tolerant/resistant plant cultivars to powdery mildew. In addition, powdery mildew can be effectively

Figure 10.11 Powdery mildew symptoms in: a. Kale (left), b. Swiss chard (middle), and c. Radicchio (right).

managed by common fungicides, and there is a wide range of fungicides with efficacy against powdery mildew pathogens. FRAC fungicide activity groups used include demethylation inhibitors (Gp3), succinate dehydrogenase inhibitors (Gp7), strobilurins (Gp11), benzimidazoles (Gp1), dithiocarbamates (Gp M3), and inorganics (Gps M1 and M2). Products containing potassium bicarbonate and petroleum or botanical oils are also used for powdery mildew control.

4.10 Sclerotinia Rot

Causes

Sclerotinia sclerotiorum (Lib.) de Bary and *S. minor* Jagger cause diseases in a wide range of plants. They cause a disease called "drop" in lettuce; while in *Brassicaceae* crops they cause the disease known as white mold, drop, or watery soft rot.

Symptoms

Water-soaked brown infected areas form on leaves and stems, which are often accompanied by white cottony fungal growth and black globular or irregular-shaped resting sclerotia (Figure 10.12). The two *Sclerotinia* spp. are distinguished by the different sizes of their sclerotia: *S. minor* having smaller match-head-sized ones while those of *S. sclerotiorum* are larger (10 mm or more in length). The soft watery rot can further develop and spread during transit and storage.

Disease Cycle

Sclerotia survive in soil for varying lengths of time depending on climatic conditions and cultural practices (Wu et al., 2008). Those of *S. sclerotiorum* can survive in soil for 3–4 yr. Both species infect plants at ground level through fungal mycelium from germinating sclerotia. The fungus invades through senescing plant tissue, through

Figure 10.12 Sclerotinia rot symptoms. a. *Sclerotinia sclerotiorum* forming black sclerotia of cabbage head (top left). b. *Sclerotinia sclerotiorum* causing a watery rot of lettuce petioles (top right). c. *Sclerotinia minor* producing fluffy white mold on leafy lettuce in storage (lower).

wounds, or directly through epidermal cells. Wet and humid conditions favor sclerotial germination and infection. Sclerotia, particularly those of *S. sclerotiorum*, also reproduce sexually by forming mushroom-like structures called apothecia, which release ascospores into the air.

Management

Crop rotations of three or more years with nonhost crops reduce soil populations of sclerotia. Cover cropping with certain cereals (such as oat, barley, or millet) has been shown to reduce Sclerotinia rot in subsequent crops. Members of the following fungicide chemical activity groups are used for control (FRAC codes in parentheses): benzimidazoles (Gp1); dicarboximides (Gp2); demethylation inhibitors (Gp3); succinate dehydrogenase inhibitors (Gp7); strobilurins (Gp11); pyrimidines (Gp9); phenylpyrrols (Gp12); and dithiocarbamates (Gp M3). Microbial biological controls are also commonly used. There are numerous formulated products containing hyperparasitic fungal strains of *Trichoderma* spp. and *Coniothyrium minitans* W.A. Campb. or bacterial strains of *Bacillus, Pseudomonas*, or *Streptomyces* spp.

Most leafy brassicas are less susceptible to these pathogens than their heading-type relatives.

5 Physiological Disorders

5.1 Introduction

Physiological disorders involve the breakdown of tissue that is not directly caused either by pests and diseases or by mechanical damage. These disorders are briefly described and included in this chapter, as the symptoms of physiological disorders can sometimes be confused with those from microbial infection, and these disorders are often sites for secondary infections.

There are literally hundreds of described disorders and injuries in lettuce, spinach, and Brassicaceae crops. The names and classification of these disorders are very descriptive and reflect their often indeterminate and seasonal nature. In addition, disorders are generally nondescript, for example, occurrence of "general tissue browning/discoloration." Because of the physiological and nonpathological nature of these disorders, the nomenclature of physiological disorders is often clouded and confused. For example, rib discoloration in lettuce has been known as rib breakdown, brown rib, red rib, and rib blight. Each of these descriptors may or may not be the same disorder. Indeed, the biochemical and biophysical mechanisms that give rise to physiological disorders are further complex and are not well understood. While the description of most disorders might be considered nonscientific, describing and understanding these disorders is a critically important step toward optimizing postharvest storage and handling practices (Wills and Golding, 2016).

Physiological disorders may develop in response to various preharvest growing and postharvest storage conditions, such as nutrient deficiency that occurs in the field and low temperature stress during storage. Indeed, many of the pre- and postharvest factors interact leading to complex interactions and expressions of physiological disorders. The effect of the preharvest growing environment on the development of many disorders cannot be overstated where often unknown preharvest growing conditions affect postharvest storage life and quality.

This section focuses on some of the common physiological disorders of lettuces and describes the symptoms, causes, and management of russet spotting, tip burn, rib discoloration, and pink rib. These disorders were selected as representative physiological disorders where the main causative agents are the preharvest growing conditions such as nutrition (e.g., tip burn and gomasho) and in-field heat stresses (e.g., rib discoloration), or the result of adverse storage atmosphere [e.g., ethylene-russet spotting and pink rib-high carbon dioxide (CO_2)]. The inter-relationship between pre- and postharvest factors in the development of these disorders is highlighted.

5.2 Russet Spotting

Symptoms

Russet spotting is a physiological disorder of lettuce, mainly Iceberg lettuce, that is characterized by many small brown oval spots, mostly on the midribs, but also on the green leaf tissue, if severe (Figure 10.13).

Figure 10.13 Russet spotting symptoms. a. Severe russet spotting symptoms on iceberg lettuce after cold storage (left). b. Close-up of russet spotting symptoms on the leaf blade (right).

Causes

Russet spotting can occur in lettuces that have been harvested overmature or stored at too high a temperature, but the primary cause is exposure to ethylene. Ethylene is the natural ripening hormone in fruit and vegetables, but exposure to low levels of ethylene at storage temperatures around 5°C induces russet spot development. Ethylene has been shown to promote the synthesis, accumulation, and oxidation of phenolic compounds in the leaf (Ke and Saltveit, 1989b), which leads to the classic russet symptoms (Figure 10.13). Susceptibility to russet spotting depends on storage temperature, ethylene concentration, oxygen (O_2) and CO_2 levels, maturity, and cultivar (Saltveit, 2016), and maybe related to the phenylpropanoid pathway with the subsequent synthesis and accumulation of *o*-diphenols (Saltveit, 2018).

Management

Russet spotting can be controlled with correct harvesting, cooling, and storage. Ensure that lettuces are stored below 2°C and kept away from any sources of ethylene gas, such as other high ethylene producing fruit (e.g., apples, melons, peaches, etc.) or machinery emitting ethylene (such as forklifts with internal combustion engines). There is also a strong genetic component in the susceptibility to russet spotting and tolerant cultivars are available (Hunter et al., 2017). Controlled atmosphere storage has been shown to reduce russet spotting, but it is expensive and other simpler postharvest management is available, such as reducing ethylene accumulation.

5.3 Tip Burn

Symptoms

Tip burn is a preharvest disorder which is often observed after harvest. The symptoms of tip burn are necrotic browning of the edge of young developing leaves in the inner part of the head. This not only occurs in lettuce (Figure 10.14a) but also in other Brassicas such as white cabbage, Chinese cabbage (Figure 10.14b), and

Figure 10.14 Tip burn symptoms on: a. Lettuce (left) and b. Chinese cabbage (right).

Brussels sprouts. Leaves with tip burn have brown, often necrotic leaf margins which become thin and brown, later turning dark brown to black. The disorder generally only affects the leaf edge but may extend to half the leaf. The symptoms of tip burn develop in the field but are often not observed as they generally develop on the inner leaves, and these only become visible when the lettuce head is cut or the outer leaves are removed. However, the symptoms of tip burn can become more severe after harvest when the leaves lose water and browning reactions have developed. A major issue with tip burn is that the injured necrotic tissue is predisposed to attack by soft rot bacteria.

Causes

Calcium is a critical element in the growth of vegetable tissues where it is important for cell membrane integrity and cell wall strength (Wills and Golding, 2016). Growing leaves have a high requirement of calcium for the formation and expansion of cell walls, but calcium transport to these growing leaves within the lettuce head is restricted leading to localized deficiency and necrotic symptoms resulting in typical tip burn symptoms. Saure (1998) reviewed the relationships among different external factors (such as growing temperature, light, and soil conditions) and the variability in its incidence.

Management

Due to the variable nature of tip burn, it is difficult to manage in the field. While minimizing stress (such as water stress and high temperatures) can reduce the incidence of tip burn, its incidence is sporadic and variable and therefore good soil and water management is essential. There have been reports of different lettuce cultivars varying in their tolerance to tip burn (Collier and Tibbitts, 1982), but all lettuces can be susceptible so careful agronomic management is essential.

5.4 Rib Discoloration

Symptoms

The symptoms of rib discoloration start with a creamy-yellow or light brown area on the inner surface of the midrib or lateral ribs of the leaf. The discoloration is greatest on the inner surface of outer head leaves and often occurs at the curvature of the prominent midrib where the affected area is less than 20 mm. The affected tissues are firm and not sunken, pitted, or slimy. As the disorder develops, rib discoloration becomes reddish-green or blackish-brown in color and extends to larger and smaller veins and finally into the leaf, where it becomes visible on both surfaces of the leaf (Jenni, 2005). The affected tissues become sunken and occasionally cracked. This severely damaged tissue is often then affected by secondary soft rots.

Causes

Rib discoloration disorder is caused by heat stress in the field, 2 wk after the heading stage (Jenni, 2005). High daytime temperatures are thought to result in cellular membrane damage leading to disruption of cell function, oxidation of phenolic compounds, and localized browning.

Management

Managing the outside field temperature is difficult, but reducing other plant stresses, such as water stress, can assist in reducing the incidence of heat stresses. This disorder can ultimately be managed by growing lettuces to ensure that harvest occurs before the risk of heat stress.

5.5 Pink Rib

Symptoms

The first symptoms of pink rib start on the main ribs (veins) at the base of the leaves of mature lettuce heads and assume a faint, diffuse, pink color that intensifies with time. Most commonly, the pink color only extends from the point of leaf attachment to that area where the main rib branches into several veins. In severe cases, the veins that extend into the leaf margin are affected. Although it is at its most intense on the inner (upper) surface, it is often visible externally.

Causes

Typical of most physiological disorders, the cause of pink rib is unknown. A number of factors have been found to affect the incidence of pinking in lettuce, such as time of transplanting, head maturity, and lettuce type or genotype (Monaghan et al., 2017).

And while the development of pinking symptoms in harvested lettuce can occur as physiological response to wounding/cutting, rib pinking can often occur in whole lettuces. Deficit irrigation for a period of 2–3 wk in the middle or end of the growth period has been shown to reduce the incidence of pink rib (Monaghan et al., 2017). Pink rib symptoms can also develop during storage, where storing lettuces at above recommended temperatures has been shown to increase the incidence of the disorder.

Management

Due to the unknown causes of pink rib, its control and management are difficult. Withholding irrigation at specific times during growth can reduce the level of pink rib, but it also negatively affects yields and may not be considered an economically viable solution (Monaghan et al., 2017). The use of good postharvest techniques, such as rapid cooling and correct storage, as well as avoiding prolonged storage times, help minimize the disorder.

5.6 Brown Stain

Symptoms

Brown stain is characterized by slightly sunken oval spots with margins that are often darker than the slightly sunken centers of the lesion. The symptoms are large, irregularly shaped brown spots or streaks and tend to be on or near the midrib, generally toward the base. If the dark borders are absent, the brown stain can resemble russet spotting (Section 5.2) but may be distinguished from it by frequent association with heart-leaf injury, another symptom of CO_2 damage (Snowdon, 1992).

Causes

Brown stain is caused by exposure to high CO_2 levels (>2% CO_2) especially during extended shipment or storage, and combining high CO_2 with low O_2 can further increase the disorder (Ke and Saltveit, 1989a). Sensitivity to brown staining varies not only with cultivar and growing area but also with the time of day at harvest. Forney and Austin (1988) showed that lettuces picked in the afternoon were more tolerant of high CO_2 levels than were heads picked in the morning. As with many other physiological disorders, the development of the typical brown stain symptoms is not always visible when lettuce is held under CO_2-enriched atmospheres but become more obvious after the tissue is transferred to air.

Management

The simplest solution to control brown stain is not to expose lettuces to high CO_2 levels. With high respiration rates of lettuces, it is important to ensure there is no accumulation of CO_2 within the storage atmosphere. Development of brown stain is highly dependent on storage temperature, with the highest incidence at 0°C and being negligible at 10°C (Brecht et al., 1973). This is a problem, as lettuces need to be stored at temperatures as low as possible and highlights the complexity of managing physiological disorders.

5.7 Gomasho

Symptoms

Gomasho is a physiological disorder in Chinese cabbage types (*Brassica rapa*), particularly *Brassica rapa* subsp. *pekinensis*. The disorder is also known as petiole spot, pepper spot, or black fleck and affects the petiole and white midrib tissue of Chinese cabbage types (wombok, napa cabbage, etc.). Gomasho is often observed in the field and can also develop during cold storage. The classic symptoms are dark spots or flecks about 1–2 mm in diameter on the petioles (Figure 10.15). The first symptoms of dark spots are small, dark circular or elongated spots that first appear on the white midribs of the outer leaves and then spread to the middle inner leaves. Initial darkening occurs at the juncture of two or more epidermal cells, spreading to include 20 or more cells that collapse to form the typical pepper spot lesion.

Causes

As with most physiological disorders, the cause of gomasho is not clear. Genetics and in-field nutrition have been shown to affect the severity of symptoms. For example, the use of high levels of nitrogen fertilizers, particularly nitrate, has been shown to be associated with the disorder (Phillips and Gersbach, 1989). In addition, high soil pH, high

Figure 10.15 Symptoms of gomasho or "pepper spot" on the rib of Chinese cabbage.

levels of copper, and low levels of boron have been associated with gomasho (Klieber, 2001). The development of the disorder generally increases during cold storage. If mild symptoms are present at harvest, these will generally worsen, becoming severe after 10 to 12 d in cold storage (Simonne et al., 2007). It has been reported that the gomasho symptoms develop faster at 5°C than at either 0 or 10°C (Brecht et al., 1987).

Management

There is wide genetic diversity in the tolerance of Chinese cabbages to gomasho, and a range of tolerant cultivars are commercially available. Growers should also avoid excessive use of fertilizers, particularly N as ammonium-nitrate and the soil pH should be maintained between pH 6.0 and 6.5. Research has also shown that overmature heads are more likely to develop symptoms of the disorder and therefore correct harvesting times should be used (Simonne et al., 2007). In addition to the correct cultivar and growing practices, good postharvest practices such as rapid cooling and storage at 0°C are recommended to reduce the expression of gomasho.

References

Adeolu, M., Seema, A., Naushad, S. and Gupta, R.S. 2016. Genome-based phylogeny and taxonomy of the "*Enterobacteriales*": Proposal for *Enterobacterales* ord. nov. divided into the families *Enterobacteriaceae, Erwiniaceae* fam. nov., *Pectobacteriaceae* fam. nov., *Yersiniaceae* fam. nov., *Hafniaceae* fam. nov., *Morganellaceae* fam. nov., and *Budviciaceae* fam. nov. *International Journal of Systematic Evolutionary Microbiology* 66, 5575–5599.

Ambuko, J., Wanjiru, F., Chemining'wa, G.N., Owino, W.O. and Mwachoni, E. 2017. Preservation of postharvest quality of leafy Amaranth (*Amaranthus* spp.) vegetables using evaporative cooling. *Journal of Food Quality*. Article ID 5303156.

Barak, J.D., Koike, S.T. and Gilbertson, R.L. 2001. The role of crop debris and weeds in the epidemiology of bacterial leaf spot of lettuce in California. *Plant Disease* 85, 169–178.

Brecht, J.K., Sherman, M., Bergsma, K., Stall, W.M. and Shuler, K.D. 1987. Influence of postharvest conditions on black speck of Chinese cabbage. *HortScience* 22, 1128.

Brecht, P.E., Kader, A.A. and Morris, L.L. 1973. Influence of postharvest temperature on brown stain of lettuce. *Journal of the American Society for Horticultural Science* 98, 399–402.

Bull, C.T., Goldman, P.H., Smith, R.F. and Koike, S.T. 2004. *Pseudomonas syringae* pv. *alisalensis* and *Pseudomonas syringae* pv. *maculicola* cause disease on crucifers used in cover crop mixtures. *Phytopathology* 94, S150.

Choudhury, R.A., Koike, S.T., Fox, A., Anchieta, A.G., Subbarao, K.V., Klosterman, S. J. and McRoberts, N. 2017. Spatiotemporal patterns in the airborne dispersal of spinach downy mildew. *Phytopathology* 107, 50–58.

Cianciotto, N.P. and White, R.C. 2017. Expanding role of Type II secretion in bacterial pathogenesis and beyond. *Infection and Immunity* 85, e00014–17.

Cintas, N.A., Koike, S.T. and Bull, C.T. 2002. A new pathovar, *Pseudomonas syringae* pv. *alisalensis* pv. nov., proposed for the causal agent of bacterial blight of broccoli and broccoli raab. *Plant Disease* 86, 992–998.

Coates, L.M., Johnson, G.I. and Dale, M. 1997. Postharvest pathology of fruit and vegetables. pp. 533–547. *In*: J. Brown and H. Ogle (eds.). *Plant Pathogens and Plant Diseases*. Rockvale Publications, Armidale, Australia.

Collier, G.F. and Tibbitts, T.W. 1982. Tipburn of lettuce. *Horticultural Reviews* 4, 49–65.

Correll, J.C., Morelock, T.E. and Guerber, J.C. 1993. Vegetative compatibility and virulence of the spinach anthracnose pathogen *Colletotrichum dematium*. *Plant Disease* 77, 688–691.

Davidsson, P.R., Kariola, T., Niemi, O. and Palva, E.T. 2013. Pathogenicity of and plant immunity to soft rot pectobacteria. *Frontiers in Plant Science* 4, 191.

Davis, R.M., Subbarao, K.V., Raid, R.N. and Kurtz, E.A. 1997. *Compendium of Lettuce Diseases*. APS Press, St. Paul, MN.

du Toit, L.J. and Hernandez-Perez, P. 2005. Efficacy of hot water and chlorine for eradication of *Cladosporium variabile, Stemphylium botryosum*, and *Verticillium dahliae* from spinach seed. *Plant Disease* 89, 1305–1312.

Fayette, J., Jones, J., Pernezny, K., Roberts, P. and Raid, R. 2017. Survival of *Xanthomonas campestris* pv. *vitians* on lettuce in crop debris, irrigation water, and weeds in south Florida. *European Journal of Plant Pathology* 151, 341–353.

Forney, C.F. and Austin, R.K. 1988. Time of day at harvest influences carbohydrate concentration in crisphead lettuce and its sensitivity to high CO_2 levels after harvest. *Journal of the American Society for Horticultural Science* 113, 581–583.

Grogan, R.G., Misaghu, I.J., Kimble, K.A., Greathead, A.S., Ririe, D. and Bardin, R. 1977. Varnish spot, destructive disease of lettuce in California caused by *Pseudomonas cichorii*. *Phytopathology* 67, 957–960.

Gruda, N. 2005. Impact of environmental factors on product quality of greenhouse vegetables for fresh consumption. *Critical Reviews in Plant Sciences* 24, 227–247.

Hernandez-Perez, P. and du Toit, L.J. 2006. Seedborne *Cladosporium variabile* and *Stemphylium botyrosum* in spinach. *Plant Disease* 90, 137–145.

Hunter, P.J., Atkinson, L.D., Vickers, L., Lignou, S., Oruna-Concha, M.A., Pink, D., Hand, P., Barker, G., Wagstaff, C. and Monaghan, J.M. 2017. Oxidative discolouration in whole-head and cut lettuce: Biochemical and environmental influences on a complex phenotype and potential breeding strategies to improve shelf-life. *Euphytica* 213, 180.

Jenni, S. 2005. Rib discoloration: A physiological disorder induced by heat stress in crisphead lettuce. *HortScience* 40, 2031–2035.

Ke, D. and Saltveit, M.E. 1989a. Carbon dioxide-induced brown stain development as related to phenolic metabolism in iceberg lettuce. *Journal of the American Society for Horticultural Science* 114, 789–794.

Ke, D. and Saltveit, M.E. 1989b. Wound-induced ethylene production, phenolic metabolism and susceptibility to russet spotting in iceberg lettuce. *Physiologia Plantarum* 76, 412–418.

Kim, J.O., Shin, J.H., Gumilang, A., Chung, K., Choi, K.Y. and Kim, K.S. 2016. Effectiveness of different classes of fungicides on *Botrytis cinerea* causing gray mold on fruit and vegetables. *The Plant Pathology Journal* 32, 570–574.

Klieber, A. 2001. Chinese cabbage; pre-harvest and harvest practices. Access to Asian Vegetables. *New South Wales Agriculture* 43, 1–2.

Koike, S.T., Gladders, P. and Paulus, A.O. 2007. *Vegetable Diseases. A Color Handbook*. Manson Publishing Ltd., London, UK.

Koike, S.T., Henderson, D.M. and Butler, E.E. 2001. Leaf spot disease of spinach in California caused by *Stemphylium botryosum*. *Plant Disease* 85, 126–130.

Koukounaras, A. 2009. Senescence and quality of green leafy vegetables. *Stewart Postharvest Review* 5, 1–5.

Lebeda, A. and Mieslerova, B. 2011. Taxonomy, distribution and biology of lettuce powdery mildew (*Golovinomyces cichoracearum sensu stricto*). *Plant Pathology* 60, 400–415.

Meena, M., Gupta, S.K., Swapnil, P., Zehra, A., Dubey, M.K. and Upadhyay, R.S. 2017. *Alternaria* toxins: Potential virulence factors and genes related to pathogenesis. *Frontiers in Microbiology* 8, 1451.

Monaghan, J.M., Vickers, L.H., Grove, I.G. and Beacham, A.M. 2017. Deficit irrigation reduces postharvest rib pinking in wholehead Iceberg lettuce, but at the expense of head fresh weight. *Journal of the Science of Food and Agriculture* 97, 1524–1528.

Morris, C.E., Glaux, C., Latour, X., Gardan, L., Samson, R. and Pitrat, M. 2000. The relationship of host range, physiology and genotype to virulence on cantaloupe in *Pseudomonas syringae* from cantaloupe blight epidemics in France. *Phytopathology* 90, 636–646.

Nabhan, S., De Boer, S.H., Maiss, E. and Wydra, K. 2013. *Pectobacterium aroidearum* sp. nov, a soft rot pathogen with preference for monocyledonous plants. *International Journal of Systematic Microbiology* 63, 2520–2525.

Narayanasamy, P. 2005. *Postharvest Pathogens and Disease Management*. John Wiley, Somerset, NJ.

Nowicki, M., Nowakowska, M., Niezgoda, A. and Kozik, E.U. 2012. Alternaria black spot of Crucifers: Symptoms, importance of disease, and perspectives of resistance breeding. *Vegetable Crops Research Bulletin* 76, 5–19.

Nunes, C. 2008. Impact of environmental conditions on fruit and vegetable quality. *Stewart Postharvest Review* 4, 1–14.

Oskiera, M., Kaluzna, M., Kowalska, B. and Smolinska, U. 2017. *Pectobacterium carotovorum* subsp. *odorierum* on cabbage and Chinese cabbage: Identification, characterization and taxonomic relatedness of bacterial soft rot causal agents. *Journal of Plant Pathology* 99, 172–185.

Patterson, C.L. and Grogan, R.G. 1991. Role of microsclerotia as primary inoculum for *Microdochium panattonianum*, incitant of lettuce anthracnose. *Plant Disease* 75, 134–138.

Pauwelyn, E., Vanhouteghem, K., Cottyn, B., De Vos, P., Maes, M., Bleyaert, P. and Höfte, M. 2010. Epidemiology of *Pseudomonas cichorii*, the cause of lettuce midrib rot. *Journal of Phytopathology* 159, 298–305.

Pernezny, K., Raid, R.N., Jones, J.B. and Dickstein, E. 2007. First report of a leaf spot disease of wild rocket (*Diplotaxis tenuifolia*) in Florida caused by *Xanthomonas campestris* pv. *raphani*. *Plant Disease* 91, 1360.

Pérombelon, M.C.M. 2002. Potato diseases caused by soft rot erwinias: An overview of pathogenesis. *Plant Pathology* 51, 1–12.

Phillips, D.R. and Gersbach, N.B. 1989. Factors influencing petiole spotting ("gomasho") in Chinese cabbage. *Acta Horticulturae* 247, 117–121.

Pogson, B. and Morris, S.C. 2004. Postharvest senescence of vegetables and its regulation. pp. 319–329. *In*: L. Noodén (ed.). *Plant Cell Death Processes*. Academic Press, San Diego, CA.

Saltveit, M. 2016. Lettuce. pp. 386–389. *In*: K.C. Gross, C.Y. Wang and M. Saltveit (eds.). *The Commercial Storage of Fruits, Vegetables, and Florist and Nursery Stocks*. Agricultural Handbook 66. United States Department of Agriculture - Agricultural Research Service, Washington, DC.

Saltveit, M. 2018. Pinking of lettuce. *Postharvest Biology and Technology* 145, 41–52.

Saure, M.C. 1998. Causes of the tipburn disorder in leaves of vegetables. *Scientia Horticulturae* 76, 131–147.

Simonne, E., Brecht, J., Liu, G. and Ozores-Hampton, M. 2007. Pepper spot ("Goma-sho") on Napa cabbage. HS1101. UF/IFAS Extension. University of Florida, USA.

Snowdon, A.L. 1992. *A Colour Atlas of Post-Harvest Diseases and Disorders of Fruits and Vegetables, Vol. 2*: Vegetables. CRC Press, Boca Raton, FL.

Subbarao, K.V., Hubbard, J.C. and Schulbach, K.F. 1997. Comparison of lettuce diseases and yield under subsurface drip and furrow irrigation. *Phytopathology* 87, 877–883.

Toussaint, V., Benoit, D.L. and Odile, C. 2012. Potential of weed species to serve as a reservoir for *Xanthomonas campestris* pv. *vitians*, the causal agent of bacterial leaf spot of lettuce. *Crop Protection* 41, 64–70.

Tsuge, T., Harimoto, Y., Akimitsu, K., Ohtani, K., Kodama, M., Akagi, Y., Egusa, M., Yamamoto, M. and Otani, H. 2013. Host-selective toxins produced by the plant pathogenic fungus *Alternaria alternata*. *FEMS Microbiology Reviews* 37, 44–66.

USDA-ARS (United States Department of Agriculture - Agricultural Research Service). 2016. *The Commercial Storage of Fruits, Vegetables, and Florist and Nursery Stocks*. *In*: K.C. Gross, C.Y. Wang and M. Saltveit (eds.). Agricultural Handbook 66. United States Department of Agriculture - Agricultural Research Service, Washington, DC.

Waleron, M., Waleron, K. and Lojkowska, E. 2013. Occurrence of *Pectobacterium wasabiae* in potato field samples. *European Journal of Plant Pathology* 137, 149–158.

Wicks, T.J., Hall, B. and Pezzaniti, P. 1994. Fungicidal control of anthracnose (*Microdochium panattonianum*) on lettuce. *Australian Journal of Experimental Agriculture* 34, 277–283.

Wills, R.B.H. and Golding, J.B. 2016. *Postharvest: An Introduction to the Physiology and Handling of Fruit and Vegetables*. NewSouth Publishing, Sydney, Australia.

Wu, B.M., Subbarao, K.V. and Liu, Y.-B. 2008. Comparative survival of sclerotia of *Sclerotinia minor* and *S. sclerotiorum*. *Phytopathology* 98, 659–665.

Wu, F., Groopman, J.D. and Pestka, J.J. 2014. Public health impacts of foodborne mycotoxins. *Annual Review of Food Science and Technology* 5, 351–372.

Xie, H., Li, X.Y., Ma, Y.L. and Tian, Y. 2018. First report of *Pectobacterium aroidearum* causing soft rot of Chinese cabbage in China. *Plant Disease* 102, 674.

Section II

GENERAL ASPECTS OF INFECTION CAUSING POSTHARVEST DISEASE

Chapter 11

Molecular Insights into the Pathogenicity of Necrotrophic Fungi Causing Postharvest Diseases

Luis González-Candelas and Ana-Rosa Ballester

Institute of Agrochemistry and Food Technology (IATA), Spanish National Research Council (CSIC), Paterna, Valencia, Spain

Abbreviations

CCR	Carbon catabolite repression
CWDEs	Cell wall-degrading enzymes
CHS	Chitin synthases
OH^-	Hydroxyl radical
H_2O_2	Hydrogen peroxide
HSTs	Host-specific toxins
HR	Hypersensitive response
MS	Mass-spectrometry
MAPK	Mitogen-activated protein kinase
NGS	Next-generation sequencing
NMR	Nuclear magnetic resonance
PAMPs	Pathogen-associated molecular patterns
PHI	Pathogen-Host Interactions
PLs	Pectate lyases
PMEs	Pectin methylesterases
PGs	Polygalacturonases
ROS	Reactive oxygen species
$O_2^{\cdot -}$	Superoxide
TFs	Transcription factors

1 Introduction

Fungal pathogens are the major cause of decay during postharvest handling and storage of fresh produce. Among the plethora of different fungi causing decay, necrotrophic fungi represent the main threat to major fruit crops, such as citrus, pome, or stone fruits. Classically, plant necrotrophic fungi have been considered unsophisticated pathogens that kill host cells with the use of toxins and/or cell wall-degrading enzymes (CWDE). However, studies conducted in recent years, mainly with *Botrytis cinerea* Pers. and *Sclerotinia sclerotiorum* (Lib.) de Bary are changing this simplistic perception about necrotrophic fungi. Consequently, there is an increased interest in elucidating the pathogenicity mechanisms deployed by these fungi. A recent review covered the molecular aspects of the interactions between postharvest pathogenic fungi and fruits, attending to both the pathogen and the host (Tian et al., 2016). This chapter is more focused on the current knowledge of the molecular mechanisms deployed by necrotrophic postharvest fungi.

The reader is referred to the Pathogen–Host Interactions (PHI-base; www.phi-base.org) database, which compiles expertly curated molecular and biological information about genes proven to affect the outcome of pathogen-host interactions reported in peer-reviewed articles (Urban et al., 2017). In Table 11.1, we present the most relevant information, which has been extracted from this database, of genes from necrotrophic postharvest pathogenic fungi.

Table 11.1 Genes from necrotrophic pathogenic postharvest fungi whose deletion modifies the outcome of the pathogen-fruit interaction. Data obtained from PHI-base version 4.3 (www.phi-base.org/index.jsp)

ID	*Gene*	*Pathogen*	*Host*	*Function*	*DOI**
PHI:2280	Velvet protein family	*F. oxysporum*	apple	no data found	10.1111/J.1364-3703.2008.00512.X
PHI:2305	BcFKBP12	*B. cinerea*	apple	Rapamycin sensitivity	10.1016/j.fgb.2008.11.011
PHI:2308	BcCRZ1	*B. cinerea*	tomato	Zinc finger transcription factor	10.1128/EC.00426-07
PHI:2308	BcCRZ1	*B. cinerea*	apricot	Zinc finger transcription factor	10.1128/EC.00426-07
PHI:2368	bac	*B. cinerea*	pepper	Adenylate cyclase	10.1094/MPMI-21-11-1443
PHI:2370	bcpka1	*B. cinerea*	pepper	cAMP pathway catalytic subunit	10.1094/MPMI-21-11-1443
PHI:2371	bcpkaR	*B. cinerea*	pepper	cAMP pathway catalytic subunit	10.1094/MPMI-21-11-1443
PHI:2372	bcras2	*B. cinerea*	pepper	cAMP pathway catalytic subunit	10.1094/MPMI-21-11-1443
PHI:2409	FgMT2	*F. graminearum*	tomato	Sphingolipid C-9 methyltransferases	10.1128/EC.00255-08
PHI:2447	BcAtf1	*B. cinerea*	cucumber	bZIP transcription factor	10.1111/J.1364-3703.2011.00778.X
PHI:2476	CcpelA	*C. coccodes*	tomato	Pectate lyase	10.1111/J.1364-3703.2011.00740.X
PHI:2476	CcpelA	*C. coccodes*	tomato	Pectate lyase	10.1111/J.1364-3703.2011.00740.X
PHI:2484	Ste11	*B. cinerea*	apple	MAP triple kinase	10.1111/J.1364-3703.2009.00579.X
PHI:2485	Ste7	*B. cinerea*	apple	MAP kinase kinas	10.1111/J.1364-3703.2009.00579.X
PHI:2486	Ste50	*B. cinerea*	apple	MAP kinase adaptor protein	10.1111/J.1364-3703.2009.00579.X
PHI:2487	ste12	*B. cinerea*	apple	Transcription factor	10.1111/J.1364-3703.2009.00579.X
PHI:2487	ste12	*B. cinerea*	apple	Transcription factor	10.1111/J.1364-3703.2009.00579.X
PHI:2653	bcvel1	*B. cinerea*	grape	Regulatory protein linking light signals with development and secondary metabolism	10.1371/journal.pone.0047840

PHI:2709	AMET	*C. gloeosporioides*	avocado	Ammonium transporters	10.1094/MPMI-07-12-0170-R
PHI:2710	mepB	*C. gloeosporioides*	avocado	Ammonium transporters	10.1094/MPMI-07-12-0170-R
PHI:2728	FgERG4	*F. graminearum*	tomato	Sterol C-24 reductase	10.1111/j.1364-3703.2012.00829.x
PHI:2826	Fgb1	*F. oxysporum*	tomato	G-protein subunit	10.1016/j.fgb.2004.10.001
PHI:2827	Fmk1	*F. oxysporum*	tomato	Required for virulence	10.1016/j.fgb.2004.10.001
PHI:2826	Fgb1	*F. oxysporum*	tomato	G-protein subunit	10.1016/j.fgb.2004.10.001
PHI:2827	Fmk1	*F. oxysporum*	tomato	Required for virulence	10.1016/j.fgb.2004.10.001
PHI:2831	AKTR-1	*A. alternata*	pear	AK-toxin biosynthesis	10.1094/MPMI.2000.13.9.975
PHI:2832	AKT3-1	*A. alternata*	pear	AK-toxin biosynthesis	10.1094/MPMI.2000.13.9.975
PHI:3075	PdChsVII	*P. digitatum*	orange	Chitin synthase	10.1016/j.fgb.2014.04.002
PHI:3180	Rho1	*F. oxysporum*	tomato	Rho-type GTPases	10.1111/j.1462-5822.2008.01130.x
PHI:3226	pnl1	*P. digitatum*	orange	Pectin lyase	10.1111/mpp.12179
PHI:3292	AREB	*C. gloeosporioides*	tomato	Regulation of PACC-dependent acid-expressed genes and pathogenicity.	10.1094/MPMI-09-14-0252-R
PHI:3437	Sho1	*F. oxysporum*	tomato	Tetraspan transmembrane protein	10.1111/mpp.12217
PHI:3438	Msb2	*F. oxysporum*	tomato	Mucin-like membrane protein	10.1111/mpp.12217
PHI:3437	Sho1	*F. oxysporum*	tomato	Tetraspan transmembrane protein	10.1111/mpp.12217
PHI:3438	Msb2	*F. oxysporum*	tomato	Mucin-like membrane protein	10.1111/mpp.12217
PHI:3437	Sho1	*F. oxysporum*	apple	Tetraspan transmembrane protein	10.1111/mpp.12217
PHI:3438	Msb2	*F. oxysporum*	apple	Mucin-like membrane protein	10.1111/mpp.12217
PHI:3437	Sho1	*F. oxysporum*	apple	Tetraspan transmembrane protein	10.1111/mpp.12217
PHI:3438	Msb2	*F. oxysporum*	apple	Mucin-like membrane protein	10.1111/mpp.12217

(Continued)

Table 11.1 (Cont.)

ID	*Gene*	*Pathogen*	*Host*	*Function*	*DOI**
PHI:3458	PdGcs1	*P. digitatum*	tangerine	Biosynthesis of glucosylceramides	10.1016/j.bbrc.2014.10.142
PHI:3720	PdpacC	*P. digitatum*	mandarin	pH-signaling transcription factor	10.1007/s00253-013-5129-x
PHI:3721	Pdpg2	*P. digitatum*	mandarin	pH-signaling transcription factor	10.1007/s00253-013-5129-x
PHI:3865	GOX2	*P. expansum*	apple	Oxidation of glucose to gluconolactone	10.1094/MPMI-05-13-0138-R
PHI:3866	IDH	*P. expansum*	apple	Isoepoxyedon dehydrogenase	10.1094/MPMI-05-13-0138-R
PHI:3867	PACC	*P. expansum*	apple	pH-response transcription factor pacC/ RIM101	10.1094/MPMI-05-13-0138-R
PHI:3973	ASP	*C. gloeosporioides*	avocado	Aspartic protease	10.1094/MPMI-03-13-0080-R
PHI:4093	Bcsas1	*B. cinerea*	apple	Rab/GTPase	10.1094/MPMI-10-13-0314-R
PHI:4600	sreA	*P. digitatum*	orange	HLH transcriptional factor	10.1371/journal.pone.0117115
PHI:4865	FgVam7	*F. graminearum*	tomato	Regulatory role in cellular differentiation and virulence	10.1111/mpp.12267
PHI:4901	bcbrn1	*B. cinerea*	apple	THN reductase	10.1094/MPMI-04-15-0085-R
PHI:4902	bcpks13	*B. cinerea*	apple	Polyketide synthase	10.1094/MPMI-04-15-0085-R
PHI:4901	bcbrn1	*B. cinerea*	pepper	THN reductase	10.1094/MPMI-04-15-0085-R
PHI:4902	bcpks13	*B. cinerea*	pepper	Polyketide synthase	10.1094/MPMI-04-15-0085-R
PHI:4901	bcbrn1	*B. cinerea*	apple	THN reductase	10.1094/MPMI-04-15-0085-R
PHI:4901	bcbrn1	*B. cinerea*	pepper	THN reductase	10.1094/MPMI-04-15-0085-R
PHI:5253	FgSge1	*F. graminearum*	tomato	Transcription factor	10.1007/s11274-015-1894-2
PHI:5253	FgSge1	*F. graminearum*	tomato	Transcription factor	10.1007/s11274-015-1894-2

PHI:5315	AQP8	*B. cinerea*	tomato	Aquaporin	10.1111/nph.13721
PHI:5316	AQP9	*B. cinerea*	tomato	Aquaporin	10.1111/nph.13721
PHI:5319	NQO1	*B. cinerea*	tomato	Quinone oxidoreductase	10.1111/nph.13721
PHI:5315	AQP8	*B. cinerea*	strawberry	Aquaporin	10.1111/nph.13721
PHI:5316	AQP9	*B. cinerea*	strawberry	Aquaporin	10.1111/nph.13721
PHI:5319	NQO1	*B. cinerea*	strawberry	Quinone oxidoreductase	10.1111/nph.13721
PHI:5352	PePatL	*P. expansum*	apple	Patulin production	10.1111/mpp.12338
PHI:5352	PePatL	*P. expansum*	apple	Patulin production	10.1111/mpp.12338
PHI:5352	PePatL	*P. expansum*	apple	Patulin production	10.1111/mpp.12338
PHI:5352	PePatL	*P. expansum*	apple	Patulin production	10.1111/mpp.12338
PHI:5352	PePatL	*P. expansum*	apple	Patulin production	10.1111/mpp.12338
PHI:5352	PePatL	*P. expansum*	apple	Patulin production	10.1111/mpp.12338
PHI:5352	PePatL	*P. expansum*	apple	Patulin production	10.1111/mpp.12338
PHI:5352	PePatL	*P. expansum*	apple	Patulin production	10.1111/mpp.12338
PHI:5352	PePatL	*P. expansum*	apple	Patulin production	10.1111/mpp.12338
PHI:5352	PePatL	*P. expansum*	apple	Patulin production	10.1111/mpp.12338
PHI:5352	PePatL	*P. expansum*	apple	Patulin production	10.1111/mpp.12338
PHI:5352	PePatL	*P. expansum*	apple	Patulin production	10.1111/mpp.12338
PHI:5352	PePatL	*P. expansum*	apple	Patulin production	10.1111/mpp.12338
PHI:5374	afpB	*P. digitatum*	orange	Antifungal protein	10.1007/s00253-015-7110-3
PHI:5450	gdh2	*C. gloeosporioides*	tomato	Carbon regulation of environmental pH	10.1111/mpp.12355
PHI:5451	gox2	*C. gloeosporioides*	plum	Carbon regulation of environmental pH	10.1111/mpp.12355

(*Continued*)

Table 11.1 (Cont.)

ID	*Gene*	*Pathogen*	*Host*	*Function*	*DOI**
PHI:6159	BcATG1	*B. cinerea*	apple	Autophagy-related	10.1111/mpp.12396
PHI:6159	BcATG1	*B. cinerea*	grape	Autophagy-related	10.1111/mpp.12396
PHI:6159	BcATG1	*B. cinerea*	tomato	Autophagy-related	10.1111/mpp.12396
PHI:6159	BcATG1	*B. cinerea*	onion	Autophagy-related	10.1111/mpp.12396
PHI:6300	FgPDK1	*F. graminearum*	tomato	Pyruvate dehydrogenase kinase	10.1371/journal.pone.0158077
PHI:6404	BcMtg2	*B. cinerea*	strawberry	Assembling ribosomal subunits	10.1038/srep28673
PHI:6404	BcMtg2	*B. cinerea*	apple	Assembling ribosomal subunits	10.1038/srep28673
PHI:6404	BcMtg2	*B. cinerea*	grape	Assembling ribosomal subunits	10.1038/srep28673
PHI:6404	BcMtg2	*B. cinerea*	tomato	Assembling ribosomal subunits	10.1038/srep28673

* Digital object identifier

2 Penetration of the Cuticle and Cell Wall: Cuticle-Degrading Enzymes, Cell Wall-Degrading Enzymes, and Proteases

Some pathogenic fungi are able to penetrate the plant tissue directly without requiring a preexisting wound or open stoma. Direct penetration of the cuticle can be accomplished by formation of specialized infection structures, termed appressoria; however, other fungi are able to breach the cuticle directly by means of enzymatic degradation accomplished by cutinases. These enzymes hydrolyze the primary alcohol ester linkages of the cutin (Kolattukudy, 1985). The first direct demonstration on the role of a cutinase in pathogenicity was achieved by converting *Mycospharella* sp., a wound pathogen of papaya fruit, into a non-wound pathogen by insertion of an extracellular cutinase from *Fusarium solani* (Mart.) Sacc. (Dickman et al., 1989). However, the role of this enzyme in penetration remains controversial, as there are examples where inactivation of a cutinase encoding gene does not affect pathogenicity (Stahl and Schäfer, 1992). Germ tubes of *B. cinerea* can penetrate undamaged host tissue directly through the cuticle. The absence of appressorium-like structures suggests that this process relies on the enzymatic degradation of the cuticle. In order to study the role of cutinase on cuticle degradation, the *B. cinerea cutA* gene, coding for cutinase A, was disrupted. Pathogenicity of two transformants lacking a functional *cutA* gene on gerbera flowers and tomato fruits was unaltered with respect to the wild type strain (van Kan et al., 1997). Another cuticle-degrading lipase was purified from *B. cinerea* and antibodies raised against this lipase were able to block the penetration of the cuticle of intact tomato leaves by the germ tubes (Comménil et al., 1998). The antibodies neither affected germination of *B. cinerea* conidia nor did they inhibit the infection of wounded tissue, suggesting a role for the lipase specifically during host-surface penetration. However, gene knock-out mutants lacking the corresponding lipase gene were as virulent as the parental strain (Reis et al., 2005). Moreover, simultaneous disruption of both lipase and cutinase encoding genes did not affect virulence. Thus, no clear conclusions can be drawn about the roles of cutinase and other cuticle-degrading enzymes on pathogenicity.

The plant cell wall constitutes a barrier to pathogen penetration. To overcome this barrier, phytopathogenic fungi produce enzymes that are able to degrade the major components of the plant cell wall, namely pectin, cellulose, xyloglucan, and structural proteins. Plant CWDEs probably constitute the most extensively studied group of enzymes in relationship with pathogenicity in plant pathogenic fungi. Pectin is the major component of the middle lamella and the primary cell wall. Pectin hydrolysis not only weakens the cell wall to facilitate penetration and colonization of the host, it also provides a carbon source for pathogen growth. Pectin-degrading enzymes have long been classically considered to be putative pathogenicity determinants (Bateman and Millar, 1966) and thus have been studied in more detail than any other class of CWDEs. Pectinases can be grouped according to their catalytic mechanism into endo- and exo-polygalacturonases (PGs), endo- and exo-pectin and pectate lyases (PLs), and pectin methylesterases (PMEs). Whereas PGs hydrolyze the glyosidic bond, PLs catalyze a non-hydrolytic β-elimination reaction, generating an unsaturated hexenuronic acid residue and a new reducing end. The genome of *B. cinerea* contains at least six genes coding for endo-PGs. They differ in abundance and pattern of expression in different host tissues and maceration capability. Although there are few examples where deletion of a single PG encoding gene affected virulence, such evidence was provided for the *B. cinerea Bcpg1* gene. Gene knock-out mutants were obtained for two genes: *Bcpg1* and *Bcpg2*. Δ*Bcpg1* knock-out mutants were less virulent toward tomato and apple fruits than the parental strain (ten Have et al., 1998), demonstrating that BcPG1 is required for full virulence. *Botrytis cinerea* Δ*Bcp2* knock-out mutants were also less virulent than the parental strain on several host species, but in these experiments no assays were conducted with fruits (Kars et al., 2005a). Interestingly, both BcPG1 and BcPG2 were the proteins with the highest necrotizing activity among the five *B. cinerea* PGs heterologously expressed and purified from *Pichia pastoris* (Kars et al., 2005a). In *Penicillium digitatum* (Pers.) Sacc., the diameter of the macerated lesions caused by the knock-out Δ*Pdpg2* mutant, which lacks the polygalacturonase PG2, was about 30% smaller than that caused by the wild type strain 4 d after inoculation (Zhang et al., 2013a). Endo-PGs were also shown to be essential for virulence in *Alternaria alternata* (Fr.) Keissl. There are three major diseases in citrus caused by this species: Alternaria leaf spot of rough lemon, Alternaria brown spot of tangerines, and Alternaria black rot of many citrus fruits. An *Alternaria citri* Ellis & N. Pierce mutant for this gene resulted in a significant reduction in its ability to cause black rot symptoms in citrus fruit (Isshiki et al., 2001). However, they found no changes in the pathogenicity of an endo-PG mutant of *A. alternata*. These results indicate that a CWDE can play different roles in the pathogenicity of fungal pathogens and its role depends upon the type of disease but not the taxonomy of the fungus.

Pectin lyases and PLs differ in substrate affinity, with the former acting on partially methylated homogalacturonan and the latter acting on fully unmethylated homogalacturonan, so-called pectate. The role of PLs in virulence has been demonstrated by gene disruption in *P. digitatum* toward orange fruits (López-Pérez et al., 2015). A knock-out mutant, lacking the gene coding for pectin lyase, was less virulent than the wild-type strain. PLs are also important during the necrotrophic growth of hemibiotrophs like *Colletotrichum gloeosporioides* (Penz.) Penz. & Sacc. and *Colletotrichum coccodes* (Wallr.) S. Hughes in avocado and tomato fruits, respectively (Yakoby et al., 2001; Ben-Daniel et al., 2012). In both cases, the mutants had reduced virulence compared to their parental strains.

Pectin methylesterases demethylate pectin and make it more prone to degradation by depolymerizing enzymes such as endo-PGs. Four PMs have been found in the proteomic studies of the *B. cinerea* secretome, two of which have been knocked out, namely *Bcpme1* and *Bcpme2* (Valette-Collet et al., 2003; Kars et al., 2005b). The first gene has been mutated in two different *B. cinerea* strains, B05.10 and Bd90. *Bcpme1* knock-out mutants in the *B. cinerea* background B05.10 had the same virulence as the wild-type strain, while deletion of *Bcpme1* in the background of Bd90 reduced the virulence of the deletant strain compared to the virulence of the corresponding wild-type strain.

3 Fungal Nutrition: Acquisition of Essential Nutrients and Their Regulation

Fungi need to obtain nutrients from the host during invasion and, not surprisingly, pathogenicity genes related to fungal nutrition were identified at early stages of infection of apples by *Penicillium expansum* L. (Sánchez-Torres and González-Candelas, 2003).

Imazaki et al. (2010) have shown that mutations affecting the *A. alternata* peroxin protein AaPEX6, essential for peroxisome biogenesis, lead to a loss of pathogenicity. This phenotype is also associated with defects in fatty acid oxidation and the production of the secondary metabolite host-specific AK-toxin.

Minerals, such as iron, are essential for fungal growth and development. High-affinity iron uptake systems, classified as reductive iron assimilation (ferroxidation/permeation) and siderophore-mediated iron uptake (or non-reductive iron assimilation), enable fungi to acquire limited iron from plant hosts (Johnson, 2008). Siderophores are small iron-chelating peptides used by fungi to acquire iron. In different phytopathogenic fungi such as *Alternaria brassicicola* (Schwein.) Wiltshire and *A. alternata*, the knock-out of a gene coding for a nonribosomal peptide synthase (*nps6* gene), which is involved in the synthesis of these siderophores, leads to reduced virulence as a result of the inability of the fungus to acquire its iron needs in planta (Oide et al., 2006; Chen et al., 2013). The introduction of a wild-type AaNPS6 under the control of its endogenous promoter to a Δ*nps6* null mutant at least partially restored siderophore production and virulence of *A. alternata* in citrus, demonstrating a functional link between iron uptake and fungal pathogenesis.

Recently, the role in pathogenicity of a sucrose transporter was analyzed in *P. digitatum* (Ramón-Carbonell and Sánchez-Torres, 2017). Knock-out mutants lacking the *PdSTU1* gene did not show growth differences when compared to the parental strain using different carbon sources, including sucrose. However, the knock-out mutants showed a delayed infection progress in inoculated orange fruits.

Carbon catabolite repression (CCR) involves the transcriptional regulation of genes involved in utilization of non-easily assimilable carbon sources. CCR is driven by glucose repression and substrate induction. In *Saccharomyces cerevisiae* Meyen, the *SNF1* (sucrose non-fermenting 1) gene is required for the expression of catabolite-repressed genes. SNF1 is a kinase that phosphorylates Mig1, a DNA-binding transcriptional repressor, whose ortholog in filamentous fungi is known as CreA. Orthologues of *SNF1* have been found in several postharvest pathogens. A *P. digitatum* Δ*PdSNF1* knock-out mutant was impaired in virulence and the expression of several genes encoding CWDEs was altered in this mutant, demonstrating a role of this kinase in adaptation to alternative carbon sources (Zhang et al., 2013b).

4 Host Colonization

4.1 Fungal Infections Regulated by pH and ROS

Changes in the ambient pH may differentially regulate the synthesis of pathogenicity factors and have a significant impact on the infection process, including both spore germination and mycelial development. A proteomic analysis indicated that spore germination of *P. expansum* was inhibited at pH 2.0 and 8.0 by changing intracellular pH and regulating protein expression (Li et al., 2010). The pH stress not only reduced the germination rate, but also led to abnormal morphology of germinated spores. The ambient pH could also regulate the secreted proteins (secretome) of different postharvest pathogens. It is well known that the secretome has a critical role during the infection process. *Botrytis cinerea* is able to secrete a large set of extracellular enzymes, such as PGs, PMEs, proteases, and laccases, to degrade host tissues (Manteau et al., 2003). Some secreted proteins are encoded by gene families and are differentially expressed according to various host and environmental factors (Wubben et al., 2000; ten Have et al., 2001). These secreted proteins are considered to be potential virulence factors during the infection process. Li et al. (2012) analyzed changes in the secretome of *B. cinerea* as influenced by pH. The study showed that at pH 4, more proteins related to proteolysis were induced in *B. cinerea*, whereas most of up-accumulated proteins were CWDEs at pH 6, indicating that *B. cinerea* can adjust the protein profile of the secretome responding to different ambient pH values. This switch of secretome responding to different pH values may contribute to the wide plant host range of *B. cinerea*.

Fungi have developed a complicated regulatory system to sense and respond to ambient pH signaling within the long-term interactions with the environment. This system of expression of pH-regulated genes is mediated by the integrated functions of seven genes: *pacC, palA, palB, palC, palF, palH*, and *palI* (Peñalva et al., 2008). The transcription factor, *PacC*, is an activator of genes expressed under alkaline conditions and a repressor of those expressed in acidic conditions. The PacC encoding gene has been found in several postharvest fungal pathogens, including *P. expansum* (Barad et al., 2014), *P. digitatum* (Zhang et al., 2013a), *C. gloesporioides* (Alkan et al., 2013), and *S. sclerotiorum* (Rollins, 2003). This transcription factor contributes to the full virulence of these pathogens, as silenced or knock-out mutants are less virulent than their corresponding wild-type parentals.

Fungal plant pathogens not only sense the ambient pH, they can also modify the host pH in several ways in order to facilitate their infection (for a recent review, see Prusky et al., 2016). Some fungi, such as many *Colletotrichum* species and *A. alternata*, induce host alkalization by secreting ammonia (Eshel et al., 2002a, 2002b; Alkan et al., 2008). The alkalization of the infection site induces the expression of genes encoding virulence factors. However, other fungi such as *P. expansum* and *P. digitatum* (Prusky et al., 2004; Vilanova et al., 2014), *B. cinerea* (Wubben et al., 2000; ten Have et al., 2001; Manteau et al., 2003), *S. scletoriorum* (Rollins and Dickman, 2001), and *Monilinia fructicola* (G. Winter) Honey (De Cal et al., 2013) induce host acidification. This acidification is achieved by secretion of small organic acids, mainly gluconic (Barad et al., 2012) or oxalic (Liang et al., 2015) acids. In these pathogens, the acidification of the ambient pH leads to a higher CWDE activity at lower pH level. Recent results indicate that availability of carbon is a major factor determining whether the fungus alkalinizes or acidifies the environment. Thus, ammonia is secreted under limited carbon and gluconic acid under excess carbon (Prusky et al., 2016).

When attacked by a pathogen, the plant reacts with an oxidative burst, which is a rapid and transient production of reactive oxygen species (ROS), including superoxide ($O_2^{\cdot -}$), hydroxyl radical ($OH^{\cdot}$), and hydrogen peroxide (H_2O_2). ROS can be harmful, causing DNA damage, protein inactivation, and lipid peroxidation; and the generation of an oxidative burst can trigger hypersensitive cell death. This is the so-called hypersensitive response (HR), which is considered a major element of plant disease resistance. The HR reduces the availability of food for the pathogen and helps to restrict the progress of biotrophic pathogens in host tissues. However, necrotrophs can utilize dead tissue and HR might facilitate the colonization of plants. Necrotrophic pathogens kill host plants before colonization, and the accumulation of higher levels of ROS is thought to be important for their infection. Necrotrophs can produce ROS themselves and stimulate the host to do so. Govrin and Levine (2000) showed that the inoculation of *Arabidopsis thaliana* with *B. cinerea* and *S. sclerotiorum* induced ROS production during the oxidative burst and that both fungi utilized plant HR for rapid colonization of their hosts. Pathogenicity of both fungi was dependent on the levels of ROS generated during infection. However, intracellular accumulation of ROS in *P. expansum*, caused by antifungal chemicals, is involved in cell damage and destruction (Qin et al., 2007). The mitochondrial impairment due to functional alteration of oxidative-stress-sensitive proteins contribute to fungal death caused by H_2O_2 (Qin et al., 2011). To counteract ROS, originating either from normal physiological processes or from environmental changes, the fungus needs an efficient detoxification system, including the glutathione and thioredoxin systems, and the detoxifying enzymes, including catalases, peroxidases, and superoxide dismutases. Ballester et al. (2006) showed that the high catalase and soluble peroxidase activities in the sporulation zone, possibly derived from the pathogen, may help *P. digitatum* to cope with the H_2O_2 generated during citrus fruit infection. In addition, the catalase activity in *Cladosporium fulvum* Cooke is 25-fold higher in the spores than in the mycelium, and it has been correlated with the existence of two differentially expressed *cat* genes (Bussink and Oliver, 2001). Several redox-sensitive transcription factors are required for cellular responses to oxidative/redox conditions. Among them, the yes-activating protein 1 (YAP1) is one of the most important determinants of the yeast's response to chemical stress, responsible for transcriptional activation of genes involved in ROS detoxification as well as drug and heavy metal resistance in *S. cerevisiae* (Rodrigues-Pousada et al., 2010). A homologue of YAP1, the redox-responsive transcription factor MFAP1, was described recently in *M. fructicola* (Yu et al., 2017). *Monilinia fructicola* wild-type strain responded to oxidative stress at the infection site by activating the expression of *MfAP1* and up-regulating the genes required for ROS detoxification and fungal virulence. So, silencing of *MfAP1* caused a reduction of the lesion on rose petals during pathogenicity assays. Similar results have been observed with an *A. alternata AP1* null mutant, which did not cause any visible necrotic lesions on citrus leaves (Lin et al., 2009). Compared to the wild type, this null mutant displayed lower catalase, peroxidase, and superoxide dismutase, confirming the involvement of YAP1 homologues in the response to oxidative stress and ROS detoxification. Deletion of *B. cinerea Bap1* gene has no apparent effect on pathogenicity (Temme and Tudzynski, 2009). However, deletion of several genes involved in NADPH oxidases complexes (*BcnnoxA, BcnoxB*, and *BcnoxR*), as well as some genes involved in ROS detoxification (*Bcsod1, Bctrx1*, and *Bctrr1*) had a severe influence on pathogenicity as they exhibited strongly retarded lesion formation (Siegmund and Viefhues, 2016).

4.2 Toxins and Other Secondary Metabolites

Necrotrophs can be divided into host-specific and broad host-range species. The archetypical broad host-range fungal necrotrophs are *B. cinerea, S. sclerotiorum*, and *A. alternata*. Phytotoxins can be either non-host-specific toxins (non-HSTs) that affect a broad range of plant species, or HSTs that affect only a particular plant species or more often genotypes of that species. The filamentous fungus *A. alternata* includes seven pathogenic variants (pathotypes), which produce different host-selective toxins and cause diseases on different plants (Tsuge et al., 2016).

The authors identified the HST biosynthetic genes, called *TOX* genes, from six pathotypes. The relation between the production of the HSTs and the pathogenicity has been described in the Japanese pear pathotype of *A. alternata* that produces AK-toxin, an epoxy-decatrienoic acid ester. An *A. alternata* strain deficient in a gene coding for a peroxin protein essential for peroxisome biogenesis completely lost the production of AK-toxin, resulting in loss of pathogenicity on susceptible pear leaves (Imazaki et al., 2010). In citrus, two pathotypes of *A. alternata* are identified according to the production of HSTs. One is the tangerine pathotype, which produces ACT-toxin specific to tangerine and their hybrids, and the other is the rough lemon pathotype, which affects rough lemons and produces ARC-toxin. These HSTs are essential for host-selective infection and disease development (Tsuge et al., 2016). On the other hand, the role of non-HSTs in virulence is not as evident. For example, some *B. cinerea* knock-out mutants unable to produce botrydial were distinctly reduced in virulence, depending on the genetic background of the strain (Siewers et al., 2005; Pinedo et al., 2008). The loss of botcinic acid production did not affect virulence. However, double mutants that did not produce botcinic acid or botrydial exhibited markedly reduced virulence (Dalmais et al., 2011).

Based on their chemical structure, phytotoxins are commonly classified as polyketides, nonribosomal peptides, alkaloids, terpenes, or metabolites of mixed biosynthetic origin. Patulin is a toxic secondary metabolite mainly produced by *P. expansum*, a fungus known as the most serious apple postharvest pathogen. Several groups have investigated the potential role of patulin as a virulence factor in *P. expansum*–apple interaction. However, existing studies have sometimes had contradictory results. Both Sanzani et al. (2012) and Barad et al. (2014) reported a direct relationship between patulin production and blue mold incidence and severity on apples by obtaining a mutant with a disruption of the *patK* gene coding for 6-methyl-salicylic acid synthetase, an enzyme involved in the first committed step of patulin biosynthesis, and a RNA-i-mediated knockdown of *patN* gene (isoepoxydon dehydrogenase, *idh*) coding for one of the later enzymes in the patulin biosynthesis pathway. However, in both cases, mutants produced residual amounts of patulin. An *Agrobacterium tumefaciens*-mediated transformation approach was used by Ballester et al. (2015) and Li et al. (2015) in order to obtain knock-out mutants in *patK, patL,* and *patN* genes. Both studies clearly demonstrated that patulin is not required by *P. expansum* to successfully infect apple fruit. Although patulin is not essential for the initiation of the disease, it acts as a cultivar-dependent aggressiveness factor for *P. expansum* (Snini et al., 2016).

5 Recognition of the Host and Signaling

5.1 MAPK Cascades and Transcription Factors

Adaptation to changes in the environment is crucial for viability of all organisms. In fungi, conserved signal transduction pathways, such as fungal mitogen-activated protein kinase (MAPK) cascades, are triggered by an array of stimuli and regulate a wide range of genes whose functions are largely involved in cell cycle, reproduction, morphogenesis, stress response, and virulence. MAPK-mediated signaling pathways involve three serine/threonine protein kinases-MAPK kinase kinase

(MAPKKK), MAPK kinase (MAPKK), and MAPK. In *S. cerevisiae*, five MAPK pathways regulate mating, invasive growth, cell wall integrity, hyperosmolarity responses, and ascospore formation. Most fungal pathogens contain three MAPKs that are orthologs of the *S. cerevisiae* Fus3/Kss1, Slt2, and Hog1 MAPKs, which function in separate signaling cascades to regulate infection-related morphogenesis, cell wall remodeling, and high osmolarity stress response, respectively (Hamel et al., 2012; Turrá et al., 2014). The three MAPK pathways contribute to virulence on plants, although each one of them has distinct and sometimes even opposite functions during the infection process.

Deletion of the *B. cinerea Bmp1*, which encodes the Fus3/Kss1p ortholog, causes defects in hydrophobicity-induced germination, delayed vegetative growth, reduced size of conidia, lack of sclerotia formation, and loss of pathogenicity (Zheng et al., 2000). Deletion of the Fus3/Kss1ortholog in *P. digitatum* (denoted as *PdMpkB* gene) also reduced the growth and conidiogenesis of the fungus, and the deletant fungal strain showed elevated resistance to osmotic stress. The Δ*PdMpkB* strain induced smaller lesions than the wild type, but the major difference between both strains was the absence of aerial sporulation in fruits infected with the deletant strain. Δ*PdMpkB* hyphae could expand within citrus tissues but failed to penetrate fruit cuticles and failed to sporulate on the fruit surface (Ma et al., 2016). *Botrytis cinerea* BcSte12 is considered as one of several effector proteins of the BsSte11-BcSte7-Bmp1 module, and its deletion resulted in delayed infection due to low penetration efficiencies, loss of sclerotinia formation, and increased melanization (Schamber et al., 2010). Similar results were observed in *P. digitatum*, in which the deletion of the *PdSte12* gene hindered the green mold decay on citrus fruit, and mutants exhibited reduced growth and impaired conidiogenesis during fungal infection of citrus fruit (Vilanova et al., 2016; de Ramón-Carbonell and Sánchez-Torres, 2017). The capacity of *B. cinerea* deletants in the *bmp3* gene, orthologous to the yeast Slt2p, to penetrate and to colonize the host tissue was significantly affected (Rui and Hahn, 2007). The high-osmolarity glycerol (HOG) pathway plays an important role in eukaryotic organisms in the adaptation to the changing environmental conditions. In *S. cerevisiae*, the HOG pathway has been intensively characterized, and the MAPK Hog1p plays an important role activating the transcription of genes related to the stress response. In *B. cinerea*, the Hog1 homolog BcSak1 is involved in the response to osmotic stress, and resistance to iprodione and oxidative stress reaction caused by H_2O_2 (Segmüller et al., 2007). Moreover, BcSak1 is required for conidiation, indispensable for the formation of sclerotia, and the deletion mutants were completely non-pathogenic in unwounded plant tissue. Bos4 and Bos5 were confirmed to act upstream of BcSak1 in the same module. The Hog2 homolog (PdOs2) in *P. digitatum* is required for salt stress resistance, cell wall construction, fungicide sensitivity, and virulence (Wang et al., 2014). In contrast to *B. cinerea*, the Δ*Pdos2* mutant was not affected in the response to oxidative stress caused by H_2O_2; however, the average lesion size in citrus fruit was smaller than that caused by the wild-type strain, suggesting that Pdos2 is needed for full virulence of *P. digitatum*.

Transcription factors (TFs) play a central role in the regulation of gene expression at the transcriptional level. The MAD-box family of transcription factors is highly conserved in eukaryotic organisms and specifically regulate a wide range of cellular functions, including primary metabolism cell cycle, and cell identity. *Bcmads1* was indispensable for sclerotia production and for the full virulence potential of *B. cinerea* on apple fruit (Zhang et al., 2016). To study the impact of stress by the oxidative burst on pathogenic development, the homologue of the *S. cerevisiae yap1* gene, *bap1* was functionally analyzed in *B. cinerea*. The Δbap1 mutants were more sensitive to oxidative stress in vitro, but had normal virulence in apple fruits (Temme and Tudzynski, 2009). Another transcription factor required to cope with oxidative stress is a GATA transcription factor (BcLTF1 for light-responsive TF1). This TF was required for full virulence of *B. cinerea*, maintaining ROS homeostasis, and secondary metabolism (Schumacher et al., 2014). Calcium is a highly versatile signaling molecule that regulates numerous pathways. For example, increasing of cytosolic calcium level could activate calcineurin, which targets the calcineurin-responsive transcription factor Crz1, resulting in dephosphorylation of Crz1p. The TF *PdCrz1* of *P. digitatum* has important roles in conidiation, virulence to citrus fruit, and resistance to 14α-demethylation inhibitor fungicides (Zhang et al., 2013c).

5.2 Effectors

Effectors are molecules released/associated with an organism that alter the physiology, structure, or function of another organism. Specifically, effectors are pathogen molecules that can modify host cell structures and manipulate function, facilitating infection and/or triggering defense responses. Effectors are responsible for promoting pathogen penetration and persistence inside the host tissue, as well as suppression of immune responses, allowing access to nutrients, proliferation, and growth. Effectors of necrotrophic fungi include toxins, which have already been covered in Section 4.2, and proteins. In general, proteinaceous effectors are small proteins with a signal peptide for secretion, no trans-membrane domains, they are rich in cysteine, have no similarity with other obvious protein domains, and are mostly species-specific. Classically, it has been considered that the necrotrophic pathogens do not need to interact with the plant because they kill host cells using CWDEs or toxins. Although in silico analyses of the proteomes of necrotrophic fungi have highlighted the presence of putative effectors (Marcet-Houben et al., 2012; Heard et al., 2015), there has been little experimental evidence for the existence of interactions between proteinaceous effectors and host targets for typical necrotrophic phytopathogens. Nep1-like proteins (NLPs) induce necrosis and ethylene production in dicotyledonous plants. *Botrytis cinerea* contains two genes encoding NLPs. Although both proteins are able to induce necrosis, they are dispensable for pathogenesis, since deletant mutants in either gene were equally infective as the parental strain (Cuesta Arenas et al., 2010).

Recent studies showed that during the infection process the *B. cinerea* endoPGs not only degraded the cell wall of the cells, but also could be recognized as microbe-associated molecular patterns by the *Arabidopsis* receptor-like protein RBPG1, thereby influencing the resistance of the host (Zhang et al., 2014). Similarly, the xylanase Xyn11A contributes to virulence by virtue of its necrotizing activity and not by its catalytic activity (Noda et al., 2010). Another effector protein with a proven role in the pathogenicity of *B. cinerea* is the cerato-platanin (Frías et al., 2013).

In *S. sclerotiorum*, a cysteine-rich, small secreted protein SsSSVP1 may manipulate plant energy metabolism to facilitate fungal infection (Lyu et al., 2016), and another small secreted protein, Ss-Caf1, functions as a pathogenicity factor to trigger host cell death during the early stages of infection (Xiao et al., 2014).

Together with the typical secreted proteins, small RNAs and toxins have been considered recently as putative effectors. Some *B. cinerea* small RNAs hijack the host RNA interference machinery by binding to *Arabidopsis* Argonaute 1 and selectively silencing host immunity genes (Weiberg et al., 2013).

6 Biosynthesis and the Integrity of Fungal Cell Walls: Chitin

Chitin is a major component of the cell walls of many fungi, together with glucans, mannans, and glycoproteins, and it is required for the maintenance of cell integrity. In general, chitin oligosaccharides released from hyphae during infection can act as pathogen-associated molecular patterns (PAMPs) in innate immunity in both plants and animals. Chitin biosynthesis is catalyzed by a family of transmembrane proteins called chitin synthases. Fungal chitin synthases (CHS) play critical roles in hyphal developments and fungal pathogenicity. They have been grouped into two divisions consisting of seven classes (classed I to VII) with at least one member from each class in most filamentous fungi, being classes III, V, VI, and VII exclusive of filamentous fungi. *Botrytis cinerea* has eight CHS and several mutants constructed by reverse genetics were characterized (Morcx et al., 2013). The *Bcchs1* mutant displayed cell wall weakening and reduced virulence (Soulié et al., 2003), the disruption of *Bcchs3a* gene resulted in drastic reduction of virulence (Soulie et al., 2006), and BcCHS7 was required for pathogenicity in a host-dependent manner. Mutation in *Bcchs6* gene is probably lethal (Morcx et al., 2013). However, all pathogenicity assays were done in leaves and not in fruit. Seven different chitin synthase genes have been identified in *P. digitatum* (Gandía et al., 2012, 2014). Δ*PdchsVII* knock-out mutants were able to infect citrus fruit and produced tissue maceration, although had reduced virulence and were impaired in the production of visible mycelium and conidia on the fruit.

7 Omics of Postharvest Phytopathogenic Fungi

Omics encloses the high-throughput approaches genomics, transcriptomics, as well as metabolomics and proteomics, among others. Genomics can be defined as the generation and analysis of whole-genome sequences of DNA extracted from an organism; transcriptomics describes the entire set of coding and non-coding RNAs; and metabolomics and proteomics are employed for the identification and quantification of the collection of metabolites and proteins from a genome. In this section, we focus on how these technologies can be used in the study of postharvest necrotrophic pathogenic fungi.

7.1 Genomics

Development of next-generation sequencing (NGS) and technology for assembly of sequence data from NGS has made whole genome sequencing of phytopathogenic fungi possible. *Botrytis cinerea* has become a model for dissecting the complexity of necrotrophs and broad host-range pathogenicity. To gain an in-depth understanding of *B. cinerea*-plant interactions at whole genome level, the first genome assemblies for *B. cinerea* T4 and B05.10 strains were obtained by using Sanger sequencing technology at low coverage (Amselem et al., 2011; Staats and van Kan, 2012), although some expressed sequence tags had been generated previously (Li et al., 2004). Currently, thanks to the improvement of sequencing technologies and the extreme reduction of sequencing costs, improved genomes sequences and gene annotations and new genome sequences of different *B. cinerea* strains are accessible (Staats and van Kan, 2012; Blanco-Ulate et al., 2013; van Kan et al., 2017). Due to these improvements, also an increased number of genome sequences of other postharvest pathogens, such as *P. digitatum, P. expansum, Penicillium italicum* Wehmer, and *A. alternata* (Marcet-Houben et al., 2012; Ballester et al., 2015; Nguyen et al., 2016), have been published in recent years (Table 11.2). Comparative genomics among different species or strains is an effective approach to understand the genetic basis of fungal pathogen evolution, pathogenicity, nutrients acquisition, etc. (Plissonneau et al., 2017). Comparative analysis of the genome sequences of *B. cinerea* B05.10 and the closely related necrotrophic plant pathogenic fungus *S. sclerotiorum* revealed high sequence identity and gene arrangement similarity, but different behavior and compatibility systems between these pathogenic fungi (Amselem et al., 2011). A similar approach was used to highlight genes and functions unique to *Rhizoctonia solani* J.G. Kühn by comparing the rice and potato pathogen strains (Hane et al., 2014), or to contrast genomic diversity between two closely related postharvest pathogens, *P. digitatum* and *P. expansum* (Julca et al., 2016). The availability of sequenced genomes of necrotrophs (Table 11.2), as well as an increasing availability of sequenced plant/fruit genomes provides essential baseline information to support experiments where the interactome cycle of pathogens and hosts can be studied.

7.2 Transcriptomics

Transcriptomics is the comprehensive study of transcripts in an organism in different situations. Microarray was the most popular tool to perform transcriptome analysis until recent years. However, the implementation of NGS and the advance of bioinformatics have made RNA sequencing (RNA-Seq) to become the most popular approach for transcriptomic analysis in both model and non-model organisms. This technology considerably improves the quality of genome annotation including gene prediction, but also constitutes the current technology for global gene expression analysis: the expression of a gene is calculated based on the number of sequence reads for each gene. As far as we are aware, few articles utilizing RNA-Seq analysis have been used to identify fungal genes during postharvest fruit-pathogen interactions. Blanco-Ulate et al. (2014) investigated the expression of CAZymes-encoding genes in *B. cinerea* during its interaction with tomato fruit and grape berries, highlighting host-specific commonalities and differences. Ballester et al. (2015), by sequencing the global RNA population, identified *P. expansum* genes expressed during the colonization of apple fruits and showed that several genes classically related to virulence were induced, such as CWDEs, proteases and oxidorreductases. In the same study, a new glucose oxidase-encoding gene (*GOX3*) was discovered with a much higher induction level than the previously described *GOX2* (Barad et al., 2012).

Table 11.2 Genomic sequences of necrotrophic postharvest pathogenic fungi.

Organism/Name	*Strain*	*Size (Mb)*	*Genes*	*Proteins*	*BioProject*
Alternaria alternata	ATCC 34957	33.4996	-	-	PRJNA300540
Alternaria alternata	B2a	33.0163	-	-	PRJNA309229
Alternaria alternata	SRC1lrK2f	32.9908	13577	13466	PRJNA342680
Alternaria alternata	Z7	34.3559	12048	12048	PRJNA278916
Botrytis cinerea	B05.10	42.6301	11698	13703	PRJNA15632
Botrytis cinerea	BcDW1	42.1323	11028	11028	PRJNA188482
Botrytis cinerea	T4	41.6129	-	-	PRJNA162725
Botrytis cinerea	T4	39.5114	16353	16353	PRJNA64593
Colletotrichum acutatum	1	52.1291	-	-	PRJNA314171
Colletotrichum acutatum	C71	44.92	-	-	PRJNA314187
Colletotrichum gloeosporioides	030206	57.6035	-	-	PRJNA377876
Colletotrichum gloeosporioides	Cg-14	53.2099	16538	16538	PRJNA176412
Colletotrichum gloeosporioides	Nara gc5	55.6071	15440	15381	PRJNA225509
Fusarium oxysporum	FoMN14	48.9969	-	-	PRJNA306247
Fusarium oxysporum	JCM 11502	46.1785	-	-	PRJDB3637
Geotrichum candidum	3C	41.3845	-	-	PRJNA243259
Geotrichum candidum	CLIB 918	24.8361	7357	6799	PRJEB5752
Monilinia fructicola	LMK 125	44.6842	-	-	PRJNA277973
Penicillium digitatum	Pd01-ZJU	25.0111	-	-	PRJNA178915
Penicillium digitatum	Pd1	26.0532	8962	8961	PRJNA302315
Penicillium digitatum	PDC 102	25.2044	-	-	PRJNA176878
Penicillium digitatum	PHI26	25.9851	9134	9133	PRJNA157541
Penicillium expansum	CMP-1	31.087	10663	10663	PRJNA255744
Penicillium expansum	d1	32.065	11023	11023	PRJNA255745
Penicillium expansum	MD-8	32.356	11060	11060	PRJNA319326
Penicillium expansum	R19	31.4157	-	-	PRJNA225688
Penicillium expansum	R21	35.0461	-	-	PRJNA339168
Penicillium expansum	NRRL 62431	31.5475	-	-	PRJNA170336
Penicillium expansum	T01	33.0326	-	-	PRJNA222879
Penicillium expansum	YT02	31.252	-	-	PRJNA50009
Penicillium griseofulvum	MRI314	28.0728	-	-	PRJNA286832

(Continued)

Table 11.2 (Cont.)

Organism/Name	*Strain*	*Size (Mb)*	*Genes*	*Proteins*	*BioProject*
Penicillium griseofulvum	PG3	29.1409	9630	9630	PRJNA289974
Penicillium italicum	B3	28.5976	-	-	PRJNA238601
Penicillium italicum	GL-Gan1	31.0266	-	-	PRJNA317511
Penicillium italicum	PHI-1	30.1644	9996	9996	PRJNA255746
Rhizoctonia solani	BBA 69670	56.0285	12012	11897	PRJEB9381
Rhizoctonia solani	123E	39.4189	10993	10993	PRJNA227561
Rhizoctonia solani	AG-1 IA	36.9381	10590	10489	PRJNA51401
Rhizoctonia solani	1802/KB	29.7303	-	-	PRJNA354902
Rhizoctonia solani	AG-1 IB	46.9144	12860	12616	PRJEB7257
Rhizoctonia solani	isolate 7/3/14	48.6737	12393	12268	PRJEB40
Rhizoctonia solani	AG-3 Ben3	51.485	-	-	PRJEB19450
Rhizoctonia solani	Rhs1AP	51.7059	12737	12726	PRJNA73133
Rhizoctonia solani	AG-8 WA-C10335	39.8229	13952	13952	PRJNA187548
Sclerotinia sclerotiorum	1980 UF-70	38.4592	38.4592	14490	PRJNA15530

7.3 Proteomics

Proteomics is the comprehensive study of proteins in an organism, organ, tissue, or cell. Peptide sequencing, that is, Edman sequencing, was the popular method to identify proteins few years ago. However, currently mass-spectrometry (MS) is used for protein identification and quantification in almost all proteomic studies. To identify proteins by MS, protein sequence databases are required and databases of allied species or other organisms can be used. However, for efficient and accurate identification, protein sequence database derived from one's own genome sequence or EST should be available for usage. Normally, the number of proteins identified in one proteomics is less than 10% of the gene number, because of the sensitivity of MS and masking of minor peptides by major ones. Before MS analysis, proteins are often separated, enriched or labeled depending on purposes. In comparative proteomics, 2D-PAGE is often used. Comparing several 2D-PAGE images, differences in protein amounts among different samples can be determined. ITRAQ labeling is also one of the methods used in comparative proteomics.

In general, proteomics studies are based only on the fruit or the pathogen. There are several studies, mainly on *B. cinerea* and *P. expansum*, analyzing proteins synthesized under conditions that could affect pathogenesis, such as during conidia germination or exposure of the fungus to different pHs (Li et al., 2010, 2012), and under different in vitro growth conditions (Fernández-Acero et al., 2010; González-Fernández et al., 2015). Comparative proteomic analyses have also been used to identify proteins that play crucial roles in the differential virulence between two *B. cinerea* strains (Fernández-Acero et al., 2007). However, only Shah et al. (2012) analyzed the interacting proteome of both fruit and pathogen simultaneously. They characterized fruit and fungal proteins solubilized in the *B. cinerea*-tomato interaction. More than 45% of the detected proteins were classified as CWDEs, including pectin methyl esterases, endo-PGs, β-galactosidase, and cellulases, indicating that they could be involved in the penetration and degradation of host cell walls, contributing to *B. cinerea* pathogenicity and playing a role in the infection process.

7.4 Metabolomics

Another approach to study fruit-pathogen interactions is to measure the abundance of metabolites, which fall downstream of genomic, transcriptomic, and proteomic variations. Metabolic profiling is often performed with nuclear magnetic resonance (NMR), or MS, such as gas chromatography-MS (GC-MS) and liquid-chromatography-MS (LC-MS). Metabolic approaches have the ability to measure both primary metabolites, such as sugars and amino acids, and secondary metabolites. Most of the metabolomics studies in postharvest interactions focused on the defense mechanisms of the fruit, in particular on phenolic compounds, because these compounds are associated with an increased resistance level in the fruit. The metabolite profiling of citrus fruit inoculated with *P. digitatum* showed an induction of different flavanones, flavones, polymethoxylated flavones, and scoparone (Ballester et al., 2013b), and some of these metabolites were also induced in elicited oranges with an increased resistance (Ballester et al., 2013a). There was an association between flavonoid structure, selective scavenging ability for different free radicals, and reduced susceptibility to *B. cinerea* infection in tomato fruit (Zhang et al., 2015). However, no studies have been conducted so far in order to identify fungal metabolites during the infection process of fruits.

7.5 Random Mutagenesis

Large-scale random insertional mutagenesis via transformation (i.e., forward genetics) facilitates high-throughput uncovering of novel genes with a relevant role in pathogenicity. Despite the enormous potential of this approach, there are few collections of random mutants from phytopathogenic fungi, including *A. alternata, B. cinerea, Colletotrichum higginsianum* Sacc., *Fusarium oxysporum* Schltdl., *Gibberella zeae* (Schwein.) Petch, *Magnaporthe oryzae* B.C. Couch, and *S. sclerotiorum* (Motaung et al., 2017). Random plasmid integration (Baldwin et al., 2010), restriction enzyme-mediated integration (Sweigard et al., 1998), and T-DNA integration (Mullins et al., 2001) have been the main systems used to generate mutant collections.

In *B. cinerea*, the screening of a mutant collection obtained by random T-DNA integration revealed 68 mutants, out of 2,368 screened mutants, which showed a significant reduction in virulence (Giesbert et al., 2011). Two of them were further characterized, confirming that a type 2A phosphoprotein phosphatase and a SPT3 transcription factor were crucial for virulence. Mutant screens for pathogenicity defects in *S. sclerotiorum* identified the secreted putative Ca^{2+}-binding protein SsCAF1 and the secreted Cu/Zn superoxide dismutase SsSOD1 (Xu and Chen, 2013;

Xiao et al., 2014). A random mutation analysis of pathogenicity was performed in *A. alternata* pv. *citri* using restriction enzyme-mediated integration. Three isolates among 1,694 transformants analyzed had a loss in pathogenicity in a citrus peel assay (Katoh et al., 2006). One of these isolates identified the gene *AcIGPD*, which encodes an imidazole glycerol phosphate dehydratase, the sixth enzyme in the histidine biosynthetic pathway. Targeted gene disruption of this gene confirmed its role in citrus black rot development.

8 Conclusions and Future Prospects

Classically, necrotrophic fungi were thought to rely on toxins, either HST or non-HST, and CWDEs, together with other extracellular hydrolases, such as proteases, as the major virulence factors. However, the characterization of gene knock-out mutants indicates that in most instances all these enzymes, except the HSTS, are not key determinants of virulence. A similar situation exists with enzymes involved in ROS production/detoxification. These results suggest the complex nature of virulence in necrotrophic fungi, where the absence of an enzyme belonging to a family may be compensated by other members of the family. In fact, all mutants isolated so far show reduced virulence at best, but no non-pathogenic mutants have yet been isolated. In addition, in many cases, mutants exhibiting reduced virulence also show reduced or abnormal growth under normal in vitro growth conditions. Therefore, despite the recent advances in the study of the molecular basis of pathogenicity in necrotrophic postharvest fungi, we are still far from understanding these phenomena. Probably, we would need to adopt a more holistic approach that take into account simultaneously the different factors that play a role in pathogenicity, such as the different kind of effectors, not only of proteinaceous nature, but also the newly discovered small RNAs. Other aspects that are likely to gain relevance in the near future are those related to the regulation of gene networks, including epigenetics. In addition, modern genome editing tools, such as the use of clustered regularly interspaced short palindromic repeats (CRISPR/Cas9) technology, can be used to generate collections of mutants and to develop non-mycotoxigenic strains that can be used as biocontrol agents with the added value of being non-recombinant strains. This technology was recently used in *A. alternata* to inactivate two genes involved in the biosynthesis pathway of melanin (Wenderoth et al., 2017). Finally, in the coming years we will probably see a wider implementation of "omics" techniques to help us to unravel the different mechanisms of pathogenicity and virulence of this group of fungi. The widespread implementation of these technologies will result in an unprecedented level of knowledge that can comprise the basis for the development of new control strategies that allow us to reduce postharvest losses.

References

Alkan, N., Fluhr, R., Sherman, A. and Prusky, D. 2008. Role of ammonia secretion and pH modulation on pathogenicity of *Colletotrichum coccodes* on tomato fruit. *Molecular Plant-Microbe Interactions* 21, 1058–1066.

Alkan, N., Meng, X., Friedlander, G., Reuveni, E., Sukno, S., Sherman, A., Thon, M., Fluhr, R. and Prusky, D. 2013. Global aspects of pacC regulation of pathogenicity genes in *Colletotrichum gloeosporioides* as revealed by transcriptome analysis. *Molecular Plant-Microbe Interactions* 26, 1345–1358.

Amselem, J., Cuomo, C.A., van Kan, J.A.L., Viaud, M., Benito, E.P., Couloux, A., Coutinho, P.M., de Vries, R.P., Dyer, P.S., Fillinger, S., Fournier, E., Gout, L., Hahn, M., Kohn, L., Lapalu, N., Plummer, K.M., Pradier, J.-M., Quévillon, E., Sharon, A., Simon, A., ten Have, A., Tudzynski, B., Tudzynski, P., Wincker, P., Andrew, M., Anthouard, V., Beever, R.E., Beffa, R., Benoit, I., Bouzid, O., Brault, B., Chen, Z., Choquer, M., Collémare, J., Cotton, P., Danchin, E.G., Da Silva, C., Gautier, A., Giraud, C., Giraud, T., Gonzalez, C., Grossetete, S., Güldener, U., Henrissat, B., Howlett, B.J., Kodira, C., Kretschmer, M., Lappartient, A., Leroch, M., Levis, C., Mauceli, E., Neuvéglise, C., Oeser, B., Pearson, M., Poulain, J., Poussereau, N., Quesneville, H., Rascle, C., Schumacher, J., Ségurens, B., Sexton, A., Silva, E., Sirven, C., Soanes, D.M., Talbot, N.J., Templeton, M., Yandava, C., Yarden, O., Zeng, Q., Rollins, J.A., Lebrun, M.-H. and Dickman, M. 2011. Genomic analysis of the necrotrophic fungal pathogens *Sclerotinia sclerotiorum* and *Botrytis cinerea*. *PLOS Genetics* 7, e1002230.

Baldwin, T.K., Gaffoor, I., Antoniw, J., Andries, C., Guenther, J., Urban, M., Hallen-Adams, H.E., Pitkin, J., Hammond-Kosack, K.E. and Trail, F. 2010. A partial chromosomal deletion caused by random plasmid integration resulted in a reduced virulence phenotype in *Fusarium graminearum*. *Molecular Plant-Microbe Interactions* 23, 1083–1096.

Ballester, A.R., Lafuente, M.T., De Vos, C.H.R., Bovy, A. and González-Candelas, L. 2013a. Citrus phenylpropanoids and defense against pathogens. Part I: Metabolic profiling in elicited fruits. *Food Chemistry* 136, 178–185.

Ballester, A.R., Lafuente, M.T. and González-Candelas, L. 2006. Spatial study of antioxidant enzymes, peroxidase and phenylalanine ammonia-lyase in the citrus fruit-*Penicillium digitatum* interaction. *Postharvest Biology and Technology* 39, 115–124.

Ballester, A.R., Lafuente, M.T. and González-Candelas, L. 2013b. Citrus phenylpropanoids and defense against pathogens. Part II: Gene expression and metabolite accumulation in the response of fruits to *Penicillium digitatum* infection. *Food Chemistry* 136, 285–291.

Ballester, A.R., Marcet-Houben, M., Levin, E., Sela, N., Selma, C., Carmona, L., Wisniewski, M., Droby, S., González-Candelas, L. and Gabaldón, T. 2015. Genome, transcriptome, and functional analyses of *Penicillium expansum* provide new insights into secondary metabolism and pathogenicity. *Molecular Plant-Microbe Interactions* 28, 232–248.

Barad, S., Horowitz, S., Kobiler, I., Sherman, A. and Prusky, D.B. 2014. Accumulation of the mycotoxin patulin in the presence of gluconic acid contributes to pathogenicity of *Penicillium expansum*. *Molecular Plant-Microbe Interactions* 27, 66–77.

Barad, S., Horowitz, S.B., Moskovitch, O., Lichter, A., Sherman, A. and Prusky, D. 2012. A *Penicillium expansum* glucose oxidase–encoding gene, *GOX2*, is essential for gluconic acid production and acidification during colonization of deciduous fruit. *Molecular Plant-Microbe Interactions* 25, 779–788.

Bateman, D.F. and Millar, R.L. 1966. Pectic enzymes in tissue degradation. *Annual Review of Phytopathology* 4, 119–144.

Ben-Daniel, B.-H., Bar-Zvi, D. and Tsror, L. 2012. Pectate lyase affects pathogenicity in natural isolates of *Colletotrichum coccodes* and in *pelA* gene-disrupted and gene-overexpressing mutant lines. *Molecular Plant Pathology* 13, 187–197.

Blanco-Ulate, B., Allen, G., Powell, A.L.T. and Cantu, D. 2013. Draft genome sequence of *Botrytis cinerea* BcDW1, inoculum for noble rot of grape berries. *Genome Announcements* 1, e00252–13.

Blanco-Ulate, B., Morales-Cruz, A., Amrine, K.C.H., Labavitch, J.M., Powell, A. and Cantu, D. 2014. Genome-wide transcriptional profiling of *Botrytis cinerea* genes targeting plant cell walls during infections of different hosts. *Frontiers in Plant Science* 5, article 435.

Bussink, H.J. and Oliver, R. 2001. Identification of two highly divergent catalase genes in the fungal tomato pathogen, *Cladosporium fulvum*. *European Journal of Biochemistry* 268, 15–24.

Chen, L.-H., Lin, C.-H. and Chung, K.-R. 2013. A nonribosomal peptide synthetase mediates siderophore production and virulence in the citrus fungal pathogen *Alternaria alternata*. *Molecular Plant Pathology* 14, 497–505.

Comménil, P., Belingheri, L. and Dehorter, B. 1998. Antilipase antibodies prevent infection of tomato leaves by *Botrytis cinerea*. *Physiological and Molecular Plant Pathology* 52, 1–14.

Cuesta Arenas, Y., Kalkman, E.R., Schouten, A., Dieho, M., Vredenbregt, P., Uwumukiza, B., Ruiz, M.O. and van Kan, J.A. 2010. Functional analysis and mode of action of phytotoxic Nep1-like proteins of *Botrytis cinerea*. *Physiological and Molecular Plant Pathology* 74, 376–386.

Dalmais, B., Schumacher, J., Moraga, J., Le Pecheur, P., Tudzynski, B., Collado, I.G. and Viaud, M. 2011. The *Botrytis cinerea* phytotoxin botcinic acid requires two polyketide synthases for production and has a redundant role in virulence with botrydial. *Molecular Plant Pathology* 12, 564–579.

De Cal, A., Sandín-España, P., Martinez, F., Egüen, B., Chien-Ming, C., Lee, M.H., Melgarejo, P. and Prusky, D. 2013. Role of gluconic acid and pH modulation in virulence of *Monilinia fructicola* on peach fruit. *Postharvest Biology and Technology* 86, 418–423.

de Ramón-Carbonell, M. and Sánchez-Torres, P. 2017. The transcription factor PdSte12 contributes to *Penicillium digitatum* virulence during citrus fruit infection. *Postharvest Biology and Technology* 125, 129–139.

Dickman, M., Podila, G. and Kolattukudy, P. 1989. Insertion of cutinase gene into a wound pathogen enables it to infect intact host. *Nature* 342, 446–448.

Eshel, D., Lichter, A., Dinoor, A. and Prusky, D. 2002a. Characterization of *Alternaria alternata* glucanase genes expressed during infection of resistant and susceptible persimmon fruits. *Molecular Plant Pathology* 3, 347–358.

Eshel, D., Miyara, I., Ailing, T., Dinoor, A. and Prusky, D. 2002b. pH regulates endoglucanase expression and virulence of *Alternaria alternata* in persimmon fruit. *Molecular Plant-Microbe Interactions* 15, 774–779.

Fernández-Acero, F.J., Colby, T., Harzen, A., Carbú, M., Wieneke, U., Cantoral, J.M. and Schmidt, J. 2010. 2-DE proteomic approach to the *Botrytis cinerea* secretome induced with different carbon sources and plant-based elicitors. *Proteomics* 10, 2270–2280.

Fernández-Acero, F.J., Inmaculada, J., Calvo, E., Inmaculada, V., Carbú, M., Emilio, C., Carlos, G., López, J.A., Jorrin, J. and Cantorral, J.M. 2007. Proteomic analysis of phytopathogenic fungus *Botrytis cinerea* as a potential tool for identifying pathogenicity factors, therapeutic targets and for basic research. *Archives of Microbiology* 187, 207–215.

Frías, M., Brito, N. and González, C. 2013. The *Botrytis cinerea* cerato-platanin BcSpl1 is a potent inducer of systemic acquired resistance (SAR) in tobacco and generates a wave of salicylic acid expanding from the site of application. *Molecular Plant Pathology* 14, 191–196.

Gandía, M., Harries, E. and Marcos, J.F. 2012. Identification and characterization of chitin synthase genes in the postharvest citrus fruit pathogen *Penicillium digitatum*. *Fungal Biology* 116, 654–664.

Gandía, M., Harries, E. and Marcos, J.F. 2014. The myosin motor domain-containing chitin synthase PdChsVII is required for development, cell wall integrity and virulence in the citrus postharvest pathogen *Penicillium digitatum*. *Fungal Genetics and Biology* 67, 58–70.

Giesbert, S., Schumacher, J., Kupas, V., Espino, J., Segmüller, N., Haeuser-Hahn, I., Schreier, P.H. and Tudzynski, P. 2011. Identification of pathogenesis-associated genes by T-DNA–mediated insertional mutagenesis in *Botrytis cinerea*: A type 2A phosphoprotein phosphatase and an SPT3 transcription factor have significant impact on virulence. *Molecular Plant-Microbe Interactions* 25, 481–495.

González-Fernández, R., Valero-Galván, J., Gómez-Gálvez, F.J. and Jorrín-Novo, J. V. 2015. Unraveling the *in vitro* secretome of the phytopathogen *Botrytis cinerea* to understand the interaction with its hosts. *Frontiers in Plant Science* 6, article 839.

Govrin, E.M. and Levine, A. 2000. The hypersensitive response facilitates plant infection by the necrotrophic pathogen *Botrytis cinerea*. *Current Biology* 10, 751–757.

Hamel, L.-P., Nicole, M.-C., Duplessis, S. and Ellis, B.E. 2012. Mitogen-activated protein kinase signaling in plant-interacting fungi: Distinct messages from conserved messengers. *The Plant Cell* 24, 1327–1351.

Hane, J.K., Anderson, J.P., Williams, A.H., Sperschneider, J. and Singh, K.B. 2014. Genome sequencing and comparative genomics of the broad host-range pathogen *Rhizoctonia solani* AG8. *PLOS Genetics* 10, e1004281.

Heard, S., Brown, N.A. and Hammond-Kosack, K. 2015. An interspecies comparative analysis of the predicted secretomes of the necrotrophic plant pathogens *Sclerotinia sclerotiorum* and *Botrytis cinerea*. *PLoS One* 10, e0130534.

Imazaki, A., Tanaka, A., Harimoto, Y., Yamamoto, M., Akimitsu, K., Park, P. and Tsuge, T. 2010. Contribution of peroxisomes to secondary metabolism and pathogenicity in the fungal plant pathogen *Alternaria alternata*. *Eukaryotic Cell* 9, 682–694.

Isshiki, A., Akimitsu, K., Yamamoto, M. and Yamamoto, H. 2001. Endopolygalacturonase is essential for citrus black rot caused by *Alternaria citri* but not brown spot caused by *Alternaria alternata*. *Molecular Plant-Microbe Interactions* 14, 749–757.

Johnson, L. 2008. Iron and siderophores in fungal–host interactions. *Mycological Research* 112, 170–183.

Julca, I., Droby, S., Sela, N., Marcet-Houben, M. and Gabaldón, T. 2016. Contrasting genomic diversity in two closely related postharvest pathogens: *Penicillium digitatum* and *Penicillium expansum*. *Genome Biology and Evolution* 8, 218–227.

Kars, I., Krooshof, G.H., Wagemakers, L., Joosten, R., Benen, J.A. and Van Kan, J.A. 2005a. Necrotizing activity of five *Botrytis cinerea* endopolygalacturonases produced in *Pichia pastoris*. *The Plant Journal* 43, 213–225.

Kars, I., McCalman, M., Wagemakers, L. and Van Kan, J.A.L. 2005b. Functional analysis of *Botrytis cinerea* pectin methylesterase genes by PCR-based targeted mutagenesis: *Bcpme1* and *Bcpme2* are dispensable for virulence of strain B05.10. *Molecular Plant Pathology* 6, 641–652.

Katoh, H., Isshiki, A., Masunaka, A., Yamamoto, H. and Akimitsu, K. 2006. A virulence-reducing mutation in the postharvest citrus pathogen *Alternaria citri*. *Phytopathology* 96, 934–940.

Kolattukudy, P. 1985. Enzymatic penetration of the plant cuticle by fungal pathogens. *Annual Review of Phytopathology* 23, 223–250.

Li, B., Lai, T., Qin, G. and Tian, S. 2010. Ambient ph stress inhibits spore germination of *Penicillium expansum* by impairing protein synthesis and folding: A proteomic-based study. *Journal of Proteome Research* 9, 298–307.

Li, B., Wang, W., Zong, Y., Qin, G. and Tian, S. 2012. Exploring pathogenic mechanisms of *Botrytis cinerea* secretome under different ambient pH based on comparative proteomic analysis. *Journal of Proteome Research* 11, 4249–4260.

Li, B., Zong, Y., Du, Z., Chen, Y., Zhang, Z., Qin, G., Zhao, W. and Tian, S. 2015. Genomic characterization reveals insights into patulin biosynthesis and pathogenicity in *Penicillium* species. *Molecular Plant-Microbe Interactions* 28, 635–647.

Li, R., Rimmer, R., Buchwaldt, L., Sharpe, A.G., Séguin-Swartz, G., Coutu, C. and Hegedus, D.D. 2004. Interaction of *Sclerotinia sclerotiorum* with a resistant *Brassica napus* cultivar: Expressed sequence tag analysis identifies genes associated with fungal pathogenesis. *Fungal Genetics and Biology* 41, 735–753.

Liang, X., Liberti, D., Li, M., Kim, Y.T., Hutchens, A., Wilson, R. and Rollins, J.A. 2015. Oxaloacetate acetylhydrolase gene mutants of *Sclerotinia sclerotiorum* do not accumulate oxalic acid, but do produce limited lesions on host plants. *Molecular Plant Pathology* 16, 559–571.

Lin, C.-H., Yang, S.L. and Chung, K.-R. 2009. The YAP1 homolog–mediated oxidative stress tolerance is crucial for pathogenicity of the necrotrophic fungus *Alternaria alternata* in citrus. *Molecular Plant-Microbe Interactions* 22, 942–952.

López-Pérez, M., Ballester, A.R. and González-Candelas, L. 2015. Identification and functional analysis of *Penicillium digitatum* genes putatively involved in virulence toward citrus fruit. *Molecular Plant Pathology* 16, 262–275.

Lyu, X., Shen, C., Fu, Y., Xie, J., Jiang, D., Li, G. and Cheng, J. 2016. A small secreted virulence-related protein is essential for the necrotrophic interactions of *Sclerotinia sclerotiorum* with its host plants. *PLOS Pathogens* 12, e1005435.

Ma, H., Sun, X., Wang, M., Gai, Y., Chung, K.-R. and Li, H. 2016. The citrus postharvest pathogen *Penicillium digitatum* depends on the PdMpkB kinase for developmental and virulence functions. *International Journal of Food Microbiology* 236, 167–176.

Manteau, S., Abouna, S., Lambert, B. and Legendre, L. 2003. Differential regulation by ambient pH of putative virulence factor secretion by the phytopathogenic fungus *Botrytis cinerea*. *FEMS Microbiology Ecology* 43, 359–366.

Marcet-Houben, M., Ballester, A.-R., De la Fuente, B., Harries, E., Marcos, J.F., González-Candelas, L. and Gabaldón, T. 2012. Genome sequence of the necrotrophic fungus *Penicillium digitatum*, the main postharvest pathogen of citrus. *BMC Genomics* 13, article 646.

Morcx, S., Kunz, C., Choquer, M., Assie, S., Blondet, E., Simond-Côte, E., Gajek, K., Chapeland-Leclerc, F., Expert, D. and Soulie, M.-C. 2013. Disruption of *Bcchs4, Bcchs6* or *Bcchs7* chitin synthase genes in *Botrytis cinere*a and the essential role of class VI chitin synthase (*Bcchs6*). *Fungal Genetics and Biology* 52, 1–8.

Motaung, T.E., Saitoh, H. and Tsilo, T.J. 2017. Large-scale molecular genetic analysis in plant pathogenic fungi: A decade of genome-wide functional analysis. *Molecular Plant Pathology* 18, 754–764.

Mullins, E.D., Chen, X., Romaine, P., Raina, R., Geiser, D. and Kang, S. 2001. *Agrobacterium*-mediated transformation of *Fusarium oxysporum*: An efficient tool for insertional mutagenesis and gene transfer. *Phytopathology* 91, 173–180.

Nguyen, H.D.T., Lewis, C.T., Lévesque, C.A. and Gräfenhan, T. 2016. Draft Genome Sequence of *Alternaria alternata* ATCC 34957. *Genome Announcements* 4, e01554–01515.

Noda, J., Brito, N. and González, C. 2010. The *Botrytis cinerea* xylanase Xyn11A contributes to virulence with its necrotizing activity, not with its catalytic activity. *BMC Plant Biology* 10, 38.

Oide, S., Moeder, W., Krasnoff, S., Gibson, D., Haas, H., Yoshioka, K. and Turgeon, B.G. 2006. *NPS6*, encoding a nonribosomal peptide synthetase involved in siderophore-mediated iron metabolism, is a conserved virulence determinant of plant pathogenic ascomycetes. *The Plant Cell* 18, 2836–2853.

Peñalva, M.A., Tilburn, J., Bignell, E. and Arst, H.N. 2008. Ambient pH gene regulation in fungi: Making connections. *Trends in Microbiology* 16, 291–300.

Pinedo, C., Wang, C.-M., Pradier, J.-M., Dalmais, B., Choquer, M., Le Pêcheur, P., Morgant, G., Collado, I.G., Cane, D.E. and Viaud, M. 2008. Sesquiterpene synthase from the botrydial biosynthetic gene cluster of the phytopathogen *Botrytis cinerea*. *ACS Chemical Biology* 3, 791–801.

Plissonneau, C., Benevenuto, J., Mohd-Assaad, N., Fouché, S., Hartmann, F.E. and Croll, D. 2017. Using population and comparative genomics to understand the genetic basis of effector-driven fungal pathogen evolution. *Frontiers in Plant Science* 8, article 119.

Prusky, D., McEvoy, J.L., Saftner, R., Conway, W.S. and Jones, R. 2004. Relationship between host acidification and virulence of *Penicillium* spp. on apple and citrus fruit. *Phytopathology* 94, 44–51.

Prusky, D.B., Bi, F., Moral, J. and Barad, S. 2016. How does host carbon concentration modulate the lifestyle of postharvest pathogens during colonization? *Frontiers in Plant Science* 7, article 1306.

Qin, G., Liu, J., Cao, B., Li, B. and Tian, S. 2011. Hydrogen peroxide acts on sensitive mitochondrial proteins to induce death of a fungal pathogen revealed by proteomic analysis. *PLoS One* 6, e21945.

Qin, G., Tian, S., Chan, Z. and Li, B. 2007. Crucial role of antioxidant proteins and hydrolytic enzymes in pathogenicity of *Penicillium expansum*: Analysis based on proteomic approach. *Molecular and Cellular Proteomics* 6, 425–438.

Ramón-Carbonell, M. and Sánchez-Torres, P. 2017. Involvement of *Penicillium digitatum* PdSUT1 in fungicide sensitivity and virulence during citrus fruit infection. *Microbiological Research* 203, 57–67.

Reis, H., Pfiffi, S. and Hahn, M. 2005. Molecular and functional characterization of a secreted lipase from *Botrytis cinerea*. *Molecular Plant Pathology* 6, 257–267.

Rodrigues-Pousada, C., Menezes, R.A. and Pimentel, C. 2010. The Yap family and its role in stress response. *Yeast* 27, 245–258.

Rollins, J.A. 2003. The *Sclerotinia sclerotiorum pac1* gene is required for sclerotial development and virulence. *Molecular Plant-Microbe Interactions* 16, 785–795.

Rollins, J.A. and Dickman, M.B. 2001. pH signaling in *Sclerotinia sclerotiorum*: Identification of a *pacC/RIM1* homolog. *Applied and Environmental Microbiology* 67, 75–81.

Rui, O. and Hahn, M. 2007. The Slt2-type MAP kinase Bmp3 of *Botrytis cinerea* is required for normal saprotrophic growth, conidiation, plant surface sensing and host tissue colonization. *Molecular Plant Pathology* 8, 173–184.

Sánchez-Torres, P. and González-Candelas, L. 2003. Isolation and characterization of genes differentially expressed during the interaction between apple fruit and *Penicillium expansum*. *Molecular Plant Pathology* 4, 447–457.

Sanzani, S.M., Reverberi, M., Punelli, M., Ippolito, A. and Fanelli, C. 2012. Study on the role of patulin on pathogenicity and virulence of *Penicillium expansum*. *International Journal of Food Microbiology* 153, 323–331.

Schamber, A., Leroch, M., Diwo, J., Mendgen, K. and Hahn, M. 2010. The role of mitogen-activated protein (MAP) kinase signaling components and the Ste12 transcription factor in germination and pathogenicity of *Botrytis cinerea*. *Molecular Plant Pathology* 11, 105–119.

Schumacher, J., Simon, A., Cohrs, K.C., Viaud, M. and Tudzynski, P. 2014. The transcription factor BcLTF1 regulates virulence and light responses in the necrotrophic plant pathogen *Botrytis cinerea*. *PLOS Genetics* 10, e1004040.

Segmüller, N., Ellendorf, U., Tudzynski, B. and Tudzynski, P. 2007. BcSAK1, a stress-activated mitogen-activated protein kinase, is involved in vegetative differentiation and pathogenicity in Botrytis cinerea. *Eukaryotic Cell* 6, 211–221.

Shah, P., Powell, A.L., Orlando, R., Bergmann, C. and Gutierrez-Sanchez, G. 2012. Proteomic analysis of ripening tomato fruit infected by *Botrytis cinerea*. *Journal of Proteome Research* 11, 2178–2192.

Siegmund, U. and Viefhues, A. 2016. Reactive oxygen species in the *Botrytis*–host interaction. pp. 269–289. *In*: S. Fillinger and Y. Elad (eds.). Botrytis*–the Fungus, the Pathogen and its Management in Agricultural Systems*. Springer International Publishing, Cham, Switzerland.

Siewers, V., Viaud, M., Jimenez-Teja, D., Collado, I.G., Gronover, C.S., Pradier, J.-M., Tudzynsk, B. and Tudzynski, P. 2005. Functional analysis of the cytochrome P450 monooxygenase gene *bcbot1* of *Botrytis cinerea* indicates that botrydial is a strain-specific virulence factor. *Molecular Plant-Microbe Interactions* 18, 602–612.

Snini, S.P., Tannous, J., Heuillard, P., Bailly, S., Lippi, Y., Zehraoui, E., Barreau, C., Oswald, I.P. and Puel, O. 2016. Patulin is a cultivar-dependent aggressiveness factor favouring the colonization of apples by *Penicillium expansum*. *Molecular Plant Pathology* 17, 920–930.

Soulie, M.C., Perino, C., Piffeteau, A., Choquer, M., Malfatti, P., Cimerman, A., Kunz, C., Boccara, M. and Vidal-Cros, A. 2006. *Botrytis cinerea* virulence is

drastically reduced after disruption of chitin synthase class III gene (*Bcchs3a*). *Cellular Microbiology* 8, 1310–1321.
Soulié, M.C., Piffeteau, A., Choquer, M., Boccara, M. and Vidal-Cros, A. 2003. Disruption of *Botrytis cinerea* class I chitin synthase gene *Bcchs1* results in cell wall weakening and reduced virulence. *Fungal Genetics and Biology* 40, 38–46.
Staats, M. and van Kan, J.A.L. 2012. Genome update of *Botrytis cinerea* strains B05.10 and T4. *Eukaryotic Cell* 11, 1413–1414.
Stahl, D.J. and Schäfer, W. 1992. Cutinase is not required for fungal pathogenicity on pea. *The Plant Cell* 4, 621–629.
Sweigard, J.A., Carroll, A., Farrall, L., Chumley, F.G. and Valent, B. 1998. *Magnaporthe grisea* pathogenicity genes obtained through insertional mutagenesis. *Molecular Plant-Microbe Interactions* 11, 404–412.
Temme, N. and Tudzynski, P. 2009. Does *Botrytis cinerea* ignore H_2O_2-induced oxidative stress during infection? Characterization of *Botrytis* activator protein 1. *Molecular Plant-Microbe Interactions* 22, 987–998.
ten Have, A., Breuil, W.O., Wubben, J.P., Visser, J. and van Kan, J.A. 2001. *Botrytis cinerea* endopolygalacturonase genes are differentially expressed in various plant tissues. *Fungal Genetics and Biology* 33, 97–105.
ten Have, A., Mulder, W., Visser, J. and van Kan, J.A. 1998. The endopolygalacturonase gene *Bcpg1* is required for full virulence of *Botrytis cinerea. Molecular Plant-Microbe Interactions* 11, 1009–1016.
Tian, S., Torres, R., Ballester, A.R., Li, B., Vilanova, L. and González-Candelas, L. 2016. Molecular aspects in pathogen-fruit interactions: Virulence and resistance. *Postharvest Biology and Technology* 122, 11–21.
Tsuge, T., Harimoto, Y., Hanada, K., Akagi, Y., Kodama, M., Akimitsu, K. and Yamamoto, M. 2016. Evolution of pathogenicity controlled by small, dispensable chromosomes in *Alternaria alternata* pathogens. *Physiological and Molecular Plant Pathology* 95, 27–31.
Turrá, D., Segorbe, D. and Di Pietro, A. 2014. Protein kinases in plant-pathogenic fungi: Conserved regulators of infection. *Annual Review of Phytopathology* 52, 267–288.
Urban, M., Cuzick, A., Rutherford, K., Irvine, A., Pedro, H., Pant, R., Sadanadan, V., Khamari, L., Billal, S., Mohanty, S. and Hammond-Kosack, K.E. 2017. PHI-base: A new interface and further additions for the multi-species pathogen–host interactions database. *Nucleic Acids Research* 45, D604–D610.
Valette-Collet, O., Cimerman, A., Reignault, P., Levis, C. and Boccara, M. 2003. Disruption of *Botrytis cinerea* pectin methylesterase gene *Bcpme1* reduces virulence on several host plants. *Molecular Plant-Microbe Interactions* 16, 360–367.
van Kan, J., Van't Klooster, J., Wagemakers, C., Dees, D. and van der Vlugt-Bergmans, C. 1997. Cutinase A of *Botrytis cinerea* is expressed, but not essential, during penetration of gerbera and tomato. *Molecular Plant-Microbe Interactions* 10, 30–38.
van Kan, J.A.L., Stassen, J.H.M., Mosbach, A., van der Lee, T.A.J., Faino, L., Farmer, A.D., Papasotiriou, D.G., Zhou, S., Seidl, M.F., Cottam, E., Edel, D., Hahn, M., Schwartz, D.C., Dietrich, R.A., Widdison, S. and Scalliet, G. 2017. A gapless genome sequence of the fungus *Botrytis cinerea. Molecular Plant Pathology* 18, 75–89.
Vilanova, L., Teixidó, N., Torres, R., Usall, J., Viñas, I. and Sánchez-Torres, P. 2016. Relevance of the transcription factor *PdSte12* in *Penicillium digitatum*

conidiation and virulence during citrus fruit infection. *International Journal of Food Microbiology* 235, 93–102.

Vilanova, L., Viñas, I., Torres, R., Usall, J., Buron-Moles, G. and Teixidó, N. 2014. Acidification of apple and orange hosts by *Penicillium digitatum* and *Penicillium expansum*. *International Journal of Food Microbiology* 178, 39–49.

Wang, M., Chen, C., Zhu, C., Sun, X., Ruan, R. and Li, H. 2014. Os2 MAP kinase-mediated osmostress tolerance in *Penicillium digitatum* is associated with its positive regulation on glycerol synthesis and negative regulation on ergosterol synthesis. *Microbiological Research* 169, 511–521.

Weiberg, A., Wang, M., Lin, F.-M., Zhao, H., Zhang, Z., Kaloshian, I., Huang, H.-D. and Jin, H. 2013. Fungal small RNAs suppress plant immunity by hijacking host RNA interference pathways. *Science* 342, 118–123.

Wenderoth, M., Pinecker, C., Voß, B. and Fischer, R. 2017. Establishment of CRISPR/Cas9 in *Alternaria alternata*. *Fungal Genetics and Biology* 101, 55–60.

Wubben, J.P., ten Have, A., van Kan, J.A. and Visser, J. 2000. Regulation of endopolygalacturonase gene expression in *Botrytis cinerea* by galacturonic acid, ambient pH and carbon catabolite repression. *Current Genetics* 37, 152–157.

Xiao, X., Xie, J., Cheng, J., Li, G., Yi, X., Jiang, D. and Fu, Y. 2014. Novel secretory protein Ss-Caf1 of the plant-pathogenic fungus *Sclerotinia sclerotiorum* is required for host penetration and normal sclerotial development. *Molecular Plant-Microbe Interactions* 27, 40–55.

Xu, L. and Chen, W. 2013. Random T-DNA mutagenesis identifies a Cu/Zn superoxide dismutase gene as a virulence factor of *Sclerotinia sclerotiorum*. *Molecular Plant-Microbe Interactions* 26, 431–441.

Yakoby, N., Beno-Moualem, D., Keen, N.T., Dinoor, A., Pines, O. and Prusky, D. 2001. *Colletotrichum gloeosporioides* pelB is an important virulence factor in avocado fruit-fungus interaction. *Molecular Plant-Microbe Interactions* 14, 988–995.

Yu, P.-L., Wang, C.-L., Chen, P.-Y. and Lee, M.-H. 2017. YAP1 homologue-mediated redox sensing is crucial for a successful infection by *Monilinia fructicola*. *Molecular Plant Pathology* 18, 783–797.

Zhang, L., Kars, I., Essenstam, B., Liebrand, T.W.H., Wagemakers, L., Elberse, J., Tagkalaki, P., Tjoitang, D., van Den Ackerveken, G. and van Kan, J.A.L. 2014. Fungal endopolygalacturonases are recognized as microbe-associated molecular patterns by the Arabidopsis receptor-like protein Responsiveness to Botrytis Polygalacturonases1. *Plant Physiology* 164, 352–364.

Zhang, T., Sun, X., Xu, Q., González-Candelas, L. and Li, H. 2013a. The pH signaling transcription factor PacC is required for full virulence in *Penicillium digitatum*. *Applied Microbiology and Biotechnology* 97, 9087–9098.

Zhang, T., Sun, X., Xu, Q., Zhu, C., Li, Q. and Li, H. 2013b. PdSNF1, a sucrose non-fermenting protein kinase gene, is required for *Penicillium digitatum* conidiation and virulence. *Applied Microbiology and Biotechnology* 97, 5433–5445.

Zhang, T., Xu, Q., Sun, X. and Li, H. 2013c. The calcineurin-responsive transcription factor Crz1 is required for conidation, full virulence and DMI resistance in *Penicillium digitatum*. *Microbiological Research* 168, 211–222.

Zhang, Y., De Stefano, R., Robine, M., Butelli, E., Bulling, K., Hill, L., Rejzek, M., Martin, C. and Schoonbeek, H.J. 2015. Different reactive oxygen species scavenging properties of flavonoids determine their abilities to extend the shelf life of tomato. *Plant Physiology* 169, 1568–1583.

Zhang, Z., Li, H., Qin, G., He, C., Li, B. and Tian, S. 2016. The MADS-Box transcription factor Bcmads1 is required for growth, sclerotia production and pathogenicity of *Botrytis cinerea*. *Scientific Reports* 6, 33901.
Zheng, L., Campbell, M., Murphy, J., Lam, S. and Xu, J.-R. 2000. The *BMP1* gene is essential for pathogenicity in the gray mold fungus *Botrytis cinerea*. *Molecular Plant-Microbe Interactions* 13, 724–732.

Chapter 12

Mechanisms of Fungal Quiescence during Development and Ripening of Fruits

Dov Prusky

Department of Postharvest Science Agricultural Research Organization (ARO), the Volcani Center, Rishon LeZion, Israel

College of Food Science and Engineering, Gansu Agricultural University, Lanzhou, PR China

Carmit Ziv

Department of Postharvest Science , Agricultural Research Organization (ARO), the Volcani Center, Rishon LeZion, Israel

Abbreviations

1-MCP	1-Methylcyclopropene
ABA	Abscisic acid
CWDE	Cell wall-degrading enzymes
HK	Hexose-kinase
JA	Jasmonic acid
MEP	Methyl ammonia permease
nor	Non-ripening
Nr	Never-ripe
OA	Oxalic acid
PAMP	Pathogen-associated molecular pattern
PELB	Pectate lyase encoding gene
PG	Polygalacturonase
PR	Pathogenesis-related protein
rin	Ripening inhibited
ROS	Reactive oxygen species
SA	Salicylic acid

1 Introduction

Postharvest pathogens may penetrate their fruit host through wounds, natural openings like lenticels, or directly through the cuticle. The length of the lag period for an infecting pathogen to activate its metabolic processes, colonize the tissue, and cause disease symptoms varies among pathogens. In some cases with fruit pathogens, the stage of fungal germination and penetration is followed by a period of fungal interaction with its host that leads to a relative long period until the fruit is harvested and symptoms are clearly observed. This period from fungal infection until the initiation of symptom development is considered to be the quiescent period and it is modulated by either the fungal development and/or the host physiology.

The structure of the fungus during quiescence varies among pathogens. While *Colletotrichum* penetrates the host by a germinated appressorium that breach the fruit cuticle and become quiescence as a germinated appressorium, *Botrytis* may penetrate directly into the tissues and remain quiescent as elongated hyphae. Moreover, the quiescent structures of the pathogen may vary according to the fruit hosts. Thus, *Colletotrichum* remains quiescent in avocado as elongated hyphae between cells, but in tomato the germinated conidia that breached the unripe fruit cuticle exhibit a quiescent stage with dendritic-like structures and swollen hyphae within the fruit epidermal cells.

The length of the quiescent stage may also vary according to the type of host and its developmental stage. Specifically in fruits, the time from fungal infection to symptom expression is closely related to the stage of fruit development and maturation. In many postharvest pathogens, disease symptoms occur long after the initial stages of infection. Unripe, infected fruits may have a lag period of many months until symptoms appear. However, this lag period is significantly shortened to several days with ripened fruits.

1.1 The Modulation of Metabolic Processes during Quiescent Infections

The initial interaction between the fruit and the fungal pathogen that leads to quiescent infection was suggested to be a result of: (i) host involvement in induction of resistance; (ii) lack of the secretion of fungal pathogenicity factors; or (iii) the possible development of a mutualistic-endophytic–like adaptation of the fungus to the plant.

To gain new insights on this unique interaction, the mutual transcriptome of a quiescent infection of *Colletotrichum gloeosporioides* (Penz.) Penz. & Sacc. in unripe tomato was characterized (Alkan et al., 2015). During quiescence, 7,903 fungal genes were expressed. Of these, 178 genes could be defined as quiescent-specific. Cell cycle components were up-regulated, which have a role in the DNA synthesis and control of the cell cycle start at the G1/S transition. Histone modifiers and ATP-dependent chromatin remodeling complexes regulating DNA accessibility were also activated and those genes modulated during quiescence are crucial for gene activation/repression. Transcripts for histone modifiers were up-regulated specifically during the quiescent stage. Additionally, Cgl-Med21 (RNApolymerase II mediator subunit 21) was up-regulated 100-fold, suggesting that both histone modifiers and modification of chromatin structure could be active at the quiescent stage. However, what the contribution of these changes on the establishment and activation of quiescence is still unclear.

Nevertheless, the importance of the fungal behavior during quiescence has not been fully addressed. Is the pathogen in an endophytic-like/non-parasitic state, which shifts to parasitic during ripening? This is evidence for the presence of a biotrophic interaction by the pathogen that involves an initial penetration of the germinated hypha into various tissues of fruits and vegetables (fruit peel, receptacle, flower stems, etc.) without any specific fungal structure. This biotrophic stage may facilitate the intimate parasitic relationship with living host cells. What then are the fungal changes induced by the host that contribute to the initiation of colonization and those modulated by the pathogen? The complete absence of haustoria from the life cycle of postharvest pathogens is evidence that the germinated spore has no real biotrophic stage and the germinated spore proceeds to immediate necrotrophic development (Prusky et al., 2013). Nevertheless, in hemibiotrophs such as *Colletotrichum*, the fungus has an early asymptomatic biotrophic phase of growth that is followed by a necrotrophic second stage characterized by tissue degradation and disease symptoms. The transition from the hemibiotrophic to the necrotrophic stage occurs after a long biotrophic/quiescent stage (Prusky et al., 2013).

1.2 The Activation of Quiescent Infections

For successful colonization, a pathogen must be able to initiate attack under prevailing physiological and environmental conditions. During this period, the pathogen must trigger pathogenicity factors at any stage after penetration that macerate host tissues and release the nutrients required to sustain its development. Since both the host and the pathogen are living entities, the conditions imposed by the host during fruit ripening must be at a critical level to enable susceptibility and initiation of the pathogen quiescent stage. Fruit ripening is accompanied by significant fruit metabolic changes affecting fruit resistance as well as host changes (sugar accumulation, pH change) that may contribute to the activation of pathogenicity factors (Prusky, 1996). At the same time, nutrient signaling and nutrient acquisition by the pathogen activates fungal metabolism contributing to pathogenicity through pH modulation of the environment.

The importance of host behavior during quiescence has been addressed in several cases suggesting that host response can inhibit the activation of pathogen development. During the physiological changes during ripening of the host, quiescent biotrophic infections, resulting from penetration of the fruit directly or through wounds, become active and necrotrophic colonization occurs (Denison et al., 1995; Calvo et al., 2002; Caracuel et al., 2003b). In this case, the host physiological changes may simply allow or induce fungal colonization.

The importance of the interaction between the host and pathogen during quiescence has also been addressed in several studies. The capability of ripening processes in fruits to enhance fungal pathogenicity factors and fungal activation is also a possibility that was discussed. Early reports suggested that four factors modulate the activation of biotrophic/quiescent interactions into necrotrophic phase: (i) induced accessibility of disassembled cell wall substrates during fruit softening (Cantu et al., 2008b) and ethylene induction (Giovannoni, 2001); (ii) a decline in preformed antifungal compounds, such as polyphenols and other preformed fungitoxic substances and inducible phytoalexins (Romanazzi et al., 2014); (iii) a decline in inducible host-defense responses (Beno-Moualem and Prusky, 2000); and (iv) the pH and carbon availability in the host (Prusky et al., 2016). However, a new factor was added based on recent findings suggesting (v) the existence of endophytic-like interaction (Card et al., 2016). In this chapter, we will review and discuss the different mechanisms affecting the occurrence of quiescent infections and its activation to necrotrophic stage in developing fruits that result in rotting and loss of crops in the field and after harvest.

2 Host Factors Activating Quiescent Infections in Ripening Fruits

2.1 Host Cell Wall and Cuticle-Dependent Resistance

Fruit softening facilitates the activation of quiescent infections. Cell walls not only are physical barriers that limit pathogen access to cellular contents, but also are involved in pathogen recognition and in activation and deployment of plant responses (Vorwerk et al., 2004; Cantu et al., 2008a, 2008b). The role of such resistance in the control of quiescence is unclear. Early work by Droby and coworkers (Droby et al., 1987) showed that fruits peeled before harvest in the orchard were immediately susceptible to infection and colonization after artificial fungal inoculation, while similar unpeeled fruits were not. This indicates the importance of the cuticle in quiescent infections of fruits in the orchard. However, as the degradation of the plant cell wall and the cuticle proceeds, by the action of pathogen-derived cell-wall and cuticle-depolymerizing enzymes that are linked to pathogen-associated molecular pattern (PAMP)-triggered immunity (reviewed by Mengiste, 2012), these processes likely impact quiescence.

In tomato, the simultaneous suppression of expression of fruit-ripening-associated *LePG* and *LeExp1* reduced susceptibility to *Botrytis cinerea* Pers. infection during ripening, indicating that *PG* and *Exp* act cooperatively to support both softening (Brummell et al., 1999, 2002; Kalamaki et al., 2003; Powell et al., 2003) and resistance to *B. cinerea* (Cantu et al., 2008a, 2008b). Interestingly, *B. cinerea* infections also induce expression of the same *LePG* and *LeExp1* proteins (Gonzalez-Bosch et al., 1996; Flors et al., 2007), suggesting that specific aspects of the cell-wall disassembly induced during fruit susceptibility are similar to those occurring during ripening. This may indicate that the enzymes involved in cell-wall metabolism play a central role in modulating active infection (Cantu et al., 2009).

Furthermore, during fruit maturation the cuticle undergoes major changes, and cuticular changes, such as cutin deficiency and permeability, were directly associated with resistance of fruit (Isaacson et al., 2009; Curvers et al., 2010; L'Haridon et al., 2011). Cell-wall disassembly of the polysaccharide matrix may also affect the embedded plant proteins that modulate pathogen recognition and plant responses (Van Loon et al., 2006). Cell-wall disassembly may lead to the loss of polygalacturonase (PG)-inhibiting proteins that have a depolymerizing action on fungal PGs and may also generate signals, including pectin-derived oligosaccharides (Cervone et al., 1989; Cote and Hahn, 1994; Vorwerk et al., 2004; Cantu et al., 2008a, 2008b) that activate plant-defense responses that are inoperative or unavailable before the onset of normal ripening and activating quiescent infections (Nurnberger and Scheel, 2001). Alternatively, disassembly of the polysaccharide matrix may simply render the cell wall ineffective as a barrier. The exact impact of these mutants on quiescence and not just necrotrophic growth needs to be examined.

However, activation of quiescence does not always occur with decreased firmness of the fruits. Quiescent infections are located in the peel (cuticle and 1–2 fruit parenchyma cell layers); however, flesh softening and ripening may change independently of fruit peel senescence. Treatments of 1-methylcyclopropene (1-MCP) to avocado fruits greatly inhibited fruit flesh softening, but did not inhibit the activation of quiescent infections located in the fruit peel or the occurrence of decay. This was explained by the differential senescence of both tissues and decline of preformed antifungal dienes (Wang et al., 2005).

2.2 Hormones and Ripening-Controlled Resistance

Plant-hormone synthesis and signaling pathways impact host responses and likely modulate entry into, duration, and exit from quiescence just as they impact leaf and fruit tissues. In leaves, defenses against *B. cinerea* utilize complex signaling networks involving ethylene, oxylipins, salicylic acid, and abscisic acid (Ferrari et al., 2003; Glazebrook, 2005; Abuqamar et al., 2006; Asselbergh and Hofte, 2007). Recent metabolomic studies in leaves revealed that cell-wall modifications played a major role in ethylene-mediated resistance of vegetative tissue against *B. cinerea* (Lloyd et al., 2011). Whereas in both unripe-quiescent and ripe-susceptible fruits, *B. cinerea* induced transcription of genes in pathways associated with ethylene synthesis, the concurrent promotion of fruit ripening by this hormone might be sufficient to offset its roles in defense responses (Chague et al., 2002; Qadir et al., 2011).

Botrytis is also known to induce fruit-ripening processes by actively secreting ethylene and abscisic acid (ABA) (Siewers et al., 2006), thereby inducing host responses (Ron and Avni, 2004; Cantu et al., 2009; Alkan et al., 2012); ABA was found to have an essential role in cuticular permeability, which may influence tomato fruit resistance to *B. cinerea* (Curvers et al., 2010) and the termination of quiescence. However, not only the individual hormones, but also the cross-talk between various hormone-response pathways that modulate pathogen responses is important. *Botrytis cinerea* exploited the antagonistic interactions between pathways to enhance disease development in tomato (El Oirdi et al., 2011). The fungus produces the exopolysaccharide β-(1, 3)(1,6)-D-glucan, which is an elicitor of the salicylic acid (SA) pathway. Subsequently, the activated SA pathway antagonizes the jasmonate signaling pathway in an NPR1-dependent manner, and thereby promotes severe disease in tomato. NPR1 is central to systemic acquired resistance and recently was reported as the receptor for SA (Wu et al., 2012).

In several cases, *B. cinerea* relied on the processes and events that occur during ripening, in order to successfully infect the organ. Fruit ripening induced by *B. cinerea* can be considered a form of regulated senescence and a mechanism for quiescence release (Giovannoni, 2007), and *B. cinerea*, which is a necrotrophic pathogen, promotes senescence. The conclusion that ripening affects the way fruits respond to the pathogen is supported by the observation that expression of a β-1,3-glucanase gene, which has been associated with responses to pathogens (Cantu et al., 2009), was induced in unripe and mature wild-type tomatoes and in all ripening *rin* (ripening inhibited), *nor* (non-ripening), and *Nr* (never-ripe) mutants, but was down-regulated by *B. cinerea* in ripe wild-type fruits. Because the ripening regulators RIN and NOR control many aspects of fruit ripening, they were used to analyze the increased susceptibility of fruit tissue to *B. cinerea* (Giovannoni, 2001, 2007; Cantu et al., 2009). Mature *nor* and *rin* mutant fruits failed to soften or to synthesize lycopene or carotenoids, and evolved ethylene only slightly and at late stages (McGlasson et al., 1975). Cantu and coworkers (Cantu et al., 2009) found that increased susceptibility of ripe fruits to *B. cinerea* depended on NOR but not on RIN, and only partially on the perception of ethylene by the fruit. This conclusion was based on the observation that *rin* mutant fruits and those treated with the ethylene perception inhibitor 1-MCP did not ripen but, nevertheless, were susceptible to *B. cinerea*. Similar results were obtained in avocado fruit treated with 1-MCP where inhibition of ripening was accompanied by enhanced susceptibility to *C. gloeosporioides* as a result of the inhibition of the continuous preformed-antifungal diene synthesis by 1-MCP (Wang et al., 2004, 2005). These results indicate that, in some cases, the properties of ripe fruit that promote susceptibility and the activation of quiescence are not controlled by perception of RIN or ethylene. Similar results were also observed by the activation of quiescent *Alternaria* infections in 1-MCP-treated persimmon fruits that, while softening of the flesh was inhibited, *Alternaria* infections in the peel were activated (Prusky unpublished data).

Other indications that ripening factors are not directly related to fruit susceptibility to disease were described in relation to the role of polyamines in fruit ripening and disease resistance. Polyamines are polycationic biogenic amines involved in many biological processes, including plant growth and development. Transgenic tomato lines overexpressing yeast spermidine synthase (an enzyme involved in polyamine biosynthesis) had a longer shelf life, less shriveling, and delayed decay symptom development than wild-type fruits (Nambeesan et al., 2010). Unexpectedly, in vegetative tissue, increased polyamine was accompanied with enhanced susceptibility to *B. cinerea*, because of cross-suppression of the ethylene pathway (Nambeesan et al., 2012). Since ethylene and polyamines share common precursors, up-regulation of polyamine synthesis in the vegetative tissue interferes with ethylene synthesis, an indication that, in this case, independent responses are involved in the termination of quiescence in leaves and in fruits.

2.3 Preformed and Inducible Antifungal Resistance

Preformed and inducible antifungal compounds are key factors for modulation of the quiescent infections (Prusky, 1996). The resistance of unripe fruits was associated with the presence of preformed antifungal phenolic compounds and inducible phytoalexins, mainly in the cuticle. During fruit ripening, concentrations of preformed and inducible phytoalexins were found to decline, and this was more rapid in susceptible than in resistant cultivars (Lattanzio et al., 1994, 2001; Prusky, 1996). Furthermore, in some cases, the concentrations of antifungal compounds, such as chlorogenic acid and proanthocyanidin, were higher in the more resistant cultivars (Jersch et al., 1989; Bostock et al., 1999). Some of the classical preformed antifungal compounds include: (i) the family of mono-, di-, and triene compounds in avocado; (ii) the resorcinol derivates in mango; (iii) tannins in banana peel (reviewed by Prusky, 1996); (iv) tomatine in tomato fruits (Itkin et al., 2011), and (v) a mixture of several gallotannins with glycosidic linkages, including 1,2,3,4,6-penta-*O*-galloyl-β-D-glucopyranose, in unripe mango fruit peel (Karunanayake et al., 2011). The aqueous phase of mango latex has significant chitinase activity that can digest conidia of *C. gloeosporioides*. Thus, mango fruit with higher concentrations of anthocyanin and flavonoids in red fruits were more resistant to pathogens (Sivankalyani et al., 2015). The activity of all these antifungal compounds declined dramatically during fruit ripening, thus enabling development of penetrating mycelia (Verhoeff, 1974).

In addition, inducible antifungal compounds such as capsicannol in pepper, falcarindiol in carrots, scoparone in citrus, mermesin in celery leaves, resveratrol in grapes, and others were activated in unripe but not always in ripe produce (reviewed by Prusky, 1996). These compounds were hypothesized to be quiescence-modulating factors. Interestingly, although the preformed antifungal compounds dramatically declined during ripening, postharvest stress treatments such as heat, superatmospheric oxygen, high CO_2, UV-C irradiation, and wounding induced fruit phenolic compounds, and may enhance host resistance by preventing the shift from quiescence to necrotrophic growth (Lattanzio et al., 2006).

2.4 Host Resistance Responses during Quiescence

Perception and response to molecules secreted from pathogen have been repeatedly studied in leaves and roots (Cote and Hahn, 1994; Collado et al., 2000; Prusky and Yakoby, 2003; An et al., 2005; Staats et al., 2005; Van Kan, 2006), but only few reports addressed these phenomena in ripening fruits. During necrotrophic colonization, ammonia secretion by *Colletotrichum* directly activated the tomato fruit

NADPH oxidase. This resulted in production of reactive oxygen species (ROS) and induction of SA-dependent genes related to biotic stress, ion leakage, and cell death, leading to enhanced colonization (Alkan et al., 2009, 2012). In parallel, SA repressed jasmonic acid (JA)-dependent genes. This would support the hypothesis that SA-induced programmed cell death (PCD) is a major process in protecting against biotrophic fungi, while at the same time it induces necrotrophic colonization (Glazebrook, 2005). These processes activate necrotrophic development of the fungus and contribute to the termination of its quiescence as they are also related to pH change (Miyara et al., 2010). However, it is known that germinating *Colletotrichum* spores are copious secretors of ammonia, but in the case of their germination on unripe fruit, this does not lead to immediate growth but quiescence. Thus, ammonia secretion by itself cannot trigger necrotrophic colonization in unripe fruit.

Analysis of gene expression in unripe, pathogen resistant, inoculated fruits by means of RNA-seq technology has detected a large number of expressed genes, which suggests that previously described antifungal secondary metabolites (Prusky, 1996) represent only a portion of the antifungal factors that are activated during quiescence. Secretome analysis of *B. cinerea*-infected tomato fruits (Shah et al., 2012) revealed that quiescent infections of unripe fruits activated many more plant-defense-related proteins than infections of red ripe fruits, suggesting that impaired cellular responsiveness in ripe fruit could be a quality of decreased resistance. Importantly, irrespective of whether *B. cinerea* was capable of infecting either unripe or ripe tomato fruits, the proteins detected that were secreted by the fungus were similar.

Botrytis cinerea, a model necrotrophic fungal pathogen, causes significant postharvest rot of fresh fruits and vegetables, including tomatoes. By describing host and pathogen proteomes simultaneously in infected tissues, the plant proteins that provide resistance and allow susceptibility and the pathogen proteins that promote colonization and facilitate quiescence can be identified. When mature green, red ripe wild-type or *rin* mutant tomato fruit were infected with *B. cinerea*, 3 d after this 186 tomato proteins were identified in common between the red ripe and red ripe-equivalent *rin* mutant tomato fruit infected by *B. cinerea* (Cantu et al., 2008a, 2008b). However, the limited infections by *B. cinerea* of mature green wild-type fruit resulted in 25% and 33% fewer defense-related tomato proteins than in red and *rin* fruit, respectively. In contrast, the ripening stage of genotype of the fruit when infected did not affect the secreted proteomes by *B. cinerea*. The composition of the collected proteins and the putative functions of the identified proteins argue for their role in plant-pathogen interactions.

Secretome analysis of the inoculated unripe tomato fruit showed 114 defense-related proteins that were classified into four subcategories: (i) pathogenesis-related (PR) proteins (43 proteins); (ii) proteases (43 proteins); (iii) peroxidases (13 proteins); and (iv) protease-inhibiting proteins (15 proteins). Chitinases were prominent (12/43) among the PR proteins. Chitinases catalyze hydrolysis of chitin and constitute the second largest group of antifungal proteins (Selitrennikoff, 2001; Rep et al., 2002; Coaker et al., 2004; Kim et al., 2004; Casado-Vela et al., 2006). In addition, 1,3- and 1,4-β-endoglucanases were identified. These hydrolyze the structural 1,3-β-glucan present in the fungal cell wall, leading to cell lysis and death (Selitrennikoff, 2001), and they can effectively inhibit fungal growth (Mauch et al., 1988; Sela-Buurlage et al., 1993). Plant inhibitors of polysaccharide hydrolases produced by pathogens have been shown to impart reduced susceptibility to infections (Powell et al., 2003); expression of a xyloglucan fungal endoglucanase-inhibitor protein (Qin et al., 2003) reduced susceptibility of tomato fruits to *B. cinerea* (Powell et al., 2003). Other proteins in the PR category include a disease-response protein, an elicitor-inducible protein, a glycosyl hydrolase, three osmotin-like proteins, and thaumatin (Powell et al., 2003).

3 Fungal Factors Activating Quiescent Infections in Ripening Fruits

Quiescence of postharvest pathogens and resistance of unripe fruit to infection are considered to be dynamic processes that switch to an active state during host maturation and ripening. In light of published analyses of gene expression during quiescence and early development stages of unripe fruits (Cantu et al., 2008a, 2009; Djami-Tchatchou et al., 2012; O'Connell et al., 2012), we consider that quiescent fungi should be regarded as eliciting biotrophic or active endophytic interactions that manipulate host responses, similar to the reported endophytic interactions in leaves (Carroll, 1988). The widespread reports of fungal species (including the hemibiotrophic *Colletotrichum* and the necrotrophic *Botrytis*) showing quiescence in similar fruits suggest that host responses are not specific to a fungal colonization mechanism. Transcriptome analysis of the *Colletotrichum higginsianum* Sacc. biotrophic stage revealed a large inventory of effector candidates (Kleemann et al., 2012). These tagged effectors were found in appressoria penetration pores before host invasion, revealing subtle early levels of functional complexity for this fungal organ. Their presence indicates that secreted effectors proteins manipulate the host during penetration and quiescence indicating that quiescence is not a passive interaction (Kleemann et al., 2012; O'Connell et al., 2012). Although *C. higginsianum* growth in *Arabidopsis* leaves is continuous, without a quiescent phase, it likely reveals behavior of similar nature to other *Colletotrichum* spp. that do pause between their biotrophic and necrotrophic stages. *Colletotrichum gloeosporioides* also secretes a large variety of stage specific effectors, non-ribosomal-peptide-synthase, cutinases, and others cell killing factors during quiescence (Alkan et al., 2015). Effectors are key factors in eliciting as well as subverting host-defense responses and PCD; thus, replacement of the terms "quiescence" or "latency" with "establishment" should be considered.

3.1 Fungal Secreted Molecules and Quiescence Release

The transition from quiescence to necrotrophic colonization probably involves modulation of fungal effectors, as reported for *C. higginsianum* and exemplified by *ChNLP1* and *ChToxB* (Kleemann et al., 2012; O'Connell et al., 2012), or of pH modulators, exemplified by ammonia, gluconic and oxalic acids for *C. gloeosporioides, Penicillium expansum* L., and *B. cinerea*, respectively (Prusky and Yakoby, 2003; Prusky and Lichter, 2007; Prusky et al., 2010). Genes regulating the expression of fungal effectors are induced during quiescence and then switch to necrotrophy, which suggests that their toxic products may contribute to termination of the biotrophic phase in preparation for subsequent necrotrophic growth. These fungal effectors are plant-induced during the switch from biotrophic to necrotrophic colonization (Kleemann et al., 2012), which may indicate again that quiescence is an active process that expresses a differential set of fungal genes (O'Connell et al., 2012) and host responses (Djami-Tchatchou et al., 2012).

Fungal pathogens have been known to adjust the extracellular pH in order to increase their infectious potential (Prusky et al., 2001, 2016). The ability of the postharvest pathogen to alter the pH locally has been described for *Colletotrichum* spp., *Alternaria alternata* (Fr.) Keissl., *B. cinerea, P. expansum, Penicillium digitatum* (Pers.) Sacc., *Penicillium italicum* Wehmer, *Phomopsis mangiferae* S. Ahmad, and *Fusarium oxysporum* Schltdl. (Prusky et al., 2001, 2004; Rollins and Dickman, 2001; Eshel et al., 2002a, 2002b; Diéguez-Uribeondo et al., 2008; Miyara et al., 2008, 2010, 2012; Davidzon et al., 2010). The fungi that alkalinize the environment are termed "alkaline fungi" and those that acidify it are termed "acidic fungi."

Alkalization of the plant host was first reported in a number of fruit-infecting species, such as *Colletotrichum* spp. and *A. alternata* (Prusky et al., 2016), and more recently in the root-infecting fungus *F. oxysporum* (Fernandes et al., 2017). These fungi are able to trigger an increase of more than 2 units in the pH of the surrounding fruit tissue or the rhizosphere, respectively. Similarly, the human pathogen *Candida albicans* (C.P. Robin) Berkhout raises the pH in host macrophages by several units, resulting in neutralization of the normally acidic phagosome (Vylkova et al., 2011; Vylkova and Lorenz, 2014). The main mechanism of host alkalization reported in these fungal species is the release of ammonia, which acts as a weak base (Prusky et al., 2001; Vylkova et al., 2011). The exact mechanism that leads to extracellular accumulation of ammonia remains to be elucidated. Work in *Saccharomyces cerevisiae* Meyen, *C. albicans*, and *C. gloeosporioides* showed that this process requires the regulated uptake of amino acids via amino acid permeases or their mobilization from vacuolar stores, followed by catabolism through different routes involving steps of deamination (Miyara et al., 2012; Bi et al., 2016). In *C. gloeosporioides*, the transformation of glutamate to α-ketoglutarate and ammonium was shown to be carried out by the NAD^+-specific glutamate dehydrogenase GDH2 (Miyara et al., 2012). To protect the cell from the toxic effects, ammonia is released either by passive diffusion or through the action of transporters, such as the members of the Ato protein family (Palkova et al., 2002; Danhof and Lorenz, 2015). The precise mechanisms of ammonia extrusion during alkalization remain to be determined.

Ammonium accumulation has been detected in association with pathogenicity of many *Colletotrichum* spp., including *C. gloeosporioides, C. acutatum* J.H. Simmonds, *C. higginsianum, C. graminicola* (Ces.) G.W. Wilson, and *C. coccodes* (Wallr.) S. Hughes (Kramer-Haimovich et al., 2006; Alkan et al., 2008; Miyara et al., 2012; O'Connell et al., 2012), *A. alternata* (Eshel et al., 2002a, 2002b), and *F. oxysporum* (Miyara et al., 2012). The ammonium secreted by *Colletotrichum* spp., *F. oxysporum*, and *A. alternata* (Eshel et al., 2002a, 2002b; Kramer-Haimovich et al., 2006; Alkan et al., 2008) enhances alkalization of the host tissue, and its concentration can reach approximately 5 mM, as found in decayed fruits (Eshel et al., 2002a, 2002b; Alkan et al., 2008; Miyara et al., 2010). Increased ammonium accumulation has been related to *Colletotrichum* spp. transformation from quiescent to necrotrophic development (Alkan et al., 2008, 2009; Miyara et al., 2010).

Other pathogenic fungi, such as *P. expansum, P. digitatum, P. italicum, Phomopsis mangiferae, Aspergillus niger* Tiegh., and *B. cinerea* (Ruijter et al., 1999), use tissue acidification during activation of the quiescent infection and transition to a necrotrophic attack. Tissue acidification is enhanced by secretion of organic acids and/or H^+ excretion. *Sclerotinia sclerotiorum* (Lib.) de Bary and *B. cinerea* decrease the host pH by secreting significant amounts of oxalic acid (OA) (Rollins and Dickman, 2001; Manteau et al., 2003); gluconic acid is secreted by *P. mangiferae* (Davidzon et al., 2010), and combinations of gluconic and citric acids are secreted by *Penicillium* (Prusky et al., 2004) and *Aspergillus* (Ruijter et al., 1999).

3.2 Fungal Pathogenicity Factors that are Activated by Ripening

In the transition from quiescence it is difficult to pinpoint the crucial break in fruit resistance that signals initiation of pathogen growth, but it is clear that the pathogen plays a significant role in acceleration of host responses, as was demonstrated in *B. cinerea* (Chague et al., 2002; Siewers et al., 2006; Cantu et al., 2009; Qadir et al., 2011) and *C. coccodes* (Alkan et al., 2012). Overall, the complex network of processes associated with host ripening and pathogen activation suggests that, whereas the pathogen deploys its pathogenic arsenal, developmental conditions in the host tissue determine the outcome of the fruit–pathogen interaction. The ability of postharvest pathogens to affect the host environment by enhancing fruit ripening and/or modulating ambient pH levels are key aspects of a complex set of characteristics of a fungus that control its transition from quiescent to active infection. Whereas enhanced ripening of fruits is a late-acting factor, modulated by the postharvest pathogens (Kader, 1985), pH modulation seems to be an early-acting factor in the activation process (Prusky et al., 2016).

Although a stable intracellular pH must be maintained to provide enzymes with optimal conditions for their activity, proteins that come in contact with host environment affected by fungal penetration through wounds and/or cell death, need to be selectively expressed at pH levels conducive to their functioning. The factors that determine host pH are part of the first wave of pathogenic attack, which determines the capability of the pathogen to successfully colonize, invade, and kill the targeted host cell. Since pH is a critical factor in fungal attack strategy, the pathogen has developed environment-sensing mechanisms that enable it to modify the ambient conditions to best suit its offensive arsenal and compromise the host defense. Fungi monitor external pH by several mechanisms (Luo et al., 2001; Luo and Michailides, 2001; Peñalva and Arst, 2002; Caracuel et al., 2003a, 2003b; Rollins, 2003). Of them, a signal transduction pathway designated as *RIM* or *PAL* for yeasts or filamentous fungi, respectively, is the major pH sensing component (Arst et al., 1994; Tilburn et al., 1995; Negrete-Urtasun et al., 1999; Arst and Peñalva, 2003; Arechiga-Carvajal and Ruiz-Herrera, 2005; Peñalva et al., 2008; Cervantes-Chávez et al., 2010; Fonseca-García et al., 2012). In filamentous fungi, Pal/Rim signaling pathway requires the proteolytic activation of a zinc-finger transcription factor named PACC (Peñalva and Arst, 2002). This factor activates the transcription of some genes at alkaline pH and represses others at acidic pH (Su and Mitchell, 1993; Espeso and Arst, 2000; Errakhi et al., 2008; Miyara et al., 2010). The pathogen's ability to change the environment and activate signaling pathways enables it to control not only its own cell growth and development but ultimately to colonize the host.

Fungal secreted factors that modulate the host environmental pH promote secretion of the arsenal of pH-specific pathogenicity factors, which includes toxins, transcription factors, genes that modulate host defenses, specific transporters, redox-protective substances, cell-wall-degrading enzymes, and proteinases (Espeso et al., 1993; Calvo et al., 2002; Spielvogel et al., 2008). Simultaneously, the secreted pH-modulating factors affect the host directly, as discussed below.

3.3 Modification of Environmental pH and Pathogenicity

The extent of the pH change in the plant tissue depends on the host's buffer capacity and the initial pH. In contrast to ambient pH, intracellular pH (pHi) influences all biological processes and tends to be constant and tightly regulated (Kane, 2016).

Three major mechanisms for pHi regulation have been reported in fungi; two of them both based on conserved proton-pumping ATPases, and the third related to methyl ammonia permease (MEP) transport. The primary determinant of cytosolic pH is plasma membrane H^+-ATP (Pma1), an essential H^+-ATPase and the most abundant plasma membrane protein in *S. cerevisiae*. Pma1 homologues are found in all fungi, as well as in plants. The second mechanism is the vacuolar ATPase (V-ATPase), a multiprotein complex that mediates acidification of organelles, such as vacuoles, endosomes, or the Golgi (Kane, 2016). Pma1 and V-ATPase are often co-regulated, for example, in response to sudden shifts in ambient pH or glucose levels. In *Colletotrichum*, low sucrose levels lead to the activation of metabolic processes related to ammonia accumulation and its export by the MEP and/or AMET (an ammonia exporter) (Miyara et al., 2012). By contrast, sucrose excess results in inhibition of GDH2 reaction activity, concomitant with an acidification of the environment. As a general rule, activation of these proton-pumping ATPases leads to a pHi increase due to increased proton export, whereas their inhibition triggers intracellular acidification.

3.4 Nutritional Factors Modulating the Secretion of Host Environmental Modulators

Plant-pathogenic fungi acquire and use nutrients during infection (Fernandez et al., 2014). Sugar availability during fruit maturation and ripening is important, given that detection of the nitrogen and carbon compounds modulates the use of regulatory pathways. A second requirement for pH modulation of the environment is carbon deprivation. Presumably, a lack of carbon prevents the efficient use of ammonia for biosynthesis of amino acids and nucleotides, favoring its accumulation (Vylkova et al., 2011; Bi et al., 2016). This may explain the possible mechanisms of activation of quiescence occurring in parallel to the modulation of ripening. Newly developed fungal pathways may activate the secretion of enzymes to depolymerize and/or mobilize these compounds, and take up and metabolize the released substrates. The gene expression underlying absorption is optimized by global regulation in response to available carbon and nitrogen quality, and fine-tuned by pathway-specific regulation. Global carbon and nitrogen regulators ensure that, if available, glucose and ammonium are used first, whereas genes for utilizing less-favored alternatives, such as xylose and nitrate, are repressed. Pathway-specific regulators ensure that genes of a required catabolic pathway are expressed when the appropriate inducer is present.

How, then, is metabolic regulation involved in the different stages of fruit colonization? When do the fungal systems switch from their own stored nutritional factors in quiescent conditions to those induced in the host during active colonization? What is the signal(s) that activate fungal primary metabolism to produce the initial molecules contributing to fungal pathogenicity? What differential metabolism is activated for acquiring nutrients from the immature, developing host versus the fully mature one? What are the metabolic strategies contributing to host maceration and pathogen colonization? Transcriptional, biochemical, and functional analyses of fungal genes in biotrophic and hemibiotrophic foliar pathogens and necrotrophic fruit pathogens have attempted to answer these questions, and to characterize the contribution of fungal metabolism during plant infection.

The ability of fungal pathogens to grow in different media is due to their capacity to sense and respond to changes in nutrient availability. The transduction pathways of the nutritional signals are induced by specific nutrients, and in general these pathways bring about changes in gene expression, mRNA stability (Cereghino and Scheffler, 1996), and post-translational modifications (Ordiz et al., 1996; Rolland et al., 2000). Studies using glucose characterized several signal-transduction pathways that allow yeast to perceive the level of glucose in the medium and initiate the appropriate metabolic response (Gancedo, 1998; Rolland et al., 2001). Three fungal hexose-kinase (HK) enzymes can catalyze this first irreversible step in the intracellular metabolism of glucose in yeast. The genes that are affected by this process include, among others, those involved in the utilization of alternative carbon sources, gluconeogenesis, the glyoxylate and Krebs cycles, respiration, and peroxisomal functions. Glucose also induces the expression of genes required for its own utilization, such as those encoding glycolytic enzymes and glucose transporters (Ozcan and Johnston, 1999). Although several of the genes implicated in the pathways that control glucose repression and induction have been identified (Gancedo, 1998), a complete mechanistic picture of the phenomenon is not yet available for pathogenic fungi on either leaves or fruit. In particular, the position of each factor in the signaling cascade and the interactions among them are still not well known.

The importance of fungal HKs and their activation in pathogenic fungi has not been fully described. What makes Hxk2 a special protein is that, not only is it the enzyme responsible for the phosphorylation of glucose, but it is also implicated in glucose repression. The dual role of HK as a metabolic enzyme and regulatory protein may not be so unique in sugar metabolism, but it indicates the complexity of the carbon metabolism at initial stages of fungal colonization (Moreno and Herrero, 2002).

3.5 The Role of pH in Post-Quiescence Release

Expression of pH-regulated genes is mediated by the transcription factor PACC (Tilburn et al., 1995; Peñalva and Arst, 2002). The loss of PACC function leads to fungal miss-expression of genes normally expressed only under acidic conditions (Tilburn et al., 1995; Bruton et al., 1998). Thus, in alkaline conditions, PACC is responsible for repression of genes expressed under acidic pH (Espeso and Arst, 2000) and for preferential induction of genes expressed under alkaline pH, including PACC itself (Movahedi and Heale, 1990; Suárez and Peñalva, 1996; Mingot et al., 1999; Caracuel et al., 2003a, 2003b). Early induction of PACC is critical as it activates the first wave of pathogenicity, which modulates the transformation of quiescent infections to active necrotrophic growth. The expression of pathogenicity factors such as pectate lyase and endoglucanases in several host/pathogen interactions are examples of PACC-controlled expression (Yakoby et al., 2000a; Prusky et al., 2001, 2004; Eshel et al., 2002b).

In *C. gloeosporioides*, pectate lyase encoding gene (PELB) expression during pH alkalization followed a pattern similar to that of *PAC1*, suggesting the presence of a pH-regulatory mechanism for control of secreted proteins during pH increase (Drori et al., 2003; Miyara et al., 2008). Similarly, for acidic environments, PG expression in *P. expansum* and *F. oxysporum* were controlled by acidic pH (Caracuel et al., 2003a, 2003b; Prusky et al., 2004). *ΔpacC* mutants of *C. gloeosporioides, C. acutatum, F. oxysporum*, and *S. sclerotiorum* were less virulent compared to the wild type (Caracuel et al., 2003a; Rollins, 2003; You et al., 2007; Miyara et al., 2008), which suggests the activity of a complex virulence mechanism that controls pathogenicity factors expression under pH change.

The activation of quiescent infection is facilitated by large gene families of CWDE. This has been observed for the endo-PGs families of *B. cinerea* (van der Cruyssen et al., 1994; Wubben et al., 1999, 2000) and glucanases in *A. alternata* (Eshel et al., 2002a). Furthermore, evidence for the differential expression of each family member under varied pH conditions illustrates the ability of a fungus to fine-tune its pathogenicity to the prevailing environmental conditions (Eshel et al., 2002a, 2002b; Manteau et al., 2003; Miyara et al., 2008). During the study of differential expression of endoPG by *B. cinerea* in apple and zucchini, with low and neutral pH, respectively, it was reported that *Bcpg2* was negatively modulated by low ambient pH, which might account for its lack of expression in apple fruits, whereas *Bcpg3* expression was induced at low pH in liquid cultures, as in apples. This fine-tuning of enzyme induction and secretion in response to the ambient pH can discriminate between hosts (Manteau et al., 2003). It further illustrates the importance of the specific regulatory system, as controlled by environmental pH, and highlights its importance during the transformation of a quiescent to an active infection.

3.6 Sensitization of the Host by pH Modulators and Activation of Host Response

The modulation of pH during quiescence and necrotrophic development affects the plant host. The pH changes in various plant-cell compartments induce ion effluxes that play important roles in regulating ROS production and plant-defense responses. It was reported that apoplast alkalization induced Rboh activation, probably as a result of plasma membrane depolarization, and that the subsequent K^+/H^+ exchange, followed by Ca^{2+} influx/Cl^-_2 efflux (Simon-Plas et al., 1997; Nurnberger and Scheel, 2001; Zhao et al., 2005) led to induction of the defense response and plant PCD (Schaller and Oecking, 1999; Clarke et al., 2005; Hano et al., 2008). In this connection, changes in external pH rapidly altered plant gene expression and modulated host responses, similarly to elicitors (Wojtaszek et al., 1995; Torres et al., 2002; Arst and Peñalva, 2003; Kim et al., 2008; Alkan et al., 2009, 2012). This may indicate that the early-secreted pH modulators might affect the host response and the development of quiescent infections.

Ammonia, as a pH modulator, could itself affect host responses and the transformation of a quiescent to an active infection. The mechanisms of ammonia action include: activation of plant-membrane transporters, activation of plasma-membrane ATPase, elevation of membrane flux, changing the concentration of cytosolic protons, disruption of the biochemical pH-stat, and induction of ethylene synthesis (Mathieu et al., 1994; Britto and Kronzucker, 2002; Zhu et al., 2011). It may be that local secretion of ammonia by penetrating hyphae could induce confined cell death, from which the fungus would initiate its necrotrophic development. In a number of studies, ammonia accumulation during fungal colonization led to ROS accumulation through NADPH oxidase activation in a calcium- and pH-dependent manner, thereby playing an important role in eliciting a wide range of defense mechanisms involved in plant PCD, including activation of host SA-dependent genes (Yakoby et al., 2000a, 2000b; Alkan et al., 2008, 2009, 2012; O'Connell et al., 2012).

OA secretion by *S. sclerotiorum* and *B. cinerea* is essential for pathogenicity, and it is remarkably multifunctional (Maxwell and Lumsden, 1970; Marciano et al., 1983; Godoy et al., 1990; Dutton and Evans, 1996; Cessna et al., 2000; Bolton et al., 2006; Kim et al., 2008; Williams et al., 2011; Veluchamy et al., 2012). OA affects host responses through numerous physiological processes, such as: (i) ambient acidification; (ii) strong Ca^{2+} chelation, which weakens the plant cell wall and compromises the host-defense responses (Martell and Calvin, 1952; Cunningham and Kuiack, 1992; Cessna et al., 2000); (iii) guard-cell regulation; (iv) vascular plugging with oxalate crystals; and (v) inhibition of the host callose deposition (Williams et al., 2011). However, the importance of OA during the early quiescent stages is not clear. During activation of the necrotrophic stage, oxalate lowered the pH and inhibited the oxidative burst of the host (cultured tobacco cells) at an initial pH of 3 to 4, thereby compromising plant-defense responses (Cessna et al., 2000). The mechanisms by which OA triggers redox reduction and modulates plant-defense pathways probably affect both host and pathogen. In *Monilinia* host-redox reduction, which is driven by phenols, affected fungal intracellular antioxidant concentration, appressoria formation, CWDE activity and release from quiescence, thereby enhancing resistance (Lee and Bostock, 2007). In contrast, OA also induced plant ROS production that correlated with a pH of 5 to 6, and which later induced ethylene production, anion channel activation, and DNA laddering, all of which indicated that host PCD was necessary for necrotrophic colonization (Errakhi et al., 2008; Kim et al., 2008). Thus, OA may have a role in host sensitization and regulation of ROS production. This regulation may down-regulate and also activate the plant oxidative burst, possibly in a pH-dependent fashion, to match the pathogen's necrotrophic development.

4 Endophytic Mutualism Modulating the Activation of Quiescent Infections in Ripening Fruits

The possibility of a mutualistic/endophytic nonpathogenic-like interaction during pathogen quiescence was not considered in the past, because fungal quiescence was considered to be mainly related to the presence of preformed antifungal compounds, or to a lack of induction of pathogenicity factors. The recent broad description of endophytic fungus colonizing symptomless hosts, raises the question if during the long periods of quiescence a mutualistic endophytic fungal association may be present (Brownbridge et al., 2012).

Endophytes are ubiquitous, colonizeers in all plants, and have been isolated from almost all plants examined to date. This association can be obligate or facultative and causes no harm to the host plants. They exhibit complex interactions with their hosts that involve mutualism and antagonism. Endophytes colonize plants systemically, and this provides their continuous protection and enhanced persistence. Based on the infection process, the endophytes differ in their ability to colonize different plant parts and to persist over a crop growth cycle (Carroll, 1988).

Endophytes are closely related to virulent pathogens but with limited, if any, pathogenic effects by themselves. "Constitutive mutualism" is the relatively faithful association, usually with grasses, of endophytes that infect host ovules and are propagated in host seed; substantial fungal biomass with probable high metabolic cost develops throughout the aerial parts of the host plant as stem-end rots. "Inducible mutualist" endophytes are those pathogens that are disseminate independently through air or in water. They infect vegetative parts of the host and remain metabolically inactive for long periods with relatively little fungal biomass (Carroll, 1988). However, these inducible mutualists grow rapidly and may produce pathogenicity factors if they are present in ripening host.

The mechanism of growth of constitutive mutualistic pathogens by stem-end causing diseases as *A. alternata* and *Lasiodiplodia theobromae* (Pat.) Griffon & Maubl. that penetrate during flowering is different than inducible mutualistic endophytes that are the result of direct penetration of the fruit cuticle by germinating *Colletotrichum* spores (Diskin et al., 2017). While in the first case, the mycelium is growing through the plant, in the inducible mutualist the fungal development is fully inhibited. In order to maintain stable mutualistic interaction, endophytes produce several compounds that promote growth of plants and help them adapt better to the environment (Khan et al., 2016; Ali et al., 2017).

Tanaka et al. (2006, 2008) demonstrated that ROS bursts originating from a mutualistic endophyte are required to inactivate plant defense responses against the fungus, thereby maintaining the mutualism. Whether the suppression of plant defense is the result of fungal, plant, or symbiotic metabolism is poorly understood. Because ROS plays a mechanistic role in PCD, general stress responses, and systemic signaling, they can have diverse effects on the success of fungal infection or endophyte colonization and the plant responses, that is, resistance, acceptance, or no effect. Moreover, antioxidants can serve to transmit stress signals through the oxidant–antioxidant interaction. This may facilitate the chemical communication between a host and an avirulent pathogen or asymptomatic endophyte enabling the host to react quickly to pathogenesis and differentiate a pathogen from a mutualist.

However, in other cases, endophytic microorganisms may be inhibited by a second interaction. The report of the presence of bacterial endophytes of the genus *Bacillus* in melons of the group 'Dulce' that exhibited antagonistic properties against the fruit pathogens may indicate that quiescence of *Fusarium* as postharvest pathogens may be a more complex process (Glassner et al., 2015). If this set of interactions is confirmed in many fruits, its ability to modulate quiescence activation by naturally occurring bacteria in fruits may be possible. Diskin et al. (2017) suggested that increase in pathogenicity was correlated with increased abundance of chitin-degrading Chitinophagaceae bacteria, suggesting that various conditions modify the microbial community at the stem. In this case, the identified bacteria may be useful for protecting plants, not only in the field, but also after harvest.

5 Concluding Remarks

During ripening, fruits undergo physiological changes, such as activation of ethylene biosynthesis, cuticular changes, cell-wall loosening, and decline of antifungal compounds, which release the fungus from its quiescent state and promote a necrotrophic and pathogenic life style. In the long run, detailed knowledge of host pathways that affect postharvest disease will enable us to design control or monitoring measures that focus on alleviating the consequences of active infection or to extend the duration of quiescence in order to enhance fruit shelf life (Prusky, 2006). The roles of the various branches of the plant immune-response pathways and the transition from the quiescent to the necrotrophic phase have not been examined, particularly in fruit tissue. The plant's perception of PAMP molecules early in the plant–pathogen encounter, and how these are activated in both mature and unripe fruit tissues need to be examined. Further studies on host genes and their patterns of temporal and spatial regulation relative to the quiescent or active stages of infection, and interactions with other fruit-specific developmental processes need to be pursued. The use of genome-wide transcriptome, proteome, and metabolome approaches will help the identification and evaluation of critical factors for entry into and exit from quiescent infection, as well as elucidation of their interactions with other developmental processes. However, as the fruit ripens and the fungal pathogen switches to active secretion of effectors and pH modulators, the necrotrophic stage becomes active. The mechanisms of fungal activation are key factors needed by the pathogen to cope effectively with the hostile environment found within the host. Understanding that regulation should lead to new approaches for disease control (Prusky, 2006).

References

Abuqamar, S., Chen, X., Dhawan, R., Bluhm, B., Salmeron, J., Lam, S., Dietrich, R.A. and Mengiste, T. 2006. Expression profiling and mutant analysis reveals complex regulatory networks involved in Arabidopsis response to *Botrytis infection. Plant Journal* 48, 28–44.

Ali, S., Khan, A.L., Ali, L., Rizvi, T.S., Khan, S.A., Hussain, J., Hamayun, M. and Al-Harrasi, A. 2017. Enzyme inhibitory metabolites from endophytic *Penicillium citrinum* isolated from *Boswellia sacra. Archives of Microbiology* 199, 691–700.

Alkan, N., Davydov, O., Sagi, M., Fluhr, R. and Prusky, D. 2009. Ammonium secretion by *Colletotrichum coccodes* activates host NADPH oxidase activity enhancing host cell death and fungal virulence in tomato fruits. *Molecular Plant-Microbe Interactions* 22, 1484–1491.

Alkan, N., Fluhr, R. and Prusky, D. 2012. Ammonium secretion during *Colletotrichum coccodes* infection modulates salicylic and jasmonic acid pathways of ripe and unripe tomato fruit. *Molecular Plant-Microbe Interactions* 25, 85–96.

Alkan, N., Fluhr, R., Sherman, A. and Prusky, D. 2008. Role of ammonia secretion and pH modulation on pathogenicity of *Colletotrichum coccodes* on tomato fruit. *Molecular Plant-Microbe Interactions* 21, 1058–1066.

Alkan, N., Friedlander, G., Ment, D., Prusky, D. and Fluhr, R. 2015. Simultaneous transcriptome analysis of *Colletotrichum gloeosporioides* and tomato fruit pathosystem reveals novel fungal pathogenicity and fruit defense strategies. *The New Phytologist* 205, 801–815.

An, H.J., Lurie, S., Greve, L.C., Rosenquist, D., Kirmiz, C., Labavitch, J.M. and Lebrilla, C.B. 2005. Determination of pathogen-related enzyme action by mass spectrometry analysis of pectin breakdown products of plant cell walls. *Analytical Biochemistry* 338, 71–82.

Arechiga-Carvajal, E.T. and Ruiz-Herrera, J. 2005. The RIM101/pacC homologue from the basidiomycete *Ustilago maydis* is functional in multiple PH-sensitive phenomena. *Eukaryotic Cell* 4, 999–1008.

Arst, H.N., Jr., Bignell, E. and Tilburn, J. 1994. Two new genes involved in signaling ambient pH in *Aspergillus-nidulans*. *Molecular & General Genetics* 245, 787–790.

Arst, H.N., Jr. and Peñalva, M.A. 2003. pH regulation in *Aspergillus* and parallels with higher eukaryotic regulatory systems. *Trends in Genetics* 19, 224–231.

Asselbergh, B. and Hofte, M. 2007. Basal tomato defences to *Botrytis cinerea* include abscisic acid-dependent callose formation. *Physiological and Molecular Plant Pathology* 71, 33–40.

Beno-Moualem, D. and Prusky, D. 2000. Early events in the development of quiescent infection of avocado fruits against *Colletotrichum gloeosporioides*. *Phytopathology* 90, 553–559.

Bi, F., Barad, S., Ment, D., Luria, N., Dubey, A., Casado, V., Glam, N., Minguez, J.D., Espeso, E.A., Fluhr, R. and Prusky, D. 2016. Carbon regulation of environmental pH by secreted small molecules that modulate pathogenicity in phytopathogenic fungi. *Molecular Plant Pathology* 17, 1178–1195.

Bolton, M.D., Thomma, B.P.H.J. and Nelson, B.D. 2006. *Sclerotinia sclerotiorum* (Lib.) de Bary: Biology and molecular traits of a cosmopolitan pathogen. *Molecular Plant Pathology* 7, 1–16.

Bostock, R.M., Wilcox, S.M., Wang, G. and Adaskaveg, J.E. 1999. Suppression of *Monilinia fructicola* cutinase production by peach fruit surface phenolic acids. *Physiological and Molecular Plant Pathology* 54, 37–50.

Britto, D.T. and Kronzucker, H.J. 2002. NH_4^+ toxicity in higher plants: A critical review. *Journal of Plant Physiology* 159, 567–584.

Brownbridge, M., Reay, S.D., Nelson, T.L. and Glare, T.R. 2012. Persistence of *Beauveria bassiana* (Ascomycota: Hypocreales) as an endophyte following inoculation of radiata pine seed and seedlings. *Biological Control* 61, 194–200.

Brummell, D.A., Harpster, M.H., Civello, P.M., Palys, J.M., Bennett, A.B. and Dunsmuir, P. 1999. Modification of expansin protein abundance in tomato fruit alters softening and cell wall polymer metabolism during ripening. *The Plant Cell* 11, 2203–2216.

Brummell, D.A., Howie, W.J., Ma, C. and Dunsmuir, P. 2002. Postharvest fruit quality of transgenic tomatoes suppressed in expression of a ripening-related expansin. *Postharvest Biology and Technology* 25, 209–220.

Bruton, B.D., Conway, W.S., Gross, K.C., Zhang, J.X., Biles, C.L. and Sams, C.E. 1998. Polygalacturonases of a latent and wound postharvest fungal pathogen of muskmelon fruit. *Postharvest Biology and Technology* 13, 205–214.

Calvo, A.M., Wilson, R.A., Bok, J.W. and Keller, N.P. 2002. Relationship between secondary metabolism and fungal development. *Microbiology and Molecular Biology Reviews* 66, 447–459.

Cantu, D., Blanco-Ulate, B., Yang, L., Labavitch, J.M., Bennett, A.B. and Powell, A.L. T. 2009. Ripening-regulated susceptibility of tomato fruit to *Botrytis cinerea* requires NOR but not RIN or ethylene. *Plant Physiology* 150, 1434–1449.

Cantu, D., Vicente, A.R., Greve, L.C., Dewey, F.M., Bennett, A.B., Labavitch, J.M. and Powell, A.L.T. 2008a. The intersection between cell wall disassembly, ripening, and fruit susceptibility to *Botrytis cinerea*. *Proceedings of the National Academy of Sciences of the United States of America* 105, 859–864.

Cantu, D., Vicente, A.R., Labavitch, J.M., Bennett, A.B. and Powell, A.L.T. 2008b. Strangers in the matrix: Plant cell walls and pathogen susceptibility. *Trends in Plant Science* 13, 610–617.

Caracuel, Z., Casanova, C., Roncero, M.I.G., Di Pietro, A. and Ramos, J. 2003a. pH response transcription factor PacC controls salt stress tolerance and expression of the P-type Na^+-ATPase ena1 in *Fusarium oxysporum*. *Eukaryotic Cell* 2, 1246–1252.

Caracuel, Z., Roncero, M.I.G., Espeso, E.A., Gonzalez-Verdejo, C.I., Garcia-Maceira, F.I. and Di Pietro, A. 2003b. The pH signalling transcription factor PacC controls virulence in the plant pathogen *Fusarium oxysporum*. *Molecular Microbiology* 48, 765–779.

Card, S., Johnson, L., Teasdale, S. and Caradus, J. 2016. Deciphering endophyte behaviour: The link between endophyte biology and efficacious biological control agents. *FEMS Microbiology Ecology* 92, fiw114.

Carroll, G. 1988. Fungal endophytes in stems and leaves: From latent pathogen to mutualistic symbiont. *Ecology* 69, 2–9.

Casado-Vela, J., Selles, S. and Martinez, R.B. 2006. Proteomic analysis of tobacco mosaic virus-infected tomato (*Lycopersicon esculentum* M.) fruits and detection of viral coat protein. *Proteomics* 6, 196–206.

Cereghino, G.P. and Scheffler, I.E. 1996. Genetic analysis of glucose regulation in saccharomyces cerevisiae: Control of transcription versus mRNA turnover. *The EMBO Journal* 15, 363–374.

Cervantes-Chávez, J.A., Ortiz-Castellanos, L., Tejeda-Sartorius, M., Gold, S. and Ruiz-Herrera, J. 2010. Functional analysis of the pH responsive pathway Pal/Rim in the phytopathogenic basidiomycete *Ustilago maydis*. *Fungal Genetics and Biology* 47, 446–457.

Cervone, F., Hahn, M.G., Delorenzo, G., Darvill, A. and Albersheim, P. 1989. Host-pathogen interactions. XXXIII. A plant protein converts a fungal pathogenesis factor into an elicitor of plant defense responses. *Plant Physiology* 90, 542–548.

Cessna, S.G., Sears, V.E., Dickman, M.B. and Low, P.S. 2000. Oxalic acid, a pathogenicity factor for *Sclerotinia sclerotiorum*, suppresses the oxidative burst of the host plant. *The Plant Cell* 12, 2191–2199.

Chague, V., Elad, Y., Barakat, R., Tudzynski, P. and Sharon, A. 2002. Ethylene biosynthesis in *Botrytis cinerea*. *FEMS Microbiology Ecology* 40, 143–149.

Clarke, A., Mur, L.A.J., Darby, R.M. and Kenton, P. 2005. Harpin modulates the accumulation of salicylic acid by Arabidopsis cells via apoplastic alkalization. *Journal of Experimental Botany* 56, 3129–3136.

Coaker, G.L., Willard, B., Kinter, M., Stockinger, E.J. and Francis, D.M. 2004. Proteomic analysis of resistance mediated by Rcm 2.0 and Rcm 5.1, two loci controlling resistance to bacterial canker of tomato. *Molecular Plant-Microbe Interactions* 17, 1019–1028.

Collado, I.G., Aleu, J., Hernandez-Galan, R. and Duran-Patron, R. 2000. *Botrytis* species: An intriguing source of metabolites with a wide range of biological activities. Structure, chemistry and bioactivity of metabolites isolated from *Botrytis* species. *Current Organic Chemistry* 4, 1261–1286.

Cote, F. and Hahn, M.G. 1994. Oligosaccharins - structures and signal-transduction. *Plant Molecular Biology* 26, 1379–1411.

van der Cruyssen, G., de Meester, E. and Kamoen, O. 1994. Expression of polygalacturonase of *Botrytis cinerea in vitro* and *in vivo*. *Mededelingen - Faculteit Landbouwkundige en Toegepaste Biologische Wetenschappen Universiteit Gent* 59, 895–905.

Cunningham, J.E. and Kuiack, C. 1992. Production of citric and oxalic acids and solubilization of calcium-phosphate by *Penicillium-bilaii*. *Applied and Environmental Microbiology* 58, 1451–1458.

Curvers, K., Seifi, H., Mouille, G., De Rycke, R., Asselbergh, B., Van Hecke, A., Vanderschaeghe, D., Hofte, H., Callewaert, N., Van Breusegem, F. and Hofte, M. 2010. Abscisic acid deficiency causes changes in cuticle permeability and pectin composition that influence tomato resistance to *Botrytis cinerea*. *Plant Physiology* 154, 847–860.

Danhof, H.A. and Lorenz, M.C. 2015. The *Candida albicans* ATO gene family promotes neutralization of the macrophage phagolysosome. *Infection and Immunity* 83, 4416–4426.

Davidzon, M., Alkan, N., Kobiler, I. and Prusky, D. 2010. Acidification by gluconic acid of mango fruit tissue during colonization via stem end infection by *Phomopsis mangiferae*. *Postharvest Biology and Technology* 55, 71–77.

Denison, S.H., Orejas, M. and Arst, H.N., Jr. 1995. Signaling of ambient pH in *Aspergillus* involves a cysteine protease. *The Journal of Biological Chemistry* 270, 28519–28522.

Diéguez-Uribeondo, J., Förster, H. and Adaskaveg, J.E. 2008. Visualization of localized pathogen-induced pH modulation in almond tissues infected by *Colletotrichum acutatum* using confocal scanning laser microscopy. *Phytopathology* 98, 1171–1178.

Diskin, S., Feygenberg, O., Maurer, D., Droby, S., Prusky, D. and Alkan, N. 2017. Microbiome alterations are correlated with occurrence of postharvest stem-end rot in mango fruit. *Phytobiomes* 1, 117–127.

Djami-Tchatchou, A.T., Straker, C.J. and Allie, F. 2012. 454 sequencing for the identification of genes differentially expressed in avocado fruit (cv. Fuerte) infected by *Colletotrichum gloeosporioides*. *Journal of Phytopathology* 160, 449–460.

Droby, S., Prusky, D., Jacoby, B. and Goldman, A. 1987. Induction of antifungal resorcinols in flesh of unripe mango fruits and its relation to latent infection by *Alternaria alternata*. *Physiological and Molecular Plant Pathology* 30, 285–292.

Drori, N., Kramer-Haimovich, H., Rollins, J., Dinoor, A., Okon, Y., Pines, O. and Prusky, D. 2003. External pH and nitrogen source affect secretion of pectate lyase by *Colletotrichum gloeosporioides*. *Applied and Environmental Microbiology* 69, 3258–3262.

Dutton, M.V. and Evans, C.S. 1996. Oxalate production by fungi: Its role in pathogenicity and ecology in the soil environment. *Canadian Journal of Microbiology* 42, 881–895.

El Oirdi, M., Abd El Rahman, T., Rigano, L., El Hadrami, A., Rodriguez, M.C., Daayf, F., Vojnov, A. and Bouarab, K. 2011. *Botrytis cinerea* manipulates the antagonistic effects between immune pathways to promote disease development in tomato. *The Plant Cell* 23, 2405–2421.

Errakhi, R., Meimoun, P., Lehner, A., Vidal, G., Briand, J., Corbineau, F., Rona, J.P. and Bouteau, F. 2008. Anion channel activity is necessary to induce ethylene synthesis and programmed cell death in response to oxalic acid. *Journal of Experimental Botany* 59, 3121–3129.

Eshel, D., Lichter, A., Dinoor, A. and Prusky, D. 2002a. Characterization of *Alternaria alternata* glucanase genes expressed during infection of resistant and susceptible persimmon fruits. *Molecular Plant Pathology* 3, 347–358.

Eshel, D., Miyara, I., Ailing, T., Dinoor, A. and Prusky, D. 2002b. pH regulates endoglucanase expression and virulence of *Alternaria alternata* persimmon fruit. *Molecular Plant-Microbe Interactions* 15, 774–779.

Espeso, E.A. and Arst, H.N. 2000. On the mechanism by which alkaline pH prevents expression of an acid-expressed gene. *Molecular and Cellular Biology* 20, 3355–3363.

Espeso, E.A., Tilburn, J., Arst, H.N., Jr. and Peñalva, M.A. 1993. pH regulation is a major determinant in expression of a fungal penicillin biosynthetic gene. *The EMBO Journal* 12, 3947–3956.

Fernandes, T.R., Segorbe, D., Prusky, D. and Di Pietro, A. 2017. How alkalinization drives fungal pathogenicity. *PLoS Pathogens* 13, e1006621.

Fernandez, J., Marroquin-Guzman, M. and Wilson, R.A. 2014. Mechanisms of nutrient acquisition and utilization during fungal infections of leaves. *Annual Review of Phytopathology* 52, 155–174.

Ferrari, S., Plotnikova, J.M., De Lorenzo, G. and Ausubel, F.M. 2003. Arabidopsis local resistance to *Botrytis cinerea* involves salicylic acid and camalexin and requires EDS4 and PAD2, but not SID2, EDS5 or PAD4. *Plant Journal* 35, 193–205.

Flors, V., Leyva, M.D., Vicedo, B., Finiti, I., Real, M.D., Garcia-Agustin, P., Bennett, A.B. and Gonzalez-Bosch, C. 2007. Absence of the endo-beta-1,4-glucanases Cel1 and Cel2 reduces susceptibility to *Botrytis cinerea* in tomato. *Plant Journal* 52, 1027–1040.

Fonseca-García, C., León-Ramírez, C.G. and Ruiz-Herrera, J. 2012. The regulation of different metabolic pathways through the Pal/Rim pathway in *Ustilago maydis*. *FEMS Yeast Research* 12, 547–556.

Gancedo, J.M. 1998. Yeast carbon catabolite repression. *Microbiology and Molecular Biology Reviews* 62, 334–361.

Giovannoni, J. 2001. Molecular biology of fruit maturation and ripening. *Annual Review of Plant Physiology and Plant Molecular Biology* 52, 725–749.

Giovannoni, J.J. 2007. Fruit ripening mutants yield insights into ripening control. *Current Opinion in Plant Biology* 10, 283–289.

Glassner, H., Zchori-Fein, E., Compant, S., Sessitsch, A., Katzir, N., Portnoy, V. and Yaron, S. 2015. Characterization of endophytic bacteria from cucurbit fruits with potential benefits to agriculture in melons (*Cucumis melo* L.). *FEMS Microbiology Ecology* 91:1-13.

Glazebrook, J. 2005. Contrasting mechanisms of defense against biotrophic and necrotrophic pathogens. *Annual Review of Phytopathology* 43, 205–227.

Godoy, G., Steadman, J.R., Dickman, M.B. and Dam, R. 1990. Use of mutants to demonstrate the role of oxalic-acid in pathogenicity of *Sclerotinia-sclerotiorum* on *Phaseolus-vulgaris*. *Physiological and Molecular Plant Pathology* 37, 179–191.

Gonzalez-Bosch, C., Brummell, D.A. and Bennett, A.B. 1996. Differential expression of two endo-1,4-beta-glucanase genes in pericarp and locules of wild-type and mutant tomato fruit. *Plant Physiology* 111, 1313–1319.

Hano, C., Addi, M., Fliniaux, O., Bensaddek, L., Duverger, E., Mesnard, F., Lamblin, F. and Laine, E. 2008. Molecular characterization of cell death induced by a compatible interaction between *Fusarium oxysporum* f. sp linii and flax (*Linum usitatissimum*) cells. *Plant Physiology and Biochemistry* 46, 590–600.

Isaacson, T., Kosma, D.K., Matas, A.J., Buda, G.J., He, Y.H., Yu, B.W., Pravitasari, A., Batteas, J.D., Stark, R.E., Jenks, M.A. and Rose, J.K.C. 2009. Cutin deficiency in the tomato fruit cuticle consistently affects resistance to microbial infection and biomechanical properties, but not transpirational water loss. *Plant Journal* 60, 363–377.

Itkin, M., Rogachev, I., Alkan, N., Rosenberg, T., Malitsky, S., Masini, L., Meir, S., Iijima, Y., Aoki, K., De Vos, R., Prusky, D., Burdman, S., Beekwilder, J. and Aharoni, A. 2011. GLYCOALKALOID METABOLISM1 is required for steroidal alkaloid glycosylation and prevention of phytotoxicity in tomato. *The Plant Cell* 23, 4507–4525.

Jersch, S., Scherer, C., Huth, G. and Schlosser, E. 1989. Proanthocyanidins as basis for quiescence of *Botrytis-cinerea* in immature strawberry fruits. *Journal of Plant Diseases and Protection* 96, 365–378.

Kader, A.A. 1985. Ethylene-induced senescence and physiological disorders in harvested horticultural crops. *HortScience* 20, 54–57.

Kalamaki, M.S., Harpster, M.H., Palys, J.M., Labavitch, J.M., Reid, D.S. and Brummell, D.A. 2003. Simultaneous transgenic suppression of LePG and LeExp1 influences rheological properties of juice and concentrates from a processing tomato variety. *Journal of Agricultural and Food Chemistry* 51, 7456–7464.

Kane, P.M. 2016. Proton transport and pH control in fungi. *Advances in Experimental Medicine and Biology* 892, 33–68.

Karunanayake, L.C., Adikaram, N., Kumarihamy, B.M.M., Bandara, B.M.R. and Abayasekara, C. 2011. Role of antifungal gallotannins, resorcinols and chitinases in the constitutive defence of immature mango (*Mangifera indica* L.) against *Colletotrichum gloeosporioides*. *Journal of Phytopathology* 159, 657–664.

Khan, A.L., Al-Harrasi, A., Al-Rawahi, A., Al-Farsi, Z., Al-Mamari, A., Waqas, M., Asaf, S., Elyassi, A., Mabood, F., Shin, J.H. and Lee, I.J. 2016. Endophytic fungi from Frankincense tree improves host growth and produces extracellular enzymes and indole acetic acid. *PLoS ONE* 11, e0158207.

Kim, K.S., Min, J.Y. and Dickman, M.B. 2008. Oxalic acid is an elicitor of plant programmed cell death during *Sclerotinia sclerotiorum* disease development. *Molecular Plant-Microbe Interactions* 21, 605–612.

Kim, S.T., Kim, S.G., Hwang, D.H., Kang, S.Y., Kim, H.J., Lee, B.H., Lee, J.J. and Kang, K.Y. 2004. Proteomic analysis of pathogen-responsive proteins from rice leaves induced by rice blast fungus, *Magnaporthe grisea*. *Proteomics* 4, 3569–3578.

Kleemann, J., Rincon-Rivera, L.J., Takahara, H., Neumann, U., Van Themaat, E.V.L., Van Der Does, H.C., Hacquard, S., Stuber, K., Will, I., Schmalenbach, W., Schmelzer, E. and O'connell, R.J. 2012. Sequential delivery of host-induced virulence effectors by appressoria and intracellular hyphae of the phytopathogen *Colletotrichum higginsianum*. *PLoS Pathogens* 8, e1002643.

Kramer-Haimovich, H., Servi, E., Katan, T., Rollins, J., Okon, Y. and Prusky, D. 2006. Effect of ammonia production by *Colletotrichum gloeosporioides* on pelB activation, pectate lyase secretion, and fruit pathogenicity. *Applied and Environmental Microbiology* 72, 1034–1039.

L'Haridon, F., Besson-Bard, A., Binda, M., Serrano, M., Abou-Mansour, E., Balet, F., Schoonbeek, H.J., Hess, S., Mir, R., Leon, J., Lamotte, O. and Metraux, J.P. 2011. A permeable cuticle is associated with the release of reactive oxygen species and induction of innate immunity. *PLoS Pathogens* 7, e1002148.

Lattanzio, V., Cardinali, A., Divenere, D., Linsalata, V. and Palmieri, S. 1994. Browning phenomena in stored artichoke (*Cynara-Scolymus* L) heads - enzymatic or chemical-reactions. *Food Chemistry* 50, 1–7.

Lattanzio, V., Di Venere, D., Linsalata, V., Bertolini, P., Ippolito, A. and Salermo, M. 2001. Low temperature metabolism of apple phenolics and quiescence of *Phlyctaena vagabunda*. *Journal of Agricultural and Food Chemistry* 49, 5817–5821.

Lattanzio, V., Lattanzio, V.M.T. and Cardinale, A. 2006. Role of phenolics in the resistance mechanisms of plants against fungal pathogens and insects. pp. 23–67. *In*: F. Imperato (ed.). *Phytochemistry: Advances in Research*. Research Signpost, Trivandrum, Kerala, India.

Lee, M.H. and Bostock, R.M. 2007. Fruit exocarp phenols in relation to quiescence and development of *Monilinia fructicola* infections in *Prunus* spp.: A role for cellular redox? *Phytopathology* 97, 269–277.

Lloyd, A.J., Allwood, J.W., Winder, C.L., Dunn, W.B., Heald, J.K., Cristescu, S.M., Sivakumaran, A., Harren, F.J.M., Mulema, J., Denby, K., Goodacre, R., Smith, A.R. and Mur, L.A.J. 2011. Metabolomic approaches reveal that cell wall modifications play a major role in ethylene-mediated resistance against *Botrytis cinerea*. *Plant Journal* 67, 852–868.

Luo, Y., Ma, Z.H. and Michailides, T.J. 2001. Analysis of factors affecting latent infection and sporulation of *Monilinia fructicola* on prune fruit. *Plant Disease* 85, 999–1003.

Luo, Y. and Michailides, T.J. 2001. Factors affecting latent infection of prune fruit by *Monilinia fructicola*. *Phytopathology* 91, 864–872.

Manteau, S., Abouna, S., Lambert, B. and Legendre, L. 2003. Differential regulation by ambient pH of putative virulence factor secretion by the phytopathogenic fungus *Botrytis cinerea*. *FEMS Microbiology Ecology* 43, 359–366.

Marciano, P., Dilenna, P. and Magro, P. 1983. Oxalic-acid, cell wall-degrading enzymes and pH in pathogenesis and their significance in the virulence of 2

Sclerotinia-sclerotiorum isolates on sunflower. *Physiological Plant Pathology* 22, 339–345.

Martell, A.E. and Calvin, M. 1952. *Chemistry of the Metal Chelate Compounds.* Prentice-Hall, New York, NY.

Mathieu, Y., Jouanneau, J.P., Thomine, S., Lapous, D. and Guern, J. 1994. Cytosolic protons as secondary messengers in elicitor-induced defense responses. *Biochemical Society Symposium* 60, 113–130.

Mauch, F., Mauch-Mani, B. and Boller, T. 1988. Antifungal hydrolases in pea tissue. II. Inhibition of fungal growth by combinations of chitinase and β-1,3-glucanase. *Plant Physiology* 88, 936–942.

Maxwell, D.P. and Lumsden, R.D. 1970. Oxalic acid production by *Sclerotinia sclerotiorum* in infected bean and in culture. *Phytopathology* 60, 1395–1398.

McGlasson, W.B., Poovaiah, B.W. and Dostal, H.C. 1975. Ethylene production and respiration in aging leaf segments and in disks of fruit tissue of normal and mutant tomatoes. *Plant Physiology* 56, 547–549.

Mengiste, T. 2012. Plant immunity to necrotrophs. *Annual Review of Phytopathology* 50, 267–294.

Mingot, J.M., Tilburn, J., Diez, E., Bignell, E., Orejas, M., Widdick, D.A., Sarkar, S., Brown, C.V., Caddick, M.X., Espeso, E.A., Arst, H.N., Jr. and Peñalva, M.A. 1999. Specificity determinants of proteolytic processing of *Aspergillus* PacC transcription factor are remote from the processing site, and processing occurs in yeast if pH signalling is bypassed. *Molecular and Cellular Biology* 19, 1390–1400.

Miyara, I., Shafran, H., Davidzon, M., Sherman, A. and Prusky, D. 2010. pH regulation of ammonia secretion by *Colletotrichum gloeosporioides* and its effect on appressorium formation and pathogenicity. *Molecular Plant-Microbe Interactions* 23, 304–316.

Miyara, I., Shafran, H., Haimovich, H.K., Rollins, J., Sherman, A. and Prusky, D. 2008. Multi-factor regulation of pectate lyase secretion by *Colletotrichum gloeosporioides* pathogenic on avocado fruits. *Molecular Plant Pathology* 9, 281–291.

Miyara, I., Shnaiderman, C., Meng, X., Vargas, W.A., Diaz-Minguez, J.M., Sherman, A., Thon, M. and Prusky, D. 2012. Role of nitrogen-metabolism genes expressed during pathogenicity of the alkalinizing *Colletotrichum gloeosporioides* and their differential expression in acidifying pathogens. *Molecular Plant-Microbe Interactions* 25, 1251–1263.

Moreno, F. and Herrero, P. 2002. The hexokinase 2-dependent glucose signal transduction pathway of Saccharomyces cerevisiae. *FEMS Microbiology Reviews* 26, 83–90.

Movahedi, S. and Heale, J.B. 1990. Purification and characterization of an aspartic proteinase secreted by *Botrytis-cinerea* pers ex pers in culture and in infected carrots. *Physiological and Molecular Plant Pathology* 36, 289–302.

Nambeesan, S., Abuqamar, S., Laluk, K., Mattoo, A.K., Mickelbart, M.V., Ferruzzi, M.G., Mengiste, T. and Handa, A.K. 2012. Polyamines attenuate ethylene-mediated defense responses to abrogate resistance to *Botrytis cinerea* in tomato. *Plant Physiology* 158, 1034–1045.

Nambeesan, S., Datsenka, T., Ferruzzi, M.G., Malladi, A., Mattoo, A.K. and Handa, A. K. 2010. Overexpression of yeast spermidine synthase impacts ripening, senescence and decay symptoms in tomato. *Plant Journal* 63, 836–847.

Negrete-Urtasun, S., Reiter, W., Diez, E., Denison, S.H., Tilburn, J., Espeso, E.A., Peñalva, M.A. and Arst, H.N., Jr. 1999. Ambient pH signal transduction in *Aspergillus*: Completion of gene characterization. *Molecular Microbiology* 33, 994–1003.

Nurnberger, T. and Scheel, D. 2001. Signal transmission in the plant immune response. *Trends in Plant Science* 6, 372–379.

O'Connell, R.J., Thon, M.R., Hacquard, S., Amyotte, S.G., Kleemann, J., Torres, M.F., Damm, U., Buiate, E.A., Epstein, L., Alkan, N., Altmuller, J., Alvarado-Balderrama, L., Bauser, C.A., Becker, C., Birren, B.W., Chen, Z.H., Choi, J., Crouch, J.A., Duvick, J.P., Farman, M.A., Gan, P., Heiman, D., Henrissat, B., Howard, R.J., Kabbage, M., Koch, C., Kracher, B., Kubo, Y., Law, A.D., Lebrun, M.H., Lee, Y.H., Miyara, I., Moore, N., Neumann, U., Nordstrom, K., Panaccione, D.G., Panstruga, R., Place, M., Proctor, R.H., Prusky, D., Rech, G., Reinhardt, R., Rollins, J.A., Rounsley, S., Schardl, C.L., Schwartz, D.C., Shenoy, N., Shirasu, K., Sikhakolli, U.R., Stuber, K., Sukno, S.A., Sweigard, J.A., Takano, Y., Takahara, H., Trail, F., Van Der Does, H.C., Voll, L.M., Will, I., Young, S., Zeng, Q.D., Zhang, J. Z., Zhou, S.G., Dickman, M.B., Schulze-Lefert, P., Van Themaat, E.V.L., Ma, L.J. and Vaillancourt, L.J. 2012. Lifestyle transitions in plant pathogenic *Colletotrichum* fungi deciphered by genome and transcriptome analyses. *Nature Genetics* 44, 1060–1065.

Ordiz, I., Herrero, P., Rodicio, R. and Moreno, F. 1996. Glucose-induced inactivation of isocitrate lyase in *Saccharomyces cerevisiae* is mediated by the cAMP-dependent protein kinase catalytic subunits Tpk1 and Tpk2. *FEBS Letters* 385, 43–46.

Ozcan, S. and Johnston, M. 1999. Function and regulation of yeast hexose transporters. *Microbiology and Molecular Biology Reviews* 63, 554–569.

Palkova, Z., Devaux, F., Icicova, M., Minarikova, L., Le Crom, S. and Jacq, C. 2002. Ammonia pulses and metabolic oscillations guide yeast colony development. *Molecular Biology of the Cell* 13, 3901–3914.

Peñalva, M.A. and Arst, H.N., Jr. 2002. Regulation of gene expression by ambient pH in filamentous fungi and yeasts. *Microbiology and Molecular Biology Reviews* 66, 426–446.

Peñalva, M.A., Tilburn, J., Bignell, E. and Arst, H.N., Jr. 2008. Ambient pH gene regulation in fungi: Making connections. *Trends in Microbiology* 16, 291–300.

Powell, A.L.T., Kalamaki, M.S., Kurien, P.A., Gurrieri, S. and Bennett, A.B. 2003. Simultaneous transgenic suppression of LePG and LeExp1 influences fruit texture and juice viscosity in a fresh market tomato variety. *Journal of Agricultural and Food Chemistry* 51, 7450–7455.

Prusky, D. 1996. Pathogen quiescence in postharvest diseases. *Annual Review of Phytopathology* 34, 413–434.

Prusky, D., Alkan, N., Mengiste, T. and Fluhr, R. 2013. Quiescent and necrotrophic lifestyle choice during postharvest disease development. *Annual Review of Phytopathology* 51, 155–176.

Prusky, D.B., Bi, F., Moral, J. and Barad, S. 2016. How does host carbon concentration modulate the lifestyle of postharvest pathogens during colonization? *Frontiers in Plant Science* 7, Article 1306.

Prusky, D., Alkan, N., Miyara, I., Barad, S., Davidzon, M., Kobiler, I., Brown-Horowitz, S., Lichter, A., Sherman, A. and Fluhr, R. 2010. Mechanisms modulating postharvest pathogen colonization of decaying fruits. pp. 43–55. *In: D. Prusky and M.L. Gullino (eds.). Postharvest Pathology in the 21st Century.* Springer, Dordrecht, The Netherlands.

Prusky, D., Kobiler, I., Akerman, M. and Miyara, I. 2006. Effect of acidic solutions and acidic prochloraz on the control of postharvest decay caused by *Alternaria alternata* in mango and persimmon fruits. *Postharvest Biology and Technology* 42, 134–141.

Prusky, D. and Lichter, A. 2007. Activation of quiescent infections by postharvest pathogens during transition from the biotrophic to the necrotrophic stage. *FEMS Microbiology Letters* 268, 1–8.

Prusky, D., McEvoy, J.L., Leverentz, B. and Conway, W.S. 2001. Local modulation of host pH by *Colletotrichum* species as a mechanism to increase virulence. *Molecular Plant-Microbe Interactions* 14, 1105–1113.

Prusky, D., McEvoy, J.L., Saftner, R., Conway, W.S. and Jones, R. 2004. Relationship between host acidification and virulence of *Penicillium* spp. on apple and citrus fruit. *Phytopathology* 94, 44–51.

Prusky, D. and Yakoby, N. 2003. Pathogenic fungi: Leading or led by ambient pH? *Molecular Plant Pathology* 4, 509–516.

Qadir, A., Hewett, E.W., Long, P.G. and Dilley, D.R. 2011. A non-ACC pathway for ethylene biosynthesis in *Botrytis cinerea. Postharvest Biology and Technology* 62, 314–318.

Qin, Q., Bergmann, C.W., Rose, J.K.C., Saladie, M., Kolli, V.S.K., Albersheim, P., Darvill, A.G. and York, W.S. 2003. Characterization of a tomato protein that inhibits a xyloglucan-specific endoglucanase. *Plant Journal* 34, 327–338.

Rep, M., Dekker, H.L., Vossen, J.H., De Boer, A.D., Houterman, P.M., Speijer, D., Back, J.W., De Koster, C.G. and Cornelissen, B.J. 2002. Mass spectrometric identification of isoforms of PR proteins in xylem sap of fungus-infected tomato. *Plant Physiology* 130, 904–917.

Rolland, F., De Winde, J.H., Lemaire, K., Boles, E., Thevelein, J.M. and Winderickx, J. 2000. Glucose-induced cAMP signalling in yeast requires both a G-protein coupled receptor system for extracellular glucose detection and a separable hexose kinase-dependent sensing process. *Molecular Microbiology* 38, 348–358.

Rolland, F., Wanke, V., Cauwenberg, L., Ma, P., Boles, E., Vanoni, M., De Winde, J. H., Thevelein, J.M. and Winderickx, J. 2001. The role of hexose transport and phosphorylation in cAMP signalling in the yeast *Saccharomyces cerevisiae. FEMS Yeast Research* 1, 33–45.

Rollins, J.A. 2003. The *Sclerotinia sclerotiorum* pac1 gene is required for sclerotial development and virulence. *Molecular Plant-Microbe Interactions* 16, 785–795.

Rollins, J.A. and Dickman, M.B. 2001. PH signaling in *Sclerotinia sclerotiorum*: Identification of a pacC/RIM1 homolog. *Applied and Environmental Microbiology* 67, 75–81.

Romanazzi, G., Feliziani, E., Bautista-Baños, S. and Sivakumar, D. 2014. Shelf life extension of fresh fruit and vegetables by chitosan treatment. *Critical Reviews of Food Science and Nutrition* 57, 579–601.

Ron, M. and Avni, A. 2004. The receptor for the fungal elicitor ethylene-inducing xylanase is a member of a resistance-like gene family in tomato. *The Plant Cell* 16, 1604–1615.

Ruijter, G.J.G., Van De Vondervoort, P.J.I. and Visser, J. 1999. Oxalic acid production by *Aspergillus niger*: An oxalate-non-producing mutant produces citric acid at pH 5 and in the presence of manganese. *Microbiology* 145, 2569–2576.

Schaller, A. and Oecking, C. 1999. Modulation of plasma membrane H^+-ATPase activity differentially activates wound and pathogen defense responses in tomato plants. *The Plant Cell* 11, 263–272.

Sela-Buurlage, M.B., Ponstein, A.S., Bres-Vloemans, S.A., Melchers, L.S., Van Den Elzen, P. and Cornelissen, B. 1993. Only specific tobacco (*Nicotiana tabacum*) chitinases and [beta]-1,3-glucanases exhibit antifungal activity. *Plant Physiology* 101, 857–863.

Selitrennikoff, C.P. 2001. Antifungal proteins. *Applied and Environmental Microbiology* 67, 2883–2894.

Shah, P., Powell, A.L.T., Orlando, R., Bergmann, C. and Gutierrez-Sanchez, G. 2012. Proteomic analysis of ripening tomato fruit infected by *Botrytis cinerea*. *Journal of Proteome Research* 11, 2178–2192.

Siewers, V., Kokkelink, L., Smedsgaard, J. and Tudzynski, P. 2006. Identification of an abscisic acid gene cluster in the grey mold *Botrytis cinerea*. *Applied and Environmental Microbiology* 72, 4619–4626.

Simon-Plas, F., Rusterucci, C., Milat, M.L., Humbert, C., Montillet, J.L. and Blein, J.P. 1997. Active oxygen species production in tobacco cells elicited by cryptogein. *Plant Cell and Environment* 20, 1573–1579.

Sivankalyani, V., Feygenberg, O., Maorer, D., Zaaroor, M., Fallik, E. and Alkan, N. 2015. Combined treatments reduce chilling injury and maintain fruit quality in avocado fruit during cold quarantine. *PLoS ONE* 10, e0140522.

Spielvogel, A., Findon, H., Arst, H.N., Araujo-Bazán, L., Hernández-Ortíz, P., Stahl, U., Meyer, V. and Espeso, E.A. 2008. Two zinc finger transcription factors, CrzA and SltA, are involved in cation homoeostasis and detoxification in *Aspergillus nidulans*. *Biochemical Journal* 414, 419–429.

Staats, M., Van Baarlen, P. and Van Kan, J.A. 2005. Molecular phylogeny of the plant pathogenic genus *Botrytis* and the evolution of host specificity. *Molecular Biology and Evolution* 22, 333–346.

Su, S.S.Y. and Mitchell, A.P. 1993. Molecular characterization of the yeast meiotic regulatory gene Rim1. *Nucleic Acids Research* 21, 3789–3797.

Suárez, T. and Peñalva, M.A. 1996. Characterization of a *Penicillium chrysogenum* gene encoding a PacC transcription factor and its binding sites in the divergent pcbAB-pcbC promoter of the penicillin biosynthetic cluster. *Molecular Microbiology* 20, 529–540.

Tanaka, A., Christensen, M.J., Takemoto, D., Park, P. and Scott, B. 2006. Reactive oxygen species play a role in regulating a fungus-perennial ryegrass mutualistic interaction. *The Plant Cell* 18, 1052–1066.

Tanaka, A., Takemoto, D., Hyon, G.S., Park, P. and Scott, B. 2008. NoxA activation by the small GTPase RacA is required to maintain a mutualistic symbiotic association between *Epichloe festucae* and perennial ryegrass. *Molecular Microbiology* 68, 1165–1178.

Tilburn, J., Sarkar, S., Widdick, D.A., Espeso, E.A., Orejas, M., Mungroo, J., Peñalva, M.A. and Arst, H.N., Jr. 1995. The *Aspergillus* Pacc zinc-finger transcription factor mediates regulation of both acid-expressed and alkaline-expressed genes by ambient pH. *The EMBO Journal* 14, 779–790.

Torres, M.A., Dangl, J.L. and Jones, J.D.G. 2002. Arabidopsis gp91(phox) homologues AtrbohD and AtrbohF are required for accumulation of reactive oxygen intermediates in the plant defense response. *Proceedings of the National Academy of Sciences of the United States of America* 99, 517–522.

Van Kan, J.A. 2006. Licensed to kill: The lifestyle of a necrotrophic plant pathogen. *Trends in Plant Science* 11, 247–253.

Van Loon, L.C., Rep, M. and Pieterse, C.M.J. 2006. Significance of inducible defense-related proteins in infected plants. *Annual Review of Phytopathology* 44, 135–162.

Veluchamy, S., Williams, B., Kim, K. and Dickman, M.B. 2012. The CuZn superoxide dismutase from *Sclerotinia sclerotiorum* is involved with oxidative stress tolerance, virulence, and oxalate production. *Physiological and Molecular Plant Pathology* 78, 14–23.

Verhoeff, K. 1974. Latent infections by fungi. *Annual Review of Phytopathology* 12, 99–110.

Vorwerk, S., Somerville, S. and Somerville, C. 2004. The role of plant cell wall polysaccharide composition in disease resistance. *Trends in Plant Science* 9, 203–209.

Vylkova, S., Carman, A.J., Danhof, H.A., Collette, J.R., Zhou, H. and Lorenz, M.C. 2011. The fungal pathogen *Candida albicans* autoinduces hyphal morphogenesis by raising extracellular pH. *MBio* 2, e00055–11.

Vylkova, S. and Lorenz, M.C. 2014. Modulation of phagosomal pH by *Candida albicans* promotes hyphal morphogenesis and requires Stp2p, a regulator of amino acid transport. *PLoS Pathogens* 10, e1003995.

Wang, X., Beno-Moualem, D., Kobiler, I., Leikin-Frenkel, A., Lichter, A. and Prusky, D. 2004. Expression of Delta(12) fatty acid desaturase during the induced accumulation of the antifungal diene in avocado fruits. *Molecular Plant Pathology* 5, 575–585.

Wang, X., Kobiler, I., Lichter, A., Leikin-Frenkel, A., Pesis, E. and Prusky, D. 2005. 1-MCP prevents ethylene-induced accumulation of antifungal diene in avocado fruit. *Physiological and Molecular Plant Pathology* 67, 261–267.

Williams, B., Kabbage, M., Kim, H.J., Britt, R. and Dickman, M.B. 2011. Tipping the balance: *Sclerotinia sclerotiorum* secreted oxalic acid suppresses host defenses by manipulating the host redox environment. *PLoS Pathogens* 7, e1002107.

Wojtaszek, P., Trethowan, J. and Bolwell, G.P. 1995. Specificity in the immobilization of cell-wall proteins in response to different elicitor molecules in suspension-cultured cells of french bean (*Phaseolus-vulgaris* L). *Plant Molecular Biology* 28, 1075–1087.

Wu, Y., Zhang, D., Chu, J.Y., Boyle, P., Wang, Y., Brindle, I.D., De Luca, V. and Despres, C. 2012. The Arabidopsis NPR1 protein is a receptor for the plant defense hormone salicylic acid. *Cell Reports* 1, 639–647.

Wubben, J.P., Mulder, W., Ten Have, A., Van Kan, J.A. and Visser, J. 1999. Cloning and partial characterization of endopolygalacturonase genes from *Botrytis cinerea*. *Applied and Environmental Microbiology* 65, 1596–1602.

Wubben, J.P., Ten Have, A., Van Kan, J.A. and Visser, J. 2000. Regulation of endopolygalacturonase gene expression in *Botrytis cinerea* by galacturonic acid, ambient pH and carbon catabolite repression. *Current Genetics* 37, 152–157.

Yakoby, N., Freeman, S., Dinoor, A., Keen, N.T. and Prusky, D. 2000a. Expression of pectate lyase from *Colletotrichum gloeosporioides* in *C. magna* promotes pathogenicity. *Molecular Plant-Microbe Interactions* 13, 887–891.

Yakoby, N., Kobiler, I., Dinoor, A. and Prusky, D. 2000b. pH regulation of pectate lyase secretion modulates the attack of *Colletotrichum gloeosporioides* on avocado fruits. *Applied and Environmental Microbiology* 66, 1026–1030.

You, B.J., Choquer, M. and Chung, K.R. 2007. The *Colletotrichum acutatum* gene encoding a putative pH-responsive transcription regulator is a key virulence determinant during fungal pathogenesis on citrus. *Molecular Plant-Microbe Interactions* 20, 1149–1160.

Zhao, J., Davis, L.C. and Verpoorte, R. 2005. Elicitor signal transduction leading to production of plant secondary metabolites. *Biotechnology Advances* 23, 283–333.

Zhu, Y., Lian, J., Zeng, H., Gan, L., Di, T., Shen, Q. and Xu, G. 2011. Involvement of plasma membrane H+ ATPase in adaption of rice to ammonium nutrient. *Rice Science* 18, 335–342.

Chapter 13

Detection and Control of Postharvest Toxigenic Fungi and Their Related Mycotoxins

Simona Marianna Sanzani and Antonio Ippolito
Department of Soil, Plant and Food Sciences, University of Bari Aldo Moro, Bari, Italy

Abbreviations

ADM	*Aspergillus* differential medium
AFB1	AflatoxinB1
AFPA	*A. flavus* and *A. parasiticus* agar
ALT	Altenuene
AME	Alternariol monomethyl ether
AOAC	Association of Official Analytical Chemists
AOH	Alternariol
CDC	Conditionally dispensable chromosome
DCMA	Dichloran-chloramphenicol malt agar
DON	Deoxynivalenol
EF-1α	Elongation factor 1-alpha
EFSA	European Food Safety Authority
ELISA	Enzyme-linked immunosorbent assay
EO	Essential oil
EU	European Union
FAO	Food and Agricultural Organization of the United Nations
FB1	Fumonisin B1
GC	Gas chromatography
GIPSA	Grain Inspection, Packers, and Stockyards Administration
HPLC/MS	High-performance liquid chromatography-mass spectrometry
HPLC	High-performance liquid chromatography
HRM	High resolution melting
ITS	Internal transcribed spacer
LODs	Limits of detection
MS	Mass spectrometry
NIR	Near-infrared
NIV	Nivalenol
OTA	Ochratoxin A
PCR	Polymerase chain reaction
PCR-RFLP	PCR-restriction fragment length polymorphism
PMA	Propidium monoazide
PRYES	Pentachloronitrobenzene, rose bengal in yeast extract-sucrose agar
qPCR	Quantitative PCR
rRNA	Ribosomal RNA
SELEX	Systematic evolution of ligands
SIDAs	Stable isotope dilution assays
SPR	Surface plasmon resonance
SSH	Suppression subtractive hybridization
TeA	Tenuazonic acid
TEN	Tentoxin
TLC	Thin layer chromatography

1 Introduction

Fungi produce a wide array of compounds classified as secondary metabolites, since they are not a fundamental requirement for their subsistence. Secondary metabolites synthesized by microorganisms, including fungi, are usually associated with sporulation processes (Calvo et al., 2002). In particular, they can be divided into three broad categories: (i) metabolites that activate sporulation, such as the linoleic acid-derived compounds of *Aspergillus nidulans* (Eidam) G. Winter; (ii) pigments required for sporulation structures, such as the melanins necessary for the formation/integrity of sexual/asexual spores and overwintering bodies; and (iii) mycotoxins and other toxic metabolites secreted by growing colonies at the approximate time of sporulation. Mycotoxins are defined as low molecular weight compounds that are toxic to vertebrates. They are produced by several fungal genera but mainly associated to *Alternaria, Aspergillus, Fusarium,* and *Penicillium.* Numerous studies have been conducted since the beginning of the 20th century, and thus a huge amount of data on mycotoxin synthesis and regulation are now available (Reverberi et al., 2010; Gallo et al., 2013). It is now known that sequences of genes encoding mycotoxins are largely shared, even among phylogenetically distant fungal taxa.

It has been estimated that more than 50% of fruits and vegetables are lost all over the world mainly because of mycotoxin contamination (Sanzani et al., 2016b), which consequently are present in the diet of a substantial portion of the world's population (Wild and Gong, 2010). Especially in the developing countries, most of these losses occur after harvest during storage and processing, when the product has a higher value due to the accumulated costs of production and the postharvest handling chain.

Although the mycotoxin chemical structures vary considerably, they share some common aspects that harm human and animal health. They interfere with several cellular and physiological functions and can cause harm as carcinogens, teratogens, mutagens, and immunosuppressants (Peraica et al., 1999). These toxic effects on animals and humans are referred to as mycotoxicosis, the severity of which depends on the toxicity of the mycotoxin, the extent of exposure, age and nutritional status of the individual, and possible synergistic effects with other chemicals to which the individual is exposed. In developing countries, where agricultural practices and regulations to control mycotoxins are not commonly adopted, the risk of exposure is continuous and often at elevated levels. In industrialized countries, despite occasional acute poisoning outbreaks, the mycotoxin exposure is mostly chronic and due to the continuous consumption of lightly contaminated food. Moreover, the consumption of products derived from animals fed by contaminated feed can cause secondary mycotoxicosis. For instance, aflatoxin B1, one of the most dangerous hepatocarcinogenic compounds found in nature, is converted by the mammalian rumen into its hydroxylated form, known as aflatoxin M1, and released in the milk (Galvano et al., 1996). Moreover, because of their persistence, mycotoxins can spread independently from their synthesizing organism and remain intact while they undergo processing into food products, thus representing a threat for not only for the consumers of the intact product but also for consumers of processed products. Attention to this topic has resulted in stringent import regulations implemented to limit mycotoxin contamination in foods in the industrialized nations. For example, the European Commission (2006) established maximum limits for mycotoxin contamination in food and feed, and regulators in countries outside Europe have done the same.

The production of these compounds by mycotoxigenic fungi within crops is highly influenced by environmental factors before and/or after harvest (e.g., temperature and available moisture) (Paterson and Lima, 2010; Battilani and Camardo Leggieri, 2014). Therefore, mycotoxin production might be affected by climate changes occurring worldwide. Large-scale clearing of forests, burning of fossil fuel, and other human activities have changed the global climate. Concentrations of methane, carbon dioxide, nitrous oxide, and chlorofluorocarbons in the atmosphere have increased, resulting in environmental warming (Chakraborty et al., 1998). As such, conditions conducive to contamination might occur in unexpected areas. Moreover, mycotoxins are influenced by noninfectious factors, such as the bioavailability of micronutrients, insects, and infection or colonization by other pests, which are in turn driven by climatic conditions. In warm regions, rain occurring at or near harvest, and/or high daily temperatures, are conducive to the growth of these fungi and may lead to unacceptable concentrations of mycotoxins in many crops (Cotty and Jaime-Garcia, 2007). Similarly, tropical, semiarid, and arid conditions, as well as drought, may be associated with contamination (Lewis et al., 2005). Indeed, the increase in average temperature in certain latitude ranges can affect the species composition in this area. For example, *Aspergillus* spp. and consequent aflatoxin contamination of maize, typically seen in tropical and subtropical regions, has become increasingly present in Europe since the 2000s, concurrent with an increase in hot and dry summers (Medina et al., 2015). In 2003, 2008, and 2015, outbreaks occurred in Italy, and in 2013 in the Balkan regions (De Rijk et al., 2015). Fortunately, changes in climate may cause a risk that is somewhat predictable, although the predictive models for mycotoxins have been validated in only one or a few countries, due to limited data availability and accessibility (Van der Fels-Klerx et al., 2018). Moreover, the impact of climate change on mycotoxin risk has focused only on some selected mycotoxin-crop combinations.

Since the discovery of mycotoxins, the scientific community has attempted to find out strategies to limit contamination in foods and feeds. However, only elucidation of the mechanisms underlying the biosynthesis of mycotoxins and their biological role will permit identification of efficient solutions to minimize them. Nevertheless, the biological role of many mycotoxins remains elusive. Only recently, mycotoxicology has become a branch of plant pathology, a discipline best suited to study biological roles of mycotoxins. Indeed, the apparent lack of specificity of mycotoxins has hindered the acceptance of their contribution to plant pathogenesis. Critical analysis of their roles was delayed until molecular tools were developed that specifically eliminated them from a model biological system (Desjardins and Hohn, 1997). For instance, it has been proved that DON, NIV, and FB1 have well-established roles in the onset of diseases on cereals by *Fusarium graminearum* Schwabe, *F. verticillioides* (Sacc.) Nirenberg, and *F. culmorum* (W.G. Sm.) Sacc. (Boddu et al., 2007; Desmond et al., 2008; Scherm et al., 2011). Moreover, Sanzani et al. (2012) recorded a reduced pathogenicity/virulence in *Penicillium expansum* L. mutants deficient in patulin production. Finally, Reverberi et al. (2008) proposed that aflatoxins could protect *Aspergillus parasiticus* Speare from the oxidative stress produced as defensive response by the host.

2 Main Postharvest Mycotoxigenic Genera

The main mycotoxin producing fungal genera commonly involved in the postharvest spoilage of fruits and vegetables are *Alternaria, Aspergillus*, and *Penicillium*.

2.1 *Alternaria*

The genus *Alternaria* contains various species pathogenic to fruits and vegetables. However, in a recent study by Woudenberg et al. (2015), a reclassification of the 26 *Alternaria* sections in 11 phylogenetic species and 1 species complex was proposed. Thirty-five morphospecies, undistinguishable even on a multigene phylogeny, were synonymized as *Alternaria alternata* (Fr.) Keissl., which therefore is the most important species within the genus, even as far as mycotoxin contamination concerns. Several host-specific pathotypes (able to infect apples, strawberries, pears, tomatoes, tangerines, etc.) exist, as well as saprotrophic forms that usually infect harvested commodities (Logrieco et al., 2009). A difference between the two forms is reported to be a host-specific pathogenicity factor (AAL-, ACT-, AM-toxin, etc.) carried by a small CDC. However, saprotrophic strains of *A. alternata* can acquire pathogenicity by transfer of the CDC chromosome (Akagi et al., 2009). These factors are defined phytotoxins and are considered primary determinants of pathogenicity. Phytotoxins usually alter the chloroplasts and the cell wall/membrane of the host, thus supporting infection by the fungus (Johnson et al., 2000). They seem not to be a significant problem from a food safety perspective (Mamgain et al., 2013), albeit the AAL toxin resembles in structure the fumonisins (Desjardins and Proctor, 2007). However, *Alternaria* is also able to produce reasonable amounts of certain mycotoxins, principally TeA, AOH and AME, TEN, and ALT, which contaminate harvested plant products and can exert poisonous effects after consumption by humans. Indeed, although there are no specific regulatory limits yet, the European Food Safety Authority suggested that *Alternaria* toxins are of high concern for public health (EFSA, 2011). As such, the European Standing Committee recommended to EU member states to collect data on their occurrence. For example, the Netherlands found AOH, AME, TeA, and TEN in at least one food commodity, while ALT was not found in any samples (López et al., 2016). TeA was found at high concentrations in 27% of analyzed cereals, tomato sauces, figs, wine, and sunflower seeds, whereas it was only incidentally found in apples, citrus fruits, tomatoes, and olives. The biosynthetic pathways of AOH and AME have been elucidated (Graf et al., 2012; Saha et al., 2012).

2.2 *Aspergillus*

The genus *Aspergillus* contains several plant saprotrophic and pathogenic species, including the mycotoxin producers *Aspergillus flavus Link* and *A. parasiticus* (sect. *Flavi*), *A. niger* Tiegh. and *A. carbonarius* (Bainier) Thom (sect. *Nigri*), *A. ochraceus* K. Wilh., and *A. westerdijkiae* Frisvad & Samson (sect. *Circumdati*). They usually produce high amounts of several mycotoxins, such as aflatoxins, ochratoxins, fumonisins, and patulin, during the entire cycle of infection, but mostly during the storage of seeds and fruits (Sanzani et al., 2016b). Indeed, although *A. flavus*, *A. parasiticus*, and *A. carbonarius* usually infect the commodity in the field, they are more harmful, in terms of mycotoxin production, during the postharvest phase (Aldars-García et al., 2016). In fact, undetected latent or quiescent field infections later develop into severe postharvest rots. Furthermore, at lower water activity, these fungi can outcompete other crop-infecting fungi (Georgianna and Payne, 2009). Thus, in the field, where the water activity of plant tissues may exceed 0.90, for example, *Fusarium* spp. can compete better on cereals than *Aspergilli* whereas at seed maturity or at postharvest, when fruits or seeds tend to dehydrate, the latter are favored (Magan and Medina, 2016). Secondary metabolites have been very valuable for *Aspergillus* chemotaxonomy and are often included in species descriptions (Kocsubé et al., 2016). Independent analysis of *Aspergillus* species, identified using either morphology plus physiology or DNA sequencing, showed that the secondary metabolite profile is species-specific, while individual secondary metabolites may occur in closely related species, in less closely related species within a genus, or even in completely unrelated species (Frisvad and Larsen, 2015). Several studies were conducted to elucidate the biosynthetic pathways of the main mycotoxins associated with this genus (Amare and Keller, 2014; Gerin et al., 2016; Susca et al., 2016).

2.3 Penicillium

Penicillium is one of the most widespread genera, comprising 150 recognized species of anamorphic fungi occurring in a diverse range of habitats: soil, vegetation, air, indoor environments, and various foodstuffs (Visagie et al., 2014). As such, it has a worldwide distribution and a large economic impact in the food industry as a decomposer of organic material and cause of destructive rots. Indeed, it produces a wide array of secondary metabolites with multiple functions. Wounds on fresh products, occurring at harvest or during postharvest handling processes, are the major site of infection, although it can penetrate through natural openings such as stem end, open calyx tube, or lenticels (Amiri and Bompeix, 2005). Postharvest contamination may come from various sources including contact with soil carried on bins, decayed fruits, air, drenching solutions, and washing water, where the spore concentration increases with the number of soaked fruit lots (Fallanaj et al., 2013). Particularly relevant within the genus, and thus elected as type species, is *P. expansum*, causal agent of blue mold, a common postharvest disease of pome fruits in particular, but present on several other commodities (Sanzani et al., 2013). *Penicillium expansum* produces several mycotoxins (e.g., patulin, citrinin, chaetoglobosins, etc.) that can affect the quality and safety of juices (Andersen et al., 2004); however, the lactone patulin is considered the most dangerous and widespread (Moake et al., 2005). The contamination is of concern during both conventional and organic production. For instance, surveys carried out on fruity foodstuffs, including baby foods, reported significantly higher levels of patulin in organic products compared to conventional ones (Piemontese et al., 2005; Sarubbi et al., 2016). Several studies were conducted to provide information on the genomic basis of the production of secondary metabolites in *P. expansum* (Ballester et al., 2015; Li et al., 2015; Julca et al., 2016). It was demonstrated that, despite a major genome contraction, *P. expansum* is the species with the highest potential to produce secondary metabolites within the genus. Indeed, 55 secondary metabolite clusters were identified, including the one for patulin. Moreover, patterns of polymorphism were observed in the species, suggesting the existence of recombination events. In accordance with putative sexual recombination and heterothallism, still unknown for this species, two alternative mating types were found (Julca et al., 2016).

3 Detection

3.1 Sampling Methods

Obtaining representative samples for a reliable analysis of mycotoxin contamination remains a major issue, since official sampling methods have been established only for a limited number of compounds. For example, sampling methods for aflatoxins, OTA, and *Fusarium* toxins in foodstuffs intended for human consumption were defined by the European Community (EC, 2006) in Commission Directives 98/53/EC, 2004/43/EC, and 2005/38/EC. Mycotoxin sampling for agricultural commodities was also defined in EC No. 519/2014 amending EC No. 401/2006. In the United States, sampling guidelines were developed by the Grain Inspection, Packers, and Stockyards Administration (GIPSA). Because of the heterogeneous distribution of mycotoxins, especially those produced by *Aspergillus* species for which "hot spots" may be found in batches of most foods and feeds (Champeil et al., 2004), large quantities of sample are needed in validated sampling procedures. Handling and preparation of samples with weights of up to 30 kg are also required, which render implementation of the sampling plans difficult (Spanjer et al., 2006). The problem is less pronounced for *Fusarium* toxins, which are regarded as less heterogeneously distributed than aflatoxins (Larsen et al., 2004). Nevertheless, the lack of scientific studies describing mycotoxin distribution and variability has limited the development of alternative sampling plans. These plans would be used depending on the mycotoxins and the type and size of the sampled commodities, and they would enable a reduction in the total variability and estimation of errors in the quantification of mycotoxin concentrations (Krska and Molinelli, 2007). In September 2014, FAO released a Mycotoxin Sampling Tool (available at www.fstools.org/mycotoxins) to provide support in analyzing the performance of sampling plans, and determining the most appropriate plan to meet user's defined objectives. The tool facilitates evaluation of the effect of varying design parameters on the performance of the sampling plan, thus determining the most appropriate mycotoxin-sampling plan to minimize risk of misclassifying lots according to needs and resources. The tool currently provides guidance for the sampling of 26 mycotoxin-commodity combinations, and it is constructed in a way to allow for inclusion of additional mycotoxin-commodity combinations, as well as new functions in future versions.

3.2 Traditional Identification Procedures

In absence of specific signs and symptoms, traditional detection relies on the use of moist chambers, which can promote the growth and the sporulation of the pathogen from the host tissues, or the isolation of the pathogen on culture media

(Lane et al., 2012). This latter technique is mostly restricted to facultative parasites (necrotrophs) and is well suited for pathogens confined in the host tissues, because contaminating microorganisms can be physically avoided. However, in complex environmental samples, faster growing or saprophytic microrganisms can conceal the presence of the primary pathogen. Under these conditions, the use of culture media with different levels of selectivity to reduce the range of saprophytic species isolated is necessary, but these are currently available only for a limited number of genera (Sanzani et al., 2014). In 1974, Bothast and Fennell developed ADM, whose differential ingredient was ferric citrate (0.05%), which, reacting with kojic acid produced only by *A. flavus* and closely related species, produced a bright orange-yellow pigment on the reverse side of the colony. Pitt et al. (1983) improved ADM and developed AFPA. This medium, recommended to detect internal contamination in nuts, corn, spices, and soil, employed incubation at 30°C for 42–48 hr to develop the orange-yellow pigment. *Aspergillus ochraceus* and *A. niger* were found to produce a yellow reverse color, but they could be easily differentiated from *A. flavus* by spore color after a further 24–48 hr of incubation. Frisvad (1983) reported that pentachloronitrobenzene, rose bengal in PRYES, consistently indicated the presence of OTA and citrinin-producing *Penicillium viridicatum* Westling with a typical violet brown reverse color. DCMA was reported to be a good differential medium for *A. alternata*; the differentiation was based on colony morphology and the shape and size of conidia (Gourama and Bullerman, 1995). Cultural methods essentially provide qualitative data, even though quantitative information can be obtained, for example, the number of colonies in relation to a certain quantity of samples (Sanzani et al., 2014). One rather obvious advantage of culturing methods is that successful isolation yields objective proof of the presence of the pathogen and a culture that is available for further characterization. However, most of the above-mentioned methods are time- and labor-consuming, and results are not always conclusive, for instance, when closely related organisms need to be discriminated from each other.

3.3 Immunological Assays

Being heat-resistant, antigens can survive after the death of the fungi that produced them. Since the 1980s, several ELISA assays for pathogens were developed. For instance, Notermans and Heuvelman (1985) developed an ELISA assay to detect, among others, *Penicillium verrucosum* Dierckx in foods. The antigens were found to be genus-specific and not present in nonmoldy food samples. The identified fungal antigens were mainly extracellular polysaccharides with immunogenic properties. However, they can also be bound to the fungal cell wall, and thus in order to be detected immunologically, they have to be released from the mycelial wall (Gourama and Bullerman, 1995).

In previous 20 yr, ELISA has become very popular due to the relatively low cost and easy application as an analytical tool for the detection of mycotoxins (Anfossi et al., 2016). ELISA employs different formats, such as microtiter plates and bead-based assays, or membranes, and can also be linked with other techniques, such as electrochemical sensors and SPR (Meneely and Elliott, 2014).

Commercially available kits are normally based on a competitive assay that uses either a primary antibody specific for the target molecule or a conjugate of an enzyme and the required target (Stanker et al., 2008). The complex formed will then interact with a chromogenic substrate to give a measurable result. They can be portable, rapid, easy-to-use, and specific. However, being intended for a single use, they might be not affordable for bulk screening. Additionally, competitive ELISA suffers from having a limited detection range due to the narrow sensitivity of the antibodies, whether they are mono- or polyclonal (Turner et al., 2009). Moreover, due to their small size, mycotoxins require conjugation to a carrier molecule, usually a protein (e.g., bovine serum albumin), to achieve immunogenicity. However, the conjugation process can decrease assay selectivity. Thus, both direct and indirect ELISA employed in analysis of mycotoxins have their advantages and limitations. Direct ELISA is quick, since only one antibody is used and the cross-reactivity of a possible secondary antibody is completely eliminated, but it suffers from the fact that immunoreactivity of the primary antibody may be reduced as a result of labeling, causing signal capture to become difficult. Indirect ELISA has several advantages, for example, it is more sensitive, since each primary antibody contains several epitopes that can be bound by the labeled secondary antibody, thereby enhancing signal amplification; as such, a wide range of labeled secondary antibodies are commercially available. However, the cross-reactivity may occur with the secondary antibody, resulting in a nonspecific signal (Zachariasova et al., 2014).

3.4 Molecular Methods

The invention of PCR in 1984 by Kary Mullis has revolutionized basic and applied studies in all biological fields, including plant pathology. Since then several PCR assays for the detection of mycotoxigenic fungi were developed. For example, Konstantinova et al. (2002) described a PCR assay for carrot contaminants *A. alternata, A. radicina* Meier, and *A. dauci* (J.G. Kühn) J.W. Groves & Skolko, targeting the ITS regions. Despite the usual low specificity of these regions, they were able to detect and differentiate the three species from carrot samples. Similarly, several methods for early detecting aflatoxin-producer fungi into figs were set, starting from 1990s (Färber et al., 1997). An improved PCR-RFLP assay allowed distinguishing the five species of *Aspergillus* present on grapes: *A. carbonarius, A. tubingensis, A. niger, A. aculeatus* Iizuka, and *A. japonicus* Saito (Spadaro et al., 2012). Some conventional PCR assays have been reported for the detection of patulin-producing *Penicillium* spp. (Paterson et al., 2000; Dombrink-Kurtzman and Engberg, 2006).

The potential of PCR increased greatly in the early 1990s with the development of real-time quantitative amplification technologies. Indeed, qPCR can be utilized to detect and quantify mycotoxin-producing fungi present as infecting pathogens or just as contaminants in foodstuffs (Sanzani et al., 2014). Several studies described a correlation between fungal biomass and specific mycotoxin contamination (Gil-Serna et al., 2009; Sanzani et al., 2013). However, as the toxigenic ability of each strain needs to be considered, various mycotoxin genotyping assays have been developed to directly quantify genes responsible for mycotoxin synthesis, from both fungal culture and plant material (Kulik et al., 2011). The availability of toxin-specific qPCR assays can help in studying the pathogen population structure, competition between toxin-producing and nonproducing subpopulations, and the effects of disease management strategies to reduce toxin contamination (Luo et al., 2009; Sanzani et al., 2009b).

Pavón et al. (2012a) developed a qPCR method based on ITS sequences for the quantification of *Alternaria* sp. in raw and processed vegetables that was able to detect up to 1 CFU/g. The measured counts were in good correlation with the counts determined by plating. The same group developed a reverse transcriptase qPCR method, based on the rRNA to detect viable *Alternaria* cells in foodstuffs (Pavón et al., 2012b). In order to distinguish between viable and nonviable *Alternaria* spp. conidia, a promising technique based on the pretreatment of samples with nucleic acid intercalating dyes, as PMA, prior to qPCR was used (Crespo-Sempere et al., 2013). PMA selectively penetrated damaged cells, inhibiting DNA amplification. The assay achieved a detection limit of 10^2 conidia/g of tomato. Artificially inoculated tomato samples, treated with 65 μM of PMA, showed a reduction in the qPCR signal by almost seven cycles between live and heat-killed *Alternaria* spp. conidia.

The *aflD* (nor-1) gene is a recognized target for obtaining species-specific amplicons from aflatoxigenic strains of *Aspergillus* (Mayer et al., 2003). A study was conducted to establish a correlation between levels of aflatoxins and *aflD* gene expression in South African feeds. Results indicated that aflatoxins levels in positive samples ranged from 0.7 to 33.0 ppb. However, they generally did not correlate with those of *aflD* gene expression in similar samples (Iheanacho et al., 2014).

Both wine and table grapes represent a target for *Aspergillus* sect. *Nigri* contamination. Several methods have been recently demonstrated suitable for amplifying specific DNA fragments from different cultivars of grapes. Notably, primers on polyketide synthase related to OTA synthesis were used for quantifying *A. carbonarius* in table grapes (Ayoub et al., 2010).

The use of qPCR provided a new useful tool for detecting as well as quantifying *Penicillium* DNA in a wide variety of food products. For instance, in the study by Sanzani et al. (2012), a qPCR method to monitor *P. expansum* growth on apples was proposed. Moreover, a further step toward the rapid identification/detection of *Penicillium* spp. was reported by using the HRM method (Sanzani et al., 2013), which permitted the discrimination of *P. expansum* from other species (*Penicillium chrysogenum* Thom and *P. crustosum* Thom) and, within the same species, among different hosts of their origin. Indeed, *Penicillium* has a wide host range, including sweet cherries and grapes. Wine and fermenting musts are greatly compromised by mycotoxigenic fungi, with an increasing need for "user-friendly" assays. Recently, a single-tube nested real-time PCR approach was developed to trace *Penicillium* spp. in musts and wines (Sanzani et al., 2016a). The method consisted of two sets of primers, targeting the beta-tubulin gene, simultaneously applied with the aim of lowering the detection limit of conventional qPCR. The assay was able to detect up to 1 fg of *Penicillium* DNA. Most of analyzed wines/musts resulted contaminated by patulin at >50 ppb and a 76% accordance with the molecular assay was observed. Moreover, assays specifically targeting the patulin toxin biosynthetic gene in terms of presence and expression were proposed (Rodríguez et al., 2011; Tannous et al., 2015).

Finally, multiplex qPCR methods to quantify different toxins (e.g., aflatoxin, OTA, patulin, and trichothecene) from toxin-producing molds in foods were developed using specific genes involved in the biosynthesis of these toxins (Rodríguez et al., 2011; Luque et al., 2013).

In the last decade, microarray-based immunoassay technology has permitted quantitative measurement of multiple samples simultaneously for multiple analytes, thus showing its great potential as monitoring system for the rapid assessment of food samples (Ngundi et al., 2005). For instance, a toxin microarray for rapid and sensitive detection of AFB1 in cereal samples was reported (Beizaei et al., 2015). For fungal identification, oligonucleotide probes were designed by exploiting the sequence variations of the EF-1α coding regions and the ITS regions of the ribosomal RNA cassette. For the detection of mycotoxin pathways (aflatoxins, patulin, DON/NIV, trichothecenes), oligonucleotide probes directed toward genes leading to toxin production from different fungal strains were designed.

3.5 Current and Emerging Technologies for Mycotoxin Analysis

Mycotoxins are generally identified by chromatographic techniques such as thin layer chromatography (TLC) and high-performance liquid chromatography (HPLC), singly or coupled with mass spectrometry (HPLC-MS/MS) (Sanzani et al., 2016b). Concerning *Alternaria* toxins, Oviedo et al. (2012) used HPLC to detect AOH and AME in soya beans, with recoveries of 95–99%. Ostry et al. (2012) described a high-performance thin-layer chromatography (HPTLC) method for the detection and quantification of ALT, AOH, and AME, with a limit of detection of 0.3 mg/kg. Asam and Rychlik (2015) described SIDAs for AOH, TeA, the perylenequinones ATX I–III, and TEN. Their isotopes were very similar in their chemical properties to their natural analytes, and thus behaved similarly in a food sample. Both isotopes (the natural and the added) could be differentiated by mass spectrometry. The concentration of the natural analyte could be calculated by knowing the concentration of the added isotope. With this method, AOH and AME could be detected in fruit juices with a recovery rate of over 100%. Moreover, low LODs could be achieved, for example, 0.03 μg/kg for AOH and 0.01 μg/kg for AME. Recently, a HPLC-MS/MS-based method for the quantification of nine *Alternaria* mycotoxins in various food matrices was developed (Hickert et al., 2016). The method relies on a single-step extraction, followed by dilution of the raw extract and direct analysis. In combination with an analysis time of 12 min per sample, the sample preparation proved to be cost-effective and easy to handle. The method was validated for tomato and bakery products, sunflower seeds, fruit juices, and vegetable oils. AOH, AME, TeA, and TEN were found in quantifiable amounts in 92% of 96 analyzed samples, although at low levels.

HPLC coupled to tandem mass spectrometry is employed to detect mycotoxins produced by the genus *Aspergillus* into fruits, mainly aflatoxins, ochratoxins, and fumonisins. Often, simple extractive procedures are coupled to sophisticated analytical techniques, which detect and quantify multiple toxins in a single run (Lacina et al., 2012). Recently, an aptamer-based detection system was developed to quantify OTA in wine and grape juice. Aptamers are single-stranded oligonucleotides that fold into distinct three-dimensional conformations capable of binding strongly and selectively to a target molecule. As molecular recognition probes, aptamers have binding affinities and specificities that are comparable to, and in some cases surpass, those of monoclonal antibodies (McKeague et al., 2011). The assay was based on the high specificity recognition between a specific aptamer and OTA (Cruz-Aguado and Penner, 2008). This aptamer was modified by molecular beacon technology for performing the analysis in a real-time PCR apparatus, skipping the expensive and laborious procedures for quantifying OTA (Sanzani et al., 2015). Aptamers are selected from random-sequence nucleic acids libraries by an in vitro process known as SELEX, through iterative cycles of affinity separation and amplification (Yang et al., 2012). Targets for which aptamers can be developed are varied and range from small molecules (Huizenga and Szostak, 1995) to proteins and even whole cells. The in vitro nature of the selection process allows for the discovery of aptamers for even non-immunogenic or highly toxic substances. In addition to this advantage, high-purity aptamers can be chemically synthesized at a low cost and can be easily modified with dyes, labels, and surface attachment groups without affecting their affinities. Moreover, aptamers are more chemically stable under most environmental conditions, have a long shelf life, and can be reversibly denatured without loss of specificity.

Conventional analytical methods for patulin include TLC, HPLC, and GC; the latter two can even be coupled with MS. However, in practice most patulin analyses are conducted according to the European Norm 14177:2001, which was also recommended for adoption by the AOAC (MacDonald et al., 2000). These techniques could be used even to assess fungal presence. For example, López et al. (2015) demonstrated that Z-3-hexenyl 2-methylbutanoate could be a potential biomarker of *P. expansum* using dynamic headspace-GC, since it was quantifiable even before the disease was visible on inoculated fruit. These analytical methods applied to food or feed samples are time consuming and produce results within hours or days. Commercial competition within the food and feed industries compels them to reduce costs and deliver products promptly. Therefore, rapid analytical methods have become increasingly important. In a report by de Champdoré et al. (2007), the synthesis of two patulin derivatives was described, along with their conjugation to bovine serum albumin, for the production of polyclonal antibodies. Moreover, a fluorescence competitive immunoassay was developed for the on-line detection of patulin. More recently, they reported a competitive immune assay for the detection of patulin based on an optical technique called SPR, in which laser beam induced interactions between probe and target molecules in proximity of the gold surface of a biochip, leading to a shift in resonance and consequently in reflectivity (Pennacchio et al., 2014). Later, the detection of the toxin directly in apple juice was accomplished using a fluorescence polarization approach based on the use of near-infrared (NIR) fluorescence probes coupled to anti-patulin antibodies. This methodology relied on an increase in fluorescence polarization emission from a labeled patulin derivative after binding to specific antibodies in competition with nonfluorescent patulin (Pennacchio et al., 2015).

4 Control

4.1 Traditional Means

Controlling mycotoxin accumulation is not only necessary to ensure consumers' safety but also useful to prevent pathogen infection and rot development. Traditionally, the control of mycotoxigenic fungi employs fungicides (e.g., pyrimethanil, thiabendazole, imazalil), but their intense use over many years has led to the development of resistant populations (Baraldi et al., 2003; Li and Xiao, 2008). Moreover, progressively restrictive legislation, concerns about environmental pollution and chemical residues in the food, costs of new registration/reregistration, and disposal of used fungicide solutions have reduced or eliminated their use after harvest, and thus increased the search for alternative control means. In a study by Schmidt-Heydt et al. (2013), seven commercial fungicide formulations, containing fosethyl-Al, iprodione, boscalid, azoxystrobin, fluopyram, tebuconazole, fenamidone, and/or mancozeb, were tested for their ability to inhibit the growth of several fungi including *P. verrucosum*. Interestingly, individual fungicides were only able to inhibit the growth of the analyzed fungi to some extent, but in case of *P. verrucosum*, iprodione, an imidazole-class fungicide, greatly induced mycotoxin biosynthesis as has been previously described for the strobilurin-class fungicides (Ellner, 2005). Consequently, before using a given fungicide to protect crops and enhance storage life, the applicability of this chemical compound should be tested not only for its ability to inhibit fungal growth but also for its effect on the level of secondary metabolite biosynthesis.

4.2 Alternative Means

In recent years, the need to develop alternatives to synthetic chemicals to control qualitative and quantitative losses of foods by molds and mycotoxin contamination has become a priority for many scientists worldwide, and one approach is to use plant-based food preservatives for these purposes. In the previous decade, aromatic plant products (EOs and their major bioactive compounds) have been extensively studied throughout the world to control fungi by their fumigant action. Shukla et al. (2009) reported the ability of the EO and its major components of *Lippia alba* to inhibit aflatoxin production by three strains of *A. flavus*. EO completely inhibited AFB1 at a concentration of 0.6–1.0 μL/mL. Among its constituents, geranial caused complete inhibition of AFB1 at a concentration of 0.6–0.8 μL/mL, while nerol did so from a concentration of 0.8–1.0 μL/mL. Tian et al. (2011) reported in vitro and in vivo antifungal activity of EO from *Anethum graveolens* seeds to *Aspergillus oryzae* (Ahlb.) Cohn, *A. flavus*, *A. niger*, and *A. alternata*. In vivo 100–120 μL/mL of oil completely protected cherry tomatoes from fungal infection. In the study of Phillips et al. (2012), EO vapor of citrus reduced in vitro growth of *P. chrysogenum*, *A. niger*, and *A. alternata* by 44, 34, and 67%, respectively, and the growth of *A. niger* and *P. chrysogenum* on grain by 50–60%. Prakash et al. (2014) recommended *Boswellia carterii* EO as a plant-based fumigant to control storage fungi and aflatoxin production, and demonstrated its in vivo efficacy on food commodities. The antimycotoxigenic mode of action of plant EOs has not been clearly elucidated, although several hypotheses and some experimental studies have been reported in recent years. The antiaflatoxigenic activity of EO may be related to inhibition of the ternary steps of aflatoxin biosynthesis involving lipid peroxidation and oxygenation (Bluma et al., 2008). Indeed, oxidative stress is a prerequisite for "modeling" fungal lifecycle also by fine-tuning the secondary metabolism. In relation to this, keeping the fungus in a perennial "juvenile" stage could aid in the control of aflatoxin production (Zaccaria et al., 2015). Moreover, Tian et al. (2011) suggested that the inhibition of AFB1 production can be partly attributed to the inhibition of carbohydrate catabolism in *A. flavus*. Since mitochondria are responsible for providing acetyl-CoA, a main precursor for aflatoxin biosynthesis, disruption of the mitochondrial respiration chain may also account for the inhibitory effects of phenolics on aflatoxin production (Razzaghi-Abyaneh et al., 2011). It has been also observed that the lack of sporulation in fungal mycelia treated with EOs may explain their antimycotoxigenic activity since a correlation between secondary metabolite production and sporulation is well known (Brodhagen and Keller, 2006).

Sanzani et al. (2009a) evaluated the effectiveness of several phenolic compounds to control *P. expansum* growth and patulin accumulation. Their in vitro screening identified the flavonoid quercetin as the most effective compound to control patulin production, although fungal growth was not significantly reduced. However, in in vivo tests with "Golden Delicious" apples, the presence of quercetin caused a 92% reduction in disease incidence and severity, and a 78% reduction in patulin accumulation. These findings indicated quercetis as an inducer of resistance to blue mold. To confirm or reject the hypothesis, genes differentially expressed in quercetin-treated apples were identified by SSH (Sanzani et al., 2010). They revealed high similarities with different classes of pathogenesis-related proteins (PR10 and PR8), or with proteins expressed under stress conditions. Other transcripts had high similarity to genes coding for proteins with a role in pathogen recognition and signaling pathways. Furthermore, authors reported a combined effect between wounding and phenolic treatment on the differential expression of evaluated genes. The inhibition of patulin production recorded both in vitro and in vivo indicated that quercetin might directly act on its biosynthetic pathway. Therefore, the expression of five biosynthesis genes was evaluated using qPCR in the presence/absence of the tested phenolic compound (Sanzani et al., 2009b). The two putative cytochrome P450 monooxygenases p450-1 and p450-2 proved to be downregulated, as compared to the untreated samples. The downregulation agreed with quercetin antioxidant properties (Nijveldt et al., 2001; Repetto and Llesuy, 2002), which could interfere with the pro-oxidant activity of both monooxygenases.

In the past 30 yr, there have been extensive research activities to explore and develop strategies based on microbial antagonists to control biologically postharvest pathogens (Spadaro and Droby, 2016). Fermentation by *Saccaromyces cereviasiae* has been reported to reduce patulin contamination by degrading it into its less-toxic precursor *E*-ascladiol (Moss and Long, 2002). However, a bacterium isolated from patulin-contaminated apples proved capable of degrading patulin to ascladiol (Ricelli et al., 2007). The bacterium was identified as *Gluconobacter oxydans*, whereas ascladiol was identified by HPLC-MS and proton and carbon nuclear magnetic resonance. Degradation of up to 96% of patulin was observed in apple juices incubated with *G. oxydans*. This finding is noteworthy, considering that patulin has a broad antibacterial spectrum. However, the origin of these bacteria (i.e., from rotted apples containing patulin) could explain this characteristic, since the presence of patulin could be a selective factor on the apple surface for these strains. In a study by Dorner and Cole (2002), biocontrol based on competitive exclusion by nontoxigenic strain of *A. flavus* applied in the field had a carryover effect and reduced aflatoxin contamination in storage. A strain of marine *Bacillus megaterium*, evaluated for its activity in reducing postharvest decay caused by *A. flavus*, proved to have a significant effect in reducing the biosynthesis of aflatoxins on peanut kernels and expression of *aflR* gene and *aflS* gene (Kong et al., 2010). Sempere and Santamarina (2007) used a *Trichoderma harzianum* Rifai strain to control the growth of *A. alternata* on rice. In addition, *T. harzianum* and *Trichoderma piluliferum* J. Webster & Rifai proved to have very high inhibitory activity against *A. alternata* on Akarkara (Thakur and

Harsh, 2014). Very little is known about the overall diversity and composition of microbial communities on harvested produce and how these communities vary across produce types, their function, the factors that influence their composition after harvest and during storage, and the distribution of individual taxa (Droby et al., 2016). Thus, metagenomic technologies might give relevant information on the composition of microbial communities on fruits and vegetables.

Sanitation after harvest is critically important for all fresh products, where it can reduce spoilage losses by 50% or more (Sargent et al., 2007). This occurs primarily by the sanitation of wash water, produce surfaces, equipment, and storage rooms rather than direct control of infections by the decay pathogens within the produce (Feliziani et al., 2016). Ozone (O_3) application to stored fruits and cereals can control or inhibit growth, germination and sporulation of fungi, thus preventing the production of their toxins (Giordano et al., 2012; Hansen et al., 2013; Feliziani et al., 2014). However, these effects are very dependent on the fungal species, growth stage, O_3 concentration and exposure time (Freitas-Silva and Venâncio, 2010). On the other hand, to promote the mycotoxin degradation, high concentrations of O_3 are needed. The exposure time, type of food, moisture content, and temperature are also factors that directly affect its efficacy (Ikeura et al., 2011; Li et al., 2014). McDonough et al. (2011) evaluated the use of ozonation on a commercial scale by applying O_3 at 4.7% in air to corn kernel using a continuous-flow system. It was delivered into a screw conveyor with a retention time for the grains moving through the system of 1.8 min. Under these conditions, there was an approximate reduction of 30% in aflatoxins.

5 Concluding Remarks and Future Prospectives

The mycotoxigenic fungi remain one of the worrisome issues for agriculture and food safety worldwide. Besides cultural practices, other protective measures applied after harvest are needed to effectively control plant diseases and mycotoxin accumulation. As a long-term strategy, both scientists and industry personnel strongly support the use of alternative control means (Nguyen et al., 2017). Over the past 30 yr, alternatives have moved from the simple idea of postharvest application of high concentrations of alternative means, as biocontrol agents, to the use of a wide array of other alternatives, integrating them into a systematic approach based on the multiple decrement or multiple hurdle concept (Wisniewski et al., 2016). In this perspective, also the rapid and early detection of mycotoxigenic fungi can greatly contribute to the success of an alternative control strategy, directed to both the producing fungus and the toxin itself. Furthermore, these strategies could greatly benefit by a modern understanding of the biosynthesis of mycotoxins and their biological roles based on the use of innovative technologies.

References

Akagi, Y., Akamatsu, H., Otani, H. and Kodama, M. 2009. Horizontal chromosome transfer, a mechanism for the evolution and differentiation of a plant pathogenic fungus. *Eukaryotic Cell* 8, 1732–1738.

Aldars-García, L., Ramos, A.J., Sanchis, V. and Marín, S. 2016. Modeling postharvest mycotoxins in foods: Recent research. *Current Opinion in Food Science* 11, 46–50.

Amare, M.G. and Keller, N.P. 2014. Molecular mechanisms of *Aspergillus flavus* secondary metabolism and development. *Fungal Genetics and Biology* 66, 11–18.

Amiri, A. and Bompeix, G. 2005. Diversity and population dynamics of *Penicillium* spp. on apples in pre- and postharvest environments: Consequences for decay development. *Plant Pathology* 54, 74–81.

Andersen, B., Smedsgaard, J. and Frisvad, J.C. 2004. *Penicillium expansum*: Consistent production of patulin, chaetoglobosins, and other secondary metabolites in culture and their natural occurrence in fruit products. *Journal of Agricultural and Food Chemistry* 52, 2421–2428.

Anfossi, L., Giovannoli, C. and Baggiani, C. 2016. Mycotoxin detection. *Current Opinion in Biotechnology* 37, 120–126.

Asam, S. and Rychlik, M. 2015. Recent developments in stable isotope dilution assays in mycotoxin analysis with special regard to *Alternaria* toxins. *Analytical and Bioanalytical Chemistry* 407, 7563–7577.

Ayoub, F., Reverberi, M., Ricelli, A., D'Onghia, A.M. and Yaseen, T. 2010. Early detection of *Aspergillus carbonarius* and *A. niger* on table grapes: A tool for quality improvement. *Food Additives & Contaminants* 27, 1285–1293.

Ballester, A.R., Marcet-Houben, M., Levin, E., Sela, N., Selma-Lázaro, C., Carmona, L., Wisniewski, M., Droby, S., González-Candelas, L. and Gabaldón, T. 2015. Genome, transcriptome, and functional analyses of *Penicillium expansum* provide new insights into secondary metabolism and pathogenicity. *Molecular Plant-Microbe Interactions* 28, 232–248.

Baraldi, E., Mari, M., Chierici, E., Pondrelli, M., Bertolini, P. and Pratella, G.C. 2003. Studies on thiabendazole resistance of *Penicillium expansum* of pears: Pathogenic fitness and genetic characterization. *Plant Pathology* 52, 362–370.

Battilani, P. and Camardo Leggieri, M. 2014. Predictive modelling of aflatoxin contamination to support maize chain management. *World Mycotoxin Journal* 8, 161–170.

Beizaei, A., O'Kane, S.L., Kamkar, A., Misaghi, A., Henehan, G. and Cahill, D.J. 2015. Highly sensitive toxin microarray assay for aflatoxin B1 detection in cereals. *Food Control* 57, 210–215.

Bluma, R., Amaiden, M.R., Daghero, J. and Etcheverry, M. 2008. Control of *Aspergillus* section *Flavi* growth and aflatoxin accumulation by plant essential oils. *Journal of Applied Microbiology* 105, 203–214.

Boddu, J., Cho, S. and Muehlbauer, G.J. 2007. Transcriptome analysis of trichothecene-induced gene expression in barley. *Molecular Plant-Microbe Interactions* 20, 1364–1375.

Bothast, R.J. and Fennell, D.I. 1974. A medium for rapid identification and enumeration of *Aspergillus flavus* and related organisms. *Mycologia* 66, 365–369.

Brodhagen, M. and Keller, N.P. 2006. Signalling pathways connecting mycotoxin production and sporulation. *Molecular Plant Pathology* 7, 285–301.

Calvo, A.M., Wilson, R.A., Bok, J.W. and Keller, N.P. 2002. Relationship between secondary metabolism and fungal development. *Microbiology and Molecular Biology Reviews* 66, 447–459.

Chakraborty, S., Murray, G.M., Magarey, P.A., Yonow, T., O'Brien, R.G. and Croft, B.J. 1998. Potential impact of climate change on plant diseases of economic significance to Australia. *Australasian Plant Pathology* 27, 15–35.

Champeil, A., Fourbet, J.-F. and Doré, T. 2004. Effects of grain sampling procedures on *Fusarium* mycotoxin assays in wheat grains. *Journal of Agricultural and Food Chemistry* 52, 6049–6054.

Cotty, P.J. and Jaime-Garcia, R. 2007. Influences of climate on aflatoxin producing fungi and aflatoxin contamination. *International Journal of Food Microbiology* 119, 109–115.

Crespo-Sempere, A., Estiarte, N., Marín, S., Sanchis, V. and Ramos, A.J. 2013. Propidium monoazide combined with real-time quantitative PCR to quantify viable *Alternaria* spp. contamination in tomato products. *International Journal of Food Microbiology* 165, 214–220.

Cruz-Aguado, J.A. and Penner, G. 2008. Determination of ochratoxin a with a DNA aptamer. *Journal of Agricultural and Food Chemistry* 56, 10456–10461.

de Champdoré, M., Bazzicalupo, P., De Napoli, L., Montesarchio, D., Di Fabio, G., Cocozza, I., Parracino, A., Rossi, M. and D'Auria, S. 2007. A new competitive fluorescence assay for the detection of patulin toxin. *Analytical Chemistry* 79, 751–757.

De Rijk, T.C., Van Egmond, H.P., Van der Fels-Klerx, H.J., Herbes, R., De Nijs, M., Samson, R., Slate, A.B. and Van der Spiegel, M. 2015. A study of the 2013 Western European issue of aflatoxin contamination of maize from the Balkan area. *World Mycotoxin Journal* 8, 641–651.

Desjardins, A.E. and Hohn, T.M. 1997. Mycotoxins in plant pathogenesis. *Molecular Plant-Microbe Interactions* 10, 147–152.

Desjardins, A.E. and Proctor, R.H. 2007. Molecular biology of *Fusarium* mycotoxins. *International Journal of Food Microbiology* 119, 47–50.

Desmond, O.J., Manners, J.M., Stephens, A.E., Maclean, D.J., Schenk, P.M. and Gardiner, D.M. 2008. The *Fusarium* mycotoxin deoxynivalenol elicits hydrogen peroxide production, programmed cell death and defence responses in wheat. *Molecular Plant Pathology* 9, 435–445.

Dombrink-Kurtzman, M.A. and Engberg, A. 2006. *Byssochlamys nivea* with patulin-producing capability has an isoepoxydon dehydrogenase gene (idh) with sequence homology to *Penicillium expansum* and *P. griseofulvum*. *Mycological Research* 110, 1111–1118.

Dorner, J.W. and Cole, R.J. 2002. Effect of application of nontoxigenic strains of *Aspergillus flavus* and *A. parasiticus* on subsequent aflatoxin contamination of peanuts in storage. *Journal of Stored Products Research* 38, 329–339.

Droby, S., Wisniewski, M., Teixidó, N., Spadaro, D. and Jijakli, M.H. 2016. The science, development, and commercialization of postharvest biocontrol products. *Postharvest Biology and Technology* 122, 22–29.

EC (European Community). 2006. Commission Regulation (EC) No 776/2006 of 23 May 2006 amending Annex VII to Regulation (EC) No 882/2004 of the European Parliament and of the Council as regards Community reference laboratories. *Official Journal of the European Union* 136, 3–8.

EFSA (European Food Safety Authority). 2011. Scientific opinion on the risks for animal and public health related to the presence of *Alternaria* toxins in feed and food. *EFSA Journal* 9, 2407–2413.

Ellner, F.M. 2005. Results of long-term field studies into the effect of strobilurin containing fungicides on the production of mycotoxins in several winter wheat varieties. *Mycotoxin Research* 21, 112–115.

Fallanaj, F., Sanzani, S.M., Zavanella, C. and Ippolito, A. 2013. Salt addition improves the control of citrus postharvest diseases using electrolysis with conductive diamond electrodes. *Journal of Plant Pathology* 95, 373–383.

Färber, P., Geisen, R. and Holzapfel, W.H. 1997. Detection of aflatoxinogenic fungi in figs by a PCR reaction. *International Journal of Food Microbiology* 36, 215–220.

Feliziani, E., Lichter, A., Smilanick, J.L. and Ippolito, A. 2016. Disinfecting agents for controlling fruit and vegetable diseases after harvest. *Postharvest Biology and Technology* 122, 53–69.

Feliziani, E., Romanazzi, G. and Smilanick, J.L. 2014. Application of low concentrations of ozone during the cold storage of table grapes. *Postharvest Biology and Technology* 93, 38–48.

Freitas-Silva, O. and Venâncio, A. 2010. Ozone applications to prevent and degrade mycotoxins: A review. *Drug Metabolism Reviews* 42, 612–620.

Frisvad, J.C. 1983. A selective and indicative medium for groups of *Penicillium viridicatum* producing different mycotoxins in cereals. *Journal of Applied Bacteriology* 54, 409–416.

Frisvad, J.C. and Larsen, T.O. 2015. Extrolites of *Aspergillus fumigatus* and other pathogenic species in *Aspergillus* section. *Fumigati. Frontiers in Microbiology* 6, 1485.

Gallo, A., Ferrara, M. and Perrone, G. 2013. Phylogenetic study of polyketide synthases and non-ribosomal peptide synthetases involved in the biosynthesis of mycotoxins. *Toxins* 5, 717–742.

Galvano, F., Galofaro, V. and Galvano, G. 1996. Occurrence and stability of aflatoxin M1 in milk and milk products: A worldwide review. *Journal of Food Protection* 59, 1079–1090.

Georgianna, D.R. and Payne, G.A. 2009. Genetic regulation of aflatoxin biosynthesis: From gene to genome. *Fungal Genetics and Biology* 46, 113–125.

Gerin, D., De Miccolis Angelini, R.M., Pollastro, S. and Faretra, F. 2016. RNA-Seq reveals OTA-related gene transcriptional changes in *Aspergillus carbonarius. PLoS ONE* 11, e0147089.

Gil-Serna, J., González-Salgado, A., González-Jaén, M.T., Vázquez, C. and Patino, B. 2009. ITS-based detection and quantification of *Aspergillus ochraceus* and *Aspergillus westerdijkiae* in grapes and green coffee beans by real-time quantitative PCR. *International Journal of Food Microbiology* 131, 162–167.

Giordano, B.N.E., Nones, J. and Scussel, V.M. 2012. Susceptibility of the in-shell Brazil nut mycoflora and aflatoxin contamination to ozone gas treatment during Storage. *Journal of Agricultural Science* 4, 1–10.

Gourama, H. and Bullerman, L.B. 1995. Detection of molds in foods and feeds: Potential rapid and selective methods. *Journal of Food Protection* 58, 1389–1394.

Graf, E., Schmidt-Heydt, M. and Geisen, R. 2012. HOG MAP kinase regulation of alternariol biosynthesis in *Alternaria alternata* is important for substrate colonization. *International Journal of Food Microbiology* 157, 353–359.

Hansen, L.S., Hansen, P. and Jensen, K.M.V. 2013. Effect of gaseous ozone for control of stored product pests at low and high temperature. *Journal of Stored Products Research* 54, 59–63.

Hickert, S., Bergmann, M., Ersen, S., Cramer, B. and Humpf, H.U. 2016. Survey of *Alternaria* toxin contamination in food from the German market, using a rapid HPLC-MS/MS approach. *Mycotoxin Research* 32, 7–18.

Huizenga, D.E. and Szostak, J.W. 1995. A DNA aptamer that binds adenosine and ATP. *Biochemistry* 34, 656–665.

Iheanacho, H.E., Dutton, M.F., Steenkamp, P.A., Steenkamp, L., Makun, H.A., Swart, A. and Mthombeni, J.Q. 2014. Real time PCR of Nor~ 1 (aflD) gene of aflatoxin producing fungi and its correlative quantization to aflatoxin levels in South African compound feeds. *Journal of Microbiological Methods* 97, 63–67.

Ikeura, H., Kobayashi, F. and Tamaki, M. 2011. Removal of residual pesticides in vegetables using ozone microbubbles. *Journal of Hazardous Materials* 186, 956–959.

Johnson, R.D., Johnson, L., Ito, Y., Kodama, M., Otani, H. and Kohmoto, K. 2000. Cloning and characterization of a cyclic peptide synthetase gene from *Alternaria alternata* apple pathotype whose product is involved in AM-toxin synthesis and pathogenicity. *Molecular Plant-Microbe Interactions* 13, 742–753.

Julca, I., Droby, S., Sela, N., Marcet-Houben, M. and Gabaldón, T. 2016. Contrasting genomic diversity in two closely related postharvest pathogens: *Penicillium digitatum* and *Penicillium expansum*. *Genome Biology and Evolution* 8, 218–227.

Kocsubé, S., Perrone, G., Magistá, D., Houbraken, J., Varga, J., Szigeti, G., Hubka, V., Hong, S.B., Frisvad, J.C. and Samson, R.A. 2016. *Aspergillus* is monophyletic: Evidence from multiple gene phylogenies and extrolites profiles. *Studies in Mycology* 85, 199–213.

Kong, Q., Shan, S., Liu, Q., Wang, X. and Yu, F. 2010. Biocontrol of *Aspergillus flavus* on peanut kernels by use of a strain of marine *Bacillus megaterium*. *International Journal of Food Microbiology* 139, 31–35.

Konstantinova, P., Bonants, P., van Gent-Pelzer, M., van der Zouwen, P. and van Den Bulk, R. 2002. Development of specific primers for detection and identification of *Alternaria* spp. in carrot material by PCR and comparison with blotter and plating assays. *Mycological Research* 106, 23–33.

Krska, R. and Molinelli, A. 2007. Mycotoxin analysis: State-of-the-art and future trends. *Analytical and Bioanalytical Chemistry* 387, 145–148.

Kulik, T., Jestoi, M. and Okorski, A. 2011. Development of TaqMan assays for the quantitative detection of *Fusarium avenaceum, Fusarium tricinctum*, and *Fusarium poae* esyn1 genotypes from cereal grain. *FEMS Microbiology Letters* 314, 49–56.

Lacina, O., Zachariasova, M., Urbanova, J., Vaclavikova, M., Cajka, T. and Hajslova, J. 2012. Critical assessment of extraction methods for the simultaneous determination of pesticide residues and mycotoxins in fruits, cereals, spices and oil seeds employing ultra-high-performance liquid chromatography-tandem mass spectrometry. *Journal of Chromatography A* 1262, 8–18.

Lane, C.R., Beales, P.A. and Hughes, J.D. 2012. *Fungal Plant Pathogens*. CABI, Wallingford, UK.

Larsen, J.C., Hunt, J., Perrin, I. and Ruckenbauer, P. 2004. Workshop on trichothecenes with a focus on DON: Summary report. *Toxicology Letters* 153, 1–22.

Lewis, L., Onsongo, M., Njapau, H., Schurz-Rogers, H., Luber, G. and Kieszak, S. 2005. Aflatoxin contamination of commercial maize products during an outbreak of acute aflatoxicosis in eastern and central Kenya. *Environmental Health Perspectives* 113, 1763–1767.

Li, B., Zong, Y., Du, Z., Chen, Y., Zhang, Z., Qin, G., Zhao, W. and Tian, S. 2015. Genomic characterization reveals insights into patulin biosynthesis and pathogenicity in *Penicillium* species. *Molecular Plant-Microbe Interactions* 8, 635–647.

Li, H.X. and Xiao, C.L. 2008. Characterization of fludioxonil-resistant and pyrimethanil-resistant phenotypes of *Penicillium expansum* from apple. *Phytopathology* 98, 427–435.

Li, M.M., Guan, E.Q. and Bian, K. 2014. Effect of ozone treatment on deoxynivalenol and quality evaluation of ozonized wheat. *Food Additives & Contaminants: Part A* 32, 544–553.

Logrieco, A., Mulè, G., Moretti, A. and Bottalico, A. 2009. *Alternaria* toxins and plant diseases: An overview of origin, occurrence and risks. *World Mycotoxin Journal* 2, 129–140.

López, L., Echeverria, G., Usall, J. and Teixidó, N. 2015. The detection of fungal diseases in the "Golden Smoothee" apple and "Blanquilla" pear based on the volatile profile. *Postharvest Biology and Technology* 99, 120–130.

López, P., Venema, D., de Rijk, T., de Kok, A., Scholten, J.M., Mol, H.J.G. and de Nijs, M. 2016. Occurrence of *Alternaria* mycotoxins in food products in the Netherlands. *Food Control* 60, 196–204.

Luo, Y., Gao, W., Doster, M. and Michailides, T.J. 2009. Quantification of conidial density of *Aspergillus flavus* and *A. parasiticus* in soil from almond orchards using real-time PCR. *Journal of Applied Microbiology* 106, 1649–1660.

Luque, M.I., Córdoba, J.J., Rodríguez, A., Núñez, F. and Andrade, M.J. 2013. Development of a PCR protocol to detect ochratoxin A producing moulds in food products. *Food Control* 29, 270–278.

MacDonald, S., Long, M., Gilbert, J. and Felgueiras, I. 2000. Liquid chromtographic method for determination of patulin in clear and cloudy apple juices and apple puree: Collaborative study. *Journal of AOAC International* 83, 1387–1394.

Magan, N. and Medina, A. 2016. Integrating gene expression, ecology and mycotoxin production by *Fusarium* and *Aspergillus* species in relation to interacting environmental factors. *World Mycotoxin Journal* 9, 673–684.

Mamgain, A., Roychowdhury, R. and Tah, J. 2013. *Alternaria* pathogenicity and its strategic controls. *Research Journal of Biological Sciences* 1, 1–9.

Mayer, Z., Bagnara, A., Färber, P. and Geisen, R. 2003. Quantification of the copy number of nor-1, a gene of the aflatoxin biosynthetic pathway by Real-time PCR, and its correlation to the CFU of *Aspergillus flavus* in foods. *International Journal of Food Microbiology* 82, 143–151.

McDonough, M.X., Campabadal, C.A., Mason, L.J., Maier, D.E., Denvir, A. and Woloshuk, C. 2011. Ozone application in a modified screw conveyor to treat grain for insect pests, fungal contaminants, and mycotoxins. *Journal of Stored Products Research* 47, 249–254.

McKeague, M., Giamberardino, A. and DeRosa, M.C. 2011. Advances in aptamer-based biosensors for food safety. pp. 17–42. *In*: V. Somerset (ed.). *Environmental Biosensors*. InTech, Rijeka, Croatia.

Medina, Á., Rodríguez, A. and Magan, N. 2015. Climate change and mycotoxigenic fungi: Mpacts on mycotoxin production. *Current Opinion in Food Science* 5, 99–104.

Meneely, J.P. and Elliott, C.T. 2014. Rapid surface plasmon resonance immunoassays for the determination of mycotoxins in cereals and cereal-based food products. *World Mycotoxin Journal* 7, 491–505.

Moake, M., Padilla-Zakour, O.I. and Worobo, R.W. 2005. Comprehensive review of patulin control methods in foods. *Comprehensive Reviews in Food Science and Food Safety* 4, 8–21.

Moss, M.O. and Long, M.T. 2002. Fate of patulin in the presence of the yeast. *Saccharomyces Cerevisiae. Food Additives & Contaminants* 19, 387–399.
Ngundi, M.M., Shriver-Lake, L.C., Moore, M.H., Lassman, M.E., Ligler, F.S. and Taitt, C.R. 2005. Array biosensor for detection of ochratoxin A in cereals and beverages. *Analytical Chemistry* 77, 248–254.
Nguyen, P.A., Strub, C., Fontana, A. and Schorr-Galindo, S. 2017. Crop molds and mycotoxins: Alternative management using biocontrol. *Biological Control* 104, 10–27.
Nijveldt, R.J., van Nood, E., van Hoorn, D.E.C., Boelens, P.G., van Norren, K. and van Leeuwen, P.A.M. 2001. Flavonoids: A review of probable mechanisms of action and potential applications. *American Journal of Clinical Nutrition* 74, 418–425.
Notermans, S. and Heuvelman, C.J. 1985. Immunological detection of moulds in food by using the enzyme-linked immunosorbent assay (ELISA); preparation of antigens. *International Journal of Food Microbiology* 2, 247–258.
Ostry, V., Skarkova, J. and Ruprich, J. 2012. Densitometric high-performance thin-layer chromatography method for toxigenity testing of *Alternaria alternata* strains isolated from foodstuffs. *JPC – Journal of Planar Chromatography – Modern TLC* 25, 388–393.
Oviedo, M.S., Barros, G.G., Chulze, S. and Ramirez, M.L. 2012. Natural occurrence of alternariol and alternariol monomethyl ether in soya beans. *Mycotoxin Research* 28, 169–174.
Paterson, R.R.M., Archer, S., Kozakiewicz, Z., Lea, A., Locke, T. and O'Grady, E. 2000. A gene probe for the patulin metabolic pathway with potential for use in patulin and novel disease control. *Biocontrol Science and Technology* 10, 509–512.
Paterson, R.R.M. and Lima, N. 2010. How will climate change affect mycotoxins in food? *Food Research International* 43, 1902–1914.
Pavón, M.Á., González, I., Martín, R. and García, T. 2012b. A real-time reverse-transcriptase PCR technique for detection and quantification of viable *Alternaria* spp. in foodstuffs. *Food Control* 28, 286–294.
Pavón, M.Á., Luna, A., de la Cruz, S., González, I., Martín, R. and García, T. 2012a. PCR-based assay for the detection of *Alternaria* species and correlation with HPLC determination of altenuene, alternariol and alternariol monomethyl ether production in tomato products. *Food Control* 25, 45–52.
Pennacchio, A., Ruggiero, G., Staiano, M., Piccialli, G., Oliviero, G., Lewkowicz, A., Synak, A., Bojarski, P. and D'Auria, S. 2014. A surface plasmon resonance based biochip for the detection of patulin toxin. *Optical Materials* 10, 1670–1675.
Pennacchio, A., Varriale, A., Esposito, M.G., Staiano, M. and D'Auria, S. 2015. A near-infrared fluorescence assay method to detect patulin in food. *Analytical Biochemistry* 481, 55–59.
Peraica, M., Radicâ, B., Lucicâ, A. and Pavlovicâ, M. 1999. Toxic effects of mycotoxins in humans. *Bulletin of World Health Organization* 77, 754–766.
Phillips, C.A., Laird, K. and Allen, S.C. 2012. The use of Citri-V™® – An antimicrobial citrus essential oil vapour for the control of *Penicillium chrysogenum, Aspergillus niger* and *Alternaria alternata* in vitro and on food. *Food Research International* 47, 310–314.
Piemontese, L., Solfrizzo, M. and Visconti, A. 2005. Occurrence of patulin in conventional and organic fruit products in Italy and subsequent exposure assessment. *Food Additives & Contaminants* 22, 437–442.

Pitt, J.I., Hocking, A.D. and Glenn, D.R. 1983. An improved medium for the detection of *Aspergillus flavus* and *A. parasiticus*. *Journal of Applied Bacteriology* 54, 109–114.

Prakash, B., Mishra, P.K., Kedia, A. and Dubey, N.K. 2014. Antifungal, antiaflatoxin and antioxidant potential of chemically characterized *Boswellia carterii* Birdw essential oil and its in vivo practical applicability in preservation of *Piper nigrum* L. fruits. *LWT-Food Science and Technology* 56, 240–247.

Razzaghi-Abyaneh, M., Shams-Ghahfarokhi, M. and Chang, P.K. 2011. Aflatoxins: Mechanisms of inhibition by antagonistic plants and microorganisms. pp. 285–304. *In*: R.G. Guevara-Gonzalez (ed.). *Aflatoxins: Biochemistry and Molecular Biology*. InTech, Rijeka, Croatia.

Repetto, M.G. and Llesuy, S.F. 2002. Antioxidant properties of natural compounds used in popular medicine for gastric ulcers. *Brazilian Journal of Medical and Biological Research* 35, 523–534.

Reverberi, M., Ricelli, A., Zjalic, S., Fabbri, A.A. and Fanelli, C. 2010. Natural functions of mycotoxins and control of their biosynthesis in fungi. *Applied Microbiology and Biotechnology* 87, 99–911.

Reverberi, M., Zjalic, S., Ricelli, A., Punelli, F., Camera, E., Fabbri, C., Picardo, M., Fanelli, C. and Fabbri, A.A. 2008. Modulation of antioxidant defense in *Aspergillus parasiticus* is involved in aflatoxin biosynthesis: A role for the Ap*yapA* gene. *Eukariotic Cell* 7, 988–1000.

Ricelli, A., Baruzzi, F., Solfrizzo, M., Morea, M. and Fanizzi, F.P. 2007. Biotransformation of patulin by *Gluconobacter oxydans*. *Applied and Environmental Microbiology* 73, 785–792.

Rodríguez, A., Luque, M.I., Andrade, M.J., Rodríguez, M., Asensio, M.A. and Córdoba, J.J. 2011. Development of real-time PCR methods to quantify patulin-producing molds in food products. *Food Microbiology* 28, 1190–1199.

Saha, D., Fetzner, R., Burkhardt, B., Podlech, J., Metzler, M., Dang, H., Lawrence, C. and Fischer, R. 2012. Identification of a polyketide synthase required for alternariol (AOH) and alternariol-9-methyl ether (AME) formation in *Alternaria alternata*. *PLoS ONE* 7, e40564.

Sanzani, S.M., De Girolamo, A., Schena, L., Solfrizzo, M., Ippolito, A. and Visconti, A. 2009a. Control of *Penicillium expansum* and patulin accumulation on apples by quercetin and umbelliferone. *European Food Research and Technology* 228, 381–389.

Sanzani, S.M., Li Destri Nicosia, M.G., Faedda, R., Cacciola, S.O. and Schena, L. 2014. Use of quantitative PCR detection methods to study biocontrol agents and phytopathogenic fungi and oomycetes in environmental samples. *Journal of Phytopathology* 162, 1–13.

Sanzani, S.M., Miazzi, M.M., Di Rienzo, V., Fanelli, V., Gambacorta, G., Taurino, M. R. and Montemurro, C. 2016a. A rapid assay to detect toxigenic *Penicillium* spp. contamination in wine and musts. *Toxins* 8, 235.

Sanzani, S.M., Montemurro, C., Di Rienzo, V., Solfrizzo, M. and Ippolito, A. 2013. Genetic structure and natural variation associated with host of origin in *Penicillium expansum* strains causing blue mould. *International Journal of Food Microbiology* 165, 111–120.

Sanzani, S.M., Reverberi, M., Fanelli, C., and Ippolito, A. 2015. Detection of ochratoxin A using molecular beacons and real-time PCR thermal cycler. *Toxins* 7, 812-820.

Sanzani, S.M., Reverberi, M. and Geisen, R. 2016b. Mycotoxins in harvested fruits and vegetables: Insights in producing fungi, biological role, conducive conditions, and tools to manage postharvest contamination. *Postharvest Biology and Technology* 122, 95–105.

Sanzani, S.M., Reverberi, M., Punelli, M., Ippolito, A. and Fanelli, C. 2012. Study on the role of patulin on pathogenicity and virulence of *Penicillium expansum*. *International Journal of Food Microbiology* 153, 323–331.

Sanzani, S.M., Schena, L., De Girolamo, A., Ippolito, A. and González-Candelas, L. 2010. Characterization of genes associated to induced resistance against *Penicillium expansum* in apple fruit treated with quercetin. *Postharvest Biology and Technology* 56, 1–11.

Sanzani, S.M., Schena, L., De Girolamo, A., Nigro, F. and Ippolito, A. 2009b. Effect of quercetin and umbelliferone on the transcript level of *Penicillium expansum* genes involved in patulin biosynthesis. *European Journal of Plant Pathology* 125, 223–233.

Sargent, S.A., Ritenour, M.A., Brecht, J.K. and Bartz, J.A. 2007. Handling, Cooling and Sanitation Techniques for Maintaining Postharvest Quality. University of Florida, Institute of Food and Agriculture Sciences, Florida Cooperative Extension Service, HS719. Available at: www.ksre.k-state.edu/foodsafety/produce/gaps/docs/UFsanitationguidelines.pdf

Sarubbi, F., Formisano, G., Auriemma, G., Arrichiello, A. and Palomba, R. 2016. Patulin in homogenized fruits and tomato products. *Food Control* 59, 420–423.

Scherm, B., Orrù, M., Balmas, V., Spanu, F., Azara, E., Delogu, G., Hammond, T.M., Keller, N.P. and Migheli, Q. 2011. Altered trichothecene biosynthesis in TRI6-silenced transformants of *Fusarium culmorum* influences the severity of crown and foot rot on durum wheat seedlings. *Molecular Plant Pathology* 12, 759–771.

Schmidt-Heydt, M., Stoll, D. and Geisen, R. 2013. Fungicides effectively used for growth inhibition of several fungi could induce mycotoxin biosynthesis in toxigenic species. *International Journal of Food Microbiology* 166, 407–412.

Sempere, F. and Santamarina, M.P. 2007. *In vitro* biocontrol analysis of *Alternaria alternata* (Fr.) Keissler under different environmental conditions. *Mycopathologia* 163, 183–190.

Shukla, R., Kumar, A., Singh, P. and Dubey, N.K. 2009. Efficacy of *Lippia alba* (Mill.) N.E. Brown essential oil and its monoterpene aldehyde constituents against fungi isolated from some edible legume seeds and aflatoxin B1 production. *International Journal of Food Microbiology* 135, 65–170.

Spadaro, D. and Droby, S. 2016. Development of biocontrol products for postharvest diseases of fruit: The importance of elucidating the mechanisms of action of yeast antagonists. *Trends in Food Science & Technology* 47, 39–49.

Spadaro, D., Patharajan, S., Lorè, A., Garibaldi, A. and Gullino, M.L. 2012. Ochratoxigenic black species of Aspergilli in grape fruits of northern Italy identified by an improved PCR-RFLP procedure. *Toxins* 4, 42–54.

Spanjer, M.C., Scholten, J.M., Kastrup, S., Jörissen, U., Schatzki, T.F. and Toyofuku, N. 2006. Sample comminution for mycotoxin analysis: Dry milling or slurry mixing? *Food Additive and Contaminants* 23, 73–83.

Stanker, L.H., Merrill, P., Scotcher, M.C. and Cheng, L.W. 2008. Development and partial characterization of high-affinity monoclonal antibodies for botulinum toxin type A and their use in analysis of milk by sandwich ELISA. *Journal of Immunological Methods* 336, 1–8.

Susca, A., Proctor, R.H., Morelli, M., Haidukowski, M., Gallo, A., Logrieco, A.F. and Moretti, A. 2016. Variation in fumonisin and ochratoxin production associated with differences in biosynthetic gene content in *Aspergillus niger* and *A. welwitschiae* isolates from multiple crop and geographic origins. *Frontiers in Microbiology* 7, 1412.

Tannous, J., Atoui, A., El Khoury, A., Kantar, S., Chdid, N., Oswald, I.P., Puel, O. and Lteif, R. 2015. Development of a real-time PCR assay for *Penicillium expansum* quantification and patulin estimation in apples. *Food Microbiology* 50, 28–37.

Thakur, S. and Harsh, N.S.K. 2014. Phylloplane fungi as biocontrol agent against Alternaria leaf spot disease of (Akarkara). *Spilanthes Oleracea. Bioscience Discovery* 5, 139–144.

Tian, J., Ban, X., Zeng, H., Huang, B., He, J. and Wang, Y. 2011. *In vitro* and *in vivo* activity of essential oil from dill (*Anethum graveolens*) against fungal spoilage of cherry tomatoes. *Food Control* 22, 1992–1999.

Turner, N.W., Subrahmanyam, S. and Piletsky, S.A. 2009. Analytical methods for determination of mycotoxins: A review. *Analytica Chimica Acta* 632, 168–180.

Van der Fels-Klerx, H.J., Van Asselt, E.D., Raley, M., Poulsen, M., Korsgaard, H., Bredsdorff, L. and Nauta, M. 2018. Critical review of methods for risk ranking of food-related hazards, based on risks for human health. *Critical Reviews in Food Science and Nutrition* 58, 178–193.

Visagie, C.M., Houbraken, J., Frisvad, J.C., Hong, S.-B., Klaassen, C.H.W., Perrone, G., Seifert, K.A., Varga, J., Yaguchi, T. and Samson, R.A. 2014. Identification and nomenclature of the genus. *Penicillium. Studies in Mycology* 78, 343–371.

Wild, C.P. and Gong, Y.Y. 2010. Mycotoxins and human disease: A largely ignored global health issue. *Carcinogenesis* 31, 71–82.

Wisniewski, M., Droby, S., Norelli, J., Liu, J. and Schena, L. 2016. Alternative management technologies for postharvest disease control: The journey from simplicity to complexity. *Postharvest Biology and Technology* 122, 3–10.

Woudenberg, J.H.C., Seidl, M.F., Groenewald, J.Z., de Vries, M., Stielow, J.B., Thomma, B.P.H.J. and Crous, P.W. 2015. *Alternaria* section *Alternaria*: Species, formae speciales or pathotypes? *Studies in Mycology* 82, 1–21.

Yang, C., Lates, V., Prieto-Simón, B., Marty, J.L. and Yang, X. 2012. Aptamer-DNAzyme hairpins for biosensing of Ochratoxin A. *Biosensors & Bioelectronics* 32, 208–212.

Zaccaria, M., Ludovici, M., Sanzani, S.M., Ippolito, A., Cigliano, R.A., Sanseverino, W., Scarpari, M., Scala, V., Fanelli, C. and Reverberi, M. 2015. Menadione-induced oxidative stress re-shapes the oxylipin profile of *Aspergillus flavus* and its lifestyle. *Toxins* 7, 4315–4329.

Zachariasova, M., Dzuman, Z., Veprikova, Z., Hajkova, K., Jiru, M., Vaclavikova, M., Pospichalova, M., Florian, M. and Hajslova, J. 2014. Occurrence of multiple mycotoxins in European feeding stuffs, assessment of dietary intake by farm animals. *Animal Feed Science and Technology* 193, 124–140.

Section III

NOVEL TECHNOLOGIES TO CONTROL POSTHARVEST DECAY OF FRUITS AND VEGETABLES

Chapter 14

Biocontrol of Postharvest Diseases with Antagonistic Microorganisms

Samir Droby

Department of Postharvest Science, Institute of Postharvest and Food Sciences, Agricultural Research Organization (ARO), the Volcani Center, Rishon LeZion, Israel

Michael Wisniewski

USDA-ARS, Appalachian Fruit Research Station, Kearneysville, USA

Neus Teixidó

Institute of Agrifood Research and Technology (IRTA), XaRTA-Postharvest, Edifici Fruitcentre, Parc Científic i Tecnològic Agroalimentari de Lleida, Parc de Gardeny, Lleida, Catalonia

Davide Spadaro

Department of Agricultural, Forestry and Food Sciences (DISAFA) and AGROINNOVA Centre of Competence for the Innovation in the Agroenvironmental Sector, University of Torino, Largo Braccini, Grugliasco, Torino, Italy

M. Haïssam Jijakli

Integrated and Urban Plant Pathology Laboratory, Gembloux Agro-Bio Tech, ULg, Passage des Déportés, Gembloux, Belgium

Abbreviations

a_w	Water activity
BTH	Benzo-thiadiazole-7-carbothioic acid S-methyl ester
CA	Controlled atmosphere
EU	European Union
GR	Granules
GRAS	Generally recognized as safe
IAA	Indole-3-acetic acid
MA	Modified atmosphere
MeJA	Methyl jasmonate
PCR	Polymerase chain reaction
RH	Relative humidity
ROS	Reactive oxygen species
SCAR	Sequence characterized amplified region
UV	Ultraviolet
VOCs	Volatile organic compounds

1 Introduction: Driving Forces for Alternative Management Strategies

Sustainable agricultural production systems, organic agriculture, regional food production, home gardening, public aversion to genetically modified crops, and environmental stewardship are all topics that over the last 20–30 yr have fueled a greater interest in finding alternative approaches to plant disease management that do not rely on the use of synthetic, chemical pesticides. On the basis of these public concerns and legitimate health and safety concerns, government regulatory agencies have become more restrictive about the materials and products that can be used in agricultural production. This has been especially true for the use of postharvest fungicides and these restrictions have had a great impact on the export and shipment of harvested produce to foreign markets. At present, the European Union (EU) is taking the greatest initiative to reduce the allowed use and/or residues of synthetic pesticides on produce sold to consumers (Wisniewski et al., 2016).

Many large supermarket chains and wholesale fruit suppliers have also started to set their own standards for chemical residues and the number of active ingredients allowed to be present on harvested commodities, or even used during the growing season. In addition, safety concerns about mycotoxins and foodborne pathogens have also increased the need to find viable alternatives (Sanzani et al., 2016). Restrictions on the use of some preharvest fungicides are also resulting in increased inoculum levels of postharvest pathogens and the number of latent infections present at harvest.

The search to identify alternative approaches to postharvest disease management must be viewed in relation to a complex regulatory environment, the need to address disease problems in a wide array of commodities and conditions, industry and consumer acceptance, and, last but not least, commercial viability. One of the major initiatives that has been conducted over the past several decades is the identification and use of microbial antagonists applied to harvested produce as biocontrol agents to inhibit and manage a variety of postharvest pathogens. Bacterial and yeast antagonists have been isolated from fruit surfaces and demonstrated to exhibit high levels of efficacy in lab and pilot tests conducted under semi-commercial conditions. In addition, biocontrol agents, mainly yeast, have been isolated from unique environments, such as salt brines, and cold geographical areas (Antarctica) with the idea that their ability to grow in these environments will provide an advantage to their use in the postharvest arena, where various abiotic stresses are prevalent. These include cold storage, exposure to high or low pH conditions, and exposure to other postharvest chemicals. Several products, based on the use of these bacterial and yeast antagonists, have been developed and commercialized. In fact, many labs throughout the world have identified unique antagonists or specific strains of previously identified antagonists as interest in this research has expanded and grown. The present chapter presents an overview of the various aspects of using microbial antagonists as postharvest biocontrol agents, including the procedures used to identify potential antagonists, what is known about the various mechanisms that contribute to biocontrol activity, the various parameters associated with the commercialization and use of postharvest biocontrol products, and integrating the use of antagonists with other alternative approaches. It should be noted that many studies and reviews on this topic have been previously published (Wilson and Wisniewski, 1989; Droby et al., 2009, 2016; Wisniewski et al., 2016).

The present contribution is only an overview and cannot adequately give credit to the many scientists who have contributed to this field of research and whose studies have shaped its direction and impact.

2 Discovery of Biocontrol Agents: Isolation, Screening, and Identification of Antagonists

The use of fungal and bacterial species to either modify or preserve food has been an integral part of human civilization. To extend this concept into a scientific approach for managing postharvest decay is perhaps very logical rather than farfetched. Regarding postharvest biocontrol, there has been an underlying hypothesis that there are species of microbes present on fruit and vegetable surfaces, as well as on harvested grain, that are antagonistic to decay fungi. By isolating these species, and reapplying them to the surface of harvested commodities in high numbers, one could extend the shelf life of the commodity without the use of a synthetic chemical.

Most postharvest pathogens are considered necrotrophic and largely dependent on the presence of surface injuries for the infection of the produce. The nature of surface wounds may vary from cracks in the cuticle and underlying tissues created during fruit growth, maturation and ripening, as well as cut surfaces in the stem end to mechanical damage inflicted during harvesting and subsequent handling of the fruit.

Protecting surface wounds has been the main theme of the research on biocontrol of postharvest diseases in the past three decades. The focus is on isolating epiphytic microorganisms that can rapidly colonize and grow in wounded tissue. Typically, pathogen spores germinate very rapidly (within 24 hr) and colonize wounds that are rich in sugars and other nutrients. Therefore, it is necessary to interfere with spore germination and/or germ tube growth in a rapid time frame to prevent or inhibit infections. To protect wounds against pathogen colonization, several features of an ideal antagonist were defined by Wilson and Wisniewski (1989). Rapid growth and colonization of fresh wounds by the biocontrol agent was one of the main features indicated. In addition, the antagonist must be genetically stable, effective at low concentrations, able to survive adverse environmental conditions, be effective against a wide range of pathogens, not fastidious in its nutrient requirements, amendable to production in inexpensive growth media, non-detrimental to human health, compatible with commercial processing practices, and possess a formulation that provides a long shelf life.

In an early attempt to develop postharvest biocontrol agents, Pusey and Wilson (1984) reported the ability of a strain (B-3) of *Bacillus subtilis* to control brown rot on peach caused by *Monilinia fructicola* (G. Winter) Honey. It was subsequently discovered that the antibiotic iturin, produced by the bacterium, was the main factor responsible for the control of brown rot. The use of an antibiotic-producing microorganism on food raised significant concerns. To avoid the isolation of antibiotic-producing antagonists, Wilson et al. (1993) suggested a strategy to utilize fruit wounds to screen organisms from unidentified microbial populations on fruit surfaces for potential yeast antagonists against postharvest rots. This strategy for the rapid selection of a number of potential antagonists for the control of postharvest diseases of fruit with a minimal expenditure of time and expense has been subsequently used in many postharvest biocontrol programs throughout the world. Rather than in vitro screening of organisms in Petri plates, which favors the identification of antibiotic-producing organisms, this method involves placing washing fluids obtained from the surface of fruit into fruit wounds that are subsequently inoculated with a rot pathogen. Organisms are then isolated from wounds that do not develop infections. These are plated out and isolated in pure culture. Pure cultures of potential antagonists are then screened individually in fruit wounds to assess their potential as a biocontrol agent. Organisms with antagonistic activity to a broad spectrum of postharvest pathogenic fungi on a wide variety of crops are identified using morphological and physiological characterization and/or by DNA sequencing of conserved regions of ribosomal DNA (Kurtzman and Droby, 2001). This procedure has led to the identification of numerous yeast antagonists. A comprehensive list of identified antagonists was published by Sharma et al. (2009) and Wisniewski et al. (2016).

Since the method of screening has a major impact on the type and properties of the antagonist that are identified, perhaps it is important to evaluate the consequence of the methods for screening that are presently being utilized and appraise whether or not they can be improved. As indicated, present methods largely favor the selection of microbial antagonists that have protective activity in small (3–5 mm deep) puncture wounds rather than curative activity against established infections, or demonstrated activity in a wide array of wounds (bruises, scrapes, broken stems, broken epidermal hairs, etc.). For postharvest biocontrol agents to work under commercial conditions, screening methods need to better reflect the "real world" when potential antagonists are being evaluated.

Another shortcoming of this strategy is that it favors the selection of antagonists that are generally fast growers with the ability to colonize a specific niche (surface wounds) rich in nutrients, and appear to have little effect on latent infections (Droby et al., 1989; El-Ghaouth et al., 2000b). This may partially explain the lack of correlation between laboratory tests with host/parasite systems and the performance of biocontrol agents/products under more varied commercial conditions (Droby et al., 1993, 2000; Wisniewski et al., 2001).

Droby et al. (2009) raised several reservations about the relevance of the existing paradigm for identifying antagonists that are expected to perform under "real world" situations where a wide range of wounds exist that serve as infection courts. In the current postharvest biocontrol paradigm, it is expected that a single antagonist isolated from one commodity will be effective on other commodities that vary in their genetic background, physiology, postharvest handling, and pathogen susceptibility. Perhaps this expectation or paradigm is inappropriate, given the knowledge of microbial ecology and plant microbiota that has accumulated, especially in the last few years, fostered by metagenomic approaches.

The use of antagonistic yeasts has been especially emphasized since the production of toxic secondary metabolites (antibiotics) is generally not involved in their inhibitory activity (Wisniewski and Wilson, 1992). The original source (fruit surfaces) for the isolation of yeast antagonists (Wilson and Wisniewski, 1989; Wilson et al., 1993) has expanded to other environments, such as sea water (Wang et al., 2008; Hernández-Montiel et al., 2010) and Antarctic soil samples (Vero et al., 2013). The main intent in exploring these new sources has been to identify yeast that can thrive in stressful environments and discover isolates with novel modes of action.

3 Mechanisms of Action of Antagonistic Microorganisms

Understanding the mechanism of action of postharvest biocontrol agents is essential for biofungicide product development and is relevant for marketing purposes, as it permits improving biocontrol performance through the development of appropriate formulations and methods of application, and facilitates the registration process registration (Spadaro et al., 2010a). Biocontrol agents show four major mechanisms of action: (i) antibiosis; (ii) competition for limiting nutrients and space; (iii) parasitism or direct interaction with the pathogen; and (iv) induced host resistance (Janisiewicz and Korsten, 2002; Liu et al., 2013; Spadaro and Droby, 2016). Often, more than one mechanism is implicated. For example, strains of the two sister species *Metschnikowia pulcherrima* and *M. fructicola* can compete for nutrients, release hydrolases, induce host resistance, and produce and induce reactive oxygen species (ROS), but can also biodegrade mycotoxins produced by the fungal pathogens (Spadaro et al., 2008, 2013; Macarisin et al., 2010; Hershkovitz et al., 2012, 2013). A successful biocontrol agent is equipped with several attributes, which work in concert. Some features of successful antagonists could be adhesion, release of extracellular hydrolases, induction of resistance, regulation of population density, biofilm formation, and secretion of antimicrobial substances. In addition, oxidative stress resistance and the ability to produce ROS are also associated with biocontrol efficacy.

The performance of a biocontrol agent can be seen as the result of complex mutual interactions between all the biotic and abiotic components of the system (Droby et al., 2009). Most studies about the mechanism of action of antagonists focused on a single mechanism, but a system approach should be rather employed to characterize the network of interactions involved in postharvest biocontrol (Droby et al., 2016).

One of the main modes of action of antagonists is competition for nutrients and space, a niche overlap, resulting from the simultaneous demand for the same resource by the pathogen and the antagonist. In competition for space, yeasts and bacteria have the advantage of rapid multiplication and colonization (Droby et al., 1989; Spadaro et al., 2010b), and they can form an extracellular polysaccharide capsule that promotes adhesion to fruit surface. On the inner surface of wounds, the ability to form biofilms, where microorganisms are enclosed in a hydrated matrix, favors competition (Lutz et al., 2013). Biofilm-forming yeasts were effective against *Penicillium expansum* Link on apple, but only yeast cells collected from the biofilm phase were able to efficiently colonize the inner surface of wounds (Ianiri et al., 2013).

In fruit wounds, competition for nutrients is particularly relevant for iron. Iron is essential for the fungal growth and pathogenesis, and iron sequestration by non-pathogenic microbes can be exploited in postharvest biocontrol systems. Under iron starvation conditions, fungi have lower catalase activity, and consequently, lower resistance to ROS. Rhodotorulic acid, a siderophore produced by *Rhodotorula glutinis*, improved the control of blue mold of apple (Calvente et al., 1999). *Metschnikowia pulcherrima* and *M. fructicola* are able to produce the red pigment pulcherrimin, formed from pulcherriminic acid and ferric ions, which is involved in the control of apple diseases (Saravanakumar et al., 2008; Spadaro and Droby, 2016).

Determining the role, components, and factors involved in competition for nutrients and space is crucial for enhancing biocontrol efficacy during upscale production and formulation. The supplementation of a limiting factor or essential nutrient for improved growth of the antagonist may significantly improve its effectiveness.

Antibiosis is the inhibition or destruction of a microorganism by diffusible or volatile antibiotics produced by another microorganism. Certain microorganisms start producing antibiotics when the substrate availability decreases, preventing other microorganisms from using the remaining quantity of substrate. Bacteriocins, antibacterial proteins produced by bacteria, induce the formation of pores in the membrane of target cells and deplete transmembrane potential, resulting in cell material leakage. The antagonist *Bacillus subtilis* produces a range of antimicrobial cyclic lipopeptides, including iturins, fengycins, and surfactins (Ongena and Jacques, 2008; Yánez-Mendizábal et al., 2012). *Aureobasidium pullulans* can produce aurebasidin, a cyclic depsipeptide, with antifungal and antibiotic properties. The main concern, related to the use of antibiotic-producing microorganisms in food products, is the risk of development of fruit microorganisms resistant to these compounds and the possible transfer of resistance to human pathogens.

Besides diffusible compounds, many fungal species are known to produce volatile organic compounds (VOCs) with antifungal properties, which can be used for biological fumigation, or biofumigation. *Muscodor albus* is an endophytic fungus that produces a mixture of VOCs with antimicrobial activity for the control of postharvest diseases of fruit (Mercier and Smilanick, 2005). VOC-producing microorganisms open new possibilities for postharvest biocontrol, as biofumigation does not require physical contact with the fruit to be treated, and reduces labor relative to spraying.

Antagonists and pathogens can also interact through direct parasitism, resulting in direct destruction or lysis of fungal mycelia and propagules composed of complex polymers of glucans, mannoproteins, and chitin. Mycoparasites use cell wall-degrading enzymes, such as β-1,3-glucanase, chitinase, and proteases, to dissolve cell walls and penetrate fungal cells. *Candida saitoana* cells, when cultivated together with mycelium of the pathogen *Botrytis cinerea* Pers., attached to fungal hyphae, inducing damage to the cell wall (El-Ghaouth et al., 1998). Two exo-ß-1,3-glucanase genes of *Pichia anomala* were separately and sequentially disrupted, resulting in reduced efficacy of the mutant strains against *B. cinerea* (Friel et al., 2007). *Metschnikowia fructicola* chitinase overexpressed in *Pichia pastoris* significantly controlled *Monilinia laxa* (Aderh. & Ruhland) Honey and *M. fructicola* in vitro and on peaches (Banani et al., 2015). Similarly, an alkaline serine protease of *A. pullulans*, expressed in *Escherichia coli* (Zhang et al., 2012) and *P. pastoris* (Banani et al., 2014), controlled different postharvest pathogens on apples.

Indirect antagonism implies the induction of resistance in the fruit tissue by the presence of beneficial microorganisms. Some antagonists can interact with host tissue, increasing cicatrisation and lignification of cell walls. Biocontrol agents interact with wounded tissue, inducing β-1,3-glucanase, chitinase, and peroxidase activity, the accumulation of phytoalexins, and the formation of structural barriers. Application of *Candida oleophila* induced phytoalexin production in grapefruit, ethylene biosynthesis, and the accumulation of chitinase and β-1,3-glucanase enzymes in grapefruit (Droby et al., 2002). When *Cryptococcus laurentii* was applied to jujube, β-1,3-glucanase activity increased and the expression of the *Glu-1* gene was highly induced in a defense response against postharvest pathogens (Tian et al., 2007). All of the results on the induction of resistance following antagonist treatment are correlative. Direct evidence for the ability of induced substances to inhibit pathogen infection and development, however, has not yet been established.

Oxidative stress plays a crucial role in biocontrol and may be involved in signaling pathways associated with the activation of fruit resistance response (Chan and Tian, 2005). The ability to survive and proliferate in wounded host tissues and to tolerate oxidative stress is essential for postharvest antagonists (Castoria et al., 2003). Fruit wounding is associated with the accumulation of ROS, which can affect host response, pathogen virulence, and yeast efficacy. Macarisin et al. (2010) demonstrated that yeast antagonists have the ability to produce relatively high amounts of superoxide anions. The ability of yeast antagonists to self-generate and possibly stimulate an oxidative response in the host tissue could be a major biocontrol feature. The application of *M. fructicola* and *C. oleophila* into fruit wounds caused an increase in H_2O_2 accumulation in fruit tissue (Macarisin et al., 2010). In oranges, *Penicillium digitatum* (Pers.) Sacc. suppressed H_2O_2 production, as well as superoxide dismutase and catalase activities, in response to infection (Torres et al., 2011). In contrast, *Pantoea agglomerans* triggered H_2O_2 production and enzymatic activity, which probably serves to prevent citrus fruit from being infected by subsequent challenges by *P. digitatum* (Macarisin et al., 2007).

Information on the mechanisms of action of biocontrol agents is still incomplete because of the difficulties involved in the study of the complex interactions between host, pathogen, antagonist, and others microorganisms present at the site of interaction. Cost-efficient, high-throughput DNA/RNA and proteomic technologies, along with bioinformatics, are offering new opportunities to obtain deeper insights into the mechanisms and interactions that have already been demonstrated (Hershkovitz et al., 2013; Kwasiborski et al., 2014). Developments in genomics, transcriptomics, proteomics, metagenomics, as well as comparative and functional genomics, will increase the ability to determine changes in the physiological status of antagonists, and the effect of environmental stress on their physiology and activity (Massart et al., 2015; Sui et al., 2015). Changes in the level of expression of "biocontrol genes" during mass production, formulation, and storage, or in response to exposure and contact with host plant tissues after application will be the object of future investigations about the mode of action of biocontrol agents.

4 Ecological Fitness of Antagonistic Microorganisms

Ecological studies are focused on the influence of environmental parameters on the growth and biocontrol properties of a biocontrol agent (Schena et al., 2000; Massart and Jijakli, 2014). The development of biocontrol agent monitoring tools is a prerequisite to any ecological fitness study. These tools must insure the specific detection of the targeted biocontrol agent strain on different environments, and also its quantification.

Colonization of wounded or intact surfaces of commodities by the biocontrol agents and their dispersal in the environment can be assessed using a plating method. This method has been used successfully and contributed to a better understanding of the population dynamics of antagonistic microorganisms on fruits (Teixidó et al., 1999; Karabulut et al., 2004; Calvo-Garrido et al., 2014). Nevertheless, the plating method has significant shortcomings that have led to the development of more specific and more rapid identification methods mainly based on DNA amplification. The yeast *Pichia membranifaciens* strain FY-101 is an antagonist against *B. cinerea*, a pathogen capable of infecting several fruit crops, including grapevine, apple, and peaches. Polymerase chain reaction (PCR) coupled with restriction fragment length polymorphism analysis of PCR products has been useful for the identification of *P. membranifaciens* (Masih et al., 2001). A specific fragment of DNA (L4) of *A. pullulans*, an antagonist that is effective against postharvest rots of sweet cherries and table grapes, was cloned, sequenced, and employed to design two sequence characterized amplified region (SCAR) primers and a 242-bp riboprobe. Both SCAR primers and the 242-bp digoxigenin-labeled riboprobe were highly specific for the L-47 strain of *A. pullulans* (Schena et al., 2002).

In spite of their specificity, these molecular tools are not able to assess the population of a biocontrol agent in a specific niche or environment. For this reason, they have been combined with dilution plating methods. For example, SCAR markers based on random amplification of polymorphic DNA fragments combined with the use of a semi-selective medium provided a valuable monitoring tool for two strains of *A. pullulans* (Ach 1–1 and 1113–5), effective against *B. cinerea* and *P. expansum* on apples (El Hamouchi et al., 2008). Similarly, combining a plating technique on a semi-selective medium, with direct strain K-SCAR amplification without DNA extraction, was used to quantify the colonies of *P. anomala* strain K on apples 24 hr after treatment. A decrease in population density was observed starting after 1 wk after application under cold storage conditions (De Clercq et al., 2003). A combination of a SCAR marker with the use of a semi-selective medium was also successfully used to quantify a bacterium, *P. agglomerans* CPA-2, which has antagonistic properties against postharvest fungal diseases of citrus (Nunes et al., 2008). Propidium monoazide, combined with quantitative PCR, was used to successfully track the presence and persistence of *P. agglomerans* CPA-2 in citrus in a field environment (Soto-Muñoz et al., 2015b) and in the packinghouse (Soto-Muñoz et al., 2015a).

Techniques used to identify and quantify biocontrol agents, based solely on molecular tools, were also developed for an antagonistic yeast. The sensitivity threshold of quantitative competitive enzyme-linked oligosorbent assay permitted the detection of less than 10^3 cells of *P. anomala* per apple. The technique was found to be very specific and sufficiently accurate to reliably monitor strain K populations of *P. anomala* (Pujol et al., 2004). A real-time PCR assay using a 3′-minor groove binding probe was developed for the specific detection and monitoring of *C. oleophila* (strain O), a biocontrol agent effective against *B. cinerea* and *P. expansum*, on harvested apples. The calculated population densities using this technique under applied conditions were similar to those obtained by plating on semi-selective media.

Biocontrol of postharvest diseases can be accomplished by postharvest but also by preharvest application of biocontrol agents (Jijakli, 2011). Ecological studies must take into account both pre- and postharvest treatment possibilities and their impact on biocontrol agent survival under different environmental conditions. Ecological studies should be conducted to determine if the biocontrol agent is adapted to packinghouse and storage conditions. An antagonist is applied shortly before harvest when biocontrol agents are used before harvest. This allows the biocontrol agent to colonize the fruit surface and any wounds inflicted during harvesting before they are colonized by wound pathogens. Researchers have highlighted the practical problems associated with promoting the effective establishment of antagonists in a natural environment. Antagonist survival can limit of the benefits of using biocontrol under field conditions and the commercialization of biocontrol agents. Fluctuations in abiotic factors, such as temperature, water availability, pH, relative humidity (RH), and ultraviolet (UV) radiation, can have an impact on the biological proprieties of biocontrol agents.

Exposure to stressful temperatures is inevitable when biocontrol agents are applied to plant surfaces before harvest. High temperatures can markedly reduce the viability of biocontrol agents, especially when preharvest applications are administered under field conditions (Teixidó et al., 1999). Hence, tolerance to high or low temperatures is regarded as a desirable feature of an efficient biocontrol agent (Vero et al., 2013; Sangorrín et al., 2014). Several attributes of yeast species make them suitable for use as biocontrol agents, including their ability to withstand osmotic stress. Many antagonistic yeast species can grow under conditions of low water activity (a_w) (Fredlund et al., 2002). This property partially explains the relatively high number of yeasts offering effective biocontrol efficacy on postharvest commodities in comparison to bacteria. It was suggested that oxidative stress resistance could also be an important component of the fitness and suitability of a yeast species for their use as a postharvest biocontrol agent. Indeed, wounded host tissues can be associated with the accumulation of ROS (Torres et al., 2003). The response of *M. fructicola*, *C. oleophila*, and *Cystofilobasidium infirmominiatum* PL1 to oxidative stress was examined by Liu et al. (2011a, 2011b, 2012). They demonstrated that tolerance to relatively high concentrations of ROS is essential for biocontrol efficacy. In addition to oxidative stress, antagonistic yeasts also need to survive at low oxygen levels associated with controlled or modified atmosphere storage (Petersson et al., 1999) and be able to adapt to changes in pH in the macro- and microenvironment (Jijakli, 2011). Both parameters, however, have not received much attention until recently. UV-B radiation (280–320 nm) is also a major environmental factor that can affect the viability of biocontrol agents when they are applied under field conditions. The effect of UV-B radiation on *P. anomala* strain K and *C. oleophila* strain O was evaluated in vitro and on apple fruit (Lahlali et al., 2009, 2011b). Both strains were relatively sensitive to such UV, relative to other biocontrol agents, including bacteria and fungi (Lahlali et al., 2011a).

More recent studies examined the effect of the simultaneous application of several stresses on biocontrol agents. Gotor-Vila et al. (2017c) reported that *Bacillus amyloliquefaciens* strain CPA-8, an effective bacterial biocontrol agent to control brown rot in stone fruit, had a wide tolerance to different pH–temperature levels and a_w–temperature profiles. They suggested that this should enable this strain to grow actively under a wide range of environmental conditions. Ecological studies on biocontrol agents should also be conducted on the targeted pathogens to allow a better understanding of the complex relationship between both microorganisms. In vitro and in vivo studies were undertaken to develop models predicting the combined effect of RH and temperature on two antagonistic strains of yeasts (*P. anomala* strain K and *C. oleophila* strain O) and one postharvest pathogen (*P. expansum*) of apple using the Box-Behnken method (Jijakli and Lhalali, 2015). Both antagonistic strains were sensitive to low RH. In contrast, the growth of *P. expansum* was limited at RH close to saturation. These results indicated that it is important to maintain a high RH under storage conditions, which has negative influence on *Penicillium* but not on the biocontrol strains K and O in storage room conditions.

5 Up-Scale Production and Formulation

The processes of formulation and production of biocontrol agents are critical aspects for the commercial applicability and success of microbial biocontrol products. Usually, these processes are conducted in association or directly by private industries under an agreement with the researcher or entity with rights to the biocontrol agent.

Biocontrol agents based on yeasts and bacteria can be mass-produced using a deep-tank liquid fermentation process (liquid production) or, in some cases, using semi-solid or solid state fermentation. Nutrient components of the growth medium and the growth conditions are critical to both, biomass and secondary metabolite production if the latter is required as part of the final formulation. Components of the growth medium should provide the necessary nutrients, be inexpensive, and readily available. The use of commercial products and by-products from other commercial industries is a good option to achieve cost and availability requirements because they are usually cheap and available. By-products, however, are not standardized and may contain impurities (Stanbury et al., 1995). Their composition may also vary with the season and place of origin. All these aspects have limited their use in industrial processes. Some of the most commonly used by-products are animal protein concentrates, lactoserum, molasses, and cereal by-products.

Optimization of the production process consists in defining the appropriate composition of the growth medium, selecting the best fermentation system, optimizing growth conditions (temperature, pH, aeration, agitation, initial inoculum, and culture duration), and evaluating the biocontrol efficacy of the final product. Some examples of biocontrol optimized mass production microorganisms are: yeasts, such as *Candida sake* CPA-1 using molasses (Abadias et al., 2003a), *Rhodotorula minuta* (Patiño-Vera et al., 2005), and *A. pullulans* (Mounir et al., 2007); bacteria, such as *P. agglomerans* CPA-2 (Costa et al., 2001), *P. agglomerans* PBC-1 (Manso et al., 2010), and *B. amyloliquefaciens* (Yánez-Mendizábal et al., 2012c); and finally, filamentous fungi, such as *Penicillium frequentans* 909 (De Cal et al., 2002) and *Epicoccum nigrum* using solid systems (Larena et al., 2004).

One of the bottlenecks in the production of biocontrol products is the formulation process because of the profound effect it has on efficacy and performance under large-scale commercial conditions. The final formulation should have a long shelf life, the ability to grow and survive after application, be effective in disease control, easy to handle, distribute and apply, and be cost-effective (Fravel et al., 1998). Formulations for biocontrol agents may be either liquid or solid. Liquid formulations include those that are oil-based, aqueous-based, polymer-based, or combinations thereof. These types of formulations require additional measures, other than the ability to store biocontrol cells in a solution with different additives, to stabilize survival during an extended period of time and enhance product application. The additives used are stabilizers, adherents, surfactants, colorants, additional nutrients, or other types of adjuvants. Oil-based products reduce the evaporation of droplets and allow for ultra-low-volume aerial application. Some examples of microorganisms formulated in liquid are *C. sake* CPA-1 (Abadias et al., 2003b; Torres et al., 2003), *C. laurentii* (Liu et al., 2009), and *P. anomala* (Melin et al., 2011). The main shortcoming of this kind of formulated products is that they usually need to be stored at low temperature.

Dry formulations, including wettable powders, water dispersible granules, soluble powders, and granules (GR), can be produced through different drying processes, such as freeze drying, spray drying, fluidized bed drying, and fluid-bed spray drying. The extra processing steps required in producing a dry formulation increases manufacturing cost, but reduces shipping costs due to reduced weight. Most dry formulations include an inert carrier, such as alginate, fine clay, peat, vermiculite, talc, cellulose derivatives (e.g., carboxymethyl cellulose), and other polymers, such as xanthan gum.

Compared to liquid formulations, dried products are more beneficial due to their extended storage capability, transportation convenience, and the ability to produce large amounts of dried product at a relatively low cost. Dried products, however, frequently have low cell viability because of the thermal and dehydration stress administered during the drying process. Nevertheless, microbial cell viability and biocontrol efficacy can be improved by the addition of certain stabilizing/protective substances (e.g., polymers, sugars, albumin, milk, salts, honey, polyols, or amino acids) to the formulation medium. Components of the formulation media have two main functions in preserving the viability of formulated cells: to provide a dry residue that acts as a receptor in the rehydration process and to protects the cells against biological injury during the drying process (Abadias et al., 2001; Melin et al., 2011). Some of these additives also help the microorganism to survive exposure to variable and frequently adverse environmental conditions, such as UV light, low water potential, nutrient limitation, extreme temperatures, and rainfall (Köhl and Fokkema, 1998; Cañamás et al., 2008).

A general recipe for formulating microbial biocontrol agents does not exist and different methods are established empirically for each organism or strain. Critical formulation requirements are determined by features of the organisms themselves and their prospective host environment. An overriding feature is the mode of action that dictates the formulator's ultimate objectives together with the environmental conditions in which the organism will be applied, as well as the function that the biocontrol agent is expected to perform (Burges and Jones, 1998).

5.1 Freeze Drying

Lyophilization is a drying system that has been employed for long time to dry microbial cells. Operationally, freeze drying can be defined as a controlled method of dehydrating liquid/semi-liquid products by vacuum desiccation. The liquid sample must be cooled until the final conversion of the solution into ice, the crystallization of the solutes, and the formation of an amorphous matrix comprising the non-crystallizing solutes are completed. The ice is then sublimated under vacuum-controlled conditions and the water from the amorphous matrix is evaporated. Finally, the moisture content of the product is desorbed (Adams, 2007).

While this technology is being widely used in the formulation of biocontrol agents, the process causes severe damage to microbial cells. To avoid negative effects on cell survival and viability (i.e., denaturation of proteins and decrease in cell viability of many cell types), cryoprotective agents are usually added to the formulation. Freeze drying is time-consuming and involves elevated costs in large-scale production systems (Strasser et al., 2009). Freeze drying has been used to formulate Biosave 11 LP (Jet Harvest Solutions) based on *Pseudomonas syringae*. Other examples of studies that used this drying technology are *P. agglomerans* CPA-2 (Costa et al., 2000), *C. sake* CPA-1 (Abadias et al., 2001), *C. laurentii* (Li and Tian, 2006; Navarta et al., 2014), *M. pulcherrima* (Spadaro et al., 2010a), *P. anomala* (Melin et al., 2011), and *Pseudomonas* spp. (Stephan et al., 2016).

5.2 Spray Drying

This process consists of transforming a product from a fluid to a solid in the form of powder by the removal of moisture from liquid droplets. Low-humidity, hot gas (drying gas/medium) is mixed with the dispersed droplets inside a chamber. The process involves the use of a rotary wheel/disc atomizer, pressure nozzle, or pneumatic-type atomizer. The moisture, in the form of a vapor, quickly evaporates from the suspended droplets due to heat and mass transfer processes. The separation of the dried particles from the drying gas, and their subsequent collection, takes place in external equipment, such as cyclones and/or bag-filter houses (Bhandari et al., 2008; Costa et al., 2015). The main disadvantage of spray drying is the extent of the destruction or injury to the microbial cells during the drying process due to the high temperatures used with this system. Survival of the biocontrol agents *P. agglomerans* CPA-2 (Costa et al., 2002a) and *C. sake* CPA-1 (Abadias et al., 2005) subjected to spray drying was found to be very low, with less than 8 and 10% of the cells surviving, respectively. Only microorganisms resistant to high temperatures, such as *Bacillus* spp., which are able to produce thermal-resistant endospores, can be mass-produced using this technique. One example of a biocontrol agent dried by spray drying is the bacterium *B. amyloliquefaciens* CPA-8 developed to control brown rot in stone fruit (Yánez-Mendizábal et al., 2012a, 2012b).

5.3 Fluid Bed Drying

This process was developed in the 1950s by the pharmaceutical industry. Other industries, including the agrochemical industry, have adopted this technology for producing products on a large scale. This technology is traditionally used for drying, granulation and coating of powders, and producing GR and spheres (Strasser, 2008). It involves extrusion of particles in which the product is included and then cut into small pellets. Filtered and optionally dehumidified inlet air enters the product through a perforated plate, fluidizing the product that is loaded or sucked into a granulator. The small particles are then carried by the airflow to a cylindrical expansion chamber after which they fall back into a conical product chamber. This cycle continues throughout the process, and the specifics are primarily determined by the variation of the product temperature and fluidization airflow (Strasser, 2008).

Fluidized bed drying is extensively used to dry heat-sensitive biological material as low inlet air temperatures are required. Several biocontrol agents have been successfully formulated with this system, including *E. nigrum* (Larena et al., 2003a), *Penicillium oxalicum* (Larena et al., 2003b), *P. frequentans* (Guijarro et al., 2006), *A. pullulans* (Mounir et al., 2007), and *C. sake* (Carbó et al., 2017a).

5.4 Fluid-Bed Spray Drying

This process combines spray drying with fluid bed drying. Both methods atomize liquid droplets into a hot air chamber to facilitate drying. By utilizing a large air volume, liquid can be sprayed into the chamber at an inlet air temperature much lower than what is used in traditional drying systems. The equipment allows one to obtain larger quantities of dry product with better dispersability and flowability and a narrow particle-size distribution (Srivastava and Mishra, 2010). Furthermore, compared to conventional fluidized-bed systems, this technology does not require previous extrusion and pelletization of the sample, thus offering lower operating costs and shorter processing times (Santivarangkna et al., 2007; Strasser et al., 2009). The suitability of fluid-bed spray drying makes this technology an attractive alternative to freeze, spray, and traditional fluidized-bed drying systems. This technology is relatively new and is largely used in the pharmaceutical industry. Recently, however, it has been used in the production of biocontrol products with very good results (Carbó et al., 2017b; Gotor-Vila et al., 2017a, 2017b).

Another issue involved in the commercial development of biocontrol products is their shelf life. A biofungicide should have a shelf-life period of at least 6 mon and preferably 2 yr (Usall et al., 2016). The biocontrol agent must be packaged and stored in a way that maintains a viable cell count and biocontrol efficacy. This may require the careful selection of packaging materials to control gas exchange, prevent the loss or gain of moisture, and avoid contamination of the product (Powell, 1992). Few studies have addressed the effect of packaging on a formulated biocontrol product. Some examples are studies on *P. agglomerans* CPA-2 (Costa et al., 2002b; Torres et al., 2014).

6 Enhancement of the Survival and Efficacy of Antagonistic Microorganisms

One strategy to enhance the efficacy of biocontrol agents involves increasing their tolerance to abiotic and biotic stresses that occur during production, application, and use. As previously indicated, the biocontrol agent, during fermentation and formulation, is exposed to various stresses, such as high and low temperature, desiccation, and oxidative stress. Teixidó et al. (1998a) reported that increasing total polyols and sugars in *C. sake* cells provided increased resistance to water stress and consequently better survival under field conditions (Teixidó et al., 1998b). The increase in stress tolerance was obtained by the addition of glycerol, glucose, or trehalose to the growth medium. *Cryptococcus laurentii* cultivated in a trehalose-containing medium improved its viability and biocontrol efficacy at low temperature (Li and Tian, 2006). Exposure of *Rhodosporidium paludigenum* to high salts concentrations resulted in greater viability under low a_w and freezing stress than non-exposed cells (Wang et al., 2010). Similarly, transient exposure to a heat shock treatment improved the tolerance of *M. fructicola* to subsequent high temperature and oxidative stress (Liu et al., 2011a).

When the ecological fitness of a biocontrol agent is well characterized, appropriate adjuvants can be added to the final formulation to protect it against unfavorable environmental factors occurring after its application. For example, an improvement in the tolerance of *C. infirmominiatum* to oxidative stress and an enhancement of biocontrol efficacy occurred when glycine betaine was added to the growth media (Liu et al., 2011b). In other work, addition of lignin or folic acid reduced mortality of the yeast *P. anomala* strain K on apple fruit surfaces caused by UV-B radiation and significantly increased its ability to control *P. expansum* (Lahlali et al., 2011a). Similarly, riboflavin and uric acid were used to protect *C. oleophila* strain O from the detrimental effects of UV light (Lahlali et al., 2011b). *Candida sake* strain O was also shown to be sensitive to low RH. In this regard, the addition of skimmed milk significantly improved the population density of this biocontrol agent on the apple fruit surface and increased its biocontrol efficacy at low RH compared to strain O alone (Jijakli and Lahlali, 2015).

A full characterization and understanding of the modes of action involved in the antagonistic activity of microorganisms will contribute to enhancing their efficacy. In this regard, *P. anomala* strain K was shown to be affected by the inactivation of two genes coding exo-β-1,3-glucanases (*PAEXG1* and *PAEXG2*) (Friel et al., 2007). On the basis of the demonstration in which exo-β-1,3-glucanases were involved in the biocontrol of *P. anomala*, a formulation including YGT (adjuvant containing β-1,3-glucans) was developed. This led to a higher level of efficacy against *B. cinerea* and *P. expansum* on apples along with a higher exo-β-1,3-glucanase activity in wound sites (Jijakli and Lepoivre, 1998; Jijakli, 2011).

Nutrients have been frequently reported to enhance the performance of biocontrol agents against postharvest pathogens (Janisiewicz, 1994). For instance, L-asparagine, L-proline, galactose, mannitol, ribose, sorbitol, or 2-deoxy-D-glucose were able to enhance the efficacy of a biocontrol agent and/or inhibit a pathogen (Harper et al., 1981; Janisiewicz et al., 1992; Jijakli et al., 1993).

7 Commercial Development of Postharvest Biocontrol Products

Significant changes in food safety regulations, as well as environmental concerns, have increased the demand for biological control products. As a result, multinational chemical companies and microbial industries (such as yeast producers) have been heavily investing in projects designed to expand their footprint in biological control technologies, including those for postharvest use. This is reflected in the number of acquisitions of small- and medium-size companies specializing in the development of green technologies for controlling plant diseases by larger, multinational companies, traditionally associated with chemical pesticides. In the case of industries involved in microbial products used in the bakery, brewery, and wine industries, the application of their microorganisms and expertise to the development of biocontrol products is obvious. Considering external constraints (growers, consumers, and legislators), large, multinational chemical companies are driven by two important parameters: preventing pesticide resistance, and the need for zero residues on marketed commodities. It is their further aim to offer a full portfolio of protection tools, including conventional and "green tools," to their clients (distributors and growers). If a product aimed primarily for postharvest use is also able to control preharvest infections or is suitable for use as a complementary treatment to existing conventional products for the purpose of managing resistance biotypes and residue issues, then companies will naturally explore the opportunity. The most important driving force, however, is minimizing or eliminating chemical residues in fresh harvested produce.

Numerous microbial antagonists (yeasts and bacteria) of postharvest pathogens have been identified in laboratory, semi-commercial, and commercial studies (Droby et al., 2009). Several of these antagonists reached advanced levels of development and commercialization. Among the first generation of biocontrol products registered and made commercially available were *C. oleophila* (Aspire, Ecogen, Langhorne, PA, USA) (Droby et al., 2016), *Cryptococcus albidus* (YieldPlus, Lallemand, Montreal, Canada), *C. sake* (Candifruit, IRTA, Lleida, Catalonia, Spain) (Teixidó et al., 2011), and *P. syringae* (BioSave, JET Harvest, Longwood, FL, USA) (Janisiewicz and Jeffers, 1997; Janisiewicz and Korsten, 2002). Aspire, Yieldplus, and Candifruit were commercialized for some years but discontinued due to business and marketing-related shortcomings. Biosave, however, is still available for use in the USA market for several postharvest uses (Janisiewicz and Peterson, 2004). *Bacillus subtilis* (Avogreen, University of Pretoria, Pretoria, South Africa) was introduced in South Africa for the control of Cercospora spot, a postharvest disease of avocado, but did not achieve commercial success due to inconsistent results (Demoz and Korsten, 2006). A strain of *C. oleophila* (Nexy, Lesaffre, Lille, France) was also developed in Belgium, and submitted for regulatory approval in 2005 for postharvest application to control wound pathogens on pome fruits, citrus, and banana (Lahlali et al., 2011). Nexy received registration approval throughout the EU in 2013 (Massart and Jijakli, 2014). A preharvest application of *Aureobasidium pullulans* (BoniProtect, Bio-Ferm, Tulln, Austria) was suggested to control wound pathogens that develop on pome fruit during storage (Lima et al., 2015). Another product based on *P. agglomerans* CPA-2, (Pantovital, Domca, Granada, Spain), effective against the major postharvest pathogens of pome and citrus fruits (Teixidóal., 2001; Plaza et al., 2004; Cañamás et al., 2008), was formulated but never commercialized (Torres et al., 2014). *Metschnikowia fructicola* (Shemer, Bayer, Leverkusen, Germany), registered in Israel for both pre- and postharvest application on various fruits and vegetables, including apricots, citrus fruits, grapes, peaches, peppers, strawberries, and sweet potatoes, represents a more successful example of a postharvest biocontrol product. Shemer was acquired by Bayer CropScience (Germany) and sublicensed to Koppert Biological Systems (The Netherlands) (Spadaro and Droby, 2016).

Commercial success of biocontrol products largely depends on the versatility of their use and the range of the target diseases they can control. A narrow range of activity is a limitation for reaching broad markets. In the case of specialized postharvest biocontrol products, this limitation exists because of the limited size of the postharvest market and the high number of different commodities and diseases that need to be managed. One possible approach to expand the uses of a specific biocontrol agent is to develop different formulations for different uses/targets. For example, Boni Protect, Blossom Protect, and Botector are different formulations of the same biocontrol agent, *A. pullulans.* These different formulations have been introduced for the control of postharvest and preharvest diseases of pome fruit, including fire blight on apples and gray mold, caused by *B. cinerea*, on grapes.

Commercialization and penetration of the market are the most critical requirements for the success of any biological control product. Effective implementation of a step-by-step approach is needed to coordinate the gathering of information, the establishment of a project plan, decision making, and the translating of this information into resources. This phase is long and fraught with a variety of difficulties, including scientific, regulatory, business, and marketing hurdles. For companies to effectively manage all these issues, they require ample data and information on a variety of aspects, including market demand, market size, profit margin, time to market, and time to volume of sales (Bailey et al., 2009). Nicot et al. (2012) highlighted a report published by ENDURE, a working group established within an EU project, which highlighted the factors associated with the success of biocontrol of arthropod pests, diseases, and weeds under field conditions. The report indicated that profit after taxes, provisions, and amortization were 18 and 2% of the sales value for chemical pesticide and biocontrol products, respectively. In the case of the postharvest market, the figures may be even lower than 2%. The size of the market in Europe for microbial biocontrol products targeting the management of plant diseases was estimated in 2012 to be about €52 million. Globally, biopesticides currently represent a US$ 1.5–2.5 billion market, compared to US$ 60 billion for the traditional pesticide market.

It is expected that the size of the biopesticide market will continue to increase, and that the highest rate of growth is expected to be in Europe (Anonymous, 2014). The projected worldwide annual market growth (2012–2020) for biopesticides is estimated to reach 12.3 versus 5% for chemical pesticides. Among the biocontrol agents that have been in advanced phases of the registration process are *M. fructicola* strain 277 (Shemer), *A. pullulans* strains DSM 14940 and DSM 14941 (Boni Protect), and *C. oleophila* strain O (Nexy). This development provides a strong indication as to the direction and prospects for the use of new biological control products in the coming years.

The discovery and development of new biological approaches for the control of postharvest diseases have been mainly done by small companies. Limited financial resources and the lack of established and well-developed marketing networks, however, have been major obstacles in commercializing new biocontrol products. The entrance of large multinational companies into this field may improve the prospects for the biopesticide industry.

Another major hurdle to the commercialization of biopesticides is the registration process required by government regulatory agencies prior to commercial sales of the product. Although the registration process is not as expensive or time-consuming as it is for synthetic chemical fungicides, this requirement poses a major hurdle for companies as it requires people with expertise in this field. The registration package must contain a clean record of safety (for both humans and the environment) for the biocontrol agent, basic toxicological tests on the formulated product (eye and skin irritation, ingestion), detailed data about mechanisms of action, detailed information about the strain identification, and origin and efficacy data, including semi-commercial and commercial tests using relatively large quantities of fruit treated under conditions that resemble commercial practices. The registration of biocontrol products for postharvest use in some countries (e.g., USA through the Environmental Protection Agency) has been straightforward and several products have received registration. In Europe, however, the situation until recently has been considerably more complex.

Another aspect related to the commercialization of biocontrol agents used on agricultural commodities grown for human consumption are health and safety issues related to the introduction of living microorganisms into the human diet. Although this may represent an obstacle to public acceptance of this technology, most of postharvest biocontrol agents were originally isolated from fruits and vegetables, and are generally indigenous to agricultural commodities. Humans are exposed to them daily when consuming fresh vegetables and fruit. Even though these antagonists are introduced in large numbers to the surface of a commodity, they survive and grow only in very restricted sites on the fruit surface (e.g., surface wounds). After their introduction on intact fruit surfaces, antagonist populations usually diminish within a very short period of time to the level of the natural epiphytic microflora.

Ultimately, the acceptance and use of a product by the industry will depend on its performance under commercial conditions. Commercial interest and industry acceptance of biological control products for postharvest use are largely dependent on the development of low-cost, stable products, which provide consistent efficacy. Solutions to key technical problems and the implementation of optimized strategies to make this happen will require research contributions from a variety of disciplines, including formulation science and technology.

8 Shortcomings of Postharvest Biocontrol

Despite hundreds of reports documenting the potential and commercial viability of antagonists, the widespread use of a single product has not occurred. Several products initially reached the market but were later discontinued, while others have achieved success in limited niche markets (Droby et al., 2016). This has been due to several factors including inconsistent performance, lack of industry acceptance, cost compared to synthetic fungicides, registration hurdles, and formulation problems (Droby et al., 2009, 2016). What started as a simple idea, using microbial antagonists to manage postharvest diseases, slowly evolved in complexity, as strategies to overcome various problems have been explored. The postharvest environment, including packinghouse treatments, storage, shipping, and all other aspects of the supply chain, are now being viewed as a system requiring a broad, holistic view to address a wide range of issues.

Droby et al. (2009) discussed the major obstacles facing the commercialization of postharvest biocontrol products in detail. The greatest obstacle is the need for consistent and reliable performance under a wide array of conditions (different packing lines, different inoculum levels, and use of other technologies used to maintain quality). The costs associated with product development, marketing, and industry acceptance are other factors that have limited the adoption of alternative technologies. There is a wealth of information on the use of biocontrol agents to control postharvest diseases, yet none of these approaches have been implanted on a large scale. There is a critical need to find a way to bridge the gap between research and industry so that these alternatives can become commercial successes.

Several registered postharvest biocontrol products have been developed jointly by researchers working with commercial companies (Droby et al., 2016). Although success was achieved in developing a product, commercial success, as measured by acceptance and widespread use, has been elusive. Efficacy of these products must be similar to that of chemical fungicides, which is in the range of 98–100% disease control. This level of effectiveness is seldom attained with biological control products when they are used as a stand-alone treatment. New ideas and approaches are, thus, needed if biological control of postharvest diseases is going to be a commercial success and implemented on a wide scale.

9 Integration of Biocontrol and Other Alternative Methods

The use of biocontrol agents for postharvest disease control usually has limitations because their efficacy is influenced by various factors that are difficult to control. As previously indicated, these factors include temperature, RH, the physiology of the host, existing infections, and prior preharvest treatment history. Biocontrol agents typically do not provide adequate control of established infections. Therefore, research has explored enhancing their efficacy by using a combination of two or more postharvest treatments in a cascade similar to the hurdle technology strategy used in the food industry.

9.1 Combination with Physical Treatments

Many attempts to combine physical treatments with biocontrol agents have been reported. The most successful and well-studied is the integration of microbial antagonists with heat treatments (hot air and hot water). Hot air or hot water treatments are simple techniques that can be easily implemented in packinghouses to reduce postharvest diseases. They are safe for humans and the environment and do not require registrations (Jijakli and Lepoivre, 2004; Usall et al., 2016).

The exposure of fruits for several hours or days to an air atmosphere heated to temperatures higher than 30°C and RH higher than 90% is known as hot air curing. This system has been used alone or in combination with other alternative control approaches, such as biocontrol. One example of an effective combination was described in citrus by Plaza et al. (2004). They demonstrated that *P. agglomerans* CPA-2 could effectively reduce the incidence of green mold on lemons inoculated a few hours prior to the application of the biocontrol agent, but this failed to control infections from inoculations made 24 hr or more prior to the application of the agent. When *P. agglomerans* was combined with a curing treatment at 33°C for 65 hr; however, 24-hr-old infections on artificially inoculated lemons stored at 20°C for 14 d was completely controlled, as were natural infections of fruit stored at 10°C for 3 wk plus 7 additional days at 20°C. Combined curing and biocontrol treatment was also used to successfully control bitter rot caused by *Colletotrichum acutatum* J.H. Simmonds on apples (Janisiewicz et al., 2003; Conway et al., 2005). A similar approach was also used to improve the ability of *P. syringae* to protect 'Gala' apples against future infections (Leverentz et al., 2000). The combination of strain BIO126 of *M. pulcherrima* applied after treatment with hot water at 50°C for 3 min and/or with some chemical treatments used to control *P. expansum* and *B. cinerea* in apple, improved the efficacy of *M. pulcherrima* (Spadaro et al., 2004). Good results were also obtained in pears when *R. glutinis* was applied after hot water treatment (Zhang et al., 2008).

Bacillus subtilis CPA-8 combined with a curing treatment at 50°C for 2 hr and 95–99% RH improved brown rot control in peaches compared to the application of these treatments alone (Casals et al., 2012). In that study, heat treatment alone effectively controlled existing infections in the fruit but did not protect fruit against future infections. When curing was combined with the application of the biocontrol agent, however, fruit were protected against new infections (Casals et al., 2012). Another example was reported by Zhang et al. (2007) who found that a heat treatment at 37°C for 48 hr combined with the biocontrol agent *C. laurentii* effectively controlled decay caused by *P. expansum* and *Rhizopus stolonifer* (Ehrenb.) Vuill. in peaches. A pretreatment of peaches with hot water at 55°C for 10 s also improved the efficacy of *C. oleophila* to control *B. cinerea* and *P. expansum* (Karabulut and Baykal, 2004). Similarly, a hot water treatment at 60°C for 40 s followed by the application of *B. subtilis* CPA-8 provided superior control of *Monilinia* spp. on peaches and nectarines to that obtained with each treatment applied separately (Casals et al., 2010). A hot air treatment (38–40°C for 24 hr) of cherry tomato fruit, combined with the biocontrol agent *Pichia guilliermondii*, effectively reduced decay caused by *B. cinerea, R. stolonifer*, and *Alternaria alternata* (Fr.) Keissl. (Zhao et al., 2010).

Controlled (CA) and modified atmospheres have also been combined with biocontrol agents to improve their efficacy. The ability of the yeast *C. sake* CPA-1 to control blue mold on apples was improved under CA of 3% O_2 and 3% CO_2 (Usall et al., 2000). Similar results were obtained with the bacterium *P. agglomerans* CPA-2 against most of the important fungal diseases of apples (Nunes et al., 2002)

9.2 Combination with Chemical Products

Several studies have been conducted to evaluate the value of combining biocontrol agents with generally recognized as safe (GRAS) substances, natural products, or resistance inducers (Spadaro and Gullino, 2004; Sharma et al., 2009). GRAS compounds, such as salt additives, can improve the efficacy of microbial antagonists in controlling postharvest decay. Among the salt additives, calcium chloride, calcium propionate, sodium carbonate, sodium bicarbonate, potassium metabisufite, and ammonium molybdate have all been reported to improve the performance of microbial antagonists. The improvement in the efficacy of the microbial antagonist depends upon the concentration of the antagonist, the concentration of salt additives, their mutual compatibility, and the duration and timing of application. Carbonic acid salts are good candidates for use with biocontrol agents for the integrated control of postharvest diseases of citrus (Usall et al., 2008). These compounds are primarily fungistatic, inexpensive, readily available, not very persistent, and can be used with a minimal risk of fruit injury.

The application of inorganic salts, such as calcium chloride (Jijakli et al., 1999; Tian et al., 2002), ammonium molybdate, (Nunes et al., 2002), and sodium bicarbonate (Teixidó et al., 2001; Karabulut et al., 2003; Wisniewski et al., 2016), has been reported to enhance the efficacy of antagonists to control postharvest diseases. Infiltration of fruit with calcium chloride, combined with the application of a biocontrol agent, improved the control of *P. expansum* on apples during 3–6 mon of storage at 1°C, compared to the biocontrol treatment alone (Janisiewicz, 1998; Conway et al., 1999). In addition to fungal pathogen control, the use of this integrated treatment had the added benefit of alleviating physiological disorders, such as bitter pit.

Some organic compounds are able to stimulate defense-related enzymes and/or antimicrobial substances in the host. The addition of such compounds (e.g., salicylic acid, methyl jasmonate (MeJA), or chitosan and its derivatives) successfully enhanced the biocontrol activity of yeasts to control various postharvest diseases of fruits (El-Ghaouth et al., 2000a; Jijakli et al., 2002; Qin et al., 2003; Zhang et al., 2009, 2010a; 2010b; Ebrahimi et al., 2012). MeJA is a naturally occurring plant growth regulator that modulates many physiological processes, including responses to environmental stress. Treatment of peaches with *C. laurentii* alone or in combination with MeJA reduced the diameter of lesions of brown rot and blue mold rot in peaches (Yao and Tian, 2005). MeJA also improved the efficacy of *Pichia membranaefaciens* against *C. acutatum* in loquat (Cao et al., 2009). *Pichia membranaefaciens* in combination with MeJA induced the higher activity of two defense-related enzymes, chitinase and glucanase, in loquat fruit. MeJA was also used together with *M. pulcherrima* to improve control of blue mold on apple (Ebrahimi et al., 2013). Salicylic acid is a plant phenolic compound involved in the induction of local and systemic resistance to fungal pathogens. Exogenous application of salicylic acid has been found to enhance the efficacy of *R. glutinis* in cherries by increasing host resistance to fungal pathogens (Qin et al., 2003). Integrating *R. glutinis* with salicylic acid also improved the control of gray mold on strawberry (Zhang et al., 2010a). Among natural products, chitin and its derivatives have been applied together with biocontrol agents to control postharvest diseases. For example, the antagonistic activity of *C. laurentii* increased after its cultivation in a growth medium supplemented with chitin, and the increase was attributed to a more rapid increase in the antagonist population in pear fruit wounds. Moreover, a cell-free filtrate of the chitin-supplemented culture medium, in which the yeast was incubated for 24 hr, inhibited growth of *P. expansum* (Yu et al., 2008b). Combining 0.2% glycolchitosan with the antagonist *C. saitoana* was also shown to be more effective in controlling green mold of oranges and lemons, caused by *P. digitatum*, and gray and blue molds, caused by *B. cinerea* and *P. expansum*, of apples than either treatment alone (El-Ghaouth et al., 2000a). Chitosan and its derivatives, including glycolchitosan, were reported to inhibit fungal growth and induce host-defense responses in treated fruit.

Essential oils and their components are gaining increasing interest due to their bioactivity in the vapor phase, relatively safe status, wide acceptance by consumers, and biodegradable properties (Lopez-Reyes et al., 2013). *Bacillus amyloliquefaciens* was used together with essential oils of lemongrass and thyme to control *B. cinerea, P. expansum*, and *R. stolonifer* on peach fruit (Arrebola et al., 2010). Tea polyphenol inhibited *B. cinerea* and improved the biocontrol efficacy of *Hanseniaspora uvarum* (Liu et al., 2010). The inhibitory effect of the tea polyphenol on spore germination and mycelial growth of *B. cinerea* might be the underlying cause of the improved efficacy. Phytic acid, due to its antioxidant potential, has been used in combination with biocontrol agents, such as *R. glutinis* or *Pichia caribbica*, to control gray mold of strawberry (Zhang et al., 2013) or blue mold of apple (Mahonu et al., 2016).

Benzo-thiadiazole-7-carbothioic acid S-methyl ester (BTH), a functional analogue of salicylic acid, can suppress disease by eliciting systemic acquired resistance. BTH was effective in improving the efficacy of *P. membranaefaciens* to control postharvest blue mold in peach fruit (Cao et al., 2011). The combined treatment had a synergistic effect on the induction of superoxide dismutase, catalase, ascorbate peroxidase, chitinase, and glucanase activity, causing a greater induction of resistance reactions and improved disease control compared to either BTH or *P. membranaefaciens* used alone.

Plant hormones have also enhanced biocontrol activity. The cytokinin N6-benzyladenine is a plant hormone involved in the inhibition of ripening and senescence in plants. The combination of *C. laurentii* with N6-benzyladenine was more effective in suppressing *P. expansum* growth in apple fruit wounds than *C. laurentii* alone (Yu et al., 2008a). Moreover, treatment of apple fruit with *C. laurentii* and N6-benzyladenine stimulated superoxide dismutase activity. Auxins, another class of plant hormones, regulate many plant growth- and development-associated processes, including a delay in ripening and senescence in harvested produce and the induction of disease resistance in plants. In addition to its physiological regulatory activity, recent studies have reported that auxins, such as indole-3-acetic acid (IAA), have direct antifungal activity. A combined treatment of *C. laurentii* and IAA was shown to be an effective approach to reduce gray mold in apple (Yu et al., 2009). Application of IAA greatly reduced gray mold development in apples when IAA was applied 24 hr prior to inoculation with *B. cinerea*. Moreover, the combination of IAA and *C. laurentii* stimulated superoxide dismutase, catalase, and peroxidase activity in apples.

Combining the use of biocontrol agents with commonly used fungicides is another approach that has been explored (Chand-Goyal and Spotts, 1996). The successful integration of biocontrol antagonists with fungicide treatments requires that the fungicide–antagonist combination not harm antagonist survival and efficacy. In some cases, the combination of a biological antagonist with a chemical fungicide allows the concentration of the chemical to be reduced (Droby et al., 1993).

Other approaches have tried to combine three different strategies to effectively control postharvest diseases. Applying sodium bicarbonate and a mixture of two antagonists (*M. pulcherrima* and *C. laurentiii*), in conjunction with controlled atmospheres (1.4% O_2 and 3% CO_2) resulted in superior control of postharvest pathogens of apple (Conway et al., 2007). The combination of a preharvest application of *E. nigrum* and a postharvest sodium bicarbonate hot treatment (1% sodium bicarbonate at 60°C for 20 s) significantly reduced the incidence of *Monilinia* spp. on stone fruits (Mari et al., 2007).

10 Summary and Conclusions

After more than three decades of intensive research on the use of antagonistic microbial organisms for the control of postharvest pathogens, the interest in finding alternative approaches that do not rely on the use of synthetic chemical pesticides is still valid.

Public and health and safety concerns, as well as more restrictive regulations about the materials and products that can be used in agricultural production and subsequent postharvest handling, have been the driving forces for finding alternative control strategies that are safe and effective. Alternative control strategies have moved from the simple idea of applying high concentrations of biocontrol agents to a harvested commodity, to using a wide array of other approaches and integrating them together into a systems approach based on the multiple decrement or multiple hurdle concept.

Significant progress has been made in the discovery, development, and improving the efficacy of a wide variety of microbial antagonists. All indications clearly show that the biopesticide industry will continue to grow and eventually become a mainstream approach to disease control. We are definitely moving into an "age of biology" and away from the "age of chemicals." In this new age, biological solutions will be used to solve disease and production problems in agriculture.

Commercial application of biocontrol products of postharvest pathogens comprise very small share of the potential market. For a biocontrol product to be viable, however, it must perform effectively and reliably, be widely accepted, have intellectual property protection (patent), safe for use, and profitable to the company that has invested the money in its development, registration, and marketing.

Understanding the mechanism of action of microbial antagonists has been a key issue in developing a successful biocontrol agent. Advances in omic technologies (genomics, transcriptomics, metabolomics, proteomics, and metagenomics) have greatly enhanced our understanding of the interactions between microbial antagonists, host tissues, pathogens, biotic and abiotic elicitors of defense mechanisms, and the environment. Available information indicates that there is no single universal mechanism of action common to all of the reported antagonists.

The microbiome is an integral and active component of harvested fruits and vegetables that is being influenced by various biotic and abiotic factors. Understanding all the factors involved in the assembly and composition of a specific microbiome is important to unraveling the multitrophic interactions involved in postharvest biocontrol systems. To overcome the scientific and technical challenges associated with developing novel biocontrol technologies based on a holistic or multifaceted approach, collaboration among a wide variety of scientific disciplines is imperative.

References

Abadias, M., Teixidó, N., Usall, J., Benabarre, A. and Viñas, I. 2001. Viability, efficacy and storage stability of freeze-dried biocontrol agent *Candida sake* using different protective ad rehydration media. *Journal of Food Protection* 64, 856–861.

Abadias, M., Teixidó, N., Usall, J., Solsona, C. and Viñas, I. 2005. Survival of the postharvest biocontrol yeast *Candida sake* CPA-1 after dehydration by spray-drying. *Biocontrol Science and Technology* 15, 835–846.

Abadias, M., Teixidó, N., Usall, J. and Viñas, I. 2003a. Optimization of growth conditions of the postharvest biocontrol agent *Candida sake* CPA-1 in a lab-scene fermenter. *Journal of Applied Microbiology* 95, 301–309.

Abadias, M., Usall, J., Teixidó, N. and Viñas, I. 2003b. Liquid formulation of the postharvest biocontrol agent *Candida sake* CPA-1 in isotonic solutions. *Phytopathology* 93, 436–442.

Adams, G. 2007. The principles of freeze-drying. pp. 15–38. *In*: J.G. Day and G. N. Stacey (eds.). Cryopreservation and Freeze-Drying Protocols. Humana Press Inc., Totowa, NJ.

Anonymous, 2014. Biopesticides - Global Strategic Business Report. Global Industry Analyst, Inc., p. 215.

Arrebola, E., Sivakumar, D., Bacigalupo, R. and Korsten, L. 2010. Combined application of antagonist *Bacillus amyloliquefaciens* and essential oils for the control of peach postharvest diseases. *Crop Protection* 29, 369–377.

Bailey, K.L., Boyetchko, S.M. and Längle, T. 2009. Social and economic drivers shaping the future of biological control: A Canadian perspective on the factors affecting the development and use of microbial biopesticides. *Biological Control* 52, 221–229.

Banani, H., Spadaro, D., Zhang, D., Matic, S., Garibaldi, A. and Gullino, M.L. 2014. Biocontrol activity of an alkaline serine protease from *Aureobasidium pullulans* expressed in *Pichia pastoris* against four postharvest pathogens on apple. *International Journal of Food Microbiology* 182–183, 1–8.

Banani, H., Spadaro, D., Zhang, D., Matic, S., Garibaldi, A. and Gullino, M.L. 2015. Postharvest application of a novel chitinase cloned from *Metschnikowia fructicola* and overexpressed in *Pichia pastoris* to control brown rot of peaches. *International Journal of Food Microbiology* 199, 54–61.

Bhandari, B.R., Patel, K.C. and Chen, X.D. 2008. Spray-drying of food materials-process and product characteristics. pp. 113–159. *In*: X.D. Chen and A. S. Majumdar (eds.). Drying Technologies in Food Processing. Blackwell Publishing Ltd., Oxford, UK.

Burges, H.D. and Jones, K.A. 1998. Formulation of bacteria, viruses and protozoa to control insects. pp. 31–128. *In*: H.D. Burges (ed.). Formulation of Microbial Biopesticides. Kluwer Academic Publishers, Dordrecht, The Netherlands.

Calvente, V., Benuzzi, D. and de Tosetti, M.I.S. 1999. Antagonistic action of siderophores from *Rhodotorula glutinis* upon the postharvest pathogen *Penicillium expansum*. *International Biodeterioration and Biodegradation* 43, 167–172.

Calvo-Garrido, C., Viñas, I., Usall, J., Rodríguez-Romera, M., Ramos, M.C. and Teixidó, N. 2014. Survival of the biological control agent *Candida sake* CPA-1 on grapes under the influence of abiotic factors. *Journal of Applied Microbiology* 117, 800–811.

Cañamás, T.P., Viñas, I., Torres, R., Usall, J., Solsona, C. and Teixidó, N. 2011. Field applications of improved formulations of *Candida sake* CPA-1 for control of *Botrytis cinerea* in grapes. *Biological Control* 56, 150–158.

Cañamás, T.P., Viñas, I., Usall, J., Casals, C., Solsona, C. and Teixidó, N. 2008. Control of postharvest on citrus fruit by preharvest application of the biocontrol agent *Pantoea agglomerans* CPA-2. Part I. Study of different formulation strategies to improve survival of cells in unfavourable environmental conditions. *Postharvest Biology and Technology* 49, 86–95.

Cao, S., Yang, Z., Hu, Z. and Zheng, Y. 2011. The effects of the combination of *Pichia membranefaciens* and BTH on controlling of blue mould decay caused by *Penicillium expansum* in peach fruit. *Food Chemistry* 124, 991–996.

Cao, S., Zheng, Y., Wang, K., Tang, S. and Rui, H. 2009. Effect of yeast antagonist in combination with methyl jasmonate treatment on postharvest anthracnose rot of loquat fruit. *Biological Control* 50, 73–77.

Carbó, A., Torres, R., Usall, J., Fons, E. and Teixidó, N. 2017a. Dry formulations of the biocontrol agent *Candida sake* CPA-1 using fluidized bed drying to control the main postharvest diseases on fruits. *Journal of the Science of Food and Agriculture* 97, 3691–3698.

Carbó, A., Torres, R., Usall, J., Solsona, C. and Teixidó, N. 2017b. Fluidised-bed spray-drying formulations of *Candida sake* CPA-1by adding biodegradable coatings to enhance their survival under stress conditions. *Applied Microbiology and Biotechnology* 101, 7865–7876.

Casals, C., Elmer, P.A.G., Viñas, I., Teixidó, N., Sisquella, M. and Usall, J. 2012. The combination of curing with either chitosan or *Bacillus subtilis* CPA-8 to control brown rot infections caused by *Monilinia fructicola*. *Postharvest Biology and Technology* 64, 126–132.

Casals, C., Teixidó, N., Viñas, I., Silvera, E., Lamarca, N. and Usall, J. 2010. Combination of hot water, *Bacillus subtilis* CPA-8 and sodium bicarbonate treatments to control postharvest brown rot on peaches and nectarines. *European Journal of Plant Pathology* 128, 51–63.

Castoria, R., Caputo, L., De Curtis, F. and De Cicco, V. 2003. Resistance of postharvest biocontrol yeasts to oxidative stress: A possible new mechanism of action. *Phytopathology* 93, 564–572.

Chan, Z. and Tian, S. 2005. Interaction of antagonistic yeasts against postharvest pathogens of apple fruit and possible mode of action. *Postharvest Biology and Technology* 36, 215–223.

Chand-Goyal, T. and Spotts, R.A. 1996. Post-harvest biological control of blue mold of apple and brown rot of sweet cherry by natural saprophytic yeasts alone or in combination with low doses of fungicides. *Biological Control* 6, 253–259.

Conway, W.S., Janisiewicz, W.J., Klein, J.D. and Sams, C.E. 1999. Strategy for combining heat treatment, calcium infiltration, and biological control to reduce postharvest decay of 'Gala' apples. *HortScience* 34, 700–704.

Conway, W.S., Janisiewicz, W.J., Leverentz, B., Saftner, R.A. and Camp, M.J. 2007. Control of blue mold of apple by combining controlled atmosphere, an antagonist mixture, and sodium bicarbonate. *Postharvest Biology and Technology* 45, 326–333.

Conway, W.S., Leverentz, B., Janisiewicz, W.J., Saftner, R.A. and Camp, M.J. 2005. Improving biocontrol using antagonist mixtures with heat and/or sodium bicarbonate to control postharvest decay of apple fruit. *Postharvest Biology and Technology* 36, 235–244.

Costa, E., Teixidó, N., Usall, J., Atarés, E. and Viñas, I. 2001. Production of the biocontrol agent *Pantoea agglomerans* strain CPA-2 using commercial products and by-products. *Applied Microbiology and Biotechnology* 56, 367–371.

Costa, E., Teixidó, N., Usall, J., Fons, E., Gimeno, V., Delgado, J. and Viñas, I. 2002a. Survival of *Pantoea agglomerans* strain CPA-2 in a spray-drying process. *Journal of Food Protection* 65, 185–191.

Costa, E., Usall, J., Teixidó, N., García, N. and Viñas, I. 2000. Effect of protective agents, rehydration media and initial cell concentration on viability of *Pantoea agglomerans* strain CPA-2 subjected to freeze-drying. *Journal of Applied Microbiology* 89, 793–800.

Costa, E., Usall, J., Teixidó, N., Torres, R. and Viñas, I. 2002b. Effect of package and storage conditions on viability and efficacy of the freeze-dried biocontrol agent *Pantoea agglomerans* strain CPA-2. *Journal of Applied Microbiology* 92, 873–878.

Costa, S.S., Machado, B.A.S., Martin, A.R., Bagnara, F., Ragadalli, S.A. and Alves, A.R.C. 2015. Drying by spray drying in the food industry: Micro-encapsulation, process parameters and main carriers used. *African Journal of Food Science* 9, 462–470.

De Cal, A., Larena, I., Guijarro, B. and Melgarejo, P. 2002. Mass production of conidia of *Penicillium frequentans*, a biocontrol agent against brown rot of stone fruits. *Biocontrol Science and Technology* 12, 715–725.

De Clercq, D., Cognet, S., Pujol, M., Lepoivre, P. and Jijakli, M.H. 2003. Development of a SCAR marker and a semi-selective medium for specific quantification of *Pichia anomala* strain K on apple fruit surfaces. *Postharvest Biology and Technology* 29, 237–247.

Demoz, B.T. and Korsten, L. 2006. *Bacillus subtilis* attachment, colonization, and survival on avocado flowers and its mode of action on stem-end rot pathogens. *Biological Control* 37, 68–74.

Droby, S., Chalutz, E., Hofstein, R., Wilson, C.L., Wisniewski, M.E., Fridlender, B., Cohen, L., Weiss, B. and Daus, A. 1993a. Pilot testing of *Pichia guilliermondii*: A biocontrol agent of postharvest diseases of citrus fruit. *Biological Control* 3, 47–52.

Droby, S., Chalutz, E., Horev, B., Cohen, L., Gaba, V., Wilson, C.L. and Wisniewski, M.E. 1993b. Factors affecting UV-induced resistance in grapefruit against the green mould decay caused by *Penicillium digitatum*. *Plant Pathology* 42, 418–424.

Droby, S., Chalutz, E., Wilson, C.L. and Wisniewski, M.E. 1989. Characterization of the biocontrol activity of *Debaryomices hansenii* in the control of *Penicillium digitatum* on grapefruit. *Canadian Journal of Microbiology* 35, 794–800.

Droby, S., Vinokur, V., Weiss, B., Cohen, L., Daus, A., Goldschmidt, E.E. and Porat, R. 2002. Induction of resistance to *Penicillium digitatum* in grapefruit by the yeast biocontrol agent *Candida oleophila*. *Phytopathology* 92, 393–399.

Droby, S., Wilson, C., Wisniewski, M. and El-Ghaouth, A. 2000. Biologically based technology for the control of postharvest diseases of fruits and vegetables. pp. 187–206. *In*: C. Wilson and S. Droby (eds.). Microbial Food Contamination. CRC Press, Boca Raton, FL.

Droby, S., Wisniewski, M., Macarisin, D. and Wilson, C. 2009. Twenty years of postharvest biocontrol research: Is it time for a new paradigm? *Postharvest Biology and Technology* 52, 137–145.

Droby, S., Wisniewski, M., Teixidó, N., Spadaro, D. and Jijakli, M.H. 2016. The science, development, and commercialization of postharvest biocontrol products. *Postharvest Biology and Technology* 122, 22–29.

Ebrahimi, L., Etebarian, H.R., Aminian, H. and Sahebani, N. 2012. Enhancement of biocontrol activity of *Torulaspora delbrueckii* with methyl jasmonate against apple blue mould disease. *Archives of Phytopathology and Plant Protection* 45, 2355–2363.

Ebrahimi, L., Etebarian, H.R., Aminian, H. and Sahebani, N. 2013. Effect of *Metschnikowia pulcherrima* and methyl jasmonate on apple blue mold disease and the possible mechanisms involved. *Phytoparasitica* 41, 515–519.

El Hamouchi, A., Bajji, M., Friel, D., Najimi, B., Achbani, E.H., El Jaafari, S., Durieux, A. and Jijakli, MH. 2008. Development of SCAR markers and a semi-selective medium for the quantification of strains Ach 1-1 and 1113-5, two *Aureobasidium pullulans* potential biocontrol agents. *Postharvest Biology and Technology* 50, 216–223.

El-Ghaouth, A., Smilanick, J.L. and Wilson, C.L. 2000a. Enhancement of the performance of *Candida saitoana* by the addition of glycolchitosan for the control of postharvest decay of apple and citrus fruit. *Postharvest Biology and Technology* 19, 103–110.

El-Ghaouth, A., Smilanick, J.L., Wisniewski, M. and Wilson, C.L. 2000b. Improved control of apple and citrus fruit decay with a combination of *Candida saitoana* and 2-deoxy-d-glucose. *Plant Disease* 84, 249–253.

El-Ghaouth, A., Wilson, C. and Wisniewski, M. 1998. Ultrastructural and cytochemical aspects of the biological control of *Botrytis cinerea* by *Candida saitoana* in apple fruit. *Phytopathology* 88, 282–291.

Fravel, D.R., Connick, W.J. and Lewis, J.A. 1998. Formulation of microorganisms to control plant diseases. pp. 187–202. *In*: H.D. Burges (ed.). Formulation of Microbial Biopesticides. Kluwer Academic Publishers, Dordrecht, The Netherlands.

Fredlund, E., Druvefors, U., Boysen, M.E., Lingsten, K.J. and Schnurer, J. 2002. Physiological characteristics of the biocontrol yeast *Pichia anomala* J121. *FEMS Yeast Research* 2, 395–402.

Friel, D., Gomez Pessoa, N.M., Vandenbol, M. and Jijakli, M.H. 2007. Separate and combined disruptions of two exo-b-1,3-glucanases genes decrease the efficiency of *Pichia anomala* (strain K) biocontrol against *Botrytis cinerea* on apples. *Molecular Plant-Microbe Interactions* 20, 371–379.

Gotor-Vila, A., Teixidó, N., Sisquella, M., Torres, R. and Usall, J. 2017c. Biological characterization of the biocontrol agent *Bacillus amyloliquefaciens* CPA-8: The effect of temperature, pH and water activity on growth, susceptibility to antibiotics and detection of enterotoxic genes. *Current Microbiology* 74, 1089–1099.

Gotor-Vila, A., Usall, J., Torres, R., Abadias, M. and Teixidó, N. 2017a. Formulation of the biocontrol agent *Bacillus amyloliquefaciens* CPA-8 using different approaches: Liquid, freeze-drying and fluid-bed spray-drying. *BioControl* 62, 545–555.

Gotor-Vila, A., Usall, J., Torres, R., Solsona, C. and Teixidó, N. 2017b. Biocontrol products based on *Bacillus amyloliquefaciens* CPA-8 using fluid-bed spray-drying process to control postharvest brown rot in stone fruit. *LWT-Food Science and Technology* 82, 274–282.

Guijarro, B., Larena, I., Melgarejo, P. and De Cal, A. 2006. Effect of drying on viability of *Penicillium frequentans*, a biological control agent against brown rot disease caused by *Monilinia* spp. *Biocontrol Science and Technology* 16, 257–269.

Harper, A.M., Strange, R.N. and Langcake, P. 1981. Characterization of the nutriments required by *Botrytis cinerea* to infect broad bean leaves. *Physiological Plant Pathology* 19, 153–167.

Hernández-Montiel, L.G., Ochoa, J.L., Troyo-Diéguez, E. and Larralde-Corona, C.P. 2010. Biocontrol of postharvest blue mold (*Penicillium italicum* Wehmer) on Mexican lime by marine and citrus *Debaryomyces hanseii* isolates. *Postharvest Biology and Technology* 56, 181–187.

Hershkovitz, V., Ben-Dayan, C., Raphael, G., Pasmanik-Chor, M., Liu, J., Belausov, E., Aly, R., Wisniewski, M. and Droby, S. 2012. Global changes in gene expression of grapefruit peel tissue in response to the yeast biocontrol agent *Metschnikowia fructicola*. *Molecular Plant Pathology* 13, 338–349.

Hershkovitz, V., Sela, N., Taha-Salaime, L., Liu, J., Rafael, G., Kessler, C., Aly, R., Levy, M., Wisniewski, M. and Droby, S. 2013. De-novo assembly and characterization of the transcriptome of *Metschnikowia fructicola* reveals differences in gene expression following interaction with *Penicillium digitatum* and grapefruit peel. *BMC Genomics* 14, 168.

Ianiri, G., Idnurm, A., Wright, S.A.I., Duran-Patron, R., Mannina, L., Ferracane, R., Ritieni, A. and Castoria, R. 2013. Searching for genes responsible for patulin degradation in a biological control yeast provides insights into the basis for resistance to this mycotoxin. *Applied and Environmental Microbiology* 79, 3101–3115.

Janisiewicz, W., Usall, J. and Bors, B. 1992. Nutritional enhancement of biocontrol of blue mold on apples. *Phytopathology* 82, 1364–1370.

Janisiewicz, W.J. 1994. Enhancement of biocontrol of blue mold with the nutrient analog 2-deoxy-D-glucose on apples and pears. *Applied and Environmental Microbiology* 60, 2671–2676.

Janisiewicz, W.J. 1998. Biocontrol of postharvest diseases of temperate fruits - challenges and opportunities. pp. 171–198. *In*: G.J. Boland and L.D. Kuykendall (eds.). Plant-Microbe Interactions and Biological Control. Marcel Dekker, New York, NY.

Janisiewicz, W.J. and Jeffers, S.N. 1997. Efficacy of commercial formulation of two biofungicides for control of blue mold and gray mold of apples in cold storage. *Crop Protection* 16, 629–633.

Janisiewicz, W.J. and Korsten, L. 2002. Biological control of postharvest diseases of fruits. *Annual Review of Phytopathology* 40, 411–441.

Janisiewicz, W.J., Leverentz, B., Conway, W.S., Saftner, R.A., Reed, A.N. and Camp, M.J. 2003. Control of bitter rot and blue mold of apples by integrating heat and antagonist treatments on 1-MCP treated fruit stored under controlled atmosphere conditions. *Postharvest Biology and Technology* 29, 129–143.

Janisiewicz, W.J. and Peterson, D.L. 2004. Susceptibility of the stem pull area of mechanically harvested apples to blue mold decay and its control with a biocontrol agent. *Plant Disease* 88, 662–664.

Jijakli, M.H. 2011. *Pichia anomala* in biocontrol for apples: 20 years of fundamental research and practical applications. *Antonie Van Leeuwenhoek* 99, 93–105.

Jijakli, M.H., De Clercq, D., Dickburt, C. and Lepoivre, P. 2002. Pre- and post-harvest practical application of *Pichia anomala* strain K, β-1,3-glucans and calcium chloride on apples: Two years of monitoring and efficacy against post-harvest diseases. *IOBC/WPRS Bulletin* 25, 29–32.

Jijakli, M.H. and Lepoivre, P. 1998. Characterization of an Exo-β-1,3-glucanase produced by *Pichia anomala* strain K, antagonist of *Botrytis cinerea* on apples. *Phytopathology* 88, 335–343.

Jijakli, M.H. and Lepoivre, P. 2004. State of the art and challenges of postharvest diseases management in apples. pp. 59–94. *In*: K.G. Mukerji (ed.). Disease Management of Fruits and Vegetables, Vol.1: Fruit and Vegetable Diseases. Kluwer Academic Publishers, Dordrecht, The Netherlands.

Jijakli, M.H., Lepoivre, P. and Grevesse, C. 1999. Yeast species for biocontrol of apple postharvest diseases: An encouranging case of study for practical use. pp. 31–49. *In*: K.G. Mukerji, B.P. Chamolaand and R.H. Upadhyay (eds.). Biotechnological Approaches in Biocontrol of Plant Pathogens. Kluwer Academic Publishers, Dordrecht, The Netherlands.

Jijakli, M.H., Lepoivre, P., Tossut, P. and Thonart, P. 1993. Biological control of *Botrytis cinerea* and *Penicillium* sp. on post-harvest apples by two antagonistic yeasts. *Medicine Faculty Landbouwkundige University of Gent* 58, 1349–1358.

Karabulut, O.A. and Baykal, N. 2004. Integrated control of postharvest disease of peaches with a yeast antagonist, hot water and modified atmosphere packaging. *Crop Protection* 23, 431–435.

Karabulut, O.A., Smilanick, J.L., Gabler, F.M., Mansour, M. and Droby, S. 2003. Near-harvest applications of *Metschnikowia fructicola*, ethanol, and sodium bicarbonate to control postharvest diseases of grape in central California. *Plant Disease* 87, 1384–1389.

Karabulut, O.A., Tezcan, H., Daus, A., Cohen, L., Wiess, B. and Droby, S. 2004. Control of preharvest and postharvest fruit rot in strawberry by *Metschnikowia fructicola*. *Biocontrol Science and Technology* 14, 513–521.

Köhl, J. and Fokkema, N.J. 1998. Strategies for biological control of necrotrophic fungal foliar pathogens. pp. 49–88. *In*: G.J. Boland and L.D. Kuykendall (eds.). Plant Microbe Interactions and Biological Control. Marcel Dekker Press, New York, NY.

Kurtzman, C.P. and Droby, S. 2001. *Metschnikowia fructicola*, a new ascosporic yeast effective for biocontrol of postharvest fruit rots. *Systematics Applied Microbiology* 24, 395–399.

Kwasiborski, A., Bajji, M., Renaut, J., Delaplace, P. and Jijakli, M.H. 2014. Identification of metabolic pathways expressed by *Pichia anomala* Kh6 in the presence of the pathogen *Botrytis cinerea* on apple: New possible targets for biocontrol improvement. *PLoS ONE* 9, e91434.

Lahlali, R., Brostaux, Y. and Jijakli, M.H. 2011a. Control of apple blue mold by the antatonistic yeast *Pichia anomala* strain K: Screening of UV protectants for preharvest application. *Plant Disease* 95, 311–316.

Lahlali, R., Bunatesta, R. and Jijakli, M.H. 2011b. UV protectants for *Candida oleophila* (strain O), a biocontrol agent of postharvest fruit diseases. *Plant Pathology* 60, 288–295.

Lahlali, R., De Clercq, D., Serrhini, M.N., Creemers, P. and Jijakli, M.H. 2009. Assessment of *Pichia anomala* (strain K) efficacy against blue mould of apples when applied pre- or post-harvest under laboratory conditions and in orchard trials. *Plant Pathology* 123, 37–45.

Larena, I., De Cal, A., Liñan, M. and Melgarejo, P. 2003a. Drying of *Epicoccum nigrum* conidia for obtaining a shelf-stable biological product against brown rot disease. *Journal of Applied Microbiology* 94, 508–514.

Larena, I., De Cal, A. and Melgarejo, P. 2004. Solid substrate production of *Epicoccum nigrum* conidia for biological control of brown rot on stone fruits. *International Journal of Food Microbiology* 94, 161–167.

Larena, I., Melgarejo, P. and De Cal, A. 2003b. Drying of conidia of *Penicillium oxalicum*, a biological control agent against Fusarium wilt of tomato. *Journal of Phytopathology* 151, 600–606.

Leverentz, B., Janisiewicz, W., Conway, W.S., Saftner, R.A., Fuchs, Y., Sams, C.E. and Camp, M.J. 2000. Combining yeasts or a bacterial biocontrol agent and heat treatments to reduce postharvest decay of 'Gala' apples. *Postharvest Biology and Technology* 21, 87–94.

Li, B. and Tian, S.P. 2006. Effects of trehalose on stress tolerance and biocontrol efficacy of *Cryptococcus laurentii*. *Journal of Applied Microbiology* 100, 854–861.

Lima, G., Sanzani, S.M., De Curtis, F. and Ippolito, A. 2015. Biological control of postharvest diseases. pp. 65–81. *In*: R.B.H. Wills and J.B. Golding (eds.). Advances in Postharvest Fruit and Vegetable Technology. CRC Press, Boca Raton, FL.

Liu, H.M., Guo, J.H., Cheng, Y.J., Liu, P., Long, C.A. and Deng, B.X. 2010. Inhibitory activity of tea polyphenol and *Hanseniaspora uvarum* against *Botrytis cinerea* infections. *Letters in Applied Microbiology* 51, 258–263.

Liu, J., Sui, Y., Wisniewski, M., Droby, S. and Liu, Y. 2013. Utilization of antagonistic yeasts to manage postharvest fungal diseases of fruit. *International Journal of Food Microbiology* 167, 153–160.

Liu, J., Tian, S.P., Li, B.Q. and Qin, G.Z. 2009. Enhancing viability of two biocontrol yeasts in liquid formulation by applying sugar protectant combined with antioxidant. *Biological Control* 54, 817–824.

Liu, J., Wisniewski, M., Droby, S., Norelli, J., Hershkovitz, V., Tian, S. and Farrell, R. 2012. Increase in antioxidant gene transcripts, stress tolerance and biocontrol efficacy of *Candida oleophila* following sublethal oxidative stress exposure. *FEMS Microbiology Ecology* 80, 578–590.

Liu, J., Wisniewski, M., Droby, S., Tian, S., Hershkovitz, V. and Tworkoski, T. 2011a. Effect of heat shock treatment on stress tolerance and biocontrol efficacy of *Metschnikowia fructicola*. *FEMS Microbiology Ecology* 76, 145–155.

Liu, J., Wisniewski, M., Droby, S., Vero, S., Tian, S. and Hershkovitz, V. 2011b. Glycine betaine improves oxidative stress tolerance and biocontrol efficacy of the antagonistic yeast *Cystofilobasidium infirmominiatum*. *International Journal of Food Microbiology* 146, 76–83.

Lopez-Reyes, J.G., Spadaro, D., Prelle, A., Garibaldi, A. and Gullino, M.L. 2013. Efficacy of plant essential oils on postharvest control of rots caused by fungi on different stone fruits in vivo. *Journal of Food Protection* 76, 631–639.

Lutz, M.C., Sosa, M.C., Rodriguez, M.E., Lopez, C.A. and Sangorrín, M.P. 2013. Efficacy and putative mode of action of native and commercial antagonistic yeasts against postharvest rots of pear pathogens. *International Journal of Food Microbiology* 164, 166–172.

Macarisin, D., Cohen, L., Eick, A., Rafael, G., Belausov, E., Wisniewski, M.E. and Droby, S. 2007. *Penicillium digitatum* suppresses production of hydrogen peroxide in host tissue during infection of citrus fruit. *Phytopathology* 97, 1491–1500.

Macarisin, D., Droby, S., Bauchan, G. and Wisniewski, M. 2010. Superoxide anion and hydrogen peroxide in the yeast antagonist-fruit interaction: A new role for reactive oxygen species in postharvest biocontrol? *Postharvest Biology and Technology* 58, 194–202.

Mahonu, G.K., Zhang, H., Yang, Q., Zhang, X., Li, D. and Zhou, Y. 2016. Improving the biocontrol efficacy of *Pichia caribbica* with phytic acid against postharvest blue mold and natural decay in apples. *Biological Control* 92, 172–180.

Manso, T., Nunes, C., Raposo, S. and Lima-Costa, M.E. 2010. Caro pulp as raw material for production of the biocontrol agent *Pantoea agglomerans* PBC-1. *Journal of Industrial Microbiology and Biotechnology* 37, 1145–1155.

Mari, M., Torres, R., Casalini, L., Lamarca, N., Mandrin, J.F., Lichou, J., Larena, I., De Cal, A., Melgarejo, P. and Usall, J. 2007. Control of post-harvest brown rot on nectarine by *Epicoccum nigrum* and physico-chemical treatments. *Journal of Science of Food and Agriculture* 87, 1271–1277.

Masih, E.I., Slezack-Deschaumes, S., Marmaras, I., Barka, E.A., Vernet, G., Charpentier, C., Adholeya, A. and Paul, B. 2001. Characterisation of the yeast *Pichia membranifaciens* and its possible use in the biological control of *Botrytis cinerea*, causing the grey mould disease of grapevine. *FEMS Microbiology Letters* 202, 227–232.

Massart, S., De Clercq, D, Salmon, M., Dickburt, C. and Jijakli, M.H. 2005. Development of real-time PCR using Minor Groove Binding probe to monitor the biological control agent *Candida oleophila* (strain O). *Journal of Microbiological Methods* 60, 73–82.

Massart, S. and Jijakli, M.H. 2014. *Pichia anomala* and *Candida oleophila* in biocontrol of postharvest diseases of fruits: 20 years of fundamental and practical research. pp. 111–122. *In*: D. Prusky and M. Gullino (eds.). Plant Pathology in the 21st Century. Springer International Publishing, Cham, Switzerland.

Massart, S., Perazzolli, M., Höfte, M., Pertot, I. and Jijakli, M.H. 2015. Impact of the omic technologies for understanding the modes of action of biological control agents against plant pathogens. *BioControl* 60, 725–746.

Melin, P., Schnürer, J. and Hakansson, S. 2011. Formulation and stabilization of the biocontrol yeast *Pichia anomala*. *Antonie Van Leeuwenhoek* 99, 107–112.

Mercier, J. and Smilanick, J.L. 2005. Control of green mold and sour rot of stored lemon by biofumigation with *Muscodor albus*. *Biological Control* 32, 401–407.

Mounir, R., Durieux, A., Bodo, E., Allard, C., Simon, J.P., Achbani, E.H., El-Jaafari, A., Douira, A. and Jijakli, M.H. 2007. Production, formulation and antagonistic activity of the biocontrol like-yeast *Aureobasidium pullulans* against *Penicillium expansum*. *Biotechnology Letters* 29, 553–559.

Navarta, L.G., Calvo, J., Posetto, P., Cerutti, S., Raba, J., Benuzzi, D. and Sanz, M.I. 2014. Postharvest control of gray mold in apples with lyophilized formulations of *Cryptococcus laurentii*: The effect of cold stress in the survival and effectiveness of the yeast. *Food Process and Technology* 7, 2962–2968.

Nicot, P.C., Alabouvette, C., Bardin, M., Blum, B., Köhl, J. and Ruocco, M. 2012. Review of factors influencing the success or failure of biocontrol: Technical, industrial and socio-economic perspectives. *IOBC-WPRS Bulletin* 78, 95–98.

Nunes, C., Stépien, V., Manso, T., Torres, R., Usall, J., Bajji, M. and Jijakli, M.H. 2008. Development and application of a SCAR marker for the specific monitoring of the biocontrol agent *Pantoea agglomerans* CPA-2. *Postharvest Biology and Technology* 47, 422–428.

Nunes, C., Usall, J., Teixidó, N., Abadias, M. and Viñas, I. 2002. Improved control of postharvest decay of pears by the combination of *Candida sake* (CPA-1) and ammonium molybdate. *Phytopathology* 92, 281–287.

Ongena, M. and Jacques, P. 2008. *Bacillus* lipopeptides: Versatile weapons for plant disease control. *Trends in Microbiology* 16, 115–125.

Patiño-Vera, M., Jiménez, B., Balderas, K., Ortiz, M., Allende, R., Carrillo, A. and Galindo, E. 2005. Pilot-scale production and liquid formulation of *Rhodotorula minuta*, a potential biocontrol agent of mango anthracnose. *Journal of Applied Microbiology* 99, 540–550.

Petersson, S., Jonsson, N. and Schnürer, J. 1999. *Pichia anomala* as a biocontrol agent during storage of high-moisture feed grain under airtight conditions. *Postharvest Biology and Technology* 15, 175–184.

Plaza, P., Usall, J., Smilanick, J.L., Lamarca, N. and Viñas, I. 2004. Combining *Pantoea agglomerans* (CPA-2) and curing treatments to control established infections of *Penicillium digitatum* on lemons. *Journal of Food Protection* 67, 781–786.

Powell, K.A. 1992. Biocontrol product fermentation, formulation and marketing. pp. 381–387. *In*: E.S. Tjamos, G.C. Papavizas and R.J. Cook (eds.). Biological Control of Plant Diseases. NATO ASI Series (Series A: Life Sciences), Vol. 230. Springer, Boston, MA.

Pujol, M., De Clercq, D., Cognet, S., Lepoivre, P. and Jijakli, M.H. 2004. Monitoring system for the biocontrol agent *Pichia anomala* strain K using quantitative competitive PCR-ELOSA. *Plant Pathology* 53, 103–109.

Pusey, P.L. and Wilson, C.L. 1984. Postharvest biological control of stone fruit brown rot by *Bacillus subtilis*. *Plant Disease* 68, 753–756.

Qin, G.Z., Tian, S.P., Xu, Y. and Wan, Y.K. 2003. Enhancement of biocontrol efficacy of antagonistic yeasts by salicylic acid in sweet cherry fruit. *Physiological and Molecular Plant Pathology* 62, 147–154.

Sangorrín, M.P., Lopes, C.A., Vero, S. and Wisniewski, M. 2014. Cold-adapted yeasts as biocontrol agents: Biodiversity, adaptation strategies and biocontrol potential. pp. 441–464. *In*: P. Buzzini and R. Margesin (eds.). Cold-Adapted Yeasts. Biodiversity, Adaptation Strategies and Biotechnological Significance. Springer-Verlag Berlin Heidelberg, Germany.

Santivarangkna, C., Kulozik, U. and Foerst, P. 2007. Alternative drying processes for the industrial preservation of lactic acid starter cultures. *Biotechnology Progress* 23, 302–315.

Saravanakumar, D., Ciavorella, A., Spadaro, D., Garibaldi, A. and Gullino, M.L. 2008. *Metschnikowia pulcherrima* strain MACH1 outcompetes *Botrytis cinerea, Alternaria alternata* and *Penicillium expansum* in apples through iron depletion. *Postharvest Biology and Technology* 49, 121–128.

Schena, L., Ippolito, A., Zahavi, T., Cohen, L. and Droby, S. 2000. Molecular approaches to assist the screening and monitoring of postharvest biocontrol yeasts. *European Journal of Plant Pathology* 106, 681–691.

Schena, L., Sialer, M.F. and Gallitelli, D. 2002. Molecular detection of strain L47 of *Aureobasidium pullulans*, a biocontrol agent of postharvest diseases. *Plant Disease* 86, 54–60.

Sharma, R.R., Singh, D. and Singh, R. 2009. Biological control of postharvest diseases of fruits and vegetables by microbial antagonists: A review. *Biological Control* 50, 205–221.

Soto-Muñoz, L., Teixidó, N., Usall, J., Viñas, I., Abadias, M. and Torres, R. 2015a. Molecular tools applied to identify and quantify the biocontrol agent *Pantoea agglomerans* CPA-2 in postharvest treatments on oranges. *Postharvest Biology and Technology* 100, 151–159.

Soto-Muñoz, L., Torres, R., Usall, J., Viñas, I., Dashevskaya, S. and Teixidó, N. 2015b. Environmental monitoring of the biocontrol agent *Pantoea agglomerans* CPA-2 applied to citrus fruit at preharvest. *Annals of Applied Biology* 167, 250–261.

Spadaro, D., Ciavorella, A., Lopez, G., Garibaldi, A. and Gullino, M.L. 2010a. Effect of culture age, protectants, and initial cell concentration on viability of freeze-dried cells of *Metschnikowia pulcherrima*. *Canadian Journal of Microbiology* 56, 809–815.

Spadaro, D., Ciavorella, A., Zhang, D., Garibaldi, A. and Gullino, M.L. 2010b. Effect of culture media and pH on the biomass production and biocontrol efficacy of a *Metschnikowia pulcherrima* strain to be used as a biofungicide for postharvest disease control. *Canadian Journal of Microbiology* 56, 128–137.

Spadaro, D. and Droby, S. 2016. Development of biocontrol products for postharvest diseases of fruit: The importance of elucidating the mechanisms of action of yeast antagonists. *Trends in Food Science and Technology* 47, 39–49.

Spadaro, D., Garibaldi, A. and Gullino, M.L. 2004. Control of *Botrytis cinerea* and *Penicillium expansum* on apple combining a biocontrol agent with hot water dipping and acibenzolar-S-methyl, baking soda, or ethanol application. *Postharvest Biology and Technology* 33, 141–151.

Spadaro, D. and Gullino, M.L. 2004. State of art and future perspectives of biological control of postharvest fruit diseases. *International Journal of Food Microbiology* 91, 185–194.

Spadaro, D., Lorè, A., Garibaldi, A. and Gullino, M.L. 2013. A new strain of *Metschnikowia fructicola* for postharvest control of *Penicillium expansum* and patulin accumulation on four cultivars of apple. *Postharvest Biology and Technology* 75, 1–8.

Spadaro, D., Sabetta, W., Acquadro, A., Portis, E., Garibaldi, A. and Gullino, M.L. 2008. Use of AFLP for differentiation of *Metschnikowia pulcherrima* strains for postharvest disease biological control. *Microbiological Research* 163, 523–530.

Srivastava, S. and Mishra, G. 2010. Fluid bed technology: Overview and parameters for process selection. *International Journal of Pharmaceutical Science and Drug Research* 2, 236–246.

Stanbury, P.F., Whitaker, A. and Hall, S.J. 1995. Media for industrial fermentation. pp. 93–121. *In*: P. Stanbury, A. Whitaker and S. Hall (eds.). Principles of Fermentation Technology. Pergamon Press, Oxford, UK.

Stephan, D., Matos Da Silva, A.P. and Bisutti, I.L. 2016. Optimization of a freeze-drying process for the biocontrol agent *Pseudomonas* spp. and its influence on viability, storability and efficacy. *Biological Control* 94, 74–81.

Strasser, S., Neureiter, M., Geppl, R., Braun, R. and Danner, H. 2009. Influence of lyophilisation, fluidized bed drying, addition of protectant, and storage on the viability of lactic acid bacteria. *Journal of Applied Microbiology* 107, 167–177.

Strasser, S. 2008. Innovative Product Formulations Applying the Fluidized Bed Technology. Ph.D. Dissertation. University of Natural Resources and Life Sciences, Vienna, Austria.

Sui, Y, Wisniewski, M., Droby, S. and Liu, J. 2015. Responses of yeast biocontrol agents to environmental stress. *Applied and Environmental Microbiology* 81, 2968–2975.

Teixidó, N., Torres, R., Viñas, I., Abadias, M. and Usall, J. 2011. Biological control of postharvest diseases in fruit and vegetables. pp. 364–402. *In*: C. Lacroix (ed.). Protective Cultures, Antimicrobial Metabolites and Bacteriophages for Food and Beverage Biopreservation. Series in Food Science, Technology and Nutrition No. 201. Woodhead Publishing Ltd., Cambridge, UK.

Teixidó, N., Usall, J., Palou, L., Asensio, A., Nunes, C. and Viñas, I. 2001. Improving control of green and blue molds on oranges by combining *Pantoea agglomerans*

(CPA-2) and sodium bicarbonate. *European Journal of Plant Pathology* 107, 685–694.

Teixidó, N., Usall, J. and Viñas, I. 1999. Efficacy of preharvest and postharvest *Candida sake* biocontrol treatments to prevent blue mold on apples during cold storage. *International Journal of Food Microbiology* 50, 203–210.

Teixidó, N., Viñas, I., Usall, J. and Magan, N. 1998a. Improving ecological fitness and environmental stress tolerance of the biocontrol yeast *Candida sake* by manipulation of intracellular sugar alcohol and sugar content. *Mycological Research* 102, 1409–1417.

Teixidó, N., Viñas, I., Usall, J. and Magan, N. 1998b. Control of blue mold of apples by preharvest application of *Candida sake* grown in media with different water activity. *Phytopathology* 88, 960–964.

Tian, S.P., Fan, Q., Xu, Y., and Jiang, A.L. 2002. Effect of calcium on biocontrol activity of yeast antagonists against the postharvest fungal pathogen *Rhizopus stolonifer*. *Plant Pathology* 51, 352–358.

Tian, S.P., Yao, H.J., Deng, X., Xu, X.B., Qin Guo, Z. and Chan, Z.L. 2007. Characterization and expression of β-1,3-glucanase genes in jujube fruit induced by the microbial biocontrol agent *Cryptococcus laurentii*. *Phytopathology* 97, 260–268.

Torres, R., Solsona, C., Viñas, I., Usall, J., Plaza, P. and Teixidó, N. 2014. Optimization of packaging and storage conditions of a freeze-dried *Pantoea agglomerans* formulation for controlling postharvest diseases in fruit. *Journal of Applied Microbiology* 117, 173–184.

Torres, R., Teixidó, N., Usall, J., Abadias, M., Mir, N., Larrigaudiere, C. and Viñas, I. 2011. Anti-oxidant activity of oranges after infection with the pathogen *Penicillium digitatum* or treatment with the biocontrol agent *Pantoea agglomerans* CPA-2. *Biological Control* 57, 103–109.

Torres, R., Usall, J., Teixidó, N., Abadias, M. and Viñas, I. 2003a. Liquid formulation of the biocontrol agent *Candida sake* by modifying water activity or adding protectants. *Journal of Applied Microbiology* 94, 330–339.

Torres, R., Valentines, M.C., Usall, J. and Larrigaudiere, C. 2003b. Possible involvement of hydrogen peroxide in the development of resistance mechanisms in 'Golden Delicious' apple fruit. *Postharvest Biology and Technology* 27, 235–242.

Usall, J., Smilanick, J.L., Palou, L., Denis-Arrue, N., Teixidó, N., Torres, R. and Viñas, I. 2008. Preventive and curative activity of combined treatments of sodium carbonate and *Pantoea agglomerans* CPA-2 to control postharvest green mold of citrus fruit. *Postharvest Biology and Technology* 50, 1–7.

Usall, J., Torres, R. and Teixidó, N. 2016. Biological control of postharvest diseases on fruit: A suitable alternative? *Current Opinion in Food Science* 11, 51–55.

Vero, S., Garmendia, G., González, M.B., Bentancur, O. and Wisniewski, M. 2013. Evaluation of yeasts obtained from Antarctic soil samples as biocontrol agents for the management of postharvest diseases of apple (*Malus* x *domestica*). *FEMS Yeast Research* 13, 189–199.

Wang, Y., Bao, Y., Shen, D., Feng, W., Yu, T., Zhang, J. and Dong Zheng, X. 2008. Biocontrol of *Alternaria alternata* on cherry tomato fruit by use of marine yeast *Rhodosporidium paludigenum* Fell & Tallman. *International Journal of Food Microbiology* 123, 234–239.

Wang, Y., Wang, P., Xia, J., Yu, T., Luo, B., Wang, J. and Dong Zheng, X. 2010. Effect of water activity on stress tolerance and biocontrol activity in antagonistic yeast

Rhodosporidium paludigenum. International Journal of Food Microbiology 143, 103–108.
Wilson, C.L., Wisniewski, M., Droby, S. and Chalutz, E. 1993. A selection strategy for microbial antagonist to control postharvest diseases of fruits and vegetables. *Scientia Horticulturae* 53, 183–189.
Wilson, C.L. and Wisniewski, M.E. 1989. Biological control of postharvest diseases of fruits and vegetables: An emerging technology. *Annual Review of Phytopathology* 27, 425–441.
Wisniewski, M., Droby, S., Norelli, J., Liu, J. and Schena, L. 2016. Alternative management technologies for postharvest disease control: The journey from simplicity to complexity. *Postharvest Biology and Technology* 122, 3–10.
Wisniewski, M., Wilson, C.L., El-Ghaouth, A. and Droby, S. 2001. Non-chemical approaches to postharvest disease control. *Acta Horticulturae* 553, 407–412.
Wisniewski, M.E. and Wilson, C.L. 1992. Biological control of postharvest diseases of fruits and vegetables: Recent advances. *HortScience* 27, 94–98.
Yánez-Mendizábal, V., Viñas, I., Usall, J., Cañamás, T. and Teixidó, N. 2012a. Endospore production allows using spray-drying as a possible formulation system of the biocontrol agent *Bacillus subtilis* CPA-8. *Biotechnology Letters* 34, 729–735.
Yánez-Mendizábal, V., Viñas, I., Usall, J., Torres, R., Solsona, C., Abadias, M. and Teixidó, N. 2012b. Formulation development of the biocontrol agent *Bacillus subtilis* strain CPA-8 by spray-drying. *Journal of Applied Microbiology* 112, 954–965.
Yánez-Mendizábal, V., Viñas, I., Usall, J., Torres, R., Solsona, C. and Teixidó, N. 2012c. Production of the postharvest biocontrol agent *Bacillus subtilis* CPA-8 using low cost commercial products and by-products. *Biological Control* 60, 280–289.
Yao, H.J. and Tian, S.P. 2005. Effects of a biocontrol agent and methyl jasmonate on postharvest diseases of peach fruit and the possible mechanisms involved. *Journal of Applied Microbiology* 98, 941–950.
Yu, T., Chen, J., Lu, H. and Zheng, X. 2009. Indole-3-acetic acid improves postharvest biological control of blue mold rot of apple by *Cryptococcus laurentii*. *Phytopathology* 99, 258–264.
Yu, T., Wang, L., Yin, Y., Feng, F. and Zheng, X. 2008a. Suppression of postharvest blue mould of apple fruit by *Cryptococcus laurentii* and N6-benzyladenine. *Journal of the Science of Food and Agriculture* 88, 1266–1271.
Yu, T., Wang, L., Yin, Y., Wang, Y. and Zheng, X. 2008b. Effect of chitin on the antagonistic activity of *Cryptococcus laurentii* against *Penicillium expansum* in pear fruit. *International Journal of Food Microbiology* 122, 44–48.
Zhang, H., Ma, L., Jiang, S., Lin, H., Zhang, X., Ge, L. and Xu, Z. 2010a. Enhancement of biocontrol efficacy of *Rhodotorula glutinis* by salicylic acid against gray mold spoilage of strawberries. *International Journal of Food Microbiology* 141, 122–125.
Zhang, H., Ma, L., Turner, M., Xu, H., Dong, Y. and Jiang, S. 2010b. Methyl jasmonate enhances biocontrol efficacy of *Rhodotorual glutinis* to postharvest blue mold decay of pears. *Food Chemistry* 117, 621–626.
Zhang, D., Spadaro, D., Valente, S., Garibaldi, A. and Gullino, M.L. 2012. Cloning, characterization, expression and antifungal activity of an alkaline serine protease of *Aureobasidium pullulans* PL5 involved in the biological control of postharvest pathogens. *International Journal of Food Microbiology* 153, 453–464.

Zhang, H., Wang, S., Huang, X., Dong, Y. and Zheng, X. 2008. Integrated control of postharvest blue mold decay of pears with hot water treatment and *Rhodotorula glutinis*. *Postharvest Biology and Technology* 49, 308–313.

Zhang, H., Wang, L., Zheng, X. and Dong, Y. 2007. Effect of yeast antagonist in combination with heat treatment on postharvest blue mold decay and *Rhizopus* decay of peaches. *International Journal of Food Microbiology* 115, 53–58.

Zhang, H., Yang, Q., Lin, H., Ren, X., Zhao, L. and Hou, J. 2013. Phytic acid enhances biocontrol efficacy of *Rhodotorula glutinis* against postharvest gray mold spoilage and natural spoilage of strawberries. *LWT–Food Science and Technology* 52, 110–115.

Zhao, Y., Tu, K., Tu, S., Liu, M., Su, J. and Hou, Y.P. 2010. A combination of heat treatment and *Pichia guilliermondii* prevents cherry tomato spoilage by fungi. *International Journal of Food Microbiology* 137, 106–110.

Chapter 15

Toward Probiotic Postharvest Biocontrol Antagonists

Appraisal of Obstacles

Anjani M. Karunaratne and Buddhie S. Nanayakkara

Department of Botany, Faculty of Science, University of Peradeniya, Peradeniya, Sri Lanka

Abbreviations

FAO	Food and Agriculture Organization of the United Nations
F&V	Fruits and vegetables
IDF	International Dairy Federation
NCD	Non-communicable diseases
OIE	World Organization for Animal Health
WHO	World Health Organization

1 Introduction

In any type of diet at a global level, fruits and vegetables (F&V) are promoted because of their health-promoting properties, especially with regard to controlling non-communicable diseases (NCD). The World Health Organization (WHO) has specified the amount of F&V that should be consumed on a daily basis as not less than 400 g per day (WHO, 2013). One reason that the Mediterranean diet is upheld as health promoting is because F&V compose a significant proportion of the meal. As most diets around the world do not contain substantial proportions of F&V, it is envisaged that the health authorities of many countries would take decisions to promote them in the local diets during the forthcoming years. This, in turn, will make the task of the postharvest technologists more challenging.

While handling techniques from field to the consumer has a major impact on the quality of F&V, appropriate harvesting techniques, optimal handling practices, and proper storage alone are not sufficient for most F&V. Over the years, in spite of the fact that various agrochemicals are registered in different countries, some previously approved chemicals have undergone changes, such as being banned in certain countries, due to environmental and health issues (Bautista-Baños et al., 2014). Therefore, finding alternative means of pathogen control on F&V is felt more than ever before.

Consequently, with the increasing demand for F&V, methods used to extend shelf life during long transshipments should keep pace by being both effective and safe. While agrochemical usage has been linked to various problems (Wisniewski et al., 2016), from the consumers' standpoint, F&V should be fresh and appealing without use of postharvest agrochemicals. This may be envisaged as an important reason for the popularity of the more organically grown F&V, in spite of their expense and shorter shelf life.

Of the different microbial groups, those that are classified as probiotics, which are microorganisms with beneficial qualities, have earned a good name in possessing both prophylactic and therapeutic effects (Karunaratne, 2018a). If the antagonistic potential of probiotics could be harnessed, the benefits will be twofold; while probiotic antagonists could combat pathogens and control postharvest diseases of F&V, their very presence on F&V will be an attraction for the health-seeking consumer. Such F&V may also qualify as synbiotic foods, with both probiotics and prebiotics (specific fiber in F&V that promote the growth of probiotics within the gut). Synbiotics are superior in terms of their health-promoting properties than either probiotics or prebiotics. It is well known that F&V are excellent sources of the latter. We focus on the feasibility of using probiotics to serve as biocontrol agents, which, in turn, act as antagonists to pathogens, and discuss issues on the quest for probiotic antagonists.

2 The Need for the Change of Attitudes

Lobbying for an attitudinal change is important to move forward. Biocontrol stands out in pathogen control, because of its natural method of antagonism toward pathogens. For this reason, many private establishments, especially in the technologically developed countries, have been working together with researchers in the pursuit for biocontrol agents. In spite of that, the success rate of effective biocontrol agents reaching the market has been limited due to several factors, especially because there is a comparison between the efficacy and the convenience of agrochemicals in pathogen control (Droby et al., 2016). However, with the global interest on green technology and the quest to reduce carbon footprint, Usall et al. (2016) discussed the recent changes on the attitude in favor of acceptance of biocontrol products by mega companies. This change of attitude is obviously connected with the increased awareness of the consumer about the benefits of consuming F&V and also the change in perception by the consumer looking for fresh produce that is not tainted with agrochemicals. An important message to be carried to the consumer is that when probiotics serve as biocontrol agents, the benefits for the consumer will be twofold, control of pathogens and thereby extended shelf life and, additionally, enhanced health beneficial effects due to the presence of both probiotics and prebiotics, the latter originating from the F&V (Karunaratne, 2018b). Therefore, selecting F&V with probiotic biocontrol agents will be an excellent opportunity for adding probiotics, prebiotics, and their synbiotic effects to the diet. This will be an attractive feature for companies to be interested in developing probiotic biocontrol products, in spite of some shortcomings when compared with the efficacy and ease of application of agrochemicals. From the consumer's standpoint, the need to educate the consumer of the enhanced benefits of probiotic treated F&V is of paramount importance.

3 Appraisal of Basic Research

Working toward the goal of probiotic antagonists is not an unachievable task if a systematic approach is practiced. Already, records have shown that many probiotics in veterinary applications have potential as biocontrol agents (Hossain et al., 2017). Many probiotics have antagonistic effects as well, and specifically, the antibacterial activity of certain probiotics of the genera *Lactobacillus* and *Bifidobacterium* have been documented (Angelakis et al., 2013). What is important is to know the ideal way of harnessing these antimicrobial effects to enable this process to extend shelf life of F&V. For this, as fungi constitute a major portion of postharvest pathogens, the antifungal effects of probiotics should also be assessed.

The quest for alternative means of controlling postharvest pathogens has a long history, as it is well documented in Section III (current section) of this book. These alternative management technologies used in postharvest disease control can be divided into three broad categories as the use of antagonists (biocontrol), natural compounds, and physical means (Wisniewski et al., 2016). Apart from biocontrol *per se*, such alternatives as discussed in different reviews range from inorganic means of control (Karunaratne and Gunasinghe, 2011), use of volatiles (Mari et al., 2016), use of coatings (Ncama et al., 2018), heat and other physical treatments (Sui et al., 2016), and coupling biocontrol with other demonstrably safe and agrochemical-free means (Zhang et al., 2018). Among all the alternative treatment methods, biocontrol seems to stand out, with a few biocontrol products reaching commercial applications (Gardener and Fravel, 2002; Spadaro and Droby, 2016; Zhang et al., 2018). In spite of that, the extent of basic and applied research done, which includes screening for antagonists and determining their efficacy in relation to that of an agrochemical used, as well as formulating them into a suitable commercial product to be dispensed conveniently, does not seem to justify the achievements in the discovery of effective antagonists (Spadaro and Droby, 2016; Wisniewski et al., 2016; Karunaratne, 2018b). Spadaro and Droby (2016) appraised 879 published biocontrol reports to assess the progress made in the development, registration, and commercialization of biocontrol products and they reported on only nine different products that were commercialized in different countries. Similarly, Zhang et al. (2018) listed 11 commercial biocontrol products. In its totality, 13 commercially available biocontrol products have been documented in different studies.

Managing communicable diseases became easy with the advent of antibiotics, and currently, control of the NCDs has become the focus in the medical sphere. Many clinical trials to combat NCDs have been performed, some with the use of plant extracts and various foods of plant origin, leading to the concept of "functional foods," which are health beneficial foods. The term functional foods was coined by the International Food Information Council Foundation (Food Insight, 2009) for "foods that provide a health benefit beyond basic nutrition." A forerunner to functional foods was the group of foods termed "probiotic foods." Originally, these were fermented foods (Karunaratne, 2018a); however, with time, innovative ways of introducing specific probiotics are being sought. In such a scenario using fresh F&V as carriers of probiotics would be truly innovative. The focus on the edible microorganisms present in fermented foods led to the identification of a range of microorganisms with prophylactic and therapeutic effects, especially with regard to curbing the occurrence of NCDs (Karunaratne, 2018a).

However, a similar predicament as for the biocontrol agents has been observed with regard to the numerous research investigations on probiotics followed by the introduction of few commercial products. It is apparent that in probiotic research, clinical trials have not kept pace with laboratory level investigations on organisms with the potential of being developed as probiotics. There were up to 1476 research papers per year on this subject from 2000 to 2014, yet only 24 examples of different types of probiotic products, in the form of capsules, powders, fruit drinks, or lozenges, have been reported (McFarland, 2015). Information describing the commercialized probiotic products was summarized by Karunaratne (2018b).

4 A Concerted Effort to Move Forward Toward Commercialization

In biocontrol, the antagonistic microbes employed inhibit pathogen growth through antibiosis, competition for nutrients and space, parasitism, and induction of resistance within the host (Francesco et al., 2016; Rahman et al., 2018). Substances that affect the control of pathogens include enzymes, antifungal molecules, and volatile organic compounds (Salas et al., 2017). Biocontrol methods also encompass the use of various plant exudates, and more sophisticated breeding and bioengineering to develop "microbe-optimized crops" (Rahman et al., 2018). In fact, Louis Pasteur's (1822–1895) experiments almost two centuries ago, which finally paved the path to the observations of Alexander Fleming (1881–1955), that, in turn, led to the discovery of antibiotics, may be considered as some of the initial official records of biocontrol research.

The term paraprobiotics, which received attention more recently, introduces the concept to use non-viable probiotics in place of viable probiotics. In fact, paraprobiotics have shown promise with regard to the use of non-viable microbial debris (De Almada et al., 2016), as in the case of the early vaccines. This is an area that calls for much basic research. The fact that the debris of microbes has the same or subdued effects of microbes, points toward the fact that one has to be careful not to abuse their use and to treat such microbial debris as potentially having significant biological activity.

It is well known that microorganisms are an integral part of the composition of F&V. Therefore, manipulation of the microbial ecosystems on the surfaces of F&V using organisms already proven to be beneficial to the consumer health would facilitate their regulatory approval. This, in turn, would make this approach of biocontrol as a strategy to be commercially feasible. For this to occur, a better understanding of the microbes of interest will help to move forward. In this regard, the current attention describing the fruit microbiome, as reviewed by Droby and Wisniewski (2018), will facilitate the selection of promising microbes.

A successful biocontrol agent on F&V should be resilient to face events and environmental changes through conditions encountered during farm to fork pipeline. In other words, such organisms should be stable and persist or increase under stressful conditions of the surroundings and circumstances. Most importantly, their antagonistic effects toward pathogens should be consistent. These are the aspects that need to be studied in depth to find out what factors influence the effectiveness of each particular biocontrol agent.

Figure 15.1 highlights the major events involved in the commercialization of a biocontrol agent. The arrow on the left depicts the current flow of events as signified by the tapering shape of the arrow, and that on the right indicates an ideal situation where most, if not all, laboratory experiments promote commercialization. As most research investigations done are at lower rungs of the scale regarding the development of a biocontrol agent, focus of investigations on the level of application is rare.

The reason for such an inclination of research toward the lower rungs of the ladder may be envisaged as barriers encountered at higher rungs with regard to field testing and surmounting regulatory approval barriers. Ideally, external factors, such as lack of regulatory protocols and legislation, should be considered at the beginning of the study to eliminate organisms that do not meet the basic criteria, so this aspect would not be a concern to arise unexpectedly when substantial discovery and development has been completed. For this to be a reality, firm guidelines should be established at national and international levels. In an ideal situation, the only reason to eliminate prospective organisms should be due to the fact that

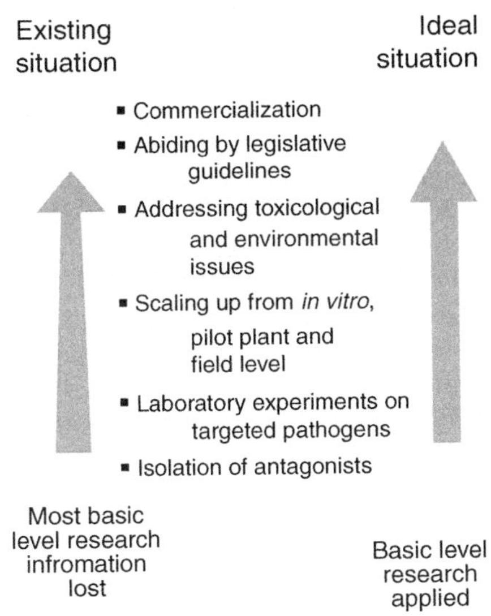

Figure 15.1 Major events in the commercialization of a biocontrol agent; current trend on the left and ideal situation on the right.

they do not meet the criteria established to qualify as potentially acceptable biocontrol agents and/or as probiotic organisms.

It is obvious when looking back that many basic level investigations having shown promise toward commercialization stopped abruptly without being able to carry out field testing, probably due to the lack of proper guidelines. This idea is highlighted by Droby et al. (2016); having reviewed literature on biocontrol, they indicate that a gap still exists between basic research involving the discovery of a biocontrol agent and its development and implementation under commercial conditions, and correctly state that a considerable volume of published research articles fall under the category of "re-inventing the wheel." At present, unless research is done at an international level, with several countries involved with large grants to handle any trying circumstance, most promising biocontrol agents tested at research level cannot reach commercialization due to various commercial or regulatory barriers. Legislation within countries has been and continues to be a major barrier (Droby et al., 2016; Wisniewski et al., 2016; Sarrocco and Vannacci, 2017; Karunaratne, 2018b). Such obstacles are created by excessively rigorous and unrealistic regulatory requirements and/or absence of a responsible body to make the decision.

It is important for research to probe into reasons as to why certain commercial products have failed, and address such issues. Basic research at higher rungs of the scale (Figure 15.1) is rare. A new research investigation does not have to commence by looking for microbes of interest, but could work on the existing problems of commercialized products that have shown promise, but have failed due to myriads of reasons. Issues related to secrecy in patenting may be one contributing factor for this seemingly reduced research output at the higher rungs. For these, however, patenting issues that are handled under secrecy, may have to be re-looked at a legislative level, and guidelines governing such issues may have to be changed to facilitate progress of scientific investigations.

A joint venture of FAO-WHO (2002) has set up guidelines to enable a systematic approach for the evaluation of probiotics in food leading to substantiating health claims provided for probiotics. According to these guidelines, even when considered as generally recognized as safe organisms (GRAS organisms), they should exhibit that they are safe with regard to certain metabolic activities of concern (Figure 15.2). The quest for biocontrol agents too has been on par with such guidelines set out to guide probiotic research.

Turning toward criteria for proper use of biocontrol agents, apart from country level guidelines, there are some broad guidelines set up by the Food and Agriculture Organization (FAO, 1995). These are wide definitions covering all aspects of biocontrol, that is, prey control of all types covering mainly fauna, and are not focused toward postharvest biocontrol. These guidelines are mainly to protect the environment from alien organisms (biocontrol agents) introduced and does not address specific issues of microbial interactions.

The use of microbial-derived products, such as antibiotics, has a long history, although their abuse rather than their use has led to various disaster generating effects, such as antibiotic resistance, dosage used not being effective, and so on. This can be the case even with the use of probiotic biocontrol agents if implemented blindly. Therefore, the story of the abuse of antibiotics could serve as a means of direction for the use of microbial products in disease control and prevention. With regard to the introduction of any biocontrol agent, environmental issues are of paramount importance. Especially with regard to F&V, any tendency to upset the balance of the phytosphere by the introduced biocontrol agent, and thereby, having a domino effect on the ecosphere at large is of concern.

A potential probiotic should be charaterized at a minimum with regard to

- •D-lactate production, bile salt deconjugation
- •Antibiotic resistance
- •Side effects during human studies
- •Epidemiological surveillance of adverse incidents in consumers
- •Toxicity effects
- •Hemolytic activity

Figure 15.2 Summary of guidelines provided by FAO-WHO (Food and Agriculture Organization - World Health Organization) (2002) on the identification of probiotics.

To work systematically, it is necessary to know the material that one is working with. Several organizations have come forward with regard to documenting data on microbes of interest. Three Bodies; the World Organization for Animal Health (OIE), WHO, and FAO, have come to a tripartite partnership on the use of antimicrobials in food animals (OIE/WHO/FAO, 2016). The European Food Safety Authority is another such authority that has decided to keep records (BIOHAZ Panel; EFSA, 2017). Also, lists of microorganisms and microbial-derived ingredients used in foods can be found at the United States Food and Drug Administration website (US FDA, 2001). Such endeavors give direction for research communities around the world as to which organisms they should concentrate on to develop biocontrol products. The International Dairy Federation (IDF) has systematically documented microbes of interest in the fermentation of dairy products that may be important in the present quest for probiotic antibiotics (IDF, 2012). These specifications encompass the criteria for developing biocontrol agents, particularly regarding the safety of the product. Taking this idea forward, the senior author reviewed the information available and proposed the idea of an umbrella organization to serve as a ready source of information on probiotic antagonists just like the set up of the Centers for Disease Control and Prevention in the USA (Karunaratne, 2018b), which would be a valuable long-term solution to existing concerns.

Both disciplines (biocontrol and probiotic) have developed independently, although the material for both are beneficial microorganisms to humans; one to extend shelf life of F&V and the other to serve as prophylactics or therapeutics. If we are to compare the pipeline from research to commercialization, there are common areas as depicted for both disciplines (Figure 15.3).

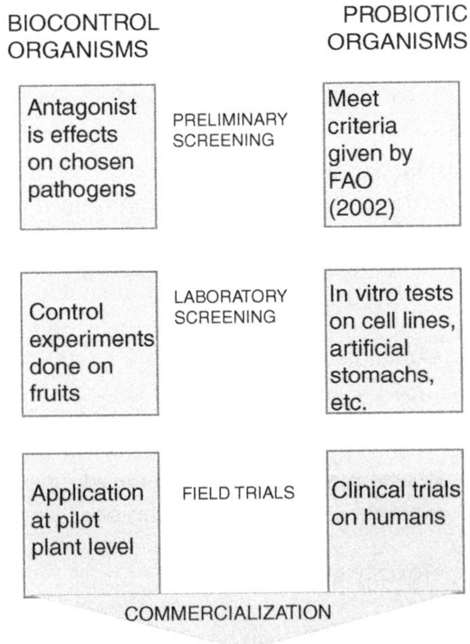

Figure 15.3 Progress of research from laboratory level to field trials and application with regard to biocontrol (left) and probiotic (right) research.

While in biocontrol research, after initial isolation, the ability of the organisms to control postharvest diseases is tested, and in probiotic research, isolation is followed by the investigations carried out to see whether the organisms meet the criteria listed by the FAO-WHO (2002) as highlighted in Figure 15.2.

When a biocontrol product is used on edible surfaces of F&V, the probiotic labeling guidelines set up by the FAO-WHO (2002) are helpful for researchers. In the present endeavor, with the initial qualifying characteristics as probiotics, potential probiotic organisms could also be tested on harvested F&V of interest (biocontrol), and on cell lines as well as artificial stomachs. When organisms show the initial qualifying criteria, field trials could follow on selected plots (biocontrol) and human volunteers (probiotic). If such field trials are successful, then additional investigation is needed to test the F&V treated with biocontrol agents on human volunteers in clinical trials, thus connecting both disciplines. Case studies to determine probiotic effects on humans with non-treated and biocontrol-treated fruits will provide valuable information concerning the feasibility of such treatments. Referring to applications related to probiotics, Bourdichon et al. (2012) discussing the history of safe use of fermented foods stated that for the safe use of microorganisms in food it is necessary to document not just the occurrence of a microorganism in a fermented food product, but also to provide evidence whether the presence of the microorganism is beneficial, fortuitous, or undesired for humans. Such an approach seems to be appropriate in keeping records of testing probiotic biocontrols as there is a manipulation of the microbial populations on the surfaces of the treated F&V, and an alteration of the microbial numbers could occur.

The problems in the application of biocontrol have been highlighted in several recent investigations. Usall et al. (2016) highlighted technical shortcomings as well as economic and regulatory constraints. Droby et al. (2016) described the process from pertinent laboratory investigations to commercial development of products and appraised the issues along this pipeline. Lack of consistency and hence unreliable performance at commercial level are major pitfalls of biocontrol agents (Usall et al., 2016). Safety is an issue even with regard to the most beneficial probiotic as illustrated by De Simone (2018). When the issue of legislation of the country is not a barrier, that is, if the legislation regarding the use of microbes for human consumption is well laid out, then the major barriers for the research not to progress beyond a certain point would be due to inherent characteristics of biocontrol organisms. Such issues may turn out to be specific safety concerns, microbial competition favoring specific undesirable organisms, lack of stability of the organisms, and so on. These are important factors to be evaluated in research investigations at higher rungs. With proper guidelines, once they go through relevant experiments on the ladder through in vitro experiments to human volunteers with regard to probiotic effectiveness, the rest of the path to commercialization should be smooth provided the guidelines are well laid out. The major issues at this point of a well carried out investigation should be the designing of the final formulation for application, and determining its degree of efficacy and stability in comparison to an effective agrochemical, which is popular among the handlers.

Among the microbes used in biocontrol are those from the phyllosphere, carposphere, flower, and rhizosphere. There is a growing interest in the microbiota of these different organs (Abdelfattah et al., 2018) and specific interest on the fruit microbiome (Droby and Wisniewski, 2018). It is interesting to note that certain elicitors that work against pathogens are compounds of interest serving as prebiotics in probiotic research. For instance, xanthan gum obtained from *Xanthomonas campestris* is used in probiotic delivery through microencapsulation (Kim et al., 2016; Shori, 2017). The role of the gum serving as a prebiotic is established, but still, will the organism from which xanthan gum is obtained, have a probiotic role as well? *X. campestris* is a well-known pathogen causing black rot in cabbage, and its control has been investigated using *Bacillus* spp. (Massomo et al., 2004). Zhang et al. (2018) reported elicitor functions for several chemicals known to have prebiotic effects, such as fructooligosaccharide, pectin, and chitosan. This is very much attractive in the fresh F&V market as both prebiotics and probiotics *per se* are considered healthy additions to the diet.

Making future biocontrol research more feasible for field application, by choosing organisms among probiotic organisms that have proven to be beneficial, will facilitate their registration. Collaboration among postharvest researchers and those involved in probiotic investigations is envisioned as a wise move to realize the goal.

Although these two areas have developed independently, there seems to be a common ground in important breakthroughs. Table 15.1 summarizes common ground on organisms found in biocontrol agents available commercially, and investigations related to probiotic research on such organisms.

Table 15.1 Active organisms in certain biocontrol agents available commercially, and reports from probiotic research on these organisms

Biocontrol agents	*Names of commercial preparations of biocontrol agents*	*Reports of same organism as probiotics or with desirable effects*
Bacillus subtilis	Avogreen®	Many commercial preparations alone or in combination
Pseudomonas syringae	Biosave®	None
Aureobasidium pullulans	Boniprotect®	Enriched culture fluid effective on mice as prophylactic against influenza (Muramatsu et al., 2012)
Ulocladium sp.	BOTRY-zen®	Evidence that it is part of the microbiome (Huffnagle and Noverr, 2013)
Candida sake	Candifruit®	None
Candida oleophila	Nexy® and Aspire®	None
Pantoea agglomerans	Pantovital®, PomaVita®, and Blossombless®	Enriched culture fluid effective on mice as probiotic (Kobayashi et al., 2018)
Metchnikowia fructicola	Shemer®	Another species, *M. pulcherrima*, is sought after in wine industry for its aroma and flavor (Barbosa et al., 2018)
Cryptococcus albidus	Yield plus®	Promising probiotic characteristics (El-Baz et al., 2018)

Let us focus our attention on some commercialized biocontrol products (Table 15.1). *Bacillus subtilis* marketed as Avogreen® (Spadaro and Droby, 2016) is also considered a probiotic that can form biofilms. A prolongevity effect of *B. subtilis* on *Caenorhabditis elegans* was associated with down regulation of the insulin-like signaling system (Ayala et al., 2017). Slightly reduced functionality of the insulin signaling was correlated with the healthy longevity of human centenarians. Also, *Aureobasidium pullulans* has been shown to prevent influenza in mice (Muramatsu et al., 2012). *A. pullulans* is a commercially used yeast bearing the name Boniprotect® in the biocontrol of important postharvest pathogens of F&V, such as *Penicillium, Botrytis,* and *Monilinia* (Spadaro and Droby, 2016). To what extent the beneficial effects shown in animal health experiments can be extrapolated to human studies is yet to be ascertained, although results tend to vary in such extrapolations (Shanks et al., 2009).

Additionally, *Ulocladium oudemansii,* which is marketed as BOTRY-zen®, is used for the control of *Botrytis cinerea* in several crops, including grapes and blackcurrants (Zhang et al., 2018), and *Candida* spp. have been reported as a part of the human microbiome, although little information is available on the fungi that constitute the microbiome (Huffnagle and Noverr, 2013). Two *Candida* spp., *C. sake* and *C. oleophila,* are already commercialized biocontrol agents bearing commercial names Candifruit® and Nexi®, respectively, the latter is also called Aspire® (Spadaro and Droby, 2016). However, it is noteworthy that some strains of *C. sake* were pathogenic to humans (Juneja et al., 2011).

5 Organisms of Interest

Investigations carried out during a long span of time show that there are species of microorganisms common to both disciplines (biocontrol and probiotic) as recorded in independent studies (Table 15.2). However, strain specificity with regard to these effects should be solved at a level of molecular characterization. For a detailed discussion on this aspect one may refer to Karunaratne (2018b).

Collating information on common organisms in both disciples is useful, as it may be possible to look at microbial volatiles and secretions that could be of value, especially if using the viable organism itself may have problems. The story, however, is not so simple not only because there may be strain variations but also because other related factors, such as storage conditions, ambient environment, the target populations, and so on, could have a bearing on the effects. Over time, while biocontrol research has used bacteria, yeasts, and filamentous fungi in the quest for biocontrol antagonists, with regard to probiotic research, bacteria have been the main focus.

However, examination of recent literature indicates that the scenario is changing rapidly. Of the commercialized biocontrol agents, there is an inclination toward the use of yeasts. The fruit microbiome constitutes an abundance of fungal genera, including several yeasts (Droby and Wisniewski, 2018). Among the reasons for this partiality toward yeasts are their tolerance to extreme environmental conditions prevailing before and after harvest (low and high temperatures, desiccation, wide range of relative humidity, low oxygen levels, pH fluctuations, and UV radiation), their unique adaptability to the fruit micro environment (high sugar

Table 15.2 A representative range of research studies over a long span of time that have indicated the use of the same species in biocontrol and probiotic research

Organism	*Experiments in biocontrol research*	*Experiments in probiotic research*	*Notes*
Bacillus cereus	Has shown much promise as a biocontrol agent (Emmert and Handelsman, 1999)	Commercial product Bactisubtil® (Hong et al., 2005)	Is also a notorious enteropathogen, and hence, knowing the strain is important*
Metarhizium anisopliae	Important fungal biocontrol agent of insect pests (Pattemore et al., 2014)	The mycobiome is still to be explored (Bourdichon et al., 2012)	As a fungus adapted to the insect intestines, what role it may have in the human intestine should be assessed
Pseudomonas spp.	*Pseudomonas syringae* formulated as frozen pellets (Emmert and Handelsman, 1999)	Probiotic and antimicrobial effects in aquatic studies (Ibrahem, 2015)	Reported to have antiviral properties as well (Ibrahem, 2015)
Streptomyces spp.	Its biocontrol includes insect biocontrol (Emmert and Handelsman, 1999)	Marine *Streptomyces* is designated as a drug to control diseases of aquaculture (Hariharan and Dharmaraj, 2018)	Numerous studies on probiotic effects of *Streptomyces* mentioned in shrimps
Pseudomonas cepacia Burkholderia cepacia and related species	Biocontrol of cotton seedling damping off (Heydari and Misaghi, 1998)	A well-known gut organism (Senan et al., 2015)	Weak pathogen affecting immunocompromised individuals (Baylan, 2012)
Lactobacillus spp.	Control toxin production and growth of fungi (Karunaratne et al., 1990; Sarrocco and Vannacci, 2017)	Well-known probiotic, with numerous citations	The genus is recorded to have antimicrobial properties in several investigations*
Enterobacter cloacae	Biocontrol of nematodes (Duponnois et al., 1999)	Inhibits human norovirus in gnotobiotic pigs (Lei et al., 2016)	Is known to be resident in the oral cavity (Leao et al., 2011)
Azotobacter spp.	Both biocontrol and biofertilizer properties (Chauhan et al., 2012)	Has been suggested as a probiotic in aquaculture of fish (Sayeda et al., 2011)	A well-known genus capable of atmospheric nitrogen fixation*
Saccharomyces cerevisiae	Used against phytopathogen *Colletotrichum acutatum* (Lopes et al., 2015)	Used against gastrointestinal tract disorders in humans (Palma et al., 2015)	Commonly used in wine making and baking*

* Well-established information.

concentration, high osmotic pressure, and low pH), they can grow rapidly on inexpensive substrates in fermenters, thus enabling production in large quantities; additionally, yeasts do not produce allergenic spores or mycotoxins, and they have simple nutritional requirements that enable them to colonize dry surfaces for long periods of time (Spadaro and Droby, 2016). It is interesting to note that even in the sphere of probiotic research, yeasts and fungi have received increasing interest recently, and there is interest in the role of the mycobiome (Bourdichon et al., 2012; Huffnagle and Noverr, 2013).

Various *Bacillus* spp. have been of interest in both disciplines, with many being commercialized (Hong et al., 2005; Cutting, 2011). Although the safety of certain *Bacillus* spp., particularly *B. subtilis* and *B. indicus*, has been questioned (Hong et al., 2008), particularly *B. subtilis* can be found as a commercial probiotic product currently. *Bacillus* spp. have been incorporated in biocontrol products, which also includes the well-known *B. thuringiensis*. Particularly, *B. subtilis* is the active ingredient in the commercial biocontrol product Avogreen®, used to control pathogens mainly on avocado. Similarly, lactobacilli and bifidobacteria, well-known probiotics (Vlasova et al., 2016), have been investigated as potential biocontrol agents.

Besides the sole use of biocontrol agents, combining biocontrol organisms with certain chemicals, that serve as elicitors in biocontrol strategies, has enhanced pathogen control (Zhang et al., 2018). As it is apparent that an integrated approach improves control of pathogens, such safe integrated means of control can be tailor-made at the level of testing on specific F&V.

It is interesting to see how organisms of interest have changed in the sphere of biocontrol. Among the bacteria, lactobacilli have been of interest in the early days of biocontrol to prevent mycotoxin effects (Karunaratne et al., 1990). With time, the need to focus on microbes on the phytosphere and thereafter to look at the plant growth-promoting microbes (Karunaratne, 2011) attracted attention. The interest later changed toward yeasts for several reasons (Spadaro and Droby, 2016). In that background of events, it seems appropriate to mention the role of the well-known biocontrol fungus, *Metarhizium anisopliae*, used for parasitic insect control (Tiago et al., 2014) (Table 15.1). Perhaps it may have a role in biocontrol, especially on fruits that need controlled atmosphere storage, and also the fungus or its debris may have a positive role in the human gut.

Also, there are numerous studies on the probiotic effects of *Streptomyces* on shrimp, one of the latest reviews mentioning this organism is by Hariharan and Dharmaraj (2018), where the focus has been on marine *Streptomyces* for drug usage. Also, *Paenibacillus polymyxa* has been shown to inhibit the growth of *Escherichia coli* and is suggested as a probiotic on bovine feed (Naghmouchi et al., 2013). As certain strains of *E. coli* are well-known human pathogens as well, it is worth investigating its effect as a probiotic on humans too. *P. polymyxa* is known as a plant growth-promoting rhizobacterium and has been used as a soil inoculant for biocontrol of plant pathogenic fungi (Hong et al., 2016).

Those microorganisms that have the potential to serve as probiotics are deemed to be of value in postharvest applications because to confer antagonistic effects, cell numbers should be adequate, and as such large numbers are needed if they are to benefit the host by serving as probiotics (Karunaratne, 2018b). However, there is always a risk involved in working with large numbers of viable microorganisms, even if they are considered to be harmless.

As antibiotic capsules are made using substances extracted from microbes in disease control, now with the evidence that microbial debris or paraprobiotics having biological effects, parts of microbes, and/or microbial metabolites may be put into a capsule to get their beneficial effects. With that background knowledge, it will be interesting to assess the effects of many of the microbial metabolites and debris on the phytosphere. For instance, xanthan gum and its utility as an elicitor in plant defense are aspects that could be investigated. It will be productive and interesting to explore the paraprobiotic role of those organisms that have served as elicitors in plant defense.

With time, data generated at research level have shown that combining biocontrol methods with physical means of control, such as with heat treatments (Sui et al., 2016) or with unconventional chemical compounds (Zhang et al., 2018), provide better control, thus extending shelf life of produce. Also, the use of volatile organic compounds in controlling postharvest pathogens has been reviewed, and it is shown that certain biocontrol agents exert their antagonistic effects through the use of such volatiles (Mari et al., 2016). Once we know where the barriers are, and the material as well as the conditions to make antagonists work effectively, the opportunities are numerous.

6 Areas Needing Attention

The barriers at the upper end of the pipeline in the quest for developing biocontrol strategies are obvious. Spadaro and Droby (2016) report from a survey on research papers carried out in 2015 that most investigations on biocontrol describe the ground work, such as isolation, efficacy, and mode of action, and less papers describe the large-scale production, formulation, and packaging to prolong viability and preservation of the biocontrol agents and the result of field testing. Inconsistent performance at commercial level is recorded as a major problem. It is reported that the cost associated with the development of a biocontrol product is very high (Usall et al., 2016). Those cost adding factors include biocontrol product development, registration, production, and marketing. Even a new agrochemical will have to go through these stages of the pipeline for commercialization and therefore, these aspects are not restricted to biocontrol products. Concerns about food safety, including chemical residues and environmental impact, over the past 20 yr have resulted in substantial regulatory changes in the use of pesticides. Regulatory restrictions are increasing on the use of a variety of chemical fungicides used to curb postharvest diseases. Several products have been lost from the market due to the unwillingness of companies to maintain registration (Droby et al., 2016).

This information points toward the fact that basic research has to continue at the upper end of the pipeline, thus making those organisms that have shown potential to cut through legislation barriers. Droby et al. (2016) explain the necessities of mass production process of antagonists concentrating on a suitable culture medium that is economical but still provides an adequate supply of nutrients and energy needed for cellular metabolism, growth, and population stability. The necessities for optimization of growth conditions (temperature, agitation, aeration, and pH), based on whether it is a solid- or liquid-phase fermentation, are other aspects needing investigation.

Probiotic research leading to probiotic commercialization has to follow these same steps. Naturally, as expected with the introduction of a new microbe, there is concern even regarding the use of probiotics (Didari et al., 2014). In both disciplines (biocontrol and probiotic), we are dealing with live organisms, which keep multiplying. Obviously, product stability, purity, and aspects of their mechanisms of action are at stake as pointed out by Usall et al. (2016). It is encouraging that in both disciplines the dead organisms have shown promising results. If such residues are used instead of live organisms, the process could be better managed and the effects could be expected to be more stable. When biopesticides are not alive, that will make a huge difference in terms of the ease of handling, and in turn making it much more practical for the users.

Even volatile organic compounds (Mari et al., 2016) and some microbial-based coatings, such as chitosan-based coatings (Ncama et al., 2018), present in fungal walls are used in postharvest research indicating that it is not necessary to have live microbes for pathogen control but microbial-derived products are sufficient. Interestingly, chitosan microencapsulation is used in probiotic work to protect probiotics *Lactobacillus bulgaricus* and *L. rhamnosus* during gastric transit (Shori, 2017). Even antimicrobial compounds from various microbes, such as the yeast *Aureobasidium pullulans*, as well as the bacteria *Lactobacillus paracasei* and *Bacillus* sp. (Seal et al., 2018), are used. Besides on humans, probiotic usage is popular to control pathogens on animals, such as fish, chicken, and many mammals (Hossain et al., 2017).

Secreted metabolites of certain microbes confer protection to crop plants. *Pseudomonas* and rhizobacteria are known to have this capability to protect plants against microbial pathogens (Rahman et al., 2018). Also, secreted antimicrobial compounds, enzymes, and exopolysaccharides can suppress pathogen establishment (Hossain et al., 2017). As several modes of action in biocontrol do not require viable organisms at the time of action, biocontrol research could investigate for such modes of action in known probiotics with antagonistic effects. Perhaps organisms having such modes of action may be termed parabiocontrol agents.

At postharvest level, manipulating the microenvironment of the F&V to provide optimum conditions to prolong their quality after harvest, could also include determination of the optimum conditions where biocontrol antagonist performance is optimized. Inconsistent performance is a problem at commercial level (Usall et al., 2016) and managing the microenvironment may help to overcome this problem. Such requirements should be determined at the time of scaling up. Questions of species purity are better handled at this time too. Shelf-life determination at the time of scaling up perhaps may be different than when the organisms are on the surface of F&V because the requirements of microorganisms could be complex and it may not be always possible to ascertain what all their requirements are. Wisniewski et al. (2016) correctly emphasize on the need for multiple interventions.

Figure 15.4 highlights a feasible path in achieving the goal of probiotic antagonists. Screening can proceed on organisms that have shown promising results. In-vitro and in vivo experimentation can proceed with both groups (scientists who have been pursuing biocontrol work and those on probiotics) collaborating to achieve a common goal. Such an endeavor will have more chance for success, especially when handling environmental and toxicological issues. The current situation, depicted in Figure 15.4, is that basic investigations often face barriers in both

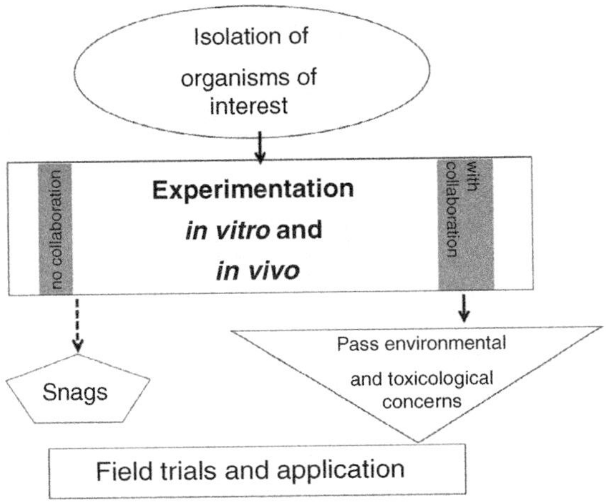

Figure 15.4 An ideal situation for testing probiotic antagonists.

disciplines. With productive collaboration, these problems could be handled collectively, thus making solution of regulatory issues easier.

An important consideration on the part of the lawmakers will be to provide appropriate guidelines for the use and commercialization of probiotic biocontrol agents. As large numbers of microbes are included in the edible covers of F&V, it may be necessary to have a special channel for probiotic treated F&V, rather than assuming that such products are beneficial for all members of a population. It is possible that the immunocompromised individuals will not benefit by the large numbers of organisms present on edible covers of F&V, and such products may not be suitable for this group of individuals. Also, considering that biocontrol agents with probiotic potential are expected to be health promoting, certain guidelines for their approval may have more lenient sanctions.

7 Concluding Remarks

While developing probiotic biocontrol agents is an achievable goal, it should not suffer the same predicaments as the current status of antibiotics (i.e., pathogen resistance and safety). To clarify further, being biological entities, unlike chemicals, they are not static but tend to evolve. The consequences of certain microbial interactions occurring on the surfaces of F&V may not be desirable. Also, probiotics may be specific to certain populations, ethnic groups, and communities. Many factors need to be considered in this endeavor, based on the understanding of who the consumer is. What the typical composition of F&V in the diet is, amount of microbial cells that may be ingested by an individual, tolerance of the high numbers of microbes even if they are labeled as probiotics, and the number of individuals who may become allergic to the organisms are some issues to be considered in developing and popularizing probiotic antagonists. Some of the difficulties that should be addressed in biocontrol research are mass production to meet the commercial demand, and having a stable formulation of organisms with a reasonable shelf life similar to that of agrochemicals. With this regard, if paraprobiotics show biocontrol effects, such effects can be monitored better at the level of mass production as non-viable microbes and microbial debris are easier to manipulate than living organisms, and certainly would be a better choice. There may be more opportunities to discover probiotic antagonists if some of the future research investigations could employ information generated in the older research that reported positive outcomes, but never progressed due to factors related to regulatory and legislative barriers. If that is possible, many efforts in prior biocontrol research will not be in vain. We will end this with the quotation of Louis Pasteur who is considered as the Father of Microbiology "In the field of observation, chance favors only the prepared mind." If all researchers involved in biocontrol and probiotic investigations could have a prepared mind to extend their sphere of observation, the goal of probiotic antagonists will not be that far.

Acknowledgments

We wish to thank Prof. C.L. Abayasekara for critically reading the manuscript.

References

Abdelfattah, A., Malacrino, A., Wisniewski, M., Cacciola, S.O. and Schena, L. 2018. Metabarcoding: A powerful tool to investigate microbial communities and shape future plant protection strategies. *Biological Control* 120, 1–10.

Angelakis, E., Merhej, V. and Raoult, D. 2013. Related actions of probiotics and antibiotics on gut microbiota and weight modification. *Lancet Infectious Diseases* 13, 889–899.

Ayala, F.R., Bauman, C., Cogliati, S., Lenini, C., Bartolini, M. and Grau, R. 2017. Microbial flora, probiotics, *Bacillus subtilis* and the search for a long and healthy human longevity. *Microbial Cell* 4, 133–136.

Barbosa, C., Lage, P., Esteves, M., Chambel, L., Mendes-Faia, A. and Mendes-Ferreira, A. 2018. Molecular and phenotypic characterization of *Metschnikowia pulcherrima* strains from Douro wine region. *Fermentation* 4, 8.

Bautista-Baños, S., Bosquez-Molina, E. and Barrera-Necha, L.L. 2014. *Rhizopus stolonifer* (soft rot). pp. 1–44. *In*: S. Bautista-Baños (ed.). *Postharvest Decay. Control Strategies.* Academic Press, Elsevier Inc., London, UK.

Baylan, O. 2012. An opportunistic pathogen frequently isolated from immunocompromised patients: *Burkholderia cepacia* complex. *Mikrobiyoloji Bülteni* 46, 304–318.

Bourdichon, F., Casaregola, S., Farrokh, C., Frisvad, J.C., Gerds, M.L., Hammes, W. P., Harnett, J., Huys, G., Laulund, S., Ouwehand, A., Powell, I.B., Prajapati, J.B., Seto, Y., Ter Schure, E., Van Boven, A., Vankerckhoven, V., Zgoda, A., Tuijtelaars, S. and Hansen, E.B. 2012. Food fermentations: Microorganisms with technological beneficial use. *International Journal of Food Microbiology* 154, 67–97.

Chauhan, S., Wadhwa, K., Vasudeva, M. and Narula, N. 2012. Potential of *Azotobacter* spp. as biocontrol agents against *Rhizoctonia_solani* and *Fusarium oxysporum* in cotton (*Gossypium hirsutum*), guar (*Cyamopsis tetragonoloba*) and tomato (*Lycopersicum esculentum*). *Archives of Agronomy and Soil Science* 58, 1365–1385.

Cutting, S.M. 2011. *Bacillus* probiotics. *Food Microbiology* 28, 214–220.

De Almada, C.N., Almada, C.N., Martinez, R.C.R. and Sant'Ana, A.S. 2016. Paraprobiotics: Evidences on their ability to modify biological responses, inactivation methods and perspectives on their application in foods. *Trends in Food Science and Technology* 58, 56–114.

De Simone, C. 2018. The unregulated probiotic market. *Clinical Gastroenterology and Hepatology* 17, 809–817.

de Vrese, M. and Schrezenmeir, J. 2008. Probiotics, prebiotics, and synbiotics. *Advances in Biochemical Engineering* 111, 1–66.

Didari, T., Solki, S., Mozaffari, S., Nikfar, S. and Abdollahi, M. 2014. A systematic review of the safety of probiotics. *Expert Opinion on Drug Safety* 13, 227–239.

Droby, S. and Wisniewski, M. 2018. The fruit microbiome: A new frontier for postharvest biocontrol and postharvest biology. *Postharvest Biology and Technology* 140, 107–112.

Droby, S., Wisniewski, M., Teixidó, N., Spadaro, D. and Jijakli, M.H. 2016. The science, development, and commercialization of postharvest biocontrol products. *Postharvest Biology and Technology* 122, 22–29.

Duponnois, R., Amadou, M.B. and Mateille, T. 1999. Beneficial effects of *Enterobacter cloacae* and *Pseudomonas mendocina* for biocontrol of *Meloidogyne incognita* with the endospore-forming bacterium *Pasteuria penetrans*. *Nematology* 1, 95–101.

EFSA (European Food safety Authority). 2017. BIOHAZ Panel. Scientific opinion on the update of the list of QPS-recommended biological agents intentionally added to food or feed as notified to EFSA. *EFSA Journal* 5, 4664.

El-Baz, A.F., El-Enshasy, H.A., Shetaia, Y.M., Mahrous, H., Othman, N.Z. and Yousef, A.E. 2018. Semi-industrial scale production of a new yeast with probiotic traits, *Cryptococcus* sp. YMHS, isolated from the Red Sea. *Probiotics and Antimicrobial Proteins* 10, 77–88.

Emmert, E.A.B. and Handelsman, J. 1999. Biocontrol of plant disease: A (Gram-) positive perspective. *FEMS Microbiological Letters* 171, 1–9.

FAO (Food and Agriculture Organization). 1995. Code of conduct for the import and release of exotic biological control agents. Available at: www.fao.org/docrep/x5585E/x5585e0h.htm

FAO-WHO (Food and Agriculture Organization-World Health Organization). 2002. Guidelines for the Evaluation of Probiotics in Food. Joint FAO/WHO Workgroup on Drafting Guidelines for the Evaluation of Probiotics in Food. London, Ontario, Canada, April 30 and May 1, 2002.

Food Insight. 2009. Functional Foods Fact Sheet. Available at: www.foodinsight.org/Content/3842/Final%20Functional%20Foods%20Backgrounder.pdf

Francesco, A.D., Martini, C. and Mari, M. 2016. Biological control of postharvest diseases by microbial antagonists: How many mechanisms of action? *European Journal of Plant Pathology* 145, 711–717.

Gardener, M.B.B. and Fravel, D.R. 2002. Biological control of plant pathogens: Research, commercialization, and application in the USA. *Plant Health Progress*. https://doi.org/10.1094/PHP-2002-0510-01-RV. Available at: www.apsnet.org/publications/apsnetfeatures/Pages/BioControl.aspx

Hariharan, S. and Dharmaraj, S. 2018. Selection of new probiotics: The case of *Streptomyces*. pp. 27–54. *In*: A. Grumezescu and A.M. Holban (eds.). *Therapeutic, Probiotic, and Unconventional Foods*. Academic Press, Amsterdam, The Netherlands.

Heydari, A. and Misaghi, I.J. 1998. Biocontrol activity of *Burkholderia cepacia* against *Rhizoctonia solani* in herbicide-treated soil. *Plant and Soil* 202, 109–116.

Hong, C.E., Kwon, S.Y. and Park, J.M. 2016. Biocontrol activity of *Paenibacillus polymyxa* AC-1 against *Pseudomonas syringae* and its interaction with *Arabidopsis thaliana*. *Microbiological Research* 185, 13–21.

Hong, H.A., Duc, L.H. and Cutting, S.M. 2005. The use of bacterial spore formers as probiotics. *FEMS Microbiological Reviews* 29, 813–835.

Hong, H.A., Huang, J.M., Khaneja, R., Hiep, L.V., Urdaci, M.C. and Cutting, S.M. 2008. The safety of *Bacillus subtilis* and *Bacillus indicus* as food probiotics. *Journal of Applied Microbiology* 105, 510–520.

Hossain, M.I., Sadekuzzaman, M. and Ha, S.D. 2017. Probiotics as potential alternative biocontrol agents in the agriculture and food industries: A review. *Food Research International* 100, 63–73.

Huffnagle, G.B. and Noverr, M.C. 2013. The emerging world of the fungal microbiome. *Trends in Microbiology* 21, 334–341.

Ibrahem, M.D. 2015. Evolution of probiotics in aquatic world: Potential effects, the current status in Egypt and recent prospective. *Journal of Advanced Research* 6, 765–791.

IDF (International Dairy Federation). 2012. Safety demonstration of the microbial food cultures in fermented food products. *Bulletin of the International Dairy Federation* 255, 2–68.

Juneja, D., Borah, A.K., Nasa, P., Singh, O., Javeri, Y. and Dang, R. 2011. *Candida sake* candidaemia in non-neutropenic critically ill patients: A case series. *Critical Care Resuscitation* 13, 187–191.

Karunaratne, A.M. 2011. Biocontrol mechanisms employed by PGPR and strategies of microbial antagonists in disease control on the post-harvest environment of fruits. pp. 131–164. *In*: D. Maheshwari (ed.). *Plant Growth & Health Promoting Bacteria, Bacteria in Agrobiology: Crop Ecosystems*. Springer-Verlag Berlin, Heidelberg, Germany.

Karunaratne, A.M. 2018a. Probiotic foods: Benefits to the cereal based Sri Lankan diet. *Ceylon Journal of Science* 47, 105–123.

Karunaratne, A.M. 2018b. A multifaceted approach to harness probiotics as antagonists on plant based foods, for enhanced benefits to be reaped at a global level. *Journal of the Science of Food and Agriculture* 98, 5189–5196.

Karunaratne, A.M. and Gunasinghe, W.K.R.N. 2011. Agrochemical free treatments to improve postharvest shelf-life of banana. pp. 1–32. *In*: A.E. Cohen (ed.). *Bananas: Nutrition, Diseases and Trade Issues*. Nova Publishers, New York, NY.

Karunaratne, A.M., Wezenberg, E. and Bullerman, L.B. 1990. Inhibition of mold growth and aflatoxin production by *Lactobacillus* spp. *Journal of Food Protection* 53, 230–236.

Kim, J., Muhammad, N., Jhun, B.H. and Yoo, J.-W. 2016. Probiotic delivery systems: A brief overview. *Journal of Pharmaceutical Investigations* 46, 377–386.

Kobayashi, Y., Inagawa, H., Kohchi, C., Kazumura, K., Tsuchiya, H., Miwa, T., Okazaki, K. and Soma, G. 2018. Oral administration of *Pantoea agglomerans*-derived lipopolysaccharide prevents development of atherosclerosis in high-fat diet-fed apoE-deficient mice via ameliorating hyperlipidemia, pro-inflammatory mediators and oxidative responses. *PLoS ONE* 13, e0195008.

Leao, M.V., Cassia, R.C., Santos, S.S., Silva, C.R. and Jorge, A.O. 2011. Influence of consumption of probiotics on presence of enterobacteria in the oral cavity. *Brazil Oral Research* 25, 401–406.

Lei, S., Samuel, H., Twitchell, E., Bui, T., Ramesh, A., Wen, K., Weiss, M., Li, G., Yang, X., Jiang, X. and Yuan, L. 2016. *Enterobacter cloacae* inhibits human norovirus infectivity in gnotobiotic pigs. *Scientific Reports* 6, 25017.

Lopes, M.R., Klein, M.N., Ferraz, L.P., da Silva, A.C. and Kupper, K.C. 2015. *Saccharomyces cerevisiae*: A novel and efficient biological control agent for *Colletotrichum acutatum* during pre-harvest. *Microbiological Research* 175, 93–99.

Mari, M., Bautista-Baños, S. and Sivakumar, D. 2016. Decay control in the postharvest system: Role of microbial and plant volatile organic compounds. *Postharvest Biology and Technology* 122, 70–81.

Massomo, S.M.S., Mortensen, C.N., Mabagala, R.B., Newman, M.A. and Hockenhull, J. 2004. Biological control of black rot (*Xanthomonas campestris* pv. *campestris*) of cabbage in Tanzania with *Bacillus* strains. *Journal of Phytopathology* 152, 98-105.

McFarland, L.V. 2015. From yaks to yogurt: The history, development, and current use of probiotics. *Clinical Infectious Diseases* 60, S85–S90.

Muramatsu, D., Iwai, A., Aoki, S., Uchiyama, H., Kawata, K., Nakayama, Y., Nikawa, Y., Kusano, K., Okabe, M. and Miyazaki, T. 2012. β-glucan derived from *Aureobasidium pullulans* is effective for the prevention of influenza in mice. *PLoS ONE* 7, e41399.

Naghmouchi, K., Baah, J., Cudennec, B. and Drider, D. 2013. Required characteristics of *Paenibacillus polymyxa* JB-0501 as potential probiotic. *Archives of Microbiology* 195, 537–543.

Ncama, K., Magwaza, L.S., Mdishwa, A. and Tesfay, S.Z. 2018. Plant based edible coatings for managing postharvest quality of fresh horticultural produce: A review. *Food Packaging and Shelf Life* 16, 157–167.

OIE/WHO/FAO (World Organization for Animal Health/United States Food and Drug Administration/World Health Organization). 2016. The OIE strategy on antimicrobial resistance and the prudent use of antimicrobials. Available at: www.oie.int/fileadmin/Home/eng/Media_Center/docs/pdf/PortailAMR/EN_OIE-AMRstrategy.pdf

Palma, M.L., Zamith-Miranda, D., Martins, F.S., Bozza, F.A., Nimrichter, L., Montero-Lomeli, M., Marques, E.T., Jr. and Douradinha, B. 2015. Probiotic *Saccharomyces cerevisiae* strains as biotherapeutic tools: Is there room for improvement? *Applied Microbiology and Biotechnology* 99, 6563–6570.

Pattemore, J.A., Hane, J.K., Williams, A.H., Bree, A.L., Ben, W., Stodart, J. and Ash, G.J. 2014. The genome sequence of the biocontrol fungus *Metarhizium anisopliae* and comparative genomics of *Metarhizium* species. *BMC Genomics* 15, 660.

Rahman, S.F.S.A., Singh, E., Pieterse, C.M.J. and Schenk, P.M. 2018. Emerging microbial biocontrol strategies for plant pathogens. *Plant Science* 267, 102–111.

Salas, M.L., Mounier, J., Valence, F., Coton, M., Thierry, A. and Coton, E. 2017. Antifungal microbial agents for food biopreservation – A review. *Microorganisms* 5, 37.

Sarrocco, S. and Vannacci, G. 2017. Preharvest application of beneficial fungi as a strategy to prevent postharvest mycotoxin contamination. *Crop Protection* 110, 166–170.

Sayeda, M.A., Wafa, M.I.A. and Abbas, W.T. 2011. Evaluation of *Azotobacter* and *Azospirillum* biofertilizers as a probiotics in *Oreochromis niloticus* aquaculture. *Journal of Fisheries and Aquatic Science* 6, 535–544.

Seal, B.S., Drider, D., Oakley, B.B., Brüssow, H., Bikard, D., Rich, J.O., Miller, S., Devillard, E., Kwan, J., Bertin, G., Reeves, S., Swift, S.M., Raicek, M. and Gay, C.G. 2018. Microbial-derived products as potential new antimicrobials. *Veterinary Research* 49, 66.

Senan, S., Prajapati, J.B., Joshi, C.G., Sreeja, V., Gohel, M.K., Trivedi, S., Patel, R.M., Pandya, H., Singh, U.S., Phatak, A. and Patel, H.A. 2015. Geriatric respondents and non-respondents to probiotic intervention can be differentiated by inherent gut microbiome composition. *Frontiers in Microbiology* 6, 944.

Shanks, N., Greek, R. and Greek, J. 2009. Are animal models predictive for humans? *Philosophy, Ethics, and Humanities in Medicine* 4, 2.

Shori, A.B. 2017. Microencapsulation improved probiotics survival during gastric transit. *HAYATI Journal of Biosciences* 24, 1–5.

Spadaro, D. and Droby, S. 2016. Development of biocontrol products for postharvest diseases of fruit: The importance of elucidating the mechanisms of action yeast antagonists. *Trends in Food Science and Technology* 47, 39–49.

Sui, Y., Wisniewski, M., Droby, S., Norelli, J. and Liu, J. 2016. Recent advances and current status of the use of heat treatments in postharvest disease management systems: Is it time to turn up the heat? *Trends in Food Science and Technology* 51, 34–40.

Tiago, V.P., de Oliveira, N.T. and Lima, E.A.L.A. 2014. Biological insect control using *Metarhizium anizopliae*: Morphological, molecular, and ecological aspects. *Ciencia Rural, Santa Maria* 44, 645–651.

US FDA (United States Food and Drug Administration). 2001. Microorganisms and microbial-derived ingredients that are used in foods (Partial list). Available at: www.fda.gov/food/ingredientspackaginglabeling/gras/microorganismsmicrobialderivedingredients/default.htm

Usall, J., Torres, R. and Teixidó, N. 2016. Biological control of postharvest diseases of fruit: A suitable alternative? *Current Opinion in Food Science* 11, 51–55.

Vlasova, A.N., Kandasamy, S., Chattha, K.S., Rajashekara, G. and Saif, L.J. 2016. Comparison of probiotic *lactobacilli* and *bifidobacteria* effects, immune responses and rotavirus vaccines and infection in different host species. *Veterinary Immunology and Immunopathology* 172, 72–84.

WHO (World Health Organization). 2013. Global strategy on diet, physical activity and health: Promoting fruit and vegetable consumption around the world. Available at: www.who.int/dietphysicalactivity/fruit/en/index2.html

Wisniewski, M., Droby, S., Norelli, J., Liu, J. and Schena, L. 2016. Alternative management technologies for postharvest disease control: The journey from simplicity to complexity. *Postharvest Biology and Technology* 122, 3–10.

Zhang, H., Mahunu, G.K., Castoria, R., Yang, Q. and Apaliya, M.T. 2018. Recent developments in the enhancement of some postharvest biocontrol agents with unconventional chemicals compounds. *Trends in Food Science and Technology* 78, 180–187.

Chapter 16

Control of Postharvest Decay of Fresh Produce by Heat Treatments; the Risks and the Benefits

Elazar Fallik
Department of Postharvest Science, Agricultural Research Organization (ARO), the Volcani Center, Rishon LeZiyyon, Israel

Zoran Ilic'
Faculty of Agriculture in Lešak, University of Pristina in Kosovska Mitrovica, Lešak, Serbia

Abbreviations

1-MCP	1-Methylcyclopropene
CAT	Catalase
CBS	Citrus black spot
HA	Hot air
HAT	Hot air treatment
HT	Heat treatment
HWD	Hot water dip
HWR	Hot water rinse
HWRB	Hot water rinsing and brushing
HWT	Hot water treatment
IMZ	Imazalil
NaOCl	Sodium hypochlorite
QHWT	Quarantine hot water treatment
RF	Radio frequency
RH	Relative humidity
ROS	Reactive oxygen species
SA	Salicylic acid

1 Introduction

Freshly harvested fruits and vegetables have become an increasingly important part of the human diet in the decade, especially since about 2005, when their nutritional and health-improving properties became increasingly known (Villa-Rodríguez et al., 2015). Numerous epidemiological studies have shown an inverse correlation between fruit and vegetable consumption, on the one hand, and, on the other hand, chronic diseases, including various types of cancer and cardiovascular diseases (Havens et al., 2012). Quality attributes of fruits or vegetables that are associated with consumer acceptance include their colorful skin, attractive fragrance, and appetizing flavors, tastes, and textures. However, approximately 1.3 billion tons of food are wasted or lost every year, of which one-third comprises fresh produce that was rendered unfit for consumption because of physiological and pathological deterioration (ISHS, 2012; Prusky et al., 2013). Apart from these, some postharvest pathogens also produce metabolites, that is, mycotoxins, which pose a health hazard to humans and can contaminate processed fruit products (Sarubbi et al., 2016). Synthetic fungicides can be applied to fresh produce to reduce pathological deterioration during handing and prolonged storage, but consumers are demanding chemical-free produce or lower presence of residues (Mahajan et al., 2016).

For many years postharvest heat treatments (HTs) have been known to be effective in managing postharvest diseases and physiological disorders. These treatments are completely safe for humans and the environment, that is, they are residue-free and environment-friendly, and feasibly can be used without constraints imposed by regulators (Usall et al., 2016). These treatments can also enhance resistance of fresh produce to environmental stress and help to preserve fruit and vegetable quality during prolonged storage and extended shelf life (Fallik, 2010;

Sivakumar and Fallik, 2013; Sui et al., 2016). However, heat-based techniques expose fruits and vegetables to the hazards of physiological disorders (Mittler et al., 2012), which can be affected by many parameters, such as the initial quality of the fruits, their physiological maturity stage, and their exposure to physical and chemical agents in the field (Woolf and Ferguson, 2000; Rodoni et al., 2016). Furthermore, the incidence of damage increased with increasing temperature and treatment duration, and with an increasing period in subsequent cold storage (Sivakumar and Fallik, 2013).

HT can directly control decay caused by agents that are found on skin surface or within two or three epidermal cell layers under the cuticle of fresh produce, but it can also control decay indirectly by inducing defense mechanisms and triggering physiological and pathological responses that enable the host to inhibit the growth of pathogens and to withstand stressful conditions encountered during storage (Fallik, 2010). Pathogens vary considerably in their sensitivity to prestorage high-temperature treatments (Sivakumar and Fallik, 2013). Pathogen mortality is not always proportional to the temperature–time product of the treatment, although reports have indicated an inverse log–linear relationship between the time to inactivate pathogens and the heat-treatment temperature (Fallik, 2010). The vegetative cells and conidia of most fungi are inactivated by exposure to 60°C for 5–10 min in vitro. Spore germination and germ tube elongation were found to be more sensitive to HTs than dormant spores, which were unaffected by hot water at the temperatures evaluated (Sivakumar and Fallik, 2013). Fruit responses to HT depend on their maturity stage at harvest, fruit size and weight, the cultivar, and the parameters of the heat-treatment treatment itself, including temperature, duration, and mode of application. On the other hand, the physiological responses of different cultivars to HTs can vary according to season, cultural conditions, growing location, and other prestorage practices (Fallik, 2010). HTs can be applied alone or in combination with other means of controlling decay on fresh fruits and vegetables; Section 2 will summarize the use of HTs during the last 5 yr, and their conditions for beneficial or deleterious effects on freshly harvested produce (Table 16.1).

2 Types of Heat Treatment

There are various physical means of applying heat, such as hot water dipping, brief hot water rinsing and brushing (HWRB), hot air (HA), steam or vapor heating, and radio frequency (RF) heating. These treatments provide quarantine security, reduce decay, enhance resistance of cold-sensitive cultivars to chilling injury, and maintain fresh produce quality during prolonged cold storage and shelf life. Effects of HTs on fruit and vegetable crops were thoroughly reviewed during the last decade (Lurie, 2008; Fallik, 2010; Schirra et al., 2011; Sivakumar and Fallik, 2013; Sui et al., 2016).

Basically, there are four types of HT, whose use depends upon the time of exposure of the fresh or fresh-cut produce; in general, the produce is exposed to a temperature of 37–65°C for periods ranging from a few seconds to several days. HTs can be applied by means of various systems: batch, continuous, or drainage (Sivakumar and Fallik, 2013).

Table 16.1 Beneficial and harmful protocols of various physical treatments on selected fresh harvested produce

Crop	*Physical treatment*	*Beneficial*	*Damage (external and internal)*	*Reference*
Apple	HWD	50–54°C, 3 min		Maxin et al. (2012)
	HWR	55°C, 20–25 s		Maxin et al. (2012)
	HWR		65°C, 20 s	Maxin et al. (2012)
	HWD		45°C, 10 min	Moscetti et al. (2013)
Banana	HWD	50°C, 20 min		Alvindia (2012)
	HWD		>50°C, 20 min	Alvindia (2012)
Broccoli	HWD	50°C, 3 min		Perini et al. (2017)
	HWV		>45°C	Duarte-Sierra et al. (2016)
Date	HA	55°C, 30 min; 60°C, 15 min; and 60°C, 20 min		Ben-Amor et al. (2016)
	RF	27.12 MHz, 6 min		Supapvanich and Promyou (2017)
	HWD		60°C, 3 min	Ben-Amor et al. (2016)
Citrus fruit	HWD	53°C, 3 min		García et al. (2016)
	HWD+	50°C, 3 min		D'Aquino et al. (2017)
	HWD	56°C, 20 s		Strano et al. (2014)
	HWD	56°C, 2 min		Yan et al. (2016)
	HWD		55°C, 5 min	Obenland and Neipp (2005)
	HWD		≥47.5°C, 2 min	Ghasemnezhad et al. (2008)
Mango	HWD	55°C, 3 min		Chiangsin et al. (2016)
	HWD	60°C, 1 min		Wang et al. (2017)
	HA		47°C, >40 min	Hoa et al. (2010)
	HWD		46.1°C, 90 min	García et al. (2016)
Papaya	HWD+	48°C, 20 min		Ayón-Reyna et al. (2017)
	HWD+	42°C, 40 min + 49°C, 20 min		Supapvanich and Promyou (2017)

(Continued)

Table 16.1 (Cont.)

Crop	Physical treatment	Beneficial	Damage (external and internal)	Reference
	HWD		42°C, 30 min + 49° C, 20 min	Armstrong and Mangan (2007)
Peach	HA	38°C, 3 h		Huan et al. (2017)
	HWD	48°C, 10 min		Huan et al. (2017)
	HWD	60°C, 1 min		Spadoni et al. (2014)
	HWD	48°C, 12 min		Jemric et al. (2011)
	HWD+	50°C, 30 s		Pazolini et al. (2016)
	HWV		>37°C	
	HWD		≥65°C	Karabulut et al. (2010)
Pepper	HWD	45°C, 3 min		Rodoni et al. (2016)
	HWD		55°C, 5 min	Rodoni et al. (2016)
Pomegranate	HWD		65°C	Mirdehghan and Rahemi (2005)
Spinach	HWD		>45°C, 60 s	Gómez et al. (2008)
	HWD		>50°C, 30 s	Glowacz et al. (2013)
Strawberry	HWD		45°C, >5 min	Caleb et al. (2016)
Tomato	HWD	40°C, 30 min		Pinheiro et al. (2014)
	HA+	38°C, 12 hr		Wei et al. (2016)
Rocket	HWD		>50°C	Koukounaras et al. (2009)
Wolfberry	HWD+	42°C, 30 min		Ban et al. (2015)
Yam	HA	29°C, 3–5 d		Lee and Park (2013)

HA, hot air; HWD, hot water dip; HWR, hot water rinse; HWV, hot water vapor; RF, radio frequency. + indicates combined treatment.

Hot air treatments (HATs), whether applied as forced air or vapor, that is, dry or wet heat, respectively, are considered to be long-exposure treatments; they are carried out in a high-airflow, controlled-temperature facility. In dry HTs, heat is applied without adding steam or water, and, in general, they use higher temperatures and longer exposure periods than steam-based treatments. However, the time and temperature requirements depend on the properties and size of the fruit or vegetable being treated and the targeted pathogen.

Hot water dips (HWDs) are medium-duration treatments lasting for several minutes to hours. Most HWD treatment facilities in commercial use form part of either a batch or continuous system. In a batch system, produce is placed in a bin or plastic crate and loaded onto a platform, which then is lowered into a hot water tank where the produce remains at the prescribed temperature for a designated period and is then taken out. In a continuous system, produce is carried, either loose or in a plastic basket, on a conveyor belt that moves slowly through hot water from one end of a tank to the other. The belt speed is set to ensure that the fruits or vegetables are submerged for a designated length of time. In this category, RF heating, which generates dry heat for up to several hours, is also considered a medium-duration treatment.

HWRB technology is considered a very short-duration treatment; it lasts for 15–25 s at temperatures between 45 and 63°C (Fallik, 2004). The apparatus comprises a stainless steel machine equipped with parallel revolving brushes, a thermostatically controlled hot water tank through which the fresh produce passes and then dried while passing along a tunnel equipped with several air fans for about 2 min, before subsequent sorting, grading, and packing.

3 Benefits of Heat Treatments

HTs, directly or indirectly, can benefit fresh produce quality by killing decay-causing agents while maintaining produce quality during prolonged storage.

Apple fruits (*Malus* × *domestica* Borkh.) are particularly vulnerable to fungal diseases because they are often stored for up to a year under controlled atmosphere. *Alternaria alternata* (Fr.) Keissl., *Botrytis cinerea* Pers., and *Penicillium expansum* Link are the main postharvest pathogens of apples; they are responsible for the development of black, gray, and blue molds, respectively (Moscetti et al., 2013). Significant reductions in apple fruit rot incidence were achieved by dipping the fruits in water for 3 min at 50–54°C, or rinsing them for 20–25 s at 55°C, followed by up to 100 d of cold storage at 2°C and 14 d at 18°C (Maxin et al., 2012). Pathogens were controlled better by dipping the fruits than by rinsing them. The effects of HA at 38°C for 3 hr or a HWD at 48°C for 10 min on peach (*Prunus persica* L. Batsch.) fruit quality and activities of reactive oxygen species (ROS) and antioxidants during storage at 4°C were investigated by Huan et al. (2017). They found that HA or HWD treatment maintained fruit quality, decreased ROS activity, and enhanced antioxidant activity. HWD was more effective than HA in alleviating internal browning symptoms. Similarly, Spadoni et al. (2014) obtained a good reduction in brown rot, caused by *Monilinia laxa* (Aderh. & Ruhland) Honey, in naturally infected peaches by HWD treatment at 60°C for 1 min. In an earlier study, Jemric et al. (2011) found a significant reduction in brown rot incidence in peach fruits after dipping in 48°C water for 12 min. In another investigation, Pazolini et al. (2016) reported that a treatment with canola and Indian mustard extracts combined with HWD at 50°C for 30 s controlled brown rot on peach effectively and that this combined treatment was as effective as treatment with the fungicide azoxystrobin (Amistar, Syngenta, Switzerland). In kiwifruit, another deciduous fruit, Chen et al. (2015) found that hot water treatment (HWT) at 45°C for 10 min controlled gray and blue mold in fruits stored at 4 and 25°C, without impairing fruit quality.

Citrus fruits also were found to benefit from HTs. Mandarins (cvs. 'Fortune,' 'Ortanique,' 'Ellendale,' 'Clemenules,' and 'Hernandina') and oranges (cvs. 'Navelate,' 'Navelina,' 'Lanelate,' 'Salustiana,' and 'Valencia') were inoculated with the fungi *Penicillium digitatum* (Pers.) Sacc. and *Penicillium italicum* Wehmer and then subjected to postharvest HWD (García et al., 2016); treatment at 53°C for 3 min was found to be the most effective regarding reduction of decay incidence without affecting juice content, soluble solids, pH, titratable acidity, or sensory quality. Strano et al. (2014) found that HWD of 'Tarocco' orange fruits (*Citrus sinensis* L. Osbeck) at 56°C for 20 s significantly reduced decay incidence without affecting external or internal quality. In a more recent study, D'Aquino et al. (2017) showed that sequential treatment with sodium hypochlorite (NaOCl) at 200 mg/L and the synthetic fungicide imazalil (IMZ) at 50 mg/L, both at 50°C, was as effective as IMZ at 1000 mg/L at 20°C in controlling Penicillium decay in lemon fruits; they concluded that decay control and quality preservation of fruits stored at room temperature for several weeks could be accomplished by a sequential treatment with NaOCl and a heated water emulsion of IMZ at 50 mg/L, when fruits were wrapped with highly gas-permeable plastic films (D'Aquino et al., 2017). Citrus black spot (CBS), caused by *Guignardia citricarpa* Kiely, is a fungal disease first described in Australia in the 1890s and first found in southwest Florida in 2010. On asymptomatic fruits, although the treatments did not significantly reduce CBS lesion development, fruits dipped in water at 56°C for 120 s showed significantly less severe disease than the controls after 2 wk of storage; none of the treatments caused peel scalding or impaired fruit quality (Yan et al., 2016).

Beneficial effects of HTs, used alone or in combination with other treatments, were reported for tropical and exotic fruits by Sivakumar and Fallik (2013). The efficacy of HWT, as an alternative to fungicide treatment to control crown rot and maintain postharvest quality in banana cv. 'Buñgulan' (*Musa* genome AAA) was studied by Alvindia (2012). The optimum combination of temperature and exposure time to control crown rot was 50°C for 20 min. This treatment also delayed ripening, improved the overall appearance and firmness of the fruits, and prolonged their green life. Hydrothermal treatment, composed of a water treatment at 48°C for 20 min followed by immersion for 20 min in 1% Ca, prepared from calcium chloride, was shown to effectively control anthracnose. It has also been evaluated to control anthracnose (*Colletotrichum gloeosporioides* [Penz.] Penz. & Sacc) in papaya (*Carica papaya* L.). An HT–Ca treatment reduced anthracnose incidence and severity compared to the control (Ayón-Reyna et al., 2017). Supapvanich and Promyou (2017) investigated the effects of dipping of 'Holland' papaya fruits in hot water containing salicylic acid (SA) on the quality of the fruits immersed in 0, 1.0, 2.0, and 3.0 mM SA solutions at 42°C for 40 min, followed by 20 min at 49°C, with the controls untreated. Both hot water and hot SA dips reduced the incidence of decay during storage, delayed ripening slightly, did not harm the fruit, and enhanced contents of carotenoids, ascorbic acid, and antioxidants. Chiangsin et al. (2016) compared postharvest dipping of mango fruits (*Mangifera indica* L.) for 3 min in the fungicides, such as prochloraz and azoxystrobin, or in hot water at 55° C, and found that hot water was the most effective treatment to control anthracnose. Ban et al. (2015) investigated the synergistic application of a HWD at 42°C for 30 min and a 1% chitosan coating on postharvest quality traits and microbe populations on wolfberry fruits (*Lycium barbarum* L.). This fruit is also called Goji berry and is well known as a medicinal fruit. They found that the combination of prestorage HT and chitosan coating both controlled decay and maintained higher levels of ascorbic acid, total phenolic contents, and antioxidant capacity. Dipping 'Ivory' mango in water at 50°C for 10 min, 60°C for 1 min, or 70°C for 5 s could improve fruit quality parameters, such as the nutritional content; however, 60°C for 1 min yielded the best improvement of fruit quality (Wang et al., 2017).

Immersing tomato (*Solanum lycopersicum* L.) in water at 40°C for 30 min was very effective in reducing decay and it delayed physiochemical changes during storage, especially at the turning maturity stage (Pinheiro et al., 2014). Wei et al. (2016) investigated the combination of treatments with HA at 38°C for 12 hr and *Cryptococcus laurentii* (Kuff.) C.E. Skinner for controlling postharvest decay caused by *B. cinerea* in cherry tomato at various maturity stages. They found that the combined treatment significantly reduced the incidence of gray mold at both mature green and pink stages, irrespective of whether fruits were artificially inoculated with *B. cinerea* before or after treatment. However, for a given treatment, disease incidence on mature green fruits was lower than that on pink fruits (Wei et al., 2016). Lee and Park (2013) observed that a treatment of Chinese yams (*Dioscorea polystachya* Turcz.) consisting of 29°C HA for 3–5 d followed by storage at 0.5°C, which was phased in to reduce chilling injury, had the best quality after 7 mon of storage.

Fresh-cut products can also benefit from physical treatments. Rodoni et al. (2016) subjected both green and ripe fresh-cut peppers (*Capsicum annuum* L.) to HWD at 45°C for 3 min and obtained lower spoilage than in the controls; the treatments markedly reduced soft rots. This treatment also inhibited shriveling, weight loss, and color changes, and reduced fruit respiration during storage. The treatments did not alter sugar content, acidity, or antioxidant capacity.

Broccoli (*Brassica oleracea* L., Italica group) is a vegetable used worldwide with high nutritional value and health benefits. Its inflorescences are harvested while the floral heads, branchlets, and florets are immature. Perini et al. (2017) found that dipping the first 5 cm of broccoli stems at 50°C for 3 min enhanced the overall quality of the product during storage, and Duarte-Sierra et al. (2016) established a methodology to determine the hormetic dose of heat to be applied to broccoli florets to improve their shelf life and maintain their quality during storage. HT at 41°C for 180 min was considered to be the hormetic heat dose for this fresh produce with regard to promoting color retention without creating excessive anaerobic conditions or damage to cells.

HT can be used to disinfest fresh or dry fruits of insects. The date palm (*Phoenix dactylifera* L.) is considered a symbol of life in the desert because it tolerates high temperatures, drought, and salinity (Haouel et al., 2010). Ben-Amor et al. (2016) studied the effect of three HATs, 55°C for 30 min, 60°C for 15 min, or 60°C for 20 min, on 'Deglet Noor' date fruits that were then stored for 45 d at 2°C followed by a retail period of 4 d at 23°C. The results showed that the use of HAT caused 100% mortality of larvae of *Ectomyelois ceratoniae* Zeller in naturally infested dates, without any harm to fruit quality. Another HT, which used RF heating to control insect infestation, was reported by Garbati Pegna et al. (2017). An exposure to date fruits of 6 min to 27.12 MHz RF caused mortality of the larvae, pupae, and adults of the dried fruit beetle *Carpophilus hemipterus* L.

4 Disadvantages of Heat Treatments

Although HTs can benefit fresh produce quality, use of high temperatures in combination with excessive exposure periods could injure the fresh produce. Symptoms of heat injury that are visible immediately or after a period of storage usually involve external damage, such as skin browning, surface pitting, rind black spots, or stem browning. However, internal damage, such as flesh darkening, water loss, and failure to ripen normally, can also occur (Fallik, 2010).

Dipping apples in water at 45°C for 10 min elicited peel damage, that is, surface browning and internal breakdown disorders, and led to undesirable early fruit senescence after only 2 mon of storage at 2°C and 90% relative humidity (RH) (Moscetti et al., 2013). In contrast, Maxin et al. (2012) observed no heat damage on 'Ingrid Marie' apples subjected to HWD at temperatures up to 50°C, or to a hot water rinse at up to 58°C. However, following rinsing at 65°C for 20 s, incidences of *Psilocybe washingtonensis* A.H. Sm., *P. expansum, Mucor* spp., and *Phoma exigua* Desm. were higher than in untreated control fruits or apples rinsed at lower temperatures, and it caused injury to the fruit skin. Karabulut et al. (2010) evaluated brief (30 or 60 s) immersion in water at 24, 50, 55, 60, 65, or 70°C for control of brown rot, caused by *Monilinia fructicola* (G. Winter) Honey, on peaches, nectarines, and plums. Immersion in water at 55°C for 60 s or at 60°C for 30 or 60 s significantly reduced both incidence and severity of decay, but water temperatures of 65°C or higher were phytotoxic and caused moderate-to-severe surface injuries. Pomegranate (*Punica granatum* L.) is one of the oldest known edible fruits and its culture is associated with ancient civilizations of the Middle East (Holland et al., 2009). It is an important fruit in the tropical and subtropical regions of the world and is cultivated mainly in the Mediterranean countries, because of their moderate climates (Opara et al., 2015). One of the major problems associated with pomegranate fruits stored for 3–4 mon is excessive weight loss, which may result in hardening of the husk and browning of the rind and arils (Pareek et al., 2015). Mirdehghan and Rahemi (2005) reported that HWT of whole 'Malas Yazdi' pomegranate fruits at 45°C for 2 or 5 min significantly reduced skin browning, but after treatment at 65°C, they observed slight heat injury and increased percentage of browning in fruits stored for 3 mon.

Rind disorders of citrus fruits are a common occurrence that reduces fruit marketability. One type of disorder, which is characterized by darkened sunken areas or pits on the rind, can be induced by exposure to high temperatures. The water status of the peel is believed to be a key factor in its development; lemon fruits that were immersed in hot water at 55°C for 5 min developed these symptoms (Obenland and Neipp, 2005). Ghasemnezhad et al. (2008) found that the optimal HWD protocol for avoiding these symptoms in 'Satsuma' mandarins (*Citrus unshiu* Marc.) was 47.5–50°C for 2 min; temperatures greater than 47.5 and 50°C for 2 and 5 min, respectively, caused heat damage in the form of rind browning.

Mangoes (*Mangifera indica* L.) are among the most popular tropical fruits, but they are susceptible to various diseases. Hoa et al. (2010) subjected 'Cat Hoa loc' mangoes at the commercially mature stage to HAT at 47°C and 90% RH, for 20, 40, 60, 90, 120, or 180 min; this treatment for 20 min did not affect the ripening process, biochemical composition, or organoleptic quality of the fruit, but application for 40–180 min significantly increased skin burns; the main symptom was appearance of brown spots. Fruits treated for 60–180 min had overripe flesh with little natural aroma and a fermented taste (Hoa et al., 2010). Kim et al. (2009) reported that darkened lenticels in mangoes might be induced by heat when treatment conditions were outside the optimum range. Osuna-García et al. (2015) treated 'Kent' mango fruits with aqueous 1-methylcyclopropene (1-MCP), combined or not combined with quarantine hot water treatment (QHWT), that is, 46.1°C for 90 min. They evaluated the effects on delay of the ripening process, extension of shelf-life period, and maintenance of fruit quality, and found that QHWT had a negative impact on fruit quality, causing surface spots and lenticel blackening to develop during simulated shipping and marketing. They were harmed less when 1-MCP was applied before QHWT, and harm was most severe when 1-MCP was applied after QHWT.

Papaya is one of the major dessert-fruit crops; cultivated in the tropical and subtropical regions of the world. A two-stage hot water treatment was developed for disinfecting Hawaiian papayas of fruit fly eggs by immersing one-quarter-ripe papayas in water at 42°C for 30 min, then in water at 49°C for 20 min, and finally, hydrocooling them by spraying with water at ambient temperature (Armstrong and Mangan, 2007). Depending upon maturity at harvest and seasonal changes, the two-stage hot water quarantine treatment caused uneven ripening when applied at some maturity stages, with surface scalding and abnormal fruit softening (Armstrong and Mangan, 2007). Poor flavor, lumpiness, and "hard-core" texture were also common quality problems in hot water-treated papayas (Sivakumar and Wall, 2013).

The efficacy of HWT, as an alternative to chemical treatment, in controlling crown rot and maintaining postharvest quality in banana cv. 'Buñgulan' was studied by Alvindia (2012), who found that HWT application protocols exceeding 50°C for 20 min caused fruit injuries, such as scalding, shriveling, and failure to soften.

Strawberry (*Fragaria* × *ananassa* Duch.) is one of the most important soft-fruit crops consumed globally. Strawberries are highly desirable for their unique flavor attributes and nutritional benefits; they are a rich source of health-promoting compounds, such as essential nutrients, vitamins, anthocyanins, and organic acids, and they exhibit high antioxidant capacity (Musto and Satriano, 2010). Caleb et al. (2016) evaluated the effects of HWDs on the physiochemical quality of 'Sonata' strawberries by dipping fruits in hot water at 35 or 45°C for 5 or 10 min, respectively, and storing them in open trays at 4°C for 9 d and then for an additional 3 d at 16°C. Overall, HWD treatment of 45°C for 5 min had no detrimental effects and was best for maintaining strawberry quality attributes, and prevented decay; however, HWD at 45°C for longer than 5 min impaired fruit quality and should, therefore, be avoided (Caleb et al., 2016).

Ben-Amor et al. (2016) found that HWT used for quarantine purposes in 'Deglet Noor' date fruits slightly reduced their quality when applied at 60°C for 3 min, although the fruits did not show any visible heat damage.

The influence of ripening stage on pepper heat tolerance was studied by Rodoni et al. (2016). They found that green fresh-cut peppers were more tolerant of HT than ripe fruits. Slight heat damage symptoms appeared only after 55°C for 5 min, the most intense application applied.

Leafy vegetables can also benefit from physical treatments. Gómez et al. (2008) reported that HWT at 40°C for 3.5 min reduced tissue breakdown and chlorophyll loss in spinach, by retaining the green color by retarding yellowing and delaying leaf senescence. Glowacz et al. (2013) found that the maximal HWT, with respect to temperature and duration, that did not cause a significant reduction in spinach leaf quality was 45°C for 60 s. Treatment at 50°C for 30 s or longer resulted in significant color change, 20% greater membrane damage, and significantly higher solute leakage and respiration after 10 d, compared with leaves treated at 40°C for up to 120 s, 45°C for up to 60 s, or 50°C for less than 30 s. Koukounaras et al. (2009) treated rocket (*Eruca sativa* Mill.) leaves with water at 50, 52.5, or 55°C and found that HWT at 50°C for 30 s delayed leaf yellowing, but higher temperatures and/or longer treatment caused visible heat injuries.

Broccoli is one of the most heavily consumed Brassica crops, because of its contents of bioactive compounds, such as glucosinolates and flavonoids. However, the edible parts, the florets, deteriorate very rapidly. Duarte-Sierra et al. (2016) treated broccoli florets with hot humidified air at temperatures of 32–52°C for periods of 5–1440 min. Subsequently, after 10 d of storage at 10°C, they found HT above 45°C led to severe anaerobic conditions and tissue damage.

5 Possible Modes of Action of Heat Disorders

When high temperatures are used in combination with long exposure times to reduce or eliminate infections by pathogens, HT can have negative impacts, which include changes in membrane fluidity and lipid structure, and activation of ion channels (Mittler et al., 2012). In plants, cell membranes are among the first components affected by stress; alterations in membrane structure may cause modification of cellular compartmentalization. Heat can cause alterations in plant tissue microstructure that influence texture: tissue softening is caused by loss of turgor pressure, purging of occluded air, thermal degradation of middle lamella pectins and other cell wall polysaccharides, and gelatinization of starch (Mittler et al., 2012). Cell membrane damage may be manifested in many different ways including changes in composition or structure, a loss of protein functionality, and changes in functionality, such as in fluidity or permeability. Injury caused by heat stress could result in either increased efflux caused by impaired semipermeability of the plasmalemma, or decreased influx because of damage to the active transport system (González and Barrett, 2010).

A combined biochemical and proteomic approach was successfully employed by Wu et al. (2015) in an investigation of the responses of grape berries to postharvest HWT during subsequent cold storage. They found that HWT was effective to retard fruit ripening and maintain fruit quality for up to 45 d, depending upon the temperature and exposure duration. Proteomic analysis revealed that under optimal heat-treatment conditions, proteins associated with the defense response remained at levels similar to those observed in untreated fruits, which may indicate the molecular basis for retarded fruit ripening and fruit-quality maintenance in heat-treated grape berries (Wu et al., 2015). On the other hand, heat stress is also well known to induce production of ROS, which cause cell damage that results in oxidative stress. Thus, the antioxidant defense system protects plants against oxidative stress by activating the main ROS-scavenging enzymes, such as ascorbate peroxidase, Mn-superoxide dismutase, glutathione reductase, catalase (CAT), and reduced nicotinamide adenine dinucleotide phosphate oxidase (Wang et al., 2012). Optimal HT was reported to increase superoxidase dismutase and CAT activities, which decreased hydrogen peroxide (a known causative factor of cell membrane damage) and protected cells from oxidation damage (Rabiei et al., 2011). In contrast, however, heat stress caused free radicals to be generated in plant cells and these caused severe damage to detached leaves of cinnamon myrtle (Sommano, 2015).

HT induced increases in free putrescine and spermidine during storage of pomegranate. These higher polyamine levels, as well as maintenance of the unsaturated:saturated fatty acid ratio, could account for the prolonged retention of membrane integrity and fluidity within these fruits during storage (Pareek et al., 2015).

In contrast, heat-treatment damage that could reduce biosynthesis of polyamine and the antioxidative system enzymes would render membrane integrity susceptible to heat injury and, in turn, increase the occurrence of both internal and external disorders in the fruit (Khademi et al., 2013). Another possible mechanism of heat injury was reported by Ghasemnezhad et al. (2008). They observed that the browning of Satsuma mandarin rind damaged by HT at 55°C for 2 min could be attributed in part to loss of membrane integrity that was associated with significant increases in vacuolar pyrophosphatase and vacuolar ATPase activities in the fruit peel, which probably occurred to compensate for a loss of tonoplast integrity.

Internal breakdown (darkening of the fruit flesh) after HWT at higher temperatures and longer exposures may also be due to a combination of unfavorable temperature and gas composition within the fruit. This injury might be attributed to an imbalance between oxygen and carbon dioxide concentrations within the fruit during HT (Duarte-Sierra et al., 2012). In addition, high respiration rates at elevated temperatures can lead to anaerobic conditions because of accelerated tissue oxygen depletion, and also because of membrane disruption and progressive leakage (Duarte-Sierra et al., 2012). Depletion of oxygen from plant tissue under high-temperature treatment can be expected, because of reduction in gas solubility with increasing temperature (degassing). Consequently, anaerobic conditions may prevail, leading to increased anaerobic respiration (Duarte-Sierra et al., 2016).

6 Conclusions

Recent investigations have shown that more than one-third of harvested fruits and vegetables are lost (Buzby et al., 2014; OECD, 2014). Most losses are due to senescence and pathogen infections in the field or after harvest, which lead to postharvest decay.

Numerous reports are available describing physical treatments that control postharvest decay in agricultural products without harming their quality. However, not all freshly harvested produce, especially heat-sensitive fruits and vegetables, can tolerate the physical conditions required to control decay-causing agents. Also, energy efficiency and cost requirements are two very important considerations in the development of commercially practical and effective physical treatments for controlling decay in fresh or fresh-cut products (Shahkoomahally and Ramezanian, 2015). The search for safe and effective alternatives to synthetic chemical fungicides for reducing postharvest losses of harvested commodities without impairing their quality is still very challenging. Environmental factors, genetic background, fruit maturity, fruit size, mineral nutrition deficiencies during production, or other pre- and/or postharvest conditions are the most likely reasons for the varied results obtained (Fallik, 2010; Sivakumar and Fallik, 2013; Usall et al., 2016). In addition, many technical and engineering factors should be considered in selecting the most appropriate treatment. The main factors generally considered are availability and cost of commercial-scale systems. By comparing the physical treatments, it is possible to take into account several other factors, and before trying to develop a new treatment, a decision should be taken as to which technology is the most appropriate for potential commercial-scale application. Most of the problems associated with each physical treatment probably can be solved, but the solutions may not always be economically or technically feasible.

The complexity of the mode of action, which involves both direct effects on the pathogen and indirect effects on the host, is still not well understood. A better understanding of both direct and indirect modes of action of HTs on pathogens and fresh produce tissue will facilitate the development of optimal, successful, and relatively inexpensive hot water dipping or rinsing treatments. Equipment applying these treatments will control pathogens without impairing the overall quality of the fruits or vegetables. However, the new tools available in molecular biology, the metabolomics and proteomic technologies, should enable exploration of possible modes of action, and ease the task of elucidating alternative physiological and biochemical means to ensure that the chosen physical treatment will be the most efficient and beneficial for postharvest produce management.

Acknowledgments

Contribution No. 782/17 from the Agricultural Research Organization, the Volcani Center, Rishon LeZiyyon, Israel.

References

Alvindia, D. 2012. Revisiting hot water treatments in controlling crown rot of banana cv. Buñgulan. *Crop Protection* 33, 59–64.

Armstrong, J.W. and Mangan, R.L. 2007. Commercial quarantine heat treatments. pp. 311–340. *In*: J. Tang, E. Mitcham, S. Wang and S. Lurie (eds.). *Heat Treatments for Postharvest Pest Control.* CABI, Wallingford, UK.

Ayón-Reyna, L.E., González-Robles, A., Rendón-Maldonado, J.G., Báez-Flores, M.E., López-López, M.E. and Vega-García, M.O. 2017. Application of a hydrothermal-calcium chloride treatment to inhibit postharvest anthracnose development in papaya. *Postharvest Biology and Technology* 124, 85–90.

Ban, Z., Wei, W., Yang, X., Feng, J., Guan, J. and Li, L. 2015. Combination of heat treatment and chitosan coating to improve postharvest quality of wolfberry (*Lycium barbarum*). *International Journal of Food Science and Technology* 50, 1019–1025.

Ben-Amor, R., Dhouibi, M.H. and Aguayo, E. 2016. Hot water treatments combined with cold storage as a tool for *Ectomyelois ceratoniae* mortality and maintenance of Deglet Noor palm date quality. *Postharvest Biology and Technology* 112, 247–255.

Buzby, J.C., Wells, H.F. and Hyman, J. 2014. The estimated amount, value, and calories of postharvest food losses at the retail and consumer levels in the United States. Report no. EIB-121, US Department of Agriculture, Economic Research Service. Available at: www.ers.usda.gov/publications/eib-economic-information-bulletin/eib-xxx.aspx.

Caleb, O.J., Wegner, G., Rolleczek, C., Herppich, W.B., Geyer, M. and Mahajan, P.V. 2016. Hot water dipping: Impact on postharvest quality, individual sugars, and bioactive compounds during storage of 'Sonata' strawberry. *Scientia Horticulturae* 210, 150–157.

Chen, H., Cheng, Z., Wisniewski, M., Liu, Y. and Liu, J. 2015. Ecofriendly hot water treatment reduces postharvest decay and elicits defense response in kiwifruit. *Environmental Science and Pollutant Research* 22, 15037–15045.

Chiangsin, R., Wanichkul, K., Guest, D.I. and Sangchote, S. 2016. Reduction of anthracnose on ripened mango fruits by chemicals, fruit bagging, and postharvest treatments. *Australasian Plant Pathology* 45, 629–635.

D'Aquino, S., Dai, S., Deng, Z., Gentile, A., Angioni, A., De Pau, L. and Palma, A. 2017. A sequential treatment with sodium hypochlorite and a reduced dose of imazalil heated at 50°C effectively control decay of individually film-wrapped lemons stored at 20°C. *Postharvest Biology and Technology* 124, 75–84.

Duarte-Sierra, A., Corcuff, R., Angers, P. and Arul, J. 2012. Effect of heat treatment using humidified air on electrolyte leakage in broccoli florets: Temperature-time relationships. *Acta Horticulturae* 945, 149–155.

Duarte-Sierra, A., Corcuff, R. and Arul, J. 2016. Methodology for the determination of hormetic heat treatment of broccoli florets using hot humidified air: Temperature–Time relationships. *Postharvest Biology and Technology* 117, 118–124.

Fallik, E. 2004. Prestorage hot water treatments (immersion, rinsing and brushing). *Postharvest Biology and Technology* 32, 125–134.

Fallik, E. 2010. Hot water treatments of fruits and vegetables for postharvest storage. *Horticultural Reviews* 38, 191–212.

Garbati Pegna, F., Sacchetti, P., Canuti, V., Trapani, S., Bergesio, C., Belcari, A., Zanoni, B. and Meggiolaro, F. 2017. Radio frequency irradiation treatment of dates in a single layer to control *Carpophilus hemipterus*. *Biosystems Engineering* 155, 1–11.

García, J.F., Olmo, M. and García, J.M. 2016. Decay incidence and quality of different citrus varieties after postharvest heat treatment at laboratory and industrial scale. *Postharvest Biology and Technology* 118, 96–102.

Ghasemnezhad, M., Marsh, K., Shilton, R., Babalar, M. and Woolf, A. 2008. Effect of hot water treatments on chilling injury and heat damage in 'satsuma' mandarins: Antioxidant enzymes and vacuolar ATPase, and pyrophosphatase. *Postharvest Biology and Technology* 48, 364–371.

Glowacz, M., Mogren, L.M., Reade, J.P.H., Cobb, A.H. and Monaghan, J.M. 2013. Can hot water treatments enhance or maintain postharvest quality of spinach leaves? *Postharvest Biology and Technology* 81, 23–28.

Gómez, F., Fernández, L., Gergoff, G., Guiamet, J.J., Chaves, A. and Bartoli, C.G. 2008. Heat shock increases mitochondrial H_2O_2 production and extends postharvest life of spinach leaves. *Postharvest Biology and Technology* 49, 229–234.

González, M.E. and Barrett, D.M. 2010. Thermal, high pressure, and electric field processing effects on plant cell membrane integrity and relevance to fruit and vegetable quality. *Journal of Food Science* 75, R121–R130.

Haouel, S., Mediouni-Ben Jemâa, J. and Khouja, M.L. 2010. Postharvest control of the date moth *Ectomyelois ceratoniae* using eucalyptus essential oil fumigation. *Tunisian Journal of Plant Protection* 5, 201–212.

Havens, E.K., Martin, K.S., Yan, J., Dauser-Forrest, D. and Ferris, A.M. 2012. Federal nutrition program changes and healthy food availability. *American Journal of Preventive Medicine* 43, 419–422.

Hoa, T.T., Hien, D.M., Self, G. and Ducamp, M.-N. 2010. Effects of hot air treatment on postharvest quality of 'cat Hoa loc' mangoes. *Fruits* 65, 237–244.

Holland, D., Hatib, K. and Bar-Yaakov, I. 2009. Pomegranate: Botany, horticulture, breeding. *Horticultural Reviews* 35, 127–191.

Huan, C., Han, S., Jiang, L., An, X., Yu, M., Xu, Y., Ma, R. and Yu, Z. 2017. Postharvest hot air and hot water treatments affect the antioxidant system in peach fruit during refrigerated storage. *Postharvest Biology and Technology* 126, 1–14.

ISHS (International Society for Horticultural Science). 2012. Growing economically. pp. 41–50. *In*: A. Aitken, E. Hewett, I. Warrington, C. Hale, D. McCaffrey (contributors). *Harvesting the Sun, a Profile of World Horticulture.* Scripta Horticulturae No. 14. Leuven, Belgium.

Jemric, T., Ivic, D., Fruk, G., Matijas, H.S., Cvjetkovic, B., Bupic, M. and Pavkovic, B. 2011. Reduction of postharvest decay of peach and nectarine caused by *Monilinia laxa* using hot water dipping. *Food Bioprocessing and Technology* 4, 149–154.

Karabulut, O.A., Smilanick, J.L., Crisosto, C.H. and Palou, L. 2010. Control of brown rot of stone fruit by brief heated water immersion treatments. *Crop Protection* 29, 903–906.

Khademi, O., Salvador, A., Zamani, Z. and Besada, C. 2013. Effects of hot water treatments on antioxidant enzymatic system in reducing flesh browning of persimmon. *Food Bioprocess Technology* 6, 3038–3046.

Kim, Y., Lounds-Singleton, A.J. and Talcott, S.T. 2009. Antioxidant phytochemical and quality changes associated with hot water immersion treatment of mangoes (*Mangifera indica* L). *Food Chemistry* 115, 989–993.

Koukounaras, A., Siomos, A.S. and Sfakiotakis, E. 2009. Impact of heat treatment on ethylene production and yellowing of modified atmosphere packaged rocket leaves. *Postharvest Biology and Technology* 54, 172–176.

Lee, D.-S. and Park, Y.-M. 2013. Optimization of curing treatment and storage temperature of Chinese yam. *Korean Journal of Horticultural Science and Technology* 31, 289–298.

Lurie, S. 2008. Heat treatment for enhancing postharvest quality. pp. 246–259. P. Paliyath, D.P. Murr, A.V. Handa and S. Lurie (eds.). *Postharvest Biology and Technology of Fruits, Vegetables, and Flowers.* Wiley-Blackwell, Ames, IA.

Mahajan, P.V., Caleb, O.J., Singh, Z., Watkins, C.B. and Geyer, M. 2016. Postharvest treatments of fresh produce. *Philosophical Transactions of the Royal Society A* 372, 20130309.

Maxin, P., Weber, R.W.S., Pedersen, H.L. and Williams, M. 2012. Control of a wide range of storage rots in naturally infected apples by hot-water dipping and rinsing. *Postharvest Biology and Technology* 70, 25–31.

Mirdehghan, S.H. and Rahemi, M. 2005. Effects of hot water treatment on reducing chilling injury of pomegranate (*Punica granatum*) fruit during storage. *Acta Horticulturae* 682, 887–892.

Mittler, R., Finka, A. and Goloubinoff, P. 2012. How do plants feel the heat? *Trends in Biochemistry and Science* 37, 118–125.

Moscetti, R., Carletti, L., Monarca, D., Cechini, M., Stella, E. and Massantini, R. 2013. Effect of alternative postharvest control treatments on the storability of 'Golden Delicious' apples. *Journal of the Science of Food and Agriculture* 93, 2691–2697.

Musto, M. and Satriano, M.L. 2010. Fruit responses to postharvest heat treatment time: Characterisation of heat-treated strawberry (*Fragaria* × *ananassa*) cv. 'Candonga' fruits. *Agronomy Research* 8, 815–826.

Obenland, D. and Neipp, P. 2005. Chlorophyll fluorescence imaging allows early detection and localization on lemon rind injury following hot water treatment. *HortScience* 40, 1821–1823.

OECD (Organization for Economic Co-operation and Development). 2014. Market and Trade Impacts of Food Loss and Waste Reduction. K. Okawa (Ed.). Paris, France. Available at: www.oecd.org/officialdocuments/publicdisplaydocumentpdf/?cote = TAD/CA/APM/WP (2014) 35/FINAL&docLanguage=En.

Opara, U.L., Atukuri, J. and Fawole, O.A. 2015. Application of physical and chemical postharvest treatments to enhance storage and shelf life of pomegranate fruit – A review. *Scientia Horticulturae* 197, 41–49.

Osuna-García, J.A., Brecht, J.K., Huber, D.J. and Nolasco-González, Y. 2015. Aqueous 1-methylcyclopropene to delay ripening of 'Kent' mango with or without quarantine hot water treatment. *HortTechnology* 25, 349–357.

Pareek, S., Valero, D. and Serrano, M. 2015. Postharvest biology and technology of pomegranate. *Journal of the Science of Food and Agriculture* 95, 2360–2379.

Pazolini, K., dos Santos, I., Giaretta, R.D., Marcondes, M.M., Reiner, D.A. and Citadin, I. 2016. The use of brassica extracts and thermotherapy for the postharvest control of brown rot in peach. *Scientia Horticulturae* 209, 41–46.

Perini, M.A., Sin, I.N., Reyes Jara, A.M., Gómez Lobato, M.E., Civello, P.M. and Martínez, G.A. 2017. Hot water treatments performed in the base of the broccoli stem reduce postharvest senescence of broccoli (*Brassica oleracea* L. Var. italic) heads stored at 20°C. *LWT - Food Science and Technology* 77, 314–322.

Pinheiro, J., Alegria, C., Abreu, M., Sol, M., Gonçalves, E.M. and Silva, C.L.M. 2014. Postharvest quality of refrigerated tomato fruit (*Solanum lycopersicum*, cv. Zinac) at two maturity stages following heat treatment. *Journal of Food Processing and Preservation* 39, 697–707.

Prusky, D., Alkan, N., Mengiste, T. and Fluhr, R. 2013. Quiescent and necrotrophic lifestyle choice during postharvest disease development. *Annual Review of Phytopathology* 51, 155–176.

Rabiei, V., Es-haghi, S., Aazami, M.A. and Sharafi, Y. 2011. Combined effects of hot air and calcium chloride on quality and antioxidant enzymes activity in 'Red Delicious' apple fruits. *Journal of Medicinal Plants Research* 5, 4954–4961.

Rodoni, L.M., Hasperue, J.H., Ortiz, C.M., Lemoine, M.L., Concellon, A. and Vicente, A.R. 2016. Combined use of mild heat treatment and refrigeration to extend the postharvest life of organic pepper sticks, as affected by fruit maturity stage. *Postharvest Biology and Technology* 117, 168–176.

Sarubbi, F., Formisano, G., Auriemma, G., Arrichiello, A. and Palomba, R. 2016. Patulin in homogenized fruit and tomato products. *Food Control* 59, 420–423.

Schirra, M., D'Aquino, S., Cabras, P. and Angioni, A. 2011. Control of postharvest diseases of fruit by heat and fungicides: Efficacy, residue levels, and residue persistence. A review. *Journal of Agricultural and Food Chemistry* 59, 8531–8542.

Shahkoomahally, S. and Ramezanian, A. 2015. Hot water combined with calcium treatment improves physical and physicochemical attributes of kiwifruit (*Actinidia deliciosa* cv. Hayward) during storage. *HortScience* 50, 412–415.

Sivakumar, D. and Fallik, E. 2013. Influence of heat treatments on quality retention of fresh and fresh-cut produce. *Food Reviews International* 29, 294–320.

Sivakumar, D. and Wall, M.M. 2013. Papaya fruit quality management during the postharvest supply chain. *Food Reviews International* 29, 24–48.

Sommano, S. 2015. Physiological and biochemical changes during heat stress induced browning of detached *Backhousia myrtifolia* (Cinnamon Myrtle) tissues. *Tropical Plant Biology* 8, 31–39.

Spadoni, A., Guidarelli, M., Sanzani, S.M., Ippolito, A. and Mari, M. 2014. Influence of hot water treatment on brown rot of peach and rapid fruit response to heat stress. *Postharvest Biology and Technology* 94, 66–73.

Strano, M.C., Calandra, M., Aloisi, V., Rapisarda, P., Strano, T. and Ruberto, G. 2014. Hot water dipping treatments on Tarocco orange fruit and their effects on peel essential oil. *Postharvest Biology and Technology* 94, 26–34.

Sui, Y., Wisniewski, M., Droby, S., Norelli, J. and Liu, J. 2016. Recent advances and current status of the use of heat treatments in postharvest disease management systems: Is it time to turn up the heat? *Trends in Food Science and Technology* 51, 34–40.

Supapvanich, S. and Promyou, S. 2017. Hot water incorporated with salicylic acid dips maintaining physicochemical quality of 'Holland' papaya fruit stored at room temperature. *Emirates Journal of Food and Agriculture* 29, 18–24.

Usall, J., Ippolito, A., Sisquella, M. and Neri, F. 2016. Physical treatments to control postharvest diseases of fresh fruits and vegetables. *Postharvest Biology and Technology* 122, 30–40.

Villa-Rodríguez, J.A., Palafox-Carlos, H., Yahia, E.M., Ayala-Zavala, J.F. and González-Aguilar, G.A. 2015. Maintaining antioxidant potential of fresh fruits and vegetables after harvest. *Critical Reviews in Food Science and Nutrition* 55, 806–822.

Wang, H., Yang, Z., Song, F., Chen, W. and Zhao, S. 2017. Effects of heat treatment on changes of respiration rate and enzyme activity of ivory mangoes during storage. *Journal of Food Processing and Preservation* 41, e12737.

Wang, H., Zhang, Z., Xu, L., Huang, X. and Pang, X. 2012. The effect of delay between heat treatment and cold storage on alleviation of chilling injury in banana fruit. *Journal of the Science of Food and Agriculture* 92, 2624–2629.

Wei, Y., Xu, M., Wu, H., Tu, S., Pan, L. and Tu, K. 2016. Defense response of cherry tomato at different maturity stages to combined treatment of hot air and *Cryptococcus laurentii*. *Postharvest Biology and Technology* 117, 177–186.

Woolf, A.B. and Ferguson, I.B. 2000. Postharvest responses to high fruit temperatures in the field. *Postharvest Biology and Technology* 21, 7–20.

Wu, Z., Yuan, X., Li, H., Liu, F., Wang, Y., Li, J., Cai, H. and Wang, Y. 2015. Heat acclimation reduces postharvest loss of table grapes during cold storage – Analysis of possible mechanisms involved through a proteomic approach. *Postharvest Biology and Technology* 105, 26–33.

Yan, J., Dewdney, M.M., Roberts, P.D. and Ritenour, M.A. 2016. The effect of postharvest hot water and fungicide treatments on *Guignardia citricarpa* growth and the development of citrus black spot symptoms on 'Valencia' orange fruit. *HortScience* 51, 1555–1560.

Chapter 17

UV-C Hormesis

A Means of Controlling Diseases and Delaying Senescence in Fresh Fruits and Vegetables during Storage

Arturo Duarte-Sierra
Department of Biotechnology and Food Sciences, Sonora Technological Institute, Ciudad Obregón, Sonora, Mexico

Marie Thérèse Charles
Saint-Jean-sur-Richelieu Research and Development Centre, Agriculture and Agri-Food Canada, Saint-Jean-sur-Richelieu, Quebec, Canada

Joseph Arul
Department of Food Science and Horticultural Research Centre, Laval University, Quebec, Quebec, Canada

Abbreviations

λ	Wavelength
ν	Frequency
ε	Molar absorptivity
6-MM	6-Methoxymellein
ABA	Abscisic acid
ACC	1-Aminocyclopropane-1-carboxylic acid
APX	Ascorbate peroxidase
BER	Base excision repair
c	Speed of light in vacuum
CA	Controlled atmosphere
CAT	Catalase
CHS	Chalcone synthase
CPDs	Cyclobutene pyrimidine dimers
CWSZ	Cell-wall stacking zone
E	Energy
EM	Electromagnetic
EP	Epicarp cell
ER	Endoplasmic reticulum
ERF	Ethylene response factors
eV	Electron-volt
FW	Fresh weight
GR	Glutathione reductase
h	Planck's constant
HCW	Host cell wall
HRGP	Hydroxyproline-rich glycoproteins
HR	Hypersensitive response
LED	Light-emitting diodes
LPM	Low pressure mercury lamps
MAA	Mycosporine-like amino acids
MAP	Modified atmosphere packaging
MDA	Malondialdehyde
MDS	Magnesium-dechelatase
MEP	2-C-Methyl-D-erythritol 4-phosphate
MPM	Medium pressure mercury lamps
NCI	National Cancer Institute
NER	Nucleotide excision repair
NPAAs	Nonprotein amino acids
OD	Optical density
OM	Osmiophilic Vesicle
PAL	Phenylalanine ammonia-lyase
PAR	Photosynthetically active radiation
PG	Polygalacturonase
PM	Plasma membrane
PME	Pectin methyl esterase

PR	Pathogenesis-related
PUFAs	Polyunsaturated fatty acids
RH	Relative humidity
ROS	Reactive oxygen species
SA	Salicylic acid
SGR	Stay-green
SMs	Secondary metabolites
SOD	Superoxide dismutase
UV	Ultraviolet
UVR	Ultraviolet radiation
V	Vacuole

1 Introduction

There are more than 2000 plant species contributing to our diet, of which only about 1% represents the most important fruits and vegetable crops in the world. The classification of these crops can be done by botanical systematics, by plant organ (flower, leaf, tuber, etc.), or by their end-uses and economic considerations (Haard, 1984). Fruits and vegetables have been part of our diet since ancient times, probably because of their delightful taste and aroma. Their contribution to health and general well-being has been recognized over the last century. One of the most illustrative examples is the identification of vitamin C as an essential nutrient present in fruits by Albert Szent-Gyorgyi, for which he was honored with the Nobel Prize in medicine in 1932 (Terry and Thompson, 2011). The American National Academy of Sciences emphasized the importance of fruits and vegetables in the diet in 1982, highlighting carotene-rich fruits and cruciferous vegetables (*Brassicas*) for reducing risk of cancer (Liu, 2003). Consumers have been increasingly eager to adopt a healthier diet since the National Cancer Institute (NCI) of the U.S. Department of Health and Human Services promoted in 1991 the '5-a-Day' consumption of fruits and vegetables as a measure to prevent cancer (Bartz and Brecht, 2002). Lower risk of coronary artery disease and stroke has been correlated with the increased dietary intake of plant-based foods such as fruits and vegetables (Hu, 2003).

1.1 Factors Limiting the Storability and Quality of Fresh Produce

Commodities are perishable and keep an active metabolism during the postharvest phase (Arul, 1994). Biochemical and metabolic events such as the loss of chlorophyll, the solubilization of pectin, and starch degradation will continue until the complete degradation of fruits and vegetables occurs. Respiratory activity, transpiration, and ethylene biosynthesis and action are the most important physiological factors that control senescence (Phan, 1987). In addition, as senescence advances, the tissue becomes increasingly susceptible to diseases and decay caused by fungal and bacterial infections, which is a major factor limiting the storable life and contributing to postharvest losses of fresh produce. Finally, abusive storage conditions involving temperature and relative humidity (RH) can negatively influence the postharvest storage of produce. Appropriate handling and storage are of vital importance for preserving quality, nutritional, and phytochemical content of fresh produce throughout storage, transportation, and distribution. Therefore, research to develop new technologies to improve postharvest handling of fresh fruits, vegetables, and minimally processed produce is critical (Sitbon and Paliyath, 2011; Graça et al., 2013; Shen et al., 2013; Park and Kim, 2015; Huang et al., 2017; Scott et al., 2017).

1.2 Strategies Being Pursued

There are three main strategies to cope with postharvest losses: (1) the use of genotypes that have longer postharvest life; (2) appropriate agricultural practices to maintain the quality of the commodities; and (3) suitable postharvest handling practices to keep the quality and safety of the produce (Kader and Rolle, 2004). These strategies are of equal importance; however, the first two are beyond the scope of this chapter. The appropriate postharvest handling of produce includes: (1) postharvest management procedures that are critical to maintain the quality, the integrity, and the safety of commodities; (2) postharvest treatments designed to manipulate the environment around the produce; and (3) postharvest treatments to minimize decay.

Temperature and RH are by far the most fundamental factors to prolong storage life and maintain the quality of produce. Appropriate temperature management is the simplest and the most important aspect to retard senescence, which is marked by color change and pigment development, tissue softening, loss of membrane integrity, and loss of proteins. Low temperatures decrease both the respiration rate and the physiological activity of the produce as well as the activity of microorganisms causing spoilage (Nunes do Nascimento and Emond, 2002). Lower the storage temperature, within limits, is generally associated with slower respiration rate and longer storage life. Most temperate and subtropical commodities have an optimal shelf life at temperatures of 0.0–1.0°C. However, most tropical crops are susceptible to chilling injury and the lowest safe temperature varies with commodities. For instance, the lowest safe temperature for avocado, pineapple, pomegranate, and okra is 7°C, while for tropical fruits such as banana, mango, breadfruit, and jicama is 13°C (Kader and Rolle, 2004). RH, on the other hand, can influence both water loss by evaporation and decay development caused by condensation of moisture on the surface of the crops that is often aggravated by storage temperature fluctuations. Reduced weight loss at high RH contributes to retain the tissue turgidity, resulting in firmer, crispier, and higher quality vegetables (Van den Berg, 1981). Proper relative humidity storage of fruits is 85–95%, while for most vegetables the range of RH varies from 90 to 98% (Kader and Rolle, 2004). In addition, many horticultural crops are sensitive to ethylene action, and hence, ethylene scrubbers that catalytically oxidize ethylene are often used in many commercial facilities. Besides, mixed loads of ethylene-producing commodities with ethylene-sensitive produce should be avoided (Kader and Rolle, 2004).

The second strategy to reduce postharvest losses during storage is by using postharvest treatments that modify the environment, achieved by controlled atmosphere (CA) and modified atmospheres packaging (MAP), as an adjutant to low temperature–high humidity atmosphere. The principle of CA/MAP consists of altering the concentrations of oxygen (O_2) and carbon dioxide (CO_2) in the storage room or in the package. Low levels of O_2 decrease the respiration rate and high levels of CO_2 inhibit the development of fungal growth. CA is mostly used for large volumes and long-term storable commodities. MAP, on the other hand, is used for fresh cut and ready-to-eat highly perishables fruits and vegetables. The concentration of the gases in MAP can be altered passively or actively. In the first case, the modification of the atmosphere inside the package is created by respiration rate of the produce and the gas transfer capacity of the packaging material. In the latter, levels of O_2 and CO_2 will be altered by replacing the atmosphere in the package with a desired gas mixture. The composition of the atmosphere can be further adjusted by the addition of ethylene scavengers or absorbers to establish a desired concentration (Kader and Watkins, 2000; Caleb et al., 2012).

Reduction of decay using postharvest treatments is the third strategy to extend the storable life of produce. When washing is required, the use of clean water with appropriate concentration of sanitizers is important to reduce the transition of pathogens from water to the produce. Curing treatment to allow wound healing is used to prolong the storage life of certain commodities such as tubers and bulbs. Short exposure to heat at relatively high temperature, that is, 50°C, is used to reduce decay in a few crops, including mangoes. Also, fungicides, such as imazalil and/or thiabendazole, are utilized to reduce spoilage in various crops. Biological control alone or in combination with fungicides has been shown to be an effective method to reduce decay by microorganisms (Kader and Rolle, 2004).

Recently, more attention is being paid to control diseases and decay by the induction of natural defenses of produce by biological, chemical, or physical elicitors. Induced resistance consists in the activation of multiple defense mechanisms, such as reinforcement of cell wall, activation of pathogenesis-related (PR) proteins, the de novo synthesis of phytoalexins as well as increased levels of constitutive secondary metabolites (SMs) (Dann, 2003). Some examples of elicitors are chitosan (Reddy et al., 2000), salicylic acid (SA) (Yalpani et al., 1991), and ultraviolet C (UV-C) irradiation (Arul et al., 2001).

Beneficial low doses, also called *hormetic* doses, of harmful abiotic stresses such as UV-C radiation have been deliberately applied not only to induce disease resistance of produce but also to slow down ripening and senescence, and to enhance phytochemical content of the treated crops. Hormesis (hormetic, the adjective of hormesis) is a biological phenomenon, where a biological system stimulates beneficial responses at low doses of stressors that are otherwise harmful to that system; and it is often characterized by biphasic dose–response relationship. For instance, UV-C irradiation induces enzymes such as phenylalanine ammonia-lyase (PAL), chalcone synthase (CHS), and the synthesis of several chemical compounds, including phenylpropanoids (Pombo et al., 2011), and delayed yellowing of broccoli (Costa et al., 2006). UV-C irradiation has been employed after harvest to induce disease defense responses as well as to delay senescence (Charles and Arul, 2007; Shama, 2007). UV-C at 3.7 kJ/m^2 induced resistance to *Botrytis cinerea* Pers. in tomato fruit by the induction of phytoalexins, ultrastructural modification in the pericarp, induction of PR-proteins, and the accumulation of phenolic compounds (Charles et al., 2008a, 2008b, 2008c, 2008d, 2009).

2 UV Radiations: Physics and Biology

Ultraviolet radiation (UVR) is part of the electromagnetic spectrum considered as nonionizing radiation. It is divided into three regions: UV-A, UV-B, and UV-C. These regions were proposed for the first time in 1932 at the Second International Congress of light at Copenhagen as follows: UV-A 400–315 nm, UV-B 315–280 nm, and UV-C 280–100 nm (Bintsis et al., 2000; Diffey, 2002). Since vacuum V-UV at 100–200 nm can be absorbed by molecular oxygen in the air, actual UV-C range is referred from 200 to 280 nm (Meulemans, 1987; Koutchma et al., 2009c). Along with the previously proposed regions, other terms such as near UV (300–380 nm), far UV (200–300 nm), and extreme UV (< 200 nm) are also common (Shama, 1999). Of the solar radiation, UV-B radiation reaching earth surface is 0.1% of the total spectrum, that of UV-A is around 5%, and those of visible and infrared radiation correspond to 39 and 56%, respectively (Hollósy, 2002; Holick, 2016). UV-C is the most damaging radiation of the three UV regions, but it is not relevant to biological action under the normal atmospheric condition, as most of this radiation (<280 nm) is filtered out in the stratospheric by the ozone layer. To study the impact of UV-C on biological systems, it must be generated artificially.

UV-C radiation has been commonly referred as germicidal UV, since it can be absorbed by the nucleic acids of microorganisms resulting in their inactivation or destruction (Stevens et al., 1998a; Shama, 2007). However, the correct definition of this term is the band of UV wavelengths between 200 and 280 nm (Kowalski, 2009b). UVR in the range of 250 ± 260 nm, with a maximum effect at 254 nm, is lethal to most microorganisms, including bacteria, viruses, protozoa, mycelial fungi, yeasts, and algae (Bintsis et al., 2000). The maximum germicidal activity is observed at about 265 nm, which decreases with either the longer or the shorter wavelengths, with the peak emission band of low-pressure lamps at 254 nm (Meulemans, 1987). Due to the germicidal effect of this radiation, one of the first applications of UV-C was air and surface sterilization processes in hospitals (Andersen et al., 2006; Nerandzic et al., 2010). Eventually, the food industry took note of the potential of short-wave radiations for the reduction of food-borne pathogens. There are three main categories of applications of UVR in foods: surface decontamination of microorganisms, destruction of microorganisms in the air, and sterilization of liquids (Bintsis et al., 2000). Most efforts on surface disinfection with UV-C in foods have been addressed to inactivate *Escherichia coli* (Gayán et al., 2011; Guan et al., 2012), *Salmonella* (Chun et al., 2009, 2010), and *Listeria monocytogenes* (Chun et al., 2009; Koutchma et al., 2009b). Disinfection of air is used to improve the quality of fresh-cut processing facilities (Bintsis et al., 2000). Nonthermal processing of fruit juices using UV-C is emerging as a promising technology for sterilization of liquids. Nevertheless, UV-C has a major limitation in this context that the lack of deep penetration into the medium limits its utilization to nonopaque fluids.

2.1 General Notions about Electromagnetic (EM) Waves

All EM waves consist of an electric field component and a magnetic field component vibrating in phase and perpendicular to the direction of the wave (Edwards and Alexander, 2017). They differ by their frequency (ν) or their wavelength (λ), which are related by the expression:

$$\lambda = \frac{c}{v} \tag{1}$$

EM waves carry energy, and it is emitted in the form of small discrete quantities, called photons, transporting energy (E) proportional to the frequency (Andrews, 1999) as follows:

$$E = hv \tag{2}$$

where E is energy (J), v is the frequency (Hz), and h is the Planck's constant (6.62×10^{-34} J s). Since Joule is a large energy unit to express the energy of a photon, either electron-volt (eV) 1.6×10^{-19} J, or Einstein, 6.02×10^{23} photons is preferred.

Combining (1) and (2) yields:

$$E = \frac{hc}{\lambda} \tag{3}$$

where c is the speed of light (299,792,458 m/s), and it can be deduced that photons with shorter wavelengths are more energetic compared to those with longer wavelengths.

The energy of 1 photon at 254 nm can be estimated from Equation (1), where v can be calculated as follows:

$$V = \frac{c}{l} \tag{4}$$

where l is the wavelength (m), thus v equals to 1.18×10^{15} Hz and multiplying this value by Planck's constant, results in 7.81×10^{-19} J or 4.88 eV.

In addition, from Equations (2) and (4), 1 eV corresponds to a wavelength of 1240 nm, thus the energy in eV of a wavelength can be easily estimated by the following expression:

$$E\,(\text{eV}) = \frac{1240}{\lambda\,(\text{nm})} \tag{5}$$

According to Harm (1980), sometimes it is useful to express photon energies in terms of kilocalories per Einstein, since the absorption of 1 Einstein can excite (exc.) 1 mole of absorbing substance with the following expression:

$$E\left(\frac{kcal}{Einstein}\right) = E_{exc}\left(\frac{kcal}{mole}\right) = \frac{28590}{\lambda(nm)} \tag{6}$$

The energy fluence of UVR is expressed as (J/m^2). The photon fluence can be estimated, which is equal to the number of photons per unit area, and the following expression can be used:

$$\text{Photon fluence}\left(\frac{\text{photon}}{m^2}\right)=\frac{\text{Fluence}\left(\frac{J}{m^2}\right)\times\lambda(m)}{h(J_S)\times c\left(\frac{m}{s}\right)} \quad (7)$$

For 1 J/m^2 of radiant energy at 254 nm, the photon fluence would be equivalent to 1.279×10^{18} photons/m^2.

Since photon counting using photochemical devices or photomultipliers has limitations for use as an independent calibration tool, the commonly used calibration method is the measure of radiation energy or power (energy per time). The latter is referred to as irradiance or fluence rate with a distinction. Irradiance is the radiation power incident on a flat surface, whereas fluence rate is the radiation power incident on a sphere of unit cross section. Because of the three-dimensional nature of plants and plant crops, the use of the term, fluence rate, is acceptable. Nonetheless, both these terms can be expressed in photon terms. The energy fluence or dose is the energy fluence rate integrated over the exposure time, which can be as follows:

$$\text{Dose}\left(\frac{J}{m^2}\right)=\text{Fluence rate}\left(\frac{W}{m^2}\right)\times\text{Exposure time(s)} \quad (8)$$

2.2 Source Lamps

UVR is naturally produced by the sun due to its ultrahigh temperature. It can be also produced artificially by passing electric current through a gas or vapor such as vaporized mercury and the vapor can be activated at relatively low temperatures (Diffey, 2002). The vapor discharge lamps are sealed tubes of silica glass, filled with mercury and an inert gas powered by alternating current. When voltage is applied, electrons and ions of the gas are accelerated by the electric field, producing collisions and the emission of spectral power (Koutchma et al., 2009c). Lamps can be identified by a defined nomenclature compromised of six parts: (1) lamp nominal wattage; (2) lamp nominal length; (3) bulb diameter; (4) lamp shape; (5) lamp base code; and (6) circuit or special description. Thus, 30 W/36T8/PH-B refers to a 30 Watt, 36 inches long, 26 mm diameter, preheat start bactericidal lamp (Kowalski, 2009c). The most common types of mercury lamps are high-intensity discharge lamps, medium pressure mercury lamps (MPM) (1000 torr), and low-pressure UV lamps (LPM) containing mercury gas (10 torr) (Kowalski, 2009c).

Koutchma et al. (2009d) have reviewed some important technical characteristics to be considered when choosing UV lamps, including operating temperature, electrical power input (W/cm), germicidal UV output (W/cm), electrical-to-germicidal UV conversion efficiency (%), and lifetime rating (h). LPM lamps operate at 40°C, the specific electrical loading is between 0.4 and 0.6 W/cm, the 254 nm band represents 85% of the total UV radiation emitted, the electrical-to-germicidal UV conversion efficiency is approximately 40%, the other 60% is mainly heat and an initial drop in emission yield occurs at 100–200 hr. Operational temperature in MPM is more elevated than LPM lamps, ranging from 400 to 800°C, the specific electrical loading is also higher between 5 and 30 W/cm. One of the main difference between LPM and MPM lamps is that the latter are polychromatic, and the conversion efficiency is directly proportional to voltage, for instance, at 160 V the efficiency is about 30%, while at 250 V, that efficiency is nearly 100%. Another important factor differentiating LPM and MPM is the aging, that is, the drop-in emission yields of MPM lamps start after 4000 hr.

The optimal operational temperature of LPM and MPM lamps hampers treatments conducted in chambers at low temperatures, where the produce is normally irradiated. Furthermore, these lamps are fragile and contain mercury, which represents a significant risk to human health in the event of any rupture. Thus, to cope with these issues, new technologies have been developed. One of the most important is the advance on light-emitting diodes (LEDs), which can also emit in the UV-C range. One important conclusion was made by Kim et al. (2016): By adjusting the wavelength of LEDs to obtain the maximum absorption by microbial DNA at 266 and 270 nm, the inactivation rates are superior to those of conventional LPM lamps emitting at 254 nm having similar intensities. However, it was also clear that intensities of LEDs are very low, compared to those of LPM lamps.

2.3 Interaction of UV Radiation with Matter

UVR interacts with the materials it encounters through absorption, reflection, refraction, and scattering depending on the structure of the matter and on the wavelength of the radiation (Koutchma et al., 2009c; Flammer et al., 2013). When UVR, light, or infrared radiation comes across with matter, there is an expected interaction between the radiation and the atoms of the matter. If absorbed, the electron of an atom, or molecule uses a photon to jump into an energetically higher state; if emitted, an electron can fall into a state of lower energy, with the energy difference being sent out as a photon (Flammer et al., 2013). Thus, radiation may be reemitted at a different wavelength, and energy absorption may result in molecular vibrations that translate into heat (Kowalski, 2009b).

Absorption of light is a fundamental process in life, and it is the basis for photosynthesis in plants as well as the physical basis for vision (Flammer et al., 2013). It consists basically of the transformation of photon energy to other types of energy (Koutchma et al., 2009c). The absorption of UVR by molecules is concomitant with changes in their electronic configuration, vibration, rotation, and energy. Two terms are important when describing absorption. The absorption spectrum is the capacity of a molecule to absorb UVR at a specific wavelength. The other term is the absorbance or optical density (OD), which is related to the probability that some change would occur when a photon of specific wavelength interacts with a molecule (Kowalski, 2009b). If UVR absorption occurs in molecules, photochemical reactions will take place only when photons have sufficient energy to promote a reaction (Koutchma et al., 2009c). Two kinds of bonding molecular orbitals may be involved in photochemical reactions, σ (sigma) orbitals and π (Pi) orbitals (Kowalski, 2009b).

Absorption of UVR corresponds to low energy transitions from nonbonded orbitals or π orbitals, being $\sigma \rightarrow \sigma^*$ and $\pi \rightarrow \pi^*$, the formally allowed transitions. Those transitions involving σ-bonds require high energy with transitions occurring below 150 nm; and on the other hand, the lower energy gaps between π-bonds allow electronic transitions above 200 nm (Edwards and Alexander, 2017). The π orbitals are common in conjugated structures. In proteins, the molecules responsible for absorption are amino acids with absorption peaks at 280 nm, and in nucleic acids the chromophores are purines and pyrimidines with absorption peaks at 265 nm (Kowalski, 2009b). In addition to absorbance, the molar absorptivity ε, also known as molar absorptivity coefficient or molar extinction coefficient, is a useful parameter to determine how strong a substance absorbs UVR at a specific wavelength. In fact, absorbance (*A*) or OD is defined as:

$$A = \varepsilon l c \tag{9}$$

where ε is the molar absorptivity (L/mole cm); l is the thickness of a solution (cm); and c is the concentration of solute (moles/L). In general, strong absorption of UVR correlates with molecules where conjugated double bonds are present (Kowalski, 2009b).

2.4 Reflection and Scattering

Most ground surfaces reflect less than 10% of UVR, with some exceptions, including gypsum, sand (15–30%), and snow (up to 90%) (Diffey, 2002). Reflection is considered as the change in the direction of EM radiation deflected by an interface that can be specular when the reflection occurs from smooth surface or diffuse if it occurs from a rough surface (Koutchma et al., 2009c). According to Hollósy (2002), reflectance as protection mechanism in plants is of secondary importance in most cases; however, in a few species such as *Dudleya brittonia* and *Argyroxiphium sandwicense*, surface reflectance can mirror 20–70% of radiation in the UV region. In *Dudleya*, the accumulation of glaucescence, a powdery wax coating, increased the reflection of UV-B radiation by 25% more than the photosynthetically active radiation (PAR) (Jansen et al., 1998). Like reflection, scattering of EM radiation relates to the change of direction, but in this case, there is interaction between the radiation and the surface. In plants, the latter occurs by leaf orientation as well as by other tissue structures (Ruban, 2009).

2.5 Penetration

For radiation to interact with matter, it must penetrate and propagate into the matter. UVR, often used for surface treatments, is strongly absorbed by most materials but it cannot penetrate beyond the surface layers of solid matter (Koutchma et al., 2009a). Also, there is an inverse relationship between the wavelength of the EM wave and the energy it can dissipate. Thus, it could be logical to think that UV-C radiation with a higher energy content compared to UV-A would have deeper penetration into biological surfaces. Nonetheless, this is not the case in human skin, because some macromolecules such as DNA and proteins would absorb shorter wavelength allowing longer wavelengths to penetrate deeper into the tissue (Holick, 2016). Evidence of blue light and UV-A radiation penetration has been reported in plant tissue, where the penetration of blue light (460 nm) was deeper than UV-A radiation (360 nm) that can be explained by the presence of screening pigments and leaf anatomy (Bornman and Vogelmann, 1988). The penetration of

UV-B radiation (285 nm) appeared to be reduced in the leaves of *Brassica napus* L. cv. 'Ceres' by the presence of the flavonoids, such as quercetin and kaempferol, that are concentrated within the upper layers to the depth of 40 μm, leading to the protection of susceptible structures of the leaf and the prevention of UV-B damage (ÅLenius et al., 1995). The data for UV-C penetration into plant tissues are not well documented, but penetration depths of 30–50 μm could be expected. Most of this radiation is likely absorbed in the surface layers by membrane lipids with conjugated dienes and trienes that absorb in the UV-C range at 220–230 nm and 265–270 nm, respectively (Vieira and Regitano-d'Arce, 1998), and proteins that have high concentration of chromophores such as tryptophan, tyrosine, cysteine, and phenylalanine (Kowalski, 2009b). While these absorptions would likely reduce the penetration depths of UV-C radiation into the tissue, that is, a few cell layers in the periderm, the dose applied would have certain impact on the potential biomolecular targets, and consequently, on the responses of the tissue to UV-C. Since lipids and proteins are at higher concentrations than DNA, it can be expected that, at low doses of UV-C radiation, they would be target structures, unless higher doses are applied to target DNA. Low dose of UV-C triggers lipid peroxidation in tomato following its application, suggesting membrane lipids are targets for low-dose UV-C, provoking oxidative stress in the tissue (Ait-Barka et al., 2000a).

2.6 Cellular Damages Caused by UV Radiation: A Harmful Stressor

Stapleton (1992) classified damages caused by UVR into two categories: direct DNA damage (strong absorption at 260 nm) and damages to physiological processes. Nucleic acids can absorb very effectively at 254 nm which is the main wavelength obtained from germicidal lamps (Bornman et al., 1983). The absorption of UVR by DNA can result in the cross- linking of the pyrimidine, thymine, and cytosine bases (Holick, 2016). The most important damage to DNA is the production of pyrimidine photoproducts, such as cyclobutene pyrimidine dimers (CPDs), pyrimidine 6-4 pyrimidone photoproducts (6-4PPs), and their Dewar isomers (Jansen et al., 1998; Hollósy, 2002). The formation of CPDs represents approximately 75% of the total DNA damage by UVR (Nawkar et al., 2013). Although guanine and cytosine bases are less susceptible to photochemical damage than thymine bases (Cockell, 1998), adenine residues undergoing photocycloaddition reactions with other adenine or thymine can induce purine photoproducts in the presence of UVR (Rastogi et al., 2010). The dimer inhibits the formation of new DNA or RNA chains in the process of cell replication, resulting in the inactivation or inability to replicate in organisms affected by UVR (Bolton and Linden, 2003).

The impact of UVR on plant physiology is concomitant with physical size, molecular weight of DNA or RNA, content of pyrimidine and purine bases, and the presence of chromophores (Kowalski, 2009a). Amino acids that may contribute to photoreactivity in DNA involving cross-linking with proteins include cysteine, cystine, tyrosine, serine, methionine, lysine, arginine, histidine, tryptophan, and phenylalanine (Kowalski, 2009b). Of these amino acids, tyrosine absorbs at 245 nm, tryptophan at 242 nm, and phenylalanine at 236 nm (Beaven and Holiday, 1952). UVR also causes photochemical reactions in proteins and peptide bonds within proteins at wavelengths at about 280 nm and below 240 nm, respectively, causing enzymatic damages by broadband UV wavelengths other than 254 nm (Kowalski, 2009b). Lipids with isolated or conjugated double bounds may also undergo photooxidation under UVR, and the consequence is the oxidation of polyunsaturated fatty acids (PUFAs) (Hollósy, 2002; Martysiak-Żurowska et al., 2017). The peroxidation of lipids leads to breakdown of their structure, thus affecting the function of cell membranes as well as the membranes in chloroplasts (Hollósy, 2002; Nawkar et al., 2013) and other organelles. Inhibition of photosynthesis, especially by UV-C, has been investigated, because of the reduction in chlorophyll contents and chloroplast size (Bornman et al., 1983; Najeeb et al., 2011). Common responses in plants exposed to UVR include increased leaf thickness and decreased leaf area, presumably to avoid further penetration of the radiation (Jansen et al., 1998; Hollósy, 2002). In addition, UV-C radiation showed a greater damage than UV-B on leaf tissues (Bornman et al., 1983).

2.7 Protection and Repair Mechanisms

On the Archean Earth, UV-C and other UV bands would have reached the Earth without much reduction due to the lack of oxygen and ozone, and early life forms would have been subject to severe damage to their DNA, proteins, and lipids. Thus, UV survivability would have mainly depended on avoidance, protection, and repair mechanisms of those organisms (Cockell, 1998). Being sessile, plants have developed diverse photoprotection mechanisms including leaf movement, chloroplast movement, screening of photoirradiation, reactive oxygen species (ROS) scavenging, thermal energy dissipation of absorbed energy, modification of cyclic electron flow, and adjustments to the photorespiratory pathway (Takahashi and Badger, 2011). Of these mechanisms, UV-screening compounds are particularly important for UVR protection in plants (Koutchma et al., 2009b). UV-screening compounds are mainly pigments that can absorb light at certain wavelengths. Organic pigments share many double bonds, that is, they are multi-unsaturated, allowing the formation of an electron cloud that can absorb UVR and light (Flammer et al., 2013). Plant pigments are divided mainly in four groups: mycosporine-like amino acids (MAA), carotenoids, phenolic compounds (phenylpropanoids, flavonols, and anthocyanins), and nitrogen-containing heterocyclic betalains (Solovchenko, 2010b). UV-screening in plants is mainly due to phenolic compounds, especially flavonoids and MAA. The latter are SMs found in algae, and other organisms such as lichens and fungi, with UV-absorbing potential due to their high molar extinction coefficients (≅50,000) (Hartmann et al., 2015). Certain carotenoids and betalains screen (i.e., block) light in the blue part of the electromagnetic spectrum (Solovchenko, 2010b).

Pigments are concentrated in superficial structures such as the cuticle and epidermis and/or are distributed within tissues and cells (Stapleton, 1992). Solovchenko (2010a) has further categorized the localization of pigments. Most common adopting intracellular structures include vacuoles for polar compounds (anthocyanins, flavonols, and phenolic acids), as well as plastoglobuli and lipid bodies for nonpolar pigments (carotenes, fatty acids, and xanthophyll esters). Extracellular adopting structures, including the cuticle and cell wall, accumulate nonpolar pigments (acylated or bound flavonoids) and polar pigments (flavonols, phenolic acids, and phenylpropanoids), respectively. Flavonoid synthesis in the epidermis is often proposed as the main mechanism for preventing UV-B radiation from reaching mesophyll and affecting photosynthesis, which can be depressed due to the loss of chlorophyll that is adversely damaged by large amounts of UV-B (Hollósy, 2002). Spectroscopy studies have shown that flavones and flavonols have two major absorbance bands: Band I at 320–385 nm and Band II at 250–285 nm (Kumar and Pandey, 2013). However, UV-screening pigments are unable to avoid UVR completely from reaching DNA in the superficial tissue, and additional repair mechanisms are deployed in the defense against UVR.

In the event of UVR reaching DNA, plants are capable of exerting a number of highly conserved repair mechanisms including plant-specific photoreactivation pathway, light-independent repair and homologous recombination (Koutchma et al., 2009b; Rastogi et al., 2010; Nawkar et al., 2013; Gill et al., 2015). Photoreactivation, which is thought to be the major DNA repair pathway in several higher plants, consists of light reverse of DNA damage by monomerizing CPDs (Kowalski, 2009b; Rastogi et al., 2010). Photoreactivation is a process that is achieved by enzymes known as photolyases, which bind specifically to CPDs or 6-4PPs and directly monomerize the cyclo-butane ring, using the energy of one photon of visible/blue light at 450–370 nm, to split approximately one dimer (Stapleton, 1992; Rastogi et al., 2010). Although photoreactivation is an effective process to reverse thymine dimers, it may not completely reverse UVR damage, but it can effectively limit it (Kowalski, 2009b). In addition to photoreactivation, excision repair is a multistep, dark repair pathway, where an abnormal or damaged base is removed by two major subpathways: base excision repair (BER) and nucleotide excision repair (NER) (Rastogi et al., 2010). In BER pathway, DNA glycosylases remove damaged bases, followed by the removal of the apurinic/apyrimidinic (AP)-site by AP-endonuclease or AP-lyase, and subsequently, phosphodiesterase removes the deoxyribose phosphate residues with the resulting gap filled by DNA polymerase, and finally, the strand is sealed by DNA ligase (Seeberg et al., 1995). Homologous recombination mechanism has been used in plants for maintaining genome integrity and viability by using a homologous chromosome as template to recover the lost information (Gill et al., 2015).

3 Hormesis: A Biological Phenomenon

Hormesis is an expression used by toxicologists and biologists to describe an adaptive response of organisms to moderate stresses (Mattson, 2008). The term, *hormesis*, is derived from the Greek and can be interpreted as 'to urge on, to impel, and to excite' (Shama, 2007). Although in a more recent review by Poschenrieder et al. (2013), two other meanings were included: 'setting into motion' or 'to boost something'. Edward Calabrese, one of the prolific authors on hormesis, considers the phenomenon as

> an adaptive response characterized by biphasic dose responses of generally similar quantitative features with respect to amplitude and range of the stimulatory responses that are either directly induced or the result of compensatory biological processes following an initial disruption in homeostasis
>
> *(Calabrese and Baldwin, 2001, 2002, 2003; Chapman, 2002; Calabrese and Blain, 2005).*

A shorter definition was proposed by Mattson (2008) and Luckey (1980), which describes hormesis as the process in which the exposure of organisms to any agent or stressor that is otherwise harmful induces adaptive and beneficial effects at low doses in that system. A more explicit definition is due to Kendig et al. (2010): Hormesis is a dose–response relationship for a single end point that is characterized by reversal of response between low and high doses of chemicals, biological molecules, physical stressors, or any other initiators of the response.

Hormesis can be obtained by either overcompensation or direct stimulation. Hormesis through overcompensation is an adaptive response to mild stresses resulting in enhanced resistance in biological systems for a finite period, which is a consequence of a short-time disruption of homeostasis, and it follows a pattern of returning to homeostasis as well as protection against subsequent and possibly more massive stresses (Calabrese and Baldwin, 2002). Even if no subsequent damaging event occurs, the overutilization of resources to face the initial damage may be employed for other useful functions. This is what is typically measured in studies assessing hormesis (Calabrese and Baldwin, 2002). Qualitatively, hormesis through overcompensation displays a U or an inverted U-shaped dose–response relationship, where the response is the measured end point (e.g., plant growth, yield, storage time, etc.) (Calabrese and Baldwin, 2002). The quantitative features of the overcompensation are such that the magnitude of the stimulatory response is modest, and not greater than twofold (typically, 130–160%) as compared with the control group and a width of the stimulatory response, usually of less than 10-fold in dose range (Calabrese and Baldwin, 2003; Calabrese and Blain, 2009; Calabrese, 2015a). However, any response greater than 10-fold is likely to represent a different phenomenon, not related to hormesis (Calabrese and Baldwin, 2002). Direct stimulation of hormesis responses generally occurs quickly after exposure to stressful conditions of a biological system and does not imply a disruption of homeostasis, but displays similar qualitative and quantitative features to the overcompensation response with respect to amplitude and dose range of the stimulatory response (Calabrese and Baldwin, 2002; Poschenrieder et al., 2013; Calabrese, 2015b).

3.1 UV-C Radiation Hormesis in Postharvest Crops: Responses and Hormetic Doses

The phenomenon of UV-C hormesis in a variety of postharvest crops is firmly established and well documented with respect to the beneficial responses of induced disease resistance (Table 17.1) as well as delayed ripening and senescence (Table 17.2) , leading to their extended shelf life and maintenance of quality during storage (Stevens et al., 1996; Arul et al., 2001; Shama and Alderson, 2005; Charles and Arul, 2007). UV-C doses ranging from 0.01 to 39 kJ/m^2 have been reported as hormetic (adjective of hormesis) for a number of fruits and vegetables, the lowest being for banana (Mohamed et al., 2017) and the highest for pineapple (Sari et al., 2016). Given to the lack of consistency between research reports regarding treatment application and radiation measurement, the effective doses could be found in a more restricted range of 0.125 to 9 kJ/m^2, with lower levels showing little beneficial effect and higher levels beyond that range showing physiological damages. Increase disease resistance has been observed in UV-C-treated tomato at 3.7 kJ/m^2 (Charles et al., 2008a, 2008c, 2008d). UV-C treatments have delayed the yellowing of broccoli florets with doses between 1.2 and 10 kJ/m^2 (Costa et al., 2006; Duarte-Sierra et al., 2012). In addition to these parameters, 1.2 kJ/m^2 of UV-C increased the content of glucosinolates in broccoli florets (Nadeau et al., 2012).

In many studies, a single UV-C dose has been characterized as hormetic for most crops, although a range of doses is considered beneficial for some. The hormetic dose required for a specific response also differs between crops. For example, the hormetic dose required for controlling fungal rots (namely, those caused by *Monilinia* spp., *Alternaria* spp.), and *Colletotrichum gloeosporioides* (Penz.) Penz. & Sacc. in apple (Stevens et al., 1996) was different from the dose required in grapefruit against *Penicillium digitatum* (Pers.) Sacc. (Rodov et al., 1992). In addition, seasonal differences (e.g., for grapefruit) (Droby et al., 1993) and varietal differences (e.g., for sweet potato) (Stevens et al., 1990) affected the hormetic doses. In some instances, only a minor difference in the hormetic dose requirement was found by independent groups (e.g., in mature green tomato) (Liu et al., 1993; Maharaj et al., 1999), while others reported significantly different dose (Liu et al., 2009). In addition, the fractionation of the hormetic doses in two subdoses was found to be less effective in bell pepper than the entire dose in a single application (Mercier et al., 2001).

From a few studies that investigated the application of hormetic doses of UV-C at varying fluence rates, it appears that a threshold fluence rate may exist to achieve a maximum level of a specific response. To achieve the maximum in the disease control response, applied fluence rates were above 6.8 W/m^2; but a fluence rate of 20.5 W/m^2 was the threshold to achieve the maximum in the accumulation of the phytoalexin, rishitin, in UV-C-treated tomato fruits and stored at 13°C and 95% RH (Charles, 1998). In another work, Cote et al. (2013) examined the effect of fluence rates, 3.0 and 33 W/m^2, in the control of diseases in UV-C-treated tomato and strawberry fruits. The higher intensity was more effective in reducing decay in both the crops as well as in the delay of senescence in strawberry, but the delay in the ripening of the treated tomato was essentially the same at both the fluence rates. Thus, this aspect needs to be taken into consideration in designing treatment protocols, in addition to the primary consideration, the hormetic dose.

UV-C radiation induces defense responses in the exposed crops that include both disease resistance responses as well as delay in the progress of senescence (Charles and Arul, 2007; Shama, 2007). The latter is also important because their vulnerability to diseases increases with the advances in senescence due to a diminished ability to maintain their natural defenses. Tables 17.1 and 17.2 show the disease resistance and related markers, and senescence-related markers, respectively, in some fruits and vegetables. UV-C at the hormetic dose of 3.7 kJ/m^2 induced resistance to *B. cinerea* in tomato fruit by the accumulation of phytoalexins, phenolic compounds, ultrastructural modification in the pericarp, and the induction of PR-proteins (Charles et al., 2008a, 2008b, 2008c, 2008d, 2009). UV-C doses of 2.4–4.9 kJ/m^2 maintained overall appearance, decreased decay percentage, and increased the shelf life of mango (González-Aguilar et al., 2007). UV-C treatment delayed yellowing and chlorophyll a and b degradation, reduced tissue damage and disruption, and maintained antioxidant capacity of broccoli (Costa et al., 2006). UV-C treatment contributed to the preservation of tissue integrity, as indicated by lower electrolyte leakage and reduced respiratory activity in treated broccoli florets (Lemoine et al., 2007). The resistance of UV-C-treated tomato fruit against *B. cinerea* and control of natural infection in the treated tomato and bell pepper are illustrated in Figures 17.1–17.3.

UV-C radiation can induce biochemical responses in plants that are very similar to those induced by pathogens such as the increase in SA by increase in the activity of benzoic acid 2-hydroxylase that catalyzes SA biosynthesis (Yalpani et al., 1994) and increased activity of PAL (González-Aguilar et al., 2007). One of the first responses of plants exposed to UV-C radiation is the synthesis of phytoalexins. These compounds vary among different crops exposed to UV-C, including 6-methoxymellein (6-MM) in carrot roots (Mercier et al., 2000), δ-viniferin in grapes (Pezet

Table 17.1 Effect of UV-C radiation on disease control and related markers in some fruits and vegetables

Commodity	*Treatment*	*Response*	*Reference*
Banana	0.01 kJ/m^2 (0.336 W/m^2 during 30 s)	Cell wall reinforcement by lignification	Mohamed et al. (2017)
Strawberry	0.25 kJ/m^2	Reduction in the incidence of disease by about 77% in UV-C exposed strawberries and stored at 4°C for 12 d	Baka et al. (1999)
Carrot	0.88 kJ/m^2	Twofold higher in endo-chitinases in UV-C-treated carrots compared to the unexposed carrots after 6 d of storage at 10°C and 90% RH	Ojaghian et al. (2017)
Mango	2.46 and 4.93 kJ/m^2 (8.22 W/m^2 during 300 and 600 s)	UV-C treatments reduced fungal disease severity by 55%.	González-Aguilar et al. (2007)
Minimally processed lily bulb	4.5 kJ/m^2	Total viable bacterial counts in the UV-C-treated bulbs were 32% lower compared to unexposed bulbs after 5 d of storage at 4°C.	Huang et al. (2017)
Minimally processed apple	1.0 kJ/m^2	Effective reduction in the counts of *Escherichia coli*, *Listeria innocua*, and *Salmonella enterica* on the surface of cut apples exposed to UV-C	Graça et al. (2013)
Pear	3.1 kJ/m^2 (approx. 21 W/m^2)	A 2.7 log reduction of *Penicillium expansum* on wounded pear disks exposed to UV-C	Syamaladevi et al. (2014)
Tomato fruit	3.7 kJ/m^2	A reduction in the lesion of 42.0% in UV-C-treated tomatoes, challenged with *Botrytis cinerea*, 7 d of treatment	Charles et al. (2008a)
	(20 W/m^2 during 30 s)	A reduction 38.0% in the lesion area in UV-C-treated tomatoes and challenged with *B. cinerea*	Scott et al. (2017)

Table 17.2 Effect of UV-C radiation on senescence and related markers in some fruits and vegetables

Commodity	*Treatment*	*Response*	*Reference*
Amaranth leaves	1.7 kJ/m^2	Twofold reduction in fresh weight loss (reduced transpiration) in UV-C-treated leaves, after 14 d of storage	Gogo et al. (2017)
Minimally processed lily bulb	4.5 kJ/m^2	Reduced tissue browning (delayed senescence) in UV-C-treated bulbs by about 37%, after 5 d of storage at 4°C	Huang et al. (2017)
Pepper	7 kJ/m^2	Delayed tissue softening and reduced chilling injury in UV-C-exposed peppers	Vicente et al. (2005)
Strawberry	0.25 kJ/m^2	Reduced respiration rate, delayed tissue softening, and retention of membrane integrity in UV-C exposed strawberry fruits and stored at 4°C	Baka et al. (1999)
Cherry tomato	4.2 kJ/m^2 (approx. 8.75 W/m^2 during 480 s)	Delay in tissue softening in UV-C-exposed cherry tomatoes, concomitant with reduced ethylene production	Bu et al. (2013)
Tomato	3.7 kJ/m^2 (approx. 3.85 W/m^2 during 960 s)	Maintenance of green color (delay in ripening) in UV-C exposed tomatoes, despite the accumulation of ethylene	Tiecher et al. (2013)
	3.7 kJ/m^2 (20 W/m^2 during 30 s)	A delay in the ripening of UV-C exposed tomatoes by about 43.0%	Scott et al. (2017)

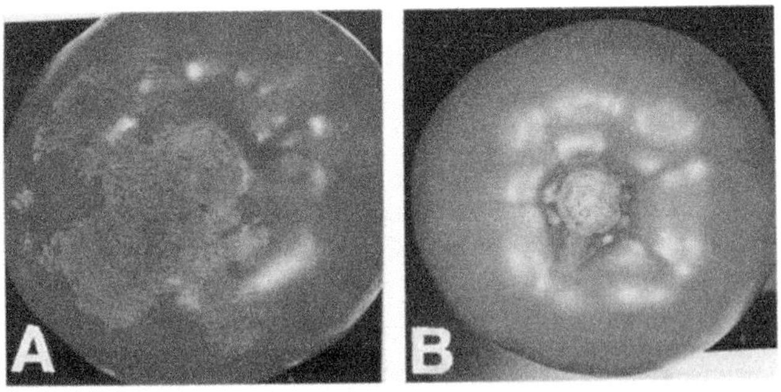

Figure 17.1 Disease resistance of UV-C-treated tomato fruits (cv. 'Trust'), inoculated with spores of *Botrytis cinerea* 10 d after treatment, and stored for 30 d at 13°C. A: Nontreated tomato; B: UV-C-treated tomato with the hormetic dose of 3.7 kJ/m^2 (Charles, 1998).

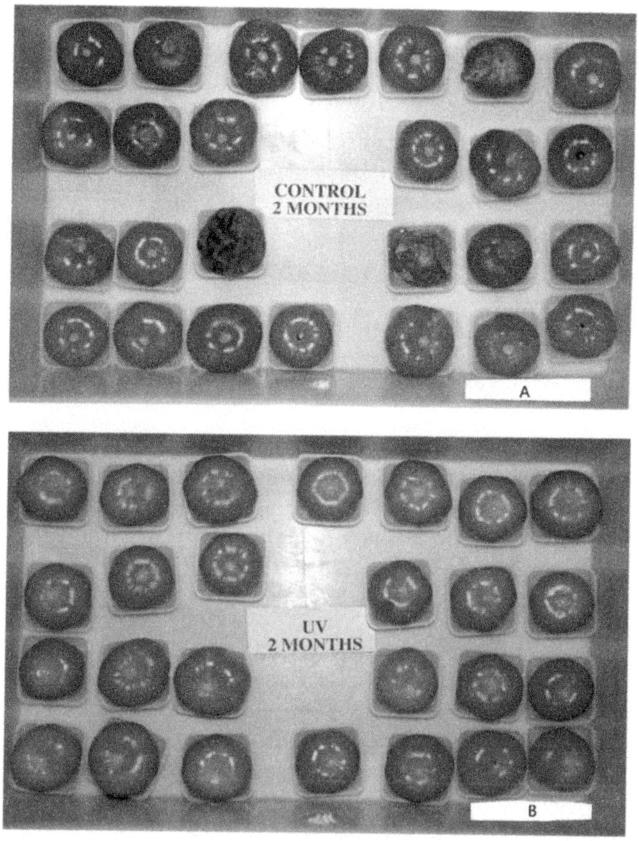

Figure 17.2 Control of natural infection in tomato fruits (cv. 'Trust') treated with the hormetic dose of UV-C (3.7 kJ/m^2) and stored at 13°C for 60 d. A: Nontreated tomatoes; B: UV-C-treated tomatoes. Fruits were surface sterilized with sodium hypochlorite and then assigned randomly to Control or UV-C groups (Charles and Arul, unpublished data).

et al., 2003), scoparone and scopoletin in oranges (D'Hallewin et al., 1999), and rishitin in tomato (Charles et al., 2008d). Similarly, UV-C can enhance the production of PR-proteins in exposed tobacco plants (Yalpani et al., 1994) and induce β-1,3-glucanases and chitinases in UV-C-treated tomato fruit (Charles et al., 2009). There is evidence that UV-C-induced resistance in produce follows changes in a chronological mode involving multiple defense mechanisms. Tomato exposed to UV-C showed first the accumulation of phytoalexins as the earliest defense response, followed by the formation of physical barriers, and ultimately by increased production of constitutive and inducible β-1,3-glucanases and chitinases (Charles et al., 2011).

The physiological age of the produce and UV-C dose are important factors since younger tissue appears to exhibit higher active responses (Arul et al., 2001). Optimal UV-C doses to achieve beneficial effects in the produce may range from 0.03 kJ/m^2 in banana (Pongprasert et al., 2011) to 39.6 kJ/m^2 in pineapple (Sari et al., 2016). It also

Figure 17.3 Control of natural infection in bell peppers (cv. 'Billy Boy') treated with the hormetic dose of UV-C (0.88 kJ/m^2) and stored at 13°C for 18 d. A: Nontreated peppers; B: UV-C-treated peppers (Baka, 1997).

appears that the induced disease resistance and the accumulation of some phytoalexins by UV-C action is a characteristic localized response, and it is not systemic (Arul et al., 2001). UV treatments of postharvest crops must take these factors into consideration to attain prolonged shelf life, and to increase valuable health-beneficial compounds.

4 Induced Resistance in Postharvest Crops by UV-C Radiation

Plants possess several natural mechanisms to protect themselves against aggression by pathogens (Benhamou, 2009; Walters, 2010). These mechanisms can be classified into two groups: preformed and induced defense systems. The preformed or constitutive defense mechanisms are related to the normal anatomy and physiology of plant organs. These defense mechanisms are also termed passive defense as they are a consequence of temporal adaptation to climatic and environmental conditions. They are structural barriers, such as the cuticle (cutin and wax), the stomata, and the trichomes, or preformed chemical compounds with antimicrobial activity, such as phenolics, alkaloids, lactones, essential oils, and enzymes. In response to biotic or abiotic stresses, the plant genes are activated and defense responses are mounted to a higher level. This active defense is also of structural and/or of biochemical nature. The structural active defense mechanisms involve reinforcement of the plant cell wall through lignification, suberization, callose apposition, incorporation of hydroxyproline-rich glycoproteins (HRGP) to the cell wall, and papilla formation.

The inducible biochemical defenses include the accumulation of phytoalexins and PR-proteins (chitinase, β-1,3-glucanase, and other proteins). Until recently, there were not many reports addressing the activation of resistance response in postharvest organs; and the diseases in postharvest crops were controlled mainly by the application of pesticides at the pre- or postharvest stage. A brief search to a web database with the key words 'postharvest and disease and resistance' would illustrate this context. It is only during the recent years that stimulation of natural host defenses is being considered as a preferred strategy for the management of postharvest disease as an alternative to the use of chemical programs (Arul et al., 2001; Terry and Joyce, 2004; Romanazzi et al., 2006; Stevens et al., 2006). UV-C hormesis has been shown to reduce disease incidence and severity in horticultural crops at the postharvest stage (Charles and Arul, 2007). This approach has been shown effective in controlling pathogenic bacteria and fungi in climacteric and non-climacteric crops. The action modes of UV-C treatment are described as either direct or indirect. The direct action mode is likely through surface decontamination, while the indirect action is related to the priming of the defense mechanisms intrinsic to the plant organs (Syamaladevi et al., 2014). At low doses, the germicidal effect of UV-C is small, and the induction of disease resistance is the dominant mode of action (Mercier et al., 2001). Although the phenomenon of UV-induced protection against pathogens is adequately established, in some instances discrepancies in the magnitude of the responses have been observed. The fact is that UV hormesis is dependent on several factors, ranging from the preharvest conditions, harvest season, the physiological status of the crops, the genotypes, and post-treatment storage conditions. The following sections outline the effect of UV-C hormesis on the preformed and the inducible defense mechanisms.

4.1 Effect of UV-C on the Constitutive Defense Mechanisms

The plant surface, namely, the cuticle through its physical, biochemical, and mechanical properties, is the first obstacle to the invading fungi and bacteria in postharvest crops. Changes in the cuticle structures through UV-C action could potentially interfere with the outcome of the interaction between the disease-causing microorganisms and their host at the postharvest stage. To the best of our knowledge, there is no research report on the effect of postharvest UV-C hormesis on wax physicochemical properties. However, the ability of UV-C irradiation to modify the wax layer of the cuticle of harvested plant tissue was highlighted recently that showed an increase in the hydrophobicity of cuticular wax and improvement in the barrier properties against water loss (Mihaly-Cozmuta et al., 2016). The precited study compared with classical ones reporting the impact of postharvest UV-C hormesis contained important differences in the methodology used: UV-C was applied daily over a 12 hr period for a duration of 6 wk, which is beyond the short exposure time usually reported in postharvest studies and key information such as lamp intensity, treatment duration, and distance were not provided. Nonetheless, due to the paucity of information on the impact of UV-C hormesis on crop surface and wax composition, consideration was given to the overall data reported by Mihaly-Cozmuta et al. (2016) with pepper, since they are in line with observations made by others showing that postharvest UV-C treatment modified surface wax apposition (Charles et al., 2008b, 2008c; D'Hallewin et al., 2009; Fava et al., 2011), but differently depending on the species.

Fava et al. (2011) reported UV-C treatment caused wax excrescences of irregular shape in grapes; Charles et al. (2008b) observed that treated tomato had a smoother surface likely covered with amorphous wax platelets; and D'Hallewin et al. (2009) reported more wax and cuticular cracks occurring during storage on treated oranges. It is likely that these ultrastructural modifications interfere with adhesion of pathogenic propagules and their ability to resume growth upon inoculation. In fact, Charles et al. (2008c) reported sparse surface colonization and fewer adhesion structures formed by *B. cinerea* on UV-C-treated tomato, suggesting that structural modification of the epicuticular wax induced by UV may have affected the ability of the fungus to attach to the treated fruit surface. Furthermore, hypersensitive reaction-like response of the epicarp cells in both grape (Fava et al., 2011) and tomato (Charles et al., 2008c) may likely have played a role in hindering pathogen colonization. In the report by D'Hallewin et al. (2009), it is likely that following UV-C exposure, antimicrobial principles contained in the oil glands were released favoring the inhibition of *P. digitatum* growth.

The other constitutive line of defense in plant organs is of chemical nature and is represented by the preformed antimicrobial compounds. These compounds innately biosynthesized by the plant tissue even in the absence of biotic or abiotic stressors are termed 'phyto-anticipins', 'prohibitins', or 'inhibitins' (Pedras and Yaya, 2015). They belong to several chemical classes with diverse enzymatic or catalytic properties, namely, phenylpropanoids, alkaloids, lactones, cyanogenic glycosides, essential oils, saponins, and proteins. They are generally typical of a plant family or genotype (species and cultivars). Although their concentrations in the plant tissue could vary with the general growth conditions, they are often present at higher titers in the juvenile tissues. It is likely that postharvest interventions such as UV-C hormesis that slow down senescence would maintain the preformed compounds at respectable levels during post-treatment storage. However, only a few studies have looked onto this aspect in horticultural crops subjected to UV-C hormesis.

SMs of the phenylpropanoid pathway are well recognized for their antimicrobial activities. A large body of studies has addressed the changes in phenolic contents in UV-C-treated crops, and several of them have established direct or circumstantial evidence of their involvement in contributing to disease resistance. Both activity and gene expression of PAL, the key enzyme catalyzing the first step in phenylpropanoid pathway, increased in response to UV-C (El Ghaouth et al., 2003; Pombo et al., 2011; Sripong et al., 2015), concomitant with disease reduction caused by necrotrophic pathogens such as *B. cinerea* in strawberry fruit (Pombo et al., 2011), *P. digitatum* in grapefruit (Droby et al., 1993), *C. gloeosporioides* in mango (Sripong et al., 2015), and *Sclerotinia sclerotiorum* (Lib.) de Bary in carrot (Ojaghian et al., 2017). Although not always assessed simultaneously in the cited reports, the trends for PAL activity and gene expression increments are generally in line with the increase in total phenol assessed by the Folin-Ciocalteau method or measurement of the content of the major individual phenols that are characteristic of the species. A case in point is chlorogenic acid in tomato (Liu et al., 2012; Severo et al., 2015b) or in carrot (Formica-Oliveira et al., 2017). In other studies, only relatively slower rate of degradation of chlorogenic acid and its derivatives was observed, depending on the UV-C application conditions (Bravo et al., 2012). Finally, the potential involvement of phenolic compounds in the repression of *B. cinerea* infection of tomato was established, using histopathological techniques (Charles et al., 2008b).

In immature and unripe tomato fruit, the resistance to *B. cinerea* is associated with the presence of the glycoalkaloid tomatine that decreases as ripening progresses, concomitant with increase in the susceptibility of the fruit to the pathogen (Eltayeb and Roddick, 1984). Tomato fruits of the cultivars 'Floradade' and 'Better Boy' that have been subjected to the 3.7 kJ/m^2 hormetic dose, maintained a higher level of tomatine than the control (Stevens et al., 1998b). Besides tomatine in tomato, the edible part of other plant species also contained preformed antimicrobial compounds. To name a few, citral in citrus fruit (Rodov et al., 1995); 5-pentadecyl resorcinol and 5-(heptadeca-8,11-dienyl) benzene1,3-diol in mango (Krishnapillai and Wilson Wijeratnam, 2017); stilbenoids in grapes (Ficke et al., 2004); polyacetylenes, falcarindiol, and falcarinol in carrot (Mercier et al., 1993a). Although Mercier et al. (1993c) reported on the polyacetylenes content of several carrot cultivars, none of the precited studies give clear input as to the possible involvement of the preformed compounds in disease resistance.

4.2 Effect of UV-C on the Inducible Defense Mechanisms

Of the cascade of events put in place in the induction of defenses, the early and transient burst of ROS, the hallmark of stressed plant tissues (Doke, 1997), is observed in UV-C-treated plant materials (Balestrazzi et al., 2010). ROS act in a dose-dependent manner; they are signaling molecules at low levels that regulate key cellular functions such as those involved in plant defense; but at high levels, they exert severe oxidative stress (Tatone et al., 2010) that may cause extensive cellular damages. Ait-Barka et al. (2000a) observed a significant and transient increase in hydrogen peroxide, a stable ROS, immediately after UV-C treatment in tomato pericarp tissue, along with increases in lipid peroxidation markers. The timing in the generation of ROS appears to be directly linked to an effective defense against disease (Király et al., 2013).

The ROS may contribute to the hypersensitive response (HR) or host cell death that would lead to the confinement of the pathogen in the necrotized cells (Mengiste, 2012). The formation of a zone of stacked cell walls (CWSZ) involving the epicarp cells and the top cell layers of the pericarp tissue, and the confinement of fungal cells in the CWSZ were observed in UV-C-treated tomato (Figure 17.4) (Charles et al., 2008a). These cell layers presumably underwent plasmolysis leading to host cell death, akin to a form of HR. Besides, contributing to HR-like reaction, ROS are also known to promote physical reinforcement of the plant cell wall by cross-linking of cell wall glycoproteins and lignin via oxidative cross-linking (Künstler et al., 2015). Reinforcement of tomato cell wall through lignification was observed to modify the outcome of the interaction in this crop with *B. cinerea* (Charles et al., 2008c). Nonetheless, it is worth mentioning that evolution of ROS content in UV-C-treated commodity appears to be biphasic (Ait-Barka et al., 2000a), as the first burst is scavenged by enzymatic (Yang et al., 2014) and nonenzymatic antioxidant molecules (Maharaj et al., 2014) preventing oxidative damages to cellular components. Among these antioxidant molecules, the phenols seem to be key players in plant resistance. However, it is difficult to enunciate the multiple roles of ROS in UV-C-induced resistance phenomenon in the absence of direct evidence at this time, and further investigations would be required to assess their role.

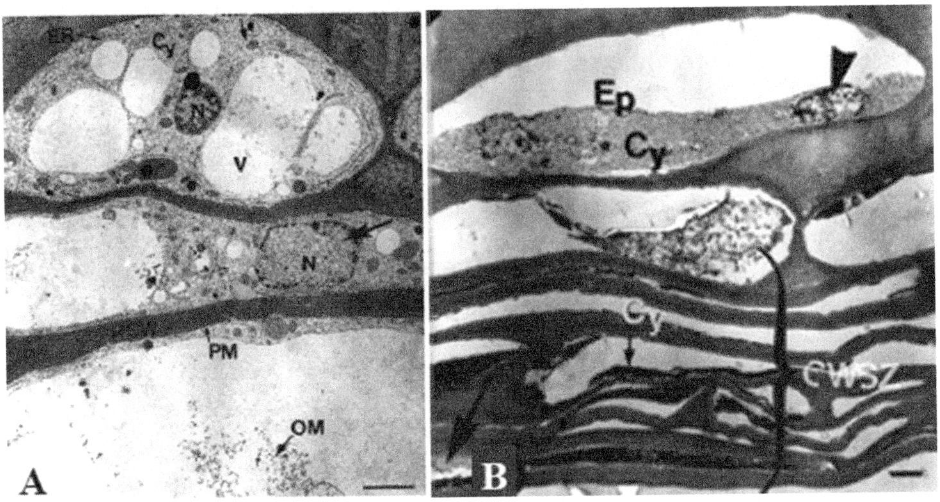

Figure 17.4 Transmission electron micrographs of non-inoculated tomato pericarp. (A) Control fruit 10 d after treatment; (B) UV-C-treated fruit 10 d after treatment. Cy: cytoplasm; CWSZ: cell-wall stacking zone; Ep: epicarp cell; ER: endoplasmic reticulum; HCW: host cell wall; N: nucleus; OM: osmiophilic vesicle; PM: plasma membrane; V: vacuole. Scale bars: 2 μm. Source of data from Charles et al. (2008a, 2011) and Charles (1998).

4.3 Phytoalexins

Phytoalexins are plant SMs with fungitoxic and fungistatic properties. They are low molecular mass compounds of various chemical structures: phenolics, terpenoids, furano-acetylenes, steroid glycoalkaloids, sulfur-containing compounds, and indoles derived from specific metabolic routes depending on the plant family and species. Their expression is also organ and developmental stage specific. Their metabolism and their role in disease resistance have been reviewed by several authors (Bailey, 1982; Purkayastha, 1994; Smith, 1996; Guest, 2017). They are absent in naturally healthy or unstressed plant tissues and their synthesis and accumulation are activated in a nonspecific manner by biotic elicitors (pathogenesis, elicitors from pathogenic and nonpathogenic organisms) and abiotic elicitors (UVR, heavy metals).

The indications from the literature suggest UV-C irradiation is an effective elicitor of phytoalexins in various crops. The information summarized in Table 17.3 indicates that UV-C induces specific chemical classes of phytoalexins in several plant foods. It should be noted however, that the plant parts investigated were not always edible tissues, and that the range of UV-C radiation doses and the conditions of its application were not in the context of postharvest preservation. To the best of our knowledge, studies on the elicitation of phytoalexins in response to UV-C hormesis as one of its modes of action in building resistance against diseases in stored crops have been conducted in a few crops, including citrus fruits, carrot, and tomato.

Table 17.3 Phytoalexins identified in some edible crops

Family	*Plants*	*Phytoalexins*	*Chemical classes*
Solanaceae	Potato	Solavetivone, Lubimin, Rishitin	Sesquiterpene
	Tomato	Rishitin	
	Sweet pepper	Capsidiol	
Leguminosae	Peas	Pisatin	Flavonoid
	Haricot	Kievitone, phaseollin	Isoflavonoid
	Soybean	Glyceollin	
	Cowpea	Vignafuran	Benzofuran
	Peanut	Resveratrol	Stilbene
		Arachidins	Flavonoïd
Rosaceae	Apple	Malusfuran	Benzofuran
	Strawberry	Resveratrol	Stilbene
Rutaceae	Lime		Methoxycoumarin
	Kumquat	Scoparone	
	Orange		
Umbelliferae	Carrot	6-methoxymellein	Isocoumarin
Convolvulaceae	Sweet potato	Ipomeamarone	Furanosesquiterpene
Brassicaceae	Rutabaga, Turnip, Cauliflower	Isalexin, S-(−)- spirobrassisinin, methylbrassitin, brassicanal C, caulilexin A–C, brassinin, cyclobrassin, rutalexin	Indole
Vitaceae	Grape	Resveratrol	Stilbene
Poaceae	Rice	Momilactones, phytocassanes, oryzalexins	Diterpenoid
		Sakuranetin	Flavonoid
	Corn	Kauralexins, zealexins	
Amaryllidaceae	Garlic	Allixin	Pyrone
Moraceae	Mulberry	Moracins (C, N, and chalcomoracin)	Benzofuran
Ericaceae	Blueberry	Resveratrol	Stilbene

An obvious relationship between improved disease resistance and phytoalexin accumulation by UV-C treatment was established in citrus fruits by a group from Volcani Center, Israel (Ben-Yehoshua et al., 1992; Rodov et al., 1992, 1994). Scoparone (6,7-dimethoxycoumarin) and its precursor scopoletin (6-methoxy, 7-hydroxycoumarin) were identified in the flavedo of various citrus fruits (kumquat, orange, and lemon) treated with UV-C. Their phytoalexin status was confirmed by their absence in the tissue of nontreated fruits and their antifungal activities, both in vivo and in vitro, against microorganisms such as *Cladosporium cladosporioides* (Fresen.) G.A. de Vries and *P. digitatum*, the latter being one of the most important diseases that causes major economical losses in citrus fruits. The reduction of disease incidence and severity tended to be greater than 50% compared to the control fruits, depending on the date at which inoculation was performed after the treatment (Rodov et al., 1992). Recently, Kuniga et al. (2006) reported on the inhibitory activity of scoparone against *B. cinerea* and *C. gloeosporioides* with EC_{50} value of about 60 and 75 ppm, respectively.

The effective range for UV-C-induced phytoalexins accumulation in citrus fruits appears to be between 1.5 and 5 kJ/m^2. The phytoalexin response to UV-C in citrus fruits was organ and harvest period dependent, with higher content found in leaf blades than in fruit albedo (Kuniga et al., 2006; Kuniga and Nesumi, 2011), and in organs harvested in August than those in December (Kuniga et al., 2006). The seasonal effect of UV-C is likely related to tissue age or maturation stage (D'Hallewin et al., 2000; Kuniga et al., 2006). Among the citrus families, kumquat flavedo (Rodov et al., 1992) and lemon leaves harvested in August (Kuniga et al., 2006) showed the highest capacity to accumulate the phytoalexins. In addition, there were differences between orange cultivars in their capacity to accumulate the antifungal compounds (Rodov et al., 1992; Kuniga et al., 2006). Grapefruit was found to be a preferential accumulator of scopoletin (D'Hallewin et al., 2000) compared to the other citrus fruits, which accumulated scoparone as the principal phytoalexin. Post-treatment storage temperatures did not affect the phytoalexin response in UV-C-treated 'Verna' lemons and 'Jaffa' oranges (D'Hallewin et al., 2009).

Carrot is known to accumulate the phytoalexin, 6-methoxymellein (6-MM), an isocoumarin, in response to infection (Goodliffe and Heale, 1978). The knowledge on the dynamics of phytoalexin accumulation in response to UV-C was gained from the work on carrots (Mercier et al., 1993a, 2000). They observed an inhibition of *B. cinerea* and *S. sclerotiorum* (>60%) in UV-C-treated carrot root tissue, concomitant with the accumulation of 6-MM (Mercier et al., 1993b). Although the rate of accumulation of 6-MM in the treated carrots was faster at 20°C than either at 4 or 1°C, the lower temperatures allowed the maintenance of higher levels for longer periods, since the rapid degradation of 6-MM was apparent at 20°C after a certain maximum level was reached (Mercier et al., 1993a,b; Kouassi et al., 2012); presumably, it is toxic to plant cells as well (Darvill and Albersheim, 1984). UV-C dose-6-MM accumulation as well as disease resistance relationships were biphasic, and the hormetic doses were 0.88 and 5.4 kJ/m^2 for wounded and healthy carrots, respectively. In addition, the capacity to accumulate 6-MM decreased with the age of the carrot (time after harvest). There were differences between carrot cultivars in their ability to accumulate 6-MM in the range of 57 to 100 µ/g fresh weight (Mercier et al., 1993a,c). Furthermore, disease resistance and 6-MM accumulation in response to UV-C were not systemic but localized effects that would require exposure of the entire surface of the produce to radiation (Mercier et al., 2000; Ojaghian et al., 2017). Besides, the polyacetylenes, 6-MM may also contribute to bitterness of carrot root (Lafuente et al., 1989). But it may not be an issue, since peeling and cooking operations remove much of this compound, and the taste is not affected (Mercier et al., 1994).

The stilbene phytoalexin, resveratrol, found in grape (Hasan and Bae, 2017), has attracted much attention due to its beneficial impact on human health (Shishodia and Aggarwal, 2005). This phytoalexin has been found in over 70 plant species (Sebastià et al., 2017) with higher levels generally found in grapes and peanuts, and they are the most investigated in relation to the ability of UV-C to trigger the accumulation of the phytoalexin and the development of defenses. Often the use of

UV-C treatment as a postharvest treatment for grapes was not aimed at extending their shelf life through the activation of disease resistance, but at elevating resveratrol level with various doses of UV-C for its utilization as a source for the extraction of this compound (Cantos et al., 2001, 2003) and to produce enriched grape juice (González-Barrio et al., 2009) and other products (Hasan and Bae, 2017).

UV-C-treated table grapes had displayed significantly greater resistance to gray mold caused by *B. cinerea* (Romanazzi et al., 2006; D'Hallewin et al., 2012). The accumulation of resveratrol was UV-C dose (0.5–3.0 kJ/m^2) and cultivar (5 cultivars) dependent (D'Hallewin et al., 2012). UV-C doses above the hormetic dose of 0.5 kJ/m^2 had a negative impact in the treated grapes, particularly at the highest dose of 3.0 kJ/m^2 that exhibited dark discoloration during storage. Moreover, grape berries in the middle of the bunch, not exposed directly to UV-C, had lower level of induced resveratrol and were apparently not protected from mold development, suggesting that the mode of action of this treatment is not systemic (D'Hallewin et al., 2012). Nigro et al. (1998) reported on the induction of resistance in *B. cinerea* inoculated grape berries by doses between 0.25 and 0.5 kJ/m^2 but they did not evaluate the elicitation of resveratrol.

The sesquiterpenes are the phytoalexins found in Solanaceae. Among them, the phytoalexin, rishitin, was induced in potato tuber slides by UV-C radiation (Cheema and Haard, 1978) and in tomato fruit (Charles et al., 2008d) by UV-C radiation. A role for this molecule in disease resistance could be drawn from the observation that a compatible race of *Phytophthora infestans* (Mont.) de Bary did not induce rishitin accumulation in potato slices, while the incompatible race of this fungus elicited significant level of the phytoalexin (Currier and Kuć, 1975). UV-C at a dose of 3.7 kJ/m^2 improved resistance of postharvest tomato to *B. cinerea* concomitantly with the accumulation of rishitin (Charles et al., 2008d). Furthermore, the accumulation of rishitin was significantly higher upon infection of the treated fruit, suggesting that UV-C had primed the tissue for this response (Figure 17.5). There was clear indication that rishitin accumulation in UV-C-treated tomato was affected by post-treatment storage conditions (Charles, 1998), and was also cultivar-dependent as some commercial tomato cultivars did not accumulate this phytoalexin to inhibitory levels (Charles et al., 2019).

In the *Brassicacea* crops, sulfur containing phytoalexins, namely, isalexin, S-(−)- spirobrassisinin, methylbrassitin, brassicanal C, caulilexin A–C, brassinin, cyclobrassin, and rutalexin were detected in UV-C-treated cauliflower, turnip, and rutabaga (Pedras et al., 2006, 2008). These phytoalexins exhibited antimicrobial properties against postharvest pathogens such as *Rhizoctonia solani* J.G. Kühn and *S. sclerotiorum*. Unfortunately, UV-C application characteristics in these studies, such as the doses, were not clearly described.

4.4 Reinforcement of the Cell Wall

The ability of a pathogen to breach the cell wall may be a determining factor in pathogenicity. Therefore, cell wall reinforcement through apposition of structural materials such as callose, lignin, or suberin is another form of protection that is believed to check pathogen ingress. Accumulation of the polysaccharide callose (β-1,3-glucan) is one of the earlier response observed in plants responding to both biotic and abiotic stress (Piršelová and Matušíková, 2013). It is believed that callose

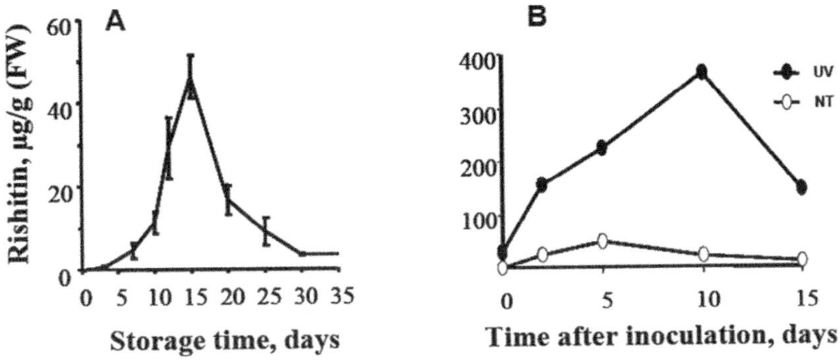

Figure 17.5 Accumulation of the phytoalexin rishitin in postharvest tomato (A) during storage of UV-C-treated tomato and (B) in response to inoculation with *Botrytis cinerea*. The data for (B) were from control and UV-C-treated fruit inoculated 12 d after UV-C treatment. FW: fresh weight; UV: UV-C-treated fruit; NT: non-UV-C-treated control. Source of data from Charles et al. (2008d, 2011) and Charles (1998).

plays its effective role not only through mechanical strengthening but also by preventing the leakage of assimilates soon after the imposition of stress. There is no report on callose apposition in postharvest crops treated with hormetic UV-C doses. However, Mintoff et al. (2015) reported that in *Arabidopsis thaliana* sublethal UV-C doses of 0.25 to 1.0 kJ/m^2 induced callose apposition. The ability for postharvest crops to develop callose apposition was shown in citrus fruit, primed with chitosan (Benhamou, 2004), and in muskmelon after hot water dips (Yuan et al., 2013).

It is well established that complex and polymeric phenolic compounds, lignin and suberin, contain plant pathogens ingress in plant tissue and restrict their access to water and nutrients (Vance et al., 1980; Bostock and Stermer, 1989). Increases in the activity of PAL, the first enzyme catalyzing the biosynthesis of phenolic monomers for lignin synthesis, were found in several UV-C-treated crops including grapefruit (Chalutz et al., 1992), oyster mushroom (Wang et al., 2017), carrot (Ojaghian et al., 2017), blueberry (Xu et al., 2016), mango (Sripong et al., 2015), and others. Increased PAL activity was linked to enhanced *FaPAL* gene transcription and expression in strawberry fruit (Pombo et al., 2011).

Likewise, a fourfold enhancement in the transcription of C′3H (*p*-coumaroyl ester 3′-hydroxylase) compared to the control was observed in globe artichoke leaves (Moglia et al., 2009). In the lignin biosynthesis pathway, the C′3H enzyme family catalyzes the conversion of *p*-coumaroyl-quinate to 5-caffeoylquinic acid (chlorogenic acid), or *p*-coumaroyl-shikimate to caffeoyl-shikimate. The resistance to anthracnose was concomitant with UV-C-induced increment in PAL activity in mango fruit (Sripong et al., 2015). Using histochemical and ultrastructural techniques, it was shown that the ingression of *B. cinerea* in UV-C-treated tomato was hampered by the reduced ability of the pathogen to deconstruct the fruit cell wall, which were likely less prone to the action of the cell wall degrading enzymes secreted by the pathogen (Figures 17.6, 17.7) (Charles et al., 2008a, 2008b, 2008c). Lignification in UV-C-treated tissues appears to be modulated variously based on the dose, plant species, and storage temperature. Gogo et al. (2017) examined the

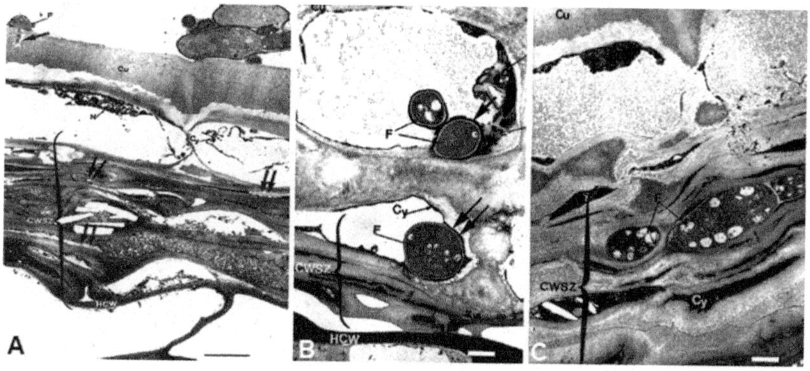

Figure 17.6 Transmission electron micrographs of UV-C-treated tomato fruit inoculated 10 d after treatment. Observations were made (A) 4 d or (B) and (C) 15 d after inoculation. Cu: cuticle; CWSZ: cell-wall stacking zone; F: fungal cell; Cy: cytoplasm; HCW: host cell wall; N: nucleus. Scale bars: (A) 5 μm; (B) and (C) 2 μm. Source of data from Charles et al. (2008a, 2011) and Charles (1998).

acid detergent lignin fraction in two leafy vegetables and observed a higher titer in the treated African nightshade that had been stored at 5°C in comparison to the control. In amaranth, they did not find any effect of either the storage temperature (20 or 5°C) or the dose (1.7 or 3.4 kJ/m^2). The lignin content decreased in UV-C-treated asparagus (1.0 kJ/m^2) after storage at 20°C (Huyskens-Keil et al., 2011).

Suberization involves the deposition of a polymeric material composed of aromatic domains similar to those in lignin and of aliphatic polyester domains similar to those in cutin. The polymeric material forms a hydrophobic barrier associated with waxes (Kolattukudy, 1984), and it is induced by wounding as well as pathogenesis (Mohan and Kolattukudy, 1990). Specific staining for suberin in tomato tissue treated with the hormetic UV-C dose revealed that this compound likely provides additional protection against tissue maceration by fungal enzymes, thereby impeding *B. cinerea*, a necrotrophic pathogen known to be good secretor of cell wall degrading enzymes (Charles et al., 2008a). Lignification, suberization, and ultrastructural modifications within the CWSZ (cell wall stacking zone) in UV-C-treated tomato may have comprised a barrier even difficult for *B. cinerea* to breach (Charles et al., 2008a).

Another strategy developed by plants to quench pathogen ingress is the strengthening of cell wall architecture through cross-linkage of cell wall polysaccharides with the HRGP and extensins by the action of peroxidase and H_2O_2 (Jackson et al., 2001; Barceló et al., 2003), which may play a significant role in the biotic and abiotic stress responses (Deepak et al., 2010). The specific wavelength of UV used in postharvest hormesis studies was shown to stimulate the production of H_2O_2 (Murphy Terence and Huerta Alfredo, 1990). The increase in H_2O_2 production, in turn, is a driver of cell wall lignification (Kuźniak and Urbanek, 2000) as observed in UV-C-treated tomato fruit (Charles et al., 2008a). A clearer contribution of HRGP and extensins cross-linking to the defense reactions elicited in UV-C-treated postharvest crops remains to be established. The lack of information in that regard may be

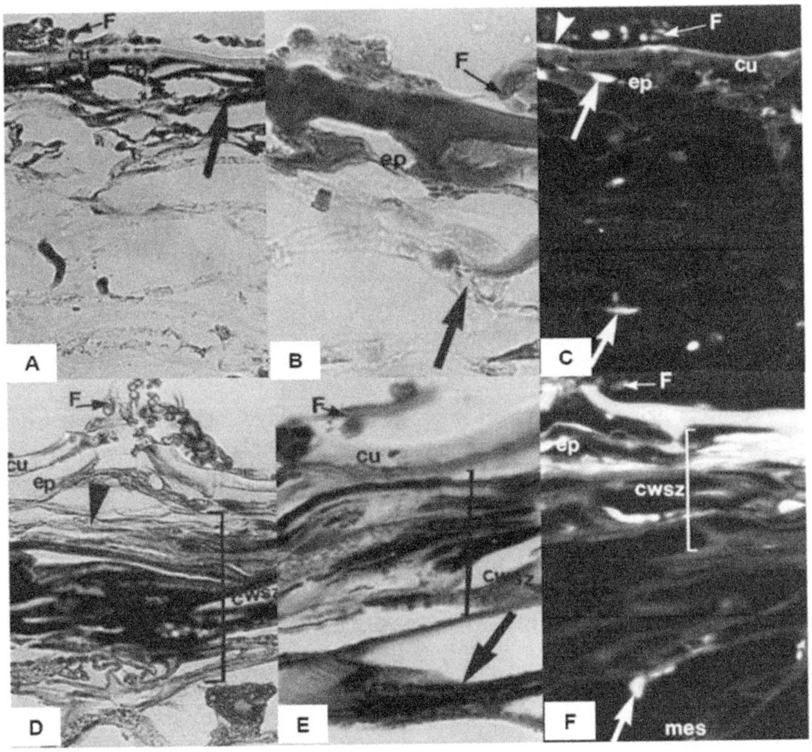

Figure 17.7 Histochemical test for (A) and (D) phenolic compounds, (B) and (E) lignin, and (C) and (F) suberin in tomato fruit inoculated with *Botrytis cinerea*. Upper plates: control tissue; lower plates: UV-C-treated tissue. Inoculation was performed 10 d after UV-C treatment, and observations were made 4 d after inoculation. cu: cuticle; cwsz: cell-wall stacking zone; ep: epicarp; F: fungal cell; mes: mesocarp. Source of data from Charles et al. (2008b, 2011) and Charles (1998).

because UV-C-induced H_2O_2 burst is an early event and transient, and studies that assess UV-C hormesis focused on the shelf life of crops monitor changes during storage, with little attention to early events.

4.5 Defense-Related Proteins

PR-proteins, also termed defense-related proteins, are induced de novo in plant tissues, and at times may also be enhanced in response to pathogen attack, ethylene, and physical and chemical stresses (Linthorst and van Loon, 1991; Van Loon et al., 2006). These proteins are classified based on their putative function in host–pathogen interactions, their cellular localization, their electrophoretic mobility under native conditions, and according to their amino acid composition (Van Loon, 1985). Defense-related proteins are classified into 17 distinct families (Van Loon et al., 2006). The most studied families in plant tissues are the β-1,3-glucanases (PR-2) and chitinases (PR-3, PR-4, PR-8, and PR-11) characterized for their capacity to hydrolyze fungal cell wall constituents. Among them, PR-2 and PR-3 activities are the only ones often studied in postharvest crops as well as in UV-treated crops. In some cases, their gene expression was the methodology embraced. The ability of UV-C treatment to affect PR-proteins is apparently modulated by the maturity stage (Petit et al., 2009) as well as by tissue type (Petit et al., 2009; Colas et al., 2012). Colas et al. (2012) observed, using in situ localization, a marked induction of *CHI4D* (a chitinase) transcripts in the exocarp of grape berries in response to UV-C, and this induction was associated with the vascular bundles in the mesocarp. On the other hand, mRNAs of *TL3* belonging to the PR-5 family (thaumatin-like protein with β-1,3-glucanase activity) increased by UV-C treatment in the whole mesocarp. Both chitinase and glucanase activities and their relative gene expression were at higher levels in UV-C-treated grape berries approaching veraison or onset of ripening (Petit et al., 2009). The apparent involvement of PR-proteins, in particular, the chitinases and glucanases, in improving the disease resistance of UV-C-treated crop was highlighted in several studies.

A study by Porat et al. (1999) suggested the involvement of PR-proteins, namely, chitinase and β-1,3-glucanase, in the resistance of UV-irradiated grapefruit to *P. digitatum*. While UV treatment alone could induce chitinase expression, a combination of both wounding and UV treatment was necessary to trigger β-1,3-glucanases in grapefruit flavedo tissue. UV-C-induced antifungal hydrolases, some of which are true PR-proteins, were also reported in bell pepper (Baka, 1997), tomato (Charles et al., 2009), mangoes (Srepong et al., 2013), and carrot roots (Mercier et al., 2000; Ojaghian et al., 2017). In carrot roots, a chitinase was induced not only by UV-C application but also by infection with *B. cinerea* (Mercier et al., 2000), a finding that confirms the PR status of this hydrolase. In peach fruit extracts, increased chitinase and β-1,3-glucanase activities were detected as early as 6 hr after UV treatment, reaching a maximum level after 96 hr (El Ghaouth et al., 2003). A 33.1 kDa β-1,3-glucanase induced by UV-C and its activity was maintained until the end of the storage period of 35 d in tomato fruit. However, it was apparent that it is not a true PR protein since it was detected neither in the control tissue nor in the tissue inoculated with *B. cinerea* (Charles et al., 2009). On the other hand, two chitinases of 37.1 and 20.6 kDa were induced by both UV-C and pathogenesis, as such they could be classified as PR proteins.

5 Delayed Senescence and Ripening in Postharvest Crops by UV-C Radiation

The susceptibility of fruits and vegetables to postharvest diseases typically increases with the senescence of the tissue. This is associated with diminishing ability of tissue to synthesize preformed and inducible inhibitors, which provide disease resistance to young tissue but degrade with age; and consequently, the latent infections resume. Thus, the maintenance of constitutive and inducible defense responses of the host by slowing down senescence could be of significance in the extension of shelf life and in the reduction of postharvest decay. The understanding of the physiological and metabolic changes during postharvest storage of produce would enable adopting appropriate strategies to delay the senescence process, especially using hormetic doses of UV-C irradiation. The effectiveness of practices such as storage temperature, atmosphere, and treatments to delay senescence can be measured by several markers including respiration rate, antioxidant systems, color change, tissue softening, membrane integrity, protein loss, hormones, and gene expression.

5.1 Respiration of Commodities and UV-C Irradiation

An effective strategy to delay the progress of senescence and to reduce postharvest losses is to decrease the respiration rate of the crops. Respiration is a metabolic process which not only maintains biochemical processes but also hastens the senescence of produce (Yang et al., 2014). Reduction of respiration rates by UV-C treatments has been reported in different commodities. Pongprasert et al. (2011) observed a reduction in both respiration rate and ethylene production in banana exposed to 0.03 kJ/m^2. This pattern was also observed by Maharaj et al. (1999) in tomatoes exposed to 3.7 kJ/m^2 and in strawberries exposed to 0.25 kJ/m^2 (Baka et al., 1999). However, UV-C treatments have also increased the respiration rates of certain commodities such as pumpkins (Erkan et al., 2001) and banana (Sheik Mohamed et al., 2017). Thus, the effect of UV-C radiation on respiration rate appears to be variable, even with the same dose and with similar crops. Despite the prior, Civello et al. (2006) grouped the effects of UV-C on respiration rates as three different patterns: (1) very low doses do not affect respiration rate; (2) a transient increase in respiration rate (stress respiration), followed by a reduction of respiration rates during the storage that occurs with increasing the dose of UV-C; and (3) very high UV-C doses may damage the tissue with a concomitant increase in respiration rates. Ultimately, the transient rise of respiration rate in the crops exposed to UV-C is an acclimatization process that leads to the synthesis of compounds with protective and defense functions.

5.2 Induction of Antioxidant Systems and Alleviation of Oxidative Stress

UV-C radiation is an abiotic stress causing oxidative events leading to the production of ROS in plant cells (Mittler, 2002). At high levels, ROS can damage DNA, cell membranes lipids, and proteins and eventually hastening senescence (Ait-Barka et al., 2000a; González-Aguilar et al., 2007). Thus, delay of senescence often reported in UV-C-treated crops must be linked to the ability of the stressed tissues to scavenge ROS. The remodeling of the balance between reduced and oxidized glutathione, and increase in phenolic biosynthesis, are among the mechanisms described to explain how UV-C-treated produces cope with oxidative stress through the production of antioxidant compounds (Erkan et al., 2008; Jiang et al., 2010; Martínez-Hernández et al., 2011). In addition to nonenzymatic compounds, exposure of produce to UV-C enhanced the activity of antioxidant enzymes including superoxide dismutase (SOD) in strawberry fruits and peaches (Erkan et al., 2008; Yang et al., 2014), ascorbate peroxidase (APX) and glutathione reductase (GR) in shitake mushrooms (Jiang et al., 2010), and catalase (CAT) in peaches (Yang et al., 2014).

5.3 Changes in Color and Pigments

Color of fresh fruits and vegetables is related to their specific pigment composition. The development of pigments is desirable in many fruits. The change in color of tomato, for example, from green to red is due to degradation of chlorophyll and synthesis of lycopene. In green leafy and flower vegetables, chlorophyll degradation and accumulation of yellow carotenoids, a sign of senescence, must be retarded. UV-C radiation has been used as a means of delaying the yellowing of broccoli florets (Costa et al., 2006; Martínez-Hernández et al., 2011; Duarte-Sierra et al., 2012), as well as to affect the development of other pigments such as anthocyanins in strawberry (Cote et al., 2013) and carotenoids in tomato (Maharaj et al., 1999; Liu et al., 2009). Ethylene action accelerates chlorophyll degradation by stimulating enzymes such as magnesium-dechelatase (MDS), chlorophyllase, and chlorophyll-peroxidase (Costa et al., 2006). Any negative impact of UV-C on ethylene production and its action could contribute to the delay in chlorophyll degradation. The exposure of fruits and vegetables to UV-C radiation is known to induce the production of polyamines (Maharaj et al., 1999; Tiecher et al., 2013; Mortazavi et al., 2014), which can delay senescence in plant tissue, and perform multiple functions: stabilize DNA and cell membranes, scavenge free radicals, suppress ethylene production, and inhibit RNAse and proteases (Galston and Sawhney, 1990). Polyamines (spermine, spermidine, and putrescine) inhibit the conversion of 1-aminocyclopropane-1-carboxylic acid (ACC) to ethylene (Abeles et al., 1992). Recently, it has been proposed that downregulation of genes in UV-C-treated broccoli coding for a chloroplast-located protein, also known as Stay-Green (SGR), may be another mechanism contributing to the delay in chlorophyll degradation (Gómez-Lobato et al., 2014).

5.4 Delay in Tissue Softening

The cell wall structure of crops undergoes modifications during senescence, which is evident by the loss of tissue firmness or softening, particularly with extended storage periods (Lers, 2012). Loss of firmness is mainly the result of the action of three cell wall degrading enzymes: pectin methyl esterase (PME), polygalacturonase (PG), and cellulase. It has been reported that the exposure of tomato fruit to either 3.7 or 4.2 kJ/m^2 of UV-C radiation reduced the activity of these enzymes as well as the expression of genes encoding for cell wall degrading enzymes in the fruit during storage (Ait-Barka et al., 2000b; Bu et al., 2013). Furthermore, polyamines may also protect the cell wall in a manner similar to that of calcium by cross-linking with pectic acid to limit the access to hydrolase enzymes (Kramer et al., 1989; Maharaj et al., 1999). The scavenging of free radicals by polyamines may also contribute to reducing the derangement of the intact supramolecular structures of pectin, thereby reducing their exposure to the endogenous pectic enzymes.

Cell wall reinforcement has been reported in response to exposure of produce to UV-C. Charles et al. (2008c), for example, reported the lignification and suberization of tomato fruit cuticle exposed to 3.7 kJ/m^2. Although lignification may lead to increases in tissue firmness, its accumulation may negatively affect the digestibility and texture (Lers, 2012). For some crops such as green asparagus, the tissue toughening is not a desirable texture attribute. A lower dose of 2.4 kJ/m^2 prevented the toughening of green asparagus, but a higher dosage of 3.8 kJ/m^2 had no effect on that characteristic (Poubol et al., 2010). In contrast, even a very low UV-C dosage (1.0 kJ/m^2) significantly enhanced tissue toughness of white asparagus (Huyskens-Keil et al., 2011).

5.5 Maintenance of Membrane Integrity

The production of ROS due to UVR may cause injury because membrane lipids are susceptible to ROS, leading to membrane disorganization (Wu et al., 2016). Although excessive doses of UV-C radiation can lead to membrane disorganization (Kovács and Keresztes, 2002), low doses of UV-C radiation appear to improve membrane integrity. Low dose of 0.03 kJ/m^2 reduced malondialdehyde (MDA) levels of banana peel (Pongprasert et al., 2011), and a dose of 3.0 kJ/m^2 increased the cell membrane fluidity in peaches (Yang et al., 2014). Such protective impact of UV-C on membranes appears to cross protect tropical and subtropical crops against chilling injury. Effectively, chilling injury was alleviated by UV-C in oyster mushrooms (Wang et al., 2017), bamboo shoots (Zeng et al., 2015), pepper (Vicente et al., 2005; Rodoni et al., 2015), cucumber (Kasim and Kasim, 2008), and broccoli florets (Lemoine et al., 2007). The increased tolerance to chilling injury in UV-C-treated crops has been attributed to increased influx of calcium into the cells as in apple (Hemmaty et al., 2007) or the increased activity of ROS scavenging enzymes such as such as CAT, APX, and GR in UV-C-treated tissues (Zeng et al., 2015). Yang et al. (2014) suggested that induction of antioxidant enzymes by UV-C might be related to the maintenance of membrane integrity, especially to SOD, which also appears to increase the tolerance to chilling injury (Lers, 2012).

5.6 Reduction of Protein Loss in Crops by UV-C Irradiation

The complex physiological changes occurring in senescing plant organs are governed by the interplay of biosynthesis and degradation of catalytic and structural proteins. The overall protein balance of senescing tissue is expected to decline. UV-C treatment was shown in several instances to reduce the rate of protein decline in postharvest commodities during storage. The attenuation of protein loss has been shown in broccoli (Khalili et al., 2017), spinach, leek, and cabbage (Liao et al., 2016), some African leafy vegetables (Gogo et al., 2017), and tomato (Charles et al., 2009). Charles et al. (2009) observed an initial increase, followed by a slower degradation of protein, in the extracts from UV-C-treated tomato. They suggested that UV-C presumably induced the synthesis of protease inhibitors, leading to retention of constitutive and induced proteins caused by a lower protease activity observed in UV-C-treated tomato (Ait-Barka et al., 2000b). On the other hand, the lower protease activity in UV-C-treated tomato may also have slowed protein loss (Ait-Barka et al., 2000b). George Dominic et al. (2015), using proteomic approach, observed reductions in the expression of specific senescence, ripening proteins (namely, AAC oxidase, AAC synthase), and stress responses-related proteins (PG inhibiting proteins, RuBisCO complex) in UV-C-treated fresh-cut mango, and attributed the reductions to its improved shelf life.

The de novo synthesis of PR-proteins might be another mechanism that could underscore the higher protein content in UV-C-treated fruits during storage. Charles et al. (2009) observed an initial increase in protein content in UV-C-treated tomato concomitant with the enhancement of constitutive proteins as well as the induction of PR-proteins The latter has also been observed in strawberries, where the accumulation of Fra a1, a PR-10 protein, was enhanced, compared with non-treated fruits treated with a dose of 4.35 kJ/m^2 (Severo et al., 2015a). Moreover, authors of this study have speculated that Fra a1 protein may also enhance the color of strawberries as a result of facilitated translocation of flavonoid molecules, which confer for the red color to this fruit. It also worthy to note that the initial increase in protein content not long after UV-C treatment (Charles et al., 2009) is in line with de novo synthesis of proteins.

5.7 Interaction of Plant Hormones with UV-C Radiation Treatments

Hormones have a central role on senescence, with the well-established function of ethylene as the key hormone controlling ripening and senescence. Ethylene effect in crops can be summarized as three main effects: the acceleration of chlorophyll loss, the promotion of ripening, and the stimulation of phenylpropanoid metabolism (Saltveit, 1999). There are mainly four strategies to delay the senescence induced by this hormone: (1) the suppression of ethylene synthesis; (2) the inhibition of its perception; (3) the enhancement of antagonistic hormones against its action; and (4) the augmented increasing perception of those inhibiting hormones (Lers, 2012). Inhibition of ethylene synthesis by polyamines has been a common observation with produce irradiated with UV-C, as stated previously in this section. Inhibition of ethylene perception by UV-C has also been hypothesized. Severo et al. (2015a) proposed that the delay of tomato ripening may be due to the activation of ethylene response factors (ERF). Abscisic acid (ABA) can be considered as another ripening factor besides ethylene that it accumulates preceding ethylene release in climacteric fruits (Leng et al., 2014). The ABA peak was suppressed by 67% in UV-C-treated tomato pericarp discs with a dose of 1.4 kJ/m^2 (Kalantari et al., 2001). The application of low doses of UV-C (0.1 and 0.3 kJ/m^2d) to pea seedlings reduced level of ABA (Katerova et al., 2009). While UV-C appears to have a more significant effect on polyamines synthesis and delay in ethylene climacteric peak and ABA, other signal molecules such as jasmonic acid can also be involved in the transduction network. For instance, Kondo et al. (2011) observed that the accumulation of polyamines in apple seedling exposed to UV-C was accompanied with increases in jasmonate. Thus, the delay of senescence of crops in response to UV-C may depend not only on ethylene but also may involve its interactions with other signal molecules and phytohormones.

6 Maintenance of Quality, and Enhancement of Secondary Metabolites in Postharvest Crops by UV-C Radiation

Fruits and vegetables add flavor and variety to the human diet and they supply essential nutrients, antioxidants, and health-promoting phytocompounds. The quality of fresh produce is a combination of characteristics and attributes that give value to a specific crop. The components of quality include visual appearance free of defects, texture, flavor (taste and aroma), nutritional value, and safety (microbial and chemical residues). Any intervention or treatment to control diseases and improve storability of produce should not compromise their sensory and nutritional attributes, which are fundamental in the acceptability of a specific crop. In addition to improved storability and reduction in losses due to decay, enhancement in health-beneficial phytocompounds in the treated produce could improve the exposure of the population to those compounds, contributing to their health and well-being.

6.1 UV-C and Quality Attributes of Fruits and Vegetables

Consumer acceptance of fresh or minimally processed fruits and vegetables is generally assessed by attributes such as color, flavor, aroma, taste, texture, and freshness, among others. Visual assessment of freshness is often restricted to the edible portion of some produce, while for some others the appearance of the nonedible part such as the peel is also important. Although many studies have reported on the effect of UV-C treatment on quality factors such as color and pigments, soluble solids, acidity, and texture, only a few have evaluated the sensory attributes using a panel. Using taste as a discriminant, the panelists could not separate UV-C-treated tomato (Charles et al., 2005), carrot (Mercier et al., 1994), and onion (Lu et al., 1988) from their respective untreated control, while they successfully identified UV-C-treated pears (Syamaladevi et al., 2014), apple (Manzocco et al., 2011a), and melon (Manzocco et al., 2011b). The observed differences may be traced to the ability of UV-C radiation to modify the components associated with taste, flavor, and aroma. The sugar/acid ratio is often used as a measure of the storability and quality of fresh crops. In different tomato cultivars treated with the hormetic UV-C dose, the general effect was higher acid and lower sugar contents (Charles et al., 2016). However, based on the sugar and acid detection thresholds, known in the literature, this effect would not impact negatively consumer preference (Charles et al., 2016).

In general, studies have shown that sensorial scores are not affected after exposing various produce to UV-C irradiation. For example, the sensory attributes of fresh-cut dragon fruit were unaffected by a treatment of 3.2 kJ/m^2 (Nimitkeatkai and Kulthip, 2016), and UV-C had a positive effect on the bracts of dragon fruit (Kowitcharoen et al., 2010). Strawberries exposed to a two-step UV-C treatment at 2 kJ/m^2 had higher sensory scores in calix color, freshness, and acceptability, compared to the nonexposed fruits (Ortiz Araque et al., 2018), and a dose of 0.5 kJ/m^2 improved sensorial quality of 'Kurdistan' strawberries (Darvishi et al., 2012). Allende and Artés (2003) reported that UV-C (2.44–8.14 kJ/m^2) improved overall sensory quality of minimally processed lettuce, and the produce did not score below the acceptance limit after storage for 8 d, and the total aerobic count of the treated lettuce had not reached the limit of the Spanish legislation.

However, discoloration has occurred after UV-C treatment during storage of the treated crops. For example, Ben-Yehoshua et al. (1992) noted brown or bronze discolorations in the rind of UV-C-treated citrus fruits. Panelists noted poorer appearance and color of UV-C-treated tomato fruits with hormetic dose of 3.7 kJ/m^2 compared to control fruits (Charles et al., 2005); arguably the treated fruits had not reached the same ripeness as the control fruits. Also, the exposure of lettuce to the highest UV-C dose (8.14 kJ/m^2) became glossy (Allende and Artés, 2003). The poorer appearance and loss of gloss may also be attributable to light reflection characteristics arising from surface modifications (Charles et al., 2008c).

The induction of 6-MM, a phytoalexin, in carrot peel after exposure to UV-C radiation is known to contribute to its bitterness. However, the taste of cooked carrot taste was not impaired, presumably 6-MM was eliminated by peeling operation or leached into the boiling water (Mercier et al., 1994). Bitterness of produce may also be associated with certain phenolic compounds (e.g., naringin, quercetin, catechin, and genistin); terpenoids (e.g., limonoids such as limonin and nomilin) as well as glucosinolates (e.g., sinigrin, progoitrin, and glucobrassicin) (Drewnowski and Gomez-Carneros, 2000). While such SMs do play a role in disease resistance, they are often concentrated in the peel, and this may not be of great concern as many fruits and vegetables are peeled before consumption. Since UV-C radiation induces biochemical changes in the treated crops that are keys to disease resistance, it is imperative to use appropriate doses that not only contribute to the immunity of the crops but also minimize undesirable characteristics such as appearance to favor consumer acceptability.

6.2 Enhancement of Secondary Metabolites on Produce by UV-C Hormetic Doses

Plants produce SMs for different functions, including defenses against microbial attack and environmental stresses, pollination, and growth regulation, and some unknown roles. These compounds are divided in two main classes: nitrogen-containing SMs and non-nitrogen containing SMs. Alkaloids, nonprotein amino acids (NPAAs), amines, cyanogenic glycosides, glucosinolates, alkamides, lectins, peptides, and polypeptides are all nitrogen-containing SMs, while terpenoids, steroids, saponins, phenolic acids, flavonoids, coumarins, stilbenes, lignin, acetate-derived fatty acids, waxes, and polyketides are SMs without nitrogen (Wink, 2010). It is increasingly clear that the SMs of different chemical classes in fruits and vegetables are beneficial to human health, which can reduce the risk of chronic diseases such as cancer, cardiovascular disease, and diabetes. It would be of great interest to evaluate the potential of UV-C radiation to enhance the health beneficial compounds in fruits and vegetables, in addition to its potential to control diseases in the crops and reduce postharvest losses as an alternative to chemical use.

On this account, prior studies examined the possibility of elevating the level of resveratrol, a stilbene, in grapes. UV-C-irradiated grapes presented a greater potential to produce wine with high contents of resveratrol and piceatannol than UV-B radiation (Cantos et al., 2003; Venditti and D'Hallewin, 2014), where resveratrol is considered a protective agent against several human diseases such as cancer and cardiovascular disease (Charles and Arul, 2007). Many of the studies on UV-C treatment of crops report on antioxidant capacity and the accumulation of phenolic compounds in the treated produce (Costa et al., 2006; Lemoine et al., 2007; Perkins-Veazie et al., 2008; Wu et al., 2017; Xu et al., 2017), and only a few studies have reported on specific SMs such as phytoalexins. Wei et al. (2013) reported a 3.4-, 4.3-, and 4.8-fold induction of naringenin, luteolin, and apigenin, respectively, in pigeon pea leaves irradiated by UV-C for

8 hr, which were significantly higher than compared to either nonexposed leaves or those exposed to UV-B radiation. Some examples of enhancement in SMs in various crops are listed in Table 17.4.

Anthocyanins are a class of flavonoids, which also have been targets in some studies due to the multiple beneficial effects on human health, including vision health, oxidative stress protection, prevention of obesity and diabetes, as well as their capacity to modulate cognitive and motor function. In addition, like many other bioactive flavonoids, genes responsible of anthocyanin synthesis are highly inducible, and thus, controlled dosage of abiotic stresses is an effective strategy to increase their concentrations (Lila, 2004). It was shown recently that UV-C radiation at a dose of 3.0 kJ/m^2 induces the expression of genes and the induction of anthocyanins in red cabbage (Wu et al., 2017). UV-C radiation has also been applied to enhance the content of anthocyanins in fruits, especially in those with intense purple and red colors such as blueberries and strawberries. In blueberries, for example, the concentration of individual anthocyanins, including malvidin-3-galatoside,

Table 17.4 Effect of UV-C radiation on different mechanisms related to the induction of secondary metabolites on fruits and vegetables

Crop	*Treatment*	*Effect on secondary metabolite*	*Reference*
Fresh-cut Bimi broccoli	4.5 and 9.0 kJ/m^2 (approx. 40.5 W/m^2 during 111 and 222 s)	Increased levels of hydroxy-cinnamoyl acid derivatives (4.8 and 4.5-folds), in UV-C-exposed vegetables at 4.5 and 9.0 kJ/m^2, respectively	Martínez-Hernández et al. (2011)
Garlic	2.0 kJ/m^2	Increase in quercetin and apigenin by fivefold in exposed UV-C garlic	Park and Kim (2015)
Mango	2.5 and 4.9 kJ/m^2 (8.22 W/m^2 during 300 and 600 s)	Increase in the total phenol content in UV-C-treated fruits	González-Aguilar et al. (2007)
Minimally processed lily bulb	4.5 kJ/m^2	Higher total phenolic content in UV-C-treated bulbs (13%) after 5 d of storage at 4°C	Huang et al. (2017)
Mushroom	1 kJ/m^2 (1 W/m^2 during 100 s)	Enhancement of total phenolic compounds in UV-C exposed mushroom during 21 d of storage at 4°C, concomitant with reduction of tyrosine and total phenolic compounds	Wu et al. (2016)
Satsuma mandarin	3.00 kJ/m^2 (7.65 W/m^2 during 392 s)	Increase in narirutin and hesperidin in UV-C-exposed fruits (21 and 8%, respectively)	Shen et al. (2013)
Tomato fruit	3.7 kJ/m^2 (approx. 3.85 W/m^2 during 960 s)	Increased accumulation of polyamines (putrescine and spermidine) in UV-C-treated fruits	Tiecher et al. (2013)

delphinidin-3-galactoside, delphinidin-3-arabinoside, and petunidin-3-arobinoside, was enhanced by UV-C expose up to 2.15 kJ/m^2, but a decline in these anthocyanins was observed with higher doses higher than 2.15 kJ/m^2 (Wang et al., 2009).

Carotenoids (e.g., lycopene, β-carotene and lutein) are tetra-terpenoids biosynthesized from 2-C-methyl-D-erythritol 4-phosphate (MEP) pathways, which are essential compounds for plant development as well as for normal health of animals (i.e., precursors of retinol, potent antioxidants, among other functions) (Saini and Keum, 2018). The effect of UV-C radiation on carotenoid content of produce has been previously described as being either not significant or negatively affected (Mditshwa et al., 2017). Bravo et al. (2012), in contrast to the report of Jagadeesh et al. (2011), reported an increase in lycopene content in UV-C-treated tomato. Differences in the methodology of these two studies may account for the discrepancies. Although indicating the additional effect of storage temperature, the report by Jagadeesh et al. (2011) is in agreement with the results from others such as Maharaj et al. (2010) and Bu et al. (2014).

UV-C radiation has also been used to induce the concentration of nitrogen-containing SMs the glucosinolates, which are amino acid-derived compounds present in *Brassicas*, including broccoli, cabbage, cauliflower, and others. Glucosinolates are substrates for myrosinase enzyme (E.C. 3.2.3.1), and the reaction leads to the synthesis of isothiocyanates, thiocyanates, oxazolidine-2-thiones, epithio-nitriles, or nitriles that serve as chemical defense against pathogens (Textor and Gershenzon, 2009). The isothiocyanates, especially sulforaphane, derived from glucoraphanin, have been related to anticarcinogenic action in humans by the induction of detoxification enzymes (Zhang et al., 1992). Thus, the enhancement of glucosinolates has become a strategy of interest not only to improve the storage of *Brassicas* by reducing decay but also as a means of increasing commercial value by enhancing nutraceutical content. Nadeau et al. (2012) observed that irradiation of broccoli florets tended to increase the levels of 4-methoxyglucobrassicin, 4-hydroxyglucobrassicin, and glucoraphanin with a hormetic UV-C dose of 1.2 kJ/m^2. Higher titers of glucosinolates in broccoli florets treated with the hormetic dose of UV-C at 1.2 kJ/m^2 have been correlated with the decreased concentration of source amino acids (Duarte-Sierra et al., 2012).

7 Challenges

It is evident that UV-C radiation at hormetic doses has much potential to induce disease resistance and reduce decay in several postharvest crops. The effect of UV-C in delaying ripening and senescence adds an advantage to the control of diseases. In addition, the possibility of enhancement of health-beneficial phytocompounds by hormetic doses of UV-C adds another attractive dimension to this approach. Initial reports on the application of UV-C radiation appeared in the early 1990s and a growing number of articles published on this topic since then, mostly on the effect of UV-C in extending the shelf life of produce and in the maintenance of their quality factors.

The concept of UV-C hormesis in the intensification of natural defenses that gives a head start in fighting the infection has not been fully embraced by the community. The hormetic doses for various crops would be clearly defined with a good appreciation of it. The determination of hormetic doses should take into account not only of disease control response but also of acceptability of the treated produce, potential formation of color defects such as discolorations that affect visual appearance or flavor defects, and potential accumulation of undesirable SMs such as terpenoids and glycoalkaloids, and bitter substances. The accumulation of such compounds in the peel is less of a concern since many fruits and vegetables are peeled before consumption.

Second, the intrinsic resistance is variable among the crops. Some crops such as tomato express multiple defenses and the protective effect persists for longer periods. In some cases, the expression of defense mechanisms may be limited, and consequently, the protection may not persist for longer periods. In the latter, it is imperative that the induced mechanisms, if only a few, should be maintained for a reasonable duration under suitable post-treatment storage conditions. Thus, it would involve a good understanding of defenses induced by UV-C and their dynamics through the course of storage.

Third, the induction of disease resistance is not systemic, rather localized. This warrants that the entire surface of the crops be exposed to radiation, which come in different shapes and sizes. This is an engineering problem with certain challenges, in addition to scaling-up to handle large volumes.

In conclusion, UV-C hormesis has potential to be an effective technology to control diseases in stored crops as an alternative to the use of fungicides, accomplished through exploiting the physiological capacity of the crops themselves. It is a low-cost treatment and can be easily adapted into the current handling and storage practices. UV-C treatment may appear to be an easy technology, but it requires a significant understanding of the physiological basis of induced resistance as well as engineering solutions to transfer into practice.

References

Abeles, F.B., Morgan, P.W. and Saltveit, M.E. 1992. *Ethylene in Plant Biology,* 2nd ed. Academic Press, Elsevier, London, UK.

Ait-Barka, E.A., Kalantari, S., Makhlouf, J. and Arul, J. 2000a. Effects of UV-C irradiation on lipid peroxidation markers during ripening of tomato (*Lycopersicon esculentum* L.) fruits. *Australian Journal of Plant Physiology* 27, 147–152.

Ait-Barka, E.A., Kalantari, S., Makhlouf, J. and Arul, J. 2000b. Impact of UV-C irradiation on the cell wall-degrading enzymes during ripening of tomato (*Lycopersicon esculentum* L.) fruit. *Journal of Agricultural and Food Chemistry* 48, 667–671.

ÅLenius, C.M., Vogelmann, T.C. and Bornman, J.F. 1995. A three-dimensional representation of the relationship between penetration of u.v.-B radiation and u.v.-screening pigments in leaves of *Brassica napus*. *New Phytologist* 131, 297–302.

Allende, A. and Artés, F. 2003. UV-C radiation as a novel technique for keeping quality of fresh processed 'Lollo Rosso' lettuce. *Food Research International* 36, 739–746.

Andersen, B.M., Banrud, H., Boe, E., Bjordal, O. and Drangsholt, F. 2006. Comparison of UV C light and chemicals for disinfection of surfaces in hospital isolation units. *Infection Control and Hospital Epidemiology* 27, 729–734.

Andrews, D.L. 1999. Electromagnetic radiation A2. pp. 397–40. *In*: J.C. Lindon (ed.). *Encyclopedia of Spectroscopy and Spectrometry.* Elsevier, Oxford, UK.

Arul, J. 1994. Emerging technologies for the control of postharvest diseases of fresh fruits and vegetables. pp. 1-10. *In*: C.L. Wilson and M. Wisniewski (eds.). *Control of Postharvest Diseases: Theory and Practice.* CRC Press, Boca Raton, FL.

Arul, J., Mercier, J., Charles, M.T., Baka, M. and Maharaj, R. 2001. Photochemical treatment for control of post-harvest diseases in horticultural crops. pp. 146–161. *In*: C. Vicent, B. Panneton and F. Fleurat-Lessard (eds.). *Physical Control Methods in Plant Protection.* INRA Editions, Paris, France.

Bailey, J.A. 1982. Mechanisms of phytoalexin accumulation. pp. 289–318. *In*: J. W. Mansfield and J.A. Bailey (eds.). *Phytoalexins.* John Wiley and Sons, New York.

Baka, M. 1997. Photochemical therapy in the preservation of fresh fruits and vegetables: Phenomenoligal aspects of disease resistance and delayed senescence. Ph.D. Thesis, Université Laval, Quèbec, Canada.

Baka, M., Mercier, J., Corcuff, R., Castaigne, F. and Arul, J. 1999. Photochemical treatment to improve storability of fresh strawberries. *Journal of Food Science* 64, 1068–1072.

Balestrazzi, A., Locato, V., Bottone, M.G., De Gara, L., Biggiogera, M., Pellicciari, C., Botti, S., Di Gesu, D., Dona, M. and Carbonera, D. 2010. Response to UV-C radiation in topo I-deficient carrot cells with low ascorbate levels. *Journal of Experimental Botany* 61, 575–585.

Barceló, A.R., Pomar, F., López-Serrano, M. and Pedreño, M.A. 2003. Peroxidase: A multifunctional enzyme in grapevines. *Functional Plant Biology* 30, 577–591.

Bartz, J.A. and Brecht, J.K. 2002. *Postharvest Physiology and Pathology of Vegetables.* CRC Press, Boca Raton, FL.

Beaven, G.H. and Holiday, E.R. 1952. Ultraviolet absorption spectra of proteins and amino acids. *Advances in Protein Chemistry* 7, 319–386.

Benhamou, N. 2004. Potential of the mycoparasite, *Verticillium lecanii*, to protect citrus fruit against *Penicillium digitatum*, the causal agent of green mold: A comparison with the effect of chitosan. *Phytopathology* 94, 693–705.

Benhamou, N. 2009. *La Résistance Chez Les Plantes. Principes de la Stratégie Défensive et Applications Agronomiques.*TEC & DOC Lavoisier, Paris, France.

Ben-Yehoshua, S., Rodov, V., Kim, J.J. and Carmeli, S. 1992. Preformed and induced antifungal materials of citrus fruits in relation to the enhancement of decay resistance by heat and ultraviolet treatments. *Journal of Agricultural and Food Chemistry* 40, 1217–1221.

Bintsis, T., Litopoulou-Tzanetaki, E. and Robinson, R.K. 2000. Existing and potential applications of ultraviolet light in the food industry – A critical review. *Journal of the Science of Food and Agriculture* 80, 637–645.

Bolton, J.R. and Linden, K.G. 2003. Standardization of methods for fluence (UV dose) determination in bench-scale UV experiments. *Journal of Environmental Engineering* 129, 209–215.

Bornman, J.F., Evert, R.F. and Mierzwa, R.J. 1983. The effect of UV-B and UV-C radiation on sugar beet leaves. *Protoplasma* 117, 7–16.

Bornman, J.F. and Vogelmann, T.C. 1988. Penetration of blue nad UV radiation measured by fiber optics in spruce and fir needles. *Physiologia Plantarum* 72, 699–705.

Bostock, R.M. and Stermer, B.A. 1989. Perspectives on wound healing in resistance to pathogens. *Annual Review of Phytopathology* 27, 343–371.

Bravo, S., García-Alonso, J., Martín-Pozuelo, G., Gómez, V., Santaella, M., Navarro-González, I. and Periago, M.J. 2012. The influence of post-harvest UV-C hormesis on lycopene, β-carotene, and phenolic content and antioxidant activity of breaker tomatoes. *Food Research International* 49, 296–302.

Bu, J., Ni, Z., Aisikaer, G., Jiang, Z., Khan, Z.U., Mou, W. and Ying, T. 2014. Postharvest ultraviolet-C irradiation suppressed Psy 1 and Lcy-β expression and altered color phenotype in tomato (*Solanum lycopersicum*) fruit. *Postharvest Biology and Technology* 89, 1–6.

Bu, J., Yu, Y., Aisikaer, G. and Ying, T. 2013. Postharvest UV-C irradiation inhibits the production of ethylene and the activity of cell wall-degrading enzymes during softening of tomato (*Lycopersicon esculentum L.*) fruit. *Postharvest Biology and Technology* 86, 337–345.

Calabrese, E.J. 2015a. Hormesis within a mechanistic context. *Homeopathy* 104, 90–96.

Calabrese, E.J. 2015b. Hormesis: Principles and applications. *Homeopathy* 104, 69–82.

Calabrese, E.J. and Baldwin, L.A. 2001. Hormesis: A generalizable and unifying hypothesis. *Critical Reviews in Toxicology* 31, 353–424.

Calabrese, E.J. and Baldwin, L.A. 2002. Defining hormesis. *Human & Experimental Toxicology* 21, 91–97.

Calabrese, E.J. and Baldwin, L.A. 2003. HORMESIS: The dose-response revolution. *Annual Review of Pharmacology and Toxicology* 43, 175–197.

Calabrese, E.J. and Blain, R. 2005. The occurrence of hormetic dose responses in the toxicological literature, the hormesis database: An overview. *Toxicology and Applied Pharmacology* 202, 289–301.

Calabrese, E.J. and Blain, R.B. 2009. Hormesis and plant biology. *Environmental Pollution* 157, 42–48.

Caleb, O., Mahajan, P., Al-Said, F. and Opara, U. 2012. Modified atmosphere packaging technology of fresh and fresh-cut produce and the microbial consequences—A review *Food and Bioprocess Technology* 5, 1–27.

Cantos, E., Espín, J.C., Fernández, M.J., Oliva, J. and Tomás-Barberán, F.A. 2003. Postharvest UV-C-irradiated grapes as a potential source for producing stilbene-enriched red wines. *Journal of Agricultural and Food Chemistry* 51, 1208–1214.

Cantos, E., Espín, J.C. and Tomás-Barberán, F.A. 2001. Postharvest induction modeling method using UV irradiation pulses for obtaining resveratrol-enriched table grapes: A new "functional" fruit? *Journal of Agricultural and Food Chemistry* 49, 5052–5058.

Chalutz, E., Droby, S., Wilson, C.L. and Wisniewski, M.E. 1992. UV-induced resistance to postharvest diseases of citrus fruit. *Journal of Photochemistry and Photobiology B: Biology* 15, 367–371.

Chapman, P.M. 2002. Defining hormesis: Comments on Calabrese and Baldwin (2002). *Human & Experimental Toxicology* 21, 99–101.

Charles, M.T. 1998. Bases physiologiques de la résistance aux maladies post-récolte induite par les rayons UV chez la tomate. Ph.D. Thesis, Université Laval, Québec, Canada.

Charles, M.T. and Arul, J. 2007. UV treatment of fresh fruits and vegetables for improved quality: A status report. *Stewart Postharvest Review* 3, 1–8.

Charles, M.T., Arul, J. and Benhamou, N. 2011. UV-C-induced disease resistance in tomato fruit is a multi-component and time-dependent system. *Acta Horticulturae* 905, 251–260.
Charles, M.T., Arul, J., Charlebois, D., Yaganza, E.-S., Rolland, D., Roussel, D. and Merisier, M.J. 2016. Postharvest UV-C treatment of tomato fruits: Changes in simple sugars and organic acids contents during storage. *LWT - Food Science and Technology* 65, 557–564.
Charles, M.T., Arul, J., Yaganza, E.-S., Rolland, D. and Roussel, D. 2019. Differential accumulation of rishitin and stigmasterol in various tomato genotypes in response to UV-C treatment. *Acta Horticulturae*. In press.
Charles, M.T., Benhamou, N. and Arul, J. 2008a. Physiological basis of UV-C induced resistance to *Botrytis cinerea* in tomato fruit. III. Ultrastructural modifications and their impact on fungal colonization. *Postharvest Biology and Technology* 47, 27–40.
Charles, M.T., Goulet, A. and Joseph, A. 2008b. Physiological basis of UV-C induced resistance to *Botrytis cinerea* in tomato fruit. IV. Biochemical modification of structural barriers. *Postharvest Biology and Technology* 47, 41–53.
Charles, M.T., Kalantari, S., Corcuff, R. and Arul, J. 2005. Postharvest quality and sensory evaluation of UV-treated tomato fruit. *Acta Horticulturae* 682, 537–542.
Charles, M.T., Makhlouf, J. and Arul, J. 2008c. Physiological basis of UV-C induced resistance to *Botrytis cinerea* in tomato fruit: II. Modification of fruit surface and changes in fungal colonization. *Postharvest Biology and Technology* 47, 21–26.
Charles, M.T., Mercier, J., Makhlouf, J. and Arul, J. 2008d. Physiological basis of UV-C-induced resistance to *Botrytis cinerea* in tomato fruit: I. Role of pre- and post-challenge accumulation of the phytoalexin-rishitin. *Postharvest Biology and Technology* 47, 10–20.
Charles, M.T., Tano, K., Asselin, A. and Arul, J. 2009. Physiological basis of UV-C induced resistance to *Botrytis cinerea* in tomato fruit. V. Constitutive defence enzymes and inducible pathogenesis-related proteins. *Postharvest Biology and Technology* 51, 414–424.
Cheema, A.S. and Haard, N.F. 1978. Induction of rishitin and lubimin in potato tuber discs by non-specific elicitors and the influence of storage conditions. *Physiological Plant Pathology* 13, 233–240.
Chun, H., Kim, J., Chung, K., Won, M. and Song, K.B. 2009. Inactivation kinetics of *Listeria monocytogenes, Salmonella enterica serovar Typhimurium*, and *Campylobacter jejuni* in ready-to-eat sliced ham using UV-C irradiation. *Meat Science* 83, 599–603.
Chun, H.H., Kim, J.Y., Lee, B.D., Yu, D.J. and Song, K.B. 2010. Effect of UV-C irradiation on the inactivation of inoculated pathogens and quality of chicken breasts during storage. *Food Control* 21, 276–280.
Civello, P.M., Vicente, A.R. and Martínez, G.A. 2006. UV-C technology to control postharvest diseases of fruits and vegetables. pp. 71–102. *In*: R. Troncoso-Rojas, M. E. Tiznado-Hernández and A. González-León (eds.). *Recent Advances in Alternative Postharvest Technologies to Control Fungal Diseases in Fruits & Vegetables*. Research Signpost, Trivandum, Kerala, India.
Cockell, C.S. 1998. Biological effects of high ultraviolet radiation on early earth—A theoretical evaluation. *Journal of Theoretical Biology* 193, 717–729.

Colas, S., Afoufa-Bastien, D., Jacquens, L., Clement, C., Baillieul, F., Mazeyrat-Gourbeyre, F. and Monti-Dedieu, L. 2012. Expression and in situ localization of two major PR proteins of grapevine berries during development and after UV-C exposition. *PLoS ONE* 7, e43681.

Costa, L., Vicente, A.R., Civello, P.M., Chaves, A.R.A. and Martínez, G.A. 2006. UV-C treatment delays postharvest senescence in broccoli florets. *Postharvest Biology and Technology* 39, 204–210.

Cote, S., Rodoni, L., Miceli, E., Concellón, A., Civello, P.M. and Vicente, A.R. 2013. Effect of radiation intensity on the outcome of postharvest UV-C treatments. *Postharvest Biology and Technology* 83, 83–89.

Currier, W.W. and Kuć, J. 1975. Effect of temperature on rishitin and steroid glycoalkaloid accumulation in potato tuber. *Phytopathology* 65, 1194–1197.

D'Hallewin, G., Ladu, G., Pani, G., Dore, A., Molinu, M.G. and Venditti, T. 2012. Postharvest physio-pathological disorders in table grapes as affected by UV-C light. *Communications in Agricultural and Applied Biological Sciences* 77, 515–525.

D'Hallewin, G., Molinu, G.M., Dore, A., Venditti, T. and Rodov, V. 2009. Abiotic stresses sequentially applied enhance natural resistance and reduce postharvest decay. *Communications in Agricultural and Applied Biological Sciences* 74, 659–665.

D'Hallewin, G., Schirra, M., Manueddu, E., Piga, A. and Ben-Yehoshua, S. 1999. Scoparone and scopoletin accumulation and ultraviolet-C induced resistance to postharvest decay in oranges as influenced by harvest date. *Journal of the American Society for Horticultural Science* 124, 702–707.

D'Hallewin, G., Schirra, M., Pala, M. and Ben-Yehoshua, S. 2000. Ultraviolet C irradiation at 0.5 $kJ.m^{-2}$ reduces decay without causing damage or affecting postharvest quality of star ruby grapefruit (*C. paradisi* Macf.). *Journal of Agricultural and Food Chemistry* 48, 4571–4575.

Dann, E. 2003. Induced resistance: Potential for control of postharvest diseases of horticultural crops. *Proceedings of Autralian Postharvest Horticulture* 2003, 144–148.

Darvill, A.G. and Albersheim, P. 1984. Phytoalexins and their elicitors-A defense against microbial infection in plants. *Annual Review of Plant Physiology* 35, 243–275.

Darvishi, S., Fatemi, A. and Davari, K. 2012. Keeping quality of use of fresh 'Kurdistan' strawberry by UV-C radiation. *World Applied Sciences Journal* 17, 826–831.

Deepak, S., Shailasree, S., Kini, R.K., Muck, A., Mithöfer, A. and Shetty, S.H. 2010. Hydroxyproline-rich glycoproteins and plant defence. *Journal of Phytopathology* 158, 585–593.

Diffey, B.L. 2002. Sources and measurement of ultraviolet radiation. *Methods* 28, 4–13.

Doke, N. 1997. The oxidative burst: Roles in signal transduction and plant stress. pp. 785–813. *In*: J.G. Scandalios (ed.). *Oxidative Stress and the Molecular Biology of Antioxidant Defenses.* Cold Spring Harbor Laboratory Press Monograph Archive, Vol. 34, 785-814, Cold Spring Harbor, New York.

Drewnowski, A. and Gomez-Carneros, C. 2000. Bitter taste, phytonutrients, and the consumer: A review. *The American Journal of Clinical Nutrition* 72, 1424–1435.

Droby, S., Chalutz, E., Horev, B., Cohen, L., Gaba, V., Wilson, C.L. and Wisniewski, M. 1993. Factors affecting UV-induced resistance in grapefruit against the green mould decay caused by *Penicillium digitatum*. *Plant Pathology* 42, 418–424.

Duarte-Sierra, A., Corcuff, R., Angers, P., Arul, J. and Forney, C. 2012. Influence of UV-C on color development and free amino acid profile in broccoli florets during postharvest storage. *Acta Horticulturae* 945, 97–103.

Edwards, A.A. and Alexander, B.D. 2017. UV-visible absorption spectroscopy, organic applications A2. pp. 511–519. *In*: J.C. Lindon, G.E. Tranter and D.W. Koppenaal (eds.). *Encyclopedia of Spectroscopy and Spectrometry,* 3rd ed. Academic Press, Oxford.

El Ghaouth, A., Wilson, C.L. and Callahan, A.M. 2003. Induction of chitinase, β-1,3-glucanase, and phenylalanine ammonia lyase in peach fruit by UV-C treatment. *Phytopathology* 93, 349–355.

Eltayeb, E.A. and Roddick, J.G. 1984. Changes in the alkaloid content of developing fruits of tomato (*Lycopersicon esculentum Mill.*): I. analyses of cultivars and mutants with different ripening characteristics. *Journal of Experimental Botany* 35, 252–260.

Erkan, M., Wang, C.Y. and Krizek, D.T. 2001. UV-C irradiation reduces microbial populations and deterioration in *Cucurbita pepo* fruit tissue. *Environmental and Experimental Botany* 45, 1–9.

Erkan, M., Wang, S.Y. and Wang, C.Y. 2008. Effect of UV treatment on antioxidant capacity, antioxidant enzyme activity and decay in strawberry fruit. *Postharvest Biology and Technology* 48, 163–171.

Fava, J., Hodara, K., Nieto, A., Guerrero, S., Alzamora, S.M. and Castro, M.A. 2011. Structure (micro, ultra, nano), color and mechanical properties of *Vitis labrusca* L. (grape berry) fruits treated by hydrogen peroxide, UV-C irradiation and ultrasound. *Food Research International* 44, 2938–2948.

Ficke, A., Gadoury, D.M., Seem, R.C., Godfrey, D. and Dry, I.B. 2004. Host barriers and responses to uncinula necator in developing grape berries. *Phytopathology* 94, 438–445.

Flammer, J., Mozaffarieh, M. and Bebie, H. 2013. The interaction between light and matter. pp. 21–39. *In*: J. Flammer, M. Mozaffarieh and H. Bebie (eds.). *Basic Sciences in Ophthalmology. Physics and Chemistry.* Springer, Heidelberg, Berlin, Germany.

Formica-Oliveira, A.C., Martínez-Hernández, G.B., Díaz-López, V., Artés, F. and Artés-Hernández, F. 2017. Effects of UV-B and UV-C combination on phenolic compounds biosynthesis in fresh-cut carrots. *Postharvest Biology and Technology* 127, 99–104.

Galston, A.W. and Sawhney, R.K. 1990. Polyamines in plant physiology. *Plant Physiology* 94, 406.

Gayán, E., Monfort, S., Álvarez, I. and Condón, S. 2011. UV-C inactivation of *Escherichia coli* at different temperatures. *Innovative Food Science & Emerging Technologies* 12, 531–541.

George Dominic, S., Razali, Z., Santhirasegaram, V. and Somasundram, C. 2015. Effect of postharvest ultraviolet-C treatment on the proteome changes in fresh cut mango (*Mangifera indica* L. cv. Chokanan). *Journal of the Science of Food and Agriculture* 96, 2851–2860.

Gill, S.S., Anjum, N.A., Gill, R., Jha, M. and Tuteja, N. 2015. DNA damage and repair in plants under ultraviolet and ionizing radiations. *Scientific World Journal* 2015, 250158.

Gogo, E.O., Opiyo, A.M., Hassenberg, K., Ulrichs, C. and Huyskens-Keil, S. 2017. Postharvest UV-C treatment for extending shelf life and improving nutritional quality of African indigenous leafy vegetables. *Postharvest Biology and Technology* 129, 107–117.

Gómez-Lobato, M.E., Mansilla, S.A., Civello, P.M. and Martínez, G.A. 2014. Expression of Stay-Green encoding gene (BoSGR) during postharvest senescence of broccoli. *Postharvest Biology and Technology* 95, 88–94.

González-Aguilar, G.A., Zavaleta-Gatica, R. and Tiznado-Hernández, M.E. 2007. Improving postharvest quality of mango 'Haden' by UV-C treatment. *Postharvest Biology and Technology* 45, 108–116.

González-Barrio, R., Vidal-Guevara, M.L., Tomás-Barberán, F.A. and Espín, J.C. 2009. Preparation of a resveratrol-enriched grape juice based on ultraviolet C-treated berries. *Innovative Food Science & Emerging Technologies* 10, 374–382.

Goodliffe, J.P. and Heale, J.B. 1978. The role of 6-methoxy mellein in the resistance and susceptibility of carrot root tissue to the cold-storage pathogen. *Botrytis Cinerea. Physiological Plant Pathology* 12, 27–43.

Graça, A., Salazar, M., Quintas, C. and Nunes, C. 2013. Low dose UV-C illumination as an eco-innovative disinfection system on minimally processed apples. *Postharvest Biology and Technology* 85, 1–7.

Guan, W., Fan, X. and Yan, R. 2012. Effects of UV-C treatment on inactivation of *Escherichia coli* O157: H7, microbial loads, and quality of button mushrooms. *Postharvest Biology and Technology* 64, 119–125.

Guest, D.I. 2017. Phytoalexins, natural plant protection. pp. 124–128. *In*: B. Thomas, B.G. Murray and D.J. Murphy (eds.). *Encyclopedia of Applied Plant Sciences*, 2nd ed. Academic Press, Oxford, UK.

Haard, N.F. 1984. Postharvest physiology and biochemistry of fruits and vegetables. *Journal of Chemical Education* 61, 277.

Harm, W. 1980. *Biological Effects of Ultraviolet Radiation*. Cambridge University Press, Cambrige, London, UK.

Hartmann, A., Becker, K., Karsten, U., Remias, D. and Ganzera, M. 2015. Analysis of mycosporine-like amino acids in selected algae and cyanobacteria by hydrophilic interaction liquid chromatography and a novel MAA from the red alga catenella repens. *Marine Drugs* 13, 6291–6305.

Hasan, M. and Bae, H. 2017. An overview of stress-induced resveratrol synthesis in grapes: Perspectives for resveratrol-enriched grape products. *Molecules* 22, 294.

Hemmaty, S., Hemmaty, S., Moallemi, N. and Naseri, L. 2007. Effect of UV-C radiation and hot water on the calcium content and postharvest quality of apples. *Spanish Journal of Agricultural Research* 5, 559–568.

Holick, M.F. 2016. Biological effects of sunlight, ultraviolet radiation, visible light, infrared radiation and vitamin D for health. *Anticancer Research* 36, 1345–1356.

Hollósy, F. 2002. Effects of ultraviolet radiation on plant cells. *Micron* 33, 179–197.

Hu, F.B. 2003. Plant-based foods and prevention of cardiovascular disease: An overview. *The American Journal of Clinical Nutrition* 78, 544S–551S.

Huang, H., Ge, Z., Limwachiranon, J., Li, L., Li, W. and Luo, Z. 2017. UV-C treatment affects browning and starch metabolism of minimally processed lily bulb. *Postharvest Biology and Technology* 128, 105–111.

Huyskens-Keil, S., Hassenberg, K. and Herppich, W.B. 2011. Impact of postharvest UV-C and ozone treatment on textural properties of white asparagus (*Asparagus officinalis* L.). *Journal of Applied Botany and Food Quality* 84, 229–234.

Jackson, P.A., Galinha, C.I., Pereira, C.S., Fortunato, A., Soares, N.C., Amancio, S.B. and Pinto Ricardo, C.P. 2001. Rapid deposition of extensin during the elicitation of grapevine callus cultures is specifically catalyzed by a 40-kilodalton peroxidase. *Plant Physiology* 127, 1065–1076.

Jagadeesh, S.L., Charles, M.T., Gariepy, Y., Goyette, B., Raghavan, G.S.V. and Vigneault, C. 2011. Influence of postharvest UV-C hormesis on the bioactive components of tomato during post-treatment handling. *Food and Bioprocess Technology* 4, 1463–1472.

Jansen, M.A.K., Gaba, V. and Greenberg, B.M. 1998. Higher plants and UV-B radiation: Balancing damage, repair and acclimation. *Trends in Plant Science* 3, 131–135.

Jiang, T., Jahangir, M.M., Jiang, Z., Lu, X. and Ying, T. 2010. Influence of UV-C treatment on antioxidant capacity, antioxidant enzyme activity and texture of postharvest shiitake (*Lentinus edodes*) mushrooms during storage. *Postharvest Biology and Technology* 56, 209–215.

Kader, A.A. and Rolle, R.S. 2004. *The Role of Post-Harvest Management in Assuring the Quality and Safety of Horticultural Produce.* Food and Agriculture Organization of the United Nations, Rome, Italy.

Kader, A.A. and Watkins, C.B. 2000. Modified atmosphere packaging - Toward 2000 and beyond. *HortTechnology* 10, 483–486.

Kalantari, S., Turcotte, C., Fliss, I., Makhlouf, J. and Arul, J. 2001. A new competitive enzyme immunoassay for the detection of abscisic acid. Proceedings of the Annual Meeting of the ASHS, Sacramento, CA. pp. 20–25.

Kasim, R. and Kasim, M.U. 2008. The effect of ultraviolet irradiation (UV-C) on chilling injury of cucumbers during cold storage. *Journal of Food, Agriculture & Environment* 6, 50–54.

Katerova, Z., Ivanov, S., Prinsen, E., Van Onckelen, H., Alexieva, V. and Azmi, A. 2009. Low doses of ultraviolet-B or ultraviolet-C radiation affect phytohormones in young pea plants. *Biologia Plantarum* 53, 365–368.

Kendig, E.L., Le, H.H. and Belcher, S.M. 2010. Defining hormesis: Evaluation of a complex concentration response phenomenon. *International Journal of Toxicology* 29, 235–246.

Khalili, F., Shekarchi, M., Razavi, K. and Rastegar, H. 2017. Postharvest UV-C irradiation delays senescence and maintains nutritional properties of broccoli florets. *International Journal of Vegetable Science* 23, 158–170.

Kim, S.J., Kim, D.K. and Kang, D.H. 2016. Using UVC light-emitting diodes at wavelengths of 266 to 279 nanometers to inactivate foodborne pathogens and pasteurize sliced cheese. *Applied and Environmental Microbiology* 82, 11–17.

Király, L., Künstler, A., Bacsó, R., Hafez, Y. and Király, Z. 2013. Similarities and differences in plant and animal immune systems — What is inhibiting pathogens? *Acta Phytopathologica Et Entomologica Hungarica* 48, 187–205.

Kolattukudy, P.E. 1984. Biochemistry and function of cutin and suberin. *Canadian Journal of Botany* 62, 2918–2933.

Kondo, S., Fiebig, A., Okawa, K., Ohara, H., Kowitcharoen, L., Nimitkeatkai, H., Kittikorn, M. and Kim, M. 2011. Jasmonic acid, polyamine, and antioxidant levels in apple seedlings as affected by Ultraviolet-C irradiation. *Plant Growth Regulation* 64, 83–89.

Kouassi, N., Corcuff, R., Arul, J. and Tweddell, R.J. 2012. Effect of storage temperature and age after harvest on the accumulation of the phytoalexin 6-methoxymellein in UV-C treated carrots. *Acta Horticulturae* 945, 135–138.

Koutchma, T.N., Forney, L.J. and Moraru, C.I. 2009a. Characterization of foods in relation to UV treatment. pp. 53–67. *In*: Ultraviolet Light in Food Technology. CRC Press, Boca Raton, FL.

Koutchma, T.N., Forney, L.J. and Moraru, C.I. 2009b. Microbial inactivation by UV light. pp. 69–101. *In*: Ultraviolet Light in Food Technology. CRC Press, Boca Raton, FL.

Koutchma, T.N., Forney, L.J. and Moraru, C.I. 2009c. Principles and applications of UV technology. pp. 1–31. *In*: Ultraviolet Light in Food Technology. CRC Press, Boca Raton, FL.

Koutchma, T.N., Forney, L.J. and Moraru, C.I. 2009d. Sources of UV light. pp. 33–51. *In*: Ultraviolet Light in Food Technology. CRC Press, Boca Raton, FL.

Kovács, E. and Keresztes, Á. 2002. Effect of gamma and UV-B/C radiation on plant cells. *Micron* 33, 199–210.

Kowalski, W. 2009a. UV rate constants. pp. 73–117. *In*: Ultraviolet Germicidal Irradiation Handbook: UVGI for Air and Surface Disinfection. Springer, Heidelberg, Berlin, Germany.

Kowalski, W. 2009b. UVGI disinfection theory. pp.17–50. *In*: Ultraviolet Germicidal Irradiation Handbook: UVGI for Air and Surface Disinfection. Springer, Heidelberg, Berlin, Germany.

Kowalski, W. 2009c. UVGI lamps and fixtures. pp.119–137. *In*: Ultraviolet Germicidal Irradiation Handbook: UVGI for Air and Surface Disinfection. Springer, Heidelberg, Berlin, Germany.

Kowitcharoen, L., Kammapana, L. and Srilaong, V. 2010. UV-C treatment delays chlorophyll degradation in the bract of dragon fruit cultivar 'Vietnam'. *Acta Horticulturae* 875, 105–110.

Kramer, G.F., Wang, C.Y. and Conway, W.S. 1989. Correlation of reduced softening and increased polyamine levels during low-oxygen storage of 'McIntosh' apples. *Journal of the American Society for Horticultural Science* 114, 942–946.

Krishnapillai, N. and Wilson Wijeratnam, R.S. 2017. Sap volatile components in relation to susceptibility of anthracnose and Aspergillus rot of mangoes (*Mangifera indica L.*). *The Journal of Horticultural Science and Biotechnology* 92, 206–213.

Kumar, S. and Pandey, A.K. 2013. Chemistry and biological activities of flavonoids: An overview. *The Scientific World Journal* 2013, 16.

Kuniga, T. and Nesumi, H. 2011. UV-C irradiation affects accumulation of scoparone in citrus. *Acta Horticulturae* 907, 81–85.

Kuniga, T., Tsumura, T., Matsuo, Y. and Matsumoto, R. 2006. Changes in scoparone concentrations in citrus cultivars after ultraviolet radiation. *Journal of the Japanese Society for Horticultural Science* 75, 328–330.

Künstler, A., Bacsó, R., Hafez, Y.M. and Király, L. 2015. Reactive oxygen species and plant disease resistance. pp. 269–303. *In*: D.K. Gupta, J.M. Palma and F. J. Corpas (eds.). *Reactive Oxygen Species and Oxidative Damage in Plants under Stress.* Springer International Publishing, Cham, Switzeland.

Kuźniak, E. and Urbanek, H. 2000. The involvement of hydrogen peroxide in plant responses to stresses. *Acta Physiologiae Plantarum* 22, 195–203.

Lafuente, M.T., Cantwell, M., Yang, S.F. and Rubatzky, V. 1989. Isocoumarin content of carrots as influenced by ethylene concentration, storage temperature and stress conditions. *Acta Horticulturae* 258, 523–534.

Lemoine, M.L., Civello, P.M., Martínez, G.A. and Chaves, A.R. 2007. Influence of postharvest UV-C treatment on refrigerated storage of minimally processed broccoli (*Brassica oleracea* var. Italica). *Journal of the Science of Food and Agriculture* 87, 1132–1139.

Leng, P., Yuan, B. and Guo, Y. 2014. The role of abscisic acid in fruit ripening and responses to abiotic stress. *Journal of Experimental Botany* 65, 4577–4588.

Lers, A. 2012. 27 - Potential application of biotechnology to maintain fresh produce postharvest quality and reduce losses during storage. pp. 425–441. *In*: A. Arie and P.M. Hasegawa (eds.). *Plant Biotechnology and Agriculture. Prospects for the 21st Century.* Academic Press, San Diego, CA.

Liao, C., Liu, X., Gao, A., Zhao, A., Hu, J. and Li, B. 2016. Maintaining postharvest qualities of three leaf vegetables to enhance their shelf lives by multiple ultraviolet-C treatment. *LWT - Food Science and Technology* 73, 1–5.

Lila, M.A. 2004. Anthocyanins and human health: An in vitro investigative approach. *Journal of Biomedicine and Biotechnology* 5, 306–313.

Linthorst, H.J.M. and van Loon, L.C. 1991. Pathogenesis-related proteins of plants. *Critical Reviews in Plant Sciences* 10, 123–150.

Liu, C.H., Cai, L.Y., Lu, X.Y., Han, X.X. and Ying, T.J. 2012. Effect of postharvest UV-C irradiation on phenolic compound content and antioxidant activity of tomato fruit during storage. *Journal of Integrative Agriculture* 11, 159–165.

Liu, J., Stevens, C., Khan, V.A., Lu, J.Y., Wilson, C.L., Adeyeye, O., Kabwe, M.K., Pusey, P.L., Chalutz, E., Sultana, T. and Droby, S. 1993. Application of ultraviolet-C light on storage rots and ripening of tomatoes. *Journal of Food Protection* 56, 868–873.

Liu, L.H., Zabaras, D., Bennett, L.E., Aguas, P. and Woonton, B.W. 2009. Effects of UV-C, red light and sun light on the carotenoid content and physical qualities of tomatoes during post-harvest storage. *Food Chemistry* 115, 495–500.

Liu, R.H. 2003. Health benefits of fruit and vegetables are from additive and synergistic combinations of phytochemicals. *The American Journal of Clinical Nutrition* 78, 517S–520S.

Lu, J.Y., Stevens, C., Yakubu, P., Loretan, P.A. and Eakin, D. 1988. Gamma, electron beam and ultraviolet radiation on control of storage rots and quality of walla walla onions. *Journal of Food Processing and Preservation* 12, 53–62.

Luckey, T.D. 1980. *Hormesis with Ionizing Radiation.* CRC Press, Boca Raton, FL.

Maharaj, R., Arul, J. and Nadeau, P. 1999. Effect of photochemical treatment in the preservation of fresh tomato (*Lycopersicon esculentum* cv. Capello) by delaying senescence. *Postharvest Biology and Technology* 15, 13–23.

Maharaj, R., Arul, J. and Nadeau, P. 2010. UV-C irradiation of tomato and its effects on color and pigments. *Advances in Environmental Biology* 4, 308–315.

Maharaj, R., Arul, J. and Nadeau, P. 2014. UV-C irradiation effects on levels of enzymic and non-enzymic phytochemicals in tomato. *Innovative Food Science & Emerging Technologies* 21, 99–106.

Manzocco, L., Da Pieve, S., Bertolini, A., Bartolomeoli, I., Maifreni, M., Vianello, A. and Nicoli, M.C. 2011a. Surface decontamination of fresh-cut apple by UV-C light exposure: Effects on structure, colour and sensory properties. *Postharvest Biology and Technology* 61, 165–171.

Manzocco, L., Da Pieve, S. and Maifreni, M. 2011b. Impact of UV-C light on safety and quality of fresh-cut melon. *Innovative Food Science & Emerging Technologies* 12, 13–17.

Martínez-Hernández, G.B., Gómez, P.A., Pradas, I., Artés, F. and Artés-Hernández, F. 2011. Moderate UV-C pretreatment as a quality enhancement tool in fresh-cut Bimi® broccoli. *Postharvest Biology and Technology* 62, 327–337.

Martysiak-Żurowska, D., Puta, M., Kotarska, J., Cybula, K., Malinowska-Pańczyk, E. and Kołodziejska, I. 2017. The effect of UV-C irradiation on lipids and selected biologically active compounds in human milk. *International Dairy Journal* 66, 42–48.

Mattson, M.P. 2008. Hormesis defined. *Ageing Research Reviews* 7, 1–7.

Mditshwa, A., Magwaza, L.S., Tesfay, S.Z. and Mbili, N.C. 2017. Effect of ultraviolet irradiation on postharvest quality and composition of tomatoes: A review. *Journal of Food Science and Technology* 54, 3025–3035.

Mengiste, T. 2012. Plant immunity to necrotrophs. *Annual Review of Phytopathology* 50, 267–294.

Mercier, J., Arul, J. and Julien, C. 1993a. Effect of UV-C on phytoalexin accumulation and resistance to *Botrytis cinerea* in stored carrots. *Journal of Phytopathology* 139, 17–25.

Mercier, J., Arul, J. and Julien, C. 1994. Effect of food preparation on the isocoumarin, 6-methoxymellein, content of UV-treated carrots. *Food Research International* 27, 401–404.

Mercier, J., Arul, J., Ponnampalam, R. and Boulet, M. 1993b. Induction of 6-methoxymellein and resistance to storage pathogens in carrot slices by UV-C. *Journal of Phytopathology* 137, 44–54.

Mercier, J., Baka, M., Reddy, B., Corcuff, R. and Arul, J. 2001. Shortwave Ultraviolet irradiation for control of decay caused by *Botrytis cinerea* in bell pepper: Induced resistance and germicidal effects. *Journal of the American Society for Horticultural Science* 126, 128–133.

Mercier, J., Ponnampalam, R., Bérard, L.S. and Arul, J. 1993c. Polyacetylene content and UV-induced 6-methoxymellein accumulation in carrot cultivars. *Journal of the Science of Food and Agriculture* 63, 313–317.

Mercier, J., Roussel, D., Charles, M.T. and Arul, J. 2000. Systemic and local responses associated with UV- and pathogen-induced resistance to *Botrytis cinerea* in stored carrot. *Phytopathology* 90, 981–986.

Meulemans, C.C.E. 1987. The basic principles of UV-disinfection of water. *Ozone Science and Engineering Journal* 9, 299.

Mihaly-Cozmuta, A., Mihaly-Cozmuta, L., Peter, A. and Nicula, C. 2016. The comparative effect of far-red light and UV-C light on some physical-chemical attributes in red bell peppers (*Capsicum annuum* L.) during storage. *Analele Universitatii Din Oradea: Fascicula Biologie* 23, 71–77.

Mintoff, S.J., Rookes, J.E. and Cahill, D.M. 2015. Sub-lethal UV-C radiation induces callose, hydrogen peroxide and defence-related gene expression in *Arabidopsis thaliana*. *Plant Biology* 17, 703–711.

Mittler, R. 2002. Oxidative stress, antioxidants and stress tolerance. *Trends in Plant Science* 7, 405–410.

Moglia, A., Comino, C., Portis, E., Acquadro, A., Vos, C.H.D., Beekwilder, M.J. and Lanteri, S. 2009. Isolation and mapping of a C3′H gene (CYP98A49) from globe artichoke, and its expression upon UV-C stress. *Plant Cell Reports* 28, 963–974.

Mohamed, N.T.S., Ding, P., Ghazali, H.M. and Kadir, J. 2017. Biochemical and cell wall ultrastructural changes in crown tissue of banana (Musa AAA 'Berangan') fruit as mediated by UVC irradiation against crown rot fungal infection. *Postharvest Biology and Technology* 128, 144–152.

Mohan, R. and Kolattukudy, P.E. 1990. Differential activation of expression of a suberization-associated anionic peroxidase gene in near-isogenic resistant and susceptible tomato lines by elicitors of *Verticillium albo-atratrum*. *Plant Physiology* 92, 276–280.

Mortazavi, S.M.H., Siruie, B., Moalemi, N. and Eshghi, S. 2014. The effects of polyamines and UV-C irradiation on postharvest quality of strawberry fruit. *Acta Horticulturae* 1049, 749–754.

Murphy Terence, M. and Huerta Alfredo, J. 1990. Hydrogen peroxide formation in cultured rose cells in response to UV-C radiation. *Physiologia Plantarum* 78, 247–253.

Nadeau, F., Gaudreau, A., Angers, P. and Arul, J. 2012. Changes in the level of glucosinolates in broccoli florets (*Brassica oleracea* var. Italic) during storage following postharvest treatment with UV-C. *Acta Horticulturae* 945, 145–148.

Najeeb, U., Xu, L., Ahmed, Z.I., Rasheed, M., Jilani, G., Naeem, M.S., Shen, W. and Zhou, W. 2011. Ultraviolet-C mediated physiological and ultrastructural alterations in *Juncus effusus* L. shoots. *Acta Physiologiae Plantarum* 33, 481–488.

Nawkar, G.M., Maibam, P., Park, J.H., Sahi, V.P., Lee, S.Y. and Kang, C.H. 2013. UV-Induced cell death in plants. *International Journal of Molecular Sciences* 14, 1608–1628.

Nerandzic, M.M., Cadnum, J.L., Pultz, M.J. and Donskey, C.J. 2010. Evaluation of an automated ultraviolet radiation device for decontamination of *Clostridium difficile* and other healthcare-associated pathogens in hospital rooms. *BMC Infectious Diseases* 10, 197.

Nigro, F., Ippolito, A. and Lima, G. 1998. Use of UV-C light to reduce *Botrytis* storage rot of table grapes. *Postharvest Biology and Technology* 13, 171–181.

Nimitkeatkai, H. and Kulthip, J. 2016. Effect of sequential UV-C irradiation on microbial reduction and quality of fresh-cut dragon fruit. *International Food Research Journal* 23, 1818–1822.

Nunes do Nascimento, M.C. and Emond, J.P. 2002. *Storage Temperature, Postharvest Physiology and Pathology of Vegetables*. CRC Press, Boca Raton, FL.

Ojaghian, M.R., Zhang, J.-Z., Xie, G.-L., Wang, Q., Li, X.-L. and Guo, D.-P. 2017. Efficacy of UV-C radiation in inducing systemic acquired resistance against storage carrot rot caused by *Sclerotinia sclerotiorum*. *Postharvest Biology and Technology* 130, 94–102.

Ortiz Araque, L.C., Rodoni, L.M., Darré, M., Ortiz, C.M., Civello, P.M. and Vicente, A.R. 2018. Cyclic low dose UV-C treatments retain strawberry fruit quality more effectively than conventional pre-storage single high fluence applications. *LWT - Food Science and Technology* 92, 304–311.

Park, M.-H. and Kim, J.-G. 2015. Low-dose UV-C irradiation reduces the microbial population and preserves antioxidant levels in peeled garlic (*Allium sativum* L.) during storage. *Postharvest Biology and Technology* 100, 109–112.

Pedras, M.S., Sarwar, M.G., Suchy, M. and Adio, A.M. 2006. The phytoalexins from cauliflower, caulilexins A, B and C: Isolation, structure determination, syntheses and antifungal activity. *Phytochemistry* 67, 1503–1509.

Pedras, M.S.C., Gadagi, R.S., Zheng, Q.-A. and Rimmer, S.R. 2008. Selective elicitation of the phytoalexin rutalexin in rutabaga and turnip roots by a biotrophic plant pathogen. *Natural Product Communications* 3, 1239–1243.

Pedras, M.S.C. and Yaya, E.E. 2015. Plant chemical defenses: Are all constitutive antimicrobial metabolites phytoanticipins? *Natural Product Communications* 10, 209–218.

Perkins-Veazie, P., Collins, J.K. and Howard, L. 2008. Blueberry fruit response to postharvest application of ultraviolet radiation. *Postharvest Biology and Technology* 47, 280–285.

Petit, A.-N., Baillieul, F., Vaillant-Gaveau, N., Jacquens, L., Conreux, A., Jeandet, P., Clément, C. and Fontaine, F. 2009. Low responsiveness of grapevine flowers and berries at fruit set to UV-C irradiation. *Journal of Experimental Botany* 60, 1155–1162.

Pezet, R., Perret, C., Jean-Denis, J.B., Tabacchi, R., Gindro, K. and Viret, O. 2003. δ-Viniferin, a resveratrol dehydrodimer: One of the major stilbenes synthesized by stressed grapevine leaves. *Journal of Agricultural and Food Chemistry* 51, 5488–5492.

Phan, C.-T. 1987. Biochemical and physiological changes during the harvest period. pp. 9–22. *In*: J. Weichmann (ed.). *Postharvest Physiology of Vegetables.* Marcel Dekker, New York, NY.

Piršelová, B. and Matušíková, I. 2013. Callose: The plant cell wall polysaccharide with multiple biological functions. *Acta Physiologiae Plantarum* 35, 635–644.

Pombo, M.A., Rosli, H.G., Martínez, G.A. and Civello, P.M. 2011. UV-C treatment affects the expression and activity of defense genes in strawberry fruit (*Fragaria×ananassa*, Duch.). *Postharvest Biology and Technology* 59, 94–102.

Pongprasert, N., Sekozawa, Y., Sugaya, S. and Gemma, H. 2011. A novel postharvest UV-C treatment to reduce chilling injury (membrane damage, browning and chlorophyll degradation) in banana peel. *Scientia Horticulturae* 130, 73–77.

Porat, R., Lers, A., Dori, S., Cohen, L., Weiss, B., Daus, A., Wilson, C.L. and Droby, S. 1999. Induction of chitinase and β-1,3-endoglucanase proteins by UV irradiation and wounding in grapefruit peel tissue. *Phytoparasitica* 27, 233.

Poschenrieder, C., Cabot, C., Martos, S., Gallego, B. and Barceló, J. 2013. Do toxic ions induce hormesis in plants? *Plant Science* 212, 15–25.

Poubol, J., Lichanporn, I., Puthmee, T. and Kanlayanarat, S. 2010. Effect of ultraviolet-C irradiation on quality and natural microflora of asparagus spears. *Acta Horticulturae* 875, 257–262.

Purkayastha, R.P. 1994. Progress in phytoalexin research during the past 50 years. pp. 1–40. *In*: M. Daniel and R.P. Purkayastha (eds.). *Handbook of Phytoalexin Metabolism and Action.* Marcel Dekker, New York, NY.

Rastogi, R.P., Richa, Kumar, A., Tyagi, M.B. and Sinha, R.P. 2010. Molecular mechanisms of ultraviolet radiation-induced DNA damage and repair. *Journal of Nucleic Acids* 2010, 592980.

Reddy, B.M.V., Belkacemi, K., Corcuff, R., Castaigne, F. and Arul, J. 2000. Effect of pre-harvest chitosan sprays on post-harvest infection by *Botrytis cinerea* and quality of strawberry fruit. *Postharvest Biology and Technology* 20, 39–51.

Rodoni, L.M., Zaro, M.J., Hasperué, J.H., Concellón, A. and Vicente, A.R. 2015. UV-C treatments extend the shelf life of fresh-cut peppers by delaying pectin solubilization and inducing local accumulation of phenolics. *LWT - Food Science and Technology* 63, 408–414.
Rodov, V., Ben-Yehoshua, S., Fang, D., D'Hallewin, G. and Castia, T. 1994. Accumulation of phytoalexins scoparone and scopoletin in citrus fruits subjected to various postharvest treatments. *Acta Horticulturae* 381, 517–525.
Rodov, V., Ben-Yehoshua, S., Fang, D.Q., Kim, J.J. and Ashkenazi, R. 1995. Preformed antifungal compounds of lemon fruit: Citral and its relation to disease resistance. *Journal of Agricultural and Food Chemistry* 43, 1057–1061.
Rodov, V., Ben-Yehoshua, S., Kim, J.J., Shapiro, B. and Ittah, Y. 1992. Ultraviolet illumination induces scoparone production in kumquat and orange fruit and improves decay resistance. *Journal of the American Society for Horticultural Science* 117, 788–792.
Romanazzi, G., Gabler, F.M. and Smilanick, J.L. 2006. Preharvest chitosan and postharvest UV irradiation treatments suppress gray mold of table grapes. *Plant Disease* 90, 445–450.
Ruban, A.V. 2009. Plants in light. *Communicative & Integrative Biology* 2, 50–55.
Saini, R.K. and Keum, Y.S. 2018. Significance of genetic, environmental, and pre- and post-harvest factors affecting carotenoid contents in crops: A review. *Journal of Agricultural and Food Chemistry* 66, 5310–5324.
Saltveit, M.E. 1999. Effect of ethylene on quality of fresh fruits and vegetables. *Postharvest Biology and Technology* 15, 279–292.
Sari, L.K., Setha, S. and Naradisorn, M. 2016. Effect of UV-C irradiation on postharvest quality of 'Phulae' pineapple. *Scientia Horticulturae* 213, 314–320.
Scott, G., Rupar, M., Fletcher, A.G.D., Dickinson, M. and Shama, G. 2017. A comparison of low intensity UV-C and high intensity pulsed polychromatic sources as elicitors of hormesis in tomato fruit. *Postharvest Biology and Technology* 125, 52–58.
Sebastià, N., Montoro, A., León, Z. and Soriano, J.M. 2017. Searching trans-resveratrol in fruits and vegetables: A preliminary screening. *Journal of Food Science and Technology* 54, 842–845.
Seeberg, E., Eide, L. and Bjørås, M. 1995. The base excision repair pathway. *Trends in Biochemical Sciences* 20, 391–397.
Severo, J., de Oliveira, I.R., Tiecher, A., Chaves, F.C. and Rombaldi, C.V. 2015a. Postharvest UV-C treatment increases bioactive, ester volatile compounds and a putative allergenic protein in strawberry. *LWT - Food Science and Technology* 64, 685–692.
Severo, J., Tiecher, A., Pirrello, J., Regad, F., Latché, A., Pech, J.-C., Bouzayen, M. and Rombaldi, C.V. 2015b. UV-C radiation modifies the ripening and accumulation of ethylene response factor (ERF) transcripts in tomato fruit. *Postharvest Biology and Technology* 102, 9–16.
Shama, G. 1999. Ultra-violet light. pp. 2208–2214. *In*: K.R. Richard (ed.). *Encyclopedia of Food Microbiology*. Elsevier, Oxford, UK.
Shama, G. 2007. Process challenges in applying low doses of ultraviolet light to fresh produce for eliciting beneficial hormetic responses. *Postharvest Biology and Technology* 44, 1–8.
Shama, G. and Alderson, P. 2005. UV hormesis in fruits: A concept ripe for commercialisation. *Trends in Food Science & Technology* 16, 128–136.

Sheik Mohamed, N., Ding, P., Kadir, J. and Hasanah, M.G. 2017. Potential of UVC germicidal irradiation in suppressing crown rot disease, retaining postharvest quality and antioxidant capacity of Musa AAA "Berangan" during fruit ripening. *Food Science & Nutrition* 5, 967–980.

Shen, Y., Sun, Y., Qiao, L., Chen, J., Liu, D. and Ye, X. 2013. Effect of UV-C treatments on phenolic compounds and antioxidant capacity of minimally processed Satsuma mandarin during refrigerated storage. *Postharvest Biology and Technology* 76, 50–57.

Shishodia, S. and Aggarwal, B.B. 2005. A polyphenol of all seasons. p. 712. *In*: B. B. Aggarwal and S. Shishodia (eds.). *Resveratrol in Health and Disease*. CRC Press, Boca raton, FL.

Sitbon, C. and Paliyath, G. 2011. Pre- and postharvest treatments affecting nutritional quality. pp. 349–357. *In*: M.Y. Murray (ed.). *Comprehensive Biotechnology*, 2nd ed. Academic Press, Burlington, NJ.

Smith, C.J. 1996. Tansley Review Accumulation of phytoalexins: Defence mechanism and stimulus response system. *New Phytologist* 86(132), 1–45.

Solovchenko, A. 2010a. Localization of screening pigments within plant cells and tissues, photoprotection in plants: optical screening-based mechanisms. pp. 67–88. *In*: Photoprotection in Plants. Optical Screening-based Mechanisms. Springer, Heidelberg, Berlin, Germany.

Solovchenko, A. 2010b. Screening pigments: general questions, photoprotection in plants: optical screening-based mechanisms. pp. 9–31. *In*: Photoprotection in Plants. Optical Screening-based Mechanisms. Springer, Heidelberg, Berlin, Germany.

Srepong, K., Jitareerat, P., Uthairatanakij, A., Srilaong, V., Wongs-Aree, C. and Tsuyumu, S. 2013. Induction of defense mechanisms on harvested mangoes by UV-C irradiation. *Acta Horticulturae* 973, 89–95.

Sripong, K., Jitareerat, P., Tsuyumu, S., Uthairatanakij, A., Srilaong, V., Wongs-Aree, C., Ma, G., Zhang, L. and Kato, M. 2015. Combined treatment with hot water and UV-C elicits disease resistance against anthracnose and improves the quality of harvested mangoes. *Crop Protection* 77, 1–8.

Stapleton, A.E. 1992. Ultraviolet radiation and plants: Burning questions. *The Plant Cell* 4, 1353–1358.

Stevens, C., Khan, V.A., Lu, J.Y., Wilson, C.L., Pusey, P.L., Kabwe, M.K., Igwegbe, E. C.K., Chalutz, E. and Droby, S. 1998a. The germicidal and hormetic effects of UV-C light on reducing brown rot disease and yeast microflora of peaches. *Crop Protection* 17, 75–84.

Stevens, C., Khan, V.A., Tang, A.Y. and Lu, J.Y. 1990. The effect of Ultraviolet radiation on mold rots and nutrients of stored sweet potatoes. *Journal of Food Protection* 53, 223–226.

Stevens, C., Khan, V.A., Wilson, C.L., Lu, J.Y., Pusey, L., Bassett, C.L., Igwegbe, E.C. K., Wisniewski, M., Chalutz, E., Droby, S. and El Ghaouth, A. 2006. The use of low dose UV-C light technology to control postharvest storage decay and delayed ripening and senescence of fruits and vegetables. pp. 195–237. *In*: B. Noureddine and S. Norio (eds.). *Advances in Postharvest Technology of Horticultural Crops*. Research Signpost, Kerala, India.

Stevens, C., Liu, J., Khan, V.A., Lu, J.Y., Wilson, C.L., Igwegbe, E.C.K., Kabwe, M.K., Chalutz, E. and Droby, S. 1998b. Application of hormetic UV-C for delayed ripening and reduction of Rhizopus soft rot in tomatoes: The effect of tomatine on storage rot development. *Journal of Phytopathology* 146, 211–221.

Stevens, C., Wilson, C.L., Lu, J.Y., Khan, V.A., Chalutz, E., Droby, S., Kabwe, M.K., Haung, Z., Adeyeye, O., Pusey, L.P., Wisniewski, M.E. and West, M. 1996. Plant hormesis induced by ultraviolet light-C for controlling postharvest diseases of tree fruits. *Crop Protection* 15, 129–134.

Syamaladevi, R.M., Lupien, S.L., Bhunia, K., Sablani, S.S., Dugan, F., Rasco, B., Killinger, K., Dhingra, A. and Ross, C. 2014. UV-C light inactivation kinetics of *Penicillium expansum* on pear surfaces: Influence on physicochemical and sensory quality during storage. *Postharvest Biology and Technology* 87, 27–32.

Takahashi, S. and Badger, M.R. 2011. Photoprotection in plants: A new light on photosystem II damage. *Trends in Plant Science* 16, 53–60.

Tatone, C., Di Emidio, G., Vento, M., Ciriminna, R. and Artini, P.G. 2010. Cryopreservation and oxidative stress in reproductive cells. *Gynecological Endocrinology* 26, 563–567.

Terry, L.A. and Joyce, D.C. 2004. Elicitors of induced disease resistance in postharvest horticultural crops: A brief review. *Postharvest Biology and Technology* 32, 1–13.

Terry, L.A. and Thompson, A.K. 2011. Introduction. pp. 1–4. *In*: L.A. Terry (ed.). *Health-Promoting Properties of Fruit and Vegetables.* Cabi Publishing, Wallingford, Oxford, UK.

Textor, S. and Gershenzon, J., 2009. Herbivore induction of the glucosinolate–myrosinase defense system: major trends, biochemical bases and ecological significance. *Phytochemistry Reviews* 8, 149-170.

Tiecher, A., de Paula, L.A., Chaves, F.C. and Rombaldi, C.V. 2013. UV-C effect on ethylene, polyamines and the regulation of tomato fruit ripening. *Postharvest Biology and Technology* 86, 230–239.

Van den Berg, L. 1981. The role of humidity, temperature, and atmospheric composition in maintaining vegetable quality during storage. pp. 95–70. *In*: R. Teranishi and H. Barrera-Benitz (eds.). *Quality of Selected Fruits and Vegetables of North America.* ACS Symposium Series, vol. 170. American Chemical Society, Washington.

Van Loon, L.C. 1985. Pathogenesis-related proteins. *Plant Molecular Biology* 4, 111–116.

Van Loon, L.C., Rep, M. and Pieterse, C.M. 2006. Significance of inducible defense-related proteins in infected plants. *Annual Review of Phytopathology* 44, 135–162.

Vance, C.P., Kirk, T.K. and Sherwood, R.T. 1980. Lignification as a mechanism of disease resistance. *Annual Review of Phytopathology* 18, 259–288.

Venditti, T. and D'Hallewin, G. 2014. Use of ultraviolet radiation to increase the health-promoting properties of fruits and vegetables. *Stewart Postharvest Review* 10, 1–3.

Vicente, A.R., Pineda, C., Lemoine, L., Civello, P.M., Martínez, G.A. and Chaves, A.R. 2005. UV-C treatments reduce decay, retain quality and alleviate chilling injury in pepper. *Postharvest Biology and Technology* 35, 69–78.

Vieira, T.M.F.S. and Regitano-d'Arce, M.A.B. 1998. Stability of oils heated by microwave: UV - spectrophotometric evaluation. *Food Science and Technology (Campinas)* 18, 433–437.

Walters, D.R. 2010. What defenses do plants use? pp. 15–76. *In*: D.R. Walters (ed.). *Plant Defense: Warding off Attack by Pathogens, Herbivores, and Parasitic Plants.* Blackwell Publishing Ltd., Chichester, UK.

Wang, C.Y., Chen, C.-T. and Wang, S.Y. 2009. Changes of flavonoid content and antioxidant capacity in blueberries after illumination with UV-C. *Food Chemistry* 117, 426–431.

Wang, Q., Chu, L. and Kou, L. 2017. UV-C Treatment maintains quality and delays senescence of oyster mushroom (*Pleurotus ostreatus*). *Scientia Horticulturae* 225, 380–385.

Wei, Z.F., Luo, M., Zhao, C.J., Li, C.Y., Gu, C.B., Wang, W., Zu, Y.G., Efferth, T. and Fu, Y.J. 2013. UV-induced changes of active components and antioxidant activity in postharvest pigeon pea [*Cajanus cajan* (L.) Millsp.] leaves. *Journal of Agricultural and Food Chemistry* 61, 1165–1171.

Wink, M. 2010. Introduction: Biochemistry, physiology and ecological functions of secondary metabolites. pp. 1–19. *In*: M. Wink (ed.). *Annual Plant Reviews Volume 40: Biochemistry of Plant Secondary Metabolism.* Blackwell Publishing Ltd., Chichester, UK.

Wu, J., Liu, W., Yuan, L., Guan, W.-Q., Brennan, C.S., Zhang, Y.Y., Zhang, J. and Wang, Z.-D. 2017. The influence of postharvest UV-C treatment on anthocyanin biosynthesis in fresh-cut red cabbage. *Scientific Reports* 7, 5232.

Wu, X., Guan, W., Yan, R., Lei, J., Xu, L. and Wang, Z. 2016. Effects of UV-C on antioxidant activity, total phenolics and main phenolic compounds of the melanin biosynthesis pathway in different tissues of button mushroom. *Postharvest Biology and Technology* 118, 51–58.

Xu, F., Wang, S., Xu, J., Liu, S. and Li, G. 2016. Effects of combined aqueous chlorine dioxide and UV-C on shelf-life quality of blueberries. *Postharvest Biology and Technology* 117, 125–131.

Xu, Y., Charles, M.T., Luo, Z., Mimee, B., Veronneau, P.Y., Rolland, D. and Roussel, D. 2017. Preharvest ultraviolet C irradiation increased the level of polyphenol accumulation and flavonoid pathway gene expression in strawberry fruit. *Journal of Agricultural and Food Chemistry* 65, 9970–9979.

Yalpani, N., Enyedi, A., León, J. and Raskin, I. 1994. Ultraviolet light and ozone stimulate accumulation of salicylic acid, pathogenesis-related proteins and virus resistance in tobacco. *Planta* 193, 372–376.

Yalpani, N., Silverman, P., Wilson, T.M., Kleier, D.A. and Raskin, I. 1991. Salicylic acid is a systemic signal and an inducer of pathogenesis-related proteins in virus-infected tobacco. *The Plant Cell* 3, 809–818.

Yang, Z., Cao, S., Su, X. and Jiang, Y. 2014. Respiratory activity and mitochondrial membrane associated with fruit senescence in postharvest peaches in response to UV-C treatment. *Food Chemistry* 161, 16–21.

Yuan, L., Bi, Y., Ge, Y., Wang, Y., Liu, Y. and Li, G. 2013. Postharvest hot water dipping reduces decay by inducing disease resistance and maintaining firmness in muskmelon (*Cucumis melo L.*) fruit. *Scientia Horticulturae* 161, 101–110.

Zeng, F., Jiang, T., Wang, Y. and Luo, Z. 2015. Effect of UV-C treatment on modulating antioxidative system and proline metabolism of bamboo shoots subjected to chilling stress. *Acta Physiologiae Plantarum* 37, 244.

Zhang, Y., Talalay, P., Cho, C.G. and Posner, G.H. 1992. A major inducer of anticarcinogenic protective enzymes from broccoli: Isolation and elucidation of structure. *Proceedings of the National Academy of Sciences* 89, 2399.

Chapter 18

Reducing or Replacing Conventional Postharvest Fungicides with Low Toxicity Acids and Salts

Salvatore D'Aquino and Amedeo Palma

Institute of Sciences of Food Production, National Research Council, Sassari, Italy

Abbreviations

AA	Acetic acid
DMI	Demethylation inhibitors
EU	European Union
FRAC	Fungicide Resistance Action Committee
GRAS	Generally recognized as safe
HOPP	Undissociated *o*-phenylphenol
HPMC	Hydroxypropyl methylcellulose
IMZ	Imazalil
LTC	Low toxicity compounds, GRAS and low toxicity inorganic and organic acids and salts
OPP^-	Dissociated *o*-phenylphenate
PAL	Phenylalanine ammonia-lyase
PDA	Potato dextrose agar
Phi	Phosphonic acid
Pi	Phosphoric acid
PS	Potassium sorbate
QoIs	Quinone outside inhibitor fungicides
RH	Relative humidity
SA	Sorbic acid
SOPP	Sodium *o*-phenylphenate tetrahydrate
TBZ	Thiabendazole
USA	Unites States of America

1 Introduction

In recent decades, many countries have made marked changes in the regulation of pesticides to minimize their residues in horticultural products. These measures involve more restrictions on the use of pesticides and in the future these will have a revolutionary impact on the standard practices concerning the use of fungicides as protectants of fresh produce from decay causing fungi. In this context, low toxicity inorganic and organic acids and salts, many of them classified as generally recognized as safe (GRAS) compounds by the US Food and Drug Administration and other regulators, and other low toxicity compounds (LTCs) already in general use for other food applications, will play an increasing role not only due to legislative issues but also above all to respond to public concerns about their real or presumed risks to human health and the environment. The worldwide increasing demand for safer food and the public's skeptical position toward chemical residues on food constantly compels regulatory authorities as well as the agricultural industries to respond to both consumers' demands and the priority of guarantying effective means to control fungal spoilage. Furthermore, in an attempt to gain a commercial advantage by satisfying consumers' concerns, international buyers and large-scale retail distribution companies impose their own safety standards, generally with stricter conditions than the legal requirements, typically by allowing a reduced number of detectable active ingredients or imposing residue limits lower than those set by regulatory authorities.

It is conceivable that the use of synthetic fungicides could be completely banned in the future, as they are today for fresh products classified as 'organic'. In recent years, intensive research work has been done on the use of LTCs as means to protect fresh fruits and vegetables from pathogens causing postharvest decay. Unfortunately, overall results have been inconsistent and not always satisfactory and their transfer to commercial implementation has been modest. Therefore, decay management continues to rely on applications of synthetic fungicides, at least for all those products for which their use is still legally authorized. However, their use is probably more widespread than data from literature indicates, because of unresolved regulatory issues about their residues. Some countries do not have a classification established for them to be used on fruit and may require official use labels and residue tolerances (like Japan), while in others this is rarely an issue (like the USA).

This chapter reviews the research to evaluate LTCs as alternative means to control postharvest decay of horticultural products and describes their main chemical and physical properties.

2 Conventional Synthetic Fungicides for Postharvest Use

2.1 Fungicide Resistance

In contrast to sanitizers, which are primarily nonselectively toxic biocides applied to reduce the total microbiological load in process water and on the fruit surface, postharvest fungicides inhibit fungal growth specifically and protect fruits and vegetables from infections by fungal postharvest pathogens.

Fungicides are classified according to their mobility in the plant as contact, cytotropic, or systemic fungicides. The former are not able to penetrate the tissue and they remain on fruit surface and prevent infections by killing or inhibiting fungi or fungal spores before they can grow and develop, while the latter two, being absorbed by tissues, may stop incipient and quiescent infections that were already present both before and after the time of the fungicide application. Therefore, the cytotropic and systemic fungicides may possess curative, protectant, and eradicant properties (Lyr, 1995; Brent and Hollomon, 1998, 2007). Fungicides can inhibit one or more sites in one or more metabolic pathways. Generally, multisite fungicides are contact fungicides, whereas single-site fungicides are usually cytotropic or systemic. Fungicides sharing a target site belong to the same chemical group (Brent and Hollomon, 1998, 2007; Hann, 2014).

Despite that a large number of fungicides has been tested to control postharvest diseases of several crops, only few are registered for postharvest use. The first generation of postharvest fungicides, including sodium *o*-phenylphenate and dichloran, was effective against wound pathogens but had little activity against latent, deep-seated infections (Eckert and Sommer, 1967, 1985). From middle 1960s, a new generation of fungicides, more active against latent infections and with greater protective activity against a wide range of diseases, was developed (Morton and Staub, 2008). Unfortunately, repeated use of the same fungicides soon led to a loss of effectiveness to control disease because of the development of resistant strains among populations of these fungi. Problems of fungicide resistance lead to failures to control *Penicillium digitatum* (Pers.) Sacc. and *Penicillium italicum* Wehmer in citrus fruits. These fungicides included biphenyl, sodium *o*-phenylphenate tetrahydrate (SOPP), thiabendazole (TBZ), benomyl, and *sec*-butylamine, with resistance detected just after a few years after their introduction (Bollen and Scholten, 1971; Houck, 1977;

McDonald et al., 1979; Dave et al., 1980; El-Goorani et al., 1983; Eckert and Sommer, 1985; Adaskaveg and Förster, 2015). Fungal isolates that are resistant to one fungicide are usually also resistant to structurally related compounds, for example, biphenyl and SOPP; and TBZ, benomyl, carbendazim, and thiophanate-methyl (Eckert and Sommer, 1985; Hann, 2014). The loss of fungicide effectiveness is an issue not limited to their specific use as postharvest treatments, but includes any application where fungicides are employed, such as the protection of textiles, painted surfaces, paper pulp, as well as the protection of crops in the field, seeds, and of fruits and vegetables during storage and shipment.

The Fungicide Resistance Action Committee (FRAC), created in 1998, is a technical group whose specific mission is to protect the efficacy of fungicides susceptible to the development of resistance in pathogen populations and managing emerging resistance problems to limit the damage they cause to crops. FRAC classifies all fungicides into groups wherein products with the same target site have the same FRAC code number (Brent and Hollomon, 1998, 2007). Fungi are more likely to develop resistance to single-site mode of action fungicides than those with multiple sites of inhibition. Based on how the decline in effectiveness of a fungicide group evolves, resistance risk for a new fungicide of the same group can be high or low. A sudden loss in effectiveness is referred as 'qualitative' or 'single step' resistance and once developed it tends to be stable. This form of resistance is due to one point mutation that causes a single amino acid change in the protein targeted in its mode of action and is responsible for the resistance. As a result, the fungal population falls into only two classes: the resistant strains and the sensitive strains. Benzimidazoles, phenylamides, dicarboximides, and methoxyacrylate or quinone outside inhibitor fungicides (QoIs) share this kind of resistance (Brent and Hollomon, 1998, 2007; Ma and Michailides, 2005). In contrast, when a decline in disease control effectiveness and fungicide sensitivity of pathogen populations occur gradually, resistance that develops is referred as 'quantitative', 'multistep' or 'polygenic', and its expression is the combined result of different genes, each contributing partially to resistance. Polygenic resistance is typical of fungicides that inhibit demethylation in the synthesis of ergosterol (DMI), such as imazalil and propiconazole (Brent and Hollomon, 1998).

Basic strategies recommended to reduce the risk of resistance development include: a) minimize the use of single-site fungicides and to rotate their application with others of different modes of action; b) combine two or more active ingredients with different modes of action; c) avoid application rates lower than those recommended by the product manufacturer; and d) implement effective sanitation actions to minimize pathogen populations. Using the highest label rates is expected to minimize the selection of strains with intermediate fungicide sensitivity when resistance involves several genes (Brent and Hollomon, 1998, 2007). Unfortunately, the current worldwide legislative trend of reducing maximum levels of residues and demands by buyers and consumers, based on concerns about the presence of chemical residues on food, and to reduce the number of detectable active ingredients of pesticides to few or none, is in conflict with the resistance management recommendations. This complicates the implementation of strategies to minimize the development of fungicide resistance that would prolong fungicide effectiveness.

2.2 Approved Fungicides and Their Mode of Action

SOPP is one of the oldest fungicides still in use in many countries. It is a wide spectrum fungicide and bactericide that, once dissolved in water, generates undissociated *o*-phenylphenol (HOPP) and dissociated *o*-phenylphenate (OPP^-). HOPP is very active and able to penetrate the fruit tissue, but it is phytotoxic to many fruits at 200–400 ppm. OPP^- is less active but much less phytotoxic. After rinsing, the OPP^- anion coming in contact with acidic-wounded tissue changes to HOPP and becomes very active. To reduce the risk of phytotoxicity, the commercial formulation of SOPP contains hexamine and sodium hydroxide. Hexamine precipitates free *o*-phenylphenol before it reaches phytotoxic levels and buffers the solution at about pH 11.8. The use of SOPP has been in decline over recent years due to disposal and human safety concerns (Eckert and Eaks, 1989).

Benzimidazoles (TBZ, benomyl, carbendazim, thiophanate-methyl), introduced in the late 1960s, are potent inhibitors of β-tubulin polymerization, which prevents cell division in many species of fungi. They can affect all the development stages of the fungi, inhibiting spore germination, germ tube elongation, cellular multiplication, and mycelial growth (Eckert and Sommer, 1985). They are active against several fungi causing decay in many crops, including *Colletrotichum* spp., *Verticillium* spp., *Botryodiplodia* spp., and *Penicillium* spp., but not against other important postharvest fungi such as *Geotrichum candidum* Link, *Alternaria* spp., *Rhizopus stolonifer* (Ehrenb.)Vuill., *Mucor* spp., and *Phytophthora* spp. Benzimidazoles, being insoluble in water, are applied as water suspension or in mixture with wax. They are single-site mode of action fungicides (FRAC Code 1). Their success was due to their systemic properties, which enabled them to control both wound pathogens as well and quiescent infections such as those causing anthracnose, brown rot of stone and pome fruits (*Monilinia* spp.), or stem-end fungi of citrus fruits [*Lasiodiplodia theobromae* (Pat.) Griffon & Maubl. (syn.: *Diplodia natalensis* P. Evans) and *Phomopsis citri* H.S. Fawc.]. Despite the decreased activity of benzimidazoles due to the development of resistant fungal strains, TBZ, the only benzimidazole currently in use, is still one of the most popular fungicides for control of postharvest diseases worldwide. It is weakly soluble in water; 160 ppm at 20°C and pH 4; 25–30 ppm at 20°C within a pH range of 7–10. Its solubility increases in combination with lactic acid. The pKa value is 4.73; at lower pH it is prevalent as cation; at higher pH the anionic form prevails (Eckert and Eaks, 1989). TBZ is compatible with SOPP and chlorine as sanitizer agents and was reported to alleviate chilling injury in cold stored sensitive citrus species (Schirra et al., 2011; Smilanick, 2011).

Several ergosterol biosynthesis inhibitor fungicides (FRAC Code 3), which block the synthesis of ergosterol in higher fungi (Siegel and Ragsdale, 1978), have been used to control postharvest diseases of several horticultural crops. Imazalil

(IMZ) was the first fungicide of this family registered for postharvest purposes, and it is still the most popular postharvest fungicide worldwide (CCQC, 1980; Schirra et al., 2011; Smilanick, 2011). It is produced as a water emulsion or as a sulfate salt. IMZ was introduced in the USA in 1981 (Eckert et al., 1994). It is systemic and its curative, protective, and eradicant activities are superior to those of all other available fungicides. It also retards sporulation of pathogens on the surface of infected fruit, especially when in combination with wax, thus reducing the production of inoculum, and soiling of adjacent healthy fruits. The water solubility of IMZ is 184 ppm at 20°C and its pKa is 7.2. The undissociated form, whose ratio increases with pH, is more active than the dissociated one. At pH 7 IMZ, activity is about 50% higher than at pH 4. IMZ residues also increase as the pH increases (Smilanick et al., 2005; Erasmus et al., 2013). IMZ is very active against *Penicillium* spp., but also moderately inhibitory to *Alternaria* spp. and other important decay causing fungi such as *Colletotrichum, Lasiodiplodia, Phomopsis*, or *Botrytis*, although it does not inhibit decay by *Geotrichum citri-aurantii* (Ferraris) E.E. Butler (Brown and Miller, 1999; Barkai-Golan, 2001). Propiconazole, which like IMZ is an inhibitor of ergosterol biosynthesis, controls a wide range of fungi, including *G. citri-aurantii* and *G. candidum* that cause sour rot (McKay et al., 2012). Propiconazole is registered in the USA for stone fruit, citrus fruit, tomato, and pepper and in the European Union (EU) for citrus fruit (Adaskaveg and Förster, 2010). Tebuconazole and prochloraz are two other ergosterol biosynthesis inhibitors registered in the USA and few other countries. The first is effective against several important decays, including those caused by species of *Penicillium, Monilinia, Rhizopus*, and *Mucor*. The latter is similar to IMZ, but it is registered only in a few countries, so its use is generally limited to local trade (Barkai-Golan, 2001; Adaskaveg and Förster, 2010).

Guazatine, a multisite contact (FRAC Code M7) broad-spectrum fungicide, is very active against sour rot and Penicillium decay, but it is not able to control other important postharvest diseases (Eckert and Eaks, 1989; Barkai-Golan, 2001). Its use is still approved in some countries such as South Africa and Australia, but it is not currently registered in other important citrus producing countries such as those of the EU, the USA, and Canada. Therefore, its application is mainly limited to products destined for local markets since it cannot be used on fruits destined for export to many markets (Adaskaveg and Förster, 2010). It is compatible with sodium bicarbonate and a synergistic effect in efficacy was observed when the two compounds were combined (Horuz and Kmay, 2010). One of the main constraints regarding guazatine registration is due to its multiple derivate components when it is synthesized, which makes residue quantification difficult (Dreassi et al., 2007).

The hydroxyanilide fenhexamid, very active against brown rot and gray mold, is registered in some countries for stone fruit, pome fruit, and kiwifruit (Förster et al., 2007). This fungicide specifically inhibits the 3-ketoreductase involved in the C-4 demethylation during ergosterol biosynthesis (FRAC Code 17) (Adaskaveg and Förster, 2010). Although fenhexamid is not systemic, it prevents the fungal penetration into the plant tissue because it is a potent inhibitor of germ tube growth (Debieu et al., 2001). Unfortunately, resistant strains of *Botrytis cinerea* Pers. have been recently identified (Billard et al., 2012).

Fludioxonil, pyrimethanil, and azoxystrobin are three relatively new postharvest fungicides, each with a different mode of action, registered first in the USA at the beginning of this century and gradually in other countries for postharvest applications in a variety of crops (Adaskaveg and Förster, 2010; Schirra et al., 2011).

Fludioxonil, belonging to the phenylpyrrole chemical group (FRAC Code 12), is very active on a large number of fungi causing decay and registered for stone fruit, pome fruit, pomegranate, kiwifruit, pineapple, and citrus fruit to control gray and brown rots, Rhizopus rot, and Penicillium decay (Förster et al., 2007; Kanetis et al., 2007, 2008a, 2008b). Fludoxonil interferes with the osmoregulatory signal transduction pathway, causing dysfunction in glycerol synthesis (Rosslenbroich and Stuebler, 2000; Kanetis et al., 2008a, 2008b). As a contact fungicide, fludioxonil has good protective activity, inhibiting spore germination, germ tube elongation, and mycelial growth. In citrus fruits, it has also shown a good curative activity to control Penicillium decay when treatments were done within 12–24 hr after infection (D'Aquino et al., 2013b). Azoxystrobin is a synthetic derivate of a naturally occurring strobilurin; it belongs to the QoI class of fungicides (FRAC Code 11) and inhibits respiration by binding at the quinone 'outer' (Qo) binding site of the cytochrome bc1 complex within the mitochondrion. Azoxystrobin inhibits all four major groups of fungi, the Ascomycota, Deuteromycota, Basidiomycota, and Oomycota. In citrus fruits, it had good activity against Penicillium decay (Kanetis et al., 2007). Although azoxystrobin is 'locally' systemic, its curative activity is quite weak, while it is very effective as a protectant fungicide, acting as a potent inhibitor of spore germination and zoospore motility (Bartlett et al., 2002). The anilinopyriminedine pyrimethanil (FRAC Code 9) is very active against gray mold and Penicillium decay. It interferes with the biosynthesis of the amino acid methionine and is involved in the inhibition of various fungal hydrolytic enzymes (Heye et al., 1994). Pyrimethanil inhibits mycelial elongation and the secretion of hydrolytic enzymes, with only slight activity on spore germination. Because of its translaminar activity, it displays both curative and preventive properties (D'Aquino et al., 2006; Kanetis et al., 2007), although in citrus fruits it was slightly more effective in controlling *P. digitatum* when applied before inoculation (Smilanick et al., 2006).

3 Inorganic Acids and Salts

3.1 Carbonate and Bicarbonate Salts

Carbonic acid salts, such as carbonates and bicarbonates, are widely used in the food industry for leavening, pH-control, and their influence on taste and texture properties (Lindsay, 1985). They are classified as food additives and are allowed with no restrictions for many applications in the EU and the USA.

Carbonic acid is a diprotic acid from which two series of salts can be formed when carbon dioxide (CO_2) is dissolved in water, either the hydrogen carbonates (or bicarbonates, HCO_3^-) or carbonates (CO_3^{2-}). The solubility of carbon dioxide in water depends on temperature and pressure. An increase in temperature decreases the carbon dioxide solubility and the equilibrium moves to the right:

$$HCO_3^- + H_3O^+ \leftrightharpoons H_2CO_3 + H_2O \leftrightharpoons CO_2 + 2H_2O$$

In aqueous solution, bicarbonate anion, carbon dioxide, and carbonic acid exist together in a dynamic equilibrium. In strongly basic conditions, the carbonate ion predominates, while in weakly basic conditions the bicarbonate ion is prevalent.

Therefore, solutions of sodium carbonate are basic while those of sodium bicarbonate are weakly basic. The solubility of bicarbonates is generally higher than carbonates in water. However, the solubility of potassium and sodium carbonate, the most used carbonates, occurs rapidly to effective levels in warm water.

Among carbonic acids salts, sodium, potassium, and calcium are the main carbonates and bicarbonates used for controlling postharvest diseases as aqueous solutions. Sodium bicarbonate and sodium carbonate have been used in California since the early decades of the 20th century with many variations in concentrations, temperature, and contact time (Palou et al., 2008). Potassium and sodium bicarbonate have also been widely used to control powdery mildew and other foliar and fruit diseases of several species although, due to the numerous factors affecting their efficacy, results sometimes have been irregular (Deliopoulos et al., 2010).

The activity of bicarbonates and carbonates to inhibit fungi relies on the alkaline pH of the solutions, on the proportion of dissociated HCO_3^- and CO_3^{-2} anions, with the latter more toxic than the former, and the cationic moiety (Corral et al., 1988; Palmer et al., 1997; Smilanick et al., 2005). In vitro studies showed that the activity of carbonate salts is markedly influenced by the pH of the medium, the temperature, the salt concentration, the pathogen organ, and the susceptibility of the target microorganism, resulting in some cases with fungicidal activity and in other cases with fungistatic activity. Olivier et al. (1998) and Ricker and Punja (1991) found carbonate salts fungitoxic on mycelial growth of *Helminthosporium solani* Durieu & Mont and *Rhizoctonia carotae* Rader, respectively, while only fungistatic activity was observed on spore germination of *P. digitatum* (Smilanick et al., 1999) and *P. italicum* (Palou et al., 2002a). Since the growth of several fungi that acidify host tissue they infect and colonize, such as *Penicillium* spp., *B. cinerea, Sclerotinia sclerotiorum* (Lib.) de Bary, and *Fusarium* spp., were inhibited by alkaline pH (Prusky and Yakoby, 2003), one mode of action of carbonates and bicarbonates could be due to the high pH of their solutions. Sodium carbonate strongly inhibited mycelial growth and spore germination of *B. cinerea* (Palmer et al., 1997) and *Fusarium sambucinum* Fuckel in a medium at pH 11, whereas it displayed almost no inhibition at pH 6. However, as growth of *F. sambucinum* is not affected by the pH in the range 5–12, it was supposed that the pH could not be the only factor affecting the treatment toxicity and that the pH of the medium could mediate the toxicity of the salts. In fact, the proportion of CO_3^{2-}, which inhibits spore germination, increases with the pH and is very high at pH 11.4, while at pH 8.2 and 6.5 the concentration of CO_3^{2-} is very low and the HCO_3^- anion predominates, which is less toxic (Punja and Grogan, 1982; Punja and Gaye, 1993; Mecteau et al., 2002). The toxicity of carbonate and bicarbonate salts in vitro is not always an indicator of their effectiveness to control disease when used on produce because the interaction between the salts and the host tissue may significantly affect the growth of the pathogen. For example, in in vitro studies, sodium carbonate was more toxic than sodium bicarbonate to conidia of *P. digitatum*, but they were equally effective in controlling green mold in oranges immersed for 2 min at 40–45°C in 3% solutions of sodium carbonate or 2% solutions of sodium bicarbonate (Smilanick et al., 1999). The acidic pH of the tissue tends to lower the pH of the solution; in contrast the salt tends to increase the pH of the wounded tissue. Consequently, conidial germination and germ tube growth, which are greater at lower pH, will depend on the final pH of the infection site, which in turn is mediated by the alkaline pH and concentration of the salts and the acidic pH of the tissue, which tends to reduce the active anions CO^{3-} and HCO^{3-} (Smilanick et al., 1999). Smilanick et al. (2005) found that the pH within 2-mm-deep wounds in lemons increased from 5.3 to 6.7 after fruit immersion in 3% sodium bicarbonate. Similarly, the pH of the albedo of different citrus species when treated with 5% sodium carbonate increased from 5.3–6.2 to 8.9–10.1 and showed no significant change during the next 6 d (Venditti et al., 2005).

Carbonate salts may also stimulate the host natural defense mechanisms. When citrus fruits were wounded and treated with 5% sodium carbonate, scoparone content in the albedo of wounded fruits increased markedly during the next 15 d following treatment (Venditti et al., 2005). The mechanical strength, the structure of the tissue, and the changes induced by carbonates may also affect host resistance to pathogens. Venditti et al. (2005) employed scanning electronic microscopy to examine the structural changes induced by sodium bicarbonate. Mandarin fruits, with a thin irregular layer of albedo, showed fewer pronounced structural changes than oranges, lemons, or grapefruits, and they attributed this in part to the superior control of decay by bicarbonate on treated lemons, oranges, and grapefruits than that on mandarins. A lower effectiveness in controlling decay in mandarins by bi- and carbonate salts with respect to other citrus fruits was also reported by Palou et al. (2002a). Increasing the concentration of salts and solution temperature, although still within levels that did not harm the fruit, generally increased the treatment efficacy (Smilanick et al., 1997; Palou et al., 2001, 2002b).

3.2 Silicate Salts

Silicon (Si) is the second most important element in earth, after oxygen, accounting for about 28% by mass of Earth's crust. In nature, Si is mainly found as silicon dioxide, otherwise known as silica, or as silicates, in combination with oxygen and metals. In the soil, Si concentration is significantly affected by the nature of the different forms of Si of the solid phase and is represented by orthosilicic acid (H_4SiO_4), as well as polymerized or complexed silicic acid (Epstein, 1999). Orthosilicic acid is stable in water at room temperature as long as its concentration remains below 100 ppm, but above this concentration individual orthosilicic acid molecules condensate to form a range of small oligomers that eventually evolve in stable and complex aggregates (Belton et al., 2012). Si polymerization to form a silica gel is also favored by pH value below 10 (Nordstrom et al., 2011). Although considering Si as a nutrient is still a topic of discussion, it is universally recognized the very important role played by Si in plant physiology, being effective in alleviating heavy metal toxicity, improving nutrient imbalances, and increasing salt tolerance and cell wall lignification (Fateux et al., 2005; Coskun et al., 2016).

The capability of Si to prevent plant disease was first demonstrated by Onodera (1917) at the beginning of the 20th century: He found a higher susceptibility to rust (*Pyricularia oryzae* Cavara) in rice plants with lower levels of Si. Studies conducted in greenhouses with added silicates in irrigation or hydroponic water definitely confirmed the evidence of a close relationship between Si levels in plant tissue and disease severity (Samuels et al., 1991; Deliopoulos et al., 2010; Rodrigues et al., 2015). Si supply in plant, either as soil amendment or fertilization, has shown to effectively control various diseases including powdery mildew (Miyaki and Takahashi, 1983); Pythium crown and root rot, and Fusarium wilt in cucumbers and other cucurbits; Fusarium crown root rot and wilting in tomatoes, gummy stem light in watermelons, Fusarium wilt in lettuce and Phytophthora blight in bell peppers (Datnoff and Rodrigues, 2015).

The main silicates used in agriculture, as silicate salts, are potassium (K_2SiO_3) and sodium (Na_2SiO_3) silicates. They are dense, soluble in water, and highly viscous. The pH of potassium and sodium silicate solutions depends on concentration, but they have always an alkaline reaction with pH values ranging from approximately 10 to 13 (Liang. et al., 2015). Soluble Si sprayed onto foliage is not absorbed by foliage but polymerized on leaves forming a barrier to pathogens penetration. This physical barrier seems less efficient against pathogens than protection offered by Si taken up by roots, which, once absorbed from the soil solution, follows the transpiration stream and polymerizes, when exceeds 100 ppm, in cell walls, cell lumen, intercellular spaces, and in the subcuticular layer forming an 'endogenous' physical barrier (Sangster et al., 2001). The role of Si in plant disease suppression is associated with different modes of action, including increased mechanical strength, direct activity against pathogens, and regulation of plant defense mechanisms, while the high pH of potassium and sodium silicates could directly inhibit the growth of several pathogens (Liang. et al., 2015). Sodium metasilicate at 3 g/L completely inhibited spore germination and significantly reduced mycelial growth of *Monilinia fructicola* (G. Winter) Honey in in vitro studies, whereas field treatments with 6 g/L completely suppressed brown rot disease on peaches at harvest and reduced decay incidence in fruits stored for 2 or 4 d at 20°C after harvest (Pivotto Pavanello et al., 2016). A persistent protective effect after preharvest treatment with Si was also reported for tomatoes. After soil fertilization with sodium silicate (100 mg/L) at flowering or fruits growth, less anthracnose developed and higher percentage of appressoria of *Colletotrichum gloeosporioides* (Penz.) Penz. & Sacc. were present among fruits inoculated after harvest. The number of appressoria remained high on fruit surface when the penetration was hindered and infection was unsuccessful. They proposed infection was unsuccessful because Si reinforced the physical structure of the peel in fruits of treated plants (Weerahewa and David, 2015).

The activity of Si varies with its chemical form. In in vitro tests, sodium silicate added to potato dextrose agar (PDA) suppressed the radial growth of *Trichothecium roseum* (Pers.) Link, whereas silicon oxide was ineffective. However, when melons were dipped in Si solutions, both silicon oxide and sodium silicate reduced the severity of pink rot caused by *T. roseum* with lesion diameters up to fivefold smaller than in control fruits (Guo et al., 2007). The enhanced peroxidase and phenylalanine ammonia-lyase (PAL) activity observed in sodium silicate-treated melons with respect to those treated with silicon oxide suggested that the mode of action of sodium silicate and silicon oxide might involve different mechanisms. The capacity of sodium silicate to control decay was lower in fruits inoculated before the treatment with respect to those inoculated 24 hr after the treatment (Guo et al., 2007). Similar results were reported by Bi et al. (2006) in melons inoculated with conidia of *Alternaria alternata* (Fr.) Keissl., *Fusarium semitectum* Berk. & Ravenel, and *T. roseum*. In melons, sodium silicate was more effective at concentrations of between 100 and 200 mg/L, although 200 mg/L was phytotoxic to some cultivars (Bi et al., 2006; Guo et al., 2007). Postharvest dips of papaya fruits in 2500 or 5000 mg/L of sodium silicate for 20 min followed by inoculation with *C. gloeosporioides* effectively controlled anthracnose decay and increased papaya shelf life up to 4–5 d (Bandara et al., 2015).

Potassium silicate is preferred in agricultural applications due to the limited solubility of sodium silicate in water and the tendency of its commercial formulations to polymerize. Tarabih et al. (2014) evaluated the effect of a 5-min dip treatment in 0.1, 0.2, or 0.3% potassium silicate solutions on the physiology and incidence of decay caused by *Penicillium expansum* L. in 'Anna' apples stored at 0° C for 60 d plus 6 d at 28°C to simulate marketing conditions and found that treatments were effective in maintaining postharvest fruit quality and controlling decay. The best results were achieved in fruits treated with 0.3% potassium silicate. These fruits lost less weight (approximately one-third less than untreated fruits) and developed less decay (9%) than the control fruits (26%). Potassium silicate applied at 90 mM had both significant protective and preventive activity in the control of *P. digitatum* and *P. italicum* in citrus fruits (Moscoso-Ramírez and Palou, 2014). As an alternative to synthetic fungicides, potassium silicate offers interesting potential to be applied commercially to control postharvest diseases of several crops. Moreover, in the USA it is exempted from requirement of tolerance for residues on all food commodities when used as fungicide, insecticide, or miticide at concentrations not exceeding 1% (US EPA, 2006).

The low toxicity shown by Si toward different biological control agents such as *Cryptococcus laurentii* and *Rhodotorula glutinis* (Tian et al., 2005), and *Candida membranifaciens* (Farahani et al., 2012) has stimulated the interest of scientists to develop integrated treatments combining biological control agents with silicates. Results of several studies showed synergistic effects and increased disease control (Qin and Tian, 2005; Tian et al., 2005; Farahani et al., 2012).

3.3 Phosphite Salts

Phosphonates were tested as an alternative to phosphate fertilizers in the 1930s, although their effectiveness was not comparable with other sources of phosphorous (P) and even delayed plant growth. This happened when supplies of rock phosphate to farmers were threatened by the Second World War. Later, they were introduced as fungicides in the mid-1970s (Guest and Grant, 1991). The term phosphonate describes the salts of phosphonic acid [$(OH)_2HPO$] (Phi) or the tautomer phosphorous acid [$(OH)_3P$], a reduced form of phosphoric acid [$(OH)_3PO$] (Pi). However, in water solutions almost all phosphorous acid changes into phosphonic acid, which with bases forms phosphonates. The term phosphonates is also used when the phosphonic anion combines with alcohols or other organic compounds with a C-P bond (Guest and Grant, 1991). Phi, therefore, differs from Pi only because a hydrogen atom has replaced an oxygen atom. However, because P is a component of key molecules such as nucleic acids, phospholipids, ATP, and enzymes regulating numerous metabolic pathways (Schachtman et al., 1998), this small difference changes the shape and the charge distribution of the molecule, with a marked effect on the biological behavior of the two acids. Most enzymes catalyzing uptake and transportation of Pi, active both in rot and in leaf tissues, do not discern between Pi and Phi. Consequently, mobility of Phi both in xylem and phloem relays on the same transporters, but being Phi more soluble than Pi, it is more rapidly absorbed and translocated than Pi. In contrast, enzymes that catalyze the transfer of phosphate groups can discriminate between phosphate and phosphonate (Guest and Grant, 1991). Thus, phosphonates do not participate in the same biochemical cycles as phosphates and cannot be metabolized by the plant. For this reason, Phi remains stable for long periods in plant tissues and its effects are long lasting. Nevertheless, Phi, by mimicking Pi in various signaling pathways, blocks important metabolic functions that make it highly phytotoxic. Toxicity by Phi is particularly evident in Pi-starved plants, where it suppresses their response to P deficiencies (Carswell et al., 1996, 1997; Förster et al., 1998; Ticconi et al., 2001; Varadarajan et al., 2002). Despite numerous studies that claim Phi as a source of P for plants, Phi is not a plant nutrient (Tao and Yamakawa, 2009).

Phi is able to effectively control many plant diseases caused by species of pseudofungi belonging to the order Oomycetes, particularly *Phytophthora* spp. (Guest and Grant, 1991). The fungicide aluminum tris-*O*-ethyl phosphonate (fosetyl aluminum or fosetyl-Al, Aliette) is an alkyl phosphonate released in 1977. This compound very effectively controls diseases caused by oomycetes, particularly downy mildew of grape caused by *Plasmopara viticola* (Berk. & M.A. Curtis) Berl. & De Toni. In water, fosetyl-Al hydrolyzes to phosphonic acid and ethanol and subsequently the phosphonic acid ionizes to the active constituent, the phosphonate anion. The aluminum component of fosetyl-Al not absorbed by the plant remains on the plant surface and provides some protectant activity (Guest and Grant, 1991). Phi can affect directly the fungal pathogen and/or act indirectly through stimulation of plant defense response against pathogens. When used directly against fungi, Phi is much less effective than when applied to the host. It is likely that in the fungus Phi acts as in plants by competing with phosphate, thereby causing P starvation (Smillie et al., 1989; Tao and Yamakawa, 2009). The capacity of potassium phosphite to inhibit conidial germination of *P. digitatum* decreased when the pH and the phosphate content of the medium were increased (Cerioni et al., 2013). The most important mode of action of Phi is due to its capacity to induce a series of changes in plants, such as hypersensitive cell death, activated defense-related biosynthetic pathways, or accumulation of higher levels of phytoalexins. Activities of enzymes involved in defense, including PAL, and the synthesis of phenolic compounds and lignin deposition following inoculation with pathogens are higher in plants treated with Phi (Smillie et al., 1989; Tao and Yamakawa, 2009; Gómez-Merino and Trejo-Téllez, 2015).

Despite the numerous studies regarding field application, relatively few studies have been conducted on the use of Phi to control postharvest diseases. Phi induced resistance in oranges inoculated with *P. digitatum* (Forbes-Smith et al., 1998) and reduced decay due to *P. digitatum* in wound inoculated apples when applied at 1 mg/L (Wild et al., 1998). In in vitro studies, mycelial growth and conidial germination of *P. expansum* were completely inhibited by Phi at 2 or 4 mg/L, respectively, while blue mold on 'Elstar' apples wounded and inoculated with a TBZ-resistant isolate of *P. expansum* was reduced by a curative treatment with Phi at 2 mg/mL (Amiri and Bompeix, 2011). Phi was also effective in reducing decay incidence in noninoculated apples stored for 6 mon at 2°C (Amiri and Bompeix, 2011). Cerioni et al. (2013) tested heated solutions (46°C) of potassium phosphite and calcium phosphite at 15 g/L as a drench treatment lasting 15 s to control green mold in 'Atwood' oranges, 'Eureka' lemons, and 'W. Murcott' mandarins inoculated with *P. digitatum* 24 hr before treatment. Both phosphites reduced decay after 3 wk storage at 20°C, with calcium phosphite being more effective than potassium phosphite. However, in citrus fruits, Phi was more effective in controlling brown rot than Penicillium decay: applied at 15 g/L after 18 hr from inoculation with *Phytophthora citrophthora* (R.E. Sm. & E.H. Sm.) Leonian, potassium phosphite reduced decay by 96% (Adaskaveg et al., 2015). In the USA, potassium phosphite was registered for postharvest use to control brown rot of citrus fruits and is exempt from residue tolerance in the USA as well as in many other countries (Adaskaveg et al., 2015).

3.4 Calcium Salts

Calcium (Ca) is a key plant nutrient that has a significant role in cell functions, including reducing softening and senescence of fruits (Conway and Sams, 1984; Conway et al., 1991). Ca, as Ca-polygalacturonates, increases stability of the cell wall by contributing to the structure of the middle lamella. It also plays a very important role at the cellular level as a secondary messenger by binding to a wide range of proteins, which as a result change their conformation and catalytic activities (White and Broadley, 2003). Although Ca scarcity in soil is rare, many deficiency symptoms may occur in young expanding leaves, inner tissue of leafy vegetables, and tissues such as bulky organs. This occurs because Ca cannot be mobilized from older tissues and redistributed via phloem. Because Ca is supplied via xylem its distribution is dependent on transpiration, which can be quite low in young leaves and fruits, localized calcium deficiencies occur in these organs and are frequently associated with cell death (White and Broadley, 2003).

Foliar applications of Ca controlled various fungi causing storage rots when applied either as organic or inorganic salts (Biggs, 1999). Preharvest application of calcium chloride 90 and 30 d before harvest reduced preharvest and postharvest gray mold incidence in table grapes from 23.4 to 9.5% and from 63.8 to 22.5%, respectively (Nigro et al., 2006), and was effective even when the grapes were picked at a late harvest date (Chervin et al., 2009). High Ca content in tissues is generally associated with greater resistance to pathogens causing preharvest as well as postharvest decay. However, while repeated sprays over the growing season lead to significant higher levels of Ca, postharvest treatments are less effective and according to species treated, Ca treatment may not increase Ca in tissue but rather cause phytotoxicity (Eaks, 1985).

Ca salts may inhibit pathogens in different ways. Firstly, by direct effects on the growth and development of the fungus, and, secondly, by indirectly enhancing host defense mechanisms by delaying ripening processes and senescence (Eckert and Eaks, 1989; Biggs, 1999; Wood et al., 2013). High external cellular concentrations of Ca^{2+} may increase the Ca^{2+} concentration in the cytosol of the pathogen to toxic levels. Since maintenance of low basal concentrations of internal Ca^{2+} is essential for normal cell functions, organisms unable to regulate intracellular Ca^{2+} may exhibit compromised growth and development, while no toxic effect would occur in those that can regulate intracellular Ca^{2+} (Biggs et al., 1997). This could be the reason why some fungal species as well as isolates of the same species can display different susceptibility to Ca treatments (Biggs et al., 1997). In apples, uptake of Ca solutions primarily occurs through lenticels and calyx openings, but postharvest dips in Ca solutions generally do not lead to significant increase in tissues, although environmental conditions, maturity stage, or postharvest treatments that cause cracks in the fruit cuticle or epidermal cells, can significantly increase

Ca uptake after pressure infiltration or a simple dip (Scott and Wills, 1979, Sebti and Tantaoui-Elaraki, 1994; Saftner et al., 1997a, 1997b). A 2-min immersion into 0, 2, 4, 6, or 8% calcium chloride solutions did not alter Ca concentration in apple tissue, while pressure infiltration at 103 kPa improved Ca uptake and effectively reduced decay caused by *P. expansum*. Furthermore, Ca uptake into apple fruits dipped into calcium chloride solutions increased with maturity (Conway, 1982).

Ca directly affected the growth and enzymatic activity of several pathogens. Although the bulk of studies on Ca have been done with calcium chloride, other calcium salts including calcium oxide, calcium hydroxide, calcium sulfate, and calcium nitrate were shown to be equally or more effective than calcium chloride (Biggs et al., 1997; Brackman et al., 2001). In in vitro studies, polygalacturonase activity of *M. fructicola* was reduced in descending order by calcium propionate, calcium sulfate, tribasic calcium phosphate, calcium gluconate, and calcium succinate, while dibasic calcium phosphate and calcium tartrate had no inhibitory activity and the solution pH had no effect (Biggs et al., 1997). In the same study, the incidence of brown rot among fruits inoculated by spraying the inoculum or by placing it in wounds was effectively controlled by all of these salts. Although in most cases inhibition of pathogen was attributed only to the Ca cation, the anionic moiety of the salts can by itself inhibit pathogens. For example, in the case of calcium chloride, it is likely that the Cl^- anion may suppress fungal diseases, as was shown for other Cl inorganic salts (KCl, NaCl, $AlCl_3$, $MnCl_2$, and NH_4Cl), regardless the cationic component (Deliopoulos et al., 2010). Calcium chloride, calcium propionate, and calcium silicate had no effect on the germination of conidia of *Colletotrichum acutatum* J.H. Simmonds and *C. gloeosporioides*, casual agents of bitter rot in apple. However, the first two salts reduced germ tube growth by 41 and 50%, respectively, compared with the control, while calcium silicate did not reduce it. When the calcium salt solutions were applied to wounded apples prior to inoculation, fruits treated with calcium chloride and calcium propionate had 30% smaller lesions and formation of acervuli was delayed compared to those treated with calcium silicate or the control, which were similar. Calcium chloride at a concentration of 0.75–1.5% was also very effective in reducing the incidence of green mold, caused by *P. digitatum*, in large-scale tests with citrus fruits (Droby et al., 1997). The efficacy of Ca salts varied among pathogenic fungi and among isolates of the same species. For example, in contrast to the results of Biggs et al. (1997), calcium chloride inhibited spore germination of *C. gloeosporioides* in vitro but did not effectively control anthracnose disease in papayas when applied as a dip treatment at 2.5% (Al Eryani-Raqeeb et al., 2009). In another study, calcium propionate reduced the in vitro growth of *Allantophomopsis cytisporea* (Fr.) Petr., *Allantophomopsis lycopodina* (Höhn.) Carris, *Coleophoma empetri* (Rostr.) Petr., *Fusicoccum putrefaciens* Shear, and *Physalospora vaccinii* (Shear) Arx & E. Müll., while calcium chloride and calcium nitrate inhibited only the growth of *C. empetri* and *F. putrefaciens*, and stimulated the growth of *P. vaccinii* (Blodgett et al., 2002). Both field sprays as well as postharvest Ca treatments have in general induced resistance to decay in many fruits. However, there have been cases when Ca did not reduce decay, as in strawberry (*B. cinerea*) and mango (*Dothiorella dominicana* Petr. & Cif.) or, as in guava, where Ca stimulated the growth of the fungus *C. gloeosporioides* (Johnson, 1977; Ellis et al., 1996).

Most Ca salts that inhibited pathogenic fungi do not inhibit yeasts, making it feasible to combine applications of Ca salts with biological control agents. When calcium chloride was applied in combination with *Pichia guillermondii* (Droby et al., 1997), *Candida guilliermondii* and *Pichia membranefaciens* (Tian et al., 2002), or *Rhodosporidium paludigenum* (Wang et al., 2010), better control of *P. digitatum* in citrus, *R. stolonifer* in nectarines, and *A. alternata* in cherry tomatoes was observed.

3.5 Other Compounds

Some inorganic compounds have shown an ability to control a wide range of postharvest pathogens. These include sodium and ammonium molybdate (Nunes et al., 2002; Palou et al., 2002b), aluminum chloride, aluminum sulfate, and other aluminum salts (Mecteau et al., 2002, 2008; Kolaei et al., 2013), potassium iodide and ammonium acetate (Elsherbiny and El-Khateeb, 2012), zinc oxide (He et al., 2011), and calcium polysulfide (Smilanick and Sorenson, 2001).

4 Organic Acids and Salts

4.1 Sorbic Acid and Salts

The first to discover the antimicrobial activity of sorbic acid (SA) (EU food additive number E-200) and potassium sorbate (PS) (E-202) were Miller and Gooding in the late 1930s (Stopforth et al., 2005). They are approved worldwide as food additives. In the EU, sorbic acid and potassium sorbate at concentrations up to 2000 mg/kg are added as preservatives to a large number of commonly consumed foods (EPC, 1995), and are also listed in the EU Register of feed additives for all animal species without restrictions (EFSA, 2014).

Sorbic acid crystallizes in needles or plates with a weak characteristic odor and a slightly acidic taste. SA solubility in water is 0.16 g/100 mL at 20°C and increases to 3.9 g/100 mL at 50°C. In contrast to SA, PS is more water soluble, with a solubility of 58.20 g/100 mL at 20°C and 61 g/100 mL at 50°C. SA is a weak acid, with a dissociation constant of 1.73×10^{-5} at 25°C and a pKa of 4.76. Within the pH range of 3.70–4.75, the percentage of undissociated acid decreases from 93 to 50%, but as the pH increases beyond the pKa, the undissociated form declines rapidly, with values of only 7% at pH 5.8 and 0.6% at pH 7 (Dharmadhikari, 1992). Because the undissociated molecule is the active form of SA, the activity of both SA and PS decreases as the pH increases, thus they should be applied at low pH. Nevertheless, although the undissociated acid is 10–600 times more effective than the dissociated form, the latter accounts for more than 50% of the growth inhibition at pH levels above 6 for most microorganisms (Eklund, 1983). SA is active against the microorganism cell wall and membranes, hindering absorption of amino acids and increasing proton flow into the cell (Bell et al., 1959; Eklund, 1980). Because in most cases postharvest studies were conducted with PS without controlling the pH of the solutions they applied, it is likely that inconsistent results could in part be due to variations in solution pH.

The first studies to evaluate the potential of PS to control decay of citrus fruits were conducted by Smoot and McCornack (1978). In in vitro tests, PS at 1000 mg/L with the medium pH adjusted at 4.5 completely inhibited the mycelial growth of *Alternaria citri* Ellis and Pierce, *C. gloeosporioides, L. theobromae, P. citri, G. citri-aurantii*, and strains resistant to benzimidazoles of *P. digitatum* and *P. italicum*. In in vivo tests with various citrus fruits, PS was as effective as SOPP or even more effective, but generally inferior to TBZ or benomyl, with the exception of a test where fruit decay was mainly caused by a population of *P. digitatum* resistant to benzimidazoles. However, results by Wild (1987) showed that PS was less effective than SOPP to control decay by *P. digitatum*, and that PS activity could be improved when the dip solution was heated to 47°C. In contrast, Hall (1988) found that a 2-min dip of 'Valencia' oranges in a PS solution at 2000 mg/L at pH 7.0–7.5 were as effective as TBZ in fruits inoculated with a *P. digitatum* strain sensitive to benzimidazoles. PS and sodium sorbate inhibited the in vitro growth of *H. solani* (Oliver et al., 1999), the causal agent of potato silver scurf in in vitro studies and reduced disease severity when applied 2 or 4 d after artificial inoculation (Hervieux et al., 2002). In cherries dipped for 1 min in a 2% PS solution, decay incidence was significantly reduced after a 30 d storage period at 0°C plus 4 d of shelf life at room temperature (Karabulut et al., 2001). PS treatment controlled green mold and sour rot in citrus fruits even when the solution concentration was reduced to 1000 mg/L and the dip time to 30 s, with better results with a solution heated to 50°C (Smilanick et al., 2008). With the objective of reducing immersion times, Montesinos-Herrero et al. (2009) tested several combinations of treatment time-temperature combinations from 20 to 68°C. They found that applications at 62°C for 30 or 60 s in 3% PS were the most effective treatments on several citrus species and cultivars artificially inoculated with *P. digitatum* or *P. italicum* 24 hr before the treatment and held at 20°C for 7 d or stored at 5°C for 60 d. However, the reduction of molds was markedly affected by the species and cultivar, ranging from 20% on 'Clemenules' and 'Nadorcott' mandarins to 95% on 'Valencia' oranges. The treatment effectiveness was further improved when a 60 s dip treatment in 3% PS solution heated at 60°C was followed by a curing treatment of 24–48 hr at 20 or 33°C either in ambient atmosphere or in an atmosphere of 15 kPa CO_2 or 30 kPa O_2 (Montesinos-Herrero and Palou, 2016).

As for other food additives or GRAS compounds, the potent activity of SA and salts observed in in vitro studies has given inconsistent and modest results in in vivo tests. For example, while 1000 or 2000 mg/L PS added to media reduced the mycelial growth of two strains (one sensitive and one resistant to TBZ) of *P. digitatum* by 50–60%, 1 min dip of lemons artificially inoculated with the same strains into 1% PS solution caused only a slight reduction of decay. Moreover, the performance of PS was only marginally improved when the solution temperature was raised to 53°C. The low efficacy of PS was attributed to the rapid degradation of PS and to its uneven distribution on the fruit surface (D'Aquino et al., 2013b). The low persistence of PS and the irregular spatial distribution of the salt on fruit surface were confirmed in another study with apples where the capacity of *P. expansum* to degrade the salt was also demonstrated (Fadda et al., 2015). In contrast, in other studies, field applications of PS in table grapevines were effective in reducing natural gray mold incidence at harvest and even during the postharvest phase, although fluctuations in the results occurred from year to year (Feliziani and Romanazzi, 2013a).

The status of food additive makes PS an eligible ingredient to incorporate into waxes and edible coatings for fresh fruits and vegetables. On citrus fruits, the combination with different waxes containing PS at a concentration of 2–5% moderately reduced green mold incidence, but in all cases they increased weight loss, by up to as much as 65%, compared to wax alone (Parra et al., 2014). In contrast, edible coatings formulated with PS effectively reduced green and blue molds on oranges and mandarins with no adverse effects on fruit quality (Valencia-Chamorro et al., 2009; Palou et al., 2015). Likewise, when pistachios were coated with a carboxymethyl cellulose based edible film containing PS at 1, 0.5, or 0.25 g/100 mL, no mold development was observed at all concentrations (Sayanjali et al., 2011). A reduction in decay incidence, a delay in postharvest ripening, a decrease in water loss, and better preservation of vitamin C and titratable acidity were observed on mangoes coated with a combination of bentonite and PS (Liu et al., 2014).

4.2 Short Chain Organic Acids and Salts

Acetic acid (AA), a stable compound manufactured in large amounts for numerous industrial processes (Lodal, 1993), could offer promise in the vapor phase to control postharvest pathogens. AA is a weak monoprotic acid. In aqueous solution, AA exists in a pH-dependent equilibrium between the undissociated and dissociated state with a pKa value of 4.76. A 1.0 M solution has a pH of 2.4 with about 0.4% of it as dissociated molecules (Brul and Coote, 1999). As a food additive, AA is approved in many countries and finds wide use in the food industry to adjust the acidity, flavor food, and, thanks to its antimicrobial activity, hinder the growth of pathogenic microorganisms (Levine and Fellers, 1940). In fact, it is an effective antiseptic when used as a 1% solution, with broad spectrum of activity against molds and bacteria such as streptococci, staphylococci, pseudomonads, enterococci, and others (Busta and Foegeding, 1983). As is true for all weak organic acids, AA is more active as preservative at low pH, since the undissociated form is able to cross the plasma membrane. The higher cytoplasmic pH within the cell favors its dissociation in the cationic and anionic components, which are not able to pass back through the plasma membrane, thus resulting in their accumulation and toxicity in the cytosol (Brul and Coote, 1999). The mode of action attributed to AA and other weak acids includes membrane disruption, inhibition of essential metabolic reactions, harmful changes in the cytosolic pH homeostasis, and accumulation of toxic anions (Brul and Coote, 1999).

Like other short chain organic compounds, AA can be used as a fumigant, owing to its high volatility at ambient temperature. This mode of application offers the advantage of minimal fruit handling and absence of wetting (Spadoni et al., 2015; Mari et al., 2016), which is particularly valuable for fruits that do not tolerate wetting well, such as strawberries or peaches. The first studies on the use of AA as fumigant on peach fruit reported germination of conidia of *M. fructicola* was completely inhibited, but the treatment blackened the fruit within a few minutes (Roberts and Dunegan, 1932). Later, Stadelbacher and Prasad (1974) found that conidia of *P. expansum* fumigated with 4.0 mg/L (0.15%) AA vapor for 60 min lost the capacity to infect apple fruit and that the concentration of acetaldehyde necessary to achieve the same result was 2%, about 13 times higher than that of AA. Wilson et al. (1987) reported significant inhibition of *M. fructicola* and *B. cinerea* by AA. In vitro studies conducted by Sholberg and Gaunce (1995) confirmed the inhibitory activity of low concentrations of AA vapor (2.7–10.8 mg/L) on conidial germination of *P. expansum* and demonstrated that the same treatment conditions even more effectively inhibited conidia of *B. cinerea*. When apples, pears, kiwifruits, table grapes, and tomatoes were inoculated with *P. expansum* or *B. cinerea* and fumigated for 1 hr with AA vapors at up to 4 mg/L, decay was completely prevented. AA vapors were also effective in inhibiting decay on navel oranges inoculated with *P. italicum*. Results showed that AA vapor was equally effective at a treatment temperature of 5 or 20°C and that treatment effectiveness increased when the relative humidity (RH) was raised from 17 to 98% (Sholberg and Gaunce, 1995). AA fumigation at 1.4 or 2.7 mg/L for 60 min destroyed fungal spores of *M. fructicola* and *R. stolonifer* present on peaches, nectarines, apricots, and cherries and prevented decay. However, fumigation with 2.7 mg/L AA caused slight injuries, which became severe with higher AA concentrations (Sholberg and Gaunce, 1996). This treatment was also effective in reducing decay on 'Lambert' cherries, primarily caused by *Alternaria* spp., but small pits developed on the peel of the treated fruits during storage at 1°C (Sholberg and Gaunce, 1996). In a comparative trial, fruits of eight cherry, 14 apple, and three citrus cultivars were fumigated with vapors of AA at 1.9 or 2.5 μL/L, formic acid at 1.2 μL/L, or propionic acid at 2.5 μL/L. Although the treatment duration was reduced to 30 min, all three fumigants completely inhibited decay caused by *M. fructicola, P. expansum*, and *R. stolonifer* on cherries, and significantly reduced blue mold caused by *P. expansum* on pome fruits and green mold caused by *P. digitatum* on oranges. In this study, it was observed that while AA and propionic acid did not cause any phytotoxicity, formic acid induced pitting on six of the eight cultivars of cherry and increased blackening of the stem, compared to the control, in all eight cultivars, while in citrus fruits browning of the peel occurred (Sholberg, 1998). However, in another experiment, 20 min fumigations with AA at 5 mg/L of plums and 8 mg/L of apricots, both previously inoculated with *M. fructicola*, modestly controlled decay (Liu et al., 2002). Sholberg et al. (2004) conducted a semicommercial test with repeated treatments of AA vapor on 'd'Anjou' pears inoculated with *B. cinerea* and fumigated within a 27 m^3 room at 2°C. The exposure time varied from 1 to 4 hr and the amount of liquid AA applied during fumigation within the room, between 48 and 100 mL, produced a vapor concentration of AA acid of 112–626 μL/L. AA vapors almost completely eliminated both stem and fruit surface microflora when assessed soon after the treatments had been applied and halved decay incidence in fruits stored for 4 mon. Two consecutive fumigations of AA vapor at 200 μL/L×h to naturally contaminated fruits decreased stem infections by 43% without damage to the pears. However, to increase the efficiency of the treatments and to reduce the risk of phytotoxicity, the authors recommended fumigation of the fruits as soon as possible after harvest and not to exceed 200 μL/L×h at a treatment temperature of 1°C. Likely, besides temperature, the risk of injury due to toxicity of AA vapor is also strongly dependent on treatment duration. Venditti et al. (2009) fumigated 'Fremont' and 'Fairchild' mandarins for 15 min that had been inoculated with *P. digitatum* with AA at 0, 5, 15, 25, 50, 75, or 100 μL/L after a curing treatment at 36°C and 95% RH for 36 hr. All treatments reduced decay, but the combined treatments were superior to the individual ones, with the best control on 'Fremont' fruits observed after an application of 75 μL/L, and on 'Fairchild' mandarins after an application of 50 μL/L. In assessment of AA concentrations that were markedly higher than those normally used, only treatments of 100 μL/L caused rind injuries. Similarly, in 'Taloppo' table grapes stored at 5°C for 8 wk plus 3 d at 20°C to simulate marketing conditions, a prestorage fumigation with 50 μL/L for 15 min followed by two other identical fumigations after 4 and 8 wk storage, or a prestorage fumigation with 30 μL/L followed by five other identical fumigations during storage at 2 wk intervals, reduced gray mold incidence by 63.6 and 57.1%, respectively. Fruit weight loss was significantly reduced by all treatments, while fruit quality was not affected by any of the treatments (Venditti et al., 2017). Fumigation with AA and other volatile compounds was also effective in controlling black spot, silver scarf, and soft rot in potatoes (Wood et al., 2013).

From the commercial point of view the use of vinegar could be a very interesting alternative to AA, considering its countless uses in food industries and consumer prepared foods, its low cost, worldwide popularity, and the absence of legal restriction to its use. Vinegar was familiar in ancient Egypt and in China, where rice vinegar was first produced about 3000 years ago. In Greece and ancient Rome, where it was prepared from wine, dates, figs, and other fruits, vinegar was used to flavor foods and beverages and as a medicine. Romans soldiers and Japanese samurais drank vinegar diluted in water as an energizing beverage and tonic (Budak et al., 2014). Sholberg et al. (2000) tested the potential of several commercially available vinegars, containing from 4 to 6% AA, to control postharvest decay of stone fruits, strawberries, and apples. Vinegar was vaporized at a rate of 0.08–0.24 mL/L for periods of 0.5 or 16 hr. All vinegars at 0.08 mL/L were effective in controlling decay and in most cases almost completely inhibited the growth of *P. expansum, M. fructigena, and B. cinerea*. Vinegar vapors were also reported to control anthracnose in tomatoes (Tzortzakis, 2010).

In contrast with fumigation, results of both AA and vinegar used in solution as dip treatments were modest. Immersion of 'Sanny' peppers in 16 mL/L vinegar (5% AA) did not prevent gray mold (Tzortzakis et al., 2016). Similarly, immersion of 'Red Delicious' apples in hot AA solutions at 2% for 3 min or 3% AA for 2 min partially controlled decay only on fruits stored for short periods, while the effect was insignificant on fruits stored for a long period (Radi et al., 2010).

4.3 Other Compounds

Benzoic acid is one of the oldest and most commonly used chemical preservatives (Theron and Lues, 2011). It has been detected as a natural constituent in cranberries, raspberries, plums, prunes, cinnamon, and cloves (Barbosa-Cánovas et al., 2003). It has limited solubility in water (1.7 g/100 mL at 0°C and 5 g/100 mL at 40°C), a pKa of 4.2 and, as is true with other organic acids, it is more active in the undissociated form, which, thanks to its lipophilic properties, crosses the fungi membranes. The reduction of the cytoplasmatic pH by benzoic acid would alter the glycolysis pathway at pyruvate kinase and glyceraldehyde dehydrogenase-phosphoglycerate kinase steps, lower the ATP availability, and in turn inhibit cell growth (Krebs et al., 1983; Warth, 1991). Its activity is strongly pH-dependent and more effective under acidic conditions, where the protonated form of the acid is predominant (Eklund, 1980, 1985).

Sodium benzoate is generally preferred in water solutions to benzoic acid because it is much more soluble, approximately 63 g/100 mL at 0–30°C (Theron and Lues, 2011). In in vitro tests at a concentration of 0.1–0.2 M, sodium benzoate completely inhibited conidial germination, mycelial growth, and sporulation of *Fusarium solani* (Mart.) Sacc. and *F. sambucinum*, causal agents of potato dry rot, and *H. solani*, the causal agent of silver scurf of potatoes (Hervieux et al., 2002; Mecteau et al., 2002). Sodium benzoate was effective in controlling green mold in oranges (Hall, 1988) and green mold and blue mold in several citrus species and cultivars, although it was more effective in oranges than in mandarins (Palou et al., 2002a, 2002b; Montesinos-Herrero et al., 2016). Palou et al. (2009) evaluated in a study the potential of more than 20 food additives and GRAS compounds to control the major postharvest diseases of stone fruits (*M. fructicola, B. cinerea, G. candidum, A. alternata, P. expansum, Mucor piriformis* A. Fisch., and *R. stolonifer*), sodium benzoate was among the most effective compounds. However, when tested in small-scale trials, sodium benzoate lacked sufficient effectiveness and persistence to control brown rot on stone fruits.

Propionic acid, another organic acid widely used in the food industry as preservative and additive (Theron and Lues, 2011), effectively inhibits fungal growth, especially at low pH, by affecting fungal membranes and inhibiting amino acids uptake at pH values below 4.5 (Eklund, 1985). The salts of propionic acid, such as sodium propionate and ammonium propionate, show similar effects against yeasts and filamentous molds at a low pH (Schnurer and Magnusson, 2005) and are preferred to propionic acid. Propionic acid salts showed some activity against a wide range of fungi causing decay in a variety of horticultural crops (Theron and Lues, 2011). Calcium propionate reduced mycelial growth of *M. fructicola* in in vitro studies, inhibited fungal polygalacturonase activity, and controlled brown rot on fruits dipped into 60 or 100 mg/L solutions (Biggs et al., 1997). On apples inoculated with *B. cinerea* or *P. expansum*, calcium propionate provided both protective and curative effects against infections caused by the two fungi when examined after 7 or 21 d at 20–22°C, although control of gray mold was superior (Droby et al., 2003). However, on potatoes, suppression of silver scurf by calcium propionate was inconsistent and its activity did not improve when it was acidified (Oliver et al., 1999). Disappointing results were also reported by Mecteau et al. (2002), who found that calcium propionate reduced in vitro mycelial growth of *F. sambucinum* but stimulated conidiation, whereas in vivo it did not control dry rot on potato tubers that had been artificially inoculated.

Parabens and some of their salts are among the most commonly used preservatives in food, cosmetic, or pharmaceutical products and they are approved for use in foods in the USA as well as in the EU. The mode of action of parabens has not been completely elucidated. It is thought they cause disruption of membrane transport processes in cells or inhibit synthesis of DNA and RNA (Jay et al., 2005; Soni et al., 2005). Parabens seem to be more effective against molds than yeasts. In contrast to other organic acids, the pKa of parabens is around 8.5 and their activity is not diminished at pH values of up to 8 (Jay et al., 2005). Parabens inhibit spore germination more than mycelial growth (Watanabe and Takesue, 1976). Propylparaben showed a high inhibitory activity on mycelial growth and conidia germination of *B. cinerea* (Yildirm and Yapici, 2007). In in vitro tests, disks of hydroxypropyl methylcellulose (HPMC)-lipid edible composite films containing parabens effectively inhibited *P. digitatum* and *P. italicum* (Valencia-Chamorro et al., 2008).

Sodium ethylparaben (Moscoso-Ramírez et al., 2013a) and sodium methylparaben (Moscoso-Ramírez et al., 2013b) at 80 mM and 200 mM, respectively, were effective in reducing decay in various citrus cultivars inoculated with *P. digitatum* or *P. italicum*. In all tests, parabens were more effective on oranges than in mandarins. When both parabens were combined with low doses of IMZ, the combined treatments were more effective than the single ones. Sodium methylparaben performance did not improve when the solution temperature was raised from 20 to 50 or 62°C; this is very important for citrus packinghouses, since implementation and application costs of nonheated solutions are lower (Moscoso-Ramírez et al., 2013a).

5 Low Toxicity Compounds: Strengths, Weaknesses, and Strategies to Improve Their Efficacy

LTCs show a wide range of modes of action interfering with various metabolic pathways occurring in target microorganisms, so the risk that resistant strains of a pathogen develop is quite low. Another important feature of LTCs is their broad spectrum of activity, which includes most of the important pathogens that cause decay. The large number of available compounds, each with specific physical and chemical properties, offers wide flexibility for their formulation and methods of their application. For example, short chain organic acids can be applied as fumigants while sorbates, benzoates, or propionates can be applied as dip treatments or sprays on fruit packinglines.

However, these compounds are generally less active than synthetic fungicides, so higher concentrations are needed to achieve comparable levels of decay control. Moreover, most of them are only fungistatic, lack preventive activity, have limited persistence, pose a higher risk of fruit injury, and can cause increased weight and firmness losses during long-term storage if treated commodities are not rinsed after treatment (Palou et al., 2008). Furthermore, LTCs, particularly organic acids, degrade very rapidly, and may have a stimulatory effect on the pathogen growth when applied at low concentrations (Theron and Lues, 2011), because they can comprise a nutritional substrate for pathogens. This is true for sorbic acid, which can be metabolized by *P. expansum* (Fadda et al., 2015) or *Penicillium roqueforti* Thom (Marth et al., 1966).

Another important shortcoming of LTCs is that their capacity to reduce fungal growth is frequently more inconsistent or modest in in vivo tests with fruits and vegetables than in in vitro tests (Palou et al., 2002b, 2008, 2009). It appears, therefore, that their capacity to control decay is mediated by the way they interact with the host. For example, changes in the solution pH induced by the fruit tissue may enhance or reduce their toxicity. The acidic pH of many fruit tissues may reduce the activity of bi- and carbonate salts, while it could enhance that of organic acids such as sorbic, benzoic, or propionic acids. However, the potential capacity of some compounds to enhance natural defense mechanisms of the host may fail if they do not come in contact with the host tissues. Highly systemic products, such as phosphites, normally give consistent results, whereas compounds such as silicates or carbonates are more effective in wounded fruits, because they are poor inducers of resistance-responses in sound, nonwounded tissues (Venditti et al., 2005). The lipophilic nature of epicuticular waxes hampers an even distribution of dissociated salts moieties on fruit surface and may produce an over/underconcentration of salts on fruit surface, even if the compounds are applied at low concentrations (D'Aquino et al., 2013a), with the result that many areas of fruit surface could remain untreated, thus unprotected, while other areas could accumulate the salts to phytotoxic levels. The efficacy of carbonate salts was improved when surfactants or lecithin were added to the solution (Homma et al., 1981a, 1981b; Horst et al., 1992; Lurie and Klein, 1992; Alvindia and Natsuaki, 2007; Deliopoulos et al., 2010). On bananas, bicarbonate as well as carbonate salts were effective in controlling crown rot, but their crystallization when dried led to large variations in their efficacy (Alvindia et al., 2004).

6 Strategies to Improve the Treatment Efficacy of Low Toxicity Compounds

It is likely that the use of LTCs will play an increasingly important role for the control of postharvest diseases commercially in the future, because there are few regulatory restrictions concerning their use and a more favorable perception by consumers of them compared to synthetic chemicals, regardless of the real hazard they may pose for human health. Implementation and optimization of LTCs to control postharvest diseases is a key objective for the postharvest industry in view of further limitations of the use of pesticides. On the other hand, in many countries where synthetic fungicides are banned from postharvest use on a number of horticultural products (i.e., stone fruits or tomatoes in the EU countries) or in products certified as 'organic', LTCs already represent the only available chemical means to control postharvest diseases.

Considering the lower overall toxicity of LTCs, new control strategies should shift from the 'silver bullet' concept where a single treatment is applied to control decay to a holistic approach, involving several integrated treatments acting together additively or synergistically to bring about an acceptable level of decay control (Wisniewshi et al., 2016). In agreement with Ippolito and Nigro (2005), because LTCs are less toxic to animals and are environmentally friendly, treatments could be repeated several times along the distribution chain or during long-term storage. Their effectiveness could further be improved by their use as preharvest treatments, where several studies showed their effectiveness extended after harvest to control decay (Youssef et al., 2012; Feliziani and Romanazzi, 2013a). Preharvest treatments with GRAS compounds reduced decay on strawberries (Romanazzi et al., 2013), table grapes (Nigro et al., 2006; Chervin et al., 2009; Li et al., 2009; Feliziani et al., 2013b), apples (Biggs et al., 1993), stone fruits (Biggs et al., 1997), mangoes (Singh et al., 1993), and other horticultural crops (Brown et al., 1996; Penter and Stassen, 2000; Plitch and Wójcik, 2002; Sivakumar et al., 2002; Nolla et al., 2006; Goutam et al., 2010; Bosse et al., 2011; Wójcik et al., 2014). The lack of protective activity is one of the major limitations of LTCs for commercial applications, especially when treated products are rinsed after treatment to minimize phytotoxicity. Their combination with coating matrixes could mitigate their toxicity while adding antifungal properties to fruit coatings, particularly conventional commercial waxes or edible coatings. For example, decay incidence in 'Comune' clementine mandarins and 'Tarocco' and 'Valencia' oranges was significantly reduced when sprayed with a commercial wax formulation containing 6% sodium carbonate, potassium carbonate, potassium bicarbonate, ammonium bicarbonate, or PS (Youssef et al., 2012). Edible HPMC-lipid-based edible coatings formulated with PS, sodium benzoate, or sodium propionate inhibited the in vitro growth of *P. digitatum* and *P. italicum* and reduced green and blue molds on several citrus cultivars cold-stored for long periods (Palou et al., 2015). Fundamentals and recent advances in the development and use of antifungal edible coatings for postharvest disease control are described in detail in Chapter 10. Increasing the solution temperature is another way to increase LTCs effectiveness. Additive or synergistic effects were reported for PS, sodium carbonate, sodium bicarbonate, sodium benzoate (Barkai-Golan and Appelbaum, 1991; Schirra et al., 2008; Kanetis et al., 2008b), and other GRAS compounds when the solution temperature was raised from 40 to 50°C and even above 60°C (Kitagawa and Kagawa, 1984; Wild, 1987; Smilanick et al., 1997, 1999; Smilanick and Sorenson, 2001; Palou et al., 2001, 2002b; Montesinos-Herrero et al., 2009; D'Aquino et al., 2012, 2015).

The use of LTCs is gaining increasing interest even for conventional production where fungicides are routinely applied. Combining LTCs with synthetic fungicides widens the mode of action of the treatments and reduces the risk of development of resistance to fungicides. Moreover, the efficacy of the treatments may be markedly enhanced, allowing significant reductions in the rates of applied fungicides and a reduction in their residues, and provide some control of pathogens if fungicide resistance does develop. Bicarbonates, carbonates, or PS are compatible with many approved fungicides, and these mixtures show in most cases additive or synergistic increases in efficacy (Schirra et al., 2011; Smilanick, 2011; Palou et al., 2016). However, since an increase in treatment temperature can increase the deposition of fungicide residues, care is needed not to exceed maximum residue levels (Schirra et al., 2011). Integrated disease management programs based on the combination of LTCs (particularly PS and sodium bicarbonate) are in commercial use and becoming increasingly common practices in some fruit packinghouses (Montesinos-Herrero and Palou, 2010; Smilanick, 2011; Palou et al., 2016).

7 Concluding Remarks

LTCs, including food additives and a large number of inorganic and organic acids and salts widely used for several purposes in the food industry as well as for other industrial applications, may represent a valid alternative to control postharvest diseases of horticultural crops. Several of these compounds are known from ancient times in many civilizations and some, as vinegar, find numerous uses in consumers' homes as well as in industrial applications.

Although most of them are effective in preventing food spoilage and have proven fungicidal properties in in vitro studies, none of the LTCs have been able to provide a level of control comparable to that achieved with synthetic fungicides. The high rates needed to arrest the growth of the pathogens, their specificity, the potential phytotoxicity at effective rates, and the narrow range of optimal conditions within which they are effective are some of the reasons for the low performance.

Although it is unlikely that LTCs will be widely adopted to control postharvest diseases commercially while the use of synthetic fungicides is still allowed, it is also true that the LTCs alternatives are currently the only postharvest decay control chemicals available for either certified 'organic' produce or the large number of fresh products for which no synthetic fungicides are registered for postharvest uses. On the other hand, the overall trend of our society is to develop a food production system that is socially, ecologically, and economically acceptable and sustainable, based on economic crop production with a minimal use of agrochemicals. In this scenario, research of strategic importance is what will make alternative treatments more cost-effective, focusing on mechanisms underlying the interactions among LTCs, the fruit host, and the targeted pathogens.

References

Adaskaveg, J.E. and Förster, H. 2010. New developments in postharvest registrations for edible horticultural crops and use strategies in the United States. pp. 107–117. *In*: D. Prusky and M.L. Gullino (eds.). *Post-Harvest Pathology. Plant Pathology in the 21st Century*. Springer, New York, NY.

Adaskaveg, J.E. and Förster, H. 2015. Resistance in postharvest pathogens of citrus in the United States. pp. 449–466. *In*: H. Ishii and D.W. Hollomon (eds.). *Fungicide Resistance in Plant Pathogens*. Springer, Tokyo, Japan.

Adaskaveg, J.E., Hao, W. and Förster, H. 2015. Postharvest strategies for managing Phytophthora brown rot of citrus using potassium phosphite in combination with heat treatments. *Plant Disease* 99, 1477–1482.

Al Eryani-Raqeeb, A., Mahamud, T.M.M., Syed Omar, S.R., Mohamed, A.R. and Al Eryani, A.R. 2009. Effect of calcium and chitosan treatments on controlling anthracnose and postharvest quality of papaya (*Carica papaya* L.). *International Journal of Agricultural Research* 4, 53–68.

Alvindia, D.G., Kobayashi, Y., Natsuaki, K.T. and Tanda, S. 2004. Inhibitory influence of inorganic salts on banana postharvest pathogens and preliminary application to control crown rot. *Journal of General Plant Pathology* 70, 61–65.

Alvindia, D.G. and Natsuaki, K.T. 2007. Control of crown rot-causing fungal pathogens of banana by inorganic salts and a surfactant. *Crop Protection* 26, 1667–1673.

Amiri, A. and Bompeix, G. 2011. Control of *Penicillium expansum* with potassium phosphite and heat treatment. *Crop Protection* 30, 222–227.

Bandara, W.M.K.I., Perera, O.D.A.N. and Weerahewa, H.L.D. 2015. Enhancing disease resistance and improving quality of papaya (*Carica papaya* L.) by postharvest application of silicon. *International Journal of Agriculture, Forestry and Plantation* 1, 24–27.

Barbosa-Cánovas, G.V., Fernández-Molina, J.J., Alzamora, S.M., Tapia, M.S., López-Malo, A. and Chanes, J.W. 2003. General considerations for preservation of fruits and vegetables. pp. 39–54. *In*: The FAO Agency (ed.). *Handling and Preservation of Fruits and Vegetables by Combined Methods for Rural Areas*. FAO Agricultural Services Bulletin No. 149. FAO, Rome, Italy.

Barkai-Golan, R. 2001. *Postharvest Diseases of Fruits and Vegetables: Development and Control*. Elsevier Science B.V., Amsterdam, The Netherlands.

Barkai-Golan, R. and Appelbaum, A. 1991. Synergistic effects of heat and sodium o-phenylphenate treatments to inactivate *Penicillium* spores and suppress decay in citrus fruits. *Tropical Science* 31, 229–233.

Bartlett, D.W., Clough, J.M., Godwin, J.R., Hall, A.A., Hamer, M. and Parr-Dobrzanski, B. 2002. The strobilurin fungicides. *Pest Management Science* 58, 649–662.

Bell, T.A., Etchells, J.L. and Borg, A.F. 1959. Influence of sorbic acid on the growth of certain species of bacteria, yeasts, and filamentous fungi. *Journal of Bacteriology* 77, 573–580.

Belton, J.B., Deschaume, O. and Perry, C.C. 2012. An overview of the fundamentals of the chemistry of silica with relevance to biosilicification and technological advances. *FEBS Journal* 279, 1710–1720.

Bi, Y., Tian, S.P., Guo, Y.R., Ge, Y.H. and Qin, G.Z. 2006. Sodium silicate reduces postharvest decay on Hami melons induced resistance and fungistatic effects. *Plant Disease* 90, 279–283.

Biggs, A.R. 1999. Effects of calcium salts on bitter rot caused by two *Colletotrichum* spp. *Plant Disease* 83, 1001–1005.

Biggs, A.R., El-Kholi, M.M., El-Neshawy, S. and Nickerson, R. 1997. Effects of calcium salts on growth, polygalacturonase activity and infection of peach by *Monilinia fructicola*. *Plant Disease* 81, 399–403.

Biggs, A.R., Ingle, M. and Solihati, W.D. 1993. Control of *Alternaria* infection on fruit of apple cultivar Nittany with calcium chloride and fungicides. *Plant Disease* 77, 976–980.

Billard, A., Fillinger, S., Leroux, P., Lachaise, H., Beffa, R. and Debieu, D. 2012. Strong resistance to the fungicide fenhexamid entails a fitness cost in *Botrytis cinerea*, as shown by comparisons of isogenic strains. *Pest Management Science* 68, 684–691.

Blodgett, A.B., Caldwell, R.W. and McManus, P.S. 2002. Effects of calcium salts on the cranberry fruit rot disease complex. *Plant Disease* 86, 747–752.

Bollen, G.J. and Scholten, G. 1971. Acquired resistance to benomyl and some other systemic fungicides in a strain of *Botrytis cinerea* in cyclamen. *Netherlands Journal of Plant Pathology* 77, 83–90.

Bosse, R.J., Bower, J.P. and Bertling, I. 2011. Pre-and post-harvest treatments on Fuerte avocados to control anthracnose (*Colletotrichum gloeosporioides*) during ripening. *South African Avocado Growers' Association* 34, 65–69.

Brackman, A., Ceretta, M. and Vizzotto, M. 2001. Use of calcium chloride and lime to control post-harvest rot in apples. *Revista Brasileira de Fruticultura* 23, 298–301.

Brent, K.J. and Hollomon, D.W. 1998. *Fungicide Resistance: The Assessment of Risk*. FRAC Monograph No. 2. Global Crop Protection Federation, Brussels, Belgium.

Brent, K.J. and Hollomon, D.W. 2007. *Fungicide Resistance in Crop Pathogens: How Can It Be Managed?* FRAC Monograph No. 1, 2nd ed. Fungicide Resistance Action Committee, Brussels, Belgium.

Brown, G.E. and Miller, W.R. 1999. Maintaining fruit health after harvest. pp. 175–187. *In*: L.W. Timmer and L.W. Duncan (eds.). *Citrus Health Management*. American Phytopathological Society Press, St. Paul, MN.

Brown, G.S., Kitchener, A.E., McGlasson, W.B. and Barnes, S. 1996. The effect of copper and calcium foliar sprays on cherry and apple fruit quality. *Scientia Horticulturae* 67, 219–227.

Brul, S. and Coote, P. 1999. Preservative agents in foods – Mode of action and microbial resistance mechanisms. *International Journal of Food Microbiology* 50, 1–17.

Budak, N.H., Aykin, E., Seydim, A.C., Greene, A.K. and Guzel-Seydim, Z.B. 2014. Functional properties of vinegar. *Journal of Food Science* 79, 757–764.

Busta, F.F. and Foegeding, P.M. 1983. Chemical food preservatives. pp. 656–688. *In*: S.S. Block (ed.). *Disinfection, Sterilization, and Preservation*, 3rd ed. Lea & Febiger, Philadelphia, PA.

Carswell, M.C., Grant, B.R., Plaxton, W.C., 1997. Disruption of thephosphate-starvation response of oilseed rape suspension cells by thefungicide phosphonate. *Planta* 203, 67–74.

Carswell, M.C., Grant, B.R., Theodorou, M.E., Harris, J., Niere, J.O. and Plaxton, W. D. 1996. The fungicide phosphonate disrupts the phosphate starvation response in *Brassica nigra* seedlings. *Plant Physiology* 110, 105–110.

CCQC (California Citrus Quality Council). 1980. Imazalil: A new weapon in the fruit decay battle. *Citrograph* 65, 95–96.

Cerioni, L., Rapisarda, V.A., Doctor, J., Fikkert, S., Uits, T., Fassel, R. and Smilanick, J.L. 2013. Use of phosphites salts in laboratory and semicommercial tests to control citrus postharvest decay. *Plant Disease* 97, 201–212.

Chervin, C., Lavigne, D. and Westercamp, P. 2009. Reduction of gray mold development in table grapes by preharvest sprays with ethanol and calcium chloride. *Postharvest Biology and Technology* 54, 115–117.

Conway, W.S. 1982. Effect of post-harvest calcium treatment on decay of Delicious apples. *Plant Disease* 66, 402–403.

Conway, W.S. and Sams, C.E. 1984. Possible mechanisms by which postharvest calcium treatment reduces decay in apples. *Phytopathology* 74, 208–210.

Conway, W.S., Sams, C.E., Abbott, J.A. and Bruton, B.D. 1991. Postharvest calcium treatment of apple fruit to provide broad-spectrum protection against postharvest pathogens. *Plant Disease* 75, 620–622.

Corral, L.G., Post, L.S. and Montville, T.J. 1988. Antimicrobial activity of sodium bicarbonate. *Journal of Food Science* 53, 981–982.

Coskun, D., Britto, D.T., Huynh, W.Q. and Kronzucker, H.J. 2016. The role of silicon in higher plants under salinity and drought stress. *Frontiers in Plant Science* 7, article 1072.

D'Aquino, S., Barberis, A., Continella, A., La Malfa, S., Gentile, A. and Schirra, M. 2012. Individual and combined effects of postharvest treatments with hot water at 50°C, soy lecithin and sodium carbonate on cold stored cactus pear fruits. *Communications in Agricultural and Applied Biological Sciences* 77, 207–217.

D'Aquino, S., Fadda, A., Barberis, A., Palma, A., Angioni, A. and Schirra, M. 2013a. Combined effects of potassium sorbate, hot water and thiabendazole against green mould of citrus fruit and residue levels. *Food Chemistry* 141, 858–864.

D'Aquino, S., Palma, A., Angioni, A. and Schirra, M. 2013b. Residue levels and efficacy of fludioxonil and thiabendazole in controlling postharvest green mold decay in citrus fruit when applied in combination with sodium bicarbonate. *Journal of Agricultural and Food Chemistry* 61, 296–306.

D'Aquino, S., Palma, A., Schirra, M., Continella, A., La Malfa, S. and Gentile, A. 2015. Decay control of cactus pear by pre- and post-cold storage dip treatments with fludioxonil, sodium bicarbonate and water at 50°C. *Acta Horticulturae* 1067, 119–125.

D'Aquino, S., Schirra, M., Palma, A., Angioni, A., Cabras, P. and Migheli, Q. 2006. Residue levels and effectiveness of pyrimethanil vs imazalil when using heated postharvest dip treatments for control of Penicillium decay on citrus fruit. *Journal of Agricultural and Food Chemistry* 54, 4721–4726.

Datnoff, L.E. and Rodrigues, F.A. 2015. History of silicon and plant disease. pp. 1–6. *In*: F.A. Rodrigues and L.E. Datnoff (eds.). *Silicon and Plant Diseases*. Springer International Publishing, Cham, Switzerland.

Dave, B.A., Kaplan, H.J. and Petri, J.F. 1980. The isolation of *Penicillium digitatum* Sacc. strains tolerant to 2-AB, SOPP, TBZ and benomyl. *Proceedings of the Florida State Horticultural Society* 93, 344–347.

Debieu, D., Bach, J., Hugon, M., Malosse, C. and Leroux, P. 2001. The hydroxyanilide fenhexamid, a new sterol biosynthesis inhibitor fungicide efficient against the plant pathogenic fungus *Botryotinia fuckeliana* (*Botrytis cinerea*). *Pest Management Science* 57, 1060–1067.

Deliopoulos, T., Ketlewell, P.S. and Hare, M.C. 2010. Fungal disease suppression by inorganic salts: A review. *Crop Protection* 29, 1059–1075.

Dharmadhikari, M. 1992. Sorbic acid. *Vineyard Vintage View Newsletter* 7, 1–5.

Dreassi, E., Zizzari, A.T., Arezzo, S.D., Visca, P. and Botta, M. 2007. Analysis of guazatine mixture by LC and LC-MS and antimycotic activity determination of principal components. *Journal of Pharmaceutical and Biomedical Analysis* 43, 1499–1506.

Droby, S., Wisniewski, M., El-Gaouth, A. and Wilson, C. 2003. Influence of food additives on the control of postharvest rots of apples and peach and efficacy of the yeast-based biocontrol product Aspire. *Postharvest Biology and Technology* 27, 127–135.

Droby, S., Wisniewski, M.E., Cohen, L., Weiss, B., Touitou, D., Eilam, Y. and Chalutz, E. 1997. Influence of $CaCl_2$ on *Penicillium digitatum*, grapefruit peel tissue, and biocontrol activity of *Pichia guilliermondii*. *Phytopathology* 87, 310–315.

Eaks, I.L. 1985. Effect of calcium on ripening, respiratory rate, ethylene production, and quality of avocado fruit. *Journal of the American Society for Horticultural Science* 110, 145–148.

Eckert, J.W. and Eaks, I.L. 1989. Postharvest disorders and diseases of citrus fruits. pp. 179–260. *In*: W. Reuther, E.C. Calavan and G.E. Carman (eds.). *The Citrus Industry*, vol. 5. University of California, Division of Agriculture and Natural Resources, Oakland.

Eckert, J.W., Sievert, J.R. and Ratnayake, M. 1994. Reduction of imazalil effectiveness against citrus green mold in California packinghouses by resistant biotypes of *Penicillium digitatum*. *Plant Disease* 78, 971–974.

Eckert, J.W. and Sommer, N.F. 1967. Control of diseases of fruits and vegetables by postharvest treatment. *Annual Review of Phytopathology* 5, 391–428.

Eckert, J.W. and Sommer, N.F. 1985. The chemical control of postharvest diseases: Subtropical and tropical fruits. *Annual Review of Phytopathology* 23, 421–454.

EFSA (European Food Safety Authority, Panel on Additives and Products or Substances used in Animal Feed). 2014. Scientific opinion on the safety and efficacy of sorbic acid and potassium sorbate when used as technological additives for all animal species. *EFSA Journal* 12, 3792–3810.

Eklund, T. 1980. Inhibition of growth and uptake processes in bacteria by some chemical food preservatives. *Journal of Applied Bacteriology* 48, 423–432.

Eklund, T. 1983. The antimicrobial effect of dissociated and undissociated sorbic acid at different pH levels. *Journal of Applied Bacteriology* 54, 383–389.

Eklund, T. 1985. Inhibition of microbial growth at different pH levels by benzoic and propionic acids and esters of p-hydroxybenzoic acid. *International Journal of Food Microbiology* 2, 159–167.

El-Goorani, M.A., El-Kasheir, H.M., Kabeel, M.T. and Shoeib, A. 1983. Resistance to benzimidazole fungicides of *Penicillium italicum* and *P. digitatum* isolated from packinghouses and orchards in Egypt. *Plant Disease* 68, 100–102.

Ellis, M.E., Madden, L.V. and Erincik, O. 1996. Evaluation of calcium chloride for control of Botrytis fruit rot in strawberry. Ohio Integrated Pest Management, Block Grant Reports. Available at: https://ipm.osu.edu/sites/ipm/files/imce/1996%20ipm%20reports.pdf

Elsherbiny, A.E. and El-Khateeb, Y. 2012. Effect of organic and inorganic salts on mycelial growth, sporulation and spore germination of potato postharvest pathogens. *Journal of Plant Protection and Pathology, Mansoura University* 3, 1353–2364.

EPC (European Parliament and Council). 1995. Directive No. 95/2/ED of 20 February 1995 on Food Additives Other Than Colours and Sweeteners. OJ L 61, 18.3.1995.

Epstein, E. 1999. Silicon. *Annual Review of Plant Physiology and Plant Molecular Biology* 50, 641–664.

Erasmus, A., Lennox, C.L., Smilanick, J.L., Lesar, K. and Fourie, P.H. 2013. Imazalil residue loading and green mould control on citrus fruit as affected by formulation, solution pH and exposure time in aqueous dip treatments. *Postharvest Biology and Technology* 77, 43–49.

Fadda, A.A., Barberis, A., D'Aquino, S., Palma, A., Angioni, A., Lai, F. and Schirra, M. 2015. Residue levels and performance of potassium sorbate and thiabendazole and their co-application against blue mold of apples when applied as water dip treatments at 20 or 53°C. *Postharvest Biology and Technology* 106, 33–43.

Farahani, L., Etebarian, H.R., Sahebani, N. and Aminian, H. 2012. Biocontrol of blue mold of apple by *Candida membranifaciens* in combination with silicon. *Archives of Phytopathology and Plant Protection* 45, 310–317.

Fateux, F., Rémus-Borel, W., Menzies, J. and Bélanger, R.R. 2005. Silicon and plant disease resistance against pathogenic fungi. *FEMS Microbiology Letters* 249, 1–6.

Feliziani, E. and Romanazzi, G. 2013a. Preharvest application of synthetic fungicides and alternative treatments to control postharvest decay of fruit. *Stewart Postharvest Review* 3, 4.

Feliziani, E., Smilanick, J.L., Margosan, D.A., Mansour, M.F., Romanazzi, G., Gu, S., Gohil, H.L. and Rubio Ames, Z. 2013b. Preharvest fungicide, potassium sorbate, or chitosan use on quality and storage decay of table grapes. *Plant Disease* 97, 307–314.

Forbes-Smith, M., Paton, J. and Mu, S. 1998. Induced resistance in oranges to *Penicillium digitatum*. Available at: http://bspp.org.uk/icpp98/1.9/8.html.

Förster, H., Adaskaveg, J.E., Kim, D.H. and Stanghelleni, M.E. 1998. Effect of phosphite on tomato and pepper plants and on susceptibility of pepper to

Phytophthora root and crown rot in hydroponic cultures. *Plant Disease* 82, 11165–11170.
Förster, H., Driever, G.F., Thompson, D.C. and Adaskaveg, J.E. 2007. Postharvest decay management for stone fruit crops in California using the "reduced-risk" fungicides fludioxonil and fenhexamid. *Plant Disease* 91, 209–215.
Gómez-Merino, F.C. and Trejo-Téllez, L.I. 2015. Biostimulant activity of phosphite in horticulture. *Scientia Horticulturae* 196, 82–90.
Goutam, M., Dhaliwal, H.S. and Mahajan, B.V.C. 2010. Effect of pre-harvest calcium sprays on post-harvest life of winter guava (*Psidium guajava* L.). *Journal of Food Science and Technology* 47, 501–506.
Guest, D. and Grant, B. 1991. The complex action of phosphonates as antifungal agents. *Biological Reviews* 66, 159–187.
Guo, Y., Liu, L., Zhao, J. and Bi, Y. 2007. Use of silicon oxide and sodium silicate for controlling *Trichothecium roseum* postharvest rot in Chinese cantaloupe (*Cucumis melo* L.). *International Journal of Food Science and Technology* 42, 1012–1018.
Hall, D.J. 1988. Comparative activity of selected food preservatives as citrus postharvest fungicides. *Proceedings of the Florida State Horticultural Society* 101, 184–187.
Hann, M. 2014. The rising threat of fungicide resistance in plant pathogenic fungi: Botrytis as a case study. *Journal of Chemical Biology* 7, 133–141.
He, L., Liu, Y., Mustapha, A. and Lin, M. 2011. Antifungal activity of zinc oxide nanoparticles against *Botrytis cinerea* and *Penicillium expansum*. *Microbiological Research* 166, 207–215.
Hervieux, V., Yaganza, E.S., Arul, J. and Tweddell, R.J. 2002. Effect of organic and inorganic salts on the development of *Helminthosporium solani*, the causal agent of potato silver scurf. *Plant Disease* 86, 1014–1018.
Heye, U.J., Speich, J., Siegle, H., Wohlhauser, R., Hubele, A. 1994. CGA-219417 A novel broad-spectrum fungicide. *Brighton Crop Protection Conference Pests and Diseases* 2, 501–508.
Homma, Y., Arimoto, Y. and Misato, T. 1981a. Effects of emulsifiers and surfactants on the protective values of sodium bicarbonates. *Journal of Pesticide Science* 6, 145–153.
Homma, Y., Arimoto, Y. and Misato, T. 1981b. The control of citrus storage disease by sodium bicarbonate formulation. *Proceedings of the International Society of Citriculture* 2, 823–825.
Horst, R.K., Kawamoto, S.O. and Porter, L.L. 1992. Effect of sodium bicarbonate and oils on the control of powdery mildew and black spot of roses. *Plant Disease* 76, 247–251.
Horuz, S. and Kmay, P. 2010. The effects of some new postharvest fungicides and combination of hot water with sodium bicarbonate against *Geotrichum citri-aurantii* on citrus. *Acta Horticulturae* 877, 1551–1558.
Houck, L.G. 1977. Problems of resistance to citrus fungicides. *Proceedings of the International Society of Citriculture* 1, 263–269.
Ippolito, A. and Nigro, F. 2005. Natural antimicrobials for preservation of fresh fruit and vegetables. pp. 513–555. *In*: W. Jongen (ed.). *Improving the Safety of Fresh Fruit and Vegetables*. Woodhead Publishing Limited, CRC Press, Boca Raton, FL.

Jay, J.M., Loessner, M.J. and Golden, D.A. 2005. Bioassay and related methods. pp. 285–298. *In*: J.M. Jay, M.J. Loessner and D.A. Golden (eds.). *Modern Microbiology*. Springer Science + Business Media Inc., New York, NY.

Johnson, D.S. 1977. A note on the effects of calcium and magnesium sprays on the short-term storage life of raspberries. *Horticultural Research* 17, 51–56.

Kanetis, L., Förster, H. and Adaskaveg, J.E. 2007. Comparative efficacy of the new postharvest fungicides azoxystrobin, fludioxonil, and pyrimethanil for managing citrus green mold. *Plant Disease* 91, 1502–1511.

Kanetis, L., Förster, H. and Adaskaveg, J.E. 2008a. Optimizing efficacy of new postharvest fungicides and evaluation of sanitizing agents for managing citrus green mold. *Plant Disease* 92, 261–269.

Kanetis, L., Förster, H., Jones, C.A., Borkovick, K.A. and Adaskaveg, J.E. 2008b. Characterization of biochemical and genetic mechanisms of fludioxonil and pyrimethanil resistance in field isolates of *Penicillium digitatum*. *Phytopathology* 98, 205–214.

Karabulut, O.A., Lurie, S. and Droby, S. 2001. Evaluation of the use of sodium bicarbonate, potassium sorbate and yeast antagonists for decreasing postharvest decay of sweet cherries. *Postharvest Biology and Technology* 23, 233–236.

Kitagawa, H. and Kagawa, K. 1984. Effect of sorbic acid and potassium sorbate on the control of sour rot of citrus fruit. *Proceedings of the Florida State Horticultural Society* 97, 133–135.

Kolaei, E.A., Cenatus, C., Tweddell, R.J. and Avis, T.J. 2013. Antifungal activity of aluminium-containing salts against the development of carrot cavity spot and potato dry rot. *Annals of Applied Biology* 163, 311–317.

Krebs, H.A., Wiggins, D., Stubbs, M., Sols, A. and Bedoya, F. 1983. Studies on the antifungal action of benzoate. *Biochemical Journal* 214, 657–663.

Levine, A.S. and Fellers, C.R. 1940. Action of acetic acid on food spoilage microorganisms. *Journal of Bacteriology* 39, 499–514.

Li, Y.C., Bi, Y., Ge, Y.H., Sun, X.J. and Wang, Y. 2009. Antifungal activity of sodium silicate on *Fusarium sulphureum* and its effect on dry rot of potato tubers. *Journal of Food Science* 74, 213–218.

Liang., Y., Nikolic, M., Bèlanger, R., Gong, H. and Song, A. 2015. *Silicon in Agriculture – From Theory to Practice*. Springer International Publishing, Dordrecht, The Netherlands.

Lindsay, R.C. 1985. Food additives. pp. 632–675. *In*: O.R. Fennema (ed.). *Food Chemistry*. Marcel Dekker, Inc., New York.

Liu, K., Wang, X. and Young, M. 2014. Effect of bentonite/potassium sorbate coatings on the quality of mangos in storage at ambient temperature. *Journal of Food Engineering* 137, 16–22.

Liu, W.T., Chu, C.L. and Zhou, T. 2002. Thymol and acetic acid vapors reduce postharvest brown rot of apricots and plums. *HortScience* 37, 151–156.

Lodal, P. 1993. Key properties and hazards. pp. 361–436. *In*: V.H. Greda and J.R. Zoeller (eds.). *Acetic Acid and Its Derivates*. Marcel Dekker, Inc., New York, NY.

Lurie, S. and Klein, J.D. 1992. Calcium and heat treatments to improve storability of 'Anna' apples. *HortScience* 27, 36–39.

Lyr, H. 1995. *Modern Selective Fungicides: Properties, Applications, Mechanisms of Action*, 2 ed. VCH Publishers, Deerfield Beach, FL.

Ma, Z.H. and Michailides, T.J. 2005. Advances in understanding molecular mechanisms of fungicide resistance and molecular detection of resistant genotypes in phytopathogenic fungi. *Crop Protection* 24, 853–863.

Mari, M., Bautista-Baños, S. and Sivakumar, D. 2016. Decay control in the postharvest system: Role of microbial and plant volatile organic compounds. *Postharvest Biology and Technology* 122, 70–81.

Marth, E.H., Capp, C.M., Hasenzahl, L., Jackson, H.W. and Hussong, R.V. 1966. Degradation of potassium sorbate by *Penicillium* species. *Journal of Dairy Science* 49, 1197–1205.

McDonald, R.E., Risse, L.A. and Hillebrand, B.M. 1979. Resistance to thiabendazole and benomyl of *Penicillium digitatum* and *P. italicum* isolates from citrus fruit from several countries. *Journal of the American Society for Horticultural Science* 1, 333–335.

McKay, A.H., Förster, H. and Adaskaveg, J.E. 2012. Efficacy and application strategies for propiconazole as a new postharvest fungicide for managing sour rot and green mold of citrus fruit. *Plant Disease* 96, 235–242.

Mecteau, M.R., Arul, J. and Tweddell, R.J. 2002. Effect of organic and inorganic salts on the growth and development of *Fusarium sambucinum*, a causal agent of potato dry rot. *Mycological Research* 106, 688–696.

Mecteau, M.R., Arul, J. and Tweddell, R.J. 2008. Effect of different salts on the development of *Fusarium solani* var. *coeruleum*, a causal agent of potato dry rot. *Phytoprotection* 89, 1–6.

Miyaki, Y. and Takahashi, E. 1983. Effect of silicon on the growth of solution-cultured cucumber plants. *Japanese Journal of Soil Science and Plant Nutrition* 29, 71–83.

Montesinos-Herrero, C., del Río, M.A., Pastor, C., Brunetti, O. and Palou, L. 2009. Evaluation of brief potassium sorbate dips to control postharvest Penicillium decay on major citrus species and cultivars. *Postharvest Biology Technology* 52, 117–125.

Montesinos-Herrero, C., Moscoso-Ramírez, P.A. and Palou, L. 2016. Evaluation of sodium benzoate and other food additives for the control of citrus postharvest green and blue molds. *Postharvest Biology and Technology* 115, 72–80.

Montesinos-Herrero, C. and Palou, L. 2010. Combination of physical and low-toxicity chemical postharvest treatments for the management of citrus fruit: A review. *Stewart Postharvest Review* 6:1, 1–11.

Montesinos-Herrero, C. and Palou, L. 2016. Synergism between potassium sorbate dips and brief exposure to high CO_2 or O_2 at curing temperature for the control of citrus postharvest green and blue molds. *Crop Protection* 81, 43–46.

Morton, V. and Staub, T. 2008. A short history of fungicides. Available at: APSnet Features, doi:10.1094/APSnetFeature2008-0308.

Moscoso-Ramírez, P.A., Montesinos-Herrero, C. and Palou, L. 2013a. Control of citrus postharvest penicillium molds with sodium ethylparaben. *Crop Protection* 46, 44–51.

Moscoso-Ramírez, P.A., Montesinos-Herrero, C. and Palou, L. 2013b. Characterization of postharvest treatments with sodium methylparaben to control citrus green and blue mold. *Postharvest Biology and Technology* 77, 128–137.

Moscoso-Ramírez, P.A. and Palou, L. 2014. Preventive and curative activity of postharvest potassium silicate treatments to control green and blue molds on orange fruit. *European Journal of Plant Pathology* 138, 721–732.

Nigro, F., Schena, L., Ligorio, A., Pentimone, I., Ippolito, A. and Salerno, M.G. 2006. Control of table grape storage rots by pre-harvest application of salts. *Postharvest Biology and Technology* 42, 142–149.

Nolla, A., Korndörfer, G.H. and Coelho, L. 2006. Efficiency of calcium silicate and carbonate in soybean disease control. *Journal of Plant Nutrition* 29, 2049–2061.

Nordstrom, J., Nilsson, E., Jarvol, P., Nayeri, M., Palqvist, A., Bengenholtz, J. and Matic, A. 2011. Concentration and pH-dependence of highly alkaline sodium silicate solutions. *Journal of Colloid and Interface Science* 356, 37–45.

Nunes, C., Usall, J., Teixidó, N., Abadias, M. and Viñas, I. 2002. Improved control of postharvest decay of pears by the combination of *Candida sake* (CPA-1) and ammonium molybdate. *Phytopathology* 92, 281–287.

Olivier, C., Halseth, D.E., Mizubuti, E.S.G. and Loria, R. 1998. Postharvest application of organic and inorganic salts for suppression of silver scurf on potato tubers. *Plant Disease* 82, 213–217.

Olivier, C., MacNeil, C.R., Loria, R. 1999. Application of organic and inorganic salts to filed-grown potato tubers can suppress silver scurf during potato storage. *Plant Disease* 83, 814-818.

Onodera, I. 1917. Chemical studies on rice blast. *Journal of Agriculture and Social Sciences* 180, 606–617.

Palmer, C.L., Horst, R.K. and Langhans, R.W. 1997. Use of bicarbonates to inhibit in vitro colony growth of *Botrytis cinerea*. *Plant Disease* 81, 1432–1438.

Palou, L., Ali, A., Fallik, E. and Romanazzi, G. 2016. GRAS, plant- and animal-derived compounds as alternatives to conventional fungicides for the control of postharvest diseases of fresh horticultural produce. *Postharvest Biology and Technology* 122, 41–52.

Palou, L., Smilanick, J.L. and Crisosto, C.H. 2009. Evaluation of food additives as alternative or complementary chemicals to conventional fungicides for the control of major postharvest diseases of stone fruit. *Journal of Food Protection* 72, 1037–1046.

Palou, L., Smilanick, J.L. and Droby, S. 2008. Alternatives to conventional fungicides for the control of citrus postharvest green and blue moulds. *Stewart Postharvest Review* 2:2, 1–16.

Palou, L., Smilanick, J.L., Usall, J. and Viñas, I. 2001. Control of postharvest blue and green molds of oranges by hot water, sodium carbonate, and sodium bicarbonate. *Plant Disease* 85, 371–376.

Palou, L., Usall, J., Muñoz, J.A., Smilanick, J.L. and Viñas, I. 2002a. Hot water, sodium carbonate, and sodium bicarbonate for the control of postharvest green and blue molds of clementine mandarin. *Postharvest Biology and Technology* 24, 93–96.

Palou, L., Usall, J., Smilanick, J.L., Aguilar, M.J. and Viþas, I. 2002b. Evaluation of food additives and low-toxicity compounds as alternative chemicals for the control of *Penicillium digitatum* and *Penicillium italicum* on citrus fruit. *Pest Management Science* 58, 459–466.

Palou, L., Valencia-Chamorro, S.A. and Pérez-Gago, M.B. 2015. Antifungal edible coatings for fresh citrus fruit: A review. *Coatings* 5, 962–986.

Parra, J., Ripoll, G. and Orihuel-Iranzo, B. 2014. Potassium sorbate effects on citrus weight loss and decay control. *Postharvest Biology and Technology* 96, 7–13.

Penter, M.G. and Stassen, P.J.C. 2000. The effect of pre- and postharvest calcium applications on the postharvest quality of Pinkerton avocados. *South African Avocado Growers' Association Yearbook* 23, 1–7.

Pivotto Pavanello, E., Brackmann, A., Dressler da Costa, I.F., Both, V. and Ludwig, V. 2016. Use of sodium metasilicate for management of peach brown rot. *Pesquisa Agropecuária Tropical* 46, 245–253.

Plitch, H. and Wójcik, P. 2002. The effect of calcium and boron foliar application on postharvest plum fruit quality. *Acta Horticulturae* 594, 445–451.

Prusky, D. and Yakoby, N. 2003. Pathogenic fungi: Leading or led by ambient pH?. *Molecular Plant Pathology* 4, 509–516.

Punja, Z.K. and Gaye, M.M. 1993. Influence of postharvest handling practices and dip treatments on development of black root on fresh market carrots. *Plant Disease* 77, 989–995.

Punja, Z.K. and Grogan, R.G. 1982. Effects of inorganic salts, carbonate-bicarbonate anions, ammonia, and the modifying influence of pH on sclerotial germination of *Sclerotium rolfsii*. *Phytopathology* 72, 635–639.

Qin, G.Z. and Tian, S.P. 2005. Enhancement of biocontrol activity of *Cryptococcus laurentii* by silicon and the possible mechanisms involved. *Phytopathology* 95, 69–75.

Radi, M., Jouybari, H.A., Mesbahi, G., Farahnaky, A. and Amiri, S. 2010. Effect of hot acetic acid solutions on postharvest decay caused by *Penicillium expansum* on Red Delicious apples. *Scientia Horticulturae* 126, 421–425.

Ricker, M.D. and Punja, Z.K. 1991. Infuence of fungicide and chemical salt dip treatments on crater rot caused by *Rhizoctonia carotae* in long-term storage. *Plant Disease* 75, 470–474.

Roberts, J.W. and Dunegan, J.C. 1932. Peach Brown Rot. U.S. Department of Agriculture, Technical Bulletin No. 328, Washington, DC.

Rodrigues, F.A., Dallagnol, L.J., Duarte, H.S.S. and Datnoff, L.E. 2015. Silicon control of foliar diseases in monocots and dicots. pp. 67–109. *In*: F.A. Rodrigues and L. E. Datnoff (eds.). *Silicon and Plant Diseases*. Springer International Publishing, Cham, Switzerland.

Romanazzi, G., Feliziani, E., Santini, M. and Landi, L. 2013. Effectiveness of postharvest treatment with chitosan and other resistance inducers in the control of storage decay of strawberry. *Postharvest Biology and Technology* 75, 24–27.

Rosslenbroich, H.J. and Stuebler, D. 2000. *Botrytis cinerea* - history of chemical control and novel fungicides for its management. *Crop Protection* 19, 557–561.

Saftner, R.A., Buta, J.G. and Conway, W.S. 1997a. Effect of surfactants on pressure infiltration of calcium chloride solutions into 'Golden Delicious' apples. *Journal of the American Society for Horticultural Science* 122, 386–391.

Saftner, R.A., Conway, W.S. and Sams, C.E. 1997b. Effects of some polyamine biosynthesis inhibitors and calcium chloride on in vitro growth and decay development in apples caused by *Botrytis cinerea* and *Penicliium expansum*. *Journal of the American Society for Horticultural Science* 122, 380–385.

Samuels, A.L., Glass, A.D.M., Ehret, D.L. and Menzies, J.G. 1991. Mobility and deposition of silicon in cucumber plants. *Plant, Cell & Environment* 14, 485–492.

Sangster, A.G., Hodson, M.J. and Tubb, H.J. 2001. Silicon deposition in higher plants. pp. 85–114. *In*: G.H. Datnoff, G.H. Snyder and G.H. Korndöfer (eds.). *Silicon in Agriculture*. Elsevier, Amsterdam, The Netherlands.

Sayanjali, S., Ghanbarzadeh, B. and Ghiassifar, S. 2011. Evaluation of antimicrobial and physical properties of edible film based on carboxymethyl cellulose containing potassium sorbate on some mycotoxigenic *Aspergillus* species in fresh pistachios. *LWT - Food Science and Technology* 44, 1133–1138.
Schachtman, D.P., Reid, R.J. and Ayling, S.M. 1998. Phosphorus uptake by plants: From soil to cell. *Plant Physiology* 111, 447–453.
Schirra, M., D'Aquino, S., Cabras, P. and Angioni, A. 2011. Control of postharvest diseases of fruit by heat and fungicides: Efficacy, residue levels, and residue persistence. A review. *Journal of Agricultural and Food Chemistry* 59, 8531–8542.
Schirra, M., D'Aquino, S., Palma, A., Angioni, A. and Cabras, P. 2008. Factors affecting the synergy of thiabendazole, sodium bicarbonate, and heat to control postharvest green mold of citrus fruit. *Journal of Agricultural and Food Chemistry* 56, 10793–10798.
Schnurer, J. and Magnusson, J. 2005. Antifungal lactic acid bacteria as biopreservatives. *Trends in Food Science and Technology* 16, 70–78.
Scott, K.J. and Wills, R.B.H. 1979. Effect of vacuum and pressure infiltration of calcium chloride; a method for reducing bitter pit and senescence of apples during storage at ambient temperatures. *Horticultural Science* 10, 75–76.
Sebti, F. and Tantaoui-Elaraki, A. 1994. In vitro inhibition of fungi isolated from 'Pastilla' papers by organic acids and cinnamon. *LWT - Food Science and Technology* 27, 370–374.
Sholberg, P., Haag, P., Hocking, R. and Bedford, K. 2000. The use of vinegar vapor to reduce postharvest decay of harvested fruit. *HortScience* 35, 898–903.
Sholberg, P.L. 1998. Fumigation of fruit with short-chain organic acids to reduce the potential of postharvest decay. *Plant Disease* 82, 689–693.
Sholberg, P.L. and Gaunce, A.P. 1995. Fumigation of fruit with acetic acid to prevent postharvest decay. *HortScience* 30, 1271–1275.
Sholberg, P.L. and Gaunce, A.P. 1996. Fumigation of stone fruit with acetic acid to control postharvest decay. *Crop Protection* 15, 681–686.
Sholberg, P.L., Shephard, T., Randall, P. and Moyls, L. 2004. Use of measured concentrations of acetic acid vapour to control postharvest decay in d'Anjou pears. *Postharvest Biology and Technology* 32, 89–98.
Siegel, M.R. and Ragsdale, N.N. 1978. Antifungal mode of action of imazalil. *Pesticide Biochemistry and Physiology* 9, 48–56.
Singh, B.P., Tandon, D.K. and Kalra, S.K. 1993. Changes in postharvest quality of mangoes affected by preharvest application of calcium salts. *Scientia Horticulturae* 54, 211–219.
Sivakumar, D., Hewarathgamagae, N.K., Wilson Wijeratnam, R.S. and Wijesundera, R.L.C. 2002. Effect of ammonium carbonate and sodium bicarbonate on anthracnose of papaya. *Phytoparasitica* 30, 486–492.
Smilanick, J.L. 2011. Integrated approaches to postharvest disease management in California citrus packinghouses. *Acta Horticulturae* 905, 145–148.
Smilanick, J.L., Mackey, B.E., Reese, R., Usall, J. and Margosan, D.A. 1997. Influence of concentration of soda ash, temperature, and immersion period on the control of postharvest green mold of oranges. *Plant Disease* 81, 379–382.
Smilanick, J.L., Mansour, M.F., Margosan, D.A. and Mlikota Gabler, F. 2005. Influence of pH and $NaHCO_3$ on effectiveness of imazalil to inhibit germination of

Penicillium digitatum and to control postharvest green mold on citrus fruit. *Plant Disease* 89, 640–648.

Smilanick, J.L., Mansour, M.F., Mlikota Gabler, F. and Goodwine, W.R. 2006. The effectiveness of pyrimethanil to inhibit germination of *Penicillium digitatum* and to control citrus green mold after harvest. *Postharvest Biology and Technology* 42, 75–85.

Smilanick, J.L., Mansour, M.F., Mlikota Gabler, F. and Sorenson, D. 2008. Control of citrus postharvest green mold and sour rot by potassium sorbate combined with heat and fungicides. *Postharvest Biology and Technology* 47, 226–238.

Smilanick, J.L., Margosan, D.A., Mlikota Gabler, F., Usall, J. and Michael, I.F. 1999. Control of citrus green mold by carbonate and bicarbonate salts and the influence of commercial postharvest practices on their efficacy. *Plant Disease* 83, 139–145.

Smilanick, J.L. and Sorenson, D. 2001. Control of postharvest decay of citrus fruit with calcium polysulfide. *Postharvest Biology and Technology* 21, 157–168.

Smillie, R., Grant, B.R. and Guest, D. 1989. The mode of action of phosphite: Evidence for both direct and indirect modes of action on three *Phytophthora* spp. in plants. *Phytopathology* 79, 921–926.

Smoot, J.J. and McCornack, A.A. 1978. The use of potassium sorbate for citrus decay control. *Proceedings of the Florida State Horticultural Society* 9, 119–122.

Soni, M.G., Carabin, I.G. and Burdock, G.A. 2005. Safety assessment of esters of p-hydroxybenzoic acid (parabens). *Food and Chemical Toxicology* 43, 985–1015.

Spadoni, A., Neri, F. and Mari, M. 2015. Physical and chemical control of postharvest diseases. pp. 89–115. *In*: R.B.H. Wills and J.B. Golding (eds.). *Advances in Postharvest Fruit and Vegetables Technology.* CRC Press, Taylor & Francis Group, Boca Raton, FL.

Stadelbacher, G.J. and Prasad, K. 1974. Postharvest decay control of apple by acetaldehyde vapor. *Journal of the American Society for Horticultural Science* 99, 364–368.

Stopforth, J., Sofos, J.N. and Busta, F.F. 2005. Sorbic acid and sorbates. pp. 49–75. *In*: P.M. Davidson, J.N. Sofos and A.L. Branen (eds.). *Antimicrobials in Food*, 3rd ed. CRC Press, Taylor & Francis Group, Boca Raton, FL.

Tao, H.T.B. and Yamakawa, T. 2009. Phosphite (phosphorous acid): Fungicide, fertilizer or bio-stimulator? *Soil Science and Plant Nutrition* 55, 228–234.

Tarabih, M.E., EL-Eryan, E.E. and EL-Metwally, M.A. 2014. Physiological and pathological impacts of potassium silicate on storability of Anna apple fruits. *American Journal of Plant Physiology* 9, 52–67.

Theron, M.M. and Lues, J.F.R. 2011. Mechanism of microbial inhibition. pp. 117–149. *In*: M.M. Theron and J.F.R. Lues (eds.). *Organic Acids and Food Preservation.* CRC Press, Taylor & Francis Group, Boca Raton, FL.

Tian, S., Qin, G. and Xu, Y. 2005. Synergistic effects of combining biocontrol agents with silicon against postharvest diseases of jujube fruit. *Journal of Food Protection* 68, 544–550.

Tian, S.P., Fan, Q., Xu, Y. and Jiang, A.L. 2002. Effects of calcium on biocontrol activity of yeast antagonists against the postharvest fungal pathogen *Rhizopus stolonifer. Plant Pathology* 51, 352–358.

Ticconi, C.A., Delatorre, C.A. and Abel, S. 2001. Attenuation of phosphate starvation responses by phosphite in Arabidopsis. *Plant Physiology* 127, 963–972.

Tzortzakis, N. 2010. Ethanol, vinegar and *Origanum vulgare* oil vapour suppress the development of anthracnose rot in tomato fruit. *International Journal of Food Microbiology* 142, 14–18.

Tzortzakis, N., Chrysargyris, A., Sivakumar, D. and Loulakakis, K. 2016. Vapour or dipping applications of methyl jasmonate, vinegar and sage oil for pepper fruit sanitation towards grey mould. *Postharvest Biology and Technology* 118, 120–127.

US EPA (United States Environmental Protection Agency). 2006. 40 CFR 180.1268. Potassium silicate; exemption from the requirement of a tolerance. Final rule. *Federal Register* 71(114), 34267–34272.

Valencia-Chamorro, S.A., Palou, L., del Río, M.A. and Pérez-Gago, M.B. 2008. Inhibition of *Penicillium digitatum* and *Penicillium italicum* by hydroxypropyl methylcellulose-lipid edible composite films containing food additives with antifungal properties. *Journal of Agricultural and Food Chemistry* 56, 11270–11278.

Valencia-Chamorro, S.A., Pérez-Gago, M.B., del Río, M.A. and Palou, L. 2009. Curative and preventive activity of hydroxypropyl methylcellulose-lipid edible composite coatings containing antifungal food additives to control citrus postharvest green and blue molds. *Journal of Agricultural and Food Chemistry* 57, 2770–2777.

Varadarajan, D.K., Karthikeyan, A.S., Matilda, P.D. and Raghothama, K.G. 2002. Phosphite, an analog of phosphate, suppresses the coordinated expression of genes under phosphate starvation. *Plant Physiology* 129, 1232–1240.

Venditti, T., Dore, A., Molinu, M.G., Agabbio, M. and D'hallewin, G. 2009. Combined effect of curing followed by acetic acid vapour treatments improves postharvest control of *Penicillium digitatum* on mandarins. *Postharvest Biology and Technology* 54, 111–114.

Venditti, T., Ladu, G., Cubaiu, L., Myronycheva, O. and D'hallewin, G. 2017. Repeated treatments with acetic acid vapors during storage preserve table grapes fruit quality. *Postharvest Biology and Technology* 125, 91–98.

Venditti, T., Molinu, G.M., Dore, A., Agabbio, M. and D'Hallewin, G. 2005. Sodium carbonate treatment induces scoparone accumulation, structural changes, and alkalinization in the albedo of wounded citrus fruits. *Journal of Agricultural and Food Chemistry* 53, 3510–3518.

Wang, Y., Ren, X., Song, X., Yu, T., Lu, H., Wang, P., Wang, J. and Zheng, X.D. 2010. Control of postharvest decay on cherry tomatoes by marine yeast *Rhodosporidium paludigenum* and calcium chloride. *Journal of Applied Microbiology* 109, 651–656.

Warth, A.D. 1991. Mechanism of action of benzoic acid on *Zygosaccharomyces bailii*: Effects on glycolytic metabolite levels, energy production, and intracellular pH. *Applied and Environmental Microbiology* 57, 3410–3414.

Watanabe, K. and Takesue, S. 1976. Selective inhibition of the germination of *Bacillus megaterium* spores by alkyl p-hydroxybenzoates. *Chemical and Pharmaceutical Bulletin* 24, 224–229.

Weerahewa, D. and David, D. 2015. Effect of silicon and potassium on tomato anthracnose and on the postharvest quality of tomato fruit (*Lycopersicon esculentum* Mill.). *Journal of the National Science Foundation of Sri Lanka* 43, 273–280.

White, P.J. and Broadley, M.R. 2003. Calcium in plants. *Annals of Botany* 92, 487–511.

Wild, B.L. 1987. Fungicidal activity of potassium sorbate against *Penicillium digitatum* as affected by thiabendazole and dip temperature. *Scientia Horticulturae* 32, 41–48.

Wild, B.L., Wilson, C.L. and Winley, E.L. 1998. Apple host defence reactions as affected by cycloheximide, phosphonate, and citrus green mould, *Penicillium digitatum*. *ACIAR Proceedings Series* 80, 155–161.

Wilson, C.L., Franklin, J.D. and Otto, B.E. 1987. Fruit volatiles inhibitory to *Monilinia fructicola* and *Botrytis cinerea*. *Plant Disease* 71, 316–319.

Wisniewshi, M., Droby, S., Norelli, J., Liu, J. and Schena, L. 2016. Alternative management technologies for postharvest disease control: The journey from simplicity to complexity. *Postharvest Biology and Technology* 122, 3–10.

Wójcik, P., Skorupińska, A. and Filipczak, J. 2014. Impacts of preharvest fall sprays of calcium chloride at high rates on quality and 'Conference' pear storability. *Scientia Horticulturae* 168, 51–57.

Wood, E.M., Miles, T.D. and Wharton, P.S. 2013. The use of natural plant volatile compounds for the control of the potato postharvest diseases, black dot, silver scurf and soft rot. *Biological Control* 64, 152–159.

Yildirm, I. and Yapici, B.M. 2007. Inhibition of conidia germination and mycelial growth of *Botrytis cinerea* by some alternative chemicals. *Pakistan Journal of Biological Sciences* 10, 1294–1300.

Youssef, K., Ligorio, A., Sanzani, S.M., Nigro, F. and Ippolito, A. 2012. Control of storage diseases of citrus by pre- and postharvest application of salts. *Postharvest Biology and Technology* 72, 57–63.

Chapter 19

Extracts and Plant-Derived Compounds as Natural Postharvest Fungicides

Rosalba Troncoso-Rojas, Martín Ernesto Tiznado-Hernández, Tania Elisa González-Soto and Alberto González-León

Food and Development Research Center (CIAD, A.C.), Hermosillo, Sonora, Mexico

Abbreviations

ABC	Transporters type ATP-binding cassette
AITC	Allyl isothiocyanate
BITC	Benzyl isothiocyanate
CE	Capillary electrophoresis
DMAPP	Dimethylallyldiphosphate
EITC	Ethyl isothiocyanate
GC	Gas chromatography
GL	Glucosinolates
GST	Glutathione transferases
IPP	Isopentenyl diphosphate
ITC	Isothiocyanates
LC	Liquid chromatography
MAE	Microwave-assisted extraction
MFS	Major facilitator superfamily
Mpa	Mega Pascal
MS	Mass spectrometry
NMR	Nuclear magnetic resonance
PEE	Pomegranate exocarp extract
PLE	Pressurized liquid extraction
SFE	Supercritical fluid extraction
SPE	Solid-phase extraction
SWE	Subcritical water extraction
UAE	Ultrasound-assisted extraction

1 Introduction

Fungal diseases are one of the most important factors that decrease the postharvest quality of fresh horticultural produce, resulting in considerable economic losses. For many years, the principal method to control fungal diseases during postharvest has been the use of synthetic fungicides. In 2014, the total quantity of pesticides utilized in the world was close to 5.6 million tons of active ingredients, of which 10% were fungicides (FAOSTAT, 2017). Although synthetic fungicides have provided effective control of major postharvest diseases, there are concerns that their application may be harmful to human health and the environment (Nicolopoulou-Stamati et al., 2016). Also, their prolonged use has induced resistance in fungi, one of the biggest problems in agriculture. Only in the past 10 yr (Lamichhane et al., 2016), several studies have reported the evolution of resistance to fungicide products among several plant pathogenic fungi and oomycetes. Therefore, the presence of synthetic fungicide residues on horticultural produce and fungal resistance to fungicides continue to be a significant challenge, with a genuine need for developing alternatives that guarantee the safety of fresh horticultural produce and the environment. As part of programs for integrated management of postharvest diseases, alternatives for the reduction of produce decay should be implemented in conjunction with elements such as sanitation, good temperature management, and careful handling. A promising strategy is the use of plant extracts and bioactive compounds derived from plants.

The antifungal activity of plant extracts and plant-derived compounds against fungi responsible for postharvest diseases was reported beginning in the last century. Due to new trends in producing safer and healthier foods, research to discover new plants and/or plant-derived compounds with antifungal activity has grown considerably. The effectiveness of these plant extracts and plant-derived compounds has been widely demonstrated, under in vitro conditions, on a wide variety of plant-pathogenic fungi of importance in agriculture. On the other hand, studies demonstrating their effectiveness to control fungi infecting fruits after harvest are limited. Plant-derived compounds with well-demonstrated antifungal activity include phenolic compounds, flavonoids, terpenoids, alkaloids, saponins, and nitrogen-containing compounds, such as the isothiocyanates (Savithramma et al., 2011). Successful results obtained with plant extracts to control postharvest decay suggest the possible development of natural fungicides. However, due to variations in fruit physiology, environmental conditions, and methods and solvents used for the extraction, plant extracts and plant-derived compounds have shown reduced and inconsistent efficacy, as compared with synthetic fungicides. In order to increase the effectiveness of natural fungicides, several studies have been conducted in which the combined use of plant extracts and plant-derived compounds with other postharvest treatments as an integrated control strategy has been reported (Palou et al., 2016). Therefore, intensive research has been undertaken on the isolation and characterization of a wide variety of plant compounds with antifungal activity, alone or in combination with different treatments, to find additive, synergistic, or complementary activity with the goal to find natural fungicides with greater effectiveness.

In this chapter, we describe the most recent advances in the science and technology of the evolving area of plant extracts and plant-derived compounds with antifungal activity.

2 Occurrence and Biosynthesis of Plant Fungicides

Since ancestral times, plants were used to obtain compounds that have made great contributions to medicine, health, and human wellness. Plants contain a large variety of secondary metabolites, some of which have antimicrobial properties, so their contribution to the control of plant diseases has been widely demonstrated.

The discovery of plant extracts has gained widespread popularity. New plants are continually investigated worldwide as sources of antifungal compounds (Zhu et al., 2011). Recent reports include plants with antifungal activity, such as *Aloe vera* (Patel et al., 2012), *Glycyrrhiza glabra* (licorice) (Roshan et al., 2012), *Allium sativum* (garlic) (Thangavelu et al., 2013), *Zingiber officinale* (Kubra et al., 2013), Brassicaceae (Troncoso-Rojas and Tiznado-Hernández, 2015), *Punica granatum* (Romeo et al., 2015; Elsherbiny et al., 2016; Li Destri Nicosia et al., 2016), and *Solanum nigrum* (Musto et al., 2014), as well as native plants collected from southern Italy (Gatto et al., 2011), south of Morocco (Askarne et al., 2012; Talibi et al., 2012a); eastern Ethiopia (Sing, 2011), Jimma,

Ethiopia (Deressa et al., 2015), West Bengal, India (Bhutia et al., 2016), northwest Argentina (Sayago et al., 2012), and Brazil (Breda et al., 2016), and some mangrove species such as *Laguncularia racemose* and *Rhizophora mangle* (Cerqueira et al., 2016), among others.

The antimicrobial activity of plant extracts is attributed to secondary metabolites that have an ecological impact on organisms during its interaction with biotic and abiotic stresses (Ullah et al., 2016). Plants have metabolic pathways leading to tens of thousands of secondary metabolites capable of effectively responding to stress situations imposed by biotic or abiotic factors. Secondary metabolites act as a defense against herbivores, microbes, viruses, or competing plants, and as signal compounds, as well as a plant protection from ultraviolet radiation and oxidants (Kabera et al., 2014). They can be divided into three principal groups according to their biosynthetic origin: phenylpropanoids known as phenolic compounds, nitrogen-containing compounds (alkaloids, glucosinolates, and cyanohydrins), and terpenoids (Savithramma et al., 2011).

2.1 Phenolic Compounds

Phenolic compounds are extensively distributed in nature as secondary metabolites that are synthesized through different pathways in plants. More than 8,000 phenolic compounds with distinct structures and functions have been found and reported from various plant sources (Croteau et al., 2000). Phenolic compounds are classified into three broad categories: simple phenols (phenolic acids), polyphenols, and a diverse group.

In general, most phenolic compounds are biosynthesized by either shikimic acid pathway or malonate/acetate pathway. In shikimic acid pathway, aromatic amino acids are derived from carbohydrate precursors that are derived from the pentose phosphate pathway (D-erythrose-4-phosphate) and from glycolysis (phosphoenolpyruvic acid), which are then directed to the phenylpropanoid pathway to generate phenolic compounds. The basic structural unit of all phenolic compounds is a phenyl ring bearing a hydroxyl group (Valcarcel et al., 2015).

The most important phenolic acids in plants are substituted derivatives of hydroxybenzoic acids and hydroxycinnamic acids. The most common hydroxycinnamic acids are caffeic, *p*-coumaric, and ferulic acids, whereas the most common forms of hydroxybenzoic acid are *p*-hydroxybenzoic, vanillic, and protocatechuic acids (Yeo and Shahidi, 2015). Some phenolic compounds are stored in plant cells as inactive bound forms but are readily converted into biologically active compounds by plant glycosidases in response to pathogen attack. Since these compounds do not involve transcription of genes, they are considered as phytoanticipins. The concentrations of preformed phytoanticipins and inducible phytoalexins, in general, were found to decline during fruit ripening, and this occurred more rapidly in susceptible cultivars than in resistant cultivars (Prusky et al., 2013). In this sense, many phenolic acids are directly implicated in the response of plants to different types of stresses. These phytochemicals show an increase in concentration in response to pathogen infection and contribute to healing by lignification of damaged zones (Wang et al., 2015).

Flavonoids and their conjugates are a part of a vast group of natural products. Their biosynthesis combines the products of the shikimate metabolic pathway and polyketide pathway when 4-coumaroyl CoA ligase enzyme activates the metabolite

4-coumarate by transforming it into its CoA derivatives. Thereafter, chalcone synthase catalyzes the addition of three malonyl-CoA units (synthesized in the polyketide pathway) and removal of $3CO_2$ to form a group of substances called chalcones that constitute the skeleton for the biosynthesis of all flavonoids (Giweli et al., 2013). The flavonoids are found in many plant tissues, where they are present inside the cells or on the surfaces of different plant organs. The chemical structures of this class of compounds are based on a C6–C3–C6 skeleton. Depending on the linkage position of the aromatic ring to the benzopyran moiety, this group of natural products may be divided into three main classes: flavonoids (2-phenylbenzopyrans), isoflavonoids (3-benzopyrans), and neoflavonoids (4-benzopyrans). Among the flavonoids, the most widespread skeletons are flavones (luteolin, apigenin, and scutellarin) and then flavanones (eriodyctiol and naringin). All these groups usually share a common chalcone precursor (Lattanzio, 2013). Plants synthesize these compounds in response to microbial infection, and they have been frequently found effective as antimicrobial substances against a broad range of microorganisms during experiments in vitro and in vivo. For instance, Ouattara et al. (2011) showed the presence of the two main groups of secondary metabolites in the genus *Thymus*, namely volatile terpenoids and polyphenolic compounds. Both these groups of compounds are mainly responsible for the biological effects in this genus. Among the polyphenolic constituents, especially flavonoids and phenolic acids have been reported in *Thymus* plants. Besides, different phenolic acids like caffeic acid and rosmarinic acids have frequently been found in the *Thymus* species.

2.2 Terpenoids

Terpenoids constitute a large family and the most diverse class of phytochemicals produced by plants (more than 40,000 known structures) and represent the oldest group of small molecules (Aqil et al., 2010). They play a variety of roles, many of which are essential for the development and differentiation of all living organisms including plants, and they have been shown to have positive effects on human health (Giweli et al., 2013). Terpenoid roles include primary functions such as structural components of membranes, growth regulators (hormones), and photoprotectors (pigments). They also act as secondary metabolites such as pathogen inhibitory molecules, attractants, odors, and flavors, among others (Singh and Sharma, 2015; Tholl, 2015).

Terpenoids are polymeric isoprene derivatives synthesized from acetate via mevalonic acid pathway. All terpenoids are derived from the C_5-units 3-isopentenyl diphosphate (IPP) or dimethylallyldiphosphate (DMAPP). They are classified into hemiterpenes (C_5), monoterpenes (C_{10}), sesquiterpenes (C_{15}), diterpenes (C_{20}), triterpenes (C_{30}), tetraterpenes (C_{40}), and polyterpenes, according to the number of linked C_5-units. Steroids, carotenoids, and gibberellic acid are some of their members (Mithöfer and Boland, 2012; Thimmappa et al., 2014).

Plants use the majority of terpenoids for protection against the harmful effects of the abiotic and biotic environments. Loreto et al. (2014) reported that volatile or semivolatile low-molecular-weight terpenoids (isoprene, monoterpenoids, sesquiterpenoids, and diterpenoids) are implicated in the protection of plants against abiotic and biotic stress. Steroids such as β-sitosterol and stigmasterol have been previously shown to be involved in plant–pathogen interactions (Griebel and Zeier, 2010). Betulinic acid, oleanolic acid, and ursolic acid (pentacyclic triterpenoids) were detected in the plant extract of *Lantana camara* (Sing and Srivastava, 2012), which showed antifungal activity against the pathogen *Alternaria alternata* (Fr.) Keissl. In cotton, Hall et al. (2011) observed a greater accumulation of terpenoids in the cells surrounding the point of infection, significantly limiting the progress of the pathogen in the xylem vessels of a resistant cotton variety inoculated with *Fusarium oxysporum* f.sp. *vasinfectum* (G.F. Atk.) W.C. Snyder & H.N. Hansen.

2.3 Alkaloids

Alkaloids are a large class of organic heterocyclic chemical compounds that mostly contain basic nitrogen atoms. Alkaloids in their structure contain carbon, hydrogen, and nitrogen, and may also include oxygen, sulfur, and other less common elements such as chlorine, bromine, and phosphorus. Many alkaloids have complex chemical structures and contain multiple asymmetric centers, complicating structure elucidation (Croteau et al., 2000). These compounds are grouped into three main classes, depending on the precursors and final structure. These classes include right alkaloids, which are basic and contain nitrogen in the heterocyclic ring, such as nicotine. The second class is pseudoalkaloids, which are also basic but they are not derived from amino acids, such as caffeine. And the third class is protoalkaloids, which are synthesized from amino acids but does not contain nitrogen in the heterocyclic ring, such as mescaline (Agbafor et al., 2011). Among the alkaloids classes, we can find phenyl alkyl amines, pyrrolidines, tropane alkaloids, pyrrolizidines, and purine alkaloids (Carson and Hammer, 2010). Some examples of alkaloids include caffeine, cocaine, morphine, nicotine, and theobromine, and these can exhibit antimicrobial activity against pathogens (Mazid et al., 2011; Garba and Okeniyi, 2012).

The antifungal activity of alkaloids has been reported widely since the past century. A recent study carried out by Musto et al. (2014) showed that the aqueous extract of *S. nigrum* leaves contained alkaloids among other secondary metabolites such as flavonoids, saponins, steroids, glycosides, terpenoids, and tannins, although alkaloids were not present at higher levels. The aqueous extract of *S. nigrum* exhibited antifungal activity against *Penicillium digitatum* (Pers.) Sacc. Lin et al. (2011) also showed that an ethanol extract of *S. nigrum* inhibited spore germination of *Alternaria brassicicola* (Schwein.) Wiltshire, the causal agent of cabbage black leaf spot disease. Rabêlo et al. (2014) reported that the genus *Annona* is a rich source of isoquinoline, particularly aporphine alkaloids. Recently, the authors found that the leaf extract of the plant *Croton aromaticus* significantly inhibited the pathogens *Colletotrichum musae* (Berk. & M.A. Curtis) Arx, *Pestalotiopsis mangiferae* (Henn.) Steyaert, *A. alternata*, and *Colletotrichum gloeosporioides* (Penz.) Penz. & Sacc. According to the authors, phytochemical analysis revealed the presence of alkaloids, terpenoids, quinones, phytosterols, and flavonoids in leaf extracts of plants (Wijesundara et al., 2016).

2.4 Glucosinolate-Derived Isothiocyanates

Glucosinolates (GL) are a group of glycosides stored within cell vacuoles of dicotyledonous angiosperms. They are specific compounds present in about 3500 plants species of the Brassicaceae. When plant tissues are damaged, glucosinolates are hydrolyzed by myrosinase enzyme (thioglucosidase EC 3.2.3.1), catalyzing the formation of bioactive compounds, of which the isothiocyanates (ITC) are the most potent biocide compounds (Troncoso-Rojas and Tiznado-Hernández, 2015). They are well-known natural fungicides (Tiznado-Hernández and Troncoso-Rojas, 2006), and their activity against a large number of fungi, bacteria, and oomycetes has been shown (Smith and Kirkegaard, 2002). Furthermore, they can inhibit fungi that infect fruit (Troncoso et al., 2005; Troncoso-Rojas et al., 2005, 2009). It was reported that allyl isothiocyanate (AITC) showed the greatest antifungal activity (Troncoso et al., 2005).

The chemical properties of the ITCs depend upon the structure of glucosinolate side chain, plant species, and reaction conditions. Depending upon the R-group, these compounds can be classified into aliphatic straight-chain, aliphatic branched-chain, aliphatic straight- and branched-chain alcohols, ketones, sulfur-containing side chains, olefins, alcohols, aromatic, indole, benzoates, and with multiple glycosylations (Agerbirk and Olsen, 2012). The biosynthesis of GLs takes place via three independent steps. The first step is amino acid chain elongation, in which additional methylene groups are inserted into side chains; the second step is the conversion of amino acid residues to the core structure of GLs; and the third step is subsequent amino acid side chain modification (Sønderby et al., 2010). Based on these reactions, it has been concluded that a diverse range of GLs with different chemical structures is formed. Agerbirk and Olsen (2012) documented about 130 GL structures.

In general, members of the Brassicaceae plant family constitutively accumulate around 1% of dry weight of glucosinolates, but in some seeds they can accumulate at levels up to 10% of dry weight. The ITC content varies widely across individual vegetables with an average level of 16.2 μmol/100 g fresh weight, ranging from 1.5 μmol in raw cauliflower to 61.3 μmol in raw mustard greens (Tang et al., 2013). Recently, Lola-Luz et al. (2014) reported a total ITC content in broccoli between 1.8 and 5.5 μmol/100 g fresh weight. According to Slusarenko et al. (2008), *Allium sativum* can yield around 2 mg/g of allicin; however, quantities as low as sevenfold to tenfold are sufficient to inhibit the growth of numerous fungal pathogens.

3 Methodological Approach to Discover Novel Plant Fungicides

In order to evaluate the antifungal activities of natural products derived from plants, several conventional and non-conventional methods have been utilized. The method selected depends on the characteristics of the extracts and compounds tested. The complexity of plant metabolism is due to a wide number of compounds, and the extracts from the same plant are more complex, with the molecular composition highly varying from one extraction to another extraction (Regnault-Roger, 2012). Therefore, the successful discovery of new active natural products from plants has traditionally been low due to the inherent difficulties and costs associated with solvent extraction, isolation, purification, and structure elucidation of individual components of plant metabolites. In agreement to Gurjar et al. (2012), it is important to point out the need to standardize the methods of extraction, so that the search for new biological active phytochemicals could be more systematic and focused.

The analytical methodologies to discover and study the natural fungicides include extraction, isolation, and characterization of bioactive compounds with antifungal activity present in plant extracts (Das et al., 2010). Extraction is the crucial first step in the plant analysis to extract the chemical components of interest from plant materials for further separation and characterization. The basic operation includes several steps such as pre-washing, drying of plant materials or freeze drying, and grinding to obtain a homogenous sample with a goal to improve the kinetics of analytic extraction and increase the contact of the sample surface with the solvent system (Sasidharan et al., 2011). The selection of the solvent system depends primarily on the specific nature of the bioactive compound to be extracted. Different solvents, including aqueous and organic solutions, have been described. Anwar and Przybylski (2012) extracted two different batches of flax seed with solvents of different polarities, and the different extracts were tested for antioxidant activity with different methods. The highest yield of extracts was achieved with 80% methanol, but the extract did not contain large amounts of phenolic compounds and flavonoids; however, when 80% ethanol was used for extraction, higher level of flavonoids was detected. Quezada and Cherian (2012) showed that the yield of total phenolic compounds extracted from flax seeds was affected by solvent polarity; for example, solvents with low polarities, such as ethyl acetate, are less efficient than more polar solvents such as ethanol. For instance, a study carried out by Arya et al. (2012) reported that the extraction of *Psidium guajava* leaves with ethanol and hydro-alcohol had the highest yield. This plant extract showed the greatest content of alkaloids, saponins, carbohydrates, tannins, and flavonoids, compared to the extraction with petroleum ether, chloroform, and water. Methanol extracts of *Garcinia atriviridis* exhibited higher antioxidant activity compared to the aqueous extract (Al-Mansoub et al., 2014). Vongsak et al. (2013) reported that the maceration of *Moringa oleifera* with 70% ethanol showed highest phenolic and flavonoid contents as compared to Soxhlet extraction and percolation methods.

Various methods such as sonication, heating under reflux, Soxhlet extraction, maceration, and percolation of fresh green plants or dried powdered plant material are commonly used. But other modern extraction techniques include supercritical fluid extraction, pressurized liquid extraction, microwave-assisted extraction, ultrasound-assisted extraction, solid-phase micro-extraction, solid-phase extraction, and surfactant-mediated techniques. These methods use less extracting organic solvent and have advantages regarding sample degradation. Additional steps such as clean-up and concentration are eliminated before chromatographic analysis, improving in extraction efficiency and selectivity (Azwanida, 2015).

Supercritical fluid extraction (SFE) has been used for many years for the extraction of volatile components, for example essential oils and aroma compounds from plant materials on an industrial scale. Recently, the application of this technique at analytical scale has attracted widespread interest for preparing samples before chromatographic analysis. The potential advantages include the ability to perform rapid (often less than 30 min) extractions, to reduce the use of hazardous solvents, and to couple the extraction step with gas, liquid, or supercritical fluid chromatography (Pereira et al., 2010). A significant advantage of using SFE to extract active compounds from plants is that degradation due to prolonged exposure to high temperatures and/or atmospheric oxygen is reduced or avoided (Capuzzo et al., 2013). The use of water under high temperature (between 100 and 374°C) and pressure (between 5 and 22 Mpa) to maintain the water in a liquid state during extraction processes is usually referred to as subcritical water extraction (SWE). This technique has emerged as a useful tool to replace traditional extraction methods. SWE has significant advantages over the conventional extraction methods; it is faster, produces high yields, reduces the use of solvents, produces extracts of great quality, lowers the costs of the extracting agent, and it is an environmentally friendly technique (Plaza et al., 2010).

Pressurized liquid extraction (PLE) is gaining popularity and it has been widely employed for the extraction of bioactive compounds from plant extracts (Miron et al., 2010). The PLE is referred to as accelerated solvent extraction (SE) and pressurized SE. This method uses organic liquid solvents at high temperature (50 to 200°C) and pressure (1450 to 2175 psi) to enable rapid extraction of the compounds (Dunford et al., 2010; Li et al., 2012).

In contrast to conventional liquid–solid extraction methods (e.g., Soxhlet extraction), in which a relatively long extraction time (typically 3–48 hr) is required, the use of microwave energy for solution heating results in a significant reduction in the extraction time (to usually less than 30 min). This technique is known as microwave-assisted extraction (MAE). The microwaves heat the solvent or solvent mixture directly, thus accelerating the speed of heating. Besides having the advantage of super extraction speed, MAE also allows a significant reduction in the consumption of organic solvent to typically less than 40 mL, compared with the 100–500 mL required in Soxhlet extraction (Zhang et al., 2011).

Ultrasound-assisted extraction (UAE) has been proposed as an alternative to conventional extraction, providing greater recovery of targeted compounds. It also reduces solvent consumption and allows a faster recovery of bioactive compounds. Its better extraction efficiency is related to the phenomenon called acoustic cavitation. When the ultrasound intensity is sufficient, the expansion cycle can generate microbubbles in the liquid. Once formed, the bubbles will absorb the energy from the sound waves and grow during the expansion cycles and recompress during the compression cycle. Later, the bubbles may start another cycle or collapse creating shock waves of high pressure and temperature (several hundred atmospheres and around 5000°K of temperature) (Soria and Villamiel, 2010). In ultrasonically assisted extraction, the use of an aqueous surfactant solution of 10% Triton X-100 as the extraction solvent resulted in faster extraction kinetics and a higher recovery compared to methanol and water (Esclapez et al., 2011).

Solid-phase extraction (SPE) is a simple preparation technique based on the principles used in liquid chromatography (LC), in which solubility and functional group interactions of sample, solvent, and adsorbent are optimized to get sample fractionation and concentration. A wide range of chemically modified adsorbent materials (silica gel or synthetic resins) enables precise group separation by different types of physicochemical interaction, i.e., reversed-phase (C_2, C_8, C_{18}), cation- and anion-exchange and so on. It should, in particular, be noted that SPE is well suited for the treatment of sample matrices with high water content, e.g., extracts of herbal materials (Rajabi et al., 2014).

Craige and Rochfor (2010) presented an overview of the instrumentation and data management tools required for metabolomics. The analysis of all the metabolites synthesized by a biological system is particularly challenging because of the diverse chemical nature of the metabolites. The techniques such as high-field nuclear magnetic resonance (NMR), spectroscopy and mass spectrometry (MS), either alone or in combination with gas chromatography (GC), LC, and capillary electrophoresis (CE), are very useful to measure the contents of low-molecular-weight metabolites in biological systems. Advanced statistical and bioinformatics tools are then used to maximize the recovery of information and interpret the large data sets generated. The scientific development of chromatographic separation and identification techniques has advanced enough to be able to elucidate the chemical structure and full identification of a single component within extracts. The double separation methods and rigorous structure elucidation techniques such as chromatography (gas or liquid) coupled with mass spectrometry, and high-field NMR will facilitate the transition from preparative-scale extraction to commercial extraction processes to obtain large quantities of molecules of interest to conduct in vitro, in vivo, and toxicological experiments.

4 Plant Fungicides for the Control of Postharvest Diseases of Fresh Horticultural Produce

Numerous scientific studies have focused on the investigation of plant-derived fungicides motivated by the need to replace synthetic fungicides with environmentally friendly and safe treatments to control postharvest fungal decay of horticultural produce. Some studies have focused on assessing the antifungal activity of native plants around the world. For example, extracts from plants from southern Italy were evaluated to control fungal diseases of apricots, nectarines, table grapes, and oranges under cold storage conditions (Gatto et al., 2011). On nectarines and apricots, *Monilinia laxa* (Aderh. & Ruhland) Honey was completely inhibited by *Sanguisorba minor* extract after 6 d at 15°C. On table grapes, plant extracts such as those from *S. minor* and *Orobanche crenata* caused a significant reduction (23 to 53%) in gray mold caused by *Botrytis cinerea* Pers. after 6 d at 20°C. Extracts of neem have been extensively tested to control diseases. Wang et al. (2010) found that decay caused by *Monilinia fructicola* (G. Winter) Honey was significantly reduced (30.6%) on plum fruits treated with 100 mg/mL of neem seed kernel extract (NSKE) 48 hr after inoculation. Furthermore, on Yali pear treated with NSKE at the concentration of 200 mg/mL, the growth of *A. alternata, Trichothecium roseum* (Pers.) Link, and *Penicillium expansum* Link 96 hr after inoculation was reduced by 42, 80, and 61%, respectively. Postharvest treatments with extracts from two wild edible plants (*O. crenata* and *S. minor*), aqueous solutions of calcium chloride ($CaCl_2$), and sodium bicarbonate ($NaHCO_3$), and the combination of the extracts with added $CaCl_2$ or $NaHCO_3$, were evaluated to control postharvest diseases of sweet cherries. Extracts from *S. minor* and *O. crenata* inhibited rot development on stored fruits by 89 and 76%, respectively, while the combined application of plant extracts and salts did not improve the level of fungal control. Chromatographic analysis of *O. crenata* extract showed that verbascoside was the main phenolic compound, comprising about 95% of the total phenolic compounds (Gatto et al., 2016). Usall et al. (2015) described some plant extracts with antifungal activity against *M. laxa* and other important postharvest pathogens causing diseases in stone fruits.

Similar studies were carried out to control citrus postharvest diseases. Aqueous extracts of several plants collected in southern Morocco were used to inhibit blue mold caused by *Penicillium italicum* Wehmer in 'Valencia' oranges during storage at 20°C and 95% relative humidity for 10 d. Disease severity on treated fruits was reduced with extracts of *Halimium umbellatum* (95%) and *Inula viscosa* (75%) at the concentration of 500 mg/mL. Furthermore, while no visible symptoms of phytotoxicity were detected on oranges treated with *H. umbellatum, Ceratonia siliqua, Bubonium odorum*, and *Asteriscus graveolens* aqueous extracts, fruits treated with aqueous extracts of *I. viscosa, Anvillea radiate*, and *Hammada scoparia* showed dehydration and browning around the treated rind lesions, indicating they were phytotoxic (Askarne et al., 2012). In another work with oranges, extracts of *S. minor* caused a significant reduction (92%) of *P. digitatum* after 25 d at 7°C. The potent antifungal activity could be ascribed to a possible synergic effect of the phenolic mixture composition of caffeic acid derivatives, quercetin-3-glucoside, and kaempferol-3-glucoside. The last two compounds were the most abundant flavonoids (Gatto et al., 2011). On clementine mandarins, the methanolic extracts of the Moroccan plants *Cistus villosus, H. umbellatum*, and *C. siliqua* successfully controlled citrus sour rot caused by *Geotrichum citri-aurantii* (Ferraris) E.E. Butler. Among them, the extract of *C. villosus* at 50 mg/mL completely controlled the disease (Talibi et al., 2012a). According to Talibi et al. (2012b), aqueous extracts of the same plant species were less effective in controlling citrus sour rot than methanolic extracts. The authors discussed that the differences observed in antifungal activity were due to the polarity of each solvent, which affected the type of molecules that were extracted. Methanol is a polar solvent that can extract several antimicrobial compounds such as alkaloids, triterpene glycosides, tannins, sesquiterpene lactones, and phenolic compounds. In another study, aqueous leaf extracts of *A. vera* effectively controlled *P. digitatum* and *P. italicum* on 'Kinnow' mandarins after 30 d at 5°C. Furthermore, the onset of fungal decay caused by both pathogens was significantly delayed, which increased the storage life by 9 d. The positive effect of this extract could be attributed to aloin and aloe emodin, the most abundant anthraquinones present in *A. vera*. All the extracts did not influence the postharvest quality parameters of mandarin fruit, which included contents of total soluble solids, titratable acidity, and ascorbic acid (Jhalegar et al., 2014). On lemon fruits, an extract of the Argentine plant *Parastrephia lepidophylla* at 600 mg/L showed curative and protective effects against decay caused by *P. digitatum*, due to its high phenolic content (Sayago et al., 2012). Recently, an aqueous extract of *S. nigrum* applied at the concentration of 0.066 mg/mL was found to be more effective as a preventive treatment rather than as a curative treatment for the control of *P. digitatum* on lemon fruit (Musto et al., 2014). The antifungal activity of the plant extracts could be attributed to phenolic compounds; nevertheless, the type of phenolic molecule was more relevant than the concentration. In this sense, the antifungal activity of the phenolic compounds quercetin, scopoletin, and scoparone was evaluated against *P. digitatum* on 'Navelina' oranges. The three phenolic compounds reduced disease severity by 36–47% up to 12 d post-inoculation, whereas, at 14 d post-inoculation, scoparone proved to be the most effective compound (37% reduction), although its possible mode of action is unknown (Sanzani et al., 2014).

Alternaria brassicicola (Schwein.) Wiltshire causes cabbage black leaf spot, one of the most deteriorative postharvest diseases of cabbage. Lin et al. (2011) reported that an aqueous extract of *S. nigrum* completely inhibited spore germination of this pathogen. According to Atanu et al. (2011), this effect could be due to compounds such as alkaloids, flavonoids, saponins, and steroids, among others, which were present in aqueous extracts of *S. nigrum*.

Several plant-based extracts have been assessed to control anthracnose development in some tropical fruits. On banana (*Musa paradisiaca*), aqueous and methanolic extracts of *Momordica charantia* inhibited by 70 and 80%, respectively, the development of the disease caused by *C. musae* when applied 2 d before fungal inoculation (Celoto et al., 2011). Cruz et al. (2013) reported that an aqueous extract of *Azadirachta indica* at concentrations of 2 and 4% (v/v) significantly reduced (41.73 and 57.83%, respectively) banana anthracnose caused by *C. musae*. A similar study was conducted with an extract of *Acacia albida*, an Ethiopian plant, which significantly reduced anthracnose incidence (29.8%). Considering the low effectiveness of the aqueous plant extract, it was combined with hot water treatment at 45 and 50°C, and lower anthracnose incidence (12.3%) and severity (23.4%) were observed when the extract was applied in combination with hot water at 50°C (Bazie et al., 2014). These results suggest that the heat treatment enhanced the fungicide activity of plant extracts, which agrees with reports from other studies (Sing, 2011). In another study with bananas, extracts of the plants *Z. officinale, Polyalthia longifolia*, and *Clerodendrum inerme* (from Bengal, India) reduced disease severity by more than 90%. In this study, the rhizome extract of *Z. officinale* was the most highly effective in controlling anthracnose on stored fruits (Bhutia et al., 2016). The antifungal activity of the extract of *Z. officinale* could be related to the presence of alpha-cucurmene and zingerone, whose antifungal activities were previously reported by other researchers (Kubra et al., 2013). On mangoes (*Mangifera indica* L.), Alemu et al. (2014) observed greatly reduced anthracnose caused by *C. gloeosporioides* when fruits were dipped in aqueous extract of *Ruta chalepensis*. The extract reduced disease incidence to below 36% during the experimental period, in contrast with the untreated fruits, in which disease reached more than 93.4%. Deressa et al. (2015) reported a higher anthracnose reduction in mangoes treated with *Proposis juliflora* and *L. camara* leaf extracts, at 0 and 48 hr after inoculation, than on control fruits. The aqueous leaf extract of *L. camara* had greater inhibitory activity than the solvent (methanol or acetone 70%) leaf extract, suggesting that the active compounds such as pentacyclic triterpenoids, betulinic acid, oleanolic acid, and ursolic acid were more soluble in water. On the other hand, the antifungal activity of pentacyclic triterpenoids has been reported in previous studies (Sing and Srivastava, 2012). In the case of papaya, the application of plant extracts to control anthracnose has been reported for more than 20 yr. Bautista-Baños et al. (2013) reported that extracts from various plants such as *Acharas sapota, Chrysophyllum cainito, Pouteria sapota, Carica papaya, Pachyrrizus erosus, Phythecellobium dulce, Cestrum nocturnum*, and *Lantana camara* provided significant control of diseases caused by *C. gloeosporioides, Rhizopus* sp., *Aspergillus* sp., and *Mucor* sp. Ademe et al. (2014) found that the methanolic extract of the Ethiopian plant *Echinops* sp. at a concentration of 25% significantly reduced anthracnose on papaya fruit, 14 d after inoculation. In the case of pineapple fruit, de Souza et al. (2015) found that the application of extracts of *Mormodica charantia* significantly reduced the development of black rot caused by the fungus *Chalara paradoxa* (De Seynes) Sacc., without adversely affecting the postharvest quality of the fruit.

Recently, pomegranate exocarp has gained attention as a source of alternative antifungal substances. The efficacy of pomegranate exocarp extract (PEE) in controlling different postharvest diseases of citrus, apples, table grape berries, and sweet cherries was evaluated. Romeo et al. (2015) found that a concentrated extract of PEE (80% ethanol/water) completely inhibited the development of *B. cinerea* on table grapes. The ethanolic extract of PEE was a very effective preventive treatment to control some of the most important postharvest pathogens including *P. digitatum* and *P. italicum* on lemons, *P. digitatum* on grapefruits, *P. expansum* on apples, and *M. laxa* and *B. cinerea* on sweet cherries (Li Destri Nicosia et al., 2016). The use of PEE at a concentration of 12 g/L completely inhibited decay caused by *P. digitatum* and *P. italicum* on lemons and *P. digitatum* on grapefruits, without harming the commercial quality of the fruit throughout the experiment. At the same concentration, decay caused by *P. expansum* on apples decreased by 80% by PEE. Lower efficacy against *P. expansum* was observed with lower concentrations of PEE (1.2 and 0.12 g/L). Additionally, PEE was applied under semi-commercial conditions and caused a significant reduction of natural decay on sweet cherry cvs. 'B. Moreau' (61%) and 'Giorgia' (95.7%), and lemons (88%) (Li Destri Nicosia et al., 2016). On potato tubers, a methanolic PEE applied at 20 mg/mL greatly reduced the growth of *Fusarium sambucinum* Fuckel (Elsherbiny et al., 2016). The antimicrobial capacity of PEE has been associated with its relatively high phenolic content with the prevalence of punicalagin A and B, a group of ellagitannins, granatin B, and ellagic acid (Akhtar et al., 2015; Romeo et al., 2015). Chlorogenic acid was the major phenolic compounds detected by HPLC in the methanolic PEE (42.4 mg/g), followed by catechin at a concentration of 30.4 mg/g. However, the antifungal and antimicrobial activities depended on the extraction method and the type of tested microorganism and fruit host evaluated.

Other plant-derived compounds that have drawn attention due to their effectiveness to control postharvest diseases of fruits and vegetables are the isothiocyanates (ITC) compounds produced by *Brassica* spp. Studies were recently conducted to control fungal diseases on apple, blueberries, and strawberries. Wang et al. (2010) reported a significant reduction of fungal decay in blueberry (*Vaccinium corymbosum* cv. 'Duke') exposed to allyl isothiocyanate (AITC) (or 2-propenyl isothiocyanate) at 5 μL/L, during storage at 10°C, and that fungal growth was retarded throughout 12 d. In a similar study, AITC was evaluated for its efficacy to control *B. cinerea* on strawberries cv. 'Monterey'. In this study, synthetic AITC and glucosinolate-derived AITC reduced gray mold by 91.5 and 86.5%, respectively (Ugolini et al., 2014). On apples, the effect of AITC and ethyl ITC (EITC), each alone or in combination, against *P. expansum* and *B. cinerea* was demonstrated. The combination of AITC and EITC at a ratio of 3:1 completely inhibited decay after 3–4 d of incubation at 20°C, suggesting a synergistic effect of both ITCs evaluated (Wu et al., 2011). Recently, the antifungal activity of garlic cloves to inhibit *B. cinerea, P. expansum*, and *Neofabraea alba* (E.J. Guthrie) Verkley was demonstrated on apple cvs. 'Granny Smith', 'Golden Delicious', and 'Pink Lady'. Disease caused by *B. cinerea* and *P. expansum* was inhibited by 85 and 95%, respectively, in all the apple cultivars evaluated after 7 d at 20°C (Daniel et al., 2015). Shinde et al. (2016) reported similar results on grapes. All these studies clearly showed the potent activity of ITCs to control postharvest diseases. Furthermore, their residues on the fruit surface were very low, as demonstrated with apples in which the skin contained <10 μg/g of AITC and <5 μg/g of EITC, and no detectable residue of either ITCs in the pulp (Wu et al., 2011). Similarly, the concentration of AITC in strawberries treated for 4 hr with 0.1 mg/L was found to be lower than 1 mg/kg after 7 d (Ugolini et al., 2014). According to the Council of Europe (CE, 2000), this concentration is lower than that reported as the tolerable human daily intake of 0.06 mg kg/bw (body weight) per day. Strawberry fruit taste was not altered by ITC treatments (Ko et al., 2012).

According to the experiments described above, the potential capacity of plant extracts and plant-derived compounds to reduce or control postharvest diseases of fresh horticultural produce is evident. However, the vast majority of studies conducted so far focused on the study of the antifungal effect of compounds at the laboratory level, and studies at a larger scale or commercial level are scarce. Therefore, it is essential to conduct more studies to confirm the efficacy of the compounds under commercial conditions, since the effectiveness of many postharvest treatments has proven to be higher under laboratory conditions than on larger scale trials (Bautista-Baños et al., 2013).

4.1 Mechanisms of Action

Recent work on the antifungal activity of plant extracts and plant-derived compounds against postharvest fungal diseases of fresh horticultural produce is currently at very early stages, and the information about their mode of action is scarce. However, some data suggest the possible mode of action of some of these compounds.

Plant extracts and plant-derived compounds have the characteristic of being hydrophobic, which allows them to break the lipids of the cell membrane and the mitochondria, altering the cellular structures and making them more permeable. The uncontrollable leakage of critical molecules and ions from the cell leads to its death (Deressa et al., 2015). In this sense, phenolic compounds are known to alter the membrane functionality of pathogens. Endo et al. (2010) suggested that the antifungal activity may be due to reaction with sulfhydryl groups, more nonspecific interactions with the proteins, and unavailability of substrates to microorganisms. Elsherbiny et al. (2016) reported that the presence of high molecular weight phenolics, tannins (punicalagin, pedunculagin, pubicalagin), ellagitannins, proanthocyanidins, complex polysaccharides, and flavonoids (catechin, epicatechin, rutin) in methanolic PEE produced extensive damage on the hypha of *F. sambucinum* that include highly vacuolated cytoplasm, migration of the protoplasm from the cell wall, and disintegration of cytoplasmic organelles. These changes might be due to the disruption of cell permeability, which resulted in an imbalance in internal osmotic pressure, subsequent disintegration of cell organelles, leakage of cytoplasmic components, and finally cell death.

In the case of ITCs, the mode of action on fungal pathogens is not yet well elucidated. Some hypotheses suggest that the mechanism of action of ITC in living organisms can be due to non-specific reactions with any protein of the organism. It is possible that ITC activity could be due to possible non-specific and covalent interaction of the functional group (–N = C = S) with sulfhydryl groups (–SH) and amino residues of the lateral amino acid group as well as with the disulfide bonds of proteins (Troncoso-Rojas and Tiznado-Hernández, 2015). During pathogenesis, fungal effectors are secreted, which are cysteine-rich proteins. In *Fusarium oxysporum* f. sp. *lycopersici* (Sacc.) W.C. Snyder & H.N. Hansen, a protein was found to be secreted in the xylem, Six 1 or Avr3, which is a small cysteine-rich protein that is required by *F. oxysporum* for full virulence on tomato (de Sain and Rep, 2015). In this sense, assuming that the ITC reacts with sulfhydryl groups, it is possible to suggest that they can react with the cysteine-proteins secreted by the fungal effector, disabling the ability of the pathogen to infect the host plant and avoiding the disease, although more experimental evidence is needed to prove this statement. Additionally, it has been reported that 2-propenyl isothiocyanate can induce the synthesis of glutathione S-transferase, which has been shown to have anticancer activity (Zhang, 2010; Geng et al., 2011; Fimognari et al., 2012), while sulforaphane and 6-methylsulfinyl hexyl ITC have shown inhibitory effects on cancer cell growth (Mikkelsen et al., 2010). It is clear from this that ITC could have positive effects on human health and could be safe to use.

5 Molecular Mechanisms of Fungal Response to Plant Fungicides

Experiments performed to elucidate the fungal responses to compounds derived from plants with fungicidal activity are described in this section. The molecule stilbene has been reported to have fungicidal activity in vitro against several species of pathogenic fungi (Celimene et al., 2001; Kumar and Nambisan, 2014). Furthermore, in grapes this molecule acts as a defense system against the attack of *B. cinerea* (Bavaresco et al., 2015). Also, it was shown that inhibition of enzyme stilbene oxidase in *B. cinerea* by the phenolic compounds present in grapefruit inhibited the infection development (Goetz et al., 1999). Besides, part of the molecular mechanism of *B. cinerea* during grapefruit infection is the induction of the enzyme stilbene oxidase to metabolize stilbene into a larger molecular weight compound with dehydrodimeric structure and more hydrophobicity characteristics (Breuil et al., 1999). In this way, this fungus can successfully infect fruit berries and avoid the effect of a plant extract containing this natural fungicide.

Several reports were published describing experiments to elucidate the molecular response of the fungal species such as *A. alternata* and *A. brassicicola* to the effect of several ITCs. These experiments are probably the most serious attempts to elucidate the molecular mechanism of the fungal response and adaptation to a natural fungicide from plants. An experiment was carried out to isolate the genes differentially expressed in spores of *A. brassicicola* germinating either in the leaf of *Arabidopsis thaliana* or in water by using a subtractive suppressive hybridization protocol. RNA was extracted after about 80% of the spores were germinated. Regulation of genes showing differential expression between the two different conditions was studied by using a protocol of dot-blot macro-arrays and virtual northern blots. *Arabidopsis thaliana* produces ITC to suppress fungal infection and, therefore, the gene induced in the fungus is related with the fungal response to the ITC. With this protocol, 47 cDNAs differentially expressed were isolated. It was found that these genes play a role in transport, cell defense/stress, metabolism, protein destination, protein synthesis, transcription, and energy. Out of this, it was found that five genes were expressed in spores germinating on the leaf surface but not in spores germinating in water. These genes encode for a cyanide hydratase, arsenic ATPase which is a type of ABC transporter, formate dehydrogenase, *Alternaria* allergen, and one of them did not show any significant hit in the database. The expression analysis of these five cDNAs was carried out again with spores germinating in nutrient media, on the leaf of cabbage (*Brassica oleraceae* cv. 'Capitata') and on leaf of *A. thaliana*. It was found that only the cDNAs with an unknown homology were expressed in spores infecting plant leaves (Cramer and

Lawrence, 2004). Efforts by the authors of this chapter to elucidate the protein encoding for this gene included communication with Cramer and Lawrence and BLAST analysis against the NCBI nucleotide database looking for genes of bacteria, fungi, and *A. brassicicola* using the sequence published in the Genebank (ID: CA405405.1). Also, a BLAST analysis against the *Alternaria* genome database (http://alternaria.vbi.vt.edu/index.html) was carried out, but all the efforts were unsuccessful and the protein encoded by the EST mentioned by Cramer and Lawrence (2004) remains unknown.

Alternaria brassicicola was treated with benzyl isothiocyanate (BITC) and the genes induced were isolated by a non-radioactive differential display reverse transcription technique. Sellam et al. (2006) showed the induction of a glutathione-S-transferase gene by the BITC treatment. The 5′ end of the gene was obtained by the PCR walking technique and designated *AbGst1*. *AbGst1* gene was also found to be expressed during fungal infection of the *A. thaliana* leaves. The recombinant protein of the gene was obtained in bacteria, and it was found that the enzyme showed activity to degrade 2-propenyl isothiocyanate and BITC. It was suggested that the glutathione-S-transferase gene isolated most likely plays a role in the defense of *A. brassicicola* against the toxic effect of ITCs. In another experiment by Sellam et al. (2006), *A. brassicicola* was exposed to 2-propenyl isothiocyanate and the phytoalexin camalexin. The genes induced by the treatments were isolated by a suppressive subtractive hybridization protocol. In the case of 2-propenyl isothiocyanate, a total of 76 non-redundant expressed sequence tags were isolated and found to play different functions such as oxidative burst, stress and defense, membrane transporters, transcription, and cell cycle as well as metabolism. Interestingly, four of the isolated ESTs showed similarity to putative glutathione-S-transferase enzymes, which agrees with the results of the experiment described above. Several genes induced by 2-propenyl isothiocyanate such as glutathione transferases (GST), glutathione peroxidase, γ-glutamylcysteine synthetases, thioredoxins, thioredoxin reductase, and oxidoreductases play a role in the oxidative stress response. Sellam et al. (2007) suggested that the oxidative stress induced by ITC is due to the depletion of glutathione, which reacts with ITCs to inactivate their toxic effects. In agreement with this, the presence of unstable reactive oxygen species in the fungal cytoplasm was shown by demonstrating the synthesis of the fluorescent compound 2′,7′-dichlorofluorescein in fungi treated for 4 hr with either allyl- or BITC. The induction of the membrane transporters type ATP-binding cassette (ABC) or major facilitator superfamily (MFS) by AITC clearly indicates that the ITC alone or covalently bound to glutathione is transported outside of the cytoplasm to eliminate its toxic effect.

According to Glawischnig (2007), camalexin is the most important phytoalexin of *A. thaliana*, and it is induced by different plant pathogens. In the case of fungi exposed to camalexin, the procedure allowed the isolation of 112 unique ESTs. Out of these, 50 ESTs showing homology to known or putative proteins were classified into different categories from the functional point of view, namely, transcription, cellular transport, oxidative burst, stress and defense, protein fate, cell wall, and metabolism. Some of the genes induced in the fungi play a role in ergosterol biosynthesis, which is a major component of the plasma membrane. The induction of these genes suggests that camalexin alters the plasma membrane integrity of the fungi. In agreement with this, it was found that camalexin treatment alters the permeability of the fungal plasma membrane using a green fluorescent dye, which can only penetrate and bind to the DNA when there is membrane damage. Furthermore, green fluorescence was only observed in hyphae treated for

24 hr with camalexin. They recorded the induction of two genes related to the melanin biosynthesis that could be playing a role decreasing cell wall permeability to avoid the intake of camalexin into the fungal cytoplasm. In agreement with this, it was observed that the induction of genes plays a role in maintaining the structure of cell walls. Furthermore, they reported the induction of two MFS membrane transporters that may play a role in transporting the camalexin out of the cytoplasm. From this description, it can be concluded that part of the fungal response to the camalexin toxic effect is to induce genes that play a role in transporting the compound outside the cytoplasm, reducing the intake of camalexin by strengthening the cell wall, or repairing the camalexin-induced damage in the fungal plasma membrane.

Experiments by our research group showed that *A. alternata* can adapt to a toxic level of 2-propenyl isothiocyanate if the fungus is exposed to higher concentrations of the ITC with small differences between the different ITC concentrations instead of being exposed to a lethal concentration from the beginning. Analysis of these fungal strains with the protocol of inter simple sequence repeats showed differences in the similarity coefficients between the resistant fungal strains and the untreated fungal strains suggesting extensive changes in the genome of the fungus due to the treatment with the ITC (Troncoso-Rojas et al., 2013). In agreement with this, it was reported that the 2-propenyl isothiocyanate can induce mutations in *Drosophila melanogaster* (Auerbach and Robson, 1944). Because of this, we decided to study the *A. alternata* mechanism of response and adaptation to the toxic effect of the 2-propenyl isothiocyanate. A suppressive subtractive hybridization protocol was carried out using the cDNAs from the fungi under the ITC treatment as a tester and the cDNAs from the untreated fungus as a driver. It was possible to isolate 50 ESTs divided into 17 contigs and 33 singletons. The annotation of the ESTs showed molecular functions related to binding, structural molecule activity, translation regulator activity, transporter activity, molecular transducer activity, and catalytic activity. In the case of the biological process, ESTs were found to play a role in the multicellular organismal process, localization, multiorganism process, the establishment of localization, biological regulation, developmental process, immune system process, response to stimulus, and metabolic process. Six ESTs putatively playing a role in the fungal response to the ITC were chosen to study the regulation of expression by real-time PCR. As a result, four of them were found to be upregulated under the ITC treatment: Ca^{++}ATPase, opsin, EF-hand protein (calmodulin), and an ABC transporter. It is interesting that the ABC transporter was found to be upregulated in the previous studies described above. However, in this experiment, gene encoding glutathione-S-transferase was not found as it was the case in the other studies described above (Báez-Flores et al., 2011). Further analysis of the subtracted cDNA library created in this study allowed the isolation and sequencing of 141 additional ESTs. After assembly, there were 124 reads remaining, consisting of 40 contigs and 84 singletons. The annotation of these ESTs showed diverse molecular functions such as antioxidant activity, electron carrier activity, nucleic acid binding transcription factor activity, transporter activity, structural molecule activity, catalytic activity, and binding activity. It was interesting that nucleic acid binding transcription factor, antioxidant activity, and transporter activity were among the molecular functions. In this work, an ABC multidrug CDR4 transporter and a MFS transporter were isolated. Furthermore, a Zn finger transcription factor was located, which seems to play a role in activating the transcription of a gene encoding an ABC multidrug transporter in *Candida albicans* (C.P. Robin) Berkhout (Coste et al., 2007). Those findings agree with the results of other experiments described above. Electron carrier activity can be related to energy to counter the toxic stress induced by the ITC (García-Coronado et al., 2015). Based on all this information, it seems that the fungal response to the toxic effect of the ITCs involves the conjugation with glutathione by the enzyme glutathione-S-transferase and elimination of the complex through an ABC multidrug CDR4 transporter. The genes encoding the ABC transporter can be induced by the Zn finger transcription factor. The depletion of the glutathione can induce the antioxidant stress, which can be controlled by enzymes such as SNOG proteins, thioredoxin reductases, glutathione peroxidases, glutamylcysteine synthetases, aldehyde dehydrogenases, and oxidoreductases. Perhaps the involvement of the calcium ATPase is related to the release of calcium acting as a second messenger or activating the calmodulin EF-hand protein.

Although few studies characterizing the molecular basis of fungal responses to natural fungicides isolated from plants have been carried out, it is clear from the reports available and from those described above that the elucidation of the phenomena will help to make informed decisions to optimize current treatments or design new ones with the goal to control fungal infections on fresh horticultural products.

6 Conclusions and Future Trends

Extract and plant-derived compounds provide excellent potential leads for the development of postharvest fungicides because of their chemical diversity and biological functionality. According to the literature, the antifungal activity of these compounds has been clearly demonstrated against a broad spectrum of fungi infecting fresh horticultural produce. Moreover, most of these compounds present little toxicological and environmental hazard, and often may have not an adverse impact on the fruit taste and the overall postharvest quality of the commodity. However, they can have reduced and inconsistent efficacy due to variability in the method of extraction, fruit physiology, ripening stage, and storage conditions. These critical disadvantages need to be solved if the plant fungicides are to be available as a cost-effective postharvest technology.

Additionally, notwithstanding the scientific advances so far achieved demonstrating the potential of plant compounds for the control of postharvest diseases, large-scale or commercial application is limited. Until now, the research has focused on studying the antifungal effect of plant extracts and plant compounds at the laboratory scale, and the few studies that have been done at the semi-commercial level reported a lower efficacy of the compounds compared to that obtained in the laboratory. Therefore, it is essential to conduct more studies to confirm the efficacy of the compounds under commercial conditions.

In conclusion, for plant extracts and plant-derived compounds to be considered as an alternative to the synthetic fungicides, these compounds must have greater effectiveness at the postharvest level. Therefore, more studies are needed to enhance their effectiveness, increase the efficiency of their extraction methods, improve their applicability, minimize their costs, confirm their commercial efficacy, and demonstrate that they do not induce the development of resistance strains of pathogenic fungi.

References

Ademe, A., Ayalew, A. and Woldetsadik, K. 2014. *In vitro* and *in vivo* activity of selected plant extracts against papaya (*Carica papaya* L.) anthracnose (*Colletotrichum gloeosporioides*). *Journal of Horticulture* 1, 1–4.

Agbafor, K.N., Akubugwo, E.I., Ogbashi, M.E., Ajah, P.M. and Ukwandu, C.C. 2011. Chemical and antimicrobial properties of leaf extracts of *Zapoteca portoricensis*. *Research Journal of Medicinal Plants* 5, 605–612.

Agerbirk, N. and Olsen, C.E. 2012. Glucosinolate structures in evolution. *Phytochemistry* 77, 16–45.

Akhtar, S., Ismail, T., Fraternale, D. and Sestili, P. 2015. Pomegranate peel and peel extracts: Chemistry and food features. *Food Chemistry* 174, 417–425.

Alemu, K., Ayalew, A. and Woldetsadik, K. 2014. Antifungal activity of plant extracts and their applicability in extending the shelf life of mango fruits. *Food Science and Quality Management* 33, 47–54.

Al-Mansoub, M.A., Asmawi, M.Z. and Murugaiyah, V. 2014. Effect of extraction solvents and plant parts used on the antihyperlipidemic and antioxidant effects of *Garcinia atroviridis*: A comparative study. *Journal of the Science of Food and Agriculture* 94, 1552–1558.

Anwar, F. and Przybylski, R. 2012. Effect of solvents extraction on total phenolics and antioxidant activity of extracts from flaxseed (*Linum usitatissimum* L.). *Acta Scientiarum Polonorum, Technologia Alimentaria* 11, 293–301.

Aqil, F., Zahin, M., Ahmad, I., Owais, M., Khan, M.S.A., Bansal, S.S. and Farooq, S. 2010. Antifungal activity of medicinal plant extracts and phytocompounds: A review. pp. 449–484. *In*: I. Ahmad, M. Owais, M. Shahid and F. Aqil (eds.). Combating Fungal Infections. Springer-Verlag Berlin Heidelberg, Berlin, Germany.

Arya, V., Thakur, N.M. and Kashyap, C. 2012. Preliminary phytochemical analysis of the extracts of *Psidium* leaves. *Journal of Phytochemistry* 1, 1–5.

Askarne, L., Talibi, I., Boubaker, H., Boudyach, E.H., Msanda, F., Saadi, B., Seghini, M.A. and Ait Ben Aoumar, A. 2012. *In vitro* and *in vivo* antifungal activity of several Moroccan plants against *Penicillium italicum*, the causal agent of citrus blue mold. *Crop Protection* 40, 53–58.

Atanu, F.O., Ebiloma, U.G. and Ajayi, E.I. 2011. A review of the pharmacological aspects of *Solanum nigrum* Linn. *Biotechnology Molecular Biology Review* 6, 1–7.

Auerbach, C. and Robson, J.M. 1944. Production of mutations by allyl isothiocyanate. *Nature* 154, 81.

Azwanida, N.N. 2015. A review on the extraction methods use in medicinal plants, principle, strength and limitation. *Medicinal & Aromatic Plants* 4, 1–6.

Báez-Flores, M.E., Troncoso-Rojas, R., Osuna, M.A.I., Domínguez, M.R., Pryor, B. and Tiznado-Hernández, M.E. 2011. Differentially expressed cDNAs in *Alternaria alternata* treated with 2-propenyl isothiocyanate. *Microbiological Research* 166, 566–577.

Bautista-Baños, S., Sivakumar, D., Bello-Pérez, A., Villanueva-Arce, R. and Hernández-López, M. 2013. A review of the management alternatives for controlling fungi on papaya fruit during the postharvest supply chain. *Crop Protection* 49, 8–20.

Bavaresco, L., Petegolli, D., Cantu, E., Fergoni, M., Chiusa, G. and Trevisan, M. 2015. Elicitation and accumulation of stilbene phytoalexins in grapevine berries infected by *Botrytis cinerea. VITIS-Journal of Grapevine Research* 36, 77–83.

Bazie, S., Ayalew, A. and Woldetsadik, K. 2014. Integrated management of postharvest banana anthracnose (*Colletotrichum musae*) through plant extracts and hot water treatment. *Crop Protection* 66, 14–18.

Bhutia, D.D., Zhimo, Y., Kole, R. and Saha, J. 2016. Antifungal activity of plant extracts against *Colletotrichum musae*, the postharvest anthracnose pathogen of banana cv. Martaman. *Nutrition & Food Science* 16, 2–15.

Breda, C.A., Gasperini, A.M., Garcia, V.L., Monteiro, K.M., Bataglion, G.A., Eberlin, M.N. and Teixeira, D.M.C. 2016. Phytochemical analysis and antifungal activity of extracts from leaves and fruit residues of Brazilian Savanna plants aiming its use as safe fungicides. *Natural Products and Bioprospecting* 6, 195–204.

Breuil, A.C., Jeandet, P., Adrian, M., Chopin, F., Pirio, N., Meunier, P. and Bessis, R. 1999. Characterization of a pterostilbene dehydrodimer produced by laccase of *Botrytis cinerea.*. *Phytopathology* 89, 298–302.

Capuzzo, A., Maffei, M. and Occhipinti, A. 2013. Supercritical Fluid Extraction of Plant Flavors and Fragrances. *Molecules* 18, 7194–7238.

Carson, C.F. and Hammer, K.A. 2010. Chemistry and bioactivity of essential oils. pp. 203–238. *In*: H. Thormar (ed.). Lipids and Essential Oils as Antimicrobial Agents. John Wiley & Sons, New York, USA.

CE (Council of Europe). 2000. European Landscape Convention. Available at: www.coe.int

Celimene, C.C., Smith, D.R., Young, R.A. and Stanosz, G.R. 2001. *In vitro* inhibition of *Sphaeropsis sapinea* by natural stilbenes. *Phytochemistry* 56, 161–165.

Celoto, M.I.B., Papa, M.F.S., Sacramento, I.V.S. and Celoto, F.J. 2011. Atividade antifúngica de extratos de *Momordica charantia* L. sobre *Colletotrichum musae. Revista Brasileira de Plantas Medicinais* 13, 337–341 (In Portuguese with English abstract).

Cerqueira, S.M.D., Bacellos, C.H., Bueno, F.P.M., Aires, V.J. and Dummer, M.D. 2016. Antifungal activity of plant extracts with potential to control plant pathogens pineapple. *Asian Pacific Journal of Tropical Biomedicine* 6, 26–31.

Coste, A., Selmecki, A., Forche, A., Diogo, D., Bougnoux, M.E., d'Enfert, C., Berman, J. and Sanglard, D. 2007. Genotypic evolution of azole resistance mechanisms in sequential *Candida albicans* isolates. *Eukaryotic Cell* 6, 1889–1904.

Craige, T.V. and Rochfor, S.J. 2010. Natural products research and metabolomics. *Comprehensive Natural Products II* 9, 595–628.

Cramer, R.A. and Lawrence, C.B. 2004. Identification of *Alternaria brassicicola* genes expressed in planta during pathogenesis of *Arabidopsis thaliana. Fungal Genetics and Biology* 41, 115–128.

Croteau, R., Kutcha, T. and Lewis, N.G. 2000. Natural products (secondary metabolites). pp. 1250–1318. *In*: B. Buchanan, W. Gruissem and R. Jones (eds.). Biochemistry & Molecular Biology of Plants. Wiley-Blackwell, NJ.

Cruz, M.E.S., Schwan-Estrada, K.R.F., Clemente, E., Itako, A.T., Jr, S. and Cruz, M.J. S. 2013. Plant extracts for controlling the post-harvest anthracnose of banana fruit. *Revista Brasileira de Plantas Medicinais* 15, 727–733.

Daniel, C.K., Lennox, C.L. and Vries, F.A. 2015. *In vivo* application of garlic extract in combination with clove oil to prevent postharvest decay caused by *Botrytis cinerea, Penicillium expansum* and *Neofabraea alba* on apples. *Postharvest Biology and Technology* 99, 88–92.

Das, K., Tiwari, R.K.S. and Shrivastava, D.K. 2010. Techniques for evaluation of medicinal plant products as antimicrobial agent: Current methods and future trends. *Journal of Medicinal Plants Research* 4, 104–111.

de Sain, M. and Rep, M. 2015. The role of pathogen-secreted proteins in fungal vascular wilt diseases. *International Journal of Molecular Science* 16, 23970–23993.

de Souza, W.C.O., Do Nascimento, L.C., Vieira, D.L., Dos Santos, T.S. and de Assis Filho, F.M. 2015. Alternative control of *Chalara paradoxa*, causal agent of black rot of pineapple by plant extract of *Mormodica charantia*. *European Journal of Plant Pathology* 142, 481–488.

Deressa, T., Lemessa, F. and Wakjira, M. 2015. Antifungal activity of some invasive alien plant leaf extracts against mango (*Mangifera indica*) anthracnose caused by *Colletotrichum gloeosporioides*. *International Journal of Pest Management* 61, 99–105.

Dunford, N., Irmak, S. and Jonnala, R. 2010. Pressurized solvent extraction of policosanol from wheat straw, germ and bran. *Food Chemistry* 119, 1246–1249.

Elsherbiny, E.A., Amin, B.H. and Baka, Z.A. 2016. Efficiency of pomegranate (*Punica granatum* L.) peels extract as a high potential natural tool towards Fusarium dry rot on potato tubers. *Postharvest Biology and Technology* 111, 256–263.

Endo, E.H., Cortéz, D.A.G., Ueda-Nakamura, T., Nakamura, C.V. and Filho, B.P.D. 2010. Potent antifungal activity of extracts and pure compounds isolated from pomegranate peels, and synergism with fluconazole against *Candida albicans*. *Research Microbiology* 161, 534–540.

Esclapez, M.D., García-Pérez, J., Mulet, A. and Cárcel, J. 2011. Ultrasound-assisted extraction of natural products. *Food Engineering Reviews* 3, 108–120.

FAOSTAT (Food and Agriculture Organization of the United Nations, Statistics Division). 2017. Food and Agriculture Data. Available at: www.fao.org/faostat/

Fimognari, C., Turrini, E., Ferruzzi, L., Lenzi, M. and Hrelia, P. 2012. Natural isothiocyanates: Genotoxic potential versus chemoprevention. *Mutation Research* 750, 107–131.

Garba, S. and Okeniyi, S.O. 2012. Antimicrobial activities of total alkaloids extracted from some Nigerian medicinal plants. *Journal of Microbiology and Antimicrobials* 4, 60–63.

García-Coronado, H., Troncoso-Rojas, R., Tiznado-Hernández, M.E., de la Cruz, M.D. C., Díaz-Camacho, S.P. and Báez-Flores, M.E. 2015. Analysis of a suppressive subtractive hybridization library of *Alternaria alternata* resistant to 2-propenyl isothiocyanate. *Electronic Journal of Biotechnology* 18, 320–326.

Gatto, M., Sergio, L., Ippolito, A. and Di Venere, D. 2016. Phenolic extracts from wild edible plants to control postharvest diseases of sweet cherry fruit. *Postharvest Biology and Technology* 20, 180–187.

Gatto, M.A., Ippolito, A., Linsalata, V., Cascarano, N.A., Nigro, F., Vanadia, S. and Di Venere, D. 2011. Activity of extracts from wild edible herbs against postharvest fungal diseases of fruit and vegetables. *Postharvest Biology and Technology* 61, 72–82.

Geng, F., Tang, L., Li, Y., Yang, L., Chois, K.S., Kazim, A.L. and Zhang, Y. 2011. Allyl isothiocyanate arrest cancer cells in mitosis, and mitotic arrest in turn leads to apoptosis via Bcl-2 protein. *Journal of Biological Chemistry* 286, 32259–32267.

Giweli, A.A., Džamić, A.M., Soković, M., Ristić, M., Janaćković, P. and Marin, P. 2013. The chemical composition, antimicrobial and antioxidant activities of the essential oil of *Salvia fruticosa* growing wild in Libya. *Archives of Biological Sciences* 1, 321–329.

Glawischnig, E. 2007. Camalexin. *Phytochemistry* 68, 401–406.

Goetz, G., Fkyerat, A., Métais, N., Kunz, M., Tabacchi, R., Pezet, R. and Pont, V. 1999. Resistance factors to grey mould in grape berries: Identification of some phenolics inhibitors of *Botrytis cinerea* stilbene oxidase. *Phytochemistry* 52, 759–767.

Griebel, T. and Zeier, J. 2010. A role of beta-sitosterol to stigmasterol conversion in plant-pathogen interactions. *Plant Journal* 63, 254–268.

Gurjar, M., Ali, S., Akhtar, M. and Singh, K. 2012. Efficacy of plant extracts in plant disease management. *Agricultural Sciences* 3, 425–433.

Hall, D.E., Robert, J.A., Keeling, C.I., Domanski, D., Quesada, A.L., Jancsik, S., Kuzyk, M.A., Hamberger, B., Borchers, C.H. and Bohlmann, J. 2011. An integrated genomic, proteomic and biochemical analysis of (+)-3-carene biosynthesis in Sitka spruce (*Picea sitchensis*) genotypes that are resistant or susceptible to white pine weevil. *Plant Journal* 65, 936–948.

Jhalegar, Md.J., Sharma, R.R. and Singh, D. 2014. Antifungal efficacy of botanicals against major postharvest pathogens of Kinnow mandarin and their use to maintain postharvest quality. *Fruits* 69, 223–237.

Kabera, J., Semana, E., Mussa, A. and He, X. 2014. Plant secondary metabolites: Biosynthesis, classification, function and pharmacological properties. *Journal of Pharmacy and Pharmacology* 2, 377–383.

Ko, J.A., Kim, W.Y. and Park, H.J. 2012. Effects of microencapsulated Allyl isothiocyanate (AITC) on the extension of the shelf-life of Kimchi. *International Journal of Food Microbiology* 53, 92–98.

Kubra, L.R., Murthy, P.S. and Rao, L.J. 2013. *In vitro* antifungal activity of dehydro-zingerone and its fungi toxic properties. *Journal of Food Science* 78, 64–69.

Kumar, S.N. and Nambisan, B. 2014. Antifungal activity of diketopiperazines and stilbenes against plant pathogenic fungi *in vitro*. *Applied Biochemistry and Biotechnology* 172, 741–754.

Lamichhane, J.R., Dachbrodt-Saaydeh, S., Kudsk, P. and Messéan, A. 2016. Toward a reduced reliance on conventional pesticide in European Agriculture. *Plant Disease* 100, 10–24.

Lattanzio, V. 2013. Phenolic compounds: Introduction. pp. 1543–1580. *In*: K. G. Ramawat and J.W. Mérillon (eds.). Natural Products. Springer-Verlag, Berlin Heidelberg, Berlin, Germany.

Li Destri Nicosia, M.G., Pangallo, S., Raphael, G., Romeo, F.V., Strano, M.C., Rapisarda, P., Droby, S. and Schena, L. 2016. Control of postharvest fungal rots on citrus fruit sweet cherries using a pomegranate peel extract. *Postharvest Biology and Technology* 114, 54–61.

Li, L.H., Zhang, H.F., Hu, S., Bai, X.H. and Li, S. 2012. Dispersive liquid-liquid microextraction coupled with high-performance liquid chromatography for determination of coumarin compounds in *Radix Angelicae Dahuricae*. *Chromatographia* 75, 131–137.

Lin, T.C., Fan, M.C., Wang, S.Y. and Huang, J.W. 2011. Identification of the *Solanum nigrum* extract component involved in controlling cabbage black leaf spot disease. *Journal of Agricultural and Food Chemistry* 59, 1667–1672.

Lola-Luz, T., Hennequart, F. and Gaffney, M. 2014. Effect on yield, total phenolic, total flavonoid and total isothiocyanate content of two broccoli cultivars (*Brassica oleraceae* var *Italica*) following the application of a commercial brown seaweed extract (*Ascophyllum nodosum*). *Agricultural and Food Science* 23, 28–37.

Loreto, F., Dicke, M., Schnitzler, J.P. and Turlings, T.C.J. 2014. Plant volatiles and the environment. *Plant Cell Environment* 37, 1905-1908.

Mazid, M., Khan, T.A. and Mohammad, F. 2011. Role of secondary metabolites in defense mechanisms of plants. *Biology and Medicine* 3, 232–249.

Mikkelsen, M.D., Olsen, C.E. and Halkier, B.A. 2010. Production of the cancer-preventive glucoraphanin in tobacco. *Molecular Plant* 3, 751–759.

Miron, T., Plaza, M., Bahrim, G., Ibáñez, E. and Herrero, M. 2010. Chemical composition of bioactive pressurized extracts of Romanian aromatic plants. *Journal of Chromatography* 1218, 4918–4927.

Mithöfer, A. and Boland, W. 2012. Plant defense against herbivores: Chemical aspects. *Annual Review of Plant Biology* 63, 431–450.

Musto, M., Potenza, G. and Cellini, F. 2014. Inhibition of *Penicillium digitatum* by a crude extract from *Solanum nigrum* leaves. *Biotechnology Agronomy Society and Environment* 18, 174–180.

Nicolopoulou-Stamati, P., Maipas, S., Kotampasi, C., Stamatis, P. and Hens, L. 2016. Chemical pesticides and human health: The urgent need for a new concept in agriculture. *Frontiers in Public Health* 4, 148.

Ouattara, L., Koudou, J., Zongo, C., Barro, N., Savadogo, A., Bassole, I.H.N., Ouattara, A. S. and Traore, A.S. 2011. Antioxidant and antibacterial activities of three species of *Lannea* from Burkina Faso. *Journal of Applied Sciences* 11, 157–162.

Palou, L., Ali, A., Fallik, E. and Romanazzi, G. 2016. GRAS, plant-and animal-derived compounds as alternatives to conventional fungicides for the control of postharvest diseases of fresh horticultural produce. *Postharvest Biology and Technology* 122, 41–52.

Patel, D., Patel, K., Tahilyani, K. and Barbaloin, V. 2012. A concise report of its pharmacological and analytical aspects. *Asian Pacific Journal of Tropical Biomedicine* 2, 835–838.

Pereira, C.G., Angela, M. and Meireles, A. 2010. Supercritical fluid extraction of bioactive compounds: Fundamentals, applications and economic perspectives. *Food Bioprocess Technology* 3, 340–372.

Plaza, M., Amigo-Benavent, M., del Castillo, M., Ibáñez, E. and Herrero, M. 2010. Facts about the formation of new antioxidants in natural samples after subcritical water extraction. *Food Research International* 43, 2341–2348.

Prusky, D., Alkan, N., Mengiste, T. and Fluhr, R. 2013. Quiescent and necrotrophic lifestyle choice during postharvest disease development. *Annual Review of Phytopathology* 51, 55–76.

Quezada, N. and Cherian, G. 2012. Lipid characterization and antioxidant status of the seeds and meals of *Camelina sativa* and flax. *European Journal of Lipid Science and Technology* 114, 974–982.

Rabêlo, S.V., Araújo, C.S., Costa, V.C.O., Tavares, J.F., Silva, M.S., BarbosaFilho, J.M. and Almeida, J.R.G.S. 2014. Occurrence of alkaloids in species of the genus *Annona* L. (Annonaceae): A review. pp. 41–60. *In*: S. Kaur Brar, S. Kaur and G. S. Dhillon (eds.). Nutraceuticals and Functional Foods: Natural Remedy. Nova Science Publishers, New York, NY.

Rajabi, M., Ghanbari, H., Barfi, B., Asghari, A. and Haji-Esfandiari, S. 2014. Ionic liquid-based ultrasound-assisted surfactant-emulsified microextraction for simultaneous determination of three important flavoring compounds in plant extracts and urine samples. *Food Research International* 62, 761–770.

Regnault-Roger, C. 2012. Essential oils in insect control: Low-risk products in a high-stakes world. *Annual Review of Entomology* 57, 405–424.

Romeo, F.V., Ballistreri, G., Fabroni, S., Pangallo, S., Li Destri Nicosia, M.G., Schena, L. and Rapisarda, P. 2015. Chemical characterization of different sumac and pomegranate extracts effective against *Botrytis cinerea* rots. *Molecules* 20, 11941–11958.

Roshan, A., Verma, N.K., Kumar, C.S., Chandra, V., Singh, D.P. and Panday, M.K. 2012. Phytochemical constituent, pharmacological activities and medicinal uses through the millenia of *Glycyrrhiza glabra* Linn: A review. *International Research Journal Pharmacy* 3, 45–55.

Sanzani, S.M., Schena, L. and Ippolito, A. 2014. Effectiveness of phenolic compounds against citrus green mold. *Molecules* 19, 12500–12508.

Sasidharan, S., Chen, Y., Saravanan, D., Sundram, K.M. and Yoga, L.L. 2011. Extraction, isolation and characterization of bioactive compounds from plant extracts. *African Journal of Traditional, Complementary and Alternative Medicines* 8, 1–10.

Savithramma, N., Linga Rao, M. and Suhrulatha, D. 2011. Screening of medicinal plants for secondary metabolites. *Middle-East Journal of Scientific Research* 8, 579–584.

Sayago, J.E., Ordoñez, R.M., Kovacevich, L.N., Torres, S. and Isla, M.I. 2012. Antifungal activity of extracts of extremophile plants from the Argentine puna to control citrus postharvest pathogens and green mold. *Postharvest Biology and Technology* 67, 19–24.

Sellam, A., Dongo, A., Guillemette, T., Hudhomme, P. and Simoneau, P. 2007. Transcriptional responses to exposure to the brassicaceous defence metabolites camalexin and allyl-isothiocyanate in the necrotrophic fungus *Alternaria brassicicola*. *Molecular Plant Pathology* 8, 195–208.

Sellam, A., Poupard, P. and Simoneau, P. 2006. Molecular cloning of *AbGst1* encoding a glutathione transferase differentially expressed during exposure of *Alternaria brassicicola* to isothiocyanates. *FEMS Microbiology Letters* 258, 241–249.

Shinde, R., Barhate, B. and Bandgar, M. 2016. Efficacy of garlic extract and yeast for the control of postharvest diseases of grape. *International Journal of Plant Protection* 9, 434–438.

Sing, P. 2011. Integrated management of storage anthracnose of mango. *Journal of Mycology and Plant Pathology* 41, 63–66.

Sing, P. and Srivastava, D. 2012. Biofungicidal or biocontrol activity of *Lantana camara* against phytopathogenic *Alternaria alternata*. *International Journal of Pharmaceutical Science Research* 3, 4818–4821.

Singh, B. and Sharma, R.A. 2015. Plant terpenes: Defense responses, phylogenetic analysis regulation and clinical applications. *3 Biotech* 5, 129–151.

Slusarenko, A.J., Patel, A. and Portz, D. 2008. Control of plant diseases by natural products: Allicin from garlic as a case study. *European Journal of Plant Pathology* 121, 313–322.

Smith, B.J. and Kirkegaard, J.A. 2002. *In vitro* inhibition of soil microorganisms by 2-phenylethyl isothiocyanate. *Plant Pathology* 51, 585–593.

Sønderby, I.E., Geu-Flores, F. and Halkier, B.A. 2010. Biosynthesis of glucosinolates – Gene discovery and beyond. *Trends in Plant Science* 15, 283–290.

Soria, A.C. and Villamiel, M. 2010. Effect of ultrasound on the technological properties and bioactivity of food: A review. *Trends in Food Science & Technology* 21, 323–331.

Talibi, I., Askarne, L., Boubaker, H., Boudyach, E.C., Msanda, F., Saadi, B. and Ait Ben Aoumar, A. 2012a. Antifungal activity of Moroccan medicinal plants against citrus sour rot agent *Geotrichum candidum*. *Letters in Applied Microbiology* 55, 155–161.

Talibi, I., Askarne, L., Boubaker, H., Boudyach, E.C., Msanda, F., Saadi, B. and Ait Ben Aoumar, A. 2012b. Antifungal activity of some Moroccan plants against *Geotrichum candidum*, causal agent of postharvest citrus sour rot. *Crop Protection* 35, 41–46.

Tang, L., Paonessa, J.D., Zhang, Y., Ambrosone, S.B. and McCann, S.E. 2013. Total isothiocyanate yield from raw cruciferous vegetables commonly consumed in the United States. *Journal of Functional Foods* 5, 1996–2001.

Thangavelu, R., Devi, P.G., Gopi, M. and Mustaffa, M.M. 2013. Management of Eumusae leaf spot disease of banana caused by *Mycosphaerella eumusae* with Zimmu (*Allium sativum* × *Allium cepa*) leaf extract. *Crop Protection* 46, 100–105.

Thimmappa, R., Geisler, K., Louveau, T., O'Maille, P. and Osbourn, A. 2014. Triterpene biosynthesis in plants. *Annual Review of Plant Biology* 65, 225–257.

Tholl, D. 2015. Biosynthesis and biological functions of terpenoids in plants. *Advances in Biochemical Engineering/Biotechnology* 148, 63–106.

Tiznado-Hernández, M.E. and Troncoso-Rojas, R. 2006. Control of fungal diseases with isothiocyanates. *Stewart Postharvest Review* 1:4, 1–14.

Troncoso, R., Espinoza, C., Sánchez-Estrada, A., Tiznado, M.E. and García, H.S. 2005. Analysis of the isothiocyanates present in cabbage leaves extract and their potential application to control Alternaria rot in bell peppers. *Food Research International* 38, 701–708.

Troncoso-Rojas, R., Báez-Flores, M.E., Pryor, B., García, H.S. and Tiznado-Hernández, M.E. 2013. Inter simple sequence repeat polymorphism in *Alternaria* genomic DNA exposed to lethal concentrations of isothiocyanates. *African Journal of Microbiology Research* 7, 838–852.

Troncoso-Rojas, R., Corral, A.Y., Sánchez-Estrada, A., García, E.R., Aguilar-Valenzuela, A., Ojeda-Contreras, A.J. and Tiznado-Hernández, M.E. 2009. Postharvest treatment of isothiocyanates to control *Alternaria* rot in netted melon. *Phytoparasitica* 37, 445–451.

Troncoso-Rojas, R., Sánchez-Estrada, A., Ruelas, C., García, H.S. and Tiznado-Hernández, M.E. 2005. Effect of benzyl isothiocyanate on tomato fruit infection development by *Alternaria alternata*. *Journal of the Science of Food and Agriculture* 85, 1427–1434.

Troncoso-Rojas, R. and Tiznado-Hernández, M.E. 2015. Plant isothiocyanates as an alternative for sustainable disease control of horticultural crops. pp. 70–90. *In*:

V. Kurucheve, G. Sangeetha and J. Jayaraman (eds.). Natural Products for Sustainable Crop Diseases Management. CAB International, Wallingford, UK.

Ugolini, L., Martini, C., Lazzeri, L., D'Avino, L. and Mari, M. 2014. Control f postharvest grey mold (*Botrytis cinerea* Pers.: Fr.) on strawberries by glucosinolate-derived allyl-isothiocyanate treatments. *Postharvest Biology and Technology* 90, 34–39.

Ullah, F., Ayaz, M., Sadiq, A., Hussain, A., Ahmad, S., Imran, M. and Zeb, A. 2016. Phenolic, flavonoid contents, anticholinesterase and antioxidant evaluation of *Iris germanica* var; *florentina*. *Natural Products Research* 30, 1440–1444.

Usall, J., Casals, C., Sisquella, M., Palou, L. and De Cal, A. 2015. Alternative technologies to control postharvest diseases of stone fruits. *Stewart Postharvest Review* 4:2, 1–6.

Valcarcel, J., Reilly, K., Gaffney, M. and O'Brien, N. 2015. Antioxidant activity, total phenolic and total flavonoid content in sixty varieties of potato (*Solanum tuberosum* L.) grown in Ireland. *Potato Research* 58, 221–244.

Vongsak, B., Sithisam, P., Mangmool, S., Thongpraditchote, S., Wongkrajang, Y. and Gritsanap, W. 2013. Maximizing total phenolics, total flavonoids contents and antioxidant activity of *Moringa oleifera* leaf extract by the appropriate extraction method. *Industrial Crops and Products* 44, 566–571.

Wang, J., Jiankang, C.J.L. and Jiang, W. 2010a. Antifungal activities of neem (*Azadiracha indica*) seed kernel extracts on postharvest diseases in fruits. *African Journal of Microbiology Research* 4, 1100–1104.

Wang, J., Li, J., Cao, J. and Jiang, W. 2011. Antifungal activities of neem (*Azadirachta indica*) seed kernel extracts on postharvest diseases in fruits. *African Journal of Microbiology Research* 4, 1100–1104.

Wang, Q., Cao, Y., Zhou, L., Jiang, C., Feng, Y. and Wei, S. 2015. Effects of postharvest curing treatment on flesh colour and phenolic metabolism in fresh-cut potato products. *Food Chemistry* 169, 246–254.

Wang, S.Y., Chen, C. and Yin, J. 2010b. Effect of allyl isothiocyanate on antioxidants and fruit decay of blueberries. *Food Chemistry* 120, 199–204.

Wijesundara, S.A.D.T.L., Kannangara, B.T.S.D.P. and Abeywickrama, K. 2016. Antifungal activity of *Croton aromaticus* L. *in vitro*, against post-harvest fungal pathogens isolated from tropical fruits. *The Journal of Agricultural Science* 11, 105–117.

Wu, H., Zhang, X., Zhang, G., Zeng, S. and Lin, K. 2011. Antinfungal vapour-phase activity of a combination of allyl isothiocyanate and ethyl isothiocyanate against *Botrytis cinerea* and *Penicillium expansum* infection on apples. *Journal of Phytopathology* 159, 450–455.

Yeo, J. and Shahidi, F. 2015. Critical evaluation of changes in the ratio of insoluble bound to soluble phenolics on antioxidant activity of lentils during germination. *Journal of Agricultural and Food Chemistry* 63, 379–381.

Zhang, H.F., Yang, X.H. and Wang, Y. 2011. Microwave assisted extraction of secondary metabolites from plants: Current status and future directions. *Trends in Food Science and Technology* 22, 672–688.

Zhang, Y. 2010. Allyl isothiocyanate as a cancer chemopreventive phytochemical. *Molecular Nutrition and Food Research* 54, 127–135.

Zhu, Y., Zhou, H.T., Hu, Y.H., Tang, J.Y., Su, M.X., Guo, Y.J., Chen, Q.X. and Liu, B. 2011. Antityrosinase and antimicrobial activities of 2-phenylethanol, 2-phenylacetaldehyde and 2-phenylacetic acid. *Food Chemistry* 124, 298–302.

Chapter 20

Use of Essential Oils to Improve Postharvest Quality and Control Postharvest Decay of Tropical, Subtropical, and Temperate Fruits

Dharini Sivakumar

Phytochemical Food Network Group, Department of Crop Sciences, Tshwane University of Technology, Pretoria West, Pretoria, South Africa

Gianfranco Romanazzi

Department of Agricultural, Food and Environmental Science, Marche Polytechnic University, Ancona, Italy

Abbreviations

GRAS	Generally recognized as safe
LEDP	Low-density polyethylene
LOX	Lipoxygenases
MAP	Modified atmosphere packaging
MIC	Minimum inhibitory concentration
PAL	Phenylalanine ammonia-lyase
PMF	Proton motive force
PR	Pathogenesis-related
ROS	Reactive oxygen species
TA	Titratable acidity
TSS	Total soluble solids
US FDA	United States Food and Drug Administration

1 Introduction

Globally, an increase in postharvest losses has been attributed to the susceptibility of the fruit to fungal pathogens, which shorten the shelf life and decrease the commercial value of the fruit in the supply chain. In developing countries, food losses can reach up to 50%, and most of the postharvest losses due to fungal decay occur during storage and transportation. In order to reduce such losses, synthetic fungicides such as imazalil or prochloraz are applied as the first postharvest fungicide immediately after harvest in packinghouses, especially for citrus and avocados, respectively (Everett et al., 2005). Thiabendazole, fludioxonil, and pyrimethanil are also widely used in the citrus industry to control postharvest diseases, mainly blue and green molds, caused by *Penicillium italicum* Wehmer and *Penicillium digitatum* (Pers.) Sacc., respectively, during storage and transportation (Bus et al., 1991; Ismail and Zhang, 2004; Smilanick et al., 2005; Palou et al., 2008). Although control of the major postharvest pathogens can be achieved with synthetic fungicides, there are growing concerns about human and environmental health risks associated with pesticides. Furthermore, coupled with the development of fungicide-resistant strains such as resistance of *Penicillia* and *Botrytis cinerea* Pers. populations to thiabendazole and other fungicides (Bus et al., 1991) and the banning of certain other fungicides has prompted the industry to search for alternative approaches (Vilanova et al., 2013). The mango fruit industry uses a prochloraz dip or spray in packinghouses to control anthracnose and stem-end rots. The commercial fungicide iprodione is also used to control postharvest decay caused by *Monilinia* spp. in stone fruits (Cindi et al., 2016). The deposit of the fungicide residues on fruit has raised concerns about their potential negative effects on the environment and their risks to consumers' health (Tyagi et al., 2013). Furthermore, in Europe, with a few exceptions, the application of fungicides after harvest is prohibited in response to consumers' demands for pesticide-free fruits arising from their concerns regarding food safety issues and human health (Zhang et al., 2010). Moreover, the use of synthetic fungicides is prohibited by the organic fruit industry regulations. Food suppliers and food catering agents are also expected to support sustainable agriculture and adopt food safety practices that avoid or minimize the use of synthetic pesticides (Popp et al., 2013; Frankova et al., 2016). All these concerns have encouraged the pursuit of research to develop alternative approaches to control postharvest diseases in the fruit industry. Other means of physical control cannot be easily applied while treatments like heat or hot water require a high usage of energy, which are often not economically feasible for the producer. Alternative control strategies aim to replace or reduce the use of synthetic fungicides to control postharvest diseases by employing treatments that are less harmful to human health, environmentally friendly, and cost-effective (Viuda-Martos et al., 2008; Simas et al., 2017).

Many essential oils are listed as Generally Recognized as Safe (GRAS) at doses normally used in foods by the United States Food and Drug Administration (US FDA) for use in foods and beverages (US FDA, 2016). The use of essential oils has attracted considerable interest as a possible natural substitute for conventional synthetic fungicides (Elshafie et al., 2015). Essential oils are highly volatile compounds found in many plant species that play a role in plant defense mechanisms against several phytopathogenic microorganisms. They are known as natural antioxidants that possess antimicrobial and biodegradable properties, and do not leave residues on fresh produce (Sivakumar and Bautista-Baños, 2014).

2 Essential Oils

Over the past two decades, there has been much research into plant extracts due to their less harmful effect on human health and the environment (Sayago et al., 2012). The interest in the possible antimicrobial properties of plant extracts is relatively new. Essential oils (also called volatile or ethereal oils) are concentrated liquids of a complex mixture of volatile compounds and can be extracted from various plant organs. Plants belonging to Lamiaceae are responsible for producing a large number of various types of essential oils (Burt, 2004; Tongnuanchan and Benjakul, 2014). Essential oils are synthesized and kept in complex secretory structures such as glandular trichomes, secretory cavities, and resin ducts (Pavela, 2015). Roots, peels, leaves, seeds, and fruits are among the plant parts that contain higher concentrations of essential oils (Tongnuanchan and Benjakul, 2014; Pavela, 2015). Although it is said that essential oils have been used since ancient times, notable work on essential oils began during the enlightenment era with Berthelot characterizing many natural substances. The most significant work was conducted to isolate the individual oil constituents and evaluate their basic properties (Baser and Buchbauer, 2010).

Essential oils are extracted from various fragrant plants, generally situated in warm temperate countries, such as the Mediterranean and tropical countries (Patrignani et al., 2015). Essential oils are composed of a large group of chemical constituents with bioactivity. These oils can be extracted from the plant tissue via distillation, expression, solvent extracts, and supercritical carbon dioxide (Lihandra, 2007; Patrignani et al., 2015). The extraction methods are important for the concentration of active ingredients of the essential oils; for example, variation in the chemical constitution of oils was observed when they were obtained by means of solvent extraction and hydro distillation, and this affected their antimicrobial properties (Burt, 2004). The essential oils are referred to as 'green pesticides' because of their eco-friendly, cost-effectiveness, and target-specific, biodegradable nature.

Essential oils are secondary metabolites that include terpenoids, shikimates, and polyketides (Baser and Buchbauer, 2010). The terpenes are unsaturated hydrocarbons that are based on simple isoprene molecules. When terpenes are modified chemically, such as by oxidation or rearrangement of the carbon skeleton, the resulting compounds are generally referred to as terpenoids. Terpenoids are the most significant group of the natural products of essential oils. Typically, there would be 60 individual compounds in an essential oil, with the major compounds called active compounds (Adeyinka and Richard, 2015). Essential oils include phenolics and phenolic-based compounds as the major active compounds that are responsible for their antioxidant and antimicrobial properties. Essential oils with aldehydes or phenols as major components (cinnamaldehyde, thymol, eugenol, carvacrol) have controlled postharvest decay most effectively, followed by essential oils containing terpene alcohols, while those oils that contain ketones or esters as their major compounds are less effective (Perricone et al., 2015). Moreover, the chemical components of the essential oils differ according to the agro-climatic conditions during cultivation, the growth stage of the plant part harvested, harvest season, soil characteristics, and fertilizer applications (Bakkali et al., 2008; Viuda-Martos et al., 2008).

3 Effects of Essential Oils on *Monilinia* spp., *Penicillium* spp., *Botrytis cinerea*, and *Colletotrichum* spp.

Essential oils of *Citrus sinensis* containing limonene as the major active compound had potent toxicity to conidia of *Penicillium expansum* L. at 500 ppm, while the mycelial growth of the pathogen was inhibited at 600 ppm under similar conditions in vitro (Sharma and Tripathi, 2006). Beetle leaf (*Piper betle* var. 'Tamluk Mitha') essential oil composed of chavibetol (22.0%), stragol (15.8%), β-cubebene (13.6%), chavicol (11.8%), and caryophyllen (11.3%) completely inhibited germination of spores of *P. expansum* at 0.5 μL/mL, while mycelial growth was inhibited at 0.6 μL/mL in vitro (Bask and Guha, 2015). Xing (personal communication) reported that the minimum inhibitory concentration (MIC) of cinnamon oil to spores of *P. expansum* was 0.3% in vitro. The fungistatic inhibitory effect of thyme oil on the radial mycelial growth of *P. expansum* and *Monilinia laxa* (Aderh. & Ruhland) Honey isolated from peaches was reported by Sellamuthu et al. (2013b). Thyme oil (*Thymus vulgaris*) at 5 μL/plate was more effective than citronella or peppermint oils at similar concentrations in vitro (Sellamuthu et al., 2013b). However, essential oils of *Eugenia caryophyllata, Cinnamomum zeylanicum*, or *Carum copticum* effectively controlled *Monilinia* spp. infection in stone fruits (Hassani et al., 2012). The mycelial growth of *P. digitatum* was progressively inhibited by increasing the concentrations of lime, thyme, and camphor oils (1–10%). At 10%, lime oil significantly reduced the in vitro growth of *P. digitatum* more so than the other oils (Badawy et al., 2011).

Lemon myrtle oil completely inhibited the spore germination and the mycelial growth of *Monilinia fructicola* (G. Winter) Honey isolated from nectarines at 1000 μL/L compared to the tea tree, valerian, clove bud, oregano, and cinnamon bark oils in vitro (Lazar-Baker et al., 2011). Furthermore, four *Thymus* species (*T. leptobotrys, T. satureioides* subsp. pseudomastichina, *T. broussonnetii* subsp.

hannois, and *T. riatarum*) were tested for their ability to inhibit *P. digitatum* and *P. italicum* in vitro. The active chemical components in *T. leptobotrys* were borneol (27.71%) and thymol (18.47%), and *T. satureioides* contained camphor (46.17%) and alpha-terpineol (7.69%). *Thymus broussonnetii* contained carvacrol (32.24%) and gamma-terpinene (19.60%), while *T. riatarum* had *p*-cymene (13.52%) (Boubaker et al., 2016). The response of spore germination and mycelial growth varied with the essential oils obtained from the different *Thymus* spp. Essential oils of *T. leptobotrys* demonstrated the lowest MIC (<500 μL/L) for *P. digitatum* and *P. italicum* (Boubaker et al., 2016).

A combination of essential oil fractions of *Cymbopogan citratus* and *Ocimum gratissimum* had greater fungicidal activity against mycotoxin producing strains of *P. expansum* than the stand-alone treatments (Nguefack et al., 2012). The synergistic effect of essential oils can be used to improve their antifungal activity and to minimize the concentration of the oils needed in order to reduce costs or impacts on the sensory properties of fruit (Nguefack et al., 2012). Synergy was observed in the effectiveness of the application of oregano oil (*Origanum vulgare*; carvacrol 64%, *p*-cymene, 5.16%, thymol 2.93%) and cinnamon oil (*C. zeylanicum*; Z-cinnamaldehyde 73.06%, limonene 4.96%; cinnamyl acetate 3.70%; eugenol 3.54%) to control *P. digitatum* and *P. expansum* (Frankova et al., 2016). The MIC of oregano oil for the control of radial mycelial growth of *P. digitatum* was 128 μL/L, whereas for *P. expansum* it was 64 μL/L. Similarly, the MIC for the control of radial mycelial growth of *P. digitatum* and *P. expansum* was 256 and 32 μL/L, respectively. The observed differences in the MIC to control the two pathogens could be due to the influence of the minor components of the essential oils that may play a major role in the mode of action of these oils (Burt, 2004; Frankova et al., 2016).

Essential oil vapor of *Solidago canadensis* leaves containing α-pinene, germacrene D, and 6-*epi*-β-cubebene as the major components exhibited antifungal activity against *B. cinerea* and altered the cellular morphology of the pathogen (Liu et al., 2016). The vapor phase of thyme oil at a concentration of 5 μL/plate inhibited the growth of *Colletotrichum gloeosporioides* (Penz.) Penz. & Sacc. isolated from mango (Perumal et al., 2016) and avocado (Sellamuthu et al., 2013a; Bill et al., 2016) fruits. The surface of the hypha was altered and collapsed when spore germination was inhibited by essential oils (Perumal et al., 2016).

A stand-alone treatment with lemongrass (0.05%) or cinnamon oil (0.4%) inhibited the mycelial growth of the anthracnose pathogens (*C. gloeosporioides* and *Colletotrichum musae* (Berk. & M.A. Curtis) Arx) in vitro (Maqbool et al., 2011). Lemongrass is an essential oil with a high citral content, which inhibits the growth of various fungal pathogens. The active compound (citral) of lemongrass inhibited stem-end rot caused by *Lasiodiplodia theobromae* (Pat.) Griffon & Maubl. on avocado fruit after 14 d of storage at 8°C. Also, basil (*Ocimum basilicum*) oil at the concentrations of 0.15–0.2% (v/v) completely inhibited the banana-isolated pathogen *C. musae* after 7 d of incubation at 25°C (Idris et al., 2015).

In vivo inoculation investigations showed that thyme oil spray (600 μL/L) or a vapor-phase application significantly reduced brown rot incidence in apricot cv. 'Asgarabad' (caused by *M. fructicola*) (Hassani et al., 2012) and in peaches (caused by *M. laxa*) (Cindi et al., 2016). In later investigations by Lazar-Baker et al. (2011), the preventative application of lime myrtle oil (1000 μL/L) reduced the incidence of decay caused by *M. fructicola* by 34.2% compared to the control fruit in artificially inoculated 'Summer Bright' nectarines for up to 7 d at 12°C. Furthermore, exposure to the essential oil vapors was effective in controlling the decay as a preventative treatment, rather than as a curative treatment, by affecting the penetration of spores that resided on the fructoplane rather than killing spores that had penetrated deeply into the fruit tissue. Therefore, the application of essential oils as vapors was more effective as a preventative treatment than a curative one (Lazar-Baker et al., 2011). Thyme oil vapors (960 µL in a 10 L volume) effectively controlled brown rot caused by *M. laxa* in peaches more so than the cinnamon oil vapors did (Cindi et al., 2016). Thyme (56.2% cyme) and vervain (*Verbena officinalis*) oils (citral 44.5%; isobornyl formate 45.4%) were tested to control decay caused by *Monilinia fructigena* (Aderh. & Ruhland) Honey, *M. laxa*, and *M. fructicola* in vivo (Elshafie et al., 2015). Vervain oils at 1000 ppm reduced brown rot severity (measured by lesion diameter) to a remarkable extent on peaches inoculated with *M. laxa* or *M. fructicola* (Elshafie et al., 2015). The effectiveness of thyme oil at 500 ppm for the control of brown rot caused by *M. fructicola* was similar to that of the vervain oils at 1000 ppm in vivo (Elshafie et al., 2015). The concentrations of monoterpenes and phenolic compounds, the major active chemical components in the essential oils, were responsible for the differences in the effectiveness of the essential oils to inhibit the three *Monilinia* spp. that were tested (Elshafie et al., 2015).

The application of rosemary essential oil vapor for 24 hr reduced gray mold infections among table grapes stored for 9 d at room temperature (Servili et al., 2017). Rosemary and mint essential oils, applied in the same environment, but under hypobaric conditions (50 Kpa, 0.5 atm) in a gas-tight tank, effectively reduced gray mold during storage at room temperature. Of the table grapes exposed to 24 hr of hypobaric fumigation with the aforementioned essential oils, then cold stored at 4°C for 7 d, and thereafter exposed to shelf conditions, only the rosemary essential oil consistently reduced gray mold, while the mint and thyme oils were not consistently effective (Servili et al., 2017). These trials suggested that thyme oil is less volatile than rosemary and mint oils.

Some of the essential oil components can interfere with the host recognition by the pathogen. Essential oils with a higher concentration of monoterpene volatiles such as limonene in citrus demonstrated that they act as signaling compounds involved in recognizing the host during the infection by *P. digitatum* and *P. italicum* (Droby et al., 2008; Simas et al., 2017). Thyme oil vapors were not effective in controlling imazalil-resistant strains of *P. digitatum* and *P. italicum* when tested preventatively and curatively in the soft citrus cv. 'Empress' at 20°C for 7 d (Mothopa et al., 2015). Treatment with lemon oil after harvest significantly reduced the severity of green mold caused by *P. digitatum* on oranges (Badawy et al., 2011). Recent information on the effective control of postharvest diseases using essential oils on selected commercial fruits is given in Table 20.1.

The effectiveness of essential oils in a vapor phase to control postharvest diseases depends on various factors such as the volatilizing temperature of the essential oils and the exposure time (Frankova et al., 2016). Their effectiveness to inhibit pathogens also depends on vapor pressure, which is positively correlated with the temperature. The efficacy of the essential oil treatment depends on the active component, application method, duration of the treatment, storage time of the fruit, type of cultivar or variety, and the virulence of the pathogens. Thyme oil vapors effectively controlled avocado anthracnose during simulated shipping conditions (Figure 20.1).

Table 20.1 Recent information on the effective control of decay caused by postharvest pathogens using essential oils on selected commercial fruits

Fruit	*Decay causal agent*	*Essential oil*	*References*
Peach	*Monilinia laxa* and *M. fructicola*	Thyme oil vapors	Cindi et al. (2016) Elshafie et al. (2015)
Nectarine	*M. fructicola*	Lemon myrtle fumigant	Lazar-Baker et al. (2011)
Table grapes	*Botrytis cinerea*	Rosemary oils under hypobaric conditions Fennel (*Foeniculum vulgare*)	Servili et al. (2017) Pedrotti et al. (2017)
Avocado	*Colletotrichum gloeosporioides*	Thyme oil vapors	Bill et al. (2016)
Strawberry	*B. cinerea*	Lemon essential oils	Vitoratos et al. (2013)
		Solidago canadensis essential oil	Liu et al. (2016)
		Tea tree oil	Shao et al. (2013)

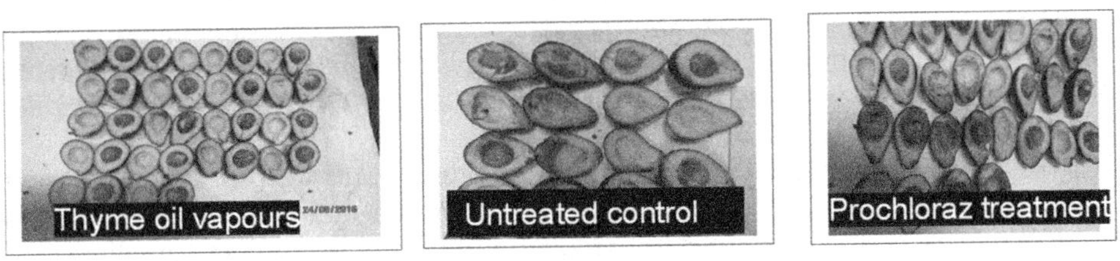

Figure 20.1 Control of anthracnose on avocado cv. 'Hass' after exposure to thyme oil vapors prior to storage for 28 d at 7°C, followed by a simulated shelf-life period of 5 d at 20°C.

4 Essential Oil Treatment and Differential Responses among Cultivars

Different responses to essential oil treatments were observed among the various cultivars of peaches, especially with respect to the control of brown rot (Cindi et al., 2016). Brown rot caused by *M. laxa* was effectively controlled by thyme oil vapors (960 μL in a 10 L volume) in yellow flesh 'Spring Princess' peaches, but not in yellow flesh 'Sonnet' fruits, after postharvest storage at 0°C for 12 d followed by 2 d at 15°C. Variations in the effectiveness of thyme oil vapor treatment between yellow and red skin peach cultivars were recorded (Khumalo et al., 2017). Thyme oil vapors (960 μL in a 10 L volume) remarkably reduced brown rot incidence in red skinned or yellow flesh peach cultivars such as 'Jim Dandy', 'Novadonna',

'Scarlet Rich', and 'Transvalia' more so than in the yellow skin and yellow flesh cultivars 'Early Gold', 'Kakamas', 'Kesie', and 'Summer Sun' after postharvest storage at 0°C for 14 d and 6 d at 15°C (Khumalo et al., 2017). Although thyme oil vapors reduced the incidence of *P. expansum* in red-skinned peach cultivars, the efficacy was inferior to the control of brown rot (Cindi et al., 2016).

The effectiveness of essential oils to control decay on apple cultivars ('Golden Delicious', 'Royal Gala', 'Granny Smith', and 'Red Chief') that had been inoculated with *P. expansum* differed. The essential oils tested were *O. basilicum* (basil), *Foeniculum sativum* (fennel), *Lavandula officinalis* (lavender), *Origanum majorana* (marjoram), *O. vulgare* (oregano), *Mentha piperita* (peppermint), *Rosemarinus officinalis* (rosemary), *Salvia officinalis* (sage), *Satureja montana* (savory), *T. vulgaris* (thyme), and *Mentha arvensis* (wild mint) (Lopez-Reyes et al., 2010). The essential oils of fennel (10%), sage, or oregano (1%) significantly reduced blue mold lesion diameter by approximately less than 5 mm on 'Golden Delicious' fruits inoculated with *P. expansum* during 15 d of storage at 1°C. However, the severity of decay increased greatly, with lesions of more than 20 mm in diameter, when storage was prolonged to 30 d (Lopez-Reyes et al., 2010). With cv. 'Granny Smith', the essential oils of thyme (10%), rosemary (10%), oregano (10%), peppermint (10%), sage (10%), marjoram (1%), and lavender (1%) significantly reduced blue mold severity to less than 5 mm during the 15 d of storage at 1°C. However, basil (1 and 10%), thyme (1%), and peppermint (1%) essential oils reduced decay to less than 10 mm during the 30 d of low temperature storage. With cv. 'Red Chief', essential oils of thyme (10%), sage (10%), and lavender (10%) controlled blue mold as well as a treatment with the synthetic fungicide tebuconazole (25%) after 15 d at 1°C. Essential oils of basil completely controlled blue mold in cv. 'Red Chief' and were more effective than tebuconazole after 15 d storage at 1°C, and both were similar in effectiveness after 30 d storage at 1°C. With cv. 'Royal Gala', essential oils of thyme at 1 or 10%, or marjoram, fennel, or basil at 10% concentration controlled blue mold as effectively as tebuconazole after 15 d storage at 1°C. The essential oils of marjoram and fennel at 10% and thyme oil at 1% reduced the decay severity of blue mold to lesions of slightly more than 10 mm in diameter and were less effective than tebuconazole (nearly 2 mm) (Lopez-Reyes et al., 2010).

The incidence of gray mold was higher in cv. 'Golden Delicious' compared to cvs. 'Royal Gala', 'Granny Smith', and 'Red Chief' stored at 4 ± 1°C. Thyme oil at 10% was similar in effectiveness to tebuconazole (25%) after 15 d storage at 4°C (Lopez-Reyes et al., 2010). In cv. 'Granny Smith', thyme oil, sage oil, oregano, peppermint, and basil essential oils at 10% concentration completely inhibited gray mold during the 15 d storage at 4°C. In cv. 'Red Chief', essential oils of thyme, sage, savory, oregano, marjoram, or basil at 10% concentration controlled gray mold after 15 d of storage at 4°C, and were similar in effectiveness to tebuconazole. Control of gray mold by the essential oils marjoram or basil at 10% was similar in effectiveness to tebuconazole after 30 d storage at 4°C. In cv. 'Royal Gala', thyme oil at 10% concentration was more effective than tebuconazole after 15 d of storage at 4°C. However, after 30 d of storage at 4°C, thyme oil at 1 or 10% were similar in effectiveness to tebuconazole after 15 d storage at 4°C (Lopez-Reyes et al., 2010). Variations in the effectiveness of essential oil vapors to control blue mold in apples may be associated with factors such as antifungal (phytochemicals) substances present in the various cultivars and the chemical components of the essential oils that were tested. Also, *B. cinerea* and *Rhizopus stolonifer* (Ehrenb.) Vuill. were effectively controlled with lemongrass oil (Tzortzakis and Economakis, 2007).

Thyme oil vapor controlled anthracnose in avocado cv. 'Hass' as effectively as the synthetic fungicide prochloraz. However, the effectiveness of the thyme oil vapor treatment to control anthracnose was superior in cv. 'Ryan' than in cv. 'Hass' (Bill, personal communication). Similarly, the application of lemongrass vapor reduced the incidence of anthracnose in avocados. However, the effectiveness of this lemon grass vapor treatment was different in various avocado cultivars: 'Ryan' > 'Hass' > 'Fuerte' (Mafune et al., 2013).

5 Mechanisms of Action of Essential Oils

Essential oils are lipophilic and can easily penetrate cells and disrupt the membrane and/or its permeability (Perricone et al., 2015). The hydrophobicity of essential oils allows them to partition the lipid layer of the membrane of the mitochondrion, causing it to be more permeable. This process leads to the leakage of ions and other cell components to a level of loss of these crucial cell contents to cause lysis and death (Adeyinka and Richard, 2015; Vergis et al., 2015). Vergis et al. (2015) reported that this mechanism exerts an impact on the cytoplasmic membrane by disrupting the proton motive force (PMF), electron flow, and active transport, with eventual coagulation of the cell contents. Adeyinka and Richard (2015) suggested that the hydroxyl (OH) group of eugenol, an active component in many essential oils, strongly adheres to proteins, rendering them inert and thus disrupting enzymatic functions and reactions. The authors also described the occurrence of a breakdown in the cell wall and cytolysis of pathogens due to the antimicrobial properties of eugenol. Tongnuanchan and Benjakul (2014) stated that the effectiveness of an essential oil lies in its ability to reduce the antioxidant activity. They described that harm to antioxidant activities could include the prevention of chain initiation; decrease in free radical scavenger functions; impaired destruction of peroxides; prevention of continued quenching of singlet oxygen formation; and reduced binding of transitional metal ion catalysts. They claimed that the antioxidant properties of various essential oils emanate from plant phenols (Tongnuanchan and Benjakul, 2014). Thyme oil contains phenolic monoterpenes such as thymol, terpinolene, and *o*-cymol as the main compounds that have a significant role in its antifungal activity against pathogens (Bill et al., 2016). The presence and number of hydroxyl groups on the phenolic rings of thymol, the major active ingredient in thyme oil, describes the increased antifungal activity of thyme oil to control postharvest pathogens and, ultimately, to control postharvest decay. The cell membrane of the pathogen is the main target of the phenolic compounds in essential oil. Hence, it disrupts the permeability barrier of the cell membrane and obstructs respiration. The fungitoxicity of lemongrass oil was attributed to its active component citral, which comprises 79% of lemongrass oil, because causes permanent cell membrane disruption of fungi, while other components, including camphene and limonene, diffuse into the cell membrane and cause its deformation (Wuryatmo et al., 2003).

Thyme oil was reported to induce the activities of chitinase, β-1,3 glucanase, and peroxidase in avocados (Sellamuthu et al., 2013a; Sivakumar and Bautista-Baños, 2014; Bill et al., 2016; Cindi et al., 2016). In avocados, exposure to thyme oil vapors induced the gene expression of chitinase and β-1,3-glucanase activities following long-term storage at 7 or 7.5°C for 21 d followed by up to 5 d at 20°C, and provided a residual protection against subsequent anthracnose (Bill et al., 2016). Furthermore, thyme oil vapors caused down-regulation of lipoxygenases (LOX) gene expression in two avocado cultivars. Thus, simultaneous up-regulation of phenylalanine ammonia-lyase (PAL) genes and down-regulation of LOX genes in avocados exposed to thyme oil vapors prior to cold storage were associated with a decrease in lipoxygenase activity and a higher epicatechin content in the fruit, which enhanced anthracnose disease resistance. Increased epicatechin content in the fruit skin was also involved in slowing the rate of decline of the antifungal diene compound via lowering/delaying the expression of the LOX genes (Prusky et al., 1988, 1996). They hypothesized that down-regulation of the LOX gene expression in 'Hass' and 'Ryan' avocados is the main mode of action of the thyme oil vapor to enhance resistance of these fruits to anthracnose.

In apples, thyme oil induced the accumulation of the pathogenesis-related protein 8 (PR-8), which codes for a class III chitinase, and improved the resistance against *B. cinerea*. Thyme oil treatment provided long-lasting residual protection to the fruit to gray mold (Spadaro et al., 2017).

Phenolic compounds improved the antioxidant capacity and reactive oxygen species (ROS) scavenging activity, thus they can also contribute to an enhanced resistance to pathogen challenge (Mittler, 2002). Most essential oils also reduced the incidence of brown rot and enhanced the total phenolic content of peach fruit (Khumalo et al., 2017). Sellamuthu et al. (2013a) and Bill et al. (2016) reported that thyme oil activated the synthesis of phenols by stimulating higher PAL activity in avocados. In peaches, the effectiveness of thyme oil vapor on brown rot control was greater in red skin cultivars due to higher levels of catechins, chlorogenic acids, and caffeic acids that are responsible for the antioxidant capacity. Moreover, higher concentrations of these compounds suppressed chitinase activity, the formation of appressoria by *Monilinia*, and inhibited the infection process (Bostock

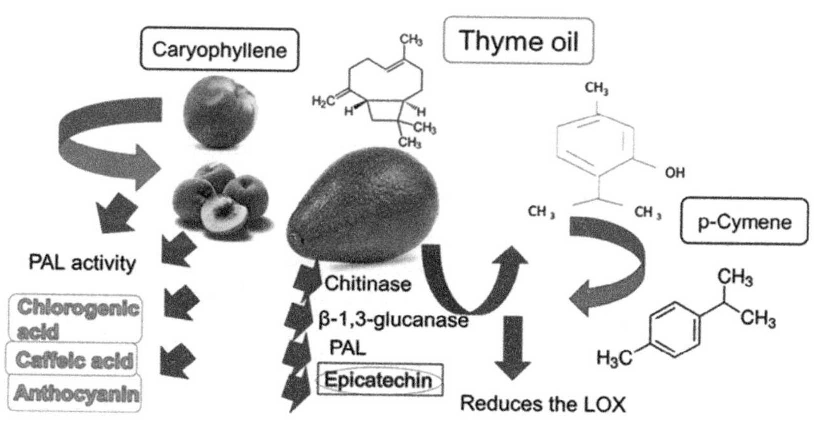

Figure 20.2 Thyme oil vapors act as elicitors to prevent anthracnose in avocados and brown rot in peaches.

et al., 1999; Lee and Bostock, 2007; Villarino et al., 2011; De Cal et al., 2013; Khumalo et al., 2017). In general, essential oils are among the most widespread natural compounds that elicit host defenses that cause a reduction in postharvest decay on fruit and vegetables (Romanazzi et al., 2016). The signaling effects of thyme oil vapors to induce defense mechanisms in avocado and peach are summarized in Figure 20.2.

6 Influence of Essential Oil Treatment on Fruit Quality and Nutritional Compounds

Essential oils caused a reduction in the rate of weight loss and improved retention of vitamin C and carotenoid content during the storage of apples (Spadaro et al., 2017). Thyme oil vapor treatment caused slightly higher retention of the non-structural carbohydrate C7 sugar, mannoheptulose, improved the shelf life of avocado fruit (Bill et al., 2016), and prolonged the retention of fruit firmness during storage (Bill et al., 2016; Khumalo et al., 2017). Thyme oil vapor enhanced the accumulation of major phenolic compounds including *p*-coumaric, ferulic and caffeic acids, and catechin and epicatechin, and improved retention of oleic (18:1) and palmitic (16:0) fatty acids in avocados. Strawberries exposed to *S. canadensis* vapors were firmer, had higher total soluble solids (TSS) and titratable acidity (TA) contents, and a reduced rate of weight loss (Liu et al., 2016). Postharvest treatment with 10% lime oil significantly increased TSS in oranges during 14 wk of cold storage (Badawy et al., 2011).

One of the possible constraints linked to the application of the essential oils may reside in the potential changes in the sensory properties and phytotoxic injury in treated fruits. In trials run on table grapes exposed to essential oils of rosemary and peppermint, Servili et al. (2017) reported that a triangulated test sensory panel perceived these essential oils on the grapes soon after application and 24 hr later, but not 48 hr after treatment. The results were similar with table grapes kept at either room temperature or in cold storage at 4°C for 48 hr.

7 Integrated Application of Essential Oils with Other Postharvest Treatments

Integrated treatments as alternatives to synthetic fungicides are becoming popular due to the restricted use of pesticides and trade regulations enforced by the importing countries (Arrebola et al., 2010). The application of essential oils is feasible in the vapor phase, and commodities that are sensitive to aqueous sanitation can be treated under pallet covers or exposed to the vapors within modified atmosphere packaging (MAP). Integrated application of lemongrass oil with the biocontrol agent *Bacillus amyloliquefaciens* within MAP (3% O_2 and 5% CO_2) at 0°C followed by a shelf-life period of 3–5 d at 20°C helped to improve the biofilm formation of the biocontrol agent and reduced gray mold incidence in peaches (Arrebola et al., 2010). The combined application of lemongrass oil with the biocontrol agent *B. amyloliquefaciens* within MAP retained the peach firmness and the ratio TSS/TA without causing phytotoxicity. Fruit flavor and the aroma were not altered by the lemongrass oil treatment (Arrebola et al., 2010).

In another study, a combined application of *Lactobacillius plantanum* A7 and ajwain (50 μL) or cinammon oils (100 μL) caused a reduction of 40 and 60% in the incidence of gray mold in strawberries infected with *B. cinerea*. The combined application did not affect the pH, TSS, TA, or ascorbic acid content of the fruit (Zamani-Zadeh et al., 2013).

A combination of lemongrass oil (1.5 mL/L) and $CaCl_2$ (1.5 g/L) reduced Rhizopus rot by 70% and was synergistic in its effectiveness to control the decay of peaches. Both treatments altered the growth, sporulation, and morphology of *R. stolonifer* (Salem et al., 2016). The integrated application of thyme oil vapors in MAP (8% CO_2 and 2% O_2) was evaluated in 'Fuerte' and 'Hass' avocados stored for 18 d at 10°C followed by ripening at 25°C. The treatment reduced anthracnose incidence and physiological disorders. The integrated application of thyme oil vapors and MAP increased the phenolic compound content and the antioxidant activity in the mesocarp (Sellamuthu et al., 2013a).

8 Use of Microencapsulated Essential Oil and Essential Oil Impregnated Films in Packaging to Control Postharvest Decay

Disadvantages of essential oils, such as insufficient volatility, poor water solubility, phytotoxicity risks, and strong aromas, currently limit their application in some food industry applications. At present, encapsulation of essential oils has overcome some of these disadvantages and improved their delivery for pathogen control (Feng et al., 2017). For example, microencapsulated thyme and rosemary oils controlled postharvest decay in strawberry fruits. The combination of two different oils (thyme and rosemary oils) in a microcapsule significantly improved retention of fruit quality after treatment, mainly due to the pattern of release of the oils from the microcapsules during the application process (Alikhani and Garmakhany, 2012). The application of microencapsulated rosemary oil from *Opuntia mucilage* helped to prevent browning and maintained the overall quality of fresh-cut mangoes packed in tray packs for up to 9 d at 6°C (Alikhani, 2014).

The use of essential oil-impregnated 'antifungal film' packaging offers advantages because it uses small volumes of essential oil during the preparation of the film, which could help preventing interferences with the flavor or aroma of the fruit (Feng et al., 2017). The use of antifungal packaging in the food industry is beneficial because it enables an extension of postharvest life by the use of minimal amounts of a preservative, ensuring safety while maintaining the fruit health promoting compounds (Alikhani, 2014). The incorporation of thyme oil into LEDP films modified the environment within the packaging and controled the decay-causing pathogens by inducing defense mechanisms (Bill et al., 2016). Furthermore, investigations into the incorporation of thyme oil into the LEDP matrix during the blown film extrusion created a film that inhibited mycelial growth of *C. gloesporioides* (Bill et al., 2016). Therefore, LEDP-thyme oil film showed promise as an antifungal packaging to minimize fungal decay and to improve the fruit quality during postharvest storage by retaining the fruit firmness at ready to eat stage.

9 Conclusions

Increasing consumer concerns regarding microbial food safety and the safety of chemical residues on fruit create a need to discover natural, environmentally friendly alternative disease control strategies and products. Alternative postharvest disease control strategies have been developed to provide the industry with a more feasible and environmentally friendly product that is also safer for both operators and consumers. The application of an essential oil postharvest treatment in a vapor phase is an efficient and cost-effective method of application. A vapor-phase application also helps to minimize both phytotoxicity and the strong odor of the essential oils. However, the method of application depends on the duration and temperature during application and the cultivar responses that influence their effectiveness to control postharvest diseases.

Acknowledgments

The authors wish to thank the South African Avocado Growers Association for their postharvest innovation financial support and the South African Research Chair program (SARChI) for Phytochemical Food Network to Improve Nutritional Quality for Consumers. Grant No. 98352 is greatly acknowledged.

References

Adeyinka, A. and Richard, F. 2015. Application of phytochemical extracts and essential oils in food products: A review. *International Journal of Biotechnology and Food Science* 3, 31–35.

Alikhani, M. 2014. Enhancing safety and shelf life of fresh-cut mango by application of edible coatings and microencapsulation technique. *Food Science and Nutrition* 2, 210–217.

Alikhani, M. and Garmakhany, A.D. 2012. Effect of microencapsulated essential oils on storage life and quality of strawberry (*Fragaria ananassa* cv. Camarosa). *Quality Assurance and Safety of Crops and Foods* 4, 106–112.

Arrebola, E., Sivakumar, D., Bacigalupo, R. and Korsten, L. 2010. Combined application of antagonist *Bacillus amyloliquefaciens* and essential oils for the control of peach postharvest diseases. *Crop Protection* 29, 369–377.

Badawy, I.B., Sallam, N.M.A., Ibrahim, A.R. and Asran, M.R. 2011. Efficacy of some essential oils on controlling green mold of orange and their effects on postharvest quality parameters. *Plant Pathology Journal* 10, 168–174.

Bakkali, F., Averbeck, S. and Idaomar, M. 2008. Biological effects of essential oils – A review. *Food and Chemical Toxicology* 46, 446–475.

Baser, K.H.C. and Buchbauer, G. 2010. *Handbook of Essential Oils Science, Technology and Applications*. CRC Press, Boca Raton, FL.

Bask, S. and Guha, P. 2015. Modelling the effect of essential oil of betel leaf (*Piper betle* L.) on germination, growth, and apparent lag time of *Penicillium expansum* on semi-synthetic media. *International Journal of Food Microbiology* 215, 171–178.

Bill, M., Sivakumar, D., Beukes, M. and Korsten, L. 2016. Expression of pathogenesis-related (PR) genes in avocados fumigated with thyme oil vapours and control of anthracnose. *Food Chemistry* 194, 938–943.

Bostock, R.M., Wilcox, S.M., Wang, G. and Adaskaveg, J.E. 1999. Suppression of *Monilinia fructicola* cutinase production by peach fruit surface phenolic acids. *Physiological and Molecular Plant Pathology* 54, 37–50.

Boubaker, H., Karim, H., Hamdaoui, E.L., Msanda, F., Leach, D., Bombarda, I., Vanloot, P., Abbad, A., Boudyach, E.H. and Ait Ben Aoumar, A. 2016. Chemical characterization and antifungal activities of four *Thymus* species essential oils against postharvest fungal pathogens of citrus. *Industrial Crops and Products* 86, 95–101.

Burt, S. 2004. Essential oils: Their antibacterial properties and potential applications in foods—a review. *International Journal of Food Microbiology* 94, 223–253.

Bus, V.G., Bongers, A.J. and Risse, L.A. 1991. Occurrence of *Penicillium digitatum* and *P. italicum* resistant to benomyl, thiabendazole, and imazalil on citrus fruit from different geographic origins. *Plant Disease* 75, 1098–1100.

Cindi, M.D., Soundy, P., Romanazzi, G. and Sivakumar, D. 2016. Different defense responses and brown rot control in two *Prunus persica* cultivars to essential oil vapours after storage. *Postharvest Biology and Technology* 119, 9–17.

De Cal, A., Sandín-Espana, P., Martínez, F., Eguen, B., Chien-Ming, C., Lee, M.H., Melgarejo, P. and Prusky, D. 2013. Role of gluconic acid and pH modulation in virulence of *Monilinia fructicola* on peach fruit. *Postharvest Biology and Technology* 86, 418–421.

Droby, S., Eick, A., Macarisin, D., Cohen, L., Rafael, G., Strange, R., McColum, G., Dudai, N., Nasser, A., Wisniewski, M. and Shapira, R. 2008. Role of citrus volatiles in host recognition, germination and growth of *Penicillium digitatum* and *Penicillium italicum*. *Postharvest Biology and Technology* 49, 386–396.

Elshafie, H.S., Mancini, E., Camele, I., De Martino, L. and De Feo, V. 2015. In vivo antifungal activity of two essential oils from Mediterranean plants against postharvest brown rot disease of peach fruit. *Industrial Crops and Products* 66, 11–15.

Everett, K.R., Owen, S.G. and Cutting, J.G.M. 2005. Testing efficacy of fungicides against postharvest pathogens of avocado (*Persea americana* var Hass.). *New Zealand Plant Protection* 58, 89–95.

Feng, K., Wen, P., Yang, H., Li, N., Lou, W.Y., Zong, M.H. and Wu, H. 2017. Enhancement of the antimicrobial activity of cinnamon essential oil-loaded electrospun nanofilm by the incorporation of lysozyme. *RSC Advances* 7, 1572–1580.

Frankova, A., Smid, J., Bernardos, A., Finkousova, A., Marsik, P., Novotny, D., Legarova, V., Pulkrabek, J. and Kloucek, P. 2016. The antifungal activity of essential oils in combination with warm air flow against postharvest phytopathogenic fungi in apples. *Food Control* 68, 62–68.

Hassani, A., Ghosta, Y. and Abdolahi, A. 2012. Evaluation of plant essential oils for control of postharvest brown and gray mold rots on apricot. *Journal of Food Safety* 32, 94–101.

Idris, F.M., Ibrahim, A.M. and Forsido, S.F. 2015. Essential oils to control *Colletotrichum musae* in vitro and in vivo on banana fruits. *American-Eurasian Journal of Agricultural and Environmental Sciences* 15, 291–302.

Ismail, M. and Zhang, J. 2004. Post-harvest citrus disease and their control. *Out-looks on Pest Management* 15, 29–35.

Khumalo, K., Tinyane, P., Soundy, P., Romanazzi, G., Glowacz, M. and Sivakumar, D. 2017. Effect of thyme oil vapour exposure on the brown rot infection, phenylalanine ammonia-lyase (PAL) activity, phenolic content and antioxidant activity in red and yellow skin peach cultivars. *Scientia Horticulturae* 214, 195–199.

Lazar-Baker, E.E., Hetherington, S.D., Ku, V.V. and Newman, S.M. 2011. Evaluation of commercial essential oil samples on the growth of postharvest pathogen *Monilinia fructicola* (G.Winter) honey. *Letters in Applied Microbiology* 52, 227–232.

Lee, M.H. and Bostock, R.M. 2007. Fruit exocarp phenols in relation to quiescence and development of *Monilinia fructicola* infections in *Prunus* spp: A role for cellular redox? *Phytopathology* 97, 269–277.

Lihandra, E.K. 2007. Assessment of ethanol, honey, milk and essential oils as potential postharvest treatments of New Zealand grown fruit. MSc. Dissertation. University of Technology. Auckland, New Zealand.

Liu, S., Shao, X., Wei, Y., Li, Y., Xu, F. and Wang, H. 2016. *Solidago canadensis* L. essential oil vapor effectively inhibits *Botrytis cinerea* growth and preserves postharvest quality of strawberry as a food model system. *Frontiers in Microbiology* 7, 1179.

Lopez-Reyes, J.G., Spadaro, D., Gullino, M.L. and Garibaldi, A. 2010. Efficacy of plant essential oils on postharvest control of rot caused by fungi on four cultivars of apples in vivo. *Flavour and Fragrance Journal* 25, 171–177.

Mafune, M., Sivakumar, D., Sellamuthu, P.S. and Bautista-Baños, S. 2013. Use of lemongrass oil and modified atmosphere packaging on control of anthracnose and quality maintenance in avocado cultivars. *Journal of Fruit Quality* 36, 198–208.

Maqbool, M., Ali, A., Alderson, P.G., Mohamed, M.T.M., Siddiqui, Y. and Zahid, N. 2011. Postharvest application of gum arabic and essential oils for controlling anthracnose and quality of banana and papaya during cold storage. *Postharvest Biology and Technology* 62, 71–76.

Mittler, R. 2002. Oxidative stress, antioxidants and stress tolerance. *Trends in Plant Sciences* 7, 405–410.

Mothopa, M.M., Moremi, M., Sivakumar, D., Tzortzakis, N. and Korsten, L. 2015. Integrated treatment with edible coatings and thyme oil on the control of Imazalil resistant *Penicillium digitatum* or *P. italicum* strains and *Geotrichum citri-aurantii*. Abstracts Book of the III International Symposium on Postharvest Pathology. 7-11 June 2015, Bari, Italy. p. 195.

Nguefack, J., Tamgue, O., Lekagne, J.B., Dongmo, L., Dakole, C.B., Leth, V., Vismer, H.F., Amvam Zollo, P.H. and Nkengfack, A.E. 2012. Synergistic action between fractions of essential oils from *Cymbopogon citratus, Ocimum gratissimum* and *Thymus vulgaris* against *Penicillium expansum. Food Control* 23, 377–383.

Palou, L., Smilanick, J.L. and Droby, S. 2008. Alternatives to conventional fungicides for the control of citrus postharvest green and blue moulds. *Stewart Postharvest Review* 4:2, 1–16.

Patrignani, F., Siroli, L., Serrazane, D.I., Gardini, F. and Lanciotti, R. 2015. Innovative strategies based on the use of essential oils and their components to improve safety, shelf-life and quality of minimally processed fruits and vegetables. *Trends in Food Science and Technology* 20, 1–12.

Pavela, R. 2015. Essential oils for the development of eco-friendly mosquito larvicides: A review. *Industrial Crops and Products* 76, 174–187.

Pedrotti, C., Ribeiro, R.T.S. and Schwambach, J. 2017. Control of postharvest fungal rots on grapes using essential oil of *Foeniculum vulgare* Mill. *Journal of Agricultural Science* 9, 205–216.

Perricone, M., Arace, E., Corbo, M.R., Sinigaglia, M. and Bevilacqua, A. 2015. Bioactivity of essential oils: A review on their interaction with food components. *Frontiers in Microbiology* 6, 1–7.

Perumal, A.B., Sellamuthu, P.S., Nambiar, R.B. and Sadiku, E.R. 2016. Antifungal activity of five different essential oils in vapour phase for the control of *Colletotrichum gloeosporioides* and *Lasiodiplodia theobromae* in vitro and on mango. *International Journal of Food Science and Technology* 51, 411–418.

Popp, J., Petö, K. and Nagy, J. 2013. Pesticide productivity and food security. A review. *Agronomy for Sustainable Development* 33, 243–255.

Prusky, D., Hamdan, H., Ardi, R. and Keen, N.T. 1996. Induction of biosynthesis of epicatechin in avocado suspension cells treated with an enriched CO_2 atmosphere. *Physiological and Molecular Plant Pathology* 48, 171–178.

Prusky, D., Kobiler, I. and Jacoby, B. 1988. Involvement of epicatechin in cultivar susceptibility of avocado fruits to *Colletotrichum gloeosporioides* after harvest. *Journal of Phytopathology* 123, 140–146.

Romanazzi, G., Sanzani, S.M., Bi, Y., Tian, S., Gutiérrez-Martínez, P. and Alkan, N. 2016. Induced resistance to control postharvest decay of fruit and vegetables. *Postharvest Biology and Technology* 122, 82–94.

Salem, E., Youssef, K. and Sanzani, S.M. 2016. Evaluation of alternative means to control postharvest Rhizopus rot of peaches. *Scientia Horticulturae* 198, 86–90.

Sayago, J.E., Ordonez, R.M., Kovacevich, L.N., Torres, S. and Isla, M.I. 2012. Antifungal activity of extracts of extremophile plants from the Argentine Pune to control citrus postharvest pathogens and green mold. *Postharvest Biology and Technology* 67, 19–24.

Sellamuthu, P.S., Mafune, M., Sivakumar, D. and Soundy, P. 2013a. Thyme oil vapour and modified atmosphere packaging reduce anthracnose incidence and maintain fruit quality in avocado. *Journal of the Science of Food and Agriculture* 93, 3024–3031.

Sellamuthu, P.S., Sivakumar, D. and Soundy, P. 2013b. Antifungal activity and chemical composition of thyme, peppermint and citronella oils in vapor phase against avocado and peach postharvest pathogens. *Journal of Food Safety* 33, 86–93.

Servili, A., Feliziani, E. and Romanazzi, G. 2017. Exposure to volatiles of essential oils alone or under hypobaric treatment to control postharvest gray mold of table grapes. *Postharvest Biology and Technology* 133, 36–40.

Shao, X., Wang, H., Xu, F. and Cheng, S. 2013. Effects and possible mechanisms of tea tree oil vapor treatment on the main disease in postharvest strawberry fruit. *Postharvest Biology and Technology* 77, 94–101.

Sharma, N. and Tripathi, A. 2006. Fungitoxicity of the essential oil of *Citrus sinensis* on post-harvest pathogens. *World Journal of Microbiology and Biotechnology* 22, 587–593.

Simas, D.L.R., Silesia, H.B.M., de Amorim, H.B.M., Goulart, F.R.V., Alviano, C.S., Alviano, D.S. and da Silva, A.J.R. 2017. Citrus species essential oils and their components can inhibit or stimulate fungal growth in fruit. *Industrial Crops and Products* 98, 108–115.

Sivakumar, D. and Bautista-Baños, S. 2014. A review on the use of essential oils for postharvest decay control and maintenance of fruit quality during storage. *Crop Protection* 64, 27–37.

Smilanick, J.L., Mansour, M.F., Margosan, D.A., Mlikota Gabler, F. and Goodwine, W.R. 2005. Influence of pH and $NaHCO_3$ on effectiveness of imazalil to inhibit germination of *Penicillium digitatum* and to control postharvest green mold on citrus fruit. *Plant Disease* 89, 640–648.

Spadaro, D., Banani, H., Santoro, K., Garibaldi, A. and Gullino, M.L. 2017. Essential oils to control postharvest diseases of apples and peaches: Elucidation of the mechanism of action. Abstracts Book of the IV International Symposium on Postharvest Pathology. May 28-June 3, 2017. Skukuza, South Africa. p. 36.

Tongnuanchan, P. and Benjakul, S. 2014. Essential oils: Extraction, bioactivities, and their uses for food preservation. *Journal of Food Science* 79, 1231–1249.

Tyagi, A.K., Gottardi, D., Malik, A. and Guerzoni, M.E. 2013. Anti-yeast activity of mentha oil and vapours through in vitro and in vivo real fruit juices assays. *Food Chemistry* 137, 108–114.

Tzortzakis, N.G. and Economakis, C.D. 2007. Antifungal activity of lemongrass (*Cymbopogon citrates* L.) essential oil against key postharvest pathogens. *Innovative Food Science and Emerging Technologies* 8, 253–258.

US FDA (Unites States Food and Drug Administration). 2016. Part 182 – Substances Generally Recognized As Safe. *Code of Federal Regulations Title* 13, 21. Available at: www.fda.gov/Food/IngredientsPackagingLabeling/GRAS/2018.

Vergis, J., Gokulakrishna, P., Agarwal, R.K. and Kumar, A. 2015. Essential oils as natural food antimicrobial agents: A review. *Critical Reviews in Food Science and Nutrition* 55, 1320–1323.

Vilanova, L., Torres, R., Viñas, I., González-Candelas, L., Usall, J., Fiori, S., Solsona, C. and Teixidó, N. 2013. Wound response in orange as a resistance mechanism against *Penicillium digitatum* (pathogen) and *P. expansum* (non-host pathogen). *Postharvest Biology and Technology* 78, 113–122.

Villarino, M., Sandín-España, P., Melgarejo, P. and De Cal, A. 2011. High chlorogenic and neochlorogenic acid levels in immature peaches reduce *Monilinia laxa* infection by interfering with fungal melanin biosynthesis. *Journal of Agricultural and Food Chemistry* 59, 3205–3213.

Vitoratos, A., Bilalis, D., Karkanis, A. and Efthimiadou, A. 2013. Antifungal activity of plant essential oils against *Botrytis cinerea, Penicillium italicum* and *Penicillium digitatum. Notulae Botanicae Horti Agrobotanici* 41, 86–92.

Viuda-Martos, M., Ruiz-Navajas, Y., Fernández-López, J. and Pérez-Álvarez, J.A. 2008. Antifungal activity of lemon (*Citrus lemon* L.), mandarin (*Citrus reticulata* L.), grapefruit (*Citrus paradisi* L.) and orange (*Citrus sinensis* L.) essential oils. *Food Control* 19, 1130–1138.

Wuryatmo, E., Klieber, A. and Scott, E.S. 2003. Inhibition of citrus postharvest pathogens by vapour of citral and related compounds in culture. *Journal of Agricultural and Food Chemistry* 51, 2637–2640.

Zamani-Zadeh, M., Soleimanian-Zad, S. and Sheikh-Zeinoddin, M. 2013. Biocontrol of gray mold disease on strawberry fruit by integration of *Lactobacillus plantarum* A7 with ajwain and cinnamon essential oils. *Journal of Food Science* 78, 1582–1588.

Zhang, D., Spadaro, D., Garibaldi, A. and Gullino, M.L. 2010. Efficacy of the antagonist *Aureobasidium pullulans* PL5 against postharvest pathogens of peach, apple and plum and its mode of action. *Biological Control* 54, 172–180.

Chapter 21

Chitosan and Other Edible Coatings for Fresh Fruit Postharvest Disease Control

Evandro Leite de Souza and Lúcia Raquel Ramos Berger

Laboratory of Food Microbiology, Department of Nutrition, Federal University of Paraíba, João Pessoa, Paraíba, Brazil

Anna Marín

Research Institute of Food Engineering for Development (IIAD), Department of Food Technology, Polytechnical University of Valencia (UPV), Valencia, Spain

María B. Pérez-Gago and Lluís Palou

Laboratory of Pathology, Postharvest Technology Center (CTP), Valencian Institute of Agrarian Research (IVIA), Montcada, Valencia, Spain

Abbreviations

2-α-AAPCS	2-(α-Arylamino phosphonate) chitosan
ABC	Ammonium bicarbonate
AC	Ammonium carbonate
APh	Ammonium phosphate
CAT	Catalase
CHI	Chitinase
CMC	Carboxymethyl cellulose
CMCS	Carboxymethyl chitosan
DD	Deacetylation degree
EO	Essential oil
GlcNAc	N-acetylglucosamine
GRAS	Generally recognized as safe
GLU	β-1,3-Glucanase
HPMC	Hydroxypropyl methylcellulose
IMZ	Imazalil
LbL	Layer-by-layer
MDA	Malondialdehyde
MW	Molecular weight
PAL	Phenylalanine ammonia lyase
PC	Potassium carbonate
POD	Peroxidase
PS	Potassium sorbate
ROS	Reactive oxygen species
SB	Sodium benzoate
SEP	Sodium ethylparaben
SMP	Sodium methylparaben
SOD	Superoxide dismutase
SP	Sodium propionate
SSC	Soluble solids concentration
TA	Titratable acidity
TBZ	Thiabendazole
UDP-GlcNAc	Uridine diphosphate N-acetylglucosamine
WL	Weight loss

1 Introduction

In recent years, there has been an increasing awareness among consumers regarding the importance of adopting a healthier diet, which has increased the consumption of fresh fruits. Fresh fruits have a relatively short shelf life due to water loss and microbial decay that decrease their market value and consumer acceptance (Poverenov et al., 2014). Therefore, producers and processors are interested in adopting effective technologies to prolong the storage life and nutritional benefits of fresh fruit, including increasing fiber contents, antioxidants, and a variety of vitamins and minerals (Rojas-Graü et al., 2007; Poverenov et al., 2014; Palou et al., 2016).

Pathogenic fungi cause decay, accelerated ripening, and in some cases accumulation of mycotoxins in fruit, resulting in significant postharvest losses (Tripathi and Dubey, 2004; Palou et al., 2016). Different fungi cause a variety of postharvest diseases in fruit, each with specific symptomatology. While some of them, such as *Alternaria alternata* (Fr.:Fr.) Keissl., *Aspergillus niger* Tiegh., *Botrytis cinerea* Pers., *Colletotrichum gloeosporioides* (Penz.) Penz. & Sacc., *Lasiodiplodia theobromae* (Pat.) Griffon & Maubl., *Penicillium expansum* Link, and *Rhizopus stolonifer* (Ehrenb.: Fr.) Vuill., have a wide range of hosts, including many different fruit species, others such as *Penicillium digitatum* (Pers.) Sacc, *Penicillium italicum* Wehmer, *Geotrichum citri-aurantii* (Ferraris) E.E. Butler, *Monilinia laxa* (Aderh. & Ruhland) Honey, or *Monilinia fructicola* (G. Winter) Honey are much more host specific and only attack particular fruit families (Ali et al., 2013; Feliziani et al., 2013; Ma et al., 2013; Romanazzi et al., 2013; Kewsuwan et al., 2014; Wang et al., 2014, 2017; Oliveira et al., 2014a, 2014b; Deng et al., 2015).

Synthetic fungicides remain the primary means to control postharvest diseases of fresh fruits (Tripathi and Dubey, 2004; Romanazzi et al., 2016). Although most of the synthetic fungicides do not adversely affect the appearance or quality of fruit and have a high degree of specificity toward the target organisms, pathogens resistant to these chemicals have emerged (Palou et al., 2016). Furthermore, consumers increasingly demand non-chemically treated foods to minimize potential health problems or environmental problems (Guerra et al., 2015). Some synthetic fungicides commonly used to control postharvest decay of fresh fruits have shortcomings that restrict their use, such as causing risks to human health through carcinogenicity, teratogenicity, and high and acute residual toxicity, or having long degradation periods and causing environmental pollution (Tripathi and Dubey, 2004; Romanazzi et al., 2016). For these reasons, some chemicals (e.g., methyl bromide, iprodione, dichloran, and guazatine) used to control fruit postharvest decay were banned in many countries (Ramos-García et al., 2012; Aloui et al., 2014). These actions have driven the search for naturally occurring and other presumably safe substances with desirable properties to be used in the formulation of edible coatings, which are cited as emerging postharvest technologies to preserve fresh fruit (Oliveira et al., 2014a, 2014b; Kerch, 2015; Guerra et al., 2015, 2016; Palou et al., 2016; Romanazzi et al., 2016).

Edible coatings are characterized as being thin layers composed mainly of proteins, lipids, or polysaccharides, which are deposited directly on the surface of fruit (Azarakhsh et al., 2014; Fagundes et al., 2014; Guerra et al., 2015). Coatings play an important primary function in fresh fruit preservation by decreasing moisture loss and respiration rate (Maqbool et al., 2011; Sánchez-González et al., 2011; Fagundes et al., 2014; Guerra et al., 2015). Edible coatings were shown to extend the shelf life of fruits without affecting their quality or fresh-like characteristics (Ponce et al., 2008; Santos et al., 2012; Oliveira et al., 2014a, 2014b; Guerra et al., 2015). They offer additional advantages: increase mechanical and textural properties, decrease flavor losses in coated fruits, posses the potential to be carriers for other additives (e.g., colorants, flavors, antioxidants, nutraceuticals, and antimicrobials) that can enhance fruit appearance and preservation, and provide new or improved functionalities (Campos et al., 2011; Maqbool et al., 2011; Valencia-Chamorro et al., 2011b; Azarakhsh et al., 2012, 2014; Aquino et al., 2015; Pérez-Gago and Palou, 2016). In the particular case of edible coatings designed for the control of postharvest diseases, the additional ingredients are food-grade antimicrobial agents or compounds that decrease the survival and retard the growth of spoilage microorganisms that cause postharvest diseases. According to their nature, these antimicrobial agents can be classified into three main categories (Palou et al., 2016): i) synthetic food preservatives or generally recognized as safe (GRAS) compounds, such as inorganic and organic salts, metal-based nanoparticles, and nanocomposites, ii) natural GRAS compounds, such as essential oils and plant extracts, and iii) antagonistic microorganisms that act as biocontrol agents (bacteria, yeast, yeast-like fungi, and filamentous fungi).

The polysaccharide chitosan has attracted attention as an edible coating material to be applied on fruit and has the advantage of being classified as GRAS by the United States Food and Drug Administration (Feliziani et al., 2013; US FDA, 2015). Chitosan-based coatings are cited as inexpensive, easily handled, non-phytotoxic, and environmentally friendly materials, with promising results that include their efficacy to delay fruit senescence and fungal growth, in addition to induce desirable quality characteristic in fresh fruit (Raybaudi-Massilia et al., 2008; Sánchez-González et al., 2011; Santos et al., 2012; Aloui et al., 2014; Azarakhsh et al., 2014; Xing et al., 2015). Chitosan has shown efficacy to control postharvest losses in fresh fruits when used alone to form edible coatings (Chien et al., 2007; Meng et al., 2008; Wang et al., 2014, 2017; Oliveira et al., 2014a, 2014b), as well as when used in combination with other antimicrobials, such as plant extracts, essential oils and their individual constituents (Sánchez-González et al., 2011; Guerra et al., 2015; Souza et al., 2015; Athayde et al., 2016), and antagonistic microorganisms (Zhou et al., 2016). Other common materials that have been used as coating matrices for fresh fruits include sodium alginate (Raybaudi-Massilia et al., 2008; Azarakhsh et al., 2014), gum arabic (Maqbool et al., 2011), mesquite gum (Bosquez-Molina et al., 2010), sodium caseinate (Ponce et al., 2008), carboxymethyl cellulose (CMC), and hydroxypropyl methylcellulose (HPMC) (Ponce et al., 2008; Valencia-Chamorro et al., 2011b).

In this chapter, information on the rationale behind the use of edible coatings formulated with chitosan and other edible coating-forming materials to preserve fresh fruits, their mode of action against postharvest pathogenic fungi, and their properties to preserve the physicochemical characteristics of the coated fruit are presented.

2 Chitosan as an Edible Antimicrobial Coating Material for Fresh Fruits

2.1 Sources, Molecular Structures, and Physicochemical Characteristics

Chitin is a homopolysaccharide formed by repeated units of β-(1→4) *N*-acetyl-D-glucosamine synthesized by a variety of living organisms, and is considered the second most abundant polymer in nature after cellulose (Rinaudo, 2006). Chitin occurs partially deacetylated forming structural components in the exoskeleton of crustaceans, mollusks, annelids, coelenterates, and insects, as well as forming the fungal cell wall, particularly in fungi belonging to the Zygomycetes class (Chatterjee et al., 2005; Rinaudo, 2006). Because of its high crystallinity, chitin is insoluble in aqueous solutions and organic solvents (Tsigos et al., 2000).

Chitosan is derived from the partial deacetylation of chitin, being characterized as a linear heteropolysaccharide formed by β-1,4-D-glucosamine bound by N-acetyl-D-glucosamine residues. The molecular chain of chitosan has a three-dimensional helical configuration stabilized by intramolecular hydrogen bonds (Kas, 1997). The chitosan molecule can have different deacetylation degrees (DD, expressed as molar percentage) and molecular weights (MW) (Synowiecki and Al-Khateeb, 2003; Rinaudo, 2006). DD higher than 50 molar percent make chitosan soluble in organic and inorganic acids because of the protonation of $-NH_2$ group at the C-2 position of glucosamine units (Rinaudo, 2006; Younes et al., 2014).

Despite the widespread occurrence of chitin, the exoskeletons of marine crustaceans have been the main sources of commercially available chitosan (Pochanavanich and Suntornsuk, 2002). The crustacean chitosan is produced by partial thermochemical deacetylation of crustacean chitin, but this method has some drawbacks. Few drawbacks are as follows: the source of chitosan (crabs, shrimp, and crayfish shells) is seasonal and limited in supply, use of high temperatures results in considerable energy costs, production of large amounts of concentrated alkaline solution wastes can cause environmental pollution if inadequately discarded, protein residues in the final product can cause allergic reactions in humans, and aggressive process of demineralization and deproteinization can cause chemical modifications in the obtained chitosan that may present variable physicochemical characteristics that are difficult to standardize (Pochanavanich and Suntornsuk, 2002; Berger et al., 2014; Naghdi et al., 2014; Tasar et al., 2016).

Enzymatic deacetylation of chitin results in the formation of chitosan with standardized physicochemical characteristics (Kafetzopoulos et al., 1993; Chatterjee et al., 2005). Unlike the production of crustacean chitosan by thermochemical deacetylation, fungal chitin and chitosan are synthesized via the action of enzymes chitin synthetase (EC 2.4.1.16) and chitin deacetylase (EC 3.5.1.41), respectively, operating in tandem. The synthesis of fungal chitosan occurs by the transfer of N-acetylglucosamine (GlcNAc) from uridine diphosphate N-acetylglucosamine (UDP-GlcNAc) into chitin, followed by chitin polymerization that is catalyzed by chitin synthetase. Fungal chitosan may also be synthesized from UDP-GlcNAc, but this requires the combined action of two enzymes: i) chitin synthetase that transfers glycosyl groups to form chitin chains, and ii) chitin deacetylase that causes the deacetylation of nascent chains to produce chitosan (Davis and Bartnicki-Garcia, 1984a, 1984b).

The production of chitosan from fungal mycelial biomass is an alternative method with the following advantages: (i) simultaneous extraction of both chitin and chitosan is possible, (ii) it is a technically easy and economically viable process, (iii) protein contamination that could cause allergic reactions in individuals with shellfish allergies can be avoided, (iv) it does not depend on seasonal factors, and (v) it is relatively easy to control the pH and nutrient concentration in the fermentation medium, which makes large-scale production feasible (Cardoso et al., 2012; Berger et al., 2014; Oliveira et al., 2014a, 2014b; Souza et al., 2015; Tasar et al., 2016). Despite the use of synthetic culture media for fungal growth and chitosan production (Amorim et al., 2001; Pochanavanich and Suntornsuk, 2002; Silva et al., 2006, 2007; Cardoso et al., 2010), a variety of low-cost alternative substrates have been tested with promising results for scaling up fungal chitosan production (Amorim et al., 2006; Kleekayai and Suntornsuk, 2010; Fai et al., 2011).

Chitosan when dissolved in diluted acid solutions forms dispersions with viscosity sufficient to form thin layers on fruit surfaces when applied by dipping, brushing, or spraying (Oliveira et al., 2014a; Petriccione et al., 2015; Jongsri et al., 2016). The MW of chitosan directly influences the viscosity of chitosan dispersions in diluted acids (Rabea et al., 2003; Jongsri et al., 2016). Dispersions from chitosan of greater MW have a higher viscosity (Jongsri et al., 2016). Furthermore, coating formulations composed of chitosan with lower MW may have a disadvantage of presenting reduced tensile strength without an increase in coating permeability (Oliveira et al., 2014b).

2.2 Inhibitory Effects of Chitosan on Postharvest Pathogenic Fungi

Chitosan has a dual effect on pathogenic fungi of fresh fruits: a direct inhibition on pathogenic fungi and an indirect inhibition by initiating defense responses in fruit tissues (Oh et al., 1998; Badawy and Rabea, 2009; Younes et al., 2014). The antifungal effects of chitosan are influenced by its concentration and physicochemical characteristics (e.g., MW and DD), as well as by natural constituents in the host matrix and environmental conditions (Song et al., 2002; Bautista-Baños et al., 2005; Guerra et al., 2015, 2016). In general, most of the available studies have observed that the antifungal effects of chitosan increase in a dose-dependent manner (Bautista-Baños et al., 2005; Kewsuwan et al., 2014; Wang et al., 2014). Concerning the DD, the higher the DD, the higher the number of deacetylated amino groups in the chitosan molecule that become protonated in acidic solutions; these positive charges enable the interaction of chitosan with the target fungal cells to achieve antifungal activity (Rabea et al., 2003).

Low-MW chitosan increased the inhibition of the mycelial growth of *B. cinerea* (Badawy and Rabea, 2009) and *R. stolonifer* (Hernández-Lauzardo et al., 2008). However, high-MW chitosan inhibited the sporulation of *R. stolonifer* more compared to low or medium MW chitosan (Hernández-Lauzardo et al., 2008). A study that evaluated chitosan of low, medium, and high MW reported that MW did not influence the inhibition of *C. gloeosporioides*; however, the inhibition increased when the chitosan concentration increased (Bautista-Baños et al., 2005).

The application of chitosan reduced blue mold caused by *P. expansum* in winter jujube fruit (*Ziziphus jujube* cv. 'Dongzao'). Direct inhibition of the pathogen by chitosan was confirmed by inhibition of spore germination, germ tube length, and fungal mycelial growth. Furthermore, chitosan caused invagination and disruption of the plasma membrane and cytoplasm degradation in spores of *P. expansum* (Wang et al., 2014). Chitosan of high MW inhibited mycelial growth of *Fusarium oxysporum* f. sp. *vasinfectum* (Atk.) W.C. Snyder & H.N. Hans., *Alternaria solani* Sorauer, and *Valsa mali* Miyabe & G. Yamada (Guo et al., 2006).

The effect of MW on the physical properties and antifungal effects of chitosan membranes (mimicking coating materials) was also reported. Membranes prepared with low-MW chitosan had greater permeability than those prepared with higher MW chitosan (Chen and Hwa, 1996). In addition, low-MW chitosan inhibited mycelial growth of postharvest pathogenic fungi more than high-MW chitosan (Chien et al., 2007; Wang et al., 2017). Wang et al. (2017) reported that low-MW chitosan strongly inhibited the mycelial growth of *Botryosphaeria* sp., in addition to cause ultrastructural damage and reduce both ergosterol synthesis and mitochondrial membrane potential. In the same study, low-MW chitosan delayed postharvest decay in pear fruit (*Pyrus bretschneideri* cv. 'Huangguan'). Similar results were observed by Chien et al. (2007) with 'Murcott' tangor fruits coated with low- or high-MW chitosan to control infections caused by *P. digitatum, P. italicum, Botroydiplodia lecanidion* (Speg.) Petr. & Syd., and *B. cinerea*. The antifungal efficacy of either low- or high-MW chitosan increased when their concentrations increased, although the former inhibited the tested fungi more, even exceeding the inhibition caused by application of the synthetic fungicide thiabendazole (TBZ).

The influence of particle size on chitosan antifungal activity has been evaluated. Ali et al. (2013) reported that treatments with conventional chitosan and submicron chitosan dispersions decreased anthracnose caused by *C. gloeosporioides* in dragon fruit (*Hylocereus polyrhizus*). The submicron chitosan dispersion with 600 nm droplets was the most effective in inhibiting the growth of this fungus in dragon fruit, and its inhibitory effects were associated with its greater ability to penetrate inside the fungal cell where chitosan may act on target sites. However, the lowest concentration of submicron chitosan dispersion had the highest release of chitosan from the emulsifier matrix. They stated that the release of chitosan from dispersion depends on the dispersion viscosity, because the higher viscosity dispersion may result in a stronger matrix after interaction with emulsifiers.

The efficacy of chitosan dissolved in acetic, glutamic, formic, and hydrochloric acids, and of a water-soluble chitosan, to control gray mold and blue mold caused by *B. cinerea* and *Penicillium* spp., respectively, in 'Camarosa' strawberries was compared by Romanazzi et al. (2013). The treatments with chitosan dissolved in all tested acids (chitosan acetate, formate, glutamate, and chloride) controlled strawberry decay similarly when compared with the water-soluble commercial chitosan formulation. However, only the water-soluble chitosan formulation controlled decay in strawberry fruits better than commercial resistance inducers (e.g., benzothiadiazole, calcium, and organic acids). Feliziani et al. (2013) reported that water-soluble chitosan also decreased the mycelial growth of *B. cinerea* and *R. stolonifer* and completely inhibited the growth of *M. laxa* and *A. alternata*. Additionally, water-soluble chitosan decreased the severity of brown rot, gray mold, Rhizopus rot, Alternaria rot, and blue mold in sweet cherry fruit and reduced disease incidence when assessed before or after harvest similar to that caused by application of the synthetic fungicide fenhexamid.

In an attempt to enhance the antifungal efficacy of chitosan, researchers have prepared some chitosan derivatives, such as 2-(α-arylamino phosphonate)-chitosan (2-α-AAPCS). The efficacy of 20 different 2-α-AAPCS to inhibit *A. solani, F. oxysporum* f. sp. *vasinfectum, C. gloeosporioides*, and *Phyllisticta zingiberi* T.S. Ramakr. was determined. All of the tested 2-α-AAPCS inhibited the mycelial growth of target fungi in a dose-dependent manner, but the chitosan derivatives with the highest MW had the greater inhibitory activity. The antifungal efficacy of 2-α-AAPCS was associated with their ability to form a film around fungal cells, disturbing the transport of nutrients into the cells. The 2-α-AAPCS with higher MW were reported to have better capability of forming thicker films around fungal cells (Zhong et al., 2009).

The antifungal effects of chitosan extracted from fungal cell walls have also been evaluated against postharvest fungal pathogens (Oliveira et al., 2014a, 2014b). The application of chitosan from *Mucor circinelloides* Tiegh. inhibited the mycelial growth and spore germination and induced marked ultrastructural alterations in spores of *A. niger* and *R. stolonifer* (Oliveira et al., 2014a). A coating comprising chitosan from *M. circinelloides* delayed the growth of *A. niger* and *R. stolonifer* in artificially inoculated table grapes, and reduced the natural mycoflora in fruits stored at either room or low temperatures. In general, the antifungal activity of chitosan was higher in fruits stored at low temperatures than at room temperatures. Similar results were observed with table grapes when chitosan from *Cunninghamella elegans* Lendner was tested to control *B. cinerea* and *P. expansum* (Oliveira et al., 2014b). Physiological processes leading to fruit senescence and growth of pathogenic fungi are delayed at low temperatures, which results in a reduced incidence of postharvest decay (Meng et al., 2008; Sánchez-González et al., 2011).

Oligochitosan is obtained through the enzymatic hydrolysis of deacetylated chitosan and it has been considered as a biodegradable plant disease vaccine with increased antifungal activity. Application of oligochitosan by spraying on navel oranges before harvest reduced the incidence and severity of anthracnose caused by *C. gloeosporioides*. This treatment activated defense-related enzymes [e.g., pectin methylesterase (EC 3.1.1.11), peroxidase (POD, EC 1.11.1), chitinase (CHI, EC 3.2.1.14), and phenylalanine ammonia lyase (PAL, EC 4.3.1.24)] and contributed to the maintenance of the cross-linking between cells and the accumulation of hidroxyproline-rich glycoprotein in fruit (Deng et al., 2015).

2.3 Mode of Action on Postharvest Pathogenic Fungi

The mode of action of chitosan to inhibit fungi has been primarily associated with the positive charge on the C-2 of the glucosamine monomer of chitosan molecule dissolved in acidic solution ($pH < 6$) (Rabea et al., 2003). The antifungal effects of chitosan occur when the positive charges of this polymer react with the negatively charged residues of macromolecules or the negative charges of plasma membrane phospholipids exposed on the fungal cell surface forming polyelectrolytic complexes (Benhamou, 1992; Laflamme et al., 1999; Oliveira et al., 2014b). This interaction disturbs the permeability of plasma membrane (Badawy and Rabea, 2009; Wang et al., 2014), leads to the leakage of intracellular electrolytes and proteinaceous constituents (Young et al., 1982; Zakrzewska et al., 2005; Oliveira et al., 2014b), inhibits the transport of compounds across the plasma membrane, and consequently, promotes an internal osmotic imbalance (Assis and de Britto, 2011; Wang et al., 2014). Additionally, some researchers have proposed that chitosan, once inside the fungal cytoplasm, is able to enter the nuclei of fungi and interfere with mRNA and protein synthesis (Rabea et al., 2003; Zhang and Li, 2011; Wang et al., 2017). The uptake of chitosan into fungal cells seems to be energy dependent rather than to involve endocytosis or passive diffusion (Rabea et al., 2003; Palma-Guerrero et al., 2008).

2.4 Edible Coatings Comprising Chitosan and Other Antimicrobial Compounds

Researchers have suggested the use of edible coatings composed of chitosan and other antimicrobials to increase the inhibitory activity against postharvest fungal pathogens and extend the shelf life of fresh fruits. Most literature has focused on the combined use of chitosan and essential oils (EOs), which was commonly found to be synergistic (Santos et al., 2012; Guerra et al., 2015; Grande-Tovar et al., 2018; Oliveira et al., 2018). EOs are obtained from natural raw materials of plant origin (e.g., flowers, buds, seeds, leaves, twigs, bark, herbs, wood, fruits, and roots) by steam distillation, mechanical processing of the epicarp of citrus fruits, or dry distillation after the separation of the aqueous phase (if present) by physical processes (ISO, 2013). Antimicrobial properties of EOs are primarily associated with their natural role in plant defense systems (Maldonado et al., 2015).

The practical application of EOs alone as antimicrobial preservatives for fresh fruits could be limited because of their volatility and strong flavor that may reduce acceptance by consumers (Bakkali et al., 2008; Guerra et al., 2016). The incorporation of EOs into chitosan-based edible coatings can be an effective approach to overcome these limitations by reducing their volatility and maintaining high concentrations of active molecules on the fruit surface. In addition, this can enhance and prolong the antimicrobial activity through synergistic interactions that allow the use of lower effective doses of chitosan and EOs in combination (Aloui et al., 2014). EOs incorporated into chitosan coatings may be continuously released to the fruit surface maintaining effective concentrations of antimicrobial components over time (Outtara et al., 2000). The incorporation of EOs into edible coatings requires the use of emulsifying agents, but chitosan may ensure their homogenous dispersion by forming a thin and translucent coating when applied on the fruit surface (Sánchez-González et al., 2010).

The enhanced antifungal effects provided by the combination of chitosan and EOs were associated with the capability of chitosan to alter the permeability of fungal membranes, enabling the partition of EO constituents into fungal cells, where they can act on target structures (Santos et al., 2012). The antifungal activity of EOs is related to their ability to attack the cell wall, causing retraction of the hyphal cytoplasm and disturbance of enzymatic reactions responsible for the synthesis of cell wall components, thus affecting fungal morphogenesis, growth, and survival (Santos et al., 2012; Souza et al., 2015).

Researchers have consistently observed enhanced antifungal effects of the combined application of chitosan and different EOs on fresh fruits, such as oregano EO and chitosan on table grape (Santos et al., 2012) and pomegranate (Munhuweyi et al., 2017), mint EO and chitosan on cherry tomato (Guerra et al., 2015) and table grape (Guerra et al., 2016), bergamot EO, bitter orange EO, and chitosan on Tunisian palm date (Aloui et al., 2014), cinnamon EO and chitosan on sweet pepper (Xing et al., 2011), bergamot EO, thyme EO, tea tree EO, and chitosan on orange (Cháfer et al., 2012), and *Lippia gracilis* EO or lemon EO and chitosan on strawberry (Perdones et al., 2012; Azevedo et al., 2014). Coatings formed by chitosan and individual constituents of EOs (e.g., carvacrol, cinnamaldehyde, eugenol, *trans*-cinnamaldehyde, benzaldehyde, and eucalyptol) effectively controlled fungal decay on blueberry (Sun et al., 2014) and cherry tomato (Souza et al., 2015). Conversely, a few other studies reported no benefits from the combination of chitosan and EOs. The addition of clove EO to chitosan dispersion did not improve the efficacy to control green mold of mandarin caused by *P. digitatum*, despite the fact that the combination of clove EO and chitosan increased the activity of fruit defense enzymes (Shao et al., 2015).

Edible coatings composed of chitosan and bergamot EO or bitter EO controlled decay caused by *Aspergillus flavus* Link on Tunisian dates over 12 d of cold storage, without inducing off-flavors and off-odors on the fruit (Aloui et al., 2014). The application of a coating composed of chitosan and lemongrass (*Cymbopogon citratus*) EO decreased Rhizopus soft rot on cherry tomato fruits stored at room temperature and preserved their physicochemical and sensory attributes (Athayde et al.,

2016). In recent work, chitosan coatings containing EO of lemongrass effectively inhibited anthracnose caused by *Colletotrichum* spp. on guava, mango, and papaya (Oliveira et al., 2018). The efficacy of chitosan and mint EO coatings was evaluated to control decay caused by *A. niger, B. cinerea, P. expansum*, and *R. stolonifer* on cherry tomatoes stored at either room or low temperature. The combination of chitosan and mint EOs inhibited mycelial growth and spore germination of these fungi, delayed the appearance of the first disease symptoms in the fruit and reduced the number of infected fruits (Guerra et al., 2015). Coatings containing chitosan and either mint EO or oregano EO effectively controlled infections caused by *A. niger, B. cinerea, P. expansum*, and *R. stolonifer* on table grapes (Santos et al., 2012; Guerra et al., 2016). Coatings composed of chitosan from *M. circinelloides* and carvacrol (a phenolic compound) controlled the growth of *A. flavus*, as well as the native fungal microflora on cherry tomato fruits stored at either room or low temperature (Souza et al., 2015). Likewise, coatings composed of chitosan and oregano EO effectively delayed the appearance of the first signs of black mold and soft rot on cherry tomatoes stored at either room or low temperature (Barreto et al., 2016).

2.5 Effects on Physiological/Metabolic Responses in Fruit

In addition to the inhibitory effects on pathogenic fungi, chitosan-based edible coatings can induce some distinct physiological/metabolic responses in fresh fruits that may positively influence their quality parameters and storability (Ali et al., 2011). Most of these effects are related with specific biochemical properties of chitosan that delay senescence of fresh fruits by reducing respiration, metabolism, and dehydration (No et al., 2007; Ali et al., 2011). In this manner, edible coatings composed of chitosan or chitosan-derivatives have shown efficacy to extend the storability of a variety of fresh fruits including papaya (Ali et al., 2011), peach (Elbarbary and Mostafa, 2014), 'pearl' guava fruit (Hong et al., 2012), strawberry (Devlieghere et al., 2004), mango (Jongsri et al., 2016), litchi (Lin et al., 2011), red kiwi (Kaya et al., 2016), loquat (Petriccione et al., 2015), tomato (Guerra et al., 2015; Souza et al., 2015; Athayde et al., 2016), and grape (Oliveira et al., 2014a, 2014b; Guerra et al., 2016).

Chitosan coatings reduce the fruit respiration rate by creating a modified internal atmosphere within the fruit by altering the permeability to carbon dioxide, and oxygen. The reduction in gas exchange delays fruit senescence due to reduction in ripening rate, water loss, respiration rate, and ethylene production (Devlieghere et al., 2004; Elbarbary and Mostafa, 2014; Jongsri et al., 2016). Additionally, chitosan coatings can maintain fruit quality by improving appearance (e.g., higher luminosity and brightness) (Jongsri et al., 2016). Higher luminosity in chitosan-coated fruit is associated with the luminosity, and high transparency of chitosan films (Oliveira et al., 2014b). Researchers have reported that chitosan coatings do not affect negatively the formation of pigments that are important for fruit color development, such as lycopene and pheophytins in tomato and table grape, respectively (Oliveira et al., 2014a, 2014b; Souza et al., 2015; Guerra et al., 2015, 2016; Athayde et al., 2016).

A chitosan coating improved the physicochemical characteristics of papaya fruits and prolonged their storage life. The coating reduced weight loss (WL), maintained firmness, delayed changes in peel color and soluble solids concentration (SSC), and reduced titratable acidity (TA) during a 5-wk storage period. Minimum WL and increase in fruit firmness were observed in fruits coated with the dispersions with higher chitosan concentrations, which formed thicker films on fruit surface (Ali et al., 2011). Chitosan coating forms a smooth, semi-transparent layer on fruit pericarp that comprises a protective barrier that reduces water loss from transpiration through the fruit surface (Hong et al., 2012). Chitosan coating maintains fruit firmness probably due to its capacity to cover the cuticle and lenticels of the fruit, thereby reducing respiration and other ripening processes during storage. In infected fruit, this effect can also reduce disease development (Ali et al., 2011; Hong et al., 2012). The antifungal effects of chitosan may also contribute to the maintenance of firmness because the coating may protect the fruit against fungal cell wall–degrading enzymes that are part of the infection and colonization processes (Oliveira et al., 2014a; Barreto et al., 2016).

A chitosan coating was also effective in retarding color development and reduce SSC in papaya fruits. These effects were related to the ability of chitosan coating to reduce fruit metabolic activity by the creation of an internal atmosphere with reduced oxygen, increased carbon dioxide, and suppression of ethylene evolution that delayed ripening and senescence. An accelerated decrease in ascorbic acid content was also observed in papaya fruit coated with chitosan. The retention of ascorbic acid was attributed to low oxygen permeability into coated fruit that delayed deteriorative oxidation reactions (Ali et al., 2011).

A chitosan coating induced increased antioxidant properties and delayed ripening in 'Pearl' guava fruits. It delayed changes in chlorophyll and malondialdehyde (MDA) contents, increased the activities of peroxidase (POD), superoxide dismutase (SOD), and catalase (CAT), and inhibited superoxide free radical production ($O_2^{\bullet -}$) (Hong et al., 2012). These findings are noteworthy because SOD protects cells from oxidant stresses caused by conversion of super oxide ($O_2^{\bullet -}$) anion to H_2O_2. SOD and POD scavenge excessive reactive oxygen species (ROS), and protect the tissues from injury. ROS induces lipid peroxidation that causes a loss of membrane integrity in plant organs through a series of redox reactions, where $O_2^{\bullet -}$ is the primary cause of plant cell damage (Scandalios, 1993; Mittler, 2017). Chitosan and oligochitosan coatings have also stimulated the transcript expression of POD and β-1,3-glucanase (GLU) in peach fruit, and enhanced the synthesis of the defense-related enzymes CAT, POD, GLU, and chitinase (CHI) (Ma et al., 2013). Some chitosan derivatives, such as carboxymethyl chitosan (CMCS) originated through the introduction of hydrophilic groups, induced higher antioxidant activity in coated peach fruit, with marked reductions in MDA content over time (Elbarbary and Mostafa, 2014). MDA is a toxic product formed by ROS during storage of peach fruit and is used as an indicator of fruit damage progress (Liu et al., 2012). Antioxidants are capable of delaying, retarding, or preventing auto-oxidation caused by ROS, which is associated with fruit resistance to disease (Mittler, 2017).

Chitosan extracted from the ephippia of the planktonic crustacean *Daphnia longispina* O.F. Müller was used as an edible coating for red kiwifruit. Contents of polyphenols in chitosan-coated red kiwifruits decreased more slowly and ascorbic acid was higher than in uncoated fruits (Kaya et al., 2016). Phenolic compounds are synthesized in fruit during maturation as secondary metabolites; however, they are also metabolized during ripening of stored fruit. The internally modified atmosphere induced by chitosan coatings may affect the biosynthesis and catabolism of these secondary metabolites (Barreto et al., 2016). The capability of chitosan coating to enhance phenolic content in fresh fruit was an interesting finding because these compounds are naturally occurring antioxidants and can benefit human health. Some studies have also shown that chitosan coatings decreased sugar losses in fruit, as well as retarded increase in organic acid contents (Barreto et al., 2016). These effects are also related to maturation delay in chitosan-coated fruit as a result of decreased respiration and delayed metabolism (Romanazzi et al., 2015). Changes in sugars and organic acids contents in fruit occur naturally during storage and are primarily associated with metabolic pathways in vegetative tissues (Singh et al., 2009).

A common concern about the use of chitosan dispersions to coat fresh fruits has been their potential negative impacts on fruit sensory acceptance, owing to the development of an acidic or even slightly bitter and astringent taste due to an increase in protonated amine groups related to the use of diluted acids to dissolve chitosan (Perdones et al., 2012; Santos et al., 2012). Nonetheless, some studies have consistently reported no negative impact of postharvest treatments with chitosan on the sensory characteristics of a variety of fresh fruits (Ali et al., 2011; Santos et al., 2012; Oliveira et al., 2014a, 2014b; Romanazzi et al., 2015; Souza et al., 2015; Athayde et al., 2016; Guerra et al., 2016; Kaya et al., 2016).

3 Other Edible Coating Materials to Control Fruit Postharvest Diseases

In contrast to natural coatings based on chitosan or *Aloe vera* gels that have inherent antimicrobial activity against a wide range of microorganisms, there are other types of edible coatings whose antimicrobial activity is typically obtained by the incorporation of specific antimicrobial compounds to the basic coating matrix components.

In general, matrix components for fruits and vegetables are selected to formulate edible coatings that could replace the natural waxes removed during postharvest produce washing, and handling in the packinghouse. In some cases (e.g., in citrus fruit industry), these biodegradable coatings are designed to substitute non-edible conventional commercial waxes. These waxes are applied before long-term storage or marketing to reduce fruit transpiration and WL, regulate gaseous exchange, enhance peel gloss and external appearance, or reduce the incidence of postharvest physiological disorders (e.g., chilling injury) (Baldwin et al., 1995; Grierson and Miller, 2006). Conventional waxes are anionic microemulsions containing different types of resins and/or waxes, e.g., paraffin wax, carnauba wax, polyethylene, petroleum waxes, shellac, beeswax and candelilla wax. Commonly, these commercial waxes are formulated in mixtures with synthetic chemical fungicides, such as imazalil (IMZ) and TBZ (Palou et al., 2015). In some cases, studies focused on the replacement of these conventional fungicides with antifungal GRAS salts or natural compounds in order to reduce the residual toxicity of commercial waxes (Kouassi et al., 2012; Youssef et al., 2012; Castillo et al., 2014; Regnier et al., 2014; Tao et al., 2014), but since the composition of the commercial waxes is not changed, the resulting coatings are not edible. The most common matrix components of substitutive edible coatings are of natural (plant or animal) origin and include hydrocolloids, namely polysaccharides or proteins, and lipids, mainly edible waxes. These two basic components are generally combined to form composite coatings (Han, 2014). Typically, polysaccharides form polymeric chains associated through hydrogen bonds. These chains comprise good barriers to gas exchange but are poor moisture barriers due to their hydrophilic properties. Cellulose and derivatives, such as CMC, HPMC, starch, alginate, pectin, pullulan, and different gums, are polysaccharide materials commonly used as matrix components of edible coatings (Pérez-Gago et al., 2010; Valencia-Chamorro et al., 2011b; Zhang et al., 2014). Similarly, coating matrices based on proteins are also of hydrophilic character and are good gas barriers. Proteins frequently used as components of biodegradable coatings include wheat gluten, soy protein, corn zein, casein, rice bran protein, whey protein, and keratin. It is noteworthy that some people may have allergies or protein intolerances, e.g., to wheat gluten (celiac disease) or milk protein, which could restrict the use of protein-based coatings (Krochta, 2002; Lacroix and Vu, 2014). On the other hand, lipids or resins are hydrophobic substances that comprise good barriers to water exchange when used in composite coatings, in addition to improving the gloss and external appearance of coated produce. Nevertheless, when applied at high rates, the resulting coatings may show poor mechanical properties and objectionable opacity. Different substances can be used as hydrophobic fraction of composite edible coatings, including various animal and vegetal oils and fats (e.g., palm, peanut, and coconut), fractionated, concentrated, and/or reconstituted oils and fats (e.g., fatty acids, mono-, di-, and triglycerides, and cocoa butter substitutes), hydrogenated and/or transesterified oils (e.g., margarine, and shortenings), natural vegetal and animal waxes (e.g., beeswax, candelilla, and carnauba), and natural resins (e.g., benjoin, shellac resins, and guarana) (Debeaufort and Voilley, 2009; Pérez-Gago and Rhim, 2014; Palou et al., 2015).

Composite coatings containing both hydrophilic and hydrophobic constituents can also be formulated as bilayer emulsions. In this case, the lipid forms a second layer over the hydrocolloid layer (Pérez-Gago and Krochta, 2005). In other cases, bilayer coatings comprising only hydrocolloids have been applied using a layer-by-layer (LbL) approach. For example, Arnon et al. (2015) developed a novel coating for preservation of citrus fruits based on an internal layer of CMC plus an external layer of chitosan. Other ingredients that are commonly used as coating components are plasticizers, such as sucrose, sorbitol, glycerol, water, propylene glycol, polyethylene glycol, and fatty acids. These compounds are incorporated to improve the coating flexibility and extensibility. Emulsifiers or surfactants, such as fatty acids, monostearates, polysorbates (tweens), esters of fatty acids, sucrose ester, and lecithin, are ingredients of composite matrices that are frequently added to reduce surface tension between the dispersed and continuous phases and, consequently, increase the emulsion stability (Pérez-Gago et al., 2010; Han, 2014).

Edible coating matrices formulated with these ingredients can be directly applied to fresh fruit to provide gas and moisture barrier functionalities, and physical and mechanical protection. These coatings influence produce physiology, and they can substantially improve quality maintenance, and prolong postharvest storage life. Nevertheless, additional antifungal ingredients for these coatings with the ability to adequately control postharvest diseases of fresh fruits are needed.

3.1 Coatings Formulated with Salts or Metal-Based Nanoparticles

Potassium sorbate (PS), sodium benzoate (SB), sodium propionate (SP), sodium methylparaben (SMP), and some mixtures of these food additives have been selected in both in vitro and in vivo tests among a number of organic and inorganic acids and salts as the most suitable antifungal ingredients for HPMC-beeswax edible composite coatings to control citrus green and blue molds caused by *P. digitatum* and *P. italicum*, respectively (Valencia-Chamorro et al., 2008, 2009a). Further research confirmed the feasibility of these coatings to reduce green and blue mold on long-term cold-stored 'Valencia' oranges (Valencia-Chamorro et al., 2009b), 'Ortanique' mandarins (Valencia-Chamorro et al., 2010), and 'Clemenules' mandarins (Valencia-Chamorro et al., 2011a). The efficacy of the coatings to reduce decay of fruits artificially inoculated with these pathogens, and their effects on external, internal, and sensory quality of fruit greatly depended on the citrus species and cultivar. In general, the performance of antifungal coatings was superior on orange than on mandarin, probably because the latter commonly has a shorter storage life and greater susceptibility to postharvest decay caused by *Penicillium* spp. Similar composite HPMC-beeswax coatings containing SB or sodium ethylparaben (SEP) were recently selected as effective in controlling citrus Diplodia stem-end rot caused by the fungus *Lasiodiplodia theobromae* (Guimarães et al., 2019). Likewise, HPMC-beeswax coatings formulated not only with PS but also with sodium paraben salts such as SMP or SEP, ammonium carbonate (AC), or ammonium bicarbonate (ABC) were effective in reducing postharvest brown rot of plums caused by *M. fructicola* (Karaca et al., 2014). PS has also been evaluated as antifungal ingredient in edible coatings formulated with polysaccharide matrices of guar gum and pea starch. These coatings reduced the incidence of postharvest molds caused by *Penicillium* spp., *Cladosporium* spp., and *Aspergillus* spp. on apple, cucumber, and tomato (Mehyar et al., 2011). A study developed novel composite HPMC-beeswax coatings containing different GRAS organic and mineral salts, and evaluated their efficacy to control gray mold and black spot caused by *B. cinerea* and *A. alternata*, respectively, in cherry tomato (Fagundes et al., 2013). While gray mold incidence and severity on fruits artificially inoculated with the fungus and incubated at 20°C was better reduced by coatings containing SP, potassium carbonate (PC), ammonium phosphate (APh), or AC, coatings containing SB, SMP, or SEP were the most effective in reducing black spot on inoculated fruits. These selected coatings were then tested in trials with cold-stored cherry tomatoes. SP-based coatings were the most effective in preventing gray mold development on cherry tomatoes artificially inoculated with *B.* cinerea, but AC-based coatings were the most effective in decreasing WL, and maintain firmness of intact cherry tomatoes during cold storage at 5°C, as well as to prolong the shelf life at 20°C (Fagundes et al., 2014). On the other hand, SB-based coatings better controlled the development of black spot in fruits artificially inoculated with *A. alternata* and preserved the internal and sensory quality of the fruit during cold storage. In contrast, paraben-based coatings adversely affected the WL and firmness of coated fruits (Fagundes et al., 2015).

Recent developments in nanotechnology have led to the synthesis of novel compounds with improved biological activities. The greater surface area-to-volume ratio and surface reactivity of these nanosized materials typically enhance their performance compared to larger-sized materials (>100 nm) with similar physicochemical characteristics. Metal nanoparticles with proven antimicrobial activity include elements and oxides, such as silver (Ag), gold (Au), zinc oxide (ZnO), silica (SiO_2), titanium dioxide (TiO_2), alumina (Al_2O_3), and iron oxides (Fe_3O_4, Fe_2O_3) (Pérez-Gago and Palou, 2016). Although the potential toxicity and impact on human health of new nanostructured materials are constantly being revised by safety agencies (Jain et al., 2016), some of these nanomaterials, including metal-based nanoparticles, have been approved for specific food applications, allowing their incorporation into biodegradable polymeric matrices to form edible nanocomposites (Corrales et al., 2014). In the case of fresh fruits, nanocomposites containing metals as antimicrobial ingredients have been developed mainly to prevent bacterial growth, especially of human pathogenic bacteria (Fayaz et al., 2009; Azeredo, 2013; Rhim and Kim, 2014), and very few studies have focused on the control of fungal pathogens causing postharvest diseases. In the available literature, only one study has evaluated the in vivo effects of Ag-substituted zeolites alone, or incorporated into HPMC-based edible coatings on oranges artificially inoculated with *P. digitatum* (Cerrillo et al., 2017). The tested composite coatings were effective in inhibiting citrus green mold, but the inhibitory effects varied with the amount of Ag^+ contained in zeolites. Nanomaterials containing low Ag^+ amounts were effective, but less than materials with high Ag^+ content, which were also phytotoxic, and stained the fruit surface. Among the different types tested, the best zeolites were those with the largest pores and highest Si/Al proportion. In general, due to the broad spectrum antimicrobial activity of Ag, Ag-substituted zeolites have been the most widely used nanomaterials for food applications. In the USA, zeolite-based technologies are listed under the US FDA Inventory of Effective Food Contact Substance Notifications for use in food-contact polymers (US FDA, 2007). In Europe, the EFSA also released a positive opinion concerning the use of two zeolites containing Ag^+ ions in food contact surfaces, with Ag migration into food matrices being restricted to 50 μg Ag^+/kg of food (Corrales et al., 2014). Other antimicrobial nanomaterials that have been developed for fruit coating are organically modified nanoclays, such as Cloisite 30B and 20A, and some nanopolymers formulated with metal oxides that are classified as GRAS compounds, such as ZnO and TiO_2 (Rhim and Kim, 2014; Klangmuang and Sothornvit, 2016; Sogvar et al., 2016). However, to our knowledge, none of these materials have been directly evaluated for the control of specific fungal postharvest diseases. Another very recent technology that has not yet been tested for this purpose is the application of nanolaminate edible coatings. These coatings are formed from a wide range of materials (e.g., proteins, polysaccharides, and lipids) and generally deposited through the LbL technique into two or more layers of nanometer dimension physically or chemically bounded to each other (Flores-López et al., 2016; Acevedo-Fani et al., 2017). Examples of nanolaminate coating applications are multilayers of pectin and chitosan or k-carrageenan and lysozyme to preserve the quality and microbial safety of whole pear and mango (Medeiros et al., 2012a, 2012b), and pectin and chitosan or pectin and alginate amended with a microencapsulated beta-cyclodextrin and trans-cinnamaldehyde antimicrobial complex to prolong the shelf life of fresh-cut fruits (Brasil et al., 2012; Mantilla et al., 2013).

3.2 Coatings Formulated with Natural Compounds

The discovery and evaluation of the antimicrobial activity of a wide variety of natural compounds from plants, herbs, or spices has increased interest in recent years to incorporate them as additional active ingredients into edible coating formulations for fresh fruits. Many plant extracts, essential oils, or their individual constituents have been evaluated as stand-alone treatments (gaseous, or liquid) for decay control, but their incorporation into coatings is increasingly addressed as a feasible option to facilitate their practical application. The main limitations for application of coating formulations containing these plant-based ingredients are the possible induction of strong odor or flavor in fruit, phytotoxicity risks, and interference with fumigants or liquids applied commercially (Palou et al., 2016). Depending on the nature and proportions of coating matrix constituents, incorporated compounds, and other ingredients, the microstructure and the physical (optical, tensile, moisture, and gas barrier) and functional (antimicrobial and antioxidant) properties of the formulated coating material can be positively or negatively affected. Desirable modifications are those that allow the regulation of the diffusion process of volatile compounds for improved and prolonged antimicrobial and physiological functions of the coating in fruit, especially during long-term cold storage (Atarés and Chiralt, 2016). Nevertheless, the effectiveness of applications of amended edible coatings to fresh fruit can be compromised by the high concentrations of natural compounds needed to effectively inhibit postharvest pathogens, and the amendments may negatively impact the sensory quality of the coated fruit (Valencia-Chamorro et al., 2011b).

The number of studies on the development and evaluation of novel edible coatings incorporating natural compounds to control postharvest diseases of fresh fruit has been increasing, especially in the case of fresh fruits commercially waxed before cold storage and marketing. On citrus fruits, different EOs or plant extracts have been added to edible hydrocolloid-lipid formulations in attempts to reduce postharvest decay, particularly that caused by green and blue molds. For instance, substantial reductions were obtained with CMC-based coatings containing EOs from *Impatiens balsamina* (Zeng et al., 2013) or clove (Chen et al., 2016b), HPMC-lipid coatings containing ethanolic extract of propolis (Pastor et al., 2010), and alginate coatings containing extract from *Ficus hirta* (Chen et al., 2016a). Electrospun zein-nanofiber mats loaded with the natural GRAS colorant curcumin effectively reduced the incidence of gray and blue molds caused in apple by *B. cinerea* and *P. expansum*, respectively (Yilmaz et al., 2016). Plant extracts from *Satureja hortensis*, meadowsweet flower, or *Bergenia crassifolia* incorporated into pullulan edible coatings also decreased bacterial counts and inoculum levels of *A. niger* on apple. These coatings were also satisfactorily tested on peppers (Krasniewska et al., 2014, 2015). The postharvest application of methylcellulose-based edible coatings amended with oregano EO effectively reduced decay caused by *R. stolonifer* on tomato (Perdones et al., 2016). Oregano EO was effective in reducing black spot of tomato caused by *A. alternata* when applied as an ingredient of pectin-based edible coatings (Rodríguez-García et al., 2016). In in vitro studies, cellulose acetate films containing oregano EO inhibited, especially in vapor phase, the growth of taxonomically diverse and important postharvest pathogens, such as *A. alternata, Geotrichum candidum* Link, and *R. stolonifer* (Pola et al., 2016). Alginate coatings containing the antimicrobial and flavoring agent vanillin, applied both before and after harvest, significantly decreased yeast and mold counts and reduced postharvest decay while maintaining table grape quality during long-term cold storage (Takma and Korel, 2017). Alginate coatings formulated with carvacrol and methyl cinnamate showed potential for the control of gray mold caused by *B. cinerea* on strawberry (Peretto et al., 2014). Edible coatings containing alginate or pectin matrices enriched with the EO constituents citral or eugenol effectively reduced microbial growth and extended strawberry shelf life (Guerreiro et al., 2015). The incorporation of thyme EO into a coating formulated with mesquite gum and candelilla wax reduced postharvest decay of papaya caused by *C. gloeosporioides* and *R. stolonifer* (Bosquez-Molina et al., 2010). The incidence of anthracnose caused by *Colletotrichum musae* (Berk. & M.A. Curtis) Arx and *C. gloeosporioides* was reduced on bananas and papayas, respectively, treated with gum Arabic coatings containing cinnamon EO. The antifungal activity of these coatings was ascribed to the presence of cinnamon EO because the application of gum Arabic alone did not affect fungal growth (Maqbool et al., 2011). Anthracnose on cold-stored papaya fruits was inhibited with gum Arabic coatings amended with ginger oil or ginger extract as antifungal ingredients (Ali et al., 2016). Anthracnose was also controlled on papayas artificially inoculated with *C. gloeosporioides* and coated with cassava starch gels before storage. The cassava-based coatings also delayed papaya ripening (Oliveira et al., 2016).

3.3 Coatings Formulated with Antagonistic Microorganisms

Edible coatings can be used as carriers of different antagonistic microorganisms as biocontrol agents for controlling postharvest decay of fresh fruits. When compatibility between the biocontrol agent and ingredients of coating formulation exists, this approach integrates the advantages of the physiological fruit preservation caused by the coating matrix constituents, and the inhibition of postharvest pathogens caused by antagonistic microorganisms in the coating formulation (Marín et al., 2017c). If coating material with inherent antimicrobial capacity is used as coating matrix (e.g., chitosan), the incorporation of antagonistic microorganisms into the formulation can provide enhanced antimicrobial activity and boost the capability of the coating to control fruit postharvest diseases (Droby et al., 2009; Palou et al., 2015).

In some cases, microbial antagonists intended for the control of postharvest diseases can be applied in the field before fruit harvest. Preharvest application of edible coatings can protect the antagonistic microorganisms against ultraviolet (UV) radiation, desiccation, rain, and/or temperature extremes (Rhodes, 1993; Potjewijd et al., 1995; Cañamás et al., 2008, 2011). Edible coatings can also act as binding elements to the fruit surface, improving the adhesion of the antagonistic cells and the colonization of the fruit surface (Potjewijd et al., 1995). The presence of different coating compounds in cell suspensions may improve their stability and dispersability, allowing for a more homogeneous spatial distribution on the fruit surface. This is especially important for antagonistic microorganisms in coating formulations whose main mode of action is competition for nutrients and space, since a sufficient number of the antagonist cells is needed on the fruit surface to successfully compete with the target pathogen (El-Ghaouth et al., 2004; Sharma et al., 2009).

The constituents of coatings may also be selected and formulated to act as sources of nutrients for the antagonistic microorganisms, permitting longer survival periods (Marín et al., 2016). A study selected biopolymers that promoted both the adhesion and survival time of the antagonist *Candida sake* (Saito & Oda) van Uden & H.R. Buckley strain CPA-1 when applied to grapes in the field (Marín et al., 2016). The results showed that the best coating-forming agents were those that better satisfied the nutritional requirements of the tested antagonistic yeast. Independently of the application time (before or after harvest), the coating-antagonist interaction may exert additional direct or indirect effects to inhibit the pathogen via either the intrinsic antifungal properties, or acting as a physical or mechanical barrier protecting the fruit from fungal infection (Chien et al., 2007; Meng et al., 2010). In contrast, similar coatings containing other specific ingredients may influence negatively the viability and performance of antagonistic microorganisms. This may occur in coatings with inherent antimicrobial activity, as well as in coatings containing food preservatives or natural antimicrobial compounds that are toxic to the antagonistic microorganisms used in coating formulations. Therefore, a comprehensive study of the ingredients and their compatibility with the antagonist, as well as an optimization of the formulations to maximize the performance under practical conditions, is necessary for the design of coatings carrying microbial antagonists (Marín et al., 2016, 2017a).

The most common strategy for integration of coatings with biocontrol antagonists has been the direct incorporation and dispersion of the microorganism cell suspension into the coating prior to the application on the fruit. Regardless of application timing (pre- or postharvest), this is the simplest existing procedure because once the coating formulation is prepared, it can be applied to the fruit surface in only one step with no required additional handling (McGuire and Baldwin, 1994; Aloui et al., 2015). This approach was employed in studies that incorporated the yeast *Wickerhamomyces anomalus* (E.C. Hansen) Kurtzman, Robnett & Basehoar-Powers into sodium alginate and locust bean gum edible coatings and evaluated their application to orange fruit previously inoculated with *P. digitatum* (Aloui et al., 2015). Alginate-based coatings that contained the antagonist *Cryptococcus laurentii* (Kuff.) C.E. Skinner were also formulated for postharvest preservation of strawberry (Fan et al., 2009). Multiple coating formulations based on different proteins and polysaccharides with the incorporation of the antagonistic yeast *C. sake* were formulated for the control of grape fungal decay (Marín et al., 2016). In a number of investigations both an increase in cell viability of antagonistic microorganisms on fruit surface and an increase in the effectiveness of biocontrol antagonists incorporated into edible coating formulations were observed, even in field applications (McGuire and Baldwin, 1994; Potjewijd et al., 1995; Cañamás et al., 2011; Calvo-Garrido et al., 2013; Parafati et al., 2016).

A variant of this approach is the use of hydrogel spheres as supports for biocontrol yeasts to generate antifungal volatile organic compounds. The yeasts *W. anomalus* and *Aureobasidium pullulans* (De Bary) G. Arnaud ex Cif., Ribaldi & Corte were able to survive in these hydrophilic polymeric materials and produce volatiles with antifungal activity against important postharvest pathogens of mandarin and strawberry fruit (Parafati et al., 2017). Another strategy consists in application of the antagonistic microorganism and the coating in separate steps. In this approach, the antagonist may be applied before or after the coating because the coating acts primarily as a protective or supporting layer for the incorporated antagonist. The immersion of strawberries in suspensions of the bacterial antagonist *Bacillus subtilis* (Ehrenberg) Cohn followed by drying and application of candelilla wax-based coatings reduced the incidence of Rhizopus rot and prolonged the fruit shelf life (Oregel-Zamudio et al., 2017). Improved control of papaya anthracnose was achieved by the postharvest application of the antagonist *Burkholderia cepacia* W.H. Burkholder followed by fruit coating with a chitosan dispersion (Rahman et al., 2009).

In some cases, the antagonist application was made in the field and the coating was applied after harvest. A reduction in the incidence of postharvest diseases, and improvements in quality of table grapes sprayed in the field with the biocontrol agent *C. laurentii* and further coated after harvest with chitosan were observed (Meng et al., 2010). Another approach was the use of edible coatings as carriers in the dry formulation of microbial antagonists, where the coating matrix constituent (e.g., polymer) acts as support of the biocontrol agents in the drying process and during the product storage. Just before the application to fruit, the dry formulation is dispersed in water to form the coating dispersion containing the live antagonist cells. This approach was successfully applied in a fluidized bed drying system to obtain formulations of *C. sake* supported in different starch derivatives with coating-forming properties (Carbó et al., 2017; Marín et al., 2017b). These formulations were considered easy to pack, store, and transport and maintained the biocontrol efficacy of yeast for up to 12 mon of refrigerated storage.

Although some bacteria, such as *Pseudomonas* spp. (McGuire, 2000), *Pantoea agglomerans* Ewing and Fife (Cañamás et al., 2008), and *B. subtilis* (Oregel-Zamudio et al., 2017), have been tested as antagonists, most of the studies dealing with coatings formulated with antagonistic microorganisms have focused on the utilization of yeasts. Yeasts show several characteristics that make them especially suitable as microbial antagonists of postharvest pathogens and, therefore, as candidates to be incorporated into carrier coatings. Most of yeasts with potential activity as biocontrol agents have been isolated from the fruit surface and are capable of rapidly colonizing fruit peel wounds, which are required characteristics for effective competition with pathogens for space and nutrients. Other yeasts, commonly known as killer yeasts, show a different mode of action and exhibit direct antifungal capacity, sometimes due to their ability to release volatile organic compounds with antifungal activity (Di Francesco et al., 2015; Parafati et al., 2017).

Although intensive research has been devoted in recent years to determine the efficacy of many antagonistic yeasts and bacteria against a wide variety of pathogenic fungi causing postharvest diseases on fresh fruit (Liu et al., 2013; Spadaro and Droby, 2016; Droby et al., 2016), the literature available on the compatibility of antagonists with edible coatings and the influence of combined applications on the survival and biocontrol activity is still scarce (Marín et al., 2017c). However, this is currently an active field of research and an increasing number of investigations are devoted to explore the possibilities and potential benefits of the use of different coating matrices as carriers of biocontrol agents.

4 Final Remarks

The available literature shows that chitosan edible coatings and other edible coating materials are effective technologies to protect fresh fruits from postharvest decay through the inhibition of a variety of well-known postharvest pathogens, as well as by retarding dehydration, suppressing respiration rate and ethylene production, in addition to induce improved texture, color, volatile characteristics, and specific metabolic responses in fruit. The functionalities of coating materials to prolong the shelf life and add value to coated fruits attend the increasing demand of consumers for fresh fruits minimally handled after harvest and free of synthetic fungicides. These aspects increase the importance of coating formulations presenting antimicrobial properties as technologies for postharvest preservation of fresh fruits, even in replacement of traditional hydrocolloid or composite coatings. Further studies could consider the effects of constituents (and their concentrations) comprising the edible coating dispersions on the adhesion and duration of coatings on fresh fruit surface, as well as on their influence on physicochemical characteristics, including fruit respiration, metabolism, and dehydration. Moreover, investigations describing the effects of edible coatings on some specific metabolites (e.g., sugars, organic acids, pigments, and phenolics), sensory characteristics, and consumer acceptance of fresh fruits are needed.

Despite the increasing research work devoted to develop novel edible coatings with antifungal properties, their commercial implementation is hindered mainly by the current availability of convenient and relatively inexpensive conventional fungicides in many markets and by limited decay control performance of many of the low-toxicity edible coatings that incorporate antifungal ingredients. Commercial postharvest decay control treatments require a very high-level efficacy (typically more than 90% disease reduction), which is difficult to achieve with control methods that employ constituents with both limited toxicity and fungicidal activity. Since the performance of these treatments depends more on the fruit host (e.g., species, cultivar, physical, and physiological conditions) than that of conventional fungicides, the development of antifungal edible coatings needs to be tailored for each specific pathosystem, which may considerably narrow the spectrum of action of these treatments. Particularly considering the use of coatings containing antagonistic microorganisms, the application timing (before, or after harvest) is a crucial factor for edible coating selection and optimization, as well as to determine the mechanisms whereby the coatings will permit or enhance the biocontrol ability of the microbial antagonist. In general, besides convenient improvement of existing coatings and development of new effective formulations, further research should also focus on the combined application of these technologies with other kinds of postharvest antifungal treatments in a multifaceted integrated approach.

Acknowledgments

The 'Conselho Nacional de Desenvolvimento Científico e Tecnológico' (CNPq – Brazil), 'Coordenação de Aperfeiçoamento de Pessoal de Nível Superior' (CAPES – Brazil), 'Ministerio de Ciencia e Innovación' (MICINN – Spain), and 'Instituto Nacional de Investigación y Tecnología Agraria y Alimentaria' (INIA – Spain) are acknowledged for financial support on the research topic covered in this chapter.

References

Acevedo-Fani, A., Soliva-Fortuny, R. and Martín-Belloso, O. 2017. Nanostructured emulsions and nanolaminates for delivery of active ingredients: Improving food safety and functionality. *Trends in Food Science and Technology* 60, 12–22.

Ali, A., Hei, G.K. and Keat, Y.W. 2016. Efficacy of ginger oil and extract combined with gum arabic on anthracnose and quality of papaya fruit during cold storage. *Journal of Food Science and Technology-Mysore* 53, 1435–1444.

Ali, A., Muhammad, M.T.M., Sijam, K. and Siddiqui, Y. 2011. Effect of chitosan coatings on the physicochemical characteristics of Eksotika II papaya (*Carica papaya* L.) fruit during cold storage. *Food Chemistry* 124, 620–626.

Ali, A., Zahid, N., Manickam, S., Siddiqui, Y., Alderson, P.G. and Maqbool, M. 2013. Effectiveness of submicron chitosan dispersions in controlling anthracnose and maintaining quality of dragon fruit. *Postharvest Biology and Technology* 86, 147–153.

Aloui, H., Khwaldia, K., Licciardello, F., Mazzaglia, A., Muratore, G., Hamdi, M. and Restuccia, C. 2014. Efficacy of the combined application of chitosan and Locust Bean Gum with different citrus essential oils to control postharvest spoilage caused by *Aspergillus flavus* in dates. *International Journal of Food Microbiology* 170, 21–28.

Aloui, H., Licciardello, F., Khwaldia, K., Hamdi, M. and Restuccia, C. 2015. Physical properties and antifungal activity of bioactive films containing *Wickerhamomyces anomalus* killer yeast and their application for preservation of oranges and control of postharvest green mold caused by *Penicillium digitatum*. *International Journal of Food Microbiology* 200, 22–30.

Amorim, R.V.S., Pedrosa, R.P., Kazutaka, F., Martínez, C.R., Ledingham, W.M. and Campos-Takaki, G.M. 2006. Alternative carbon sources from sugar cane process for submerged cultivation of *Cunninghamella bertholletiae* to produce chitosan. *Food Technology and Biotechnology* 44, 519–523.

Amorim, R.V.S., Souza, W., Fukushima, K. and Campos-Takaki, G.M. 2001. Faster chitosan production by Mucoralean strains in submerged culture. *Brazilian Journal of Microbiology* 32, 20–23.

Aquino, A.B., Blank, A.F. and Santana, L.C.L.A. 2015. Impact of edible chitosan-cassava starch coatings enriched with *Lippia gracilis* Schauer genotype mixtures on the shelf life of guavas (*Psidium guajava* L.) during storage at room temperature. *Food Chemistry* 171, 108–116.

Arnon, H., Granit, R., Porat, R. and Poverenov, E. 2015. Development of polysaccharides-based edible coatings for citrus fruits: A layer-by-layer approach. *Food Chemistry* 166, 465–472.

Assis, O.B.G. and de Britto, D. 2011. Evaluation of the antifungal properties of chitosan coating on CUT apples using a non-invasive image analysis technique. *Polymer International* 60, 932–936.

Atarés, L. and Chiralt, A. 2016. Essential oils as additives in biodegradable films and coatings for active food packaging. *Trends in Food Science and Technology* 48, 51–62.

Athayde, A.J.A.A., Oliveira, P.D.L., Guerra, I.C.D., Conceição, M.L., Lima, M.A.B., Arcanjo, N.M.O., Madruga, M.S., Berger, L.R.R. and Souza, E.L. 2016. A coating composed of chitosan and *Cymbopogon citratus* (Dc. Ex Nees) essential oil to control Rhizopus soft rot and quality in tomato fruit stored at room temperature. *The Journal of Horticultural Science and Biotechnology* 91, 582–591.

Azarakhsh, N., Osman, A., Ghazali, H.M., Tan, C.P. and Adzahan, N.M. 2012. Optimization of alginate and gellan-based edible coating formulations for fresh-cut pineapples. *International Food Research Journal* 19, 279–285.

Azarakhsh, N., Osman, A., Ghazali, H.M., Tan, C.P. and Adzahan, N.M. 2014. Lemongrass essential oil incorporated into alginate-based edible coating for shelf-life extension and quality retention of fresh-cut pineapple. *Postharvest Biology and Technology* 88, 1–7.

Azeredo, H.M.C. 2013. Antimicrobial nanostructures in food packaging. *Trends in Food Science and Technology* 30, 56–69.

Azevedo, A.N., Buarque, P.R., Cruz, E.M.O., Blank, A.F., Alves, P.B., Nunes, M.L. and Santana, L.C.L.A. 2014. Response surface methodology for optimisation of edible chitosan coating formulations incorporating essential oil against several foodborne pathogenic bacteria. *Food Control* 43, 1–9.

Badawy, M.E.I. and Rabea, E.I. 2009. Potential of the biopolymer chitosan with different molecular weights to control postharvest gray mold of tomato fruit. *Postharvest Biology and Technology* 51, 110–117.

Bakkali, F., Averbeck, S., Averbeck, D. and Idaomar, I. 2008. Biological effects of essential oils – A review. *Food and Chemical Toxicology* 46, 446–475.

Baldwin, E.A., Nisperos-Carriedo, M.O. and Baker, R.A. 1995. Edible coatings for lightly processed fruits and vegetables. *HortScience* 30, 35–38.

Barreto, T.A., Andrade, S.C.A., Maciel, J.F., Arcanjo, N.M.O., Madruga, M.S., Meireles, B., Cordeiro, A.M.T., Souza, E.L. and Magnani, M.A. 2016. Chitosan coating containing essential oil from *Origanum vulgare* L. to control postharvest mold infections and keep the quality of cherry tomato fruit. *Frontiers in Microbiology* 7, 1–14.

Bautista-Baños, S., Hernández-López, M., Hernández-Lauzardo, A.N., Trejo-Espino, J.L., Bautista-Cerón, M. and Melo-Giorgana, G.E. 2005. Effect of chitosan on in vitro development and morphology of two isolates of *Colletotrichum gloeosporioides* (Penz.) Penz. and Sacc. *Mexican Journal of Phytopathology* 23, 62–67.

Benhamou, N. 1992. Ultrastructural and cytochemical aspects of chitosan on *Fusarium oxysporum* f. sp. *radicis-lycopersici*, agent of tomato crown and root rot. *Phytopathology* 82, 1185–1193.

Berger, L.R.R., Stamford, T.C.M., Stamford-Arnaud, T.M., Franco, L.O., Nascimento, A.E., Cavalcante, H.M.M., Macedo, R.O. and Campos-Takaki, G. M. 2014. Effect of corn steep liquor (CSL) and cassava wastewater (CW) on chitin and chitosan production by *Cunninghamella elegans* and their physicochemical characteristics and cytotoxicity. *Molecules* 19, 2771–2792.

Bosquez-Molina, E., Ronquillo-de Jesús, E., Bautista-Baños, S., Verde-Calvo, J.R. and Morales-López, J. 2010. Inhibitory effect of essential oils against *Colletotrichum gloeosporioides* and *Rhizopus stolonifer* in stored papaya fruit and their possible application in coatings. *Postharvest Biology and Technology* 57, 132–137.

Brasil, I.M., Gomes, C., Puerta-Gómez, A., Castell-Pérez, M.E. and Moreira, R.G. 2012. Polysaccharide-based multilayered antimicrobial edible coating enhances quality of fresh-cut papaya. *LWT -Food Science and Technology* 47, 39–45.

Calvo-Garrido, C., Elmer, P.A.G., Viñas, I., Usall, J., Bartra, E. and Teixidó, N. 2013. Biological control of botrytis bunch rot in organic wine grapes with the yeast antagonist *Candida sake* CPA-1. *Plant Pathology* 62, 510–519.

Campos, C.A., Gerschenson, L.N. and Flores, S.K. 2011. Development of edible films and coatings with antimicrobial activity. *Food and Bioprocess Technology* 4, 849–875.

Cañamás, T.P., Viñas, I., Torres, R., Usall, J., Solsona, C. and Teixidó, N. 2011. Field applications of improved formulations of *Candida sake* CPA-1 for control of *Botrytis cinerea* in grapes. *Biological Control* 56, 150–158.

Cañamás, T.P., Viñas, I., Usall, J., Torres, R., Anguera, M. and Teixidó, N. 2008. Control of postharvest diseases on citrus fruit by preharvest applications of biocontrol agent *Pantoea agglomerans* CPA-2. Part II. Effectiveness of different cell formulations. *Postharvest Biology and Technology* 49, 96–106.

Carbó, A., Torres, R., Usall, J., Fons, E. and Teixidó, N. 2017. Dry formulations of the biocontrol agent *Candida sake* CPA-1 using fluidised bed drying to control the main postharvest diseases on fruits. *Journal of the Science of Food and Agriculture* 97, 3691–3698.

Cardoso, A., Lins, C.I.M., Santos, E.R., Silva, M.C.F. and Campos-Takaki, G.M. 2012. Microbial enhance of chitosan production by *Rhizopus arrhizus* using agroindustrial substrates. *Molecules* 17, 4904–4914.

Cardoso, A., Silva, M.C.F., Batista, A.C.L., Santos, V.A. and Campos-Takaki, G.M. 2010. Submerged fermentation for chitin and chitosan production by *Rhizopus arrhizus* UCP 402. *Asian Chitin Journal* 6, 17–22.

Castillo, S., Pérez-Alfonso, C.O., Martínez-Romero, D., Guillén, F., Serrano, M. and Valero, D. 2014. The essential oils thymol and carvacrol applied in the packing lines avoid lemon spoilage and maintain quality during storage. *Food Control* 35, 136–136.

Cerrillo, J.L., Palomares, A.E., Rey, F., Valencia, S., Palou, L. and Pérez-Gago, M.B. 2017. Ag-zeolites as fungicidal material: Control of citrus green mold caused by *Penicillium digitatum*. *Microporous and Mesoporous Materials* 254, 69–76.

Cháfer, M., Sánchez-González, L., González-Martínez, C. and Chiralt, A. 2012. Fungal decay and shelf life of oranges coated with chitosan and bergamot, thyme, and tea tree essential oils. *Journal of Food Science* 77, 182–187.

Chatterjee, S., Adhya, M., Guha, A.K. and Chatterjee, B.P. 2005. Chitosan from *Mucor rouxii*: Production and physico-chemical characterization. *Process Biochemistry* 40, 395–400.

Chen, C., Peng, X., Zeng, R., Chen, M., Wan, C. and Chen, J. 2016a. Effect of *Ficus hirta* fruits extract incorporated into an alginate-based edible coating for Nanfeng mandarin preservation. *Scientia Horticulturae* 202, 41–48.

Chen, C., Zheng, J., Wan, C., Chen, M. and Chen, J. 2016b. Effect of carboxymethyl cellulose coating enriched with clove oil on postharvest quality of 'Xinyu' mandarin oranges. *Fruits* 71, 319–327.

Chen, R.H. and Hwa, H.D. 1996. Effect of molecular weight of chitosan with the same degree of deacetylation on the thermal, mechanical, and permeability properties of the prepared membrane. *Carbohydrate Polymers* 29, 353–358.

Chien, P.-J., Sheu, F. and Lin, H.-R. 2007. Coating citrus (*Murcott tangor*) fruit with low molecular weight chitosan increases postharvest quality and shelf life. *Food Chemistry* 100, 1160–1164.

Corrales, M., Fernández, A. and Han, J.J. 2014. Antimicrobial packaging systems. pp. 133–170. *In*: J.H. Han (ed.). *Innovations in Food Packaging*, 2nd ed. Academic Press, Waltham, MA.

Davis, L.L. and Bartnicki-Garcia, S. 1984a. The co-ordination of chitosan and chitin synthesis in *Mucor rouxii*. *Journal of General Microbiology* 130, 2095–2102.

Davis, L.L. and Bartnicki-Garcia, S. 1984b. Chitosan synthesis by the tandem action of chitin synthetase and chitin deacetylase from *Mucor rouxii*. *Biochemistry* 23, 1065–1073.

Debeaufort, F. and Voilley, A. 2009. Lipid-based edible films and coatings. pp. 135–168. *In*: M. Embuscado and K.C. Huber (eds.). *Edible Films and Coatings for Food Applications*. Springer-Verlag, New York, NY.

Deng, L., Zhou, Y. and Zeng, K. 2015. Pre-harvest spray of oligochitosan induced the resistance of harvested navel oranges to anthracnose during ambient temperature storage. *Crop Protection* 70, 70–76.

Devlieghere, F., Vermeulen, A. and Debevere, J. 2004. Chitosan: Antimicrobial activity, interactions with food components and applicability as a coating on fruit and vegetables. *Food Microbiology* 21, 703–714.

Di Francesco, A., Ugolini, L., Lazzeri, L. and Mari, M. 2015. Production of volatile organic compounds by *Aureobasidium pullulans* as a potential mechanism of action against postharvest fruit pathogens. *Biological Control* 81, 8–14.

Droby, S., Wisniewski, M., Teixidó, N., Spadaro, D. and Jijakli, M.H. 2016. The science, development, and commercialization of postharvest biocontrol products. *Postharvest Biology and Technology* 122, 22–29.

Droby, S., Wisniewski, M.E., Macarisin, D. and Wilson, C.L. 2009. Twenty years of postharvest biocontrol research: Is it time for a new paradigm? *Postharvest Biology and Technology* 52, 137–145.

Elbarbary, A.M. and Mostafa, T.B. 2014. Effect of γ-rays on carboxymethyl chitosan for use as antioxidant and preservative coating for peach fruit. *Carbohydrate Polymers* 104, 109–117.

El-Ghaouth, A., Wilson, C.L. and Wisniewski, M.E. 2004. Biologically based alternatives to synthetic fungicides for the postharvest diseases of fruit and vegetables. pp. 511–535. *In*: S.A.M.H. Naqvi (ed.). *Diseases of Fruits and Vegetables, Diagnosis and Management, Vol. II.* Kluwer Academic Publishers, Dordrecht, The Netherlands.

Fagundes, C., Palou, L., Monteiro, A.R. and Pérez-Gago, M.B. 2014. Effect of antifungal hydroxypropyl methylcellulose-beeswax edible coatings on gray mold development and quality attributes of cold-stored cherry tomato fruit. *Postharvest Biology and Technology* 92, 1–8.

Fagundes, C., Palou, L., Monteiro, A.R. and Pérez-Gago, M.B. 2015. Hydroxypropyl methylcellulose-beeswax edible coatings formulated with antifungal food additives to reduce alternaria black spot and maintain postharvest quality attributes of cold-stored cherry tomatoes. *Scientia Horticulturae* 193, 249–257.

Fagundes, C., Pérez-Gago, M.B., Monteiro, A.R. and Palou, L. 2013. Antifungal activity of food additives in vitro and as ingredients of hydroxypropyl methylcellulose-lipid edible coatings against *Botrytis cinerea* and *Alternaria alternata* on cherry tomato fruit. *International Journal of Food Microbiology* 166, 391–398.

Fai, A.E.C., Stamford, T.C.M., Stamford-Arnaud, T.M., Santa-Cruz, P.A., Silva, M.C.F., Campos-Takaki, G.M. and Stamford, T.L.M. 2011. Physicochemical characteristics and functional properties of chitin and chitosan produced by *Mucor circinelloides* using yam bean as substrate. *Molecules* 16, 7143–7154.

Fan, Y., Xu, Y., Wang, D., Zhang, L., Sun, J., Sun, L. and Zhang, B. 2009. Effect of alginate coating combined with yeast antagonist on strawberry (*Fragaria × ananassa*) preservation quality. *Postharvest Biology and Technology* 53, 84–90.

Fayaz, A.M., Balaji, K., Girilal, M., Kalaichelvan, P.T. and Venkatesan, R. 2009. Mycobased synthesis of silver nanoparticles and their incorporation into sodium alginate films for vegetable and fruit preservation. *Journal of Agricultural and Food Chemistry* 57, 6246–6252.

Feliziani, E., Santini, M., Landi, L. and Romanazzi, G. 2013. Pre- and postharvest treatment with alternatives to synthetic fungicides to control postharvest decay of sweet cherry. *Postharvest Biology and Technology* 78, 133–138.

Flores-López, M.L., Cerqueira, M.A., de Rodríguez, D.J. and Vicente, A.A. 2016. Perspectives on utilization of edible coatings and nano-laminate coatings for extension of postharvest storage of fruits and vegetables. *Food Engineering Reviews* 8, 292–305.

Grande-Tovar, C.D., Chaves-Lopez, C., Serio, A., Rossi, C. and Paparella, A. 2018. Chitosan coatings enriched with essential oils: Effects on fungi involved in fruit decay and mechanisms of action. *Trends in Food Science and Technology* 78, 61–71.

Grierson, W. and Miller, W.M. 2006. Storage of citrus fruit. pp. 547–581. *In*: W. F. Wardowski, W.M. Miller, D.J. Hall and W. Grierson (eds.). *Fresh Citrus Fruits*, 2nd ed. Florida Science Source, Longboat Key, FL.

Guerra, I.C.D., Oliveira, P.D.L., Pontes, A.L.S., Lúcio, A.S.S.C., Tavares, J.F., Barbosa-Filho, J.M., Madruga, M.S. and Souza, E.L. 2015. Coatings comprising chitosan and *Mentha piperita* L. or *Mentha* x *villosa* Huds essential oils to prevent common postharvest mold infections and maintain the quality of cherry tomato fruits. *International Journal of Food Microbiology* 214, 168–178.

Guerra, I.C.D., Oliveira, P.D.L., Santos, M.M.F., Lúcio, A.S.S.C., Tavares, J.F., Barbosa-Filho, J.M., Madruga, M.S. and Souza, E.L. 2016. The effects of composite coatings containing chitosan and *Mentha* (*piperita* L. or *x villosa* Huds) essential oil on postharvest mold occurrence and quality of table grape cv. Isabella. *Innovative Food Science and Emerging Technologies* 34, 112–121.

Guerreiro, A.C., Gago, C.M.L., Faleiro, M.L., Miguel, M.G.C. and Antunes, M.D.C. 2015. The use of polysaccharide-based edible coatings enriched with essential oils to improve shelf-life of strawberries. *Postharvest Biology and Technology* 110, 51–60.

Guimarães, J.E.R., de la Fuente, B., Pérez-Gago, M.B., Andradas, C., Carbó, R., Mattiuz, B. and Palou, L. 2019. Antifungal activity of GRAS salts against *Lasiodiplodia theobromae* in vitro and as ingredients of hydroxypropyl methylcellulose-lipid composite edible coatings to control Diplodia stem-end rot and maintain postharvest quality of citrus fruit. *International Journal of Food Microbiology* 301, 9–18.

Guo, Z., Chen, R., Xing, R., Liu, S., Yu, H., Wang, P., Li, C. and Pengcheng, L. 2006. Novel derivatives of chitosan and their antifungal activities in vitro. *Carbohydrate Research* 341, 351–354.

Han, J.H. 2014. Edible films and coatings: A review. pp. 213–255. *In*: J.H. Han (ed.). *Innovations in Food Packaging*, 2nd ed. Academic Press, Elsevier, London, UK.

Hernández-Lauzardo, A.N., Bautista-Baños, S., Velázquez-del Valle, M.G., Méndez-Montealvo, M.G., Sánchez-Rivera, M.M. and Bello-Pérez, L.A. 2008. Antifungal effects of chitosan with different molecular weights on *in vitro* development of *Rhizopus stolonifer* (Ehrenb: Fr.) Vuill. *Carbohydrate Polymers* 73, 541–547.

Hong, K., Xie, J., Zhang, L., Sun, D. and Gong, D. 2012. Effects of chitosan coating on postharvest life and quality of guava (*Psidium guajava* L.) fruit during cold storage. *Scientia Horticulturae* 144, 172–178.

ISO (International Organization for Standardization). 2013. Aromatic natural raw materials-Vocabulary. ISO 9235:2013. Available at: www.iso.org/iso/catalogue_detail.htm?csnumber=51017.

Jain, A., Ranjan, S., Dasgupta, N. and Ramalingam, C. 2016. Nanomaterials in food and agriculture: An overview on their safety concerns and regulatory issues. *Critical Reviews in Food Science and Nutrition* 58, 297–317.

Jongsri, P., Wangsomboondee, T., Rojsitthisak, P. and Seraypheap, K. 2016. Effect of molecular weights of chitosan coating on postharvest quality and physicochemical characteristics of mango fruit. *LWT – Food Science and Technology* 73, 28–36.

Kafetzopoulos, D., Martinou, A. and Bouriotis, V. 1993. Bioconversion of chitin to chitosan: Purification and characterization of chitin deacetylase from *Mucor rouxii*. *Proceedings of the National Academy of Sciences of the United States of America* 90, 2564–2568.

Karaca, H., Pérez-Gago, M.B., Taberner, V. and Palou, L. 2014. Evaluating food additives as antifungal agents against *Monilinia fructicola* in vitro and in hydroxypropyl methylcellulose-lipid composite edible coatings for plums. *International Journal of Food Microbiology* 179, 72–79.

Kas, H.S. 1997. Chitosan: Properties, preparations and application to microparticulate systems. *Journal of Microencapsulation* 14, 689–711.

Kaya, M., Cesoniene, L., Daubaras, R., Leskauskaite, D. and Zabulione, D. 2016. Chitosan coating of red kiwifruit (*Actinidia melanandra*) for extending of the shelf life. *International Journal of Biological Macromolecules* 85, 355–360.

Kerch, G. 2015. Chitosan films and coatings prevent losses of fresh fruit nutritional quality: A review. *Trends in Food Science and Technology* 46, 159–166.

Kewsuwan, P., Rujitanapanich, S., Bhasabutra, T., Puripunyavanich, V., Busamongkol, A., Injarean, U. and Pichetpong, P. 2014. Irradiated oligochitosan against *Colletotrichum gloeosporioides* in chili. *Energy Procedia* 56, 274–279.

Klangmuang, P. and Sothornvit, R. 2016. Barrier properties, mechanical properties and antimicrobial activity of hydroxypropyl methylcellulose-based nanocomposite films incorporated with Thai essential oils. *Food Hydrocolloids* 61, 609–616.

Kleekayai, T. and Suntornsuk, W. 2010. Production and characterization of chitosan obtained from *Rhizopus oryzae* grown on potato chip processing waste. *World Journal of Microbiology and Biotechnology* 27, 1145–1154.

Kouassi, K.S., Bajji, M. and Jijakli, H. 2012. The control of postharvest blue and green molds of citrus in relation with essential oil wax formulations, adherence and viscosity. *Postharvest Biology and Technology* 73, 122–128.

Krasniewska, K., Gniewosz, M., Synowiec, A., Przybyl, J.L., Baczek, K. and Weglarz, Z. 2014. The use of pullulan coating enriched with plant extracts from *Satureja hortensis* L. to maintain pepper and apple quality and safety. *Postharvest Biology and Technology* 90, 63–72.

Krasniewska, K., Gniewosz, M., Synowiec, A., Przybyl, J.L., Baczek, K. and Weglarz, Z. 2015. The application of pullulan coating enriched with extracts from *Bergenia crassifolia* to control the growth of food microorganisms and improve the quality of peppers and apples. *Food Bioproducts Processing* 94, 422–433.

Krochta, J.M. 2002. Proteins as raw materials for films and coatings: Definitions, current status, and opportunities. pp. 1–41. *In*: A. Gennadios (ed.). *Protein-Based Films and Coatings*. CRC Press, Boca Raton, FL.

Lacroix, M. and Vu, K.D. 2014. Edible coating and film materials: Proteins. pp. 277–304. *In*: J.H. Han (ed.). *Innovations in Food Packaging*, 2nd ed. Academic Press, Elsevier, London, UK.

Laflamme, P., Benhamou, N., Bussières, G. and Dessureault, M. 1999. Differential effect of chitosan on root rot fungal pathogens in forest nurseries. *Canadian Journal of Botany* 77, 1460–1468.

Lin, B.F., Du, Y.M., Liang, X.Q., Wang, X.Y., Wang, X.H. and Yang, J.H. 2011. Effect of chitosan coating on respiratory behavior and quality of stored litchi under ambient temperature. *Journal of Food Engineering* 102, 94–99.

Liu, J., Sui, Y., Wisniewski, M., Droby, S. and Liu, Y. 2013. Review: Utilization of antagonistic yeasts to manage postharvest fungal diseases of fruit. *International Journal of Food Microbiology* 167, 153–160.

Liu, J., Wisniewski, M., Droby, S., Norelli, J., Hershkovitz, V., Tian, S. and Farrell, R. 2012. Increase in antioxidant gene transcripts, stress tolerance and biocontrol efficacy of *Candida oleophila* following sublethal oxidative stress exposure. *FEMS Microbiology Ecology* 80, 578–590.

Ma, Z., Yang, L., Yan, H., Kennedy, J.F. and Meng, X. 2013. Chitosan and oligochitosan enhance the resistance of peach fruit to brown rot. *Carbohydrate Polymers* 94, 272–277.

Maldonado, A.F.S., Schieber, A. and Gänzle, M.G. 2015. Plant defense mechanisms and enzymatic transformation products and their potential applications in food preservation: Advantages and limitations. *Trends in Food Science & Technology* 46, 49–59.

Mantilla, N., Castell-Perez, M.E., Gomes, C. and Moreira, R.G. 2013. Multilayered antimicrobial edible coating and its effect on quality and shelf-life of fresh-cut pineapple (*Ananas comosus*). *LWT -Food Science and Technology* 51, 37–43.

Maqbool, M., Ali, A., Alderson, P.G., Mohamed, M.T.M., Siddiqui, Y. and Zahid, N. 2011. Postharvest application of gum arabic and essential oils for controlling anthracnose and quality of banana and papaya during cold storage. *Postharvest Biology and Technology* 62, 71–76.

Marín, A., Atarés, L., Cháfer, M. and Chiralt, A. 2017a. Properties of biopolymer dispersions and films used as carriers of the biocontrol agent *Candida sake* CPA-1. *LWT - Food Science and Technology* 79, 60–69.

Marín, A., Atarés, L., Cháfer, M. and Chiralt, A. 2017b. Stability of biocontrol products carrying *Candida sake* CPA-1 in starch derivatives as a function of water activity. *Biocontrol Science and Technology* 27, 268–287.

Marín, A., Atarés, L. and Chiralt, A. 2017c. Improving function of biocontrol agents incorporated in antifungal fruit coatings: A review. *Biocontrol Science and Technology* 27, 1220–1241.

Marín, A., Cháfer, M., Atarés, L., Chiralt, A., Torres, R., Usall, J. and Teixidó, N. 2016. Effect of different coating-forming agents on the efficacy of the biocontrol agent *Candida sake* CPA-1 for control of *Botrytis cinerea* on grapes. *Biological Control* 96, 108–119.

McGuire, R.G. 2000. Population dynamics of postharvest decay antagonists growing epiphytically and within wounds on grapefruit. *Phytopathology* 90, 1217–1223.

McGuire, R.G. and Baldwin, E.A. 1994. Compositions of cellulose coatings affect populations of yeasts in the liquid formulation and on coated grapefruits. *Proceedings of the Florida State Horticultural Society* 107, 293–297.

Medeiros, B.G. de S., Pinheiro, A.C., Carneiro-da-Cunha, M.G. and Vicente, A.A. 2012a. Development and characterization of a nanomultilayer coating of pectin and chitosan e evaluation of its gas barrier properties and application on 'Tommy Atkins' mangoes. *Journal of Food Engineering* 110, 457–464.

Medeiros, B.G. de S., Pinheiro, A.C., Teixeira, J.A., Vicente, A.A. and Carneiro-da-Cunha, M.G. 2012b. Polysaccharide/protein nanomultilayer coatings: Construction, characterization and evaluation of their effect on 'Rocha' pear (*Pyrus communis* L.) shelf-life. *Food and Bioprocess Technology* 5, 2435–2445.

Mehyar, G.F., Al-Qadiri, H.M., Abu-Blan, H.A. and Swanson, B.G. 2011. Antifungal effectiveness of potassium sorbate incorporated in edible coatings against spoilage molds of apples, cucumbers, and tomatoes during refrigerated storage. *Journal of Food Science* 76, M210–M217.

Meng, X., Li, B., Liu, J. and Tian, S. 2008. Physiological responses and quality attributes of table grape fruit to chitosan preharvest spray and postharvest coating during storage. *Food Chemistry* 106, 501–508.

Meng, X.H., Qin, G.Z. and Tian, S.P. 2010. Influences of preharvest spraying *Cryptococcus laurentii* combined with postharvest chitosan coating on postharvest diseases and quality of table grapes in storage. *LWT - Food Science and Technology* 43, 596–601.

Mittler, R. 2017. ROS are good. *Trends in Plant Science* 22, 11–19.

Munhuweyi, K., Caleb, O.J., Lennox, C.L., van Reenen, A.J. and Opara, U.L. 2017. *In vitro* and *in vivo* antifungal activity of chitosan-essential oils against pomegranate fruit pathogens. *Postharvest Biology and Technology* 129, 9–22.

Naghdi, M., Zamani, A. and Karimi, K. 2014. A sulfuric-lactic acid process for efficient purification of fungal chitosan with intact molecular weight. *International Journal of Biological Macromolecules* 63, 158–162.

No, H.K., Meyers, S.P., Prinyawiwatkul, W. and Xu, Z. 2007. Applications of chitosan for improvement of quality and shelf life of foods: A review. *Journal of Food Science* 72, 87–100.

Oh, S.K., Cho, D. and Yu, S.H. 1998. Development of integrated pest management techniques using biomass for organic farming (I). Suppression of late blight and fusarium wilt of tomato by chitosan involving both antifungal and plant activating activities. *Korean Journal of Plant Pathology* 14, 278–285.

Oliveira, B.F., Cruz, A.F. and Alves, E. 2016. Cassava starch coatings for postharvest control of papaya anthracnose. *Phytopathologia Mediterranea* 55, 276–284.

Oliveira, C.E.V., Magnani, M., Sales, C.V., Pontes, A.L.S., Campos-Takaki, G.M., Stamford, T.C.M. and de Souza, E.L. 2014a. Effects of post-harvest treatment using chitosan from *Mucor circinelloides* on fungal pathogenicity and quality of table grapes during storage. *Food Microbiology* 44, 211–219.

Oliveira, C.E.V., Magnani, M., Sales, C.V., Pontes, A.L.S., Campos-Takaki, G.M., Stamford, T.C.M. and de Souza, E.L. 2014b. Effects of chitosan from *Cunninghamella elegans* on virulence of post-harvest pathogenic fungi in table grapes (*Vitis labrusca* L.). *International Journal of Food Microbiology* 171, 54–61.

Oliveira, P.D.L., de Oliveira, K.A.R., dos Santos Vieira, W.A., Camara, M.P.S. and de Souza, E.L. 2018. Control of anthracnose caused by *Colletotrichum* species in guava, mango and papaya using synergistic combinations of chitosan and *Cymbopogon citratus* (D.C. ex Nees) Stapf. essential oil. *International Journal of Food Microbiology* 266, 87–94.

Oregel-Zamudio, E., Angoa-Pérez, M.V., Oyoque-Salcedo, G., Aguilar-González, C.N. and Mena-Violante, H.G. 2017. Effect of candelilla wax edible coatings combined with biocontrol bacteria on strawberry quality during the shelf-life. *Scientia Horticulturae* 214, 273–279.

Outtara, B., Simard, R.E., Piette, G., Begin, A. and Holley, R.A. 2000. Diffusion of acetic and proprionic acids from chitosan-based antimicrobial packaging films. *Journal of Food Science* 65, 768–773.

Palma-Guerrero, J., Jansson, H.–.B., Salinas, J. and Lopez-Llorca, L.V. 2008. Effect of chitosan on hyphal growth and spore germination of plant pathogenic and biocontrol fungi. *Journal of Applied Microbiology* 104, 541–553.

Palou, L., Ali, A., Fallik, E. and Romanazzi, G. 2016. GRAS, plant- and animal-derived compounds as alternatives to conventional fungicides for the control of postharvest disease of fresh horticultural produce. *Postharvest Biology and Technology* 122, 41–52.

Palou, L., Valencia-Chamorro, S.A. and Pérez-Gago, M.B. 2015. Antifungal edible coatings for fresh citrus fruit: A review. *Coatings* 5, 962–986.

Parafati, L., Vitale, A., Restuccia, C. and Cirvilleri, G. 2016. The effect of locust bean gum (LBG)-based edible coatings carrying biocontrol yeasts against *Penicillium digitatum* and *Penicillium italicum* causal agents of postharvest decay of mandarin fruit. *Food Microbiology* 58, 87–94.

Parafati, L., Vitale, A., Restuccia, C. and Cirvilleri, G. 2017. Performance evaluation of volatile organic compounds by antagonistic yeasts immobilized on hydrogel spheres against gray, green and blue postharvest decays. *Food Microbiology* 63, 191–198.

Pastor, C., Sánchez-González, L., Cháfer, M., Chiralt, A. and González-Martínez, C. 2010. Physical and antifungal properties of hydroxypropylmethylcellulose based films containing propels as affected by moisture content. *Carbohydrate Polymers* 82, 1174–1183.

Perdones, A., Sánchez-González, L., Chiralt, A. and Vargas, M. 2012. Effect of chitosan-lemon essential oil coatings on storage-keeping quality of strawberry. *Postharvest Biology and Technology* 70, 32–41.

Perdones, A., Tur, N., Chiralt, A. and Vargas, M. 2016. Effect on tomato plant and fruit of the application of biopolymer–Oregano essential oil coatings. *Journal of the Science of Food and Agriculture* 96, 4505–4513.

Peretto, G., Du, W.-X., Avena-Bustillos, R.J., Berrios, J.D.J., Sambo, P. and McHugh, T.H. 2014. Optimization of antimicrobial and physical properties of alginate coatings containing carvacrol and methyl cinnamate for strawberry application. *Journal of Agricultural and Food Chemistry* 62, 984–990.

Pérez-Gago, M.B., González-Aguilar, G.A. and Olivas, G.I. 2010. Edible coatings for fruits and vegetables. *Stewart Postharvest Review* 6:3, 1–14.

Pérez-Gago, M.B. and Krochta, J.M. 2005. Emulsion and bi-layer edible films. pp. 384–402. *In*: J.H. Han (ed.). *Innovations in Food Packaging*, 1st ed. Academic Press, Elsevier, Amsterdam, The Netherlands.

Pérez-Gago, M.B. and Palou, L. 2016. Antimicrobial packaging for fresh and fresh-cut fruits and vegetables. pp. 403–452. *In*: S. Pareek (ed.). *Fresh-Cut Fruits and Vegetables: Technology, Physiology, and Safety*. CRC Press, Taylor and Francis Group, Boca Raton, FL.

Pérez-Gago, M.B. and Rhim, J.W. 2014. Edible coating and film materials: Lipid bilayers and lipid emulsions. pp. 325–350. *In*: J.H. Han (ed.). *Innovations in Food Packaging*, 2nd ed. Academic Press, Elsevier, London, UK.

Petriccione, M., Pasquariello, M.S., Mastrobuoni, F., Zampella, L., Di Patre, D. and Scortichini, M. 2015. Influence of a chitosan coating on the quality and nutraceuticals traits of loquat fruit during postharvest life. *Scientia Horticulturae* 197, 287–296.

Pochanavanich, P. and Suntornsuk, W. 2002. Fungal chitosan production and its characterization. *Letters in Applied Microbiology* 35, 17–21.

Pola, C.C., Medeiros, E.A.A., Pereira, O.L., Souza, V.G.L., Otoni, C.G., Camilloto, G. P. and Soares, N.F.F. 2016. Cellulose acetate active films incorporated with oregano (*Origanum vulgare*) essential oil and organophilic montmorillonite clay control the growth of phytopathogenic fungi. *Food Packaging and Shelf Life* 9, 67–78.

Ponce, A.G., Roura, S.I., del Valle, C.E. and Moreira, M.R. 2008. Antimicrobial and antioxidant activities of edible coatings enriched with natural plant extracts: *In vitro* and *in vivo* studies. *Postharvest Biology and Technology* 49, 294–300.

Potjewijd, R., Nisperos, M.O., Burns, J.K., Parish, M. and Baldwin, E.A. 1995. Cellulose-based coatings as carriers for *Candida guillermondii* and *Debaryomyces* sp. in reducing decay of oranges. *HortScience* 30, 1417–1421.

Poverenov, E., Zaitsev, Y., Arnon, H., Granit, R., Alkalai-Tuvia, S., Perzelan, Y., Weinberg, T. and Fallik, E. 2014. Effects of a composite chitosan-gelatin edible coating on postharvest quality and storability of red bell peppers. *Postharvest Biology and Technology* 96, 106–109.

Rabea, E.I., Badawy, M.E., Stevens, C.V., Smagghe, G. and Steurbaut, W. 2003. Chitosan as antimicrobial agent: Applications and mode of action. *Biomacromolecules* 4, 1457–1465.

Rahman, M.A., Mahmud, T.M.M., Kadir, J., Abdul Rahman, R. and Begum, M.M. 2009. Enhancing the efficacy of *Burkholderia cepacia* B23 with calcium chloride and chitosan to control anthracnose of papaya during storage. *The Plant Pathology Journal* 25, 361–368.

Ramos-García, M., Bosquez-Molina, E., Hernández-Romano, J., Zavala-Padilla, G., Terrés-Rojas, E., Alia-Tejacal, I., Barrera-Necha, L., Hernández-López, M. and Bautista-Baños, S. 2012. Use of chitosan-based edible coatings in combination with other natural compounds, to control *Rhizopus stolonifer* and *Escherichia coli* DH5α in fresh tomatoes. *Crop Protection* 38, 1–6.

Raybaudi-Massilia, R.M., Mosqueda-Melgar, J. and Martín-Belloso, O. 2008. Edible alginate-based coating as carrier of antimicrobials to improve shelf-life and safety of fresh-cut melon. *International Journal of Microbiology* 121, 313–327.

Regnier, T., Combrinck, S., Veldman, W. and Du Plooy, W. 2014. Application of essential oils as multi-target fungicides for the control of *Geotrichum citri-aurantii* and other postharvest pathogens of citrus. *Industrial Crops and Products* 61, 151–159.

Rhim, J.-W. and Kim, Y.-T. 2014. Biopolymer-based composite packaging materials with nanoparticles. pp. 413–442. *In*: J.H. Han (ed.). *Innovations in Food Packaging*, 2nd ed. Academic Press, Elsevier, London, UK.

Rhodes, D.J. 1993. Formulation of biological control agents. pp. 411–439. *In*: D. G. Jones (ed.). *Exploitation of Microorganisms*. Chapman & Hall, London, UK.

Rinaudo, M. 2006. Chitin and chitosan: Properties and applications. *Progress in Polymer Science* 31, 603–632.

Rodríguez-García, I., Cruz-Valenzuela, M.R., Silva-Espinoza, B.A., González-Aguilar, G.A., Moctezuma, E., Gutiérrez-Pacheco, M.M., Tapia-Rodríguez, M.R., Ortega-Ramírez, L.A. and Ayala-Zavala, J.F. 2016. Oregano (*Lippia graveolens*) essential oil added within pectin edible coatings prevents fungal decay and increases the antioxidant capacity of treated tomatoes. *Journal of the Science of Food and Agriculture* 96, 3772–3778.

Rojas-Graü, M.A., Raybaudi-Massilia, R.M., Soliva-Fortuny, R.C., Avena-Bustillos, R. J., McHugh, T.H. and Martín-Belloso, O. 2007. Apple puree-alginate edible coating as carrier of antimicrobial agents to prolong shelf-life of fresh-cut apples. *Postharvest Biology and Technology* 45, 254–264.

Romanazzi, G., Feliziani, E., Bautista-Baños, S. and Sivakumar, D. 2015. Shelf life extension of fresh fruit and vegetables by chitosan treatment. *Critical Reviews in Food Science and Nutrition* 11, 579–601.

Romanazzi, G., Feliziani, E., Santini, M. and Landi, L. 2013. Effectiveness of postharvest treatment with chitosan and other resistance inducers in the control of storage decay of strawberry. *Postharvest Biology and Technology* 75, 24–27.

Romanazzi, G., Sanzani, S.M., Bi, Y., Tian, S., Martínez, P.G. and Alkan, N. 2016. Induced resistance to control postharvest decay of fruit and vegetables. *Postharvest Biology and Technology* 122, 82–94.

Sánchez-González, L., Cháfer, M., Chiralt, A. and González-Martínez, C. 2010. Physical properties of edible chitosan films containing bergamot essential oil and their inhibitory action on *Penicillium italicum*. *Carbohydrate Polymers* 82, 277–283.

Sánchez-González, L., Pastor, P., Vargas, M., Chiralt, A., González-Martínez, C. and Cháfer, M. 2011. Effect of hydroxypropylmethylcellulose and chitosan coatings with and whitout bergamot essential oil on quality and safety of cold-stored grapes. *Postharvest Biology and Technology* 60, 57–63.

Santos, N.S.T., Aguiar, A.J.A.A., Oliveira, C.E.V., Sales, C.V., Silva, S.M., Silva, R.S., Stamford, T.C.M. and de Souza, E.L. 2012. Efficacy of the application of a coating composed of chitosan and *Origanum vulgare* L. essential oil to control *Rhizopus stolonifer* and *Aspergillus niger* in grapes (*Vitis labrusca* L.). *Food Microbiology* 32, 345–353.

Scandalios, J.G. 1993. Oxygen stress and superoxide dismutase. *Plant Physiology* 107, 7–12.

Shao, X., Cao, B., Xu, F., Xie, S., Yu, D. and Wang, H. 2015. Effect of postharvest application of chitosan combined with clove oil against citrus green mold. *Postharvest Biology and Technology* 99, 37–43.

Sharma, R.R., Singh, D. and Singh, R. 2009. Biological control of postharvest diseases of fruits and vegetables by microbial antagonists: A review. *Biological Control* 50, 205–221.

Silva, F.M.C., Stamford, T.C.M., Franco, L.O. and Campos-Takaki, G.M.C. 2006. Effect of salinity and glucose on chitin and chitosan production by *Cunninghamella elegans*. *Asian Chitin Journal* 2, 29–38.

Silva, M.C.F., Barros Neto, B., Stamford, T.C.M. and Campos-Takaki, G.M. 2007. Effect of environmental conditions on chitin and chitosan production by *Cunninghamella elegans* UCP 542 using factorial design. *Asian Chitin Journal* 3, 15–22.

Singh, S.P., Singh, Z. and Swinny, E.E. 2009. Sugars and organic acids in Japanese plums (*Prunus salicina* Lindell) as influenced by maturation, harvest date, storage temperature and period. *International Journal of Food Science and Technology* 44, 1973–1982.

Sogvar, O.B., Saba, M.K., Emamifar, A. and Hallaj, R. 2016. Influence of nano-ZnO on microbial growth, bioactive content and postharvest quality of strawberries during storage. *Innovative Food Science and Emerging Technologies* 35, 168–176.

Song, Y., Babiker, E.E., Usui, M., Saito, A. and Kato, A. 2002. Emulsifying properties and bactericidal action of chitosan-lysozyme conjugates. *Food Research International* 35, 459–466.

Souza, E.L., Sales, C.V., Oliveira, C.E.V., Lopes, L.A.A., da Conceição, M.L., Berger, L.R.R. and Stamford, T.C.M. 2015. Efficacy of a coating composed of chitosan from *Mucor circinelloides* and carvacrol to control *Aspergillus flavus* and the quality of cherry tomato fruits. *Frontiers in Microbiology* 6, 1–9.

Spadaro, D. and Droby, S. 2016. Development of biocontrol products for postharvest diseases of fruit: The importance of elucidating the mechanisms of action of yeast antagonists. *Trends in Food Science and Technology* 47, 39–49.

Sun, X., Narciso, J., Wang, Z., Ference, C., Bai, J. and Zhou, K. 2014. Effects of chitosan-essential oil coatings on safety and quality of fresh blueberries. *Journal of Food Science* 79, 955–960.

Synowiecki, J. and Al-Khateeb, N.A.A. 2003. Production, properties, and some new applications of chitin and its derivatives. *Critical Reviews in Food Science and Nutrition* 43, 144–171.

Takma, D.K. and Korel, F. 2017. Impact of preharvest and postharvest alginate treatments enriched with vanillin on postharvest decay, biochemical properties, quality and sensory attributes of table grapes. *Food Chemistry* 221, 187–195.

Tao, N., Fan, F., Jia, L. and Zhang, M. 2014. Octanal incorporated in postharvest wax of Satsuma mandarin fruit as a botanical fungicide against *Penicillium digitatum*. *Food Control* 45, 56–61.

Tasar, O.C., Erdal, S. and Taskin, M. 2016. Chitosan production by psychrotolerant *Rhizopus oryzae* in non-sterile open fermentation conditions. *International Journal of Biological Macromolecules* 89, 428–433.

Tripathi, P. and Dubey, N.K. 2004. Exploitation of natural products as an alternative strategy to control postharvest fungal rotting of fruit and vegetables. *Postharvest Biology and Technology* 32, 235–245.

Tsigos, I., Martinou, A., Kafetzopoulos, D. and Bouriotis, V. 2000. Chitin deacetylases: New, versatile tools in biotechnology. *Trends in Biotechnology* 18, 305–312.

US FDA (United States Food and Drug Administration). 2007. *Inventory of Effective Food Contact Substance (FCS) Notifications: Zeolite A in Which Silver and Zinc Ions Have Been Exchanged for Sodium Ions*, FCS No. 697. U.S. Food and Drug Administration, Washington, DC.

US FDA (United States Food and Drug Administration). 2015. Title 21 – Food and Drugs. Part 182 – Substances generally recognized as safe. Essential oils, oleoresins (solvent-free), and natural extractives (including distillates); final rule, Federal Register. Published 1 April 2015. Available at: www.accessdata.fda.gov/scripts/cdrh/cfdocs/cfcfr/CFRSearch.cfm?fr=182.20

Valencia-Chamorro, S.A., Palou, L., del Río, M.A. and Pérez-Gago, M.B. 2008. Inhibition of *Penicillium digitatum* and *Penicillium italicum* by hydroxypropylmethylcellulose-lipid edible composite films containing food additives with antifungal properties. *Journal of Agricultural and Food Chemistry* 56, 11270–11278.

Valencia-Chamorro, S.A., Palou, L., del Río, M.A. and Pérez-Gago, M.B. 2011a. Performance of hydroxypropyl methylcellulose (HPMC)-lipid edible coatings with antifungal food additives during cold storage of 'Clemenules' mandarins. *LWT - Food Science and Technology* 44, 2342–2348.

Valencia-Chamorro, S.A., Pérez-Gago, M.B., del Río, M.A. and Palou, L. 2009a. Curative and preventive activity of hydroxypropyl methylcellulose-lipid edible composite coating antifungal food additives to control citrus postharvest green and blue molds. *Journal of Agricultural and Food Chemistry* 57, 2770–2777.

Valencia-Chamorro, S.A., Pérez-Gago, M.B., del Río, M.A. and Palou, L. 2009b. Effect of antifungal hydroxypropyl methylcellulose (HPMC)-lipid edible composite coatings on postharvest decay development and quality attributes of cold-stored 'Valencia' oranges. *Postharvest Biology and Technology* 54, 72–79.

Valencia-Chamorro, S.A., Pérez-Gago, M.B., del Río, M.A. and Palou, L. 2010. Effect of antifungal hydroxypropyl methylcellulose (HPMC)-lipid edible composite coatingson penicillium decay development and postharvest quality of cold-stored 'Ortanique' mandarins. *Journal of Food Science* 75, 418–426.

Valencia-Chamorro, S.A., Pérez-Gago, M.B., del Río, M.A. and Palou, L. 2011b. Antimicrobial edible films and coatings for fresh and minimally processed fruits and vegetables: A review. *Critical Reviews in Food Science and Nutrition* 51, 872–900.

Wang, L., Wu, H., Qin, G. and Meng, X. 2014. Chitosan disrupts *Penicillium expansum* and controls postharvest blue mold of jujube fruit. *Food Control* 41, 56–62.

Wang, Y., Li, B., Zhang, X., Peng, N., Mei, Y. and Liang, Y. 2017. Low molecular weight chitosan is an effective antifungal agent against *Botryosphaeria* sp. and preservative agent for pear (*Pyrus*) fruits. *International Journal of Biological Macromolecules* 95, 1135–1143.

Xing, Y., Li, X., Xu, Q., Yun, J., Lu, Y. and Tang, Y. 2011. Effects of chitosan coating enriched with cinnamon oil on qualitative properties of sweet pepper (*Capsicum annuum* L.). *Food Chemistry* 124, 1443–1450.

Xing, Y., Lin, H., Cao, D., Xu, Q., Han, W., Wang, R., Che, Z. and Li, X. 2015. Effect of chitosan coating with cinnamon oil on the quality and physiological attributes of china jujube fruits. *BioMed Research International* 2015, article 835151.

Yilmaz, A., Bozkurt, F., Cicek, P.K., Dertli, E., Durak, M.Z. and Yilmaz, M.T. 2016. A novel antifungal surface-coating application to limit postharvest decay on coated apples: Molecular, thermal and morphological properties of electrospun zein–Nanofiber mats loaded with curcumin. *Innovative Food Science and Emerging Technologies* 37, 74–83.

Younes, I., Sellimi, S., Rinaudo, M., Jellouli, K. and Nasri, M. 2014. Influence of acetylation degree and molecular weight of homogeneous chitosans on antibacterial and antifungal activities. *International Journal of Food Microbiology* 185, 57–63.

Young, D.H., Köhle, H. and Kauss, H. 1982. Effect of chitosan on membrane permeability of suspension-cultured *Glycine max* and *Phaseolus vulgaris* cells. *Plant Physiology* 70, 1449–1454.

Youssef, K., Ligorio, A., Nigro, F. and Ippolito, A. 2012. Activity of salts incorporated in wax in controlling postharvest diseases of citrus fruit. *Postharvest Biology and Technology* 65, 39–43.

Zakrzewska, A., Boorsma, A., Brul, S., Hellingwerf, K.J. and Klis, F.M. 2005. Transcriptional response of *Saccharomyces* to the plasma membrane-perturbing compound chitosan. *Eukaryotic Cell* 4, 703–715.

Zeng, R., Zhang, A., Chen, J. and Fu, Y. 2013. Impact of carboxymethyl cellulose coating enriched with extract of *Impatiens balsamina* stems on preservation of 'Newhall' navel orange. *Scientia Horticulturae* 160, 44–48.

Zhang, H.Y. and Li, W.M. 2011. Effects of chitin and its derivative chitosan on postharvest decay of fruits: A review. *International Journal of Molecular Science* 12, 917–934.

Zhang, Y., Rempel, C. and McLaren, D. 2014. Edible coating and film materials: Carbohydrates. pp. 305–323. *In*: J.H. Han (ed.). *Innovations in Food Packaging*, 2nd ed. Academic Press, Elsevier, London, UK.

Zhong, Z., Li, P., Xing, R., Chen, X. and Liu, S. 2009. Preparation, characterization and antifungal properties of 2-(α-arylamino phosphonate)-chitosan. *International Journal of Biological Macromolecules* 45, 255–259.

Zhou, Y., Zhang, L. and Zeng, K. 2016. Efficacy of *Pichia membranaefaciens* combined with chitosan against *Colletotrichum gloeosporioides* in citrus fruits and possible modes of action. *Biological Control* 96, 39–47.

Chapter 22

Aloe spp.

Gels to Reduce Fruit Disease and Maintain Quality Properties

Domingo Martínez-Romero, Fabián Guillén, Salvador Castillo, Pedro Javier Zapata, and Juan Miguel Valverde

Department of Food Technology, University Miguel Hernández, Orihuela, Alicante, Spain

María Serrano

Department of Food Biology, University Miguel Hernández, Orihuela, Alicante, Spain

Daniel Valero

Department of Food Technology, University Miguel Hernández, Orihuela, Alicante, Spain

Abbreviations

1-MCP	1-Methylcyclopropene
ACC	1-aminocyclopropane-1-carboxylic acid
AG	*Aloe* gel
AVG	*Aloe vera* gel
BC	Before Christ
β-Gal	β-galactosidase
BI	Browning index
CAT	Catalase
Cel	Cellulase
CFU	Colony-forming units
CTR	Carbon dioxide transmission rate
OD	Optical density
OTR	Oxygen transmission rate
PAL	Phenylalanine ammonia-lyase
PDA	Potato dextrose agar
PE	Pectinsterase
PG	Polygalacturonase
PL	Pectin lyase
PPO	Polyphenol oxidase
POD	Peroxidase
POX	Proline oxidase
PME	Pectin methylesterase
RH	Relative humidity
SAM	S-adenosil methionine
SEM	Scanning electron microscopy
SOD	Superoxide dismutase
TA	Total acidity
TP	Total polyphenols
TSS	Total soluble solids
UV	Ultraviolet light
WVTR	Water vapor transmission rate

1 Botany of *Aloe* spp. with Commercial Potential for the Food Industry

The botanical name of the most important *Aloe* spp. is *A. vera* (L.) Burm. f., and it is a synonym of *A. barbadensis* Mill. (Anonymous, 2013). It belongs to the Asphodelaceae (Liliaceae) family and is a shrubby or arborescent, perennial, xerophytic, succulent, pea-green color plant. It grows mainly in the dry regions of Africa, Asia, Europe, and America. The plant has triangular, fleshy leaves with serrated edges, yellow tubular flowers, and fruits that contain numerous seeds. Each leaf is composed of three layers: (1) an outer thick rind of 15–20 cell layers, which has a protective function and synthesizes carbohydrates and proteins; (2) a middle layer of latex, which is the bitter yellow sap that contains anthraquinones and glycosides. Inside the rind there are vascular bundles that transport substances such as water (xylem) and starch (phloem); and (3) an inner clear gel that contains 99% water with the balance consisting of glucomannans, amino acids, lipids, sterols, and vitamins.

The first physical record of an *Aloe* plant occurs on a stone tablet, written in Sumerian, from about 2100 B.C., although there is evidence to suggest that it was used long before this time to treat a wide variety of illnesses and ailments. It was a favorite plant of the Egyptians, Romans, and Greeks, as well as the Chinese and Indian cultures (Reynolds, 2004). While the earliest record is from 2100 B.C., it is likely that these cultures were using *Aloe* materials for thousands of years earlier, although no one recorded the use. In 1552 B.C., *Aloe* was mentioned as a laxative in the Egyptian *Papyrus Ebers* (Manvitha and Bidya, 2014). Discordes, a Greek physician, stated in the book *De Materia Medica* that *Aloe* could treat wounds, heal skin infections, diminish hair loss, and eliminate hemorrhoids (Shelton, 1991). In the 4th century B.C., Greeks found *Aloe* on the island of Socotra in the Indian Ocean. Alexander the Great was persuaded by his mentor Aristotle to capture this island to obtain supplies of *Aloe* to heal his wounded soldiers. *Aloe* was used by Cleopatra (69–30 B.C.) and Nefertiti as part of their beauty treatments. *Aloe* was included in the cargo sent along with Columbus when he discovered the new world. It may very well have been these voyages, along with other explorers that followed his lead, which took this plant to the new world, where it found new root and new purpose. These plants were transported in pots across the ocean and were used for their healing purposes on ships where gastrointestinal discomfort, malnutrition, and minor wounds and burns would have been common. When the plants were brought to the Mayan and Aztec civilizations, they regarded them just as the earliest humans and Egyptians did—a gift from the gods. In 1920, the species *Aloe vera* (L.) Burm. f. was first cultivated for pharmaceutical distribution (Shelton, 1991). The commercial use of *A. vera* gel (AVG) started in the 1950s, and in the 1960s pharmacist Bill C. Coates of Dallas, Texas (USA), succeeded in extracting the gel of *A. vera* while preserving its healing properties (Reynolds, 2004). This stabilized gel opened new fields of application. In fact, AVG is being used nowadays commercially as a beverage or functional ingredient in foods (Moore and MacAnalley, 1995; He et al., 2005), due to its proven health benefits and alleviation of symptoms of several diseases such as diabetes, inflammatory illnesses, cancer, and others (Reynolds and Dweck, 1999; Eshun and He, 2004). Currently, Mexico is the main producer of AVG worldwide (Pal et al., 2013; Sánchez-Machado et al., 2017).

2 Leaf Harvesting, Composition, Processing, and Gel Yields

Processing of *A. vera* leaves to obtain a gel derived from the leaf pulp has become a large industry worldwide due to its applications in food industries. AVG has been utilized as a component in health drinks and other beverages, including a tea containing AVG that has no laxative effects. It is also used in other food products, for example, milk, ice cream confectionery, and others (Ahlawat and Khatkar, 2011). A considerable amount of AVG finds its application in the pharmaceutical industry in the production of topical ointments, gel preparations, tablets, and capsules.

Two methods of processing *A. vera* leaves are utilized to obtain the juice ingredient commonly used in commercial AVG products: (1) a leaf processing, where *A. vera* leaf juice is obtained by grinding or macerating the entire leaf followed by purification to remove the phenolic compounds found in the latex. This purification step is usually accomplished via activated carbon filtration in a process known as decolorization; and (2) an inner leaf processing, where *A. vera* leaf juice is obtained by stripping away the outer leaf rind, rinsing or washing away the latex, and processing the remaining inner leaf material. Decolorization is also sometimes employed with this method.

Aloe vera leaf processing begins with the removal of the leaf from the plant and washing it, after which the leaf is cut into sections and ground into slurry to produce a material of soup-like consistency colloquially known as 'guacamole'. This in-process material may be treated with the enzyme cellulase in order to obtain a less viscous product (Upton et al., 2012). Afterwards, the resulting liquid is subjected to a series of filtration steps to remove remaining rind particles and the undesirable phenolic constituents. The resulting liquid is pumped into sanitized stainless steel holding tanks, before advancing to a depulping extractor that removes the remaining pulp and rind particles generated by the initial grinding process. The liquid is then passed through a series of press filters with carbon-coated plates that remove the phenolic constituents, including aloin, and any microscopic pieces of leaf, sand, or other particles. The *Aloe* juice is continually passed through the filter press until 99.9% or more of the aloin is removed. Filtration is performed as a final purification step before the liquid is ready for stabilization. A complete leaf processing procedure was described in detail by Ramachandra and Rao (2008).

There are several factors that affect *Aloe* gel (AG) yield and composition, such as the plant species and the environment during leaf growth, apart from other agronomical factors such as irrigation and fertilization. In view of the large number of *Aloe* spp. (more than 400 have been described), it is surprising that only *A. vera* and *A. ferox* Mill. have been studied in depth, and to a lesser extent *A. arborescenes* Mill. Recently, Zapata et al. (2013b) reported an investigation of eight *Aloe* spp. (three named previously plus *A. aristata* Haw., *A. claviflora* Burch., *A. mitriformis* Mill., *A. saponaria* Ait., and *A. striata* Haw.) that described their leaf characteristics and the properties and chemical composition of their AGs. It was shown that leaves of the *Aloe* spp. differed significantly in size and weight. *Aloe mitriformis* and *A. aristata* had the smallest leaves (~14–21 cm in length and ~32–34 g of weight), while *A. ferox* had the largest leaves (~66 cm in length and ~725 g of weight). Gel yields were evaluated through three harvest seasons, and yield generally decreased from winter to summer for all *Aloe* spp., although the decrease depended on the species. The greatest yield decrease (22%) occurred with *A. saponaria*, while non-significant decreases were observed with *A. claviflora, A. striata*, and *A. vera*. The percentage of obtained AG was also dependent on the species, with the highest gel yield obtained from leaves of *A. vera* followed by *A. claviflora*, with the lowest gel yield from *A. arborescens*.

3 Physicochemical Composition of *Aloe* spp. Gels

One of the main biologically active constituents of *Aloe* extracts is aloin or barbaloin (10-glucopyranosyl-1.8-dihydroxy-3-(hydroxymethyl)-9 (10H)-anthracenone), which is found in nature as a mixture of two diastereosiomers, aloin A (10R) and aloin B (10S). These two compounds are generally used as key components for the quality control of this plant and its derivatives. Aloin is generally contained in the bitter, smelly exudate seeping out from freshly cut leaves, while very low amounts of aloin exist in AVG obtained from the internal mass of *A. vera* leaf (Fanalli et al., 2010). Aloin concentrations were 1 to 9 mg/100 g depending on *Aloe* spp. and harvest season (Zapata et al., 2013b). Interestingly, harvest season had a strong influence on aloin content, since aloin significantly increased from winter to summer for all *Aloe* spp., which was in accordance with the concentration of total phenolics and total antioxidant activity. This could be due to the higher total radiation, which are mainly ultraviolet (UV) rays of summer season.

The anthraquinones (aloin and/or barbaloin) have a laxative effect when ingested and can cause irritation or allergic reactions when applied to the skin. The anthraquinones must be eliminated from the gels (decolorization) to achieve a concentration lower than 10 ppm, according to International Aloe Science Council standard for oral products (IASC, 2015). When *Aloe* gels are applied as edible coatings for postharvest treatment of horticultural products, the amount of anthraquinones that can be ingested with respect to the total food is negligible.

The main chemical compounds found in AGs are shown in Table 22.1.

Table 22.1 Chemical composition of *Aloe* spp. gels

Type	*Chemical compounds*
Carbohydrates	Pure mannan, acetylated mannan, acetylated glucomannan, glucogalactomannan, galactan, galactogalacturan, arabinogalactan, galactoglucoarabinomannan, pectic substance, xylan, cellulose
Proteins	Lectins
Aminoacids	Alanine, arginine, aspartic acid, glutamic acid, glycine, histidine, hydroxyproline, isoleucine, leucine, lysine, methionine, phenylalanine, proline, threonine, tyrosine, valine
Saccharides	Mannose, glucose, L-rhamnose, aldopentose
Lipids	Arachidonic acid, γ-linolenic acid, steroids (campestrol, cholesterol, β-sitosterol), triglicerides
Vitamins	B1, B2, B6, C, β-carotene, choline, folic acid, α-tocopherol
Anthraquinones	Aloe-emodin, aloetic-acid, anthranol, aloin A and B (or collectively known as barbaloin), isobarbaloin, emodin, ester of cinnamic acid
Chromones	8-C-glucosyl-(2′-O-cinnamoyl)-7-O-methylaloediol A, 8-C-glucosyl-(S)aloesol, 8-C-glucosyl-7-O-methyl-(S)-aloesol, 8-C-glucosyl-7-O-methylaloediol, 8-C-glucosyl-noreugenin, isoaloeresin D, isorabaichromone, neoaloesin A
Minerals	Calcium, chlorine, chromium, copper, iron, magnesium, manganese, potassium, phosphorous, sodium, zinc
Enzymes	Alkaline phosphatase, amylase, carboxypeptidase, catalase, cyclooxidase, cyclooxygenase, lipase, oxidase, phosphoenol-pyruvate carboxylase, superoxide dismutase

Source: Dagne et al. (2000), Reynolds (2004), Chen et al. (2012).

4 Antimicrobial Activity of *Aloe* spp. Gels

Since 1995, the activity of different AGs to inhibit some fungi that cause plant diseases has been analyzed (Saks and Barkai-Golan, 1995; Ali et al., 1999; Jasso de Rodríguez et al., 2005). Diseases such as those caused by the pathogens *Botrytis cinerea* Pers., *Penicillium digitatum* (Pers.) Sacc., *Penicillium expansum* L., and *Alternaria alternata* (Fr.) Keissl. have a great economic impact because they can cause substantial losses during the postharvest storage and marketing of fruits and vegetables (Table 22.2).

The effect of AGs has been studied in vitro on spore survival and mycelium radial growth and in vivo by using AGs as preventive or curative treatments applied to horticultural products. In these experiments, mycelial growth inhibition and minimum AGs inhibitory concentrations were determined and, in most cases, it was established whether the treatments with AGs were fungistatic or fungicidal (Table 22.2).

The inhibitory activity of AVG differed depending on the fungus genera. For example, Ortega-Toro et al. (2017) found that AVGs inhibited *Fusarium oxysporum* Schltdl. more than *A. alternata, Colletotrichum gloeosporioides* (Penz.) Penz. & Sacc., *Bipolaris spicifera* (Bainier) Subram., *Curvularia hawaiiensis* Manamgoda, L. Cai, K.D Hyde, or *B. cinerea.* On the other hand, Vieira et al. (2016) indicated that *B. cinerea* was more sensitive to coatings with AVG than *P. expansum* or *Aspergillus niger* Tiegh. Castillo et al. (2010) reported that *B. cinerea* was 3-fold more inhibited than *P. digitatum* to a similar AVG concentration. In addition, in all cases, the AVG inhibition of fungi was always dose-dependent.

The magnitude of AVG inhibition of fungi varies, even within the same species. Vieira et al. (2016) reported a 90% inhibition of *B. cinerea* growth with an AVG concentration of 500 mL/L, while Castillo et al. (2010) reported a similar level of inhibition with 100 mL/L, a 5-fold lower concentration). These differences indicate that the environmental and agronomical conditions in which the *A. vera* plant grows as well as the procedures used for obtaining the gels might be important factors that influence its antifungal efficacy. In this sense, Zapata et al. (2013b) verified that AGs obtained from different *Aloe* spp. varied in composition depending on the species and the date of harvest within seasons. AGs from leaves collected in summer (Northern hemisphere) were richer in aloin than those obtained from leaves harvested in other seasons and had greater antifungal activity. In addition, the leaves of *A. ferox, A. mitriformis*, and *A. saponaria* had the highest concentrations of aloin and were the most effective to control the infections by *B. cinerea, P. digitatum, P. expansum*, and *Penicillium italicum* Wehmer (Zapata et al., 2013b).

Other individual components found in AVG, such as saponins, acemannan, and anthraquinone derivatives, are also known to have antibiotic activity and could be responsible for its antibacterial properties (Serrano et al., 2006). In addition, when the antifungal activities of the gel (parenchyma) or liquid (latex) fractions were measured separately, the liquid fraction had the greatest inhibitory activity due to its higher polyphenol and aloin contents and greater antioxidant activity (Jasso de Rodríguez et al., 2005; Zapata et al., 2013b; Flores-López et al., 2016; Vieira et al., 2016). Moreover, UV-A light treatment for 36 d of *A. vera* plants led to an increase in the concentrations of aloin in the gel, epidermis, and liquid fraction (latex), and the growth of *B. cinerea* on artificially inoculated *A. vera* leaf surface decreased significantly in the leaves treated with UV-A (Martínez-Romero et al., 2013b).

Table 22.2 Effect of *Aloe* spp. gels on fungi causing diseases in fruits and vegetables.

Fungal species	Aloe *gel composition*	*Assay*	*Effects*	*Reference*
Penicillium digitatum, *Penicillium expansum*, *Botrytis cinerea*, and *Alternaria alternata*	Pure AVG prepared from fresh *Aloe vera* (L.) Burm. f. leaves.	**In vitro. Spore survival.** 0.1–0.2 mL of spores 10^3–10^4 CFU/mL spread in PDA with 0–10^5 µL AVG. **In vitro. Mycelium radial growth.** Growth of 5 mm fungal mycelium was measured in PDA containing 0–10^5 µL AVG. **In vivo. Preventive.** Untreated and treated grapefruit (1–10 µL AVG) were inoculated with 20 µL of *P. digitatum* (10^4 CFU/mL).	Spore survival and mycelium radial growth of *P. digitatum*, *A. alternata*, and *B. cinerea* were reduced. Incidence of infection on *P. digitatum* inoculated grapefruit was reduced by 54–77% with 1–10^3 µL.	Saks and Barkai-Golan (1995)
Rhizoctonia solani, *Fusarium oxysporum*, and *Colletotrichum coccodes*	Two types of pulp or liquid gel prepared from fresh *A. vera* leaves: Manually scraped out and by using a laboratory roll processor.	**In vitro. Mycelium radial growth.** Growth of 0.4 mm fungal mycelium was measured in PDA containing 0–10^5 µL AVG.	For the two types of *Aloe* fractions the activities were similar. AVG reduced the rate of colony growth at a concentration of 10^5 µL.	Jasso de Rodríguez et al. (2005)
Botrytis gladiolorum, *Fusarium oxysporum F. gladioli*, *Heterosporium pruneti*, and *Penicillium gladioli*	*A. vera* fresh leaves with the water-ethanol extract.	**In vitro. Mycelium radial growth.** Growth of 5 mm fungal mycelium was measured in Czapek-agar containing 0 (control), 20, 40, 80, and 100 µL/mL AVG.	AVG inhibited *B. gladiolorum* mycelial growth at 80 µL/mL and *F. oxysporum* f. *gladioli*, *H. pruneti* and *P. gladioli* growth at 100 µL/mL.	Rosca-Casian et al. (2007)
P. digitatum and *B. cinerea*	AVG gel (food grade) was purchased	**In vitro. Mycelium radial growth.** Fungi were placed in	Radial mycelial growth of *B. cinerea* and *P. digitatum*	Castillo et al. (2010)

(*Continued*)

Table 22.2 (Cont.)

Fungal species	Aloe *gel* *composition*	*Assay*	*Effects*	*Reference*
	from AVISA (Aloe Vera Internacional Sociedad Anónima, Fuerteventura, Canary Islands, Spain). Pure AVG was used for in vitro treatment.	PDA containing 0 (control), 1, 5, 25, 50, and 100 mL/L AVG.	was reduced. The dose of 250 mL/L led to 4- and 2-log reductions of mycelial growth for *P. digitatum* and *B. cinerea*, respectively.	
Aspergillus niger, Aspergillus flavus, A. alternata, Drechslera hawaiensis, and *P. digitatum*	Pure AVG prepared from fresh *A. vera* leaves.	**In vitro. Mycelium radial growth.** Radial growth of the inoculated Petri plates was measured in PDA containing 0, 0.15, 0.25, and 0.35% AVG.	An AVG concentration of 0.35% completely inhibited the growth of *D. hawaiensis* and *A. alternata*, while 70, 90, and 95% inhibition was found for *A. niger, A. flavus*, and *P. digitatum*, respectively.	Sitara et al. (2011)
Rhizopus stolonifer, B. cinerea, and *P. digitatum*	Pure AVG prepared from fresh *A. vera* leaves. AVG and AVG plus thymol (1 mL/L).	**In vivo. Preventive.** Untreated and treated nectarines (*Prunus persica* L. cvs. Flavela and Flanoba) (1–10 µL AVG) were inoculated with 20 µL of the corresponding fungus (10^4 CFU/mL).	Reduced decay incidence, severity, and growth of *R. stolonifer, B. cinerea*, and *P. digitatum*. Inoculation of coated fruits reduced postharvest ripening, ethylene production, and respiration rate.	Navarro et al. (2011)
Colletotrichum gloeosporioides	Different combinations of AVG (2%) and thyme oil (1%) were	**In vitro. Mycelium radial growth.** Growth of 6 mm fungal mycelium was measured in PDA	Reduced radial mycelial growth of *C. gloeosporioides*. AVG coating alone showed fungistatic	Bill et al. (2014)

(*Continued*)

Table 22.2 (Cont.)

Fungal species	Aloe *gel composition*	*Assay*	*Effects*	*Reference*
	prepared [1:1 v/v], [3:1 v/v]. AVG dried powder was purchased from Aloway Natural Health Products Pty., Limpopo, South Africa.	containing AVG. Fungicidal and fungistatic activity was determined. **In vivo. Preventive and curative.** Coatings were applied before or after the avocado inoculation with 20 μL of 10^5 CFU/mL of *C. gloeosporioides*.	effect while AVG + thyme oil showed fungicidal effect. Incidence and severity of anthracnose caused by *C. gloeosporioides* were reduced in both preventive and curative experiments. The activity of defense-related and antioxidant enzymes increased in coated avocados.	
P. expansum and *B. cinerea*	Gel and liquid fractions at 3 doses (0.1, 1.0, and 50%, v/v) and bagasse extracts (ethanol or distilled water) at 3 doses (50, 100, and 500 ppm, w/v) prepared from fresh *A. vera* leaves.	**In vitro. Fungal growth.** 100 μL (10^4 CFU/mL) of *P. expansum* and *B. cinerea* in gel, liquid bagasse extracts were monitored spectrophotometrically at 530 nm by measuring optical density (OD).	For all the treatments, the antifungal effect was concentration dependent and varied according to the fungal species. The liquid fraction showed the best inhibition effect.	Flores-López et al. (2016)
B. cinerea, *P. expansum*, and *A. niger*	AVG was obtained from fresh *A. vera* leaves. In vitro AVG fractions (pulp and liquid), at 0.5, 5, 20, and 50%. In vivo coatings with	**In vitro. Fungal growth.** 100 μL (10^4 CFU/mL) of *B. cinerea* in pulp or liquid fractions were monitored spectrophotometrically at 530 nm by measuring OD. **In vivo. Preventive.** Coatings were applied before the cherry	The liquid fraction showed the best inhibitory effect. *A. vera* liquid fractions at 0.5% showed a growth inhibition of 84.86 ± 1.44% for *B. cinerea*, 49.59 ± 8.56% for *P. expansum*, and	Vieira et al. (2016)

(Continued)

Table 22.2 (Cont.)

Fungal species	Aloe *gel composition*	*Assay*	*Effects*	*Reference*
	0.5% (w/v) chitosan + 0.5% (w/v) glycerol + 0.1% (w/v) Tween 80 + 0.5% (v/v) AVG liquid.	tomato inoculation (10 μL of *B. cinerea* at 10^4 CFU/mL).	15.91 ± 0.94% for *A. niger*. In coated fruit the growth of *B. cinerea* was inhibited.	
F. oxysporum, *A. alternata*, *C. gloeosporioides*, *Bipolaris spicifera*, *Curvularia hawaiiensis*, and *B. cinerea*	AVG was obtained from fresh *A. vera* leaves. AVG was mixed (in 1:1 w/W ratio) with Potato Dextrose Agar (PDA).	**In vitro. Mycelium radial growth.** Growth of 8 mm fungal mycelium was measured in PDA containing 50% AVG. **In vivo. Preventive and curative.** Coatings were applied before or after inoculation of cherry tomatoes with 5 μL of *F. oxysporum* (10^6 CFU/mL).	Fungal growth decreased when *A. vera* was present in the medium. Higher inhibition was observed for *F. oxysporum*, *B. spicifera*, and *C. hawaiiensis*. After preventive and curative coating application, fruit showed 30 and 40% of decay, respectively, while it was 50% in control fruit.	Ortega-Toro et al. (2017)
P. digitatum	AVG was obtained from fresh *A. vera* leaves. AVG coating alone or supplemented with *Pichia guilliermondii* BCC5389 (10^8 CFU/mL).	**In vitro. Fungal growth.** AVG coating alone or supplemented with *P. guilliermondii* at 1:1 ratio. The fungal growth inhibition was determined using spherical concavity culture slides. **In vivo. Preventive.** Coatings were applied before mandarin inoculation with 5 μL of 10^4 CFU/mL of *P. digitatum*. Wounded fruits were immersed for 5 min in AVG.	Inhibition of mycelial growth and reduced incidence of green mold caused by *P. digitatum*.	Jiwanit et al. (2018)

5 Effect of Preharvest Application of *Aloe* spp. Gels on Fruit Quality

Fruit and vegetable quality management starts in the field and continues during postharvest storage, transport, and distribution, ending when they reach the consumers. From harvest to consumer, fruits and vegetables can undergo severe quality losses attributable to over-ripening accompanied in most cases by a high incidence of postharvest diseases. Fortunately, postharvest quality and storage life potential of fruits can be manipulated by preharvest treatments. Thus, preharvest foliar applications of different minerals, growth regulators, ethylene inhibitors, fungicides, or edible coatings are among practices that enhance quality and storage potential of various fruit crops. However, the efficacy of a particular preharvest treatment to increase storage life of horticultural products can vary among different species and/or even cultivars of the same fruit species (Khan and Ali, 2018). To control postharvest diseases, several fungicides applied as preharvest treatments are available, although their use is regulated and the approved maximum residue levels can be very low and limit marketing opportunities. In this sense, our research group was the first to study the effect of AG applications before harvest to reduce fungal decay and maintain fruit quality during storage. In general, results have indicated positive economic benefits of this practice since field infections can be controlled so postharvest losses are reduced.

5.1 Quality and Safety at Harvest

Our research group has shown that preharvest AVG treatment delayed the on-tree fruit ripening process in several fruit species, such as table grape, sweet cherry, peach, and nectarine. The respiration rate at harvest was lower in all of these fruits when sprayed with AVG before harvest. This reduction could be associated with a diminution of the general fruit metabolism (Valero and Serrano, 2010). Another two parameters related to fruit quality, color, and firmness were not significantly affected by AVG treatments. On the other hand, fruit maturity index was not affected in treated table grape while an increase in the retention of total acidity (TA) occurred in sweet cherry, peach, and nectarine, causing a reduction of the maturity index at harvest on the treated fruits (Zapata et al., 2013a).

With respect to microbial spoilage, AVG treatments were effective in reducing the microbial population at harvest compared to control fruits on table grapes and stone fruits. In table grapes, AVG was especially effective to reduce mold and yeast populations, and all AVG-treated fruits had also lower mesophilic aerobics counts (Castillo et al., 2010; Zapata et al., 2013a). Ali et al. (1999) reported that the antifungal activity of AVG was suppression of spore germination and inhibition of mycelial growth, which could be attributed to the presence of more than one active compound with antifungal activity. There were also findings that AVG acts as an elicitor for natural plant disease resistance mechanisms such as the biosynthesis of antioxidant enzymes and total phenolic compounds (Bill et al., 2014; Hassanpour, 2015).

5.2 Postharvest Quality Maintenance

Fruit quality maintenance during postharvest storage after preharvest treatment with AGs has been also observed, and a delay in the decline of quality previously observed during on-tree ripening was even greater during storage. In fact, AVG was more effective on maintaining table grape quality during postharvest storage when the treatment was applied to the vines twice (1 and 7 d before harvest) instead of just once, especially for parameters such as fruit firmness and color (Hue angle). These parameters were significantly higher in treated berries in comparison to control fruit during storage. AVG applied before harvest up-regulated anthocyanin production by stimulating gene expression of enzymes involved in the anthocyanin biosynthetic pathway, such as PAL (Castillo et al., 2010). Weight loss and respiration rates were higher in control table grapes compared to the AGV-treated ones, and a reduction of microbial spoilage was observed. The antifungal activity of AVG could be attributed to the presence of more than one active compound with antifungal activity, as commented previously (Castillo et al., 2010).

6 Effect of Postharvest *Aloe* spp. Gel (AG) Coatings on Fruit Quality and Disease Reduction

Valverde et al. (2005) first reported the use of AVG as an edible coating for fruits and postulated that it could be a good tool for improving quality attributes and controlling microbial spoilage of tables grapes (*Vitis vinifera* L. cv. 'Crimson Seedless'). Valverde et al. (2005) stated that AVG could be an alternative to the use of conventional chemical fungicides. These authors reported that postharvest senescence was delayed in AVG-coated grapes by reducing respiration rate and weight loss and maintaining quality parameters such as total soluble solids (TSS), TA, color, firmness, and rachis condition. Moreover, microbiological populations (total mesophilic aerobic and yeast and mold counts) were lower in coated grapes than those of the controls during storage.

Since 2005 until the present, about 50 scientific papers on the use of AGs as edible coatings of intact and fresh-cut horticultural products have been published (Tables 22.3 and 22.4). Among them, stone fruits (*Prunus* spp.) such as plum, peach, and nectarine have been the most studied commodities. These papers reported positive effects of AGs (alone or in combination with other compounds) to improve produce quality during postharvest storage.

6.1 Preparation of AG Coatings

The source of *Aloe* extracts to be used as coatings is quite variable (Table 22.3). Early reports revealed that in most cases AG coatings, all of them 100% pure extracts, were liquid gels, freeze-dried powders (Ahmed et al., 2009; Dang et al., 2008), or spray-dried powders (Bill et al., 2014). In addition, in most of the studies, AGs were purchased either from pharmaceutical Valverde et al., 2005; Martínez-Romero et al., 2006; Serrano et al., 2006; Ahmed et al., 2009; García et al., 2014) or agricultural companies (Castillo et al., 2010; Padmaja and Bosco, 2014). Most of the commercial *Aloe* extracts (gel or powder) were analyzed to check their purity, authenticity, or adulteration as well as their degradation rate over time. Several authors reported that AG composition may vary due to processing techniques (Ramachandra and Rao, 2008), AGs can be adulterated with artificial preservatives, or be extremely diluted, as well as have excessive concentration of compounds related to both fermentative or degradative processes due to poor conservation or absence of stabilization (Plaskett, 1997; Lachenmeier et al., 2005; Bozzi et al., 2007). Accordingly, as shown in Table 22.3, coatings based on *Aloe* spp. from different sources did not show the same effects on different fruits. For this reason, since 2011, the authors of the present chapter prepared their own AGs from parenchyma obtained from freshly harvested leaves (Table 22.3). In some cases, our results were similar to those obtained with other edible coatings such as carnauba wax (Dang et al., 2008), gum tragacanth (Mohebbi et al., 2012), chitosan (Bill et al., 2014; de Bruin et al., 2016; Vieira et al., 2016), gum Arabic (Bill et al., 2014; Ullah et al., 2017), cinnamon oil (Ullah et al., 2017), or commercial coatings such as Semperfresh™ (Dang et al., 2008) or Shellac™ (Chauhan et al., 2015).

Most authors have followed a protocol very similar to the one described by Navarro et al. (2011) when the gels were made by the researchers themselves (Marpudi et al., 2011; Mohebbi et al., 2012; Guillén et al., 2013; Zapata et al., 2013a; Nasution et al., 2015; Sogvar et al., 2016; Vieira et al., 2016; Martínez-Romero et al., 2017; Jiwanit et al., 2018). Normally, fresh *Aloe* spp. leaves are harvested from 3- to 4-yr-old plants about 3 hr after sunrise. The plant basal leaves are cut with a knife and immediately cleaned in water with 0.03–2% sodium hypochlorite for 2–5 min to remove dirt from the surface. Then, the yellow substance or latex is extracted by cutting the base of the leaves to detach them and then standing them upright to drain these substances from them. Thereafter, the leaves spines and edges are cut. The leaves are split longitudinally in half. The top and bottom epidermis layers are carefully separated from the parenchyma by using a sharp knife. The colorless parenchyma fillets are crushed and uniformly mixed to yield a mucilaginous gel. Then, the gel is filtered and the fibrous fraction discarded. By using this procedure, the yield of AVG is 75–80%.

The gel characteristics of this AVG are: pH 4.7–5.3, °Brix 0.98–1.15, and TA 0.038–0.053 g 100/g citric acid equivalent. The gel is stabilized by lowering the pH to 3–4 with phosphoric, citric, or ascorbic acid. In some works, the AGs were heated between 65 and 80°C for 10 s to 45 min, and immediately cooled to 5–20°C before use. Chauhan et al. (2015) analyzed the physicochemical characteristics of AVG, and they were compared with those from Shellac™ and the combination of Shellac™ + AVG. AVG had a viscosity 9.8±0.1 (cp at 30°C); drying time 3.6±0.1 (min); oxygen transmission rate (OTR) 1560±25 (cc/m^2/d/atm), carbon dioxide transmission rate (CTR) 1200±4 (cc/m^2/d/atm), and water vapor transmission rate (WVTR) 78.2±1.4 (g/m^2/d/atm), all at 75% relative humidity (RH) and 30°C. The addition of AVG to Shellac™ improved its physical properties, permeability to O_2 and CO_2, and transparency while the drying time was not affected. In addition, as AVG concentration increased (from 0 to 15%) in a solution of 0.5% carboxymethyl cellulose (0.5%), peanut oil (5%), and glycerol monostearate (2% as emulsifier), the values of the physical properties (OTR, CTR, and WVTR) of the AVG composites diminished (Panwar et al., 2016).

6.2 Improving AG Coating Barrier Properties

Many AG coating treatments were performed manually (Dang et al., 2008) or most frequently by immersion or dipping for different periods (from 3 to 10 min or with time as a variable; Padmaja and Bosco, 2014). In other cases, the coating treatment was repeated twice (Marpudi et al., 2013). AVG was applied either undiluted or diluted in distilled water to 50, 33, 25, or 1% concentration, where the later was the lowest concentration used (Shahkoomahally and Ramezanian, 2014).

As shown in section 3 of this chapter, pure AGs are characterized as having low dried matter and lipids and being rich in long-chain polysaccharides such as acemannan derivatives. Guillén et al. (2013) compared the application of two AG coatings, one from *A. vera* and the other from *A. arborecens*, to 'Santa Rosa' plums and 'Red Heaven' peaches, and showed that *A. arborescens* coating was more effective than *A. vera* coating in terms of modulating the ripening and quality-related parameters, which were attributed to the higher lipid content of the former and its capacity to increase the barrier properties (Zapata et al., 2013b).

Pure AGs have been modified with different compounds to enhance their antimicrobial, antioxidant, and preservation activities (Tables 22.3 and 22.4) or to improve their physicochemical properties to form emulsions or to increase their plasticity, permeability, or mechanical properties. In this sense, AGs were assessed in whole fruit to increase their antifungal capacity by the addition of thyme oil or thymol (Navarro et al., 2011; Bill et al., 2014), cinamaldehyde (Athmaselvi et al., 2013), chitosan (Vieira et al., 2016), or the antagonistic yeast *Pichia guilliermondii* (Jiwanit et al., 2018). Ascorbic acid was added to AGs to increase their antioxidant properties (Athmaselvi et al., 2013; Sogvar et al., 2016). To improve texture and firmness, added substances included calcium chloride (Shahkoomahally and Ramezanian, 2014) and plasticizers such as glycerol and oleic acid (Athmaselvi et al., 2013). Rosehip oil improved the barrier properties of AGs to water vapor diffusion (Paladines et al., 2014; Martínez-Romero et al., 2017). In other cases, AGs were enriched with polysaccharides (Padmaja and Bosco, 2014; Vieira et al., 2016).

In the case of fresh-cut fruits and vegetables (Table 22.4), pure AGs (Benítez et al., 2013; Supapvanich et al., 2016; Alberio et al., 2017) without the addition of other compounds have been used. However, in most cases, more complex formulations than those used for intact fruit have been developed to control tissue browning by using organic acids (Martínez-Romero et al., 2013a; Nasution et al., 2015; Kuwar et al., 2015), salicylic acid (Riaie et al., 2017), and cysteine (Song et al., 2013). Calcium chloride was added to improve firmness (Kuwar et al., 2015; Nasution et al., 2015), while chitosan (Treviño-Garza et al., 2017), vainillin (Kuwar et al., 2015), and potassium sorbate (Nasution et al., 2015) were added to control microbial spoilage. Finally, glycerol was added to increase plasticity (Kuwar et al., 2015; Nasution et al., 2015; Treviño-Garza et al., 2017).

6.3 Effect of AG Coatings on Physiological and Quality Parameters of Whole Fruit and Fungal Spoilage

6.3.1 Physiological Parameters: Respiration Rate, Ethylene Production, Ion Leakage, and Weight Loss

Ethylene production and respiration rate are the most important physiological parameters related to quality and perishability of harvested fruits (Valero and Serrano, 2010). Fresh plant products are characterized by the fact that they continue metabolic processes after harvest. These are intimately related to their respiration where the plant tissues consume oxygen (O_2) and release carbon dioxide (CO_2). The majority of edible coatings applied to fruits and vegetables reduce the inherent permeability of tissues of plant products to these gases to reduce their respiration rates (Lee et al., 2003; Maftoonazad and Ramaswamy, 2005).

It has been reported that the respiration rate during storage (temperatures of 1–20°C including refrigeration and shelf-life periods) of whole fruit coated with pure AGs was lower than that of control fruit for a wide range of fruits, including table grapes (Valverde et al., 2005), cherries, nectarines, peaches, plums (Ahmed et al., 2009; Martínez-Romero et al., 2006, 2017; Navarro et al., 2011; Guillén et al., 2013; Paladines et al., 2014; Ravanfar et al., 2014), tomato (Chauhan et al., 2015), and mango (Muangdech, 2016). The degree of respiration rate reduction was variable depending on the plant species, cultivar, and the concentration of AG in the coatings. In the case of mangoes treated with pure AVG or AVG at a 1:1 water dilution, no significant differences in respiration rate were observed (Dang et al., 2008). However, up to 50% reduction in respiration rate in some fruits has been observed with AG coatings (Table 22.3). As previously noted (Section 3 of this chapter), AGs are very rich in polysaccharides and poor in lipids and proteins; for this reason, they have relatively low permeability to gases, but little resistance to transmission of water vapor (Baldwin, 2002).

AGs, due to their barrier properties, have a selective permeability to O_2 and CO_2 on the surface of the fruits that causes an increase the internal concentration of CO_2 and a reduction in the internal concentration of O_2 (de Wild et al., 2005). This could be responsible for the decrease in the metabolism of fruits, such as their respiration and ethylene production rates. Ethylene production, like respiration, is a process that requires O_2. Generally, low O_2 (below 8%) and high CO_2 (above 5%) concentrations slow down respiration and retard ethylene production and therefore fruit ripening (Kader, 1986).

Another factor to consider is the *Aloe* species used to make the gel. Guillén et al. (2013) observed that gel of the species *A. arborescens* was more effective than that of *A. vera* when plums and nectarines were coated. These results were attributed to the different composition of these gels. Specifically, gels of *A. arborecens* are richer in lipids, which would reduce the gas permeability of these coatings. Taking into account this circumstance, Paladines et al. (2014) observed that with the incorporation of different fractions of rosehip oil (0, 1, 2, and 5%) to AVG, the respiration rate of coated cherries, peaches, nectarines, and apricots decreased during storage at 20°C. The increasing concentrations of rosehip oil proportionally reduced the respiration rate. These results were attributed to the reduction of coating permeability and the creation of a micro-modified internal atmosphere, rich in CO_2 and depleted of O_2, capable of reducing fruit respiration rate. In particular, Park and Chinnan (1990) observed that by adding acetylated and monoglycerides lipids to hydroxymethylcellulose, the permeability to O_2 and CO_2 significantly decreased by 65 and 90%, respectively.

Ethylene is the hormone responsible for triggering the ripening process in climacteric fruits (Valero and Serrano, 2010), affecting all changes related to fruit quality. To date, different methods have been described to inhibit and/or control ethylene production (Martínez-Romero et al., 2007). Among them, 1-MCP (1-methylcyclopropene) is the most effective commercial method and has been used successfully for many *Prunus* spp. (Valero et al., 2005), including plums (Martínez-Romero et al., 2003). In addition, some edible coatings, such as alginate applied to tomatoes (Zapata et al., 2008), four varieties of plum (Valero et al., 2013), or chitosan applied to pear (Li and Yu, 2001), reduced ethylene emission rates. Accordingly, as occurred for respiration, ethylene emission was reduced and/or delayed in fruits of *Prunus* spp. (Ahmed et al., 2009; Navarro et al., 2011; Guillén et al., 2013) or tomatoes (Chauhan et al., 2015) treated with pure AVG compared to control fruits. In addition, when AVG was modified with rosehip oil and applied to different stone fruits stored either at 20 or 1–2°C, the ethylene inhibition rate increased proportionally with the added oil concentration (Paladines et al., 2014; Martínez-Romero et al., 2017). This coating, due to its barrier properties, had a selective permeability to O_2 and CO_2 (Banks et al., 1993; de Wild et al., 2005), which could be responsible for the decrease in ethylene production in fruits coated with AVG. According to the results obtained with tomatoes coated with alginate and zein, which showed a decrease in the rate of ethylene production due to a decrease in the concentration of ACC (1-aminocyclopropane-1-carboxylic acid) (Zapata et al., 2008), this could be attributed to the effect of CO_2 on the inhibition of the conversion of SAM (S-adenosyl methionine) to ACC by means of the enzyme ACC synthase (de Wild et al., 2005).

At harvest time, the water content of plant products is very high, but once they are harvested, the transpiration process continues and water is lost continuously. The quality of dehydrated fruits and vegetables is poor, and the consumer is not willing to accept a dehydrated, soft, and withered fruit. The deficit of water during storage affects the turgor of plant tissues, and its loss diminishes the perception of freshness of horticultural products. Thus, maintaining a high proportion of water in plant products is essential to maintain quality (Gómez-Galindo et al., 2004). Transpiration rates can vary according to many parameters related to external conditions (such as temperature and RH), characteristics of fruit surface (epidermal and subepidermal cell layers, cuticle composition, and fruit surface/volume ratio), and presence of external barriers such as edible coatings (composition and structure of the coating). The composition and structure of edible films and coatings affect the mechanism of water transfer and, therefore, their barrier behavior (Embuscado and Huber, 2009).

A reduction of weight loss during the storage of AVG-coated fruits occurred with many commodities (Table 22.3). For example, when pure AVG was applied, weight loss with respect to control fruit decreased between 5% in papayas stored at 30°C and 60% RH for 10 d (Marpudi et al., 2011) and 30% in figs stored at 29°C and 45% RH for 6 d (Marpudi et al., 2013). Exceptions were reported by García et al. (2014) working with tomatoes (cv. 'Charleston') and Dang et al. (2008) working with mangos (cv. 'Kensington Pride'), who observed no significant control of weight loss on AVG-coated fruits.

In general, differences in the ability of edible coatings to reduce weight loss are attributed to the different water vapor permeability of the polysaccharides used in the formulations (Vargas et al., 2008), the addition of other compounds such as glycerol (Zapata et al., 2008; Valero et al., 2013) or even lipids (oils or fatty acids) (Embuscado and Huber, 2009) that act as plasticizers. Individually, water vapor permeability of coatings based on lipids, fatty acids, or monoglycerides is lower than that of coatings based on polysaccharides (Dhall, 2013). Lipids are used for their hydrophobic nature as water vapor barriers, although their efficiency depends not only on their chemical structure, degree of saturation, and physical state, but also on the homogeneity of the film (Callegarin et al., 1997). The length of fatty acid chains plays a very important role in the rate of water vapor transmission since it increases with the number of carbon atoms (from 14 to 18). The relative proportion of the lipophilic part of the molecule increases with longer fatty acid chains, minimizing water solubility and moisture transfer. Fatty acids chains of 18 carbons are those with the lowest permeability. However, when the carbon chain contains more than 18 carbon atoms, they have higher moisture permeability (Debeaufort and Voilley, 2009). In this sense, the weight loss of whole fruits coated with AVG plus the addition of lipids was lower than that of fruits coated with pure AVG, as observed for coatings with AVG + oleic acid in tomato (Athmaselvi et al., 2013), AVG + rosehip oil in *Prunus* spp. (Paladines et al., 2014; Martínez-Romero et al., 2017), and AVG + cinnamon oil in pepper (Ullah et al., 2017). Moreover, AG coatings also cover the stomata resulting in a decrease in respiration rate, as shown by Athmaselvi et al. (2013). They reported this for tomato 'Ruchi 618' coated with AVG and observed by scanning electron microscopy (SEM) that a thin and uniform semipermeable membrane for gas exchange was present.

During storage, the integrity of cell membranes is reduced as a consequence of processes linked to maturation, senescence, and even the appearance of physiological disorders such as chilling injury. Thus, AVG coating treatments, storage period, and their interaction significantly affected the cell membranes in nectarine (Ahmed et al., 2009) and bell pepper (Ullah et al., 2017). In both cases, the electrolyte leakage in control fruits was 20–34% higher than in coated fruits. In addition, bell peppers coated with AVG exhibited a lower percentage of fruits affected by chilling injury than control fruits (Ullah et al., 2017).

6.3.2 Quality and Sensory Parameters

Color, firmness, total soluble solids (TSS), and titratable acidity (TA) are the usual quality parameters evaluated in fruits (Table 22.3). In many reports, the use of AVG as edible coating delayed the loss of color, maintained firmness and acidity levels within whole fruits, and delayed the accumulation of sugars. The magnitude of these AVG effects was variable, mainly depending on the plant species, cultivar, and the fruit storage conditions. In addition, when AVG was incorporated with other polysaccharides or plasticizers, the results improved considerably. These effects could be attributed to the reduction of ethylene production, respiration rate, and weight loss (transpiration) in coated fruits, which are the mechanisms with the greatest influence on fruit quality losses after harvest (Valero and Serrano, 2010).

External color is a very important parameter for marketing of fruit and vegetables, since consumers find produce appearance to be a decisive characteristic for their acquisition. Color gives an idea of the maturity stage and conservation period of the fruit (Crisosto et al., 1995). The fruit color changes during storage and maturation for different reasons, including degradation of chlorophylls, accumulation of pigments (mainly anthocyanins and carotenoids), and the appearance of browning (Valero and Serrano, 2010). The synthesis and degradation of colored compounds are directly related to fruit metabolism as well as to the action of ethylene.

Fruit softening involves structural as well as compositional changes in various components of the cell wall carbohydrates, partly as a result of the action of fruit softening enzymes. The delayed softening observed in AVG-treated fruits could be due to the inhibitory effect of AVG on cell wall degrading-enzymes, mainly polygalacturonase (PG), pectinsterase (PE), β-galactosidase (β-Gal), pectin lyase (PL), and cellulase (Cel) (Valero and Serrano, 2010; Yoshioka et al., 2011; Bustamante et al., 2012). The majority of the enzymes that act on the degradation of the middle lamella and the primary cell wall are controlled by the action of ethylene (Martínez-Romero et al., 2007). Fruits that were coated with AVG had lower ethylene production and in general lower softening (Table 22.3).

Lower TSS in AVG-coated fruit might be due to a delay in fruit ripening, which indicates that control fruit was more mature than coated fruit. This could be related to higher respiration, ethylene production, and weight loss observed in uncoated fruit. Reduced TA in control fruit may be due to a higher respiration rate leading to degradation of organic acids. This was confirmed by reports that organic acids act as substrates for the enzymatic reactions of respiration that result in a reduction of the fruit acidity. Thus, most of the fruits showed a reduced rate of increase in the TSS:TA ratio when AVG was used as coating (Table 22.3).

There are some reports describing the effect of AVG coatings on sensory quality of fruits such as jujube (Padmaja and Bosco, 2014), different species and cultivars of *Prunus* (Martínez-Romero et al., 2006; Paladines et al., 2014; Ravanfar et al., 2014; Hazrati et al., 2017), table grape (Valverde et al., 2005; Ali et al., 2016), papaya (Marpudi et al., 2011), and fig (Marpudi et al., 2013). In most of these reports, a sensory analysis was conducted by 10 to 30 trained judges. Scales of 5 or 6 points have been used, where the assessment was made of different attributes of the fruits (color, crunchiness, juiciness, sweetness, sourness, chewing ability, and global quality) or for the appearance of physiological disorders, browning, or dehydration. None of the fruits was considered to have an objectionable appearance (brightness, appearance of strange color, etc.) after treatment with pure AVG. In the case of a complex coating where rosehip oil was added at 2 to 10% concentration, the fruits were shinier and had a greasier appearance (Paladines et al., 2014). In all cases, when the judges analyzed the whole fruits, those covered with AVG were always rated highest. Also, AVG coatings were effective in reducing dehydration and browning of the rachis in cherries (Martínez-Romero et al., 2006) and table grapes (Valverde et al., 2005; Ali et al., 2016), according to the visual aspect evaluated by panelists.

Edible coatings also affect the content of bioactive compounds with antioxidant potential in fruits, such as ascorbic acid, phenolics, and carotenoids. Ahmed et al. (2009) indicated that during the ripening of nectarine at different temperatures there was an accumulation of ascorbic acid. However, AVG-coated fruit had less ascorbic acid compared to uncoated fruit during ripening. They stated the lower concentration of ascorbic acid in coated nectarines could be due to an increase in the enzymatic activity of cytochrome oxidase, ascorbic oxidase, and peroxidase. Conversely, in other fruits such as strawberry (Sogvar et al., 2016), table grape (Serrano et al., 2006; Shahkoomahally and Ramezanian, 2014), and mandarin (Jiwanit et al., 2018), a continuous loss of vitamin C and antioxidant activity was found in control fruit during storage, while the rate of their loss in fruits coated with AVG was lower. The delay of vitamin C loss could be due to a minor decrease in the autoxidation of ascorbic acid in the presence of O_2. The reduced loss of antioxidant activity could have been due to the reduced accumulation of O_2 inside the fruit, which was attributed to the lower oxygen permeability of the coatings (Ayranci and Tunc, 2003). The antioxidant activity of fruits and vegetables has been correlated with the presence and accumulation of efficient oxygen radical scavengers such as polyphenols, carotenoids, and ascorbic acid, among others, and the activity of antioxidant enzymes. On the other hand, fruits coated with AVG had higher antioxidant activity than uncoated fruits, as observed in grape (Serrano et al., 2006) and raspberry (Hassanpour, 2015). However, in other fruits coated with pure AVG, the total antioxidant activity was not affected, although it increased with the incorporation of ascorbic acid to the coatings (Sogvar et al., 2016). The increase of antioxidant activity in these coated fruits was a consequence of the stimulation of the gene expression of enzymes involved in the biosynthetic pathway of anthocyanin or phenolic compounds, such as PAL in raspberry (Hassanpour, 2015) or avocado (Bill et al., 2014). In addition, raspberry (Hassanpour, 2015) or avocado (Bill et al., 2014) fruits treated with AVG had higher activities of the antioxidant enzymes glutathione-peroxidase, glutathione reductase, superoxide dismutase (SOD), ascorbate peroxidase, catalase (CAT), and guaiacol peroxidase.

6.3.3 Microbial Populations and Fungal Spoilage of Whole Fruit after Harvest

AVG coatings reduced populations of total mesophilic aerobic bacteria and yeast and molds (in CFU/g or CFU/cm^2) during storage of tables grapes (Valverde et al., 2005), sweet cherry (Martínez-Romero et al., 2006), and zucchini (de Bruin et al., 2016). Valverde et al. (2005) observed in table grapes that coatings were more effective in reducing the proliferation of yeast and molds (3-fold reduction compared to control fruit, <2 log CFU/g) than that of mesophilic aerobic bacteria (2-fold reduction compared to control fruit, <3 log CFU/g). However, in sweet cherry (Martínez-Romero et al., 2006), AVG treatments were more effective in reducing populations of mesophilic aerobic bacteria (2.5-fold reduction compared to control fruit, <2 log CFU/g) than yeast and molds (2-fold reduction compared to control fruit, <1.2 log CFU/g).

Fungal decay incidence was reduced after harvest by AVG coatings by 60% in papaya (Marpudi et al., 2011), 45% in fig (Marpudi et al., 2013), 18% in jujube (Padmaja and Bosco, 2014), 12% in raspberry (Hassanpour, 2015), 20% in mango (Muangdech, 2016), and 37% in pepper (Ullah et al., 2017). These reductions have been associated with the gel ability to suppress germination and inhibit mycelial growth due to the presence of more than one bioactive compound (Nabigol and Asghari, 2013).

In addition, it was shown that AVG application increased the resistance of tissues to decay by enhancing their antioxidant systems and free radical scavenging capability. Thus, fruits treated with AVG had higher total phenolic contents and higher activities of the antioxidant enzymes glutathione-peroxidase, glutathione reductase, SOD, ascorbate peroxidase, and guaiacol peroxidase (Hu et al., 2005; Bill et al., 2014). Recently, Jiwanit et al. (2018) reported that the increased activities of these enzymes in mandarin were associated with a stimulation of their gene expression leading to induction of natural defense mechanisms.

6.4 AG Coatings for Minimally Processed Fruit

Production of minimally processed fresh fruit has increased recently due to increasing consumer demand. However, fresh-cut fruits are more difficult to preserve than whole fruits because preservation of internal tissues exposed by processing

Table 22.3 *Aloe* spp. gel coating characteristics and their effects on quality of entire fruits

Plant species and cultivar	*Aloe gel composition*	*Treatment*	*Quality, physiological, and microbiological parameters improved*	*Reference*
Avocado *Persea americana* Mill. cv. 'Hass'	AVGs were prepared at different concentrations. *Aloe vera* (L.) Burm. f. (2%) and thyme oil (1%) [1:1 v/v], [3:1 v/v], AVG spray dried powder from Aloway Natural Health Products Pty, Limpopo, South Africa.	Fruits were dipped into coating solutions for 5 min.	Firmness, color, total phenolic compounds, defense enzymes, antioxidant enzymes, and antioxidant activity.	Bill et al. (2014)
Blueberry *Vaccinium corymbosum* L. cv. 'Duke'	AVG was obtained from fresh *Aloe* leaves. Evaluation of postharvest quality of cold-stored blueberries coated with chitosan-based coating containing AVG fractions. Solution containing 0.5% (w/v) chitosan; 0.5% (v/v) glycerol; 0.1% (w/v) Tween 80; and 0.5% (v/v) AVG liquid fraction.	Fruits were dipped into the coating for 5 min.	Weight loss, total soluble solids (TSS), total acidity (TA), and total phenolics.	Vieira et al. (2016)
Fig *Ficus carica* L.	Pure AVG prepared from fresh *A. vera* leaves.	Fresh fig fruits were dipped into the coating solutions at room temperature for	Weight loss, firmness, color, TSS, TA, and sensory analysis.	Marpudi et al. (2013)

(Continued)

Table 22.3 (Cont.)

Plant species and cultivar	Aloe gel composition	Treatment	Quality, physiological, and microbiological parameters improved	Reference
		5 min. At regular intervals, the fruits were stirred to increase the coating efficiency.	The number of spoiled fruits was controlled.	
Indian jujube *Zizyphus mauritiana* Lam.	Pectin with *Aloe* juice was prepared at three different concentrations: distilled water-AVG at proportions 1:1, 1:2, and 1:3. *Aloe* gel was obtained from PSSGT EXPORT, Tuticorin, Tamil Nadu, India.	Fruits were dipped into different coating solutions for 3, 5, or 7 min.	Weight loss, firmness, color, TSS, TA, ascorbic acid, and sensory analysis. The number of spoiled fruits was controlled. Mesophilic aerobic and yeast and mold colony-forming units (CFU) were reduced.	Padmaja and Bosco (2014)
Mandarin *Citrus reticulata* Blanco cv. 'Shogun'	AVG was obtained from fresh *Aloe* leaves. AVG coating alone or supplemented with *Pichia guilliermondii* BCC5389 (10^8 CFU/mL).	Fruits immersed for 5 min in: water, AVG, AVG + BCC5389, or BCC5389 (10^8 cells/mL).	Weight loss, TSS, TA, and ascorbic acid. Total phenolics, levels of gene transcripts associated to natural defense mechanisms.	Jiwanit et al. (2018)
Mango *Mangifera indica* L.	AVG was obtained from fresh *Aloe* leaves. Different *Aloe* gel coatings at 10, 20, and	Fruits were dipped into coating solutions for 30 s.	Weight loss, firmness, color, TSS, TA, and respiration rate. The number of spoiled fruits was controlled.	Muangdech (2016)

(Continued)

Table 22.3 (Cont.)

Plant species and cultivar	Aloe *gel composition*	*Treatment*	*Quality, physiological, and microbiological parameters improved*	*Reference*
	40% were compared with chitosan coatings at 0.5, 1.0, and 1.5% and with carnauba coatings at 2, 4, and 6%.			
Mango *M. indica* cv. 'Kensington Pride'	AVG (1:1, v/v) or AVG (100%). AVG was obtained as freeze-dried powder from Yunnan Yuanjiang Evergreen Biological Industry (Group) Co., Ltd. (Northern Section of Kunming High-Tech Industrial Development Zone, Kunming, China).	Fruits were manually coated.	Color, lower concentrations of total aroma volatiles.	Dang et al. (2008)
Mushroom *Agaricus bisporus* (J.E. Lange) Imbach	AVG from *A. vera* leaves (30% w/w).	Mushrooms were immersed in a solution of diluted AVG with distilled water (1:3) for 5 min.	Weight loss, firmness, color, and carbohydrate content.	Mohebbi et al. (2012)
Nectarine *Prunus persica* L. Batsch cv. 'Arctic Snow'	AVG freeze dried powder (200:1) was obtained from Yunnan Yuanjiang, Evergreen Biological Industry (Group) Co.,	Fruits were coated with AVG by immersing in an aqueous solution containing AVG for 5 min.	Weight loss, firmness, TSS, and TA. Ascorbic acid, total antioxidants. Reduction of electrolyte leakage, respiration	Ahmed et al. (2009)

(*Continued*)

Table 22.3 (Cont.)

Plant species and cultivar	*Aloe gel composition*	*Treatment*	*Quality, physiological , and microbiological parameters improved*	*Reference*
	Ltd. Solution of 2.5 g AVG freeze dried powder + 0.05% Tween-20 (surfactant) per L of distilled water.		rate, and ethylene production.	
Nectarine *P. persica* cvs. 'Flavela' and 'Flanoba'	Pure AVG prepared from fresh *A.* vera leaves. AVG and AVG plus thymol (1 mL/L).	Fruits were dipped into coating AVG solutions for 10 min.	Weight loss, firmness, TSS, TA, respiration rate, and ethylene production.	Navarro et al. (2011)
Papaya *Carica papaya* L.	Pure gel prepared from fresh *A. vera* leaves. 50% AVG diluted, 50% AVG plus papaya leaf extract (1:1) and 2.5% chitosan.	Fruits were dipped into solutions for 15 min.	Weight loss, firmness, TSS, TA, and sensory parameters. The number of spoiled fruits was controlled.	Marpudi et al. (2011)
Peach *P. persica*	AVG origin not indicated. AVG (which in terms of medicinal quality was pure) diluted with distilled water (1:3).	Fruits were immersed for 5 min.	Weight loss, firmness, color, TSS, TA, and sensory evaluation.	Hazrati et al. (2017)
Peach *P. persica* cv. 'Red Heaven' **Plum** *Prunus salicina* Lindl. cv. 'Santa Rosa'	Pure *A vera* gel and *A. arborescens* gels were obtained from fresh *Aloe* leaves.	Fruits were dipped into the corresponding solution for 10 min.	Weight loss, firmness, color, TSS, TA, respiration rate, and ethylene production.	Guillén et al. (2013)
Peach	AVG was obtained from	Treatments were performed by	Weight loss, firmness, color,	Paladines et al. (2014)

(*Continued*)

Table 22.3 (Cont.)

Plant species and cultivar	Aloe *gel composition*	*Treatment*	*Quality, physiological, and microbiological parameters improved*	*Reference*
P. persica cvs. 'Roma' and 'B-424-16' flat type) **Plum** *P. salicina* cvs. 'Red Beauty' and 'Songria' **Nectarine** *P. persica* cv. 'Garofa' **Sweet cherry** *Prunus avium* L. cv. 'Brooks'	fresh *Aloe* leaves. AVG at 100% or AVG + rosehip oil (2 or 10%).	dipping the fruits into the corresponding solution for 10 min.	TSS, TA, and sensory analysis. Reduced respiration rate.	
Pepper *Capsicum annuum* L. cv. 'Yolo Wonder'	AVG origin not indicated. AVG (4, 5, and 6%).	Fruits were dipped into AVG at different concentrations (4, 5, and 6%), gum arabic (6, 9, and 12%), and cinnamon oil (0.5, 0.75, and 1%) for 5 min.	Weight loss, firmness, color, TSS, TA, ascorbic acid, and chilling injury. Reduced electrolyte leakage and decay incidence.	Ullah et al. (2017)
Plum *Prunus domestica* L. cv. 'President'	*Aloe* gels were obtained from fresh *Aloe* leaves. *A. vera* or *A. arborescens* gel and rosehip oil at 2%.	Fruits were dipped for 10 min.	Weight loss, firmness, color, TSS/TA, bioactive compounds (phenolics, anthocyanins, and carotenoids), and antioxidant activity. Respiration and ethylene production rates were controlled.	Martínez-Romero et al. (2017)

(*Continued*)

Table 22.3 (Cont.)

Plant species and cultivar	Aloe *gel composition*	*Treatment*	*Quality, physiological, and microbiological parameters improved*	*Reference*
Raspberry *Rubus* spp.	AVG origin not indicated. AVG diluted at 1:3, 1:1, and 3:1 with distilled water.	Treatment was performed at 20°C by immersing the fruits for 5 min in the solution of AVG.	TSS, TA, total anthocyanin, total phenolics, antioxidant capacity, and antioxidant enzymes. Reduced decay incidence.	Hassanpour (2015)
Sapota *Manilkara zapota* L. (P. Royen)	AVG mixed with 1.5% pectin at 60°C for gelation was prepared at three different concentrations (1:1, 1:2, and 1:3). *Aloe* gel was obtained from PSSGT EXPORT.	For each concentration, fruits were immersed for two different periods (5 and 7 min).	Weight loss, firmness, color, TSS, TA, and ascorbic acid.	Padmaja et al. (2015)
Sour cherry *Prunus cerasus* L.	Pure, untreated AVG obtained from fresh *Aloe* leaves.	Fruits were dipped.	Weight loss, firmness, color, TSS, TA, ascorbic acid, and sensory analysis. Reduced respiration rate.	Ravanfar et al. (2014)
Strawberry *Fragaria x ananassa* Duch. cv. 'Parous'	AVG was obtained from fresh *Aloe* leaves. AVG was diluted (1:3) with distilled water in combination with 0, 1, 3, or 5% ascorbic acid.	Fruits were dipped at 20°C for 5 min in AVG diluted with distilled water (1:3) in combination with ascorbic acid.	Weight loss, firmness, TSS, TA, ascorbic acid, and total anthocyanin, phenolics, and antioxidant activity. Mesophilic aerobic and yeast and mold CFU were reduced.	Sogvar et al. (2016)

(*Continued*)

Table 22.3 (Cont.)

Plant species and cultivar	Aloe *gel composition*	*Treatment*	*Quality, physiological, and microbiological parameters improved*	*Reference*
Sweet cherry *P. avium* cv. 'StarKing'	AVG (pharmaceutical quality, 100% purity) manufactured by Roig Farma S.A. (Terrassa, Barcelona, Spain).	Treatment was performed by immersion of fruits for 5 min in a solution of AVG diluted with distilled water (1:3).	Weight loss, firmness, color, TSS, TA, sensory evaluation (whole fruit and stem). Mesophilic aerobic and yeast and mold CFU were reduced.	Martínez-Romero et al. (2006)
Table grape *Vitis vinifera* L. cv. 'Thompson Seedless'	AVG was obtained from fresh *Aloe* leaves and applied at 10, 20, and 40%.	Grapes were dipped for 5 min.	Weight loss, color, rachis appearance, and sensorial quality.	Ali et al. (2016)
Table grape *V. vinifera* cv. 'Askari'	AVG origin not indicated. AVG combined with calcium chloride (2%) or citric acid (1%).	Fruits were immersed for 5 min in a solution of AVG diluted with distilled water (1:3).	Weight loss, TSS, TA, ascorbic acid, and rachis browning. Reduced decay index.	Shahkoomahally and Ramezanian (2014)
Table grape *V. vinifera* cv. 'Autum Royal'	AVG (food grade) was purchased from AVISA (Aloe Vera Internacional Sociedad Anónima, Fuerteventura, Canary Islands, Spain). AVG was diluted at 1:3, for preharvest treatment.	Preharvest spray treatments were conducted once (1 d before harvest) or twice (1 and 7 d before harvest).	Firmness, color, respiration rate, and ethylene production. Mesophilic aerobic and yeast and mold CFU were reduced.	Castillo et al. (2010)
Table grape *V. vinifera* cv. 'Crimson Seedless'	AVG (pharmaceutical quality, 100% purity) manufactured by Roig Farma S.A.	Treatment was performed by dipping for 5 min into AVG diluted with	Weight loss, firmness, color, TSS, TA, sensory quality (whole fruit and raquis),	Serrano et al. (2006); Valverde et al. (2005)

(*Continued*)

Table 22.3 (Cont.)

Plant species and cultivar	Aloe *gel composition*	*Treatment*	*Quality, physiological , and microbiological parameters improved*	*Reference*
		distilled water (1:3).	ascorbic acid, and total antioxidant activity, phenolics, and anthocyanins. Reduced respiration rate. Mesophilic aerobic and yeast and mold CFU were reduced.	
Tomato *Lycopersicon sculentum* Mill.	AVG was obtained from fresh *Aloe* leaves Shellac and AVG	Fruits were dipped into the AVG solution for 5 min at 25°C	Weight loss, firmness, color, TSS, respiration rate, and ethylene production.	Chauhan et al. (2015)
Tomato *L. sculentum* cv. 'Charleston'	AVG (pharmaceutical quality, 100% purity) manufactured by Pharmaceutical Laboratories Mario Muñoz (La Havana, Cuba) Aqueous extract (100% purity) or diluted (50%).	The treatment was performed at 22°C by double immersion for 2 min in film forming solutions.	No effect on weight loss, firmness, TSS, or TA. Growth of *Rhizoctonia solani, Alternaria alternata*, and *Phytophthora parasitica* was reduced.	García et al. (2014)
Tomato *L. esculentum* cv. 'Ruchi 618'	AVG based edible coating was formed by mixing AVG (500 ml), 0.3% antioxidant rich herb, thickening agent (20 g), glycerol (2%) for plasticizing effect, oleic acid (3 ml) and cinnamaldehyde	Tomatoes were coated by dipping them into the coating solutions.	Weight loss, firmness, color, TSS, TA, ascorbic acid, and lycopene.	Athmaselvi et al. (2013)

(*Continued*)

Table 22.3 (Cont.)

Plant species and cultivar	Aloe *gel composition*	*Treatment*	*Quality, physiological, and microbiological parameters improved*	*Reference*
	(0.2 ml) dissolved in tween 80 at the ratio of 1:1.			
Wax apple *Syzygium samarangense* (Blume) Merr. & L.M. Perry	AVG was obtained from fresh *Aloe* leaves and diluted to 50, 75, and 100% with distilled water.	Fruits were dipped into the coating solutions for 10 min.	Weight loss, TA, ascorbic acid, and total anthocyanins.	Mubarak and Rao Engakanah (2017)
Zucchini *Cucurbita pepo* L.	Commercial AVG concentrate was purchased from a local pharmaceutical retail store and diluted at 1:3 ratio with distilled water.	Fruits were dipped into the AVG solution for 5 min at 25°C.	Weight loss, firmness, color, TSS, and total anthocyanins, phenolics, and antioxidant activity.	de Bruin et al. (2016)

requires a combination of appropriate strategies that extend shelf life while maintaining fruit sensory and organoleptic properties. Fresh-cut fruits are expected to maintain fresh-like quality during their shelf life. However, processing operations such as washing, sanitizing, peeling, cutting, dicing, and slicing can cause serious damage to plant tissues. The rupture of cell components induces biochemical reactions related to changes in color (browning and chlorophyll degradation), texture loss, vitamin oxidation, and/or accelerated metabolism (Treviño-Garza et al., 2017). Furthermore, fruit juice leakage stimulates microbial growth, which may accelerate spoilage or the creation of foodborne microbial hazards (Rico et al., 2007). For these reasons, the shelf life of minimally processed fruit tends to be short, and extending it by even few days could represent a considerable marketing advantage (Manzocco et al., 2011).

Regarding whole fruits, the use of edible coatings can extend the shelf life by forming a gas barrier, causing a reduction in respiration and weight loss (Baldwin, 2002). These coatings also play an important role in preserving other fruit properties such as texture, color, soluble solids, and acid content, as well as in avoiding enzymatic browning, loss of vitamin C, and microbial spoilage (Oms-Oliu et al., 2008; Falguera et al., 2011).

Considering the increasing demand of fresh-cut fruits by consumers and processors, as well as dietary safety and environmental concerns, the use of safe technologies and additives based on natural compounds could be an alternative for the preservation of fresh-cut fruits and vegetables. Thus, the application of edible coatings possessing inherent antioxidant and/or antimicrobial properties would be a very promising approach for safety and quality improvement of minimally processed fruits and vegetables.

Researchers claim that AVG is a good potential edible coating due to its transparency, eco-friendly characteristics, and tasteless nature. The effectiveness of AVG as a coating material for minimally processed fruit has been evaluated in a wide range of fruit species (Table 22.4). Usually, all the studies were performed by applying AVG obtained from fresh *A. vera* leaves in concentrations from 5 to 100%, the lowest concentrations of *A. vera* being used when it was obtained from commercial AVG products (Benítez et al., 2013). Glycerin, glycerol, and oleic acid appear to be important additives to increase plasticizer properties of AVG coatings when used at concentrations between 0.5 and 2%, depending on the product (Panwar et al., 2016; Treviño-Garza et al., 2017). On the other hand, to control non-enzymatic and enzymatic browning in fresh-cut fruits, it seems that AVG benefits from the addition to the gel of low concentrations of anti-browning agents such as ascorbic acid, citric acid, or cysteine (Table 22.4). Additions of organic acids have been used to maintain the pH of AVG formulations (Martínez-Romero et al., 2013a; Pérez et al., 2016).

6.4.1 Physiological Parameters: Weight Loss, Respiration Rate, Ethylene Emission, and Ion Leakage

Fresh-cut fruits can easily suffer rapid weight loss due to the removal of skin that exposes the fruit internal tissues (Watada and Qi, 1999). However, fresh-cut fruits coated with AVG that lost less weight after storage include apple (Song et al., 2013; Amiri et al., 2018), guava (Nasution et al., 2015), mango (Pérez et al., 2016), and papaya (Treviño-Garza et al., 2017). This effect could be due to the hygroscopic properties of AVG, which forms a water vapor barrier between the fruit tissues and the environment to retard water loss (Valverde et al., 2005).

Application of AVG edible coatings alone or in combination with other additives influenced respiratory behavior of fruit slices. Cutting increases the respiration rate of fresh commodities and, therefore, this reduction could extend shelf life (Chauhan et al., 2011). AVG coatings have significantly suppressed respiration compared to uncoated fruit slices (Benítez et al., 2013; Pérez et al., 2016). Similar results were obtained with minimally processed table grape berries and pomegranate arils that were coated with AVG (Martínez-Romero et al., 2013a; Alberio et al., 2017).

On the other hand, the application of AVG surface coatings on fresh-cut apples significantly influenced the pattern of ethylene synthesis, causing a delay in the peak period of ethylene synthesis compared to uncoated samples. This resulted in longer retention of the quality of coated apple slices (Chauhan et al., 2011).

Fresh-cut commodities tend to lose electrolytes because their cell membranes senescence at faster rates compared to those of undamaged tissues (Chauhan et al., 2006a). In this sense, when apple slices were coated with AVG, lower electrolyte leakage was observed as compared to uncoated ones (Chauhan et al., 2011), and a reduction in the rate of senescence of the apple slices occurred. The reduced electrolyte leakage from the surfaces of the samples could be ultimately attributed to the reduced respiratory outputs caused by the application of surface coatings, as previously described in this chapter (Chauhan et al., 2006b).

6.4.2 Physical, Chemical, and Sensory Quality Parameters

The color of fruit slices is an important organoleptic factor that affects the product acceptability by consumers (Olivas and Barbosa-Canovas, 2005). Several studies indicated that AVG application to fresh-cut commodities did not substantially modify the color of uncoated fresh-cut fruits (Table 22.4).

The Browning Index (BI) was previously used as an indicator of browning intensity during storage (Olivas et al., 2007). BI of several products increased rapidly during the first hours after cutting. In general, the presence of an edible coating retards enzymatic browning by providing an oxygen barrier. Chauhan et al. (2011) reported that the color parameters L^*, a^*, and b^* in AVG-coated apple slices did not significantly change compared to control slices during 30 d of storage at 6°C, suggesting an anti-browning functionality of the AVG coating. However, the coating did not completely prevent oxidative browning (Tien et al., 2001). In several studies to inhibit enzymatic browning, cysteine and other anti-browning agents were added to AVG coating solutions (Table 22.4). Furthermore, the oxidative degradation of ascorbic acid and other non-enzymatic browning processes were reported to be major deteriorative reactions occurring during storage (Wibowo et al., 2015). Thus, the addition to AVG of anti-browning agents such as ascorbic acid, citric acid, cysteine, sodium benzoate, polysorbates, and inhibitors of polyphenol oxidase (PPO), such as calcium chloride (Chauhan et al., 2011; Song et al., 2013; Kuwar et al., 2015; Pérez et al., 2016; Amiri et al., 2018), led to better results than when AVG was applied alone. Wounding initiates enzymatic browning, caused by the mixing of PPO with phenolic compounds, which consequently impairs nutritional and sensory attributes of fresh-cut products (He and Luo, 2007). The increase in PPO activity in uncoated fresh-cut fruits during storage was concomitant with the increase in browning incidence (Chauhan et al., 2011; Supapvanich et al., 2016; Alberio et al., 2017). AVG coating greatly inhibited the activity of PPO, peroxidase (POD), and proline oxidase (POX) in all reports, which could be attributed to the respiratory gas barrier properties of AVG coatings (Martínez-Romero et al., 2006).

It is widely recognized that the evolution of total phenolic content and antioxidant activity of fruits during storage differs depending on the cultivar and climatic and environmental conditions during their growth (Kalt, 2005). A significant increase in the phenolic content in uncoated control fresh-cut wax apples and minimally processed table grapes was reported (Serrano et al., 2006; Supapvanich et al., 2016). In contrast, higher levels of total phenols were found in pomegranates arils and fresh-cut papaya coated with AVG (Martínez-Romero et al., 2013a; Kuwar et al., 2015). Also, the content of total anthocyanins in minimally processed pomegranate arils treated with AVG increased significantly during cold storage compared with that of control arils (Martínez-Romero et al., 2013a).

With respect to carotenoid content in minimally processed carrots, samples coated with AVG retained higher carotenoid content compared to uncoated peeled carrots. This might be attributed to the modified atmosphere produced by the edible coating, which was a barrier to O_2 and CO_2 diffusion. Thus, autoxidation potential of carotenes decreased with the application of AVG edible coatings (Panwar et al., 2016).

Table 22.4 *Aloe* coating characteristics and their effects on quality of fresh-cut fruits

Plant species and cultivar	Aloe *treatment composition*	*Treatment*	*Quality, physiological, and microbiological parameters improved*	*Reference*
Apple *Malus x domestica* Borkh cv. 'Red Delicious'	AVG was from fresh *Aloe vera* (L.) Burm. f. leaves. AVG coating solutions were prepared with gelatin (1%), citric acid (0.1%), calcium chloride (0.5%), with or without ascorbic acid (0.5%).	Fresh-cut apple slices were sliced in chilled water and then dipped into AVG coating solution for 1 min. Apple slices were placed in polyethylene terephthalate clamshells.	Weight loss, firmness, color, total acidity (TA), total soluble solids (TSS), and sensory parameters. Total aerobic counts were reduced.	Amiri et al. (2018)
Apple *Malus x domestica* cv. 'Hongro'	AVG was from fresh *A. vera* leaves. Apple slices were pretreated separately with anti-browning solutions at 0.5% each: cysteine, citric acid, and ascorbic acid. Coating AVG solutions (50%) including cysteine (0.5%).	Fresh-cut apple slices (2 cm) were dipped for 2 min. Then slices were packed in low density polyethylene packaging film.	Weight loss, firmness, Browning Index (BI), TA, and sensory parameters. Molds and yeasts counts and total aerobic bacteria were reduced.	Song et al. (2013)
Apple *Malus x domestica* cv. 'Fuji'	AVG was from fresh *A. vera* leaves and blended with shellac (1:1). Fruit slices were subjected to a cold shock after slicing in chilled water for 5 min and then	Apple slices were coated by dipping them into the AVG coating solution for 2 min.	Respiration, ethylene, color, electrolytic leakage, weight loss, and firmness. Polyphenol oxidase (PPO) and peroxidase (POD) were reduced. Total plate counts, coliforms, molds	Chauhan et al. (2011)

(*Continued*)

Table 22.4 (Cont.)

Plant species and cultivar	Aloe *treatment composition*	*Treatment*	*Quality, physiological, and microbiological parameters improved*	*Reference*
	ozonized. The ozonized slices were soaked in an additive solution containing ascorbic acid, citric acid, and sodium benzoate (200 mg/kg per additive). Oleic acid was used as emulsifier.		and yeasts counts were reduced.	
Carrot *Daucus carota* L.	AVG was from fresh *A. vera* leaves and small cubes of mucilaginous material were heated for 10 min at 80°C. AVG solutions (5, 10, and 15%) containing carboxymethyl cellulose (0.5%), peanut oil (5%), and glycerol (2%) were homogenized for 30 s in hot water.	Carrots were washed, peeled, coated with AVG solutions, and packed in low density polyethylene. Dipping time not specified.	TSS, total carotenoids. Permeability to O_2, CO_2, and water vapor were reduced by AVG coatings. Microflora growth was reduced.	Panwar et al. (2016)
Guava *Psidium guajava* L. cv. 'Kampuchea'	AVG was from fresh *A. vera* leaves. Flesh AVG was blanched in hot water at 100°C. After blending, AVG was filtered through activated carbon, the pH was adjusted to	Guava fruits were cut into 2–2-1.5 cm peeled cubes and coated by dipping for 5 min.	Weight loss, color, firmness, ascorbic acid, and sensory parameters. Molds and yeasts counts and aerobic plate counts were reduced.	Nasution et al. (2015)

(*Continued*)

Table 22.4 (Cont.)

Plant species and cultivar	Aloe *treatment composition*	*Treatment*	*Quality, physiological, and microbiological parameters improved*	*Reference*
	3.0 with citric acid, and pasteurized at 85°C for 1 min. Glycerine was added as plasticizer.			
Kiwifruit *Actinidia deliciosa* C.F. Liang & A.R. Ferguson cv. 'Hayward'	Commercial AVG (food grade) diluted at 1, 5, and 15%.	Fresh-cut kiwifruit slices (6 mm) were peeled, dipped for 10 min, and packed on sealed polypropylene trays.	Respiration, color, firmness and sensory parameters. Pectin depolymerization was reduced. Molds, yeasts, and total mesophilic bacterial counts were reduced.	Benítez et al. (2013)
Mango *Mangifera indica* L. cv. 'Tommy Atkins'	AVG was from fresh *A. vera* leaves (*Aloe barbadensis* Mill.)	Fresh-cut mango slices were cut into 2 cm side cubes and dipped for 30 s into a solution containing pure AVG (50%), carnauba wax (0.1%), polysorbate 80 (0.01%), and glycerol (1%). Slices were previously dipped or not into citric acid (1%) and ascorbic acid (1%) for 2 min.	Respiration, weight loss, firmness, color, TA, TSS, and sensory parameters. Molds, yeasts, and total aerobic bacteria counts were reduced.	Pérez et al. (2016)

(Continued)

Table 22.4 (Cont.)

Plant species and cultivar	Aloe *treatment composition*	*Treatment*	*Quality, physiological, and microbiological parameters improved*	*Reference*
Melon *Cucumis melo* L.	AVG solutions (33 and 20%) alone or in combination with salicylic acid at 0.1 and 2 mmol/L.	Fresh-cut melon slices were dipped for 3 min.	pH, TSS, TA, ripening index, and ascorbic acid.	Riaie et al. (2017)
Papaya *Carica papaya* L. cv. 'Pusa Delicious'	AVG was from fresh *A. vera* leaves. Pure AVG was pasteurized at 70°C for 45 min. Coating pH was adjusted at 5.6 containing 1.5% glycerol.	Papayas were cut into 1–1.5 cm side peeled cubes, with or without a previous immersion for 3 min in a chemical solution containing 2% $CaCl_2$ + 1% ascorbic acid + 0.12% vanillin.	Weight loss, BI, TSS, ascorbic acid, and sensory parameters. Total polyphenol content increased and PPO and POX activity were delayed. Molds, yeasts, and total mesophilic bacterial counts were reduced.	Kuwar et al. (2015)
Pineapple *Ananas comosus* (L.) Merr.	AVG was from fresh *A. vera* leaves. Combination of *Aloe* solution (50% *A. vera* + 0.5% glycerol) with a chitosan solution (1.5% + 0.5% glycerol).	Fresh-cut pineapple slices were cut into 2 cm side cubes and dipped for 2 min.	Weight loss, firmness, ripening index, color, decay incidence, ascorbic acid, and sensory parameters. Molds and yeasts, total aerobic and psychrotrophic counts were reduced.	Treviño-Garza et al. (2017)
Pomegranate *Punica granatum* L. cv. 'Mollar de Elche'	AVG was from fresh *A. vera* leaves. Pure AVG (*A. barbadensis*)	Minimally processed arils were dipped for 5 min into AVG solutions containing	Respiration, firmness, color, TSS, TA, ripening index, and sensory parameters. Total phenolics	Martínez-Romero et al. (2013a)

(*Continued*)

Table 22.4 (Cont.)

Plant species and cultivar	Aloe *treatment composition*	*Treatment*	*Quality, physiological, and microbiological parameters improved*	*Reference*
	diluted at 50 and 100%.	citric acid and ascorbic acid (0.5 or 1% each) and then stored into rigid polypropylene boxes.	and anthocyanins increased. Molds and yeasts and total mesophilic bacterial counts were reduced.	
Table grape *Vitis vinifera* L. cvs. 'Sugar One', 'Victoria', and 'Black Magic'	Commercial AVG was used. Pure AVG (*A. barbadensis*).	Minimally processed table grapes under MAP conditions.	BI, sensory parameters, and respiration rate. PPO, pectin methylesterase (PME), and β-galactosidase (β-Gal) activities were reduced.	Alberio et al. (2017)
Wax apple *Syzygium samarangenese* (Blume) Merrill & L.M. Perry cv. 'Taaptimjaan'	Homogenates of AVG (*A. barbadensis*). AVG diluted at 25, 50, 75, and 100%.	Fresh-cut apple wax slices were dipped for 2 min.	Color, whitening, BI, and sensory parameters. Increases in polyphenol content and PPO activity were delayed.	Supapvanich et al. (2016)

Regarding ascorbic acid content, it has been reported that AVG-coated fresh-cut pineapples, kiwifruits, and guavas had higher ascorbic acid contents than control fruits (Benítez et al., 2015; Nasution et al., 2015; Treviño-Garza et al., 2017). Accordingly, an increase of AVG concentration increased ascorbic acid content in fresh-cut oranges (Radi et al., 2017). This can be attributed to the low gas permeability of AVG coating, which reduced the respiratory rate (Hassanpour, 2015).

7 Concluding Remarks

Gels of *Aloe* ssp., applied alone or in combination with other compounds, have been employed as edible coatings to reduce postharvest decay and improve the physicochemical, sensorial, and microbiological properties of fresh produce, leading to an extension of the shelf life of either whole or fresh-cut fruits. However, a commercial coating based on *Aloe* spp. gels for postharvest use is not available yet.

In general, fruits coated with AVG lost less weight, were firmer, and had higher ascorbic acid and TA contents. Other attributes include retention of fruit color, resistance to CI, higher TSS and sugar contents, and reduced membrane leakage as they aged. Microbe populations were reduced, as was postharvest decay in both in vitro and in vivo experiments, where both preventive and curative antifungal activity were evaluated. Authors of most of the extensive research on these natural coatings stated they were promising eco-friendly postharvest tools for use by the fresh produce industry.

References

Ahlawat, K.S. and Khatkar, B.S. 2011. Processing, food applications and safety of *Aloe vera* products, a review. *Journal of Food Science and Technology* 48, 525–533.

Ahmed, M.J., Singh, Z. and Khan, A.S. 2009. Postharvest *Aloe vera* gel-coating modulates fruit ripening and quality of 'Arctic Snow' nectarine kept in ambient and cold storage. *International Journal of Food Science and Technology* 44, 1024–1033.

Alberio, G.R.A., Muratore, G., Licciardello, F., Giardina, G. and Spagna, G. 2017. *Aloe vera* extract as a promising treatment for the quality maintenance of minimally-processed table grapes. *Food Science and Technology. Campinas* 35, 299–306.

Ali, J., Pandey, S., Singh, V. and Joshi, P. 2016. Effect of coating of *Aloe vera* gel on shelf life of grapes. *Current Research in Nutrition and Food Science* 4, 58–68.

Ali, M.I.A., Shalaby, N.M.M., Elgamal, M.H.A. and Mousa, A.S.M. 1999. Antifungal effects of different plant extracts and their major components of selected *Aloe* species. *Phytotherapy Research* 13, 401–407.

Amiri, S., Akhavan, H.R., Zare, N. and Radi, M. 2018. Effect of gelatin-based edible coatings incorporated with *Aloe vera* and green tea extracts on the shelf-life of fresh-cut apple. *Italian Journal of Food Science* 30, 61–74.

Anonymous. 2013. The Plant List, Version 1.1, 2013. Available at: www.theplantlist.org

Athmaselvi, K.A., Sumitha, P. and Revathy, B. 2013. Development of *Aloe vera* based edible coating for tomato. *International Agrophysics* 27, 369–375.

Ayranci, E. and Tunc, S. 2003. A method for the measurement of the oxygen permeability and the development of edible films to reduce the rate of oxidative reactions in fresh foods. *Food Chemistry* 80, 423–431.

Baldwin, E.A. 2002. Coatings and other supplemental treatments to maintain vegetable quality respiration and gas exchange. pp. 413–435. *In*: J.A. Bartz and J.K. Brecht (eds.). *Postharvest Physiology and Pathology of Vegetables*. Marcel Dekker, New York, USA.

Banks, N.H., Dadzie, B.K. and Cleland, D.J. 1993. Reducing gas exchange of fruits with surface coatings. *Postharvest Biology and Technology* 3, 269–284.

Benítez, S., Achaerandio, I., Pujol, M. and Sepulcre, F. 2015. *Aloe vera* as an alternative to traditional edible coatings used in fresh cut fruits, A case of study with kiwifruit slices. *LWT – Food Science and Technology* 61, 184–193.

Benítez, S., Achaerandio, I., Sepulcre, F. and Pujolà, M. 2013. *Aloe vera* based edible coatings improve the quality of minimally processed 'Hayward' kiwifruit. *Postharvest Biology and Technology* 81, 29–36.

Bill, M., Sivakumar, D., Korsten, L. and Thompson, A.K. 2014. The efficacy of combined application of edible coatings and thyme oil in inducing resistance components in avocado (*Persea americana* Mill.) against anthracnose during postharvest storage. *Crop Protection* 64, 159–167.

Bozzi, A., Perrin, C., Austin, S. and Arce Vera, F. 2007. Quality and authenticity of commercial *Aloe vera* gel powders. *Food Chemistry* 103, 22–30.

Bustamante, C.A., Budde, C.O., Borsani, J., Lombardo, V.E., Lauxmann, M.A., Andreo, C.S., Lara, M.V. and Drincovich, M.F. 2012. Heat treatment of peach fruit, Modifications in the extracellular compartment and identification of novel extracellular proteins. *Plant Physiology and Biochemistry* 60, 35–45.

Callegarin, F., Quezada Gallo, J.A., Debeaufort, F. and Voilley, A. 1997. Review, lipids and biopackaging. *Journal of the American Oil Chemists' Society* 74, 1183–1192.

Castillo, S., Navarro, D., Zapata, P.J., Guillén, F., Valero, D., Serrano, M. and Martínez-Romero, D. 2010. Antifungal efficacy of *Aloe vera in vitro* and its use as a preharvest treatment to maintain postharvest table grape quality. *Postharvest Biology and Technology* 57, 183–188.

Chauhan, O.P., Nanjappa, C., Ashok, N., Ravi, N., Roopa, N. and Raju, P.S. 2015. Shellac and *Aloe vera* gel based surface coating for shelf life extension of tomatoes. *Journal of Food Science and Technology* 52, 1200–1205.

Chauhan, O.P., Raju, P.S., Ashok, N., Dasgupta, D.K. and Bawa, A.S. 2006a. Development of edible surface coatings based on shellac and Aloe vera gel. pp. 16–17. *In*: AFSTI (eds.). Proceedings of Indian Convention of Food Scientists and Technologists. Hyderabad, India.

Chauhan, O.P., Raju, P.S., Shylaja, R., Dasgupta, D.K. and Bawa, A.S. 2006b. Synergistic effects of modified atmosphere and minimal processing on the keeping quality of pre-cut papaya (*Carica papaya* L.). *Journal of Horticultural Science and Biotechnology* 81, 903–909.

Chauhan, O.P., Raju, P.S., Singh, A. and Bawa, A.S. 2011. Shellac and aloe-gel-based surface coatings for maintaining keeping quality of apple slices. *Food Chemistry* 126, 961–966.

Chen, W., Van Wykb, B.E., Vermaak, I. and Viljoen, A.M. 2012. Cape aloes - A review of the phytochemistry, pharmacology and commercialisation of *Aloe ferox*. *Phytochemistry Letters* 5, 1–12.

Crisosto, C.H., Mitchell, F.G. and Johnson, S. 1995. Factors in fresh market stone fruit quality. *Postharvest News and Information* 6, 17–21.

Dagne, E., Bisrat, D., Viljoen, A. and Van Wyk, B.E. 2000. Chemistry of *Aloe* species. *Current Organic Chemistry* 4, 1055–1078.

Dang, K.T.H., Singh, Z. and Swinny, E.E. 2008. Edible coatings influence fruit ripening, quality, and aroma biosynthesis in mango fruit. *Journal of Agricultural and Food Chemistry* 56, 1361–1370.

de Bruin, W., Rossouw, W. and Korsten, L. 2016. Comparison of safe alternative dipping treatments to maintain quality of zucchini. *Journal of Food Quality* 39, 109–115.

de Wild, H.P.J., Balk, P.A., Fernandes, E.C.A. and Peppelenbos, H.W. 2005. The action site of carbon dioxide in relation to inhibition of ethylene production in tomato fruit. *Postharvest Biology and Technology* 36, 273–280.

Debeaufort, F. and Voilley, A. 2009. Lipid-based edible films and coatings. pp. 135–169. *In*: M.E. Embuscado and K.C. Kerry Huber (eds.). *Edible Films and Coatings for Food Applications*. Springer, Dordrecht, The Netherlands.

Dhall, R.K. 2013. Advances in edible coatings for fresh fruits and vegetables, a review. *Critical Reviews in Food Science and Nutrition* 53, 435–450.

Embuscado, M.E. and Huber, K.C. 2009. *Edible Films and Coatings for Food Applications*. Springer, Dordrecht, The Netherlands.

Eshun, K. and He, Q. 2004. *Aloe vera*, a valuable ingredient for the food, pharmaceutical and cosmetic industries: A review. *Critical Reviews in Food Science and Nutrition* 44, 91–96.

Falguera, V., Quintero, J.P., Jiménez, A., Muñoz, J.A. and Ibarz, A. 2011. Edible films and coatings: structures, active functions and trends in their use. *Trends in Food Science & Technology* 22, 292–303.

Fanalli, S., Aturki, Z., D'Orazio, G., Rocco, A., Ferranti, A., Mercolini, L. and Raggi, M.A. 2010. Analysis of Aloe-based phytotherapeutic products by using nano-LC–MS. *Journal of Separation Science* 33, 2663–2670.

Flores-López, M.L., Romaní, A., Cerqueira, M.A., Rodríguez-García, R., Jasso de Rodríguez, D. and Vicente, A.A. 2016. Compositional features and bioactive properties of whole fraction from *Aloe vera* processing. *Industrial Crops and Products* 91, 179–185.

García, M.A., Ventosa, M., Díaz, R., Falco, S. and Casariego, A. 2014. Effects of *Aloe vera* coating on postharvest quality of tomato. *Fruits* 69, 117–126.

Gómez-Galindo, F., Herppich, W., Gekas, V. and Sjöholm, I. 2004. Factors affecting quality and postharvest properties of vegetables, integration of water relations and metabolism. *Critical Reviews in Food Science and Nutrition* 44, 139–154.

Guillén, F., Díaz-Mula, H.M., Zapata, P.J., Valero, D., Serrano, M., Castillo, S. and Martínez-Romero, D. 2013. *Aloe arborescens* and *Aloe vera* gels as coatings in delaying postharvest ripening in peach and plum fruit. *Postharvest Biology and Technology* 83, 54–57.

Hassanpour, H. 2015. Effect of *Aloe vera* gel coating on antioxidant capacity, antioxidant enzyme activities and decay in raspberry fruit. *LWT-Food Science and Technology* 60, 495–501.

Hazrati, S., Kashkooli, A.B., Habibzadeh, F., Tahmasebi-Sarvestani, Z. and Sadeghi, A.R. 2017. Evaluation of *Aloe vera* gel as an alternative edible coating for peach fruits during cold storage period. *Gesunde Pflanzen* 69, 131–137.

He, Q., Changhong, L., Kojo, E. and Tian, Z. 2005. Quality and safety assurance in the processing of *Aloe vera* gel juice. *Food Control* 16, 95–104.

He, Q. and Luo, Y. 2007. Enzymatic browning and its control in fresh-cut produce. *Stewart Postharvest Review* 3, 1–7.

Hu, Q., Hu, Y. and Xu, J. 2005. Free radical-scavenging activity of *Aloe vera* (*Aloe barbadensis* Miller) extracts by supercritical carbon dioxide extraction. *Food Chemistry* 91, 85–90.

IASC (International Aloe Science Council). 2015. *Aloe vera* quality standard. Available at: www.iasc.org/Portals/19/Documents/Certification/AloeVeraQualityStandard.pdf?ver=2015-09-01-113340-987.

Jasso de Rodríguez, D., Hernández-Castillo, D., Rodríguez-García, R. and Angulo-Sánchez, J.L. 2005. Antifungal activity in vitro of *Aloe vera* pulp and liquid fraction against plant pathogenic fungi. *Industrial Crops and Products* 21, 81–87.

Jiwanit, P., Pitakpornpreecha, T., Pisuchpen, S. and Leelasuphakul, W. 2018. The use of *Aloe vera* gel coating supplemented with *Pichia guilliermondii* BCC5389 for enhancement of defense-related gene expression and secondary metabolism in mandarins to prevent postharvest losses from green mold rot. *Biological Control* 117, 43–51.

Kader, A.A. 1986. Biochemical and physiological basis for effects of controlled and modified atmospheres on fruits and vegetables. *Food Technology* 40, 99–104.

Kalt, W. 2005. Effects of production and processing factors on major fruit and vegetables antioxidants. *Journal of Food Science and Technology* 47, 4639–4644.

Khan, A.S. and Ali, S. 2018. Preharvest sprays affecting shelf life and storage potential of fruits. pp. 209–255. *In*: M.W. Siddiqui (ed.). *Preharvest Modulation of Postharvest Fruit and Vegetable Quality*. Academic Press, Elsevier, London, UK.

Kuwar, U., Sharma, S. and Tadapaneni, V.R.R. 2015. *Aloe vera* gel and honey-based edible coatings combined with chemical dip as a safe means for quality maintenance and shelf life extension of fresh-cut papaya. *Journal of Food Quality* 38, 347–358.

Lachenmeier, K., Kuepper, U., Musshoff, F., Burkhard Madea, B., Helmut Reusch, H. and Lachenmeier, D.W. 2005. Quality control of *Aloe vera* beverages. *Electronic Journal of Environmental, Agricultural and Food Chemistry* 4, 1033–1042.

Lee, J.Y., Parka, H.J., Lee, C.Y. and Choi, W.Y. 2003. Extending shelf-life of minimally processed apples with edible coatings and antibrowning agents. *Lebensmittel-Wissenschaft und Technologie* 36, 323–329.

Li, H. and Yu, T. 2001. Effect of chitosan on incidence of brown rot, quality and physiological attributes of postharvest peach fruit. *Journal of the Science of Food and Agriculture* 81, 269–274.

Maftoonazad, N. and Ramaswamy, H.S. 2005. Postharvest shelf-life extension of avocados using methyl cellulose – Based coating. *Lebensmittel-Wissenschaft Und Technologie* 38, 617–624.

Manvitha, K. and Bidya, B. 2014. *Aloe vera*, a wonder plant its history, cultivation and medicinal uses. *Journal of Pharmacognosy and Phytochemistry* 2, 85–88.

Manzocco, L., Foschia, M., Tomasi, N., Maifreni, M., Dalla Costa, L., Marino, M., Cortella, G. and Cesco, S. (2011). Influence of hydroponic and soil cultivation on quality and shelf life of ready-to-eat lamb's lettuce (*Valerianella locusta* L. Laterr). *Journal of the Science of Food and Agriculture* 8, 1373–80.

Marpudi, S.L., Abirami, L.S.S., Pushkala, R. and Srividya, N. 2011. Enhancement of storage life and quality maintenance of papaya fruits using *Aloe vera* based antimicrobial coating. *Indian Journal of Biotechnology* 10, 83–89.

Marpudi, S.L., Ramachandran, P. and Srividya, N. 2013. *Aloe vera* gel coating for postharvest quality maintenance of fresh fig fruits. *Research Journal of Pharmaceutical, Biological and Chemical Sciences* 4, 878–887.

Martínez-Romero, D., Alburquerque, N., Valverde, J.M., Guillén, F., Castillo, S., Valero, D. and Serrano, M. 2006. Postharvest sweet cherry quality and safety maintenance by *Aloe vera* treatment: A new edible coating. *Postharvest Biology and Technology* 39, 93–100.

Martínez-Romero, D., Castillo, S., Guillén, F., Díaz-Mula, H.M., Zapata, P.J., Valero, D. and Serrano, M. 2013a. *Aloe vera* gel coating maintains quality and safety of ready-to-eat pomegranate arils. *Postharvest Biology and Technology* 86, 107–112.

Martínez-Romero, D., Dupille, E., Guillén, F., Valverde, J.M., Serrano, M. and Valero, D. 2003. 1-Methylcyclopropene increases storability and shelf life in climacteric and nonclimacteric plums. *Journal of Agricultural and Food Chemistry* 51, 4680–4686.

Martínez-Romero, D., Guillén, F., Pérez-Aguilar, H., Castillo, S., Serrano, M., Zapata, P.J. and Valero, D. 2013b. Is it possible to increase the aloin content of *Aloe vera* by the use of ultraviolet light? *Journal of Agricultural and Food Chemistry* 6, 2165–2170.

Martínez-Romero, D., Serrano, M., Guillén, F., Castillo, S. and Valero, D. 2007. Tools to maintain post-harvest fruit and vegetable quality through the inhibition of ethylene action, a review. *Critical Reviews in Food Science and Nutrition* 47, 543–560.

Martínez-Romero, D., Zapata, P.J., Guillén, F., Paladines, D., Castillo, S., Valero, D. and Serrano, M. 2017. The addition of rosehip oil to *Aloe* gels improves their properties as postharvest coatings for maintaining quality in plum. *Food Chemistry* 217, 585–592.

Mohebbi, M., Ansarifar, E.A.E., Hasanpour, N. and Amiryousefi, M.R. 2012. Suitability of *Aloe vera* and gum tragacanth as edible coatings for extending the shelf life of button mushroom. *Food Bioprocess Technology* 5, 3193–3322.

Moore, E.D. and MacAnalley, B.H. 1995. A drink containing mucilaginous polysaccharides and its preparation. U.S. Patent # 5,443,830.

Muangdech, A. 2016. Research on using natural coating materials on the storage life of mango fruit cv. Nam Dok Mai and technology dissemination. *Walailak Journal of Science and Technology* 13, 205–220.

Mubarak, A. and Rao Engakanah, T. 2017. *Aloe vera* edible coating retains the bioactive compounds of wax apples (*Syzygium samarangense*). *Malaysian Applied Biology* 46, 141–148.

Nabigol, A. and Asghari, A. 2013. Antifungal activity of *Aloe vera* gel on quality of minimally processed pomegranate arils. *International Journal of Agronomy and Plant Production* 4, 833–838.

Nasution, Z., Ye, J.N.W. and Hamzah, Y. 2015. Characteristics of fresh-cut guava coated with *Aloe vera* gel as affected by different additives. *Kasetsart Journal - Natural Science* 49, 1–11.

Navarro, D., Díaz-Mula, H.M., Guillén, F., Zapata, P.J., Castillo, S., Serrano, M., Valero, D. and Martínez-Romero, D. 2011. Reduction of nectarine decay caused by *Rhizopus stolonifer, Botrytis cinerea* and *Penicillium digitatum* with *Aloe vera* gel alone or with the addition of thymol. *International Journal of Food Microbiology* 151, 241–246.

Olivas, G.I. and Barbosa-Canovas, G.V. 2005. Edible coatings for fresh-cut fruits. *Critical Reviews in Food Science and Nutrition* 45, 657–670.

Olivas, G.I., Mattinson, D.S. and Barbosa-Canovas, G.V. 2007. Alginate coatings for preservation of minimally processed 'Gala' apples. *Postharvest Biology and Technology* 45, 89–96.

Oms-Oliu, G., Soliva-Fortuny, R. and Martin-Belloso, O. 2008. Edible coatings with antibrowning agents to maintain sensory quality and antioxidant properties of fresh-cut pears. *Postharvest Biology and Technology* 50, 87–94.

Ortega-Toro, R., Collazo-Bigliardi, S., Roselló, J., Santamarina, P. and Chiralt, P. 2017. Antifungal starch-based edible films containing *Aloe vera*. *Food Hydrocolloids* 72, 1–10.

Padmaja, N. and Bosco, S.J.D. 2014. Preservation of jujube fruits by edible *Aloe vera* gel coating to maintain quality and safety. *Indian Journal of Scientific Research and Technology* 2, 79–88.

Padmaja, N., Bosco, S.J.D. and Rao, J.S. 2015. Physico chemical analysis of Sapota (*Manilkara zapota*) coated by edible *Aloe vera* gel. *International Journal of Applied Sciences and Biotechnology* 3, 20–25.

Pal, S., Sahrawat, A. and Prakash, D. 2013. *Aloe vera*. Composition, processing and medicinal properties. *Current Discovery* 2, 106–122.

Paladines, D., Valero, D., Valverde, J.M., Díaz-Mula, H.M., Serrano, M. and Martínez-Romero, D. 2014. The addition of rosehip oil improves the beneficial effect of *Aloe vera* gel on delaying ripening and maintaining postharvest quality of several stone fruit. *Postharvest Biology and Technology* 92, 23–28.

Panwar, S., Mishra, B. and Goyal, P. 2016. Permeability of *Aloe vera* composite coatings and their effect on quality of peeled carrots. *Current Science* 111, 2031–2035.

Park, H.J. and Chinnan, M.S. 1990. Properties of edible coatings for fruits and vegetables. *American Society of Agriculture Engineers Meeting Presentation* 90, 6510–6529.

Pérez, T.A.F., Aristizábal, T.I.D. and Restrepo, F.J.I. 2016. Conservation of minimally processed mango Tommy Atkins by applying an Aloe vera (*Aloe barbadensis* Miller) coating. *Vitae* 23, 65–77.

Plaskett, L.G. 1997. Aloe and its quality control checking upon the genuineness of products. *Biomedical Information Services Ltd.*, Issue 11. Available at: www.desertharvest.com/physicians/documents/Plaskett%20Quality%20Control.pdf

Radi, M., Firouzi, E., Akhavan, H. and Amiri, S. 2017. Effect of gelatin-based edible coatings incorporated with *Aloe vera* and black and green tea extracts on the shelf life of fresh-cut oranges. *Journal of Food Quality* 2017, 1–10. Article ID 9764650.

Ramachandra, C.T. and Rao, P.S. 2008. Processing of *Aloe vera* leaf gel, a review. *American Journal of Agricultural and Biological Sciences* 3, 502–510.

Ravanfar, R., Niakousari, M. and Maftoonazad, N. 2014. Postharvest sour cherry quality and safety maintenance by exposure to hot- water or treatment with fresh *Aloe vera* gel. *Journal of Food Science and Technology* 51, 2872–2876.

Reynolds, T. 2004. *Aloes: The Genus* Aloe. CRC Press, Boca Raton, FL.

Reynolds, T. and Dweck, A.C. 1999. *Aloe vera* leaf gel, a review update. *Journal of Ethnopharmacology* 68, 3–37.

Riaie, S., Saadatian, M., Aghaie, M., Alizadeh, M. and Hajitaghiloo, R. 2017. Application salicylic acid and *Aloe vera* gel as edible coating layer to preserving fresh-cut melon slices in cold storage. *International Food Research Journal* 24, 2456–2459.

Rico, D., Martin-Diana, A.B., Barat, J.M. and Barry-Ryan, C. 2007. Extending and measuring the quality of fresh-cut fruits and vegetables: A review. *Trends in Food Science and Technology* 18, 373–386.

Rosca-Casian, O., Parvu, M., Vlase, L. and Tamas, M. 2007. Antifungal activity of *Aloe vera* leaves. *Fitoterapia* 78, 219–222.

Saks, Y. and Barkai-Golan, R. 1995. *Aloe vera* gel against plant pathogenic fungi. *Postharvest Biology and Technology* 6, 159–165.

Sánchez-Machado, D.I., López-Cervantes, J., Sendón, R. and Sanches-Silva, A. 2017. *Aloe vera*: Ancient knowledge with new frontiers. *Trends in Food Science and Technology* 61, 94–102.

Serrano, M., Valverde, J.M., Guillén, F., Castillo, S., Martínez- Romero, D. and Valero, D. 2006. Use of *Aloe vera* gel coating preserves the functional properties of table grapes. *Journal of Agricultural and Food Chemistry* 54, 3882–3886.

Shahkoomahally, S. and Ramezanian, A. 2014. Effect of natural *Aloe vera* gel coating combined with calcium chloride and citric acid treatments on grape (*Vitis vinifera* L. Cv. Askari) quality during storage. *American Journal of Food Science and Technology* 2, 1–5.

Shelton, R.M. 1991. *Aloe vera*. Its chemical and therapeutic properties. *International Journal of Dermatology* 30, 679–683.

Sitara, U., Hassan, N. and Naseem, J. 2011. Antifungal activity of *Aloe vera* gel against plant pathogenic fungi. *Pakistan Journal of Botany* 43, 2231–2233.

Sogvar, O.B., Saba, M.K. and Emamifar, A. 2016. *Aloe vera* and ascorbic acid coatings maintain postharvest quality and reduce microbial load of strawberry fruit. *Postharvest Biology and Technology* 114, 29–35.

Song, H.-Y., Jo, W.-S., Song, N.-B., Min, S.C. and Song, K.B. 2013. Quality change of apple slices coated with *Aloe vera* gel during storage. *Journal of Food Science* 78, 817–822.

Supapvanich, S., Mitrsang, P., Srinorkham, P., Boonyaritthongchai, P. and Wongs-Aree, C. 2016. Effects of fresh *Aloe vera* gel coating on browning alleviation of fresh cut wax apple (*Syzygium samarangenese*) fruit cv. Taaptimjaan. *Journal of Food Science and Technology* 53, 2844–2850.

Tien, G.L., Vachon, C., Matteescu, M.A. and Lacroix, M. 2001. Milk protein coatings prevent oxidative browning of apples and potatoes. *Journal of Food Science* 66, 512–516.

Treviño-Garza, M.Z., García, S., Heredia, N., Alanís-Guzmán, G. and Arévalo-Niño, K. 2017. Layer-by-layer edible coatings based on mucilages, pullulan and chitosan and its effect on quality and preservation of fresh-cut pineapple (*Ananas comosus*). *Postharvest Biology and Technology* 128, 63–75.

Ullah, A., Abbasi, N.A., Shafique, M. and Qureshi, A.A. 2017. Influence of edible coatings on biochemical fruit quality and storage life of bell pepper cv. "Yolo Wonder". *Journal of Food Quality* 2017, Article ID 2142409.

Upton, R., Axentiev, P. and Swisher, D. 2012. *Aloe Vera Leaf, Aloe Vera Leaf Juice, Aloe Vera Inner Leaf Juice, Aloe Vera (L.) Burm. F, Standards of Identity, Analysis, and Quality Control*. American Herbal Pharmacopoeia, Scotts Valley, CA.

Valero, D., Díaz-Mula, M., Zapata, P., Guillén, F., Martínez-Romero, D., Castillo, S. and Serrano, M. 2013. Effects of alginate edible coating on preserving fruit quality in four plum cultivars during postharvest storage. *Postharvest Biology and Technology* 77, 1–6.

Valero, D., Guillén, F., Valverde, J.M., Martínez-Romero, D., Castillo, S. and Serrano, M. 2005. 1-MCP use on *Prunus* spp. to maintain fruit quality and to extend shelf life during storage, a comparative study. *Acta Horticulturae* 682, 933–940.

Valero, D. and Serrano, M. 2010. *Postharvest Biology and Technology for Preserving Fruit Quality*. CRC Press, Taylor and Francis, Boca Raton, FL.

Valverde, J.M., Valero, D., Martínez-Romero, D., Guillén, F., Castillo, S. and Serrano, M. 2005. Novel edible coating based on of *Aloe vera* gel to maintain table grape quality and safety. *Journal of Agricultural and Food Chemistry* 53, 7807–7813.

Vargas, M., Pastor, C., Chiralt, A., Mc Clements, D.J. and Martínez, C.G. 2008. Recent advances in edible coatings for fresh and minimally processed fruits. *Critical Reviews in Food Science and Nutrition* 48, 496–511.

Vieira, J.M., Flores-López, M.L., Jasso de Rodríguez, D., Sousa, M.C., Vicente, A.A. and Martins, J.T. 2016. Effect of chitosan–*Aloe vera* coating on postharvest quality of blueberry (*Vaccinium corymbosum*) fruit. *Postharvest Biology and Technology* 116, 88–97.

Watada, A.E. and Qi, L. 1999. Quality of fresh-cut produce. *Postharvest Biology and Technology* 15, 201–205.

Wibowo, S., Grauwet, T., Santiago, J.S., Tomic, J., Vervoort, L., Hendrickx, M. and Van Loey, A. 2015. Quality changes of pasteurised orange juice during storage: A kinetic study of specific parameters and their relation to color instability. *Food Chemistry* 187, 140–151.

Yoshioka, H., Mase, K., Yoshioka, M., Kobayashi, M. and Asai, S. 2011. Regulatory mechanisms of nitric oxide and reactive oxygen species generation and their role in plant immunity. *Nitric Oxide* 25, 216–221.

Zapata, P.J., Guillén, F., Castillo, S., Martínez-Romero, D., Valero, D. and Serrano, M. 2013a. Preharvest application of *Aloe vera* gel exhibits antimicrobial activity by reducing yeast, mould, and aerobic counts at harvest in several *Prunus* spp. *Acta Horticulturae* 1012, 121–126.

Zapata, P.J., Guillén, F., Martínez-Romero, D., Castillo, S., Valero, D. and Serrano, M. 2008. Use of alginate or zein as edible coatings to delay postharvest ripening process and to maintain tomato (*Solanum lycopersicon* Mill.) quality. *Journal of the Science of Food and Agriculture* 88, 1287–1293.

Zapata, P.J., Navarro, D., Guillén, F., Castillo, S., Martínez-Romero, D., Valero, D. and Serrano, M. 2013b. Characterisation of gels from different *Aloe* spp. as antifungal treatment, potential crops for industrial application. *Industrial Crops and Products* 42, 223–230.

Chapter 23

Antifungal Peptides and Proteins with Activity against Fungi Causing Postharvest Decay

Jose F. Marcos, Mónica Gandía, Sandra Garrigues, and Paloma Manzanares

Institute of Agrochemistry and Food Technology (IATA), Spanish National Research Council (CSIC), Paterna, Valencia, Spain

María Coca

Centre for Research in Agricultural Genomics (CRAG), CSIC-IRTA-UAB-UB, Campus UAB, Cerdanyola del Vallès, Barcelona, Spain

Abbreviations

AMPs	Antimicrobial peptides
AFPs	Antifungal proteins
CPPs	Cell-penetrating peptides
CRPs	Cysteine-rich proteins
CSαβ	Cysteine-stabilized αβ-motif
CW	Cell wall
CWI	Cell wall integrity
GlcCer	Glucosylceramide
GPLs	Glycerophospholipids
GSLs	Glycosphingolipids
LTPs	Lipid-transfer proteins
M(IP)2C	Mannosyldiinositol phosphorylceramide
MAPK	Mitogen-activated protein kinase
MIC	Minimum inhibitory concentration
NO	Nitric oxide
PA	Phosphatidic acid
PI(4,5)P2	Phosphatidylinositol 4,5-bisphosphate
PK	Protein kinase
ROS	Reactive oxygen species
SP	Signal peptide

1 Introduction

Plant pathogens cause disease and postharvest decay and are responsible for major losses in crop production worldwide, and for the contamination of food and feed through the production of toxins that are detrimental to human health. Crop breeding programs devote efforts in obtaining resistant cultivars, but the targeted pathogens overcome resistance quickly in few generations. Therefore, control and protection rely mainly on chemical antimicrobials and pesticides, yet their use is currently subject to strong restrictions and regulatory requirements due to their long-term negative impact on health and environment (Hirooka and Ishii, 2013; Palou et al., 2016).

Antimicrobial peptides (AMPs) are a broad class of peptides and small proteins with direct antimicrobial activity. Previous publications extensively reviewed the discovery, significance, and potential applications of AMPs (Zasloff, 2002; Brogden, 2005; Hancock and Sahl, 2006). AMPs are widely distributed in nature, produced by organisms all along the phylogenetic scale from bacteria, fungi, plants, and insects to highly evolved animal species with complex immune systems. AMPs are evolutionarily selected from ancient systems of competition for nutrients or protection/attack against invaders. Depending on each AMP, they have different spectra of target organisms ranging from viruses to parasites.

AMPs are gene-encoded and most of them derive from precursor peptides through one or more proteolytic activation steps. They are short (10–50 amino acids), usually cationic (overall positive charge +2 to +9) and with a substantial proportion of hydrophobic residues, but their sequences and secondary structures vary widely (Brogden, 2005). The most prevalent structural families of AMPs are: (i) linear, α-helical peptides, such as human LL-37, amphibian magainins, and insect cecropins; (ii) small cysteine-rich proteins (CRPs) with multiple disulphide bridges and β-sheet structures, such as the defensins from plants, insects and mammals, and defensin-like antifungal proteins of fungal origin; (iii) extended structures rich in specific amino acids as glycine, proline, tryptophan, arginine, and/or histidine, such as bovine indolicidin or human histatine; and (iv) loop peptides with one disulfide bridge, for example bovine bactenecin.

AMPs have attracted an enormous attention as promising candidates for the development of novel antimicrobial compounds and therapies due to several factors. In clinical use, they can fight the rapid and alarming increase of drug-resistant microbes and to aid in the treatment of immune-compromised patients (Yeung et al., 2011). The use of AMPs in crop protection has been recognized as an alternative to chemical fungicides and antibiotics that are hazardous to the environment, toxic for human consumers, and which extensive use has resulted in the development of resistant pathogens (Rao, 1995; Montesinos, 2007; Marcos et al., 2008; López-García et al., 2012). More importantly, studies show that AMPs have a variety of mechanisms of action that are different from those of existing antimicrobial compounds, which reinforces the enormous potential that AMPs have as alternative to conventional antibiotics. Currently, there are thousands of AMPs with demonstrated antimicrobial activity (Zhao et al., 2013) and the number increases continuously.

Paradoxically, very few peptides have reached the market or are in advanced clinical trial phases (Haney and Hancock, 2013; Rautenbach et al., 2016). An example is the antifungal P113 derived from human histatin 5 (Rothstein et al., 2001), an antifungal tested against oral candidiasis (Table 23.1). The bacteriocin nisin produced by food-grade lactic acid bacteria is an example of AMP which use has been approved for dairy products and canned goods such as vegetables and soups (Cotter et al., 2005). However, similar success with other peptides in the agrifood environment has yet to be repeated on the same scale (Barbosa et al., 2017). Therefore, and despite nearly two decades of serious efforts to exploit the advantageous features of AMPs, there are still important challenges to be solved for their application, which are different in the clinic or agricultural scenarios. Among limitations of natural AMPs are their potential toxicity to humans, and their potential lability to proteases and harsh environmental conditions. These two issues are being addressed and several solutions have been validated. The peptidic nature of these biomolecules enables the changing of their properties by rational design of new peptide variants with site-specific amino acid changes (Hancock and Sahl, 2006; Marcos et al., 2008) (see section 2). An additional limitation to reach commercial use, especially in the case of agriculture, is the high cost of manufacturing peptides and proteins. There is a growing need for low-cost and effective AMP production platforms at the scale and purity required for the different applications.

In this scenario, AMPs have also been proposed for the treatment, conservation, and protection of fruits and vegetables from postharvest pathogens (López-García et al., 2000; Jones and Prusky, 2002) and this will be the subject of review in this chapter. There are more than 21,000 scientific articles that include the term "antimicrobial peptides," from which more than 3,000 are reviews (Web of Science,

Table 23.1 Representative antimicrobial peptides with activity against fungal phytopathogens

Peptide	*Amino acid sequence*	*Origin*	*Reference*
BP22	Ts-FKLFKKILKVL-NH_2	Rational design	Badosa et al. (2009)
BM0	Ac-RFWWFRRR-NH_2	Combinatorial library	Monk et al. (2005); Muñoz et al. (2007)
CecropinA	KWKLFKKIEKVGDNIRDGII-KAGPAVAVVGQATQIAK	Insect	De Lucca et al. (1997)
D4E1	FKLRAKIKVRLRAKIKL	Rational design	Cary et al. (2000)
GMA4-L	RGFRRR	Plant (fragment of MtDef4)	Sagaram et al. (2011)
Indolicidin	Ac-ILPWKWPWWPWRR-NH_2	Bovine	Muñoz et al. (2007)
LfcinB17-31	Ac-FKCRRWQWRMKKLGA-NH_2	Bovine (fragment of Lactoferrin)	Muñoz and Marcos (2006)
Melittin	GIGAVLKVLTTGLPALISS-WIKRKRQQ	Insect	Muñoz et al. (2006)
MsrA1	MALEHMKWKLFKKIGI-GAVLKVLTTGLPALKLTK	Rational design (CecA::Mel hybrid)	Osusky et al. (2000)
P113	AKRHHGYKRKFH	Human (fragment of Histatin 5)	López-García et al. (2015)
PAF26	RKKWFW	Combinatorial library	Muñoz et al. (2007)
Pep3	WKLFKKILKVL	Rational design (CecA)	Cavallarín et al. (1998)

www.webofknowledge.com/). Web sites and databases dedicated to the classification, categorization and mining of AMPs also proliferate. Therefore, it is almost impossible to compile a complete, or even a novel, review on the subject and the main goal of this chapter is to discuss results and provide ideas related to the potential application of AMPs to postharvest pathology, focused on peptides and proteins with activity against phytopathogenic fungi causing postharvest decay.

2 Antifungal Peptides and Proteins

Over the years, most of the efforts in the study and identification of AMPs concentrated on peptides showing antibacterial activity for their relevance and potential application in human diseases, which are mainly caused by bacteria and viruses (Marcos et al., 2012; van der Weerden et al., 2013). However, fungal infections also pose a severe threat to human health as a major cause of morbidity and mortality due to the emergence of novel pathogens and the growing number of immunocompromised patients (Brown et al., 2012). The situation is different in the case of crop protection. Fungi cause the majority of infections that affect crop production.

Chemical fungicides are the main method for the control of plant disease during cultivation and postharvest conservation, despite all the concerns regarding the appearance of fungicide-resistant strains, nonspecific activity, and the negative impact in human and plant health and the environment (Holmes and Eckert, 1999; Hirooka and Ishii, 2013; Palou et al., 2016). There is thus an urgent need for the development of novel antifungal agents with properties and mechanisms of action different from the existing ones, and AMPs with antifungal activity offer this potential (Duncan and O'Neil, 2013; Shah and Read, 2013).

As consequence, there has been an increase in the number of peptides and proteins with demonstrated antifungal activity, which vary in length and amino acid composition, but share a common cationic and amphipathic character. Representative examples are reviewed in this section.

2.1 Short Linear Antifungal Peptides

Examples of short and linear naturally occurring AMPs with antifungal activity include: cecropins and melittin from insects, magainins and dermaseptins from amphibians, or indolicidin from mammals (Table 23.1). Some short AMPs of natural origin have undesirable nonspecific toxicity towards plant or animal cells, for example melittin obtained from bee venom that is very active but also highly hemolytic.

The increasing number of biotechnological companies with competitive synthesis facilities has boosted the diversity of rationally designed short linear peptides that can be tested in the search for novel AMPs. A more detailed review of successful examples of AMP design and sequence modifications is available (Marcos et al., 2008); we will only discuss selected examples in this chapter. The strategies performed in order to obtain novel antifungal peptides with improved properties include fusions of antimicrobial fragments into peptide hybrids; sequence point modifications; de novo rational design based on known AMPs; blind combinatorial approaches; the identification of minimal motifs that retain full activity; or the incorporation of non-natural amino acids or derivatizations. A classic example is the generation of cecropin::melittin hybrids and the many variants and modifications derived from them. Cecropins are susceptible to degradation by plant proteases while melittin is highly hemolytic and toxic. An early work demonstrated that shorter versions of cecropin A and cecropin A::melittin hybrids inhibited the in vitro growth of filamentous fungal plant pathogens with minimum inhibitory concentration (MIC) of 10–60 μM depending on the fungus (Cavallarín et al., 1998). Further modifications of these chimeras (MsrA1, Table 23.1) were designed to be expressed in tobacco (Osusky et al., 2000). A step forward was the combinatorialization of selected positions of the Pep3 cecropin A::melittin hybrid to lower the MIC below 6.25 μM (Badosa et al., 2009). The peptide synthesis also facilitates the design and testing of shorter versions of natural AMPs and proteins of interest in order to define the minimal bioactive motifs that confer antimicrobial activity, as was demonstrated in the peptides P113 from human histatin 5 (Rothstein et al., 2001), $LfcinB_{17\text{-}31}$, and $LfcinB_{20\text{-}25}$ from bovine lactoferrin (Muñoz and Marcos, 2006), or GMA4-L from the plant defensin MtDef4 (Sagaram et al., 2011), all of them with antifungal activity in the low micromolar range. De novo design of peptides based on the essential features of natural AMPs has resulted in the identification of potent and widely used peptides as D4E1 (Cary et al., 2000), which has MIC values against phytopathogenic fungi of 1–25 μM depending on the fungus, and has been successfully expressed in plants.

In parallel, combinatorial peptide chemistry has allowed the preparation of synthetic libraries, which represent all the desired sequence variations of a given peptide length or sequence. Their in vitro screening for inhibition of microbial growth has led to the nonbiased identification of novel peptides with potent activity against phytopathogenic fungi (Reed et al., 1997; Gonzalez et al., 2002; López-García et al., 2002). Examples of synthetic antifungal peptides obtained by this approach are the hexapeptide PAF26, originally identified as an inhibitor of the growth of the postharvest pathogen *Penicillium digitatum* (López-García et al., 2002) or the octapeptide BM0 identified as an inhibitor of the growth of the yeast *Saccharomyces cerevisiae* (Monk et al., 2005) (Table 23.1). In these screening experiments, appropriate controls should be established to select for peptides with specific antifungal activity and low non-specific toxicity. For instance, the peptide 77–3 (Gonzalez et al., 2002) is as potent as PAF26 in terms of antifungal activity but has unspecific toxicity that results in significant hemolysis (López-García et al., 2015). The specificity also allows the discrimination between peptides with potent toxicity to fungi and bacteria and peptides that are more specific to fungi (Muñoz et al., 2007).

Synthetic libraries are very powerful because they allow the screening of a wide sequence diversity (Blondelle et al., 1996). Although the number of compounds/peptides that can be screened is very high, it is also limited because they are also costly and time-consuming. To overcome these problems, computer-aided approaches are being developed to virtually predict the activity and mode of action of a higher number of sequences of design (Fjell et al., 2012; Porto et al., 2017).

2.2 Cysteine-Rich Proteins: Plant Defensins

A class of peptides that is of particular interest for their biotechnological applications in multiple fields, including agriculture, is the small CRPs. These are peptides and small proteins containing multiple cysteine residues that form disulfide bonds and fold into compact structures, conferring a high degree of stability against adverse biochemical and biophysical conditions. Several groups of these CRPs have a defense role within their producing organisms and specific antimicrobial activity, as for instance human and plant defensins or plant thionins, lipid transfer proteins or thaumatins (Ganz, 2003; van der Weerden et al., 2013). Defensins and defensin-like proteins form by far the largest family of CRPs with antimicrobial activity. Plant defensins compose a numerous group of small cationic CRPs (45–54 amino acids in length) that typically include eight cysteines and four intramolecular disulfide bonds (van der Weerden and Anderson, 2013). They are found ubiquitously throughout the plant kingdom as part of the innate immunity against microbial infections, mostly fungal (Terras et al., 1993; Carvalho and Gomes, 2009). They are constitutively expressed in both plant storage and reproductive organs, but also induced locally as well as systemically during plant defense responses. Plant defensins are structurally and functionally related to the previously characterized defensins found in insects and mammals (Carvalho and Gomes, 2009; Vriens et al., 2014); they are structurally similar peptides, but their sequences are divergent and show different activities that include antifungal, antibacterial, or antitumoral activities (van der Weerden and Anderson, 2013) (Figure 23.1A). Examples of plant defensins with antifungal activity include RsAFP1 and RsAFP2 from *Raphanus sativus*, MsDefl and MtDef4 from *Medicago* spp., Dm-AMP1 and Dm-AMP2 from *Dahlia merckii*, Psd1 from *Pisum sativum* or NaD1 from *Nicotiana alata*. Plant defensins are very active against phytopathogenic filamentous fungi, with MIC values as low as 1–8 μM depending on the defensin and fungus (Terras et al., 1993; Sagaram et al., 2011; Bleackley et al., 2016).

The transcript of the majority of plant defensins encodes a polypeptide that contains a signal peptide (SP) that targets the protein to the extracellular space, a pro-peptide that blocks the activity of the mature peptide until necessary, and the mature antifungal protein (Carvalho and Gomes, 2009). The structure of plant defensins typically comprises a cysteine-stabilized αβ-motif (CSαβ) with a prominent α-helix and a triple-stranded antiparallel β-sheet (Figure 23.1C). Additionally, a three-dimensional peptide signature called the γ-core is present in virtually all defensins and other antimicrobial CRPs. This γ-core includes two cysteines in the conserved motif Gly-X-Cys-$X_{3\text{-}9}$-Cys, either in dextromeric or levomeric isoform (Yount and Yeaman, 2004). The antimicrobial activity of synthetic peptides comprising this motif has been demonstrated for some plant defensins. This is the case, for example, of the γ-core present in MtDef4 from *Medicago truncatula* (Sagaram et al., 2011) or BhDef2 from *Brassica hybrid* (Kaewklom et al., 2016), and is related to its cationic net charge due to the presence of basic residues. However, other γ-core peptides do not show any antimicrobial activity (Sagaram et al., 2011), suggesting that the γ-core serves as a structural scaffold to which other antimicrobial determinants (e.g., β-sheets) are attached (Yount and Yeaman, 2004).

It is interesting to note that distinct but sequence-related plant defensins may have very different antifungal potency and/or specificity for fungi, as shown for the NaD1 and NaD2 from *Nicotiana* (Bleackley et al., 2016), or the *Medicago* MsDefl and MtDef4 (Sagaram et al., 2011). This indirectly indicates that subtle amino acid changes could alter the antimicrobial properties of the protein and points towards the application of rational design approaches, similar to those used in shorter peptides, to improve the antifungal efficacy of defensins. In fact, point mutations and exchange of motifs between *Medicago* defensins result in modulation of activity and exchange of properties (Sagaram et al., 2011). A very significant improvement in antifungal activity was observed through the exchange of motifs between NaD1 and NaD2, with a resulting inhibitory concentration below the micromolar range, lower than any of the two parental proteins (Bleackley et al., 2016).

2.3 Cysteine-Rich Proteins: Fungal Antifungal Proteins

Another CRP group of interest comprises the antifungal proteins (AFPs) of fungal origin. AFPs are small (45–57 amino acids) and cationic defensin-like proteins that are produced and secreted to the culture medium by filamentous ascomycetes, mostly from the genera *Aspergillus* and *Penicillium*, and exhibit antifungal activity at low micromolar concentrations (Hegedüs and Marx, 2013; Silva et al., 2014). The first identified AFP was the one produced by *Aspergillus giganteus* (Olson and Goerner, 1965; Nakaya et al., 1990; Wnendt et al., 1994). The second so-called PAF protein was identified in a study of abundant proteins secreted by *Penicillium*

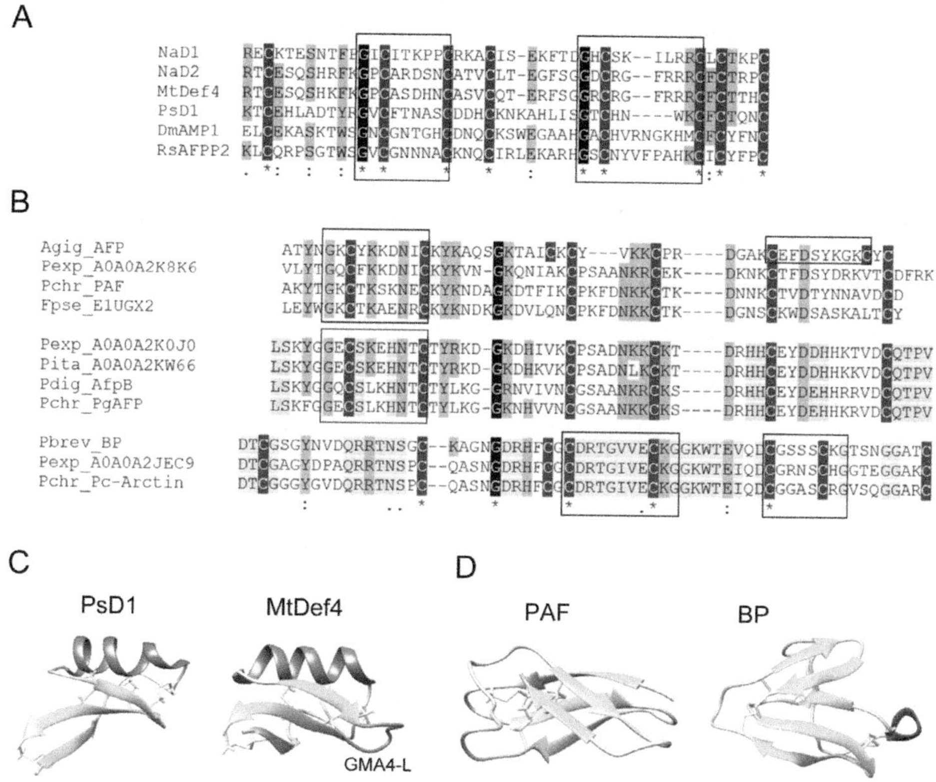

Figure 23.1 Amino acid sequence and structure of selected plant defensins and fungal antifungal proteins. (A) Sequence alignment of the plant defensins NaD1 and NaD2 from *N. alata*; MtDef4 from *M. truncatula*; PsD1 from *P. sativum*; DmAMP1 from *D. merckii*; and RsAFP2 from *R. sativus*. (B) Sequence alignment of the antifungal proteins AFP from *A. giganteus*; PAF, PgAFP and Pc-Arctin from *P. chrysogenum*; AfpB from *P. digitatum*; BP from *P. brevicompactum*; and other AFPs identified in the genomes of filamentous fungi (for additional details, see Garrigues et al., 2016). The sequence alignment is divided in three blocks to emphasize separation of classes A, B, and C (Garrigues et al., 2016). In (A) and (B), the antimicrobial γ-core motif (Yount and Yeaman, 2004) is boxed, and residues are shaded to emphasize sequence similarities within each group. (C) Visualization of the protein structure of the plant defensins PsD1 from *P. sativum* (Protein Data Bank ID 1JKZ) and MtDef4 from *M. truncatula* (PDB ID 2LR3). (D) Visualization of the protein structure of the antifungal proteins PAF from *P. chrysogenum* (PDB ID 2MHV) and BP from *P. brevicompactum* (PDB ID 1UOY). In (C) and (D), α-helixes are shown in green, β-sheets in blue and disulfide bridges in yellow. The loop containing the sequence of the peptide GMA4-L (RGFRRR) of MtDef4 is marked (see Table 23.1 and text for further details).

chrysogenum (Marx et al., 1995). These proteins are abundantly secreted by the producer fungus, with yields above 50 mg/L of culture medium. Additional AFPs are Anafp from *Aspergillus niger* (Lee et al., 1999), AcAFP from *Aspergillus clavatus* (Skouri-Gargouri and Gargouri, 2008), the bubble protein (BP) from *Penicillium brevicompactum* (Seibold et al., 2011), or NFAP and the recent NFAP2 from *Neosartorya fischeri* (Kovács et al., 2011; Tóth et al., 2016). Several of these proteins have activity towards fungal plant pathogens, with MIC values below 10 μM or even below micromolar concentrations depending on the fungus (Lee et al., 1999; Vila et al., 2001; Kaiserer et al., 2003; Skouri-Gargouri and Gargouri, 2008; Chen et al., 2013; Leiter et al., 2017). They were not toxic to plant or mammalian cells (Moreno et al., 2006; Hegedüs and Marx, 2013).

Similar to plant defensins, AFPs contain the conserved γ-core motif, are produced as SP-pro-proteins, and have to be processed in order to release the mature protein with antifungal activity, even though the functional role of the pro-sequence is still unclear. Figure 23.1B shows the annealed sequences of selected mature AFPs. The structure of AFPs differs from the structure of classical defensin-like proteins. AFPs fold into five β-strands forming two packed β-sheets that share a common interface, and have typically six cysteine residues forming three disulfide bonds (Campos-Olivas et al., 1995; Batta et al., 2009) (Figure 23.1D).

The identification of the BP protein in *P. brevicompactum* indicated the presence of an orthologous gene in the genome of *P. chrysogenum* and therefore more than one AFP-coding gene in this fungus, and the separation of AFP-related sequences in two distinct groups (Seibold et al., 2011). In parallel, it was observed that different strains of *P. chrysogenum* produced two additional AFPs: the PgAFP produced by an isolate from dry-cured ham (Rodríguez-Martín et al., 2010) and the Pc-Arctin produced by an arctic isolate (Chen et al., 2013). More recently, the availability of numerous sequenced genomes of filamentous fungi has allowed mining to search for potential novel AFPs and AFP-like sequences and indicated the presence of multiple *afp* genes in fungal genomes (Galgóczy et al., 2013; Garrigues et al., 2016; Tóth et al., 2016), as earlier shown for plant defensins. Phylogenetic analyses confirmed that there are at least three different groups/classes of AFPs and AFP-like proteins. The initial proposal identified classes A, B, and C, each including at least one AFP with experimentally proven antifungal activity (Garrigues et al., 2016). Each different AFP encoded by *P. chrysogenum* belongs to a different class: PAF from class A, PgAFP from class B, and the Pc-Artic class C protein. The representative of the distant class C phylogenetic cluster is the BP protein that folds in a slightly different arrangement with a minor α-helix contribution (Figure 23.1D). The recently identified NFAP2 from *N. fischeri* would be the first member of an additional class of cationic AFPs (Tóth et al., 2016). Fungal plant and postharvest pathogens also encode AFPs in their genomes. For instance, the citrus pathogens *P. digitatum* and *P. italicum* encode only one AFP that groups in the class B phylogenetic cluster (Marcet-Houben et al., 2012; Ballester et al., 2015). On the other hand, *Penicillium expansum*, which is the causal agent of the blue mold disease in apples (Ballester et al., 2015) encodes three different AFPs, one belonging to each class. None of these latter proteins have been either produced or studied yet.

The level of natural production of different AFPs is variable and depends on the fungal strain, the protein and the growth media. In addition to the well-known AFP from *A. giganteus* and PAF from *P. chrysogenum*, a remarkable record is the production of PgAFP from *P. chrysogenum*, for which yields up to 600 mg/L are reported (Delgado et al., 2015). In contrast, there are reports of very low (Kovács et al., 2011; Tóth et al., 2016) or even of undetectable protein production in the case of the AfpB from *P. digitatum* despite the high expression level of the corresponding *afpB* gene (Garrigues et al., 2016). Even a given fungus might produce different AFPs depending on culture broth (Kovács et al., 2011; Tóth et al., 2016).

Different synthetic peptides derived from the *P. digitatum* AfpB or the *P. chrysogenum* PAF sequence have shown modest activity, below the expected potency for any AFP (Garrigues et al., 2017). This is in contrast with a similar study with the MtDef4 plant defensin, for which synthetic peptides could be identified having antifungal potency in the same range as the native protein (Sagaram et al., 2011). Another difference with the MtDef4 defensin was that the synthetic peptide that corresponds to the γ-core of *P. digitatum* AfpB did not show activity (Garrigues et al., 2017).

The production of an antifungal protein by a fungus is, at the very least, intriguing. It is assumed that AFP proteins are not active against the producer fungus, but their biological role is still unclear. A proposed function related to the competition for nutrients is obvious although no experimental evidence is available. Null mutation of the *paf* gene in *P. chrysogenum* reduces the production of asexual spores and affects negatively the expression of genes that regulate asexual development, suggesting that AFPs could play a role beyond their antifungal activity (Hegedüs et al., 2011). A similar approach with the null mutation of the *afpB* gene in *P. digitatum* did not result, however, in any significant difference regarding production of spores, asexual development, growth or virulence to citrus fruit (Garrigues et al., 2016). The PAF protein from *P. chrysogenum* is active against *P. digitatum* when added exogenously to the culture media. However, *P. digitatum* strains engineered to express the heterologous *paf* gene produced abundant quantities of the PAF protein, comparable to the native *P. chrysogenum* fungus, without deleterious effects on the axenic growth or pathogenicity of *P. digitatum* (Sonderegger et al., 2016). Authors of a recent transcriptomic meta-analysis of *A. niger* proposed novel functions for the AnAFP produced by this fungus in nutrient recycling during starvation, autophagy, and development (Paege et al., 2016).

As demonstrated in the case of plant defensins, the small size of AFPs makes these proteins suitable to rationally design and engineer amino acid substitutions with an impact on their activity, which have been used to study structure–activity relationships (Batta et al., 2009; Sonderegger et al., 2016, 2017).

3 Modes of Antifungal Action

As with any other antimicrobial, the study of the mode of action of AMPs is essential for their future development and application. The mode of action of AMPs has been reviewed extensively, especially in the case of antibacterial peptides (Brogden, 2005; Nicolas, 2009; Nguyen et al., 2011) but also in the specific antifungal peptides (Rautenbach et al., 2016). A detailed mechanistic knowledge might also provide information to improve the properties of AMPs by rational design and residue substitution, and could led to novel microbial targets for the development of other antimicrobials, not necessarily peptides. In contrast to conventional antibiotics designed to target a specific essential microbial process/molecule with the highest possible affinity, in general AMPs seem to act at multiple low-affinity targets so hindering the development of microbial resistance. Their mechanisms of action include different strategies ranging from interaction with the cell membrane to exerting their action internally (Hancock and Sahl, 2006; Nicolas, 2009).

3.1 Binding to the Fungal Cell

The first step in the AMP mode of action is the physical interaction with the outer structures that surround the microbial cell. The cationic properties of many AMPs enable an electrostatic attraction to the negative-charged microbial envelopes, which is a plausible explanation for the specificity of most AMPs for microbes and lower toxicity to animal and plant cells (Zasloff, 2002). However, there are examples of AMPs whose interaction and binding with microbes does not rely primarily on electrostatic attraction and is aided by specific components. The cell wall (CW) and membrane of fungi have specific compositions, different from other eukaryotic cells, and these have been long recognized as targets for antifungal molecules (Munro, 2013). Most antifungal peptides target this specificity (Rautenbach et al., 2016) and interact with the fungal envelopes as a first step in their mode of action (Figure 23.2).

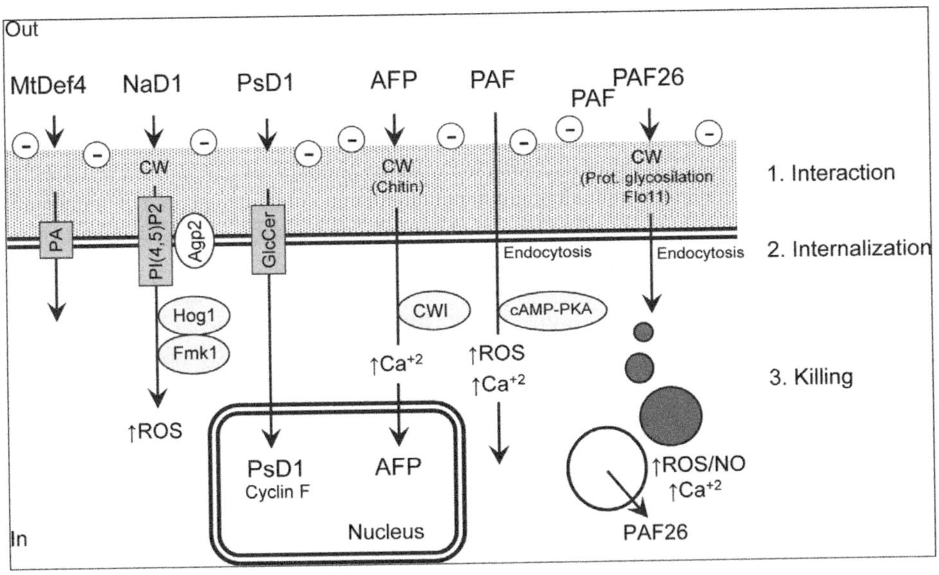

Figure 23.2 Schematic model of the non-lytic mode of action of antifungal peptides and proteins. The model is divided in three steps of (1) interaction, (2) internalization, and (3) intracellular killing, as indicated at the right. The available data are summarized for the plant defensins MtDef4 from *M. truncatula*, NaD1 from *N. alata*, and PsD1 from *P. sativum*; the fungal antifungal proteins AFP from *A. giganteus*, and PAF from *P. chrysogenum*; and the synthetic hexapeptide PAF26 (see the main text for further details and explanation).

The fungal CW is composed, from the outermost to the innermost, of a highly glycosylated cover of mannoproteins, a thick intermediate layer of β-glucans (predominantly 1,3-β-glucan) and a thin layer of chitin (a polymer of N-acetylglucosamine), after which the lipid bilayer of the cell membrane stands as the barrier to the cell interior. The enzymatic degradation of CW reduces the sensitivity of fungal cells to the defensin NaD1 and the synthetic PAF26 (van der Weerden et al., 2010; Harries et al., 2013), demonstrating that an intact CW is required for peptide interaction and activity. Reinforcement of CW is a defensive countermeasure of fungal cells exposed to antifungal peptides and proteins (López-García et al., 2010a; Ouedraogo et al., 2011). Mannosylated proteins and CW composition regulate the sensitivity to the antifungal plant osmotin (Coca et al., 2000; Ibeas et al., 2001). *S. cerevisiae* genes involved in protein glycosylation of CW mannoproteins also determine the interaction and internalization of PAF26 against the yeast (Harries et al., 2013), in a demonstration that the fungal CW is required for internalization of the peptide. These studies were partly extended to *P. digitatum* with the demonstration that mutants in the protein glycosylation gene *pmt2* were selectively more tolerant to PAF26 (Harries et al., 2015). Specific binding roles have been shown for the CW protein Ssa1/2 that mediates binding of P113/histatin 5 to *Candida* (Li et al., 2003), the plasma membrane protein Ore20/Pho36 that is required for sensitivity to osmotin (Narasimhan et al., 2005), and the hyperglycosylated CW protein Flo11 that mediates the binding to *S. cerevisiae* of PAF26 but not of P113 (Bou Zeidan et al., 2013) (Table 23.1).

Once the peptides diffuse through the CW, they encounter the fungal cell membrane. Different membrane glycosphingolipids (GSLs) and glycerophospholipids (GPLs) from fungi have been shown to determine the interaction and activity of antifungal peptides and defensins (Wilmes et al., 2011; Vriens et al., 2014; Rautenbach et al., 2016) (Figure 23.2). The genes *ipt1* and *skn1* involved in the biosynthesis of the acidic GSL mannosyldiinositol phosphorylceramide [M(IP)2C] determine the binding and sensitivity of yeast to the plant defensin DmAMP1 (Thevissen et al., 2000, 2005). This is a remarkable example since deletion of *ipt1* results in tolerance to other unrelated antifungal peptides such as the cyclic lipopeptide syringomycin E (Stock et al., 2000) or the synthetic hexapeptide PAF26 (López-García et al., 2010a). Therefore, a binding function for this class of lipids might not be the unique explanation for the observed tolerant phenotype of *S. cerervisae* (Thevissen et al., 2010). The involvement of GSLs in apoptosis, response to oxidative stress, or autophagy in fungi (Aerts et al., 2006; Thevissen et al., 2010) could indicate a role in the intracellular cell response/signaling to peptide exposure (see next section). The defensins RsAFP2 and PsD1 bind a different GSL, the neutral glucosylceramide (GlcCer), and *S. cerevisiae* mutants deficient in the gene *gcs1* that is involved in the biosynthetis of GlcCer are resistant to RsAFP2 (Thevissen et al., 2004; de Medeiros et al., 2010). *Fusarium graminearum* mutants in the homologous GlcCer synthase *gcs1* gene are depleted in GlcCer and they are more tolerant to MsDef1 and also to RsAFP2, but interestingly not to the related MtDef4 (Ramamoorthy et al., 2007a), which shows a remarkable degree of specificity.

Conversely, this latter MtDef4 binds in vitro to the GPL phosphatidic acid (PA), a signaling lipid and precursor of membrane phospholipids (Sagaram et al., 2013). Mutation of the RGFRRR motif in the loop of the γ-core of MtDef4 (Figure 23.1) abolishes the capacity of the protein to bind PA and also to enter fungal cells (see next section). Interestingly, the *Nicotiana* NaD1 and NaD2 differ in the same loop; being NaD2 identical to MtDef4 and also capable of binding PA, while NaD1 has a preferred binding for another GPL, phosphatidylinositol 4,5-bisphosphate [PI(4,5)P2] (Poon et al., 2014; Bleackley et al., 2016) (Figure 23.2).

Once the interaction has occurred, the peptides can be membrane disruptive resulting in cell lysis or be transported into the cell in a non-disruptive manner, bringing them into contact with intracellular targets. The cationic and amphipathic character of most AMPs would allow the interaction, insertion and disruption of lipid cell membranes causing their rupture, loss of biophysical properties and the cell killing through a lytic mechanism (Nguyen et al., 2011). Different models to describe the membrane disruption by AMPs have been proposed such as the "barrel-stave," the "carpet," the "toroidal-pore" or the "aggregate channel" model (Brogden, 2005; Nicolas, 2009; Guilhelmelli et al., 2013). These models are not necessarily exclusive and very likely depend on the membrane mimetic used for in vitro studies, the microorganism, the peptide, or a combination of them. However, it must be taken into consideration that peptides that mostly act through a membrane disrupting mechanism are usually highly lytic and also toxic to different cell types and organisms, including plant and animal cells. Thus, AMPs with non-lytic mechanisms and specific modes of action and selectivity towards the targeted microorganism (fungi) are preferred as lead compounds for future development (Marcos and Gandía, 2009).

3.2 Cell-Penetrating Antifungal Peptides

Most AMPs show sequence and biophysical similarities with the so-called cell-penetrating peptides (CPPs) that are capable of cell internalization in a non-disruptive manner (Henriques et al., 2006; Marcos and Gandía, 2009; Nicolas, 2009; Marcos et al., 2012). CPPs are cationic and amphipathic, as occurs with AMPs, and in fact antimicrobial properties have been demonstrated for peptides initially known as CPPs and vice versa. CPPs have two general pathways to penetrate into the cells: direct translocation or endocytosis. Both pathways can act simultaneously for the same peptide, being concentration-dependent; direct translocation across the cell membrane occurs mainly at high peptide concentrations, while endocytosis would occur at low concentrations (Muñoz et al., 2012; Guidotti et al., 2017). After translocation into the cell interior, these types of peptides would be able to bind to inner targets and interfere functions inside the cell. We will focus in this and the following section on examples of antifungal CPPs for which a linkage between specificity for fungi, cell internalization and antifungal killing has been established: the synthetic PAF26, plant defensins (MtDef4 and NaD1), and fungal antifungal proteins (AFP and PAF) (Figure 23.2). These examples are also representatives of AMPs for which there is experimental evidence of fruit postharvest diseases control (see last section).

The hexapeptide PAF26 (RKKWFW) (Table 23.1) has been proposed as a model to investigate the mode of action and specificity of antifungal CPPs (Muñoz et al., 2013a). It was early demonstrated that PAF26 is internalized by conidia and mycelium of *P. digitatum* at very low, sub-inhibitory concentrations (in the nanomolar range) in a mechanism that preceded any cell lytic effect (Muñoz et al., 2006). In the model fungus *Neurospora crassa*, PAF26 is internalized inside the cells following two different pathways: an energy-dependent endocytic-like mechanism when the peptide is applied at low concentrations and an energy-independent translocation at high concentrations (Muñoz et al., 2012). PAF26 follows an endocytic entry route in which it is localized first at the outer CW, from where it is later observed inside small endocytic vesicles that coalesce and grow bigger with time to fill large central vacuoles, and the intracellular killing correlates with the active extrusion of the peptide from the vacuoles to the cytosol (Muñoz et al., 2006, 2012; Harries et al., 2013) (see diagram in Figure 23.2).

By changing specific amino acids in the PAF26 sequence, different peptides with distinct activity were designed and characterized (Muñoz et al., 2013a,b). Substitution of the C-terminal aromatic motif by alanines (RKKAAA) resulted in a non-active peptide that remained at the cell envelope and was not internalized. On the other end, substitution of the cationic N-terminal motif (AAAWFW) also resulted in an inactive peptide that weakly interacts and is internalized by *N. crassa* cells, in which it remains in the vacuoles without causing cell lysis. These studies on PAF26 define a model framework of three separated steps of: (1) interaction, (2) internalization, and (3) intracellular killing (Muñoz et al., 2013b) (Figure 23.2), which has been further supported with the identification of mutants of *S. cerevisiae* and *N. crassa* that specifically block each stage (Carmona et al., 2012; Muñoz et al., 2012, 2013b; Bou Zeidan et al., 2013; Harries et al., 2013). This general model is also applicable to more complex antifungal peptides such as the defensin NaD1 (Bleackley et al., 2014). Thus, the PAF26 peptide with just six amino acids could help us to understand the mode of action of other more complex antifungal CPPs, which cause cell death by a non-lytic mechanism.

The plant defensins Psd1, NaD1, and MtDef4 are also internalized by fungi (Lobo et al., 2007; van der Weerden et al., 2008; Sagaram et al., 2013); Psd1 co-localized with the nuclei of *N. crassa*, while the other two defensins accumulated mainly in the cytosol of *Fusarium* species. Substitution for alanines of residues of the MtDef4 RGFRRR motif that binds PA (see previous section) in the loop of the γ-core abolishes the cell penetration and antifungal activity of the defensin, linking cell internalization to activity as well (Sagaram et al., 2013). The yeast *agp2* gene, which encodes a plasma membrane regulator of polyamine uptake, also mediates the internalization of the NaD1 defensin into yeast and its yeast killing activity (Bleackley et al., 2014).

Two different fungal AFPs were shown to also enter fungal cells as part of their fungal inhibition mechanism (Oberparleiter et al., 2003; Moreno et al., 2006). In the case of the *P. chrysogenum* PAF, the protein accumulated in the cytosol and its internalization was energy-dependent and prevented in the presence of inhibitors of actin polymerization, which suggested an endocytosis-like mechanism (Oberparleiter et al., 2003). The PAF^{D19S} mutant that has decreased antifungal activity has also a reduced uptake in *N. crassa* cells (Sonderegger et al., 2017). The AFP from *A. giganteus* also entered *Magnaporthe oryzae* cells, but co-localized within nuclei (Moreno et al., 2006).

3.3 Intracellular Action of Antifungal Peptides

Once inside the cell, intracellular AMPs may target multiple processes. Again, intracellular bacterial targets have been much more extensively studied and generally include binding to DNA, RNA and proteins, inhibition of cell wall DNA, RNA or protein synthesis, or interference with bacterial cytokinesis or division (Brogden, 2005; Hancock and Sahl, 2006).

Although less studied, there are also examples of antifungal peptides that exert their action intracellularly, usually by disrupting the homeostasis of cellular functions (Marcos and Gandía, 2009; Wilmes et al., 2011; Marcos et al., 2012; Guilhelmelli et al., 2013; Hegedüs and Marx, 2013; Hayes et al., 2014) (Figure 23.2). Maybe the most specific example described so far is that of defensin PsD1 that localizes within the nucleus of *N. crassa*, interacts in vitro with cyclin F, and interferes with cell division in vivo after DNA duplication (Lobo et al., 2007). Common signatures of several antifungal peptides and proteins whose mechanism of action has been studied are the induction of reactive oxygen species (ROS) and the disruption of intracellular Ca^{+2} homeostasis. Examples of the production of ROS include the plant defensins RsAFP2 (Aerts et al., 2007) and NaD1 (van der Weerden et al., 2008), the fungal PAF (Leiter et al., 2005), and the synthetic hexapeptide PAF26 (Carmona et al., 2012). The problem is that ROS induction is an apoptotic marker and it is very difficult to determine whether it is the cause of the cell killing induced by the peptides or just a marker of the cell death. In the particular case of PAF26, ROS production is linked to the production of nitric oxide (NO) that causes nitrosative stress in *S. cerevisiae* (Carmona et al., 2012), and mutants and treatments that block NO production also confer tolerance to the peptide indicating that the antifungal activity of PAF26 correlates with NO production.

The fungal PAF (Binder et al., 2010) and AFP proteins (Binder et al., 2011) disrupted cytosolic Ca^{+2} homeostasis in sensitive fungi, and PAF resistant mutants in the catalytic subunit of protein kinase A (PKA) do not exhibit an increase in Ca^{+2} levels after PAF treatment (Binder et al., 2015). Upon treatment with PAF26, *N. crassa* cells also suffer a dose-dependent increase of cytosolic Ca^{+2} that is blocked by treatments that impede the peptide internalization and the killing activity (Muñoz et al., 2012).

There are intracellular mechanisms that do not necessarily require the internalization of the peptide but rather the activation of signalling cascades after binding to the cell surface. Most plant pathogenic fungi have at least four protein kinase (PK) signalling pathways that govern pathogenicity, growth, and response to adverse conditions (Turrà et al., 2014): the cAMP-PKA cascade that regulates differentiation and morphogenesis; and the three mitogen-activated PK (MAPK) Hog1, Kss1/Fmk1/Fus3, and Slt2/Mpk1 that regulate response to osmotic stress (Hog1), invasive growth, and CW integrity (CWI), respectively. These pathways could be either hijacked by the antifungal peptides to exert the killing activity or, alternatively, be part of the fungal defense response to counteract the damage produced by the peptides. The first alternative seems to operate in the case of the plant osmotin, which subverts the Kss1/Fus3 and cAMP-PKA pathways to induce ROS increase and apoptosis in *S. cerevisiae* (Yun et al., 1998; Narasimhan et al., 2001). Analogously, the cAMP-PKA pathway is also an active determinant of the PAF toxicity to *A. niger* (Binder et al., 2015). The binding of defensins to signalling lipids at the membrane, such as MtDef4 to PA (Sagaram et al., 2013) or NaD1 to PI(4,5)P2 (Poon et al., 2014), might alter the normal function of these lipids. Several cellular proteins with a signalling role are recruited to the membrane by their binding to PA, although it is not known how this is related to the antifungal action of these defensins (Sagaram et al., 2013).

On the other hand, the activation of the stress signalling pathways is a broad defensive response that enhances tolerance to antifungal peptides and proteins, as well as to fungicides (Ouedraogo et al., 2011; Hayes et al., 2014). However, there are differences in the different antifungal proteins and pathways activated. Mutations at the invasive growth and CWI pathways, but not at the osmotic stress pathway, result in increased susceptibility of *F. graminearum* to the MsDef1 defensin, but not to the related MtDef4 (Ramamoorthy et al., 2007b). On the contrary, the Hog1 osmotic stress pathway protects *Candida albicans* and *F. oxysporum* against the defensin NaD1 (Hayes et al., 2013; Dracatos et al., 2016).

4 Application of Antifungal Peptides in Crop Protection

AMPs can be biotechnologically produced through genetic engineering of the host plant to confer disease resistance. Multiple reports on the expression of genes encoding AMPs in model and crop plant species conferring different degrees of protection against phytopathogens can be found in the literature, and are the subject of comprehensive reviews (López-García et al., 2012; Holásková et al., 2014). Here, we present only the main different approaches and transgenes used.

Strategies based on the heterologous expression of genes encoding plant AMPs have demonstrated efficacy to enhance resistance in crop and ornamental plants towards fungal pathogens (Terras et al., 1995; Gao et al., 2000; Chan et al., 2005). Among these AMPs are CRPs such as thionins, defensins, LTPs, hevein-like peptides, and snakins.

AMPs from origins other than plants could potentially be more effective against plant pathogens than plant AMPs due to coevolutionary aspects. A successful example is the AFP from the fungus *A. giganteus* which conferred enhanced resistance to transgenic rice plants against the blast fungus *M. oryzae* (Coca et al., 2004), and to transgenic wheat plants against the powdery mildew fungus *Erysiphe graminis* f.sp. *tritici* and the leaf rust fungus *Puccinia recondite* f.sp. *tritici* (Oldach et al., 2001). Other peptides used include linear peptides with broad-spectrum and potent antimicrobial activity as the insect AMPs cecropins, attacins, and apidecins, or the frog peptides magainins and temporins. Consistently, the transgenic plants producing these AMPs exhibited high levels of protection not only against fungal but also bacterial pathogens (Ko et al., 2002; Coca et al., 2006; Rahnamaeian et al., 2009; Bundó et al., 2014). Early attempts to express these genes in plants produced contradictory results regarding pathogen resistance, mainly attributed to differential susceptibility of these peptides to host proteases and lack of accumulation (Mills et al., 1994). Strategies targeting the AMPs to specific subcellular compartments to be protected from host proteases resulted in their accumulation, and consistently in improved plant disease resistance (Coca et al., 2006; Rahnamaeian et al., 2009; Bundó et al., 2014). Remarkably, the protection provided by the attacin E to the 'Royal Gala' apple tree against the fire blight pathogen has proven to be stable for a long period of time, remaining effective throughout a 12-year period of orchard growth of the transgenic apple trees (Borejsza-Wysocka et al., 2010). These results demonstrate the efficacy and durability of the strategy of producing AMPs by the own plant for protection against phytopathogens.

Genes encoding several synthetic AMPs have been also transferred to plants leading to improved fungal resistance (López-García et al., 2012; Yevtushenko and Misra, 2012). A successful example is the synthetic peptide MsrA1 that conferred potato plants with broad-spectrum resistance to fungal, oomycete, and bacterial pathogens (Osusky et al., 2000). Interestingly, the tubers retained their resistance to challenge inoculations for over one year during postharvest storage, and were not harmful when fed to mice.

Simultaneous expression in plants of different AMPs, preferably with different modes of action, is an attractive approach to attain resistance against a broad range of pathogens, and to minimize the probability of pathogens to overcome the engineered trait. Different approaches for gene-stacking include crossing individual transformants, sequential transformation, or co-transformation with different transgenes. Such strategies are, however, time-consuming and with limited probability of success, and are associated with gene silencing problems when gene promoters or marker genes are used more than once. To circumvent such difficulties, the expression of genes encoding polyproteins is a good alternative (Marcos and Beachy, 1994). A successful example is the production in rice plants of a chimeric polyprotein containing two plant defensins DmAMP1 and RsAFP2 fused through the IbAMP linker peptide, which is proteolytically cleaved in plant cells leading to the accumulation of the two independent AMPs (Jha and Chattoo, 2009). The plants transformed with the polyprotein construct exhibited improved disease resistance to two important fungal pathogens of rice, *M. oryzae* and *Rhizoctonia solani.*

In addition to the AMP properties, the choice of a suitable promoter to drive strong and convenient spatio-temporal transgene expression is an important factor to attain AMP-based resistance. In most of the studies, constitutive promoters such as the CaMV35S (Cauliflower Mosaic Virus) or the maize *Ubiquitin1* promoter have been used. Even though this strategy is suitable to assess the effectiveness of the transgene expression, it presents a number of potential drawbacks for actual use in genetically improved crops. Among them, a potential negative impact on fitness and yield of the host plant, or the selection of resistant populations of target pathogens due to the permanent exposure to AMPs. Instead, a controlled production of the AMP at the site where it is needed and only when needed is a more desirable strategy, which can be accomplished by driving the expression of *AMP* genes by pathogen-inducible promoters. There are several reports on the controlled production of AMPs in genetically modified plants. As an example, the pathogen-inducible expression of the *afp* gene from *A. giganteus* in rice plants resulted in superior protection against rice blast disease compared to that observed in the plants constitutively expressing the same gene (Moreno et al., 2005). Not only temporal but also spatial control of the transgene expression can be achieved using tissue-specific promoters, resulting in the accumulation of AMPs in specific organs such as seeds or tubers.

The engineered plant resistance against fungal pathogens by AMP transgene expression has been assessed mainly in controlled inoculation experiments, at most under greenhouse conditions. Hence, field trials are necessary to evaluate these resistances to natural infections. One of the first examples of disease protection under field conditions was reported in potato plants by transgenic expression of the alfalfa antifungal defensin (*alfAFP*) gene, increasing resistance against *Verticillium dahliae* to levels that were equal to or that exceed those obtained through current practices based on fumigants (Gao et al., 2000). More recently, the performance of transgenic cotton plants expressing the plant defensin NaD1 was evaluated under field conditions over three growing seasons (Gaspar et al., 2014). These plants showed substantial resistance to two economically important fungal pathogens, namely *F. oxysporum* and *V. dahliae*, whereas they did not exhibit any detrimental agronomic features in non-infected soils. These data suggest that this technology can be implemented in crop protection programs.

Although the benefits of AMPs in enhancing disease resistance in transgenic plants have been demonstrated repeatedly, studies on the transgene induced changes in the host plants are still scarce. Profiling technologies such as transcriptomics, proteomics, and metabolomics can be used to investigate intended and unintended differences in transgenic AMP plants. Additionally, physiological parameters of agronomic relevance should be evaluated. Compositional and toxicological analyses are also necessary to assess the substantial equivalence of transgenics compared to their non-transgenic counterparts. The introduction of the AMP gene must be demonstrated not to adversely affect the performance of the crop, nor introduce unsafe alterations in food or feed products derived from these plants. However, and despite of successful results, the most important challenges currently to the production of AMPs in plants are public concerns about transgenic technologies in general, although AMPs are natural compounds produced by all living organisms to fight pathogen infections.

5 Use of Antifungal Peptides in Postharvest Conservation

The potential use of AMPs in the control of postharvest decay as alternative to the current use of chemical fungicides has been recognized (López-García et al., 2000; Jones and Prusky, 2002; Palou et al., 2016). There is evidence that antifungal peptides are able to control postharvest decays caused by fungi under controlled experimental conditions. Table 23.2 summarizes examples of the strategies followed for the application of AMP in the control of postharvest decay of artificially inoculated fruits. In these experiments, synthetic peptides, proteins purified from natural sources or genetically modified microorganisms producing them, were either included in the fungal inoculum or applied before or soon after inoculation. Most of the examples concern citrus and apples, the two most important tree fruits worldwide.

In the case of the synthetic PAF26 hexapeptide and derivatives, both the fungicidal activity of peptides in addition to the fungistatic activity is critical to attain a good level of protection (Muñoz et al., 2007; Badosa et al., 2009). Thus, peptides that have in vitro fungistatic activity similar to PAF26 or BM0 but are not fungicidal do not perform well in these experimental assays. By analyzing a set of antifungal peptides of variable fungicidal and/or fungistatic activity, a correlation between fungicidal properties and level of protection was demonstrated. This is mostly due to the experimental setup in which the peptide is present only at the inoculation site and not in the surrounding fruit surface adjacent regions. Thus, once a few spores are able to germinate and escape from the inoculation site they can infect fruit. Nonetheless, comparison experiments with imazalil or thiabendazole demonstrated that the PAF26 perform at least as well as the fungicides under the same experimental conditions (Figure 23.3). Another parameter

Table 23.2 Overview of experimental strategies for the application of AMPs to control postharvest decay

Peptide	*Peptide origin*	*Mode of application*	*Fungus*	*Fruit*	*Reference*
CecropiA	Insect	Genetically modified *S. cerevisiae* (point co-inoculation)	*Colletotrichum coccodes*	Tomato	Jones and Prusky (2002)
PAF26 and derivatives	Peptide library	Synthetic (point co-inoculation)	*Penicillium digitatum* *Penicillium italicum*	Citrus	López-García et al. (2003); Muñoz et al. (2007)
BM0	Peptide library	Synthetic (point co-inoculation)	*P. digitatum*	Citrus	Muñoz et al. (2007)
PsD1	Plant defensin (*P. sativum*)	Genetically modified *P. pastoris* (point pre-inoculation)	*Penicillium expansum*	Apple	Janisiewicz et al. (2008)
BP22	Peptide library (N-terminal derivatization)	Synthetic (point pre-application)	*P. expansum*	Apple	Badosa et al. (2009)
CecropinA	Insect	Genetically modified *P. pastoris* (point pre-inoculation)	*P. expansum*	Apple	Ren et al. (2012a)
CecropinA	Insect	Genetically modified *P. pastoris* (point pre-inoculation)	*P. digitatum*	Citrus	Ren et al. (2012b)
Aq-AMP	Plant (*A. quitensis*)	Recombinantly produced (*E. coli*) and purified (point co-inoculation)	*P. digitatum*	Citrus	Alem et al. (2014)
AFP	Fungal (*A. giganteus*)	Purified from fungal cultures (sprayed on fruit surface)	*Alternaria alternata*	Banana	Barakat (2014)
CecA::PsD1	Fusion of modified cecropin A and defensin PsD1	Genetically modified *P. pastoris* (point pre-inoculation)	*Geotrichum citri-aurantii*	Citrus	Kong et al. (2016)

to be considered is the susceptibility of peptides to degradation on the fruit surface. Introduction of unnatural amino acids or chemical modifications into lead peptides may improve their stability on fruits while maintaining their biological activity (Ng-Choi et al., 2014). For example, the peptide BP22 (Tables 23.1, 23.2) protected

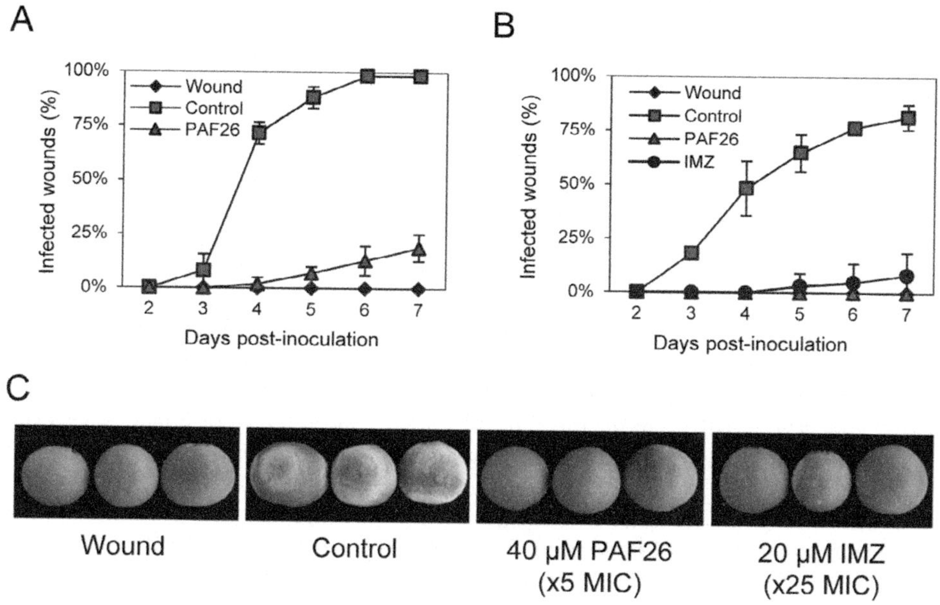

Figure 23.3 The synthetic peptide PAF26 (see Table 23.1) protects orange fruits from the green mold caused by *P. digitatum.* (A) and (B) show two independent experiments as the incidence of infection caused by the reference *P. digitatum* strain CECT20796 (2.5×10^4 conidia/mL) alone (control, squares) or in the presence of 40 μM PAF26 (triangles) and the absence of infection in the non-inoculated but wounded controls (diamonds). In (B), the activity of PAF26 is also compared with 20 μM of the fungicide imazalil (IMZ) (circles). (C) Representative images of fruits from the experiment shown in (B) pictured at 7 post-inoculation. The MIC value of PAF26 and IMZ are 8 μM and 0.8 μM, respectively, to *P. digitatum* CECT20796; therefore the experiments were conducted at 5 and 25 times the in vitro MIC value, respectively. Additional details of the inoculation procedures are found in the previously published literature (Muñoz et al., 2007).

apples against *P. expansum* decay as effectively as a commercial formulation of the fungicide imazalil (Badosa et al., 2009).

However, and despite these promising results, all of these experiments were conducted with synthetic peptides, and further pilot-scale studies that simulate commercial application remain scarce. The few field trials conducted with synthetic peptides have shown promising results, as it is the control of the brown spot of pear trees caused by the fungus *Stemphylium vesicarium* by the peptide BP15, a non-derivatized variant of BP22 (Puig et al., 2015).

5.1 The Problem of the Cost-Effective Production of AMPs

The development of AMPs for future application as antifungals requires their production in large amounts. The cost-effective production and delivery of peptides of sufficient purity is critical; this will determine if the practical use of peptides as part of postharvest conservation treatments is feasible. Even with the increasing number of peptide companies and the declining costs of synthetic procedures, production costs via chemical synthesis and purification are still too high for their application to be cost-effective for postharvest conservation, even in the case of very short peptides, and this currently precludes the use of synthetic peptides in postharvest treatments. Another alternative is the purification of AMPs from their natural sources. However, this is restricted to natural peptides, and not to those obtained by rational design or modification of natural ones. Also, usually low yields are obtained from natural sources. For example, the extraction of plant defensins from natural sources is rather complicated due to their low abundance and the presence of other compounds that coexist in the plant tissues where they are produced (Vriens et al., 2014). To circumvent this major problem, there are different alternatives to the production and use of AMPs for the control of postharvest pathogens, and they mostly rely on the biotechnological production of peptides and proteins through genetically modified organisms. One of the primary attractive attributes of AMPs is their peptidic nature that would, theoretically, enable their production through biotechnology. The heterologous production of AMPs in bacteria, fungi, or plants as biofactories has been explored and extensively reviewed in previous publications (Li, 2011; Parachin et al., 2012). It must be pointed out, however, that no AMPs in advanced clinical trials have been biotechnologically produced yet (Haney and Hancock, 2013).

5.2 AMP Production in Plants to Control Postharvest Decay

Following the approaches described in section 4, AMP expression in plants can be directed to seeds or fruits using specific gene promoters, to protect them from pathogen infection after harvesting and during postharvest storage (Bundó et al., 2014; Diaz et al., 2016; Montesinos et al., 2017). To reach this goal, the expression of AMPs in crops and fruit trees and the availability of fruit specific promoters must be validated. AMP expression associated with protection against infection has been reported in citrus fruit (Furman et al., 2013; Hao et al., 2016; Zou et al., 2017) and apples (Borejsza-Wysocka et al., 2010), although no data have been reported yet on protection against postharvest decay. On the other hand, the genomic studies and tools available for citrus or apple trees have permitted the identification of highly specific fruit, and even fruit peel, promoters that could be potentially applied (Alquézar et al., 2009; Rodrigo et al., 2013).

Engineering plants to produce AMPs can be applied not only to protect host plants against pathogen infection, but also as AMP biofactories for further uses (Abdallah et al., 2010; López-García et al., 2012). Plants represent a safe, easily scalable, and economic production system of AMPs. Although challenging due to the physicochemical properties of these peptides, several plant-based platforms have been reported to be suitable for the large-scale production of AMPs. Among them are chloroplasts-based or rice seeds-based production systems (Lee et al., 2011; Bundó et al., 2014; Montesinos et al., 2016, 2017). The success of these strategies is based on limiting the accumulation of AMPs to storage organs so as to avoid interference with plant vegetative growth; or to limit it to specialized subcellular organelles that would both protect AMPs from protease degradation and facilitate their purification (Bundó et al., 2014; Montesinos et al., 2016, 2017). Compartments that favor AMP accumulation include chloroplasts (Lee et al., 2011), protein bodies (Bundó et al., 2014; Montesinos et al., 2017), and oil bodies (Montesinos et al., 2016).

5.3 AMP Production in Microorganisms

Another alternative is the expression and in situ production of AMP by epiphytic microorganisms growing on the fruit surface, a very attractive approach that was tested early and proved feasible in proof-of-concept experiments, using genetically modified yeast laboratory strains that produced the insect peptide cecropin A (Jones and Prusky, 2002). Surprisingly, only a few reports have tested this alternative approach. Expression of cecropin A (or cecropin A-derived peptides) (Jones and Prusky, 2002; Ren et al., 2012a) or the plant defensin PsD1 (Janisiewicz et al., 2008; Kong et al., 2016), showed promising control of mostly *Colletotrichum* sp. and *Penicillium* spp. in tomato, apple, and citrus (Table 23.2).

The most widely explored approach is the use of microorganisms as biofactories for the production of AMPs (Li, 2011). In the case of AMPs, the production in microorganisms must solve the obvious problem of the intrinsic toxicity of the peptide for the producer host. Strategies such as the fusion to inactivation domains or the strict regulation/induction of gene expression have solved the toxicity issue in some cases. Production of functional plant defensins fused to histidine tags that facilitate purification in the bacterium *Escherichia coli* was reported (Kant et al., 2009). The CRP Ac-AMP2 found in *Amaranthus* (Broekaert et al., 1992) was fused to the protein thioredoxin and a histidine tag, then produced in *E. coli* and purified using metal affinity chromatography (Alem et al., 2014). Yields of 38 mg/L of the purified peptide fusion were reported. The purified peptide controlled the green mold decay of citrus under experimental inoculations (Table 23.2).

Nonetheless, eukaryotic systems are preferred for CRPs with disulphide bonds and complex folding. Some plant defensins and AFPs have been efficiently produced in the recombinant system *Pichia pastoris*, but the maximum yields of the well-processed proteins were not as high as desired and in general always lower than yields reported in bacteria (Kant et al., 2009; López-García et al., 2010b; Parachin et al., 2012; Virágh et al., 2014). High-throughput and systematic optimization of the expression and production of AMPs is being attempted (Schreiber et al., 2017), but yields of at most 20 mg/L for bacteria (*E. coli*) and 1 mg/L for yeast (*P. pastoris*) are still too low to support practical use.

In a twist of biotechnology, it is possible that the best producers of antifungal peptides and proteins are the fungi themselves. The biotechnological production and commercial use of proteins and enzymes by genetically modified filamentous fungi is well established (Meyer et al., 2016). As described in the previous sections, the AFPs of fungi are secreted and can be easily purified from the culture broth. These properties offer an opportunity for the use of these strains for the production of AFPs as biofungicides in the agricultural and food industry (Delgado et al., 2016; Leiter et al., 2017). Biotechnologically produced AFPs could be used as additional compounds for the postharvest treatment of fruits or vegetables. Proof-of-concept in the control of fungal infection of plants was initially provided by the application of solutions containing AFP purified from its natural producer, *A. giganteus* (Vila et al., 2001; Moreno et al., 2003). More recently, harvested bananas sprayed with

AFP solutions were partially protected from infection by *Alternaria alternata* (Barakat, 2014) (Table 23.2). A step forward was taken recently with the development of specific gene promoters and systems for the efficient production of a variety of heterologous AFPs, with protein yields ranging up to 100 mg/L, and the use of filamentous fungi as AFP biofactories (Sonderegger et al., 2016). Interestingly, this latter report showed that the citrus postharvest pathogen *P. digitatum* is an effective biofactory for the production of AFPs, with no self-toxicity under the production conditions even though the produced proteins (for instance, the PAF from *P. chrysogenum*) are highly toxic to the fungus when applied exogenously.

6 Concluding Remarks

AMPs have drawn considerable attention of the scientific community as alternatives to the current crisis of antibiotics and fungicides. Distinct types of antifungal peptides that include short linear peptides and CRPs have demonstrated in vitro activity against fungal phytopathogens and in vivo control of postharvest decay of fruits. Mechanistic studies are demonstrating novel modes of action and mechanisms of specificity that reinforce the interest in these antifungal peptides. However, the high cost of chemical synthesis, the low yields obtained via purification from natural sources, and the difficulties of production through biotechnology have limited so far their use in crop protection or postharvest conservation. Current research has the objective to accomplish cost-effective production of these novel biofungicides, in order to facilitate extensive pilot assays of their effectiveness and practical integration with other novel postharvest management strategies.

Acknowledgments

We apologize to all the colleagues whose research could not be appropriately cited due to space limitations. Work on antifungal peptides in our laboratories is funded by grants BIO2015-68790-C2-1-R and BIO2015-68790-C2-2-R from the "Ministerio de Economía y Competitividad" (Spain) (MINECO/FEDER Funds).

References

Abdallah, N.A., Shah, D., Abbas, D. and Madkour, M. 2010. Stable integration and expression of a plant defensin in tomato confers resistance to fusarium wilt. *GM Crops* 1, 344–350.

Aerts, A.M., François, I.E.J.A., Bammens, L., Cammue, B.P.A., Smets, B., Winderickx, J., Accardo, S., De Vos, D.E. and Thevissen, K. 2006. Level of M(IP)(2)C sphingolipid affects plant defensin sensitivity, oxidative stress resistance and chronological life-span in yeast. *FEBS Letters* 580, 1903–1907.

Aerts, A.M., François, I.E.J.A., Meert, E.M.K., Li, Q.T., Cammue, B.P.A. and Thevissen, K. 2007. The antifungal activity of RsAFP2, a plant defensin from *Raphanus sativus*, involves the induction of reactive oxygen species in *Candida albicans*. *Journal of Molecular Microbiology and Biotechnology* 13, 243–247.

Alem, D., Díaz-Dellavalle, P., Leoni, C., De-Simone, S.G., Correa, A., Oppezzo, P. and Rizza, M.D. 2014. In search of topical agricultural biofungicides: Properties of the recombinant antimicrobial peptide TrxAq-AMP Obtained from *Amaranthus quitensis*. *Journal of Microbial and Biochemical Technology* 6, 268–273.
Alquézar, B., Zacarías, L. and Rodrigo, M.J. 2009. Molecular and functional characterization of a novel chromoplast-specific lycopene β-cyclase from Citrus and its relation to lycopene accumulation. *Journal of Experimental Botany* 60, 1783–1797.
Badosa, E., Ferré, R., Francés, J., Bardají, E., Feliu, L., Planas, M. and Montesinos, E. 2009. Sporicidal activity of synthetic antifungal undecapeptides and control of *Penicillium* rot of apples. *Applied and Environmental Microbiology* 75, 5563–5569.
Ballester, A.R., Marcet-Houben, M., Levin, E., Sela, N., Selma-Lázaro, C., Carmona, L., Wisniewski, M., Droby, S., González-Candelas, L. and Gabaldón, T. 2015. Genome, transcriptome, and functional analyses of *Penicillium expansum* provide new insights into secondary metabolism and pathogenicity. *Molecular Plant-Microbe Interactions* 28, 232–248.
Barakat, H. 2014. Bio-control of *Alternaria alternata* during banana storage by purified AFP using isoelectric focusing technique. *Food and Nutrition Sciences* 5, 1482–1495.
Barbosa, A.A.T., Mantovani, H.C. and Jain, S. 2017. Bacteriocins from lactic acid bacteria and their potential in the preservation of fruit products. *Critical Reviews in Biotechnology* 37, 852–864.
Batta, G., Barna, T., Gáspári, Z., Sándor, S., Kövér, K.E., Binder, U., Sarg, B., Kaiserer, L., Chhillar, A.K., Eigentler, A., Leiter, E., Hegedüs, N., Pócsi, I., Lindner, H. and Marx, F. 2009. Functional aspects of the solution structure and dynamics of PAF - A highly-stable antifungal protein from *Penicillium chrysogenum*. *The FEBS Journal* 276, 2875–2890.
Binder, U., Benčina, M., Eigentler, A., Meyer, V. and Marx, F. 2011. The *Aspergillus giganteus* antifungal protein AFP_{NN5353} activates the cell wall integrity pathway and perturbs calcium homeostasis. *BMC Microbiology* 11, 209.
Binder, U., Benčina, M., Fizil, A., Batta, G., Chhillar, A.K. and Marx, F. 2015. Protein kinase A signaling and calcium ions are major players in PAF mediated toxicity against *Aspergillus niger*. *FEBS Letters* 589, 1266–1271.
Binder, U., Chu, M., Read, N.D. and Marx, F. 2010. The antifungal activity of the *Penicillium chrysogenum* protein PAF disrupts calcium homeostasis in *Neurospora crassa*. *Eukaryotic Cell* 9, 1374–1382.
Bleackley, M.R., Payne, J.A.E., Hayes, B.M.E., Durek, T., Craik, D.J., Shafee, T.M.A., Poon, I.K.H., Hulett, M.D., Van der Weerden, N.L. and Anderson, M.A. 2016. *Nicotiana alata* defensin chimeras reveal differences in the mechanism of fungal and tumor cell killing and an enhanced antifungal variant. *Antimicrobial Agents and Chemotherapy* 60, 6302–6312.
Bleackley, M.R., Wiltshire, J.L., Perrine-Walker, F., Vasa, S., Burns, R.L., Van der Weerden, N.L. and AnDerson, M.A. 2014. Agp2p, the plasma membrane transregulator of polyamine uptake, regulates the antifungal activities of the plant defensin nad1 and other cationic peptides. *Antimicrobial Agents and Chemotherapy* 58, 2688–2698.
Blondelle, S.E., Pérez-Payá, E. and Houghten, R.A. 1996. Synthetic combinatorial libraries: Novel discovery strategy for identification of antimicrobial agents. *Antimicrobial Agents and Chemotherapy* 40, 1067–1071.

Borejsza-Wysocka, E., Norelli, J.L., Aldwinckle, H.S. and Malnoy, M. 2010. Stable expression and phenotypic impact of attacin E transgene in orchard grown apple trees over a 12 year period. *BMC Biotechnology* 10, 41.

Bou Zeidan, M., Carmona, L., Zara, S. and Marcos, J.F. 2013. *FLO11* gene is involved in the interaction of flor strains of *Saccharomyces cerevisiae* with a biofilm-promoting synthetic hexapeptide. *Applied and Environmental Microbiology* 79, 2023–2032.

Broekaert, W.F., Marien, W., Terras, F.R.G., De Bolle, M.F.C., Proost, P., Van Damme, J., Dillen, L., Claeys, M., Rees, S.B., Vanderleyden, J. and Cammue, B. P.A. 1992. Antimicrobial peptides from *Amaranthus caudatus* seeds with sequence homology to the cysteine/glycine-rich domain of chitin-binding proteins. *Biochemistry* 31, 4308–4314.

Brogden, K.A. 2005. Antimicrobial peptides: Pore formers or metabolic inhibitors in bacteria? *Nature Reviews Microbiology* 3, 238–250.

Brown, G.D., Denning, D.W., Gow, N.A.R., Levitz, S.M., Netea, M.G. and White, T.C. 2012. Hidden killers: Human fungal infections. *Science Translational Medicine* 4, 165rv13.

Bundó, M., Montesinos, L., Izquierdo, E., Campo, S., Mieulet, D., Guiderdoni, E., Rossignol, M., Badosa, E., Montesinos, E., San Segundo, B. and Coca, M. 2014. Production of cecropin A antimicrobial peptide in rice seed endosperm. *BMC Plant Biology* 14, 102.

Campos-Olivas, R., Bruix, M., Santoro, J., Lacadena, J., Martínez del Pozo, A., Gavilanes, J.G. and Rico, M. 1995. NMR solution structure of the antifungal protein from *Aspergillus giganteus*: Evidence for cysteine pairing isomerism. *Biochemistry* 34, 3009–3021.

Carmona, L., Gandía, M., López-García, B. and Marcos, J.F. 2012. Sensitivity of *Saccharomyces cerevisiae* to the cell-penetrating antifungal peptide PAF26 correlates with endogenous nitric oxide (NO) production. *Biochemical and Biophysical Research Communications* 417, 56–61.

Carvalho, A. and Gomes, V.M. 2009. Plant defensins-Prospects for the biological functions and biotechnological properties. *Peptides* 30, 1007–1020.

Cary, J.W., Rajasekaran, K., Jaynes, J.M. and Cleveland, T.E. 2000. Transgenic expression of a gene encoding a synthetic antimicrobial peptide results in inhibition of fungal growth in vitro and in planta. *Plant Science* 154, 171–181.

Cavallarín, L., Andreu, D. and San Segundo, B. 1998. Cecropin A-derived peptides are potent inhibitors of fungal plant pathogens. *Molecular Plant-Microbe Interactions* 11, 218–227.

Chan, Y.L., Prasad, V., Sanjaya, Kuei, H.C., Po, C.L., Chan, M.T. and Cheng, C.P. 2005. Transgenic tomato plants expressing an *Arabidopsis* thionin (Thi2.1) driven by fruit-inactive promoter battle against phytopathogenic attack. *Planta* 221, 386–393.

Chen, Z., Ao, J., Yang, W., Jiao, L., Zheng, T. and Chen, X. 2013. Purification and characterization of a novel antifungal protein secreted by *Penicillium chrysogenum* from an Arctic sediment. *Applied Microbiology and Biotechnology* 97, 10381–10390.

Coca, M., Bortolotti, C., Rufat, M., Peñas, G., Eritja, R., Tharreau, D., Martínez del Pozo, A., Messeguer, J. and San Segundo, B. 2004. Transgenic rice plants expressing the antifungal AFP protein from *Aspergillus giganteus* show enhanced resistance to the rice blast fungus *Magnaporthe grisea*. *Plant Molecular Biology* 54, 245–259.

Coca, M., Damsz, B., Yun, D.J., Hasegawa, P.M., Bressan, R.A. and Narasimhan, M.L. 2000. Heterotrimeric G-proteins of a filamentous fungus regulate cell wall composition and susceptibility to a plant PR-5 protein. *The Plant Journal* 22, 61–69.

Coca, M., Penas, G., Gomez, J., Campo, S., Bortolotti, C., Messeguer, J. and Segundo, B.S. 2006. Enhanced resistance to the rice blast fungus *Magnaporthe grisea* conferred by expression of a cecropin A gene in transgenic rice. *Planta* 223, 392–406.

Cotter, P.D., Hill, C. and Ross, R.P. 2005. Bacteriocins: Developing innate immunity for food. *Nature Reviews Microbiology* 3, 777–788.

De Lucca, A.J., Bland, J.M., Jacks, T.J., Grimm, C., Cleveland, T.E. and Walsh, T.J. 1997. Fungicidal activity of Cecropin A. *Antimicrobial Agents and Chemotherapy* 41, 481–483.

de Medeiros, L.N., Angeli, R., Sarzedas, C.G., Barreto-Bergter, E., Valente, A.P., Kurtenbach, E. and Almeida, F.C.L. 2010. Backbone dynamics of the antifungal Psd1 pea defensin and its correlation with membrane interaction by NMR spectroscopy. *Biochimica et Biophysica Acta – Biomembranes* 1798, 105–113.

Delgado, J., Acosta, R., Rodriguez-Martin, A., Bermudez, E., Nunez, F. and Asensio, M.A. 2015. Growth inhibition and stability of PgAFP from *Penicillium chrysogenum* against fungi common on dry-ripened meat products. *International Journal of Microbiology* 205, 23–29.

Delgado, J., Owens, R.A., Doyle, S., Asensio, M.A. and Nunez, F. 2016. Antifungal proteins from moulds: Analytical tools and potential application to dry-ripened foods. *Applied Microbiology and Biotechnology* 100, 6991–7000.

Diaz, A.H., Kovacs, I. and Lindermayr, C. 2016. Inducible expression of the *de novo*-designed antimicrobial peptide SP1-1 in tomato confers resistance to *Xanthomonas campestris* pv. *Vesicatoria*. *PLoS ONE* 11, e0164097.

Dracatos, P.M., Payne, J., Di Pietro, A., Anderson, M.A. and Plummer, K.M. 2016. Plant defensins NaD1 and NaD2 induce different stress response pathways in fungi. *International Journal of Molecular Sciences* 17, 1473.

Duncan, V.M.S. and O'Neil, D.A. 2013. Commercialization of antifungal peptides. *Fungal Biology Reviews* 26, 156–165.

Fjell, C.D., Hiss, J.A., Hancock, R.E.W. and Schneider, G. 2012. Designing antimicrobial peptides: Form follows function. *Nature Reviews Drug Discovery* 11, 37–51.

Furman, N., Kobayashi, K., Zanek, M.C., Calcagno, J., García, M.L. and Mentaberry, A. 2013. Transgenic sweet orange plants expressing a dermaseptin coding sequence show reduced symptoms of citrus canker disease. *Journal of Biotechnology* 167, 412–419.

Galgóczy, L., Virágh, M., Kovács, L., Tóth, B., Papp, T. and Vágvölgyi, C. 2013. Antifungal peptides homologous to the *Penicillium chrysogenum* antifungal protein (PAF) are widespread among Fusaria. *Peptides* 39, 131–137.

Ganz, T. 2003. Defensins: Antimicrobial peptides of innate immunity. *Nature Reviews Immunology* 3, 710–720.

Gao, A.G., Hakimi, S.M., Mittanck, C.A., Wu, Y., Woerner, B.M., Stark, D.M., Shah, D.M., Liang, J.H. and Rommens, C.M.T. 2000. Fungal pathogen protection in potato by expression of a plant defensin peptide. *Nature Biotechnology* 18, 1307–1310.

Garrigues, S., Gandía, M., Borics, A., Marx, F., Manzanares, P. and Marcos, J.F. 2017. Mapping and identification of antifungal peptides in the putative antifungal protein AfpB from the filamentous fungus *Penicillium digitatum*. *Frontiers in Microbiology* 8, 592.

Garrigues, S., Gandía, M. and Marcos, J.F. 2016. Occurrence and function of fungal antifungal proteins: A case study of the citrus postharvest pathogen *Penicillium digitatum*. *Applied Microbiology and Biotechnology* 100, 2243–2256.

Gaspar, Y.M., McKenna, J.A., McGinness, B.S., Hinch, J., Poon, S., Connelly, A.A., Anderson, M.A. and Heath, R.L. 2014. Field resistance to *Fusarium oxysporum* and *Verticillium dahliae* in transgenic cotton expressing the plant defensin NaD1. *Journal of Experimental Botany* 65, 1541–1550.

Gonzalez, C.F., Provin, E.M., Zhu, L. and Ebbole, D.J. 2002. Independent and synergistic activity of synthetic peptides against thiabendazole-resistant *Fusarium sambucinum*. *Phytopathology* 92, 917–924.

Guidotti, G., Brambilla, L. and Rossi, D. 2017. Cell-penetrating peptides: From basic research to clinics. *Trends in Pharmacological Sciences* 38, 406–424.

Guilhelmelli, F., Vilela, N., Albuquerque, P., Derengowski, L.S., Silva-Pereira, I. and Kyaw, C.M. 2013. Antibiotic development challenges: The various mechanisms of action of antimicrobial peptides and of bacterial resistance. *Frontiers in Microbiology* 4, 353.

Hancock, R.E.W. and Sahl, H.G. 2006. Antimicrobial and host-defense peptides as new anti-infective therapeutic strategies. *Nature Biotechnology* 24, 1551–1557.

Haney, E.F. and Hancock, R.E.W. 2013. Peptide design for antimicrobial and immunomodulatory applications. *Biopolymers* 100, 572–583.

Hao, G., Stover, E. and Gupta, G. 2016. Overexpression of a modified plant Thionin enhances disease resistance to Citrus canker and Huanglongbing (HLB). *Frontiers in Plant Science* 7, 1078.

Harries, E., Carmona, L., Muñoz, A., Ibeas, J.I., Read, N.D., Gandía, M. and Marcos, J.F. 2013. Genes involved in protein glycosylation determine the activity and cell internalization of the antifungal peptide PAF26 in *Saccharomyces cerevisiae*. *Fungal Genetics and Biology* 58-59, 105–115.

Harries, E., Gandia, M., Carmona, L. and Marcos, J.F. 2015. The *Penicillium digitatum* protein *O*-mannosyltransferase Pmt2 is required for cell wall integrity, conidiogenesis, virulence and sensitivity to the antifungal peptide PAF26. *Molecular Plant Pathology* 16, 748–761.

Hayes, B.M.E., Anderson, M.A., Traven, A., van der Weerden, N.L. and Bleackley, M.R. 2014. Activation of stress signalling pathways enhances tolerance of fungi to chemical fungicides and antifungal proteins. *Cellular and Molecular Life Sciences* 71, 2651–2666.

Hayes, B.M.E., Bleackley, M.R., Wiltshire, J.L., Anderson, M.A., Traven, A. and van der Weerden, N.L. 2013. Identification and mechanism of action of the plant defensin NaD1 as a new member of the antifungal drug arsenal against *Candida albicans*. *Antimicrobial Agents and Chemotherapy* 57, 3667–3675.

Hegedüs, N. and Marx, F. 2013. Antifungal proteins: More than antimicrobials? *Fungal Biology Reviews* 26, 132–145.

Hegedüs, N., Sigl, C., Zadra, I., Pocsi, I. and Marx, F. 2011. The *paf* gene product modulates asexual development in *Penicillium chrysogenum*. *Journal of Basic Microbiology* 51, 253–262.

Henriques, S.T., Melo, M.N. and Castanho, M.A.R.B. 2006. Cell-penetrating peptides and antimicrobial peptides: How different are they? *Biochemical Journal* 399, 1–7.

Hirooka, T. and Ishii, H. 2013. Chemical control of plant diseases. *Journal of General Plant Pathology* 79, 390–401.

Holásková, E., Galuszka, P., Frébort, I. and Öz, M.T. 2014. Antimicrobial peptide production and plant-based expression systems for medical and agricultural biotechnology. *Biotechnology Advances* 33, 1005–1023.

Holmes, G.J. and Eckert, J.W. 1999. Sensitivity of *Penicillium digitatum* and *P. italicum* to postharvest citrus fungicides in California. *Phytopathology* 89, 716–721.

Ibeas, J.I., Yun, D.J., Damsz, B., Narasimhan, M.L., Uesono, Y., Ribas, J.C., Lee, H., Hasegawa, P.M., Bressan, R.A. and Pardo, J.M. 2001. Resistance to the plant PR-5 protein osmotin in the model fungus *Saccharomyces cerevisiae* is mediated by the regulatory effects of SSD1 on cell wall composition. *The Plant Journal* 25, 271–280.

Janisiewicz, W.J., Pereira, I.B., Almeida, M.S., Roberts, D.P., Wisniewski, A. and Kurtenbach, E. 2008. Improved biocontrol of fruit decay fungi with *Pichia pastoris* recombinant strains expressing Psd1 antifungal peptide. *Postharvest Biology and Technology* 47, 218–225.

Jha, S. and Chattoo, B.B. 2009. Transgene stacking and coordinated expression of plant defensins confer fungal resistance in rice. *Rice* 2, 143–154.

Jones, R.W. and Prusky, D. 2002. Expression of an antifungal peptide in *Saccharomyces*: A new approach for biological control of the postharvest disease caused by *Colletotrichum coccodes*. *Phytopathology* 92, 33–37.

Kaewklom, S., Euanorasetr, J., Intra, B., Panbangred, W. and Aunpad, R. 2016. Antimicrobial activities of novel peptides derived from defensin genes of *Brassica hybrid* cv Pule. *International Journal of Peptide Research and Therapeutics* 22, 93–100.

Kaiserer, L., Oberparleiter, C., Weiler-Gorz, R., Burgstaller, W., Leiter, E. and Marx, F. 2003. Characterization of the *Penicillium chrysogenum* antifungal protein PAF. *Archives of Microbiology* 180, 204–210.

Kant, P., Liu, W.Z. and Pauls, K.P. 2009. PDC1, a corn defensin peptide expressed in *Escherichia coli* and *Pichia pastoris* inhibits growth of *Fusarium graminearum*. *Peptides* 30, 1593–1599.

Ko, K.S., Norelli, J.L., Reynoird, J.P., Aldwinckle, H.S. and Brown, S.K. 2002. T4 lysozyme and attacin genes enhance resistance of transgenic 'Galaxy' apple against *Erwinia amylovora*. *Journal of the American Society for Horticultural Science* 127, 515–519.

Kong, Q., Liang, Z., Xiong, J., Li, H. and Ren, X. 2016. Overexpression of the bivalent antibacterial peptide genes in *Pichia pastoris* delays sour rot in citrus fruit and induces *Geotrichum citri-aurantii* cell apoptosis. *Food Biotechnology* 30, 79–97.

Kovács, L., Virágh, M., Takó, M., Papp, T., Vágvölgyi, C. and Galgóczy, L. 2011. Isolation and characterization of *Neosartorya fischeri* antifungal protein (NFAP). *Peptides* 32, 1724–1731.

Lee, D.G., Shin, S.Y., Maeng, C.Y., Jin, Z.Z., Kim, K.L. and Hahm, K.S. 1999. Isolation and characterization of a novel antifungal peptide from *Aspergillus niger*. *Biochemical and Biophysical Research Communications* 263, 646–651.

Lee, S., Li, B., Jin, S. and Daniell, H. 2011. Expression and characterization of antimicrobial peptides Retrocyclin-101 and Protegrin-1 in chloroplasts to control viral and bacterial infections. *Plant Biotechnology Journal* 9, 100–115.

Leiter, É., Gáll, T., Csernoch, L. and Pócsi, I. 2017. Biofungicide utilizations of antifungal proteins of filamentous ascomycetes: Current and foreseeable future developments. *Biocontrol* 62, 125–138.

Leiter, E., Szappanos, H., Oberparleiter, C., Kaiserer, L., Csernoch, L., Pusztahelyi, T., Emri, T., Pócsi, I., Salvenmoser, W. and Marx, F. 2005. Antifungal protein PAF severely affects the integrity of the plasma membrane of *Aspergillus nidulans* and induces an apoptosis-like phenotype. *Antimicrobial Agents and Chemotherapy* 49, 2445–2453.

Li, X.S., Reddy, M.S., Baev, D. and Edgerton, M. 2003. *Candida albicans* Ssa1/2p Is the cell envelope binding protein for human salivary Histatin 5. *Journal of Biological Chemistry* 278, 28553–28561.

Li, Y. 2011. Recombinant production of antimicrobial peptides in *Escherichia coli*: A review. *Protein Expression and Purification* 80, 260–267.

Lobo, D.S., Pereira, I.B., Fragel-Madeira, L., Medeiros, L.N., Cabral, L.M., Faria, J., Bellio, M., Campos, R.C., Linden, R. and Kurtenbach, E. 2007. Antifungal *Pisum sativum* defensin 1 interacts with *Neurospora crassa* cyclin F related to the cell cycle. *Biochemistry* 46, 987–996.

López-García, B., Gandía, M., Muñoz, A., Carmona, L. and Marcos, J.F. 2010a. A genomic approach highlights common and diverse effects and determinants of susceptibility on the yeast *Saccharomyces cerevisiae* exposed to distinct antimicrobial peptides. *BMC Microbiology* 10, 289.

López-García, B., González-Candelas, L., Pérez-Payá, E. and Marcos, J.F. 2000. Identification and characterization of a hexapeptide with activity against phytopathogenic fungi that cause postharvest decay in fruits. *Molecular Plant-Microbe Interactions* 13, 837–846.

López-García, B., Harries, E., Carmona, L., Campos-Soriano, L., López, J.J., Manzanares, P., Gandía, M., Coca, M. and Marcos, J.F. 2015. Concatemerization increases the inhibitory activity of short, cell-penetrating, cationic and tryptophan-rich antifungal peptides. *Applied Microbiology and Biotechnology* 99, 8011–8021.

López-García, B., Moreno, A.B., San Segundo, B., De los Ríos, V., Manning, J.M., Gavilanes, J.G. and Martínez del Pozo, A. 2010b. Production of the biotechnologically relevant AFP from *Aspergillus giganteus* in the yeast *Pichia pastoris*. *Protein Expression and Purification* 70, 206–210.

López-García, B., Pérez-Payá, E. and Marcos, J.F. 2002. Identification of novel hexapeptides bioactive against phytopathogenic fungi through screening of a synthetic peptide combinatorial library. *Applied and Environmental Microbiology* 68, 2453–2460.

López-García, B., San Segundo, B. and Coca, M. 2012. Antimicrobial peptides as a promising alternative for plant disease protection. pp. 263–294. *In*: K. Rajasekaran, J.W. Cary, J. Jaynes and E. Montesinos (eds.). *Small Wonders: Peptides for Disease Control*, Vol. 1095. American Chemical Society, Washington, DC.

López-García, B., Veyrat, A., Pérez-Payá, E., González-Candelas, L. and Marcos, J.F. 2003. Comparison of the activity of antifungal hexapeptides and the fungicides thiabendazole and imazalil against postharvest fungal pathogens. *International Journal of Food Microbiology* 89, 163–170.

Marcet-Houben, M., Ballester, A.-R., de la Fuente, B., Harries, E., Marcos, J.F., González-Candelas, L. and Gabaldón, T. 2012. Genome sequence of the necrotrophic fungus *Penicillium digitatum*, the main postharvest pathogen of citrus. *BMC Genomics* 13, 646.

Marcos, J.F. and Beachy, R.N. 1994. In vitro characterization of a cassette to accumulate multiple proteins through synthesis of a self-processing polypeptide. *Plant Molecular Biology* 24, 495–503.

Marcos, J.F. and Gandía, M. 2009. Antimicrobial peptides: To membranes and beyond. *Expert Opinion on Drug Discovery* 4, 659–671.

Marcos, J.F., Gandía, M., Harries, E., Carmona, L. and Muñoz, A. 2012. Antifungal peptides: Exploiting non-lytic mechanisms and cell penetration properties. pp. 337–357. *In*: K. Rajasekaran, J.W. Cary, J. Jaynes and E. Montesinos (eds.). *Small Wonders: Peptides for Disease Control*, Vol. 1095. American Chemical Society, Washington, DC.

Marcos, J.F., Muñoz, A., Pérez-Payá, E., Misra, S. and López-García, B. 2008. Identification and rational design of novel antimicrobial peptides for plant protection. *Annual Review of Phytopathology* 46, 273–301.

Marx, F., Haas, H., Reindl, M., Stöffler, G., Lottspeich, F. and Redl, B. 1995. Cloning, structural organization and regulation of expression of the *Penicillium chrysogenum paf* gene encoding an abundantly secreted protein with antifungal activity. *Gene* 167, 167–171.

Meyer, V., Andersen, M.R., Brakhage, A.A., Braus, G.H., Caddick, M.X., Cairns, T.C., de Vries, R.P., Haarmann, T., Hansen, K., Hertz-Fowler, C., Krappmann, S., Mortensen, U.H., Peñalva, M.A., Ram, A.F.J. and Head, R.M. 2016. Current challenges of research on filamentous fungi in relation to human welfare and a sustainable bio-economy: A white paper. *Fungal Biology and Biotechnology* 3, 6.

Mills, D., Hammerschlag, F.A., Nordeen, R.O. and Owens, L.D. 1994. Evidence for the breakdown of cecropin-B by proteinases in the intercellular fluid of peach leaves. *Plant Science* 104, 17–22.

Monk, B.C., Niimi, K., Lin, S., Knight, A., Kardos, T.B., Cannon, R.D., Parshot, R., King, A., Lun, D. and Harding, D.R.K. 2005. Surface-active fungicidal D-peptide inhibitors of the plasma membrane proton pump that block azole resistance. *Antimicrobial Agents and Chemotherapy* 49, 57–70.

Montesinos, E. 2007. Antimicrobial peptides and plant disease control. *FEMS Microbiology Letters* 270, 1–11.

Montesinos, L., Bundo, M., Badosa, E., San Segundo, B., Coca, M. and Montesinos, E. 2017. Production of BP178, a derivative of the synthetic antibacterial peptide BP100, in the rice seed endosperm. *BMC Plant Biology* 17, 63.

Montesinos, L., Bundó, M., Izquierdo, E., Campo, S., Badosa, E., Rossignol, M., Montesinos, E., San Segundo, B. and Coca, M. 2016. Production of biologically active cecropin A peptide in rice seed oil bodies. *PLoS ONE* 11, e0146919.

Moreno, A.B., Martínez del Pozo, A., Borja, M. and San Segundo, B. 2003. Activity of the antifungal protein from *Aspergillus giganteus* against *Botrytis cinerea*. *Phytopathology* 93, 1344–1353.

Moreno, A.B., Martínez del Pozo, A. and San Segundo, B. 2006. Biotechnologically relevant enzymes and proteins - Antifungal mechanism of the *Aspergillus giganteus* AFP against the rice blast fungus *Magnaporthe grisea*. *Applied Microbiology and Biotechnology* 72, 883–895.

Moreno, A.B., Peñas, G., Rufat, M., Bravo, J.M., Estopà, M., Messeguer, J. and San Segundo, B. 2005. Pathogen-induced production of the antifungal AFP protein from *Aspergillus giganteus* confers resistance to the blast fungus *Magnaporthe grisea* in transgenic rice. *Molecular Plant-Microbe Interactions* 18, 960–972.

Muñoz, A., Gandía, M., Harries, E., Carmona, L., Read, N.D. and Marcos, J.F. 2013a. Understanding the mechanism of action of cell-penetrating antifungal peptides using the rationally designed hexapeptide PAF26 as a model. *Fungal Biology Reviews* 26, 146–155.

Muñoz, A., Harries, E., Contreras-Valenzuela, A., Carmona, L., Read, N.D. and Marcos, J.F. 2013b. Two functional motifs define the interaction, internalization and toxicity of the cell-penetrating antifungal peptide PAF26 on fungal cells. *PLoS ONE* 8, e54813.

Muñoz, A., López-García, B. and Marcos, J.F. 2006. Studies on the mode of action of the antifungal hexapeptide PAF26. *Antimicrobial Agents and Chemotherapy* 50, 3847–3855.

Muñoz, A., López-García, B. and Marcos, J.F. 2007. Comparative study of antimicrobial peptides to control citrus postharvest decay caused by *Penicillium digitatum*. *Journal of Agricultural and Food Chemistry* 55, 8170–8176.

Muñoz, A. and Marcos, J.F. 2006. Activity and mode of action against fungal phytopathogens of bovine lactoferricin-derived peptides. *Journal of Applied Microbiology* 101, 1199–1207.

Muñoz, A., Marcos, J.F. and Read, N.D. 2012. Concentration-dependent mechanisms of cell penetration and killing by the *de novo* designed antifungal hexapeptide PAF26. *Molecular Microbiology* 85, 89–106.

Munro, C.A. 2013. Chitin and glucan, the yin and yang of the fungal cell wall, implications for antifungal drug discovery and therapy. pp. 145–172. *In*: S. Sima and M. G. Geoffrey (eds.). *Advances in Applied Microbiology*, Vol. 83. Elsevier, Academic Press, Waltham, MA.

Nakaya, K., Omata, K., Okahashi, I., Nakamura, Y., Kolekenbrock, H. and Ulbrich, N. 1990. Amino acid sequence and disulfide bridges of an antifungal protein isolated from *Aspergillus giganteus*. *European Journal of Biochemistry* 193, 31–38.

Narasimhan, M.L., Coca, M.A., Jin, J.B., Yamauchi, T., Ito, Y., Kadowaki, T., Kim, K. K., Pardo, J.M., Damsz, B., Hasegawa, P.M., Yun, D.J. and Bressan, R.A. 2005. Osmotin is a homolog of mammalian adiponectin and controls apoptosis in yeast through a homolog of mammalian adiponectin receptor. *Molecular Cell* 17, 171–180.

Narasimhan, M.L., Damsz, B., Coca, M.A., Ibeas, J.I., Yun, D.J., Pardo, J.M., Hasegawa, P.M. and Bressan, R.A. 2001. A plant defense response effector induces microbial apoptosis. *Molecular Cell* 8, 921–930.

Ng-Choi, I., Soler, M., Güell, I., Badosa, E., Cabrefiga, J., Bardají, E., Montesinos, E., Planas, M. and Feliu, L. 2014. Antimicrobial peptides incorporating non-natural amino acids as agents for plant protection. *Protein and Peptide Letters* 21, 357–367.

Nguyen, L.T., Haney, E.F. and Vogel, H.J. 2011. The expanding scope of antimicrobial peptide structures and their modes of action. *Trends in Biotechnology* 29, 464–472.

Nicolas, P. 2009. Multifunctional host defense peptides: Intracellular-targeting antimicrobial peptides. *The FEBS Journal* 276, 6483–6496.

Oberparleiter, C., Kaiserer, L., Haas, H., Ladurner, P., Andratsch, M. and Marx, F. 2003. Active internalization of the *Penicillium chrysogenum* antifungal protein

PAF in sensitive Aspergilli. *Antimicrobial Agents and Chemotherapy* 47, 3598–3601.
Oldach, K.H., Becker, D. and Lörz, H. 2001. Heterologous expression of genes mediating enhanced fungal resistance in transgenic wheat. *Molecular Plant-Microbe Interactions* 14, 832–838.
Olson, B.H. and Goerner, G.L. 1965. Alpha sarcin, a new antitumor agent. I. Isolation, purification, chemical composition, and the identity of a new amino acid. *Applied Microbiology* 13, 314–321.
Osusky, M., Zhou, G., Osuska, L., Hancock, R.E.W., Kay, W.W. and Misra, S. 2000. Transgenic plants expressing cationic peptide chimeras exhibit broad-spectrum resistance to phytopathogens. *Nature Biotechnology* 18, 1162–1166.
Ouedraogo, J.P., Hagen, S., Spielvogel, A., Engelhardt, S. and Meyer, V. 2011. Survival strategies of yeast and filamentous fungi against the antifungal protein AFP. *Journal of Biological Chemistry* 286, 13859–13868.
Paege, N., Jung, S., Schape, P., Muller-Hagen, D., Ouedraogo, J.P., Heiderich, C., Jedamzick, J., Nitsche, B.M., van den Hondel, C.A., Ram, A.F. and Meyer, V. 2016. A transcriptome meta-analysis proposes novel biological roles for the antifungal protein AnAFP in *Aspergillus niger*. *PLoS ONE* 11, e0165755.
Palou, L., Ali, A., Fallik, E. and Romanazzi, G. 2016. GRAS, plant- and animal-derived compounds as alternatives to conventional fungicides for the control of postharvest diseases of fresh horticultural produce. *Postharvest Biology and Technology* 122, 41–52.
Parachin, N.S., Mulder, K.C., Viana, A.A.B., Dias, S.C. and Franco, O.L. 2012. Expression systems for heterologous production of antimicrobial peptides. *Peptides* 38, 446–456.
Poon, I.K.H., Baxter, A.A., Lay, F.T., Mills, G.D., Adda, C.G., Payne, J.A.E., Phan, T. K., Ryan, G.F., White, J.A., Veneer, P.K., van der Weerden, N.L., Anderson, M. A., Kvansakul, M. and Hulett, M.D. 2014. Phosphoinositide-mediated oligomerization of a defensin induces cell lysis. *eLife* 3, e01808.
Porto, W.F., Pires, A.S. and Franco, O.L. 2017. Computational tools for exploring sequence databases as a resource for antimicrobial peptides. *Biotechnology Advances* 35, 337–349.
Puig, M., Moragrega, C., Ruz, L., Montesinos, E. and Llorente, I. 2015. Controlling brown spot of pear by a synthetic antimicrobial peptide under field conditions. *Plant Disease* 99, 1816-1822.
Rahnamaeian, M., Langen, G., Imani, J., Khalifa, W., Altincicek, B., Von Wettstein, D., Kogel, K.H. and Vilcinskas, A. 2009. Insect peptide metchnikowin confers on barley a selective capacity for resistance to fungal ascomycetes pathogens. *Journal of Experimental Botany* 60, 4105–4114.
Ramamoorthy, V., Cahoon, E.B., Jia, L., Thokala, M., Minto, R.E. and Shah, D.M. 2007a. Glucosylceramide synthase is essential for alfalfa defensin-mediated growth inhibition but not for pathogenicity of *Fusarium graminearum*. *Molecular Microbiology* 66, 771–786.
Ramamoorthy, V., Zhao, X.H., Snyder, A.K., Xu, J.R. and Shah, D.M. 2007b. Two mitogen-activated protein kinase signalling cascades mediate basal resistance to antifungal plant defensins in *Fusarium graminearum*. *Cellular Microbiology* 9, 1491–1506.
Rao, A.G. 1995. Antimicrobial peptides. *Molecular Plant-Microbe Interactions* 8, 6–13.

Rautenbach, M., Troskie, A.M. and Vosloo, J.A. 2016. Antifungal peptides: To be or not to be membrane active. *Biochimie* 130, 132–145.

Reed, J.D., Edwards, D.L. and Gonzalez, C.F. 1997. Synthetic peptide combinatorial libraries: A method for the identification of bioactive peptides against phytopathogenic fungi. *Molecular Plant-Microbe Interactions* 10, 537–549.

Ren, X., Kong, Q., Wang, H., Yu, T., Tang, Y.J., Zhou, W.W. and Zheng, X. 2012a. Control of apple blue mold by *Pichia pastoris* recombinant strains expressing cecropin A. *Bioprocess and Biosystems Engineering* 35, 761–767.

Ren, X., Kong, Q., Wang, H., Yu, T., Zhou, W.W. and Zheng, X. 2012b. Biocontrol of fungal decay of citrus fruit by *Pichia pastoris* recombinant strains expressing cecropin A. *Food Chemistry* 131, 796–801.

Rodrigo, M.J., Alquézar, B., Alós, E., Medina, V., Carmona, L., Bruno, M., Al-Babili, S. and Zacarías, L. 2013. A novel carotenoid cleavage activity involved in the biosynthesis of Citrus fruit-specific apocarotenoid pigments. *Journal of Experimental Botany* 64, 4461–4478.

Rodríguez-Martín, A., Acosta, R., Liddell, S., Nuñez, F., Benito, M.J. and Asensio, M. A. 2010. Characterization of the novel antifungal protein PgAFP and the encoding gene of *Penicillium chrysogenum*. *Peptides* 31, 541–547.

Rothstein, D.M., Spacciapoli, P., Tran, L.T., Xu, T., Roberts, F.D., Dalla Serra, M., Buxton, D.K., Oppenheim, F.G. and Friden, P. 2001. Anticandida activity is retained in P-113, a 12-amino-acid fragment of histatin 5. *Antimicrobial Agents and Chemotherapy* 45, 1367–1373.

Sagaram, U.S., El-Mounadi, K., Buchko, G.W., Berg, H.R., Kaur, J., Pandurangi, R.S., Smith, T.J. and Shah, D.M. 2013. Structural and functional studies of a phosphatidic acid-binding antifungal plant defensin MtDef4: Identification of an RGFRRR motif governing fungal cell entry. *PLoS ONE* 8, e82485.

Sagaram, U.S., Pandurangi, R., Kaur, J., Smith, T.J. and Shah, D.M. 2011. Structure-activity determinants in antifungal plant defensins MsDef1 and MtDef4 with different modes of action against *Fusarium graminearum*. *PLoS ONE* 6, e18550.

Schreiber, C., Muller, H., Birrenbach, O., Klein, M., Heerd, D., Weidner, T., Salzig, D. and Czermak, P. 2017. A high-throughput expression screening platform to optimize the production of antimicrobial peptides. *Microbial Cell Factories* 16, 29.

Seibold, M., Wolschann, P., Bodevin, S. and Olsen, O. 2011. Properties of the bubble protein, a defensin and an abundant component of a fungal exudate. *Peptides* 32, 1989-1995.

Shah, D.M. and Read, N.D. 2013. Antifungal peptides come of age. *Fungal Biology Reviews* 26, 107–108.

Silva, P.M., Gonçalves, S. and Santos, N.C. 2014. Defensins: Antifungal lessons from eukaryotes. *Frontiers in Microbiology* 5, 97.

Skouri-Gargouri, H. and Gargouri, A. 2008. First isolation of a novel thermostable antifungal peptide secreted by *Aspergillus clavatus*. *Peptides* 29, 1871-1877.

Sonderegger, C., Fizil, Á., Burtscher, L., Hajdu, D., Muñoz, A., Gáspári, Z., Read, N. D., Batta, G. and Marx, F. 2017. D19S Mutation of the cationic, cysteine-rich protein PAF: Novel insights into its structural dynamics, thermal unfolding and antifungal function. *PLoS ONE* 12, e0169920.

Sonderegger, C., Galgóczy, L., Garrigues, S., Fizil, Á., Borics, A., Manzanares, P., Hegedüs, N., Huber, A., Marcos, J.F., Batta, G. and Marx, F. 2016.

A *Penicillium chrysogenum*-based expression system for the production of small, cysteine-rich antifungal proteins for structural and functional analyses. *Microbial Cell Factories* 15, 192.

Stock, S.D., Hama, H., Radding, J.A., Young, D.A. and Takemoto, J.Y. 2000. Syringomycin E inhibition of *Saccharomyces cerevisiae*: Requirement for biosynthesis of sphingolipids with very-long-chain fatty acids and mannose- and phosphoinositol-containing head groups. *Antimicrobial Agents and Chemotherapy* 44, 1174–1180.

Terras, F.R.G., Eggermont, K., Kovaleva, V., Raikhel, N.V., Osborn, R.W., Kester, A., Rees, S.B., Torrekens, S., Van Leuven, F., Vanderleyden, J., Cammue, B.P.A. and Broekaert, W.F. 1995. Small cysteine-rich antifungal proteins from radish: Their role in host defense. *The Plant Cell* 7, 573–588.

Terras, F.R.G., Torrekens, S., Van Leuven, F., Osborn, R.W., Vanderleyden, J., Cammue, B.P.A. and Broekaert, W.F. 1993. A new family of basic cysteine-rich plant antifungal proteins from Brassicaceae species. *FEBS Letters* 316, 233–240.

Thevissen, K., Cammue, B.P.A., Lemaire, K., Winderickx, J., Dickson, R.C., Lester, R. L., Ferket, K.K., Van Even, F., Parret, A.H. and Broekaert, W.F. 2000. A gene encoding a sphingolipid biosynthesis enzyme determines the sensitivity of *Saccharomyces cerevisiae* to an antifungal plant defensin from dahlia (*Dahlia merckii*). *Proceedings of the National Academy of Sciences of the USA* 97, 9531–9536.

Thevissen, K., Idkowiak-Baldys, J., Im, Y.J., Takemoto, J., François, I.E.J.A., Ferket, K.K.A., Aerts, A.M., Meert, E.M.K., Winderickx, J., Roosen, J. and Cammue, B.P.A. 2005. SKN1, a novel plant defensin-sensitivity gene in *Saccharomyces cerevisiae*, is implicated in sphingolipid biosynthesis. *FEBS Letters* 579, 1973-1977.

Thevissen, K., Warnecke, D.C., François, I.E.J.A., Leipelt, M., Heinz, E., Ott, C., Zahringer, U., Thomma, B.P.H.J., Ferkel, K.K.A. and Cammue, B.P.A. 2004. Defensins from insects and plants interact with fungal glucosylceramides. *Journal of Biological Chemistry* 279, 3900–3905.

Thevissen, K., Yen, W.L., Carmona-Gutierrez, D., Idkowiak-Baldys, J., Aerts, A.M., François, I.E.J.A., Madeo, F., Klionsky, D.J., Hannun, Y.A. and Cammue, B.P.A. 2010. Skn1 and Ipt1 negatively regulate autophagy in *Saccharomyces cerevisiae*. *FEMS Microbiology Letters* 303, 164–168.

Tóth, L., Kele, Z., Borics, A., Nagy, L.G., Váradi, G., Virágh, M., Takó, M., Vágvölgyi, C. and Galgóczy, L. 2016. NFAP2, a novel cysteine-rich anti-yeast protein from *Neosartorya fischeri* NRRL 181: Isolation and characterization. *AMB Express* 6, 75.

Turrà, D., Segorbe, D. and Di Pietro, A. 2014. Protein kinases in plant-pathogenic Fungi: Conserved regulators of infection. *Annual Review of Phytopathology* 52, 267–288.

van der Weerden, N.L. and Anderson, M.A. 2013. Plant defensins: Common fold, multiple functions. *Fungal Biology Reviews* 26, 121–131.

van der Weerden, N.L., Bleackley, M.R. and Anderson, M.A. 2013. Properties and mechanisms of action of naturally occurring antifungal peptides. *Cellular and Molecular Life Sciences* 70, 3545–3570.

van der Weerden, N.L., Hancock, R.E.W. and Anderson, M.A. 2010. Permeabilization of fungal hyphae by the plant defensin NaD1 occurs through a cell wall-dependent process. *Journal of Biological Chemistry* 285, 37513–37520.

van der Weerden, N.L., Lay, F.T. and Anderson, M.A. 2008. The plant defensin, NaD1, enters the cytoplasm of *Fusarium oxysporum* hyphae. *Journal of Biological Chemistry* 283, 14445–14452.

Vila, L., Lacadena, V., Fontanet, P., Martínez del Pozo, A. and San Segundo, B. 2001. A protein from the mold *Aspergillus giganteus* is a potent inhibitor of fungal plant pathogens. *Molecular Plant-Microbe Interactions* 14, 1327–1331.

Virágh, M., Vörös, D., Kele, Z., Kovács, L., Fizil, Á., Lakatos, G., Maróti, G., Batta, G., Vágvölgyi, C. and Galgóczy, L. 2014. Production of a defensin-like antifungal protein NFAP from *Neosartorya fischeri* in *Pichia pastoris* and its antifungal activity against filamentous fungal isolates from human infections. *Protein Expression and Purification* 94, 79–84.

Vriens, K., Cammue, B.P.A. and Thevissen, K. 2014. Antifungal plant defensins: Mechanisms of action and production. *Molecules* 19, 12280–12303.

Wilmes, M., Cammue, B.P.A., Sahl, H.G. and Thevissen, K. 2011. Antibiotic activities of host defense peptides: More to it than lipid bilayer perturbation. *Natural Product Reports* 28, 1350–1358.

Wnendt, S., Ulbrich, N. and Stahl, U. 1994. Molecular cloning, sequence analysis and expression of the gene encoding an antifungal-protein from *Aspergillus giganteus. Current Genetics* 25, 519–523.

Yeung, A.T.Y., Gellatly, S.L. and Hancock, R.E.W. 2011. Multifunctional cationic host defence peptides and their clinical applications. *Cellular and Molecular Life Sciences* 68, 2161–2176.

Yevtushenko, D.P. and Misra, S. 2012. Transgenic expression of antimicrobial peptides in plants: Strategies for enhanced disease resistance, improved productivity, and production of therapeutics. pp. 445–458. *In*: K. Rajasekaran, J.W. Cary, J. Jaynes and E. Montesinos (eds.). *Small Wonders: Peptides for Disease Control*, Vol. 1095. American Chemical Society, Washington, DC.

Yount, N.Y. and Yeaman, M.R. 2004. Multidimensional signatures in antimicrobial peptides. *Proceedings of the National Academy of Sciences of the USA* 101, 7363–7368.

Yun, D.J., Ibeas, J.I., Lee, H., Coca, M.A., Narasimhan, M.L., Uesono, Y., Hasegawa, P.M., Pardo, J.M. and Bressan, R.A. 1998. Osmotin, a plant antifungal protein, subverts signal transduction to enhance fungal cell susceptibility. *Molecular Cell* 1, 807–817.

Zasloff, M. 2002. Antimicrobial peptides of multicellular organisms. *Nature* 415, 389–395.

Zhao, X., Wu, H., Lu, H., Li, G. and Huang, Q. 2013. LAMP: A database linking antimicrobial peptides. *PLoS ONE* 8, e66557.

Zou, X., Jiang, X., Xu, L., Lei, T., Peng, A., He, Y., Yao, L. and Chen, S. 2017. Transgenic citrus expressing synthesized cecropin B genes in the phloem exhibits decreased susceptibility to Huanglongbing. *Plant Molecular Biology* 93, 341–353.

Chapter 24

Induced Resistance in Fruits and Vegetables by Elicitors to Control Postharvest Diseases

Bi Yang

College of Food Science and Engineering, Gansu Agricultural University, Lanzhou, China

Xue Huali

College of Science, Gansu Agricultural University, Lanzhou, China

Wang Junjie

College of Biological Science and Engineering, North Minzu University, Yinchuan, China

Abbreviations

APX	Ascorbate peroxidase
BABA	β-aminobutyric acid
BTH	Benzothiadiazole
CAT	Catalase
CHT	Chitinase
C4H	Cinnamic acid-4-hydroxylase
4CL	4-coumarate-CoA-ligase
INA	2,6-dichloronicotinic acid
GLU	β-1,3-glucanase
GR	Glutathione reductase
HR	Hypersensitive response
ISR	Induced systemic resistance
JA	Jasmonic acid
MeSA	Methyl salicylic acid
NOX	NADPH oxidase
NO	Nitric oxide
PR	Pathogenesis-related
POD	Peroxidase
PAL	Phenylalanine ammonia-lyase
PPO	Polyphenol oxidase
ROS	Reactive oxygen species
SA	Salicylic acid
SOD	Superoxide dismutase
SAR	Systemic acquired resistance

1 Introduction

Fresh fruits and vegetables are quite perishable and susceptible to attack by various pathogens. Postharvest diseases begin by either latent infections established in the field or the infection of wounds during subsequent harvesting and handling (Bi et al., 2007; Tian et al., 2007). Recent research showed that more than one-third of harvested fruits and vegetables are lost annually because of inadequate sanitation and refrigeration chain in developing countries, while losses may be lower in developed countries (FAO, 2011; Prusky, 2011; Okawa, 2015).

Application of synthetic fungicides is generally the main strategy to control postharvest decay (Tian et al., 2007; Romanazzi et al., 2016). However, within the last decade, some fungicides have been withdrawn from the market due to the resistance of pathogens to the fungicide, toxicity to humans and the environment, increasing consumer concerns about risk of fungicide residues on products, and increasing costs of registration and re-registration (Romanazzi et al., 2016). Therefore, new strategies that are alternatives to synthetic fungicides to control postharvest decay with characteristics such as low cost, a high degree of safety, and eco-friendliness are in urgent demand (Xue et al., 2017).

Induced resistance by various biotic inducers (such as fungi, bacteria, and viruses) and abiotic stresses (such as chemical and physical inducers) is considered to be a sustainable strategy to control postharvest decay of fruits and vegetables, and it has gained increasing attention during the last years (Romanazzi et al., 2016). Induced resistance is a natural defense response triggered by pathogens or elicitors, and it includes both local resistance and systemic resistance (Kuc, 2001). Induction of local resistance implies that disease resistance is directly triggered at the site of pathogen infection, while induction of systemic resistance implies that disease resistance is triggered at an uninfected plant part and then expressed widely in the host (Ryals et al., 1996; Bi et al., 2007). Systemic resistance can be classified into systemic acquired resistance (SAR) and induced systemic resistance (ISR), based on different signal pathways. The former is dependent on the signal molecule salicylic acid (SA) and it is associated with the accumulation of pathogenesis-related (PR) proteins, while the latter is dependent on the signal molecules jasmonic acid (JA) and ethylene (Van Wees et al., 1999).

This chapter reviews the various physical, chemical, and biological inducers that have been proven useful to manage postharvest disease, and also briefly presents knowledge on the mechanisms of induced resistance in fruits and vegetables.

2 Induced Resistance by Physical Inducers

Control of postharvest diseases by induction of natural resistance by using physical elicitors has received considerable attention in recent years (Terry and Joyce, 2004; Palou et al., 2008). The primary mode of action of many postharvest physical treatments is considered to be the disinfection of the product, which avoids or stops the infectious process by fungal spores or mycelia. However, physical treatments can also elicit induced resistance to protect against subsequent infections in some commodities. Application of ultraviolet-C (UV-C) radiation, heat treatment, ionizing irradiation, and other physical means before storage at low temperatures and/or in controlled atmospheres may enhance natural disease resistance.

2.1 UV-C Irradiation

UV-C irradiation is an environmentally friendly strategy to control postharvest disease that has been studied for many different horticultural products (Charles and Arul, 2007). The first study on the effects of UV-C irradiation was conducted with carrots (Mercier et al., 1993). Low doses (0.5–8 kJ/m^2) of UV-C (wavelengths from 100 to 280 nm) can control many postharvest diseases of fruits and vegetables (Palou et al., 2008), although the effectiveness of this treatment and the risk of phytotoxicity varies with irradiation dose and duration, fruit type and maturity, harvest time, and storage conditions (Bi, 1996). Since UV-C irradiation targets the DNA of the microorganism, the treatment has been used as a germicidal or mutagenic agent. In addition to the direct germicidal effects of illumination with UV-C on conidia and mycelia of pathogens, the prevalent mode of action of this treatment to control postharvest diseases of fresh produce is the elicitation of beneficial responses in the host when applied at sub-lethal doses (hormesis) (Palou et al., 2008).

Postharvest UV-C irradiation increased the resistance of citrus fruits against infection by *Botrytis cinerea* Pers. (Ben-Yehoshua et al., 1992). Exposure of tomato to UV-C irradiation at 7.5 kJ/m^2 significantly reduced the disease caused by *Alternaria alternata* (Fr.) Keissl. (Liu et al., 1993). Exposure to UV-C irradiation at 0.5 kJ/m^2 effectively controlled gray mold of kiwifruit caused by *B. cinerea* (Nigro et al., 1998). Similar results were found by Stevens et al. (1999), who suggested that UV-C irradiation at 4.8 kJ/m^2 noticeably decreased Fusarium root rot of sweet potato tuber. Pre-storage UV-C irradiation treatments have also been used to control *B. cinerea* on bell pepper (Mercier et al., 2001). Stevens et al. (2004) showed that UV-C irradiation delayed ripening and Rhizopus soft rot development on tomato. Exposure to UV-C irradiation at a sub-lethal dose of 4.1 kJ/m^2 delayed strawberry fruit softening (Pombo et al., 2009). Defensive reactions in fruits and vegetables as a response to UV-C include alteration in levels of phytoalexins, accumulation of PR proteins, and induction of the activity of enzymes related to resistance. Pombo et al. (2011) found that the reduction of postharvest disease caused by *B. cinerea* on strawberry fruits treated with UV-C at harvest was related to the increase in the transcription and activity of enzymes and proteins involved in the defense against pathogens. During the response of tomato fruit to postharvest UV-C irradiation, 274 genes mainly involved in signal transduction, defense response, and metabolism were up-regulated, while 403 genes related to cell wall disassembly, photosynthesis, and lipid metabolism were down-regulated (Liu et al., 2011). However, there are few reports describing induced resistance stimulated by UV-C in fruits or vegetables. The physiological mechanism by which resistance to *B. cinerea* occurred in tomato fruit treated with UV-C was researched systemically by chemical analysis, electrophoresis, ultrastructure, and tissue chemical localization. Charles et al. (2008d) reported that UV-C irradiation stimulated the accumulation of the phytoalexin rishitin, caused ultrastructural modifications of the fruit surface, which impacted fungal colonization (Charles et al., 2008a, 2008c), induced biochemical modifications in structural barriers (Charles et al., 2008b), and increased the activity of constitutive defense enzymes and inducible PR proteins (Charles et al., 2009).

2.2 Heat Treatment

Heat treatment is one of the most traditional, commonly used, and promising means to control postharvest diseases of fruits and vegetables. It has been shown that heat treatment not only directly inhibits pathogen development, but it also induces natural resistance in postharvest commodities. Heat treatment has been successfully applied to manage postharvest diseases in fruits and vegetables (Schirra et al., 2000).

Heat treatments can be applied to fresh produce after harvest by application of hot water, hot dry air, or vapor heat fumigation for brief periods (Terry and Joyce, 2004). Postharvest heat treatments decreased the development of *Penicillium expansum* L. on apple fruits (Fallik et al., 1995) and enhanced the resistance of tomatoes to gray mold caused by *B. cinerea* (Lurie et al., 1997). Heat treatment at 42°C of mature-green tomato fruits reduced decay by 60% (McDonald et al., 1999).

As a physical elicitor, heat can elicit the accumulation of antifungal compounds, the enhancement of PR proteins, or the improvement in the activities of defensive enzymes in fruits and vegetables. Yuan et al. (2013) reported that the activities of phenylalanine ammonia-lyase (PAL), cinnamic acid-4-hydroxylase (C4H), 4-coumarate-CoA-ligase (4CL), peroxidase (POD), and polyphenol oxidase (PPO) were significantly enhanced in muskmelon fruit treated with hot water. Similar results were observed in peach (Jin et al., 2009), Chinese bayberry (Wang et al., 2010), and loquat (Liu et al., 2010). Chitinase (CHT) and β-1,3-glucanase (GLU) are enzymes that belong to a PR protein family involved in the degradation of the fungal cell wall. Liu et al. (2012) reported that the levels of expression of CHT and GLU genes were improved in peach fruit treated with heat. The accumulation of antifungal compounds such as phenolic compounds, flavonoid substances, and lignin has been associated with disease resistance in fruits and vegetables (Mayr et al., 1997; Shadle et al., 2003; Deepak et al., 2007; Tronchet et al., 2010; Yuan et al., 2013).

3 Induced Resistance by Chemical Elicitors

3.1 Salicylic Acid (SA) and Its Analogs

Salicylic acid, *o*-hydroxybenzoic acid, is the most widely distributed simple phenolic compound in plants. It is synthesized via the shikimate pathway from chorismate through the activity of isochorismate synthase. SA is a key signaling molecule that participates in the mechanisms of defense (it is involved in ISR), plant development, fruit ripening, and responses to various abiotic factors (Sticher et al., 1997; Barkai-Golan, 2001; Romanazzi et al., 2016). Exogenous SA application triggers the development of both local and systemic defense responses (Martinez et al., 2000; Cordelier et al., 2003; Alkan and Fortes, 2015) and the induction of the hypersensitive response (HR) and programmed cell death in plants (Mur et al., 1997; Alkan et al., 2012).

SA can also effectively induce disease resistance in fruits and vegetables after harvest, although the effectiveness of SA application greatly varies with fruit and maturation stage, mode of SA treatment, and SA concentration. Poole et al. (1998) reported that SA at 0.14 mg/mL applied as spray before harvest or as a soaking treatment before storage induced natural resistance to gray mold in kiwifruit. Preharvest or postharvest SA treatment at 2.0 mg/mL effectively reduced the incidence of mango postharvest anthracnose (Zainuri et al., 2001). SA application effectively reduced postharvest rots of citrus fruits (Iqbal et al., 2012; Zhu et al., 2016). Mangoes treated with SA at 2 mg/mL showed suppressed anthracnose caused by *Colletotrichum gloeosporioides* (Penz.) Penz. & Sacc. (Zainuri et al., 2001). SA application to muskmelons at 0.4 mg/L induced resistance to Alternaria rots and pink rot after harvest (Ge et al., 2006). Decay incidence and lesion diameter caused by *P. expansum* were effectively reduced on sweet cherry fruits treated with SA at 0.5 mM (Chan and Tian, 2006).

The mechanism of induction of defense responses by SA is related to the accumulation of reactive oxygen species (ROS), which can kill the pathogen cells (Mittler et al., 2011). ROS levels are regulated by the activities of enzymes such as NADPH oxidase (NOX) and antioxidants such as SOD, ascorbate peroxidase (APX), glutathione reductase (GR), and catalase (CAT) (Sharma et al., 2012). Moreover, SA-induced defense responses involve the expression of defense-related genes, especially those encoding PR proteins such as CHT, GLU, PPO, PAL, and POD. A combination treatment of SA and antagonistic yeast significantly increased the activities of PPO, PAL, and GLU in cherries (Qin et al., 2003). SA increased the activities of PAL, CHT, GLU, and GR and reduced the activities of CAT and APX in pears (Cao et al., 2006). The application to cherry of the SA analog methyl salicylic acid (MeSA) increased the activities of CAT, POD, APX, and SOD (Valverde et al., 2015). Therefore, the resistance induced by SA in fruits and vegetables consists of a global defense response, including activation of the phenylpropanoid pathway, induction of accumulation of ROS and antioxidants, and production of PR proteins (Liu et al., 2014; Wang et al., 2015).

3.2 Benzothiadiazole (BTH)

Benzothiadiazole [benzo (1,2,3)-thiadiazole-7-carbothioicacid S-methyl ester, BTH; syn.: acibenzolar-S-methyl, ASM] is perhaps the most potent synthetic elicitor discovered to date (Terry and Joyce, 2004; Bi et al., 2010). BTH is a functional analogue of SA that induces resistance against a broad range of phytopathogens by activating the SAR pathway in plants (Mandal et al., 2008; Boava et al., 2009).

Preharvest BTH application effectively controlled postharvest disease in fruits and vegetables, including strawberry (Terry and Joyce, 2000; Mazaro et al., 2008; Cao et al., 2011), pear (Cao and Jiang, 2005), peach (Liu et al., 2005a, 2005b), muskmelon (Ge et al., 2008; Zhang et al., 2011b; Ren et al., 2012; Liu et al., 2014), and potato (Bokshi et al., 2003) by increasing disease resistance in these hosts to control latent infections by the pathogens. The incidence of latent infections by various fungal pathogens in muskmelons significantly decreased by preharvest application of BTH at 100 mg/L (Zhang et al., 2011b). Field treatment with BTH significantly enhanced the resistance of pears to black rot and blue mold caused by *A. alternata* and *P. expansum*, respectively (Cao and Jiang, 2006). Postharvest application of BTH at the concentration of 100 mg/L significantly reduced disease caused by *Trichothecium roseum* (Pers.) Link in both artificially inoculated and naturally infected melons (Bi et al., 2006a; Ren et al., 2012). However, BTH at concentrations higher than 300 mg/L failed to promote better resistance to pathogens and was phytotoxic (Bi et al., 2006a). Postharvest BTH treatment by spraying or soaking also effectively controlled postharvest blue mold of peach (Liu et al., 2005a, 2005b) caused by *P. expansum*, anthracnose of mango caused by *C. gloeosporioides* (Lin et al., 2011; Pan and Liu, 2011), pink rot of muskmelon caused by *T. roseum* (Bi et al., 2006a; Wang et al., 2008; Ren et al., 2012), and gray mold of strawberry (Mazaro et al., 2008; Cao et al., 2011) and tomato (Iriti et al., 2007) caused by *B. cinerea*. Vacuum treatment with BTH at 0.5 mM decreased the severity of Alternaria rot and blue mold of 'Yali' pear (Cao et al., 2005). Alternaria rot, Fusarium rot, and pink rot of melons were significantly suppressed by soaking in BTH at 100 mg/L (Bi et al., 2006a; Ge et al., 2008; Wang et al., 2008; Ren et al., 2012). In addition, anthracnose of mango and gray mold of tomato were significantly suppressed by a postharvest BTH treatment (Zhu et al., 2008).

The mechanism of disease resistance induced by BTH is linked to the induction of the expression of resistance genes and an increase in the activity of a range of enzymes involved in defense reactions in fruits and vegetables. Cao and Jiang (2006) reported that the activities of PR proteins and enzymes such as POD, CHT, and GLU were significantly enhanced in pears treated with BTH. Similar results were observed in potato (Bokshi et al., 2003), peach (Liu et al., 2005a), and melon (Bi et al., 2006a; Zhang et al., 2011b). Rapid generation of ROS has been recognized as one of the earliest events correlated with plant resistance to pathogens (Mittler et al., 2011). It was reported that the levels of hydrogen peroxide (H_2O_2) and the activities of enzymes related with its metabolism, including NOX, SOD, APX, GR, and CAT, were increased in fruits treated with BTH (Liu et al., 2005a; Cao and Jiang, 2006; Zhang et al., 2011b; Ren et al., 2012). In order to strengthen the structure of the cell wall and avoid the ingress by pathogens, BTH activates phenylpropanoid pathway by enhancing the activity of PAL, C4H, 4CL, and CAD, as well as augmenting the contents of total phenols, flavonoids, and lignin in fruits (Zhang et al., 2011b; Liu et al., 2014). Therefore, resistance induced by BTH application in fruits and vegetables is related to the induction of various defensive responses, including activation of the phenylpropanoid pathway, induction of ROS accumulation, and production of PR proteins (Ge et al., 2008).

3.3 2,6-Dichloroisonicotinic Acid (INA)

The synthetic compound 2,6-dichloroisonicotinic acid (INA) is a structural and functional analogue of SA. INA can induce host resistance to a broad spectrum of pathogens by activating the SAR pathway in plants (Bi et al., 2010). Lucas (1999) suggested that the disease resistance mediated by INA is persistent.

A preharvest foliar spray of INA at 50 mg/L significantly reduced postharvest diseases of melon fruits (Bokshi et al., 2005, 2006). Preharvest treatment with INA enhanced resistance to anthracnose caused by *C. gloeosporioides* on mango (Santiago et al., 2006). The incidence and severity of decay on mango fruits were reduced by postharvest vacuum infiltration treatments of INA at 20 and 100 mg/L (Zeng and Jiang, 2009). Postharvest treatment with INA at 0.5 g/L effectively reduced the size of disease spots on the peel of bananas inoculated with a spore suspension of *Colletotrichum musae* (Berk. & M.A. Curtis) Arx (Huang et al., 2011b).

INA acts as a signal molecule and its application regulates and activates the host defense response by inducing the accumulation of ROS, which can directly kill the pathogen cells (Baker and Orlandi, 1995; Mittler et al., 2011). However, ROS metabolism relates to the activities of enzymes including NOX, SOD, APX, GR, and CAT. The increase in host resistance is closely related to the content of H_2O_2 and the activity of the enzymes responsible for resistance. INA application increased the content of H_2O_2 and decreased the activity of enzymes related to H_2O_2 scavenging, including CAT and APX, and suppressed the gene expression of both enzymes.

3.4 Jasmonic Acid (JA) and Methyl Jasmonate (MJ)

Jasmonic acid is a signal molecule naturally synthetized in plants that plays a crucially important role in diverse physiological processes in plant development and fruit maturation. JA not only stimulates the synthesis of proteins related to defense responses and the accumulation of secondary metabolites, but also induces postharvest resistance to pathogens in fruits and vegetables.

Application of JA or methyl jasmonate (MJ) has effectively controlled decay in horticultural plants. Preharvest MJ treatment at 0.2 mM reduced brown rot of sweet cherry and was more effective than if applied after harvest (Yao and Tian, 2005b). Postharvest treatment with JA at 0.01 mM significantly reduced green mold of grapefruit and orange (Droby et al., 1999; Porat et al., 2002). MJ at 0.2 mM reduced the diameter of lesions on sweet cherry fruit infected with *Monilinia fructicola* (G. Winter) Honey (Yao and Tian, 2005a). In addition, effective concentrations of JA or MJ to induce disease resistance vary greatly among different fruits infected with the same pathogen, and among different pathogens infecting the same fruit. For instance, the minimum effective concentration of MJ to control tomato anthracnose was 0.0448 mM (Tzortzakis, 2007), while it was 10 mM to control tomato gray mold (Zhu and Tian, 2012).

JA stimulated the production of signaling molecules related to resistance, the accumulation of antimicrobial compounds, and also strengthened structural barriers, which limited pathogen infection (Tian et al., 2007). MJ application promoted the accumulation of signaling molecules, such as NO and H_2O_2, induced an increase in the activities of the enzymes PAL, C4H, 4CL, and RS, and stimulated an increase in contents of the phytoalexin resveratrol and its dehydrodimer (Wang et al., 2009). The activity of GLU after preharvest application of MJ at 0.2 mM once every day for 3 d was 1.8 times higher than in the controls at harvest, and the activities of PAL and POD were also increased (Yao and Tian, 2005b). Similarly, the activities of CHT and GLU were increased in peach by MJ treatment, which induced resistance to brown rot and blue mold (Yao and Tian, 2005a). MJ treatment promoted the early accumulation of H_2O_2 and increased the expression of Cu-Zn, SOD, CAT, and APX genes. Moreover, MJ treatment also enhanced the contents of ASA and GSH, which could scavenge the excess of ROS to alleviate oxidation injury to proteins (Zhu and Tian, 2012).

3.5 Silicates

Silicon (Si) is the second most abundant element in the Earth's lithosphere, and it is as important as phosphorus and magnesium in the biota. Also, it is considered to be biologically active and trigger a faster and more extensive deployment of plant natural defenses. Silicon oxide and sodium silicate significantly reduced postharvest pink rot severity caused by *T. roseum* on muskmelons (Guo et al., 2007). Sodium silicate at 100 mM also reduced rots of melons caused by *A. alternata, F. semitectum*, and *T. roseum* (Bi et al., 2006b; Li et al., 2012). However, a higher concentration of 200 mM was not more effective and caused rind injury (Bi et al., 2006b). Results with potato tubers were similar, where sodium silicate applied at 100 mM decreased the area of dry rot lesions caused by *Fusarium sulphureum* Schltdl. (Xue et al., 2017).

Sodium silicate proved effective for suppressing pathogen growth, as well as inducing resistance to disease in fruits and vegetables. Si also inhibited the conidial germination and mycelial growth of pathogens. For example, soluble Si treatment inhibited conidial germination and appressorial formation of the strawberry powdery mildew pathogen *Sphaerotheca aphanis* var. *aphanis* (Kanto et al., 2007). Bi et al. (2006b) also found that sodium silicate significantly inhibited the mycelial growth of *A. alternata, F. semitectum*, and *T. roseum* in vitro. Although Si treatment failed to increase the activities of POD, PPO, PAL, and GLU or to enhance the contents of total phenols or flavonoids in potato tubers, a significant increase in the activities of these enzymes and the contents of total phenols or flavonoids was found after a challenge inoculation with *F. sulphureum* (Li et al., 2009a). These results indicated that Si primed resistance responses to this pathogen in potato tuber (Conrath et al., 2001). Bi et al. (2006b) reported that Si-induced protection was correlated with the activation of POD and CHT in 'Hami' melons. Si also induced a significant increase in the activities of PAL, PPO, and POD in sweet cherry (Qin and Tian, 2005). Recently, Li et al. (2012) reported that the effects of Si on postharvest disease of muskmelon fruit may be associated with the elicitation of antioxidant defense systems in the fruit.

3.6 Other Chemical Elicitors

The application of oxalic acid has been shown to induce systemic resistance to diseases (Zhang et al., 2009). Dipping artificially inoculated muskmelon fruits after harvest in an oxalic acid solution at 50 mM reduced the diameter of lesions caused by *A. alternata, F. semitectum*, and *T. roseum*, but the use of a higher concentration and longer exposure periods failed to improve resistance to these pathogens and were phytotoxic (Deng et al., 2008). Zheng et al. (2005) found that treatment with oxalic acid at 5 mM increased antioxidant capacity and PPO activity, which were beneficial in delaying ripening and enhancing disease resistance in peach fruit during cold storage.

Citric acid at 2% effectively inhibited the occurrence of decay and the appearance of superficial brown, black, and withered stems in pear fruit (Cao and Jiang, 2005). Citric acid at 45 mM effectively inhibited decay on potato tubers and slices inoculated with *F. sulphureum*. This treatment increased the activities of POD, PPO, PAL, and GLU and enhanced the contents of flavonoids and total phenolic compounds in the tubers (Zhang et al., 2009).

Although β-aminobutyric acid (BABA) is only rarely found naturally in plants, it has been proven to be a potent inducer of acquired resistance with a broad spectrum of activity to inhibit various pathogens. BABA application reduced the development of dry rot on potato tubers inoculated with *F. sulphureum* (Xue et al., 2017). Postharvest treatment of BABA (50 mmol/L) combined with hot water (45°C) effectively reduced the diameter of lesions on apples inoculated with *P. expansum* (Li et al., 2011). BABA caused brief activation of CHT and POD activities on melon after a first spray application that increased after the second spray, but these levels were lower than those induced by INA (Bokshi et al., 2005, 2006).

Nitric oxide (NO) is an important bioactive molecule with complex signal functions in plants (Beligni and Lamattina, 2001). NO, a free radical gas, is soluble in both aqueous and hydrophobic environments (Thomas et al., 2008). The NO molecule not only easily moves between hydrophilic areas of the cytoplasm, but also freely diffuses through the lipid phase of the membranes (Arasimowicz and Floryszak-Wieczorek, 2007). NO is an endogenous maturation and senescence regulating factor in higher plants (Leshem et al., 1998). NO at low concentration extended the shelf life of strawberry (Soegiarto and Wills, 2006), peach (Zhu et al., 2006), longan (Duan et al., 2007), and winter jujube (Zhu et al., 2009). Lai et al. (2011) suggested that NO treatment suppressed the production of ethylene, increased the activity of antioxidant enzymes, up-regulated related after-ripening gene expression, delayed the reddening of tomato, and prolonged the retention of fruit quality during storage. Furthermore, NO also plays a key role in the induction of resistance to pathogens. Niritake et al. (1996) showed that treatment with 1-hydroxy-2-oxo-3,3-*bis*(2-aminoethyl)-1-triazene (NOC-18), a NO-releasing compound, induced the accumulation of the phytoalexin rishitin in potato tuber tissues. Clark et al. (2000) used a variety of NO donors to demonstrate that NO inhibited tobacco CAT and APX activity. Delledonne et al. (1998) showed that NO potentiates the induction of hypersensitive cell death in soybean cells by reactive oxygen intermediates and functions independently of such intermediates to induce genes for the synthesis of protective natural products.

Many inorganic salts, such as bicarbonates, phosphites, chlorides, phosphates, and others, can suppress the growth of pathogenic fungi that cause postharvest diseases. For example, potassium borate treatment effectively controlled blue mold of apple. It inhibited both conidial germination and the mycelial growth of pathogens and also suppressed the expression of two antioxidant enzymes (glutathione S-transferase and CAT) of the pathogen *P. expansum* (Qin et al., 2007). In addition, some inorganic salts can induce resistance against some postharvest diseases of fruits and vegetables. For example, sodium carbonate treatment induced resistance against postharvest green mold of citrus fruit caused by *Penicillium digitatum* (Pers.) Sacc, through scoparone accumulation, structural changes, and alkalinization in the albedo of wounded citrus fruits (Venditti et al., 2005; Youssef et al., 2014). Sodium bicarbonate treatments induced crystalline wax generation, activated host-resistance, and increased imazalil level in rind wounds of oranges, improving the control of green mold during storage (Dore et al., 2010; Youssef et al., 2014).

4 Induced Resistance by Biological Elicitors

4.1 Protein Elicitor: Harpin

Harpin (Messenger™; Eden Bioscience Co., Bothell, WA, USA) is an acidic, heat-stable, glycine-rich, 44-kDa protein encoded by the hrpN gene that was first described in *Erwinia amylovora*, a bacterium that causes fire blight of apple, pear, and other members of Rosaceae. It can elicit HR and induce SAR in plants (Baker et al., 1995) and induce resistance to infection by pathogens in many commodities after harvest. Field spraying with harpin effectively reduced latent infections in muskmelons. Wang et al. (2011) reported that harpin treatment before harvest effectively controlled pink rot caused by *T. roseum* in harvested muskmelons, and the effect in the reduction of latent infections was in proportion to the amount of harpin applied. A similar result was found by Ge et al. (2006), who indicated that harpin application 1 wk before harvest decreased pink rot of melon. Postharvest treatment with harpin at 0.04 to 0.16 mg/L reduced the incidence of blue mold of apple caused by *P. expansum* (de Capdeville et al., 2003). Harpin at 90 mg/L decreased pink rot and Fusarium rot of muskmelon caused by *T. roseum* and *F. semitectum*, respectively. However, a higher concentration (>90 mg/L) failed to promote resistance and caused phytotoxicity to melons (Bi et al., 2005).

Bi et al. (2005) reported that harpin significantly reduced lesion diameter in muskmelons inoculated with *T. roseum*. The mechanism of action was attributed to local acquired resistance and SAR induced by harpin. Some of the defensive reactions elicited by harpin on melons were identified: alteration of the levels of preformed antifungal substances such as phenols and flavonoids, accumulation of PR proteins such as CHT and GLU, induction of the activity of enzymes such as PAL and POD, modulation of metabolism of ROS such as SOD, CAT, and hydrogen peroxide; and reinforcement of cell wall by lignin accumulation (Bi et al., 2005; Wang et al., 2008, 2011). Although harpin had no direct inhibitory effect on the pathogen, it was considered as a promising strategy for controlling postharvest decay, thanks to the activation of SAR mechanisms (Bi et al., 2005).

4.2 Chitosan

Chitin is a primary constituent of crustacean shells, insect cuticles, and fungal cell walls (Wu et al., 2005). Chitosan (poly-β-(1,4)-N-acetyl-d-glucosamine), its deacetylated soluble form, inhibits many pathogens and can elicit defensive responses in fresh fruits and vegetables at low concentrations (Palou et al., 2008; Yin et al., 2010). Chitosan and its derivatives such as oligochitosan can be used in solution, powder form, or as wettable coatings. The potential of chitosan and its derivatives to reduce disease in many plants has been confirmed repeatedly (Benhamou and Theriault, 1992; Makino and Hirata, 1997; Ren et al., 2001; Bautista-Baños et al., 2006; Meng et al., 2012). Pre- or postharvest treatment with chitosan or its derivatives tended to suppress decay in many postharvest commodities such as strawberry (Bhaskara-Reddy et al., 2000), citrus (Chien et al., 2007; Zeng et al., 2010), jujube (Yan et al., 2012), banana (Meng et al., 2012), and potato (Xue et al., 2017).

A low molecular weight (15 kDa) chitosan coating at 0.2% controlled citrus decay better than the synthetic fungicide thiabendazole (Chien et al., 2007). Similarly, oligochitosan at 20 g/L controlled anthracnose of banana nearly as well as the fungicide prochloraz at 0.5 g/L (Meng et al., 2012). Chitosan and oligochitosan application before harvest reduced postharvest disease of strawberry and jujube during storage (Bhaskara-Reddy et al., 2000; Yan et al., 2012). The rate of expansion of lesions caused by *F. sulphureum* on potato tubers was significantly reduced after treatment with chitosan at 0.5% (Xue et al., 2017). The germicidal activity of chitosan has been well documented both in vitro and in vivo (Li et al., 2009b; Yin et al., 2010; Zhang et al., 2011a).

Some studies indicated that chitosan inhibits the growth of pathogens by induction of significant morphological changes, structural alterations, and molecular disorganization of the fungal cells (Rabea et al., 2003; Bautista-Baños et al., 2006; Li et al., 2009b; Yin et al., 2010; Zhang et al., 2011a). In addition to its direct antifungal activity, other studies strongly suggested that chitosan induces a series of defense responses by increasing enzymatic activities (Bautista-Baños et al., 2006). Meng et al. (2012) reported that a significant increase in CHT and GLU activities was observed in banana and jujube treated with oligochitosan. Chitosan or oligochitosan treatment induced a significant increase in the activities of PPO (Liu et al., 2007), POD (Yan et al., 2012), and PAL (Meng et al., 2012) after harvest in various commodities. Chitosan treatment enhanced total phenolics, flavonoids, other antifungal substances (Sun et al., 2008; Badawy and Rabea, 2009; Yan et al.,

2012), and induced ROS through regulating the activity of metabolic enzymes such as SOD, CAT, and APX (Zeng et al., 2010; Yan et al., 2012). Structural defense responses including preservation of pectin binding sites and a thick and regular cellulose distribution over host walls were observed in the first tissue layers beneath the ruptured cells in bell peppers that had been treated with chitosan (El Ghaouth et al., 1994).

4.3 Other Biological Elicitors

Interest in the use of microbial antagonists to control postharvest decay of fruits and vegetables has increased in recent years (Romanazzi et al., 2016). Many antagonistic microorganisms can induce defense responses in host tissues. For example, *Candida oleophila* treatment significantly reduced decay caused by *P. digitatum* on grapefruits by induction of host resistance (Droby et al., 2002). Four antagonistic yeasts, *Pichia membranifaciens, Cryptococcus laurentii, Candida guilliermondii,* and *Rhodotorula glutinis* stimulated CAT and POD activities and reduced the levels of protein carbonylation in response to ROS caused by *M. fructicola* infection in peach fruit (Xu et al., 2008b). *Candida saitoana* induced resistance in apple fruit after harvest against *B. cinerea*, with increased activities of CHT and GLU (El Ghaouth et al., 2007). *Candida intermedia* induced strawberry fruit defense mechanisms, which included the biosynthesis of β-1,3-exoglucanase, and effectively controlled postharvest decay caused by *B. cinerea* (Huang et al., 2011a).

Essential oils recently received increasing interest for the control of postharvest decay of fruits and vegetables. The term "essential oils" encompasses a large variety of volatile compounds with antifungal activity isolated from plants. For example, postharvest treatment with thyme oil significantly reduced citrus sour rot caused by *Geotrichum citri-aurantii* (Liu et al., 2009).

In addition, plants and animals can also produce a variety of peptides and small proteins with antimicrobial activity, which are presumed to be part of constitutive or inducible defense mechanisms to inhibit infections by pathogens (Broekaert et al., 1995). Blondelle and Houghten (1992) stated that the mechanism of action was the interaction of an amphipathic cationic peptide with the target cell membrane, followed by membrane disruption. Several peptides from different origins have been identified, characterized, and tested for activity against postharvest fungi.

5 Conclusion

Induced resistance, where the mode of action is the stimulation of local and/or systemic defensive responses in fruits and vegetables, is a new and relatively safe strategy to control a broad spectrum of postharvest diseases. However, the effectiveness of induced resistance is far lower than that of conventional chemical fungicides. At this time, the categories of elicitors reported by researchers have been relatively limited. Few systematic or in-depth studies have been conducted of factors that stimulate elicitor responses or the mechanisms of induced resistance. Furthermore, the combination of elicitor and fungicidal treatments may be a better strategy to improve the effectiveness of the elicitor treatment alone (Keinath et al.,

2007). In addition, there is a need to broaden the spectrum of action, determine optimum concentrations and application methods, investigate the key factors that influence preharvest or postharvest treatments, and reveal the mechanisms of induced resistance in order to extend the induction periods. Finally, the effect of treatment on product quality and shelf life should be objectively evaluated (Bi et al., 2007). Comparing with other decay control methods, treatments with elicitors have some desired characteristics, such as simplicity, eco-friendliness, safety, sustainability, and a broad spectrum of action, that make them suitable strategies for the control of postharvest diseases and extension of the shelf life of fruits and vegetables in the future.

References

Alkan, N., Fluhr, R. and Prusky, D. 2012. Ammonium secretion during *Colletotrichum coccodes* infection modulates salicylic and jasmonic acid pathways of ripe and unripe tomato fruit. *Molecular Plant Microbe Interactions* 25, 85–96.

Alkan, N. and Fortes, A.M. 2015. Insights into molecular and metabolic events associated with fruit response to postharvest fungal pathogens. *Frontiers in Plant Science* 6, article 889.

Arasimowicz, M. and Floryszak-Wieczorek, J. 2007. Nitric oxide as a bioactive signalling molecule in plant stress responses. *Plant Science* 172, 876–887.

Badawy, M.E.I. and Rabea, E.I. 2009. Potential of the biopolymer chitosan with different molecular weights to control postharvest gray mold of tomato fruit. *Postharvest Biology and Technology* 51, 110–117.

Baker, C.J. and Orlandi, E.W. 1995. Active oxygen in plant pathogenesis. *Annual Review of Phytopathology* 33, 299–321.

Barkai-Golan, R. 2001. *Postharvest Diseases of Fruit and Vegetables: Development and Control*. Elsevier Press, Amsterdam, The Netherlands.

Bautista-Baños, S., Hernández-Lauzardo, A.N., Velázquez-del Valle, M.G., Hernández-López, M., Ait Barka, E., Bósquez-Molina, E. and Wilson, C.L. 2006. Chitosan as a potential natural compound to control pre and postharvest diseases of horticultural commodities. *Crop Protection* 25, 108–118.

Beligni, M.V. and Lamattina, L. 2001. Nitric oxide in plants: The history is just beginning. *Plant, Cell and Environment* 24, 267–278.

Benhamou, N. and Theriault, G. 1992. Treatment with chitosan enhances resistance of tomato plants to the crown and root pathogen *Fusarium oxysporum* f. sp. *radicis-lycopersici*. *Physiological and Molecular Plant Pathology* 41, 34–52.

Ben-Yehoshua, S., Rodov, S., Kim, J.J. and Carnelli, S. 1992. Preformed and induced antifungal materials of citrus fruit in relation to enhancement of decay resistance by heat and UV treatment. *Journal of Agricultural and Food Chemistry* 40, 1217–1221.

Bhaskara-Reddy, M.V., Belkacemi, K., Corcuff, R., Castaigne, F. and Arul, J. 2000. Effect of pre-harvest chitosan sprays on post-harvest infection by *Botrytis cinerea* and quality of strawberry fruit. *Postharvest Biology and Technology* 20, 39–51.

Bi, Y. 1996. UV-C irradiation on control of the postharvest disease in fruit and vegetable. *Food Science* 12, 58–61 (in Chinese with English abstract).

Bi, Y., Ge, Y.H., Li, Y.C., Wang, J.J., Miao, X.Y. and Li, X.W. 2006a. Postharvest acibenzolar-S-methyl treatment suppresses decay and induces resistance in Hami melons. *Acta Horticulturae* 712, 393–399.

Bi, Y., Li, Y.C. and Ge, Y.H. 2007. Induced resistance in postharvest fruits and vegetables by chemicals and its mechanism. *Stewart Postharvest Review* 3:6, 1–7.

Bi, Y., Li, Y.C., Ge, Y.H. and Wang, Y. 2010. Induced resistance in melons by elicitors for the control of postharvest diseases. pp. 31–41. *In*: D. Prusky and M. L. Gullino (eds.). *Post-Harvest Pathology*. Springer, Dordrecht, The Netherlands.

Bi, Y., Tian, S.P., Guo, Y.R., Ge, Y.H. and Qin, G.Z. 2006b. Sodium silicate reduces postharvest decay on Hami melons: Induced resistance and fungistatic effects. *Plant Disease* 90, 279–283.

Bi, Y., Tian, S.P., Zhao, J. and Ge, Y.H. 2005. Harpin induces local and systemic resistance against *Trichothecium roseum* in harvested Hami melons. *Postharvest Biology and Technology* 38, 183–187.

Blondelle, S.E. and Houghten, R.A. 1992. Design of model amphipathic peptides having potent antimicrobial activities. *Biochemistry* 31, 12688–12694.

Boava, L.P., Kuhn, O.J., Pascholati, S.F., Di Piero, R.M. and Furtado, E.L. 2009. Effect of acibenzolar-S-methyl and *Saccharomyces cerevisiae* on the activation of *Eucalyptus* defences against rust. *Australasian Plant Pathology* 38, 594–602.

Bokshi, A.I., Morris, S.C. and Deverall, B.J. 2003. Effects of benzothiadiazole and acetylsalicylic acid on β-1,3-glucanase activity and disease resistance in potato. *Plant Pathology* 52, 22–27.

Bokshi, A.I., Morris, S.C., McConchie, R.M. and Deverall, B.J. 2006. Pre-harvest application of 2,6-dichloroisonicotinic acid, beta-aminobutyric acid or benzothiadiazole to control post-harvest storage diseases of melons by inducing Systemic Acquired Resistance (SAR). *Journal of Horticultural Science & Biotechnology* 81, 700–706.

Bokshi, A.I., Morris, S.C., McDonald, K. and McConchie, R.M. 2005. Application of INA and BABA control pre and postharvest diseases of melons through induction of systemic acquired resistance. *Acta Horticulturae* 694, 417–419.

Broekaert, W.F., Terras, F.R., Cammue, B.P. and Osborn, R.W. 1995. Plant defensins: Novel antimicrobial peptides as components of the host defense system. *Plant Physiology* 108, 1353–1358.

Cao, J.K. and Jiang, W.B. 2005. Effect of citric acid treatment on the quality of Ya pear fruit during storage. *Food Science and Technology* 10, 84–87 (In Chinese with English abstract).

Cao, J.K. and Jiang, W.B. 2006. Induction of resistance in Yali pear (*Pyrus bretschneideri* Rehd.) fruit against postharvest diseases by acibenzolar-S-methyl sprays on trees during fruit growth. *Scientia Horticulturae* 110, 181–186.

Cao, S.F., Hu, Z., Zheng, Y., Yang, Z. and Lu, B. 2011. Effect of BTH on antioxidant enzymes, radical-scavenging activity and decay in strawberry fruit. *Food Chemistry* 125, 145–149.

Chan, Z.L. and Tian, S.P. 2006. Induction of H_2O_2-metabolizing enzymes and total protein synthesis by antagonistic yeast and salicylic acid in harvested sweet cherry fruit. *Postharvest Biology and Technology* 39, 314–320.

Charles, M.T. and Arul, J. 2007. UV treatment of fresh fruits and vegetables for improved quality: A status report. *Stewart Postharvest Review* 3:3, 1–8.

Charles, M.T., Benhamou, N. and Arul, J. 2008a. Physiological basis of UV-C induced resistance to *Botrytis cinerea* in tomato fruit: III. Ultrastructural modifications and their impact on fungal colonization. *Postharvest Biology and Technology* 47, 27–40.

Charles, M.T., Goulet, A. and Arul, J. 2008b. Physiological basis of UV-C induced resistance to *Botrytis cinerea* in tomato fruit: IV. Biochemical modification of structural barriers. *Postharvest Biology and Technology* 47, 41–53.

Charles, M.T., Makhlouf, J. and Arul, J. 2008c. Physiological basis of UV-C induced resistance to *Botrytis cinerea* in tomato fruit: II. Modification of fruit surface and changes in fungal colonization. *Postharvest Biology and Technology* 47, 21–26.

Charles, M.T., Mercier, J., Makhlouf, J. and Arul, J. 2008d. Physiological basis of UV-C-induced resistance to *Botrytis cinerea* in tomato fruit: I. Role of pre- and post-challenge accumulation of the phytoalexin-rishitin. *Postharvest Biology and Technology* 47, 10–20.

Charles, M.T., Tano, K., Asselin, A. and Arul, J. 2009. Physiological basis of UV-C induced resistance to *Botrytis cinerea* in tomato fruit. V. Constitutive defence enzymes and inducible pathogenesis-related proteins. *Postharvest Biology and Technology* 51, 414–424.

Chien, P.J., Sheu, F. and Lin, H.R. 2007. Coating citrus (*Murcott tangor*) fruit with low molecular weight chitosan increases postharvest quality and shelf life. *Food Chemistry* 100, 1160–1164.

Clark, D., Durner, J., Navarre, D.A. and Klessig, D.F. 2000. Nitric oxide inhibition of tobacco catalase and ascorbate peroxidase. *Molecular Plant-Microbe Interactions* 13, 1380–1384.

Conrath, U., Thulke, O., Katz, V., Schwindling, S. and Kohler, A. 2001. Priming as a mechanism in induced systemic resistance of plants. *European Journal of Plant Pathology* 107, 113–119.

Cordelier, S., de Ruffray, P., Fritig, B. and Kauffmann, S. 2003. Biological and molecular comparison between localized and systemic acquired resistance induced in tobacco by a *Phytophthora megasperma* glycoprotein elicitin. *Plant Molecular Biology* 51, 109–118.

de Capdeville, G., Beer, S.V., Watkins, C.B., Wilson, C.L., Tedeschi, L.O. and Aist, J. R. 2003. Pre-and postharvest harpin treatments of apples induced resistance to blue mold. *Plant Disease* 89, 39–44.

Deepak, S., Shailasree, S., Kini, R.K., Hause, B. and Shetty, S.H. 2007. Role of hydroxyproline-rich glycoproteins in resistance of pearl millet against downy mildew pathogen *Sclerospora graminicola*. *Plant* 226, 323–333.

Delledonne, M., Xia, Y.J., Dixon, R.A. and Lamb, C. 1998. Nitric oxide functions as a signal in plant disease resistance. *Nature* 394, 585–588.

Deng, J.J., Bi, Y., Xie, D.F., Ge, Y.H., Wang, Y., Sun, X.J. and Li, Y.H. 2008. Effect of oxalic acid treatment on postharvest diseases and fruit quality of muskmelons. *Journal of Gansu Agricultural University* 43, 82–86 (in Chinese with English Abstract).

Dore, A., Molinu, M.G., Venditti, T. and D'Hallewin, G. 2010. Sodium bicarbonate induces crystalline wax generation, activates host-resistance, and increases imazalil level in rind wounds of oranges, improving the control of green mold during storage. *Journal of Agricultural Food and Chemistry* 58, 7297–7304.

Droby, S., Porat, R., Cohen, L., Weiss, B., Shapiro, B., Philosoph-Hadas, S. and Meir, S. 1999. Suppressing green mold decay in grapefruit with postharvest jasmonate application. *Journal of the American Society for Horticultural Science* 124, 184–188.

Droby, S., Vinokur, V., Weiss, B., Cohen, L., Daus, A., Goldschmid, E. and Porat, R. 2002. Induction of resistance to *Penicillium digitatum* in grapefruit with the biocontrol agent *Candida oleophila*. *Phytopathology* 92, 393–399.

Duan, X., Su, X., You, Y., Qu, H., Li, Y. and Jiang, Y. 2007. Effect of nitric oxide on pericarp browning of harvested longan fruit in relation to phenolic metabolism. *Food Chemistry* 104, 571–576.

El Ghaouth, A., Arul, J., Wilson, C. and Benhamou, N. 1994. Ultrastructural and cytochemical aspects of the effect of chitosan on decay of bell pepper fruit. *Physiological and Molecular Plant Pathology* 44, 417–432.

El Ghaouth A, Wilson C., Wisniewski M. 2007. Control of postharvest decay of apple fruit with *Candida saitoana* and induction of defense responses. *Phytopathology* 93, 344–348.

Fallik, E., Grinberg, S., Gambourg, M., Klein, J.D. and Lurie, S. 1995. Prestorage heat treatment reduces pathogenicity of *Penicillium expansum* in apple fruit. *Plant Pathology* 45, 92–97.

FAO (Food and Agriculture Organization of the United Nations). 2011. Global food losses and food waste - extent, causes and prevention. Available at: www.fao.org/docrep/014/mb060e/mb060e00.pdf

Ge, Y.H., Bi, Y., Li, X. and Li, M. 2008. Induced resistance against Fusarium and pink rots by acibenzolar-S-methyl in harvested muskmelon (cv. Yindi). *Agricultural Sciences in China* 7, 58–64 (in Chinese with English abstract).

Ge, Y.H., Bi, Y. and Yang, D.M. 2006. Effect of elicitors treatment on postharvest *Trichothecium* and *Alternaria* rots in harvested muskmelon (cv. Yindi). *Food Science* 27, 246–249 (in Chinese with English abstract).

Guo, Y.R., Liu, L., Zhao, J. and Bi, Y. 2007. Use of silicon oxide and sodium silicate for controlling *Trichothecium roseum* postharvest rot in Chinese cantaloupe (*Cucumis melo* L.). *International Journal of Food Science and Technology* 42, 1012–1018.

Huang, R., Li, G.Q., Zhang, J., Yang, L., Che, H.J., Jiang, D.H. and Huang, H.C. 2011a. Control of postharvest Botrytis fruit rot of strawberry by volatile organic compounds of *Candida intermedia*. *Phytopathology* 101, 859–869.

Huang, X.M., Zhang, C., Pang, X.Q. and Zhang, Z.Q. 2011b. Early changes of reactive oxygen species in 2,6-dichloroisonicotianic acid inducing tolerance in postharvest banana fruits. *Acta Horticulturae Sinica* 38, 265–272 (in Chinese with English abstract).

Iqbal, Z., Singh, Z., Khangura, R. and Ahmad, S. 2012. Management of citrus blue and green moulds through application of organic elicitors. *Australasian Plant Pathology* 41, 69–77.

Iriti, M., Mapelli, S. and Faoro, F. 2007. Chemical-induced resistance against post-harvest infection enhances tomato nutritional traits. *Food Chemistry* 105, 1040–1046.

Jin, P., Zheng, Y.H., Tang, S.S., Rui, H.J. and Wang, C.Y. 2009. A combination of hot air and methyl jasmonate vapor treatment alleviates chilling injury of peach fruit. *Postharvest Biology and Technology* 52, 24–29.

Kanto, T., Maekawa, K. and Aino, M. 2007. Suppression of conidial germination and appressorial formation by silicate treatment in powdery mildew of strawberry. *Journal of General Plant Pathology* 73, 1–7.

Keinath, A.P., Holmes, G.J., Everts, K.L., Egel, D.S. and Langston Jr, D.B. 2007. Evaluation of combinations of chlorothalonil with azoxystrobin, harpin, and disease forecasting for control of downy mildew and gummy stem blight on melon. *Crop Protection* 26, 83–88.

Kuc, J. 2001. Concepts and directions of induced systemic resistance in plants and its application. *European Journal of Plant Pathology* 107, 7–12.

Lai, T.F., Wang, Y.Y., Li, B.Q., Qin, G.Z. and Tian, S.P. 2011. Defensive responses of tomato fruit to exogenous nitric oxide during postharvest storage. *Postharvest Biology and Technology* 62, 127–132.

Leshem, Y.A.Y., Wills, R.B.H. and Ku, V.V. 1998. Evidence for the function of the free radical gas - nitric oxide (NO•) - as an endogenous maturation and senescence regulating factor in higher plants. *Plant Physiology and Biochemistry* 36, 825–833.

Li, W.H., Bi, Y., Ge, Y.H., Li, Y.C., Wang, J.J. and Wang, Y. 2012. Effects of postharvest sodium silicate treatment on pink rot disease and oxidative stress-antioxidative system in muskmelon fruit. *European Food Research and Technology* 234, 137–145.

Li, Y.C., Bi, Y., Ge, Y.H., Sun, X.J. and Wang, Y. 2009a. Antifungal activity of sodium silicate on *Fusarium sulphureum* and its effect on dry rot of potato tubers. *Journal of Food Science* 74, 213–218.

Li, Y.C., Sun, X.J., Bi, Y., Ge, Y.H. and Wang, Y. 2009b. Antifungal activity of chitosan on *Fusarium sulphureum* in relation to dry rot of potato tuber. *Agricultural Sciences in China* 8, 597–604.

Li, Y.C., Yin, Y., Chen, S.J., Bi, Y., Shen, X.J. and Wu, Y.H. 2011. Effect of hot water treatment combined with β-aminonbutyric acid on incidence of blue mold rot disease in postharvest apple fruits. *Food Science* 32, 265–269 (in Chinese with English abstract).

Lin, J.H., Gong, D.Q., Zhu, S.J., Zhang, L.J. and Zhang, L.B. 2011. Expression of PPO and POD genes and contents of polyphenolic compounds in harvested mango fruits in relation to benzothiadiazole-induced defense against anthracnose. *Scientia Horticulturae* 130, 85–89.

Liu, C., Cai, L., Han, X. and Ying, T. 2011. Temporary effect of postharvest UV-C irradiation on gene expression profile in tomato fruit. *Gene* 486, 56–64.

Liu, F., Tu, K., Shao, X., Zhao, Y., Tu, S. and Su, J. 2010. Effect of hot air treatment in combination with *Pichia guilliermondii* on postharvest anthracnose rot of loquat fruit. *Postharvest Biology and Technology* 58, 65–71.

Liu, H.X., Jiang, W.B., Bi, Y. and Luo, Y.B. 2005a. Postharvest BTH treatment induces resistance of peach (*Prunus persica* L. cv. Jiubao) fruit to infection by *Penicillium expansum* and enhances activity of fruit defense mechanisms. *Postharvest Biology and Technology* 35, 263–269.

Liu, H.X., Wang, B.G., Luo, Y.B. and Jiang, W.B. 2005b. Improving disease resistance in peach fruit during storage using benzo-(1,2,3)-thiodiazole-7-carbothioic acid S-methyl ester (BTH). *The Journal of Horticultural Science and Biotechnology* 80, 736–740.

Liu, J., Stevens, C., Khan, V.A., Lu, J.Y., Wilson, C.L., Adeyeye, O., Kabwe, M.K., Pusey, P.L., Chalutz, E., Sultana, T. and Droby, S. 1993. Application of ultraviolet-C light on storage rots and ripening of tomatoes. *Journal of Food Protection* 56, 868–872.

Liu, J., Sui, Y., Wisniewski, M., Droby, S., Tian, S.P., Norelli, J. and Hershkovitz, V. 2012. Effect of heat treatment on inhibition of *Monilinia fructicola* and induction of disease resistance in peach fruit. *Postharvest Biology and Technology* 65, 61–68.

Liu, J., Tian, S.P., Meng, X.H. and Xu, Y. 2007. Effects of chitosan on control of postharvest diseases and physiological responses of tomato fruit. *Postharvest Biology and Technology* 44, 300–306.

Liu, X., Wang, L.P., Li, Y.C., Li, H.Y., Yu, T. and Zheng, X.D. 2009. Antifungal activity of thyme oil against. *Geotrichum citri-aurantii* in *vitro* and in *vivo*. *Journal of Applied Microbiology* 107, 1450–1456.

Liu, Y.Y., Ge, Y.H., Bi, Y., Li, Y.C., Deng, H.W., Hu, L.G. and Dong, B. 2014. Effect of postharvest acibenzolar-S-methyl dipping on phenylpropanoid pathway metabolism in muskmelon (*Cucumis melo* L.) fruits. *Scientia Horticulturae* 168, 113–119.

Lucas, J.A. 1999. Plant immunisation: From myth to SAR. *Pesticide Science* 55, 193–196.

Lurie, S., Fallik, E., Handros, A. and Shapira, R. 1997. The possible involvement of peroxidase in resistance to *Botrytis cinerea* in heat treated tomato fruit. *Physiological and Molecular Plant Pathology* 50, 141–149.

Makino, Y. and Hirata, T. 1997. Modified atmosphere packaging of fresh produce with a biodegradable laminate chitosan-cellulose and polycaprolactone. *Postharvest Biology and Technology* 10, 247–254.

Mandal, B., Mandal, S., Csinos, A.S., Martinez, N., Culbreath, A.K. and Pappu, H.R. 2008. Biological and molecular analyses of the acibenzolar-S-methyl-induced systemic acquired resistance in flue-cured tobacco against tomato spotted wilt virus. *Phytopathology* 98, 196–204.

Martinez, C., Baccou, J.C., Bresson, E., Baissac, Y., Daniel, J.F. and Jalloul, A. 2000. Salicylic acid mediated by the oxidative burst is a key molecule in local and systemic responses of cotton challenged by an avirulent race of *Xanthomonas capestris* pv *malvacearum*. *Plant Physiology* 122, 757–766.

Mayr, U., Michalek, S., Treutter, D. and Feucht, W. 1997. Phenolic compounds of apple and their relationship to scab resistance. *Journal of Phytopathology* 145, 69–75.

Mazaro, S.M., Deschamps, C., de Mio, L.L., Biasi, L.A., de Gouvea, A. and Sautter, C. K. 2008. Postharvest behavior of strawberry fruits after pre harvest treatment with chitosan and acibenzolar-S-methyl. *Revista Brasileira de Fruticultura* 30, 185–190 (In Portuguese with English abstract).

McDonald, R.E., McColluma, T.G. and Baldwinb, E.A. 1999. Temperature of water heat treatments influences tomato fruit quality following low-temperature storage. *Postharvest Biology and Technology* 16, 147–155.

Meng, X.C., Tang, Y.X., Zhang, A.Y., Huang, X.M. and Zhang, Z.Q. 2012. Effect of oligochitosan on development of *Colletotrichum musae in vitro* and *in situ* and its role in protection of banana fruits. *Fruits* 67, 147–155.

Mercier, J., Arul, J. and Cohen, C. 1993. Effect of UV-C on phytoalexin accumulation and resistance to *Botrytis cinerea* in stored carrots. *Journal of Phytopathology* 137, 44–54.

Mercier, J., Baka, M., Reddy, B., Corcuff, R. and Arul, J. 2001. Short wave ultraviolet irradiation for control of decay by *Botrytis cinerea* in bell pepper: Induced resistance and germicidal effects. *Journal of the Amercian Society for Horticultural Science* 126, 128–133.

Mittler, R., Vanderauwera, S., Suzuki, N., Miller, G., Tognetti, V.B., Vandepoele, K., Gollery, M., Shulaev, V. and Van Breusegem, F. 2011. ROS signaling: The new wave? *Trends in Plant Science* 16, 300–309.

Mur, L.A.J., Bi, Y.M., Darby, R.M., Firek, S. and Draper, J. 1997. Compromising early salicylic acid accumulation delays the hypersensitive response and increase viral dispersal during lesion establishment in TMV-infected tobacco. *Plant Journal* 12, 1113–1126.

Nigro, F., Ippolito, A., Lattanzio, V., Di Venere, D. and Salerno, M. 1998. UV-C irradiation in preventing botrytis storage rot of strawberry, table grapes and kiwifruit. pp. 209–214. *In*: P. Bertolini, P.C. Sijmons, M.E. Guerzoni and F. Serra (eds.). *Proceedings of the COST 914 and 915 Joint Workshop Non Conventional Methods for the Control of Postharvest Disease and Microbiological Spoilage*. Bologna, Italy.

Niritake, T., Kawakita, K. and Doke, N. 1996. Nitric oxide induces phytoalexin accumulation in potato tuber tissues. *Plant and Cell Physiology* 37, 113–116.

Okawa, K. 2015. Market and trade impacts of food loss and waste reduction. Organisation for Economic Co-operation and Development, Food, Agriculture and Fisheries (OECD) Papers, No. 75, OECD Publishing, Paris. Available at: www.oecd-ilibrary.org/docserver/download/5js4w29h0wr2-en.pdf?expires=1511345965&id=id&accname=guest&checksum=6C9B6BF177BE9E8E0582A394D65FF29A

Palou, L., Smilanick, J.L. and Droby, S. 2008. Alternatives to conventional fungicides for the control of citrus postharvest green and blue moulds. *Stewart Postharvest Review* 4:2, 1–16.

Pan, Y.G. and Liu, X.H. 2011. Effect of benzo-thiadiazole-7-carbothioic acid S-methyl ester (BTH) treatment on the resistant substance in postharvest mango fruits of different varieties. *African Journal of Biotechnology* 10, 15521–15528.

Pombo, M.A., Dotto, M.C., Martínez, G.A. and Civello, P.M. 2009. UV-C irradiation delays strawberry fruit softening and modifies the expression of genes involved in cell wall degradation. *Postharvest Biology and Technology* 51, 141–148.

Pombo, M.A., Rosli, H.G., Martinez, G.A. and Civello, P.M. 2011. UV-C treatment affects the expression and activity of defense genes in strawberry fruit (*Fragaria x ananassa* Duch.). *Postharvest Biology and Technology* 59, 94–102.

Poole, P.R., McLeod, L.C., Whitmore, K.J. and Whitaker, G. 1998. Postharvest control of *Botrytis cinerea* rots in stored kiwifruit. *Acta Horticulturae* 464, 71–76.

Porat, R., McCollum, T.G., Vinokur, V. and Droby, S. 2002. Effects of various elicitors on the transcription of a β-1,3-endoglucanase gene in citrus fruit. *Journal of Phytopathology* 150, 70–75.

Prusky, D. 2011. Reduction of the incidence of postharvest quality losses, and future prospects. *Food Security* 3, 463–474.

Qin, G.Z., Tian, S., Chan, Z. and Li, B.Q. 2007. Crucial role of antioxidant proteins and hydrolytic enzymes in pathogenicity of *Penicillium expansum*: Analysis based on proteomics approach. *Molecular and Cellular Proteomics* 6, 425–438.

Qin, G.Z., Tian, S.P., Xu, Y. and Wan, Y.K. 2003. Enhancement of biocontrol efficacy of antagonistic yeasts by salicylic acid in sweet cherry fruit. *Physiological and Molecular Plant Pathology* 62, 147–154.

Qin, G.Z., Tian, S.P. 2005. Enhancement of biocontrol activity of *Cryptococcus laurentii* by silicon and the possible mechanisms involved. *Phytopathology* 95, 69–75.

Rabea, E.I., Badawy, M.E.T., Stevens, C.V., Smagghe, G. and Steurbaut, W. 2003. Chitosan as antimicrobial agent: Applications and mode of action. *Biomacromolecules* 4, 1457–1465.

Ren, H., Endo, H. and Hayashi, T. 2001. Antioxidative and antimutagenic activities and polyphenol content of pesticide-free and organically cultivated green vegetable using water-soluble chitosan as a soil modifier and leaf surface spray. *Journal of the Science of Food and Agriculture* 81, 1426–1432.

Ren, Y.L., Wang, Y.F., Bi, Y., Ge, Y.H., Wang, Y., Fan, C.F., Li, D.Q. and Deng, H.W. 2012. Postharvest BTH treatment induced disease resistance and enhanced reactive oxygen species metabolism in muskmelon (*Cucumis melo* L.) fruit. *European Food Research and Technology* 234, 963–971.

Romanazzi, G., Sanzani, S.M., Bi, Y., Tian, S.P., Martínez, P.G. and Alkan, N. 2016. Induced resistance to control postharvest decay of fruit and vegetables. *Postharvest Biology and Technology* 122, 82–94.

Ryals, J.A., Neuenschwander, U.H., Willits, M.G., Molina, A., Steiner, H.Y. and Hunt, M.D. 1996. Systemic acquired resistance. *The Plant Cell* 8, 1809–1818.

Santiago, J.A., Rivera-Vargas, L.I., Rodriguez, R.D.P. and Macchiavelli, R. 2006. Resistance-inducing chemicals against *Colletotrichum gloeosporioides* in mango. *Journal of Agriculture of the University of Puerto Rico* 90, 221–235.

Schirra, M., D'hallewin, G., Ben-Yehoshua, S. and Fallik, E. 2000. Host–Pathogen interactions modulated by heat treatment. *Postharvest Biology and Technology* 21, 71–85.

Shadle, G.L., Wesley, S.V., Korth, K.L., Chen, F., Lamb, C. and Dixon, R.A. 2003. Phenylpropanoid compounds and disease resistance in transgenic tobacco with altered expression of l-phenylalanine ammonia-lyase. *Phytochemistry* 64, 153–161.

Sharma, P., Jha, A.B., Dubey, R.S. and Pessarakli, M. 2012. Reactive oxygen species, oxidative damage, and antioxidative defense mechanism in plants under stressful conditions. *Journal of Botany* 2012, 1–26.

Soegiarto, L. and Wills, R.B.H. 2006. Effect of nitric oxide, reduced oxygen and elevated carbon dioxide levels on the postharvest life of strawberries and lettuce. *Australian Journal of Experimental Agriculture* 46, 1097–1100.

Stevens, C., Khan, V.A., Lu, J.Y., Wilson, C.L., Chalutz, E., Droby, S., Kabwe, M.K., Haung, Z., Adeyeye, O., Pusey, P.L. and Tang, A.Y.A. 1999. Induced resistance of sweet potato to Fusarium root rot by UV-C hormesis. *Crop Protection* 18, 463–470.

Stevens, C., Liu, J., Khan, V.A., Lu, J.Y., Kabwe, M.K., Wilson, C.L., Igwegbe, E.C.K., Chalutz, E. and Droby, S. 2004. The effect of low-dose ultraviolet light-C treatment on polygalacturonase activity, delay ripening and Rhizopus soft rot development of tomatoes. *Crop Protection* 23, 551–554.

Sticher, L., Mauch-Mani, B. and Metraux, J.P. 1997. Systemic acquired resistance. *Annual Review of Phytopathology* 35, 235–270.

Sun, X.J., Bi, Y., Li, Y.C.I., Han, R.F. and Ge, Y.H. 2008. Postharvest chitosan treatment induces resistance in potato against *Fusarium sulphureum*. *Agricultural Sciences in China* 7, 615–621.

Terry, L.A. and Joyce, D.C. 2000. Suppression of grey mould on strawberry fruit with the chemical plant activator acibenzolar. *Pest Management Science* 56, 989–992.

Terry, L.A. and Joyce, D.C. 2004. Elicitors of induced disease resistance in postharvest horticultural crops: A brief review. *Postharvest Biology and Technology* 32, 1–13.

Thomas, D.D., Ridnour, L.A., Isenberg, J.S., Flores-Santana, W., Switzer, C.H., Donzelli, S., Hussain, P., Vecoli, C., Paolocci, N., Ambs, S., Colton, C.A., Harris, C.C., Roberts, D.D. and Wink, D.A. 2008. The chemical biology of nitric oxide: Implications in cellular signaling. *Free Radical Biology and Medicine* 45, 18–31.

Tian, S.P., Qin, G.Z., Li, B.Q., Wang, Q. and Meng, X.H. 2007. Effects of salicylic acid on disease resistance and postharvest decay control of fruits. *Stewart Postharvest Review* 3:6, 1–7.

Tronchet, M., Balagué, C., Kroj, T., Jouanin, L. and Roby, D. 2010. Cinnamyl alcohol dehydrogenases-C and D, key enzymes in lignin biosynthesis, play an essential role in disease resistance in *Arabidopsis*. *Molecular Plant Pathology* 11, 83–92.

Tzortzakis, N.G. 2007. Methyl jasmonate-induced suppression of anthracnose rot in tomato fruit. *Crop Protection* 26, 1507–1513.

Valverde, J., Giménez, M., Guillén, F., Valero, D., Martínez-Romero, D. and Serrano, M. 2015. Methyl salicylate treatments of sweet cherry trees increase antioxidant systems in fruit at harvest and during storage. *Postharvest Biology and Technology* 109, 106–113.

Van Wees, S.M., Luijendijk, M., Smoorenburg, I., van Loon, L. and Pieterse, C.J. 1999. Rhizobacteria-mediated induced systemic resistance (ISR) in *Arabidopsis* is not associated with a direct effect on expression of known defense-related genes but stimulates the expression of the jasmonate-inducible gene *Atvsp* upon challenge. *Plant Molecular Biology* 41, 537–549.

Venditti, T., Molinu, M.G., Dore, A., Agabbio, M. and D'hallewin, G. 2005. Sodium carbonate treatment induces scoparone accumulation, structural changes, and alkalinization in the albedo of wounded Citrus fruits. *Journal of Agricultural and Food Chemistry* 53, 3510–3518.

Wang, J.J., Bi, Y., Zhang, Z.K., Zhang, H.Y. and Ge, Y.H. 2011. Reducing latent infection and enhancing disease resistance in muskmelon by preharvest application of harpin. *Journal of Agricultural and Food Chemistry* 59, 12527–12533.

Wang, K., Cao, S., Jin, P., Rui, H. and Zheng, Y. 2010. Effect of hot air treatment on postharvest mould decay in Chinese bayberry fruit and the possible mechanisms. *International Journal of Food Microbiology* 141, 11–16.

Wang, K., Jin, P., Cao, S., Shang, H., Yang, Z. and Zheng, Y. 2009. Methyl jasmonate reduces decay and enhances antioxidant capacity in Chinese bayberries. *Journal of Agricultural and Food Chemistry* 57, 5809–5815.

Wang, K., Liao, Y., Cao, S., Di, H. and Zheng, Y. 2015. Effects of benzothiadiazole on disease resistance and soluble sugar accumulation in grape berries and its possible cellular mechanisms involved. *Postharvest Biology and Technology* 102, 51–60.

Wang, Y., Li, X., Bi, Y., Ge, Y.H., Li, Y.C. and Xie, F. 2008. Postharvest ASM or harpin treatment induced resistance of muskmelons against *Trichothecium roseum*. *Agricultural Sciences in China* 7, 217–223.

Wu, T., Zivanovic, S., Draughon, F.A., Conway, W.S. and Sams, C.E. 2005. Physicochemical properties and bioactivity of fungal chitin and chitosan. *Journal of Agricultural and Food Chemistry* 53, 3888–3894.

Xu, X.B., Qin, G.Z. and Tian, S.P. 2008b. Effect of microbial biocontrol agents on alleviating oxidative damage of peach fruit subjected to fungal pathogen. *International Journal of Food Microbiology* 126, 153–158.

Xue, H., Bi, Y., Zong, Y., Alejandro, C.U., Wang, H., Pu, L., Wang, Y. and Li, Y.C. 2017. Effects of elicitors on trichothecene accumulation and Tri genes expression in potato tubers inoculated with *Fusarium sulphureum*. *European Journal of Plant Pathology* 148, 673–685.

Yan, J.Q., Cao, J.K., Jiang, W.B. and Zhao, Y.M. 2012. Effects of preharvest oligochitosan sprays on postharvest fungal diseases, storage quality, and defensive responses in jujube (*Zizyphus jujuba* Mill. cv. Dongzao) fruit. *Scientia Horticulturae* 142, 196–204.

Yao, H.J. and Tian, S.P. 2005a. Effects of a biocontrol agent and methyl jasmonate on postharvest diseases of peach fruit and the possible mechanisms involved. *Journal of Applied Microbiology* 98, 941–950.

Yao, H.J. and Tian, S.P. 2005b. Effects of pre- and post-harvest application of salicylic acid or methyl jasmonate on inducing disease resistance of sweet cherry fruit in storage. *Postharvest Biology and Technology* 35, 253–262.

Yin, Y., Li, Y.C., Bi, Y., Chen, S.J., Li, Y.C., Yuan, L., Wang, Y. and Wang, D. 2010. Postharvest treatment with β-aminobutyric acid induces resistance against dry rot caused by *Fusarium sulphureum* in potato tuber. *Agricultural Sciences in China* 9, 1372–1380.

Youssef, K., Sanzani, S.M., Ligorio, A., Ippolito, A. and Terry, L.A. 2014. Sodium carbonate and bicarbonate treatments induce resistance to postharvest green mould on citrus fruit. *Postharvest Biology and Technology* 87, 61–69.

Yuan, L., Bi, Y., Ge, Y.H., Wang, Y., Liu, Y.Y. and Li, Y.C. 2013. Postharvest hot water dipping reduced decay by inducing disease resistance and maintaining firmness in muskmelon (*Cucumis melo* L.) fruit. *Scientia Horticulturae*, 161, 101–110.

Zainuri, J., Joyce, D.C., Wearing, A.H., Coates, L. and Terry, L. 2001. Effects of phosphonate and salicylic acid treatments on anthracnose disease development and ripening of 'Kensington Pride' mango fruit. *Australian Journal of Experimental Agriculture* 41, 805–813.

Zeng, K.F., Deng, Y.Y., Ming, J. and Deng, L.L. 2010. Induction of disease resistance and ROS metabolism in navel oranges by chitosan. *Scientia Horticulturae* 126, 223–228.

Zeng, K.F. and Jiang, W.B. 2009. Effects of 2,6-dichloroisonicotinic acid treatments on postharvest quality of mango fruits. *Transactions of the Chinese Society of Agricultural Engineering (CSAE)* 25, 267–271.

Zhang, H.Y., Li, R.P. and Liu, W.M. 2011a. Effects of chitin and its derivative chitosan on postharvest decay of fruits: A review. *International Journal of Molecular Sciences* 12, 917–934.

Zhang, Q.C., Li, Y.C., Bi, Y., Sun, X.J. and Wang, H.J. 2009. The inhibiting effect of post harvest citric acid treatment on dry rot of wounded inoculated potato and the activity of defense enzymes. *Journal of Gansu Agricultural University* 44, 146–150 (in Chinese with English abstract).

Zhang, Z.K., Bi, Y., Ge, Y.H., Wang, J.J., Deng, J.J., Xie, D.F. and Wang, Y. 2011b. Multiple pre-harvest treatments with acibenzolar-S-methyl reduce latent infection and induce resistance in muskmelon fruit. *Scientia Horticulturae* 130, 126–132.

Zheng, X.L., Tian, S.P., Li, B.Q. and Xu, Y. 2005. Changes in antioxidant systems and polyphenol oxidase activity in peach fruit treated with exogenous oxalic acid during storage at low temperature. *Acta Horticulturae Sinica* 32, 788–792.

Zhu, F., Chen, J., Xiao, X., Zhang, M., Yun, Z., Zeng, Y., Xu, J., Cheng, Y. and Deng, X. 2016. Salicylic acid treatment reduces the rot of postharvest citrus fruit by inducing the accumulation of H_2O_2, primary metabolites and lipophilic polymethoxylated flavones. *Food Chemistry* 207, 68–74.

Zhu, S., Liu, M. and Zhou, J. 2006. Inhibition by nitric oxide of ethylene biosynthesis and lipoxygenase activity in peach fruit during storage. *Postharvest Biology and Technology* 42, 41–48.

Zhu, S., Sun, L. and Zhou, J. 2009. Effects of nitric oxide fumigation on phenolic metabolism of postharvest Chinese winter jujube (*Zizyphus jujuba* Mill. cv. Dongzao) in relation to fruit quality. *LWT - Food Science and Technology* 42, 1009–1014.

Zhu, X., Cao, J.K., Wang, Q. and Jiang, W.B. 2008. Postharvest infiltration of BTH reduces infection of mango fruits (*Mangifera indica* L. cv. Tainong) by *Colletotrichum gloeosporioides* and enhances resistance inducing compounds. *Journal of Phytopathology* 156, 68–74.

Zhu, Z. and Tian, S.P. 2012. Resistant responses of tomato fruit treated with exogenous methyl jasmonate to *Botrytis cinerea* infection. *Scientia Horticulturae* 142, 38–43.

Index

A

B

C

D

E

F

V

W